THE NEUROLOGICAL BASIS OF PAIN

Notice

Medicine is an ever-changing science. As new research and clinical experience broaden our knowledge, changes in treatment and drug therapy are required. The authors and the publisher of this work have checked with sources believed to be reliable in their efforts to provide information that is complete and generally in accord with the standards accepted at the time of publication. However, in view of the possibility of human error or changes in medical sciences, neither the authors nor the publisher nor any other party who has been involved in the preparation or publication of this work warrants that the information contained herein is in every respect accurate or complete, and they disclaim all responsibility for any errors or omissions or for the results obtained from use of the information contained in this work. Readers are encouraged to confirm the information contained herein with other sources. For example and in particular, readers are advised to check the product information sheet included in the package of each drug they plan to administer to be certain that the information contained in this work is accurate and that changes have not been made in the recommended dose or in the contraindications for administration. This recommendation is of particular importance in connection with new or infrequently used drugs.

THE NEUROLOGICAL BASIS OF PAIN

Marco Pappagallo, MD
Associate Professor of Neurology
Albert Einstein College of Medicine

Director, Division of Chronic Pain
Department of Pain Medicine and Palliative Care
Beth Israel Medical Center
Manhattan Campus for the Albert Einstein College of Medicine

McGraw-Hill
Medical Publishing Division

New York   Chicago   San Francisco   Lisbon   London   Madrid   Mexico City   Milan
New Delhi   San Juan   Seoul   Singapore   Sydney   Toronto

The **McGraw·Hill** Companies

THE NEUROLOGICAL BASIS OF PAIN

Copyright © 2005 by The McGraw-Hill Companies, Inc. All rights reserved. Printed in the United States of America. Except as permitted under the United States Copyright Act of 1976, no part of this publication may be reproduced or distributed in any form or by any means, or stored in a data base or retrieval system, without the prior written permission of the publisher.

1 2 3 4 5 6 7 8 9 0   CCI   CCI   0 9 8 7 6 5 4 3 2 1

ISBN: 0-07-144087-9

This book was set in Garamond by International Typesetting and Composition.
The editors were Janet Foltin, Marsha Loeb, and Regina Brown.
The production supervisor was Catherine Saggese.
The cover designer was Pehrsson Design.
Courier Kendallville was printer and binder.

This book is printed on acid-free paper.

Library of Congress Cataloging-in-Publication Data

The neurological basis of pain/[edited by] Marco Pappagallo.
  p. ; cm.
Includes bibliographical references and index.
ISBN 0-07-144087-9 (alk. paper)
 1. Pain. I. Pappagallo, Marco.
 [DNLM: 1. Pain—physiopathology. 2. Neurophysiology—methods. 3. Pain—therapy.
WL 704 N4941 2004]
RB127.N455 2004
616′0472—dc22

2004042629

To those who have spent many hours with me
during the editing of this book:
my beloved Jennifer, Olivia, and baby Isabella,
and my devoted Poppy, Willie, Sparty, and Panzanella

CONTENTS

Contributors . . . . *ix*
Foreword . . . . *xv*
Preface . . . . *xvii*

PART I The Neurological Basis of Pain

1 Neurophysiology of Nociception. 3

2 Nociceptors and Pain: Correlation of Electrophysiology in Animals to Pain Perception in Humans 21

3 Pharmacology of Pain Transmission and Modulation 31
      I. Central Mechanisms 31
      II. Peripheral Mechanisms 53

4 Opioid Pharmacology 61

5 Mechanisms of Neuropathic Pain. 71

6 Mechanisms of Visceral Pain 95

7 Pain and the Autonomic Nervous System. 105

8 Mechanisms of Primary Headaches 113

9 Neuroanatomy of Pain 121

10 Brain Imaging of Pain 151

11 Psychological Aspects of Pain: A Consciousness Studies Perspective. 157

PART II The Pain Physician and the Patient with Pain

12 Ethics 171

13 The Epidemiology of Pain 179

14 Principles of Pain Assessment 195

15 Comprehensive Evaluation of the Patient with Chronic Pain. 209

16 Psychological Evaluation of the Patient with Chronic Pain. 219

17 Pain in Children 225

18 Pain in the Elderly 243

19 Pain Medicine and Chemical Dependency. 257

20 Management of the Patient with Chronic Pain: Aspects of Quality Assurance and Outcomes . . . 273

PART III Syndromes and Disorders in Pain Medicine

21 Taxonomy of Pain Syndromes 289

22 Central Neuropathic Pain 301

23 Peripheral Neuropathic Pain 321

24 Pain in Neurological Disorders. 343

25 Complex Regional Pain Syndromes 359

26 Chronic Daily Headaches. 379

27 Migraine and Cluster Headaches 391

28 Trigeminal Neuralgia and Orofacial Pains 401

29 Cervicogenic Headaches 415

30 Spinal Pain: Pathogenesis, Evolutionary Mechanisms, and Management 421

31 Chronic Pelvic Pain 453

32 Cancer Pain: Essentials of Pharmacological Management.............................467

33 Internal Medicine Aspects of Pain..................................481

34 The Myalgic Syndromes....................503

35 Assessment and Treatment of the Myofascial Trigger Point...........................513

36 Acute Pain Management....................519

37 Psychiatric, Somatic, and Behavioral/Psychological Comorbidities in Chronic Pain Patients: Diagnostic and Treatment Approaches.....................527

PART IV The Therapeutic Basis of Pain

38 Nonsteroidal Anti-Inflammatory Medications and Acetaminophen.......................545

39 Opioid Therapy..........................559

40 Antidepressants, Anticonvulsants, and Miscellaneous Agents.....................581

41 Psychological Treatments of Patients with Chronic Pain...........................599

42 Principles of Interventional Pain Medicine.....607

43 Neurostimulatory Techniques in Pain Medicine...........................621

44 Neurosurgical Approaches to the Treatment of Pain................................631

45 Principles of Palliative Care for the Pain Physician..........................641

46 Physical Therapy and Rehabilitation.........647

Index...................................661

CONTRIBUTORS

Steven B. Abramson, MD *(Chapter 38)*
NYU School of Medicine
Professor of Medicine and Pathology
Department of Rheumatology
Hospital for Joint Diseases
New York, NY

Charles Argoff, MD *(Chapter 35)*
Director
Neurological Rehabilitation
North Shore University Hospital
Manhasset, NY

Nadine Attal, MD, PhD *(Chapter 22)*
Director
Laboratory of Pathophysiology and Clinical
 Pharmacology of Pain
Ambroise Pare Hospital
Assistance Publique Hopitaux de Paris
Universite de Versailles
Saint Quentin, France

Sheena K. Aurora, MD *(Chapter 8)*
Swedish Medical Center
Director, Swedish Headache Center
Assistant Professor
Department of Neurology
University of Washington
Seattle, WA

Ralf Baron, MD, PhD *(Chapter 25)*
Professor of Neurology at the University
 of Schleswig-Holstein
Associate Director of the Department of Neurology
University of Schleswig-Holstein
Campus Kiel

Eduardo E. Benarroch, MD *(Chapter 7)*
Professor of Neurology
Department of Neurology
Mayo Clinic-Mayo Foundation
Rochester, MN

Andreas Binder *(Chapter 25)*
Physician and Scientific Assistant at the Department
 of Neurology
University of Schleswig-Holstein
Campus Kiel

John Birkness, MD *(Chapter 43)*
Resident in Neurology
Department of Neurosurgery
Thomas Jefferson University
Philadelphia, PA

Didier Bouhassira, MD, PhD *(Chapter 22)*
Centre d'Evaluation et de Traitement de la Douleur
Ambroise Pare Hospital
Assistance Publique Hopitaux de Paris
Paris, France

Brenda Breuer, PhD, MPH *(Chapter 13)*
Director of Epidemiologic Research
Department of Pain Medicine and Palliative Care
Beth Israel Medical Center
New York, NY

Stephen C. Brown, MD, FRCPC *(Chapter 17)*
Director, Chronic Pain Management Center
 and Staff Anaesthesiologist
The Hospital for Sick Children
Assistant Professor of Anaesthesia
University of Toronto
Toronto, Ontario, Canada

James N. Campbell, MD *(Chapter 44)*
Professor
Department of Neurosurgery
Director of the Blaustein Pain
 Treatment Center
The Johns Hopkins University
 School of Medicine
Baltimore, MD

C. Richard Chapman, PhD *(Chapter 11)*
Professor and Director
Pain Research Center
Department of Anesthesiology
University of Utah
Salt Lake City, UT

Beverly J. Collett, MBBS, FRCA *(Chapter 31)*
Consultant in Pain Management
  and Anesthesia
Pain Management Service
Department of Anesthesia, Critical Care,
  and Pain Management
University Hospitals of Leicester
Leicester, UK

Flaminia Coluzzi, MD *(Chapter 40)*
Resident
Anaesthesiology and Intensive Care
Department of Anaesthesia
Intensive Care and Pain Therapy
University of Rome "La Sapienza"
Rome, Italy

Christine J. Cordle, BSc, MSc, AFBPsS *(Chapter 31)*
Consultant Clinical Psychologist
Clinical Psychology Department
University of Leicester
Department of Medical Psychology
University Hospitals of Leicester
Leicester, UK

Giorgio Cruccu, MD *(Chapter 28)*
Professor of Neurology
Department of Neurological Sciences
University of Rome "La Sapienza"
Rome, Italy

Karen D. Davis, PhD *(Chapter 10)*
Associate Professor of Surgery
Canada Research Chair in Brain and Behaviour
University of Toronto
Toronto Western Hospital
Toronto Western Research Institute
Toronto, Ontario, Canada

Mellar P. Davis, MD, FCCP *(Chapter 39)*
Director of Research
The Harry R. Horvitz Center for Palliative Medicine
The Cleveland Clinic Foundation
Cleveland, OH

Robert A. Duarte, MD *(Chapter 29)*
Attending
North Shore-Long Island Jewish Health System
Pain and Headache Treatment Center
Department of Neurology
Manhasset, NY

David A. Fishbain, MD, FAPA *(Chapter 37)*
Professor of Psychiatry
Adjunct Professor of Neurological Surgery
  and Anesthesiology
University of Miami School of Medicine
University of Miami Pain Center
Miami Beach, FL

Gerald F. Gebhart, PhD *(Chapter 6)*
Professor and Head
Department of Pharmacology
Carver College of Medicine
The University of Iowa
Iowa City, IA

Robert D. Gerwin, MD *(Chapter 34)*
Pain and Rehabilitation Medicine
The Johns Hopkins University
Baltimore, MD

Christopher G. Gharibo, MD *(Chapter 36)*
Assistant Professor of Anesthesiology
Associate Director of Comprehensive Pain
  Treatment Center
New York University Medical Center
Hospital for Joint Diseases
New York, NY

Brian D. Golden, MD *(Chapter 38)*
Department of Rheumatology
Hospital for Joint Diseases
New York, NY

Debra B. Gordon, RN, MS *(Chapter 20)*
Senior Clinical Nurse Specialist
Faculty Associate
University of Wisconsin Hospitals and Clinics
Madison, WI

Stephen W. Harkins, PhD *(Chapter 18)*
Professor
Gerontology, Psychiatry and Biomedical Engineering
Virginia Commonwealth University
Richmond, VA

Karla S. Hayes, MD *(Chapter 45)*
Director of Pediatric Pain
Departments of Anesthesia and Critical Care
Massachusetts General Hospital
Boston, MA

Jennifer A. Haythornthwaite, MD *(Chapter 15)*
Associate Professor
Department of Psychiatry and Behavioral Sciences
Johns Hopkins University School of Medicine
Baltimore, MD

E. Daniela Hord, MD *(Chapters 15, 24, and 45)*
Clinical Instructor in Anesthesia and Neurology
Massachusetts General Hospital Pain Center
Interventional Pain Program
Department of Anesthesia and Critical Care
Department of Neurology
Massachusetts General Hospital
Harvard Medical School
Boston, MA

Ling Hsiu-Hsien, MD *(Chapter 32)*
Fellow
Pain and Palliative Care Service
Department of Neurology
Memorial Sloan-Kettering Cancer Center
New York, NY

Helena Knotkova, MD *(Chapter 3)*
Research Fellow
Department of Pain Medicine and Palliative Care
Beth Israel Medical Center
Clinical Psychologist
Hospic Strasburk
Prague, Czech Republic

Kristina R. Krmpotic, BA *(Chapter 17)*
Administrative Coordinator
Chronic Pain Research Center
The Hospital for Sick Children
Toronto, Ontario, Canada

Josephine Lai, PhD *(Chapter 3)*
Professor
Department of Pharmacology
University of Arizona
Tucson, AZ

Allen H. Lebovits, PhD *(Chapter 41)*
Associate Professor
Departments of Anesthesiology and Psychiatry
Co-Director
New York University Pain Management Center
New York University School of Medicine
New York, NY

Paolo Manfredi, MD *(Chapter 32)*
Fellowship Director
Neurology Department
Pain and Palliative Care Service
Memorial Sloan-Kettering Cancer Center
New York, NY

Marco Maresca, MD *(Chapter 33)*
Researcher
Department of Medical and Surgical Critical Care
University of Florence
Florence, Italy

Consalvo Mattia, MD *(Chapter 40)*
Associate Professor
Anaesthesiology and Intensive Care Medicine
Department of Anaesthesia
Intensive Care and Pain Therapy
University of Rome "La Sapienza"
Rome, Italy

Alexander Mauskop, MD, FAAN *(Chapter 27)*
Associate Professor of Clinical Neurology
SUNY—Downstate Medical Center
Director, New York Headache Center
New York, NY

Patricia A. McGrath, PhD *(Chapter 17)*
Director, Chronic Pain Research Center
The Hospital for Sick Children
Senior Associate Scientist
Brain and Behavior Program
The HSC Research Institute and Professor
  of Anaesthesia
The University of Toronto
Toronto, Ontario, Canada

Lynette A. Menefee, PhD *(Chapter 16)*
Assistant Professor
Department of Anesthesiology
Jefferson Medical College
Philadelphia, PA

Richard A. Meyer, MS *(Chapter 2)*
Professor
Department of Neurosurgery and of Biomedical
  Engineering
The Johns Hopkins University School
  of Medicine
Applied Physics Laboratory
The Johns Hopkins University
Baltimore, MD

Christine Miaskowski, RN, PhD, FAAN *(Chapter 14)*
Professor and Chair
Department of Physiological Nursing
University of California
Director of the Program in Symptom Management,
 Supportive Care, and Survivorship
Comprehensive Cancer Center
University of California
San Francisco, CA

Daniel B. Murrey, MD *(Chapter 30)*
Charlotte Orthopedic Specialists
Charlotte Spine Center
Charlotte, NC

T. J. Ness, MD, PhD *(Chapter 6)*
Department of Anesthesiology
School of Medicine
University of Alabama at Birmingham
Birmingham, AL

Michael H. Ossipov, PhD *(Chapter 3)*
Research Associate Professor
Department of Pharmacology
University of Arizona
Tucson, AZ

Marco Pappagallo, MD *(Chapters 3 and 23)*
Associate Professor of Neurology
Albert Einstein College of Medicine
Director, Division of Chronic Pain
Department of Pain Medicine and Palliative Care
Beth Israel Medical Center
Manhattan Campus for the Albert Einstein College
 of Medicine
New York, NY

Gavril W. Pasternak, MD, PhD *(Chapter 4)*
Head, Laboratory of Molecular Neuropharmacology
Memorial Sloan-Kettering Cancer Center
Professor of Neurology and Neuroscience,
 Pharmacology, and Psychiatry
Weill College of Medicine of Cornell University
New York, NY

Anca Popescu, MD *(Chapters 9 and 24)*
Assistant Professor of Neurology
Temple University School of Medicine
Temple University Hospital
Philadelphia, PA

Frank Porreca, PhD *(Chapter 3)*
Professor
Departments of Pharmacology
 and Anesthesiology
University of Arizona
Tucson, AZ

Srinivasa N. Raja, MD *(Chapter 15)*
Professor of Anesthesiology and Critical Care
 Medicine
The Johns Hopkins University School of Medicine
Baltimore, MD

Ali R. Rezai MD *(Chapter 43)*
Head, Section of Stereotactic and Functional
 Neurosurgery
Department of Neurosurgery
The Cleveland Clinic Foundation
Cleveland, OH

Andrew Rosenberg, MD *(Chapter 36)*
Chairman Department of Anesthesiology
Hospital for Joint Diseases
Associate Professor of Clinical Anesthesiology
New York University School of Medicine
New York, NY

Nathan J. Rudin, MD, MA *(Chapter 20)*
Medical Director
UW Pain Treatment and Research Center
Assistant Professor
Section of Rehabilitation Medicine
Department of Orthopedics and Rehabilitation
University of Wisconsin Medical School
Madison, WI

Paola Sandroni, MD, PhD *(Chapter 7)*
Assistant Professor of Neurology
Mayo College of Medicine
Rochester, MN

Seddon R. Savage, MD, MS *(Chapter 19)*
Associate Professor of Anesthesiology
Dartmouth Medical School
Pain Consultant
Manchester VA Medical Center
Manchester, NH

Joachim Scholz, MD *(Chapter 5)*
Instructor
Neural Plasticity Research Group
Department of Anesthesia and Critical Care
Massachusetts General Hospital and Harvard Medical School
Boston, MA

Gil Schreier, MD *(Chapter 32)*
Fellow
Pain and Palliative Care Service
Department of Neurology
Memorial Sloan-Kettering Cancer Center
New York, NY

Jacob P. Schwarz, MD *(Chapter 2)*
Resident
Department of Neurosurgery
The Johns Hopkins University School of Medicine
Baltimore, MD

Daniel M. Sciubba, MD *(Chapter 44)*
Department of Neurosurgery
The Johns Hopkins University School of Medicine
Baltimore, MD

Ashwini D. Sharan, MD *(Chapter 43)*
Assistant Professor
Department of Neurosurgery
Thomas Jefferson University
Philadelphia, PA

Maureen J. Simmonds, MSc, PhD *(Chapter 46)*
Professor and Head
School of Health Professionals and Rehabilitation Sciences
University of Southampton
Highfield, Southampton, Hants, UK

Howard S. Smith, MD *(Chapter 21)*
Academic Director of Pain Management
Department of Anesthesiology
Albany Medical College
Associate Professor, Anesthesiology
Albany Medical College
Albany, NY

Anan Srikiatkhachorn, MD *(Chapter 26)*
Associate Professor
Neuroscience Unit, Department of Physiology
King Chulalongkorn Memorial Hosptial and Faculty of Medicine
Chulalongkorn University
Bangkok, Thailand

Charles R. Stewart, MBBS, FRCOG *(Chapter 31)*
Consultant Obstetrician and Gynecologist Emeritus
University Hospitals of Leicester
Leicester, UK

Milan P. Stojanovic, MD *(Chapter 42)*
Director
Interventional Pain Program
MGH Pain Center
Department of Anesthesia and Critical Care
Massachusetts General Hospital
Assistant Professor of Anesthesia
Harvard Medical School
Boston, MA

Laura S. Stone, PhD *(Chapter 3)*
Assistant Professor
Department of Neuroscience and Anesthesiology
University of Minnesota
Minneapolis, MN

Carol R. Taylor, CSFN, RN, PhD *(Chapter 12)*
Center for Clinical Bioethics
Georgetown University
Washington, DC

Andrea Truini, MD *(Chapter 28)*
Neuropathic Pain Unit
Department of Neurological Sciences
University of Rome "La Sapienza"
Rome, Italy

Kathleen Vits, MSc, MCSP, SRP *(Chapter 31)*
Clinical Specialist Physiotherapist
Women's Health
Princess Anne Hospital
Southampton, UK

Karin N. Westlund, PhD *(Chapter 1)*
Professor
Department of Neuroscience and Cell Biology
University of Texas Medical Branch
Galveston, TX

Anthony H. Wheeler, MD *(Chapter 30)*
Pain and Orthopedic Neurology
Charlotte Spine Center
Charlotte, NC

George L. Wilcox, PhD *(Chapter 3)*
Professor
Departments of Neuroscience and Pharmacology
Graduate Program in Neuroscience
University of Minnesota
Minneapolis, MN

Clifford J. Woolf, MD, PhD *(Chapter 5)*
Professor of Anesthesia Research
Harvard Medical School
Professor of Anesthesia Research
Neural Plasticity Research Group
Department of Anesthesia and Critical Care
Massachusetts General Hospital
Boston, MA

S. Farhan Zaidi, MD *(Chapter 29)*
Resident
North Shore-Long Island Jewish Health System
Department of Neurology
Division of Physical Medicine and Rehabilitation
Manhasset, NY

Sheng Ping Zou, MD *(Chapter 36)*
Clinical Assistant Professor
Department of Anesthesiology
New York University Medical Center
Attending Physician
Director of Pain Management Center
Bellevue Hospital Center
New York, NY

FOREWORD

Half a century ago, the notion that professionals could spend a career dedicated to the management of pain, or to the study of the neural systems subserving pain perception or clinical pain states, would not have been taken seriously. The basic sciences were still focused on clarifying the essential anatomy and physiology of nociceptive systems, and the clinical sciences, much like clinicians themselves, viewed pain as an indicator of some other process that may be of interest, but not interesting in itself. There were no grant funds available to study pain, no professional societies to support the interest of investigators or clinicians, no specialty journals to disseminate information, and no standards of practice or credentialing processes for clinicians who treated pain.

The advances in both pain research and therapy since that time have been stunning. The scientific community has evolved from a rather straightforward examination of the sensory apparatus to an inchoate understanding of the neurophysiology, neurochemistry, and molecular biology of highly complex and redundant systems for sensing noxious events and processing this information before and after it elicits autonomic reactions and invades consciousness. Research into the pathogenesis of painful diseases has identified an array of complicated mechanisms that have already provided the basis for translational work. This work has influenced the development of new drugs and other treatments, and holds the promise of true mechanism-based therapy some day in the future. Studies of the psychological processes that contribute to extraordinary clinical variability have yielded concepts and principles with direct relevance to the diagnosis and treatment of patients.

Also during this period, pain management has achieved specialty status in numerous medical disciplines and other fields, such as nursing and psychology. Credentialing bodies have fashioned standards for professionals and a formal specialty of pain medicine has been accepted in the United States. There are now many professional societies devoted to pain, a large number of pain-related publications, and numerous specialty journals. Educational initiatives focused on pain can reach virtually every type of clinician, and broad education has been driven by regulatory endorsement of effective pain management as the best practice in hospitals and other institutions.

To some extent, these advances reflect a profound paradigm shift—from the view of pain as a symptom of disease to that of pain as disease, and indeed, as an illness. This new paradigm has challenged the medical community to approach the management of acute and chronic pain as a clinical imperative, necessary to recognize and treat to avoid the deranged physiology and functional loss that can accompany unrelieved pain.

With these advances, and with a prevalence of chronic pain affecting nearly one in three citizens in the United States, virtually all practitioners now recognize pain as a major problem and address pain-related concerns in practice as a matter of course. Notwithstanding, the various medical disciplines have differed in the extent to which pain education and skills-based training have been emphasized. Just a few disciplines have opted to pursue specialty status in pain medicine.

Those disciplines that have endorsed specialty status for pain medicine are likely to provide more education, offer leadership in clinical investigation, and access better and more extensive care for patients earlier. This is now happening in neurology. This acceptance of pain as a subspecialty in neurology is an extraordinary accomplishment, which recognizes both the fundamental importance of neurological diseases in the pathogenesis of pain and the potential for clinical neurology to make important contributions to the management of challenging pain syndromes.

Specialty status in neurology is still in its infancy, however, and many changes will be needed to fulfill the promise of this development. Published texts focused on pain are essential and other types of pain education and training must be developed. Access to specialty training will require the development of pain programs in academic departments and a growing cadre of mentors and role models with full-time interest in this area. This evolution will require time.

Nonetheless, organized neurology in the United States has made a strong commitment to pain medicine. This will have a positive impact on the overall discipline and yield better care for patients.

Russell K. Portenoy, MD
Chairman, Department of Pain Medicine and Palliative Care
Beth Israel Medical Center, New York, NY

Professor of Neurology
Albert Einstein College of Medicine

PREFACE

Pain management has evolved into a contemporary medical discipline, thanks to the invaluable work of a dedicated and farseeing physician, John (Giovanni) Bonica. In the mid '50s he wrote the first edition of the classic textbook, *The Management of Pain*. Subsequently, Dr. Bonica founded the world's first multidisciplinary pain center at the University of Washington, in Seattle. In 1973, he spearheaded the first international scientific meeting of pain medicine and founded the International Association for the Study of Pain (IASP), currently the largest network of professionals dedicated to the study and treatment of pain. Since then, the efforts of an ever-increasing number of outstanding individuals have led to the growth of pain medicine and its progression to a multifaceted and self-regulating medical field.

Neuroscience has always played a fundamental role in this field. The neurologists' knowledge of pain physiology and their diagnostic and pharmacological skills have greatly contributed to the expansion of pain medicine. Neurologists are often involved in the comprehensive assessment and treatment of patients with chronic pain, not only as consultants, but also as providers of long-term care. Individuals with chronic pain are the most common patients in the practice of a general neurologist. The most prevalent and expensive neurological disorder is chronic pain, affecting approximately 35% of the U.S. and European populations. In the U.S., more than $100 billion a year is spent for the management of pain. Today, pain management has become a major public health issue. Moreover, as the general population ages, needs for adequate treatments, services, and related costs are destined to increase even further.

Compiling *The Neurological Basis of Pain* has been an extraordinary professional experience. When Mark Strauss, the senior editor at McGraw-Hill, proposed the idea of developing a pain medicine textbook as a companion to the popular Adams and Victor's *Principles of Neurology*, I was both enthusiastic and perplexed. I was honored to edit a book that would complement the *Principles of Neurology*, a landmark among the textbooks of neurology. Nevertheless, I was also at a loss. I knew that in order to create a text that would command the respect of neurologists, I would need to face the challenge of gathering a panel of distinguished faculty in the field of pain medicine.

I was lucky that a group of nationally and internationally recognized experts agreed to join me and share their knowledge, experience, and skills in the making of the *Neurological Basis of Pain*. Working with them was a thoroughly enlightening and pleasant experience and I thank them all.

Part I of this book provides an in-depth and updated overview of the scientific basis of pain medicine. Part II covers a variety of issues related to the pain physician—pain patient rapport, including ethics and risk management. Part III discusses syndromes and disorders in detail, and Part IV addresses therapies. Hopefully, alongside the available comprehensive textbooks in the field, like the well-known Bonica's and the Wall and Melzack's textbooks, *Neurological Basis of Pain* will help to complement and cover the editorial field of pain medicine. While our book is addressed to neurologists, a variety of professionals, such as residents, fellows, clinicians and academicians from a range of medical fields, will benefit from it. *Neurological Basis of Pain* is intended to advance the physicians' fund of knowledge in pain medicine to a considerable degree of scientific and clinical sophistication. Thus, even pain specialists should find this textbook a source of the most recent information and references. I hope you enjoy the reading.

Marco Pappagallo, MD

PART I

The Neurological Basis of Pain

CHAPTER 1
Neurophysiology of Nociception

Karin N. Westlund

The definition of pain provided by the International Association for the Study of Pain[1] is "an unpleasant sensory and emotional experience associated with actual or potential tissue damage, or described in terms of such damage." Thus neuronally-generated input provided to higher brain centers that is described or interpreted as pain is included in this definition. This definition also includes experiences interpreted as pain irrespective of whether the patient experiencing the pain is able to express an opinion verbally about the sensation (i.e., aphasic, nonfluent, infants, animals).

Input that activates sensory pathways and leads to responses interpreted as pain can include noxious mechanical, heat, cold, chemical, and inflammatory stimuli. Many neural receptors and circuits participate in the transduction, transmission, and responses to pain[2] (Fig. 1-1). Sensory afferent fibers have receptive endings, the peripheral nociceptors, that transduce or "sense" noxious stimuli transmitted by the sensory afferent fibers to the spinal or medullary dorsal horn. Nociceptors have their cell bodies in the dorsal root or cranial nerve ganglia and extend their central axonal endings into the spinal gray matter to communicate with dorsal horn neurons. Information about a noxious event in the periphery can initiate a protective monosynaptic reflexive withdrawal event. The information is transmitted cranially by second-order projection neurons in the spinal cord or brain stem. Both excitatory and inhibitory interneuronal circuits in the spinal cord can be activated by the input. The integrated information can result in sensitization of the spinal cord neurons if the input is persistent.

Further processing of nociceptive input occurs in numerous brain structures, including the thalamus and somatosensory cortex, leading to the sensory discriminative perception of pain. Parameters encoded about the stimuli by these structures include the perceived intensity of the pain and the body map site location of the pain. Noxious sensory input also initiates a variety of other physiologic reactions to the pain, including somatic and autonomic reflexes, emotional reactions, endocrine actions, and affective-cognitive responses having an impact on learning and memory of the event.

Other brain centers receive information about nociceptive stimuli and can provide information in the form of either negative or positive feedback to the spinal cord circuitry. The integrated input derived from higher-order neurons in the brain stem and cortex can reduce or accentuate the subjective interpretation of the perception of pain.

Negative feedback provided to the spinal cord that can damp the perception of pain is mediated by descending pathways often referred to as the endogenous analgesia system. Likewise, a descending facilitation system can accentuate the perception of pain. While positive and negative feedback can be provided through neural circuitry to the spinal cord, local mechanisms also exist at many brain sites, serving as sensory processing centers for accentuation and diminution of pain and pain responses. Thus increased responsivity to pain, referred to as central sensitization, can occur at multiple levels throughout the pain system in addition to the spinal cord. While pain serves as a protective signal important

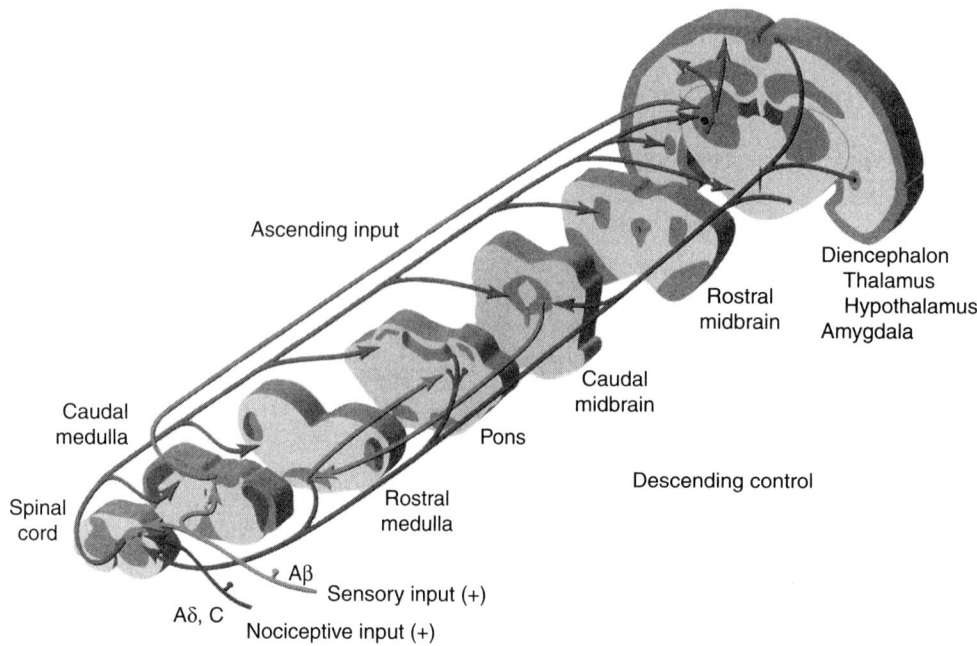

Figure 1-1. Sensory processing occurs at many levels throughout the neuraxis. Incoming noxious and non-noxious input brought in by primary afferent nerve fibers is received by the spinal cord dorsal horn and sent to multiple higher brain centers by ascending projections (*left*). Non-noxious input is relayed directly to the dorsal column nuclei before being sent to the thalamus and to the cortex. Descending projections (*right*) arising from numerous brain levels can facilitate or inhibit neural responses to sensory input.

to the prevention of further tissue damage, chronic pain produces a pathophysiologic pain state that is counterproductive and likely participates in continued tissue damage through aggravation of inflammatory and neuropathic mechanisms. The mechanisms responsible for these events are currently being explored.

▶ NOCICEPTORS

Cutaneous, Muscle, Joint, Dural, and Visceral Receptors

Peripheral nerves provide bidirectional flow of information (Fig. 1-2). Sensory input is brought into the central nervous system (CNS) by primary afferent nerve fibers over the dorsal root to the dorsal horn of the spinal cord (Fig. 1-3). The ventral root carries the axons of motor neurons located in the ventral horn that transmit motor output destined for the autonomic and somatic structures innervated. Incoming non-noxious sensory input from the periphery is important for discerning fine, discriminative touch, pressure, and position in space. The non-noxious sensory information is received by nerve endings specialized to affect the rate of adaptation to specific sensory input. The innocuous mechanical information is transduced by specialized encapsulated endings and is transmitted by large, myelinated Aβ afferent nerve fibers directly to the spinal cord and brain stem. The large fibers collateralize to innervate the dorsal horn of the spinal cord (see Fig. 1-3), as well as ascend in the dorsal column to the dorsal column nucleus at the dorsal surface of the caudal medulla.

Noxious input is transmitted to the spinal cord from free nerve endings in the skin, muscles, joints, dura, and viscera that do not have accessory specializations or myelin protection[3] (Fig. 1-4). Noxious mechanical, heat, cold, or chemical stimuli are transduced by nociceptors specialized to receive either mechanical (mechanoreceptors), mechanical and thermal (mechanothermal nociceptors), thermal and chemical (polymodal nociceptors), or cold (cold receptors) stimuli. Detailed descriptions of nociceptive endings from a variety of tissues have been provided, including those localized in muscle, fascia, and adventitia of blood vessels,[4,5] knee joint[6,7] (Fig. 1-4), dura,[8] and viscera.[9,10] The noxious information reaches the dorsal

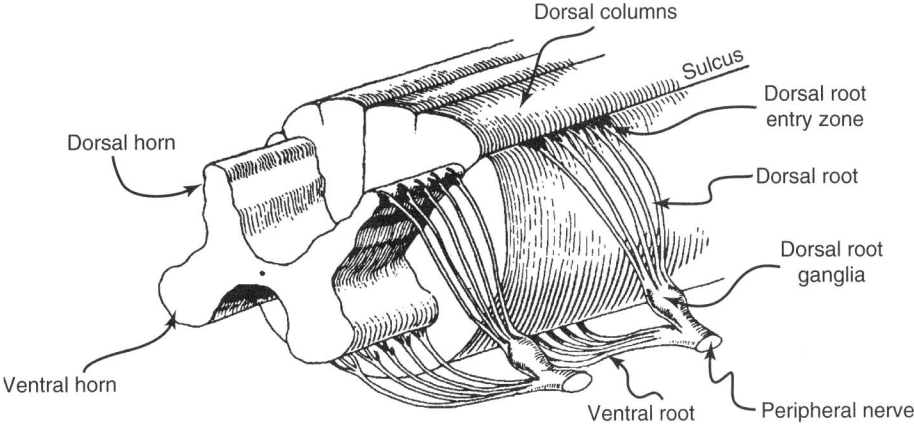

Figure 1-2. Schematized drawing of the spinal cord depicting gray matter in the foreground and white matter in the background, including the dorsal columns. Peripheral nerves contain both motor and sensory nerves. The nerves separate just outside the spinal cord into primary afferent nerve dorsal roots that enter the dorsal root entry zone and the motor nerve axons in the ventral root. The primary afferent nerve cell bodies are located in the dorsal root ganglia. (*Used with permission from Jeanmonod and Sindou.[192]*)

horn of the spinal cord or the trigeminal nucleus via two common types of nociceptive afferent nerve fiber types, the Aδ mechanoreceptive and the C polymodal nociceptive nerve fibers.[11,12]

Both similarities and differences are apparent for afferent nerve transmission from cutaneous, muscular, joint, dural, and visceral structures. The axons that relay information to the CNS about intense noxious input applied to the skin and all other tissues fall characteristically into two types: (1) small, unmyelinated C fibers with conduction velocities under 2.5 m/s for the group IV nociceptors[13] and (2) slightly larger Aδ fibers wrapped in a thin layer of myelin produced by specialized glial Schwann cells in the periphery (see Figs. 1-3 and 1-4). The myelin sheath allows nerve conduction at a rate of 4 to 30 m/s in the case of the Aδ fibers with group III nociceptive endings.[14] Primary afferent C fibers are far more numerous than myelinated primary afferent nerves. For example, in dorsal roots, the ratio of C fibers to A fibers is about 2.5:1,[15] and in joint nerves (after sympathetic postganglionic axons are removed by sympathectomy), the ratio of C to A fibers is 2.3:1.[16] In skin, the ratio for primary afferent nerve fibers is typically 70 percent C fibers and 10 percent Aδ fibers. All classes of axons can transmit non-noxious information encoded in the periphery. Under normal circumstances, only C and Aδ fibers but not Aβ fibers can transmit nociceptive information. Under normal circumstances, nociceptive primary afferent nerve fibers innervating the joint and viscera are unresponsive.[7,10,17–19] The "silent" or "sleeping" nociceptors are activated by a variety of chemical and pathologic inputs.[20–23]

Nociceptive Input to Spinal and Medullary Dorsal Horn

The central processes of primary afferent nerves enter the spinal cord as the dorsal roots (see Figs. 1-2 and 1-3). The fiber types tend to segregate functionally in the dorsal root entry zone so that thin, unmyelinated C fibers of nociceptors are situated laterally in the dorsal root entry zone and terminate superficially in the dorsal horn.[24] The larger, myelinated fibers are situated to enter the dorsal root entry zone more medially, traveling over the dorsal horn on their way to terminate in deeper layers of the medial dorsal horn.[25] Distribution patterns of C fibers originating in cutaneous tissue are more discrete than the diffuse innervation of visceral C fibers in the dorsal horn (Fig. 1-5). Collateral branches of both large and small axons carrying non-noxious sensory information enter the dorsal column, where they travel rostrally to the dorsal column nuclei in the dorsal medullary midline of the caudal brain stem (see Figs. 1-1 and 1-3). The ascending dorsal column pathway continues contralaterally to the thalamus as the medial leminiscus. Noxious input is relayed to spinothalamic tract cells situated in laminae I, IV, V, and X of the spinal cord, as described below (Figs. 1-3 and 1-6).

Neurotransmitters Used by Nociceptors

As in most central neuronal circuits, glutamate is the primary neurotransmitter substance in primary afferent nociceptors. The actions of glutamate are modulated by neuropeptides.[26,27] The dorsal root ganglia (DRG) are composed of large and small cell bodies of the primary

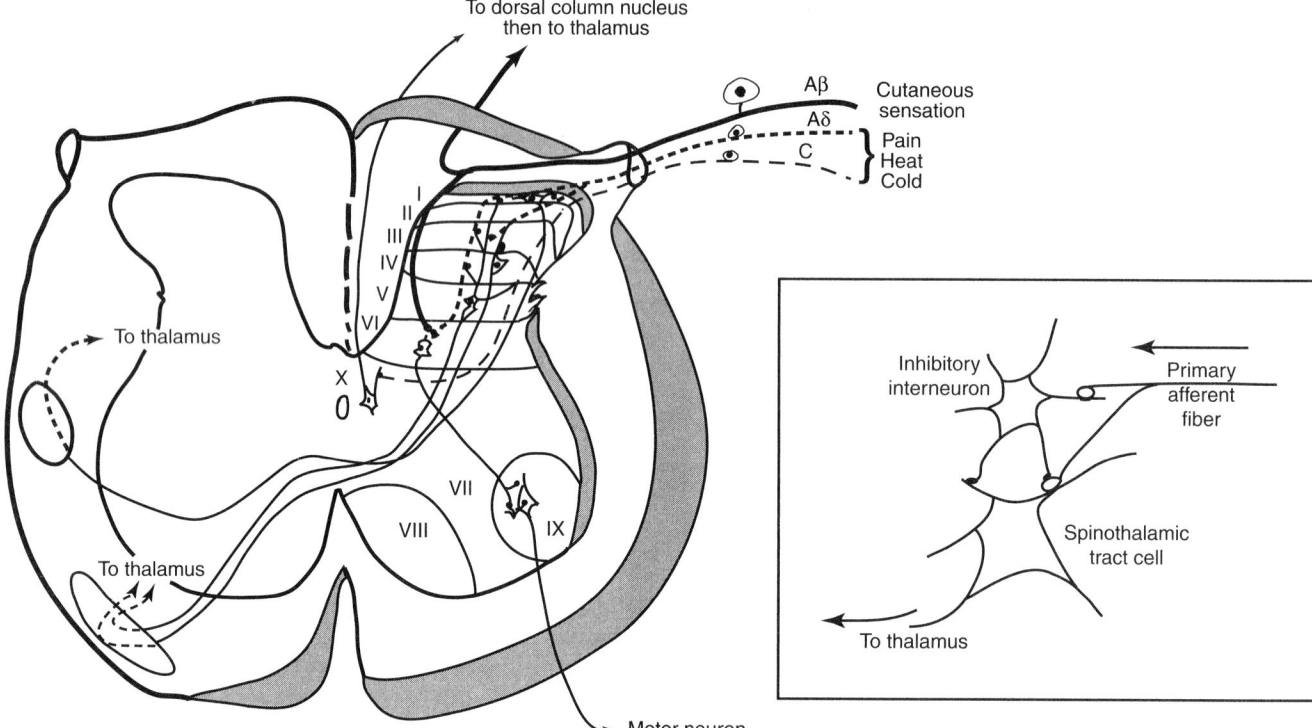

Figure 1-3. Schematized drawing of the spinal terminations of various primary afferent nerve endings onto spinal interneurons and neurons with ascending projections to the brain stem and thalamus. Cutaneous sensation is carried by large Aβ primary afferent fibers to the spinal cord and the dorsal column nucleus before transmission to the thalamus. Input perceived as painful is carried by small Aδ and C fibers to spinothalamic tract neurons located primarily in Rexed's laminae I, IV, and V. The spinothalamic tract axons cross the midline and ascend in the contralateral white matter to the thalamus. Noxious stimulation of afferent nerve fibers in visceral structures is provided to cells in and around lamina X. The postsynaptic dorsal column neurons send axonal projections in the dorsal column that then are relayed by the crossed medial leminscal pathway to the thalamus. Activation of spinal interneurons provides input to motor neurons in lamina IX for reflexive withdrawal from painful stimuli. The inset shows the synaptic arrangement whereby CGRP-containing primary afferent endings terminate on both inhibitory interneurons and spinothalamic tract cells. Through dendroaxonic contacts, the GABA interneurons inhibit the release of transmitter, effectively reducing primary afferent fiber input to the projection neurons (gate theory of pain).

afferent nerve fibers.[28] Many of the small DRG cells belong to the nociceptors.[29] Neurochemical content differences between primary afferent nerve fibers from various tissues are beginning to be apparent. In nociceptive DRG cells with cutaneous axonal endings, glutamate is often co-localized with neuropeptides such as calcitonin gene–related peptide (CGRP), substance P (SP), neurokinin A, galanin, and somatostatin.[29] Since the sole source of CGRP in the dorsal horn of the spinal cord is primary afferents, this peptide serves as a marker for primary afferent terminals.[30] The neuropeptide SP is more prominent in C fibers arising from muscle and other deep tissues as compared with skin.[31,32] Visceral afferent fibers are rich in neuropeptides such as vasoactive intestinal polypeptide (VIP), bombesin, CGRP, and/or SP.[28,29,33] While the peptides are concentrated in afferent nerve endings, glutamate and aspartate can be localized throughout the extent of primary afferent nerve fibers.[34–38]

These neurotransmitters are assumed to function in the sensory transduction process and modulation of neurotransduction. A functional role also has been described for neurotransmitters released by peripheral nerves in targeting of immune cells to peripheral damage. This includes contributions to the following clinically relevant conditions reported to involve neurogenic inflammation induced by release of neurotransmitters into peripheral sites: arthritis, nausea, emesis, respiratory disease, and urinary bladder incontinence.[39,40]

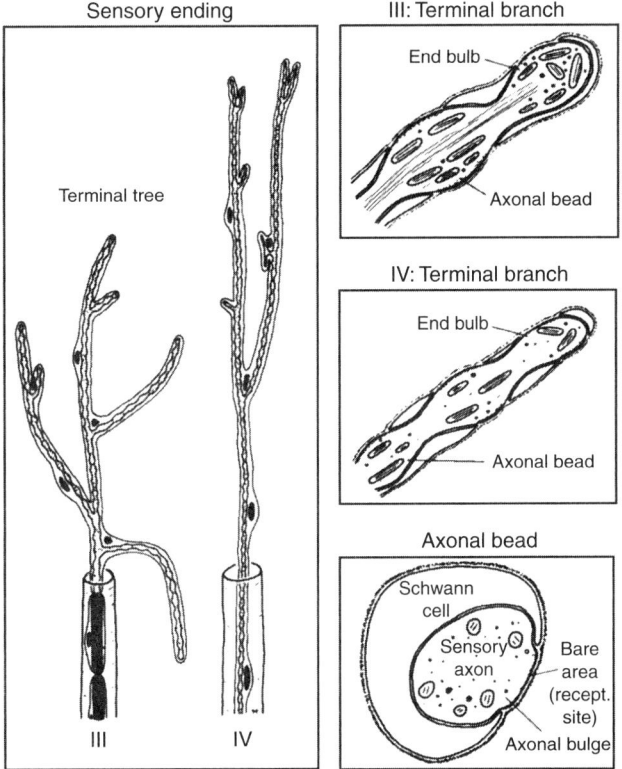

Figure 1-4. Schematic drawings of group III and IV primary afferent sensory nerve endings. The terminal endings are essentially free of myelin produced by Schwann cells in the periphery to insulate and increase conduction velocity. The beaded endings of both fiber types have bare areas that presumably are sites of chemical, mechanical, and thermal transduction. This example is reconstructed from serial electron micrographs of the articular nerve in the knee joint capsule of a cat. (*Used with permission from Hanesch et al.*[3])

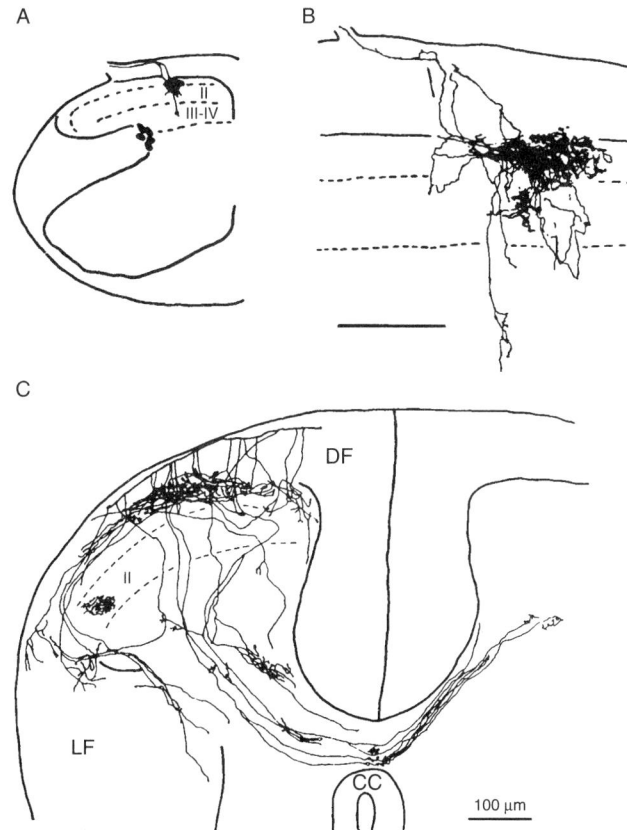

Figure 1-5. Terminal distribution of primary afferent nerve fibers in the spinal cord. Individual C fibers responsive to noxious input were characterized physiologically, and then the terminals in the dorsal horn were visualized after intracellular filling with *Phaseolus* leukoagglutinin (*A*, *B*). A cutaneous C polynociceptor ends as a tightly defined arborization in the superficial laminae, providing precise information about site localization (*C*). A visceral afferent nerve ending arborizes extensively in the superficial and deep dorsal horn, including terminations in the visceral processing areas near the central canal on both sides of the spinal cord. The terminal field distribution reflects the precise versus diffuse activation of the spinal cord by these two types of input. LF = lateral fasciculus; DF = dorsal fasciculus; CC = central canal. (*Used with permission from Sugiura et al.*[48])

Responses of peripheral nerves to noxious stimulation are mediated by both ionotropic and metabotropic receptors. The ion channels include the vanilloid, VR1, capsaicin,[41] H^+, and sodium channels.[42,43] Ionotropic and metabotropic glutamate receptors, as well as mu and delta opiate and somatostatin peptide receptors, have been identified immunohistochemically on the peripheral endings of cutaneous nerve fibers.[44–46] Neurotransmitter receptors also have been reported on peripheral visceral targets such as lungs, gastrointestinal (GI) tract, joints, and bladder.

▶ PAIN TRANSMISSION

Spinal Cord Dorsal Horn

The primary and initial response to painful stimuli is reflexive withdrawal, accomplished by primary afferent nerve activation of motor effector neurons (see Fig. 1-3).

Modulation of spinal neuronal circuits to enhance or inhibit this response is influenced by feedback from higher brain centers that have an impact on the spinal interneuronal circuitry. Signals from peripheral nociceptors are relayed through at least one spinal neuron prior to further relay to higher brain regions (see Fig. 1-3). Noxious input from the body and extremities is received by the dorsal horn of the spinal cord, which contains both local interneurons and projection neurons relaying information about noxious input to higher processing centers in the brain. The afferent fibers enter the gray matter of the spinal cord through the dorsal root entry zone and primarily

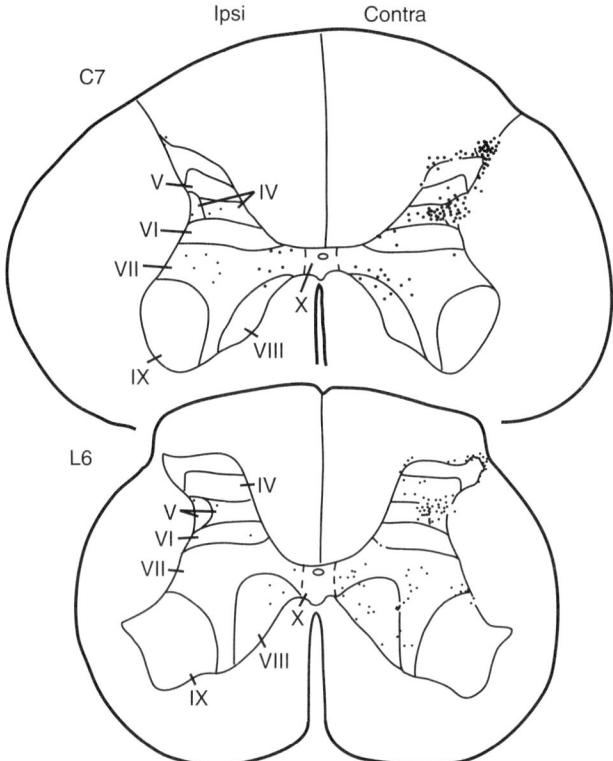

Figure 1-6. The cells of origin of the spinothalamic tract have been localized using retrograde tracers placed in the contralateral thalamus. The spinothalamic tract cells are localized primarily in Rexed's laminae I, IV, and V. A few scattered cells are also located in lamina VII and the ipsilateral spinal cord. While some cells are scattered in these locations throughout the length of the spinal cord, more spinothalamic tract cells are located in the cervical and lumbar enlargements, where cutaneous innervation of primary afferent nerves is greatest. (*Based on Apkarian and Hodge.[104]*)

innervate regions of the spinal cord within the same spinal segment matching that spinal nerve (see Fig. 1-2). The fine afferent fibers may ascend rostrally or descend caudally through Lissauer's tract, situated just dorsal to the dorsal horn. It is known that primary afferent nerves can innervate up to six spinal segments above and below their level of entry.[30]

In general, large primary afferent nerve fibers enter the spinal cord through the dorsal roots medially to travel with the dorsal column. The smaller fibers enter Lissauer's tract laterally to innervate the gray matter core of the spinal cord, where neuronal cell bodies and dendrites receive their synaptic endings (see Fig. 1-3). Ascending and descending axonal fiber connections between the spinal cord and brain travel longitudinally in the white matter. Many of the CNS axons are ensheathed in myelin formed by oligodendroglia. Somatic C nociceptive afferents are distributed mainly to laminae I and II in the same and adjacent segments, whereas visceral C afferents can extend for more than five segments, and they terminate in laminae I, II, V, and X ipsilaterally and in laminae V and X contralaterally.[47–49] Thus somatic C nociceptors end rather focally in the spinal cord gray matter, whereas terminal endings of visceral C nociceptors are widely distributed (see Fig. 1-5). Cutaneous Aδ mechanical nociceptors terminate in the ipsilateral laminae I and V, and they also may have endings in lamina X and the contralateral dorsal horn.[25]

The gray matter has been divided topographically into 10 laminae by Rexed[50] based on histologic features[51] (see Figs. 1-3 and 1-6). The dorsal horn includes laminae I through VI. The outer layer, or lamina I, of the dorsal horn contains mostly longitudinally arranged cells. Many of these are interneurons, but there are also lamina I cells that project to the brain stem, hypothalamus, and thalamus (reviewed in ref. 52). Lamina II, or the substantia gelatinosa, contains excitatory and inhibitory interneurons but few projection neurons. The interneurons take the form of stalked (limiting) cells and islet (central) cells.[53–55] Stalked cells project dorsally into lamina I. Islet cells are oriented longitudinally, and their dendritic trees tend to be flattened so that they resemble slabs oriented in the dorsoventral plane. Lamina II interneurons synthesize either inhibitory or excitatory neurotransmitters. Laminae III and IV contain interneurons and the largely tactile projection neurons of the postsynaptic dorsal column pathway and the spinocervical tract. Laminae IV through VI include interneurons and also nociceptive projection neurons with terminations in the brain stem, thalamus, and hypothalamus. There is a visceral nociceptive processing area in the central region of the spinal cord, including lamina X and adjacent parts of laminae VII and VIII. In addition to interneurons and autonomic preganglionic neurons, this area contains postsynaptic dorsal column projection cells that send axons to the dorsal column nuclei and to the thalamus, transmitting visceral pain.[56–58]

Typical of lamina II are the large glomerular synaptic complexes, sometimes referred to as *scalloped primary afferent nerve endings*, that contact multiple dendrites of dorsal horn cells simultaneously.[59] While most primary afferent nerve endings contain small, round clear vesicles, the large glomerular endings typically contain large, dense-cored vesicles as well that are believed to contain peptides. Glutamate is also localized in the glomerular endings of primary afferent nerve endings.[60] CGRP receptors have been localized on postsynaptic membranes opposite glomerular, elongated, and dome-shaped synaptic-type endings in the superficial dorsal horn.[61] Increased release of glutamate and activation of dorsal horn neurons by glutamate is enhanced by the modulatory effects of neuropeptides[62,63] such as SP, CGRP, VIP, and cholecystokinin (CCK).

Spinotrigeminal Nucleus

The incoming information received via the trigeminal nerve from head, neck, and dura projects by way of the spinal trigeminal tract to the spinal trigeminal nucleus in the medulla situated in a position equivalent to the spinal cord dorsal horn. Glutamate and glutaminase have been found in spinal trigeminal neurons.[64] Projection neurons of the spinal trigeminal nucleus send their axons to the contralateral ventromedial nucleus of the thalamus, relaying nociceptive information from the face and the dura.

Spinal Interneurons

Most of the neurons in the dorsal horn are interneurons.[65] Neurons can be found in the dorsal horn that stain for any of a large number of neuroactive substances that are presumably neurotransmitters or modulators. These substances include adenosine, choline acetyltransferase, CCK, corticotropin-releasing factor, dynorphin, enkephalin, galanin, γ-aminobutyric acid (GABA), glutamate, glycine, neurotensin, neuropeptide Y, somatostatin, SP, and thyrotropin-releasing hormone.[28] Many of these substances are thought to be involved in excitatory or inhibitory nociceptive processing by interneuronal circuits of the dorsal horn. SP seems to act directly on nociceptive projection cells in both lamina I and the deep dorsal horn, including lamina I cells with NK1 receptors.[66,67] Interestingly, neurons in lamina II, the substantia gelatinosa, do not respond to release of SP because they lack NK1 (SP) receptors.[68] The cholinergic mechanisms can reduce nociception and potentiate nonopiate analgesia.[69,70] These analgesic actions may be mediated through presynaptic terminations of laminae III through V cholinergic interneurons reported to synapse directly onto primary sensory endings.[71]

Inhibition of spinal transmission can occur through activation of either segmental or supraspinal circuitries. Weak mechanical stimulation, for example, can damp down spinal transmission of the nociceptive input.[72] While physiologists have conveniently explained "surround" type inhibition of dorsal horn synaptic transmission in terms of dorsal horn axoaxonic synaptic transmission, only a few anatomic figures of this type have been described for the dorsal horn.[55,73,74] Rather, the anatomic substrate for the gate theory of pain proposed by Melzack and Wall,[75] which can reduce neurotransmitter release from nociceptor terminals, is more likely mediated through the axodendritic arrangements of CGRP primary afferent terminals innervating the dendrites of GABA inhibitory interneurons in lamina II and dendroaxonic synapses back onto the CGRP endings (see Fig. 1-3, *inset*). Fine myelinated and unmyelinated CGRP-labeled afferent fibers are observed synapsing with GABAergic dendritic profiles of islet cells.[74] The GABA interneurons described as islet cells[54] are found in laminae I and III and have been stained for another inhibitory amino acid, glycine.[55] The GABA interneurons are uniquely qualified to provide inhibition of nociceptive input either through presynaptic afferent contacts provided by their dendrites back onto CGRP primary afferent nerve endings or through axonal contacts back onto primary afferent nerve endings or onto other dorsal horn neurons.

Projection Neurons Relaying Nociceptive Information from the Spinal Cord

Nociceptive information entering the spinal cord dorsal horn can be relayed directly by spinothalamic tract projection neurons to higher centers or modulated by both excitatory and inhibitory influences at the level of the spinal cord, presumably through local interneuronal circuitry (see Figs. 1-3 and 1-6). Several types of projection neurons in lamina I have been described by Lima and Coimbra.[76] These cell types include fusiform (elongated), multipolar, flattened, and pyramidal neurons. Different subsets of these neurons with ascending axons project to the nucleus of the solitary tract, the dorsal and lateral reticular formation of the medulla and pons, the periaqueductal gray matter, and the thalamus.[77–80] Neurons of the same morphologic types were found to stain for SP, enkephalin, dynorphin, or GABA;[81] however, it was not determined if these were interneurons or projection neurons. Similarly, three types of lamina I spinothalamic tract (STT) cells have been described in cats and monkeys: fusiform, pyramidal, and multipolar.[82,83] Evidence has been provided that the fusiform and multipolar STT cells are nociceptive, whereas the pyramidal cells are thermoreceptive.[82] Consistent with this, most fusiform and multipolar lamina I STT neurons express NK1 receptors, and most pyramidal STT cells do not.[68,84]

The spinothalamic tract projection cells that are found in deeper layers (lateral portions of laminae IV and V) of the dorsal horn are typically large, multipolar neurons with extensive dendritic arbors (see Figs. 1-3 and 1-6). The dendrites of the deep projection cells tend to arborize extensively in all directions, in contrast to the lamina I projection cells, whose dendrites are longitudinally distributed along the dorsal surface of the spinal cord gray matter. Many deep STT cells, for example, have dendrites that extend radially into the dorsal horn and even into lamina I and II, and so they can receive direct synaptic connections from the terminals of nociceptive afferent fiber endings in the superficial dorsal horn. Other STT cells deep in the dorsal horn were identified that had dendrites that extend chiefly ventrally and may receive synapses from afferent nerves supplying deep somatic and visceral structures.[85–87] Most STT neurons are believed to contain glutamate,[87] and some also contain peptides such as enkephalin, dynorphin, and VIP.[88,89] Lamina X STT cells have been shown to contain CCK, bombesin, and/or galanin.[90,91] Synapses found on

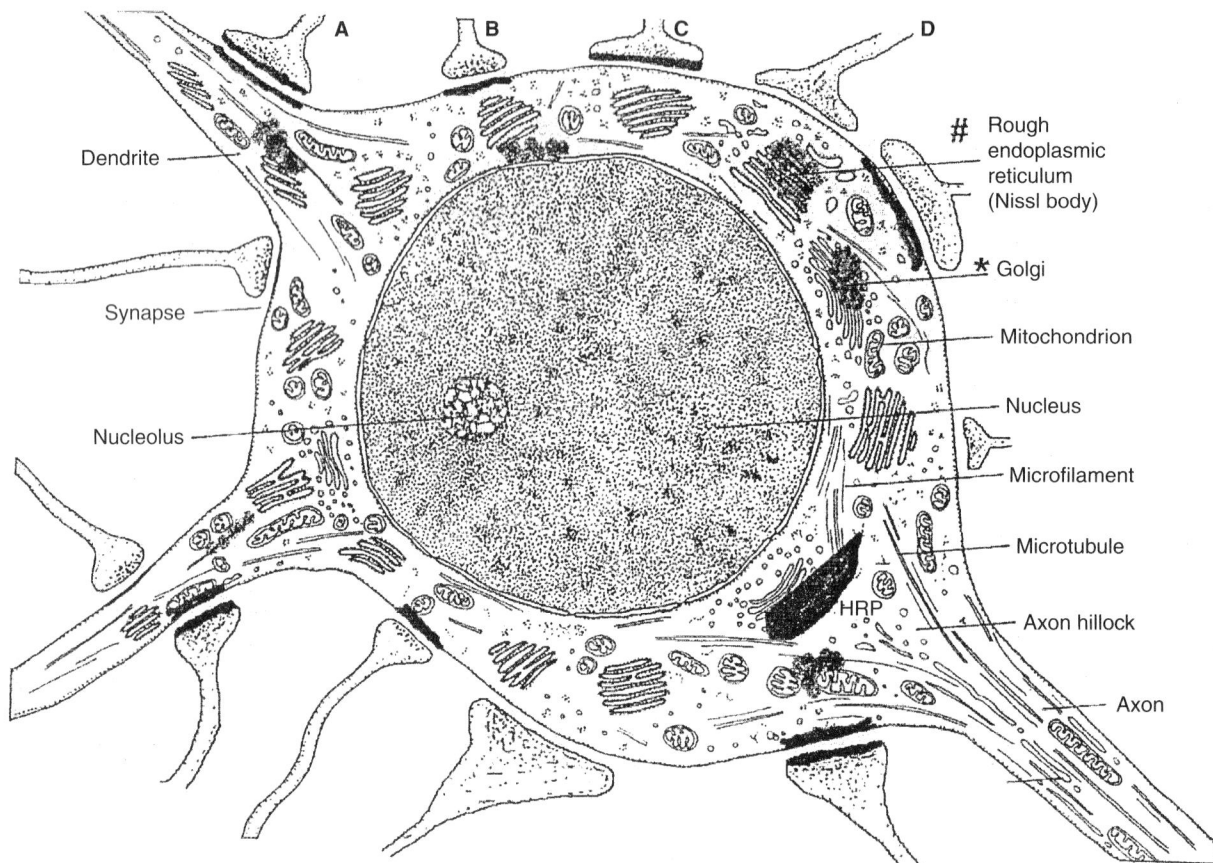

Figure 1-7. Schematic diagram of the glutamate terminal types observed contacting spinothalamic tract cells in the dorsal horn of rats by electron microscopy. The NMDA NR1, AMPA GluR1, and GluR2/3 ionotropic receptor subtypes have been observed synapsing on spinothalamic tract cells that have been identified using retrograde HRP marker. NMDA receptors were sometimes localized at both pre- and postsynaptic sites (A). Most often the receptor proteins were localized only postsynaptically (B). A few synapses were labeled presynaptically only (C), and more than half the terminals contained no glutamate receptors (D). Half the terminals contacting spinothalamic tract cells contained glutamate. Glutamate is also observed within the spinothalamic tract cells, and glutamate receptor protein was observed in sites of synthesis within the cell body. Based on these anatomic studies and many other pharmacologic and electrophysiologic studies, it is clear that glutamate is a major neurotransmitter for both the primary afferent nerve fibers and the spinothalamic relay cells of the nociceptive system. (*Based on Westlund et al.[87] and Ye and Westlund.[101]*)

the cell bodies of STT neurons of deep layers of the dorsal horn have been shown to contain glutamate (Fig. 1-7), GABA, glycine, SP, CGRP, vasopressin, norepinephrine, or serotonin,[66,92–98] indicating that these STT cells are affected by both excitatory and inhibitory influences associated with events related to nociceptive processing. Half the terminals contacting STT cells contain glutamate, and a third contain GABA. The dorsally directed dendrites of the many deep dorsal horn neurons contain neurokinin NK1 receptors for SP,[68] and these receptors are internalized following presumably painful stimulation occurring during inflammation.[99] Ultrastructural studies have revealed postsynaptic localization of glutamate receptor subunits NMDA NR1, AMPA GluR1 and Glu R2/3, and metabotropic mGluR1 and mGluR2/3 associated with terminals in contact with identified STT neurons[100,101] (see Fig. 1-7). Occasionally, the NMDA NR1 and AMPA GluR 2/3 also were localized presynaptically on the terminals contacting STT neurons.

▶ ASCENDING NOCICEPTIVE PATHWAYS

Information about noxious sensory experiences is processed in the dorsal horn and is relayed to higher brain centers to generate the conscious perception of pain. The information facilitates cognitive and emotional responses. Ascending input also is provided to brain stem sites integrating autonomic and somatic motor control. The information provided to higher brain sites may be used by autonomic and somatic motor control centers to provide appropriate feedback and to provide appropriate protective "escape" and preprogrammed autonomic and emotional responses. Neurons in the spinal cord relaying nociceptive information to the brain have axonal projections innervating the thalamus, dorsal column nuclei, reticular formation, parabrachial region, periaqueductal gray matter, anterior pretectal nucleus, hypothalamus, and amygdala (see Fig. 1-1).

Spinothalamic Tract (STT)

STT neurons transmit information to higher centers about the precise location of the noxious input on the body map. The STT neurons receiving cutaneous input are largely situated in lamina I and the lateral half of the neck of the dorsal horn in laminae IV and V[85,102–105] (see Fig. 1-6). Other STT neurons are scattered throughout the deep dorsal horn and intermediate region of the spinal cord. Some STT cells are even found medially in lamina VII in the ventral horn. The axons of the STT cells cross the midline of the spinal cord in the anterior white commissure[85] (see Figs. 1-1 and 1-3) and ascend primarily in the contralateral white matter in the lateral and ventrolateral funiculus.[106] The STT axons terminate in the posterior complex and ventroposterior and mediodorsal nuclei.[107,108] Axons of the STT cells located in lamina X near the central canal ascend in the ventromedial funiculus and terminate in the parafascicular and other medial thalamic structures.[109] Nociceptive responses also have been recorded in the posterior thalamic nuclear group[110] and intralaminar thalamic nuclei in rats.[111]

The STT terminates primarily in three regions of the thalamus—the ventroposterolateral nucleus,[112] the intralaminar nuclei (primarily the central lateral nucleus),[113] and the posterior complex.[114,115] The STT axon terminals are also found in the nucleus submedius,[116] but they are sparse.[117]

Postsynaptic Dorsal Column Pathway

In addition to collaterals of primary afferent neurons, the dorsal column contains the axons of second-order spinal cord projection neurons called *postsynaptic dorsal column neurons*. The cell bodies of many of these cells are in laminae III and IV.[118] These postsynaptic dorsal column neurons respond to a broad range of sensory input but are not believed to specifically convey cutaneous nociceptive information.[119] Postsynaptic dorsal column neurons also are found in intermediate regions of the spinal cord, most notably around lamina X[56,58,109] (Fig. 1-8). These cells transmit nociceptive information from visceral structures.[120–122] Postsynaptic dorsal column cells in the lumbosacral spinal levels terminate in the medial gracile nucleus, and those in the thoracic cord terminate medial and ventral to the cuneate nucleus. The visceral nociceptive information is then relayed from the dorsal column nucleus to sites in the thalamus[120–122] (see Fig. 1-8). The sites activated by visceral input typically are located more medially than input from skin and other somatic structures, which is transmitted to the posterior and lateral thalamus.[85,107,108]

Spinoreticular Pathways

Spinothalamic neurons send collaterals to several regions of the CNS,[123] including the medullary reticular formation,[124] the periaqueductal gray matter,[125,126] and the parabrachial area.[127] These projections innervate regions of the brain stem involved in pain-related activites, including autonomic responsiveness, the alerting response, and escape responses. Collectively, these neurons are referred to as the *spinoreticular system* (see Fig. 1-1). Most spinoreticular axons ascend in the ventrolateral white matter of the spinal cord.[28] Specific sites are innervated that are more closely related specifically to the modulation of nociceptive input. These integrative centers may either amplify or dampen the nociceptive signals. For example, a dorsal spinomedullary pathway arises from cells in laminae I, IV, and X that terminiate in the subnucleus reticularis dorsalis of the medullary reticular formation. This site is located just ventral to the cuneate and solitary nuclei of the dorsal medulla.[128] This nucleus is one of the brain stem sites active in descending inhibition of nociceptive processes termed the *diffuse noxious inhibitory control (DNIC) system*. Activation of this site produces inhibition of wide dynamic range neurons in the spinal cord by competing with afferent nerve input. The dampening effect on nociception depends on the supraspinal loop through the subnucleus reticularis dorsalis and a descending pathway back to the spinal cord.[72,129]

In the medulla, a major ascending axonal projection relaying information about pain passes through and terminates in the reticular formation of the ventrolateral medulla[79,130,131] (see Fig. 1-1). This is part of a direct spinal projection to brain stem regions that contains catecholaminergic neurons, including the C1, A1, A2, A5, A6, and A7 regions of both the medulla and pons.[100] Some of these neurons are located in the dorsolateral pons in the parabrachial and Kolliker-Fuse region (A7). Small injections of retrograde tracer in the parabrachial nucleus confirmed that this nucleus receives projections primarily from lamina I.[132] Lima and Craig and their colleagues have further characterized axonal projections of lamina I–specific morphologic subtypes of nociceptive neurons extending to the parabrachial region.[79,83] The catecholamine cells of the brain stem are involved in

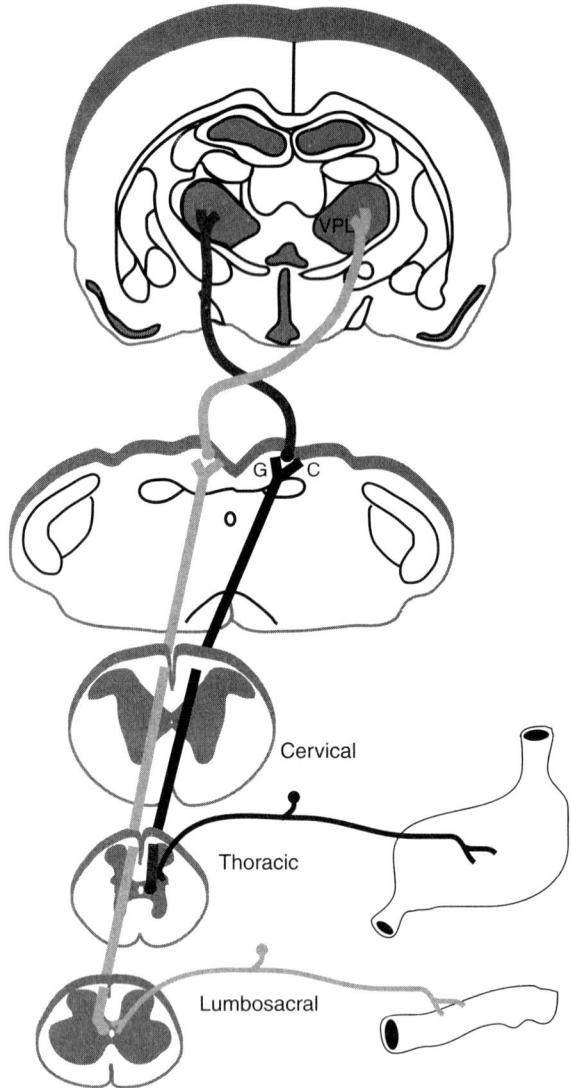

Figure 1-8. Recently, a pathway carrying information about visceral pain was discovered. Visceral afferent nerve input to the visceral processing region near the central canal is relayed by postsynaptic dorsal column neurons through the dorsal column. Postsynaptic dorsal column cells are localized in lamina X throughout the length of the spinal cord but are more prominent in thoracic and lumbosacral cord, which receives the predominant input from visceral structures. The information about visceral pain is relayed by the dorsal column nuclei through the medial lemniscus to the thalamus (VPL and other sites). While the dorsal column traditionally has been thought of as relaying only site-specific information about cutaneous sensory input such as proprioception and mechanical and vibratory sensation, new clinical and basic studies have provided information indicating that a midline dorsal column pathway carries pelvic visceral information. Visceral input arising from the thoracic regions is sent somatotopically as fibers between the gracile and the cuneate fasciculi along the dorsal intermediate septum (see Ref. 193).

diverse functions, including alerting, memory, and learning, and this has been reviewed elsewhere. Descending inhibitory input to the spinal cord limits responses to painful stimulation.[133] Recently, a direct neuronal projection from the central region of the spinal cord to the raphe magnus has been described.[108] Interconnections between the ventromedial medulla and the catecholamine cells of the pons and the descending inhibitory region of the periaqueductal gray matter also have been described.[134–136] Ascending pathways relaying information about pain also terminate in the periaqueductal gray matter and midbrain reticular formation.[108,131,137–139] All these brain stem sites are well-studied sources of input driving descending inhibition of nociceptive neuronal responses (see Fig. 1-1).

Another site of termination for ascending nociceptive pathways sending axons directly from the spinal cord and the dorsal column nuclei is the anterior pretectal nucleus.[139] The anterior pretectal nucleus in the rostral midbrain also has been found to be important as the origin of descending inhibition, with descending input interfacing with the periaqueductal gray matter. Evidence suggests that the pretectal nucleus innervates the catecholaminergic neurons of the parabrachial region.[141] It should be noted that stimulation in sites other than the pretectal nucleus result in both pro- and antinociceptive descending input, whereas stimulation of the anterior pretectal nucleus produces only antinociception.

Spinohypothalamic, Limbic, and Cortical Connections

Pain is often accompanied by motivational-affective responses, including emotional suffering, anxiety, increased attention, increased blood pressure and heart rate, and arousal, as well as changes in endocrine and autonomic responses. The neural structures involved in these integrated responses are innervated by spinal pathways that parallel those relaying information localizing the source of the noxious input on the body surface map and may indeed be collaterals of the spinothalamic tract. Some of the pathways in unique positions to assume such a role would include the spinal pathways relaying information through the solitary nuclei, the ventrolateral medulla, the parabrachial nucleus of the dorsalateral pons, the periaqueductal gray matter, the hypothalamus, and the amygdala (see Fig. 1-1). These same sites are implicated in control of descending inhibition of pain.

Spinohypothalamic and spinoamygdalar pathways have been described that may be involved in affective components of pain transmission through connections with the thalamus and with other parts of the limbic system.[117,142,143] Ascending axonal projections arise primarily from the spinal cord laminae I and X, as well as from the lateral reticulated region of the spinal cord. Axonal terminations of spinal neurons are made in the lateral hypothalamic and central nucleus of the amygdala. Both these regions are also involved in antinociceptive actions, as well as autonomic and affective responses to nociceptive input.

Spinal projections to the ventrolateral medulla innervate regions with axonal projections to the paraventricular nucleus of the hypothalamus,[144] amygdala,[145] and the medial preoptic region.[146] The direct route taken by spinal projections to the amygdala and the spinoparabrachial-amygdalar pathways[147,148] are clearly involved in integrated responses to noxious insults. In higher brain regions, a spinoreticulothalamocortical pathway is proposed as the route taken by nociceptive information to forebrain structures that presumably will be used in evoking affective responses and the subjective interpretation of a painful insult. This route is based on known anatomic projections of the reticular formation to higher brain centers.[149] Relays to cortical structures involved in somatosensory interactions have been described.[150,151] Human and experimental animal functional magnetic resonance imaging (fMRI) results indicate that the anterior cingulate cortex, thalamus, periaqueductal gray matter, raphe, and somatosensory cortex are highly activated in pain states.[152–155]

▶ DESCENDING INHIBITORY PATHWAYS

Descending inhibition of spinal nociceptive processes was described initially in behavioral experiments in which antinociceptive processes were initiated when the periaqueductal gray matter was stimulated electrically.[156] The concept of antinociception as proposed, based on behavioral, electrophysiologic, and morphologic studies, occurs through neuronal circuitry that includes a connection between the periaqueductal gray matter and the nucleus raphe magnus and adjacent reticular formation and then onto the spinal cord[137,157–167,168] (see Fig. 1-1). Subsequent findings suggest, however, that the process is mediated through more complicated brain stem circuitry and pharmacology,[169] including the anterior pretectal nucleus.[139]

The raphe spinal pathway is largely, but not entirely, serotonergic.[170–172] The serotonergic component that originates from the nucleus raphe magnus descends in the dorsolateral funiculus and terminates in the dorsal horn. Thus the serotonergic descending pathway provides the anatomic and pharmacologic criteria established for descending inhibition of pain. There are also SP and inhibitory enkephalinergic connections between the nucleus raphe magnus and the noradrenergic neurons of the dorsolateral pons.[134,136] The mu opiate receptors in the ventrolateral periaqueductal gray matter are key players in the descending inhibitory pathways mechanisms.

Descending noradrenergic input to the spinal cord from the dorsolateral pons also provides direct inhibitory modulation of spinal nociceptive mechanisms. Large populations of spinally-projecting noradrenergic neurons have been identified in the locus ceruleus, the subceruleus, and the Kolliker-Fuse nuclei.[173–175] The A5 cell group and the parabrachial nuclei also send noradrenergic projections to the spinal cord, as do adrenergic neurons of the ventrolateral medulla.[45]

The anterior pretectal nucleus also has been associated with inhibition of nociceptive somatosensory function.[140] The anterior pretectal nucleus receives direct input from laminae I and X in the spinal cord, as well as indirect innervation via input from the dorsal column nucleus.[176] Anterior pretectal nucleus stimulation evokes antinociceptive responses in the tail flick, formalin, and paw pressure tests. Indications are that these effects are mediated through anatomic connectivity with the dorsolateral pontine pathways.

Many hypothalamic nuclei, including the lateral hypothalamic nucleus, also have descending connections with the periaqueductal gray matter.[177–179] The hypothalamic input constitutes the largest input to the periaqueductal gray matter.[179] The input to the periaqueductal gray matter is topographic, aligning particular hypothalamic nuclei with particular regions of the periaqueductal gray matter.

Descending projections arise from several forebrain areas, including the anterior cingulate gyrus, the central nucleus of the amygdala, parts of the medial cerebral cortex (including the infra- and prelimbic cortex and the precentral medial cortex), and lateral portions of the cerebral cortex such as the anterior and posterior insular cortex and the perirhinal cortex.[177,180]

▶ CENTERS FOR HIGHER PROCESSING

While it is clear that the perception of pain requires the activity of higher cortical centers of the brain, the precise regional association has remained a difficult discovery until the advent of positron-emission tomography (PET) and fMRI studies in both animals and humans. These imaging studies have mapped areas of increased blood flow, indicating regions of increased activity, in response to noxious insult.[151,181–187] These studies implicate the thalamus, the SI and SII somatosensory cortex, the anterior cingulate gyrus, the insula, the prefrontal cortex, the lentiform nucleus, and the cerebellum as major players in the higher processing of pain. Many of these same regions are activated after noxious visceral stimulation.[188–190] The thalamus and the SI and SII cortex are known to be somatosensory processing regions, and evidence from imaging studies is consistent with a sensoridiscriminitive role of these structures in specific site localization of the noxious input. The anterior cingulate cortex presumably is involved in interpretation of the emotional significance of the noxious input by the limbic system,[186] whereas the lentiform nucleus and the cerebellum may be involved in the reflexive motor responsiveness to noxious input. The activity in the frontal cortex may contribute to memory and learning of events related to the unpleasant experience for future avoidance,[191] whereas the insular cortex is likely involved in cortical control of autonomic responses to the unpleasant event.

REFERENCES

1. International Association for the Study of Pain. *www.insppain.org*.
2. Hardy JD, Wolff HG, Goodell H: *Pain Sensations and Reactions*. New York: Hafner, 1967.
3. Hanesch U, Heppelmann B, Messlinger K, Schmidt RF: Nociception in normal and arthritic joints: Structural and functional aspects, in Willis WD Jr (ed), *Hyperalgesia and Allodynia*. New York: Raven Press, 1992:81–106.
4. Stacey MJ: Free nerve endings in skeletal muscle of the cat. *J Anat* 1969;105:231–254.
5. Mense S: Nociception from skeletal muscle in relation to clinical muscle pain. *Pain* 1993;54:241–289.
6. Hanesch U, Heppelmann B, Schmidt RF: Neurokinin A-like immunoreactivity in articular afferents of the cat. *Brain Res* 1992;586:332–335.
7. Schaible HG, Grubb BD: Afferent and spinal mechanisms of joint pain. *Pain* 1993;55:5–54.
8. Messlinger K: Functional morphology of nociceptive and other fine sensory endings (free nerve endings) in different tissues. *Prog Brain Res* 1996;113:273–298.
9. Ness TJ, Gebhart GF: Visceral pain: A review of experimental studies. *Pain* 1990;41:167–234.
10. Cervero F: Sensory innervation of the viscera: Peripheral basis of visceral pain. *Physiol Rev* 1994;74:95–138.
11. Perl ER: Cutaneous polymodal receptors: Characteristics and plasticity. *Prog Brain Res* 1996;113:21–37.
12. McCleskey EW: Biophysics of a trespasser, Na^+ block of Ca^{2+} channels. *J Gen Physiol* 1997;109:677–680.
13. Gasser HS: Conduction in nerves in relation to fiber types. *Proc Assoc Res Ment Dis* 1934;15:35–59.
14. Boivie J, Perl ER: Neural substrates of somatic sensation, in *MTP International Review of Science*. Baltimore: University Park Press, 1975:303–411.
15. Hulsebosch CE, Coggeshall RE: Quantitation of sprouting of dorsal root axons. *Science* 1981;213:1020–1021.
16. Langford LA, Schmidt RF: Afferent and efferent axons in the medial and posterior articular nerves of the cat. *Anat Rec* 1983;206:71–78.
17. McMahon S, Koltzenburg M: The changing role of primary afferent neurones in pain. *Pain* 1990;43:269–272.
18. McMahon SB, Koltzenburg M: Novel classes of nociceptors: Beyond Sherrington. *Trends Neurosci* 1990;13:199–201.
19. Treede RD, Davis KD, Campbell JN, Raja SN: The plasticity of cutaneous hyperalgesia during sympathetic ganglion blockade in patients with neuropathic pain. *Brain* 1992;115:607–621.
20. Reeh PW, Bayer J, Kocher L, Handwerker HO: Sensitization of nociceptive cutaneous nerve fibers from the rat's tail by noxious mechanical stimulation. *Exp Brain Res* 1987;65:505–512.
21. Sengupta JN, Gebhart GF: Characterization of mechanosensitive pelvic nerve afferent fibers innervating the colon of the rat. *J Neurophysiol* 1994;71:2046–2060.
22. Koltzenburg M: Stability and plasticity of nociceptor function and their relationship to provoked and ongoing pain. *Semin Neurosci* 1995;7:199–210.
23. Dmitrieva N, McMahon SB: Sensitisation of visceral afferents by nerve growth factor in the adult rat. *Pain* 1996;66:87–97.
24. Sugiura Y, Terui N, Hosoya Y, et al: Quantitative analysis of central terminal projections of visceral and somatic unmyelinated (C) primary afferent fibers in the guinea pig. *J Comp Neurol* 1993;332:315–325.
25. Light AR, Perl ER: Spinal termination of functionally identified primary afferent neurons with slowly conducting myelinated fibers. *J Comp Neurol* 1979;186:133–150.
26. Xu XJ, Dalsgaard CJ, Wiesenfeld-Hallin Z: Spinal substance P and N-methyl-D-aspartate receptors are coactivated in the induction of central sensitization of the nociceptive flexor reflex. *Neuroscience* 1992;51:641–648.
27. Yaksh TL, Hua XY, Kalcheva I, et al: The spinal biology in humans and animals of pain states generated by persistent small afferent input. *Proc Natl Acad Sci USA* 1999;96:7680–7686.
28. Willis WD, Coggeshall RE: *Sensory Mechanisms of the Spinal Cord*. New York: Plenum Press, 1991.
29. Lawson SN: Peptides and cutaneous polymodal nociceptor neurones. *Prog Brain Res* 1996;113:369–385.
30. Chung K, Lee WT, Carlton SM: The effects of dorsal rhizotomy and spinal cord isolation on calcitonin gene–related peptide–labeled terminals in the rat lumbar dorsal horn. *Neurosci Lett* 1988;90:27–32.
31. Lawson SN, Crepps BA, Perl ER: Relationship of substance P to afferent characteristics of dorsal root ganglion neurones in guinea pig. *J Physiol* 1997;505:177–191.
32. Lynn B: Substance P and nociceptive afferent neurones. *J Physiol* 1997;505:1.
33. Ruocco I, Cuello AC, Shigemoto R, Ribeiro-da-Silva A: Light and electron microscopic study of the distribution of substance P–immunoreactive fibers and neurokinin-1 receptors in the skin of the rat lower lip. *J Comp Neurol* 2001;432:466–480.
34. Westlund KN, McNeill DL, Patterson JT, Coggeshall RE: Aspartate immunoreactive axons in normal rat L4 dorsal roots. *Brain Res* 1989;489:347–351.
35. Westlund KN, McNeill DL, Coggeshall RE: Glutamate immunoreactivity in rat dorsal root axons. *Neurosci Lett* 1989;96:13–17.
36. Tracey DJ, De Biasi S, Phend K, Rustioni A: Aspartate-like immunoreactivity in primary afferent neurons. *Neuroscience* 1991;40:673–686.
37. Valtschanoff JG, Phend KD, Bernardi PS, et al: Amino acid immunocytochemistry of primary afferent terminals in the rat dorsal horn. *J Comp Neurol* 1994;346:237–252.

38. Larsson M, Persson S, Ottersen OP, Broman J: Quantitative analysis of immunogold labeling indicates low levels and nonvesicular localization of L-aspartate in rat primary afferent terminals. *J Comp Neurol* 2001;430:147–159.
39. Geppetti P, Holzer P: *Neurogenic Inflammation.* Boca Raton, FL: CRC Press, 1996.
40. Stein C, Schafer M, Machelska H: Attacking pain at its source: New perspectives on opioids. *Nature Med* 2003; 9:1003–1008.
41. Caterina MJ, Julius D: Sense and specificity: A molecular identity for nociceptors. *Curr Opin Neurobiol* 1999; 9:525–530.
42. Akopian AN, Souslova V, England S, et al: The tetrodotoxin-resistant sodium channel SNS has a specialized function in pain pathways. *Nature Neurosci* 1999;2:541–548.
43. Reeh PW, Kress M: Molecular physiology of proton transduction in nociceptors. *Curr Opin Pharmacol* 2001;1:45–51.
44. Sato K, Kiyama H, Park HT, Tohyama M: AMPA, KA, and NMDA receptors are expressed in the rat DRG neurones. *Neuroreport* 1993;4:1263–1265.
45. Carlton SM, Hargett GL, Coggeshall RE: Localization and activation of glutamate receptors in unmyelinated axons of rat glabrous skin. *Neurosci Lett* 1995;197:25–28.
46. Coggeshall RE, Carlton SM: Receptor localization in the mammalian dorsal horn and primary afferent neurons. *Brain Res Brain Res Rev* 1997;24:28–66.
47. Morgan C, Nadelhaft I, de Groat WC: The distribution of visceral primary afferents from the pelvic nerve to Lissauer's tract and the spinal gray matter and its relationship to the sacral parasympathetic nucleus. *J Comp Neurol* 1981; 201:415–440.
48. Sugiura Y, Terui N, Hosoya Y: Difference in distribution of central terminals between visceral and somatic unmyelinated (C) primary afferent fibers. *J Neurophysiol* 1989;62:834–840.
49. Sugiura Y: Spinal organization of C-fiber afferents related with nociception or nonnociception. *Prog Brain Res* 1996; 113:319–339.
50. Rexed B: The cytoarchitectonic organization of the spinal cord in the rat. *J Comp Neurol* 1952;96:415–466.
51. Paxinos G: *The Rat Nervous System.* San Diego: Academic Press, 1995.
52. Willis WD, Westlund KN: Neuroanatomy of the pain system and of the pathways that modulate pain. *J Clin Neurophysiol* 1997;14:2–31.
53. Cajal, SR: *Histologie du systeme nerveux de l'homme et des vertébrés.* Paris: Maloine, 1909.
54. Gobel S: Neural circuitry in the substantia gelatinosa of Rolando: Anatomical insights. *Adv Res Pain Ther* 1979; 3:175–195.
55. Spike RC, Todd AJ: Ultrastructural and immunocytochemical study of lamina II islet cells in rat spinal dorsal horn. *J Comp Neurol* 1992;323:359–369.
56. Hirshberg RM, Al Chaer ED, Lawand NB, et al: Is there a pathway in the posterior funiculus that signals visceral pain? *Pain* 1996;67:291–305.
57. Wang CC, Lu Y, Willis WD, Westlund KN: A new visceral nociceptive pathway in the dorsal column: A PHA-L study of ascending projections from the area around the central canal of the T7 or S1 spinal cord in rats. *Neurosci Abstr* 1997;23:2349.
58. Willis WD, Al Chaer ED, Quast MJ, Westlund KN: A visceral pain pathway in the dorsal column of the spinal cord. *Proc Natl Acad Sci USA* 1999;96:7675–7679.
59. Ribeiro-da-Silva A, Coimbra A: Two types of synaptic glomeruli and their distribution in laminae I–III of the rat spinal cord. *J Comp Neurol* 1982;209:176–186.
60. Maxwell DJ, Christie WM, Short AD, et al: Central boutons of glomeruli in the spinal cord of the cat are enriched with L-glutamate-like immunoreactivity. *Neuroscience* 1990; 36:83–104.
61. Ye Z, Wimalawansa SJ, Westlund KN: Receptor for calcitonin gene–related peptide: Localization in the dorsal and ventral spinal cord. *Neuroscience* 1999;92:1389–1397.
62. Murase K, Ryu PD, Randic M: Excitatory and inhibitory amino acids and peptide-induced responses in acutely isolated rat spinal dorsal horn neurons. *Neurosci Lett* 1989; 103:56–63.
63. Dougherty PM, Willis WD: Enhancement of spinothalamic neuron responses to chemical and mechanical stimuli following combined microiontophoretic application of N-methyl-D-aspartic acid and substance P. *Pain* 1991;47:85–93.
64. Magnusson KR, Larson AA, Madl JE, et al: Co-localization of fixative-modified glutamate and glutaminase in neurons of the spinal trigeminal nucleus of the rat: An immunohistochemical and immunoradiochemical analysis. *J Comp Neurol* 1986;247:477–490.
65. Chung K, Kevetter GA, Willis WD, Coggeshall RE: An estimate of the ratio of propriospinal to long tract neurons in the sacral spinal cord of the rat. *Neurosci Lett* 1984;44:173–177.
66. Carlton SM, LaMotte CC, Honda CN, et al: Ultrastructural analysis of substance P and other synaptic profiles innervating an identified primate spinothalamic tract neuron. *Neurosci Abstr* 1985;11:578.
67. Mantyh PW, DeMaster E, Malhotra A, et al: Receptor endocytosis and dendrite reshaping in spinal neurons after somatosensory stimulation. *Science* 1995;268:1629–1632.
68. Brown JL, Liu H, Maggio JE, et al: Morphological characterization of substance P receptor-immunoreactive neurons in the rat spinal cord and trigeminal nucleus caudalis. *J Comp Neurol* 1995;356:327–344.
69. Hayes RL, Katayama Y, Watkins LR, Becker DP: Bilateral lesions of the dorsolateral funiculus of the cat spinal cord: Effects on basal nociceptive reflexes and nociceptive suppression produced by cholinergic activation of the pontine parabrachial region. *Brain Res* 1984;311:267–280.
70. Bannon AW, Decker MW, Holladay MW, et al: Broad-spectrum, nonopioid analgesic activity by selective modulation of neuronal nicotinic acetylcholine receptors. *Science* 1998;279:77–81.
71. Ribeiro-da-Silva A, Cuello AC: Choline acetyltransferase–immunoreactive profiles are presynaptic to primary sensory fibers in the rat superficial dorsal horn. *J Comp Neurol* 1990;295:370–384.
72. Villanueva L, Bouhassira D, Le Bars D: The medullary subnucleus reticularis dorsalis (SRD) as a key link in both the transmission and modulation of pain signals. *Pain* 1996; 67:231–240.
73. Ralston HJ III: Dorsal root projections to dorsal horn neurons in the cat spinal cord. *J Comp Neurol* 1968; 132:303–330.

74. Hayes ES, Carlton SM: Primary afferent interactions: Analysis of calcitonin gene–related peptide–immunoreactive terminals in contact with unlabeled and GABA-immunoreactive profiles in the monkey dorsal horn. *Neuroscience* 1992;47:873–896.
75. Melzack R, Wall PD: Pain mechanisms. *Science* 1965;150:971–979.
76. Lima D, Coimbra A: A Golgi study of the neuronal population of the marginal zone (lamina I) of the rat spinal cord. *J Comp Neurol* 1986;244:53–71.
77. Lima D, Coimbra A: Morphological types of spinomesencephalic neurons in the marginal zone (lamina I) of the rat spinal cord, as shown after retrograde labelling with cholera toxin subunit B. *J Comp Neurol* 1989;279:327–339.
78. Lima D, Coimbra A: Structural types of marginal (lamina I) neurons projecting to the dorsal reticular nucleus of the medulla oblongata. *Neuroscience* 1990;34:591–606.
79. Lima D, Mendes-Ribeiro JA, Coimbra A: The spino-latereticular system of the rat: Projections from the superficial dorsal horn and structural characterization of marginal neurons involved. *Neuroscience* 1991;45:137–152.
80. Esteves F, Lima D, Coimbra A: Structural types of spinal cord marginal (lamina I) neurons projecting to the nucleus of the tractus solitarius in the rat. *Somatosens Mot Res* 1993;10:203–216.
81. Lima D, Avelino A, Coimbra A: Morphological characterization of marginal (lamina I) neurons immunoreactive for substance P, enkephalin, dynorphin and gamma-aminobutyric acid in the rat spinal cord. *J Chem Neuroanat* 1993;6:43–52.
82. Zhang ET, Han ZS, Craig AD: Morphological classes of spinothalamic lamina I neurons in the cat. *J Comp Neurol* 1996;367:537–549.
83. Zhang ET, Craig AD: Morphology and distribution of spinothalamic lamina I neurons in the monkey. *J Neurosci* 1997;17:3274–3284.
84. Yu XH, Zhang ET, Shigemoto R, et al: NK-1 receptor immunostaining of different morphological types of lamina I neurons in primate spinal cord. *Neurosci Abstr* 1998;24:390.
85. Willis WD, Kenshalo DR Jr, Leonard RB: The cells of origin of the primate spinothalamic tract. *J Comp Neurol* 1979;188:543–573.
86. Surmeier DJ, Honda CN, Willis WD Jr: Natural groupings of primate spinothalamic neurons based on cutaneous stimulation: Physiological and anatomical features. *J Neurophysiol* 1988;59:833–860.
87. Westlund KN, Carlton SM, Zhang D, Willis WD: Glutamate-immunoreactive terminals synapse on primate spinothalamic tract cells. *J Comp Neurol* 1992;322:519–527.
88. Coffield JA, Miletic V: Immunoreactive enkephalin is contained within some trigeminal and spinal neurons projecting to the rat medial thalamus. *Brain Res* 1987;425:380–383.
89. Nahin RL: Immunocytochemical identification of long ascending, peptidergic lumbar spinal neurons terminating in either the medial or lateral thalamus in the rat. *Brain Res* 1988;443:345–349.
90. Ju G, Melander T, Ceccatelli S, et al: Immunohistochemical evidence for a spinothalamic pathway co-containing cholecystokinin- and galanin-like immunoreactivities in the rat. *Neuroscience* 1987;20:439–456.
91. Leah J, Menetrey D, de Pommery J: Neuropeptides in long ascending spinal tract cells in the rat: Evidence for parallel processing of ascending information. *Neuroscience* 1988;24:195–207.
92. LaMotte CC, Carlton SM, Honda CN, et al: Innervation of identified primate spinothalamic tract neurons: Ultrastructure of serotonergic and other synaptic profiles. *Neurosci Abstr* 1988;14:852.
93. Carlton SM, Westlund KN, Zhang DX, et al: Calcitonin gene–related peptide containing primary afferent fibers synapse on primate spinothalamic tract cells. *Neurosci Lett* 1990;109:76–81.
94. Westlund KN, Carlton SM, Zhang D, Willis WD: Direct catecholaminergic innervation of primate spinothalamic tract neurons. *J Comp Neurol* 1990;299:178–186.
95. Westlund KN, Sorkin LS, Ferrington DG, et al: Serotoninergic and noradrenergic projections to the ventral posterolateral nucleus of the monkey thalamus. *J Comp Neurol* 1990;295:197–207.
96. Carlton SM, Westlund KN, Zhang D, Willis WD: GABA-immunoreactive terminals synapse on primate spinothalamic tract cells. *J Comp Neurol* 1992;322:528–537.
97. Lekan HA, Carlton SM: Glutamatergic and GABAergic input to rat spinothalamic tract cells in the superficial dorsal horn. *J Comp Neurol* 1995;361:417–428.
98. Ye Z, Westlund KN: Glycine-like immunoreactive terminals contacting spinothalamic tract neurons in rat spinal cord. *Neurosci Abstr* 1998;24:387.
99. Abbadie C, Trafton J, Liu H, et al: Inflammation increases the distribution of dorsal horn neurons that internalize the neurokinin-1 receptor in response to noxious and non-noxious stimulation. *J Neurosci* 1997;17:8049–8060.
100. Westlund KN, Craig AD: Association of spinal lamina I projections with brain stem catecholamine neurons in the monkey. *Exp Brain Res* 1996;110:151–162.
101. Ye Z, Westlund KN: Ultrastructural localization of glutamate receptor subunits (NMDAR1, AMPA GluR1 and GluR2/3) and spinothalamic tract cells. *Neuroreport* 1996;7:2581–2585.
102. Hayes NL, Rustioni A: Spinothalamic and spinomedullary neurons in macaques: A single and double retrograde tracer study. *Neuroscience* 1980;5:861–874.
103. Granum SL: The spinothalamic system of the rat: I. Locations of cells of origin. *J Comp Neurol* 1986;247:159–180.
104. Apkarian AV, Hodge CJ: Primate spinothalamic pathways: I. A quantitative study of the cells of origin of the spinothalamic pathway. *J Comp Neurol* 1989;288:447–473.
105. Burstein R, Dado RJ, Giesler GJ Jr: The cells of origin of the spinothalamic tract of the rat: A quantitative reexamination. *Brain Res* 1990;511:329–337.
106. Applebaum AE, Beall JE, Foreman RD, Willis WD: Organization and receptive fields of primate spinothalamic tract neurons. *J Neurophysiol* 1975;38:572–586.
107. Craig AD, Bushnell MC, Zhang ET, Blomqvist A: A thalamic nucleus specific for pain and temperature sensation. *Nature* 1994;372:770–773.
108. Craig AD: Distribution of brainstem projections from spinal lamina I neurons in the cat and the monkey. *J Comp Neurol* 1995;361:225–248.

109. Wang CC, Willis WD, Westlund KN: Ascending projections from the area around the spinal cord central canal: A *Phaseolus vulgaris* leukoagglutinin study in rats. *J Comp Neurol* 1999;415:341–367.
110. Guilbaud G, Peschanski M, Gautron M, Binder D: Neurones responding to noxious stimulation in Vβ complex and caudal adjacent regions in the thalamus of the rat. *Pain* 1980;8:303–318.
111. Peschanski M, Guilbaud G, Gautron M: Posterior intralaminar region in rat: Neuronal responses to noxious and nonnoxious cutaneous stimuli. *Exp Neurol* 1981;72:226–238.
112. Peschanski M, Mantyh PW, Besson JM: Spinal afferents to the ventrobasal thalamic complex in the rat: An anatomical study using wheat-germ agglutinin conjugated to horseradish peroxidase. *Brain Res* 1983;278:240–244.
113. Ma W, Peschanski M, Ralston HJ III: Fine structure of the spinothalamic projections to the central lateral nucleus of the rat thalamus. *Brain Res* 1987;414:187–191.
114. Ledoux JE, Ruggiero DA, Forest R, et al: Topographic organization of convergent projections to the thalamus from the inferior colliculus and spinal cord in the rat. *J Comp Neurol* 1987;264:123–146.
115. Dado RJ, Giesler GJ Jr: Afferent input to nucleus submedius in rats: Retrograde labeling of neurons in the spinal cord and caudal medulla. *J Neurosci* 1990;10:2672–2686.
116. Craig AD Jr., Burton H: Spinal and medullary lamina I projection to nucleus submedius in medial thalamus: A possible pain center. *J Neurophysiol* 1981;45:443–466.
117. Cliffer KD, Burstein R, Giesler GJ Jr: Distributions of spinothalamic, spinohypothalamic, and spinotelencephalic fibers revealed by anterograde transport of PHA-L in rats. *J Neurosci* 1991;11:852–868.
118. Bennett GJ, Seltzer Z, Lu GW, et al: The cells of origin of the dorsal column postsynaptic projection in the lumbosacral enlargements of cats and monkeys. *Somatosens Res* 1983;1:131–149.
119. Giesler GJ Jr., Cliffer KD: Postsynaptic dorsal column pathway of the rat: II. Evidence against an important role in nociception. *Brain Res* 1985;326:347–356.
120. Al-Chaer ED, Lawand NB, Westlund KN, Willis WD: Visceral nociceptive input into the ventral posterolateral nucleus of the thalamus: A new function for the dorsal column pathway. *J Neurophysiol* 1996;76:2661–2674.
121. Al-Chaer ED, Westlund KN, Willis WD: Potentiation of thalamic responses to colorectal distension by visceral inflammation. *Neuroreport* 1996;7:1635–1639.
122. Houghton AK, Wang CC, Westlund KN: Do nociceptive signals from the pancreas travel in the dorsal column? *Pain* 2001;89:207–220.
123. Lu GW, Willis WD: Branching and/or collateral projections of spinal dorsal horn neurons. *Brain Res Brain Res Rev* 1999;29:50–82.
124. Kevetter GA, Willis WD: Collaterals of spinothalamic cells in the rat. *J Comp Neurol* 1983;215:453–464.
125. Liu RP: Spinal neuronal collaterals to the intralaminar thalamic nuclei and periaqueductal gray. *Brain Res* 1986;365:145–150.
126. Harmann PA, Carlton SM, Willis WD: Collaterals of spinothalamic tract cells to the periaqueductal gray: A fluorescent double-labeling study in the rat. *Brain Res* 1988;441:87–97.
127. Hylden JL, Anton F, Nahin RL: Spinal lamina I projection neurons in the rat: Collateral innervation of parabrachial area and thalamus. *Neuroscience* 1989;28:27–37.
128. Raboisson P, Dallel R, Bernard JF, et al: Organization of efferent projections from the spinal cervical enlargement to the medullary subnucleus reticularis dorsalis and the adjacent cuneate nucleus: A PHA-L study in the rat. *J Comp Neurol* 1996;367:503–517.
129. Villanueva L, Bouhassira D, Bing Z, Le Bars D: Convergence of heterotopic nociceptive information onto subnucleus reticularis dorsalis neurons in the rat medulla. *J Neurophysiol* 1988;60:980–1009.
130. Menetrey D, Roudier F, Besson JM: Spinal neurons reaching the lateral reticular nucleus as studied in the rat by retrograde transport of horseradish peroxidase. *J Comp Neurol* 1983;220:439–452.
131. Chaouch A, Menetrey D, Binder D, Besson JM: Neurons at the origin of the medial component of the bulbopontine spinoreticular tract in the rat: An anatomical study using horseradish peroxidase retrograde transport. *J Comp Neurol* 1983;214:309–320.
132. Cechetto DF, Standaert DG, Saper CB: Spinal and trigeminal dorsal horn projections to the parabrachial nucleus in the rat. *J Comp Neurol* 1985;240:153–160.
133. Bernard JF, Huang GF, Besson JM: The parabrachial area: Electrophysiological evidence for an involvement in visceral nociceptive processes. *J Neurophysiol* 1994;71:1646–1660.
134. Yeomans DC, Proudfit HK: Projections of substance P–immunoreactive neurons located in the ventromedial medulla to the A7 noradrenergic nucleus of the rat demonstrated using retrograde tracing combined with immunocytochemistry. *Brain Res* 1990;532:329–332.
135. Clark FM, Proudfit HK: The projections of noradrenergic neurons in the A5 catecholamine cell group to the spinal cord in the rat: Anatomical evidence that A5 neurons modulate nociception. *Brain Res* 1993;616:200–210.
136. Holden JE, Proudfit HK: Enkephalin neurons that project to the A7 catecholamine cell group are located in nuclei that modulate nociception: Ventromedial medulla. *Neuroscience* 1998;83:929–947.
137. Basbaum AI, Fields HL: Endogenous pain control systems: Brain stem spinal pathways and endorphin circuitry. *Annu Rev Neurosci* 1984;7:309–338.
138. Zhang DX, Carlton SM, Sorkin LS, Willis WD: Collaterals of primate spinothalamic tract neurons to the periaqueductal gray. *J Comp Neurol* 1990;296:277–290.
139. Keay KA, Feil K, Gordon BD, et al: Spinal afferents to functionally distinct periaqueductal gray columns in the rat: An anterograde and retrograde tracing study. *J Comp Neurol* 1997;385:207–229.
140. Rees H, Roberts MH: The anterior pretectal nucleus: A proposed role in sensory processing. *Pain* 1993;53:121–135.
141. Christensen M, Willis WD, Westlund KN: Anterior pretectal nucleus projections to pontine catecholamine cell regions. *Neurosci Abstr* 1997;23:157.
142. Burstein R, Potrebic S: Retrograde labeling of neurons in the spinal cord that project directly to the amygdala or the orbital cortex in the rat. *J Comp Neurol* 1993;335:469–485.

143. Newman HM, Stevens RT, Apkarian AV: Direct spinal projections to limbic and striatal areas: Anterograde transport studies from the upper cervical spinal cord and the cervical enlargement in squirrel monkey and rat. *J Comp Neurol* 1996;365:640–658.
144. Sawchenko PE, Swanson LW: The organization of noradrenergic pathways from the brainstem to the paraventricular and supraoptic nuclei in the rat. *Brain Res* 1982;257:275–325.
145. Petrov T, Krukoff TL, Jhamandas JH: Branching projections of catecholaminergic brainstem neurons to the paraventricular hypothalamic nucleus and the central nucleus of the amygdala in the rat. *Brain Res* 1993;609:81–92.
146. Saper CB, Levisohn D: Afferent connections of the median preoptic nucleus in the rat: Anatomical evidence for a cardiovascular integrative mechanism in the anteroventral third ventricular (AV3V) region. *Brain Res* 1983;288:21–31.
147. Fallon JH, Koziell DA, Moore RY: Catecholamine innervation of the basal forebrain: II. Amygdala, suprarhinal cortex and entorhinal cortex. *J Comp Neurol* 1978;180:509–532.
148. Bernard JF, Dallel R, Raboisson P, et al: Organization of the efferent projections from the spinal cervical enlargement to the parabrachial area and periaqueductal gray: A PHA-L study in the rat. *J Comp Neurol* 1995;353:480–505.
149. Nauta HJ, Kuypers HG: Some ascending pathways in the brain stem reticular formation, in Jasper HH, Proctor LD, Knighton RS, et al (eds), *Reticular Formation of the Brain*. Boston: Little Brown, 1958:3–30.
150. Saporta S, Kruger L: The organization of thalamocortical relay neurons in the rat ventrobasal complex studied by the retrograde transport of horseradish peroxidase. *J Comp Neurol* 1977;174:187–208.
151. Svensson P, Minoshima S, Beydoun A, et al: Cerebral processing of acute skin and muscle pain in humans. *J Neurophysiol* 1997;78:450–460.
152. Peyron R, Laurent B, Garcia-Larrea L: Functional imaging of brain responses to pain: A review and meta-analysis. *Neurophysiol Clin* 2000;30:263–288.
153. Bentley DE, Youell PD, Jones AK: Anatomical localization and intra-subject reproducibility of laser evoked potential source in cingulate cortex, using a realistic head model. *Clin Neurophysiol* 2002;113:1351–1356.
154. Gracely RH, Petzke F, Wolf JM, Clauw DJ: Functional magnetic resonance imaging evidence of augmented pain processing in fibromyalgia. *Arthritis Rheum* 2002;46:1333–1343.
155. Borsook D, Becerra L: Pain imaging: Future applications to integrative clinical and basic neurobiology. *Adv Drug Deliv Rev* 2003;55:967–986.
156. Reynolds DV: Surgery in the rat during electrical analgesia induced by focal brain stimulation. *Science* 1969;164:444–445.
157. Mayer DJ, Wolfle TL, Akil H, et al: Analgesia from electrical stimulation in the brainstem of the rat. *Science* 1971;174:1351–1354.
158. Liebeskind JC, Guilbaud G, Besson JM, Oliveras JL: Analgesia from electrical stimulation of the periaqueductal gray matter in the cat: Behavioral observations and inhibitory effects on spinal cord interneurons. *Brain Res* 1973;50:441–446.
159. Oliveras JL, Besson JM, Guilbaud G, Liebeskind JC: Behavioral and electrophysiological evidence of pain inhibition from midbrain stimulation in the cat. *Exp Brain Res* 1974;20:32–44.
160. Oliveras JL, Woda A, Guilbaud G, Besson JM: Inhibition of the jaw-opening reflex by electrical stimulation of the periaqueductal gray matter in the awake, unrestrained cat. *Brain Res* 1974;72:328–331.
161. Oliveras JL, Redjemi F, Guilbaud G, Besson JM: Analgesia induced by electrical stimulation of the inferior centralis nucleus of the raphe in the cat. *Pain* 1975;1:139–145.
162. Proudfit HK, Anderson EG: Morphine analgesia: Blockade by raphe magnus lesions. *Brain Res* 1975;98:612–618.
163. Basbaum AI, Clanton CH, Fields HL: Opiate and stimulus-produced analgesia: Functional anatomy of a medullospinal pathway. *Proc Natl Acad Sci USA* 1976;73:4685–4688.
164. Beall JE, Martin RF, Applebaum AE, Willis WD: Inhibition of primate spinothalamic tract neurons by stimulation in the region of the nucleus raphe magnus. *Brain Res* 1976;114:328–333.
165. Basbaum AI, Clanton CH, Fields HL: Three bulbospinal pathways from the rostral medulla of the cat: an autoradiographic study of pain modulating systems. *J Comp Neurol* 1978;178:209–224.
166. Basbaum AI, Fields HL: Endogenous pain control mechanisms: Review and hypothesis. *Ann Neurol* 1978;4:451–462.
167. Besson JM, Chaouch A: Peripheral and spinal mechanisms of nociception. *Physiol Rev* 1987;67:67–186.
168. Willis WD, Haber LH, Martin RF: Inhibition of spinothalamic tract cells and interneurons by brain stem stimulation in the monkey. *J Neurophysiol* 1977;40:968–981.
169. Bajic D, Van Bockstaele EJ, Proudfit HK: Ultrastructural analysis of ventrolateral periaqueductal gray projections to the A7 catecholamine cell group. *Neuroscience* 2001;104:181–197.
170. Oliveras JL, Guilbaud G, Besson JM: A map of serotoninergic structures involved in stimulation producing analgesia in unrestrained freely moving cats. *Brain Res* 1979;164:317–322.
171. Bowker RM, Steinbusch HW, Coulter JD: Serotonergic and peptidergic projections to the spinal cord demonstrated by a combined retrograde HRP histochemical and immunocytochemical staining method. *Brain Res* 1981;211:412–417.
172. Bowker RM, Westlund KN, Coulter JD: Origins of serotonergic projections to the lumbar spinal cord in the monkey using a combined retrograde transport and immunocytochemical technique. *Brain Res Bull* 1982;9:271–278.
173. Stevens RT, Hodge CJ, Jr., Apkarian AV: Kolliker-Fuse nucleus: The principal source of pontine catecholaminergic cells projecting to the lumbar spinal cord of cat. *Brain Res* 1982;239:589–594.
174. Westlund KN, Bowker RM, Ziegler MG, Coulter JD: Origins and terminations of descending noradrenergic projections to the spinal cord of monkey. *Brain Res* 1984;292:1–16.
175. Clark FM, Proudfit HK: Projections of neurons in the ventromedial medulla to pontine catecholamine cell groups involved in the modulation of nociception. *Brain Res* 1991;540:105–115.

176. Ossipov MH, Lai J, Malan TP Jr, Porreca F: Spinal and supraspinal mechanisms of neuropathic pain. *Ann NY Acad Sci* 2000;909:12–24.
177. Beitz AJ: The organization of afferent projections to the midbrain periaqueductal gray of the rat. *Neuroscience* 1982; 7:133–159.
178. Meller ST, Dennis BJ: Afferent projections to the periaqueductal gray in the rabbit. *Neuroscience* 1986;19:927–964.
179. Venning J, Buma P, Ter Horst GJ, et al: Hypopthalamic projections to the PAG in the rat: Topographical, immunoelectromiscroscopical and functional aspects, in Depaulis A, Bandler R (eds): *The Midbrain Periaqueductal Gray Matter: Functional, Anatomical, and Neurochemical Organization.* New York: Plenum Press; 1991:387–415.
180. Shipley MT, Ennis M, Rizvi TA, Behbehani MM: Topographical specificity of forebrain inputs to the midbrain periaqueductal gray: Evidence for discrete longitudinally organized input columns, in Depaulis A, Bandler R (eds): *The Midbrain Periaqueductal Gray Matter: Functional, Anatomical, and Neurochemical Organization.* New York: Plenum Press; 1991:417–448.
181. Jones AK, Brown WD, Friston KJ, et al: Cortical and subcortical localization of response to pain in man using positron emission tomography. *Proc R Soc Lond [B]* 1991; 244:39–44.
182. Casey KL, Minoshima S, Berger KL, et al: Positron emission tomographic analysis of cerebral structures activated specifically by repetitive noxious heat stimuli. *J Neurophysiol* 1994;71:802–807.
183. Coghill RC, Talbot JD, Evans AC, et al: Distributed processing of pain and vibration by the human brain. *J Neurosci* 1994;14:4095–4108.
184. Casey KL, Minoshima S, Morrow TJ, Koeppe RA: Comparison of human cerebral activation pattern during cutaneous warmth, heat pain, and deep cold pain. *J Neurophysiol* 1996;76:571–581.
185. Davis KD, Taylor SJ, Crawley AP, et al: Functional MRI of pain- and attention-related activations in the human cingulate cortex. *J Neurophysiol* 1997;77:3370–3380.
186. Rainville P, Duncan GH, Price DD, et al: Pain affect encoded in human anterior cingulate but not somatosensory cortex. *Science* 1997;277:968–971.
187. Bingel U, Quante M, Knab R, et al: Single trial fMRI reveals significant contralateral bias in responses to laser pain within thalamus and somatosensory cortices. *Neuroimage* 2003;18:740–748.
188. Silverman DH, Munakata JA, Ennes H, et al: Regional cerebral activity in normal and pathological perception of visceral pain. *Gastroenterology* 1997;112:64–72.
189. Drossman DA: Review article: An integrated approach to the irritable bowel syndrome. *Aliment Pharmacol Ther* 1999;13:3–14.
190. Berman S, Munakata J, Naliboff BD, et al: Gender differences in regional brain response to visceral pressure in IBS patients. *Eur J Pain* 2000;4:157–172.
191. Friedman DP, Murray EA, O'Neill JB, Mishkin M: Cortical connections of the somatosensory fields of the lateral sulcus of macaques: Evidence for a corticolimbic pathway for touch. *J Comp Neurol* 1986;252:323–347.
192. Jeanmonod D, Sindou M: Somatosensory function following dorsal root entry zone lesions in patients with neurogenic pain or spasticity. *J Neurosurg* 1991;74:916–932.
193. Westlund KN: Visceral nociception. *Curr Rev Pain* 2000; 4:478–487.

CHAPTER 2

Nociceptors and Pain: Correlation of Electrophysiology in Animals to Pain Perception in Humans

Jacob P. Schwarz and Richard A. Meyer

The peripheral nervous system plays a vital role in protecting the body from injury. The study of primary afferent fibers reveals that certain stimuli elicit predictable electrophysiologic responses. If these responses are meaningful for the perception of pain, they must correlate with the experience of pain in humans. In this chapter, we review the stimulus-response properties of nociceptive fibers and compare their response patterns to the perception of pain among human subjects.

Nociceptors are receptors that respond to noxious or injurious stimuli. Nociceptors typically are classified according to unique characteristics, including conduction velocity, the stimulus modality they respond to, the unique way in which they respond, and the ion channels they express in their membranes. These characteristics make it possible to relate electrophysiologic and psychophysical responses. If a heat stimulus causes a predictable, reproducible perception, the nociceptor that is firing in response to that stimulus may be responsible for that perception. Using this approach, we consider the behaviors of unmyelinated C fibers and myelinated A fibers for their potential role in the perception of pain.

▶ NOCICEPTORS AND PAIN TO CONTROLLED HEAT STIMULI

C Fibers and Painful Heat Stimuli

Near pain threshold, two types of fibers respond to heat stimuli delivered to the glabrous skin in humans. Warm fibers are unmyelinated afferents that respond to gentle warming of the skin.[1] Electrophysiologic studies show that warm fibers have spontaneous activity at normal skin temperatures and exhibit an increase in response to slight warming of the skin. Their response reaches a peak for temperatures near 45°C and then decreases for temperatures above 45°C. Since the threshold for pain is around 45°C, the role of warm fibers for encoding warmth, but not heat pain, becomes evident. The encoding of pain more likely would fall to a fiber whose rate of response increased steadily through the temperature range that humans report to be increasingly painful. C mechanoheat-sensitive fibers (CMHs) fit this description. Humans report that thermal stimuli become steadily hotter and then painful through the range of 40 to 50°C, with the threshold for pain occurring at approximately 45°C. While both CMHs and warm fibers are active between 40 and 45°C, warm fibers are more responsive from 40 to 45°C but yield to CMHs that start responding as the temperature approaches 45°C and then increase their rate of firing from 45 to 50°C. The stimulus-response function of CMHs over the temperature range of 41 to 49°C matches exactly the pain ratings of human subjects over this temperature range.[2]

Several other observations collected from direct comparisons of human judgments of pain to heat stimuli between 41 and 49°C and the activity of CMHs also support the hypothesis that CMHs play an important role in pain sensation.[2,3] The first observation involves *stimulus interaction*. Human judgments of heat pain are subject to stimulation-interaction effects. If a heat stimulus in the painful range is given alone, humans will consider it more painful than if the same stimulus is immediately preceded by an intense stimulus. CMHs respond in a similar fashion. The same stimulus will elicit a more robust firing response from CMHs if given alone than if preceded by a stronger stimulus. Likewise, both the human judgment of pain and the CMH response will decrease in response to multiple presentations of the same stimulus.[3] The second observation involves a *latency to pain perception*. Another correlation between psychophysical measurements and CMH activity is the latency to first pain observed in glabrous skin. When a painful heat stimulus is applied to glabrous skin in humans, the onset of the perception of pain lags behind the onset of the stimulus by more than 700 ms. This cannot be explained by the time it takes the nociceptor to respond because the utilization time of CMHs is very short (<50 ms). It is explained by the slow conduction from the hand to the spinal cord in the unmyelinated afferents. In contrast, the onset of pain is much faster (around 400 ms) in hairy skin, arguing that first pain sensation in hairy skin is conveyed by activity in myelinated fibers (see below). The third observation involves *patients without unmyelinated fibers*. Another argument for the important role of C fibers in the sensation of heat pain is the observation that in some patients whose glabrous skin is insensitive to painful heat stimuli, microscopic examination of the peripheral nerves indicates absence of C fibers.[4]

Direct recordings from nerves in awake humans allow further correlation of nerve fiber activity and the conscious perception of pain. Single fibers from sensory nerves, such as the superficial radial nerve at the wrist, can be identified after percutaneous placement of microelectrodes. These electrodes can then record activity from a specific fiber, or once the receptive field of a fiber is identified, the electrodes can be used to stimulate the axon to determine the perception evoked by activity in that fiber. While the size of the electrode may limit the specificity of the results,[5] these microneurographic studies implicate CMHs in the transmission of pain. Intraneural electrical stimulation of CMHs causes pain.[6] In awake humans, the threshold temperature for the perception of pain is just slightly greater than the threshold temperature to elicit a response in CMHs.[7] If a range of temperature stimuli is given, there is a linear relationship between the CMH activity recorded in awake subjects and the magnitude of pain they report.[8]

A membrane channel responsible for encoding heat into the noxious range has been identified recently.[9] The channel, called *VR1* or *TRPV1*, is located on the terminals of unmyelinated fibers in the epidermis and also is responsive to capsaicin and protons.

A-Fiber Nociceptors and Heat Pain

When a prolonged, intense heat stimulus (e.g., 53°C for 30 s) is applied to glabrous skin, pain reaches a plateau level that persists until the stimulus is removed.[10] While CMHs respond vigorously to the onset of the heat stimulus, they adapt quickly, and their activity tapers off. CMH activity, then, cannot completely explain the human experience in response to a prolonged heat stimulus. In contrast to the CMH behavior, certain A fibers, called *type I A mechanoheat-sensitive fibers* (AMHs), are initially unresponsive to short-duration heat stimuli. With prolonged, intense heat stimulation, however, they begin to respond vigorously and continue to respond for the duration of the stimulus.[10] The activities of CMHs and type I AMHs complement each other to produce the psychophysical experience of heat pain in glabrous skin.

In contrast to glabrous skin, a dual pain sensation is perceived on hairy skin in response to a stepped heat stimulus. Human subjects report an initial pricking pain sensation followed by a slower, more sustained burning pain sensation. These two sensations are called *first pain* and *second pain* respectively. The latency to the first pain sensation is too short for unmyelinated fibers to play a role in its transduction.[11] The longer latency of the second pain sensation is consistent with it being signaled by C fibers. A second class of A-fiber nociceptors, called *type II A-fiber nociceptors*, has been described that fits important criteria to be the transmitter of first pain. First, they have a heat threshold near that described for first pain.[12] Second, the onset of receptor activation after the stimulus is applied to type II nociceptors is very short.[13] Third, the burst of activity that type II nociceptors exhibit at the onset of the stimulus correlates temporally with the perception of momentary pricking of the skin. Finally, humans do not perceive first pain in glabrous skin, and no type II A-fiber nociceptors have been isolated there. Interestingly, the mechanical thresholds for type II A-fiber nociceptors are much higher than those of type I AMHs, with many not responding to mechanical stimuli.

While much evidence exists to link activity in CMHs and AMHs to the perception of heat pain, there also is considerable evidence that activity in these fibers does not always signal pain and that pain also may be transmitted by other sources. Nociceptors may be active at low levels without correlate human perception of pain.[7,14,15] This activity likely is screened by central mechanisms that are then able to edit them or distract attention to more important stimuli. Likewise, as we will discuss later, benign activity from receptors other than nociceptors may be interpreted as pain after an injury.

▶ NOCICEPTORS AND PAIN TO CONTROLLED MECHANICAL STIMULI

Responses of Nociceptors to Controlled Mechanical Stimuli

A-fiber and C-fiber mechanically sensitive nociceptive afferents (MSAs) both respond to punctate mechanical stimuli. When a stimulus is applied to the receptive field with controlled force, the response from the fiber is greatest at stimulus onset and then adapts slowly. If the stimulus is repeated multiple times, this response will fatigue unless an adequate interval between stimuli is allowed. C fibers recover from fatigue more slowly than A fibers.[16]

The discharge rate from nociceptors is proportional to force and pressure but inversely proportional to probe size.[17] If cylindrical probes of different diameters are used, similar discharge rates are obtained if the force per unit length of circumference is constant. This observation suggests that the nociceptors are responding to the stress or strain component of the mechanical stimulus, which would be greatest at the edges of the stimulus and thus most dependent on circumference. For a given probe size, the response of A fibers increases monotonically as the force is increased and is more robust than the response of C fibers, which saturate at higher force levels.[16]

A-Fiber Nociceptors Signal Sharp Pain

When the skin is pretreated for several days with capsaicin, the pain to heat stimuli but not to mechanical stimuli is reduced significantly.[18] This finding suggests that the fibers responsible for mechanical pain sensations are different from those responsible for heat pain sensation. Further, they argue that not all the fibers that respond to mechanical stimuli are sensitive to capsaicin. These observations drive the conclusion that A fibers, and specifically capsaicin-insensitive A fibers such as type I fibers or high-threshold mechanically sensitive nociceptors (HTMs), are responsible for the sharp pain experienced in response to punctate mechanical stimuli. Further support for this conclusion comes from observations between psychophysical and electrophysiologic experiments. The time to pain perception from a mechanical stimulus is short and on the order of the A-fiber conduction time. When A fibers are stimulated, the response intensity correlates well with the intensity curve reported by human subjects over the same force range.[19] Finally, when a selective A-fiber block is applied, the perception of pain in response to sharp probes is reduced dramatically.[18]

C-Fiber Mechanically Insensitive Afferents Signal Pain to Tonic Pressure

Given that an A-fiber block does not eliminate mechanical pain completely, some role for C fibers must remain. In response to prolonged controlled-force stimuli, C- and A-fiber MSAs initially respond vigorously, but that activity adapts quickly. In contrast, human judgments of pain in response to long-duration mechanical stimuli increase throughout the stimulus,[14] and this phenomenon will persist in the presence of an A-fiber block.[20] While the adapting response of C-fiber MSAs to mechanical stimuli is not likely to support this perception, the response of other C fibers may. Recently, C-fiber mechanically insensitive afferents (C-MIAs) have been described. C-MIAs are defined as those fibers whose activity cannot be provoked by gentle pinching of the skin. The receptive field of these fibers is identified by electrocutaneous stimulation.[21] When these C-MIAs are exposed to long-duration mechanical stimulation, they develop a response that is similar in profile to the pain intensities reported by human subjects.[22] C-MIAs likely play a role in the perception of pain associated with tonic mechanical pressure.

▶ NOCICEPTORS AND COLD PAIN SENSATION

The sensation of cold pain, like that of heat pain, is served by specific fibers. However, the properties of cold pain sensation are quite different from those of heat pain. First, the cold pain threshold (14°C) is more removed from the resting skin temperature (33°C) than the threshold for heat pain (45°C).[23] Second, the stimulus-response curve for painful heat stimuli is much steeper than that for cold stimuli.[24] Finally, the lag time to response after a stimulus is greater for cold pain than for heat pain.

The Aδ fibers that encode gentle cooling sensations are electrophysiologically active at room temperatures, and their activity increases significantly in response to gentle cooling.[25] The presence of a cold pain sensation that is both different from cooling and produced at a much lower threshold temperature argues for a unique receptor for cold pain. Unmyelinated and myelinated nociceptors respond to intense cold stimuli and likely encode cold pain. Two membrane channels that respond to cold have been identified in dorsal root ganglion neurons. The TRPM8 channel responds to gentle cooling and likely is involved in cool sensation.[26] The ANKTM1 channel does not become active until the temperature of the stimulus falls below 17.5°C.[27] The onset of activity in this channel correlates well with the temperature that human subjects identify as the threshold between cool and cold pain, and the response-curve slope reflects closely that of psychophysical measures.

▶ NOCICEPTORS AND CHEMICALLY EVOKED SENSATIONS

Certain exogenous chemical agents induce painful sensations when applied topically. Intradermal injection of capsaicin, the active ingredient of hot peppers, produces intense burning pain that lasts for several minutes. Capsaicin invokes little response from the receptive fields of mechanically sensitive C fibers[28] but a robust response when injected into the receptive fields of mechanically insensitive A and C fibers,[29,30] suggesting that these fibers are responsible for signaling pain in response to capsaicin. Histamine produces long-lasting itch when administered to the skin. This response is mediated by a subpopulation of mechanically insensitive C fibers that responds vigorously and for a long duration when histamine is injected into their receptive fields.[31] Other fibers, such as mechanically sensitive C fibers, respond only briefly and weakly in response to histamine injection.[32] These observations argue that mechanically insensitive C fibers play an important role in the transduction of chemically evoked pain and itch.

▶ HYPERALGESIA: ROLE OF NOCICEPTORS

Pain in response to noxious stimuli is only one aspect of pain perception. However, pain sensations change dramatically after injury. This phenomenon, called *hyperalgesia*, leads to pain in response to stimuli that otherwise would not be painful. This plasticity is crucial to avoid harmful conditions or further tissue damage.

Hyperalgesia is defined as a leftward shift of the stimulus-response function that relates magnitude of pain to stimulus intensity. Hyperalgesia may occur at the site of injury and in the surrounding uninjured skin. Primary hyperalgesia occurs at the site of injury, and secondary hyperalgesia is a property of the surrounding uninjured skin.[33] While their mechanisms differ, both provide evidence that the nervous system is plastic and can change in response to injury.

The plastic change in the nervous system that accounts for hyperalgesia is called *sensitization*. Similar to hyperalgesia, sensitization refers to a leftward shift of the stimulus-response function that relates magnitude of neural response to stimulus intensity. Just as hyperalgesia describes enhanced pain, sensitization refers to an increase in neural response. This sensitization can occur in the peripheral or central nervous system. Carefully designed studies reveal elements of sensitization in animal models that can be correlated with human responses under similar conditions. Once the response properties of a given neuron are known, the neuron can be manipulated in a manner that usually causes hyperalgesia. In designing a study, consideration must be given to the energy of the injury, the tissue involved, the energy form of the test stimulus, and the location of the test. In the setting of these variables, one can then assess whether or not the manipulation has changed the response properties of the neuron. Using this approach, thermal injuries, chemical burns, or skin incisions can be studied, and their unique effects on hyperalgesia can be better characterized.[34]

▶ PRIMARY HYPERALGESIA

Hyperalgesia to Heat Stimuli

When the skin is burned, marked hyperalgesia to heat stimuli develops at the site of injury. Human subjects rate thermal stimuli as painful that were not painful prior to the injury. In order to understand the neurophysiologic correlates of this hyperalgesia better, studies have been performed that compare human measurements of pain with changes in nociceptor fiber activity in response to similar stimuli after an identical injury. In one study,[10] heat testing was done before and after a burn to the glabrous skin of the hand at 53°C for 30 s. After this injury, human subjects report marked hyperalgesia to heat stimuli. Likewise, the response of AMHs in the monkey increased markedly. Interestingly, CMHs responded less to the same temperatures after a burn to the glabrous skin. Thus hyperalgesia to heat after a burn to the glabrous skin is signaled by sensitization of AMH nociceptors to heat stimuli.

Although CMHs in glabrous skin do not sensitize after a burn injury, CMHs that innervate hairy skin do.[35] Thus CMHs may play a role in heat hyperalgesia after a burn in hairy but not glabrous skin.[36] These data indicate a correlation between the increase in activity among specific nociceptors in specific tissues with human judgments of pain and argue that the hyperalgesia that occurs at the site of an injury is because of sensitization of primary afferents.

Hyperalgesia to Mechanical Stimuli

Just as hyperalgesia to heat stimuli can develop in injured tissue, innocuous mechanical stimuli also may become painful. Different forms of hyperalgesia develop to different types of stimuli. *Stroking*, or *dynamic*, *hyperalgesia* occurs in response to light stroking of the skin, such as one might do with a cotton swab. *Punctate hyperalgesia* refers to enhanced pain in response to stimulation with punctate stimuli such as von Frey probes. Hyperalgesia to blunt pressure or to impact from small projectiles at constant speed also has been observed in injured tissue.[37] The presence of different forms of hyperalgesia argues for different transduction mechanisms. Stroking hyperalgesia is likely to be signaled by sensory afferents that usually signal stroking and, therefore, do not usually signal nociception. In order to transmit pain, either these fibers must intrinsically change their properties, or the signal they send must be interpreted differently. In other words, the mechanisms to support hyperalgesia could rely on changes in either the peripheral or central nervous system.

The circuitry of hyperalgesia becomes more complicated when one considers that hyperalgesia occurs both in the tissue that is injured (i.e., *primary hyperalgesia*) and in the surrounding uninjured zone (i.e., *secondary hyperalgesia*). While fibers with receptive fields within the injured tissue could become sensitized as a result of the injury, supporting a peripheral mechanism of hyperalgesia, peripheral sensitization seems less likely for fibers with receptive fields in the secondary, non-injury zone. Careful examination of the activity of different fibers in these zones lends a better understanding of the neural mechanisms supporting the sensory manifestations of primary and secondary hyperalgesia.

Nociceptor Sensitization as a Mechanism for Mechanical Hyperalgesia in the Primary Zone

Observations from a number of studies support the role of nociceptor sensitization in mechanical hyperalgesia in the primary zone of injury. After injury, both mechanically sensitive and mechanically insensitive afferents change the way they respond to stimuli in ways that could encode primary hyperalgesia.[38–41] Mechanically insensitive Aδ fibers become mechanically sensitive in the presence of certain inflammatory mediators.[42] Although the AMHs and CMHs do not change thresholds, they respond more robustly to suprathreshold stimulation.[43,44] There is also evidence that in response to burn, mechanical injury, or inflammation, the receptive fields of A- and C-fiber nociceptors can expand.[41,43,45] Because of the expanded receptive fields, a greater number of fibers is activated in response to a mechanical stimulus, and through spatial summation, a greater pain response is generated. While this explanation is attractive for certain types of hyperalgesia, the fact that mechanical thresholds in the expanded receptive field are the same as in the original fields[41] makes it less likely to be the explanation for stroking hyperalgesia, where thresholds are markedly lower than are observed for nociceptors.

▶ SECONDARY HYPERALGESIA

Secondary hyperalgesia differs from primary hyperalgesia in important ways. The zone of secondary hyperalgesia describes that region immediately surrounding injured tissue but does not include any injured tissue. Any change in pain sensation in this region must be because of sensitization spreading from the zone of injury or to changes in processing in the central nervous system.

Secondary Hyperalgesia to Mechanical but Not Heat Stimuli

Primary hyperalgesia is characterized by enhanced pain to both mechanical and heat modalities. Secondary hyperalgesia, in contrast, is characterized by hyperalgesia only to mechanical stimuli.[46–48] If a burn is induced on the glabrous skin of the hand, lightly touching the skin at the burn site and in a large region of skin surrounding the burn will produce pain within minutes of the injury. The decrease in pain threshold is similar at the site of the burn and in the surrounding tissue. Heat hyperalgesia also develops in the primary area of the burn but not in the secondary zone. When the uninjured tissue between two burns was tested, heat stimuli actually were transiently less painful than they had been prior to the burns, whereas subjects rated mechanical stimuli as markedly more painful.[49]

Spreading Sensitization of Nociceptors Does Not Occur

Activation of nociceptors is known to result in an axon-reflex flare response; stimulation of one branch of a nociceptive terminal by a noxious stimulus leads to action potential activity that antidromically invades other branches of the nociceptor, leading to the release of vasodilatory substances. Lewis speculated that sensitizing substances also could be released, and thus peripheral sensitization could spread to adjoining uninjured tissue.[50] Spreading sensitization is an attractive hypothesis to explain secondary hyperalgesia because the area of flare and the area of secondary hyperalgesia are both large and surround the injury site.

Several findings indicate that spreading sensitization does not occur. For example, if a burn is applied to only one-half of a receptive field for a given nociceptor, the sensibility of the other half is unchanged.[51] In monkeys, mechanical injury does not induce mechanical sensitization among CMHs whose receptive fields are adjacent to the injury.[38,52] Electrical stimulation of nociceptive fibers that would produce antidromic activation of terminal branches does not cause sensitization in either monkeys or rats.[52,53] Finally, a chemical burn such as that from mustard oil, when applied to only part of a receptive field, does not lead to sensitization in other parts of the field.[40]

In addition, several lines of evidence suggest that the mechanisms of flare and secondary hyperalgesia are different.[54] Flare can be induced without inducing secondary hyperalgesia (e.g., by intradermal injection of histamine). Secondary hyperalgesia can be induced without flare (e.g., when the skin is pretreated with capsaicin).[18] Finally, the zone of hyperalgesia is often larger than the zone of flare, and flare will cross the midline, whereas secondary hyperalgesia will not.

Central Mechanisms of Secondary Hyperalgesia

The preceding evidence argues against a peripheral mechanism for secondary hyperalgesia. It is intuitive, then, that the central nervous system (CNS) must play an important role in whatever transformation accounts for secondary hyperalgesia. If the CNS does play an important role, then it follows that under conditions of injury, pain may not reside completely with nociceptors. Receptors that usually carry other signals, such as touch, may acquire the capacity to evoke pain. This principle applies to all states where the sensation of pain is not explained completely by the activity in peripheral nociceptors, including neuropathic pain states. If peripheral nociceptors are not sending the pain signal, it is possible that central interneurons may be induced aberrantly to send pain signals in response to activity from low-threshold mechanoreceptors. This transformation is called *central sensitization*.

Capsaicin has been used widely as the injury stimulus to elicit secondary hyperalgesia. Capsaicin is ideal for this role for several reasons. Capsaicin selectively activates nociceptors.[55] It causes both intense pain and a large zone of secondary hyperalgesia without injuring the skin.[56] Finally, the hyperalgesia that capsaicin induces closely resembles that from heat or cut injuries. The immediate zone of application develops hyperalgesia to both heat and mechanical stimuli. The secondary zone is sensitized only to mechanical stimuli.[46]

Taking advantage of the specific properties of capsaicin, LaMotte and colleagues conducted experiments carefully designed to tease apart the role of peripheral and central mechanisms in secondary hyperalgesia.[54] To test whether peripheral nerves were sensitized, capsaicin was administered during a peripheral nerve block. When the nerve is blocked proximal to the capsaicin site, the CNS is spared the initial response to capsaicin, but the peripheral responses distal to the block should remain intact. Under these conditions, axon-reflexive flare developed, but no hyperalgesia was present after the nerve block had worn off. Thus, if the signal induced by noxious stimuli cannot reach the CNS, secondary hyperalgesia does not develop.[54,57] Other evidence that central sensitization but not peripheral sensitization plays a major role in the development of secondary hyperalgesia supports this finding: First, electrical stimulation bypasses peripheral receptor sensitization mechanisms, yet it is able to produce a large zone of secondary hyperalgesia.[58] Second, a strip of anesthetic skin will prevent the spread of flare from an injury placed adjacent to the anesthetic strip without changing the zone of secondary hyperalgesia.[59] Thus the peripheral action of the anesthetic is sufficient to stop the peripheral axon-reflexive flare but cannot disrupt the central mechanism of secondary hyperalgesia. Third, the secondary hyperalgesia that develops after injecting the territory of a single nerve appears to spread to the territory of a second nerve in a pattern that could only occur with central sensitization.[60]

Different Mechanisms for Stroking and Punctate Hyperalgesia

Hyperalgesia to mechanical stimuli develops in the secondary zone for more than punctate stimuli. Brushing or stroking stimuli also elicit hyperalgesia, but blunt pressure does not.[61] Stroking hyperalgesia, also called *allodynia*, appears to be signaled by activity in low-threshold mechanoreceptors (LTMs). Experiments using blood pressure cuffs to cause a selective large myelinated fiber block have shown that stroking hyperalgesia after injury or from neuropathic pain disappears at the same time that the skin becomes anesthetic to touch but remains sensitive to temperature.[54,61,62] In studies of awake humans, it also has been shown that intraneural microelectrical stimulation of the axons of primary afferents dedicated to touch is not painful before injury but becomes painful after secondary hyperalgesia is induced.[63]

The mechanisms of stroking and punctate hyperalgesia appear to be different. Punctate hyperalgesia appears to be signaled by nociceptors. Punctate hyperalgesia has been reported in the absence of stroking hyperalgesia in a patient who suffered from large-fiber neuropathy,[64] suggesting that punctate but not stroking hyperalgesia is mediated by small fibers, presumably nociceptors. Wool fabrics elicit pain in areas of secondary hyperalgesia in proportion to their prickliness, a property transmitted by nociceptors.[65,66] Secondary hyperalgesia to punctate stimuli occupies a larger area than that to stroking stimuli. Likewise, where stroking hyperalgesia after capsaicin injection will wear off in 1 to 2 hours, punctate hyperalgesia can persist for 12 hours or more.[54] If the area of primary hyperalgesia is cooled or anesthetized, stroking hyperalgesia goes away immediately, whereas punctate hyperalgesia persists. Thus stroking hyperalgesia appears to be more dependent on an ongoing signal from the zone of injury.[54,67]

Model for Stroking Hyperalgesia

Secondary hyperalgesia for stroking stimuli appears to occur when the CNS interprets input from LTMs as pain rather than touch. Thus, signals from LTMs gain access to the pain pathway. Enhanced responsiveness to mechanical stimuli of neurons in the dorsal horn has been reported. One model to explain this is based on the primary afferent depolarization (PAD) that normally occurs in nociceptors following peripheral stimulation of LTMs. Investigators have proposed that the injury signal causes sensitization of interneurons in the dorsal horn such that this PAD is enhanced.[68] This PAD is now sufficient to generate spike activity in nociceptors that then can be propagated by the nerve antidromically as in an axonal reflex, as well as orthodromically to pain-signaling interneurons. This activity affords the LTMs access to the central pain pathway after injury that they do not have in the absence of injury. Now light touch activates LTMs, and they, in turn, activate the interneurons of the pain-signaling pathway. Stimuli that were once innocuous become painful.

Model for Punctate Hyperalgesia

Punctate hyperalgesia appears to rely on central sensitization to nociceptor input. Most nociceptors respond to both mechanical and heat stimuli, yet there is only hyperalgesia to mechanical stimuli in the zone of secondary hyperalgesia. The dissociation of heat and mechanical responses in the zone of secondary hyperalgesia argues that transmission of these stimuli must have divergent transduction pathways. One possibility is that punctate hyperalgesia is signaled by primary afferent nociceptors that respond only to mechanical stimuli. If heat-sensitive afferents do not project to the type of interneurons that eventually sensitize but mechanical afferents do, then only punctate hyperalgesia would be observed. To investigate the role of heat-sensitive nociceptors in punctate hyperalgesia, topical capsaicin was applied to the skin to desensitize the heat-sensitive nociceptors. Secondary punctate hyperalgesia persisted in skin that had been desensitized to heat by capsaicin.[18] Punctate hyperalgesia disappeared after an Aδ-fiber block.[69] Taken together, these two observations suggest that punctate hyperalgesia depends on capsaicin-insensitive, mechanically sensitive A-fiber nociceptors.

REFERENCES

1. Johnson KO, Darian-Smith I, LaMotte C, et al: Coding of incremental changes in skin temperature by a monkey: Correlation with intensity discrimination in man. *J Neurophysiol* 1979;42(5):1332–1353.
2. Meyer RA, Campbell JN: Peripheral neural coding of pain sensation. *Johns Hopkins APL Tech Dig* 1981;2:164–171.
3. LaMotte RH, Campbell JN: Comparison of responses of warm and nociceptive C-fiber afferents in monkey with human judgements of thermal pain. *J Neurophysiol* 1978; 41:509–528.
4. Bischoff A: Congenital insensitivity to pain with anhidrosis: A morphometric study of sural nerve and cutaneous receptors in the human prepuce, in Bonica JJ, Liebeskind JC, Albe-Fessard DG (eds), *Advances in Pain Research and Therapy*, 3d ed. New York: Raven Press, 1979:53–65.
5. Wall PD, McMahon SB: Microneuronography and its relation to perceived sensation: A critical review. *Pain* 1985; 21:209–229.
6. Torebjörk E, Ochoa J: Specific sensations evoked by activity in single identified sensory units in man. *Acta Physiol Scand* 1980;110:445–447.
7. Van Hees J, Gybels JC: C nociceptor activity in human nerve during painful and nonpainful skin stimulation. *J Neurol Neurosurg Psychiatry* 1981;44:600–607.
8. Torebjörk HE, LaMotte RH, Robinson CJ: Peripheral neural correlates of magnitude of cutaneous pain and hyperalgesia: Simultaneous recordings in humans of sensory judgments of pain and evoked responses in nociceptors with C-fibers. *J Neurophysiol* 1984;51:325–339.
9. Caterina MJ, Schumacher MA, Tominaga M, et al: The capsaicin receptor: A heat-activated ion channel in the pain pathway. *Nature* 1997;389:816–824.
10. Meyer RA, Campbell JN: Myelinated nociceptive afferents account for the hyperalgesia that follows a burn to the hand. *Science* 1981;213:1527–1529.
11. Campbell JN, LaMotte RH: Latency to detection of first pain. *Brain Res* 1983;266:203–208.
12. Dubner R, Price DD, Beitel RE, et al: Peripheral neural correlates of behavior in monkey and human related to sensory-discriminative aspects of pain, in Anderson DJ, Mathews B (eds), *Pain in the Trigeminal Region*. Amsterdam: Elsevier North Holland, 1977:57–66.
13. Treede R-D, Meyer RA, Campbell JN: Myelinated mechanically insensitive afferents from monkey hairy skin: Heat-response properties. *J Neurophysiol* 1998;80: 1082–1093.
14. Adriaensen H, Gybels J, Handwerker HO, et al: Nociceptor discharges and sensations due to prolonged noxious mechanical stimulation: A paradox. *Hum Neurobiol* 1984;3:53–58.
15. Cervero F, Gilbert R, Hammond RGE, et al: Development of secondary hyperalgesia following non-painful thermal stimulation of the skin: A psychophysical study in man. *Pain* 1993;54:181–189.
16. Slugg RM, Meyer RA, Campbell JN: Response of cutaneous A- and C-fiber nociceptors in the monkey to controlled-force stimuli. *J Neurophysiol* 2000;83:2179–2191.
17. Garell PC, McGillis SLB, Greenspan JD: Mechanical response properties of nociceptors innervating feline hairy skin. *J Neurophysiol* 1996;75:1177–1189.
18. Magerl W, Fuchs PN, Meyer RA, et al: Roles of capsaicin-insensitive nociceptors in cutaneous pain and secondary hyperalgesia. *Brain* 2001;124:1754–1764.
19. Meyer RA, Ringkamp M, Campbell JN, et al: Peripheral mechanisms of cutaneous nociception, in McMahon S, Koltzenburg H, (eds), *Wall & Melzack's Textbook of Pain*, 5th ed. London: Churchill Livingstone, 2004.

20. Andrew D, Greenspan JD: Peripheral coding of tonic mechanical cutaneous pain: Comparison of nociceptor activity in rat and human psychophysics. *J Neurophysiol* 1999;82:2641–2648.
21. Meyer RA, Davis KD, Cohen RH, et al: Mechanically insensitive afferents (MIAs) in cutaneous nerves of monkey. *Brain Res* 1991;561:252–261.
22. Schmelz M, Schmidt R, Bickel A, et al: Differential sensitivity of mechanosensitive and -insensitive C-fibers in human skin to tonic pressure and capsaicin injection. *Soc Neurosci Abstr* 1997;23:1004.
23. Harrison JL, Davis KD: Cold-evoked pain varies with skin type and cooling rate: A psychophysical study in humans. *Pain* 1999;83:123–135.
24. Morin C, Bushnell MC: Temporal and qualitative properties of cold pain and heat pain: A psychophysical study. *Pain* 1998;74:67–73.
25. Darian-Smith I, Johnson KO, Dykes R: "Cold" fiber population innervating palmar and digital skin of the monkey: Responses to cooling pulses. *J Neurophysiol* 1973;36:325–346.
26. Peier AM, Moqrich A, Hergarden AC, et al: A TRP channel that senses cold stimuli and menthol. *Cell* 2002; 108: 705–715.
27. Story GM, Peier AM, Reeve AJ, et al: ANKTM1, a TRP-like channel expressed in nociceptive neurons, is activated by cold temperatures. *Cell* 2003;112:819–829.
28. Baumann TK, Simone DA, Shain CN, et al: Neurogenic hyperalgesia: The search for the primary cutaneous afferent fibers that contribute to capsaicin-induced pain and hyperalgesia. *J Neurophysiol* 1991;66:212–227.
29. Ringkamp M, Peng YB, Wu G, et al: Capsaicin responses in heat-sensitive and heat-insensitive A-fiber nociceptors. *J Neurosci* 2001;21:4460–4468.
30. Schmelz M, Schmid R, Handwerker HO, et al: Encoding of burning pain from capsaicin-treated human skin in two categories of unmyelinated nerve fibres. *Brain* 2000;123:560–571.
31. Schmelz M, Schmidt R, Bickel A, et al: Specific C-receptors for itch in human skin. *J Neurosci* 1997;17:8003–8008.
32. LaMotte RH, Simone DA, Baumann TK, et al: Hypothesis for novel classes of chemoreceptors mediating chemogenic pain and itch, in Dubner R, Gebhart GF, Bond MR (eds), *Proceedings of the Vth World Congress on Pain*. Amsterdam: Elsevier, 1988:529–535.
33. Lewis T: Experiments relating to cutaneous hyperalgesia and its spread through somatic fibres. *Clin Sci* 1935;2:373–423.
34. Treede R-D, Meyer RA, Raja SN, et al: Peripheral and central mechanisms of cutaneous hyperalgesia. *Prog Neurobiol* 1992; 38:397–421.
35. Campbell JN, Meyer RA: Sensitization of unmyelinated nociceptive afferents in the monkey varies with skin type. *J Neurophysiol* 1983;49:98–110.
36. LaMotte RH, Thalhammer JG, Robinson CJ: Peripheral neural correlates of magnitude of cutaneous pain and hyperalgesia: A comparison of neural events in monkey with sensory judgements in human. *J Neurophysiol* 1983; 50:1–26.
37. Kilo S, Schmelz M, Koltzenburg M, et al: Different patterns of hyperalgesia induced by experimental inflammation in human skin. *Brain* 1994;117:385–396.
38. Campbell JN, Khan AA, Meyer RA, et al: Responses to heat of C-fiber nociceptors in monkey are altered by injury in the receptive field but not by adjacent injury. *Pain* 1988; 32:327–332.
39. Campbell JN, Meyer RA, LaMotte RH: Sensitization of myelinated nociceptive afferents that innervate monkey hand. *J Neurophysiol* 1979;42:1669–1679.
40. Schmelz M, Schmidt R, Ringkamp M, et al: Limitation of sensitization to injured parts of receptive fields in human skin C-nociceptors. *Exp Brain Res* 1996;109:141–147.
41. Thalhammer JG, LaMotte RH: Spatial properties of nociceptor sensitization following heat injury of the skin. *Brain Res* 1982;231:257–265.
42. Davis KD, Meyer RA, Campbell JN: Chemosensitivity and sensitization of nociceptive afferents that innervate the hairy skin of monkey. *J Neurophysiol* 1993;69:1071–1081.
43. Andrew D, Greenspan JD: Mechanical and heat sensitization of cutaneous nociceptors after peripheral inflammation in the rat. *J Neurophysiol* 1999;82:2649–2656.
44. Cooper B, Ahlquist M, Friedman RM, et al: Properties of high-threshold mechanoreceptors in the goat oral mucosa: II. Dynamic and static reactivity in carrageenan-inflamed mucosa. *J Neurophysiol* 1991;66:1280–1290.
45. Reeh PW, Bayer J, Kocher L, et al: Sensitization of nociceptive cutaneous nerve fibers from the rat tail by noxious mechanical stimulation. *Exp Brain Res* 1987;65: 505–512.
46. Ali Z, Meyer RA, Campbell JN: Secondary hyperalgesia to mechanical but not heat stimuli following a capsaicin injection in hairy skin. *Pain* 1996;68:401–411.
47. Dahl JB, Brennum J, Arendt-Nielsen L, et al: The effect of pre- versus postinjury infiltration with lidocaine on thermal and mechanical hyperalgesia after heat injury to the skin. *Pain* 1993;53:43–51.
48. Warncke T, Stubhaug A, Jorum E: Ketamine, an NMDA receptor antagonist, suppresses spatial and temporal properties of burn-induced secondary hyperalgesia in man: A double-blind, cross-over comparison with morphine and placebo. *Pain* 1997;72:99–106.
49. Raja SN, Campbell JN, Meyer RA: Evidence for different mechanisms of primary and secondary hyperalgesia following heat injury to the glabrous skin. *Brain* 1984; 107:1179–1188.
50. Lewis T: *Pain*. New York: Macmillan, 1942.
51. Thalhammer JG, LaMotte RH: Heat sensitization of one-half of a cutaneous nociceptor's receptive field does not alter the sensitivity of the other half, in Bonica JJ, Lindblom U, Iggo A (eds), *Advances in Pain Research and Therapy,* Vol 5. New York: Raven Press, 1983:71–75.
52. Reeh PW, Kocher L, Jung S: Does neurogenic inflammation alter the sensitivity of unmyelinated nociceptors in the rat? *Brain Res* 1986;384:42–50.
53. Meyer RA, Campbell JN, Raja SN: Antidromic nerve stimulation in monkey does not sensitize unmyelinated nociceptors to heat. *Brain Res* 1988;441:168–172.
54. LaMotte RH, Shain CN, Simone DA, et al: Neurogenic hyperalgesia: Psychophysical studies of underlying mechanisms. *J Neurophysiol* 1991;66:190–211.
55. Szolcsányi J: Capsaicin, irritation, and desensitization: Neurophysiological basis and future perspectives, in Green BG, Mason JR, Kare MR (eds), *Chemical Senses,* Vol 2: *Irritation*. New York: Marcel Dekker, 1990:141–169.

56. Simone DA, Baumann TK, LaMotte RH: Dose-dependent pain and mechanical hyperalgesia in humans after intradermal injection of capsaicin. *Pain* 1989;38:99–107.
57. Pedersen JL, Crawford ME, Dahl JB, et al: Effect of preemptive nerve block on inflammation and hyperalgesia after human thermal injury. *Anesthesiology* 1996;84: 1020–1026.
58. Koppert W, Dern SK, Sittl R, et al: A new model of electrically evoked pain and hyperalgesia in human skin: The effects of intravenous alfentanil, S(+)-ketamine, and lidocaine. *Anesthesiology* 2001;95:395–402.
59. Klede M, Handwerker HO, Schmelz M: Central origin of secondary mechanical hyperalgesia. *J Neurophysiol* 2003; 90:353–359.
60. Sang CN, Gracely RH, Max MB, et al: Capsaicin-evoked mechanical allodynia and hyperalgesia cross nerve territories: Evidence for a central mechanism. *Anesthesiology* 1996;85:491–496.
61. Koltzenburg M, Lundberg LER, Torebjörk HE: Dynamic and static components of mechanical hyperalgesia in human hairy skin. *Pain* 1992;51:207–219.
62. Campbell JN, Raja SN, Meyer RA, et al: Myelinated afferents signal the hyperalgesia associated with nerve injury. *Pain* 1988;32:89–94.
63. Torebjörk HE, Lundberg LER, LaMotte RH: Central changes in processing of mechanoreceptive input in capsaicin-induced secondary hyperalgesia in humans. *J Physiol (Lond)* 1992;448:765–780.
64. Treede R-D, Cole JD: Dissociated secondary hyperalgesia in a subject with large fibre sensory neuropathy. *Pain* 1993;53:169–174.
65. Cervero F, Meyer RA, Campbell JN: A psychophysical study of secondary hyperalgesia: Evidence for increased pain to input from nociceptors. *Pain* 1994;58:21–28.
66. Garnsworthy RK, Gully RL, Kenins P, et al: Identification of the physical stimulus and the neural basis of fabric-evoked prickle. *J Neurophysiol* 1988;59:1083–1097.
67. Magerl W, Wilk SH, Treede R-D: Secondary hyperalgesia and perceptual wind-up following intradermal injection of capsaicin in humans. *Pain* 1998;74:257–268.
68. Cervero F, Laird JMA, García-Nicas E: Secondary hyperalgesia and presynaptic inhibition: An update. *Eur J Pain* 2003;7:345–351.
69. Ziegler EA, Magerl W, Meyer RA, et al: Secondary hyperalgesia to punctate mechanical stimuli: Central sensitization to A-fibre nociceptor input. *Brain* 1999;122:2245–2257.

CHAPTER 3

Pharmacology of Pain Transmission and Modulation

I. Central Mechanisms

George L. Wilcox, Laura S. Stone, Michael H. Ossipov, Josephine Lai, and Frank Porreca

The pharmacology of pain transmission and modulation involves receptors and mechanisms in many loci in the peripheral and central nervous systems. Of these, the central nervous system (CNS), including the brain stem and particularly the spinal cord, has received much attention over the past three decades. This chapter will focus on supraspinal and spinal receptors and mechanisms involved in pain transmission, plasticity, and modulation.

Understanding the spinal mechanisms that contribute to pain states is particularly important because of the richness of the transmitter and receptor populations that are juxtaposed there. The unique combination of neurotransmitters involved in the dorsal horn of the spinal cord presents an opportunity for selective blockade of excitatory transmission or enhancement of inhibitory transmission to produce analgesia. Identifying and understanding the transmitters that are released from primary afferent nociceptive neurons, the consequences of transmitter interaction with particular receptors, and the effects of therapeutic agents on transmitter release and receptor activation in both normal and pathologic conditions are key issues for ultimate therapeutic intervention in pain states. These transmitters include simple molecules such as glutamate, with actions in the millisecond time frame, and complex molecules such as peptides, which produce actions lasting seconds to minutes. In addition, events involving permanent remodeling of neuronal activity and circuits involving growth factors and proteins called cytokines, *both of which invoke synthesis of new proteins, will be discussed. The receptors for these neurotransmitters and modulatory proteins mediate a wide variety of cellular actions. Anesthetic and analgesic agents share sites of action at the axonal conduction, synaptic transmission, or cellular signaling phases of this neurotransmission. This chapter describes the pharmacology within the context of the cellular processes involved in neurotransmission. It is, therefore, a review focused on mechanism rather than a comprehensive survey of the pharmaceutical agents prescribed for pain relief.*

► TRANSMITTERS AND RECEPTORS MEDIATING EXCITATORY PAIN TRANSMISSION

Transmitters Released in the Spinal Cord on Noxious Stimulation

Excitatory Amino Acids (EAAs)
Glutamate is the neurotransmitter most consistently associated with fast excitatory neurotransmission between nociceptors and spinal neurons.[1] Chemical and electrical activation of nociceptors stimulates the release of glutamate from primary afferent neurons in culture and in microdialysates of the dorsal spinal cord in vivo.

Peptides
The peptides thought to be involved in nociceptive transmission, calcitonin gene–related peptide, substance P, somatostatin, vasoactive intestinal polypeptide, and galanin, most likely colocalize with glutamate in primary afferent terminals. Differential compartmentalization of peptides and small-molecule neurotransmitters in these terminals permits preferential release in response to different firing patterns of the axon: Glutamate is released in response to a brief stimulus that results in a few action potentials, whereas peptide release requires a prolonged train of action potentials, such as that which results from a persistent stimulus of high intensity.

CALCITONIN GENE–RELATED PEPTIDE (CGRP)
The majority of CGRP-containing primary sensory neurons probably are nociceptive, and immunoreactive CGRP is released from primary afferent terminals in the spinal cord following peripheral noxious stimulation.[2]

SUBSTANCE P (SP) AND NEUROKININ A (NKA)
After CGRP, SP is the most widely distributed peptide among dorsal root ganglion (DRG) neurons, occurring in almost half of dorsal root ganglion (DRG) neurons, mostly small and medium-sized cells, and many nociceptors contain and release SP on activation of small-diameter afferent fibers. NKA, which is synthesized from some of the same precursor protein that yields SP, is also released in response to the same stimuli that release SP.

SOMATOSTATIN (SOM)
SOM most likely is expressed in small DRG neurons, and its expression is largely mutually exclusive to that of SP. It is released in the spinal cord in response to noxious thermal but not noxious mechanical stimulation. Unlike most other sensory transmitters, SOM couples with inhibitory cellular signaling systems ($G_{i/o}$) yet has pronociceptive activity, suggesting a contrary mode of synaptic communication.

VASOACTIVE INTESTINAL POLYPEPTIDE (VIP)
VIP-immunoreactive primary afferent neurons are numerous in thoracic and sacral spinal nerves, as well as in cranial nerves that innervate viscera, suggesting preferential expression in visceral afferent fibers. VIP exposure results in an increase in activity in dorsal horn neurons.[3]

GALANIN (GAL)
GAL immunoreactivity is present in a minority of DRG neurons, and superficial spinal GAL-positive varicosities are about half as numerous as CGRP-positive terminals. GAL appears to facilitate the flexor reflex in response to a noxious peripheral stimulus and, at higher doses, depresses the reflex.

Modality and Target Tissue Specificity
Although peptidergic content seems related to nociceptive modality of fibers, the relationship is not completely predictive. CGRP, SP, and SOM have each been identified in primary afferent neurons responding to innocuous thermal and mechanical stimuli. There may be a stronger relationship between peptide content and target-tissue innervated by the primary afferent neuron, such as visceral afferent fibers.

Effectors Mediating Excitatory Pain Transmission

Pain transmission can be modulated or inhibited primarily at two sites in the spinal cord dorsal horn: A drug or neurotransmitter can alter excitatory transmitter release by acting at receptors on nociceptive primary afferent terminals or can interfere with the excitatory effect of the transmitter on secondary neurons. Often, a given type of modulatory receptor is located both pre- and postsynaptically. However, the intracellular coupling mechanisms may differ between these sites of action.

Cellular Effector Systems in Postsynaptic Neurons in the Dorsal Horn
The release of multiple nociceptive neurotransmitters, some co-contained in and co-released by the same terminals, spawns numerous excitatory events with distinct time courses. For example, glutamate and SP are co-released from primary afferent neurons on noxious stimulation and initiate signaling processes with widely different time courses. The cell surface receptors activated by these neurotransmitters fall into two major classes: ionotropic receptors, which contain an ion channel that opens on ligand binding, and G protein–coupled receptors (GPCRs, or metabotropic), which couple with other channel or enzyme molecules via G proteins.

There are multiple transmitter-receptor pairs in the dorsal horn, such as glutamate-AMPA, glutamate-NMDA, and glutamate-mGluR and SP-NK_1. These transmitter-receptor pairs interact to produce neuronal excitation and rostrad transmission of pain from nociceptive afferents through the dorsal horn to the brain. Ionotropic glutamate-mediated excitation—glutamate-AMPA followed by glutamate-NMDA—initiates the temporal cascade after transmitter release (1 to 100 ms). Subsequently, slow depolarization mediated by SP-NK$_1$ and metabotropic glutamate (mGluR) receptors prolongs the excitation as long as 1 s. Most analgesic neurotransmitter systems, by contrast, reduce transmission through the dorsal horn via actions at inhibitory GPCRs.

IONOTROPIC EFFECTOR SYSTEMS

The ligand-gated ion channels activated by glutamate are cation channels admitting monovalent (Na$^+$ and K$^+$, all AMPA- and kainate-operated channels) and divalent (Ca^{2+} or Mg^{2+}, some variant AMPA/kainate receptors, and all NMDA receptors). Activation of these nonselective cation channels depolarizes neural processes, producing excitatory postsynaptic potentials (EPSPs) ranging from 5 to 50 ms in duration. In addition, Ca^{2+} entry mediated by these channels can initiate several intracellular changes triggered by increasing intracellular concentrations of Ca^{2+} and activating such signaling systems as calcium-calmodulin kinase (CaM-kinase II).

G-PROTEIN COUPLING

G proteins transduce ligand-receptor binding to cellular effects by converting an inactive heterotrimeric (three different subunits α, β, and γ bound together) form to a dissociated α and a βγ heterodimer, both of which mediate some of the effects of G proteins. Activated α monomers and/or βγ dimers interact with enzymes and ion channels in neurons to alter their response to other inputs. Different subtypes of G proteins associate with and mediate the effects of different GPCRs, creating a rich array of possible effects that are detailed below.

Neurokinin (NK$_1$ and NK$_2$) and some metabotropic glutamate receptors (mGluR1, mGluR3, and mGluR5) couple through G$_q$ proteins to activate phospholipase C (PLC). On ligand-receptor binding, the α subunit takes on a molecule of GTP, dissociates from the βγ subunit, and serves as an activated, locally effective, intracellular messenger. Activation of G$_q$ causes production of inositol triphosphate (IP$_3$) and diacyl glycerol (DAG). IP$_3$-gated Ca^{2+} channels in endoplasmic reticulum release intracellular stores of Ca^{2+} into the cytosol, where it can activate CaM kinase II and/or promote neurotransmitter release. DAG may diffuse less in the cell, but its action may persist for several minutes or longer. DAG activates protein kinase C (PKC), which phosphorylates numerous targets, including NMDA receptors, an action that enhances channel sensitivity to depolarization. Appropriate temporal sequences of multiple afferent excitatory messages can initiate changes in synaptic efficacy through changes in release, receptor number, and receptor coupling; such changes may underlie progressive development of abnormalities in certain pain states, including acute postoperative pain, joint pain, cutaneous hyperalgesia, and complex regional pain syndrome (CRPS).

Spinal Ionotropic Receptors Mediating Excitatory Pain Transmission

AMPA AND KAINATE RECEPTORS: RAPID, NONCONTINGENT ACTIVATION

Although the AMPA/kainate family of glutamate receptors long has been known to be involved in synaptic transmission from nonnociceptive afferent fibers in nonpathologic states, this family of receptors is now thought to be involved in nociceptive neurotransmission, particularly in neuropathic pain states. AMPA receptors are localized in the most superficial band of the dorsal horn; although generally thought to be located predominantly on postsynaptic neurons in the dorsal horn, recent evidence suggests their presence on afferent fibers as well.[4] These receptors, when occupied by an agonist, will produce strong depolarization that is not dependent on concurrent activity, making this process *noncontingent*.

A multiplicity of receptor subunits for AMPA (GluR1–4) and kainate (GluR5–7) are expressed in the spinal cord, both dorsal and ventral horns. These receptors mediate glutamate-induced decreases in the activation threshold of nociceptive neurons; antagonist blockade of these receptors (e.g., CNQX or NBQX) reduces synaptic activation of dorsal horn neurons by all afferent fibers and most peripheral stimuli. Therefore, spinally-administered AMPA antagonists produce paralysis at doses at or near those doses with antinociceptive effects.

NMDA RECEPTORS: SLOWER ACTIVATION, CONTINGENT ON PRIOR DEPOLARIZATION

NMDA receptors are involved more directly in responses of dorsal horn neurons to intense noxious stimuli than are AMPA/kainate receptors. NMDA receptors are localized in the superficial dorsal horn, but their distribution beyond the substantia gelatinosa is more extensive than that of AMPA receptors. Although a majority of NMDA receptors are expressed on secondary neurons, some receptors are contained presynaptically on primary afferent terminals in the dorsal horn. Postsynaptic receptors mediate synaptic activation of dorsal horn neurons by primary afferent input, whereas the presynaptic receptors could promote or modulate release of transmitter from primary afferent fibers.[5] The latter role suggests the existence of glutamate-mediated feedback (positive or negative) onto the fibers that released it.

Several NMDA receptor subtypes also have been identified. NR1 subunits are thought to combine with NR2A, NR2B, NR2C, or NR2D subunits, each of which confers different temporal characteristics of desensitization and different susceptibility to modulation by kinases. In situ hybridization shows intense labeling for NR1 subunits throughout the dorsal and ventral gray matter, accompanied by light staining in superficial layers by NR2B and NR2D subunits. NMDA receptors in the spinal cord are subject to positive modulation by PKC.

The NMDA-operated channel can be blocked by dissociative anesthetics, including phencyclidine (PCP), ketamine, MK-801, and the inactive opioid congener dextrorphan. In addition, competitive antagonists such as D-2-aminophosphonovaleric acid (D-APV, or AP5) block ligand-receptor interactions. On the other hand, glycine, at nanomolar concentrations, potentiates NMDA responses. Because both APV and CNQX (antagonists at NMDA and AMPA receptors, respectively) decrease nociceptive responses, both EAA components must be intact for complete nociceptive transmission. Whereas AMPA receptor activation decreases the threshold for activation of nociceptive neurons, activation of NMDA receptors enhances responses to the most intense stimuli without altering threshold. Interestingly, mu opioid agonists block the effects of NMDA but not those of AMPA, suggesting that opioids act selectively on hyperalgesia and that hyperalgesia is a predominantly NMDA-mediated action.

Exogenously applied NMDA mimics noxious stimulation and, unlike AMPA agonists, induces a state of hyperalgesia. NMDA selectively excites nociceptive neurons, and NMDA antagonists can block behavioral and neuronal responses to intense nociceptive stimulation. This action is most evident in situations where afferent nociceptors are tonically activated by chemical stimuli and least evident in rapid reflexive tests, such as the tail-flick test. These characteristics are consistent with the biophysical character of the NMDA receptor cation channel. Blockade of the channel by Mg^{2+} at resting potential prevents ligand gating of the channel; depolarization of the plasma membrane removes this Mg^{2+} block, allowing subsequent ligand gating. Removal of this Mg^{2+} block may underlie the participation of this receptor in the "windup" phenomenon, recruited by repetitive activation of C fibers. Intense activation of primary afferent nociceptors is thought to release sufficient glutamate to depolarize postsynaptic neurons via postsynaptic AMPA receptors, enabling subsequent enhanced responses mediated by NMDA receptors. Finally, NMDA receptors are critical participants in the induction of acute (formalin) and chronic (chronic constriction injury, CCI) hyperalgesia. Interestingly, the expression of this hyperalgesia is mediated partly by AMPA receptors; this characteristic is similar to that seen in hippocampal long-term potentiation (LTP). The net result of ionotropic EAA receptors on nociception is that both NMDA and AMPA receptors are important for dorsal horn neuronal responses to noxious stimuli and that antagonism of either receptor can have profound effects on nociceptive responsiveness. Because the activity induced by occupancy of these receptors is dependent on temporally contiguous membrane activity, their activation is said to be *contingent*.

Spinal G Protein–Coupled Receptors Mediating Excitatory Pain Transmission

Neurokinin Receptors

Neurokinin receptors include NK_1, NK_2, and NK_3 receptors, which have the highest affinity for SP, neurokinin A (NKA), and neurokinin B (NKB), respectively. NK_1 receptors are densely distributed on the somata and dendrites of many superficial dorsal horn neurons, as well as on some neurons in the deeper dorsal horn. SP evokes prolonged EPSPs after a long latency in spinal neurons and only excites nociceptive neurons. Furthermore, nerve terminals that contain SP appose SP-responsive neurons, and NK_1 receptor immunoreactivity is internalized following peripheral injection of a noxious chemical stimulus. Application of exogenous SP produces behaviors suggestive of nociception and causes hyperalgesia. Antagonists of the NK_1 receptor decrease the afterdischarge that occurs in nociceptive neurons in response to an intense noxious stimulus and attenuate the hyperalgesia that occurs in response to persistent noxious heat and administration of a noxious chemical. However, NK_1 receptors may not contribute to thermal hyperalgesia in neuropathic pain, and clinical trials of NK_1 antagonists for treatment of pain failed to demonstrate efficacy.[6] The role of NK_1 receptors in hyperalgesia probably involves potentiation of NMDA receptor excitability via coupling of NK_1 receptors to PLC, activating PKC. NK_2 receptors also contribute to long latency responses and increased excitability following C-fiber activation by the same pathway involving PKC.

Metabotropic EAA Receptors

Some metabotropic receptors for glutamate (mGluR) participate in nociceptive signaling by coupling to the same excitatory intracellular signaling systems activated by NK_1 receptors. Metabotropic glutamate receptors have been shown to participate in the hyperexcitable state following inflammation of the knee joint, suggesting that they are involved in more tonic forms of nociception. It is important to recognize the heterogeneity of mGluRs and of their coupling to intracellular signaling systems. Eight subtypes have been identified (mGluR1–8), and at least one subtype of the excitatory class (mGluR5) is localized on somata and dendrites in the dorsal horn. In addition to coupling through G_q to PLC and PKC, several subtypes of mGluRs can couple via G_i to inhibit adenylyl cyclase or open K^+ channels, as do analgesic receptors such as opioid receptors. Therefore, the actions of glutamate or other nonselective agonists, such as (1S, 3R)-1-aminocyclopentane-1,3-dicarboxylic acid (ACPD), at mGluRs is expected to be complex, having both excitatory and inhibitory components.

Excitatory Effector Systems Receiving New Attention

NITRIC OXIDE (NO)

One of the enzyme isoforms that synthesizes NO from arginine, neuronal nitric oxide synthase (nNOS), is localized in some neurons in the dorsal horn and is activated by Ca^{2+} via calcium-calmodulin. NO has characteristics unique among neuronal signaling molecules in that it is a free radical that is freely permeable to aqueous and lipid barriers. Therefore, a point source of NO will affect other neurons and blood vessels within a sphere of radius governed by the diffusion coefficient of NO and by its half-life, on the order of 200 to 500 μm. Inhibition of NOS by such drugs as L-NAME blocks behavioral and electrophysiologic responses to noxious stimuli (intraplantar injections of formalin or carageenan) and exogenous NMDA. NOS activation after strong nociceptive input is likely involved in the plastic changes that follow such stimulation. nNOS may co-localize with γ-aminobutyric acid (GABA) in dorsal horn interneurons. Long-term potentiation in the spinal cord may rely on activation of NMDA receptors and NOS.[7] The action of NO is not solely facilitative; NO-induced oxidation of sulfhydryls on the extracellular surface of NMDA receptors downregulates those receptors.

PURINERGIC RECEPTORS

Adenosine triphosphate (ATP) acts at P_2 purinergic receptors to depolarize dorsal horn neurons, suggesting a postsynaptic location. P_{2X} receptors are ligand-gated cation channels, several subtypes of which have been localized on primary afferent terminals in the superficial dorsal horn.[8,9] Therefore, the mode of depolarization of postsynaptic neurons may involve the release of another excitatory transmitter, most likely glutamate.[10] Currently, there are no pharmacologic agents selective for the different P_{2X} receptors. However, recent studies using antisense oligonucleotides have suggested that the P_{2X3} subtype is involved in the development and maintenance of chronic pain,[11,12] consistent with its localization on nociceptive primary afferent neurons.[9,13] In addition to the ionotropic P_{2X} receptors, ATP also acts at metabotropic P_{2Y} receptors. Although the role of these receptors in pain is just beginning to be explored, their activation may produce long-term changes in nociceptor excitability.[14]

PROSTAGLANDINS

Strong synaptic activation of the spinal cord dorsal horn invokes prostaglandin (PG) synthesis; released prostaglandins activate PG receptors located either presynaptically or postsynaptically, thereby enhancing spinal nociceptive transmission. PGE_2 activates EP_2 receptors, which couple via G_s to activate adenylyl cyclase, elevate cyclic adenosine monophosphate (cAMP), and activate protein kinase A (PKA), enhancing nociceptive neurotransmitter release from primary afferent fibers. $PGF_{2\alpha}$ activates EP_3 receptors, which couple via G_q to activate PLC, producing DAG and activating PKC, increasing postsynaptic responsiveness. Spinal administration of inhibitors of PG synthesis (nonsteroidal anti-inflammatory drugs (NSAIDs), including nonselective inhibitors of cyclooxygenase such as aspirin or ketorolac, and selective inhibitors of COX-2, such as celecoxib) or EP receptor antagonists can attenuate nociceptive transmission. Therefore, NSAIDs, in addition to decreasing peripheral inflammation, probably have important analgesic actions within the CNS.[15]

CAPSAICIN AND VANILLOID RECEPTORS

Capsaicin and synthetic derivatives, such as resiniferatoxin, excite mostly C-polymodal nociceptors and Aδ mechanoheat-sensitive nociceptors. Capsaicin acts by binding to a nonselective cation channel cloned and identified as the vanilloid receptor, VR-1.[16,17] This channel, when activated by capsaicin, resiniferatoxin, low pH, or elevated temperature, allows calcium and other cations to enter C-fiber terminals, depolarizing them and increasing intracellular calcium levels. Depolarization leads to action potential generation signaling pain, and increased intracellular calcium leads to release of peptide neurotransmitters such as SP, eliciting neurogenic inflammation. High doses of capsaicin or resiniferatoxin desensitize nerve fibers to subsequent applications of the substance of physical or chemical stimuli sufficient to activate the afferent fibers containing the receptors. Desensitization of nociceptors following local application of capsaicin-related compounds has therapeutic value for analgesia, as well as anti-inflammatory action, by blocking excitation of the terminals and decreasing their synthesis of peptides. Resiniferatoxin applied to cultured dorsal root ganglion neurons induces cytotoxicity selectively directed to nociceptors,[18] and applied epidurally to rats, produces profound, prolonged regional analgesia.[19]

The VR-1 receptor belongs to the large superfamily of transient receptor potential (TRP) channels. Family members have been shown to respond to a wide variety of stimuli, including heat, cooling, osmolarity, cell volume, phorbol esters, H_2O_2, protons, bradykinin, and PLC.[20] The importance of this family in a plethora of biologic functions is just beginning to be elucidated, and many of them likely will be found to be involved in nociceptive processing.

Interactions Among Excitatory Receptors

Multiple neurotransmitters are released from primary afferent fibers on activation by peripheral noxious stimulation. The actions of these transmitters are not singular events because the transmitters interact with multiple receptors, and the receptor systems activated interact to alter the activity of spinal cord neurons. Interactions between excitatory transmitter-receptor systems appear to be importantly involved in the responses of dorsal horn neurons to noxious stimuli. For example, NK_1 and NMDA receptors may synergize because NK_1-driven PKC would enhance the responsiveness of NMDA receptors. In view of the high probability that glutamate and SP are co-released from primary afferent terminals on activation by nociceptive stimuli, this positive and excitatory interaction probably figures prominently in the overall responses of neurons to pain-driven input to the dorsal horn. One might predict that parallel transmission systems would render analgesic interruption of one of them ineffective; for example, NK_1 antagonists may have shown low analgesic efficacy in clinical trials because of this parallelism.

▶ TRANSMITTERS AND RECEPTORS MEDIATING INHIBITION AND ANALGESIA

Afferent processing through the spinal cord has been shown to be subject to powerful regulation by a wide variety of inhibitory spinal receptor systems. Many analgesic agents act as agonists at receptors mediating this inhibitory transmission. The World Health Organization's (WHO) analgesic ladder guides the treatment of pain: (1) mild to moderate pain often will respond to nonopioid drugs alone such as the NSAIDs, (2) pain of greater severity can be relieved by milder opioids, and (3) severe pain requires stronger opioids, sometimes with adjuvant therapy.[21]

Peptidergic Modulation of Nociception

Opioid Peptides and Receptors

Opioid receptors exert powerful regulatory effects of the processing of afferent-evoked activity at the spinal level, as well as at peripheral nociceptive endings and supraspinal structures. Early binding studies emphasized that opioid-binding sites could be identified both on primary afferent terminals and on membranes of secondary neurons. The two spinal sites and their underlying mechanisms often are generically termed *presynaptic* and *postsynaptic*, respectively. The endogenous activators of these receptors include met- and leu-enkephalin (preferring delta opioid receptors), dynorphin (preferring kappa opioid receptors, but see the discussion of nonopioid actions below), and endomorphins (preferring mu opioid receptors). Most clinically used exogenous opioid analgesic agonists are thought to target mu opioid receptors.

PRESYNAPTIC OPIOID RECEPTORS

Mu opioid receptors are expressed by a proportion of DRG neurons, mostly smaller neurons thought to give rise to pain-encoding C fibers. Immunohistochemical studies have localized mu and delta opioid receptors to small-diameter primary afferent neurons, and delta opioid receptors appear to be expressed only presynaptically on SP-containing neurons, where it is trafficked both centrally into the spinal cord and to nerve endings in the periphery.[22] Morphine applied to isolated hind-paw skin overlying inflamed tissue inhibits a similar fraction of nociceptive afferents.

INHIBITORY EFFECTS

Agonist activation of opioid receptors on the spinal terminals of primary afferent neurons or sensory nerve terminals in inflamed tissue inhibits terminal activation and the release of transmitter. Agonist action at mu and delta opioid receptors inhibits DRG neurons either by increasing potassium conductance or by directly inhibiting N- and P-type calcium channels; activation of kappa receptors suppresses only calcium currents. The result of these presynaptic actions decreases excitatory transmitter release, reducing the size of excitatory postsynaptic potentials.

EXCITATORY EFFECTS

Some concentrations of mu, delta, and kappa opioid receptor agonists also can produce excitatory effects under some circumstances. Both opioid tolerance and persistent hyperalgesia require a common chain of coupled events: activation of NMDA receptors, NOS, and PKC. The similarities between mechanisms underlying opioid tolerance and those inducing persistent pain states suggest that these low-dose effects also may invoke changes in the spinal cord similar to the cascade outlined earlier.

G-PROTEIN COUPLING

The inhibitory actions of opioids on primary afferent neurons are mediated by pertussis toxin–sensitive G proteins, i.e., either G_i or G_o. Inhibition of calcium channels probably is mediated by G_o. Opioid receptors on primary afferent neurons probably couple through G_i to inhibit adenylyl cyclase, decreasing levels of cAMP and thereby decreasing transmitter release. Another mode of G-protein coupling, activation of PKC presumably by G_q and DAG, paradoxically has been implicated in the inhibition of calcium channels by opioids in DRG neurons.

OPIOID RECEPTORS ON PERIPHERAL AFFERENT TERMINALS

Opioid inhibition of sensory transduction at the peripheral terminals of nociceptors appears to occur only when nociceptors are sensitized, as in peripheral tissue injury, perhaps due to increased expression of opioid receptors on afferent terminals in the inflamed tissue. For example, opioid agonists decrease spontaneous firing in small-diameter primary afferent neurons innervating inflamed joints, prevent the antidromic release of SP into the knee joint, and decrease stimulus-evoked firing of cutaneous afferents.[23] Opioids injected directly into inflamed tissue decrease the hyperalgesia exhibited in response to thermal and mechanical stimuli, as well as autonomic responses to compression of the inflamed knee joint.[24] Hyperalgesia in these cases can be reversed by agonists selective for mu, delta, or kappa opioid receptors, consistent with the occurrence of multiple opioid receptors on individual sensory neurons.

POSTSYNAPTIC OPIOID RECEPTORS
Opioid receptors also participate postsynaptically in neurotransmission in spinal dorsal horn, and mu and kappa receptors occur on both somata and dendrites of dorsal horn neurons; participation of mu receptors is best documented. Kappa receptors, on the other hand, seem to mediate mostly pronociceptive effects, and postsynaptic effects for delta opioid agonists rarely are seen. All three opioid receptor subtypes couple preferentially through G_i or G_o proteins to inhibitory effectors inhibiting adenylyl cyclase, reducing Ca^{2+} influx, or activating K^+ efflux. Postsynaptically, these actions hyperpolarize neurons, impairing depolarization and rostrad propagation of nociceptor-driven action potentials. However, some evidence points to excitatory coupling of some opioid receptor subtypes: mu opioid agonists also can enhance NMDA-induced currents via activation of PKC and reduce GABAergic inhibition by inhibiting GABA-containing interneurons.

Neuropeptide Y Receptors
Neuropeptide Y (NPY) receptors occur primarily on small DRG cells, but the distribution shifts to include larger cells after peripheral nerve injury. NPY receptors may mediate analgesic effects, perhaps because they block SP release by inhibiting calcium currents.

Monoaminergic Transmission, Adrenergic and Serotonergic Receptors

α_2-Adrenergic Receptors
Spinally-administered α_2-adrenergic receptor (α_2AR) agonists produce analgesia in acute and chronic pain states. Norepinephrine itself, released in spinal cord from bulbospinal pathways, has a largely inhibitory effect mediated by α_2ARs on nociceptive transmission in the spinal cord.

PRESYNAPTIC α_2-ADRENERGIC RECEPTORS
α_2-adrenergic receptors on the spinal terminals of primary sensory afferent fibers predominantly mediate inhibition of nociceptive neurotransmission by inhibition of transmitter release from primary afferent neurons via pertussis toxin–sensitive G proteins (G_i or G_o). Three subtypes of α_2ARs have been cloned in human and rat, α_{2A}, α_{2B}, and α_{2C}. The α_{2A}ARs are localized mostly on the central terminals of SP-containing, capsaicin-sensitive primary afferent neurons.[25] The lack of subtype-selective pharmacologic agents has made difficult determination of the relative contributions of the three α_2AR subtypes to spinal α_2AR agonist–mediated analgesia. Pharmacologic and genetic studies indicate that activation of α_{2A}ARs mediates the action of most α_2AR agonists,[26] although the α_{2C}AR may mediate the effects of some α_2 agonists.[27,28] Although most effects of α_2AR agonists are inhibitory, excitatory effects of norepinephrine via an α_2AR have been observed in some C-polymodal nociceptors after peripheral nerve injury.

POSTSYNAPTIC α_2-ADRENERGIC RECEPTORS
Unlike spinal α_{2A}ARs, spinal α_{2C}ARs are found on both primary afferent fibers and spinal neurons.[25] Recent immunohistochemical associations between glutamate transporters and α_{2C}ARs suggest that activation of these receptors would have an inhibitory outcome on synaptic transmission in the dorsal horn.[29,30] Therefore, postsynaptic effects of norepinephrine or exogenous α_2AR agonists in the spinal cord are likely mediated by α_{2C}AR, and functional studies indicate α_2AR-mediated inhibition of dorsal horn neuronal responses to noxious stimulation, apparently via a $G_{i/o}$ protein-mediated pathway, promoting K^+ efflux through G-protein-coupled inwardly rectifying K^+ (GIRK) channels. Whether coupling of this receptor to ion channels or adenylyl cyclase contributes most to the spinal analgesic and CNS-wide toxic side effects of α_2AR agonists is unclear at present. Whereas agents in current clinical use (e.g., clonidine) and others in development (e.g., dexmedetomidine) act largely via α_{2A}ARs and have significant CNS depressant side-effect profiles, other adrenergic agents with possible utility (e.g., moxonidine) may act at non-α_{2A}AR sites[27,28] and have been shown clinically to have reduced side effects compared with clonidine.

Serotonin Receptors
Serotonin-mediated descending inhibition of spinal pain transmission has received much attention. Although the drug classes for treatment of chronic neuropathic pain include the tricyclic antidepressants (TCAs), the mechanism responsible for the TCAs' therapeutic effect is in doubt. Blockade of reuptake of serotonin may account for some of the analgesic activity of the TCAs, such as amytriptyline. However, intrathecally-administered serotonin or activation of descending serotonin-releasing systems (e.g., nucleus raphe magnus, or NRM) can either inhibit or stimulate nociceptive reflexes depending on the dose and species tested. Three distinct serotonin receptor subtypes probably account for this multiplicity of effects: $5\text{-}HT_{1B}$ ($5\text{-}HT_{1D}$ in humans), $5\text{-}HT_2$, and $5\text{-}HT_3$

receptors. Most investigators concur that the 5-HT$_{1B/D}$ receptor subtype is most effective and selective in inhibiting nociceptive neurons via G$_i$ or G$_o$ signaling. Interestingly, new evidence indicates that the over-the-counter analgesic acetaminophen (paracetamol) acts, at least in part, at spinal 5-HT$_{1B/D}$ receptors, as well as 5-HT$_{1A}$ receptors,[31] although other groups attribute its analgesic effect to 5-HT$_3$ receptors.[32] 5-HT$_3$ receptors are excitatory receptors (ligand-gated cation channels) that mediate pronociceptive effects in the periphery and antinociceptive effects in the spinal cord involving both GABA$_B$ and GABA$_A$ receptors via activation of GABAergic interneurons. Similarly, agonist activation of spinal 5-HT$_{2A/C}$ receptors produces both pro- and antinociceptive actions[33] by a mechanism similar to SP (G$_q$-mediated activation of PLC liberating intracellular Ca^{2+} and the PKC activator DAG; see above).

Amino Acid and Purinergic Transmission Contributing to Inhibition of Nociception

Glutamate Receptors

Several of the metabotropic glutamate receptors (mGluR2, mGluR4, and mGluR6–8), similar to receptors for analgesic substances such as opioids and norepinephrine, couple to G$_{i/o}$ proteins. Activation of G$_{i/o}$ reduces the open time (α_o) of voltage-gated calcium channels, increases conductance (G$_{\beta\gamma}$) of a particular kind of potassium channel called *G-protein-coupled inwardly rectifying K$^+$ channels* (GIRK channels), and decreases synthesis (α_i) of cAMP, all of which can result in decreased transmitter release from terminals by reducing calcium influx, and reducing cAMP levels and activity of PKA. Phosphorylation of K$^+$ channels by PKA is thought to decrease their conductance or open time, whereas phosphorylation of Ca^{2+} channels increases their conductance or open time. Therefore, decreasing activity of PKA would be expected to decrease excitation-release coupling. Unlike cAMP, the effects of βγ dimers are restricted to the immediate vicinity of the receptor and G protein initiating the action.

GABA Receptors

GABA acts as an agonist at both ionotropic (GABA$_A$) and metabotropic (GABA$_B$) receptors. Presynaptic GABA$_B$ receptors may mediate inhibition of transmitter release by mechanisms similar to α_{2A}ARs and opioid receptors because the effect of GABA is mediated by a pertussis toxin–sensitive G protein. However, GABA$_A$ receptors also may participate in the effect. The general consensus is that GABA$_B$ receptors are primarily responsible for inhibition of transmitter release from primary sensory afferent fibers by inhibition of Ca^{2+} influx mediated by G$_o$. The GABA$_B$ agonist baclofen appears to be more efficacious against chronic neuropathic pain than GABA$_A$ agonists such as muscimol.

Postsynaptic GABA$_B$ receptors, which are ligand-gated anion channels selective for Cl$^-$, likely mediate moderation of dorsal horn neuronal responses to peripheral activation. Therefore, their activation would be expected to hyperpolarize dorsal horn neurons or hold them near resting potential by a shunting action (i.e., increased driving force toward the Cl$^-$ equilibrium potential). Serotonin released from axons that descend from the brain stem may stimulate release of GABA from interneurons, which in turn can inhibit dorsal horn neurons. GABA$_A$ antagonists elevate responses to EAA agonists and block the inhibition produced by agonists selective for 5-HT$_3$ serotonin receptors. It is notable, however, that anesthetics and antianxiety agents such as barbiturates and benzodiazepines, which act as agonists or potentiators of GABA at these receptors, fail to produce analgesia. It is conceivable that the rapid temporal characteristics of GABA$_A$ receptors (i.e., inhibitory potential durations of several milliseconds) reduce the analgesic efficacy of these agents.

Glycine Receptors

Glycine receptors, like GABA$_A$ receptors, are ligand-gated anion channels selective for Cl$^-$. Because their activation similarly would hold dorsal horn neurons near resting potential by a shunting action, their effects would be expected to be similar. Whereas GABA agonists (see above) and glycine do not exert strong analgesic effects, their receptors clearly play major roles in regulating the ongoing encoding of afferent activity because antagonists to GABA$_A$ (not GABA$_B$) and glycine (strychnine) receptors increase the ability of low-threshold mechanical stimuli (touch) to evoke large increases in pain behavior.

A$_1$ Adenosine Receptors

Although both A$_1$ and A$_2$ adenosine receptors appear to be involved in spinal antinociception, A$_1$ adenosine receptors are likely the dominant mediators of antinociception produced by adenosine and its analogues. Most spinal adenosine receptors are localized on intrinsic neurons, and intrathecally administered adenosine analogues inhibit nociceptive responses and moderate some neuropathic states. Opioid agonists promote release of adenosine at spinal sites, and adenosine interacts with opioid and non-opioid analgesic systems. Manipulations elevating endogenous adenosine produce analgesia, whereas intrathecally administered adenosine antagonists (e.g., caffeine and theophylline) enhance nociception.

Interactions Among Inhibitory Receptors

Endogenous analgesic systems are similarly coactivated on activation of nociceptive pathways or in situations of stress. As with the excitatory transmitters discussed earlier, the inhibitory transmitters also activate multiple receptors, and the effects of receptor co-activation underlie the observed analgesic effects. Interactions between inhibitory transmitter-receptor systems are also prominent in spinal nociceptive processing. Both clinical and mechanistic studies have documented the synergistic interaction between opioid and α_2-adrenergic agonists; the presence of low levels of one agonist class can reduce the amount of the other agonist required for full analgesic effect by tenfold or more. Furthermore, the clinically used agent, tramadol, may owe its efficacy and safety to simultaneous activation of adrenergic and opioid receptors. Mechanistic studies indicate that adenosine and opioid agonists interact similarly. The synergistic receptor pairs $\delta OR/\mu OR$ and $\alpha_{2A}AR/\mu OR$ have nonoverlapping distributions in the rat spinal cord, suggesting that these synergistic interactions result from action in multiple cellular elements. The synergistic receptor pair $\delta OR/\alpha_{2A}AR$ is highly co-localized in the spinal cord, suggesting that this synergistic interaction requires some interaction at the G-protein/effector level within common cellular elements. Exploiting such synergistic pairs of spinally active analgesic agents may yield therapeutic regimens in the future with markedly reduced side-effect profiles.

▶ OTHER TARGETS: CELLS OR MOLECULES ALTERED IN CHRONIC PAIN

Presynaptic Inhibition by Blockade of Voltage-Gated Ion Channels

Sodium Channels
Sufficient blockade of sodium channels in normal peripheral axons ultimately leads to conduction failure. Peripheral axons change their expression and trafficking of sodium channel subtypes after nerve injury, possibly contributing to chronic pain states. Altered expression and trafficking of tetrodotoxin-resistant channels (TTX-Rs, $Na_V1.8$ and $Na_V1.9$) versus TTX-sensitive channels (TTX-Ss, $Na_V1.3$) may render the axons hyperexcitable,[34] possibly explaining the efficacy of low doses of local anesthetic agents in relieving chronic pain states (neuroma or injured tissue) and in synergizing with spinal morphine. The mechanism of action of many anticonvulsant agents (e.g., carbamazepine and lamotrigine) likely also involves favoring inactivation of sodium channels, thereby discouraging rapid firing in afferent fibers. These agents are less effective blockers of sodium channels than traditional local anesthetics, giving them an improved side-effect profile that lacks local anesthetic effects at therapeutic levels.

Calcium Channels
Other targets for ion channel blockade are voltage-gated Ca^{2+} channels in terminals of primary afferent fibers. Synaptic neurotransmission from primary afferent neurons requires the arrival of action potentials at nerve terminals in the spinal dorsal horn that trigger the opening of voltage-gated Ca^{2+} channels, increasing cytoplasmic Ca^{2+} and initiating exocytotic release of transmitter. N-type Ca^{2+} channels have been implicated in the release of peptides from primary afferent neurons, and P- and Q-type channels are involved in the release of glutamate from central neurons. Intrathecal administration of N- and P-type calcium channel blockers decreases nociceptive responsiveness, and the new spinal therapeutic agent ziconotide (SNX-111) selectively blocks N-type Ca^{2+} channels. Finally, the novel anticonvulsant gabapentin binds with high affinity to a subunit of voltage-gated Ca^{2+} channels and may produce its therapeutic effects in neuropathic pain states via a similar mechanism.

Receptors for Other Peptides, Growth Factors, and Cytokines

Spinal Peptide Levels
The expression of SP and CGRP is upregulated in primary afferent neurons during peripheral inflammation. Peptides not normally present in sensory neurons of monkey and rat, such as cholecystokinin 8 (CCK-8) and NPY, appear following nerve injury (see below). Dynorphin is upregulated following nerve injury and, when applied intrathecally or released from endogenous terminals, likely participates in the induction of prolonged hyperalgesic states.[35]

Role of Growth Factors and Cytokines
The changes in the expression of peptides by primary afferent neurons in response to either inflammation or peripheral nerve injury reflect, in part, the availability of growth factors. Cells of the immune system, Schwann cells, and skin provide peripheral sources of nerve growth factor (NGF) and other growth factors. High-affinity receptors for NGF (TrkA) on primary afferent neurons expressing CGRP and SP mediate regulation of the expression of CGRP and SP in adult neurons. Conversely, decreased transport of NGF after axotomy or peripheral nerve injury may underlie decreased expression of CGRP and SP in primary afferent neurons in these conditions. In contrast, the TrkC receptor, which exhibits high affinity for the growth factor neurotrophin 3 (NT-3), is expressed most frequently among muscle afferents. Cutaneous afferents express either TrkA or TrkC receptors. Expression of neurotrophins, such as NGF, and pro-inflammatory cytokines, such as interleukin 1β (IL-1β), IL-6, and tumor necrosis factor α (TNF-α), normally expressed in low concentrations in peripheral nerve or spinal cord, increases after nerve injury and in neuropathic pain states. Growth factor receptors are being explored as potential therapeutic targets for chronic pain.[36] The importance of these novel mediators in the spinal cord to the induction of chronic

pain states is further supported by recent anti-inflammatory cytokine manipulations.[37]

Plasticity in Connections or Responsiveness of Dorsal Horn Neurons

The term *neural plasticity* generally refers to long-term changes in the nervous system. Two changes relevant to the induction and treatment of chronic pain include neuropathic pain and opioid tolerance. Both changes involve the spinal cord, and both changes involve the same excitatory cascade that we term the *NMDA/NOS cascade:* glutamate release, NMDA receptor activation, postsynaptic Ca^{2+} influx, and NOS activation (reviewed in ref. 38). With the possible exception of dextromethorphan, several NMDA receptor antagonists have failed to survive toxicity screens, suggesting that high-affinity, high-potency NMDA antagonists are inherently toxic. Recently, a less potent and less toxic endogenous blocker of both NMDA receptors and the NOS enzyme, agmatine, has been shown to defeat opioid tolerance[38] and the development of neuropathic pain.[39] The combination of mechanisms exhibited by agmatine, together with its apparent antihyperalgesic efficacy, suggests a new strategy for moderating plasticity. Perhaps agents that antagonize multiple steps in the NMDA/NOS cascade will prove to be more effective at preventing permanent plastic changes underlying chronic pain states without the side-effect profile of agents directed specifically at one step in the cascade.

▶ PAIN IN THE PATHOLOGIC STATE

Inflammatory versus Neuropathic Pain

One of the most significant health problems in the U.S. is the inadequate treatment of pain. The impact of pain is determined not only in economic terms (approximately $100 billion annually) but also in terms of human suffering.[40] Estimates suggest that as many as one-third of all Americans suffer from some form of chronic pain at some point in their lives. Pain of inflammatory origin is relatively amenable to treatments with currently available therapeutic agents. In contrast, neuropathic pain is resistant to conventional therapeutic approaches in many instances. It has been estimated that one-third of chronic pain patients have pain that is resistant to the treatment efforts of the medical community. The behavioral manifestations of inflammatory and neuropathic pain may present with similar symptoms, which include exaggerated sensitivity to nociceptive stimuli (hyperalgesia) and a perception of normally innocuous stimuli as being painful (allodynia).[41] Importantly, however, the underlying neurobiology of these forms of chronic pain differs substantially.

Nerve damage arising from either trauma or disease affecting peripheral nerves can lead to abnormal pain states. Such pain may be long lasting and continue for long periods after the initial injury has healed or in the absence of observable tissue damage. Among the many causes of neuropathic pain are disease processes such as diabetic neuropathy, herpes zoster infection (shingles), and complications from acquired immune deficiency syndrome (AIDS). Neuropathic pain also may arise from physical injuries such as nerve traction or compression and radiation or chemotherapy for lung and other cancers. The neuropathic dysesthesias frequently are characterized by burning or shooting pains, along with tingling, crawling, or electrical sensations. Painful neuropathy represents a seriously debilitating syndrome that affects millions of people, and their quality of life suffers significantly.

Sodium Channels

Tissue injury in the vicinity of sensory primary afferent neurons often results in abnormal repetitive discharge or exaggerated response to subsequent sensory stimuli that is believed to contribute to chronic inflammatory pains. Additionally, central projection neurons that relay sensory afferent signals to the sensory cortex also may become hyperresponsive, a process termed *central sensitization*. Voltage-gated sodium channels (VGSCs) play a fundamental role in the excitability of all neurons. These channels are located in the plasma membrane and mediate the influx of sodium ions into the cell in response to local membrane depolarization, resulting in generation of the action potential. Consequently, changes in the expression or function of VGSCs have a profound effect on the firing pattern of sensory primary afferent neurons. Both physiologic and pharmacologic evidence implicates a critical role for VGSCs in the development and maintenance of hyperexcitability observed in primary afferent neurons following nerve and tissue injury. Importantly, use-dependent sodium channel inhibitors, which block channels that have been activated recently, are clinically effective in the treatment of many types of chronic pain.[42]

Subtypes of Sodium Channels

The VGSCs have been characterized on the basis of their biophysical characteristics, such as stimulus-response properties and action potential conduction velocity. Immunohistochemical or molecular biologic techniques have been used to identify the different VSGCs present on different classes of sensory neurons (see ref. 42 for review). Electrophysiologic characterization of VGSCs present in sensory neurons, in combination with the neurotoxin TTX, indicates that there are two general classes of current in sensory neurons: One is blocked by TTX (TTX-sensitive, or TTX-S), and the other is insensitive to

TTX (TTX-resistant, or TTX-R). TTX-S currents are blocked by TTX at concentrations in the low nanomolar range. These VGSCs tend to have a low threshold for activation (between –55 and –40 mV), are rapidly activating, and are rapidly inactivating. Approximately 50 percent of these channels are available for activation at potentials close to resting membrane potential.[43] VGSCs that are TTX-R have been subdivided further into several different classes of ionic current on the basis of distinct biophysical properties. One of these TTX-R currents has similar biophysical properties to those of TTX-S channels, with a low threshold for activation and relatively rapid rates of activation and inactivation. This low-threshold TTX-R current has been referred to as TTX-R3,[44] or fast TTX-R, current.[45] A second TTX-R (TTX-R1) current is resistant to TTX at concentrations greater than 10 mM, has a high activation threshold, and activates and inactivates relatively slowly but reprimes rapidly.[43] This current may account for the high activation threshold observed in nociceptive afferents, and the sodium channel alpha subtype, $Na_V1.8$, likely underlies this high-threshold TTX-R current. The nuclear injection of $Na_V1.8$ cDNA into sensory neurons isolated from $Na_V1.8$ knockout mice results in the expression of a TTX-R current identical to TTX-R1.[46]

There is evidence that $Na_V1.8$ channels are present and functional in peripheral terminals of nociceptive afferents. It was found that TTX-R channels mediated action potential initiation in polymodal nociceptive afferents and that these initiation sites are very close to, if not at, the terminal endings.[47] The TTX-R currents may contribute to the release of transmitter from the central terminals of nociceptive afferents.[48] The release of ATP from primary afferent terminals facilitated additional release of glutamate from the same terminals. Importantly, the additional glutamate release was dependent on active conduction in the afferent terminal that, in turn, was dependent on the activation of TTX-R channels in the afferent terminal.[48] Thus, VGSCs are clearly linked to neuronal excitability, and changes in populations of these channels may enhance the ability of sensory neurons to respond to external stimuli and may mediate sensitized pain states.

Sodium Channels and Inflammatory Pain

It is well established that tissue injury results in local inflammation; enhanced pain is one of the cardinal signs associated with this inflammation, and this pain reflects an increase in the excitability of afferent neurons innervating the injured tissue. This increase in excitability reflects the actions of a number of inflammatory mediators, including ATP, bradykinin, serotonin, cytokines such as TNF-α, and prostaglandins. That analgesic agents such as the NSAIDs act to inhibit the production of prostaglandins from arachidonic acid by blocking cyclooxygenases (COX) and are highly efficacious in alleviating inflammatory hyperalgesia suggests that prostaglandins are critical inflammatory mediators that promote pain.[49,50] It is believed that prostaglandins produce hyperalgesia and/or nociceptor sensitization while causing little direct activation of nociceptive terminals. Prostaglandins have been shown to sensitize nociceptors to all modes of stimuli tested, including mechanical, thermal, and chemical.[51] Prostaglandin E_2 (PGE_2) modulates TTX-R sodium current in sensory neurons in a manner consistent with an underlying mechanism of sensitization; the current activates at more hyperpolarized potentials, and the magnitude of the current is increased, as are its rates of activation and inactivation.[52] As discussed earlier, TTX-R sodium currents are essential for action potential generation in the majority of nociceptive neurons. Thus, the ability of PGE_2 to modulate the activity of these channels presents a highly effective mechanism by which PGE_2 can selectively enhance the excitability of the nociceptive neurons. Support for a causal relationship between PGE_2-induced hyperalgesia and TTX-R sodium current has been provided by using an antisense deoxynucleotide (ODN) that specifically disrupts the synthesis ("knockdown") of one of the TTX-R VGSCs, $Na_V1.8$, in the DRG in vivo.[53] Antisense, but not a control mismatch, ODN treatment reduces the expression of $Na_V1.8$ by approximately 50 percent, and rats treated with the antisense ODN show a significant decrease in PGE_2-induced mechanical hyperalgesia.[53] Because $Na_V1.8$ is normally expressed predominantly in unmyelinated nociceptive C fibers, these data show that acute PGE_2-induced hyperalgesia is mediated by nociceptive C-fiber activity that is sustained by TTX-R sodium current. The data also implicate $Na_V1.8$ as the critical VGSC subtype that is necessary for initiation of hyperalgesia. To date, a number of acute inflammatory mediators, including adenosine,[54] 5-HT,[55] bradykinin,[56] endothelin 1,[57] and plasma epinephrine,[58] also have been demonstrated to directly modulate the excitability of primary afferents. Their direct action on primary afferent fibers indicates that the receptors for these mediators must be expressed on nociceptive neurons. These mediators, with the exception of bradykinin (and possibly ET-1), have been shown to modulate the activity of TTX-R sodium current in a manner similar to that seen for PGE_2.[54,55,57] Modulation of TTX-R sodium current thus appears to be a common mechanism that underlies the sensitizing effect of multiple inflammatory mediators.

Sodium Channels and Neuropathic Pain

A reduction in expression of $Na_V1.8$ by antisense administration to rats with spinal nerve ligation (SNL) injury reverses neuropathic pain, bringing the sensory thresholds to thermal and tactile stimuli back to control levels; this result indicates that the activity of this sodium channel subtype in the sensory primary afferent fiber is necessary for the expression of neuropathic pain.[34,59] These results also suggest that primary afferent fibers that contain functional $Na_V1.8$ after nerve injury may be "sensitized." The important role of $Na_V1.8$ in action potential generation and the additional observation that this channel subtype enables DRG neurons to fire repetitively on stimulation provide functional evidence to support such a hypothesis.[60]

As mentioned earlier, injured primary afferents show a significant downregulation of $Na_V1.8$; thus, the site of action of $Na_V1.8$ is not likely to be in the injured nerve fibers. Following tight ligation of the L5 and L6 spinal nerves, the level of expression of $Na_V1.8$[61] and the density and kinetics of the TTX-R current[34] in the uninjured L4 DRG cells are not different from control, suggesting that the expression of $Na_V1.8$ is maintained. However, a significant upregulation of the channel protein is apparent by day 2 after injury in the sciatic nerve.[34] The upregulation of $Na_V1.8$ immunoreactivity is correlated with an increase in the TTX-R compound action potential at C-fiber conduction velocity; a minor TTX-R A-fiber conduction velocity is also evident. These data demonstrate a functional reorganization of $Na_V1.8$ along unmyelinated fibers and in some myelinated fibers. Antisense-mediated "knockdown" of $Na_V1.8$ immunoreactivity and TTX-R current in these uninjured axons correlates with the reversal of both mechanical and thermal hypersensitivity, suggesting that this reorganization of $Na_V1.8$ activity along the uninjured axons may be necessary for expression of neuropathic pain in the injured rat.[34] These and other data[62,63] argue that abnormal activity in the uninjured primary afferent may be critical for the observed hypersensitivity to sensory input in the injured animal. A redistribution of $Na_V1.8$ along the injured sciatic nerve also has been observed in the chronic constriction injury model of neuropathic pain,[64] and $Na_V1.8$ immunoreactivity is evident in peripheral nerve tissues from patients with chronic neuropathic pain.[65,66]

The use of antisense oligonucleotides to disrupt the expression and function of $Na_V1.8$ also has been applied to other models of chronic inflammatory and visceral pain.[67,68] These findings further substantiate the role of $Na_V1.8$ in the hypersensitivity of primary afferent neurons, suggesting that the changes in $Na_V1.8$ seen following nerve injury may have wider mechanistic implications and potential clinical relevance.

Transgenic mice lacking $Na_V1.8$ provide an alternative animal model to evaluate the role of $Na_V1.8$ in neuropathic pain.[46] Nerve injury elicits thermal hyperalgesia and tactile hypersensitivity by day 3 in both wild-type and $Na_V1.8$-null mutant mice, suggesting that neuropathic pain is developed and maintained despite the lack of $Na_V1.8$.[69] An important confounding factor in the interpretation of the behavioral data in these animals, however, is the uncertain phenotype of the peripheral nervous system of these mice. The $Na_V1.8$-null mutant mice exhibit an upregulation of the TTX-S $Na_V1.7$ in the C-type DRG neurons and modified activity of the C fibers.[46] Other changes that may not be in common with the wild-type control after nerve injury also may occur. Complete elimination of the TTX-R current carried by $Na_V1.8$ has a profound effect on the conductance of other channels, including sodium channel subtypes, the emergence of non-Na^+ action potentials, and calcium channel activity.[60] How these biophysical characteristics may influence the neurons' response to nerve injury is not known, making interpretation of data from this transgenic model difficult.

Neuropeptide Y

Neuropeptide Y (NPY) is widely distributed in the CNS.[70,71] NPY has been detected in nerve terminals and varicosities throughout the dorsal horn, with the greatest concentration being in the superficial laminae, suggesting a role in the processing of sensory inputs.[72] Furthermore, manipulations that destroy afferent inputs into the spinal cord do not reduce NPY content in the spinal cord, indicating that NPY is expressed principally in spinal interneurons.[72,73] It is believed that NPY may modulate spinal sensory transmission by acting either on primary afferent terminals or on second-order neurons to inhibit sensory inputs.[73–75] The spinal administration of NPY has been reported to attenuate the release of SP evoked by electrical stimulation of C fibers, inhibit Ca^{2+} currents in cultured DRG cells, and inhibit depolarization-induced Ca^{2+} influx into DRG neurons, suggesting a possible antinociceptive function (for review, see ref. 72). Additionally, it was found that NPY reduced capsaicin-evoked release of CGRP from DRG cells.[76] Despite these observations, however, the evidence for NPY-mediated antinociception is relatively sparse. It was demonstrated that spinal NPY was effective against thermal nociception but completely without effect against mechanical pain (for review, see ref. 72). Mice lacking the gene coding for the Y1 receptor demonstrated enhanced responses in several (e.g., tail flick, hot plate) but not all (e.g., phase II formalin, stress-induced analgesia) nociceptive assays.[77] In the flexor reflex model, spinal NPY or the Y1-preferring agonist [Leu^{31}, Pro^{34}]NPY elicited a biphasic dose effect, with lower doses (below 300 ng) producing a facilitation and higher doses producing inhibition.[78] The supraspinal administration of NPY also has produced conflicting results. Supraspinal NPY has either enhanced hypoalgesia in spontaneously hypertensive rats or produced no changes in nociceptive responses in normotensive rats (for review, see ref. 72). In mice, i.c.v. NPY produced either hyperalgesia or hypoalgesia depending on the testing conditions.[79,80] Finally, excitatory effects of NPY have been observed in vitro, especially in DRG cells taken from rats with nerve injury.[81] Thus, the actions of exogenous NPY in normal conditions are presently unclear.

NPY and Inflammation

Several studies have described an upregulation of spinal NPY in response to inflammation, and it has been suggested that NPY may modulate inflammation-induced pain. Studies employing in situ hybridization histochemistry and immunohistochemistry have shown that rats with complete Freund's adjuvant (CFA)–induced inflammation demonstrate a marked increase in NPY mRNA expression in ipsilateral dorsal horn neurons, especially in lamina II. This increase in peptide message is accompanied by increased expression of Y1 receptor mRNA in laminae II and III but not in primary afferent fibers (for review, see ref. 72). In rats with carrageenan-induced inflammation, the spinal injection of NPY more than doubled paw-withdrawal latency to noxious thermal stimuli ipsilateral to the inflammation while increasing contralateral hind-paw latency only marginally,[82] indicating an antihyperalgesic rather than an antinociceptive effect. Conversely, spinal injection of the Y1 antagonist BIBO 3304 decreased paw-withdrawal latency to noxious heat ipsilateral but not contralateral to carrageenan injection,[82] suggesting that inflammation strengthens inhibitory NPY tone and that spinal Y1 receptors contribute to the inhibitory effects of NPY on thermal hypersensitivity. These findings suggest that endogenous NPY exerts a compensatory, adaptive inhibition of thermal hypersensitivity in the inflammatory state,[82] although enhanced nociception is clearly present despite this adaptation. Consistent with this interpretation, mutant mice lacking Y1 receptors (Y1−/−) show hyperalgesia to acute and inflammatory pain.[77] Curiously, the Y1−/− mice also demonstrated reduced capsaicin-evoked SP release and plasma extravasation in the periphery when compared with normal mice,[77] suggesting that, in the periphery, activation of the Y1 receptor is necessary for SP release and the subsequent development of neurogenic inflammation and plasma leakage.[77]

Evidence also exists for excitatory actions of NPY. Intrathecal NPY produced a biphasic effect on the flexor-reflex model, with low doses producing facilitation and higher doses producing inhibition.[78,83] After carrageenan-induced inflammation, the facilitatory effect of NPY was enhanced, whereas the inhibitory effect of the higher dose remained unchanged, leading to the conclusion that the spinal upregulation of NPY after inflammation is associated with an increased excitatory effect.[78,83]

NPY in Neuropathic Pain

It has been demonstrated repeatedly that peripheral nerve injury results in marked changes in NPY expression throughout the CNS.[72,84–87] A particularly impressive upregulation of NPY occurs within the DRGs after nerve injury because NPY is not normally present in the DRGs in the normal state.[72,87] Novel expression of NPY or of mRNA for NPY was detected in medium- and large-diameter DRG neurons after sciatic transection, chronic constriction injury (CCI), or SNL.[84,87] Importantly, novel NPY expression occurred ipsilateral and not contralateral to the injury and did not occur in sham-operated animals.[72,87] Additionally, peripheral nerve injury also resulted in novel expression of NPY in laminae III to V of the spinal dorsal horn ipsilateral to nerve injury.[72,87] Since only medium- to large-diameter DRG neurons expressed NPY and the spinal distribution of NPY in the dorsal horn corresponded to the terminal regions of large-diameter myelinated Aβ fibers, it was concluded that nerve injury elicits a preferential synthesis of NPY in low-threshold mechanoreceptors (Aβ and some Aδ fibers) and not in C-fiber nociceptors.[72]

The large-diameter Aβ fibers ultimately communicate with the dorsal column nuclei (n. gracilis and n. cuneatus) either through direct projections or through the postsynaptic dorsal column neurons (PSDCs). Although these nuclei do not normally contain neuropeptides, peripheral nerve injury causes an upregulation of NPY in the gracilis nerve ipsilateral but not contralateral to the insult.[86–88] Injury-induced NPY upregulation in the gracilus nerve was prevented by ipsilateral dorsal rhizotomy or dorsal column lesions,[86,87] as were behavioral signs of neuropathic pain.[89] Moreover, retrograde tracing with fluorogold in the gracilis nerve showed that nearly all the fluorogold-labeled cells in the DRG neurons expressed NPY-ir after sciatic section and that NPY was restricted to terminals and varicosities of nerve fibers, thus suggesting that NPY is derived from the primary afferent fibers and not synthesized in neurons of the gracilis nerve (for review, see ref. 72).

Experimental evidence indicates that NPY upregulation in the gracilis nerve may selectively mediate tactile hypersensitivity in response to peripheral nerve injury. The microinjection of NPY into the gracilis nerve of uninjured rats produced short-acting tactile but not thermal hypersensitivity.[87] Importantly, the microinjection of antiserum to NPY, the nonselective NPY receptor antagonist $NPY_{(18-36)}$, or the selective Y1 receptor antagonist BIBO 3304 into the gracilis nerve reversed tactile hypersensitivity but not thermal hyperalgesia in rats with experimental neuropathic pain.[87] It was concluded that nerve injury–induced tactile hypersensitivity is selectively mediated by upregulated NPY in large-diameter primary afferent fibers projecting to the gracilis nerve.

Dynorphin

The dynorphin peptide family, which includes dynorphin A(1–17), dynorphin A(1–13), and dynorphin A(1–8), was identified originally as a family of highly efficacious endogenous opioid agonists. Receptor binding studies, coupled with observations from isolated tissue preparations, led to the general consensus that dynorphin is an endogenous opioid with selectivity for the kappa opioid receptor (for review, see ref. 90). However, despite this pharmacologic profile, convincing evidence of an antinociceptive role for dynorphin has not materialized (for review, see refs. 90, 91, and 92). On the other hand, considerable evidence suggests that other actions of dynorphin, particularly those of neuronal excitotoxicity, are not mediated by opioid receptors. These nonopioid actions of dynorphin are blocked by NMDA receptor antagonist MK-801, implicating possible direct or indirect interaction with the NMDA receptors.[90–94] In this regard, spinal injection of dynorphin produced long-lasting behavioral signs of abnormal pain, and these were blocked by MK-801.[95,96] Moreover, the *des*-tyrosyl fragments of dynorphin do not bind to opioid receptors but produce similar excitotoxic effects as dynorphin. In many instances, dynorphin-induced "antinociception" was due to loss of motor function of the tail and hindlimbs.[35,90–92,97] This effect was resistant to naloxone. It has been suggested that physiologically relevant levels of spinal dynorphin may contribute to endogenous pain inhibitory systems, whereas pathologic elevations in spinal dynorphin promote nociception.[35,91,92] Moreover, the pronociceptive pathologic functions of dynorphin appear to be mediated through either a direct or indirect activation of the NMDA receptor complex. MK-801 blocked dynorphin-induced tactile hypersensitivity.[95,96] Electrophysiologic studies in vitro show that dynorphin can have inhibitory[98] or excitatory[93,99] effects on NMDA-mediated current. Similar conflicting data are also apparent in dynorphin A–mediated increase in intracellular calcium concentration. It has been reported that the effects of dynorphin A could be blocked by the NMDA antagonists AP-5 and MK-801 in cultured embryonic spinal cord neurons or spinal cord neurons,[98,100,101] whereas other reports suggest that this effect of dynorphin A is independent of opioid receptor or NMDA receptor activation.[102,103] A high-affinity binding site for dynorphin A(2–17) on the NMDA receptor has been identified,[94] however, the pharmacologic characteristics of this dynorphin-binding site implicate an inhibitory rather than an excitatory action on the NMDA receptor. Thus, while these findings support an excitatory action of dynorphin on neuronal activity, the underlying mechanism(s) remains to be elucidated.

Dynorphin and Inflammatory Pain

Many studies have shown that the inflammatory state is associated with an increase in spinal dynorphin or prodynorphin content. Rats with adjuvant-induced chronic arthritis or carrageenan-induced inflammation have consistently demonstrated increased levels of immunoreactivity for dynorphin or mRNA for prodynorphin in neurons of the dorsal horn of the spinal cord or in the trigeminal nuclei.[104–108] Furthermore, prodynorphin mRNA was elevated in the spinal cords of rats treated with formalin.[109] Microdialysis studies performed in polyarthritic rats demonstrated a 40 to 50 percent increase in dynorphin-like immunoreactivity in spinal cord perfusate collected from in vivo intrathecal catheters and a concomitant increase (300 to 550 percent) in prodynorphin mRNA in the spinal cord.[110] It was thought originally that the upregulation of spinal dynorphin in response to peripheral inflammation provided for an endogenous antinociceptive activity.[108,111–113] Accordingly, the systemic administration of nor-Binaltorphimine (nor-BNI) or spinal injection of dynorphin antiserum produced a modest enhancement of formalin-induced flinching behavior in rats.[113] However, it became clear that spinal dynorphin was more likely to mediate a pronociceptive rather than an antinociceptive function. Peripheral inflammation elicited dynorphin expression in local interneurons, as well as projection neurons, in laminae I, II, V, and VI of the spinal cord, and since these neurons receive noxious inputs, this localization suggests an important role for dynorphin in nociception.[104–108] Moreover, the time course of upregulation of spinal preprodynorphin mRNA and changes in receptive fields of spinal neurons expressing increased preprodynorphin mRNA was consistent with a pronociceptive role of spinal dynorphin with regard to inflammatory hyperalgesia (for reviews, see refs. 92 and 108). It was suggested that spinal dynorphin, along with SP and CGRP, elicits an enhancement of neuronal excitability at NMDA receptor sites, leading first to dorsal horn hyperexcitability and then to excessive depolarization and excitotoxicity.[92,108] The spinal administration of MK-801 prior to and during CFA administration prevented both hyperalgesia and spinal dynorphin upregulation.[106]

Dynorphin and Neuropathic Pain

A pronociceptive function of spinal dynorphin is more clearly established with regard to neuropathic pain states. Like inflammation, peripheral nerve injury is associated with an elevation of spinal dynorphin or of prodynorphin mRNA (for reviews, see refs. 92, 108, and 114). Rats with CCI showed significant ipsilateral increases in dynorphin-like immunoreactivity and of mRNA for preprodynorphin in laminae I and II and V through VII of the dorsal horn within 5 days of injury that lasted for over 20 days, with peak elevations at day 10 (for reviews, see refs. 92, 108, and 114). Rats with SNL also demonstrated time-dependent upregulation of spinal dynorphin, with a peak occurring 10 days after the injury.[115] Importantly, this increase was not restricted to the spinal segments corresponding to the zone of entry of the injured nerves but extended rostrally and caudally beyond these regions.[115] Likewise, mice with SNL also demonstrated increased spinal dynorphin content.[116,117] Importantly, strains of mice (129s6) that did not develop behavioral signs of neuropathic pain also did not show spinal dynorphin upregulation.[117] Immunohistochemical techniques demonstrated significant bilateral increases in spinal dynorphin content in a sciatic cryoneurolysis model of neuropathic pain, which also produces signs of neuropathic pain, including touch-evoked and tactile allodynia.[118]

As with inflammatory pain, it was thought originally that spinal dynorphin content is elevated after nerve injury in order to dampen chronic nociceptive input. However, further studies suggested that dynorphin may be involved in nociceptive transmission (for reviews, see refs. 92, 108, and 114). There is considerable evidence that dynorphin has pronociceptive properties in chronic pain states. The spinal injection of antiserum to dynorphin A(1–17) blocked thermal hyperalgesia in rats and both tactile allodynia and thermal hyperalgesia in mice with SNL.[115,116,119] Similar results were obtained after cryoneurolysis.[118] Furthermore, mice in which the prodynorphin gene was deleted failed to sustain tactile and thermal allodynia after nerve injury, whereas the wild-type control littermates developed and sustained tactile hypersensitivity.[116] Although tactile and thermal hypersensitivity occur within 2 to 3 days after SNL, peak spinal dynorphin content is not detected until 10 days after nerve injury in rats and mice.[115,116] Correspondingly, tactile and thermal hypersensitivity in mice with nerve injury was blocked by spinal MK-801 but not by antiserum to dynorphin 2 days after nerve injury, whereas both MK-801 and dynorphin antiserum blocked hypersensitivity at 10 days after injury. These data suggest that dynorphin does not contribute to the initial onset of experimental neuropathy but is critical for the maintenance of the nerve injury–induced pain state, as evidenced by the temporal correlation between dynorphin upregulation and the antiallodynic and antihyperalgesic effects of dynorphin antiserum. Importantly, manipulations that block bulbospinal pain facilitatory pathways (see below) in animals with nerve injury block both neuropathic pain behaviors and spinal dynorphin upregulation.[120,121] Thus, it appears that pathophysiologic levels of spinal dynorphin serve to maintain nociceptive inputs and an enhanced pain state.

Descending Facilitation as a Mechanism of Chronic Pain

The rostroventromedial medulla (RVM) is well recognized as an important relay in the modulation of nociceptive inputs at the level of the spinal cord. Although its function as a source of inhibitory controls has been long appreciated, it is also well established that the RVM has a pain facilitatory component (for reviews, see refs. 92, 122, and 123). Considerable evidence has been generated to indicate that descending facilitatory influences from the RVM can contribute to chronic pain states;[122–124] such contributions now have been recognized as important factors to the development and maintenance of tactile and thermal hyperalgesia in chronic pain states (see ref. 127 for review). It has been demonstrated that electrical stimulation or microinjection of excitatory neurotransmitters into the RVM can either inhibit or facilitate spinal nociceptive reflexes, depending on injection site, intensity, and concentration.[124–126] Importantly, it appears that nociceptive inputs may activate pain facilitatory neurons of the RVM, thus promoting further nociception and setting the stage for the maintenance of a chronic pain state.[127] For example, formalin injection into the tail increased responses of L4–L6 neurons to heating of the hind paw.[128] Similarly, when injected into a hind paw, formalin facilitated tail-withdrawal reflexes to thermal and mechanical noxious stimuli, and these enhanced responses were blocked by local anesthetic injection at the site of injury, indicating the need for persistent afferent input to initiate sensitization.[129] Conversely, hyperalgesia induced by a variety of means is reversed by intra-RVM lidocaine or an electrolytic lesion.[124]

These physiologically contradictory functions of the RVM are attributed to the existence of classes of neurons with opposing functions. Based on electrophysiologic responses of the neurons to noxious thermal stimulation, these cells have been described as "ON" and "OFF" cells.[122,123] OFF cells are tonically active and pause in firing immediately before the animal withdraws from the noxious thermal stimulus. In contrast, ON cells accelerate firing immediately before the nociceptive reflex occurs. An additional class, the neutral cells, were characterized initially by the absence of response to noxious thermal stimulation. It was determined subsequently that activation of OFF cells correlated with inhibition of nociceptive input and nocifensive responses.[122,123] The response characteristics of the ON cells are consistent with a role in descending facilitation of nociception.[122,123,130] Accordingly, manipulations that increase nociceptive responsiveness, thus indicating facilitation, also increase ON cell activity.[122,123,131] It is now generally accepted that a spinobulbospinal loop may be important to the development and maintenance of exaggerated pain behaviors produced by noxious (i.e., hyperalgesia) and non-noxious (i.e., allodynia) peripheral stimuli.[124]

Descending Facilitation and Inflammation

Considerable evidence has accumulated to indicate that enhanced pain associated with inflammation may be mediated through descending facilitation from the RVM. It was found that formalin-induced, noxious-evoked facilitation occurred beyond the boundaries of the nociceptive insult.[128,129] Moreover, spinal methysergide or RVM stimulation blocked formalin-induced hyperalgesia, suggesting that serotonergic bulbospinal projections from the RVM mediate, at least in part, descending facilitation of nociceptive responses.[129] Spinal transection or lidocaine in the RVM reversed behavioral and electrophysiologic signs of facilitated nociception in rats with carrageenan-induced inflammation of the hind paw.[132,133] Behavioral hyperalgesia elicited by CFA was abolished by microinjection of ibotenic acid into the RVM.[134] Likewise, bilateral lesions of the RVM made with ibotenic acid blocked behavioral signs of secondary hyperalgesia in rats with either intra-articular injection of carrageenan/kaolin into the knee or topical application of mustard oil to the hind leg.[124,135] Experiments involving chemical lesions of RVM somata and physical disruption of descending fibers have further demonstrated that hyperalgesia secondary to inflammation is mediated through descending facilitation from the RVM.[124,135] Finally, time-dependent increases in RVM activity associated with facilitation and inhibition were observed following persistent hind paw inflammation, indicating a dynamic plasticity of this region in response to persistent pain.[136] It was suggested that enhanced facilitation may be due to an upregulation of NMDA receptors in the RVM induced by persistent nociceptive inputs.[136] These studies provide evidence that prolonged noxious stimulation may cause an activation of descending facilitatory fibers arising from the RVM that, in turn, leads to enhanced pain-related behaviors.

Descending Facilitation and Neuropathic Pain

Evidence suggests that persistent nociceptive inputs result in an enhanced activity of descending facilitation from the RVM. For example, prolonged delivery of a noxious thermal stimulus produced increased ON cell activity and facilitation of nociceptive reflexes, both of which are blocked by RVM lidocaine.[122,123] Similarly, behavioral signs of neuropathic pain, which were attributed to a facilitation of spinal nociceptive input, have been blocked by lidocaine given into the RVM.[120,137–139] In behavioral studies in naive rats, the microinjection of cholecystokinin (CCK) into the RVM, where it may drive ON cells,[140] elicited reversible thermal hyperalgesia and tactile allodynia.[138] Moreover, the CCK_B antagonist L365,260 microinjected into the RVM produced a reversal of established thermal hyperalgesia and tactile allodynia induced by SNL.[138] In addition, heightened responses to mechanical or cold but not noxious thermal stimulation in rats with peripheral nerve injury was abolished by transection or ipsilateral hemisection of the thoracic spinal cord.[141,142] Moreover, the selective disruption of the dorsolateral funiculus (DLF) ipsilateral but not contralateral to SNL abolished tactile and thermal hypersensitivity, whereas DLF transection in sham-operated rats did not affect normal response thresholds.[120,143] These results suggest that behavioral manifestations of chronic pain states are dependent on descending facilitation of spinal nociceptive input from the RVM because it is the principal source of descending DLF projections.[122,123,144] The microinjection of a dermorphin-saporin conjugate into the RVM resulted in the selective ablation of mu opioid receptor–expressing neurons.[120,145] These neurons are believed to represent the likely source of descending facilitation because they are the only neurons directly hyperpolarized by morphine.[146] Rats treated with dermorphin-saporin conjugate either before or after SNL did not express behavioral signs of neuropathic pain, although normal nociceptive responses were intact.[120,145] Importantly, behavioral signs of neuropathic pain were abolished 4 or more days after SNL, but not earlier, by either RVM lidocaine, DLF lesions, or RVM dermorphin-saporin.[120] Furthermore, manipulations that blocked descending facilitation also blocked the upregulation of spinal dynorphin.[120,121] It has been suggested that persistent afferent input arising from peripheral nerve injury is likely to elicit neuroplastic changes, perhaps within the RVM itself. The ultimate consequence of these changes is the activation of facilitatory influences on further nociceptive input that perpetuates the dysesthesias associated with chronic pain states and which may be mediated, in part, through increased spinal dynorphin content.

Evoked Release from Primary Afferent Terminals

The stimulated release of neuropeptides from spinal cord preparations or cultured DRG neurons has emerged as a useful tool to assess the function of peptidergic neurons.[147] Evoked release of neuropeptides commonly associated with nociceptive transmission may be employed as a means of determining enhanced sensitivity to noxious inputs. Moreover, neuropeptide release evoked selectively by capsaicin may indicate changes in excitability of nociceptors expressing the VR-1 receptor. Capsaicin-evoked release of SP or of CGRP, therefore, has been employed commonly in the examination of enhanced neuronal activity in the inflammatory and neuropathic pain states.[147–150]

Neuropeptide Release and Inflammatory Pain

Experimentally induced inflammation is associated with hyperalgesia and increased sensitivity to noxious and non-noxious inputs. Spinal cord preparations obtained from rats with carageenan-induced inflammation exhibited increased spontaneous and capsaicin-evoked release of both SP and CGRP with a similar time course as the appearance of hyperalgesia, and thus may serve as biochemical markers for inflammation-induced hyperalgesia.[150] It was concluded that prostacyclin sensitizes sensory neurons and that prostacyclin-enhanced release of neuropeptides may contribute to inflammatory hyperalgesia.[150–152] Dose-dependent, capsaicin-evoked release of CGRP from trigeminal tissue was blocked by capsazepine, indicating that neuropeptide release was mediated through the capsaicin-sensitive VR-1 receptor.[148] Enhanced capsaicin-evoked release of SP and CGRP from spinal tissue obtained from rats with CFA-induced inflammation was observed.[150] The administration of ketorolac or (S)-ibuprofen to the rats during the inflammation blocked the enhanced release of SP and CGRP without attenuating normal evoked release from spinal tissue obtained from the contralateral side.[150] It was concluded that prostaglandins act at central terminals of primary afferent neurons to enhance nociception during inflammation and that this prostaglandin-induced enhanced release of neuropeptides is attenuated by the inhibition of cyclooxygenase.[150]

Neuropeptide Release and Neuropathic Pain

Although the relationship between hyperalgesia and enhanced primary afferent activity through increased neuropeptide release has been explored in conditions of inflammation, similar studies have not been initiated aggressively with regard to neuropathic pain until recently. Peripheral nerve injury was shown to increase release of SP in response to electrical stimulation of peripheral nerve fibers in vitro[153] or to potassium introduced through a microdialysis probe in vivo.[154] Early studies employing immunohistochemistry or in situ hybridization techniques for mRNA for CGRP suggested that peripheral nerve injury results in either no change or a downregulation in CGRP in the spinal cord.[155,156] Moreover, rats with CCI or complete transection of the sciatic nerve demonstrated either no change or a decrease in CGRP expression in DRG neurons.[156] It was found recently that peripheral nerve injury enhances the capsaicin-evoked release of CGRP from primary afferent terminals, which may provide a mechanism driving injury-induced enhanced pain.[121] Since spinal and DRG CGRP expression is decreased in these animals, the enhanced evoked release of CGRP indicates enhanced neurotransmitter outflow from primary afferents, thus indicating increased sensory inputs into the spinal cord.[121] Importantly, the addition of *des*-Tyr dynorphin to the perfusion medium enhanced the capsaicin-evoked release of CGRP.[121,157,158] This observation correlates with the fact that peripheral nerve injury results in an upregulation of spinal dynorphin. Moreover, the addition of antiserum to the perfusion medium reduced enhanced capsaicin-evoked release of CGRP from spinal tissue obtained from nerve-injured rats without altering the evoked release of CGRP from sham-operated animals.[121] In addition, the disruption of descending facilitation from the RVM through either the microinjection of dermorphin-saporin conjugate into the RVM or ablation of the DLF in rats with SNL also blocked capsaicin-evoked enhanced release of CGRP without blocking normal release in sham-operated animals.[121] These manipulations also have been shown to prevent the upregulation of spinal dynorphin and to block the maintenance of behavioral signs of neuropathic pain.[92,114,120,121,127,145]

Taken together, these observations provide firm evidence that neuropathic pain states are maintained, at least in part, by increased sensitivity of primary afferent neurons to noxious stimuli. In addition, it is hypothesized that persistent nociceptive inputs elicit supraspinal neuroplastic changes that result in the activation of a tonic descending facilitation of nociceptive inputs that is mediated by an upregulation of spinal dynorphin content. Thus, increased spinal dynorphin serves to promote excitatory transmitter release, which in turn promotes the maintenance of the tonic descending facilitatory system, thereby perpetuating a chronic pain state.[121]

▶ CONCLUSIONS

The preceding sections outlined a mechanistic and experimental view of pain transmission and modulation in the CNS. Therapeutic reality presents a different view with pain of complex origin, genomic heterogeneity among patients, and therapeutic choices limited by side effects. Whereas many laboratory models of chronic pain have sought to refine and reduce the factors contributing to the chronic pain state, some recent models have sought to model pain states akin to complex human conditions such as cancer pain. A recent series of studies from one of our laboratories[159–161] highlights this developing trend. Syngeneic fibrosarcoma cells implanted in mouse bone (femur, humerus, or calcaneus) give rise to persistent and profound nociceptive states with many of the characteristics of human cancer pain: irreversible progression, somatic components involving endogenous peptide mediators, relatively high doses of morphine required for relief, and neuropathic components in later stages of the tumor's progression. Despite their complexity, models of this kind offer opportunities to identify and differentiate somatic and neuropathic mechanisms, as well as to permit screening of multiple therapeutic strategies, e.g., comparing potency and efficacy of morphine versus clonidine. This chapter should serve as a guide to interpreting future studies of this kind.

REFERENCES

1. Wilcox GL, Seybold V: Pharmacology of spinal afferent processing, in TL Yaksh (ed), *Anesthesia: Biologic Foundations.* Philadelphia: Lippincott-Raven, 1997:557–576.
2. Zhang Y, Malmberg AB, Yaksh TL, et al: Capsaicin-evoked release of pituitary adenylate cyclase activating peptide (PACAP) and calcitonin gene–related peptide (CGRP) from rat spinal cord in vivo. *Regul Pept* 1997;69:83–87.
3. Dickinson T, Fleetwood-Walker SM, Mitchell R, et al: Evidence for roles of vasoactive intestinal polypeptide (VIP) and pituitary adenylate cyclase activating polypeptide (PACAP) receptors in modulating the responses of rat dorsal horn neurons to sensory inputs. *Neuropeptides* 1997;31:175–185.
4. Kinkelin I, Brocker EB, Koltzenburg M, et al: Localization of ionotropic glutamate receptors in peripheral axons of human skin. *Neurosci Lett* 2000;283:149–152.

5. Liu HT, Mantyh PW, Basbaum AI: NMDA-receptor regulation of substance P release from primary afferent nociceptors. *Nature* 1997;386:721–724.
6. Hill R: NK1 (substance P) receptor antagonists: Why are they not analgesic in humans? (comment). *Trends Pharmacol Sci* 2000;21:244–246.
7. Svendsen F, Tjolsen A, Hole K: AMPA and NMDA receptor-dependent spinal LTP after nociceptive tetanic stimulation. *Neuroreport* 1998;9:1185–1190.
8. Vulchanova L, Riedl MS, Shuster SJ, et al: Immunohistochemical study of the P2X2 and P2X3 receptor subunits in rat and monkey sensory neurons and their central terminals. *Neuropharmacology* 1997;36:1229–1242.
9. Vulchanova L, Riedl MS, Shuster JS, et al: P_{2X3} is expressed by DRG neurons that terminate in inner lamina II. *Eur J Neurosci* 1998;10:3470–3478.
10. Li P, Calejesan AA, Zhuo M: ATP P_{2X} receptors and sensory synaptic transmission between primary afferent fibers and spinal dorsal horn neurons in rats. *J Neurophysiol* 1998; 80:3356–3360.
11. Honore P, Kage K, Mikusa J, et al: Analgesic profile of intrathecal $P_{2X(3)}$ antisense oligonucleotide treatment in chronic inflammatory and neuropathic pain states in rats. *Pain* 2002;99:11–19.
12. Barclay J, Patel S, Dorn G, et al: Functional downregulation of P_{2X3} receptor subunit in rat sensory neurons reveals a significant role in chronic neuropathic and inflammatory pain. *J Neurosci* 2002;22:8139–8147.
13. Cook SP, Vulchanova L, Hargreaves KM, et al: Distinct ATP receptors on pain-sensing and stretch-sensing neurons. *Nature* 1997;387:505–508.
14. Molliver DC, Cook SP, Carlsten JA, et al: ATP and UTP excite sensory neurons and induce CREB phosphorylation through the metabotropic receptor, P_{2Y2}. *Eur J Neurosci* 2002;16:1850–1860.
15. Svensson CI, Yaksh TL: The spinal phospholipase-cyclooxygenase-prostanoid cascade in nociceptive processing. *Annu Rev Pharmacol Toxicol* 2002;42:553–583.
16. Tominaga M, Caterina MJ, Malmberg AB, et al: The cloned capsaicin receptor integrates multiple pain-producing stimuli. *Neuron* 1998;21:531–543.
17. Caterina MJ, Schumacher MA, Tominaga M, et al: The capsaicin receptor: A heat-activated ion channel in the pain pathway. *Nature* 1997;389:816–824.
18. Olah Z, Szabo T, Karai L, et al: Ligand-induced dynamic membrane changes and cell deletion conferred by vanilloid receptor 1. *J Biol Chem* 2001;276:11021–11030.
19. Szabo T, Olah Z, Iadarola MJ, et al: Epidural resiniferatoxin induced prolonged regional analgesia to pain. *Brain Res* 1999;840:92–98.
20. Montell C, Birnbaumer L, Flockerzi V: The TRP channels, a remarkably functional family. *Cell* 2002;108:595–598.
21. World Health Organization: *Cancer Pain Relief*. Geneva, Switzerland: WHO, 1996.
22. Wenk HN, Honda CN: Immunohistochemical localization of delta opioid receptors in peripheral tissues. *J Comp Neurol* 1999;408:567–579.
23. Yaksh TL: Substance P release from knee joint afferent terminals: Modulation by opioids. *Brain Res* 1988;458:319–324.
24. Nagasaka H, Awad H, Yaksh TL: Peripheral and spinal actions of opioids in the blockade of the autonomic response evoked by compression of the inflamed knee joint. *Anesthesiology* 1997;85:808–816.
25. Stone LS, Broberger C, Vulchanova L, et al: Differential distribution of α_{2A} and α_{2C} adrenergic receptor immunoreactivity in the rat spinal cord. *J Neurosci* 1998;18:5928–5937.
26. Stone LS, Macmillan L, Kitto KF, et al: The α_{2a}-adrenergic receptor subtype mediates spinal analgesia evoked by α_2 agonists and is necessary for spinal adrenergic/opioid synergy. *J Neurosci* 1997;17:7157–7165.
27. Fairbanks CA, Stone LS, Kitto KF, et al: $\alpha(2C)$-Adrenergic receptors mediate spinal analgesia and adrenergic-opioid synergy. *J Pharmacol Exp Ther* 2002;300:282–290.
28. Fairbanks CA, Wilcox GL: Moxonidine, an α_2-adrenergic and imidazoline receptor agonist, produces spinal antinociception in mice. *J Pharmacol Exp Ther* 1999;290:403–412.
29. Olave MJ, Maxwell DJ: Neurokinin-1 projection cells in the rat dorsal horn receive synaptic contacts from axons that possess α_{2C}-adrenergic receptors. *J Neurosci* 2003;23:6837–6846.
30. Olave MJ, Maxwell DJ: Axon terminals possessing the α_{2c}-adrenergic receptor in the rat dorsal horn are predominantly excitatory. *Brain Res* 2003;965:269–273.
31. Roca-Vinardell A, Ortega-Alvaro A, Gibert-Rahola J, et al: The role of 5-$HT_{1A/B}$ autoreceptors in the antinociceptive effect of systemic administration of acetaminophen. *Anesthesiology* 2003;98:741–747.
32. Alloui A, Pelissier T, Dubray C, et al: Tropisetron inhibits the antinociceptive effect of intrathecally administered paracetamol and serotonin. *Fundam Clin Pharmacol* 1996; 10:406–407.
33. Obata H, Saito S, Sasaki M, et al: Antiallodynic effect of intrathecally administered 5-HT(2) agonists in rats with nerve ligation. *Pain* 2001;90:173–179.
34. Gold MS, Weinreich D, Kim CS, et al: Redistribution of Na(V)1.8 in uninjured axons enables neuropathic pain. *J Neurosci* 2003;23:158–166.
35. Laughlin TM, Larson AA, Wilcox GL: Mechanisms of induction of persistent nociception by dynorphin. *J Pharmacol Exp Ther* 2001;299:6–11.
36. Sah DW, Ossipo MH, Porreca F: Neurotrophic factors as novel therapeutics for neuropathic pain. *Nature Rev Drug Discov* 2003;2:460–472.
37. Laughlin TM, Bethea JR, Yezierski RP, et al: Cytokine involvement in dynorphin-induced allodynia. *Pain* 2000; 84:159–167.
38. Fairbanks CA, Wilcox GL: Acute tolerance to spinally administered morphine compares mechanistically with chronically induced morphine tolerance. *J Pharmacol Exp Ther* 1997;282:1408–1417.
39. Fairbanks CA, Schreiber KL, Brewer KL, et al: Agmatine reverses pain induced by inflammation, neuropathy, and spinal cord injury. *Proc Natl Acad Sci USA* 2000; 97:19584–10589.
40. National Institutes of Health: *NIH Guide* 1995;24:15.
41. Chaplan SR, Sorkin LS: Agonizing over pain terminology. *Pain Forum* 1997;6:81–87.
42. Lai J, Hunter JC, Porreca F: The role of voltage-gated sodium channels in neuropathic pain. *Curr Opin Neurobiol* 2003;13:291–297.

43. Waxman SG, Cummins TR, Dib-Hajj S, et al: Sodium channels, excitability of primary sensory neurons, and the molecular basis of pain. *Muscle Nerve* 1999;22:1177–1187.
44. Rush AM, Brau ME, Elliott AA, et al: Electrophysiological properties of sodium current subtypes in small cells from adult rat dorsal root ganglia. *J Physiol* 1998;511:771–789.
45. Scholz A, Appel N, Vogel W: Two types of TTX-resistant and one TTX-sensitive Na+ channel in rat dorsal root ganglion neurons and their blockade by halothane. *Eur J Neurosci* 1998;10:2547–2556.
46. Akopian AN, Souslova V, England S, et al: The tetrodotoxin-resistant sodium channel SNS has a specialized function in pain pathways. *Nature Neurosci* 1999;2:541–548.
47. Brock JA, Pianova S, Belmonte C: Differences between nerve terminal impulses of polymodal nociceptors and cold sensory receptors of the guinea-pig cornea. *J Physiol* 2001;533:493–501.
48. Gu JG, MacDermott AB: Activation of ATP P_{2X} receptors elicits glutamate release from sensory neuron synapses. *Nature* 1997;389:749–753.
49. Samad TA, Sapirstein A, Woolf CJ: Prostanoids and pain: unraveling mechanisms and revealing therapeutic targets. *Trends Mol Med* 2002;8:390–396.
50. Samad TA, Moore KA, Sapirstein A, et al: Interleukin-1β-mediated induction of Cox-2 in the CNS contributes to inflammatory pain hypersensitivity. *Nature* 2001;410:471–475.
51. Evans AR, Junger H, Southall MD, et al: Isoprostanes, novel eicosanoids that produce nociception and sensitize rat sensory neurons. *J Pharmacol Exp Ther* 2000;293:912–920.
52. England S, Bevan S, Docherty RJ: PGE_2 modulates the tetrodotoxin-resistant sodium current in neonatal rat dorsal root ganglion neurones via the cyclic AMP-protein kinase A cascade. *J Physiol* 1996;495:429–440.
53. Khasar SG, Gold MS, Levine JD: A tetrodotoxin-resistant sodium current mediates inflammatory pain in the rat. *Neurosci Lett* 1998;256:17–20.
54. Gold MS, Reichling DB, Shuster MJ, et al: Hyperalgesic agents increase a tetrodotoxin-resistant Na+ current in nociceptors. *Proc Natl Acad Sci USA* 1996;93:1108–1112.
55. Cardenas LM, Cardenas CG, Scroggs RS: 5HT increases excitability of nociceptor-like rat dorsal root ganglion neurons via cAMP-coupled TTX-resistant Na(+) channels. *J Neurophysiol* 2001;86:241–248.
56. Millan MJ: The induction of pain: An integrative review. *Prog Neurobiol* 1999;57:1–164.
57. Zhou Z, Davar G, Strichartz G: Endothelin-1 (ET-1) selectively enhances the activation gating of slowly inactivating tetrodotoxin-resistant sodium currents in rat sensory neurons: A mechanism for the pain-inducing actions of ET-1. *J Neurosci* 2002;22:6325–6330.
58. Khasar SG, McCarter G, Levine JD: Epinephrine produces a beta-adrenergic receptor-mediated mechanical hyperalgesia and in vitro sensitization of rat nociceptors. *J Neurophysiol* 1999;81:1104–1112.
59. Lai J, Gold MS, Kim CS, et al: Inhibition of neuropathic pain by decreased expression of the tetrodotoxin-resistant sodium channel, NaV1.8. *Pain* 2002;95:143–152.
60. Renganathan M, Cummins TR, Waxman SG: Contribution of Na(v)1.8 sodium channels to action potential electrogenesis in DRG neurons. *J Neurophysiol* 2001;86:629–640.
61. Decosterd I, Ji RR, Abdi S, et al: The pattern of expression of the voltage-gated sodium channels Na(v)1.8 and Na(v)1.9 does not change in uninjured primary sensory neurons in experimental neuropathic pain models. *Pain* 2002;96:269–277.
62. Wu G, Ringkamp M, Murinson BB, et al: Degeneration of myelinated efferent fibers induces spontaneous activity in uninjured C-fiber afferents. *J Neurosci* 2002;22:7746–7753.
63. Li Y, Dorsi MJ, Meyer RA, et al: Mechanical hyperalgesia after an L5 spinal nerve lesion in the rat is not dependent on input from injured nerve fibers. *Pain* 2000;85:493–502.
64. Novakovic SD, Tzoumaka E, McGivern JG, et al: Distribution of the tetrodotoxin-resistant sodium channel PN3 in rat sensory neurons in normal and neuropathic conditions. *J Neurosci* 1998;18:2174–2187.
65. Coward K, Plumpton C, Facer P, et al: Immunolocalization of SNS/PN3 and NaN/SNS2 sodium channels in human pain states. *Pain* 2000;85:41–50.
66. Bucknill AT, Coward K, Plumpton C, et al: Nerve fibers in lumbar spine structures and injured spinal roots express the sensory neuron-specific sodium channels SNS/PN3 and NaN/SNS2. *Spine* 2002;27:135–140.
67. Porreca F, Lai J, Bian D, et al: A comparison of the potential role of the tetrodotoxin-insensitive sodium channels, PN3/SNS and NaN/SNS2, in rat models of chronic pain. *Proc Natl Acad Sci USA* 1999;96:7640–7644.
68. Yoshimura N, Seki S, Novakovic SD, et al: The involvement of the tetrodotoxin-resistant sodium channel Na(v)1.8 (PN3/SNS) in a rat model of visceral pain. *J Neurosci* 2001;21:8690–8696.
69. Kerr BJ, Souslova V, McMahon SB, et al: A role for the TTX-resistant sodium channel Na(v)1.8 in NGF-induced hyperalgesia, but not neuropathic pain. *Neuroreport* 2001;12:3077–3080.
70. Thorsell A, Heilig M: Diverse functions of neuropeptide Y revealed using genetically modified animals. *Neuropeptides* 2002;36:182–193.
71. Pedrazzini T, Pralong F, Grouzmann E: Neuropeptide Y: The universal soldier. *Cell Mol Life Sci* 2003;60:350–377.
72. Hokfelt T, Broberger C, Zhang X, et al: Neuropeptide Y: Some viewpoints on a multifaceted peptide in the normal and diseased nervous system. *Brain Res Brain Res Rev* 1998;26:154–166.
73. Polgar E, Shehab SA, Watt C, et al: GABAergic neurons that contain neuropeptide Y selectively target cells with the neurokinin 1 receptor in laminae III and IV of the rat spinal cord. *J Neurosci* 1999;19:2637–2646.
74. Mark MA, Colvin LA, Duggan AW: Spontaneous release of immunoreactive neuropeptide Y from the central terminals of large diameter primary afferents of rats with peripheral nerve injury. *Neuroscience* 1998;83:581–589.
75. Parker D, Soderberg C, Zotova E, et al: Co-localized neuropeptide Y and GABA have complementary presynaptic effects on sensory synaptic transmission. *Eur J Neurosci* 1998;10:2856–2870.
76. Gibbs JL, Flores CM, Poon A, et al: Evaluation of NPY Y1 agonist effects on exocytotic activity from central and

peripheral terminals of capsaicin-sensitive nociceptors. *Soc Neurosci Abstr* 2000;26:935.
77. Naveilhan P, Hassani H, Lucas G, et al: Reduced antinociception and plasma extravasation in mice lacking a neuropeptide Y receptor. *Nature* 2001;409:513–517.
78. Xu IS, Hao JX, Xu XJ, et al: The effect of intrathecal selective agonists of Y1 and Y2 neuropeptide Y receptors on the flexor reflex in normal and axotomized rats. *Brain Res* 1999;833:251–257.
79. Mellado ML, Gibert-Rahola J, Chover AJ, et al: Effect on nociception of intracerebroventricular administration of low doses of neuropeptide Y in mice. *Life Sci* 1996; 58:2409–2414.
80. Broqua P, Wettstein JG, Rocher MN, et al: Antinociceptive effects of neuropeptide Y and related peptides in mice. *Brain Res* 1996;724:25–32.
81. Abdulla FA, Smith PA: Nerve injury increases an excitatory action of neuropeptide Y and Y2 agonists on dorsal root ganglion neurons. *Neuroscience* 1999;89:43–60.
82. Taiwo OB, Taylor BK: Antihyperalgesic effects of intrathecal neuropeptide Y during inflammation are mediated by Y1 receptors. *Pain* 2002;96:353–363.
83. Xu IS, Luo L, Ji RR, et al: The effect of intrathecal neuropeptide Y on the flexor reflex in rats after carrageenan-induced inflammation. *Neuropeptides* 1998;32:447–452.
84. Marchand JE, Cepeda MS, Carr DB, et al: Alterations in neuropeptide Y, tyrosine hydroxylase, and Y-receptor subtype distribution following spinal nerve injury to rats. *Pain* 1999;79:187–200.
85. Zhang X, Ji RR, Arvidsson J, et al: Expression of peptides, nitric oxide synthase and NPY receptor in trigeminal and nodose ganglia after nerve lesions. *Exp Brain Res* 1996; 111:393–404.
86. Li WP, Xian C, Rush RA, et al: Upregulation of brain-derived neurotrophic factor and neuropeptide Y in the dorsal ascending sensory pathway following sciatic nerve injury in rat. *Neurosci Lett* 1999;260:49–52.
87. Ossipov MH, Zhang ET, Carvajal C, et al: Selective mediation of nerve injury-induced tactile hypersensitivity by neuropeptide Y. *J Neurosci* 2002;22:9858–9867.
88. Ma W, Bisby MA: Partial sciatic nerve ligation induced more dramatic increase of neuropeptide Y immunoreactive axonal fibers in the gracile nucleus of middle-aged rats than in young adult rats. *J Neurosci Res* 2000;60: 520–530.
89. Sun H, Ren K, Zhong CM, et al: Nerve injury-induced tactile allodynia is mediated via ascending spinal dorsal column projections. *Pain* 2001;90:105–111.
90. Naqvi T, Haq W, Mathur KB: Structure-activity relationship studies of dynorphin A and related peptides. *Peptides* 1998; 19:1277–1292.
91. Caudle RM, Mannes AJ: Dynorphin: Friend or foe? *Pain* 2000;87:235–239.
92. Ossipov MH, Lai J, Malan TP Jr, et al: Spinal and supraspinal mechanisms of neuropathic pain. *Ann NY Acad Sci* 2000; 909:12–24.
93. Lai SL, Gu Y, Huang LY: Dynorphin uses a non-opioid mechanism to potentiate N-methyl-D-aspartate currents in single rat periaqueductal gray neurons. *Neurosci Lett* 1998; 247:115–118.
94. Tang Q, Gandhoke R, Burritt A, et al: High-affinity interaction of (*des*-tyrosyl)dynorphin A(2–17) with NMDA receptors. *J Pharmacol Exp Ther* 1999;291:760–765.
95. Vanderah TW, Laughlin T, Lashbrook JM, et al: Single intrathecal injections of dynorphin A or *des*-Tyr-dynorphins produce long-lasting allodynia in rats: Blockade by MK-801 but not naloxone. *Pain* 1996;68:275–281.
96. Laughlin TM, Vanderah TW, Lashbrook J, et al: Spinally administered dynorphin A produces long-lasting allodynia: Involvement of NMDA but not opioid receptors. *Pain* 1997; 72:253–260.
97. Vanderah TW, Ossipov MH, Lai J, et al: Mechanisms of opioid-induced pain and antinociceptive tolerance: Descending facilitation and spinal dynorphin. *Pain* 2001; 92:5–9.
98. Chen L, Huang LY: Dynorphin block of N-methyl-D-aspartate channels increases with the peptide length. *J Pharmacol Exp Ther* 1998;284:826–831.
99. Brauneis U, Oz M, Peoples RW, et al: Differential sensitivity of recombinant N-methyl-D-aspartate receptor subunits to inhibition by dynorphin. *J Pharmacol Exp Ther* 1996; 279:1063–1068.
100. Hu WH, Zhang CH, Yang HF, et al: Mechanism of the dynorphin-induced dualistic effect on free intracellular Ca^{2+} concentration in cultured rat spinal neurons. *Eur J Pharmacol* 1998;342:325–332.
101. Hauser KF, Foldes JK, Turbek CS: Dynorphin A(1–13) neurotoxicity in vitro: Opioid and non-opioid mechanisms in mouse spinal cord neurons. *Exp Neurol* 1999; 160:361–375.
102. Tang Q, Lynch RM, Porreca F, et al: Dynorphin A elicits an increase in intracellular calcium in cultured neurons via a non-opioid, non-NMDA mechanism. *J Neurophysiol* 2000; 83:2610–2615.
103. Luo MC, Gardell LR, Vanderah TW, et al: Excitatory mechanisms for the pronociceptive actions of spinal dynorphin [abstract]. *Abstracts of the 10th World Congress on Pain.* Seattle: IASP Press, 2002.
104. Spetea M, Rydelius G, Nylander I, et al: Alteration in endogenous opioid systems due to chronic inflammatory pain conditions. *Eur J Pharmacol* 2002;435:245–252.
105. Calza L, Pozza M, Zanni M, et al: Peptide plasticity in primary sensory neurons and spinal cord during adjuvant-induced arthritis in the rat: An immunocytochemical and in situ hybridization study. *Neuroscience* 1998;82:575–589.
106. Zhang RX, Ruda MA, Qiao JT: Pre-emptive intrathecal MK-801, a non-competitive N-methyl-D-aspartate receptor antagonist, inhibits the up-regulation of spinal dynorphin mRNA and hyperalgesia in a rat model of chronic inflammation. *Neurosci Lett* 1998;241:57–60.
107. Imbe H, Ren K: Orofacial deep and cutaneous tissue inflammation differentially upregulates preprodynorphin mRNA in the trigeminal and paratrigeminal nuclei of the rat. *Brain Res Mol Brain Res* 1999;67:87–97.
108. Przewlocki R, Przewlocki B: Opioids in chronic pain. *Eur J Pharmacol* 2001;429:79191.
109. Li JL, Li YQ, Kaneko T, et al: Preprodynorphin-like immunoreactivity in medullary dorsal horn neurons projecting to the thalamic regions in the rat. *Neurosci Lett* 1999;264:13–16.

110. Pohl M, Ballet S, Collin E, et al: Enkephalinergic and dynorphinergic neurons in the spinal cord and dorsal root ganglia of the polyarthritic rat: In vivo release and cDNA hybridization studies. *Brain Res* 1997;749:18–28.
111. Wu H, Hung K, Ohsawa M, et al: Antisera against endogenous opioids increase the nocifensive response to formalin: Demonstration of inhibitory beta-endorphinergic control. *Eur J Pharmacol* 2001;421:39–43.
112. Wu HE, Hung KC, Mizoguchi H, et al: Roles of endogenous opioid peptides in modulation of nocifensive response to formalin. *J Pharmacol Exp Ther* 2002;300:647–654.
113. Ossipov MH, Kovelowski CJ, Wheeler-Aceto H, et al: Opioid antagonists and antisera to endogenous opioids increase the nociceptive response to formalin: Demonstration of an opioid kappa and delta inhibitory tone. *J Pharmacol Exp Ther* 1996;277:784–788.
114. Ossipov MH, Lai J, Malan TP, Jr., et al: Tonic descending facilitation as a mechanism of neuropathic pain, in PT Hansson, HL Fields, RG Hill, P Marchettini (eds), *Neuropatic Pain: Pathophysiology and Treatment*. Seattle: IASP Press, 2001:107–124.
115. Malan TP, Ossipov MH, Gardell LR, et al: Extraterritorial neuropathic pain correlates with multisegmental elevation of spinal dynorphin in nerve-injured rats. *Pain* 2000;86:185–194.
116. Wang Z, Gardell LR, Ossipov MH, et al: Pronociceptive actions of dynorphin maintain chronic neuropathic pain. *J Neurosci* 2001;21:1779–1786.
117. Gardell LR, Ibrahim M, Wang R, et al: Mouse strains that lack spinal dynorphin upregulation after peripheral nerve injury do not develop neuropathic pain. *Neuroscience* 2004;123:43–52.
118. Wagner R, Deleo JA: Pre-emptive dynorphin and N-methyl-D-aspartate glutamate receptor antagonism alters spinal immunocytochemistry but not allodynia following complete peripheral nerve injury. *Neuroscience* 1996;72:527–534.
119. Nichols ML, Lopez Y, Ossipov MH, et al: Enhancement of the antiallodynic and antinociceptive efficacy of spinal morphine by antisera to dynorphin A(1–13) or MK-801 in a nerve-ligation model of peripheral neuropathy. *Pain* 1997;69:317–322.
120. Burgess SE, Gardell LR, Ossipov MH, et al: Time-dependent descending facilitation from the rostral ventromedial medulla maintains, but does not initiate, neuropathic pain. *J Neurosci* 2002;22:5129–5136.
121. Gardell LR, Vanderah TW, Gardell SE, et al: Enhanced evoked excitatory transmitter release in experimental neuropathy requires descending facilitation. *J Neurosci* 2003;23:8370–8379.
122. Heinricher MM, Pertovaara A, Ossipov MH: Descending modulation after injury, in DO Dostrovsky, DB Carr, M Koltzenburg (eds), *Proceedings of the 10th World Congress on Pain*. Seattle: IASP Press, 2003:251–260.
123. Fields HL: Pain modulation: Expectation, opioid analgesia and virtual pain. *Prog Brain Res* 2000;122:245–253.
124. Urban MO, Gebhart GF: Supraspinal contributions to hyperalgesia. *Proc Natl Acad Sci USA* 1999;96:7687–7692.
125. Zhuo M, Gebhart GF: Biphasic modulation of spinal nociceptive transmission from the medullary raphe nuclei in the rat. *J Neurophysiol* 1997;78:746–758.
126. Urban MO, Gebhart GF: Characterization of biphasic modulation of spinal nociceptive transmission by neurotensin in the rat rostral ventromedial medulla. *J Neurophysiol* 1997;78:1550–1562.
127. Porreca F, Ossipov MH, Gebhart GF: Chronic pain and medullary descending facilitation. *Trends Neurosci* 2002;25:319–325.
128. Biella G, Bianchi M, Sotgiu ML: Facilitation of spinal sciatic neuron responses to hindpaw thermal stimulation after formalin injection in rat tail. *Exp Brain Res* 1999;126:501–508.
129. Calejesan AA, Ch'ang MH, Zhuo M: Spinal serotonergic receptors mediate facilitation of a nociceptive reflex by subcutaneous formalin injection into the hindpaw in rats. *Brain Res* 1998;798:46–54.
130. McNally GP: Pain facilitatory circuits in the mammalian central nervous system: Their behavioral significance and role in morphine analgesic tolerance. *Neurosci Biobehav Rev* 1999;23:1059–1078.
131. Heinricher MM, Roychowdhury SM: Reflex-related activation of putative pain facilitating neurons in rostral ventromedial medulla requires excitatory amino acid transmission. *Neuroscience* 1997;78:1159–1165.
132. Pertovaara A, Hamalainen MM, Kauppila T, et al: Carrageenan-induced changes in spinal nociception and its modulation by the brain stem. *Neuroreport* 1998;9:351–355.
133. Kauppila T, Kontinen VK, Pertovaara A: Influence of spinalization on spinal withdrawal reflex responses varies depending on the submodality of the test stimulus and the experimental pathophysiological condition in the rat. *Brain Res* 1998;797:234–242.
134. Wei F, Dubner R, Ren K: Nucleus reticularis gigantocellularis and nucleus raphe magnus in the brain stem exert opposite effects on behavioral hyperalgesia and spinal Fos protein expression after peripheral inflammation [published erratum appears in *Pain* 1999;81:215–219]. *Pain* 1999;80:127–141.
135. Urban MO, Zahn PK, Gebhart GF: Descending facilitatory influences from the rostral medial medulla mediate secondary, but not primary hyperalgesia in the rat. *Neuroscience* 1999;90:349–352.
136. Terayama R, Guan Y, Dubner R, et al: Activity-induced plasticity in brain stem pain modulatory circuitry after inflammation. *Neuroreport* 2000;11:1915–1919.
137. Pertovaara A, Wei H, Hamalainen MM: Lidocaine in the rostroventromedial medulla and the periaqueductal gray attenuates allodynia in neuropathic rats. *Neurosci Lett* 1996;218:127–130.
138. Kovelowski CJ, Ossipov MH, Sun H, et al: Supraspinal cholecystokinin may drive tonic descending facilitation mechanisms to maintain neuropathic pain in the rat. *Pain* 2000;87:265–273.
139. Calejesan AA, Kim SJ, Zhuo M: Descending facilitatory modulation of a behavioral nociceptive response by stimulation in the adult rat anterior cingulate cortex. *Eur J Pain* 2000;4:83–96.
140. Heinricher MM, McGaraughty S, Tortorici V: Circuitry underlying antiopioid actions of cholecystokinin within the rostral ventromedial medulla. *J Neurophysiol* 2001;85:280–286.

141. Sung B, Na HS, Kim YI, et al: Supraspinal involvement in the production of mechanical allodynia by spinal nerve injury in rats. *Neurosci Lett* 1998;246:117–119.
142. Bian D, Ossipov MH, Zhong C, et al: Tactile allodynia, but not thermal hyperalgesia, of the hindlimbs is blocked by spinal transection in rats with nerve injury. *Neurosci Lett* 1998;241:79–82.
143. Ossipov MH, Sun H, Malan TP, et al: Mediation of spinal nerve injury induced tactile allodynia by descending facilitatory pathways in the dorsolateral funiculus in rats. *Neurosci Lett* 2000;290:129–132.
144. Fields HL, Basbaum AI: Central nervous system mechanisms of pain modulation, in PD Wall, R Melzack (eds), *Textbook of Pain*. Edinburgh: Churchill Livingstone, 1999:309–329.
145. Porreca F, Burgess SE, Gardell LR, et al: Inhibition of neuropathic pain by selective ablation of brain stem medullary cells expressing the mu opioid receptor. *J Neurosci* 2001;21:5281–5288.
146. Heinricher MM, Morgan MM, Tortorici V, Fields HL: Disinhibition of off-cells and antinociception produced by an opioid action within the rostral ventromedial medulla. *Neuroscience* 1994;63:279–288.
147. Chen JJ, Barber LA, Dymshitz J, et al: Peptidase inhibitors improve recovery of substance P and calcitonin gene–related peptide release from rat spinal cord slices. *Peptides* 1996;17:31–37.
148. Ulrich-Lai YM, Flores CM, Harding-Rose CA, et al: Capsaicin-evoked release of immunoreactive calcitonin gene–related peptide from rat trigeminal ganglion: Evidence for intraganglionic neurotransmission. *Pain* 2001;91:219–226.
149. Huang H, Wu XW, Nicol GD, et al: ATP augments peptide release from rat sensory neurons in culture through activation of P2Y receptors. *J Pharmacol Exp Ther* 2003;306:1137–1144.
150. Southall MD, Michael RL, Vasko MR: Intrathecal NSAIDs attenuate inflammation-induced neuropeptide release from rat spinal cord slices. *Pain* 1998;78:39–48.
151. Flores CM, Leong AS, Dussor GO, et al: Capsaicin-evoked CGRP release from rat buccal mucosa: Development of a model system for studying trigeminal mechanisms of neurogenic inflammation. *Eur J Neurosci* 2001;14:1113–1120.
152. Kilo S, Harding-Rose C, Hargreaves KM, et al: Peripheral CGRP release as a marker for neurogenic inflammation: A model system for the study of neuropeptide secretion in rat paw skin. *Pain* 1997;73:201–207.
153. Malcangio M, Ramer MS, Jones MG, et al: Abnormal substance P release from the spinal cord following injury to primary sensory neurons. *Eur J Neurosci* 2000;12:397–399.
154. Wallin J, Schott E: Substance P release in the spinal dorsal horn following peripheral nerve injury. *Neuropeptides* 2002;36:252–256.
155. Ma W, Bisby MA: Increase of calcitonin gene–related peptide immunoreactivity in the axonal fibers of the gracile nuclei of adult and aged rats after complete and partial sciatic nerve injuries. *Exp Neurol* 1998;152:137–149.
156. Ma W, Bisby MA: Ultrastructural localization of increased neuropeptide immunoreactivity in the axons and cells of the gracile nucleus following chronic constriction injury of the sciatic nerve. *Neuroscience* 1999;93:335–348.
157. Gardell LR, Wang R, Burgess SE, et al: Sustained morphine exposure induces a spinal dynorphin-dependent enhancement of excitatory transmitter release from primary afferent fibers. *J Neurosci* 2002;22:6747–6755.
158. Claude P, Gracia N, Wagner L, et al: Effect of dynorphin on ICGRP release from capsaicin-sensitive fibers, in *Abstracts of the 9th World Congress on Pain,* Vol 9. Seattle: IASP Press, 1999:262.
159. Wacnik PW, Stone LS, Laughlin TM, et al: A practical model of cancer pain: Comparing different hind limb sites of melanoma cell implantation. *Society for Neuroscience Abstracts,* Vol 24. Washington, DC: Society for Neuroscience; 1998:628.
160. Wacnik PW, Wilcox GL, Clohisy DR, et al: Animal models of cancer pain, in M Devor, MC Rowbotham, Z Wiesenfeld-Hallin (eds), *Pain Research and Clinical Management: Proceedings of the Ninth World Congress on Pain*. Seattle: IASP Press, 2000:615–637.
161. Wacnik PW, Eikmeier LJ, Ruggles TR, et al: Functional interactions between tumor and peripheral nerve: Morphology, algogen identification, and behavioral characterization of a new murine model of cancer pain. *J Neurosci* 2001;21:9355–9366.

II. Peripheral Mechanisms

Helena Knotkova and Marco Pappagallo

The principal transducing elements involved in peripheral mechanisms of nociception are unmyelinated C-fibers and the smallest group of myelinated fibers, the Aδ. Nociceptive afferents are sensitive to different kinds of noxious stimuli. A considerable amount of evidence has been accumulated about mechanisms underlying the chemosensitivity of nociceptive afferent neurons. There are a variety of receptors and ion channels relevant to nociception that are found exclusively or largely in the periphery, and many of these receptors and channels are promising therapeutic targets. The process of validation of these potential targets has been facilitated by recent advances in molecular neurobiology and neurophysiology. It has been demonstrated that some of these targets occur and act directly on primary nociceptive neurons while others (e.g., functional bradykinin B1 receptors) are not expressed on sensory nerves and their effects are mediated via non-neuronal cells. The role of particular receptors and ion channels in the process of nociception has been studied under physiological as well as pathological conditions. For example, recent findings suggest that during some pathological conditions (e.g., rheumatological disorders, cancer, trauma) a number of diffusible factors can be involved in the development of neuropathic pain via altered gene expression. Specific proinflammatory cytokines (e.g., tumor necrosis factor alpha, IL-1beta), in combination with tissue-related growth factor, are known not only to sensitize uninjured tissue nociceptors, but also possibly generate ectopic and spontaneous nociceptor activity in otherwise "healthy" nociceptors. An intense research-interest has been directed to neurotrophic factors. There is evidence that the nerve growth factor plays a role in some forms of persistent pain. However, there is considerable hope that some neurotrophic factors can be useful in the treatment of neurodegenerative disorders. Detection of potential therapeutic targets and understanding of their role in peripheral and central mechanisms of nociception have been inevitable steps in the development of new therapeutic strategies and new approaches in pain management.

▶ POTENTIAL THERAPEUTIC TARGETS IN THE PERIPHERY

- Serotonin receptors ($5\text{-}HT_1$, $5\text{-}HT_2$)
- Bradykinin receptors (B2, B1)
- Sensory neuron specific receptors (SNSRs)
- Proteinase-activated receptor (PAR2)
- Transient receptor potential channels (TRPV1-4, TRPM8/CMR1, ANTKM1)
- Proton-gated ion channels (ASIC1, ASIC3/DRASIC)
- Calcium channels
- Sodium channels (TTX-R Na_v 1.8, 1.9)
- Potassium channels
- Purine receptors (P2X)
- Cannabinoid receptors (CB1, CB2)
- Growth factors
- Cytokines

Serotonin Receptors ($5\text{-}HT_1$, $5\text{-}HT_2$)

The activation of $5\text{-}HT_{1B/D}$ subtype is of considerable clinical significance, as these receptors are associated with trigeminal afferents innervating the dura mater. $5\text{-}HT_{1B/D}$ receptors are located on terminals of these sensory afferents, and activation inhibits adenylate cyclase via activation of $G_{i/o}$ proteins. Antimigraine drugs of the triptan class act in this way to inhibit neuropeptide release from terminals of these afferents. Acting on $5\text{-}HT_{1B/D}$ receptors, these drugs can block neurogenic vasodilatation in the meninges, thus preventing stretch-induced excitation of vascular nociceptors.[1] Serotonin is also known to produce a peripherally mediated mechanical hyperalgesia by action on 5-HT$_{1A}$ receptors,[1,2] while 5-HT$_{2A}$ receptors are involved in peripheral thermal hyperalgesia.[3]

Bradykinin Receptors (B2, B1)

B2 receptors are expressed by nociceptive sensory neurons as well as other, non-neuronal cell types. Excitation of sensory neurons by bradykinin acting on B2 receptors involves activation of a pertussis toxin-sensitive G-protein which stimulates phospholipase C that cleaves the substrate membrane lipids to generate two intracellular messengers, diacylglycerol and IP$_3$ (inositol 1,4,5-triphosphate).[4] B2 receptor has received attention as a target for many years.[5] B1 receptors are not normally expressed in tissues, but their expression is induced by tissue injury when they contribute significantly to inflammatory hyperalgesia.[6] Evidence suggests that functional B1 receptors are not expressed on sensory nerves and that the effects of B1 agonists are mediated via non-neuronal cells.[4] B1 activation evokes the release of PGE$_2$ and PGI$_2$, nitric oxide and various cytokines from non-neuronal cells, and these inflammatory mediators are probably responsible for the hyperalgesia induced by B1-receptor agonists. The fact that B1 is a substantial element in the inflammatory-hyperalgesia cascade makes B1 an interesting target for potential therapeutic interventions.

Sensory Neuron Specific Receptors (SNSRs)

Sensory neuron specific receptors (SNSRs) have recently been discovered.[7,8] These receptors are specifically located on small—presumably nociceptive sensory neurons—and are substrates for a number of naturally-occurring neuropeptides, with the highest affinity to a family of opioid peptides known as bovine adrenal medulla peptides (BAMs). However, their binding is not opioid-like as it does not respond to opioid receptor antagonists. Recently, it has been shown that an active fragment of BAM, BAM 8-22, is a potent agonist of SNSRs. BAM 8-22 increases the excitability of the nociceptive flexor reflex, suggesting that antagonists at this receptor might have analgesic properties.[9]

Proteinase-Activated Receptor (PAR2)

The proteinase-activated receptor PAR2 is one of recently discovered new potential targets. It is involved in mechanisms of hyperalgesia, and acts through sensory neuropeptide regulation.[10] The proteinases involved in the activation of the PAR family are more commonly associated with protein degradation and include trombin (PAR1, PAR3 and PAR4), trypsin (PAR2 and PAR4), and tryptase (PAR2). Sixty percent of DRG neurons express PAR2 immunoreactivity, and many of these also express calcitonin gene-related peptide (CGRP) and substance P, the two major neuropeptides contained in nociceptive C fibers. Activation of PARs causes rapid intracellular Ca^{2+} release. Trypsin, tryptase, and PAR2 selective agonists cause the release of CGRP and substance P from C fibers in peripheral tissues and in the spinal cord.[11] It already has been demonstrated that agents such as opioids and triptans acting as agonists at presynaptic GPCRs (G-protein coupled receptors), which, like the PARs, regulate the release of neuropeptides such as CGRP and substance P, are potent substances for clinical relief of pain.[12]

Transient Receptor Potential Channels (TRPV1-4, TRPM8/CMR1, ANTKM1)

TRP-family channels are grouped into three subclasses designated TRPC, TRPV and TRPM.[13] The three classes of TRP channels are distinguished according to overall similarity as well as several unique characteristics. The involvement of TRP channels in sensory functions has been evident since the first channel, TRPC, was cloned and found to be essential for sensing light.[14]

Four TRPV-class channels have been implicated in sensing heat, and one TRPM-class channel in sensing cold.

TRPV1 is expressed in a subpopulation of small and medium DRG and trigeminal neurons, most of which correspond to unmyelinated C-fiber nociceptors. TRPV1-positive cells include both peptidergic- and non-peptidergic neurons.[15,16] TRPV1 is a nonselective cation channel with high permeability to calcium ions, activated by capsaicin (a pungent ingredient that elicits tingling and burning sensation), and by heat; it shows steep temperature dependence, with the low threshold near 43°C, a temperature that most mammals perceive as noxious.[17,18] Genetic methods were used to test the roles of TRPV1 knockout in thermo- and chemonociception.[19-21] In TRPV1 knockout mice, significant deficits in heat sensitivity were observed, but these mice are not thermally analgesic, suggesting that TRPV1 channel mediates some, but not all mechanisms of acute heat sensitivity. Also, almost complete lack of thermal sensitivity during inflammation was observed in TRPV1 knockout mice. However, these mice still showed usual hypersensitivity to mechanical stimuli.[19,20] It indicates that TRPV1 is an important part of the signal transduction pathway linked to thermal hyperalgesia following tissue injury.

The three other TRPV channels with greater than 40% amino acid identity to TRPV1 have also been characterized as thermosensors.[22] These channels are activated at various heat thresholds, ranging from warm temperatures near 33°C for TRPV3, to noxious heat around 55°C for TRPV2. TRPV4, originally described as an osmo-sensor, has also been shown to be activated by warm temperatures.[23–27] The threshold of TRPM8/CMR1, a menthol- and cold-activated channel, is near 20°C.[23] The combined range of temperatures that activate TRPV1-4 and TRPM8/CMR1 channels covers a majority of the physiological spectrum sensed by most mammals, with a significant gap in the noxious cold range. Noxious cold stimulation is mediated by a distant family member of TRP channels, the ANKTM1 (ankyrin-like protein with transmembrane domains 1) channel, with very little amino acid similarity to TRPM81/CMR1. It is found in a group of nociceptive sensory neurons where it is coexpressed with TRPV1, but not TRPM8/CMR1.[22] ANKTM1 is activated at lower temperatures than TRPM8, starting near 17°C, which corresponds with the threshold of noxious cold for humans.[28] A role of ANKTM1 in noxious cold detection is also suggested by its expression pattern. ANKTM1-positive cells coexpress CGRP and TRPV1, markers for nociceptive neurons, and also, ANKTM1 is not found in neurons that contain TRPM8, the channel involved in non-noxious cold sensation.[22] It suggests that two distinct populations of cold detectors exist.

Recently, it has been recognized that other classes of ion channels are homologous to the classical TRP channels described above, and a nomenclature system has been proposed to reflect these findings.[13] The new subtypes involve TRPP for PKD2-like channels linked to polycystic kidney disease, TRPML for mucolipidin-like channels, that are responsible for some lysosomal storage disorders, and TRPN for NOMPC-like channels that are involved in mechanosensory functions.[22]

Proton-Gated Ion Channels (ASIC1, ASIC3/DRASIC)

Several types of proton sensitive channels (ASICs) have recently been discovered. Both ASIC1 and ASIC3 are activated by low pH and are found on primary nociceptive neurons.[29,30] ASIC1 is localized predominantly on small to medium primary afferent fibers and in lamina I and II in the dorsal horn. ASIC3, which is most sensitive to smaller changes in pH[31] is found in large- or small-diameter primary afferent fibers. ASIC1 co-localizes with CGRP and substance P in the population of small capsaicin-sensitive, peptide containing fibers thought to have an important role in nociception.[30] The fact that decreased pH is related to many clinically-important states, e.g, inflammation, tumor-induced acidosis, has generated intense research and clinical interest linked to acid-sensing channels. Experiments with mice lacking ASIC3 channel indicate that ASIC3, but not ASIC1, is essential for the development of chronic muscle pain.[32] Also, peptidergic primary afferent-containing ASICs can be involved in generating and maintaining bone cancer pain (Mantyh, et al. 2003). There are several cancer-related mechanisms that can influence pH in involved tissues. Tumor-associated inflammatory cells invade the neoplastic tissue and release protons that generate local acidosis, or acidosis may occur in relation to apoptosis of tumor cells.[33] Recent studies have localized peptidergic fibers in bone marrow and cortical bone, and have shown significant expression of ASICs in peptidergic primary afferents.[16,34] This evidence suggests that exposure of these peptidergic fibers to the acidic extracellular microenvironment could activate resident proton-sensitive ion channels that can result in bone-pain sensation (Mantyh, et al. 2003).

Calcium Channels

The role of Ca^{2+} channels in the control of transmitter release from terminals of nociceptive primary afferents has been well-described,[35,36] but less is known about the role of Ca^{2+} channels as targets for new analgesic drugs. It has been known for many years that some anticonvulsant drugs are useful in the treatment of pain, especially pain of neuropathic origin. However, a mechanism of action of such drugs was not fully explained. Experiments with gabapentin, an anticonvulsant that is used for treating chronic neuropathic pain, led to findings that voltage-gated Ca^{2+} channels are involved in an antinociceptive mechanism of this drug. Gabapentin binds to the $\alpha_2\delta$ subunit of the voltage-gated Ca^{2+} channels that are expressed in sensory neurons, and inhibits high-threshold Ca^{2+} currents.[37–39]

Sodium Channels (TTX-R Na_v 1.8, 1.9)

The recently cloned tetrodotoxin-resistant (TTX-r) Na^+ channel Na_v 1.8 is found almost exclusively in small, nociceptive primary afferent fibers,[40,41] and therefore might be an attractive therapeutic target. Experiments with mutant mice in which Na_v 1.8 was nonfunctional suggest that Na_v 1.8 is directly involved in mechanical and thermal nociception and in hyperalgesia induced by peripheral inflammation.[40] As the other TTX-r channel, Na_v 1.9, was still functional in the Na_v 1.8 null mutants, it was not possible to assess the full contribution of the TTX-r currents to nociceptive processes. However, findings from further experiments lead to suggestion that Na_v 1.8 is the most important nociceptive channel in sensory neurons.[42]

Potassium Channels

The crucial role of potassium channels in the regulatory process of cell excitability is reflected in a great heterogeneity of K-channel subtypes. A variety of voltage-gated K channels is found in sensory neurons.[43–46] Several delayed rectifier type K channels (I_K) have been described in small and medium DRG neurons. I_K channels are involved in the process of cell membrane repolarization that terminates the action potential. I_K determines the action potential configuration, but has little effect on the firing pattern.[47] The other major type of K current, I_A, is regarded as one of the main elements regulating the maximum firing frequency of neurons.[48] Any suppression of currents mediated by I_A channels can contribute to the type of neuronal spontaneous activity which is associated with damage of nociceptive peripheral sensory neurons.

Calcium-activated K channels occur in many types of neurons and are responsible for later phases of the afterhyperpolarization that follows an action potential.[36] Sensory neurons show two types of afterhyperpolarization[49] with different time courses, which are associated with the activation of two distinct types of calcium-activated K channels. The fast afterhyperpolarization is mediated by the "big conductance" channels (BK), and occurs in all sensory neurons,[49] while the slow afterhyperpolarization involves "small conductance" channels (SK), and is seen in a subpopulation of C neurons, but not in A neurons.[47,50] The slow afterhyperpolarization is susceptible to an inhibition by inflammatory mediators (prostaglandins, bradykinin, 5-hydroxytriptamine, histamine), which promotes repetitive firing.[45,51]

Purine Receptors (P2X)

Much recent interest has been generated by the cloning of the P2X family of purine receptors. P2X receptors differ structurally from most other types of ligand-gated ion channels as they consist of subunits with only two transmembrane domains.[52,53] P2Xs are expressed on unmyelinated primary afferent fibers, and these fibers can be depolarized by application of ATP.[54] ATP is released when cells are damaged and can also be co-released with noradrenaline and neuropeptide Y from sympathetic nerves.[55] Nerve endings of nociceptor fibers express a rapidly desensitizing, slowly recovering P2X3 receptor, but Ca2+ causes rapid recovery from desensitization and increases the current flowing through the ATP-gated channel.[56]

Cannabinoid Receptors (CB1, CB2)

Two cannabinoid receptors have not been described yet: CB1, which is mainly expressed by neurons, and CB2, mainly expressed by immune cells.[57,58] In addition to psychotropic effect of cannabinoids, an increasing body of evidence supports the hypothesis of the role of cannabinoids in central and peripheral mechanisms of nociception, particularly during inflammation and neuropathic pain. Strong evidence indicates that not only neurons located in different parts of brain, but also primary afferent neurons express CBs.[59,60]

A peripheral effect of cannabinoids was documented by numerous studies describing analgesic effects of locally delivered cannabinoids at doses that were not systemically effective.[61,62] Locally administered anandamide, one of the endogenously produced cannabinoids, attenuated thermal hyperalgesia,[61] and a similar effect was described for the synthetic cannabinoid agonist WIN 55,212-2 in a model of cutaneous inflammatory hyperalgesia.[62,63] Recent studies indicate that the local analgesic effect of cannabinoids is mediated by both CB1 and CB2 receptors,[59,60,64] while systemic effects of stress-induced analgesia (especially its nonopioid-mediated component) involves CB1 receptors only.[65]

Growth Factors

An interest in a potential therapeutic use of neurotrophic factors is focused mainly on two areas: there is considerable hope that neurotrophic factors will be useful in the treatment of neurodegenerative disorders,[66] and also there is evidence that the nerve growth factor (NGF) is an important mediator of some forms of persistent pain.[67] The neurotrophins (nerve growth factor, NGF; brain-derived neurotrophic factor, BDNF; glial derived neurotrophic factor, GDNF; neurotropin 3, NT-3; and neurotrophin-4/5, NT 4/5) exert long-term regulatory influences on the growth, survival, and phenotypic properties of a distinct subset of neurons.[68]

The family of neurotrophin receptors consists of tyrosine kinase (trk) receptors that bind to particular neurotrophins with extremely high affinity, and so-called receptor p75, with the capacity to bind all the members of the neurotrophin family with equal and relatively low affinity. The TrkA receptor interacts principally with NGF, TrkB with BDNF and NT-4/5, and TrkC with NT-3.[68–70] It is widely accepted that NGF is essential for the development of neurons mediating nociceptive functions. Small peptidergic DRG neurons expressing CGRP and substance P are lost in experimental models involving knockouts of the NGF or trkA genes.[68–70] Although the dependence of sensory neurons on neurotrophins is a developmental phenomenon, many adult nociceptive neurons continue to express TrkA, and NGF continues to exert a biological effect on these afferents. External NGF injected systemically or locally into human or animal tissues produces pain and hyperalgesia.[67,73] Sensitization of nociceptive afferent terminals after external application of NGF has been directly demonstrated for both cutaneous and visceral afferents.[74,75] Endogenously produced NGF plays a role in the development of sensory abnormalities during inflammation.[76–78] As demonstrated in various experimental models,[78–80] NGF up-regulation is a critical event in the pathogenesis of inflammatory pain.

Cytokines

Cytokines are small regulatory proteins that are produced by white blood cells and a variety of other cells including microglia in the nervous system. The action of cytokines includes numerous effects on cells of the immune system and modulation of inflammatory responses. Recently, evidence has emerged that cytokines link the immune and nervous system and that they may be involved in the generation of pain and hyperalgesia.[81–84]

Proinflammatory cytokine interleukin-1β (IL-1β) is produced under pathological conditions such as chronic inflammatory diseases, cancer, or neuropathies that are associated with increased pain and hyperalgesia. In peripheral nerves, IL-β has been found in DRG neurons and in Schwann cells.[85,86] In addition, expression of IL-1 receptor type I (IL-1RI) mRNA was detected in DRG neurons, suggesting a possible action of IL-β on sensory processing.[87] The peripheral pronociceptive action of IL-β may be mediated by a complex signaling cascade and secondary production of nitric oxide, bradykinin or prostaglandins.[88] However, it was also demonstrated that IL-β excites nociceptive fibers in vivo within 1 min, indicating a more direct action.[89] Results of a recent study[90] indicate that IL-β can act directly on sensory neurons via IL-1RI/TyrK/PKC-dependent mechanisms.[91]

The cytokine interleukin-6 (IL-6) is markedly upregulated during various pathologic situations (musculoskeletal disorders, burn injury, neuropathies, malignant tumors) associated with pain and hyperalgesia.[92] The neuronal effects of IL-6 depends on the presence of the soluble IL-6 receptor (sIL-6R). In vitro and in vivo studies[90,93] demonstrated that short exposure to the IL-6/sIL-6R complex modulates nociceptor-specific release of CGRP, suggesting that IL-6 could directly sensitize nociceptors to noxious stimuli.

Tumor necrosis factor-α (TNF-α) is widely considered the prototypic proinflammatory cytokine due to its principal role in the inflammatory response. After injury or during inflammation, TNF-α is released by a variety of cell types. Intraplantar injection of TNF-α in rats has been shown to induce mechanical allodynia and thermal hyperalgesia.[82,84] Indeed, subcutaneously injected TNF-α lowers the mechanical activation threshold in C nociceptors.[95] Recently TNF-α has also been linked to the generation and maintenance of neuropathic pain; blocking TNF-α reduces hyperalgesia in models of painful neuropathy.[85,86] TNF-α exerts its effect through two receptors, TNFR1 and TNFR2. The effects associated with experimental hyperalgesia have been dependent on TNFR1,[86] which is in line with an upregulation of TNFR1 following experimental nerve lesion.[84,96]

REFERENCES

1. Longmore J, Shaw D, Smith DW, et al: Differential distribution of 5-HT1D and 5-HT1B immunoreactivity within the human trigemino-cerebrovascular system: implication for the discovery of new antimigraine drugs. *Cephalgia* 1997;17:833–842.
2. Taiwo YO, Levine JD: Serotonin is a directly acting hyperalgesia agent in the rat. *Neurosci* 1992;48:485–590.
3. Tokunaga A, Saika M, Senba E: 5-HT receptor subtype is involved in the termal hyperalgesic mechanism of serotonin in the periphery. *Pain* 1998;76:349–355.
4. Seabrook GR, Bowery BJ, Heavens R, et al: Expression of B1 and B2 bradykinin receptor mRNA and their functional roles in sympathetic ganglia and sensory dorsal root ganglia from wild-type and B2 receptor knockout mice. *Neuropharm* 1997;36:1009–1017.
5. Hill RG: Peripheral analgesic pharmacology—an update. In: Max M. (ed) *Pain 1999—an updated review*. IASP Press: Seattle 1999;391–395.
6. Davis AJ, Perkins, MN: Induction of B1 receptors in vivo in a model of persistent inflammatory mechanical hyperalgesia in the rat. *Neuropharmacol* 1994;33:127–133.
7. Dong X, Han SK, Zylka MJ, Simon MI, Anderson DJ: A diverse family of GPCRs expressed in specific subsets of nociceptive sensory neurons. *Cell* 2001;106:619–632.
8. Lembo PMC, Grazzini E., Groblewski, T: Proenkefalin A gene products activate a new family of sensory neuron specific GPCRs. *Nat Neurosci* 2002;5:201–209.
9. Cao CQ, Dray A, Perkins MN: Tonic effecs of BAM 8-22, an agonist at the novel sensory neuron specific receptor, on the nociceptive flexor reflex in rats. In: JO Dostrovsky, DB Carr and M Koltzenburg (eds). *Proceedings of the 10th World Congress on Pain*. IASP Press, Seattle, 2002;89–99.
10. Vergnolle N, Bunnett NW, Sharkey KA, et al: Proteinase-activated receptor-2 and hyperalgesia: a novel pain pathway. *Nat Med* 2001;7:821–826.
11. Oliver KR, Hill RG: Feeling below PAR: proteinase-activated receptors and the perception of neuroinflammatory pain. *Pharmacogenomics J* 2002;2:10–11.
12. Oliver KR, Sirinathsinghji DJS, Hill RG: From basic research on neuropeptide receptors to clinical benefit. *Drug News Perspect* 2000;13:530–542.
13. Montell C, Bimbaumer L, Flockerzi V, et al: A unified nomenclature for the superfamily of TRP cation channels. *Mol Cel* 2002;9:229–231.
14. Montell C, Rubin GM: Molecular characterization of the Drosophila trplocus: a putative integral membrane protein required for photo transduction. *Neuron* 1989;2:1313–1323.
15. Tominaga M: The cloned capsaicin receptor integrates multiple pain-producing stimuli. *Neuron* 1998;21:1–20.
16. Guo A, Vulchanova L, Wang J, Li X, Elde R: Immunocytochemical localization of vanilloid receptor 1 (TRPV1): relationship to neuropeptides, the P2X3 purinoceptor and IB4 binding sites. *Eur J Neurosci* 1999;11:946–958.
17. Caterina MJ, Schumacher MA, Tominaga M, et al: The capsaicin receptor: a heat-activated ion channel in the pain pathway. *Nature* 1997;389:816–824.
18. Welch JM, Simon SA, Reinhart PH: The activation mechanism of rat vanilloid receptor 1 by capsaicin involves the pore domain and differs from the activation by either acid or heat. *Proc Natl Acad Sci USA* 2000;97:13889–13894.
19. Caterina MJ, Leffler A, Malmberg AB, et al: Impaired nociception and pain sensation in mice lacking the capsaicin receptor. *Science* 2000;288:306–313.
20. Davis JB, Gray J, Gunthorpe, MJ: Vanilloid receptor-1 is essential for inflammatory thermal hyperalgesia. *Nature* 2000;405:183–187.

21. Julius D: The molecular biology of thermosensation. In: JO Dostrovsky, DB Carr, M Koltzenburg (eds), *Proceedings of the 10th World Congress on Pain.* IASP Press: Seattle 2003;64–70.
22. Story GM, Peier AM, Reeve AJ, et al: ANKTM1, a TRP-like channel expressed in nociceptive neurons, is activated by cold temperatures. *Cell* 2003;112:819–829.
23. Peier AM, Reeve AJ, Andersson DA, et al: A heat-sensitive TRP channel expressed in keratinocytes. *Science* 2002;296:2046–2049.
24. Smith GD, Gunthorpe MJ, Keisell RE, Hayes PD, Reilly P, et al: TRPV3 is a temperature-sensitive vanilloid receptor-like protein. *Nature* 2002;418:186–190.
25. Xu H, Ramsey IS, Kotecha SA, Moran MM, et al: TRPV3 is a calcium-permeable temperature sensitive cation channel. *Nature* 2002;418:181–186.
26. Guler AD, Lee H, Lida T, Shimizu I, Tominaga M, Caterina M: Heat-evoked activation of the ion channel, TRPV4. *J Neurosci* 2002;22:6408–6414.
27. Strotmann R, Harteneck C, Nunnenmacher K, Schultz G, Plant TD: OTRPC4, a nonselective cation channel that confers sensitivity to extracellular osmolarity. *Nat Cell Biol* 2000;2: 695–702.
28. Davis KD, Pope GE: Noxious cold evokes multiple sensations with distinct time courses. *Pain* 2002;98:179–185.
29. Waldman R, Champigny G, Bassilana F: A proton-gated cation channel involved in acid-sensing. *Nature* 1997;386:173–177.
30. Olson TH, Riedl MS, Vulchanova L, et al: An acid sensing ion channel (ASIC) localizes to small primary afferent neuron in rats. *Neuroreport* 1998;9:1109–1113.
31. Benson CJ, Wemmie JA, et al: Heteromultimers of DEG/EnaC subunits from H+-gated channels in mouse sensory neurons. *Proc Natl Acad Sci* USA 2002;99:2338–2343.
32. Sluka KA, Price MP, Wemmie JA, Welsh MJ: ASIC3, but not ASIC1, channels are involved in the development of chronic muscle pain. In: Dostrovsky JO, Carr DB, Koltzenburg M (eds) *Proceedings of the 10th World Congress on Pain.* IASP Press: Seattle 2003;71–80.
33. Helmlinger G, Sckell A, Dellian M, et al: Acid production in glycolysis-impaired tumors provides new insights into tumor metabolism. *Clin Cancer Res* 2002;8:1284–1291.
34. Mach DB, Rogers SD, Sabino MC, et al: Origins of skeletal pain: sensory and sympathetic innervation of the mouse femur. *J Neurosci* 2002;113:155–166.
35. Scroggs RS, Fox AP: Calcium current variation between actually isolated adult rat dorsal root ganglion neurons of different sizes. *J Physiol* 1992;445:639–658.
36. Hille B: Modulation of ion-channel function by G-protein coupled receptors. *Trends in Neurosci* 1994;17:531–536.
37. Marais E, Klugbauer N, Hofmann F: Calcium channel $\alpha_2\delta$ subunits—structure and gabapentin binding. *Mol Pharmacol* 2001;59:1234–1248.
38. Sutton KG, Martin DJ, Pinnock RD, Lee K, Scott RH: Gabapentin inhibits high-threshold calcium channel currents in cultured rat dorsal root ganglion neurons. *Br J Pharmacol* 2002;135:257–265.
39. Martin DJ, McClelland D, Herd MB, et al: Gabapentin-mediated inhibition of voltage-activated Ca^{2+} channel currents in cultured sensory neurons is dependent on culture conditions and channel subunit expression. *Neuropharmacology* 2002; 42:353–366.
40. Akopian AN, Souslova V, England S, et al: The tetrodotoxin-resistant sodium channel SNS has a specialized function in pain pathways.
41. Catterall WA: From ionic currents to molecular mechanisms: the structure and function of voltage-gated sodium channels. *Neuron* 2000;26:13–25.
42. Lai J, Gold MS, Kim CS, et al: Inhibition of neuropathic pain by decreased expression of the tetrodotoxin-resistant sodium channel, Na_v 1.8. *Pain* 2002;95:143–152.
43. Gold MS, Shuster MJ, Levine JD: Role of a Ca dependent slow after hyperpolarization in prostaglandin E2-induced sensitization of cultures rat sensory neurons. *Neurosci Let* 1996b; 205:161–164.
44. Safranov BV, Bischoff U, Vogel W: Single voltage-gated K channels and their functions in small dorsal root neurones of the rat. *J Physiol* 1996;493:393–408.
45. Strichartz G: Targeting K channels for the treatment of inflammatory and neuropathic pain. *Proceedings of Annual Meeting of American and Canadian Pain Societies,* Vancouver 2004. Vancouver, 2004.
46. Passmore G: KCNQ/M-currents in sensory neurons: Implications for pain suppression. *Proceedings of Annual Meeting of American and Canadian Pain Societies,* Vancouver 2004.
47. Gold MS, Shuster MJ, Levine JD: Characterization of six voltage-gated K current in adult rat sensory neurons. *J Neurophysiol* 1996a;75:2629–2646.
48. McFarlane S, Cooper E: Kinetics and voltage dependence of A-type currents in neonatal rat sensory neurons. *J Physiol* 1991;66:1380–1391.
49. Fowler JC, Greene R, Weinreich D: Two calcium-sensitive spike after-hyperpolarization in visceral sensory neurones of the rabbit. *J Physiol* 1985;365:59–75.
50. Akins PT, McCleskey EW: Characterization of potassium currents in adult rat sensory neurons and modulation by opioids and cyclic AMP. *Neuroscience* 1993;56:759–769.
51. Jafri MS, Moore KA, Taylor GE, Weinrich D: Histamine H1 receptor activation blocks two classes of potassium current IK and I to excite ferret vagal afferents. *J Physiol* 1997; 503:533–546.
52. Brake AJ, Wagenbach MJ, Julius D: New structural motif for ligand-gated ion channels defined by an ionotropic ATP receptor. *Nature* 1994;371:519–523.
53. Valera S, Hussy N, Evans RJ: Cloning and expression of the P2X receptor for extracellular ATP reveals a new class of ligand-gated ion channel. *Nature* 1994;371:516–519.
54. Cesare P, McNaughton P: Peripheral pain mechanisms. *Curr Opin Neurobiol.* 1997;7:493–499.
55. Burnstock G, Wood JN: Purinergic receptors: their role in nociception and primary afferent neurotransmission. *Curr Opin Neurobiol* 1996;6:526–532.
56. Cook SP, Rodland KD, McCleskey EW: A memory for extracellular Ca by speeding recovery of P2X receptors from desensitization. *J Neurosci* 1998;18:9238–9244.
57. Matsuda LA, Lolait SJ, Brownstein MJ, et al: Structure of cannabinoid receptor and functional expression of the cloned cDNA. *Nature* 1990;346:561–564.
58. Munro S, Thomas KL, Abu Shaar M: Molecular characterization of a peripheral receptor for cannabinoids. *Nature* 1993; 365:61–65.

59. Hohman AG, Herkenham M: Cannabinoid receptors undergo axonal flow in sensory nerves. *Neurosci* 1999; 92:1171–1175.
60. Bridges D, Thompson SWN, Rice ASC: Mechanisms of neuropathic pain. *Br J Anesth* 2001;87:12–26.
61. Richardson JD, Kilo S, Hargreaves KM: Cannabinoids reduce hyperalgesia and inflammation via interaction with peripheral CB1 receptors. *Pain* 1998;75:111–119.
62. Calignano A, La Rana G, Giuffrida A, Piomelli D: Control of pain initiation by endogenous cannabinoids. *Nature* 1998;394:277–281.
63. Piomelli D, Beltramo M, Giuffrida A, Stella N: Endogenous cannabinoid signalling. *Neurobiol. Dis.* 1998;5:462–473.
64. Ross RA, Coutts AA, McFarlane SM: Actions of cannabinoid receptor ligands on rat cultured sensory neurons: Implications for antinociception. *Neuropharmacol* 2001; 40:221–232.
65. Hohman AG, Neely MH, Suplita RL: Endocannabinoid mechanisms of stress-induced analgesia. In: 2002 *Symposium on the cannabinoid*. International Cannabinoid Research Society, 2002;30.
66. McMahon SB, Priestly JV: Peripheral neuropathies and neurotrophic factors: animal models and clinical perspectives. *Curr Opin Neurol* 1995;Vol 5.
67. Apfel SC, Adomato BT, Dyck PJ, et al: Results of a double blind, placebo controlled trial of recombinant human nerve growth factor in diabetic polyneuropathy. *Ann Neurol* 1996; 40:194–198.
68. McMahon SB, Bennett DLH, Koltzenburg M: The biological effects of nerve growth factor on primary sensory neurons. In: D. Borsook (ed): *Molecular neurobiology of pain*. Seattle: IASP Press, 1997;59–78.
69. Bergman I, Priestley JV, McMahon SB, et al: Analysis of cutaneous sensory neurons in transgemic mice lacking the low affinity neurotrophin receptor p75. *Eur J Neurosci* 1997; 9:18–28.
70. Carter BD, Kaltschmidt C, Kaltschmidt B, et al: Selective activation of NF-kappa B by nerve growth factor through the neurotrophin receptor p75. *Science* 1996;272: 542–545.
71. Crowley C, Spencer SD, Nishimura MC, et al: Mice lacking nerve growth factor display perinatal loss of sensory and sympathetic neurons yet develop basal forebrain cholinergic neurons. *Cell* 1994;76:1001–1011.
72. Smeyne RJ, Klein R, Schnapp A: Severe sensory and sympathetic neuropathies in mice carrying a disrupted Trk/NGF receptor gene. *Nature* 1994;368:246–249.
73. Andreev NY, Dimitrieva N, Koltzenburg M, McMahon SB: Peripheral administration of nerve growth factor in the adult rat produces a thermal hyperalgesia that requires the presence of sympathetic post-ganglion neurones. *Pain* 1995;63:109–115.
74. Dimitrieva N, McMahon SB: Sensitisation of visceral afferents by nerve growth factor in the adult rat. *Pain* 1996;66:87–97.
75. Rueff A, Mendell LM: Nerve growth factor and NT-3 induce increased thermal sensitivity of cutaneous nociceptors in-vitro. *J Neurophysiol* 1996;76:593–596.
76. Aloe L, Tuveri MA, Levi-Montalcini R: Studies on carrageenan-induced arthritis in adult rats: presence of nerve growth factor and role of sympathetic innervation. *Rheumatol Int* 1992;12:213–216.
77. Oddiah D, McMahon SB, Rattray M: Inflammation produces up-regulation of neurotrophin messenger RNA levels in bladder. *Soc Neuroci Abstr* 1995;21:604–15.
78. Lowe EM, Anand P, Terenghi G: Increased nerve growth factor levels in the urinary bladder of women with idiopathic sensory urgency and intersticinal cystitis. *Br J Urol* 1997;79:572–577.
79. Safieh-Garabedian B, Poole S, Allchorne A, Winter J, Woolf CJ: Contribution of interleukin-1 beta to the inflammation-induced increase in nerve growth factor levels and inflammatory hyper-algesia. *Br J Pharmacol* 1995;115:1265–1275.
80. Andreev NY, Priestley JV, Rattray M: Synthesis of neurotrophins is upregulated by inflammation of urinary bladder in adult rats [abstract]. *Soc Neuroci Abstr* 1993;18:248.
81. Maier SF, Wiertelak EP, Martin D, Watkins LR: Interleukin-1 mediates behavioral hyperalgesia produced by lithium chloride and endotoxin. *Brain Res* 1993;623:324–328.
82. Cunha F, Poole S, Lorenzeni B, Ferreira, S: The pivotal role of tumor necrosis factor alpha in the development of inflammatory hyperalgesia. *Br J Pharmacol* 1992; 107:660–664.
83a. Schafers M, Lee DH, Brors D, Yaksh TL, Sorkin LS: Increased sensitivity of injured and adjacent uninjured rat primary sensory neurons to exogenous tumor necrosis factor-alpha after spinal nerve ligation. *J Neurosci* 2003; 3028–3038.
84b. Schafers M, Sorkin LS, Geis C, Shubayev VI: Spinal nerve ligation induces transient upregulation of tumor necrosis factor receptors 1 and 2 in injured and adjacent uninjured dorsal root ganglia in the rat. *Neurosci Lett* 2003;347: 179–182.
85. Watkins LR, Maier SF: Implications of immune-to-brain communication for sickness and pain. *Proc Natl Acad Sci USA* 1999;96:7710–7713.
86. Sommer C, Schmidt C, George A: Hyperalgesi in experimental neuropathy is dependent on the TNF receptor 1. *Exp Neurol* 1998;151:138–142.
87. Copray JC, Mantingh I, Brouwer N, et al: Expression of interleukin-1 beta in rat dorsal root ganglia. *J Neuroimmunol* 2001;118:203–211.
88. Poole S, Cunha FQ, Ferreira SH: Hyperalgesia from subcutaneous cytokines. In: LR Watkins, SF Maier (eds), *Cytokines and pain*, Birkhauser, Basel, 1999;59–87.
89. Fukuoka H, Kawatani M, Hisamitsu T, Takeshige C: Cutaneous hyperalgesia induced by peripheral injection of interleukin-1 beta in the rat. *Brain Res* 1994;657: 133–140.
90. Obreja O, Schmelz M, Poole S, Kress M: Interleukin-6 in combination with its soluble IL-6 receptor sensitises rat skin nociceptors to heat in vivo. *Pain* 2002;96:57–62.
91. Sommer C, Kress M: Recent findings on how proinflammatory cytokines cause pain: peripheral mechanisms in inflammatory and neuropathic hyperalgesia. *Neurosci Let* 2004;361:184–187.
92. Sommer C: Cytokines and neuropathic pain. In: Hansson P, Fields H, Hill R, Marchettini P (eds), *Neuropathic pain: pathophysiology and treatment*. IASP Press, Seattle, WA, 2001;21:37–62.

93. Opree A, Kress M: Involvement of the inflammatory cytokines tumor necrosis factor-alpha, IL-1beta, and IL-6, but not IL-8 in the development of heat hyperalgesia: effects on heat-evoked calcitonin-gene related peptide release from rat skin. *J Neurosci* 2000;20:6289–6293.
94. Perkins MN, Kelly D, Davis AJ: Bradykinin B1 and B2 receptor mechanisms and cytokine-induced hyperalgesia in the rat. *Can J Physiol Pharmacol* 1995;73:832–836.
95. Junger H, Sorkin LS: Nociceptive and inflammatory effects of subcutaneous TNF alpha. *Pain* 2000;85:145–151.
96. Shubayev VI, Myers RR: Upregulation and interaction of TNF alpha and gelatinoses A and B in painful peripheral nerve injury. *Brain Res* 2000;855:83–89.

CHAPTER 4

Opioid Pharmacology

Gavril W. Pasternak

The management of pain long has depended on the availability of opium.[1] A thick, viscous substance obtained from the poppy plant, opium contains a wide variety of alkaloids, including two that have been used for pain control—morphine and codeine. It also contains high concentrations of another alkaloid, thebaine, that has been used widely as an intermediate in the development of semisynthetic drugs. The opioids have unique properties in the management of pain. Unlike agents that abolish all sensation, such as local anesthetics, the opioids work on the perception of pain. Thus, following the administration of opioids, patients often will say that the pain is still there but that it does not "hurt." This is a unique aspect of the drugs that makes them particularly valuable.

Two major pain pathways have been defined that may help to explain the concept of "first" and "second" pain. First pain is well localized and defined. It is thought to be conducted through the neospinothalamic tract, which is a monosynaptic pathway that rapidly conducts pain from the dorsal horn of the spinal cord to the thalamus. From there, thalamic neurons project to the sensory cortex. In contrast, second pain is poorly localized and is conducted through the paleospinothalamic tract, a polysynaptic pathway involving the reticular and limbic systems. Unlike the neospinothalamic tract, the pain conducted by the paleospinothalamic tract can be modulated by opioids, perhaps helping to explain the observation from patients noted earlier.

▶ PAIN MODULATORY SYSTEMS

Opioid Peptides

The presence of pain modulatory systems in the brain has been known for many years. Electrical stimulation of very specific regions of the brain, such as the periaqueductal gray matter, produces an analgesic response in animals.[2,3] Similar results have been reported with electrical stimulation of the periaqueductal gray matter in clinical trials.[4,5] In these early studies, it was noted that these responses were blocked by the opioid antagonist naloxone, implying that the stimulation was releasing an opioid-like material that is responsible for mediating these actions. We now know that these materials are the endogenous opioid peptides.

The opioid peptides[6-9] were first identified several years after the discovery of the opioid receptors.[10-12] They are derived from three independent genes that generate precursor proteins that are then cleaved to form the actual opioid peptides.[13] The three families include the enkephalins, the dynorphins, and β-endorphin[14,15] (Table 4-1). There are two enkephalins, [met[5]]enkephalin and [leu[5]]enkephalin. These compounds are pentapeptides sharing the same first four amino acids (Tyr-Gly-Gly-Phe-) and differ only at the fifth. The precursor for the enkephalins has several copies of the enkephalins, as well as a number of other potential opioid peptides with extended structures. It is not known if these additional peptides are important. The enkephalins are localized throughout the brain and spinal cord, as well as in the adrenal medulla. They act through the delta class of opioid

▶ TABLE 4-1. OPIOID AND RELATED PEPTIDES

[Leu⁵]Enkephalin	**Tyr-Gly-Gly-Phe-Leu**
[Met⁵]Enkephalin	**Tyr-Gly-Gly-Phe-Met**
Dynorphin A	**Tyr-Gly-Gly-Phe-Leu**-Arg-Arg-Ile-Arg-Pro-Lys-Leu-Lys-Trp-Asp-Asn-Gln
Dynorphin B	**Tyr-Gly-Gly-Phe-Leu**-Arg-Arg-Gln-Phe-Lys-Val-Val-Thr
α-Neoendorphin	**Tyr-Gly-Gly-Phe-Leu**-Arg-Lys-Tyr-Pro-Lys
β-Neoendorphin	**Tyr-Gly-Gly-Phe-Leu**-Arg-Lys-Tyr-Pro
$β_h$-Endorphin	**Tyr-Gly-Gly-Phe-Met**-Thr-Ser-Glu-Lys-Ser-Gln-Thr-Pro-Leu-Val-Thr-Leu-Phe-Lys-Asn-Ala-Ile-Ile-Lys-Asn-Ala-Tyr-Lys-Lys-Gly-Glu
Endomorphin-1	Tyr-Pro-Trp-Phe-NH_2
Endomorphin-2	Tyr-Pro-Phe-Phe-NH_2

▶ TABLE 4-2. CLASSIFICATION OF OPIOID RECEPTORS

Receptor	Clone	Actions
Mu	MOR-1	Sedation
Mu_1		Supraspinal and peripheral analgesia, prolactin release, feeding, acetylcholine release in the hippocampus
Mu_2		Spinal analgesia, respiratory depression, inhibition of gastrointestinal transit, dopamine release by nigrostriatal neurons, guinea pig ileum bioassay, feeding
Kappa		
$Kappa_1$	KOR-1	Analgesia, dysphoria, diuresis, feeding
$Kappa_2$	KOR-1/DOR-1 dimer	Unknown
$Kappa_3$		Analgesia
Delta	DOR-1	Mouse vas deferens bioassay, feeding, dopamine turnover in the striatum
$Delta_1$		Supraspinal analgesia
$Delta_2$		Spinal and supraspinal analgesia

NOTE: The receptor subtype designations for mu and delta receptors are based on pharmacologic studies and have not been correlated with a specific clone or splice variant. The $kappa_3$ receptor is related to the ORL1/KOR-3 clone, but it has not yet been fully identified.

receptor, as discussed below. They are quite metabolically labile and initially were difficult to examine pharmacologically. However, the development of stable analogues has permitted extensive studies into their actions.[16] Like all the opioid peptides, the enkephalins are effective analgesics. However, their other pharmacologic profiles differ from morphine, particularly their reported decreased respiratory depressant activity.

Dynorphin A is the major member of the dynorphin family.[14] It is composed of 17 amino acids, with the first 5 corresponding to the sequence of [leu⁵]enkephalin.[1] Its precursor has a number of other putative opioid peptides that also start with the enkephalin sequence, such as dynorphin B and α-neoendorphin, but their relevance remains unknown.[13] Dynorphin A is a potent analgesic that acts through the $kappa_1$ opioid receptor, as discussed below.

The third member of the opioid peptide family is β-endorphin.[15,17] β-Endorphin is a larger peptide, composed of 31 amino acids, starting with the sequence of [met⁵]enkephalin. Its precursor is processed to produce a number of other hormones, including adrenocorticotropic hormone (ACTH).[13] β-Endorphin is produced in the pituitary, where it is coreleased with ACTH. It is present in the brain, although its extensive projections are thought to all derive from cell bodies within the arcuate nucleus.

Opioid Receptors

The opioid receptors were first proposed on the basis of structure activity studies with opioids[18-20] (Table 4-2). Detailed pharmacologic studies led Martin to propose mu and kappa receptors based on the pharmacology of morphine and ketocyclazocine, respectively.[21] The discovery of the enkephalins then led to identification of delta receptors.[22,23] The pharmacology of these three families of receptors has been well studied, both clinically and in animal models. Most of the opioids used clinically fall within the mu receptor classification, with some having mixed activity at kappa receptors as well, as noted below. The receptors were cloned approximately a decade ago. Understanding them at the molecular level has provided valuable insights into their actions.

▶ OPIOIDS

The original opioids were morphine and codeine (Fig. 4-1), components of opium.[1] They were isolated from opium over 100 years ago and have been used widely since. Prior to that, opium itself was taken orally or smoked. Both morphine and codeine are mu opioids. They are effective analgesics, although codeine is less potent, and are used primarily to relieve pain, but they have a number of other actions as well. Sedation is common, explaining why morphine was named after Morpheus, the god of sleep. Another action of mu opioids is constipation, which

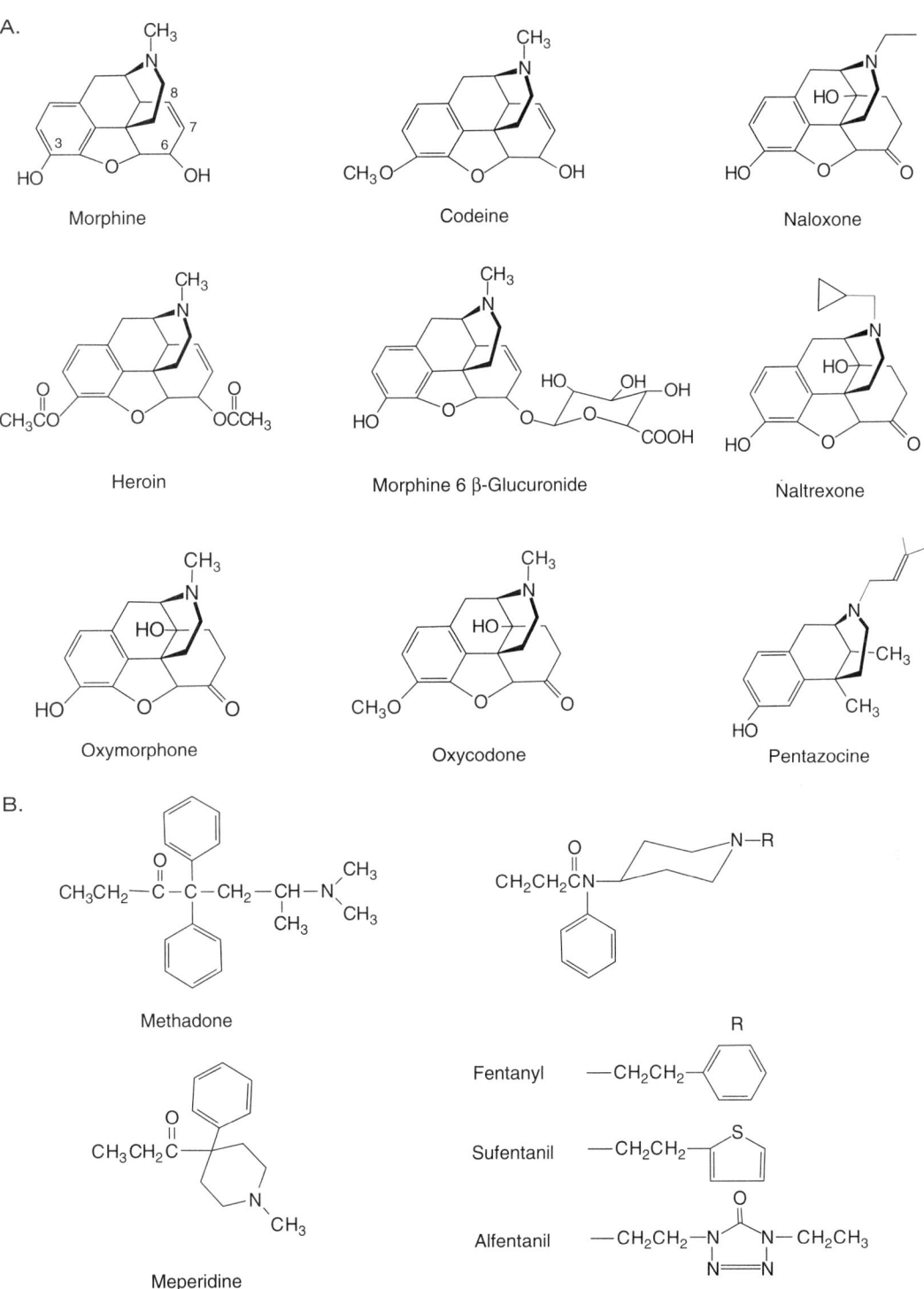

Figure 4-1. Structures of opioids.

is seen frequently as a result of inhibition of gastrointestinal transit. When treating pain, this inhibition of gastrointestinal transit can be problematic, but opioids also are used widely to treat diarrhea. Mu opioids depress respiration and must be used carefully in patients with known underlying pulmonary disease.

There have been numerous efforts over the past century to develop analgesics that lack undesirable side effects, leading to the synthesis of a vast array of compounds. The structures of these agents are quite varied[1] (see Fig. 4-1). Morphine is a naturally occurring alkaloid with a complex ring structure. The structure of codeine is very similar, differing only by the replacement of the hydroxyl group at the 3′ position with a methoxy group. Most opioids are classified as mu, or morphine-like. They include compounds such as oxycodone, hydrocodone, meperidine, methadone, fentanyl, and others. Slight structural changes also can lead to marked differences in the pharmacology of the drug. This is well illustrated with naltrexone (see Fig. 4-1). Simply replacing the *N*-methyl group with an *N*-allyl or a methyl-*c*-propyl converts the drug to a pure mu antagonist. Other changes have been associated with different pharmacologic profiles. For example, benzomorphans such as pentazocine lack the C ring of morphine and typically are mixed agonists/antagonists, interacting with both kappa and mu receptors, respectively. Although they are analgesics, they must be used carefully in tolerant patients because their antagonist activity at mu receptors can precipitate withdrawal. Thus their use should be restricted to naïve patients.

▶ CLINICAL USE OF OPIOIDS

Although they are used for a variety of purposes, the major clinical indication for the opioids involves the relief of pain. Opioids generally are used in combination with other analgesic agents, including aspirin, acetaminophen, and nonsteroidal anti-inflammatory (NSAID) drugs.[1] Most clinicians initiate therapy with NSAIDs or acetaminophen before moving to opioids. Moderate pain may be treated with codeine or its combinations. Patients with more severe pain may be given more potent agents. Although potency is often a criterion for choosing opioids, all mu opioids have similar abilities to relieve pain if the doses are sufficiently high. Mu opioids lack a ceiling effect. Thus, increasing the dose will increase pain relief. Unfortunately, they are often limited by side effects, which can vary from patient to patient. Another consideration is the potential issue of their misuse. Obviously, care must be taken when evaluating a patient for the use of an opioid, with particular attention to prior illicit drug use. However, with appropriate care in the choice of patients and continuing oversight of drug use, diversion and/or misuse is not a common problem.

The hallmark of opioid therapy is the need to individualize treatment for each patient. It is crucial to recognize that patients will respond differently to the various drugs in terms of both the potency of the drug and its effectiveness. Variability among patients can be quite profound. This can extend toward both the analgesic effects and the side effects. It is not uncommon for patients to indicate that one mu opioid is more effective than another or for a patient unable to tolerate morphine due to profound nausea and vomiting to be able to take a different mu opioid, such as methadone, without problem. Thus it is difficult to predict beforehand which drug will be best for an individual patient.

Choice of Drug

Most of the drugs used clinically are mu opioids. Some, like codeine and propoxyphene, are used more commonly for mild to moderate pain, leaving opioids such as morphine, oxycodone, hydrocodone, methadone, and fentanyl for more severe pain problems.[1] However, since there is no known ceiling effect for the analgesic action of any of the mu opioids, even codeine or propoxyphene could be used at higher doses for more severe pain.

A major consideration in the choice of an analgesic is its pharmacokinetics. Long-standing pain, such as that seen with cancer, is controlled more easily with long-acting drugs, including those whose formulations have been designed to provide twice- or thrice-daily dosing. On the other hand, when dealing with elderly patients, or patients with renal or hepatic dysfunction for whom the buildup of drug may be a problem, or for pain problems of shorter duration, such as in the postoperative setting, shorter-acting drugs may be more suitable.

In general, it is recommended that the dose of drug be escalated slowly to effect, keeping in mind that the long-duration drugs/formulations may require several days at a fixed dose to reach steady-state levels. Thus, when using long-acting agents, it is safer to use short-acting drugs as rescue treatment and to increase the dosage of the long-acting drug slowly. The objective is to achieve pain relief with a minimum of problematic side effects, such as sedation and constipation. Although respiratory depression is always a concern, it is rarely an issue in the outpatient setting in the absence of underlying pulmonary disease. In combination with anesthetics and other drugs, however, it may become more of a problem. All patients should be cautioned about the constipating effects of the opioids and should be put on a bowel regimen when they start treatment.

After initiating a drug, the dose should be escalated until the pain is controlled or until side effects prevent further escalation of dose. Again, escalating the long-action drugs should be done slowly. If one drug is not effective, or if it induces intolerable side effects, the patient can be switched immediately onto a different drug. The response of individual patients to various opioids may be very different, and it is not uncommon to find that the patient can be controlled adequately with acceptable side effects simply by switching to a second mu opioid. Unfortunately, the choice of these drugs is empirical, and several may need to be tried before finding the best one for a patient.

Tolerance

With continued use of any of the opioid drugs, patients will become both tolerant and dependent. Thus, to maintain a consistent level of pain relief, it may be necessary to escalate the dose. Clinically, it is difficult to assess tolerance. For example, the need initially to escalate the drug to achieve adequate pain control may be due, in part, to tolerance, as well as other factors. Most patients with chronic cancer pain can be maintained at a relatively constant dose for prolonged periods of time. Indeed, in this setting, the need for dose escalation more often is due to progression of disease than tolerance. However, most of these patients are already taking doses that are greater than those used in naïve patients and already may be displaying some degree of tolerance. Thus, tolerance may exist in a "steady state."

When pain is no longer well controlled, regardless of the reason, the dose of opioid should be increased. It is usually best to keep patients on the simplest regimen, so monotherapy is preferred when possible. The dose of drug should be escalated until the pain is under control or side effects intervene. However, if a patient has been on a particular drug for an extended period of time, it may be advantageous to switch the patient to a different opioid. However, switching patients who have been on opioids for a period of time differs from trying different drugs on naïve patients. Opioids show cross-tolerance to each other;[1] patients highly tolerant to one mu opioid also will be tolerant to another. However, this cross-tolerance is often incomplete. Animal models clearly illustrate incomplete tolerance.[24-26] Thus, when switching a patient from one opioid to another, it is common to find that the relative dose of the second drug is far less than the dose of the first.[27] Tables indicating the relative potencies of the opioids were developed in naïve patients, and the ratios of the drugs vary significantly in tolerant patients. Indeed, when calculating the dose of a second drug from these tables, it is common practice to start the second drug at only 25 to 35 percent of the calculated dose. In most situations, pain control can be achieved at doses far lower than those predicted from the tables. The utility of this approach has led to its widespread use, termed *opioid rotation*.[27]

Dependence

Prolonged administration of opioids results in dependence in all patients.[1] Dependence is a physiologic response to the exposure to the drug and is quite distinct from addiction, which implies drug-seeking behavior and is uncommon in most medical settings. Dependence needs to be considered both when taking a patient off an opioid or when switching a patient to a new analgesic. As patients become more dependent, they become more sensitive to antagonists. This may be problematic if they are given a mixed agonist/antagonist, such as pentazocine, because it can precipitate withdrawal. Thus, it is preferable to maintain patients on mu agonists.

Dependence also must be considered when stopping a patient's medication. In a tolerant patient, abrupt termination of an opioid or administration of an antagonist can precipitate withdrawal, which is commonly referred to as "going cold turkey." However, most patients can be taken off medication without encountering these symptoms. Except for the unusual patient, the dose of the opioid being administered can be cut in half every other day without significant side effects. The longer-acting drugs may need to be tapered more slowly due to their prolonged half-lives. However, most patients will be able to be taken off their drug in a relatively short period of time without any observable symptoms. Of course, if symptoms appear during the taper, increasing the opioid dose and slowing the taper can bring relief.

Reversing Opioid Actions

A major advantage of opioids is the availability of highly selective antagonists, the most common being naloxone.[1] Antagonists can be very helpful in cases of drug overdose, both from street drugs and from drugs administered clinically. The actions of naloxone are extremely rapid following intravenous administration and can be seen within seconds. However, it is important to be aware of several issues when using naloxone. First, the duration of action of naloxone is relatively brief and shorter than that of many agonists. It is not uncommon for a patient to respond to naloxone, only to see the patient relapse as the naloxone wears off. Clinicians should consider administering naloxone every several hours or even by infusion depending on the half-life of the drug being reversed.

It also is important to note that as patients become increasingly dependent on opioids, their sensitivity to antagonists increases. A patient who has been taking opioids for long periods of time becomes extremely more sensitive to antagonists such as naloxone. Despite the fact that the patient is taking larger doses of opioid, he or she needs far lower doses of naloxone to reverse the actions. Therefore, in patients with a history of opioid use, naloxone should be diluted and administered slowly to titrate the reversal of opioid effects without precipitating withdrawal.

▶ MOLECULAR ASPECTS OF OPIOID ACTION

Opioid receptors are members of the G protein–coupled receptor family.[28] The receptors are located within the membrane and have the standard structure that traverses the membranes seven times with an extracellular N-terminus and intracellular C-terminus (see Fig. 4-3). These transmembrane domains then arrange themselves within the membrane in a donut-like shape and provide the binding pocket for the opioid. The opioid receptors interact primarily with inhibitory G proteins located inside the cell (i.e., G_o and G_i). These G proteins comprise three components. They are the alpha subunits and the beta and gamma subunits, which are tightly bound together. On activation of the receptor, the GDP present in the binding pocket on the G_α subunit is replaced by GTP, and the G_α subunit disassociates from the $G_{\beta\gamma}$ subunits. These subunits then interact with downstream transduction systems.[1]

The three major families of opioid receptors, mu, delta, and kappa, all have been cloned, and all are members of the G protein–coupled receptor family.[1] Although they have marked similarities in their amino acid sequences, particularly in the transmembrane domains, they are each unique and products of different genes. When expressed, these receptors all show the expected selectivity for the opioids. The mu opioid receptor clone, MOR-1, is highly selective for mu drugs, whereas DOR-1 selectively binds delta ligands (see Table 4-2). The clone for the kappa$_1$ receptor, KOR-1, shows high affinity for the dynorphins and the highly selective synthetic kappa opioids. The receptors have been localized within the brain and spinal cord, where they show different regional distributions similar to those seen previously using receptor-binding (i.e., autoradiographic) approaches.

An important question arising with the cloning of these receptors was whether they represented the actual receptor responsible for the analgesic actions of the opioids in vivo. Making this correlation between the cloned protein and its behavioral pharmacology was crucial but difficult. The first studies confirming the relevance of these clones was based on antisense mapping.[29-31] In this approach, short oligodeoxynucleotides complementary to the mRNA for the specific receptor are administered to the animal. Since the oligodeoxynucleotide is complementary to the mRNA, it will bind and form a duplex structure, which signals the cell to destroy the mRNA. This leads to a downregulation of the mRNA and thus the protein. The high specificity associated with the interactions between the antisense and the mRNA make this approach extremely specific because the oligodeoxynucleotide will not bind to other mRNAs. When an oligodeoxynucleotide was administered into the periaqueductal gray matter of the rat, it blocked the analgesic activity of morphine microinjected into the same region to produce analgesia.

More recently, the importance of these receptors in the actions of the opioids was confirmed using "knockout" mice.[32-36] In this approach, the gene responsible for the protein is disrupted, leading to a loss of the mRNA and thus the protein. Again, loss of MOR-1 was associated with the loss of morphine analgesia. Similarly, disruption of the delta and kappa genes also led to the loss of enkephalin and kappa opioid analgesia, respectively.

Identifying and Characterizing MOR-1 and Its Splice Variants

As noted earlier, the need to individualize therapy with the opioids has been clearly established clinically. However, this raises the question of why drugs acting through the same receptor should show such diverse actions in different patients. Although one drug may be more potent than another, if they act through a single receptor, their relative potencies should be the same from patient to patient. Yet this is not what is seen. A similar situation exists in animal models.[25,26,37,38] For example, the *CXBK* mouse is relatively insensitive to morphine compared with standard CD-1 mice. Yet other mu drugs such as methadone, fentanyl, and even heroin show similar potencies in the two species. Antisense mapping studies suggested that there may be more than one cloned MOR-1 receptor, supporting pharmacologic studies going back over two decades proposing mu receptor subtypes.[24-26,39] In these earlier studies, novel antagonists were able to selectively block the receptors responsible for morphine analgesia without affecting those involved with either respiratory depression or the inhibition of gastrointestinal transit. At the same time, the antisense and the MOR-1 knockout studies clearly showed that the disruption of the gene encoding MOR-1 eliminated all morphine's actions. One possible explanation for these observations involves alternative splicing of the gene encoding MOR-1.

Genes are transcribed into mRNA that, in turn, is translated into proteins (Fig. 4-2). It has been estimated that there are approximately 30,000 human genes, far less than the number of proteins. One explanation for this difference is alternative splicing, in which a single gene is capable of generating a variety of proteins. Genes are composed of exons and introns. Exons correspond to the sequence that is contained within the mRNA, whereas introns are regions that are not. Most genes consist of alternating exons and introns. After the premature RNA is generated, it is processed, or spliced, so that the exons within the mRNA are coupled together, and the introns are "cut out." Some genes have several possible exons for a specific position or may have variants that contain more or fewer exons within the mRNA, leading to the generation of a variety of proteins from the same gene.

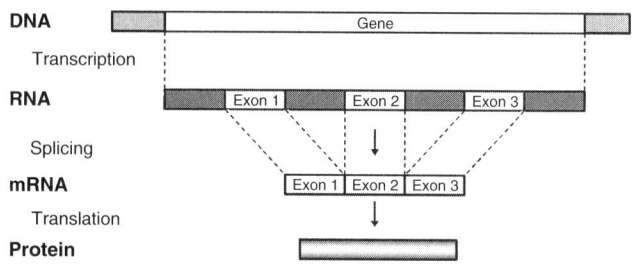

Figure 4-2. Schematic for splicing.

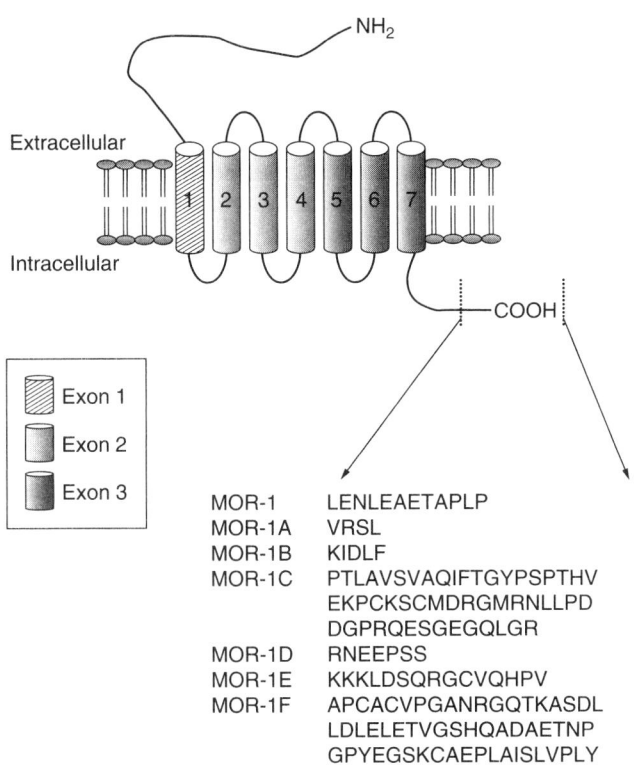

Figure 4-3. Schematic of C-terminal variants. (*From Pan et al.*[46–49])

Soon after cloning of the mu receptor MOR-1,[40–43] several splice variants were reported.[44,45] However, this was followed by identification of a large number of variants.[46–49] The variants containing the traditional G protein–coupled receptor structure involve splicing of the mRNA region important for generating the tip of the intracellular C-terminus (Fig. 4-3). They share the same transmembrane domains, which comprise the binding pocket. Thus, it is not surprising that these different MOR-1 variants are all selective for mu opioids in receptor-binding assays. Structurally, they differ at the C-terminus, a region important in transduction of the signal following receptor activation. When the variants were expressed and examined for their ability to activate G proteins, marked differences among the drugs were observed for the various splice variants.[50] Opioids that were pure agonists with one variant were partial agonists with another. Furthermore, the rank-order potency and efficacy of the drugs changed from one variant to another, showing that the drugs differed in their ability to activate these variants and providing a possible explanation for the pharmacologic differences of the mu opioids.

The MOR-1 variants also have been characterized in a number of additional studies. Overall, they show different regional distributions within the brain and spinal cord, suggesting region-specific RNA processing.[51–53] More detailed confocal microscopy shows that even when variants exist in the same brain region, there is cell-specific processing. Both MOR-1 and MOR-1C are expressed in the dorsal horn of the spinal cord, but high-power confocal microscopy shows that they are located in different cells. This suggests that there is a very complex mechanism for alternative splicing in which a cell is committed to a generation of one receptor or another.

Ultrastructural studies also illustrate differences. Whereas MOR-1 itself is located both pre- and post-synaptically, the MOR-1C receptor is located almost exclusively presynaptically. Furthermore, MOR-1C is colocalized with calcitonin gene–related peptide (CGRP), a neuropeptide involved in pain transmission, whereas MOR-1 is not. Alternative splicing has been examined in greatest detail in mice, as discussed earlier. However, there is evidence for alternative splicing in humans and rats as well. The roles of these subtypes are quite intriguing in that they may help to explain the different responses seen by patients. It is also possible that selective drugs may be developed based on their ability to activate specific subtypes within this mu opioid receptor family.

▶ CONCLUSION

Opioids play a pivotal role in the management of pain. The treatment of pain requires individualization of therapy due to the wide range of patient responses to individual drugs. There is no ceiling effect for analgesia, and doses can be escalated until limiting side effects are reached, at which point patients can be switched to an alternative medication. The presence of incomplete cross-tolerance helps to explain the utility of opioid rotation. The recent cloning of the opioid receptors has provided many insights into the actions of these drugs. Most clinical analgesics act through mu receptors. Recent studies now indicate that there are a number of splice variants of this receptor. Pharmacologic differences among these variants may help to explain variances among the drugs, despite the fact that they all are mu-selective. Furthermore, genetic differences in the generation of these variants also may help to explain the variability in response among patients to these drugs. Molecular approaches are helping to define the actions of these drugs and provide insights into why clinicians still need to individualize therapy for their patients.

REFERENCES

1. Reisine T, Pasternak GW: Opioid analgesics and antagonists, in Hardman JG, Limbird LE (eds), *Goodman & Gilman's The Pharmacological Basis of Therapeutics*. New York: McGraw-Hill, 1996:521–556.
2. Akil H, Mayer DJ, Liebeskind JC: Antagonism of stimulation-produced analgesia by naloxone, a narcotic antagonist. *Science* 1976;191:961–962.

3. Mayer DJ, Liebeskind JC: Pain reduction by focal electrical stimulation of the brain: An anatomical and behavioral analysis. *Brain Res* 1974;68:73–93.
4. Hosobuchi Y, Rossier J, Bloom FE, Guillemin R: Stimulation of human periaqueductal gray for pain relief increases immunoreactive β-endorphin in ventricular fluid. *Science* 1979;203:279–281.
5. Hosobuchi Y, Adams JE, Linchitz R: Pain relief by electrical stimulation of the central gray matter in humans and its reversal by naloxone. *Science* 1977;197:183–186.
6. Hughes J, Smith TW, Kosterlitz HW, et al: Identification of two related pentapeptides from the brain with potent opioid agonist activity. *Nature* 1975;258:577–579.
7. Pasternak GW, Simantov R, Snyder SH: Characterization of an endogenous morphine-like factor (enkephalin) in mammalian brain. *Mol Pharmacol* 1976;12:504–513.
8. Pasternak GW, Goodman R, Snyder SH: An endogenous morphine like factor in mammalian brain. *Life Sci* 1975;16:1765–1769.
9. Terenius L, Wahlstrom A: Search for an endogenous ligand for the opioid receptor. *Acta Physiol Scand* 1975;94:74–81.
10. Pert CB, Snyder SH: Opioid receptor: Demonstration in nervous tissue. *Science* 1973;179:1011–1014.
11. Simon EJ, Hiller JM, Edelman I: Stereospecific binding of the potent narcotic analgesic [^{3}H]etorphine to rat-brain homogenate. *Proc Natl Acad Sci USA* 1973;70:1947–1949.
12. Terenius L: Characteristics of the "receptor" for narcotic analgesics in synaptic plasma membrane from rat brain. *Acta Pharmacol Toxicol* 1973;33:377–384.
13. Evans CJ, Hammond DL, Frederickson RCA: The opioid peptides, in Pasternak GW (ed), *The Opioid Receptors*. Clifton, NJ: Humana Press, 1988:23–74.
14. Goldstein A, Tachibana S, Lowney LI, et al: Dynorphin(1–13), an extraordinarily potent opioid peptide. *Proc Natl Acad Sci USA* 1979;76:6666–6670.
15. Li CH, Chung D: Primary structure of human β-lipotropin. *Nature* 1976;260:622–624.
16. Pert CB, Pert A, Chang JK, Fong BTW: [D-Ala^2]-Met-enkephalinamide: A potent, long-lasting synthetic pentapeptide analgesic. *Science* 1976;194:330–332.
17. Li CH, Chung D, Doneen BA: Isolation, characterization and opioid activity of beta-endorphin from human pituitary glands. *Biochem Biophys Res Commun* 1976;72:1542–1547.
18. Beckett AH, Casy AF: Synthetic analgesics: Sterochemical considerations. *J Pharm Pharmacol* 1954;6:986–1001.
19. Portoghese PS: Stereochemical factors and receptor interactions associated with narcotic analgesics. *J Pharm Sci* 1966;55:865–887.
20. Martin WR: Opioid antagonists. *Pharmacol Rev* 1967;19:463–521.
21. Martin WR, Eades CG, Thompson JA, et al: The effects of morphine and nalorphine-like drugs in the nondependent and morphine-dependent chronic spinal dog. *J Pharmacol Exp Ther* 1976;197:517–532.
22. Lord JAH, Waterfield AA, Hughes J, Kosterlitz HW: Endogenous opioid peptides: Multiple agonists and receptors. *Nature* 1977;267:495–499.
23. Chang K-J, Cuatrecasas P: Multiple opioid receptors. *J Biol Chem* 1979;254:2610–2618.
24. Pasternak GW: Insights into mu opioid pharmacology: The role of mu opioid receptor subtypes. *Life Sci* 2001;68:2213–2219.
25. Pasternak GW: The pharmacology of mu analgesics: From patients to genes. *Neuroscientist* 2001;7:220–231.
26. Pasternak GW: Incomplete cross-tolerance and multiple mu opioid peptide receptors. *Trends Pharmacol Sci* 2001;22:67–70.
27. Cherny N, Ripamonti C, Pereira J, et al: Strategies to manage the adverse effects of oral morphine: An evidence-based report. *J Clin Oncol* 2001;19:2542–2554.
28. Uhl GR, Childers S, Pasternak GW: An opioid-receptor gene family reunion. *Trends Neurosci* 1994;17:89–93.
29. Rossi G, Pan YX, Cheng J, Pasternak GW: Blockade of morphine analgesia by an antisense oligodeoxynucleotide against the mu receptor. *Life Sci* 1994;54:PL375–PL379.
30. Pasternak GW, Standifer KM: Mapping of opioid receptors using antisense oligodeoxynucleotides: Correlating their molecular biology and pharmacology. *Trends Pharmacol Sci* 1995;16:344–350.
31. Pasternak GW, Pan Y-X: Antisense mapping: Assessing the functional significance of genes and splice variants. *Methods Enzymol* 2000;314:51–60.
32. Sora I, Takahashi N, Funada M, et al: Opioid receptor knockout mice define μ receptor roles in endogenous nociceptive responses and morphine-induced analgesia. *Proc Natl Acad Sci USA* 1997;94:1544–1549.
33. Loh HH, Liu HC, Cavalli A, et al: Mu opioid receptor knockout in mice: Effects on ligand-induced analgesia and morphine lethality. *Mol Brain Res* 1998;54:321–326.
34. Matthes HWD, Smadja C, Valverde O, et al: Activity of the delta opioid receptor is partially reduced, whereas activity of the kappa receptor is maintained in mice lacking the mu receptor. *J Neurosci* 1998;18:7285–7295.
35. Schuller AG, King MA, Zhang J, et al: Retention of heroin and morphine-6 beta-glucuronide analgesia in a new line of mice lacking exon 1 of MOR-1. *Nature Neurosci* 1999;2:151–156.
36. Law PY, Wong YH, Loh HH: Molecular mechanisms and regulation of opioid receptor signaling. *Annu Rev Pharmacol Toxicol* 2000;40:389–430.
37. Pasternak GW: Genetics and opioid pharmacology, in Pfaff DW, Berrettini WH, Joh TH, Maxson SC (eds), *Genetic Influences on Neural and Behavioral Functions*. Boca Raton, FL: CRC Press, 1999:13–29.
38. Rossi GC, Brown GP, Leventhal L, et al: Novel receptor mechanisms for heroin and morphine-6β-glucuronide analgesia. *Neurosci Lett* 1996;216:1–4.
39. Pasternak GW: The molecular biology of mu opioid analgesia, in Devor M, Rowbotham MC, Wiesenfeld-Hallin Z (eds), *Proceedings of the 9th World Congress on Pain*. Seattle: IASP Press, 2000:147–162.
40. Chen Y, Mestek A, Liu J, et al: Molecular cloning and functional expression of a μ-opioid receptor from rat brain. *Mol Pharmacol* 1993;44:8–12.
41. Eppler CM, Hulmes JD, Wang J-B, et al: Purification and partial amino acid sequence of a mu opioid receptor from rat brain. *J Biol Chem* 1993;268:26447–26451.
42. Thompson RC, Mansour A, Akil H, Watson SJ: Cloning and pharmacological characterization of a rat mu opioid receptor. *Neuron* 1993;11:903–913.

43. Wang JB, Imai Y, Eppler CM, et al: Mu opioid receptor: cDNA cloning and expression. *Proc Natl Acad Sci USA* 1993;90:10230–10234.
44. Bare LA, Mansson E, Yang D: Expression of two variants of the human mu opioid receptor mRNA in SK-N-SH cells and human brain. *FEBS Lett* 1994;354:213–216.
45. Zimprich A, Simon T, Hollt V: Cloning and expression of an isoform of the rat mu opioid receptor (rMOR-1B) which differs in agonist induced desensitization from rMOR-1. *FEBS Lett* 1995;359:142–146.
46. Pan Y-X, Xu J, Mahurter L, et al: Generation of the mu opioid receptor (MOR-1) protein by three new splice variants of the *OPRM* gene. *Proc Natl Acad Sci USA* 2001;98:14084–14089.
47. Pan Y-X, Xu J, Mahurter L, et al: Identification and characterization of two new human mu opioid receptor splice variants, hMOR-1O and hMOR-1X. *Biochem Biophys Res Commun* 2003;301:1057–1061.
48. Pan Y-X, Xu J, Bolan EA, et al: Identification and characterization of three new alternatively spliced mu opioid receptor isoforms. *Mol Pharmacol* 1999;56:396–403.
49. Pan Y-X, Xu J, Bolan E, et al: Isolation and expression of a novel alternatively spliced mu opioid receptor isoform, MOR-1F. *FEBS Lett* 2000;466:337–340.
50. Bolan EA, Pasternak GW: Functional analysis of MOR-1 splice variants of the mu opioid receptor gene, *OPRM*. *Synapse* 2004;51:11–18.
51. Abbadie C, Pan Y-X, Drake CT, Pasternak GW: Comparative immunhistochemical distributions of carboxy-terminus epitopes from the mu opioid receptor splice variants MOR-1D, MOR-1, and MOR-1C in the mouse and rat central nervous systems. *Neuroscience* 2000;100:141–153.
52. Abbadie C, Pan Y-X, Pasternak GW: Differential distribution in rat brain of mu opioid receptor carboxy-terminal splice variants MOR-1C and MOR-1-like immunoreactivity: Evidence for region-specific processing. *J Comp Neurol* 2000;419:244–256.
53. Abbadie C, Pasternak GW, Aicher SA: Presynaptic localization of the carboxy-terminus epitopes of the mu opioid receptor splice variants MOR-1C and MOR-1D in the superficial laminae of the rat spinal cord. *Neuroscience* 2001;106:833–842.

CHAPTER 5

Mechanisms of Neuropathic Pain

Joachim Scholz and Clifford J. Woolf

Neuropathic pain characterizes a large number of neurological disorders. This type of pain manifests in a complex and nonuniform way that does not correlate directly either with identifiable etiologic factors or the nature of the underlying lesion. Rather, the symptoms and signs of neuropathic pain are the result of multiple mechanisms that differ in space (peripheral versus central nervous system) and time (immediate responses versus long-term changes), but there is as yet no reliable way of predicting which mechanism is operational in patients with neurological disease.[1]

Painful disorders affecting the peripheral nervous system include metabolic diseases such as diabetes mellitus and nutritional deficits due to insufficient supply of vitamins. Painful neuropathies may develop as a side effect of drugs, following chronic alcohol abuse, or as a consequence of exposure to toxic chemicals. Neuropathic pain can occur in idiopathic inflammatory diseases such as the Guillain-Barré syndrome, in autoimmune disorders and in infections of the nervous system such as human immunodeficiency virus (HIV) or the varicella-zoster virus (VZV). Mechanical injuries leading to neuropathic pain include the traumatic severing of peripheral nerves, entrapment neuropathies, and the compression of spinal nerve roots. Neuropathic pain can be caused by cancer either through direct compression of nerves by tumor masses, as a feature of paraneoplastic disorders, or as a consequence of therapeutic interventions, including chemotherapy and radiation. Central neuropathic pain can result from a lesion or dysfunction in the central nervous system (CNS), including spinal cord and brain injuries, stroke, multiple sclerosis, and tumors. For the purposes of this chapter, we will focus only on mechanisms operating in pain following injury to the peripheral nervous system.

▶ THE DIVERSITY OF NEUROPATHIC PAIN

Neuropathic pain generally is considered a symptom of an underlying disease, and its anatomic distribution and temporal characteristics are used to establish the diagnosis of the disease. Pain may be the initial or leading symptom, for example, after the prolapse of an intervertebral disk or in neuralgic amyotrophy. In this sense, the pain still possesses a protective value, indicating the acute compression of a spinal nerve root or, in the latter case, inflammation of the brachial plexus. The location of the pain and, if present, the neurological deficit helps to identify the site of the injury. Pain can be further characterized by its sensory and affective quality, temporal characteristics, the dependency on external or internal stimuli, and any nonpainful symptoms or signs associated with the pain.

Based on the diagnosis and the nature of the underlying lesion, specific symptoms and signs of neuropathic pain have been attributed to lesions in specific compartments of the nervous system or types of injury. The quality of pain caused by an injury to the nervous system has been investigated systematically using tools such as the McGill Pain Questionnaire[2] and the Neuropathic Pain Scale.[3] However, patients with the same diagnosis, e.g., traumatic injury of a peripheral nerve, can experience pain differently. For example, "hot pain" often is felt by patients suffering radicular pain, as in postherpetic neuralgia, but a hot pain quality is also reported frequently by patients with painful diabetic polyneuropathy.[3] Therefore, the quality of pain by itself lacks the specificity to distinguish reliably between pain caused by lesions to different parts of the peripheral nervous system or different types of nerve injury. It may be of some use in the recognition of pain of central origin, which is often characterized by accompanying complex, nonpainful sensations. The LANSS Pain Scale, a recently developed tool for the assessment of neuropathic pain, uses the quality of pain as a parameter to determine pain of neuropathic origin and separate it from pain caused by other, nonneuropathic etiologies. Dysesthesia, paroxysmal pain, thermally evoked pain, mechanical allodynia, and an altered threshold for pain produced by pinprick are reportedly common features in neuropathic conditions.[4] According to this study, a hot, burning character of the pain is of high discriminative value, indicating neuropathic pain.[4]

Cranial neuralgias illustrate the diagnostic importance of the temporal pattern of pain. The key feature of these mostly idiopathic disorders is an intermittent pain that comes in paroxysmal attacks of specific duration and frequency. Complaints about ongoing pain or other continuous symptoms are atypical and alert the clinician to other diagnoses of symptomatic neuralgia. Most of the pain attacks occur spontaneously, but they can, e.g., in trigeminal neuralgia, be elicited by touch in a distinct area of the face or the mouth or by swallowing or chewing (triggered pain). Such stimuli may even evoke a series of pain paroxysms. Sometimes the temporal summation of successive nonpainful stimuli can trigger a pain attack.

Other types of evoked pain are found frequently after peripheral nerve lesions. Evoked pain can occur as an enhanced painful response to noxious stimulation (hyperalgesia) or in response to normally innocuous stimuli (allodynia). Hyperalgesia to pinprick and allodynia produced by static or moving mechanical stimuli are common. Pain also can be evoked by pressure on a nerve from certain positions or movements, e.g., in entrapment neuropathies such as carpal tunnel syndrome. Pain provoked by cold is a frequent complaint, whereas patients rarely describe pain generated by warmth. Hyperpathia is a complex form of evoked pain described mainly in patients with CNS lesions, traumatic injuries to peripheral nerves, or postherpetic neuralgia. It is reported as explosive in character, radiating, and as being associated with aftersensations, i.e., pain that may outlast the stimulus.

Nonpainful abnormal sensations (called *paresthesias* or, if characterized as unpleasant or disturbing, *dysesthesias*) can be associated with neuropathic pain. They may occur spontaneously or be evoked by stimuli. Paresthesias may be felt concomitantly with the pain or independently. Numerous qualities of paresthesias have been reported: pins and needles, tingling, etc. Paresthesias often are described by patients suffering from diabetic or alcoholic polyneuropathies or nerve entrapment. Most of these sensations are thought to be generated by overactive (or overly reactive) primary sensory neurons. More complex phenomena occur after extended deafferentation, e.g., traumatic avulsion of the brachial plexus. This injury usually affects several spinal roots whose connections with the spinal cord are severed. The resulting syndrome involves pain in almost all cases. The pain is described as burning, pounding, or electric, with episodes of an additional paroxysmal pain of high intensity. Many of these patients have a phantom limb, sometimes with imaginary movements.

Changes in skin color and temperature suggest altered cutaneous blood flow and, together with an abnormal sudomotor activity, may indicate autonomic nervous system involvement. *Sympathetically-maintained pain* (SMP) is the term used to describe pain attributable to the efferent activity of the sympathetic nervous system. In clinical practice, SMP usually is employed to describe the component of a painful syndrome that can be treated by sympatholytic intervention. SMP is variably present in complex regional pain syndrome (CRPS), a constellation of spontaneous pain, allodynia or hyperalgesia after a nerve injury, and changes in skin blood flow or sudomotor activity in the region of pain.[5] SMP and CRPS are not related uniquely to pain after injury to the nervous system; they also can occur after soft tissue injury or bone fractures.

▶ HOW TO INTERPRET SYMPTOMS AND SIGNS

The current management of patients with neuropathic pain starts with a diagnosis based on their underlying condition, such as diabetic neuropathy or postherpetic neuralgia. Clearly, patients with painful diabetic neuropathy need their diabetes control to be optimized and any other complications, such as retinopathy or peripheral vascular disease, to be treated. In considering only the treatment of their pain, however, linking the pain with the etiologic cause or disease is not always useful. Although there are certain features of neuropathic pain that are almost diagnostic, e.g., the tic douloureux in trigeminal neuralgia, in general there is considerable overlap between symptoms and signs of pain after quite different types of nervous system insults. In the absence of a past medical history and based solely on the neuropathic pain features, the underlying condition cannot be distinguished. The priority for patients with neuropathic pain is not to make a disease diagnosis but to make a diagnosis that helps to predict treatment. Current guidelines for symptom control in neuropathic pain are nonselective and consist of algorithms recommending the prescription of tricyclic antidepressants (TCAs) or gabapentin as first-line treatment, followed by trials of other antiepileptic drugs and opioids. Each subsequent step is taken after deciding that pain reduction is not sufficient or that unacceptable side effects have occurred.[6] This approach reflects the dilemma that no standardized tools are available to determine the mechanisms responsible for pain in the individual patient. Apart from such rare conditions as Guillain-Barré syndrome, in which a causative therapy such as modulation of the immune system is possible, the treatment of neuropathic pain is based on trial and error. One should refrain from disguising this shortcoming by labeling it an "individualization of therapy."[7]

Successful pain treatment needs to address the specific mechanisms that cause the painful features of the patient's neuropathic syndrome. The question then arises as to whether individual mechanisms that correlate with specific symptoms can be identified from the history, examination, and special investigations. Successful examples of how to derive pain mechanisms from symptoms and signs (or indirectly, their successful treatment) have been demonstrated in trigeminal neuralgia and postherpetic neuralgia. Since the characteristic, brief attacks of intense pain in trigeminal neuralgia respond well to the sodium channel blockers carbamazepine and phenytoin, it can be concluded that the activation of voltage-gated sodium channels contributes to the generation of this paroxysmal pain. In patients with postherpetic neuralgia, a predominantly spontaneous pain occurs in scarred skin areas with a profound loss of thermal and mechanical sensitivity. By contrast, severe mechanical allodynia often is found in skin territories characterized by minimal sensory loss. Topical lidocaine is effective against the allodynia but does not provide pain relief in areas with reduced thermal and mechanical sensitivity.[8,9] It was concluded from these observations that so-called irritable nociceptors in the skin areas with a preserved response to temperature generate a state of central sensitization (see below), which, in turn, produces the painful response to normally innocuous mechanical stimulation. On the other hand, thermal insensitivity correlates with a loss of nociceptors, demonstrated histologically in skin biopsies. In these patients, allodynia is likely caused by low-threshold mechanosensitive fibers that formed new connections to pain transmission neurons in the dorsal horn of the spinal cord. The spontaneous pain in denervated skin areas is explained by deafferentation and a subsequent hyperactivity of central transmission neurons, normally conveying nociceptive pain signals. Since this central mechanism is uncoupled from afferent input, topical administration of lidocaine predictably has no effect.[9]

This demonstrates that sensory testing can be used to help determine pain mechanisms and that this has therapeutic significance. However, quantitative testing of the sensory nervous system is time-consuming and usually performed only at institutions conducting clinical research. The urgent need for simple diagnostic tests suitable to determine sensory disorders during bedside examination is now widely recognized.[10,11] Table 5-1 tentatively assigns symptoms and signs to potential mechanisms of neuropathic pain based on results obtained from physiologic studies or laboratory research. For the purpose of a bedside examination, von Frey filaments are sufficient to determine sensitivity to punctate, static stimuli. Dynamic stimulation can be applied using a brush. A safety pin can be used to assess the sensitivity for punctate painful stimuli. Metal probes kept at nonpainful warm (>40°C) and cold (<15°C) temperatures are suitable to test for temperature discrimination and thermal hypersensitivity. Additional bedside tests can be employed to examine the sense for position, movement, and vibration.

Standard neurological investigations, including neurography and electromyography, employed to classify peripheral nerve disorders are not particularly helpful because they do not assess the function of nociceptors. Moreover, neuropathic pain does not correlate with a specific pathology; it can manifest in primarily demyelinating neuropathies as well as in neuropathies with mainly axonal degeneration. The search for suitable biomarkers of neuropathic pain mechanisms has been of limited success. There are, however, recent reports of tumor necrosis factor alpha (TNF-α) being increased in Schwann cells, macrophages, and T cells from nerve biopsies of patients with painful neuropathy.[12,13] Animal studies have shown that TNF-α facilitates neuropathic pain by sensitizing primary sensory neurons (see below). Blood levels of catecholamines have been proposed as a possible biomarker of SMP, but the concentrations of norepinephrine and its metabolite 3,4-dihydroxyphenylethyleneglycol actually were lower or did not differ in venous blood samples taken from limbs affected by CRPS compared with the contralateral side.[14,15] Adequate assessment of altered neurovascular regulation as an indicator of a changed activity level of the sympathetic nervous system is limited to specialized centers.

Functional magnetic resonance imaging (fMRI) and positron-emission tomography (PET) have been used to identify brain areas involved in the processing of pain. Activation of the right anterior cingulate cortex may relate to the affectional aspects of pain. Activation changes in the prefrontal cortex, the inferior frontal gyrus, and the sensory association cortex in Brodman areas 5 and 7 have been correlated with experimental mechanical allodynia.[16–18] Specific patterns of cortical reorganization after amputation have been associated with phantom limb pain.[19]

TABLE 5-1. MECHANISM-BASED ASSESSMENT OF SYMPTOMS AND SIGNS

Spontaneous Sensory Disorders		Mechanisms	Stimulus-Evoked Sensory Disorders	
Low-threshold Aβ-afferents	**High-threshold Aδ- or C-fiber afferents**	**Peripheral Nervous System**	**Low-threshold Aβ-afferents**	**High-threshold Aδ- or C-fiber afferents**
Decreased sensation of touch, deficits in the discrimination of position, movement, vibration	Decreased sensation of temperature and pinprick-evoked pain	**Deafferentation** through degeneration of peripheral sensory nerve fiber terminals, DRG neurons, or a lesion of ascending dorsal column fibers		
		Reduced threshold of activation in nociceptors		Pinprick hyperalgesia (mediated through Aδ-fibers) Heat hyperalgesia and cold allodynia (C-fibers)
Paresthesia, dysesthesia	Painful (burning) sensation	**Increased membrane excitability** and **ectopic activity** of the DRG neuron cell soma or along the axon		Pain caused by stretching (straight-leg-raising sign)
		Phenotypic changes including novel expression of receptors and ion channels	Paresthesia, dysesthesia on direct pressure (Tinel sign)	Painful neuroma sign
Centrally caused or mediated through low-threshold Aβ-afferents	**Centrally caused or mediated through high-threshold Aδ- or C-fiber afferents**	**Central Nervous System**	**Mediated through low-threshold Aβ-afferents**	**Mediated through high-threshold Aδ- or C-fiber afferents**
		Abnormal temporal summation	Painful sensation after repetitive application of low-intensity mechanical stimuli	
		Central sensitization: abnormal sensory excitation caused by an increased synaptic release of neurotransmitter or enhanced post-synaptic sensitivity	Allodynia on touch (static or dynamic), pressure, movement	Hyperalgesia on stimulation with pinprick, heat, or cold temperature

▶ TABLE 5-1. (CONTINUED)

Paresthesia, dysesthesia	Spontaneous pain	**Structural reorganization** of synapses	Allodynia on touch (static or dynamic)
		Loss of inhibition	Allodynia on touch (static or dynamic), pressure, movement
Paresthesia, dysesthesia Complex sensations	Spontaneous pain	**Spontaneous activity** in sensory transmission neurons of the dorsal horn or higher centers	Hyperalgesia on stimulation with pinprick, heat, or cold temperature
Sympathetic Nervous System (central or peripheral)			
Altered sweating Changes in skin blood flow, edema		**Altered level of activity**	
Paresthesia, dysesthesia	Painful sensation	**Sympathetic-sensory coupling** at a neuroma site or within the DRG	

GENETIC FACTORS

Several lines of evidence indicate that the genetic background of individuals has a significant influence on the risk to develop pain after nerve injury and the response to analgesic drugs.[20] Pain usually manifests only in a subgroup of patients affected by a disease of the nervous system. Neuropathic pain occurs in genetically-transmitted diseases such as diabetes mellitus or hereditary neuropathies. The incidence of neuropathic pain may vary with gender, e.g., in trigeminal neuralgia.

The search for genes associated with pain is advancing rapidly. High-throughput screens are being employed to identify genes regulated after nerve injury.[21] Genetically-modified animals that either lack or overexpress a candidate gene product can be produced to determine the significance of dependent molecular mechanisms for nociception and after nerve injury. Since laboratory pain research usually is performed on inbred rodent strains that are genetically well defined, comparative genetics can be used to look for heritable traits in these strains that determine differences in the responsiveness to painful stimuli of the animals.[22] Within individual mouse strains, there is a covariation of nociceptive sensitivity across different tests. Based on this cross-correlation of test results, five major types of genetically-predetermined nociception and pain hypersensitivity have been suggested.[22] Genetic background further affects the response of animals to antinociceptive drugs from a variety of pharmacologic classes,[23] which highlights the caution required in interpreting data from animal studies across strains or species.

ANIMAL MODELS OF NEUROPATHIC PAIN

Our understanding of the biologic mechanisms involved in neuropathic pain is based mainly on the results of animal studies modeling different types of nerve injury, typically in rats or mice. A criticism of animal models, however, is the lack of a direct measure of pain. The tests available are based on behavioral changes and mostly depend on the assessment of an adverse reaction to a stimulus. These reactions may involve a reflex component, but many represent complex movements, e.g., a targeted withdrawal from a pinprick, followed by shaking of the limb and licking of the paw. While it seems reasonable to assume that an immediate stimulus response, followed by an extended protection of the tested body part, corresponds to a painful sensation, the interpretation of other outcome measures is less certain, especially if a test relies on the detection of a response threshold.

A related difficulty in animal models of neuropathic pain is the absence of a valid and reliable parameter of spontaneous pain. Such a parameter is highly desirable because spontaneous pain is a predominant complaint in humans. Reduced motor activity, increased guarding of an affected extremity, asymmetric grooming, and a lack of growth have been used to measure spontaneous pain in rodents, but none of these parameters is a specific indicator of pain.

One of the first attempts to model peripheral nerve injury was made by Wall, who studied behavioral alterations in rats and mice after a complete transection of the sciatic nerve, which he considered a model of anesthesia dolorosa.[24] By contrast, partial nerve lesions allow testing of behavioral responses to defined thermal or mechanical stimuli applied to the spared or only partially denervated skin. Currently, the most widely used models of neuropathic pain include a ligation and transection of two peripheral branches of the sciatic nerve while leaving the sural nerve intact (spared nerve injury, or SNI),[25] chronic constriction injury (CCI) of the sciatic nerve by loose ligations of chromic gut or a polyethylene cuff,[26,27] a partial ligation of the sciatic nerve,[28] and a ligation of the L5 and L6 spinal nerves (SNL).[29] Chronic compression of the dorsal root ganglia (DRGs)[30] and implantation of autologous intervertebral disk tissue[31] mimic the effect of intervertebral foramen stenosis or disk herniation, respectively. Streptozotocin, a cytotoxin selectively acting on pancreatic beta cells, has been used to induce a condition reminiscent of painful diabetic peripheral polyneuropathy, but the validity and reliability of this model have been questioned.[32] Photochemically-produced ischemia of peripheral nerves has been proposed to model the vascular component of painful neuropathies caused by diabetes mellitus or entrapment.[33] Aspects of neuropathic cancer pain have been investigated after compression of the sciatic nerve caused by the progressive growth of sarcoma cells[34] or the induction of pain by chemotherapeutic drugs with neurotoxic side effects, e.g., paclitaxel, vincristine, and cisplatin.

Animal models for other, special causes of neuropathic pain such as alcohol and infections by members of the herpes virus family have been established. To explore the contribution of the immune cell response to a nerve injury, animal models of inflammatory neuropathies have been developed.

MECHANISMS OF NEUROPATHIC PAIN

Under physiologic conditions, distinct mechanisms operate in primary sensory neurons, the spinal cord, and the brain to control the flow and the integration of afferent information from peripheral receptors to the CNS so that the nature of a stimulus, its intensity, and its location can be determined with the precision necessary for an appropriate response. In conditions of neuropathic pain, this elaborate system of control derails, and a state emerges in which innocuous stimuli cause unpleasant (paresthesia) or painful sensations (allodynia), the intensity of stimuli is overestimated (hyperalgesia), the discrimination of spatial (poor localization, radiating pain) and temporal (summation, aftersensation) characteristics of the sensory information is lost, or pain is felt in areas that actually became disconnected from the sensory nervous system (deafferentation pain, phantom pain). Successful therapy requires an understanding of the pathologic mechanisms that produce this "system failure."

The following is an overview of mechanisms underlying the development and maintenance of neuropathic pain. It should be noticed that despite the recent, enormous progress in fundamental pain research, insight into several of these mechanisms is still patchy.

Changes Affecting Intact Afferents in Partially-Injured Peripheral Nerves

Most conditions of neuropathic pain develop after partial injuries to the peripheral nervous system (PNS). Complete severing of a nerve or spinal nerve root occurs by trauma but is relatively rare. Animal models of partial nerve injury reveal that both injured and neighboring, uninjured nerve fibers contribute to the generation of pain.[35]

The microenvironment of intact nerve fibers changes profoundly after injury as a result of Wallerian degeneration of lesioned fibers nearby and responses in nonneuronal Schwann cells, satellite cells around the cell bodies of primary sensory neurons in the DRGs, and different components of the immune system. In the proximal stump of a lesioned nerve, cellular and molecular changes, including the dedifferentiation and proliferation of Schwann cells, ensue to promote regrowth of surviving neurons. In the distal stump, hematogenous macrophages are recruited to remove axonal and myelin debris. During this process, neurotrophic factors, cytokines, and their receptors are expressed de novo or upregulated and serve to transmit signals between nonneuronal cells, lesioned nerve fibers, and adjacent intact neurons (Fig. 5-1). As a result, primary sensory neurons become increasingly excitable and may even exhibit spontaneous activity. Mutant mice with a delayed degeneration of nerve fibers develop only attenuated pain-like behavioral changes compared with wild-type mice after nerve injury.[36]

Neurotrophic Factors

Target-derived neurotrophins regulate the differentiation and the survival of sensory neurons during embryonic and postnatal development. Their neuroprotective function becomes relevant again after peripheral nerve injury.[37,38] Transection of peripheral nerve fibers leads to an increased expression of nerve growth factor (NGF) in skin keratinocytes within the nerve's territory. An enhanced transcription of brain-derived neurotrophic factor (BDNF), NGF, and neurotrophin 4 (NT-4) is found in Schwann cells at the site of and distal to the nerve injury. The expression of NGF increases in satellite cells of affected and neighboring DRGs, as well as in injured sensory neurons themselves.[39] Uninjured sensory neurons also show an enhanced expression of BDNF.[40] In the DRGs, neurotrophic factors possibly act as paracrine signals between sensory neurons and contribute to the pain hypersensitivity that develops after nerve injury. NGF, for example, tonically regulates the expression of substance P (SP) in nociceptors.[41] Inactivation of NGF, BDNF, and NT-3 reduces the development of mechanical allodynia.[42]

One effect of peripheral nerve injury is the creation of new contacts and chemical signaling between sensory neurons and the sympathetic nervous system (see Fig. 5-1). Following injury, efferent sympathetic fibers sprout into the DRGs and form basket-like structures around the cell bodies of large-diameter, low-threshold Aβ fibers.[43] Wallerian degeneration and the release of neurotrophins are critically involved in the creation of this abnormal connection. Sympathetic sprouting occurs late and is markedly reduced in mutant mice with a delayed type of Wallerian degeneration.[36] Blocking the endogenous levels of BDNF, NGF, and NT-3 also decreases the sprouting of sympathetic efferents.[44–46] Since these sprouts provide a potential source of norepinephrine, they may represent a morphologic basis for SMP in humans. Tyrosine hydroxylase–immunoreactive axon sprouts of noradrenergic neurons have been seen in some human DRGs obtained from patients with chronic neuropathic pain.[47] Noradrenergic receptors of the α_2 subtype are constitutively present on DRG sensory neurons. After nerve injury, the sensitivity to norepinephrine is enhanced, and the number of DRG neurons expressing α_{2A} receptors increases fivefold. Many of these neurons have peripheral axons not directly affected by the lesion.[48,49] Electrophysiologic recordings have demonstrated an abnormal sensitivity of sensory neurons to α-adrenergic agonists as a consequence of nerve injury.[50,51] The responsive cells include C- and Aδ-type nociceptors rather than large-diameter, low-threshold Aβ neurons, which are the cells surrounded primarily by the sympathetic fibers that enter the DRGs. However, since the response of sensory neurons to electrical stimulation of postganglionic sympathetic efferents sets in with a several-second delay, sympathetic-sensory coupling may occur indirectly via diffusion rather than by direct synaptic contact.[47]

The significance of sympathetic sprouting into the DRGs and its contribution to neuropathic pain are questioned by studies describing a lack of catecholaminergic modulation of pain-related behavior in animals after nerve injury or even enhanced mechanical allodynia following sympathectomy.[52] Strikingly, mice lacking functional subtypes of the α_2-adrenergic receptor do not differ

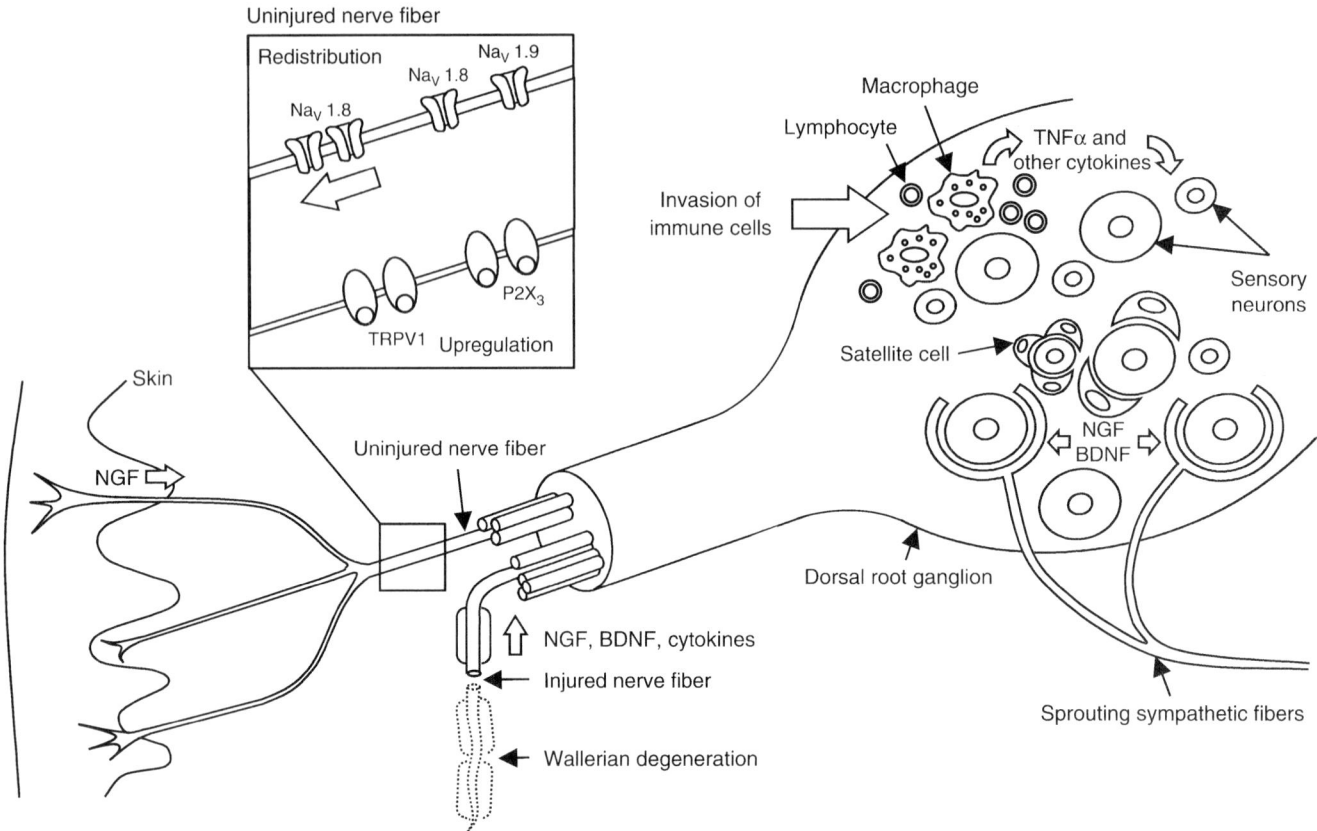

Figure 5-1. Multiple mechanisms produce changes in the activity of spared sensory afferents after nerve injury. Neurotrophic factors are released from Schwann cells at the injury site, and keratinocytes of the skin in the territory innervated by the lesioned nerve. In the DRGs, neurotrophic factors synthesized by satellite cells exercise a regulatory function through paracrine signaling. Cytokines derive from sources at the site of the nerve lesion (Schwann cells, immune cells) and from macrophages and T-lymphocytes invading the DRGs following a nerve injury. These changes in the microenvironment lead to an enhanced excitability of intact sensory afferents. The membrane excitability of spared afferents is also increased by a redistribution of the $Na_V1.8$ sodium ion channel and upregulation of receptors such as TRPV1 and $P2X_3$. Sympathetic fibers that sprout into the DRGs form baskets mainly around intact sensory neurons, which can then be further sensitized through the release of norepinephreine.

in their behavioral phenotype from wild-type mice after partial ligation of the sciatic nerve.[53] Excitatory sympathetic-sensory coupling is, in fact, only a transient phenomenon that fades over time and turns into suppression in the chronically injured nerve.[50] The extent to which the sympathetic nervous system contributes to pain in patients with nerve injury has not been sufficiently investigated by randomized, double-blind clinical studies of sympatholytic treatments.

Cytokines

Involvement in the development of pain hypersensitivity after nerve injury has been demonstrated for the cytokine TNF-α.[54] There is also an upregulation of interleukin 1β (IL-1β), IL-6, and IL-10 in the injured nerve, although with a different temporal pattern.[55] TNF-α accumulates at the injury site after release from Schwann cells in the neighborhood.[56] TNF-α concentrations in the DRGs rise within a few days of nerve injury. This increase coincides with the invasion of macrophages and T-lymphocytes.[57] Proliferating satellite cells represent another potential source for the increase of TNF-α in DRGs. The tight cell-to-cell contacts of satellite cells may support the spread of the cytokine signal to injured and uninjured neurons (see Fig. 5-1). The sensitivity to TNF-α of primary sensory neurons increases after chronic compression injury of the DRGs or a nerve lesion.[58,59] Signaling is presumably mediated through activation of TNF receptor 1 and involves the phosphorylation of the mitogen-activated protein (MAP) kinase p38.[54]

Changes in the Membrane Excitability of Intact Fibers

The functional state of neurons is defined by the electrical potential across the cell membrane. Ion channels are responsible for maintaining the resting membrane potential and for eliciting action potentials. Under resting conditions, most ion channels, except for some potassium channels, are closed; ligand binding, alterations in the membrane voltage, or mechanical stretch of the membrane are required to switch them open (gated channels). After nerve injury, multiple changes occur in the composition and the properties of ion channels in the membranes of primary sensory afferents. In addition, there is constant physiologic turnover with newly-synthesized ion channels. Following nerve injury, the expression and trafficking of ion channels change dramatically depending on the lesion site (proximal or distal to the cell body in the DRG) and whether the afferent fiber is lesioned itself or is a neighboring, intact fiber (see Fig. 5-1).

Sodium channels are critical for the generation and conduction of action potentials in sensory neurons. They consist of a principal α subunit containing the ion pore and auxiliary β subunits that are required for normal kinetics of the sodium current and modulation of the channel gating. Nociceptors preferentially express two particular α subunits of voltage-gated ion channels, $Na_V1.8$ and $Na_V1.9$.[60,61] $Na_V1.8$ (formerly known as SNS or PN_3) and $Na_V1.9$ (SNS_2 or Na_N) are resistant to tetrodotoxin (TTX-R). $Na_V1.8$ is expressed in the membrane of C fibers and some myelinated A fibers (mainly representing thinly myelinated high-threshold Aδ fibers). $Na_V1.9$ is found exclusively in unmyelinated C-fiber afferents of the DRGs and the trigeminal ganglion.[60–62]

While the expression profile of $Na_V1.8$ and $Na_V1.9$ is unchanged in the cell bodies of uninjured sensory neurons after a partial lesion of a nerve,[63] $Na_V1.8$ is redistributed and accumulates along the peripheral axons of uninjured nociceptive C-fiber afferents. The redistribution of $Na_V1.8$ is detectable as early as 2 days after the injury and parallels an increase in the TTX-R portion of the compound action potential of a partially-injured sciatic nerve.[64] Incomplete nerve injury, therefore, may result in a net augmentation of excitability depending on the proportion of uninjured nerve fibers with an increased integration of $Na_V1.8$ and possibly also $Na_V1.9$ subunits along the membrane of their distal axons. Selective knockdown of $Na_V1.8$ subunit expression by intrathecal injections of antisense oligodeoxynucleotides can reverse mechanical hypersensitivity and thermal hyperalgesia after SNL, whereas it has no effect on mechanical or thermal nociception.[65] However, knockout of $Na_V1.8$ has only a minimal effect on neuropathic pain–like behavior.[66]

Following nerve injury, neuronal activity may become uncoupled from the application of a peripheral stimulus. Ongoing activity has been demonstrated in uninjured nociceptive C fibers starting 1 day following the lesion of an adjacent dorsal root or the ventral root of the same spinal segment.[35,67] Ectopic discharges also originate in the cell bodies of uninjured myelinated muscle afferents.[68]

Upregulation of Ligand-Gated Ion Channels

Upregulation of a ligand-gated receptor, the transient receptor potential (TRP) V1 receptor (previously known as *vanilloid receptor subtype 1,* or VR1), occurs in uninjured C fibers as well as A fibers of partially lesioned nerves.[69] The TRPV1 receptor is activated by heat, low pH (protons), and the pungent ingredient of chili peppers, capsaicin. The normal activation threshold for TRPV1 is above 42°C. This threshold falls during peripheral sensitzation induced by inflammatory mediators. It is unknown if at this reduced threshold TRPV1 may contribute to the spontaneous activity that occurs after peripheral nerve injury, with body temperature potentially representing the stimulus here. It is also conceivable that endogenous ligands such as the vanilloid anandamide, a bioactive lipid and structural relative of arachidonic acid, are released after injury and activate TRPV1. However, null mutation of TRPV1 reduces inflammatory but not neuropathic pain.[70]

The proportion of sensory neurons expressing the purinoreceptor $P2X_3$, whose endogenous ligand is adenosine-5′-triphosphate (ATP), transiently increases after partial nerve injuries, and intense immunoreactivity is found at the injury site.[71,72] Knockdown of the $P2X_3$ receptor subunit by antisense oligonucleotide treatment reduces mechanical hyperalgesia and allodynia after partial ligation of the sciatic nerve,[73] indicating a possible role of the $P2X_3$ channel in tactile hypersensitivity. A remarkable effect against mechanical allodynia and thermal hyperalgesia in models of neuropathic pain also was found for an antagonist of the $P2X_3$ and $P2X_{2/3}$ receptors.[74]

Injured Nerve Fibers

The Neuroma Site

After severance of nerve fascicles, regenerating processes or neurites grow out from the proximal lesion site within hours. Some of these sprouts may successfully reinnervate their peripheral target if they are able to cross the lesion and enter into the distal myelin sheath. If growth is impeded, regenerating sprouts and Schwann cells build a tangled mass, the neuroma. A neuroma forms in continuity if only some of the nerve sprouts reconnect with the target. Aberrant sprouts also may produce scattered microneuromas along the distal nerve stump. Neuropathic pain conditions associated with neuromas encompass traumatic nerve injuries, entrapment neuropathies, and iatrogenic nerve lesions caused by surgery. All patients with stump pain after amputation have neuromas, but the presence of a neuroma does not necessarily mean that the patient will have pain.[75]

Neuromas are a significant source of spontaneous ectopic activity in injured primary afferents. These ectopic discharges are characterized by irregular firing patterns and repetitive spikes. The proportion of spontaneously active nerve fibers usually peaks within the first 3 weeks and decreases substantially thereafter. Discharges originating from Aβ and Aδ fibers are prominent during the first 2 weeks after injury and then subside, whereas spontaneous activity from C-fiber endings prevail for longer.[76–78] Mechanical and chemical stimuli, including low pH, can enhance the firing rate of spontaneously active afferents in neuromas. Inflammatory mediators present in the acute environment of a lesioned nerve may sensitize afferent fibers in a neuroma similar to the way they sensitize intact afferent peripheral terminals.[79] The heightened excitability of afferent fibers in a neuroma results in an increased sensitivity to mechanical stimulation. Paresthesias elicited by palpation are exploited by physicians in the form of the Tinel sign, which can be used to trace the progress of regrowth in regenerating nerve fibers.

The neuroma is another possible site of sympathetic-sensory coupling after nerve injury (Fig. 5-2). Sprouts of noradrenergic axons into neuromas have been identified in the first week following a nerve lesion but disappear quickly thereafter. During this period of noradrenergic fiber sprouting, spontaneously active Aβ and Aδ fibers are responsive to α_2-adrenergic agonists.[80] Application of epinephrine to long-persisting neuromas in humans produces intense burning pain,[81] suggesting that the receptor state in chronic neuromas also allows for adrenergic stimulation of sensory afferents, possibly C-fiber nociceptors. On the other hand, adrenergic-receptor antagonists do not affect the intensity of spontaneous discharges.[80] The latter observation is consistent with the finding that sympathectomy has no impact on spontaneous activity recorded from neuromas of the inferior alveolar nerve.[82] It is, therefore, unlikely that endogenous norepinephrine tonically drives the pain that originates from a neuroma.

The Cell Bodies of Injured Neurons Are the Main Source for Spontaneous, Ectopic Activity

For a limited time period (seconds to minutes) immediately following a traumatic nerve lesion, sensory neurons fire intensely, eliciting a barrage of afferent input termed the *nerve injury discharge*. A second type of spontaneous electrical activity develops later, after the first day. Most of this activity arises in myelinated Aβ and Aδ fibers and originates in the DRGs rather than directly at the injury site (see Fig. 5-2). Relatively fewer injured C fibers become spontaneously active.[83] The ectopic discharge from injured fibers can exhibit a rhythmic pattern that is caused by the emergence of sinusoidal subthreshold membrane potential oscillations and maintained by depolarizing afterpotentials.[84,85] Hyperpolarization-activated "pacemaker" channels in the membranes of the cell somata are responsible for the spontaneous activity deriving from Aβ fibers after nerve injury. These channels, which are modulated by cyclic nucleotides and permeable for sodium and potassium ions, also account for almost half the spontaneous activity of Aδ fibers.[86] Importantly, blockade of these pacemaker channels reduces mechanical allodynia, which is mediated mainly through low-threshold Aβ afferents. However, it remains to be resolved how the spontaneous ectopic activity deriving from pacemaker channels contributes to this type of stimulus-dependent behavioral change.[86]

Heightened Responsiveness to Stimuli

Sensory neurons that become spontaneously active after nerve injury may be identical to afferents that exhibit increased ectopic sensitivity to direct mechanical pressure.[87] Ectopic firing can, in fact, be exacerbated by a number of factors, including heat and endogenous inflammatory substances such as prostaglandins, cytokines, and catecholamines. The altered responsiveness of injured nerve fibers is the result of a dramatic change in the composition and the properties of ion channels, voltage-dependent and ligand-gated channels alike, in the membranes of injured afferents (see Fig. 5-2). In addition, injured neurons undergo a substantial shift in their receptor profiles and gain the ability to express new signaling molecules, contributing to the acquisition of a new phenotype.

The expression of $Na_V1.8$ and $Na_V1.9$ is largely reduced in injured nerve fibers, and this is followed by a loss of the TTX-R sodium current.[63,88,89] In contrast to the substantial redistribution of $Na_V1.8$ in uninjured C-fiber afferents, only little accumulation of $Na_V1.8$ is found proximal to the lesion site in injured nerve fibers.[64] On the other hand, injured sensory neurons start reexpressing a tetrodotoxin-sensitive (TTX-S) voltage-dependent sodium channel, $Na_V1.3$ (previously termed *brain type III*).[90] Normally, this sodium channel is only present in DRG neurons during embryonic development. The novel induction of $Na_V1.3$ and the associated increase in TTX-S sodium current produce a net augmentation of membrane excitability.[90,91] This shift in the balance of sodium channel expression depends on the interrupted supply of target-derived neurotrophic factors that is produced by the nerve injury: glial cell line–derived neurotrophic factor (GDNF) protects DRG neurons from the loss of $Na_V1.8$ and $Na_V1.9$ and the concomitant reduction of TTX-R sodium currents after injury.[89,92] At the same time, intrathecal application of NGF or GDNF prevents the induction of $Na_V1.3$ after axotomy or partial ligation of the sciatic nerve.[89,93] GDNF also reduces spontaneous firing of injured nerve fibers.[89]

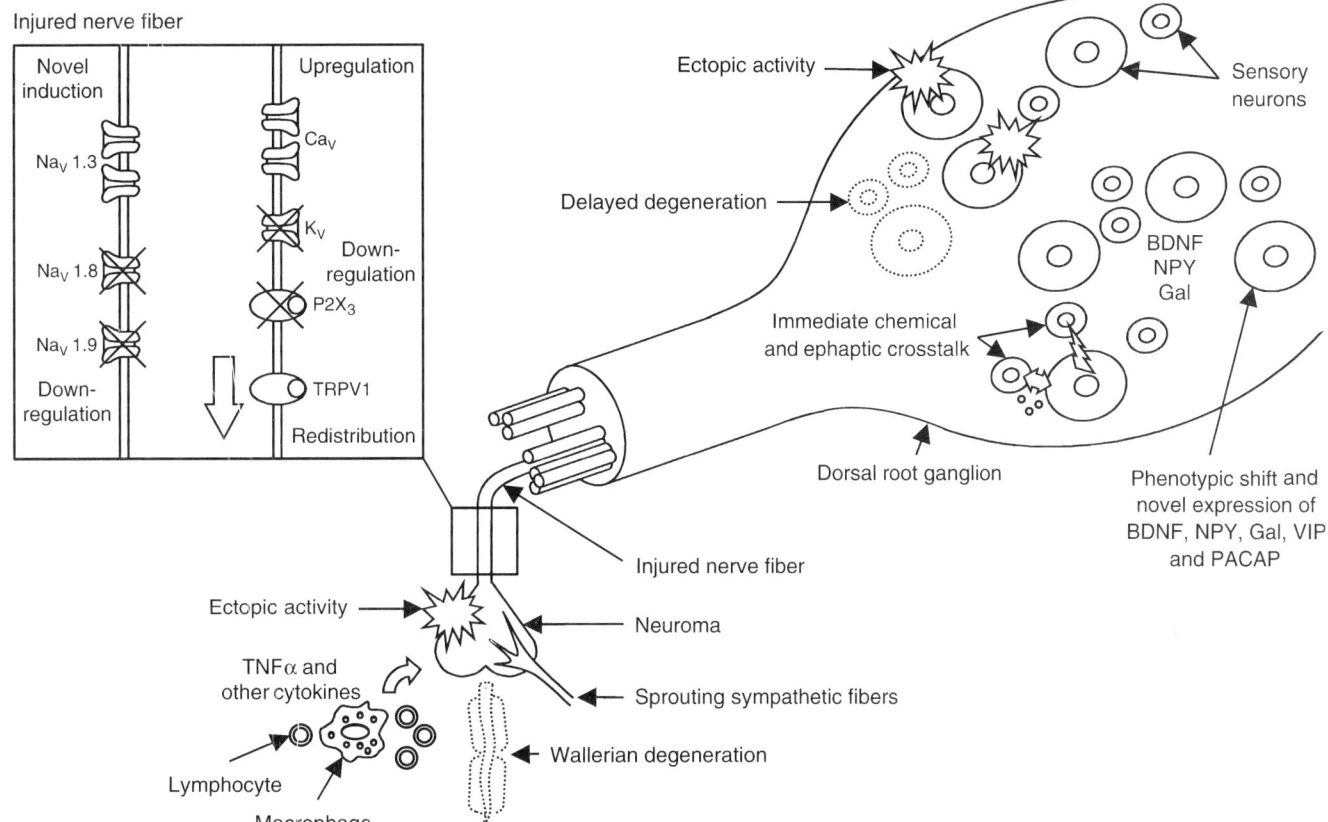

Figure 5-2. Profound changes occur in nerve fibers directly affected by the nerve lesion. Abnormal spontaneous discharges and enhanced stimulus-evoked activity are recorded from injured nerve fibers at a neuroma site. TNF-α and other cytokines released from immune cells invading the neuroma, as well as sympathetic-sensory coupling following the outgrowth of postganglionic sympathetic fibers, may underly this increase in afferent activity. While the sodium channels $Na_V1.8$ and $Na_V1.9$ are downregulated, $Na_V1.3$ is newly expressed in injured sensory afferents. The induction of $Na_V1.3$ together with the upregulation of voltage-gated calcium channels and the reduced expression of potassium channels responsible for repolarization results in a net increase in the membrane excitability of injured afferents. Furthermore, injured sensory neurons newly acquire the ability to express BDNF and peptides such as NPY and Gal, which serve paracrine functions within the DRGs. There is evidence for an immediate chemical or ephaptic (electrical) crosstalk between afferents following nerve injury. A delayed consequence of a peripheral nerve lesion is the slow degeneration of predominantly small-diameter sensory neurons.

Whether post-translational modification of the sodium channel subunits has a major role in the development of neuropathic pain is unclear, although it contributes to peripheral sensitization in inflammatory pain.[94] Phosphorylation sites on both TTX-S and TTX-R sodium channel subunits are targeted by protein kinase A (PKA) and PKC, and activation of PKA or PKC enhances TTX-R currents in primary sensory neurons.[95,96] Phosphorylation by PKA reduces the activation threshold and increases the amplitude of the TTX-R sodium current; it also influences the availability of TTX-R sodium channels by increasing the rate of current activation and inactivation.[95,96] In nociceptive, small-diameter sensory neurons, the two kinases apparently act in concert.[95]

The membrane excitability of sensory neurons is defined not only by changes in the expression and distribution of sodium channels. Potassium currents (I_K) sustained by voltage-gated rectifier type potassium channels (K_V) are responsible for membrane repolarization. Fast inactivating potassium currents (I_A) and calcium-dependent potassium currents determine the maximum frequency and pattern of firing of sensory neurons. Voltage-dependent inwardly rectifying channels are present mainly in myelinated A-type sensory neurons. Mammalian K_V1 proteins are composed of four transmembrane α subunits coassembled in various combinations with accessory, modulating subunits. In myelinated nerve fibers, K_V1 channels are expressed predominantly as $K_V1.1$, $K_V1.2$, and $K_V\beta2.1$.

Expression of $K_V1.4$ is largely confined to small-diameter sensory neurons.[97]

The expression of K_V is reduced by more than 50 percent after lesions of sensory nerve fibers.[97,98] Whole-cell patch-clamp recordings in dissociated DRG neurons after axotomy of the rat sciatic nerve have revealed decreases in I_K and I_A, whereas slow inactivating potassium currents (I_D) are preserved in large-diameter neurons.[99,100] The reduction of I_K and I_A currents remains small or is prevented after nerve crush, an incomplete lesion with a maintained supply of neurotrophic factors from the periphery. Exogenous NGF, when applied to the proximal nerve stump, rescues I_K and I_A currents after nerve transection.[101] The downregulation of potassium channels, along with the upregulation of $Na_V1.3$, leads to a shift in the membrane excitability toward a heightened activity of sensory neurons. Changes in the expression of $Na_V1.3$ are associated with spontaneous ectopic firing and, therefore, may be responsible for spontaneous pain. The significance of changes in potassium channels is less certain. I_{KD}, a 4-aminopyridine-sensitive voltage-gated potassium current, may contribute to the development of cold allodynia.[102]

Nerve injury–induced alterations also occur in a third family of voltage-gated channels that controls the entry of calcium ions into the cell. A number of voltage-activated calcium channels are expressed in DRG neurons, including high-voltage-activated (HVA) L-, N-, P/Q-, and R-type channels and low-voltage-activated (LVA) T-type calcium channels. Nociceptors exhibit mainly L- and N-type currents during activation, whereas T-type currents are recorded predominantly from medium-diameter neurons. Gabapentin, an anticonvulsant that is effective against neuropathic pain, binds to the $\alpha_2\delta$-1 subunit of these calcium channels. Upregulated transcription results in a marked increase (up to 17-fold) of the $\alpha_2\delta$-1 subunit protein 1 week after lesion to the peripheral axon of a sensory neuron but not following rhizotomy.[103,104] The exact role of the $\alpha_2\delta$-1 calcium channel subunit in the pathophysiology of nerve injury and the development of neuropathic pain is unknown and can only be deduced from the effect of gabapentin, which is most pronounced against static and dynamic mechanical allodynia.[105,106] Gabapentin may reduce transmitter release from the central terminals of primary afferents,[107] as do N-type calcium channel blockers such as the ω-conotoxin ziconitide,[108] which has good analgesic efficacy in neuropathic pain models and in patients but a low therapeutic index as it causes severe autonomic and CNS side effects.

Ligand-Gated Ion Channels

Nerve injury generally produces a downregulation of ligand-gated ion channels in sensory neurons. This has been shown for the TRPV1 receptor[109] and the purinoreceptor $P2X_3$.[110] However, TRPV1 immunoreactivity is preserved in axons of injured nerve fibers proximal to the lesion site despite a reduction in TRPV1 mRNA and immunoreactivity of their cell somata,[69] indicating a redistribution similar to that of $Na_V1.8$ in intact neurons (see Fig. 5-2). The expression level of the purinoreceptor $P2X_3$ is reduced by more than 50 percent after nerve injury, and this can be reversed by exogenous GDNF.[110]

Phenotypic Shift

Following partial nerve injury, both injured and uninjured primary sensory neurons acquire the ability to express novel genes that fundamentally change their phenotype, one example of which is the induction of catecholamine receptors described earlier. Reversal of this phenotypic shift is associated with a reduction of neuropathic pain.[38]

BDNF is one of the few molecules induced in sensory neurons after both tissue inflammation and nerve injury. Generally, molecular changes in injured neurons are the opposite of those found in inflammatory pain models.[111] Under physiologic conditions, BDNF is produced almost exclusively in nociceptors. BDNF is downregulated in these small-diameter DRG neurons after axotomy, a consequence of the deprivation of peripheral NGF.[112] In contrast, DRG neurons of medium and large diameter newly express BDNF after nerve injury.[112,113] BDNF is also upregulated in neighboring uninjured primary afferents of all diameters, including nociceptors.[40] Direct injections of exogenous BDNF or NGF into the DRGs have demonstrated their role in the development of mechanical allodynia, and antibodies directed against BDNF, NGF, or NT-3 are effective against touch-evoked pain.[42] It is, therefore, likely that BDNF exerts a paracrine signaling effect in the DRGs, acting, after its release, on mechanosensitive neurons (see Fig. 5-2). Apart from this peripheral function, BDNF is anterogradely transported to the central terminals of sensory neurons and enhances the excitability of neurons in the dorsal horn.[114] This is comparable with the principal effect of BDNF in conditions of inflammatory pain, in which the release of BDNF in the dorsal horn modulates tactile hypersensitivity.

Neuropeptide Y (NPY) is normally not expressed in the DRGs but is induced after partial or complete peripheral nerve lesions.[115,116] NPY-binding sites are found in the central projection areas of these neurons within the dorsal horn, but they are also constitutively expressed in up to 20 percent of mostly small- to medium-sized neurons in the trigeminal ganglion and the sensory neurons. The prominent receptor subtype is Y_2.[117] Therefore, as with BDNF, NPY may be involved in a paracrine modulation of nociceptor activity in the DRGs after nerve injury. Co-localization studies have suggested an additional autocrine modulatory effect of NPY through the activation of Y_2 and Y_1 receptors expressed on large-diameter neurons newly synthesizing NPY.[118]

Galanin, a 29-amino-acid peptide, is expressed in sensory neurons during development and is dramatically upregulated in nociceptors after nerve injury. A study of partial nerve injury has demonstrated that galanin expression is induced in injured and uninjured medium- and large-sized DRG neurons.[119] The new expression of galanin in primary sensory neurons is apparently under the regulatory control of leukemia inhibitory factor (LIF), which is released by nonneuronal cells at the injury site.[120] G protein–coupled receptors for galanin (Gal_1, Gal_2) are constitutively expressed on A- and C-type primary afferents. Despite downregulation of both receptors after axotomy, primary afferents develop inward currents in response to galanin only after a sciatic nerve transection. A more important effect of galanin may be its increased release from central terminals of sensory neurons after injury.[121] Possibly, galanin exerts an inhibitory function on pre- and postsynaptic Gal_1 receptors and an excitatory effect on presynaptic Gal_2 receptors in the dorsal horn.[122] However, knockout mice for the Gal_1 receptor showed only an unchanged pain hypersensitivity after partial sciatic nerve injury.[123]

Other neuropeptides that are newly expressed in subpopulations of primary sensory neurons after nerve injury are vasoactive intestinal peptide (VIP)[115] and pituitary adenylate cyclase–activating polypeptide.[124] The functional significance of these changes, however, as with most of the hundreds of genes whose expression changes after axotomy,[21] has yet to be determined.

Structural Changes: The Altered Connectivity of Sensory Afferents

New Synapse Formation in the Dorsal Horn

Mechanical allodynia after nerve injury is predominantly mediated by myelinated A-fiber afferents. Elimination of unmyelinated nociceptors by resiniferatoxin, a potent capsaicin analogue, is effective against thermal hyperalgesia, but it has no impact on mechanical hypersensitivity.[125] By contrast, differential blockade of A fibers by compression or a local anesthetic results in a loss of tactile allodynia.[126]

Following a lesion of a peripheral nerve, the central axons of injured myelinated Aβ fibers start to grow into lamina II of the dorsal horn, violating the normal laminar topography of their projections in the dorsal horn and entering areas restricted to nociceptors.[127–129] Sprouting of injured Aβ fibers is maximal at 2 weeks after nerve injury and persists over an extended period, being reversed only after 9 months.[130] The outgrowth of central Aβ-fiber terminals is prevented by NGF and GDNF treatment, suggesting an important role for neurotrophins in the regulation of this manifestation of structural plasticity.[131] Electrophysiologic recordings have shown that stimulation of Aβ fibers in injured nerves begins to activate neurons in lamina II, with a configuration of the currents resembling the postsynaptic responses normally only elicited by high-threshold afferents in intact animals.[132,133] The formation of functional synapses between Aβ terminals and dorsal horn neurons that normally receive input only from nociceptive Aδ and C fibers contributes to mechanical allodynia mediated through low-threshold afferents.

Crosstalk Between Afferent Fibers

Crosstalk between afferents through pathologic contacts is cited as underlying the triggered pain in cranial neuralgias, most notably trigeminal neuralgia. To explain the dramatic outbursts of intense pain in trigeminal neuralgia, Rappaport and Devor have introduced the hypothesis of an ignition focus within the trigeminal ganglion, a site from which a local paroxysm spreads to neighboring neurons before it is stopped by intrinsic, suppressive ion currents.[134] A key feature of this model is propagation of the initial barrage of activity to adjacent afferents. Ephaptic (electrical) coupling of individual axons has been demonstrated acutely after nerve lesions and in neuromas.[135,136] More extensive crosstalk between afferents has been observed after trains of electrical stimuli are applied to injured nerve fibers. Following tetanic excitation, neighboring, previously silent afferents start to exhibit prolonged, rhythmic firing known as *crossed afterdischarge*. Chemical instead of ephaptic transmission has been postulated for this type of cross-activation[137] (see Fig. 5-2). Pathologic changes in trigeminal ganglia from patients with trigeminal neuralgia include focal demyelination and remyelination and, indeed, close apposition of demyelinated fibers without interfering glial processes.[138] Afterdischarge following physiologic stimulation has been demonstrated in chronically-injured nerve fibers,[139] but evidence for the spread of this abnormal activation to other afferents remains elusive and has only been recorded after repetitive electrical excitation. Currently, there is no convincing way to explain how the often weak triggering stimuli are capable of eliciting outbursts of intense, painful sensations or how different types of stimuli applied to a trigger zone can provoke paroxysms of uniform character.

Delayed Degeneration of Injured Sensory Neurons

In the adult PNS, DRG neurons survive axotomy for several weeks because their viability is no longer dependent on neurotrophic factors supplied by the innervation target. After 32 weeks in the rat, however, the number of surviving neurons is decreased by 37 percent[140] (see Fig. 5-2). A similar loss of DRG neurons also is found after nerve crush.[141] What causes this delayed loss of sensory neurons is not clear. Possible factors promoting cell death may be the effect of cytokines released from macrophages and other immune cells invading the DRGs after nerve injury. TNF-α, for example, induces apoptotic neurodegeneration either directly or via silencing of survival signals.[142] There also may be a loss of intrinsic survival factors.[143] The injury-induced loss of DRG cells affects predominantly unmyelinated neurons of small diameter, the nociceptors.[140] Slow but continuous degeneration of nociceptors ultimately may result in reduced spontanous pain and hypersensitivity and contribute to the abatement of neuropathic pain that is, for example, observed in the majority of patients with postherpetic neuralgia within 2 years.

Central Processing of Sensory Input in the Spinal Cord

Sensory information from primary afferents reaches the dorsal horn of the spinal cord before being conveyed further to the brain stem or centers in the forebrain. Processing of afferent input in the dorsal horn is important for the localization of a stimulus and modality discrimination. As a key relay station, the dorsal horn regulates the flow of sensory information to the brain while integrating modulatory input from local interneurons and descending pathways. This processing of sensory information adapts to different physiologic and pathologic conditions through activity-dependent plasticity. The constant barrage of intense afferent information from injured nerves, however, can lead to maladaptive plasticity, resulting in a paradoxical enhancement of sensory input and destroying the normal relationship between primary afferent input and projection neuron output from the dorsal horn. This involves mechanisms of temporal summation, reinforced transmitter signaling, and disinhibition, that are mediated by both neuronal and non-neuronal mechanisms.

The Integration of Sensory Information Over Time: Windup, Summation, and Long-Term Potentiation

Windup is the progressive increase in the firing intensity of neurons in the dorsal horn in response to synchronous stimulation (at a frequency between 0.3 and 3 Hz) of C-fiber afferents. The neuronal hyperactivity can last for a few seconds after the stimulation has ended. Windup has been interpreted as the electrophysiologic correlate of an abnormal temporal summation of stimuli observed in some patients with neuropathic pain, particularly trigeminal neuralgia. However, temporal summation in these patients usually is elicited by initially nonpainful stimuli that would activate low-threshold Aβ fibers rather than Aδ or C fibers. Repetitive application of innocuous stimuli never produces pain under physiologic conditions. Whether C-fiber-mediated windup contributes to neuropathic pain syndromes is uncertain, and terms such as *windup-like pain* to describe temporal summation are misleading.

N-Methyl-D-aspartate (NMDA) receptor antagonists reduce action potential windup[144] and abnormal temporal summation in patients with chronic neuropathic pain. Ketamine, for example, inhibits temporal summation of nonpainful mechanical stimulation in patients with postherpetic neuralgia. In contrast, morphine is ineffective against or aggravates abnormal temporal summation.[145] NMDA receptor–independent mechanisms are thought to contribute to the phenomenon of temporal summation as well.[146]

Tetanic stimulation of Aδ- or C-fiber afferents at high frequencies, between 20 and 200 Hz, induces a long-term potentiation (LTP) of synaptic strength in neurons of dorsal horn laminae I and V. This effect may last for several hours depending on the experimental conditions. LTP requires a sufficient increase in intracellular calcium ions, which can be provoked by electrical stimulation, NMDA, or SP.[147] LTP develops only in neurons in lamina I of the dorsal horn that carry the SP receptor NK-1.[148] Selective ablation of NK-1-positive dorsal horn neurons reduces the excitability of wide dynamic range neurons of lamina V and diminishes windup.[149] While the effects of LTP alter behavioral responses in nociceptive tests and suggest the potential existence of "learning and memory in pain pathways,"[150] an involvement of LTP in neuropathic pain needs to be demonstrated.

Afferent-Dependent Central Sensitization

Enhanced nociceptive input produces a switch in spinal transmission neurons to a state of increased excitability. This process of central sensitization manifests as a reduction in the activation threshold of dorsal horn neurons and heightened responsiveness to synaptic input.[151] An increased recruitment of afferent connections furthermore leads to an expansion of the receptive field of wide-dynamic-range neurons. Central sensitization normally is evoked only by nociceptors,[152] but changes in the chemical phenotype of primary afferents after nerve injury enable low-threshold primary afferents to contribute to the sensitization of dorsal horn neurons.[153]

The NMDA glutamate receptor plays a critical role in regulating synaptic efficacy.[111] Glutamate is the major neurotransmitter released from primary afferents in the dorsal horn. Under resting conditions, glutamate activates ionotropic α-amino-3-hydroxy-5-methyl-4-isoxazole propionate (AMPA) and kainate receptors and metabotropic glutamate receptors (mGluRs). The NMDA receptor channel is blocked by a magnesium ion, which prevents the flow of sodium or calcium when glutamate binds to the receptor. Repeated postsynaptic activation, through AMPA receptors, mGluRs, or the NK-1 receptor for SP, leads to a summation of synaptic potentials and progressive membrane depolarization. This relieves the NMDA receptor of its magnesium block and, therefore, allows the influx of cations into the cell. Membrane depolarization increases further, and voltage-dependent calcium channels are opened. Activation of G protein–coupled receptors triggers intracellular signal cascades that provoke an additional release of calcium from intracellular storage sites. Calcium-induced activation of nonselective cation channels and the generation of plateau potentials finally stabilize the depolarized state of dorsal horn neurons. These changes affecting the membrane potential can be regarded as the initial phase of central sensitization.[94]

The second phase consists of the post-translational modification of AMPA and NMDA glutamate receptors (Fig. 5-3). The intracellular rise of the calcium concentration is a key regulator of PKA, calcium-calmodulin-dependent protein kinase II, and PKC. PKC is upstream of a phosphorylation cascade involving cell-adhesion kinase β (CAKβ) and the tyrosine kinase Src. Activation of PKC by calcium results in the phosphorylation of tyrosine and serine residues on the NMDA receptor and provides for an increased activation of the ion channel. Phosphorylation of the NMDA receptor by Src and a related kinase, Fyn, is further induced by activation of mGluR-5[154] or, independent of PKC, by activation of mGluR-1.[155] These first two phases in the process of central sensitization happen quickly; potentiated depolarization of transmission neurons occurs within seconds, the trafficking from intracellular stores and membrane insertion of AMPA receptors and the post-translational modification of AMPA and NMDA receptors require only minutes.

A state of central sensitization is actively maintained by ectopic or evoked activity from injured and uninjured afferents.[35,156,157] Sustained activity after nerve injury mainly originates in muscle afferents.[68] Input from these muscle afferents has an even greater central sensitizing effect than the activity of cutaneous C fibers.[158] Nevertheless, repetitive stimulation of afferents with C-fiber strength at a frequency comparable to the rate of spontaneous discharges observed after nerve injury produces mechanical hypersensitivity.[35]

NMDA antagonists and a conditional knockout of the NMDA receptor subunit NR_1 reduce behavioral changes related to central sensitization, most importantly hypersensitivity for mechanical stimuli but also cold allodynia.[159–161] Clinical studies have demonstrated the effect of NMDA antagonists on neuropathic pain, as well as their use-limiting side effects.[162] A contribution of SP to the phenomenon of central sensitization following nerve injury is suggested by the fact that ablation of spinal NK-1 receptors before spinal nerve ligation reduces mechanical allodynia,[163] but clinical studies using NK-1 antagonists have failed to show any efficacy. The significance of PKC for the induction of nerve injury–induced behavioral alterations is demonstrated by the substantially reduced mechanical allodynia found after partial sciatic nerve ligation in a knockout mouse lacking PKC.[164]

Longer-lasting mechanisms of central sensitization involve the regulation of gene expression at the transcriptional level (see Fig. 5-3). Reports of transcriptional changes related to central sensitization are based mainly, however, on evidence from models studying inflammatory pain. Only a few groups have studied transcriptional regulation in dorsal horn neurons after nerve injury. Partial sciatic nerve ligation leads to increased phosphorylation of the transcription factor cyclic AMP response element–binding protein (CREB).[165] Reduced mechanical allodynia following CCI has been found in mice lacking the downstream regulatory element antagonistic modulator (DREAM), a regulatory factor suppressing the transcription of dynorphin.[166] Ongoing research using gene expression profiling for the investigation of molecular changes in the dorsal horn can be expected to elucidate regulatory changes and their relationship to central sensitization in animal models of neuropathic pain.

Inhibition and Facilitation of Sensory Input Integrated in the Dorsal Horn

Afferent input in the dorsal horn is under tonic and phasic inhibitory control. Local inhibitory interneurons in the dorsal horn produce γ-aminobutyric acid (GABA) and glycine and are directly activated by primary sensory neurons. Dorsal horn interneurons presynaptically modulate the release of excitatory transmitters from the central terminals of primary afferents through $GABA_A$ and G protein–coupled $GABA_B$ receptors at axoaxonic contacts. Rapid postsynaptic inhibition is mediated through ligand-gated $GABA_A$ and glycine receptors. Administration of GABA or glycine antagonists in naïve animals produces painlike behavior, demonstrating the importance of inhibitory control of afferent information.[167]

Partial injuries of the sciatic nerve cause a specific reduction in the GABAergic component of afferent-evoked inhibitory postsynaptic currents, whereas the glycinergic component is mainly preserved.[168] As levels of GABA in the dorsal horn decrease following nerve injury,[169] this may be the result of decreased GABA synthesis in the dorsal horn.[168,170] On the other hand, the occurrence of apoptotic profiles in the dorsal horn ipsilateral to peripheral nerve lesions[168,171] has led to the hypothesis that interneurons die as a consequence of increased afferent input and excessive release of glutamate from injured nerves[94] (see Fig. 5-3). Such a loss of interneurons and the resulting disinhibition would be irreversible, providing a possible explanation as to why neuropathic pain in humans persists after the initial injury and becomes chronic. In addition, a shift in the transmembrane anion gradient of some neurons in lamina I of the dorsal horn causes GABA to evoke excitatory instead of inhibitory postsynaptic currents after nerve injury.[172]

Apart from controlling afferent input through local inhibitory and excitatory connections, dorsal horn neurons also integrate descending input from higher centers. The midbrain periaqueductal gray, the rostral ventromedial medulla (RVM), and the dorsolateral pontine tegmentum give rise to catecholaminergic and serotonergic spinal projections that synapse onto spinal interneurons

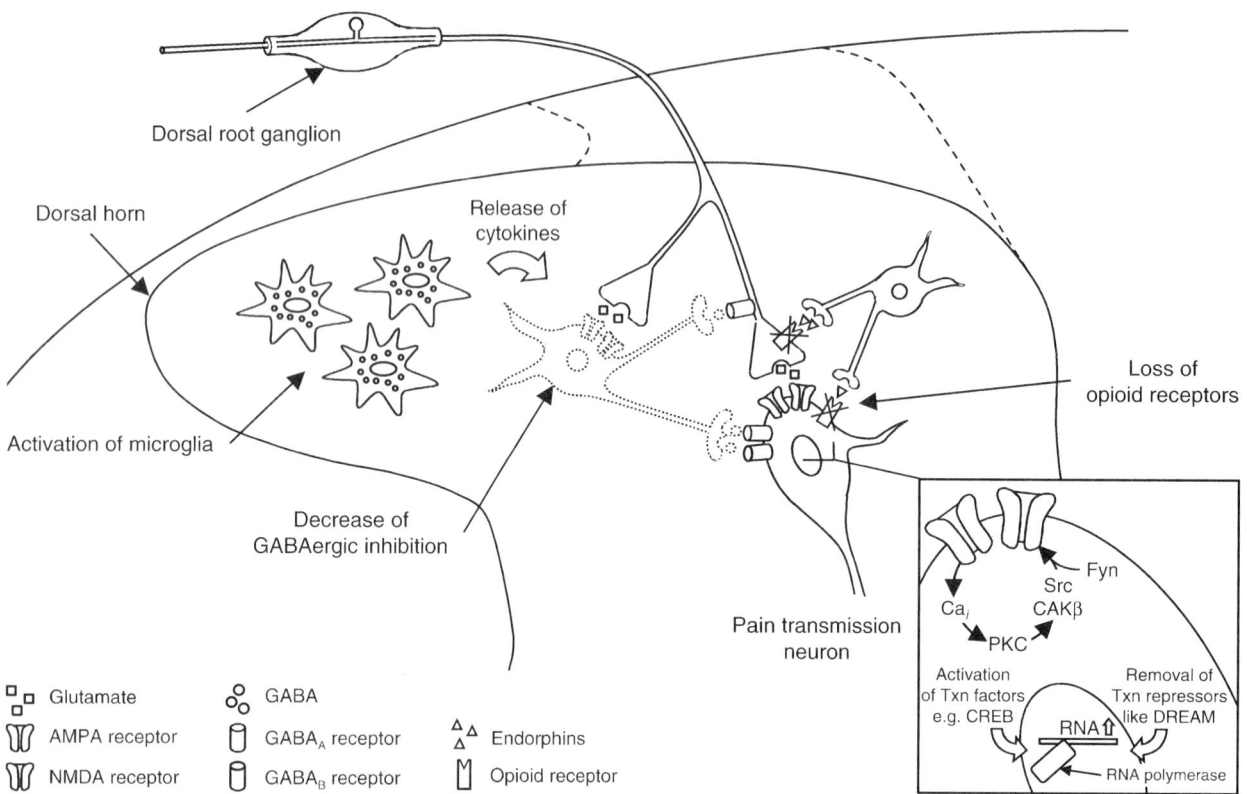

Figure 5-3. Central sensitization and loss of inhibition after nerve injury result in an altered processing of sensory information. Enhanced input from injured and uninjured afferents causes the activation of NMDA-type glutamate receptors. Activation of intracellular signaling cascades leads to phosphorylation of the NMDA receptor by the kinases Src, CAKβ, and Fyn. This post-translational modification facilitates excitatory synaptic responses. Longer-term changes in dorsal horn neurons involve altered regulation of gene expression either by activation of transcription factors such as CREB or removal of transcription repressors such as DREAM. Inhibition in the dorsal horn is diminished after nerve injury because of the reduced expression of opioid receptors, a decrease in the synthesis of GABA, and possibly a loss of GABAergic interneurons. Precisely how the activation of glial cells (astrocytes, microglia) and the release of cytokines in the dorsal horn are involved in central sensitization or interfere with spinal inhibition still needs to be elucidated. Not shown in the diagram is the contribution of descending facilitation and reduced inhibition through pathways originating in the brain stem.

and transmission neurons involved in the modulation of sensory information and pain.[173]

Activation of Spinal Glia

Injury to a peripheral nerve provokes a marked immune reaction in the ipsilateral dorsal horn, as well as around motoneurons in the ventral horn of the spinal cord. Recently, it has become clear that activated glial cells and invading hematogenic macrophages serve more functions than mere removal of degenerating terminals or redundant synaptic sites of injured nerve fibers, as it was previously assumed,[174] and, in fact, play a role in pathologic sensory processing (see Fig. 5-3).

Activation of microglia and astrocyte activation in the dorsal horn is observed after spinal nerve injury (peripheral axon) and rhizotomy (central axon) and involves upregulation of IL-1β, IL-6, IL-10, and TNF-α.[175] The immune response seems to follow a specific temporal pattern with relatively slow onset and a peak several days after the nerve lesion. An association with neuropathic pain–like behavior, predominantly mechanical allodynia, is variably reported.[176,177] Immunomodulation with fluorocitrate,[178] propentofylline,[179] or methotrexate[177] reduces pain-related behavioral responses such as mechanical allodynia following nerve injury in animals. More specifically, the development of neuropathic pain appears to be related to the spinal effect of the cytokines IL-1β, IL-6, and TNF-α[178,179] and the activation of mitogen-activated protein kinases in microglia[178,180] or astrocytes.[181] Immunosuppression has been used previously for the treatment of neuropathic pain in humans. A meta-analysis of controlled clinical trials for the treatment of peripheral neuropathic pain[182] listed corticosteroids as the only treatment consistently effective against CRPS, although a differentiation between CRPS with or without nerve injury was not made. Systematic studies on the effect of glucocorticoids or other immunosuppressive agents on neuropathic pain are hampered by the serious side effects associated with the long-term use of these drugs. Drugs interfering with specific immune-related mechanisms certainly would be desirable.

Brain Stem and Forebrain

Brain Stem Nuclei and Their Spinal Projections

The investigation of molecular changes in the brain as a consequence of peripheral nerve injury is just beginning. One of the most prominent brain stem nuclei related to nociception is the parabrachial nucleus in the pons, which receives projections through the spinoparabrachial pathway from the superficial dorsal horn and connects to the hippocampus and amygdala, brain regions involved in the affective reaction to pain. The midbrain periaqueductal gray integrates input from several cortical areas, the hypothalamus, and the amygdala and has a pivotal role in the control of descending pathways, mainly through projections to the nucleus raphe magnus in the RVM.[173]

Pharmacologic blockade or lesions of the RVM and its spinal projections have demonstrated the importance of descending facilitation for the persistence of neuropathic pain.[183] Descending facilitation originating in the RVM seems to be mediated by the effects of dynorphin in the spinal cord.[184] Knockout mice for prodynorphin exhibit early recovery of mechanical allodynia and thermal hypersensitivity after SNL at a time point where dynorphin in the spinal cord begins to rise in the wild-type mouse.[185] The role of dynorphin in pain modulation, however, is controversial. Most likely its effects are concentration-dependent; under pathologic conditions or when given exogenously, dynorphin facilitates sensory input by binding to the NMDA-type glutamate receptor, whereas physiologic levels of dynorphin exert an inhibitory effect through activation of the G_i-coupled kappa opioid receptor.[186] Endogenously elevated levels of dynorphin acting on the kappa receptor contribute to the phenotype of reduced neuropathic pain in knockout mice lacking the transcription repressor DREAM.[166]

Endomorphins (EMs) 1 and 2 have been identified recently as endogenous opioids acting on mu-type opioid receptors.[187] EM-1-immunoreactive fibers are found in the parabrachial nucleus, the periaqueductal gray, the nucleus of the solitary tract and other areas of the brain involved in the processing of pain. Endomorphins are also strongly expressed in the dorsal hypothalamic area.[188] The greatest density of EM-2-immunoreactive fiber terminals was detected in the spinal trigeminal tract and the superficial dorsal horn. These fibers may derive from primary afferents or the posterior hypothalamus. Endomorphins reduce thermal hyperalgesia and mechanical allodynia after nerve crush.[189]

▶ CONCLUSION

Multiple mechanisms operate in neuropathic pain, leading to a situation where neuronal activity becomes uncoupled from the physiologic modality- and intensity-dependent processing of stimuli and switches to a state in which spontaneous firing of neurons and the misinterpretation of stimuli produce pain even in the absence of actual or imminent tissue damage. The plasticity of the nervous system, which under normal conditions contributes to its adaptive capacity, ultimately leads to a fixed state of distorted information processing.

The widening gap between rapidly advancing laboratory research and unsatisfactory clinical progress in tackling neuropathic pain represents a challenge that will only be met if two fundamental issues are resolved. First, a simple, standardized method needs to be established to allow the identification of pain mechanisms that are operational in patients. The diversity of neuropathic pain corresponds to a multitude of possible treatment targets for pharmacologic or other interventions. However, the relationship between pain mechanisms and the clinical manifestation of neuropathic pain is complex. Many factors may produce similar symptoms or signs, and it will require a sophisticated analysis of pain-related symptoms and signs to help identify underlying mechanisms. The current approach to a patient with neuropathic pain either ignores or merely describes the manifestations of pain and fails to turn them into viable strategies for targeted treatment.

The second challenge is that once mechanisms have been identified, they need to be targeted with effective treatments. At present, pain therapy is mainly symptomatic and aims to provide global pain relief. In the face of the different mechanisms underlying neuropathic pain, it comes as no surprise that this approach is rarely successful. No treatment option will be equally effective against all the manifestations of neuropathic pain. A standardized, mechanism-based assessment of pain should be used in clinical trials to define distinct profiles of drug effects against specific features of neuropathic pain. Treatments then can be selected for patients whose pain syndromes include these pain features rather than being used globally against pain as if pain were a homogeneous state. In the absence of such standardized profiles of analgesic drugs, currently available treatment options need to be rationally assigned, wherever possible, to the mechanisms of neuropathic pain as defined by experimental and clinical evidence (Table 5-2).

TABLE 5-2. PROPOSED MECHANISM-BASED TREATMENT OF NEUROPATHIC PAIN

Mechanisms	Targets	Possible Pharmacologic Interventions
Peripheral Nervous System		
Primary sensory neuron hyperexcitability and ectopic activity	VGSC	Nonselective sodium channel blockers (local anesthetics, carbamazepine, lamotrigine, mexiletine)
	TTX-r and TTX-s VGSC	Selective blockers of TTX-r VGSC, depending on the contribution of injured versus uninjured nerve fibers
	Potassium channels	Potassium channel activators
	$P2X_3$	$P2X_3$ antagonists
Reduced threshold of nociceptor activation	TRPV1	Capsaicin
		TRPV1 antagonists
	NGF	Antibodies against NGF or the NGF receptor, TrkA
Phenotypic changes	BDNF	Antibodies against BDNF or the BDNF receptor, TrkB
	GFRα-3/RET	Artemin
	TNF receptors	Antibodies against TNFα or soluble TNF receptor-fusion proteins
Synaptic reorganization	TrkA	NGF
	GFRα-1(2)/RET	GDNF
Degeneration of peripheral sensory neurons	Neurotrophin-receptors	Neurotrophic factors
	TNFα	Antibodies against TNFα or soluble TNF receptor-fusion proteins
	Caspases	Upstream inhibitors of caspase activation, for example Hsp27
		Caspase inhibitors
Central Nervous System		
Spontaneous activity	Sodium channels	Nonselective sodium channel blockers (local anesthetics, carbamazepine, lamotrigine, mexiletine)
	VGCC, $\alpha_2\delta_1$ subunit	Gabapentin
	N-type Ca^{2+} channels	ω-Conotoxin
Abnormal temporal integration of stimuli	Wind-up	NMDA-receptor antagonists (ketamine, dextromethorphan, amantadine)
	Temporal summation	NMDA-receptor antagonists
	Long-term potentiation	NK1 antagonists
		NMDA-receptor antagonists
Abnormal excitability of sensory transmission neurons	Kainate receptor	Kainate receptor antagonists
	mGlu-R	mGlu-R antagonists
	NMDA-receptor	NMDA-receptor antagonists (ketamine, dextromethorphan, amantadine)
	PKCγ	PKCγ inhibitors
	NK-1	NK-1 antagonists
	nNOS	nNOS inhibitors
	MAPK/ERK	MAPK/ERK inhibitors
	VGCC, $\alpha_2\delta_1$ subunit	Gabapentin
	N-type Ca^{2+} channels	ω-Conotoxin
Decreased inhibition	GABA	$GABA_A$ and $GABA_B$ agonists (baclofen)
	MOR	μ-Opioid agonists (morphine, oxycodone)
	Cannabinoid-1 receptor	Cannabinoids
	Caspases	Upstream inhibitors of caspase activation, for example Hsp27
		Caspase inhibitors
	α_2-Adrenoreceptor	α_2-Adrenoreceptor agonists (clonidine)
		Tricyclic antidepressants (amitriptyline, imipramine, clomipramine)
	Adenosine receptor	Adenosine receptor agonists
Activation of spinal glia	Activated glia	Fluorocitrate
		Metothrexate
		Corticosteroids
	TNFα	Antibodies against TNFα or soluble TNF receptor-fusion proteins
Sympathetic Nervous System		
Altered level of activity	VGSC	Sympathetic blocks using local anesthetics
	α_2-Adrenoreceptor	α_2-Adrenoreceptor agonists (clonidine)
Sympathetic-sensory coupling	NGF	Antibodies against NGF or the NGF receptor, TrkA
	VGSC	Sympathetic blocks
	α_1-Adrenoreceptor	α_1-Adrenoreceptor antagonists (phentolamine, guanethidine)

ACKNOWLEDGMENT

We wish to thank Richard J. Mannion for carefully reading the manuscript and for his many helpful suggestions. We also thank Jordan M. Lawless for his assistance.

REFERENCES

1. Woolf CJ, Mannion RJ: Neuropathic pain: Aetiology, symptoms, mechanisms, and management. *Lancet* 1999;353:1959.
2. Melzack R: The McGill Pain Questionnaire: Major properties and scoring methods. *Pain* 1975;1:277.
3. Galer BS, Jensen MP: Development and preliminary validation of a pain measure specific to neuropathic pain: The Neuropathic Pain Scale. *Neurology* 1997;48:332.
4. Bennett M: The LANSS Pain Scale: The Leeds assessment of neuropathic symptoms and signs. *Pain* 2001;92:147.
5. Stanton-Hicks M, Janig W, Hassenbusch S, et al: Reflex sympathetic dystrophy: Changing concepts and taxonomy. *Pain* 1995;63:127.
6. Fields HL, Baron R, Rowbotham MC: Peripheral neuropathic pain: An approach to management, in PD Wall, R Melzack (eds), *Textbook of Pain*, 4th ed. Edinburgh: Churchill Livingstone, 1999:1523–1534.
7. Backonja M-M: Painful neuropathies, in JD Loeser (ed), *Bonica's Management of Pain*, 3d ed. Philadelphia: Lippincott, Williams & Wilkins, 2001:371–387.
8. Rowbotham MC, Fields HL: The relationship of pain, allodynia, and thermal sensation in post-herpetic neuralgia. *Brain* 1996;119:347.
9. Fields HL, Rowbotham M, Baron R: Postherpetic neuralgia: Irritable nociceptors and deafferentation. *Neurobiol Dis* 1998;5:209.
10. Woolf CJ, Decosterd I: Implications of recent advances in the understanding of pain pathophysiology for the assessment of pain in patients. *Pain* 1999;6:S141.
11. Jensen TS, Baron R: Translation of symptoms and signs into mechanisms in neuropathic pain. *Pain* 2003;92:147.
12. Empl M, Renaud S, Erne B, et al: TNF-α expression in painful and nonpainful neuropathies. *Neurology* 2001;56:1371.
13. Lindenlaub T, Sommer C: Cytokines in sural nerve biopsies from inflammatory and noninflammatory neuropathies. *Acta Neuropathol* 2003;105:593.
14. Goldstein DS, Tack C, Li ST: Sympathetic innervation and function in reflex sympathetic dystrophy. *Ann Neurol* 2000;48:49.
15. Wasner G, Schattschneider J, Heckmann K, et al: Vascular abnormalities in reflex sympathetic dystrophy (CRPS I): Mechanisms and diagnostic value. *Brain* 2001;124:587.
16. Iadarola MJ, Max MB, Berman KF, et al: Unilateral decrease in thalamic activity observed with positron emission tomography in patients with chronic neuropathic pain. *Pain* 1995;63:55.
17. Baron R, Baron Y, Disbrow E, et al: Brain processing of capsaicin-induced secondary hyperalgesia: A functional MRI study. *Neurology* 1999;53:548.
18. Witting N, Kupers RC, Svensson P, et al: Experimental brush-evoked allodynia activates posterior parietal cortex. *Neurology* 2001;57:1817.
19. Lotze M, Flor H, Grodd W, et al: Phantom movements and pain: An fMRI study in upper limb amputees. *Brain* 2001;124:2268.
20. Mogil JS, Wilson SG, Chesler EJ, et al: The melanocortin-1 receptor gene mediates female-specific mechanisms of analgesia in mice and humans. *Proc Natl Acad Sci USA* 2003;100:4867.
21. Costigan M, Befort K, Karchewski L, et al: Replicate high-density rat genome oligonucleotide microarrays reveal hundreds of regulated genes in the dorsal root ganglion after peripheral nerve injury. *BMC Neurosci* 2002;3:16.
22. Lariviere WR, Wilson SG, Laughlin TM, et al: Heritability of nociception: III. Genetic relationships among commonly used assays of nociception and hypersensitivity. *Pain* 2002;97:75.
23. Wilson SG, Smith SB, Chesler EJ, et al: The heritability of antinociception: Common pharmacogenetic mediation of five neurochemically distinct analgesics. *J Pharmacol Exp Ther* 2003;304:547.
24. Wall PD, Devor M, Inbal R, et al: Autotomy following peripheral nerve lesions: Experimental anaesthesia dolorosa. *Pain* 1979;2:103.
25. Decosterd I, Woolf CJ: Spared nerve injury: An animal model of persistent peripheral neuropathic pain. *Pain* 2000;87:149.
26. Bennett GJ, Xie YK: A peripheral mononeuropathy in rat that produces disorders of pain sensation like those seen in man. *Pain* 1988;33:87.
27. Mosconi T, Kruger L: Fixed-diameter polyethylene cuffs applied to the rat sciatic nerve induce a painful neuropathy: Ultrastructural morphometric analysis of axonal alterations. *Pain* 1996;64:37.
28. Seltzer Z, Dubner R, Shir Y: A novel behavioral model of neuropathic pain disorders produced in rats by partial sciatic nerve injury. *Pain* 1990;43:205.
29. Kim SH, Chung JM: An experimental model for peripheral neuropathy produced by segmental spinal nerve ligation in the rat. *Pain* 1992;50:355.
30. Hu SJ, Xing JL: An experimental model for chronic compression of dorsal root ganglion produced by intervertebral foramen stenosis in the rat. *Pain* 1998;77:15.
31. Obata K, Tsujino H, Yamanaka H, et al: Expression of neurotrophic factors in the dorsal root ganglion in a rat model of lumbar disc herniation. *Pain* 2002;99:121.
32. Fox A, Eastwood C, Clive Gentry, et al: Critical evaluation of the streptozotocin model of painful diabetic neuropathy in the rat. *Pain* 1999;81:307.
33. Kupers R, Yu W, Persson J, et al: Photochemically-induced ischemia of the rat sciatic nerve produces a dose-dependent and highly reproducible mechanical, heat and cold allodynia, and signs of spontaneous pain. *Pain* 1998;76:45.
34. Shimoyama M, Tanaka K, Hasue F: A mouse model of neuropathic cancer pain. *Pain* 2002;99:167.
35. Wu G, Ringkamp M, Hartke TV, et al: Early onset of spontaneous activity in uninjured C-fiber nociceptors after injury to neighboring nerve fibers. *J Neurosci* 2001;21:RC140.
36. Ramer MS, French GD, Bisby MA: Wallerian degeneration is required for both neuropathic pain and sympathetic sprouting into the DRG. *Pain* 1997;72:71.

37. Bennett DL, Michael GJ, Ramachandran N, et al: A distinct subgroup of small DRG cells express GDNF receptor components and GDNF is protective for these neurons after nerve injury. *J Neurosci* 1998;18:3509.
38. Gardell LR, Wang R, Ehrenfels C, et al: Multiple actions of systemic artemin in experimental neuropathy. *Nature Med* 2003;9:1383–1389.
39. Li L, Xian CJ, Zhong JH, et al: Lumbar 5 ventral root transection–induced upregulation of nerve growth factor in sensory neurons and their target tissues: A mechanism in neuropathic pain. *Mol Cell Neurosci* 2003;23:232.
40. Ha SO, Kim JK, Hong HS, et al: Expression of brain-derived neurotrophic factor in rat dorsal root ganglia, spinal cord and gracile nuclei in experimental models of neuropathic pain. *Neuroscience* 2001;107:301.
41. Malcangio M, Ramer MS, Boucher TJ, et al: Intrathecally injected neurotrophins and the release of substance P from the rat isolated spinal cord. *Eur J Neurosci* 2000;12:139.
42. Zhou XF, Deng YS, Xian CJ, et al: Neurotophins from dorsal root ganglia trigger allodynia after spinal nerve injury in rats. *Eur J Neurosci* 2000;12:100.
43. McLachlan EM, Janig W, Devor M, et al: Peripheral nerve injury triggers noradrenergic sprouting within dorsal root ganglia. *Nature* 1993;363:543.
44. Michael GJ, Averill S, Shortland PJ, et al: Axotomy results in major changes in BDNF expression by dorsal root ganglion cells: BDNF expression in large trkB and trkC cells, in pericellular baskets, and in projections to deep dorsal horn and dorsal column nuclei. *Eur J Neurosci* 1999;11:3539.
45. Ramer MS, Bisby MA: Adrenergic innervation of rat sensory ganglia following proximal or distal painful sciatic neuropathy: Distinct mechanisms revealed by anti-NGF treatment. *Eur J Neurosci* 1999;11:837.
46. Deng YS, Zhong JH, Zhou XF: Effects of endogenous neurotrophins on sympathetic sprouting in the dorsal root ganglia and allodynia following spinal nerve injury. *Exp Neurol* 2000;164:344.
47. Shinder V, Govrin-Lippmann R, Cohen S, et al: Structural basis of sympathetic-sensory coupling in rat and human dorsal root ganglia following peripheral nerve injury. *J Neurocytol* 1999;28:743.
48. Gold MS, Dastmalchi S, Levine JD: α_2-Adrenergic receptor subtypes in rat dorsal root and superior cervical ganglion neurons. *Pain* 1997;69:179.
49. Birder LA, Perl ER: Expression of α_2-adrenergic receptors in rat primary afferent neurones after peripheral nerve injury or inflammation. *J Physiol* 1999;515:533.
50. Michaelis M, Devor M, Janig W: Sympathetic modulation of activity in rat dorsal root ganglion neurons changes over time following peripheral nerve injury. *J Neurophysiol* 1996;76:753.
51. Petersen M, Zhang J, Zhang JM, et al: Abnormal spontaneous activity and responses to norepinephrine in dissociated dorsal root ganglion cells after chronic nerve constriction. *Pain* 1996;67:391.
52. Moon DE, Lee DH, Han C, et al: Adrenergic sensitivity of the sensory receptors modulating mechanical allodynia in a rat neuropathic pain model. *Pain* 1999;80:589–595.
53. Malmberg AB, Hedley LR, Jasper JR, et al: Contribution of alpha-2 receptor subtypes to nerve injury–induced pain and its regulation by dexmedetomidine. *Br J Pharmacol* 2001;132:1827.
54. Schafers M, Svensson CI, Sommer C, et al: Tumor necrosis factor-alpha induces mechanical allodynia after spinal nerve ligation by activation of p38 MAPK in primary sensory neurons. *J Neurosci* 2003;23:2517.
55. Okamoto K, Martin DP, Schmelzer JD, et al: Pro- and anti-inflammatory cytokine gene expression in rat sciatic nerve chronic constriction injury model of neuropathic pain. *Exp Neurol* 2001;169:386.
56. Schafers M, Geis C, Brors D, et al: Anterograde transport of tumor necrosis factor-alpha in the intact and injured rat sciatic nerve. *J Neurosci* 2002;22:536.
57. Hu P, McLachlan EM: Macrophage and lymphocyte invasion of dorsal root ganglia after peripheral nerve lesions in the rat. *Neuroscience* 2002;112:23.
58. Liu B, Li H, Brull SJ, et al: Increased sensitivity of sensory neurons to tumor necrosis factor alpha in rats with chronic compression of the lumbar ganglia. *J Neurophysiol* 2002;88:1393.
59. Schafers M, Lee DH, Brors D, et al: Increased sensitivity of injured and adjacent uninjured rat primary sensory neurons to exogenous tumor necrosis factor-alpha after spinal nerve ligation. *J Neurosci* 2003;23:3028.
60. Gold MS, Weinreich D, Kim CS, et al: Redistribution of $Na_V1.8$ in uninjured axons enables neuropathic pain. *J Neurosci* 2003;23:158.
61. Amaya F, Decosterd I, Samad TA: Diversity of expression of the sensory neuron-specific TTX-resistant voltage-gated sodium ion channels SNS and SNS_2. *Mol Cell Neurosci* 2000;15:331.
62. Novakovic SD, Eglen RM, Hunter JC: Regulation of sodium channel distribution in the nervous system. *Trends Neurosci* 2001;24:473.
63. Waxman SG, Dib-Hajj S, Cummins TR, et al: Sodium channels and pain. *Proc Natl Acad Sci USA* 1999;96:7635.
64. Decosterd I, Ji RR, Abdi S, et al: The pattern of expression of the voltage-gated sodium channels Na(v)1.8 and Na(v)1.9 does not change in uninjured primary sensory neurons in experimental neuropathic pain models. *Pain* 2002;96:269.
65. Lai J, Gold MS, Kim CS, et al: Inhibition of neuropathic pain by decreased expression of the tetrodotoxin-resistant sodium channel, $Na_V1.8$. *Pain* 2002;95:143.
66. Kerr BJ, Souslova V, McMahon SB, et al: A role for the TTX-resistant sodium channel $Na_V1.8$ in NGF-induced hyperalgesia, but not neuropathic pain. *Neuroreport* 2001;12:3077.
67. Wu G, Ringkamp M, Murinson BB, et al: Degeneration of myelinated efferent fibers induces spontaneous activity in uninjured C-fiber afferents. *J Neurosci* 2002;22:7746.
68. Michaelis M, Liu X, Janig W: Axotomized and intact muscle afferents but no skin afferents develop ongoing discharges of dorsal root ganglion origin after peripheral nerve lesion. *J Neurosci* 2000;20:2742.
69. Hudson LJ, Bevan S, Wotherspoon G, et al: VR1 protein expression increases in undamaged DRG neurons after partial nerve injury. *Eur J Neurosci* 2001;13:2105.
70. Caterina MJ, Leffler A, Malmberg AB, et al: Impaired nociception and pain sensation in mice lacking the capsaicin receptor. *Science* 2000;288:306.

71. Novakovic SD, Kassotakis LC, Oglesby IB, et al: Immunocytochemical localization of P2X$_3$ purinoceptors in sensory neurons in naive rats and following neuropathic injury. *Pain* 1999;80:273.
72. Tsuzuki K, Kondo E, Fukuoka T, et al: Differential regulation of P2X$_3$ mRNA expression by peripheral nerve injury in intact and injured neurons in the rat sensory ganglia. *Pain* 2001;91:351.
73. Barclay J, Patel S, Dorn G, et al: Functional downregulation of P2X$_3$ receptor subunit in rat sensory neurons reveals a significant role in chronic neuropathic and inflammatory pain. *J Neurosci* 2002;22:8139.
74. Jarvis MF, Burgard EC, McGaraughty S, et al: A-317491, a novel potent and selective nonnucleotide antagonist of P2X$_3$ and P2X$_{2/3}$ receptors, reduces chronic inflammatory and neuropathic pain in the rat. *Proc Natl Acad Sci USA* 2002;26:17179.
75. Loeser JD: Pain after amputation: Phantom limb and stump pain, in JD Loeser (ed), *Bonica's Management of Pain*, 3d ed. Philadelphia: Lippincott, Williams & Wilkins, 2001: 412–423.
76. Govrin-Lippmann R, Devor M: Ongoing activity in severed nerves: Source and variation with time. *Brain Res* 1978; 159:406.
77. Scadding JW: Development of ongoing activity, mechanosensitivity, and adrenaline sensitivity in severed peripheral nerve axons. *Exp Neurol* 1981;73:345.
78. Blumberg H, Janig W: Discharge pattern of afferent fibers from a neuroma. *Pain* 1984;20:335.
79. Rivera L, Gallar J, Pozo MA, et al: Responses of nerve fibres of the rat saphenous nerve neuroma to mechanical and chemical stimulation: An in vitro study. *J Physiol* 2000; 527:305.
80. Chen Y, Michaelis M, Janig W, et al: Adrenoreceptor subtype mediating sympathetic-sensory coupling in injured sensory neurons. *J Neurophysiol* 1996;76:3721.
81. Chabal C, Jacobson L, Russell LC, et al: Pain response to perineuromal injection of normal saline, epinephrine, and lidocaine in humans. *Pain* 1992;49:9.
82. Bongenhielm U, Yates JM, Fried K, et al: Sympathectomy does not affect the early ectopic discharge from myelinated fibres in ferret inferior alveolar nerve neuromas. *Neurosci Lett* 1998;245:89.
83. Liu C, Wall PD, Ben-Dor E, et al: Tactile allodynia in the absence of C-fiber activation: Altered firing properties of DRG neurons following spinal nerve injury. *Pain* 2000; 85:503.
84. Amir R, Michaelis M, Devor M: Membrane potential oscillations in dorsal root ganglion neurons: Role in the normal electrogenesis and neuropathic pain. *J Neurosci* 1999;19:8589.
85. Amir R, Michaelis M, Devor M: Burst discharge in primary sensory neurons: Triggered by subthreshold oscillations, maintained by depolarizing afterpotentials. *J Neurosci* 2002; 22:1187.
86. Chaplan SR, Guo HQ, Lee DH, et al: Neuronal hyperpolarization-activated pacemaker channels drive neuropathic pain. *J Neurosci* 2003;23:1159.
87. Chen Y, Devor M: Ectopic mechanosensitivity in injured sensory axons arises from the site of spontaneous electrogenesis. *Eur J Pain* 1998;2:165.
88. Dib-Hajj S, Black JA, Felts P, et al: Downregulation of transcripts for Na channel α-SNS in spinal sensory neurons following axotomy. *Proc Natl Acad Sci USA* 1996;93:14950.
89. Boucher TJ, Okuse K, Bennett DL, et al: Potent analgesic effects of GDNF in neuropathic pain states. *Science* 2000; 290:124.
90. Black JA, Cummins TR, Plumpton C, et al: Upregulation of a silent sodium channel after peripheral, but not central, nerve injury in DRG neurons. *J Neurophysiol* 1999;82:2776.
91. Zhang JM, Donnelly DF, Song XJ, et al: Axotomy increases the excitability of dorsal root ganglion cells with unmyelinated axons. *J Neurophysiol* 1997;78:2790.
92. Cummins TR, Black JA, Dib-Hajj SD, et al: Glial-derived neurotrophic factor upregulates expression of functional SNS and NaN sodium channels and their currents in axotomized dorsal root ganglion neurons. *J Neurosci* 2000; 20:8754.
93. Leffler A, Cummins TR, Dib-Hajj SD, et al: GDNF and NGF reverse changes in repriming of TTX-sensitive Na$^+$ currents following axotomy of dorsal root ganglion neurons. *J Neurophysiol* 2002;88:650.
94. Scholz J, Woolf CJ: Can we conquer pain? *Nature Neurosci* 2002;5:1062.
95. Gold MS, Levine JD, Correa AM: Modulation of TTX-R I_{Na} by PKC and PKA and their role in PGE$_2$-induced sensitization of rat sensory neurons in vitro. *J Neurosci* 1998; 18:10345.
96. Fitzgerald EM, Okuse K, Wood JN, et al: cAMP-dependent phosphorylation of the tetrodotoxin-resistant voltage-dependent sodium channel SNS. *J Physiol* 1999;516:433.
97. Rasband MN, Park EW, Vanderah TW, et al: Distinct potassium channels on pain-sensing neurons. *Proc Natl Acad Sci USA* 2001;98:13373.
98. Kim DS, Choi JO, Rim HD, et al: Downregulation of voltage-gated potassium channel alpha gene expression in dorsal root ganglia following chronic constriction injury of the rat sciatic nerve. *Brain Res Mol Brain Res* 2002;105:146.
99. Everill B, Kocsis JD: Reduction in potassium currents in identified cutaneous afferent dorsal root ganglion neurons after axotomy. *J Neurophysiol* 1999;82:700.
100. Abdulla FA, Smith PA: Axotomy- and autotomy-induced changes in calcium and potassium channel currents of rat dorsal root ganglion neurons. *J Neurophysiol* 2001;85:644.
101. Everill B, Kocsis JD: Nerve growth factor maintains potassium conductance after nerve injury in adult cutaneous afferent dorsal root ganglion neurons. *Neuroscience* 2000; 100:417.
102. Viana F, Pena E, Belmonte C: Specificity of cold thermotransduction is determined by differential ionic channel expression. *Nature Neurosci* 2002;5:254.
103. Luo DZ, Chaplan SR, Higuera ES, et al: Upregulation of dorsal root ganglion α$_2$δ calcium channel subunit and its correlation with allodynia in spinal nerve-injured rats. *J Neurosci* 2001;21:1869.
104. Newton RA, Bingham S, Case PC: Dorsal root ganglion neurons show increased expression of the calcium channel α$_2$δ-1 subunit following partial sciatic nerve injury. *Brain Res Mol Brain Res* 2001;95:1.
105. Field MJ, Hughes J, Singh L: Further evidence for the role of the alpha(2)delta subunit of voltage dependent calcium

channels in models of neuropathic pain. *Br J Pharmacol* 2000;131:282.
106. Luo ZD, Calcutt NA, Higuera ES, et al: Injury type–specific calcium channel $\alpha_2\delta$-1 subunit upregulation in rat neuropathic pain models correlates with antiallodynic effects of gabapentin. *J Pharmacol Exp Ther* 2002;303:1199.
107. Fehrenbacher JC, Taylor CP, Vasko MR: Pregabalin and gabapentin reduce release of substance P and CGRP from rat spinal tissues only after inflammation or activation of protein kinase C. *Pain* 2003;105:133.
108. Smith MT, Cabot PJ, Ross FB, et al: The novel N-type calcium channel blocker AM336 produces potent dose-dependent antinociception after intrathecal dosing in rats and inhibits substance P release in rat spinal cord slices. *Pain* 2002;96:119.
109. Michael GJ, Priestley JV: Differential expression of the mRNA for the vanilloid receptor subtype 1 in cells of the adult rat dorsal root and nodose ganglia and its downregulation by axotomy. *J Neurosci* 1999;1:1844–1854.
110. Bradbury EJ, Burnstock G, McMahon SB: The expression of $P2X_3$ purinoreceptors in sensory neurons: Effects of axotomy and glial-derived neurotrophic factor. *Mol Cell Neurosci* 1998;12:256.
111. Woolf CJ, Salter MW: Neuronal plasticity: Increasing the gain in pain. *Science* 2000;288:1765.
112. Karchewski LA, Gratto KA, Wetmore C, et al: Dynamic patterns of BDNF expression in injured sensory neurons: Differential modulation by NGF and NT-3. *Eur J Neurosci* 2002;16:1449.
113. Zhou XF, Chie ET, Deng YS, et al: Injured primary sensory neurons switch phenotype for brain-derived neurotrophic factor in the rat. *Neuroscience* 1999;92:841.
114. Kerr BJ, Bradbury EJ, Bennett DL, et al: Brain-derived neurotrophic factor modulates nociceptive sensory inputs and NMDA-evoked responses in the rat spinal cord. *J Neurosci* 1999;19:5138.
115. Ma W, Bisby MA: Partial and complete sciatic nerve injuries induce similar increases of neuropeptide Y and vasoactive intestinal peptide immunoreactivities in primary sensory neurons and their central projections. *Neuroscience* 1998;86:1217.
116. Ossipov MH, Zhang ET, Carvajal C, et al: Selective mediation of nerve injury–induced tactile hypersensitivity by neuropeptide Y. *J Neurosci* 2002;22:9858.
117. Mantyh PW, Allen CJ, Rogers S, et al: Some sensory neurons express neuropeptide Y receptors: Potential paracrine inhibition of primary afferent nociceptors following peripheral nerve injury. *J Neurosci* 1994;14:3958.
118. Landry M, Holmberg K, Zhang X, et al: Effect of axotomy on expression of NPY, galanin, and NPY Y1 and Y2 receptors in dorsal root ganglia and the superior cervical ganglion studied with double-labeling in situ hybridization and immunohistochemistry. *Exp Neurol* 2000;162:361.
119. Ma W, Bisby MA: Differential expression of galanin immunoreactivities in the primary sensory neurons following partial and complete sciatic nerve injuries. *Neuroscience* 1997;79:1183.
120. Kerekes N, Landry M, Hokfelt T: Leukemia inhibitory factor regulates galanin/galanin message–associated peptide expression in cultured mouse dorsal root ganglia; with a note on in situ hybridization methodology. *Neuroscience* 1999;89:1123.
121. Flatters SJ, Fox AJ, Dickenson AH: Nerve injury induces plasticity that results in spinal inhibitory effects of galanin. *Pain* 2002;98:249.
122. Liu HX, Hokfelt T: The participation of galanin in pain processing at the spinal level. *Trends Pharmacol Sci* 2002;23:468.
123. Blakeman KH, Hao JX, Xu XJ, et al: Hyperalgesia and increased neuropathic pain–like response in mice lacking galanin receptor 1 receptors. *Neuroscience* 2003;117:221.
124. Zhang Q, Shi TJ, Ji RR, et al: Expression of pituitary adenylate cyclase–activating polypeptide in dorsal root ganglia following axotomy: Time course and coexistence. *Brain Res* 1995;705:149.
125. Ossipov MH, Bian D, Malan TP Jr, et al: Lack of involvement of capsaicin-sensitive primary afferents in nerve-ligation injury induced tactile allodynia in rats. *Pain* 1999;79:127.
126. Campbell JN, Raja SN, Meyer RA, et al: Myelinated afferents signal the hyperalgesia associated with nerve injury. *Pain* 1988;32:89.
127. Woolf CJ, Shortland P, Coggeshall RE: Peripheral nerve injury triggers central sprouting of myelinated afferents. *Nature* 1992;355:75.
128. Shortland P, Woolf CJ: Chronic peripheral nerve section results in a rearrangement of the central axonal arborizations of axotomized Aβ primary afferent neurons in the rat spinal cord. *J Comp Neurol* 1993;330:65.
129. Shehab SA, Spike RC, Todd AJ: Evidence against cholera toxin B subunit as a reliable tracer for sprouting of primary afferents following peripheral nerve injury. *Brain Res* 2003;964:218.
130. Woolf CJ, Shortland P, Reynolds M, et al: Reorganization of central terminals of myelinated primary afferents in the rat dorsal horn following peripheral axotomy. *J Comp Neurol* 1995;360:121.
131. Bennett DL, Michael GJ, Ramachandran N, et al: A distinct subgroup of small DRG cells expresses GDNF receptor components and GDNF is protective for these neurons after nerve injury. *J Neurosci* 1998;15:3059.
132. Koerber HR, Mirnics K, Brown PB, et al: Central sprouting and functional plasticity of regenerated primary afferents. *J Neurosci* 1994;14:3655.
133. Kohama I, Ishikawa K, Kocsis JD: Synaptic reorganization in the substantia gelatinosa after peripheral nerve neuroma formation: Aberrant innervation of lamina II neurons by Aβ afferents. *J Neurosci* 2000;20:1538.
134. Rappaport ZH, Devor M: Trigeminal neuralgia: The role of self-sustaining discharge in the trigeminal ganglion. *Pain* 1994;56:127.
135. Seltzer Z, Devor M: Ephaptic transmission in chronically damaged peripheral nerves. *Neurology* 1979;29:1061.
136. Meyer RA, Raja SN, Campbell JN, et al: Neural activity originating from a neuroma in the baboon. *Brain Res* 1985;325:255.
137. Amir R, Devor M: Chemically mediated cross-excitation in rat dorsal root ganglia. *J Neurosci* 1996;16:4733.
138. Love S, Coakham HB: Trigeminal neuralgia: Pathology and pathogenesis. *Brain* 2001;124:2347.

139. Andrew D, Greenspan JD: Modality-specific hyperresponsivity of regenerated cat cutaneous nociceptors. *J Physiol* 1999;516:897.

140. Tandrup T, Woolf CJ, Coggeshall RE: Delayed loss of small dorsal root ganglion cells after transection of the rat sciatic nerve. *J Comp Neurol* 2000;422:172.

141. Degn J, Tandrup T, Jakobsen J: Effect of nerve crush on perikaryal number and volume of neurons in adult rat dorsal root ganglion. *J Comp Neurol* 1999;412:186.

142. Venters HD, Dantzer R, Kelley KW: A new concept in neurodegeneration: TNF-α is a silencer of survival signals. *Trends Neurosci* 2000;23:175.

143. Benn SC, Perrelet D, Kato AC, et al: Hsp27 upregulation and phosphorylation is required for injured sensory and motor neuron survival. *Neuron* 2002;36:45.

144. Woolf CJ, Thompson SW: The induction and maintenance of central sensitization is dependent on N-methyl-D-aspartic acid receptor activation: Implications for the treatment of postinjury pain hypersensitivity states. *Pain* 1991;44:293.

145. Eide PK, Jorum E, Stubhaug A, et al: Relief of post-herpetic neuralgia with the N-methyl-D-aspartic acid receptor antagonist ketamine: A double-blind, cross-over comparison with morphine and placebo. *Pain* 1994;58:347.

146. Morisset V, Nagy F: Plateau potential-dependent windup of the response to primary afferent stimuli in rat dorsal horn neurons. *Eur J Neurosci* 2000;12:3087.

147. Liu XG, Sandkuhler J: Activation of spinal N-methyl-D-aspartate or neurokinin receptors induces long-term potentiation of spinal C-fibre-evoked potentials. *Neuroscience* 1998;86:1209.

148. Ikeda H, Heinke B, Ruscheweyh R, et al: Synaptic plasticity in spinal lamina I projection neurons that mediate hyperalgesia. *Science* 2003;299:1237.

149. Suzuki R, Morcuende S, Webber M, et al: Superficial NK-1-expressing neurons control spinal excitability through activation of descending pathways. *Nature Neurosci* 2002;5:1319.

150. Sandkuhler J: Learning and memory in pain pathways. *Pain* 2000;88:113.

151. Woolf CJ: Evidence for a central component of postinjury pain hypersensitivity. *Nature* 1983;306:686.

152. Thompson SW, Woolf CJ, Sivilotti LG: Small-caliber afferent inputs produce a heterosynaptic facilitation of the synaptic responses evoked by primary afferent A-fibers in the neonatal rat spinal cord in vitro. *J Neurophysiol* 1993;69:2116.

153. Neumann S, Doubell TP, Leslie T, et al: Inflammatory pain hypersensitivity mediated by phenotypic switch in myelinated primary sensory neurons. *Nature* 1996;384:360.

154. Benquet P, Gee CE, Gerber U: Two distinct signaling pathways upregulate NMDA receptor responses via two distinct metabotropic glutamate receptor subtypes. *J Neurosci* 2002;22:9679.

155. Heidinger V, Manzerra P, Wang XQ, et al: Metabotropic glutamate receptor 1–induced upregulation of NMDA receptor current: Mediation through the Pyk2/Src-family kinase pathway in cortical neurons. *J Neurosci* 2002;22:5452.

156. Eschenfelder S, Habler HJ, Janig W: Dorsal root section elicits signs of neuropathic pain rather than reversing them in rats with L5 spinal nerve injury. *Pain* 2000;87:213.

157. Liu X, Eschenfelder S, Blenk KH, et al: Spontaneous activity of axotomized afferent neurons after L5 spinal nerve injury in rats. *Pain* 2000;84:309.

158. Wall PD, Woolf CJ: Muscle but not cutaneous C-afferent input produces prolonged increases in the excitability of the flexion reflex in the rat. *J Physiol* 1984;356:443.

159. Chaplan SR, Malmberg AB, Yaksh TL: Efficacy of spinal NMDA receptor antagonism in formalin hyperalgesia and nerve injury evoked allodynia in the rat. *J Pharmacol Exp Ther* 1997;280:829.

160. Yashpal K, Fisher K, Chabot JG: Differential effects of NMDA and group I mGluR antagonists on both nociception and spinal cord protein kinase C translocation in the formalin test and a model of neuropathic pain in rats. *Pain* 2001;94:17.

161. South SM, Kohno T, Kaspar BK, et al: A conditional deletion of the NR_1 subunit of the NMDA receptor in adult spinal cord dorsal horn reduces NMDA currents and injury-induced pain. *J Neurosci* 2003;23:5031.

162. Sindrup SH, Jensen TS: Efficacy of pharmacological treatments of neuropathic pain: An update and effect related to mechanism of drug action. *Pain* 1999;83:389.

163. Nichols ML, Allen BJ, Rogers SD, et al: Transmission of chronic nociception by spinal neurons expressing the substance P receptor. *Science* 1999;286:1558.

164. Malmberg AB, Brandon EP, Idzerda RL, et al: Diminished inflammation and nociceptive pain with preservation of neuropathic pain in mice with targeted mutation of the type I regulatory subunit of cAMP-dependent protein kinase. *J Neurosci* 1997;17:7462.

165. Ma W, Quirion R: Increased phosphorylation of cyclic AMP response element–binding protein (CREB) in the superficial dorsal horn neurons following partial sciatic nerve ligation. *Pain* 2001;93:295.

166. Cheng HY, Pitcher GM, Laviolette SR, et al: DREAM is a critical transcriptional repressor for pain modulation. *Cell* 2002;108:31.

167. Malan TP, Mata HP, Porreca F: Spinal GABA(A) and GABA(B) receptor pharmacology in a rat model of neuropathic pain. *Anesthesiology* 2002;96:1161.

168. Moore KA, Kohno T, Karchewski LA, et al: Partial peripheral nerve injury promotes a selective loss of GABAergic inhibition in the superficial dorsal horn of the spinal cord. *J Neurosci* 2002;22:6724.

169. Ibuki T, Hama AT, Wang XT, et al: Loss of GABA immunoreactivity in the spinal dorsal horn of rats with peripheral nerve injury and promotion of recovery by adrenal medullary grafts. *Neuroscience* 1997;76:845.

170. Eaton MJ, Plunkett JA, Karmally S, et al: Changes in GAD- and GABA-immunoreactivity in the spinal dorsal horn after peripheral nerve injury and promotion of recovery by lumbar transplant of immortalized serotonergic precursors. *J Chem Neuroanat* 1998;16:57.

171. Whiteside GT, Munglani R: Cell death in the superficial dorsal horn in a model of neuropathic pain. *J Neurosci Res* 2001;64:168.

172. Coull JA, Boudreau D, Bachand K, et al: Transsynaptic shift in anion gradient in spinal lamina I neurons as a mechanism of neuropathic pain. *Nature* 2003;424:938.

173. Hunt SP, Mantyh PW: The molecular dynamics of pain control. *Nature Rev Neurosci* 2001;2:83.

174. Aldskogius H, Kozlova EN: Central neuron–glial and glial-glial interactions following axon injury. *Prog Neurobiol* 1998;55:1.
175. Winkelstein BA, Rutkowski MD, Sweitzer SM, et al: Nerve injury proximal or distal to the DRG induces similar spinal glial activation and selective cytokine expression but differential behavioral responses to pharmacologic treatment. *J Comp Neurol* 2001;439:127.
176. Colburn RW, DeLeo JA, Rickman AJ, et al: Dissociation of microglial activation and neuropathic pain behaviors following peripheral nerve injury in the rat. *J Neuroimmunol* 1997;79:163.
177. Hashizume H, Rutkowski MD, Weinstein JN, et al: Central administration of methotrexate reduces mechanical allodynia in an animal model of radiculopathy/sciatica. *Pain* 2000;87:159.
178. Milligan ED, Twining C, Chacur M, et al: Spinal glia and proinflammatory cytokines mediate mirror-image neuropathic pain in rats. *J Neurosci* 2003;23:1026.
179. Sweitzer SM, Schubert P, DeLeo JA: Propentofylline, a glial modulating agent, exhibits antiallodynic properties in a rat model of neuropathic pain. *J Pharmacol Exp Ther* 2001;297:1210.
180. Jin SX, Zhuang ZY, Woolf CJ, et al: p38 mitogen-activated protein kinase is activated after a spinal nerve ligation in spinal cord microglia and dorsal root ganglion neurons and contributes to the generation of neuropathic pain. *J Neurosci* 2003;23:4017.
181. Ma W, Quirion R: Partial sciatic nerve ligation induces increase in the phosphorylation of extracellular signal-regulated kinase (ERK) and c-Jun N-terminal kinase (JNK) in astrocytes in the lumbar spinal dorsal horn and the gracile nucleus. *Pain* 2002;99:175.
182. Kingery WS: A critical review of controlled clinical trials for peripheral neuropathic pain and complex regional pain syndromes. *Pain* 1997;73:123.
183. Burgess SE, Gardell LR, Ossipov MH, et al: Time-dependent descending facilitation from the rostral ventromedial medulla maintains, but does not initiate, neuropathic pain. *J Neurosci* 2002;22:5129.
184. Gardell LR, Vanderah TW, Gardell SE, et al: Enhanced evoked excitatory transmitter release in experimental neuropathy requires descending facilitation. *J Neurosci* 2003;23:8370.
185. Wang Z, Gardell LR, Ossipov MH, et al: Pronociceptive actions of dynorphin maintain chronic neuropathic pain. *J Neurosci* 2001;21:1779.
186. Laughlin TM, Larson AA, Wilcox GL: Mechanisms of induction of persistent nociception by dynorphin. *J Pharmacol Exp Ther* 2001;299:6.
187. Zadina JE, Hackler L, Ge LJ, et al: A potent and selective endogenous agonist for the mu-opiate receptor. *Nature* 1997;386:499.
188. Martin-Schild S, Gerall AA, Kastin AJ, et al: Differential distribution of endomorphin 1– and endomorphin 2–like immunoreactivities in the CNS of the rodent. *J Comp Neurol* 1999;405:450.
189. Przewlocka B, Mika J, Labuz D, et al: Spinal analgesic action of endomorphins in acute, inflammatory, and neuropathic pain in rats. *Eur J Pharmacol* 1999;367:189.

CHAPTER 6

Mechanisms of Visceral Pain

T. J. Ness and Gerald F. Gebhart

Clinically, visceral pain is common, but only recently have scientific studies defined some of the qualities and mechanisms of this type of pain. These studies have demonstrated that visceral pain is uniquely different from pain that arises from superficial structures of the body, both in substrate and in response to environmental factors. At the same time, it has been possible to determine that there are underlying similarities in multiple visceral sensory systems such that an understanding of one particular system may improve understanding of other systems. This chapter will summarize and review what is known about visceral pain and will contrast and compare it with what is known about somatic pains.

▶ CLINICAL DIFFERENCES BETWEEN SUPERFICIAL AND VISCERAL PAIN

It is a hallmark feature of the viscera that when they are healthy, they rarely give rise to conscious sensation. However, when diseased or inflamed, they can become an overwhelming source of sensation that can stop all activity and demand all attention. In contrast, the surface of our body continuously generates conscious sensations, which also typically increase with pathology. Injury of the surface of our body inspires motion with nociceptive reflexes or "fight or flight" behavioral responses. Both visceral pain and somatic pain produce emotional responses, but it is clinical lore that visceral pain produces stronger emotional responses that may appear out of proportion to the perceived intensity of the pain. Nausea appears more commonly with visceral pain than with superficial pain. Sweating (diaphoresis), dyspnea, and other autonomic responses can be profound with visceral pain, such as angina, but are of a lesser magnitude when evoked by superficial pain. There is a poor correlation between the amount of visceral pathology and the intensity of pain. For example, very extensive inflammatory processes (e.g., ulcerative colitis) or tissue damage (e.g., gastric perforation) may produce little or no pain in some individuals, and minimally discernible pathology may produce out-of-control pain in others. There is recognized variability in individual response to any stimulus, but this variability appears greater for visceral stimuli than for somatic stimuli.

A second hallmark clinical feature of visceral pain is its poor and unreliable localization. Generally stated, visceral pain is deep and diffuse, and often the only localization possible comes with physical examination manipulations that stimulate the painful organ. This is again in contrast to superficial pain, which is easily and well localized. Depending on the site of the body surface tested, adjacent painful stimuli can be localized to within millimeters. Perhaps more important, superficial sensations from a specific site are always reliably localized to the same site and do not "migrate" to other body areas in the absence of nerve injury. The same cannot be said for visceral pain. Visceral pain can be felt in several different areas at the same time or can migrate throughout a region even though pathology is localized to a single organ. Unless highly recurrent, visceral pain is not normally perceived as localized to the organ itself but to somatic structures that receive afferent inputs at the same spinal segments as visceral afferent entry. For this reason, visceral pain classically is described as being *referred* and may have two separate components: (1) the sensation is transferred to another site (e.g., angina can be felt in the neck and arm) and/or (2) other sites become more sensitive to inputs applied directly to those other sites (e.g., flank muscle becomes sensitive to palpation when passing a kidney stone). This latter phenomenon is also described as *secondary somatic hyperalgesia*, an old clinical observation. A hypersensitivity to pinprick stimulation of the skin, develops in a dermatomal fashion that is typical for specific visceral disease processes. Head described this phenomenon in 1893, and using an artful combination of insight and reason, he was able to combine these findings with clinical data related to outbreaks of herpes zoster and generated one set of the dermatomal charts we use in clinical practice today.

▶ VISCERAL HYPERSENSITIVITY DISORDERS

The observation that pathology and symptomatology may not correlate is readily apparent in numerous visceral pain disorders. Some disorders such as chronic pancreatitis have definable pathology, but alterations in pain appear out of proportion to changes in radiographic or laboratory findings. Other disorders such as irritable bowel syndrome, noncardiac chest pain, and postcholecystectomy syndrome appear to have no histopathologic bases for the discomfort and pain. Instead, visceral discomfort and pain in such conditions are termed *functional* and are associated with altered patterns/pressures associated with motility, production of gas, and ingestion of food or beverage. Accordingly, hypersensitivity to natural visceral stimuli in the physiologic range can be associated with discomfort and pain in the absence of obvious visceral pathology. In contrast, hypersensitivity to applied stimuli in the somatic realm is always associated with tissue damage and inflammation or nerve injury.

Psychophysical studies have demonstrated evidence for hypersensitivity in virtually all clinically relevant visceral pain disorders. This includes hypersensitivity to gastric distension in patients with functional dyspepsia,[1] intestinal and rectal distension in patients with irritable bowel syndrome,[2,3] biliary and/or pancreatic duct distension in patients with postcholecystectomy syndrome or chronic pancreatitis,[4] and bladder distension in patients with interstitial cystitis.[5] In all cases, pain and/or discomfort (e.g., "bloating") is experienced at intensities of stimulation lower than required to produce the same quality and intensity of sensation in a healthy population. A more sophisticated testing of visceral sensitivity using random-order, graded distension of the rectum in irritable bowel syndrome patients suggests that the population of subjects is heterogeneous.[6] One subgroup appears to be reliably hypersensitive, and another subgroup appears to be hypervigilant. Dissociating potential psychological modifiers of sensory reports from other, more "physiologic" pathologies has proved to be a difficult and at times insurmountable methodologic problem.

Psychophysical studies related to visceral sensation in normal, healthy subjects have suggested a basis for some of the emotional factors that may affect pain reports. A hallmark study by Strigo and colleagues[7] compared sensations evoked by balloon distension of the esophagus with thermal stimulation of the midchest skin. Using graded intensities of both distending and thermal stimuli, it was possible to match the intensity of evoked sensations produced at the two different sites. Consistent with clinical lore, visceral sensations were poorly localized, and equal intensities of reported sensation produced greater emotional responses when the visceral stimulus was employed: unpleasantness ratings were higher when the esophageal stimulus was administered; word selection from the McGill Pain Questionnaire suggested a stronger affective component to the sensation evoked by esophageal distension; and greater anxiety was evoked by esophageal distension. There was a tight temporal link between the thermal cutaneous stimulus and the evoked sensations. In contrast, there was a poor temporal correlation with the esophageal stimulus; a sustained, relatively high intensity of sensation was perceived even after termination of the distending esophageal stimulus.

Other psychophysical studies of experimental visceral pain sensation in humans have indicated that a sensitization process occurs with repeated distension of the gut[8] and of the urinary bladder.[9] In normal subjects, initial visceral stimuli that are perceived as nonpainful become painful (produce discomfort) with repeated presentation. Patients with clinical visceral pain conditions such as irritable bowel syndrome typically report pain with initial stimuli at low intensities,[6] as well as sensitization when neighboring structures are stimulated.[10] Correlates to a repeated-stimulation sensitization process occur in nonhuman animals where repeated presentation of the same visceral stimulus produces increasing vigor of neuronal, cardiovascular, and visceromotor reflex responses.[11–13]

The treatment of visceral hypersensitivity (functional) disorders is poorly defined. Use of opioid analgesics has been avoided in those disorders without histopathologically-defined disease processes. Even when there is defined pathology, such as in a patient with chronic pancreatitis, there is a reluctance on the part of clinicians to use opioids due to fears of addiction or other phenomena associated with long-term opioid use (e.g., physical dependence) or patient histories of substance abuse. As a consequence, there has been a therapeutic focus on nonopioid options for medical management of visceral hypersensitivity disorders, with reported use of antidepressants, anticonvulsants, antiarrhythmics, antispastic agents, anticholinergics, and anxiolytics. A rational basis for many of these therapeutics has not been presented.

▶ ANIMAL MODELS OF VISCERAL PAIN

Application of a tissue-damaging stimulus to the surface of a healthy body evokes a localized motor response, termed a *flexion (nociceptive)-withdrawal reflex*. Apply a similar tissue-damaging stimulus to a healthy viscus, and likely nothing obvious happens. Stimuli that produce tissue damage or predict potential tissue damage (e.g., cutting, burning, pinching) universally produce reports of pain when applied to the skin but unreliably evoke reports of pain when applied to visceral structures. This has made the modeling of visceral pain in nonhuman species difficult because the definition of accepted criteria necessary for a pain model is less obvious than in somatic, cutaneous systems.

The usefulness of any visceral pain model is a function of three things: (1) the validity of the model's stimulus as a "noxious" visceral stimulus, (2) the reliability and reproducibility of responses to that stimulus, and (3) the specificity of those responses to noxious (as opposed to non-noxious) intensities of the same stimulus. Stimuli that have been employed in studies of visceral pain generally can be lumped into four groups: electrical, mechanical, chemical, and ischemic. The most commonly used models in nonhuman animals employ mechanical or chemical stimuli. Although ischemia is associated with visceral pain, it is a difficult stimulus to control, and few animal models exist. Electrical stimuli are also not used widely in animal models, although electrical stimulation of human visceral structures, typically to produce cerebral evoked potentials and thus assess central mechanisms, is used in some research trials.

To be valid, the visceral stimulus should produce pain when applied to human viscera, should produce aversive behaviors in the studied species, and should produce responses that are modified by manipulations known to modulate visceral pain in humans (e.g., inhibited by morphine). In humans, visceral pain is characterized by nonlocalized responses such as immobility, coupled with generalized increases in muscle tone, as well as vigorous but nonspecific autonomic responses such as changes in respiration, heart rate, and blood pressure. These same reflex responses to pain-producing stimuli have been termed by Sherrington[14] as *pseudaffective responses*. These include muscle contractions, pupillary dilatation, respiratory changes, and alterations in heart rate and blood pressure. Pseudaffective responses have been proven to be reliable but nonspecific measures of nociception. For example, numerous manipulations unrelated to pain can produce similar autonomic and motor reflexes. For this reason, studies using pseudaffective reflexes as the only response output must be interpreted with some element of uncertainty. Neurophysiologic responses to visceral stimuli also have proved to be reliable, but due to the invasive surgery necessary to enable recording of afferent fibers or single, central neurons, animals must be anesthetized or reduced (spinal transection and/or decerebration). Other responses include neuron early-intermediate gene induction (e.g., *c-fos*), behavioral alterations such as hypolocomotion, or secondary effects of visceral stimulation on thermal or mechanical thresholds for somatic reflexes.

Over 50 different models of visceral pain have been described, but only a few have been well characterized. These include the common pharmaceutical screening model know as the *writhing test*, which consists of the intraperitoneal injection of a chemical irritant (e.g., acetic acid, phenylquinone, or hypertonic saline), followed by counting the number of "writhes" produced. Writhes are a characteristic contraction of abdominal muscles accompanied by a hindlimb extensor motion. Variations have been described in primates, cats, dogs, and guinea pigs, but the model has been used predominantly in rats and mice. Methodologic and ethical concerns have presented significant constraints to the use of this model.

Distension of hollow organs has been employed in multiple models of visceral pain[11,15] (see also ref. 16 for review). Most commonly, distension of the distal gastrointestinal tract (cecum, colon, or rectum) has been used to evoke respiratory, cardiovascular, visceromotor, behavioral, and neurophysiologic responses in multiple species, including horses, dogs, cats, rabbits, and rats. Studies have been performed in multiple laboratories in the United States, Europe, and Asia with consistent findings between sites. Distension of the gallbladder and associated biliary system produces pathologic pain when the gallbladder is inflamed and/or associated ducts are obstructed and has been used experimentally. Distension or chemical stimulation of the urinary bladder and other urinary tract structures also has been employed commonly. Distension, compression, or traction on reproductive organs also produces nocifensive responses.

Some visceral pain models are representative of specific pathophysiologic processes. Such is the case in relation to the work of Giamberardino and colleagues,[17] who modeled urolithiasis by creating a model of artificial ureteral calculosis. Following surgical exposure, an artificial stone is placed into the upper third of the ureter by injecting 0.02 ml of dental resin cement (while still liquid) through a fine needle. Rats are then continuously observed for days for "visceral episodes" demonstrating behaviors similar to those observed in the writhing test.

Another unique difference between cutaneous and visceral sensation appears to be a differential effect of "stress" on the magnitude of responses to stimuli.

Although stress-induced analgesia (or *hypo*algesia) has been a long-recognized phenomenon associated with cutaneous sensation, it would appear that stress-induced *hyper*algesia is the correlate phenomenon associated with visceral sensation. Clinically, stressful life events have been viewed as classic "triggers" for the evocation of diffuse abdominal complaints of presumed visceral origin.[3]

There is a correlate to human behavior in animal models. Classic behavioral stressors such as a cold-water swim produce an elevation in thresholds for the evocation of responses to thermal stimuli (stress-induced analgesia), but the same animals have an increased vigor of visceromotor responses to colon distension (visceral hyperalgesia).[18] This phenomenon appears to be associated with early-life events[18] and can be modified by gonadal hormones, neurokinins, corticotropin-releasing factor, and mast cell function.[19,20]

A notable neurophysiologic correlate is the demonstration by Qin and colleagues[21–23] that the injection of glucocorticoids or aldosterone into the amygdala, a manipulation that is known to produce an increased measure of anxiety in animal subjects, also produces a hypersensitivity to visceral stimulation, as measured by an increased vigor of both visceromotor responses and responses of spinal dorsal horn neurons to colon or urinary bladder distension. This suggests a chicken-egg relation in which visceral hypersensitivity may be producing more anxiety or more anxiety may be producing more visceral sensation.

▶ THE NEUROBIOLOGY OF VISCERAL PAIN

General Organization

Sensations begin with the activation of receptors on the peripheral terminals of primary afferent nerve fibers. It is notable that visceral primary afferents differ significantly from cutaneous primary afferents in both number and pattern of distribution. Grossly, visceroceptive afferents are diffusely organized into weblike plexuses rather than forming distinct peripheral nerve entities. Afferents with endings in a specific visceral site may have cell bodies in the dorsal root ganglia (DRGs) of 10 or more spinal levels in a bilaterally distributed fashion. In contrast, cutaneous afferents arise from a limited number of unilateral DRGs. Individual visceroceptive afferent fibers have been demonstrated to branch within the spinal cord and to spread over multiple spinal segments. Individual cutaneous afferents, on the other hand, have been demonstrated to form tight "baskets" of input to localized spinal cord segments. Visceral afferents also selectively terminate in a different spinal laminar pattern than cutaneous afferents, suggesting that they also may interact with different spinal neuronal populations. When examined quantitatively, spinal dorsal horn neurons with visceral inputs have multiple, convergent inputs from other viscera, from joints, from muscle, *and* from cutaneous structures. Convergent receptive fields for these neurons, therefore, are large and have diffuse inputs. In contrast, neurons with exclusively cutaneous input commonly are identified in the spinal dorsal horn, in particular from glabrous skin. Taken together, these results suggest an imprecise organization of visceral primary inputs that would be consistent with an imprecise localization by the central nervous system (CNS). Specifics related to these differing levels of neurophysiologic processing are given below.

Primary (Sensory) Afferents

The sensory innervation of the internal organs is unlike the sensory innervation of somatic tissues in that most visceral sensory input to the CNS is not perceived consciously. For example, much of the gastrointestinal vagal afferent input is associated with mucosal endings that sample or "taste" the luminal contents and secretory and motor events that do not reach conscious appreciation.[24] When visceral sensory input is perceived consciously, it is commonly associated with mechanical stimuli (e.g., tension on the mesentery, distension or stretch of hollow organs), and the sensations are invariably discomfort and pain. Among the stimuli that produce discomfort and pain from the viscera, mechanical stimuli are best studied and understood.

It should be appreciated that we have little knowledge about the morphology of mechanosensitive (or any other) receptive endings in organs with the exception of two vagal mechanosensitive endings in the gastrointestinal tract, principally the rostral gut. Intraganglionic laminar endings (IGLEs) and intramuscular arrays (IMAs) have been described, both of which are mechanosensitive and located in hollow-organ muscle layers (see refs. 25 and 26 for reviews and commentary). Powley and Phillips[26] speculate, based on the distribution and morphology of these endings, that IGLEs transduce tension in the gut and detect rhythmic motor activity, whereas IMAs detect changes in length or stretch. These functional characteristics have not been established experimentally, but IGLEs and IMAs are the only mechanosensitive endings in the viscera for which detailed (and elegant) morphology is available. All other peripheral nerve endings in the viscera are assumed, like their somatic counterparts, to be unencapsulated "free" nerve endings. Information about visceral nerve end-organ morphology is lacking primarily because of difficulty in filling with anterograde tracer the very fine endings in visceral tissue. New experimental approaches soon may help to improve our understanding of end-organ morphology, which is critical to understanding mechanisms of energy transduction into electrical activity.[27,28]

The functional characterization of visceral afferents reveals two general populations of mechanosensitive fibers in standard in vivo–teased fiber recording preparations and at least four mechanosensitive endings in organ-nerve-attached in vitro recording preparations. In teased-fiber preparations, mechanosensitive afferent fibers with low thresholds for response (about 75 to 80 percent of the population) or high thresholds for response (20 to 25 percent of the population) have been described for all spinal visceral nerves studied in a variety of species (see ref. 29 for review). The low-threshold afferent fiber population responds to mechanical stimuli (usually balloon distension of hollow organs) in the physiologic range (e.g., 5 mmHg) but also encodes stimulus intensity well into the noxious range (60 to 80 mmHg), a characteristic that distinguishes them from low-threshold mechanoreceptors in the somatic realm (which do not encode noxious intensities of mechanical stimulation). The high-threshold afferent fiber population begins to respond at or above 30 mmHg distending pressure, an intensity established in behavioral experiments to be aversive or noxious, and then encodes noxious stimulus intensity. In contrast to the spinal visceral innervation, only low-threshold mechanosensitive afferent fibers have been found in the vagal innervation of the gut.[30] Like low-threshold spinal afferents, gastric vagal afferent fibers also encode stimulus intensity into the noxious range. Given their slowly adapting response properties and sensitivity to tension or stretch, it is assumed that the receptive endings of both spinal and vagal mechanosensors are located in the muscle layers of organs.

It has been assumed that low-threshold visceral afferents are analogous to somatic low-threshold mechanoreceptors and that high-threshold visceral afferents are analogous to somatic nociceptors. It is likely that activation of the high-threshold visceral afferent population signals acute visceral pain, but it is inappropriate to relegate the low-threshold visceral afferent population to signaling only non-noxious visceral events for several reasons. First, low-threshold visceral afferents encode mechanical stimuli well into the noxious range and, moreover, respond to noxious intensities of mechanical stimuli with greater magnitude than do high-threshold visceral afferent fibers.[29] Second, low-threshold visceral afferents sensitize when exposed to chemical mediators. Sensitization is an increase in response magnitude after tissue insult and is a distinguishing characteristic of somatic nociceptors. Low-threshold somatic mechanoreceptors do not sensitize, and the ability of both low- and high-threshold visceral mechanoreceptors to sensitize strongly suggests that both low- and high-threshold visceral mechanoreceptors contribute to discomfort and pain arising from the viscera. Indeed, because low-threshold visceral mechanoreceptors respond in the low physiologic range of mechanical stimuli, when sensitized—as may be the case in functional gastrointestinal disorders—they may play a dominant role in inappropriately giving rise to discomfort or pain when activated by stimuli in the physiologic range. Another characteristic of these two types of mechanoreceptors is that, when tested, they both have been found also to respond to chemical and/or thermal (heat) stimuli, revealing that they are polymodal in character (like polymodal somatic nociceptors). Accordingly, if sensitized, normally innocuous chemical stimuli can activate both low- and high-threshold visceral mechanoreceptors.

When characterized in vitro in organ-nerve-attached preparations, at least four types of mechanoreceptors have been described. In these preparations, the organ typically is opened and pinned flat in a perfusion chamber, thus providing direct access to the mucosal (or serosal) surface of the organ. Mucosal mechanoreceptors respond to gently stroking the mucosa of the organ and do not respond to circumferential stretch. Muscular (or tension) receptors do not respond to gentle stroking of the mucosa but respond to and encode the force of circumferential stretch of the organ. Serosal receptors respond to blunt probing of the receptive field and encode the force of probing; they do not encode circumferential stretch of the organ. All three of these mechanoreceptors have been described in different organs (esophagus, stomach, and colon) and in rats and mice. A fourth mechanoreceptor, a mucosal-muscular receptor, responds to gentle stroking of the mucosa and to circumferential stretch and has been described in the pelvic nerve supply of the distal colon of the mouse. An advantage of the in vitro preparation is the ability to determine receptive fields of the afferent fibers in the organ, which have been found to be small and spotlike (1 to 4 mm^2). Generally, only one receptive field has been found for each fiber, although multiple receptive fields have been described for some fibers in the stomach and colon. Mechanoreceptors also have been found along the mesenteric attachments of the distal gut.

Given the nature of the experimental preparation, it is also possible to study chemosensitivity of these mechanoreceptors, many of which respond to mediators such as serotonin and/or are sensitized by bradykinin, prostaglandins, and a soup of inflammatory mediators, by directly applying chemicals to the receptive field of the fiber in the organ. This is a developing field of investigation, but it is likely that mucosal mechanoreceptors will be found to be chemosensitive and perhaps selective for different chemicals, permitting subclassification and improved understanding of heretofore poorly understood mucosal receptor contributions to visceral pain mechanisms. Muscular mechanoreceptors are sensitized by chemical mediators, but whether there exist two populations of muscular receptors, low and high threshold, as described in teased-fiber preparations, remains to be determined.

Central Terminations of Visceral Afferents

Vagal afferent fibers have their cell bodies in nodose ganglia and their central terminals principally in the nucleus of the solitary tract in the dorsal medulla. A small percentage of vagal afferents either continue through the medulla or by another nonmedullary route terminate in the first and second cervical segments of the spinal cord, where they are proposed to contribute to propriospinal modulation of spinal nociceptive processing (see ref. 31 for review of vagal afferent modulation of spinal nociceptive processing). Vagal afferent input to the CNS is not considered to contribute to visceral pain, but mechanosensitive vagal afferent fibers sensitize when exposed to thermal (heat) or chemical (e.g., hydrochloric acid, bile salts, or nerve growth factor) stimuli, and there is growing evidence that vagal afferent fibers may contribute to chemonociceptive input to the CNS. For example, intragastric instillation of hydrochloric acid leads to the expression of c-fos protein in the brain stem and also produces a visceromotor response.[32,33] The visceromotor response to dilute hydrochloric acid instilled into the stomach is blocked by vagotomy but not splanchnectomy (which does block visceromotor responses to gastric balloon distension).[32] These outcomes suggest that sensations produced by gastric distension are conveyed to the CNS by spinal visceral afferents in the splanchnic nerves, whereas chemonociceptive sensations are conveyed to the CNS by vagal afferents.

Spinal visceral afferent fibers have their cell bodies in DRGs and their central terminals in the superficial dorsal horn of the spinal cord (laminae I and II), the intermediolateral cell column and sacral parasympathetic nucleus (pelvic nerve), as well as in the area around and dorsal to the central canal (often termed *lamina X*). Almost all second-order neurons in the spinal cord that receive a visceral input also receive convergent somatic input (e.g., skin and/or muscle), and this is considered the basis of referral of visceral sensation to somatic sites (e.g., myocardial ischemia typically radiates to the left shoulder and upper arm and occasionally to the jaw; the pain is not felt at the source—the heart). Interestingly, the termination of visceral afferent fibers in laminae I and II of the spinal dorsal horn overlaps with the central terminations of somatic nociceptors, perhaps explaining why the principal conscious sensations that arise from the visceral inputs are discomfort and pain. In addition to convergence of somatic and visceral inputs on second-order spinal neurons, viscerovisceral convergence is also common (e.g., bladder and colon, colon and uterus, etc.). In conjunction with the diffuse character of visceral sensations (contributed to by the relatively sparse innervation of the viscera as compared with somatic structures) and referral to somatic sites, convergence of inputs from multiple viscera onto the same spinal neurons further contributes to the difficulty both patients and physicians experience when diagnosing the source of visceral pain.

An Alternative Spinal Pathway

Traditionally, it has been taught that the primary pathways for pain-related information from the dorsal horn of the spinal cord to the brain is via the anterolateral quadrant white matter of the spinal cord. Tracts located within these sites include the spinothalamic and spinoreticular tracts, as well as the spinomesencephalic and spinohypothalamic tracts. The anterolateral quadrant of the spinal cord is clearly important for cutaneous pain sensation because lesions of those areas of white matter lead to pinprick analgesia in contralateral dermatomes below the level of the lesion. However, an important recent finding related to the neurobiology of visceral pain is evidence that a dorsal column pathway may be involved. Researchers have demonstrated that surgical lesions of the dorsal midline of the spinal cord have profound effects on visceral pain–related responses in humans, primates, and rodents. Specifically, a punctate midline myelotomy in humans has been demonstrated to relieve cancer-related pelvic and abdominal pain.[34–36] In primates, similar lesions reduce the activity of thalamic neurons evoked by colorectal distension.[37] In rats, effects of similar lesions have been demonstrated to reduce or abolish thalamic neuronal responses and/or behavioral responses to colorectal distension,[38,39] duodenal distension,[40] pancreatic stimulation,[41] and hypersensitivity following lower extremity osteotomy.[42] Whereas dorsal midline lesions did affect visceral inputs to the nucleus gracilis of the medulla,[43] these lesions did not affect inputs to the ventrolateral medulla.[39] Spinal neurons with viscerosomatic convergence and axonal extensions into the dorsal columns have been demonstrated for primates[44] and rats.[34,43] In rats, Al-Chaer and colleagues[45] demonstrated that acute inflammation of the colon, produced by the topical application of mustard oil, resulted in increased responses of postsynaptic dorsal column and thalamic neurons excited by colorectal distension.

Supraspinal Terminations of Visceral Input

Standard anterograde and retrograde anatomic tracing methods and electrophysiologic studies have established widespread distribution of visceral input in the brain. The axons of second-order spinal neurons that receive visceral input have been shown to ascend the spinal cord to the brain, with sites of termination in the medulla, pons, mesencephalon, hypothalamus, and thalamus. Neurons excited by visceral stimuli likewise have been identified at these same sites with extensive characterizations of neurons located within the ventroposterolateral, mediodorsal, and submedius nuclei of the thalamus,[46] the locus coeruleus,[47] parabrachial nucleus, ventrolateral medulla, and numerous brain stem and limbic sites.[48-52] Higher-order neurons excited by visceral stimuli also have been demonstrated to be present in the somatosensory and ventrolateral orbital cerebral cortices.[53-55] A lack of visceral sensation has been noted in neurosurgical patients who have sustained damage to their frontal lobes.[56-58]

Functional imaging of humans during visceral stimulation has revealed some consistencies but is most notable for the multitude of sites that demonstrate increased regional blood flow. Rectal distension and urinary bladder distension both produce increased blood flow in select areas of the thalamus, hypothalamus, mesencephalon, pons, and medulla.[59] Cortical sites of processing include the anterior and midcingulate cortices, the frontal and parietal cortices, and in the cerebellum.[60] The most illustrative imaging study to date comparing visceral pain sensation with cutaneous pain sensation is that of Strigo and colleagues.[61] These investigators matched the intensity of pain sensation produced by esophageal distension with that produced by heating of the skin of the midchest region. Cutaneous and esophageal pain sensations were associated with similar activation of the secondary somatosensory and parietal cortices plus the thalamus, basal ganglia, and cerebellum. Cutaneous pain evoked a higher activation of the anterior insular cortex bilaterally than did esophageal pain and also selectively activated the ventrolateral prefrontal cortex. Esophageal pain lead to activation of the inferior primary somatosensory cortex bilaterally, the primary motor cortex bilaterally, and a more anterior locus of the anterior cingulate cortex than cutaneous pain. This all suggests some shared components of sensation from same-segmental structures but also a selective activation of some structures by different types of pain.

▶ CONCLUSION

It is abundantly clear that visceral pain differs from other pain in many ways. This is not to say that there are no similarities. Primary afferent cell bodies associated with visceral nociception reside within DRGs, and the initial processing of sensory information occurs at the level of the dorsal horn of the spinal cord or in the brain stem (e.g., vagal and trigeminal inputs). Many sites of higher processing in the brain are activated by both noxious visceral and noxious somatic stimuli. Where visceral pain differs from somatic pain is in the encoding properties of visceral primary afferent transducers and in their distribution to and within the CNS. The final consequence of these dissimilarities is a difference in localization and a difference in the magnitude of emotional and autonomic responses to visceral stimuli. For this reason, the treatment of visceral pain may need to differ from that of other pain. At present, clinical practice is to use the same therapeutics for pain of any type. With additional information, it may become possible to determine treatments that are selective for the type of pain.

REFERENCES

1. Salet GA, Samsom M, Roelofs JM, et al: Responses to gastric distension in functional dyspepsia. *Gut* 1998;42:823–829.
2. Ritchie J: Pain from distension of the pelvic colon by inflating a balloon in the irritable colon syndrome. *Gut* 1973;14:125–132.
3. Mertz H: Visceral hypersensitivity. *Aliment Pharmacol Ther* 2003;17:623–633.
4. Corazziari E, Shaffer EA, Hogan WJ, et al: Functional disorders of the biliary tract and pancreas. *Gut* 1999;45:II48–54.
5. Pontari MA, Hanno PM, Wein AJ: Logical and systematic approach to the evaluation and management of patients suspected of having interstitial cystitis. *Urology* 1997; 49:114–120.
6. Naliboff BD, Munakata J, Fullerton S, et al: Evidence for two distinct perceptual alterations in irritable bowel syndrome. *Gut* 1997;41:505–512.
7. Strigo IA, Bushnell MC, Boivin M, et al: Psychophysical analysis of visceral and cutaneous pain in human subjects. *Pain* 2002;97:235–246.
8. Ness TJ, Metcalf AM, Gebhart GF: A psychophysiological study in humans using phasic colonic distension as a noxious visceral stimulus. *Pain* 1990;43:377–386.
9. Ness TJ, Richter HE, Varner RE, et al: A psychophysical study of discomfort produced by repeated filling of the urinary bladder. *Pain* 1998;76:61–69.
10. Munakata J, Naliboff B, Harraf F, et al: Repetitive sigmoid stimulation induces rectal hyperalgesia in patients with irritable bowel syndrome. *Gastroenterology* 1997;112:55–63.
11. Ness TJ, Gebhart GF: Colorectal distension as a noxious visceral stimulus: Physiologic and pharmacologic characterization of pseudaffective reflexes in the rat. *Brain Res* 1988; 450:153–169.
12. Ness TJ, Gebhart GF: Inflammation enhances reflex and spinal neuron responses to noxious visceral stimulation in rats. *Am J Physiol* 2001;280:G649–G657.
13. Ness TJ, Lewis-Sides A, Castroman P: Characterization of pressor and visceromotor reflex responses to bladder distension in rats: Sources of variability and effect of analgesics. *J Urol* 2001;165:968–974.
14. Sherrington CS: *The Integrative Action of the Nervous System*. New Haven: Yale University Press, 1906.
15. Ozaki N, Bielefeldt K, Sengupta J, et al: Models of gastric hyperalgesia in the rat. *Am J Physiol* 2002;283:G666–G676.
16. Ness TJ, Gebhart GF: Methods in visceral pain research, in Kruger L (ed), *Methods in Pain Research*. Boca Raton, FL: CRC Press, 2001:93.

17. Giamberardino MA, Valente R, de Bigontina P, et al: Artificial ureteral calculosis in rats: behavioural characterization of visceral pain episodes and their relationship with referred lumbar muscle hyperalgesia. *Pain* 1995;61:459–469.
18. Coutinho SV, Plotsky PM, Sablad M, et al: Neonatal maternal separation alters stress-induced responses to viscerosomatic nociceptive stimuli in rat. *Am J Physiol Gastrointest Liver Physiol* 2002;282:G307–G316.
19. Gue M, Del Rio-Lacheze C, Eutamene H, et al: Stress-induced visceral hypersensitivity to rectal distension in rats: Role of CRF and mast cells. *Neurogastroenterol Motil* 1997; 9:271–279.
20. Bradesi S, Eutamene H, Garcia-Villar R, et al: Stress-induced visceral hypersensitivity in female rats is estrogen-dependent and involves tachykinin NK_1 receptors. *Pain* 2003; 102:227–234.
21. Qin C, Greenwood-Van Meerveld B, Foreman RD: Visceromotor and spinal neuronal responses to colorectal distension in rats with aldosterone onto the amygdala. *J Neurophysiol* 2003;90:2–11.
22. Qin C, Greenwood-Van Meerveld B, Myers DA, et al: Corticosterone acts directly at the amygdala to alter spinal neuronal activity in response to colorectal distension. *J Neurophysiol* 2003;89:1343–1352.
23. Qin C, Greenwood-Van Meerveld B, Foreman RD: Spinal neuronal responses to urinary bladder stimulation in rats with corticosterone and aldosterone onto the amygdala. *J Neurophysiol* 2003;90:2180–2189.
24. Raybould H: Visceral perception: Sensory transduction in visceral afferents and nutrients. *Gut* 2002;51:i11–i14.
25. Phillips RJ, Powley TL: Tension and stretch receptors in gastrointestinal smooth muscle: Re-evaluating vagal mechanoreceptor electrophysiology. *Brain Res Rev* 2000; 34:1–26.
26. Powley TL, Phillips RJ: Musings on the wanderer: What's new in our understanding of vago-vagal reflexes? I. Morphology and topography of vagal afferents innervating the GI tract. *Am J Physiol* 2002;283:G1217–G1225.
27. Lynn P, Olsson C, Zagorodnyuk V, et al: Rectal intraganglionic laminar endings are transduction sites of extrinsic mechanoreceptors in the guinea pig rectum. *Gastroenterology* 2003;125:786–794.
28. Zagorodnyuk V, Bao NC, Costa M, et al: Mechanotransduction by intraganglionic laminar endings of vagal tension receptors in the guinea pig oesophagus. *J Physiol (Lond)* 2003;553:575–587.
29. Sengupta JN, Gebhart GF: Mechanosensitive afferent fibers in the gastrointestinal and lower urinary tracts, in Gebhart GF (ed), *Visceral Pain: Progress in Pain Research and Management*, Vol 5. Seattle: IASP Press, 1995:75.
30. Ozaki N, Sengupta JN, Gebhart GF: Mechanosensitive properties of gastric vagal afferent fibers in the rat. *J Neurophysiol* 1999;82:2210–2220.
31. Randich A, Gebhart GF: Vagal afferent modulation of nociception. *Brain Res Rev* 1992;17:77–99.
32. Lamb K, Kang Y-M, Gebhart GF, et al: Gastric inflammation triggers hypersensitivity to acid in awake rats. *Gastroenterology* 2003;125:1410–1418.
33. Holzer P, Danzer M, Schicho R, et al: Regulation of vagal afferent sensitivity to gastric acid, in Holtmann G, Talley NJ (eds), *Gastrointestinal Inflammation and Disturbed Gut Function.* London: Kluwer, 2003:126.
34. Hirshberg RM, Al-Chaer ED, Lawand NB, et al: Is there a pathway in the posterior funiculus that signals visceral pain? *Pain* 1996;67:291–305.
35. Nauta HJW, Hewitt E, Westlund KN, et al: Surgical interruption of a midline dorsal column visceral pain pathway. *J Neurosurg* 1997;86:538–542.
36. Nauta HJ, Soukup VM, Fabian RH, et al: Punctate midline myelotomy for the relief of visceral cancer pain. *J Neurosurg* 2000;92:125–130.
37. Al-Chaer ED, Feng Y, Willis WD: A role for the dorsal column in nociceptive visceral input into the thalamus of primates. *J Neurophysiol* 1998;79:3143–3150.
38. Al-Chaer ED, Westlund KN, Willis WD: Nucleus gracilis: An integrator for visceral and somatic information. *J Neurophysiol* 1997;78:521–527.
39. Ness TJ: Evidence for ascending visceral nociceptive information in the dorsal midline and lateral spinal cord. *Pain* 2000;87:83–88.
40. Feng Y, Cui M, Al-Chaer ED, et al: Epigastric antinociception by cervical dorsal column lesions in rats. *Anesthesiology* 1998;89:411–420.
41. Houghton AK, Wang CC, Westlund KN: Do nociceptive signals from the pancreas travel in the dorsal column? *Pain* 2001;89:207–220.
42. Houghton AK, Hewitt, E, Westlund KN: Dorsal column lesion prevents mechanical hyperalgesia and allodynia in osteotomy model. *Pain* 1999;82:73–80.
43. Al-Chaer ED, Lawand NB, Westlund KN, et al: Pelvic visceral input into the nucleus gracilis is largely mediated by the postsynaptic dorsal column pathway. *J Neurophysiol* 1996; 76:2675–2690.
44. Al-Chaer ED, Feng Y, Willis WD: Comparative study of viscerosomatic input onto postsynaptic dorsal column and spinothalamic tract neurons in the primate. *J Neurophysiol* 1999;82:1876–1882.
45. Al-Chaer ED, Westlund KN, Willis WD: Potentiation of thalamic responses to colorectal distension by visceral inflammation. *Neuroreport* 1996;7:1635–1639.
46. Bruggemann J, Shi, T, Apkarian AV: Viscerosomatic interactions in the thalamic ventral posterolateral nucleus (VPL) of the squirrel monkey. *Brain Res* 1998;787:269–276.
47. Elam M, Thoren P, Svensson TH: Locus coeruleus neurons and sympathetic nerves: Activation by visceral afferents. *Brain Res* 1986;375:117–125.
48. Traub RJ, Silva E, Gebhart GF, et al: Noxious colorectal distension induced c-fors protein in limbic brain structures in the rat. *Neurosci Lett* 1996;215:165–168.
49. Almeida A, Lima D: Activation by cutaneous or visceral noxious stimulation of spinal neurons projecting in the medullary dorsal reticular nucleus in the rat: A c-fos study. *Eur J Neurosci* 1997;9:686–695.
50. Ness TJ, Follett, KA, Piper JG, et al: Characterization of neurons in the area of the medullary lateral reticular nucleus responsive to noxious visceral and cutaneous stimuli. *Brain Res* 1998;802:163–174.
51. Lanteri-Minet M, Bon K, de Pommery J, et al: Cyclophosphamide cystitis as a model of visceral pain in rats: Model elaboation and spinal structures involved as revealed by

the expression of c-fos and Krox-24 proteins. *Exp Brain Res* 1995;105:220–232.
52. Bon K, Lanteri-Minet M, Michiels JF, et al: Cyclophosphamide cystitis as a model of visceral pain in rats: A c-fos and Krox-24 study at telencephalic levels, with a note on pituitary adenylate cyclase activating polypeptide (PACAP). *Exp Brain Res* 1998;122:165–174.
53. Follett KA, Dirks B: Characterization of responses of primary somatosensory cerebral cortex neurons to noxious visceral stimulation in the rat. *Brain Res* 1994;656:27–32.
54. Follett KA, Dirks B: Responses of neurons in ventrolateral orbital cortex to noxious visceral stimulation in the rat. *Brain Res* 1995;669:157–162.
55. Snow PJ, Lumb BM, Cervero F: The representation of prolonged and intense, noxious somatic and visceral stimuli in the ventrolateral orbital cortex of the cat. *Pain* 1992;48:89–99.
56. Andrew J, Nathan PW: Lesions of the anterior frontal lobes and disturbances of micturition and defaecation. *Brain* 1964;87:233–262.
57. Andrew J, Nathan PW, Spanos NC: Disturbances of micturition and defaecation due to aneurysms of anterior communicating or anterior cerebral arteries. *J Neurosurg* 1966;24:1–10.
58. Nathan P: Lesions of the anterior frontal lobes and disturbances of micturition and defaecation. *Neurologia* 1963;5:9–17.
59. Blok BFM: Central pathways controlling micturition and urinary continence. *Urology* 2002;59:13–17.
60. Athwal BS, Berkley KJ, Hussain I, et al: Brain responses to changes in bladder volume and urge to void in healthy men. *Brain* 2001;124:369–377.
61. Strigo IA, Duncan GH, Boivin M, et al: Differentiation of visceral and cutaneous pain in the human brain. *J Neurophysiol* 2003;89:3294–3303.

CHAPTER 7

Pain and the Autonomic Nervous System

Eduardo E. Benarroch and Paola Sandroni

There is increasing evidence that the nociceptive and autonomic systems interact at multiple levels, including the periphery, dorsal horn, brain stem, and forebrain. Both nociceptive dorsal root and autonomic ganglion neurons derive from the neural crest under the trophic influence of nerve growth factor and may be involved together in hereditary and acquired neuropathies. The interaction between these two systems also occurs centrally at the levels of the spinal cord, brain stem, and forebrain.[1] These regions contain neurons that receive convergent nociceptive and viscerosensory information and initiate autonomic, antinociceptive, and behavioral responses to noxious as well as visceral stimuli. Understanding the complexity of these interactions is important for an integrated approach to chronic pain syndromes.

▶ INTEGRATION OF PAIN AND AUTONOMIC NERVOUS SYSTEM: FUNCTIONAL ANATOMY

Integration of Nociceptive and Viscerosensitive Inputs to the Cerebral Cortex and Brain Stem

Pain should be considered a visceral sensation that monitors the integrity of all tissues and elicits visceral motor and emotional responses.[2] There is convergence between nociceptive and viscerosensory pathways at the levels of the dorsal horn,[3] brain stem, basal forebrain, thalamus, and cerebral cortex. Neurons in laminae I, IV, and V of the dorsal horn and lamina X of the intermediate gray matter receive converging nociceptive and viscerosensory inputs.[3-6] Laminae I and V convey this integrated information via the spinothalamic tract (STT). These STT axons convey pain, as well as thermal, and visceral sensation to an adjacent territory in the ventromedial nucleus of the thalamus that projects to the dorsal posterior insular cortex.[6-8] The dorsal insula contains a continuous viscerotopic and somatotopic map with an ordered rostrocaudal representation of taste, general visceral, nociceptive, and thermoreceptive information. Another cortical region receiving nociceptive inputs is the anterior cingulate cortex via a subdivision of the mediodorsal nucleus of the thalamus.[9-11] Functional neuroimaging studies show that the same insular region activated by innocuous or painful thermal stimuli is also activated by visceral stimuli, such as the Valsalva maneuver.[12-18] Functional neuroimaging studies also show that the human anterior cingulate is activated by stimuli that produce the unpleasant sensation of burning pain, such as the thermal-grill illusion.[13]

Integrated nociceptive and visceral information is also conveyed via spinobulbar pathways to catecholaminergic and serotonergic groups of the medulla, the nucleus of the tratus solitarius (NTS), the parabrachial nucleus (PBN), and the periaqueductal gray matter (PAG). The NTS and PBN provide additional pathways that relay this integrated information to the thalamus, hypothalamus, and amygdala.[2,11,19-21]

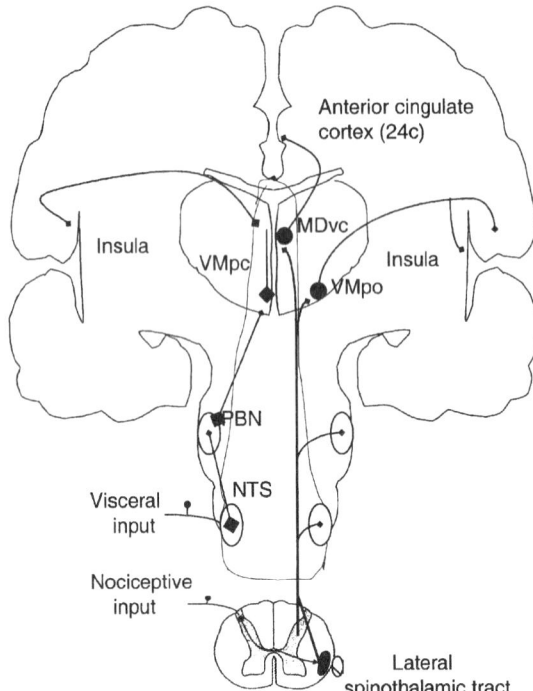

Figure 7-1. Schematic representation of the overlap of visceral, autonomic, and nociceptive pathways at the brain stem and thalamic and cortical levels. Input integration at each level allows for the complex pattern of reflex and modulatory responses elicited by painful and visceral sensations. NTS = nucleus tractus solitarius; PBN = parabrachial nucleus; VMPo = posterior ventromedial nucleus; mediodorsal thalamic nucleus (MDvc); and parvicellular ventromedial nucleus (VMPc).

Integrated Autonomic and Antinociceptive Responses to Nociceptive and Visceral Signals

All the central relay structures receiving nociceptive and visceroceptive information are reciprocally interconnected and are involved in control of autonomic function (Fig. 7-1). At the spinal level, lamina I neurons project, via interneurons, to preganglionic sympathetic neurons mediating segmental somatosympathetic and viscerosympathetic reflexes.[22–26] The ventrolateral medulla and caudal raphe nuclei, which receive input from lamina I, are critically involved in tonic maintenance of vasomotor tone and thermoregulatory mechanisms, respectively.[2,5,20,27–36] The NTS is the critical relay station for all medullary cardiovascular, respiratory, and gastrointestinal reflexes and contributes antinociceptive inputs to the dorsal horn.[21,37] For example, activation of baroreceptive NTS neurons by an increase in arterial pressure results in analgesia in experimental animals.[38,39]

The PBN, via its connections with the medulla, the hypothalamus, or the amygdala, participates in integrated homeostatic responses to nociceptive and visceral inputs.[20,36,40] The PAG, which receives nociceptive inputs from laminae I and V, contains separate columns of neurons that regulate different patterns of integrated cardiovascular and antinociceptive responses.[19,41] The lateral PAG, which receives cutaneous nociceptive inputs, initiates sympathoexcitatory "fight or flight" responses associated with opioid-independent (naloxone-insensitive) analgesia, whereas the ventrolateral PAG, which receives nociceptive inputs from muscle and visceral receptors, initiates sympathoinhibitory responses associated with opioid-dependent analgesia.[42–44] The insula and the anterior cingulate cortex are reciprocally interconnected and project to the autonomic nuclei of the hypothalamus and brain stem.[45–48] Stimulation of the anterior cingulate or the insular cortex in alert humans during surgery for epilepsy elicits a wide range of autonomic responses.[49]

▶ PAIN AND AUTONOMIC INTERACTIONS: PATHOPHYSIOLOGY AND PHARMACOLOGY

The interactions between the nociceptive and autonomic (particularly the sympathetic) systems at the level of the spinal cord and periphery are relevant for understanding the pathophysiology of neurogenic pain associated with autonomic manifestations and the rationale for pharmacologic treatment.

Spinal Cord Level: Autonomic Dysreflexia and Potential Mechanism of Syncope and Referred Pain

Due to the convergence of visceral and somatic afferents, the spinal cord is a major integration center where pain can affect the autonomic system, and vice versa.[2–5] Although supraspinal structures modulate autonomic and motor responses to stress induced by pain, some of the responses have a spinal reflex component. Like all spinal reflexes, these somatosympathetic and viscerosympathetic responses become exaggerated following lesions disconnecting the spinal cord from modulatory supraspinal influences.[42,50,51] Noxious stimuli to the skin trigger muscle vasoconstriction, sweat production, and variable skin vasomotor responses. The direct reflex causes vasodilatation, but generally, the more diffuse arousal in reflex to a painful stimulation causes vasoconstriction. Visceral noxious stimulation elicits much more powerful reflexes: skin and muscle vasoconstriction, which may be severe enough to cause hypertension, as well as excessive sweating and other manifestations of autonomic dysreflexia.

In addition, feedback protective reflexes are elicited by visceral pain, presumably arising from mechanoreceptors, and inhibit visceral motility and thus reduce the pain. One possible mechanism is stimulation of prevertebral sympathetic outflow to the gut, which is known to inhibit enteric peristaltic reflexes. A dysfunction in this reflex is thought to play a role in irritable bowel syndrome.[42,52] However, if the visceral stimulus is of high enough intensity, after the initial sympathetic surge there may be sudden withdrawal, leading to neurally-mediated syncope. Although the mechanism is undetermined, one possibility is via activation of the ventrolateral PAG, which, in response to visceral afferent input, inhibits sympathetic outflow, resulting in bradycardia and hypotension. An alternative possibility is vagally-mediated activation of the NTS, which initiates vasodepressor and bradycardic responses. These possibilities require experimental confirmation.

Cervical spinothalamic cells respond to activation of cardiothoracic vagal afferents.[3] These cervical cord neurons convey this information to supraspinal autonomic and antinociceptive regions and also participate in propriospinal pathways that modulate visceral inputs.[53–56] Propriospinal connections may allow interactions between different types of visceral inputs.[54,57–59] In animals, bladder distension can inhibit cardiothoracic STT neurons,[60] and conversely, cardiopulmonary afferents inhibit lumbosacral STT neuron activity.[59,61] These interactions require an intact cervical cord. The convergence of craniovascular, somatic, and vagal afferents from the heart and gut at the cervical cord level may contribute to referred head pain, e.g., in conditions of gastrointestinal distress.[62]

Dorsal Root Ganglia Level: Plastic Changes as a Potential Mechanism for Pain and Sympathetic Interactions Following Injury

Normally, the dorsal root ganglia (DRGs) have only sympathetic fibers innervating vessels. Following injury, sympathetic fibers sprout into the ganglion and form baskets predominantly around large (Aβ) more than small neuronal cell bodies.[63–65] Different types of injury (proximal versus distal) induce different types of basket formation. With distal lesions, only large neurons are targets for the basket formation, and the sprouts originate from perivascular sympathetic fibers.[64,65] In proximal lesions, they mainly come from spinal nerves, and synapses are formed (this has not been demonstrated with distal lesions).[63,65,66] Nerve growth factors (NGFs) stimulate basket formation, and anti-NGFs inhibit it.[66–71] P75 is upregulated following injury in satellite cells in DRGs.[71–75] NGF may be a primary trigger for sympathetic fiber sprouting; this also may have a feed-forward mechanism because beta-receptor activation enhances NGF production. Various cytokines (particularly leukemia inhibitory factor (LIF) and interleukin 6),[76–80] produced by immunocompetent cells and Schwann cells following nerve injury, also can induce sympathetic sprouting. The functional role of these baskets has not been established unequivocally. There is evidence that the sympathetic terminals have close proximity to the cell bodies, but whether such contacts represent functional synapses is still unclear.[63,64,67,72,73,81] Certainly, sympathetic fibers appear to be in a key position to modulate DRG activity, both spontaneous activity and that evoked by incoming stimuli. Such an amplification process will reverberate on the dorsal horn, augmenting the barrage of painful stimuli and leading to the windup phenomenon and central sensitization. Aβ-fiber excitation amplified by sympathetic fiber activation may account for mechanical allodynia, whereas the same amplification on C fibers will sustain the windup phenomenon. Overall, data indicate that these baskets may have a functional role in amplifying sensory inflow. However, there are no data demonstrating a direct correlation between the degree of sympathetic sprouting in DRGs and pain level.

Peripheral Level: Potential for Pain-Sympathetic Cross-Talk Following Injury

Activation of neurotransmitter release from sympathetic nerves, upregulation of expression of adrenoreceptors in nociceptive afferents, and interactions mediated by circulating and local chemical factors following injury all may contribute to sensitization of nociceptors, as well as local vasomotor, sudomotor, and presumably trophic changes[82–90] (Fig. 7-2). After axotomy, regenerating fibers may form "cross-talk" between somatic and sympathetic terminals.[91–93] Ephaptic transmission also may occur along nerve axons, and sympathetic activation will result in excitation of primary afferent somatic and nociceptive fibers.

After partial nerve injury, injured and uninjured C and Aβ fibers and axons abnormally upregulate expression of α-adrenoreceptors. This renders nociceptive axons sensitive to norepinephrine (NE) released from postganglionic sympathetic terminals, as well as to circulating catecholamines. It has been shown in experimental partial nerve injury models that there is increased responsiveness of C fibers to NE and α-adrenergic stimulation. Besides their higher density, these receptors are also more sensitive and more likely to lower nociceptor afferent threshold by altering K currents. Thus, sensory neurons become not only responsive but also hypersensitive to sympathetic activation. In these conditions, vasomotor and sudomotor phenomena accompanying the pain sensation may be prominent. Some of these vasomotor effects may reflect antidromic release of substance P (SP) or calcitonin gene–related peptide (CGRP) from the sensitized nociceptors. In humans, there is increased density of $α_1$-receptor density in hyperalgesic skin,[86] and phentolamine

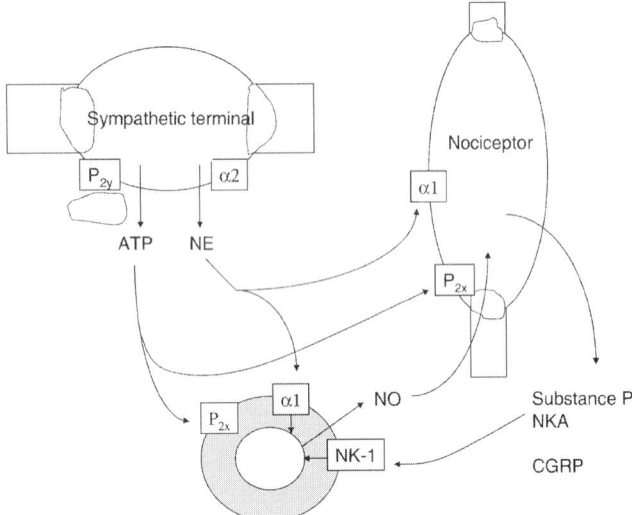

Figure 7-2. Simplified diagram depicting the basic interaction of the sympathetic terminal with primary afferent nociceptor and blood vessel. The release from the sympathetic terminal of norepinephrine (NE) and ATP activates and sensitizes the nociceptor directly and, furthermore, affects local blood vessel permeability and contractility status and in turn its release of nitric oxide (NO). The primary nociceptor in turn releases substance P, neurokinin A (NKA), and calcitonin gene–related peptide (CGRP), which can produce changes in vasomotor tone and capillary permeability. Both the sympathetic and nociceptive terminals trigger, and are affected by, mediators of inflammation, including prostaglandins, cytokines, and growth factors.

infusion relieved pain in subject patients with nerve injury but not in controls.

In addition to NE, sympathetic terminals release other mediators that may contribute to nociceptor sensitization. For example, sympathetic vasomotor fibers release adenosine triphosphate (ATP), which, via P2X purinoreceptors, increases Ca^{2+} influx and depolarizes nociceptive afferents. Sympathetic terminals also release prostaglandins, which also sensitize nociceptors and elicit inflammatory and vasomotor changes. Sympathetic sudomotor fibers release acetylcholine, which not only mediates sudomotor responses but also may act via nicotinic receptors to sensitize nociceptors. Cholinergic activation also elicits release of nitric oxide (NO), which contributes to vasodilatation and magnifies the sensitization of nociceptors via interactions with inflammatory mediators.

Sympathetic terminals also may alter their receptor expression and function following injury. Chemali and colleagues[84] recently demonstrated that in patients with complex regional pain syndrome (CRPS), adrenergic stimulation elicited a much larger sweat response in the affected than in the unaffected limb or limbs of control subjects. This suggests that regenerating sweat glands express α receptors, in addition to M3 muscarinic receptors. This situation mimics that of sweat glands during development. In these patients, resolution of CRPS symptoms was associated with return to the normal lack of adrenergically-induced sudomotor responses.

The sympathetic system participates in complex interactions with the inflammatory responses that sensitize the nociceptors. Tissue injury elicits release of several proinflammatory mediators from the damaged tissues, inflammatory cells, and abnormally permeable capillaries. These include bradykinin, prostaglandins, serotonin, histamine, serotonin, cytokines, and NGF. All these mediators may facilitate nociceptor sensitization by several mechanisms, including changes in expression of Na^+ channels or potentiation of responses mediated by vanilloid receptors. Sympathetic activity may exert an anti-inflammatory effect. For example, sympathetically-induced vasoconstriction may prevent excessive skin vasodilatation, reducing at venular plasma extravasation of inflammatory pronociceptive mediators. Catecholamines have an overall anti-inflammatory effect through β-receptor stimulation that reduces production of proinflammatory cytokines (such as tumor necrosis factor α) and increases production of anti-inflammatory cytokines.[94] On the other hand, sympathetic activation may elicit release of prostaglandins, NO, and other inflammatory and nociceptor sensitizing substances, as discussed earlier. Furthermore, sympathetically-mediated vasoconstriction may induce ischemia and prevent the removal of irritating substances.[42,92]

Visceral Pain: Nociceptive and Antinociceptive Effects of Vagal Afferents

Animal studies show that a subpopulation of vagal (myelinated) afferents may be involved in nociception. Pain mediated by vagal afferents generates protective behavior (anorexia, immobility, etc.) for the gut, but aberrant responses may contribute to persistence of abdominal pain in some chronic conditions such as irritable bowel syndrome.[3,92] Sympathectomies (surgical or due to diabetic neuropathy, for instance) abolish or markedly reduce angina pain.[3,95] Referred neck/jaw pain, however, can still be present, conveyed by vagal afferents.[3,96]

In general, activation of vagal afferents exerts antinociceptive effects, inhibiting the STT via propriospinal mechanisms (via the cervical cord) or through inputs to the NTS or other central antinociceptive nuclei. The type of vagal afferent may determine whether vagal activation elicits nociceptive or antinociceptive responses. Stimulation of type A (myelinated) fibers may elicit pain, whereas activation of type C vagal afferents may elicit antinociceptive response. Ness and colleagues[97] demonstrated that low-intensity vagal nerve stimulation performed for seizure control lowered the thermal pain threshold in humans, whereas higher-intensity stimuli were antinociceptive. These effects were independent of any hemodynamic changes.

▶ CONCLUSION

Chronic pain is a manifestation of plasticity within the central and peripheral nociceptive pathways. The evidence indicates that the interactions between the nociceptive and the autonomic systems are complex and involve a variety of peripheral, central antinociceptive, autonomic, emotional, and behavioral control mechanisms. An integrated approach to the care of patients with chronic pain syndromes requires an awareness of these multilevel interactions, which in pathologic conditions may have the opposite effect than in normal ones and which could provide the basis for innovative pharmacologic, physical, and behavioral therapy approaches.

REFERENCES

1. Benarroch EE: Pain-autonomic interactions: A selective review. *Clin Autonom Res* 2001;11:343–349.
2. Saper CB: Pain as a visceral sensation, in Mayer EA, Saper CB (eds), *Progress in Brain Research*. Amsterdam: Elsevier, 2000:237–243.
3. Foreman RD: Integration of viscerosomatic sensory input at the spinal level, in Mayer EA, Saper CB (eds), *Progress in Brain Research*. Amsterdam: Elsevier, 2000:209–221.
4. Cervero F, Foreman RD: Sensory innervation of the viscera, in Loewy AD, Spyer KM (eds), *Central Regulation of Autonomic Functions*. New York: Oxford University Press, 1990:104–125.
5. Craig AD: An ascending general homeostatic afferent pathway originating in lamina I, in Holstege G, Bandler R, Saper CB (eds), *Progress in Brain Research*. Amsterdam: Elsevier, 1996:225–242.
6. Craig AD: Distribution of brain stem projections from spinal lamina I neurons in the cat and the monkey. *J Comp Neurol* 1995;361:225–248.
7. Cechetto DF: Central representation of visceral function. *Fed Proc* 1987;46:17–23.
8. King AB, Menon RS, Hachinski V, et al: Human forebrain activation by visceral stimuli. *J Comp Neurol* 1999;413:572–582.
9. Craig AD, Dostrovsky JO: Processing of nociceptive information at supraspinal levels, in Yaksh TL (ed), *Anesthesia: Biologic Foundations*. Philadelphia: Lippincott-Raven, 1997:625–642.
10. Craig AD, Zhang ET: Anterior cingulate projection from MDvc (a lamina I spinothalamic target in the medial thalamus of the monkey). *Soc Neurosci Abstr* 1996;22:111.
11. Vogt BA, Sikes RW: The medial pain system, cingulate cortex, and parallel processing of nociceptive information, in Mayer EA, Saper CB (eds), *Progress in Brain Research*. Amsterdam: Elsevier, 2000:223–235.
12. Casey KL, Minoshima S, Berger KL, et al: Positron emission tomographic analysis of cerebral structures activated specifically by repetitive noxious heat stimuli. *J Neurophysiol* 1994;71:802–807.
13. Craig AD, Reiman EM, Evans A, et al: Functional imaging of an illusion of pain. *Nature* 1996;384:258–260.
14. Davis KD, Taylor SJ, Crawley AP, et al: Functional MRI of pain- and attention-related activations in the human cingulate cortex. *J Neurophysiol* 1997;77:3370–3380.
15. Davis KD, Wood ML, Crawley AP, et al: fMRI of human somatosensory and cingulate cortex during painful electrical nerve stimulation. *Neuroreport* 1995;7:321–325.
16. Lenz FA, Rios M, Zirh A, et al: Painful stimuli evoke potentials recorded over the human anterior cingulate gyrus. *J Neurophysiol* 1998;79:2231–2234.
17. Talbot JD, Marrett S, Evans AC, et al: Multiple representations of pain in human cerebral cortex. *Science* 1991;251:1355–1358.
18. Vogt BA, Derbyshire S, Jones AK: Pain processing in four regions of human cingulate cortex localized with co-registered PET and MR imaging. *Eur J Neurosci* 1996;8:1461–1473.
19. Bandler R, Shipley MT: Columnar organization in the midbrain periaqueductal gray: Modules for emotional expression? *Trends Neurosci* 1994;17:379–389.
20. Bernard JF, Bester H, Besson JM: Involvement of the spinoparabrachioamygdaloid and -hypothalamic pathways in the autonomic and affective emotional aspects of pain. *Prog Brain Res* 1996;107:243–255.
21. Loewy AD: Central autonomic pathways, in Loewy AD, Spyer KM (eds), *Central Regulation of Autonomic Functions*. New York: Oxford University Press, 1990:88–103.
22. Foreman RD, Blair RW: Central organization of sympathetic cardiovascular response to pain. *Annu Rev Physiol* 1988;50:607–622.
23. Janig W: Spinal cord reflex organization of sympathetic systems. *Prog Brain Res* 1996;107:43–77.
24. Janig W, McLachlan EM: Characteristics of function-specific pathways in the sympathetic nervous system. *Trends Neurosci* 1992;15:475–481.
25. Randich A: Interactions between cardiovascular and pain regulatory systems, in Kunos G, Ciriello J (eds), *Central Neural Mechanisms in Cardiovascular Regulation*. Boston: Birkhauser, 1992:297–320.
26. Sato A, Schmidt RF: Somatosympathetic reflexes: Afferent fibers, central pathways, discharge characteristics. *Physiol Rev* 1973;53:916–947.
27. Dampney RA: Functional organization of central pathways regulating the cardiovascular system. *Physiol Rev* 1994;74:323–364.
28. Day TA, Sibbald JR: A1 cell group mediates solitary nucleus excitation of supraoptic vasopressin cells. *Am J Physiol* 1989;257:R1020–R1026.
29. Day TA, Sibbald JR: Noxious somatic stimuli excite neurosecretory vasopressin cells via A1 cell group. *Am J Physiol* 1990;258:R1516–R1520.
30. Feldman JL, Ellenberger HH: Central coordination of respiratory and cardiovascular control in mammals. *Annu Rev Physiol* 1988;50:593–606.
31. Fields HL, Basbaum AI: Central nervous system mechanisms of pain modulation, in Wall PD, Melzack R (eds), *Textbook of Pain*. New York: Churchill Livingston, 1994:243–257.

32. Guyenet PG: Role of the ventral medulla oblongata in blood pressure regulation, in Loewy AD, Spyer KM (eds), *Central Regulation of Autonomic Functions.* New York: Oxford University Press, 1990:145–167.
33. Guyenet PG, Koshiya N, Huangfu D, et al: Role of medulla oblongata in generation of sympathetic and vagal outflows. *Prog Brain Res* 1996;107:127–144.
34. Morrison SF, Reis DJ: Reticulospinal vasomotor neurons in the RVL mediate the somatosympathetic reflex. *Am J Physiol* 1989;256:R1084–R1097.
35. Spyer KM: Annual Review Prize Lecture: Central nervous mechanisms contributing to cardiovascular control. *J Physiol* 1994;474:1–19.
36. Stark AM, Sawyer WB, Hughes JH, et al: A general pattern of CNS innervation of the sympathetic outflow demonstrated by transneuronal pseudorabies viral infections. *Brain Res* 1989;491:156–162.
37. Feldman JL: Neurophysiology of breathing in mammals, in Mountcastle UB, Bloom F, Geiger SR (eds), *Handbook of Physiology.* Bethesda, MD: American Physiological Society, 1986:463–524.
38. Randich A, Maixner W: The role of sinoaortic and cardiopulmonary baroreceptor reflex arcs in nociception and stress-induced analgesia. *Ann NY Acad Sci* 1986;467:385–401.
39. Rosa C, Vignocchi G, Panattoni E, et al: Relationship between increased blood pressure and hypoalgesia: Additional evidence for the existence of an abnormality of pain perception in arterial hypertension in humans. *J Hum Hypertens* 1994;8:119–126.
40. Cechetto DF, Saper CB: Neurochemical organization of the hypothalamic projection to the spinal cord in the rat. *J Comp Neurol* 1988;272:579–604.
41. Wang WH, Lovick TA: Role of medullary raphe nuclei in the inhibition of rostral ventrolateral medullary neurones evoked from the ventrolateral periaqueductal grey matter in anaesthetised rats. *J Physiol* 1992;446:521.
42. Janig W: The sympathetic nervous system in pain. *Eur J Anaesthesiol Suppl* 1995;10:53–60.
43. Lovick TA: Integrated activity of cardiovascular and pain regulatory systems: Role in adaptive behavioural responses. *Prog Neurobiol* 1993;40:631–644.
44. Van Bockstaele EJ, Aston Jones G, Pieribone VA, et al: Subregions of the periaqueductal gray topographically innervate the rostral ventral medulla in the rat. *J Comp Neurol* 1991;309:305–327.
45. Cechetto DF, Saper CB: Role of the cerebral cortex in autonomic function, in Loewy AD, Spyer KM (eds), *Central Regulation of Autonomic Functions.* New York: Oxford University Press, 1990:208–223.
46. Loewy AD: Forebrain nuclei involved in autonomic control (review). *Prog Brain Res* 1991;87:253–268.
47. Neafsey EJ: Prefrontal cortical control of the autonomic nervous system: Anatomical and physiological observations. *Prog Brain Res* 1990;85:147–165; discussion 165–166.
48. Verberne AJ, Owens NC: Cortical modulation of the cardiovascular system. *Prog Neurobiol* 1998;54:149–168.
49. Oppenheimer SM, Gelb A, Girvin JP, et al: Cardiovascular effects of human insular cortex stimulation. *Neurology* 1992;42:1727–1732.
50. Mathias CJ, Frankel HL: Autonomic disturbances in spinal cord lesions, in Bannister R, Mathias CJ (eds), *Autonomic Failure.* New York: Oxford University Press, 1992:839–881.
51. Wallin BG, Stjernberg L: Sympathetic activity in man after spinal cord injury: Outflow to skin below the lesion. *Brain* 1984;107:183–198.
52. Mayer EA, Raybould HE: *Pain Research and Clinical Management.* Amsterdam: Elsevier Science, 1993.
53. Basbaum AI, Fields HL: Endogenous pain control systems: Brain stem spinal pathways and endorphin circuitry. *Annu Rev Neurosci* 1984;7:309–338.
54. Gerhart KD, Yezierski RP, Giesler GJJ, et al: Inhibitory receptive fields of primate spinothalamic tract cells. *J Neurophysiol* 1981;46:1309–1325.
55. Jones SL: Descending control of nociception, in Light AR (ed), *The Initial Processing of Pain and Its Descending Control: Spinal and Trigeminal Systems.* New York: Karger, 1992:203–295.
56. Willis WD: Anatomy and physiology of descending control of nociceptive responses of dorsal horn neurons: Comprehensive review, in Fields HL, Besson JM (eds), *Pain Modulation: Progress in Brain Research.* Amsterdam: Elsevier, 1988:1–29.
57. Dickenson AH, Le Bars D: Diffuse noxious inhibitory controls (DNIC) involve trigeminothalamic and spinothalamic neurons in the rat. *Exp Brain Res* 1983;49:174–180.
58. Foreman RD, Hobbs SF, Oh UT: Differential modulation of thoracic and lumbar spinothalamic tract cell activity during stimulation of cardiopulmonary sympathetic afferent fibers in the primate. A new concept for visceral pain, in Dubner R, Gebhart GF, and Bond MR (eds.): *Proceedings of the 5th World Congress on Pain.* New York: Elsevier, pp. 227–231, 1988.
59. Hobbs SF, Oh UT, Chandler MJ, et al.: Evidence that C1 and C2 propriospinal neurons mediate the inhibitory effects of viscerosomatic spinal afferent input on primate spinothalamic tract neurons. *J Neurophysiol* 1992;67:852–860.
60. Brennan TJ, Oh UT, Hobbs SF, et al: Urinary bladder and hindlimb afferent input inhibits activity of primate T2–T5 spinothalamic tract neurons. *J Neurophysiol* 1989;61:573–588.
61. Zhang J, Chandler MJ, Foreman RD: Thoracic visceral inputs use upper cervical segments to inhibit lumbar spinal neurons in rats. *Brain Res* 1997;709:337–342.
62. Mavromichalis I, Zaramboukas T, Giala MM: Migraine of gastrointestinal origin. *Eur J Pediatr* 1995;154:406–410.
63. Chung K, Lee BH, Yoon YW, et al: Sympathetic sprouting in the dorsal root ganglia of the injured peripheral nerve in a rat neuropathic pain model. *J Comp Neurol* 1996;376:241–252.
64. McLachlan EM, Janig W, Devor M, et al: Peripheral nerve injury triggers noradrenergic sprouting within dorsal root ganglia. *Nature* 1993;363:543–546.
65. Ramer MS, Bisby MA: Differences in sympathetic innervation of mouse DRG following proximal or distal nerve lesions. *Exp Neurol* 1998;152:197–207.
66. Ramer MS, Bisby MA: Adrenergic innervation of rat sensory ganglia following proximal or distal painful sciatic neuropathy: Distinct mechanisms revealed by anti-NGF treatment. *Eur J Neurosci* 1999;11:837–846.

67. Davis BM, Albers KM, Seroogy KB, et al: Overexpression of nerve growth factors in transgenic mice induces novel sympathetic projections to primary sensory neurons. *J Comp Neurol* 1994;349:464–474.
68. Davis BM, Goodness TP, Soria A, et al: Overexpression of NGF in skin causes formation of novel sympathetic projections to trkA-positive sensory neurons. *Neuroreport* 1998;9:1103–1107.
69. Jones MG, Munson JB, Thompson SW: A role for nerve growth factor in sympathetic sprouting in rat dorsal root ganglia. *Pain* 1999;79:21–29.
70. Walsh GS, Kawaja MD: Sympathetic axons surround nerve growth factor–immunoreactive trigeminal neurons: Observations in mice overexpressing nerve growth factor. *J Neurobiol* 1998;34:347–360.
71. Zhou XF, Deng YS, Chie E, et al: Satellite cell–derived nerve growth factor and neurotrophin-3 are involved in noradrenergic sprouting in the dorsal root ganglia following peripheral nerve injury in the rat. *Eur J Neurosci* 1999;11:1711–1722.
72. Devor M: Abnormal excitability in injured axons, in Waxman SG, Kocsis JD, Stys PK (eds), *The Axon*. New York: Oxford University Press, 1995:530–552.
73. Ramer MS, French GD, Bisby MA: Wallerian degeneration is required for both neuropathic pain and sympathetic sprouting into the DRG. *Pain* 1997;72:71–78.
74. Walsh GS, Krol KM, Kawaja MD: Absence of the p75 neurotrophin receptor alters the pattern of sympathosensory sprouting in the trigeminal ganglia of mice overexpressing nerve growth factor. *J Neurosci* 1999;19:258–273.
75. Zhou XF, Rush RA, McLachlan EM: Differential expression of the p75 nerve growth factor receptor in glia and neurons of the rat dorsal root ganglia after peripheral nerve transection. *J Neurosci* 1996;16:2901–2911.
76. Hirota H, Kiyama H, Kishimoto T, et al: Accelerated nerve regeneration in mice by upregulated expression of interleukin (IL) 6 and IL-6 receptor after trauma. *J Exp Med* 1996;183:2627–2634.
77. Kurek JB, Austin L, Cheema SS, et al: Up-regulation of leukaemia inhibitory factor and interleukin-6 in transected sciatic nerve and muscle following denervation. *Neuromusc Disord* 1996;6:105–114.
78. Murphy PG, Grondin J, Altares M, et al: Induction of interleukin-6 in axotomized sensory neurons. *J Neurosci* 1995;15:5130–5138.
79. Satoh T, Nakamura S, Taga T, et al: Induction of neuronal differentiation in PC12 cells by B-cell stimulatory factor 2/interleukin 6. *Mol Cell Biol* 1988;8:3546–3549.
80. Thompson SW, Majithia AA: Leukemia inhibitory factor induces sympathetic sprouting in intact dorsal root ganglia in the adult rat in vivo. *J Physiol* 1998;506:809–816.
81. Devor M, Janig W, Michaelis M: Modulation of activity in dorsal root ganglion neurons by sympathetic activation in nerve-injured rats. *J Neurophysiol* 1994;71:38–47.
82. Ali Z, Ringkamp M, Hartke TV, et al: Uninjured C-fiber nociceptors develop spontaneous activity and alpha-adrenergic sensitivity following L6 spinal nerve ligation in monkey. *J Neurophysiol* 1999;81:455–466.
83. Arnold JM, Teasell RW, MacLeod AP, et al: Increased venous alpha-adrenoceptor responsiveness in patients with reflex sympathetic dystrophy. *Ann Intern Med* 1993;118:619–621.
84. Chemali KR, Gorodeski R, Chelimsky TC: Alpha-adrenergic supersensitivity of the sudomotor nerve in complex regional pain syndrome. *Ann Neurol* 2001;49:453–459.
85. Drummond PD, Finch PM, Smythe GA: Reflex sympathetic dystrophy: The significance of differing plasma catecholamine concentrations in affected and unaffected limbs. *Brain* 1991;114:2025–2036.
86. Drummond PD, Skipworth S, Finch PM: Alpha 1-adrenoceptors in normal and hyperalgesic human skin. *Clin Sci* 1996;91:73–77.
87. Sato J, Perl ER: Adrenergic excitation of cutaneous pain receptors induced by peripheral nerve injury. *Science* 1991;251:1608–1610.
88. Sato J, Suzuki S, Iseki T, et al: Adrenergic excitation of cutaneous nociceptors in chronically inflamed rats. *Neurosci Lett* 1993;164:225–228.
89. Sato K: Normal and abnormal eccrine sweat gland function, in Low PA (ed), *Clinical Autonomic Disorders: Evaluation and Management*. Philadelphia: Lippincott-Raven, 1997:97–108.
90. Scadding JW: Development of ongoing activity, mechanosensitivity, and adrenaline sensitivity in severed peripheral nerve axons. *Exp Neurol* 1981;73:345–364.
91. Chung K, Kim HJ, Na HS, et al: Abnormalities of sympathetic innervation in the area of an injured peripheral nerve in a rat model of neuropathic pain. *Neuroscience Lett* 1993;162:85–88.
92. Janig W, Levine JD, Michaelis M: Interactions of sympathetic and primary afferent neurons following nerve injury and tissue trauma. *Prog Brain Res* 1996;113:161–184.
93. Michaelis M, Devor M, Janig W: Sympathetic modulation of activity in rat dorsal root ganglion neurons changes over time following peripheral nerve injury. *J Neurophysiol* 1996;76:753–763.
94. Elenkov IJ, Wilder RL, Chrousos GP, et al: The sympathetic nerve—an integrative interface between two supersystems: The brain and the immune system. *Pharmacol Rev* 2000;52:595–638.
95. Lindgren I, Olivercrona H: The surgical treatment of angina pectoris. *J Neurosurg* 1947;4:19–39.
96. Janig W, Khasar SG, Levine JD, et al: The role of vagal visceral afferents in the control of nociception, in Mayer EA, Saper CB (eds), *Progress in Brain Research*. Amsterdam: Elsevier, 2000:273–287.
97. Ness TJ, Fillingim RB, Randich A, et al: Low-intensity vagal nerve stimulation lowers human thermal pain thresholds. *Pain* 2000;86:81–85.

CHAPTER 8

Mechanisms of Primary Headaches

Sheena K. Aurora

Headache is one of the most common presenting symptoms to a physician's office. The majority of headaches are in the category known as primary headaches *where there are no structural disturbances.* Secondary headaches *are uncommon and usually occur in less than 10% of patients. The mechanisms of secondary headaches are usually due to the underlying pathology. These are usually quite evident on neuroimaging or laboratory testing. This chapter will focus mainly on mechanisms of primary headache, i.e., migraine and cluster.*

There have been remarkable strides in the last decade in unraveling the mystery of primary headache disorders like migraine and cluster. The vascular theory has been superseded by the neurovascular phenomena that seem to be the permissive triggering factors in migraine and cluster headache. This change from the vascular theory has come about through new imaging modalities such as PET (positron imaging tomography) and fMRI (functional magnetic resonance imaging). Prior to these imaging techniques, it was impossible to study the primary headache disorders since these had no structural organic basis.

▶ MIGRAINE

The exact pathogenesis of migraine remains to be determined. The pendulum for concepts of migraine pathophysiology swung from primary vascular to primary neural mechanisms. Harold G. Wolff, a pioneer of the vascular theory of migraine, proposed that the neurological symptoms of the migraine aura were caused by cerebral vasoconstriction, and the headache by vasodilatation.[1] Lashley's[2] experience of his own visual aura led him to the concept of the spreading cortical depression (SCD) of Leao being the primary cause, thus promulgating the neural theory of migraine.[3] Newer imaging techniques have made it possible to study the very early events of migraine. There seems to be an increasing body of evidence for an inherited disorder that occurs in susceptible individuals, i.e., a phenomenon of central neuronal hyperexcitability that represents the pivotal disturbance predisposing to migraine.[4] Most recently, abnormality of calcium channels has been introduced as a potential mechanism of interictal neuronal excitability.[5] Mutant voltage-gated P/Q type calcium channel genes likely influence presynaptic neurotransmitter release, possibly of excitatory amino acid systems. It could, therefore, be hypothesized that genetic abnormalities result in a lowered threshold of response to trigger factors, since migraine is an episodic disorder involving head pain and cortical phenomena without structural abnormalities. Investigations aimed at studying the function of the brain may provide insight into migraine pathophysiology. The mechanisms of aura and head pain and then the interictal disturbances that lead to a propensity to developing migraine are discussed below.

Mechanisms of Aura

The unpredictable and elusive nature of migraine has prevented many investigators from systematically studying migraine aura. Recent studies by Cao et al, wherein migraine was reliably visually triggered in 50% of subjects, enabled the immediate early events of the migraine attack to be measured for the first time.[6] A red and green checkerboard was used for visual stimulation since migraineurs are known to be sensitive to linear stimuli. Using the recently developed fMRI based on the blood oxygen level-dependent (fMRI-BOLD) technique, the authors were able to measure, with millimeter resolution, and by second-to-second, the activation of the occipital cortex from visual stimulation in subjects with migraine. None of 6 normal controls developed a headache and displayed normal patterns of BOLD signals on visual activation. Six patients with migraine with aura (MwA) and 2 patients with migraine without aura (MwoA) experienced visually triggered headache; 2 patients also had accompanying visual change. Headache was preceded by the suppression of the initial activation that slowly propagated into contiguous occipital cortex at a rate ranging from 3 to 6 mm/min. This neuronal suppression was accompanied by an increase in baseline contrast intensity, indicative of cerebral vasodilatation and tissue hyperoxygenation. The baseline contrast increase appeared to be similar to that witnessed in experimental SCD.[7] In this study, patients were selected based on a history of visually triggered headache, so that generalizing these findings to all migraine patients must be done with caution. Nevertheless, previously hypothesized mechanisms of SCD in migraine were clarified by this study, and the previously controversial findings of ischemia accompanying migraine aura were not supported. Recently, a spontaneous migraine with aura has been described by Hadjiakani et al[8] with similar changes on fMRI-BOLD as described by Cao et al.

In a different study using PWI, another novel functional neuroimaging technique particularly suited to study short-lived events such as migraine aura, 19 patients were studied during spontaneous migraine.[9] Twenty-eight attacks were studied because some patients were imaged more than once. There was relative reduction of cerebral blood flow in the occipital cortex contralateral to the visual defect during MwA, but observed only in the occipital cortex and not in other brain regions. One subject with attacks of both MwA and MwoA demonstrated these phenomena only during MwA. No significant changes in blood flow were observed in MwoA. The hemodynamic changes were demonstrated only on PWI and not on diffusion-weighted imaging (DWI); DWI is sensitive to ischemia and thus further supports MwA not being an ischemic event.

These imaging studies, albeit favoring the neural basis of migraine, are not able to demonstrate SCD as the putative mechanism of migraine aura. To date, SCD has been recorded successfully in animal models only.[10] In animals, the SCD's band of hyperexcited neurons travels into sulci or fissures eliciting a MEG [magnetoencephalogram] signal. Using the seven channels MEG, Barkley et al reported DC shifts in spontaneous migraine.[11] A further study of a larger number of patients has not been possible because of the unpredictable nature of migraine and time of capture of these spontaneous events. Using the visual trigger modeled by Cao et al, Bowyer et al have now been able to detect DC shifts when headache or aura was precipitated.[12] These studies were performed using the whole head MEG, which permits precise localization of signals. In this study, headache was triggered in 5 of 8 migraine patients and none of 6 controls. DC-MEG shifts were observed in migraine subjects during visually triggered aura and in a patient studied during the first few minutes of spontaneous aura. No DC-MEG shifts were seen in control subjects. This is additional evidence supporting the primary neural basis migraine and confirms MEG-recorded DC shifts typical of those found during SCD, reported previously in migraine attacks.

Mechanism of Migraine Pain

The brainstem, and specifically the trigeminovascular system, has been implicated to play a large role during a migraine attack, from recent experimental and clinical data.[13] It is hypothesized that a sterile inflammatory response occurs due to the release of neuropeptides, i.e., calcitonin gene–related peptide (CGRP), neurokinin A, and substance P.[14] The development of novel antimigraine drugs for the treatment of migraine has been based predominantly on these animal models. This mechanism is further strengthened by the discovery of binding sites for the 5 HT 1B/1D agonists on brainstem structures.[14–16] The first human study to show activation in the brainstem used positron emission tomography (PET) performed in subjects during spontaneous migraine. Because PET lacks sufficient resolution for exact anatomical localization, the activation was hypothesized to be in the regions of dorsal raphe nuclei (DRN), periaqueductal gray (PAG), and locus ceruleus (LC).[17] Recently an isolated case report found red nucleus (RN) and substantia nigra (SN) to be activated in a spontaneous migraine attack.[18] The same authors also now report the RN and SN to be activated in the subjects with visually triggered migraine.[19] The RN and SN are best known for their functional roles in motor control. The RN, however, has also been associated with pain and/or nociception.[20] Numerous animal studies have documented a response of RN neurons to a variety of sensory and noxious stimuli. In a PET study performed on normal volunteers during capsaicin-induced pain, ipsilateral activation of RN was documented. It remains to be clarified whether or not the RN is involved in the pain pathways or in the motor response to pain.

Evidence of Interictal Disturbances

Electroencephalography was one of the first techniques undertaken to discern physiological differences between migraine and controls. A recent review suggests that EEG is not a valuable diagnostic tool for primary headache disorders.[21] The enhanced photic drive response on the EEG H-response, which was thought to be characteristic of migraine,[22] has recently been confirmed by spectral analysis.[23,24] The specificity of the H-response, however, has been questioned since it may occur with other primary headache disorders.[21] Abnormal steady-state response evoked by a sine-wave visual stimulus (SVEP) was seen in migraineurs, and improved after administration of propranalol.[25,26] Finally, following a repetitive pattern-reversal stimulation, migraineurs, but not controls, displayed potentiation of VEP amplitude which reached its maximum in the second to fourth blocks.[27] Similar results were seen using prolonged stimulation.[28] More recently, however, in agreement with VEP studies, strong interictal dependence of the AEPs on stimulus intensity was demonstrated in migraine.[29] Furthermore, the response was modulated by zolmitriptan.[30]

Transcranial Magnetic Stimulation (TMS) of Motor Cortex in Migraine

Several studies have investigated the motor cortex of migraineurs using TMS [transcranial magnetic stimulation]. TMS has been developed to study cortical physiology.[31,32] Three studies have been performed on the motor cortex, two of which reported increased excitability in migraineurs and suggested that this neurophysiological correlate may have a role in migraine mechanisms.[33,34] The first study that compared subjects with migraine with and without aura to controls demonstrated an increased motor threshold in classic migraine.[33] The motor threshold was increased on the side corresponding to the aura. The threshold difference could not be attributed to attack frequency. The second study was performed on menstrual migraineurs during the cycle compared to controls.[34] An increased threshold was demonstrated, similar to the first study, but in this study the patients had migraine without aura.

Following these studies, two other studies were performed. In the first study, there was a difference in amplitude of MEPs (motor evoked potentials) in migraine with aura compared to controls, but found no differences in the motor threshold.[35] The differences in this study compared to previous reports of increased threshold were explained on the basis of attack frequency, which was higher in their group of patients. In a second study performed on familial hemiplegic migraine, the threshold of motor cortex was higher on the side corresponding to the aura.[36] Using paired pulses, a recent study demonstrated reduced motor cortical excitability after administration of zolmitriptan, a centrally-acting 5 $HT_{1B/D}$ used in the treatment of migraine.[37] This technique thus provides a new opportunity to study cortical physiology and the effects of drugs in migraine.

Cortical Silent Period (CSP) in Migraine

Two studies have examined the cortical silent period (CSP). Although the results were judged to be preliminary, both reported no differences in CSP at high levels of stimulus intensity,[38,39] but at low stimulus intensity, a shorter CSP was documented in migraine with aura compared to controls.[39] Since the CSP, in part, is a measure of central inhibition of motor pathways, this shortening of the CSP suggests reduced central inhibition, inferring increased excitability.

TMS of Occipital Cortex in Migraine

Using TMS to study the occipital cortex is perhaps more relevant to migraine because enhanced excitability of the occipital cortex may underlie either spontaneous or visually triggered migraine aura.[4] Occipital cortex excitability in migraine has been evaluated by the generation of phosphenes by TMS of occipital cortex. The first study reported a low threshold for generation of phosphenes in subjects with MwA, inferring hyperexcitability of the occipital cortex.[40] In contrast, occipital cortex hypoexcitability was reported in MwA based on a lower prevalence of phosphenes stimulated by TMS.[38] Important technical differences, such as the type of stimulator or coil size, might explain these conflicting findings.[41] Since these early reports, there have been two more studies performed on the occipital cortex using TMS, both confirming the initial reports of hyperexcitability.[42,43] In one of these, hyperexcitability of the occipital cortex was associated with a propensity to visually triggered headache in the same patients.[43] Recently Battelli, and colleagues investigated the extrastriate visual area V5, which is important for the perception of motion.[44] Both migraine with and without aura groups required significantly lower magnetic field strength for the induction of moving phosphenes, as compared to the control group; this difference was significant for V5 in both left and right hemispheres. In addition the phosphenes were better defined and had clearer presentation in migraine groups, whereas in controls they tended to be more transient and ill-defined.

Repetitive TMS (rTMS) has also been used to study brain physiology in migraine. In healthy normal subjects, a few minutes of low-frequency rTMS (about 1 Hz) appears to reduce cortical excitability for a few minutes post-stimulation, whereas cortical excitability is increased after higher-frequency rTMS (more than 5 Hz).[45] Bohotin et al. used an interesting design in which a widely used measure of visual system function, the pattern-reversal visual evoked response (PR-VER), was recorded before and after both 1 Hz and 10 Hz rTMS in migraine without aura, migraine with aura, and normal control groups.[46] In both migraine groups, the PR-VEP amplitude was greater after 10 Hz rTMS (900 pulses in total), whereas PR-VEP amplitude was unaffected by 1 Hz rTMS. By contrast, in the control group, PR-VEP amplitude was decreased after 1 Hz rTMS and unaffected by 10 Hz rTMS. Bohotin et al further demonstrated that an abnormal pre-rTMS habituation of PR-VEP amplitude is normalized in both patient groups after 10 Hz rTMS; 1 Hz rTMS, by contrast, does not alter the abnormal habituation. A second study, by Brighina et al, demonstrated reduction in inhibition in migraine using the effect of rTMS on phosphene threshold.[46] They demonstrated a reduction in phosphene threshold in controls but not in migraine after 1 Hz rTMS. Thus they once again inferred cortical hyperexcitability in migraine.

Summary for Migraine Headache

Triggers of an attack initiate a cortical depolarizing neuroelectric and metabolic event similar to the spreading depression of Leao. This event activates the mechanisms of head pain and associated headache features. These mechanisms remain to be determined, but appear to involve either peripheral trigeminovascular or brainstem pathways, or both. Excitability of cell membranes, perhaps in part genetically determined, is the brain's susceptibility to attacks. Factors that increase or decrease neuronal excitability constitute the threshold for triggering attacks.

By using a model of visual stress-induced migraine or by studying spontaneous attacks, and then applying advanced imaging and neurophysiological methods, results have been obtained that support spreading cortical phenomenon as the basis of aura. This neuroelectric event is accompanied by hyperoxia of the brain, possibly associated with vasodilatation. Evidence has been obtained also that the spreading cortical event can activate subcortical centers possibly involved in nociception and associated symptoms of the migraine attack. Susceptibility to migraine attacks appears related to brain hyperexcitability. These newer techniques of functional neuroimaging have confirmed the primary neural basis of the migraine attack with secondary vascular changes, reconciling previous theories into a neurovascular mechanism.

▶ CLUSTER HEADACHE

Patients suffering from cluster headache report it as the most severe pain they have ever experienced. Women who have given birth and have cluster headaches as well rate the latter to be a much worse pain. Accompanying the pain in cluster headache are also symptoms of autonomic dysfunction. Cluster headache has previously been thought of as a disorder of the vascular system with inflammation around the cavernous sinus as the predominant event. The vascular theory, however, does not explain the circadian rhythmicity of this disorder.[47,48] Clinically, most patients have a worsening of the disorder in the spring and the fall.[49] Also, most patients are awakened during the REM period with a severe attack.[50,51] Therefore, the hypothalamus, which is the predominant player determining the circadian rhythm intuitively, plays an important role in this hitherto elusive disorder. The underlying pathogenic mechanisms of cluster headache, like other primary headache disorders, were difficult to ascertain accurately prior to new neuroimaging technology. This review will discuss the underlying mechanisms of cluster headache and the available treatment options.

Evidence for a Neural-Based Disorder

As explained earlier, the hypothalamus seems to play a pivotal role in the generation of cluster headache. The first persuasive evidence came from a PET study where the region of the hypothalamus was found to have an increase in regional cerebral blood flow (rCBF).[52] In this study, 9 subjects with cluster headache were studied in the ictal and interictal period. The cluster headache was induced by nitroglycerin, which has been shown to emulate spontaneous cluster headache.[53,54] Substantial activation ascribable to cluster headache was observed in the ipsilateral hypothalamic gray area when compared with the headache-free state. Regional CBF was not increased in the hypothalamic gray area in the 8 control subjects in whom headache was not induced after NTG. The hypothalamus areas known to be involved in pain processing, such as cingulate and insular cortex and contralateral thalamus, were also activated. Therefore, this study concluded that there is specificity found in relationship to cluster headache with the activation in the hypothalamus but there was also activation in other areas known to be involved in pain.

Since positron emission testing is not sensitive for anatomical localization, PET studies could not clarify if the increased activation was in the hypothalamic gray matter or whether dilation of blood vessels in that region could explain the finding. Recently, the same authors have shown that the structure of the hypothalamus is also different in sufferers with cluster headache, using voxel morphometric studies.[55] The voxel-based morphometry is a more objective and automated method of analyzing changes in brain structure. Voxel-based morphometric analysis of the structural T1-weighted magnetic resonance imaging (MRI) scans were performed in 25 right-handed patients, 14 in an active headache phase and 11 in a headache-free state. A significant structural difference in gray matter density between these patients and 29 right-handed healthy male volunteers was observed. There was a structural difference in the diencephalon, rostral to the aqueduct and adjacent to the third ventricle. This change seemed to be most consistent with the anatomy of the inferior posterior hypothalamus. The increase in volume was present in the entire cohort of patients. These differences existed both in the interictal and ictal period. This was identical to the area of activation seen on the PET studies. Since cluster headache is a unilateral syndrome, the investigators used a mirrored analysis. The structural changes were then noted to be lateralized and ipsilateral to the side of pain. The fact that these structural changes were seen independent of headache-state signifies an inherent dysfunction of the hypothalamus in cluster headache rather than an epiphenomenon.

Indirect evidence of a hypothalamic dysfunction also comes from hormonal studies. Low testosterone levels have been noted in males with cluster headache,[56,57] although it is not clear whether this is a primary problem or secondary to stress. Other neurohormone levels secreted by the hypothalamus such as melatonin, cortisol, beta endorphins, and prolactin levels differ during cluster attacks and in the remission periods.[58] The morning cortisol levels have been found to be high in the remission phase of CH compared to controls. This indicates an inherent hyperactivity of the hypothalamic-pituitary-adrenal system rather than a stress response. Melatonin is a surrogate marker of the circadian rhythm.[59] Melatonin prophylaxis has been reported to reduce cluster attacks in some patients.[60] The suprachiasmatic nuclei, located in the ventral hypothalamus, are responsible for the endogenous circadian rhythm reacting to the environmental light via the retinohypothalamic pathway. The aforementioned imaging studies localized the abnormality in patients with cluster headache to this precise localization.

Leone and his colleagues reported alleviation of chronic cluster headache by implantation of an electrode in the posterior hypothalamus in a single case report.[61] Topiramate has also been shown to be effective in cluster prophylaxis,[62] although indirectly it may be speculated to have an effect on hypothalamic function on the satiety center since it produces weight loss.

Similar to migraine, there may also be a generalized dysfunction of the central nervous system in cluster headache. Clinically this is manifested as premonitory symptoms similar to the aura in migraine.[63] Silberstein and his colleagues, in a recent retrospective review of 101 CH patients, found 6 with aura. Five had visual aura and one olfactory lasting in a range of 5 to 120 minutes. Others also have reported these symptoms of aura in the past. Therefore, the underrecognition of these symptoms is surprising and may be attributed to the intensity of the other symptoms such as pain.

Evidence for a Vascular Disorder

The clinical features of cluster headache definitely suggest the involvement of vascular structures, in particular the trigeminal vascular pathway. The characteristic pulsating pain and the relief with abortive agents that are vasoconstrictors also support the vasculature as playing the pivotal role. Earlier, Kudrow[64] had suggested that the peripheral chemoreceptors in the carotid body were dysfunctional and led to aberrations in the chronobiological functioning of the hypothalamus. With the neuroimaging evidence discussed earlier, the peripheral vasculature is unlikely to be the main driver. The vessels, however, are definitely involved but are most likely epiphenomena. Blood flow has been measured using SPECT; because this method is semi-quantitative, the studies have shown conflicting results. Some studies have shown an increase in blood flow,[65,66] some a decrease,[67] and some no differences at all.[68] A reduction in blood flow velocities of middle cerebral arteries indicative of vasodilation has been reported to occur in spontaneous and GTN-provoked attacks of cluster headache.[69,70]

An inflammatory process in the cavernous sinus and tributary veins is also thought to play a role in cluster headache.[71] This theory is based on findings of inflammation on orbital phlebography studies[72] and the induction of cluster headaches via NTG.[73] This theory only partially explains the clinical syndrome of cluster. An inflammatory response alone cannot explain the nature of cluster headache with regard to circadian rhythmicity. In addition, orbital phlebography studies performed in migraine, tension-type headache and cervicogenic headache, have noted similar findings as those noted in cluster headache.[74] Support of an inflammatory theory as well as unilateral blockade of venous drainage was obtained by doing craniometric measures in cluster headache patients.[75] In this study, external morphometric skull measures were obtained in 25 cluster headache patients as compared with 21 healthy controls. The cluster headache patients had significantly smaller values, which suggested a narrower anterior/middle cranial fossa. The authors inferred a narrower cavernous sinus loggia. However, this was not ascertained by a quantitative MRI study.[76] No definite pathological changes were seen on MRI in the area of the cavernous study, but quantitative measurements were not performed. Cavernous sinus inflammation, however, has been noted in cluster headache as well as headache invoked in experimental conditions. Given this new evidence, it seems that the hypothalamus may be the driver in cluster headache, with the vessel involvement with the inflammatory response in the cavernous sinus being an epiphenomenon.

Is There a Link Between the Neural and Vascular Bases?

There is strong evidence for both a neural as well as a vascular-based disorder for cluster headache. Since there is strong evidence for the hypothalamus or a more central origin as the driving force, one might question how the vessel involvement occurs. One of the molecules of interest is nitric oxide, which is generated from neurons as well as endothelium by specific enzymes known as nitric oxide synthase. Since it exists both in the neurons and endothelium, nitric oxide could serve as a very likely candidate to mediate a neural-based vasodilatation. Clinical evidence for the involvement of nitric oxide exists since nitrates can trigger cluster headache during a cluster period.[77] After NTG administration in cluster headache patients, an immediate noncluster headache is usually followed by a delay of 30 to 50 minutes by a cluster headache attack, which is similar to spontaneous attacks. This would support the notion that although the brain structures, perhaps the hypothalamus, are responsible for the initiation of attack, substances like nitric oxide may be the promoter.

Summary for Cluster Headache

Given the circadian rhythmicity of the disorder and the new evidence on imaging studies, it may be assumed that the hypothalamus plays a pivotal role in the development of CH. The hypothalamus may activate the efferent arc causing the symptoms of ptosis, miosis, and facial flushing. Peripheral release of neuropeptides may result in vasodilatation and in intense pulsating pain. Mediators like nitric oxide released from the neurons may play an important role in the vasodilatation.

Drugs that target primarily the neuronal structures, i.e., antiepileptics or calcium channel blockers, are used in the prevention of cluster headache, whereas abortive agents work at both the trigeminal level and at the vasculature.

▶ CONCLUSIONS

With the advent of noninvasive neuroimaging, the mechanisms of primary headache disorders have been elucidated. There is strong evidence for a neuronal basis for both migraine and cluster headaches. These disorders were hitherto thought to be predominantly vasculature in nature. It has now been clarified that neurons are the primary drivers for these headache disorders and the vascular structures are involved as epiphenomena.

REFERENCES

1. Wolff HG: *Headache and other head pain.* 2d ed. New York: Oxford University Press, 1963.
2. Lashley KS: Patterns of cerebral integration indicated by the scotomas of migraine. *Arch Neurol Psych* 1941;46:331–339.
3. Leao AAP: Spreading depression of activity in the cerebral cortex. *J Neurophysiol* 1944;8:379–390.
4. Welch KMA, D'Andrea TN, Barkley G, Ramadan NM: The concept of migraine as a state of central neuronal hyperexcitabiliy. *Neurol Clinics* 1990;8:817–828.
5. Ophoff RA, Terwindt GM, Vergouwe MN, et al: Familial hemiplegic migraine and episodic ataxia type-2 are caused by mutations in the Ca^{+2} channel gene CACNL1A4. *Cell* 1996; 87:543–552.
6. Cao Y, Welch KMA, Aurora SK, Vikingstad EM: Functional MRI-BOLD of visually triggered headache and visual change in migraine sufferers. *Archives of Neurology* 1999;56:548–554.
7. Gardner-Medwin AR, Bruggen NV, Williams SR, Ahier RG: Magnetic resonance imaging of propagating waves of spreading depression in the anaesthetized rat. *J of Cerebral Blood Flow and Metabolism* 1994;14:7–11.
8. Hadjikhani N, Sanchez del Rio M, Wu O, et al: Mechanisms of migraine aura revealed by functional MRI in human visual cortex. *Proc Natl Acad Sci USA* 2001 98:4687–4692.
9. Sanchez del Rio M, Bakker D, Wu O, et al: Perfusion weighted imaging during migraine spontaneous visual aura and headache.
10. Bowyer SM, Okada YC, Papuashvili N, et al: Analysis of MEG signals of spreading cortical depression with propagation constrained to a rectangular cortical strip: I. Lissencephalic rabbit model: *Brain Research* 1999;843:71–78.
11. Barkley GL, Tepley N, Nagel-Leiby S, Moran JE, Simkins RT, Welch KMA: Magnetoencephalograhic studies of migraine. *Headache* 1990;30:428–434.
12. Bowyer SM, Aurora SK, Burdette DE, Moran JE, Tepley N, Welch KMA: Neuromagnetic measurements of evoked and spontaneous migraine with aura. *Congress of the International Headache Society.* June, 1999, Barcelona Spain.
13. Raskin NH, Hosobuchi Y, Lamb S: Headache may arise from perturbation of the brain. *Headache* 1987;27:416–420.
14. Moskowitz MA: The neurobiology of vascular head pain. *Ann Neurol* 1984;15:157–168.
15. Goadsby PJ, Gundlach AL: Localization of ^{3}H-dihydroergotamine-binding sites in cat central nervous system: Relevance to migraine. *Ann Neurol* 1991;29:91–94.
16. Longmore J, Shaw D, Smith D, et al: Differential distribution of 5-HT 1D and 5-HT 1B immunoreactivity within the human trigemino-cerebrovascular system: Implications for the discovery of new anti-migraine drugs. *Cephalalagia* 1997;17:835–842.
17. Weiller C, May A, Limmroth V, et al: Brainstem activation in spontaneous human migraine attacks. *Nature Medicine* 1995;1:658–660.
18. Welch KMA, Cao Y, Aurora SK, Wiggins G, Vikingstad EM: MRI of the occipital cortex, red nucleus, and substantia nigra during visual aura of migraine. *Neurology* 1998; 51:1465–1469.
19. Cao Y, Aurora SK, Vikingstad EM, Patel SC, Welch KMA: Functional MRI of the red nucleus and occipital cortex during visual stimulation of subjects with migraine. *Neurology.* 2002;59:72–78.
20. Iadarola MJ, Berman KF, Zeffiro TA, et al: Neural activation during acute capsaicin-evoked pain and allodynia assessed with PET. *Brain* 1998;121:931–947.
21. Gronseth GS, Greenberg MK: The utility of the electroencephalogram in the evaluation of patients presenting with headache: A review of the literature. *Neurology* 1995 45:1263–1267.
22. Golla FL, Winter AL: Analysis of cerebral responses to flicker in patients complaining of episodic headache. *Electroencephalopgr Clin Neurophysiol* 1982;53:270–276.
23. Simon RH, Zimmerman AW, Tasman A, Hale MS: Spectral analysis of photic stimulation in migraine. *Electroencephalogr Clin Neurophysio* 1982;53:270–276.

24. Pechadre JC, Gibert J: Demonstration, by the cartographic test, of an unusual reaction to intermittent light stimulation in patients with migraine. *Encephale* 1987;13:245–247.
25. Nyrke T, Kangasniemi P, Lang AH: Difference of steady-state visual evoked potential in classic and common migraine. *Electroencephalopgr Clin Neurophysiol* 1989;73:284–294.
26. Nyrke T, Kangasniemi P, Lang AH: Steady-state visual evoked potentials during migraine prophylaxis by propranolol and femoxetine. *Acta Neurol Scand* 1984;69:9–14.
27. Schoenen J, Wang, W, Albert A, Delwaide PJ: Potentiation instead of habituation characterizes visual evoked potentials in migraine patients between attacks. *Eur J Neurology* 1995;2:115–122.
28. Afra J, Cecchini AP, DePasqua V, Albert A, Schoenen J: Visual evoked potentials during long periods of pattern-reversal stimulation in migraine. *Brain* 1998;121:233–241.
29. Wang W, Timsit-Berthier M, Schoenen J: Intensity dependence of auditory evoked potentials is pronounced in migraine: An indication of cortical potentiation and low serotonergic neurotransmission? *Neurology* 1996;46:1404–1409.
30. Proietti-Cecchini A, Afra J, Schoenen J: Intensity dependence of the cortical auditory evoked potentials as a surrogate marker of CNS serotonin transmission in man: Demonstration of a central effect for the 5-HT1B/1D agonist zomitriptan (311C90, Zomig). *Cephalalgia* 1997;17:1–18.
31. Barker AT, Freeston IL, Jalinous R, Jarratt JA: Magnetic stimulation of the human brain and peripheral nervous system: An introduction and the results of an initial clinical evaluation. *Neurosurgery* 1987;20:100–109.
32. Barker AT, Jalinous R, Freeston IL. Non-invasive magnetic stimulation of human motor cortex. *Lancet* 1985; 1:1106–1107.
33. Maertens de Noordhout AL, Pepin JL, Schoenen J, Delwaide PJ: Percutaneous magnetic stimulation of the motor cortex in migraine. *Electroencephalogr Clin Neurophysiol* 1992;85:110–115.
34. Bettucci D, Cantello R, Gianelli M, Naldi P, Mutani R: Menstrual migraine without aura: Cortical excitability to magnetic stimulation. *Headache* 1992;32:345–347.
35. van der Kamp W, Maassenvandenbrink A, Ferrari MD, vanDijk JG: Interictal cortical hyperexcitability in migraine patients demonstrated with transcranial magnetic stimulation. *J Neurol Sci* 1996;139:106–110.
36. van der Kamp W, Maassenvandenbrink A, Ferrari MD, vanDijk JG: Interictal cortical excitability to magnetic stimulation in familial hemiplegic migraine. *Neurology* 1997;48:1462–1464.
37. Werhahn KJ, Förderreuther S, Straube A: Effects of serotonin 1B/1D receptor agonist zolmitriptan on motor cortical excitability in humans. *Neurology* 1998;51:896–898.
38. Afra J, Mascia A, Gérard P, Maertens de Noordhout A, Schoenen J: Interictal cortical excitability in migraine: A study using transcranial magnetic stimulation of motor and visual cortices. *Ann Neurol* 1998;44:209–215.
39. Aurora SK, Al-Sayed F, Welch KMA: The cortical silent period is shortened in migraine with aura. *Cephalalgia* 1999;19:708–712.
40. Aurora SK, Al-Sayed F, Welch KMA: The threshold for magnetophosphenes is lower in migraine. *Neurology* 1999;52:A472.
41. Aurora SK, Welch KMA: Phosphene generation in migraine. [Letter to Editor]: *Ann of Neurol* 1999;45:416.
42. Aggugia M, Zibetti M, Febbraro A, Mutani R: Transcranial magnetic stimulation in migraine with aura: Further evidence of occipital cortex hyperexcitability. *Cephalalgia* 1999;19:465.
43. Aurora SK, Cao Y, Bowyer SM, Welch KMA: The occipital cortex is hyperexcitable in migraine; evidence from TMS, fMRI and MEG studies (Wolff Award 1999), *Headache* 1999;39:469–476.
44. Battelli L, Black KR, Wray SH: Transcranial magnetic stimulation of visual area V5 in migraine. *Neurology* 2002;58:1066–1069.
45. Bohotin V, Fumal A, Vandenheede M, Gérard P, Bohotin C, Maertens de Noordhout A: Effects of repetitive transcranial magnetic stimulation on visual evoked potentials in migraine. *Brain* 2002;125:912–922.
46. Brighina F, Piazza A, Daniele O, Fierro B: Modulation of visual cortical excitability in migraine with aura: Effects of 1 Hz repetitive transcranial magnetic stimulation. *Experimental Brain Research* 2002;145(2):177–181.
47. Ekbom K: A clinical comparison of cluster headache and migraine. *Acta Neurol Scand* 1970;41:1–48.
48. Ekbom K: Patterns of cluster headache with a note on the relations to angina pectoris and peptic ulcer. *Acta Neurol Scand* 1970;46:225–237.
49. Ekbom K: A clinical comparison of cluster headache and migraine. *Acta Neurol Scand* 1970;46:225–237.
50. Kudrow L: The cyclic relationship of natural illumination to cluster period frequency. *Cephalalgia* 1987;7:76-77.
51. Kudrow L, Kudrow DB: Association of sustained oxyhemoglobin desaturation and onset of cluster headache attacks. *Headache* 1994;30:400–407.
52. May A, Bahra A, Buchel C, Frackowiak RSJ, Goadsby PJ: Hypothalamic activation in cluster headache attacks. *Lancet* 1998;351:275–278.
53. Ekbom K: Nitroglycerin as a provocative agent in cluster headache. *Arch Neurol* 1968;19:487–493.
54. Fanciullacci M, Alessandri M, Figini M, Geppetti P, Michelacci S: Increases in plasma calcitonin gene-related peptide from extracerebral circulation during nitroglycerin-induced cluster headache attack. *Pain* 1995;60:119–123.
55. May A, Ashburner J, Buchel C, et al: Correlation between structural and functional changes in brain in an idiopathic headache syndrome. *Nat Med* 1999;5:836–838.
56. Facchinetti F, Nappi G, Cicoli C, et al: Reduced testosterone levels in CH, a stress-related phenomenon? *Cephalalgia* 1996;2:29–34.
57. Polleri A, Nappi G, Murialdo G, Bono G, Martignoni E, Savoldi F: Changes in the 24-hour prolactin pattern in CH. *Cephalalgia* 1982;2:1–7.
58. Waldenlind E, Gustafsson SA: Prolactin in CH, diurnal: secretion, response to thyrotropin-releasing hormone and relation to sex steroids and gonadotropins. *Cephalalgia* 1987;7:43–54.
59. Waldenlind E, Gustafsson SA, Ekbom K, Wetterberg L: Circadian secretion of cortisol and melatonin in cluster headache during active cluster period and remission. *J Neurol Neurosurg Psychiatry* 1987;50:207–213.

60. Leone M, D'Amico D, Moschiano F, Fraschini F, Bussone G: Melatonin versus placebo in the prophylaxis of CH, a double-blind pilot study with parallel groups. *Cephalalgia* 1996;16:494–496.
61. Leone M, Franzine A, D'amico D, Grazzi A, Rigamonti A, Usai S, Broggi G, Bussone G: Intractable chronic cluster headache relieved by electrode implant to posterior hypothalamus. *Cephalalgia* 2001;21:503.
62. Wheeler SD, Carrazana EJ: Topiramate-treated cluster headache. *Neurology* 1999;53:234–236.
63. Silberstein SD, Niknam R, Rozen TD, Young WB: Cluster headache with aura. *Neurology* 2000;54:219–222.
64. Kudrow L: The pathogenesis of cluster headache. *Curr Opin Nuerol* 1994;7:278–282.
65. Norris JW, Hachiniski VC, Cooper PW: Cerebral blood flow changes in cluster headache. *Acta Neurol Scand* 1976;54:371–374.
66. Sakai P, Meyer JS: Regional cerebral hemodynamics during migraine and cluster headaches measured by the 133-Xe inhalation method. *Headache* 1978;18:122–132.
67. Nelson RF, du Boulay GH, Marshall J, Russell RW, Symon L, Zilkha E: Cerebral blood flow studies in patients with cluster headache. *Headache* 1980;20:184–189.
68. Henry PY, Vernhiet J, Orgoogozo JM, Caille JM: Cerebral blood flow in migraine and cluster headache. Compartmental analysis and reactivity to anaesthetic depression. *Res Clin Stud Headache* 1978;6:81–88.
69. Krabbe AA, Henriksen L, Olesen J: Tomographic determination of cerebral blood flow during attacks of cluster headache. *Cephalalgia* 1984;4:17–23.
70. Dahl A, Russell D, Nyberg-Hansen R, Rootwelt K: Cluster headache: Transcranial Doppler ulrasound and rCBF studies. *Cephalalgia* 1990;10:87–94.
71. Tegeler CH, Davidai G, Gengo FM: Middle cerebral artery velocity correlates with nitroglycerin-induced headache onset. *J Neuroimag* 1996;6:81-86.
72. Harbedo JE: How cluster headache is explained as an intracavernous inflammatory process lesioning sympathetic fibres. *Headache* 1994;34:125–131.
73. Hannerz J, Ericson K, Bergstrand G: Orbital phlebography in patients with cluster headache. *Cephalalgia* 1987;7:207–211.
74. Bovin G, Jenssen G, Ericson K: Orbital phlebography: A comparison between cluster headache and other headaches. *Headache* 1992;32:408–412.
75. Afra J, Proietti Cecchini A, Schoenen J: Craniometric measures in cluster headache patients. *Cephalalgia* 1998;18:143–145.
76. Sjaastad O, Rinck P: Cluster headache: MRI studies of the cavernous sinus and the base of the brain. *Headache* 1990;30:350–351.
77. Drummond PD, Anthony M: Extracranial vascular responses to sublingual nitroglycerin and oxygen inhalation in cluster headache patients. *Headache* 1985;25:70–74.

CHAPTER 9

Neuroanatomy of Pain

Anca Popescu

Pain is a unique experience. Although the genotype (pain generators and pain transmission pathways) has a well-characterized structure, the phenotype (the way humans perceive pain and react to it) is the result of a number of permutations and combinations. Pain research has moved away from the mechanistic, Cartesian approach to a more holistic, multi-dimensional, dynamic outlook. The perception of the noxious stimulus triggers adaptive (endocrine, autonomic, homeostatic, somatic) and modulatory responses (neurotransmitter release and inhibition, receptor upregulation and downregulation). The brain, however, is the ultimate comparator, judge, and effector. It intervenes in all the adaptive and modulatory mechanisms via activation of attention, arousal, motivation processes, influence of the endocrine and autonomic systems, release of excitatory and inhibitory transmitters, reflex, and goal-directed behavior. The result is the pain behavior, which is finally modulated by culture, previous experiences, and reward/punishment expectation from the environment.

This chapter outlines what is known about the neuroanatomic basis of various well-defined pain syndromes. It is a paradox that we know more about the rarer pain syndromes (e.g., trigeminal neuralgia, Ramsay Hunt syndrome) than the more common ones (e.g., chronic daily headache), possibly because the former have a characteristic neurosignature as opposed to the latter, which vary according to genetic, environmental, and emotional influences.

▶ PAIN RECEPTORS

Noxious stimuli are transduced into neural signals by distal axons of the receptive neurons. Various channels on the neuronal membrane respond to different stimuli (thermal, mechanical, or chemical), and the channels are located on different types of receptors.

Nociceptors are sensory receptor neurons that are sensitive to noxious or tissue-damaging stimuli and mediate pain. *Thermoreceptors* are sensitive to cold and warmth. Nociceptors and thermoreceptors have a simple morphology; they are bare nerve endings and dendrites of small-diameter axons (Aδ and C) that are unmyelinated or lightly myelinated and, therefore, conduct slowly. Itch-sensitive neurons are also unmyelinated. All these neurons have the cell body located in the dorsal root ganglion (DRG) and provide sensory information to the anterolateral (spinothalamic) system (Fig. 9-1).

The *dorsal root ganglion* (DRG) is made of pseudounipolar neurons that receive somatic sensory information from sensory receptors located on a dermatome (skin area innervated by a single dorsal root). Their axons enter the spinal cord via the dorsal root of the spinal nerves. The dorsal roots give off segmental, ascending, and descending branches that travel in the spinal cord (Fig. 9-2).

The *dorsal horn of the spinal cord* (DH) has a laminar neuronal arrangement (Fig. 9-3). The neurons form flattened sheets (Rexed laminae) that run parallel to the long axis of the spinal cord. There are 10 laminae. The dorsal horn consists of laminae 1 through 7. Lamina I (marginal zone) receives direct input from the small sensory fibers and is the outermost in the DH; lamina II is

Modality and Submodality	Receptor Type	Fiber Diameter (μm)	Group	Myelination[1]
Touch	Mechanoreceptors	6–12	A-β (2)	Myelinated
Texture/superficial				
Pressure/deep				
Vibration	Pacinian			
Position Sense	Mechanoreceptors	13–20	A-α (1),	Myelinated
		6–12	A-β (2)	
Static				
Dynamic (kinesthesia)				
Temperature Sense	Thermoreceptors	1–5	A-δ (3),	Myelinated;
		0.2–1.5	C (4)	unmyelinated
Cold				
Warmth				
Pain	Nociceptors	1–5	A-δ (3)	Myelinated;
		0.2–1.5	C (4)	unmyelinated
Fast (pricking)				Myelinated
Slow (burning)				Unmyelinated
Itch	Histamine	0.2–1.5	C (4)	Unmyelinated

[1]A small number of thinly myelinated and unmyelinated fibers are sensitive to mechanical stimuli. These mechanoreceptors are present in the hairy skin.

Figure 9-1. Sensory modalities and their afferent fibers. (*Used with permission from Martin JH: Neuroanatomy: Text and Atlas, 3d ed. New York: McGraw-Hill, 2003, p. 108.*)

known as the *substantia gelatinosa*; laminae III and IV are known as the *nucleus proprius*; and laminae V and VI are termed the *base* of the DH. Lamina V receives direct and indirect input from small-diameter sensory fibers, as well as mechanoreceptive inputs, and contains wide dynamic range (WDR) neurons, which respond to a wide range of stimuli (from weak and mechanical to strong and noxious). Laminae VII through IX contain motor nuclei, and lamina X is the gray matter surrounding the central canal. Laminae I and V are the origins of the spinothalamic tract (STT), which conveys the pain, itch, and temperature sensations. Laminae VI through VIII also contain neurons that project to the intralaminar nuclei of the thalamus and pontomedullary reticular formation and are involved in arousal. Large-diameter sensory fibers (Aβ) enter the spinal cord medial to the Lissauer's tract (the white matter region that caps the DH) in the dorsal root entry zone (DREZ). Their axons enter the white matter of the dorsal columns but also give several segmental branches (which end in laminae V through VIII of the spinal cord).

The *spinal somatic sensory systems* convey touch, limb position, pain, itch, and temperature sensations. They also have a major role in maintenance of the arousal state and sensory regulation of limb and trunk movements. This information is collected from the cutaneous receptors, as well as from the deep receptors within muscles, joints, and viscera. The information from the large-diameter fibers ascends in the dorsal columns to the brain stem. The information from the small fibers travels both in ascending and descending nerve fascicles that form the Lissauer's tract. The axons of these neurons end in the gray matter of the spinal cord (laminae I, II, and V).

The *spinothalamic tract (anterolateral system)* receives information about the painful, pruritic, and thermal receptors. It consists of axons of second-order neurons that cross the midline in the anterior white commissure to the opposite lateral columns. They ascend toward the thalamus, reticular formation, nucleus raphe magnus, and periaqueductal gray matter (PAG). There are several tracts. The *lateral spinothalamic (neospinothalamic) tract* receives information from WDR neurons about the location, intensity, and duration of nociceptive stimuli. It has a somatotopic organization in the spinothalamic tract (leg fibers

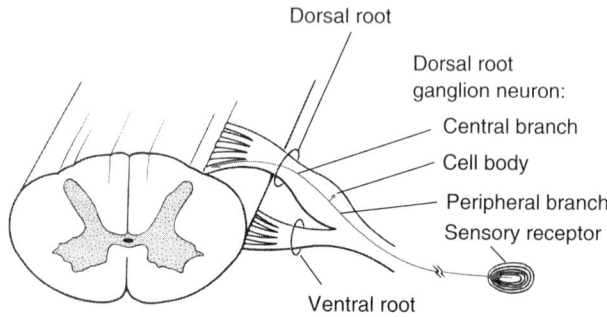

Figure 9-2. The dorsal root ganglion. (*Used with permission from Martin JH: Neuroanatomy: Text and Atlas, 3d ed. New York: McGraw-Hill, 2003, p. 110.*)

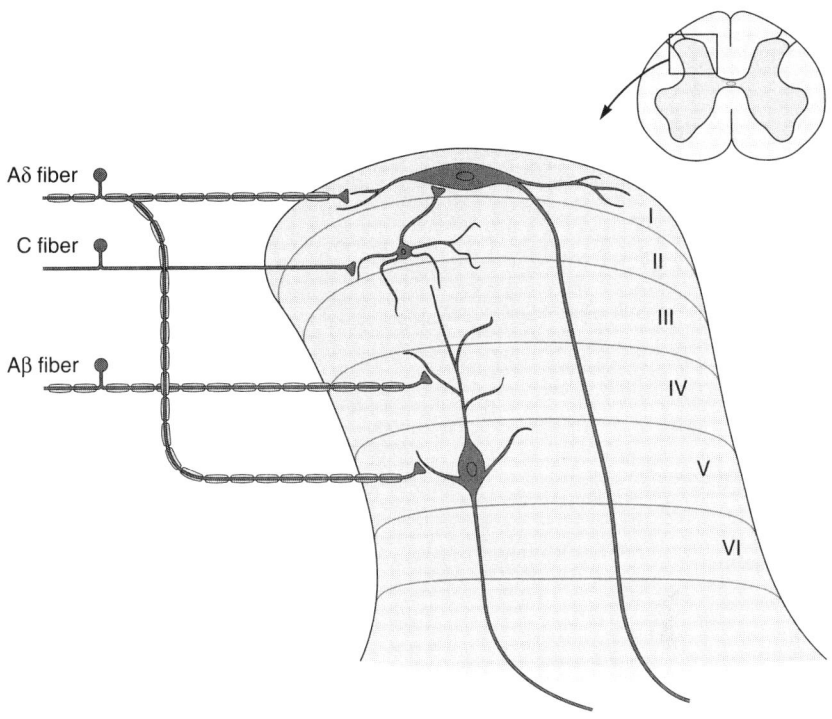

Figure 9-3. The lamination of the dorsal horn. Pain fibers (Aδ and Aβ) end in laminae 1, 2 and 5. (*Used with permission from Kandel ER, Schwartz JH, Jessell TM: Principles of Neural Science, 4th ed. New York: McGraw-Hill, 2000, p. 474.*)

are the most lateral fibers from the trunk; arm and neck are medial). It synapses in the ventral posterolateral nucleus of the thalamus. The *anterior spinothalamic (paleospinothalamic)* tract conveys information from laminae VII and VIII of the DH about location of the injury and projects to the medial and intralaminar nuclei of the thalamus. This is a polysynaptic system that contributes to the arousal to pain stimuli. Injury or electrical stimulation to the spinothalamic tract or its targets results in a severe type of pain (central pain). Lesioning of the spinothalamic tract (cordotomy) results in reduction of pain in the opposite half of the body.

Collateral fibers off the PAG constitute a link between the ascending and descending pain pathways. There are also collaterals toward the reticular ascending activating system responsible for arousal to painful stimuli. The *spinoreticular tract* receives information from laminae VII and VIII of the spinal cord and does not cross like the lateral spinothalamic tract. Some of the fibers ascend in the anterolateral and some in the dorsal part of lateral funiculus of the spinal cord. The axons end in the reticular formation of the medulla and pons. Axons of neurons in the reticular formation project further into the thalamus and superior colliculus (playing a role in orientation to somatic stimuli). The presence of these fibers explains the persistence of pain after anterolateral cordotomy. Another projection of the spinoreticular tract is to the PAG (which has a key role in modulation of the pain perception and is the origin of the descending inhibitory system). The *spinomesencephalic tract* consists of axons from neurons in laminae I and V and ascends in the anterolateral quadrant of the spinal cord to the mesencephalic reticular formation, the PAG, and the parabrachial nuclei (the latter project to the amygdala) [Fig. 9-4].

The *dorsal columns–medial lemniscal pathway* receives input from the mechanoreceptors. It is formed by first-order neurons (axons of sensory receptors, Aβ fibers). Information from the leg and lower trunk is carried by the gracile fascicle; information from the upper trunk, arm, neck, and posterior aspect of the head down to the T6 vertebra is carried by the cuneate fascicle. There is a somatotopic organization in the dorsal columns. From medial to lateral are the axons from the leg and then the lower trunk, upper trunk, arm, neck, and occiput. In the medulla, the axons of these fascicles cross toward the opposite ventral and medial parts of the medulla, just behind the medullary pyramids (forming *internal arcuate fibers*) and synapse with neurons in the dorsal column nuclei (gracile and cuneate). After they cross the midline, they ascend in the ventral brain stem as the *medial lemniscus*. The latter also has a somatotopic organization (axons from the gracile nucleus decussating ventral to the axons from the cuneate nucleus, giving the medial lemniscus the resemblance of a person standing upright). In the pons, the medial lemniscus is located more dorsally, and is oriented from medial to lateral, and it ascends toward the ventroposterior lateral nucleus of the thalamus (VPL). Here the fibers of the medial lemniscus synapse with the third-order neurons that project to the primary somatic sensory cortex (S1). Complete lesioning of this pathway leads to loss of the fine discriminatory quality of

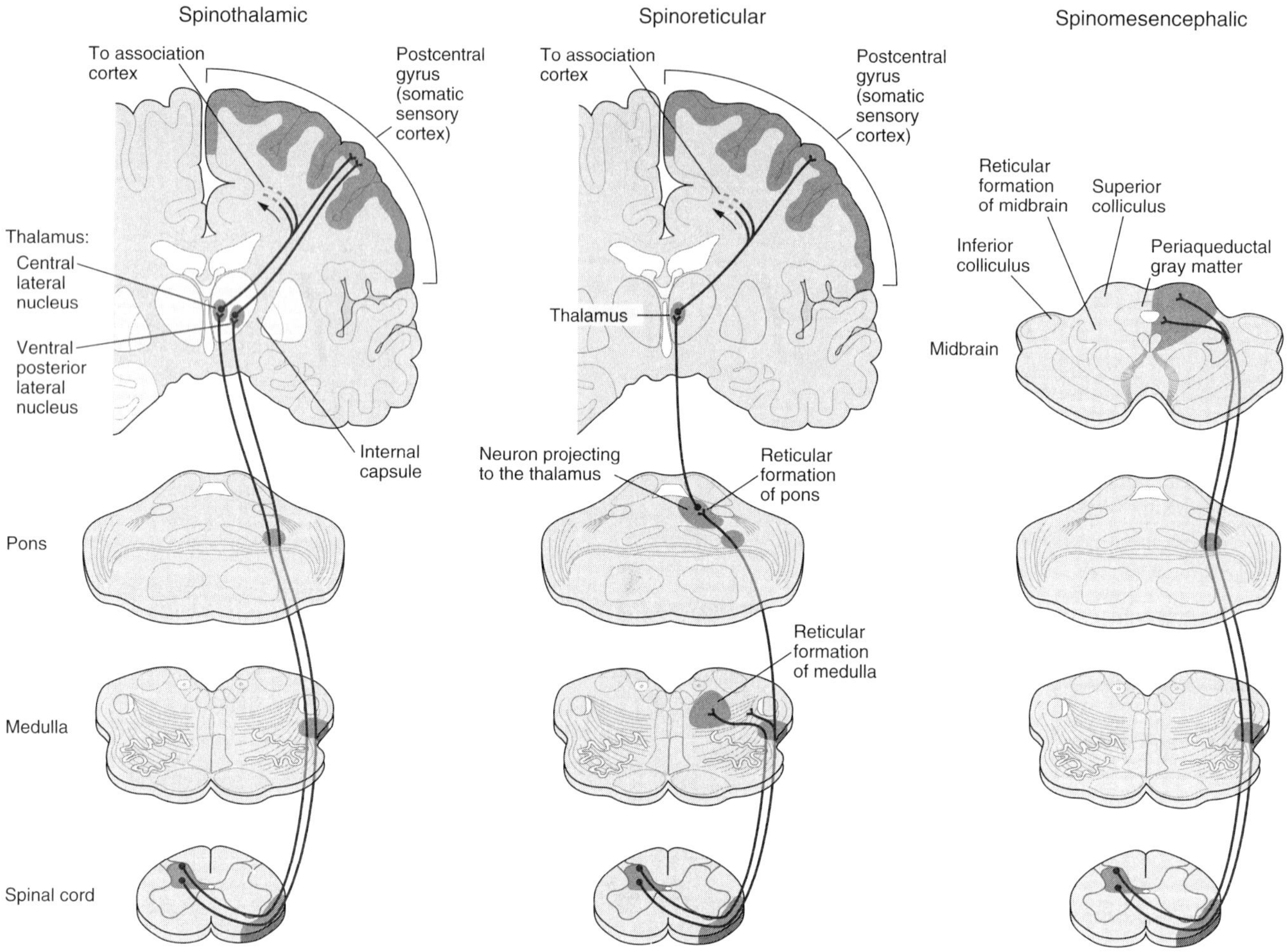

Figure 9-4. Anatomy of the anterolateral sensory systems and their cortical projections. (*Used with permission from Kandel ER, Schwartz JH, Jessell TM:* Principles of Neural Science, *4th ed. New York: McGraw-Hill, 2000, p. 482.*)

touch (i.e., to distinguish between rough and smooth), loss of position sense (i.e., loss of balance control with eyes closed), and preservation of a crude sense of touch (probably mediated by the anterolateral spinothalamic system). (Figs. 9-5 and 9-6).

The *descending nociceptive inhibitory system* originates in the midbrain in the (PAG). This represents the area surrounding the third ventricle and aqueduct of Sylvius. It receives projections from the spinoreticular tract and limbic and cortical structures involved in emotions. It also blocks the spinal withdrawal reflexes to pain. The PAG sends efferents to the nucleus raphe magnus in the rostroventral medulla (large serotonin reservoir). The *serotonergic system* suppresses pain transmission in the DH of the spinal cord by inhibiting the ascending nociceptive pathways and by stimulating the inhibitory neurons in the DH (which have enkephalin as a neurotransmitter). Thus, when stimulated, the PAG inhibits nociceptive neurons in laminae I, II, and V of the DH, leading to a profound and selective analgesia (only to pain).

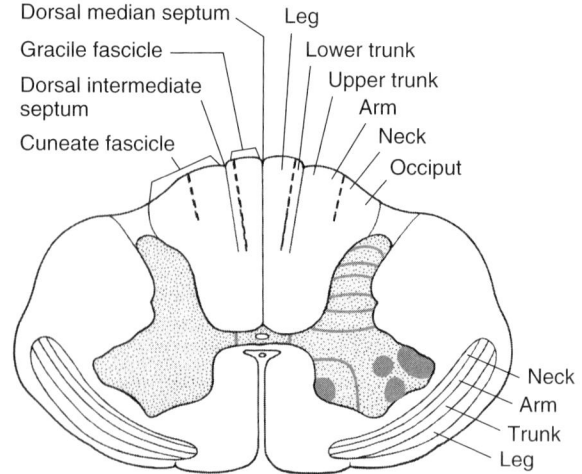

Figure 9-5. The somatotopic organization of the anterolateral and dorsal column systems. (*Used with permission from Martin JH:* Neuroanatomy: Text and Atlas, *3d ed. New York: McGraw-Hill, 2003, p. 119.*)

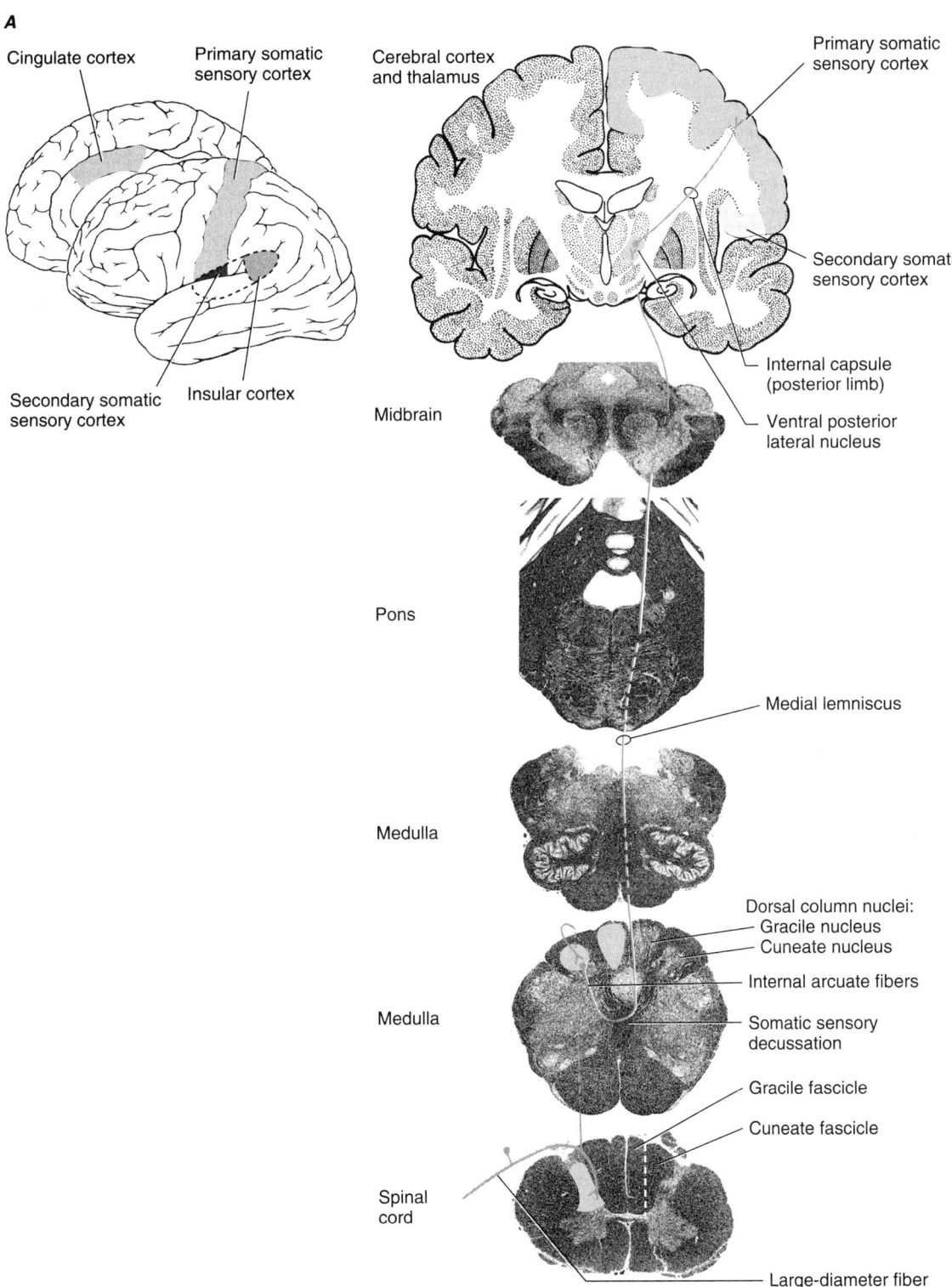

Figure 9-6. *A.* The dorsal column-medial lemniscal pathway. (*Used with permission from Martin JH:* Neuroanatomy: Text and Atlas, *3d ed. New York: McGraw-Hill, 2003, p. 112.*) *B.* The thalamic projection of the dorsal column-medial lemniscal pathway. (*Used with permission from Martin JH:* Neuroanatomy: Text and Atlas, *3d ed. New York: McGraw-Hill, 2003, p. 128.*)

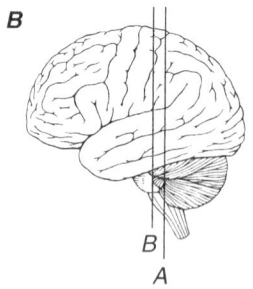
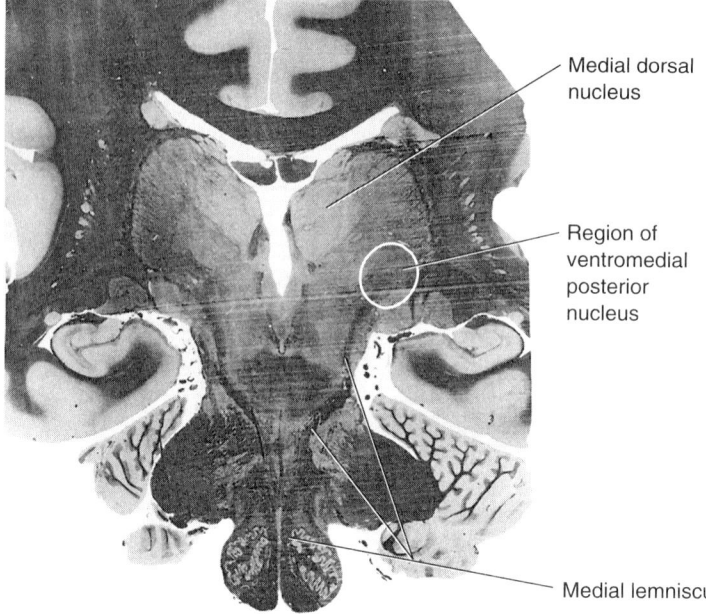

Figure 9-6. (Continued).

Another descending inhibitory system, the *noradrenergic system*, originates in the *locus coeruleus* and the lateral medullary reticular formation. Its projections suppress the output of neurons in laminae I and V and interact with opioid circuits in the DH. The two desending inhibitory systems are probably the physiologic basis of survival mechanisms that allow people to function better despite severe pain during physical combat or during childbirth. Experimentally, morphine-induced analgesia is blocked by injection of naloxone into the PAG or nucleus raphe magnus (Fig. 9-7).

The *thalamus* is the place where axons of second-order neurons synapse with third-order neurons. Axons of the third-order neurons of the lateral spinothalamic tract synapse in the lateral nuclear group of the thalamus [*ventroposterior lateral (VPL), ventroposterior medial (VPM), dorsomedial,* and *posterior*] and from there, through the posterior limb of the internal capsule, to the primary and secondary somatosensory cortex (projections from the VPL nucleus) and insular cortex (projections from the posterior nucleus). The VPL nucleus receives information from the lateral spinothalamic tract and the dorsal columns–medial lemniscus system. The VPM nucleus receives sensory information from the face. The spinoreticulothalamic tract projects to the *medial dorsal* and *intralaminar nuclei* of the thalamus and from there to the cingulate gyrus (part of the limbic system) and conveys the autonomic and unpleasant aspect of pain integrated with arousal and attention. Electrical stimulation of certain areas of the thalamus can evoke a severe pain that the patient has experienced in the past (Fig. 9-8).

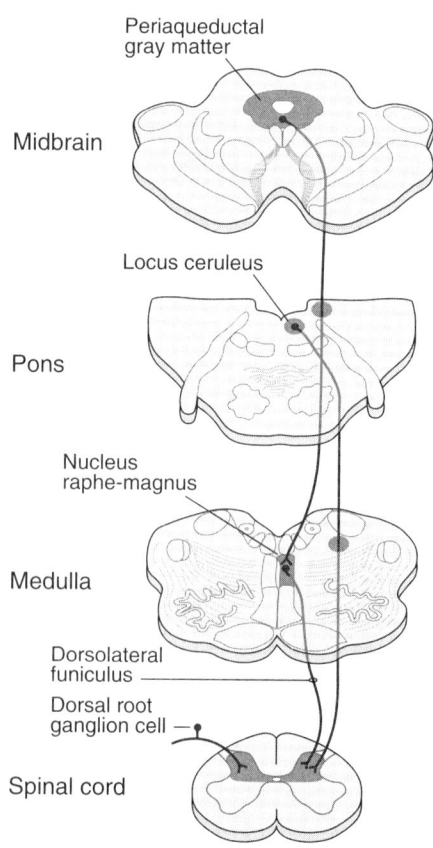

Figure 9-7. The descending nociceptive inhibitory system. *(Used with permission from Kandel ER, Schwartz JH, Jessell TM: Principles of Neural Science, 4th ed. New York: McGraw-Hill, 2000, p. 486.)*

Figure 9-8. The thalamus sectioned at the level of the ventral posterior nucleus (third-order neuron of the sensory pathway). (*Used with permission from Martin JH:* Neuroanatomy: Text and Atlas, *3d ed. New York: McGraw-Hill, 2003, p. 128.*)

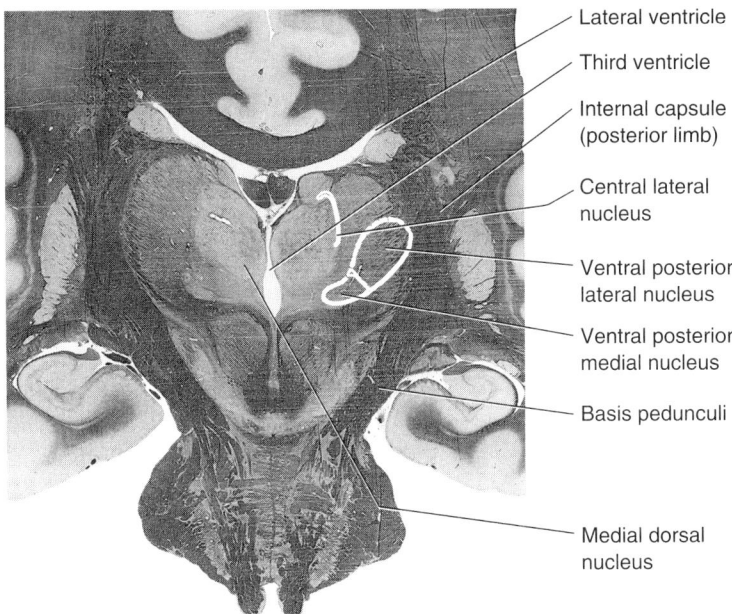

The *primary and secondary cortical somatosensory areas* are located in the postcentral gyrus of the parietal cortex and superior wall of the sylvian fissure. They receive sensory projections from the VPL and VPM nuclei of the thalamus. The cortex processes the sensory-discriminative information.

1. The *primary somatic sensory cortex (S1, sensory homunculus)* is represented by Brodmann areas 1, 2, and 3a/3b. Areas of the body used in discriminative tasks (e.g., fingers, lips, and tongue) have larger cortical representaton than those with a lower density of tactile receptors (e.g., legs and back). The body map is plastic and can undergo dynamic changes, so areas of the body that are used more for tactile exploration can enlarge (activity-dependent plasticity). The somatosensory cortex has a laminated, columnal structure. The thalamus projects mainly to layer IV, and its information is distributed to the other adjoining layers (I through VI). Most of the excitatory connections within a cortical area remain confined to a cortical column (functional unit). Areas 2 and 3a process information from mechanoreceptors located in the muscles and joints (position sense of the limbs, discrimination of shape of grasped objects, with coarser cortical representation), whereas areas 1 and 3b receive information from the skin mechanoreceptors (light touch, including texture discrimination, with very detailed cortical representation). The areas of cortical activation and the intensity of activation are weak but depend on the physiologic context in which the stimulus is applied and the person's expectation of pain.

2. The *association areas (multimodal cortical areas)* have afferent and efferent connections to S1, as well as to cortical areas that process other types of sensory information (e.g., visual, auditory, olfactory, and gustatory). They contribute to the mental construct and perception of the body image and its spatial relationships. Two well-defined such areas are the parietotemporooccipital junction and the posterior parietal cortex (Brodmann areas 5 and 7). Lesion of areas 5 and 7 in the nondominant hemisphere (usually the right in right-handed persons) leads to neglect of the contralateral part of the body (failure to dress the left half of the body or to comb the left part of the hair).

3. The *secondary somatic sensory cortex (S2)* is located on the parietal operculum and insular cortex. It is also somatotopically organized and receives tactile information necessary for object recognition. The posterior part of S2 receives information from the posterior thalamic group about high-intensity mechanical stimuli (Figs. 9-9 through 9-11).

The *parts of the limbic system* that receive information about noxious stimuli and itch and temperature modalities are the amygdaloid complex and the limbic association cortex. These are the centers that process and integrate the affective-emotional components of pain with the sensory and cognitive components. The anterior cingulate gyrus and insular cortex are consistently and intensely activated by noxious stimuli. The insular cortex processes the autonomic sensory information, thus giving the autonomic response to pain; it has direct connections to the limbic system. Resection of the prefrontal cingular cortices or the amygdala leads to isolation of pain perception from its affective-motivational

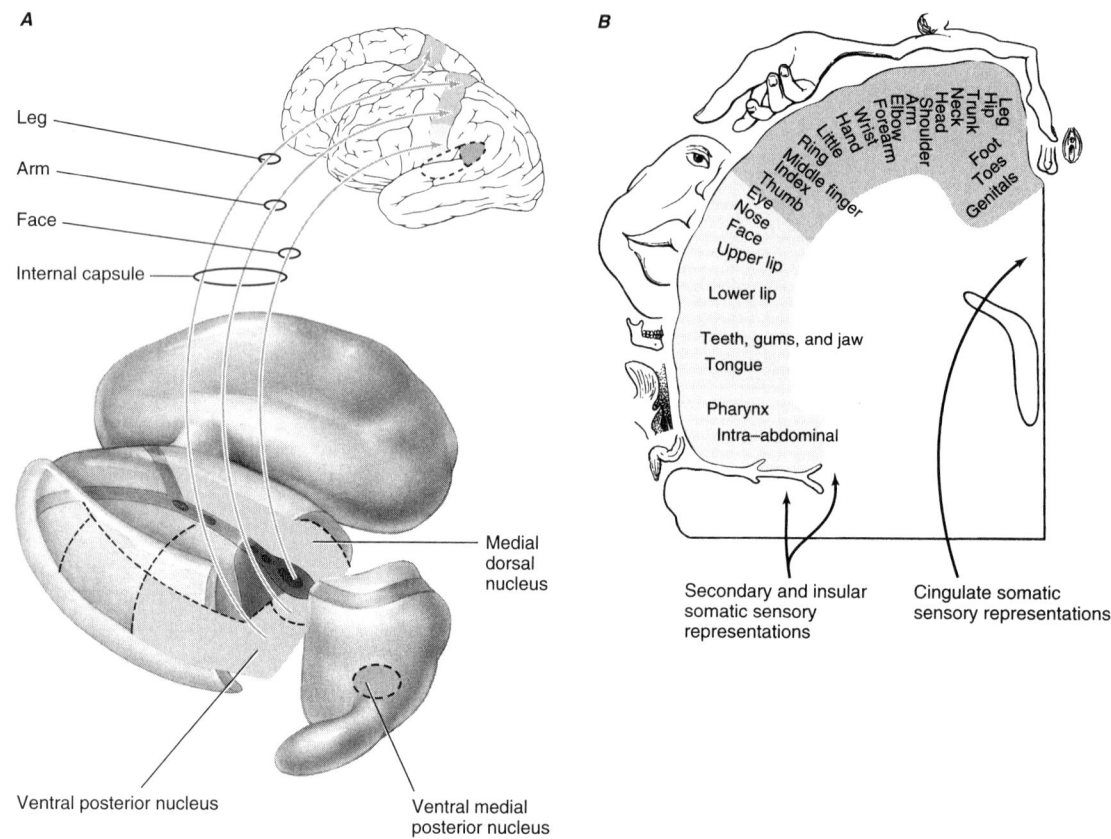

Figure 9-9. Lateral view of the cerebral hemispheres with somatotopic representation of the limbs, face, and trunk in the postcentral gyrus. (*Used with permission from Martin JH:* Neuroanatomy: Text and Atlas, *3d ed. New York: McGraw-Hill, 2003, p. 127.*)

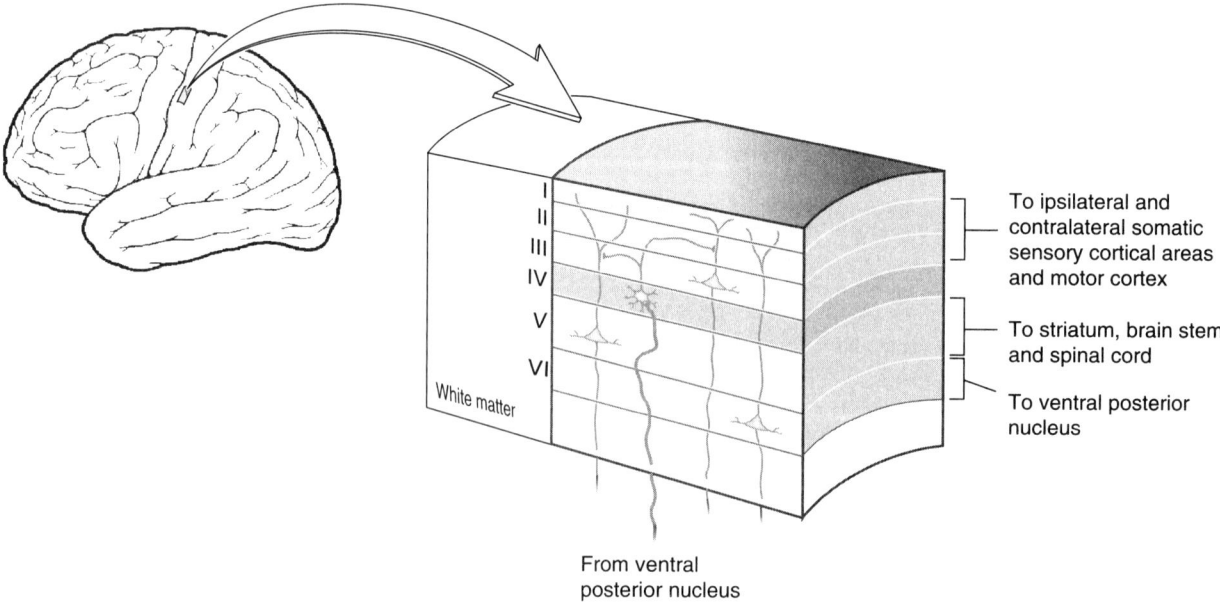

Figure 9-10. The cerebral cortex (in the postcentral gyrus) has a lamination similar to the spinal cord. Neurons in layer VI project to the thalamus, whereas those in layer IV receive input from the thalamus and connect to neurons in other cortical areas. (*Used with permission from Martin JH:* Neuroanatomy: Text and Atlas, *3d ed. New York: McGraw-Hill, 2003, p. 129.*)

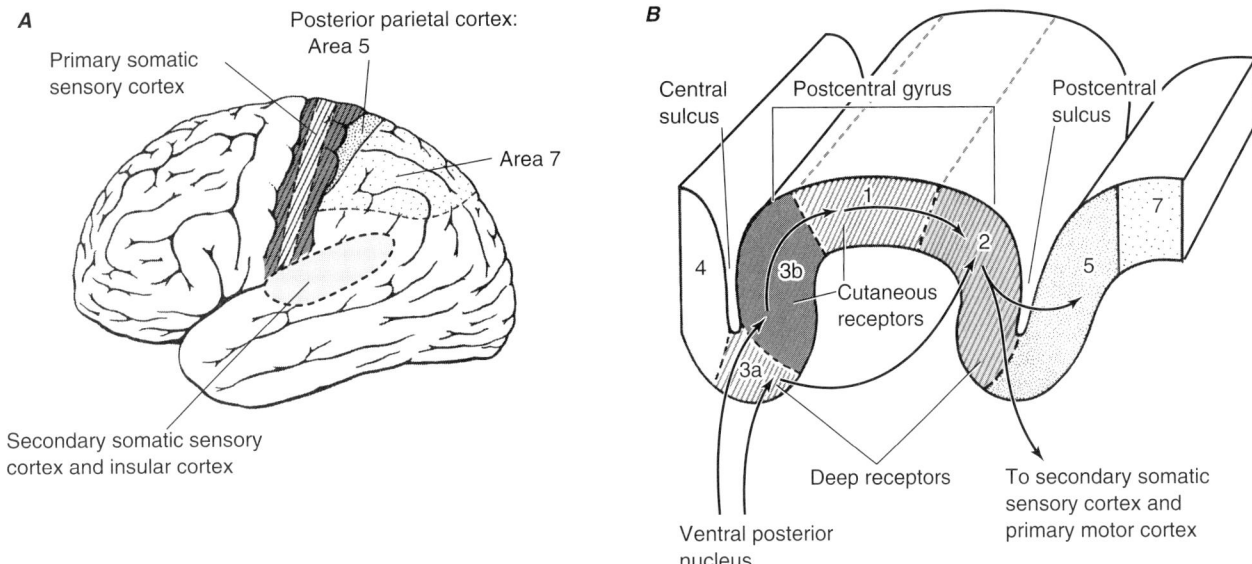

Figure 9-11. Primary and secondary somatic sensory areas in a lateral view and section. The areas in light blue represent the insular cortex and the temporoparietal operculum (which are beneath the surface). (*Used with permission from Martin JH:* Neuroanatomy: Text and Atlas, *3d ed. New York: McGraw-Hill, 2003, p. 129.*)

component and from reactions of the body to pain. Lesions of the insular cortex lead to asymbolia for pain.

▶ ANATOMY OF CEREBRAL AND BRAINSTEM CENTRAL PAIN

Central poststroke pain is one of the most difficult poststroke sequelae to treat. It has a gradual onset, with painful paresthesias in the area of the initial sensory deficit to temperature and pain. Patients with impairment of the medial lemniscal pathways but without involvement of the spinothalamic modalities also may develop pain. Most of the time, pain is in the territory of the previous sensory deficit, but it may involve only patch areas or have a pseudo-dermatomal or acral pattern (in the distal parts of the extremities). Poststroke anterior chest wall pain may mimic acute coronary syndromes. Various descriptions of pain include burning, aching, squeezing, pricking, cold, lacerating, and worsened by cold, heat, stress, and fatigue, with dysesthesias and allodynia in the painful area. A new stroke on the same side may decrease or abolish the pain, whereas a new stroke on the contralateral side may worsen it. The mechanism of pain is not well understood, but there are several theories, including lesions along the descending inhibitory pathways, hyperexcitability of the affected afferent sensory pathways, denervation hypersensitivity, and imbalance between γ-aminobutyric acid (GABA) and glutamate transmission in the central nervous system (CNS).

Thalamic pain (Dejerine-Roussy syndrome) is the classic central poststroke pain syndrome. This syndrome, however, may happen after any stroke, irrespective of the location. It also may be caused by thalamic tumors. It is an unpleasant, excruciatingly painful sensation on the body contralateral to the affected thalamus. There is significant tactile allodynia with prolonged latency between stimulus and pain and hyperpathia (pain persists after the stimulus has been removed). There is also associated sensory loss to pinprick in the affected part of the body, so this has been termed *anesthesia dolorosa*. Similar painful syndromes may occur with postherpetic neuralgia (causing damage to the DRGs) and trigeminal neuralgia after trigeminal nerve rhizotomy.

Asymbolia to pain (anosodiaphoria) represents an inappropriate response to pain in a patient with left (dominant) or bilateral parietal lobe lesions. Rarely, insular cortex lesions may cause it. Patients perceive noxious stimuli as painful and can distinguish sharp from dull pain but do not display appropriate emotional responses to pain or may even smile. Another interesting feature of this syndrome is *alloesthesia* (stimuli that are applied on the affected side are reported on the contralateral side).

Pontine strokes cause a trigeminal sensory deficit usually accompanied by decreased hearing, vertigo, and ataxia. Occasionally, the sensory symptoms are restricted to the intraoral area and may involve taste sensation. Small pontine strokes may cause symptoms similar to trigeminal neuralgia.

Lateral medullary infarct (Wallenberg syndrome) is due to infarction of a penetrating branch from the

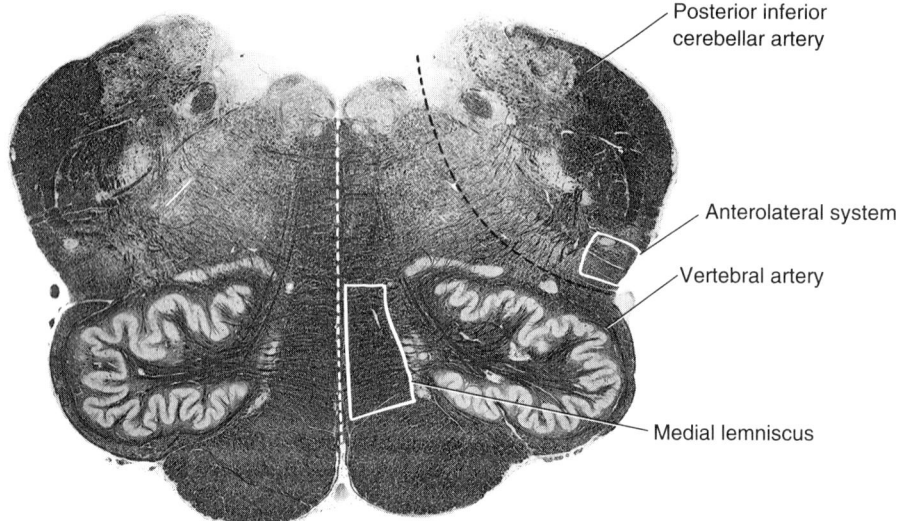

Figure 9-12. Transverse section at the level of medulla oblongata. Infarct of the territory supplied by the posteroinferior cerebellar artery gives rise to Wallenberg syndrome, affecting the anterolateral sensory system. (*Used with permission from Martin JH:* Neuroanatomy: Text and Atlas, *3d ed. New York: McGraw-Hill, 2003, p. 124.*)

posteroinferior cerebellar artery or vertebral artery. Patients present with gait ataxia, clumsiness of the ipsilateral limbs, dizziness or spinning sensation, blurred or double vision, decreased sensation to pain and temperature of the ipsilateral face and contralateral limbs, hoarseness, dysarthria, dysphagia, and hiccups. Occasionally, they may complain of headache that is continuous, unilateral, hemicranial, hemifacial (constant or paroxysmal, ipsilateral to the lesion, with an epicenter around the eye, associated with hypesthesia), or occipital (if the mechanism of stroke is vertebral artery dissection). Some patients may develop central poststroke pain in the areas affected by the sensory deficit (Fig. 9-12).

▶ ANATOMY OF CRANIAL PAIN

Anatomy of the Meninges

The meninges are three protective layers of the brain and spinal cord:

1. The *dura mater (pachymeninges)* is the outermost layer, made of dense connective tissue that is adherent to the inner surface of the skull. It is rich in blood vessels (middle meningeal artery and dural venous sinuses) and free nerve endings. The dura forms septa (falx cerebri, tentorium cerebelli, and diaphragma sellae) that divide the cranial cavity into compartments (lateral compartments for the cerebral hemispheres and posterior compartment for the cerebellum and brain stem). The dura is supplied by the middle meningeal artery (branch of the maxillary artery from the external carotid), the anterior meningeal artery from the ophthalmic artery (off the internal carotid), and posterior meningeal arteries (from the occipital and vertebral arteries). Skull fractures may lead to lacerations of the dural arteries and veins, giving rise to epidural hematomas. The supratentorial dura (of the anterior and middle cranial fossa) is innervated by branches from the trigeminal nerve. The infratentorial dura is innervated by the C2 nerve and ninth and tenth cranial nerves. The spinal dura continues the cranial dura from the foramen magnum to the level of S2. It extends caudally along the *filum terminalis* to the coccyx to form the coccygeal ligament and then becomes continuous with the periosteum. The dura also surrounds the nerve roots, forming the dural sleeves.

2. The *arachnoid* is the next inner layer, made of reticular, weblike, avascular fibers. It passes over the sulci but not between them. At the base of the brain, the subarachnoid space becomes larger than over the hemispheres and forms the subarachnoid (basal) cisterns. Cerebrospinal fluid (CSF) flows from the fourth ventricle through the foramina of Luschka and Magendie into the cisterna magna (cerebellomedullary cistern) and then upward into the prepontine, interpeduncular, magna, and suprachiasmatic cisterns. CSF also flows downward toward the conus medullaris and forms the lumbar cistern (extending from L2 to S2) and contains the filum terminale and cauda equina. CSF is obtained via spinal tap from the

lumbar cistern. The arachnoid villi and granulations are tufted extensions of the pia-arachnoid complex that protrude through the dura into the superior sagittal sinus. They are passive one-way valves where the CSF is absorbed into the venous system.

3. The *pia mater* is the innermost layer and is thin, translucent, and adherent to brain and spinal cord surface. It invaginates around the cerebral vessels. The spinal pia gives rise to fibers that attach medially to the spinal cord (midway between the dorsal and ventral roots) and laterally to arachnoid and dura. These fibers form the denticulate ligaments, which continue throughout the length of the spinal cord caudally to coat the filum terminale. The pia and arachnoid form the *leptomeninges*.

Anatomical Sources of Headache

- Intracranial structures:
  - Cerebral and dural arteries
  - Venous sinuses
  - Dura
- Extracranial structures:
  - Cervical roots
  - Cranial nerves
  - Cervical nerve roots
  - Extracranial arteries
  - Pericranial muscles
  - Skull periosteum and sinuses

Patterns of Pain Referral

- Supratentorial structures (dura and blood vessels) via trigeminal nerve into the face
- Infratentorial structures via the ninth and tenth cranial nerves and the C2 dorsal root to the occiput, neck, and over the skull convexity above the eyebrow

Pain Locations and Examples of Corresponding Pathology

- Pain above the eye: pituitary tumor
- Pain between the eyes or at the vertex: sphenoid sinus pathology
- Pain behind the eye: aneurysm of the posterior communicating or internal carotid arteries, retrobulbar (optic) neuritis
- Pain in the inner angle of the orbit: internal carotid artery pathology
- Pain in the eye: retinal detachment, glaucoma
- Anterior temporal pain: temporal arteritis, aneurysm or thrombosis of the middle cerebral artery, middle meningeal artery, occipital lobe hemorrhage
- Occipital pain: basilar artery stroke

▶ ANATOMY OF FACIAL PAIN

Optic neuritis is an inflammatory disease of the optic nerve. It may be caused by autoimmune diseases (e.g., multiple sclerosis, connective tissue diseases, after infections, or after vaccinations), granulomas (e.g., sarcoidosis), and infections (e.g., viral, bacterial, spirochetal, or parasitic). Most of the time it is associated with pain in or behind the eye, worsened by eye movements. Pain may precede the visual loss.

Eye pain (ophthalmodynia) may be a manifestation of cluster headache, migraine, paroxysmal hemicrania, hemicrania continua, or sudden-onset neuralgiform pain with conjunctival injection and tearing (SUNCT) syndrome. These conditions cause unilateral pain that rarely may be worsened by eye movements. If there is bilateral pain with eye movement, it is usually the expression of meningeal irritation.

Painful ophthalmoplegias may be caused by ophthalmoplegic migraine (headache associated with extraocular eye movement palsy). Rarely, diabetic and vasculitic third or sixth nerve palsies may be painful.

The *sphenopalatine ganglion* in the pterygopalatine fossa is associated with the maxillary nerve (V2), and receives parasympathetic afferents (from the superior salivary nucleus via the palatine, greater superficial petrosal, and vidian nerves), sympathetic branches (from the internal carotid plexus via the greater deep petrosal nerve), and sensory fibers (from V2 and the seventh and ninth cranial nerves via the tympanic plexus and vidian nerve). It gives efferents via the pharyngeal rami to the roof of the pharynx (likely from the pharyngeal rami); to the nasal cavity, uvula, tonsil, and hard palate (via the nasal and palatine rami); and to the lacrimal glands and orbital periosteum (via the orbital rami). Cluster headache consists of attacks of unilateral, often nocturnal pain of the face and eye with conjunctival congestion and tearing with nasal discharge. It may be associated with Horner's syndrome (Fig. 9-13).

The *semilunar (gasserian) ganglion* is located in Meckel's cave, in the middle cranial fossa, near the petrous apex and lateral to the cavernous sinus and represents the origin of the trigeminal sensory afferents. Inflammatory or infectious processes involving the middle cranial fossa (e.g., tumor, herpes zoster, tuberculosis, arachnoiditis, trauma, or abscess) may damage the ganglion and cause severe paroxysmal pain involving half the face in a V2/V3 distribution. Pain may be associated with numbness in the same territory and weakness of the masticatory muscles. Autoimmune/connective tissue disorders (e.g., Sjögren's disease, systemic sclerosis, or lupus) may cause facial numbness with or without pain and paresthesias in a V2 distribution (unilateral or bilateral).

Raeder's paratrigeminal neuralgia consists of aching, burning, and frontotemporal or periorbital unilateral facial pain (in trigeminal nerve distribution) associated with

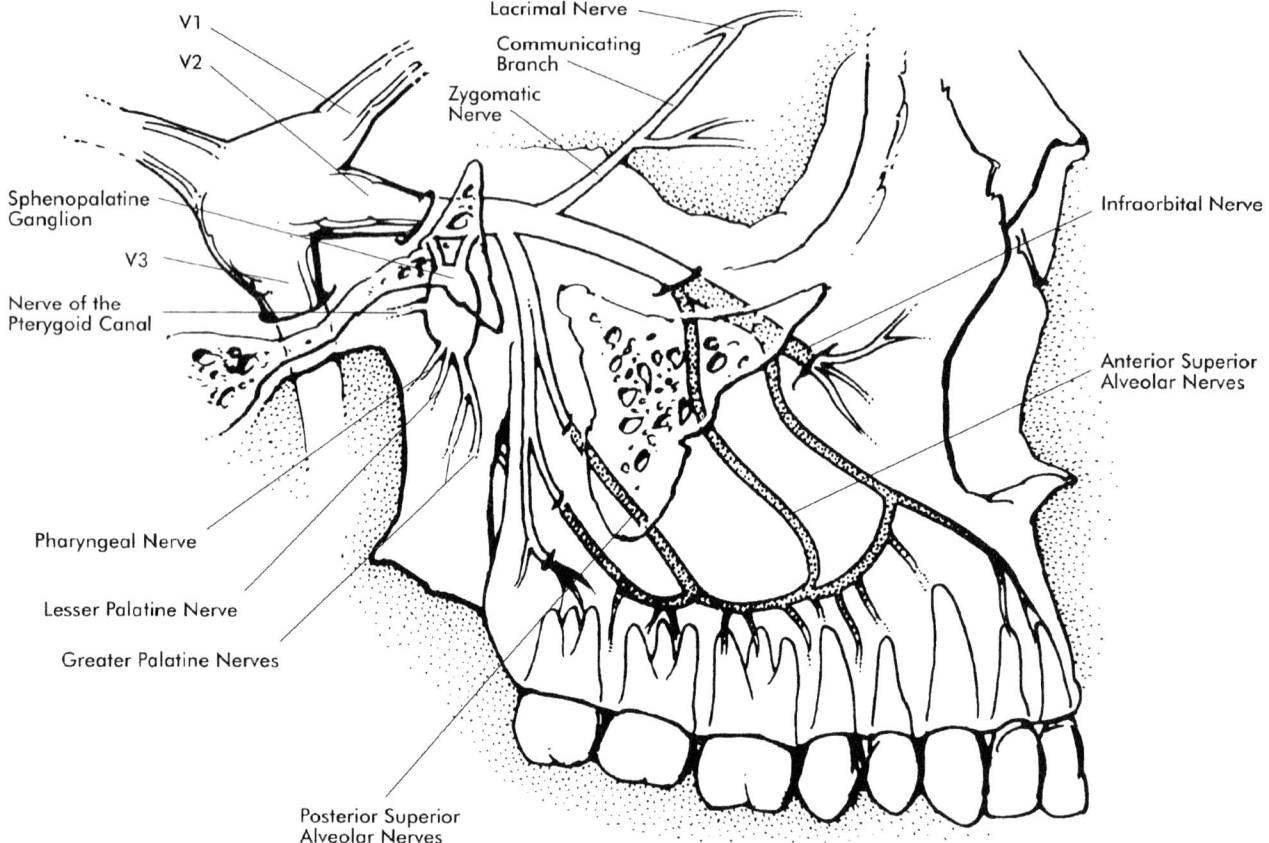

Figure 9-13. The sphenopalatine ganglion and the branches of the maxillary division of the trigeminal nerve (V2). (*Used with permission from Brazis PW, Masdeu JC, Biller J: Localization in Clinical Neurology, 3d ed. Boston: Little, Brown, 1996, p. 257.*)

Horner's syndrome (i.e., ptosis, miosis, and oculosympathetic paralysis). It may be associated with involvement of the third, fourth, and sixth cranial nerves. It is caused by lesions of the carotid siphon or middle cranial fossa at the petrous apex or around sella turcica (e.g., tumor, carotid artery aneurysm or carotid dissection, infection, or trauma). The type of pain may correlate with the underlying pathology, similar to trigeminal neuralgia (episodic pain suggests cluster headache, and constant pain is associated with the anatomic abnormalities mentioned earlier).

Tolosa-Hunt syndrome and *orbital pseudotumor* are manifested by orbital pain radiating to the eyebrow, temple, or occiput. Orbital pseudotumor may have other associated orbital signs, such as proptosis, extraocular eye muscle palsies, and orbital mass. Tolosa-Hunt syndrome frequently results from granulomas and inflammatory lesions (e.g., tuberculosis, syphilis, and vasculitides) involving the wall of the cavernous sinus or dura of the sphenoid bone.

Superior orbital fissure syndrome shares many features in common with Tolosa-Hunt and orbital pseudotumor syndromes. All three of these entities actually may represent different manifestations of a disease process affecting the same anatomic structures. The superior orbital fissure syndrome consists of pain and sensory deficit in the distribution of several branches of the trigeminal nerve (i.e., lacrimal, supraorbital, supratrochlear, zygomatic, infraorbital, and sometimes maxillary). It results from infectious, inflammatory, or infiltrative processes (e.g., cavernous sinus thrombosis, primary malignancy, or metastasis to the orbit). The patient also may have ptosis, proptosis (or, on the contrary, enophthalmos), restriction of eye movements, chemosis, and periorbital swelling.

Pituitary apoplexy is hemorrhagic infarction of the pituitary gland in patients with pituitary tumors and is a neurological emergency. It presents with sudden onset or worsening headaches, double vision (due to extraocular eye muscle palsy) or blurred vision, visual loss, visual field defects (most classic, bitemporal hemianopia), nausea, vomiting, altered mental status that may progress to coma, fever, hypotension leading to circulatory collapse, and respiratory failure.

Gradenigo syndrome consists of pain and sensory deficit in a V1 distribution, ipsilateral abducens (sixth) nerve palsy, and possibly with Horner's syndrome (without anhydrosis). It is due to infectious (e.g., osteitis or meningitis associated with otitis media) or tumoral processes involving the apex of the temporal bone.

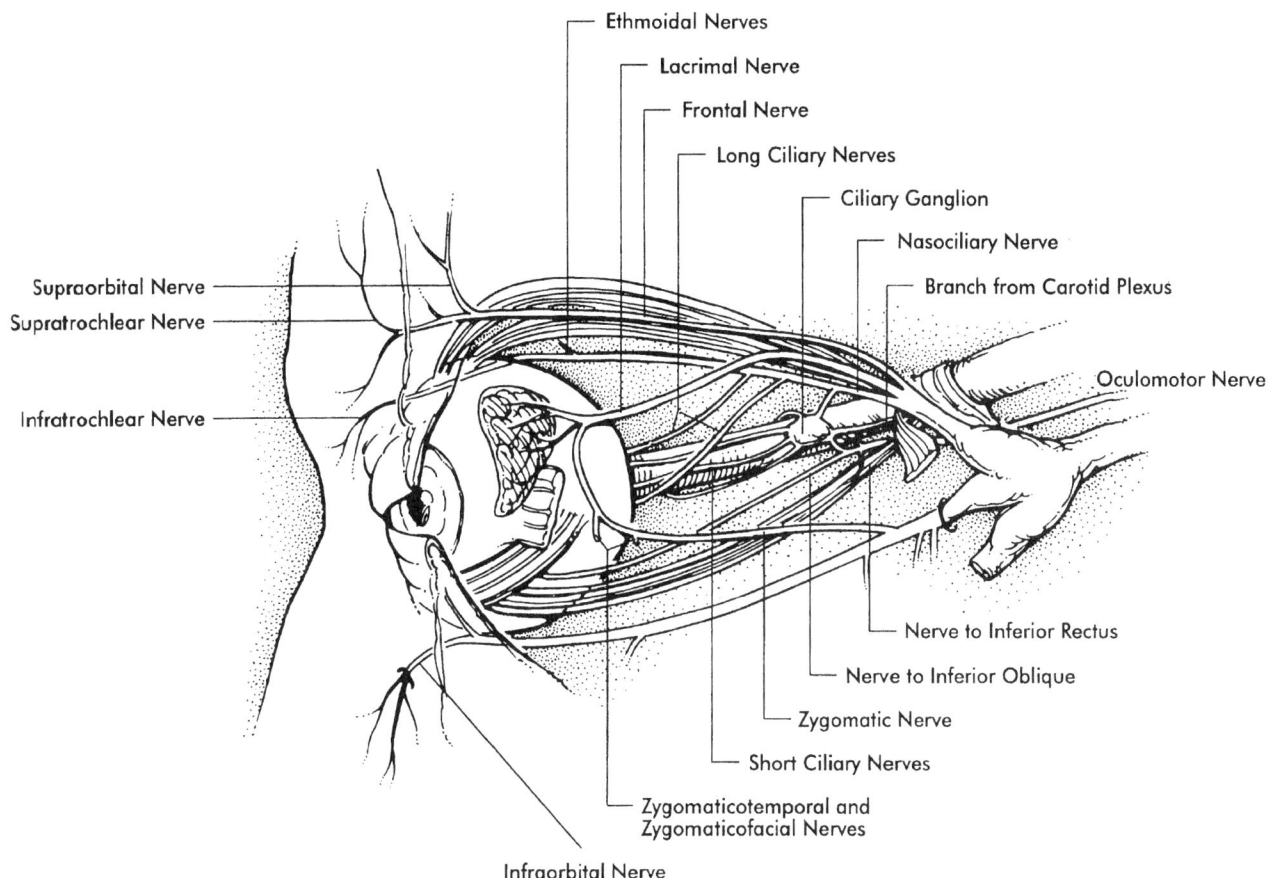

Figure 9-14. The branches of the ophthalmic division of the trigeminal nerve (V1). (*Used with permission from Brazis PW, Masdeu JC, Biller J:* Localization in Clinical Neurology, *3d ed. Boston: Little, Brown, 1996, p. 254.*)

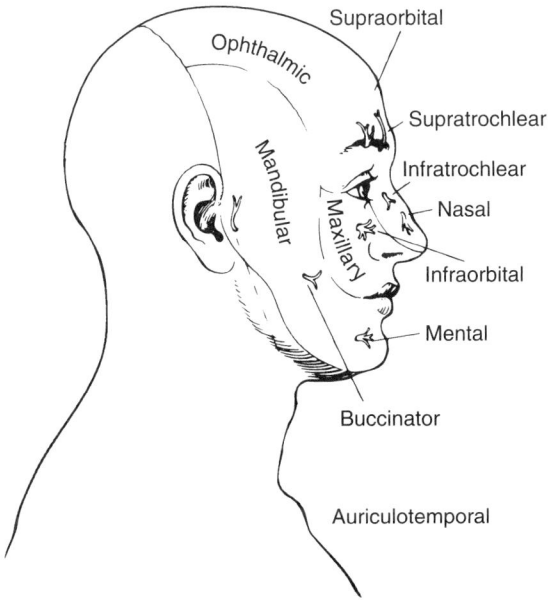

Figure 9-15. Trigeminal nerve and its branches: territories of skin innervation. (*Used with permission from Brazis PW, Masdeu JC, Biller J:* Localization in Clinical Neurology, *3d ed. Boston: Little, Brown, 1996, p. 257.*)

Trigeminal neuralgia (tic douloureux) is manifested by paroxysmal, lancinating, severe, unilateral pain in the distribution of one or more of the trigeminal nerve divisions [most commonly mandibular (lower lip) or maxillary (upper lip, nose, and cheek)]. Pain paroxysms last a few seconds up to a minute and may be triggered by tactile stimuli applied on the face, a blow of air, chewing, jaw movements, and drinking hot or cold liquids. The patient may respond to the severe pain by contraction of the facial muscles on the affected side (hemifacial spasm). It is seen mostly in individuals aged 50 to 60 years. It may be idiopathic tic douloureux or secondary due to compression, demyelination, or irritation affecting the trigeminal nerve, gasserian ganglion, trigeminal dorsal root entry zone, or brainstem. The pathology that can cause trigeminal neuralgia includes conditions such as a demyelinating plaque, an aberrant branch of the superior cerebellar artery, and a cerebellopontine angle tumor. Bilateral trigeminal neuralgia is characteristic of multiple sclerosis (Figs. 9-14 through 9-16).

Maxillary neuralgia may result from lesions affecting the lateral wall of the cavernous sinus, foramen rotundum, pterygopalatine fossa, orbital foramen, or the facial

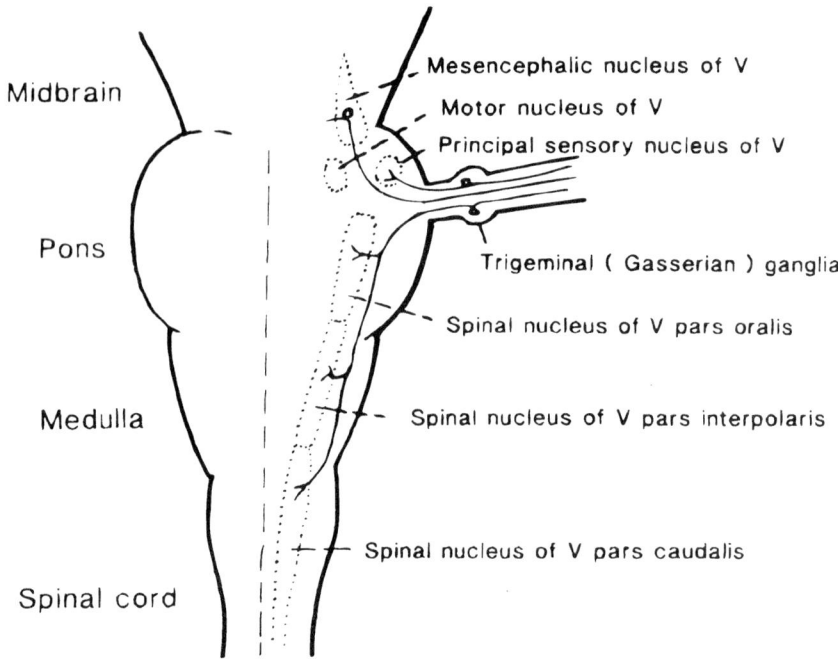

Figure 9-16. The trigeminal nerve and its origin in the brain stem. Note the locations of its sensory and motor nuclei. (*Used with permission from Brazis PW, Masdeu JC, Biller J:* Localization in Clinical Neurology, *3d ed. Boston: Little, Brown, 1996, p. 252.*)

territory of the maxillary nerve. These can be represented by nasophayngeal, cavernous sinus, or orbital tumors, as well as facial trauma. The "numb cheek" syndrome consists of hypesthesia in the cheek, upper lip, upper incisors, canines, and adjacent gum caused by lesions affecting the infraorbital foramen (most commonly squamous cell carcinoma of the skin).

Mandibullary neuralgia is manifested as pain and numbness in the lower lip, chin, and mucous membranes inside the lower lip. There may be associated weakness of the ipsilateral masticatory muscles. It results from lesions affecting the maxillary nerve in its trajectory through the foramen ovalae, zygomatic fossa, or face. The "numb chin" syndrome consists of pain and numbness in the above-mentioned distribution and usually results from trauma, metastasis, or malignant infiltration of the mental nerve. Patients with sickle cell disease may develop a burning sensation in the lower lip that is later followed by numbness as a result of infarction of the mental or inferior alveolar nerve (Fig. 9-17).

Atypical facial pain is a syndrome associated with pain that does not follow solely the distribution of a cranial nerve. Sometimes it starts after a dental procedure or facial trauma, and it may be caused by pathology of the eyes, teeth, nose, and sinuses. The location of the pain is in the mouth, around the mouth, or in the face, and it may radiate to the cervical region. It is diffuse, boring, gnawing, deep, aching, and never sharp or paroxysmal. The duration varies between hours and days. Trigger points are uncommon. There may be precipitants such as tactile or cold stimulation of the face and oral cavity (similar to trigeminal and glossopharyngeal neuralgias). As opposed to these well-defined neuralgias, it is refractory to diagnostic blocks of the trigeminal or glossopharyngeal nerve and surgical ablative procedures.

Otalgia (ear pain) can be primary or secondary. It is described as constant or intermittent, burning, aching, throbbing, stabbing, lancinating, and electric shock–like. The severity of pain does not necessarily correlate with the severity of the underlying illness. *Primary otalgia* is ear pain resulting from pathologic conditions of the ear (i.e., mastoid, external ear, ear canal, tympanic membrane, and even otitic barotraumas and barootalgia as in air travelers) and is more common in children. In these conditions, there may be mild redness of the tympanic membrane or mild swelling of the external auditory canal, facial palsy, sensorineural deafness and a sensation of fullness, and "popping" sounds in the ear. *Secondary otalgia* is pain referred to the ear from nonotologic sites (e.g., teeth, temporomandibular joint, oral cavity, tongue, pharynx, larynx, esophagus, neck and cervical spine, and aberrant/aneurysmal carotid artery in the middle ear). It is more common in adults. Pain is referred in the distributions of the fifth, sixth, ninth, and tenth cranial nerves or the cervical plexus. It may be constant or paroxysmal (similar to trigeminal neuralgia) and may involve the external auditory canal and deep structures of the face. It may be triggered by cold, noise, swallowing, or pressure on the tragus. Diseases that may include secondary otalgia are nasopharyngeal carcinoma, trigeminal neuralgia, trigeminal/glossopharyngeal/hypoglossal nerve tumors, chronic paroxysmal hemicrania, whiplash injury, cervical radiculopathy, carotidynia, and myofascial pain syndromes.

Cogan's syndrome presents with eye pain, photophobia, and lacrimation (intersitial keratitis), followed

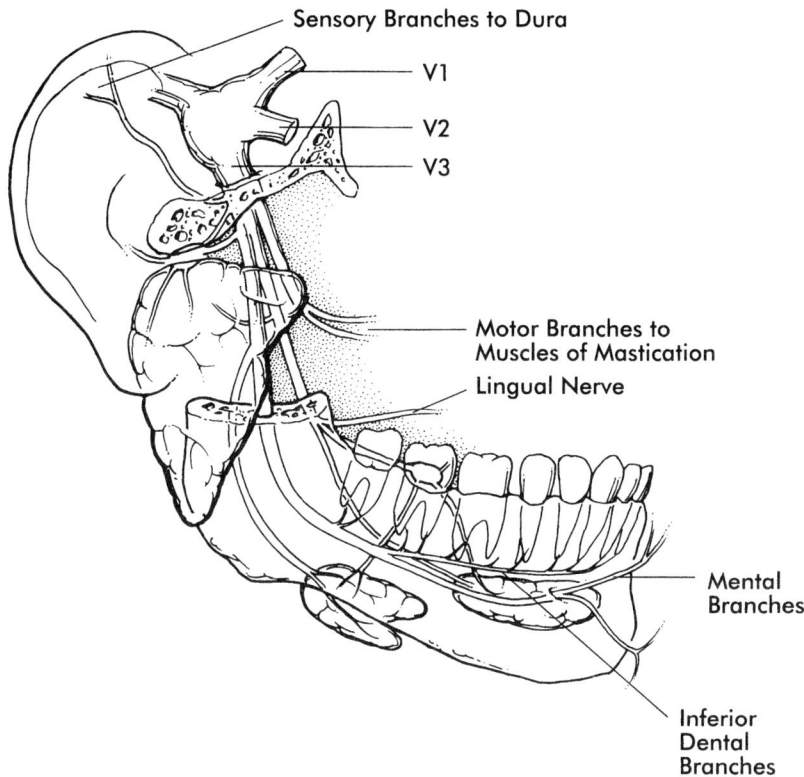

Figure 9-17. The mandibular division of the trigeminal nerve and its branches. (*Used with permission from Brazis PW, Masdeu JC, Biller J:* Localization in Clinical Neurology, *3d ed. Boston: Little, Brown, 1996, p. 258.*)

after a few months by Ménière-like attacks (i.e., vertigo, tinnitus, nausea, vomiting, oscillopsia, and hearing loss). The syndrome is usually caused by a necrotizing vasculitis that is associated at some point in its temporal course with systemic manifestations.

Geniculate neuralgia is a rare condition affecting the sensory branch of the facial (seventh) nerve. It consists of pain behind and within the ear (i.e., the external auditory meatus, walls of the auditory canal, and tympanic membrane) that may be associated with loss of taste, vertigo, tinnitus, and hearing deficit. The most common etiology is herpes zoster infection of the genicular ganglion, tympanum, or concha with peripheral facial palsy (Ramsay Hunt syndrome).

Glossopharyngeal neuralgia is an uncommon pain syndrome similar to trigeminal neuralgia. It represents pain triggered by swallowing, chewing, coughing, talking, and yawning that may be paroxysmal or constant. It radiates to the base of the tongue, throat, and ear (in the sensory distribution of the glossopharyngeal and vagus nerves) and occasionally into the larynx, tonsils, and beneath the angle of the jaw. The paroxysms last seconds to minutes and may be associated with salivation, hoarseness, syncope, or convulsions (due to bradycardia, asystole, and hypotension mediated by the baroreceptors in the carotid sinus). Occasionally, there may be a constant, dull, aching, pressure-like, or burning pain. The syndrome may be idiopathic or secondary to lesions in the posterior fossa on the trajectory of the glossopharyngeal nerve (e.g., tumors, infections, trauma, or demyelinating plaque at the dorsal root entry zone). The diagnosis can be made by relieving pain by application of topical anesthetic to the pharynx or injection of local anesthetic at the jugular foramen.

Spinal accessory nerve neuralgia may occur as a complication of surgery (e.g., lymph node biopsy, carotid endarterectomy, coronary artery bypass surgery, or radical neck resection for malignancy), cannulation of the internal jugular vein in the neck, shoulder trauma, stretch injuries caused by lifting heavy objects with concomitant head turning to the opposite side, and compression by lymphadenopathy or neoplasms. Weakness of the trapezius and sternocleidomastoid muscles may be associated with severe burning pain and allodynia in the shoulder and lateral neck area.

▶ NON-NEUROLOGICAL FACIAL PAIN MIMICKING CRANIAL NEURALGIAS

Eye Pathology

- Closed-angle glaucoma is associated with severe facial pain with diffuse radiation, edematous cornea, red eye that feels hard and tender to palpation, midriasis poorly reactive to light, blurred vision, nausea, and vomiting.
- Dysthyroid ophthalmopathy, i.e., unilateral or bilateral proptosis and sometimes diplopia, may be

associated with local signs of irritation and increased intraocular pressure.
- Orbital cellulitis: decreased vision, diplopia, local inflammation, and increased intraocular pressure.
- Endophthalmitis: photophobia, blurred vision, and local signs of inflammation.
- Eyelid disorders: palpebral edema may mimic ptosis and signs of local inflammmation.

Sinus headache (rhinosinusitis) may result from acute or chronic inflammatory and infectious processes. Pain is perceived in more than one region of the head, face, ears, or teeth and is concomitant with the onset of nasal discharge. Once the sinus infection is treated successfully and the pus drains, the headache should resolve. Sphenoid sinusitis pain may radiate to frontal, temporal, or occipital areas and only rarely to the vertex. Since it is a sphenoid closed cavity and does not have spontaneous drainage pathways, any maneuvers that increase intracranial pressure (bending, coughing, and even standing and walking) may worsen the headache.

Temporomandibular joint (TMJ) syndrome presents with pain in the masticatory muscles and preauricular area that radiates to the neck, face, and head. Pain is worse with jaw movements. Patients have jaw deviation on mouth opening, painful clicking sounds, locking of the jaw, or new changes in occlusion of teeth. The pathogenesis of TMJ pain may be myofascial and may trigger or maintain chronic or episodic headaches.

Chronic dental disease may be associated with facial pain. Atypical odontalgia represents chronic dental pain for which no local cause can be identified. To date, the relationship between atypical odontalgia and headache is unclear.

▶ ANATOMY OF MYELOPATHIC PAIN

Hemisection of the spinal cord (Brown-Sequard syndrome) can be caused by trauma or tumor encroachment of the spinal cord from the ispilateral side. It is manifested by loss of pain and temperature sensation contralateral to the hemisection (one or two segments below the level of the lesion), ipsilateral loss of position and vibration sensation, and position and ipsilateral spastic hemiparesis. At the level of the lesion there may be weakness and numbness caused by damage of the nerve roots and anterior horn cells. The anatomic organization of the two sensory systems (spinothalamic and dorsal columns) is different, and therefore, there will be two different levels of sensory impairment for touch and pain sensation. Touch sensation will be impaired at the level of the injury, whereas pain modality will be impaired starting two dermatomes below the area of spinal cord injury (Fig. 9-18).

Central cord syndrome may be seen in cavities in the center of the cord (e.g., syringomyelia, hydromyelia, and spinal cord tumors). Initial symptoms result from interruption of the decussating fibers of the spinothalamic tract (analgesia and thermoanesthesia in a "capelike distribution" with sparing of sacral sensation) by the centrally situated lesion. Therefore, sensory loss is bilaterally symmetric and spares the mechanical sensation. Body regions below the lesion are not affected (*suspended sensory deficit*). Proprioception (vibration and position) is preserved. Later, when the cavity expands more laterally and in the ventrodorsal direction, the fibers of the corticospinal tract and the dorsal columns are affected, and patients develop weakness, atrophy, and loss of proprioception. When central cord syndrome results from trauma (e.g., hyperextension injuries to the cervical spine), weakness is more severe in the arms than in the legs and distally more than proximally ("man in a barrel" syndrome) (Fig. 9-19).

Funicular (central) pain is a deep, poorly localized dysesthesia, common with intramedullary lesions. It is probably caused by abnormal functioning of the sensory pathways (spinothalamic and dorsal columns).

Dorsal columns disease (posterolateral cord syndrome) results from tabes dorsalis, B_{12} deficiency, multiple sclerosis, viral infections (such as HIV and HTLV), spondylosis, or radiation myelopathy. Patients present with impaired vibration and position sense and tactile discrimination and have gait ataxia initially in the dark (when visual control is poor). Later they develop trophic changes in the joints (Charcot or analgesic joints). There may be spontaneous lancinating sensations down the back or into the arms or legs, triggered by neck flexion (*Lhermitte's phenomenon*). Patients also have impaired control of micturition and defecation and lose their ankle and knee reflexes.

Anterior spinal artery syndrome is caused most commonly by hypotension/hypoperfusion during abdominal aortic surgery. It leads to ischemia of the anterior funiculi, anterior horns, and anterior part of lateral columns. The presentation is acute with radicular pain, followed by flaccid weakness and thermal anesthesia and analgesia below the level of the lesion. Proprioception is preserved because the dorsal columns are perfused by the posterior spinal arteries. Patients may develop burning dysesthesias below the level of the spinal cord injury. The mechanism of pain development is believed to be the sensory imbalance between the two sensory modalities: loss of temperature and pain with preservation of vibration, position, and pressure modalities.

▶ ANATOMY OF RADICULAR PAIN

Radicular pain results from irritative lesions of a dorsal root. Pain is sharp, lancinating, electric or burning, or "pins and needles" in quality in one or more dermatomes.

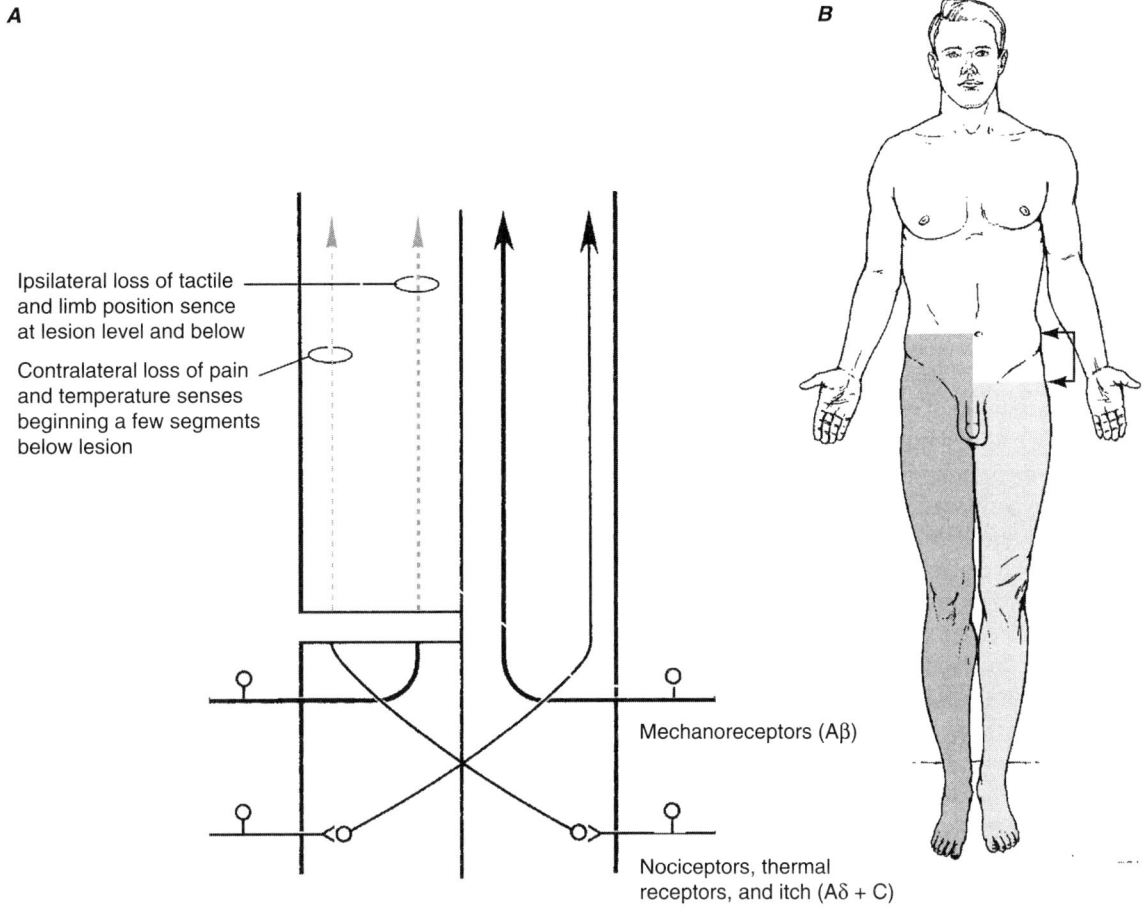

Figure 9-18. The patterns of decussation of the dorsal columns–medial lemniscal pathways. In Brown-Sequard syndrome, there is loss of pain, temperature, and itch one to two segments below the level of the lesion on the contralateral side and at the level of the lesion on the ispilateral side. (*Used with permission from Martin JH:* Neuroanatomy: Text and Atlas, *3d ed. New York: McGraw-Hill, 2003, p. 121.*)

It is worse with movements that increase the intraspinal pressure (e.g., coughing, sneezing, or straining) or stretch the nerve root (e.g., flexion/extension and lateral movements of the spine). It is more common with extradural masses (i.e., tumors, abscesses, and hematomas) and rare with intramedullary lesions. It may be the initial symptom and sign of a radiculopathy. Later it is accompanied by weakness, atrophy and sensory loss in a specific root distribution.

Painful Cervical Radiculopathies and Neuralgias: Cervical Plexus (C1–4 Ventral Rami)

Sensory Branches
- The greater occipital nerve (C2) innervates the skin of the posterior scalp.
- The lesser occipital nerve (C2) innervates the skin over the mastoid process and lateral aspect of the scalp.
- The greater auricular nerve (C2–3) innervates the skin of the lower cheek over the mandible, lower part of the external ear, and upper neck below the ear.
- The transverse colli (C2–3) innervates the skin of the anterior aspect of the neck.
- The supraclavicular nerves (C3–4) innervate the skin above the clavicle.

Motor Branches
- The ansa hypoglossi (C1–3 nerve fibers joined with the twelfth nerve) innervates the neck flexor muscles with hyoid insertion (sternohyoid, thyrohyoid, geniohyoid, and omohyoid).
- The phrenic nerve (C3–5) innervates the diaphragm.
- The scapular branches stem from the C3–4 ventral rami.
- Motor branches to the accessory (eleventh cranial) nerve innervate the sternocleidomastoid and trapezius (Fig. 9-20).

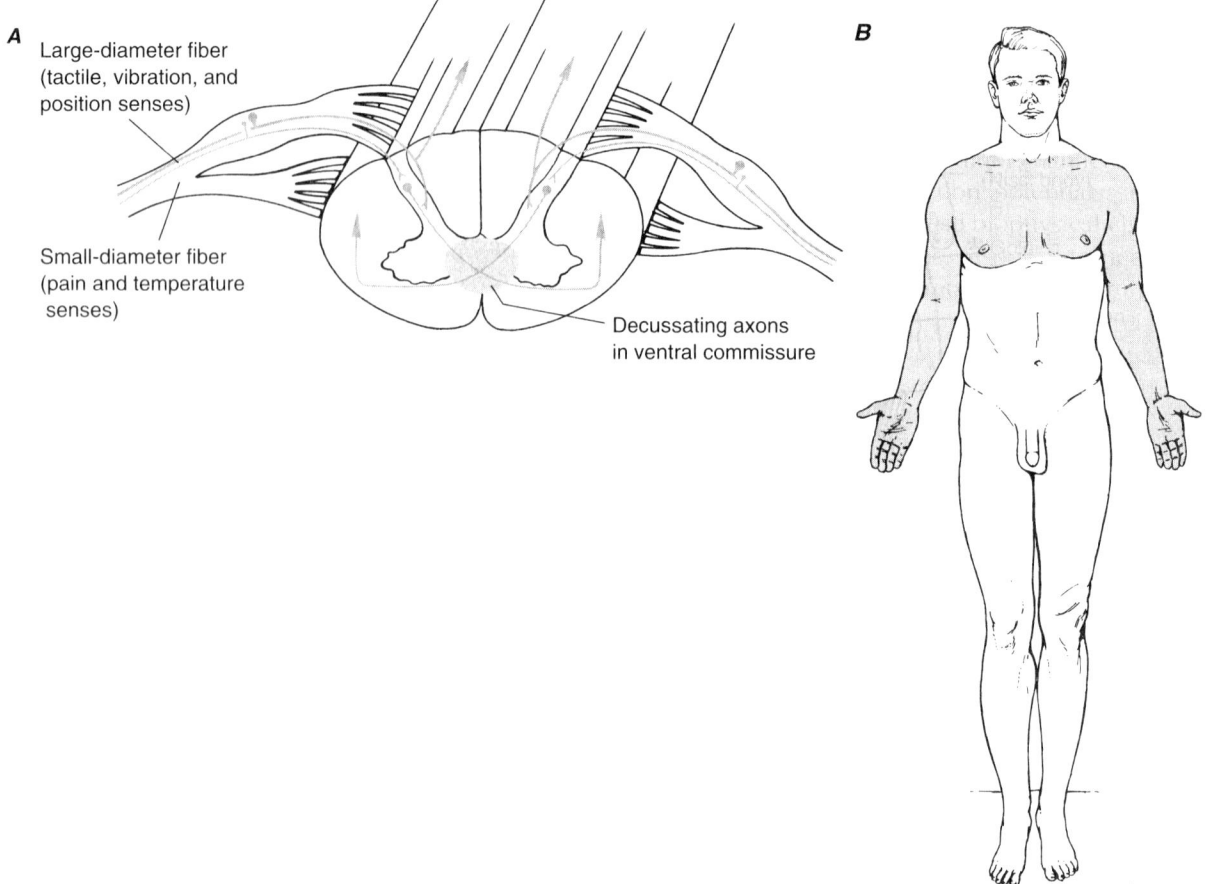

Figure 9-19. Syringomyelia. The decussating fibers of the anterolateral system are the first to be interrupted by the formation of the cavity (syrinx) represented in gray. Also, the distribution of sensory deficit is highlighted in gray on the human body. (*Used with permission from Martin JH:* Neuroanatomy: Text and Atlas, *3d ed. New York: McGraw-Hill, 2003, p. 122.*)

C2 ganglionopathy and radiculopathy are commonly due to spondylosis that may be preceded or not by previous neck trauma, whiplash injury, or a thickened ligament entrapping the C2 DRG and nerve. The C2 nerve rarely may be compressed between the C1 and C2 laminae by fractures and dislocations, congenital abnormalities of the cervical spine, subluxation of the atlantoaxial joint in rheumatoid arthritis, Paget's disease, or osteomalacia. The C2 ganglion and occipital nerve blocks have diagnostic (neuropathic versus non-neuropathic, ganglion/root/nerve lesion) as well as therapeutic value.

Greater occipital neuralgia usually is due to compression of the nerve by the semispinalis and trapezius but also may happen by external compression (such as falling asleep with the head leaned against the back of a chair or trauma to the neck, as in whiplash injuries). It presents with suboccipital pain, sometimes radiating to the vertex. It may be associated with hyperalgesia of the scalp, pinprick hypesthesia in a C2 distribution, and tenderness over the greater occipital nerve. There may be associated vertigo due to abnormal impulses from the uncovertebral joints and muscle spindles in the neck muscles to the vestibular nuclei in the brain stem.

Third occipital neuralgia usually is caused by a whiplash injury or degenerative disease of the C2–3 facet joints and is manifested by occipital and suboccipital pain. The third occipital nerve is a branch of the C3 dorsal ramus. The condition responds to third occipital nerve blocks.

Greater auricular neuralgia is due to lacerations or facial or neck surgeries (facial lift, carotid endarterectomy) where the nerve crosses the sternocleidomastoid at the angle of the jaw as it passes through the parotid gland. There may be persistent numbness in and around the ear, occasional painful paresthesias, and formation of neuromas.

The *lesser occipital nerve* may be lesioned by trauma to the lateral aspect of the neck or during surgical interventions on the posterior neck triangle (most commonly lymph node biopsy). Patients report numbness behind the ear.

Supraclavicular neuralgia may occur following clavicular fractures or neck surgery. Patients may present

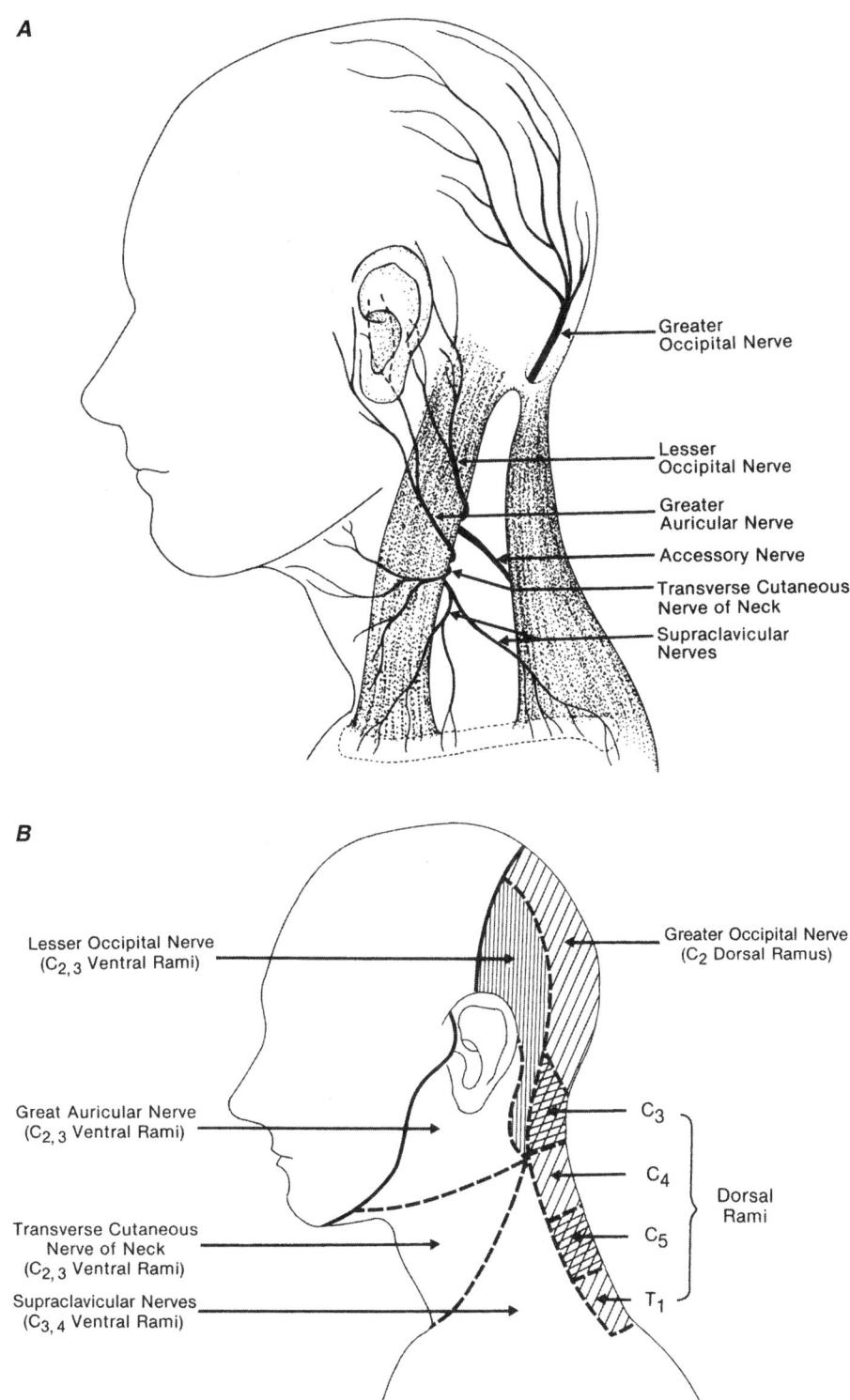

Figure 9-20. *A.* The sensory innervation of the neck: nerves and their course. (*Used with permission from Brazis PW, Masdeu JC, Biller J:* Localization in Clinical Neurology, *3d ed. Boston: Little, Brown, 1996, p. 74.*) *B.* The sensory innervation of the neck: dermatomes. (*Used with permission from Brazis PW, Masdeu JC, Biller J:* Localization in Clinical Neurology, *3d ed. Boston: Little, Brown, 1996, p. 75.*)

with severe pain or painless numbness on a small area over and below the midclavicle.

Cervical Root Avulsion Syndromes

Erb-Duchenne's paralysis (avulsion of the C5 and C6 roots) results from shoulder trauma or during delivery from forceps injuries if the shoulders are stuck and the head is pulled too hard. There is paralysis of shoulder abduction and elbow flexion, with the arm hanging limp at the side.

Klumpke's paralysis (avulsion of the C8 and T1 nerve roots) occurs during falls while the patient attempts to grab onto something and the arm is pulled up while the body continues to move. There is weakness of the intrinsic hand muscles and flexors and extensors of the fingers.

Cervical plexopathies are rare due to the deep location of the cervical plexus. They may occur following carotid endarterectomies or neck dissections for cancer. They also can occur in combination with brachial plexopathies (during motor vehicle accidents). Complete injuries present with sensory loss in the upper cervical dermatomes and diaphragmatic paralysis. Incomplete injuries without diaphragmatic involvement are difficult to detect because of the contribution of the cranial nerves to the motor innervation of the cervical muscles.

The Brachial Plexus (Ventral Rami of C5–T1) and Painful Neuropathies of the Upper Extremity

The *long thoracic nerve* (C5–7, innervating the serratus anterior), the *subclavian nerve* (C5–6, innervating the subclavius muscle), and the *dorsal scapular nerve* (C4–5, innervating the levator scapulae and rhomboids) arise from the brachial plexus before formation of the upper, middle, and lower trunks. The *upper trunk of the brachial plexus* gives rise to the suprascapular nerve (C5–6), innervating the supraspinatus and infraspinatus. The *medial* (C8–T1) and *lateral* (C5–7) *pectoralis nerves* arise from the upper/middle trunks and medial cords, respectively. The *lateral cord of the brachial plexus* gives rise to the musculocutaneous nerve (C5–7) and the lateral head of median nerve (C5–7). The *medial cord of brachial plexus* gives rise to the medial pectoralis nerve (C8–T1), the medial cutaneous nerve of the forearm (C8–T1), and the ulnar nerve (C8–T1). The *posterior cord of the brachial plexus* gives rise to the subscapular nerve (C5–7), the thoracodorsal nerve (C5–8), the axillary nerve (C5–6), and the radial nerve (C5–8) (Fig. 9-21).

Brachial plexopathy may be neoplastic, post-traumatic, autoimmune (vaccine-induced), postradiation, inflammatory/infectious, or hereditary. The brachial plexus is superficial and may be injured during anesthesia and by stretch injuries. Usually, brachial plexopathy presents as a painless motor deficit involving more than one peripheral nerve.

1. *Upper trunk (C5–6) lesions* may be caused by stab or bullet wounds to the neck. The presentation is with numbness over the lateral aspect of the arm, forearm, and hand and weakness of abduction, internal and external rotation of the shoulder, elbow flexion and wrist extension, and absent biceps and brachioradialis reflex. The rhomboid and serratus magnus will be spared (the nerves to these muscles arise proximal to the upper trunk). Neuralgic amyotrophy (Parsonage-Turner syndrome) presents

Figure 9-21. The brachial plexus, its trunk, and its cords. (*Used with permission from Stewart JD:* Focal Peripheral Neuropathies, *3d ed. Philadelphia: Lippincott, Williams & Wilkins, 2000, p. 118.*)

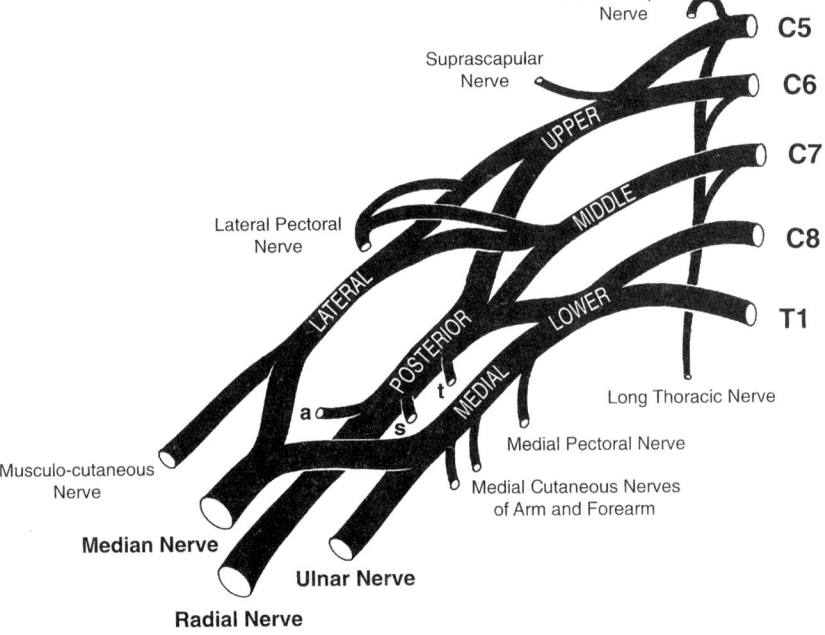

with acute onset of severe pain in the shoulder, arm, neck, and back that is worsened by movements of the arm (which is held in antalgic position, adducted at shoulder and flexed at elbow) and weakness of the shoulder muscles (deltoid, supraspinatus/infraspinatus, and serratus anterior). It may follow immunizations and become bilateral. Post–coronary bypass brachial plexopathy may occur during injury from jugular vein cannulation. Radiation-induced plexopathy most commonly involves the upper trunk of the brachial plexus, and is associated with weakness in the shoulder girdle muscles; pain is less common but may occur at the onset.

2. *Lower trunk lesions* may be caused by a cervical rib or by metastatic disease in the axilla or lung or breast malignancy. Pain is common in neoplastic lesions (primary and metastatic) and malignant infiltration of the brachial plexus. In Pancoast syndrome (tumor affecting the upper lobe of the lung), there is severe pain in the shoulder and tingling and numbness down the medial aspect of the arm, forearm, and hand (lower trunk distribution); weakness of the intrinsic hand muscles and finger flexors and extensors; and a Horner's syndrome. Pain is particularly severe at night when the patient is in a supine position and slightly better when the patient is sitting up.

3. The *thoracic outlet syndrome* results from compression of the brachial plexus or subclavian vessels between the first rib and clavicle by a cervical rib, enlarged C7 transverse process, hypertrophied scalene muscle, or fibrous bands between the C7 transverse process and the first rib or anterior scalene muscle. In neurogenic thoracic outlet syndrome, pain is intermittent in the lower trunk of the brachial plexus and ulnar nerve distribution (medial aspect of arm and forearm) associated with thenar muscle atrophy (exclusively C8 innervated) and weakness in the lower trunk innervated muscles (Fig. 9-22).

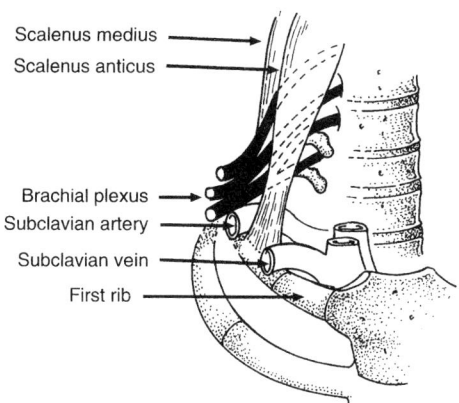

Figure 9-22. The brachial plexus in the interscalene triangle (anterior view). (*Used with permission from Stewart JD: Focal Peripheral Neuropathies, 3d ed. Philadelphia: Lippincott, Williams & Wilkins, 2000, p. 119.*)

The *suprascapular nerve* (C5–6, upper trunk of brachial plexus) may be compressed on its course in the posterior triangle of the neck, under the trapezius, at the suprascapular, and at the spinoglenoid notch. Trauma with resulting fractures or dislocations of the scapula or shoulder joint during open or arthroscopic shoulder surgery, suprascapular fossa ganglia and masses, compression from callus, and athletic activity (e.g., baseball pitchers, volleyball players, weight lifters, and wrestlers who perform acute forward rotation of the scapula). Patients may present with shoulder pain, weakness, or asymptomatic supra- and/ or infraspinatus muscle atrophy. The patient may be tender at the suprascapular notch and may experience pain when he or she performs forced arm adduction. The differential diagnosis of acute brachial plexus neuropathy is made by testing the deltoid muscle (which is spared by suprascapular neuropathy but weak in brachial plexopathy). Suprascapular nerve injury may coexist with a rotator cuff tear.

The *musculocutaneous nerve* arises from the lateral cord of the brachial plexus (C5–7), piercing the coracobrachialis, then traverses between the biceps and brachialis on the ventral aspect of the arm (giving off muscular branches to these three muscles), goes lateral to the biceps tendon at the elbow and continues as the lateral cutaneous nerve of the forearm (innervating the skin on the radial aspect of the forearm). Musculocutaneous neuropathy may be in the context of brachial plexus lesions (upper trunk or lateral cord) and after shoulder dislocations, traction maneuvers, or surgery to reduce shoulder dislocations, axillary node resection, arthroscopic shoulder surgery, malpositioning of the arm during anesthesia, and humeral fractures. Patients develop atrophy of the biceps and brachialis and sensory loss from elbow to wrist.

The median nerve (C5–T1) is formed by the medial and lateral cords of the brachial plexus. The lateral component derives from C5 and C6 ventral rami through the upper trunk and lateral cord and from C7 via the middle trunk. The medial part of the nerve derives from the C8 and T1 ventral rami (via the medial cord and lower trunk). The nerve originates in the lateral wall of the axilla next to the axillary artery, and then it descends in the medial compartment of the arm next to the brachial artery and the radial and ulnar nerves. At the elbow it crosses the antecubital fossa medial to the biceps tendon and the brachial artery underneath the bicipital aponeurosis, between the superficial and deep heads of pronator teres, then between the heads of the flexor digitorum superficialis, and between the flexor digitorum superficialis and profundus. It enters the hand through the carpal tunnel (between the carpal bones and the transverse carpal ligament, also known as the *flexor retinaculum*). Distal to the carpal tunnel it divides into a motor branch (innervating the abductor pollicis brevis, opponens pollicis, the first and second lumbricals, and sometimes the flexor pollicis brevis). The sensory palmar digital branches innervate the palmar surface of the thumb, the second and third digits, and half the fourth digit. The median nerve gives no branches in the arm; in the forearm it gives rise to motor branches to the pronator teres, flexor carpi radialis, and flexor digitorum superficialis. The anterior interosseous nerve (AIN) is a purely motor nerve that takes off from the median nerve when the latter emerges from the two heads of the pronator teres muscle. It passes under the flexor digitorum superficialis and courses down the forearm between the anterior interosseous membrane and the flexor digitorum profundus. AIN innervates the flexor pollicis longus, pronator quadratus, and flexor digitorum profundus muscles for the second and third digits. The palmar cutaneous branch arises from the median nerve above the wrist, does not pass through the carpal tunnel but through another small tunnel of its own, and innervates the skin of the thenar eminence.

1. The *pronator teres syndrome* represents compression of the median nerve between the two parts of the pronator teres. Patients present with sensory disturbances, weakness of the hand, or both. Sensory symptoms are worse with repetitive activities, especially pronation.
2. *Lesions of the anterior intra-osseous nerve* can occur by trauma to the proximal part of the median nerve (elbow/radial fractures, elbow trauma with hemorrhage into deep muscles, lacerations, elbow arthroscopy, catheterization of the brachial artery or venipuncture at the antecubital fossa), compression or inflammation of the nerve, fibrous bands of the pronator teres or flexor digitorum superficialis, or in the presence of an accessory bicipital aponeurosis. Occasionally, AIN may be affected in the context of acute brachial plexitis (patients will complain of severe pain in the shoulder and arm), during pregnancy, or after excessive forearm exercises. Patients will have weakness of the flexor pollicis longus, flexor digitorum profundus of the second and third digits, and pronator quadratus. Clinically, patients will not be able to perform the pince-finger grasp (pinching of the thumb and index finger) and forearm pronation with the elbow flexed. There is no sensory deficit.
3. *Carpal tunnel syndrome* is due to compression of the median nerve at the wrist. It is the most common neurological complication of pregnancy. Pain is in the thumb, index and middle fingers but may involve the whole hand or extend up to the medial forearm and shoulder. It is most severe at night and may be absent during the day. It may be relieved by swinging the arm or flexing/extending the wrist. Pain may be present in the absence of any sensory or motor deficit. Carpal tunnel syndrome may have four types of presentations:
   - *Pure sensory syndrome:* gradual onset without pain. Usually it is associated with repetitive movements of the wrist (knitting, sewing, or crocheting) in the presence of osteoarthritic changes of the wrist.
   - *Pure motor syndrome* (with minimal sensory symptoms) seen in carpenters, mechanics, and plumbers. It presents with weakness and atrophy in the lateral part of the thenar eminence (abductor pollics brevis).
   - *Autonomic presentation* with severe burning and stinging pain in the whole hand and a sensation of swollen, cold, and clammy hands. Pain attacks are triggered by working with the arms elevated, reading the newspaper, or driving. This syndrome occurs in pregnancy.
   - *Classic carpal tunnel syndrome.* A combination of the above.

The differential diagnosis includes cervical radiculitis.

The *ulnar nerve (C8–T1)* derives from the lower trunk of the brachial plexus and arises in the proximal axilla, travels along the lateral aspect of the axilla and medial aspect of the upper arm, close to the brachial artery and radial nerve. In the midarm it pierces the intermuscular septum that separates the flexor and extensor muscles and then turns posteriorly, close to the humerus and triceps muscle. It does not give rise to any branches in the arm. Then it passes into the ulnar (condylar) groove behind the medial epicondyle and then passes through the aponeurotic arch of the flexor carpi ulnaris between the medial epicondyle and olecranon (humeroulnar arcade). When the elbow is flexed, the distance between the olecranon and medial epicondyle increases by 1 cm, and the humeroulnar arcade tightens over the ulnar nerve, and the triceps pushes the nerve posteriorly. In extreme elbow flexion, the nerve is stretched against the medial epicondyle. After passing under the humeroulnar arcade, the nerve goes through the cubital tunnel (between the humeroulnar arcade and the flexor carpi ulnaris), and then it passes through the flexor carpi ulnaris toward the wrist. In the forearm, it gives branches to the flexor carpi ulnaris and flexor digitorum profundus. It also gives off a sensory branch, the *palmar cutaneous nerve,* at about the level of the midforearm that runs on the volar aspect of the forearm and wrist and innervates the proximal part of the ulnar aspect of the palm. There is another sensory branch, the *dorsal cutaneous branch,* that arises more distally, approximately 2 in above the wrist, and innervates the ulnar side of the dorsum of the hand and the dorsal surfaces of the 5th and the medial half of the 4th digits. At the wrist, the ulnar nerve is located in Guyon's canal (between the pisiform and the hook of hamate), together with the ulnar artery. The canal is bordered by the transverse carpal ligament and pisohamate ligament. The ulnar nerve divides into superficial and deep branches in Guyon's canal.

1. *Ulnar neuropathy in the axilla and upper arm* occurs while the arm hangs over a sharp edge, compression by crutches or a tourniquet, a misplaced injection in the posterior part of the deltoid muscle, and hematoma or aneurysmal compression of the brachial plexus (compartment syndrome). It may occur together with median and radial nerve palsy (triad neuropathy). An ischemic ulnar neuropathy may occur following the creation of an atrioventricular (AV) fistula.
2. *Supracondylar fractures of the humerus* may lesion the ulnar nerve or its branches at the time of injury or during fracture reduction or external fixation. *Tardy ulnar palsy* occurs years after a supracondylar fracture of the humerus in childhood that results in a developmental (cubitus) elbow deformity with increased carrying angle and an ulnar neuropathy with a delayed acute presentation. Patients present with tingling and numbness in fourth and fifth digits and weakness and wasting of the first dorsal interosseous muscle. The hypothenar eminence muscles, all other interossei, the flexor carpi ulnaris, and the flexor digitorum profundus may be spared.
3. *Ulnar neuropathies at the elbow* may occur in the condylar groove or cubital tunnel. The ulnar nerve is compressed easily at the elbow by repeated and trivial trauma (using a wooden armchair or driving with elbow resting on the door, or in bedridden patients). Perioperative ulnar palsy is the most frequent litigation claim against anesthesiologists in the United States; claims are made of improper positioning or inadequate padding of the arm during anesthesia. Predisposing factors for development of perioperative ulnar nerve palsy are male sex, thin or obese body habitus, duration of hospitalization (>14 days), and coronary artery bypass surgery. Rheumatoid arthritis, osteoarthritis, Paget's disease, long-standing diabetes, cubitus valgus, anterior dislocation of the radial head, shallow condylar groove, and multiple nerve traumas all may predispose to ulnar nerve palsy at the elbow.
4. *Deep ulnar nerve branch lesions at the wrist* are caused by repeated trauma to the heel of the hand that compresses the nerve against the carpal bones or by compression by a synovial cyst. The patient has weakness of all the ulnar-innervated intrinsic hand muscles without any sensory symptoms. Severe ulnar neuropathy leads to claw hand deformity. Early surgery should be undertaken to prevent muscle atrophy.

The *radial nerve (C5–T1)* is a continuation of the posterior trunk of the brachial plexus. It courses down in the lateral wall of the axilla and on the medial aspect of the humerus, winds around the humerus in the spiral groove, goes between the heads of the triceps, enters the anterior compartment of the arm below the insertion of the deltoid, and then passes into the forearm between the distal part of the biceps and proximal part of the brachioradialis. Above the elbow it gives rise to muscular branches (for the triceps, brachioradialis, and extensor carpi radialis longus and brevis) and sensory branches (the *posterior cutaneous nerve of the arm* and the *posterior cutaneous nerve of the forearm*). At about the elbow, it divides into a deep motor branch (*posterior interosseous nerve*) and a superficial sensory branch (*superficial radial nerve*). The *posterior interosseous nerve* is a pure motor branch that takes off from the radial nerve at the level of the neck of the radius, passes through the arcade of Frohse (under the tendinous edge of the supinator muscle at the level of lateral humeral epicondyle), courses between the superficial and deep parts of the supinator, and then enters the forearm. Here it innervates the extensor carpi ulnaris, extensor digitorum communis, extensor pollicis longus and brevis, abductor pollicis longus, extensor indicis, and extensor digiti minimi. The *superficial radial nerve* passes over the supinator and pronator teres and deep to brachioradialis and extensor carpi radialis longus and brevis along the lateral aspect of the radius on the dorsal aspect of the wrist and divides into terminal digital branches supplying the dorsolateral aspect of the wrist, hand, and digits one through 3 (dorsal aspect).

The radial nerve can be damaged at several sites:

1. *Axilla* (by crutches) results in weakness of all radial nerve–innervated muscles (i.e., triceps, brachioradialis, wrist extensors, and finger extensors) and pain, tingling, and numbness in the radial nerve cutaneous distribution but rarely with an objective sensory deficit (which may be present in the dorsum of the thumb and index finger). Lesion of the radial nerve spares the latissimus dorsi and deltoid muscles (as opposed to a posterior trunk of brachial plexus, which affects all radial nerve–innervated muscles, as well as the latissimus dorsi and deltoid).
2. *Trauma to the spiral groove of the humerus* (lateral aspect of upper arm) leads to wrist drop and severe pain on the anterolateral aspect of the forearm,

and sensory symptoms similar to compression in axilla.

3. *Radial nerve lesion on the medial aspect of the arm (Saturday night palsy)* occurs in patients who fall asleep with the arm hanging over the edge of the bed or over the back of a chair.
4. *Tennis elbow* is caused by trauma that leads to inflammation of the tendinous insertion of the extensors of the forearm to the lateral epicondyle. Most commonly this occurs in manual workers. Patients complain of pain in the extensor aspect of the forearm, without any muscle weakness, that is worse with supination and dorsiflexion of the fingers and wrist against resistance. It is not known whether the syndrome is caused by chronic inflammation of the tendinous insertion at the lateral epicondyle or entrapment of the posterior interosseous nerve (radial tunnel syndrome). Surgical decompression of the posterior interosseous nerve by incision of the fibromuscular structures may relieve the pain. Steroid injections, immobilization in a sling, and physical therapy are also beneficial.
5. *Fractures of the midshaft of the humerus (by the fracture itself or callus formation)* result in weakness of the wrist extensors, brachioradialis, and finger extensors.
6. *Lesions of the posterior cutaneous nerves of the arm and forearm* may be caused by blows to the arm or forearm, surgical procedures on the upper arm, injections, or tourniquet application.
7. *Posterior interosseous nerve syndrome* may be caused by twisting movements of the forearm, such as using a screwdriver or Indian bar wrestling, causing edema or hemorrhage into the supinator muscle. Patients present with weakness of the abductor policis longus and finger extensors (especially the index finger and thumb extensors), so there is finger drop as opposed to wrist drop. There is no sensory loss.
8. *Superficial radial neuropathy (Wartenberg syndrome, cheiralgia paresthetica)* is caused by entrapment of the nerve in the fascia between the tendons of the brachioradialis and extensor carpi radialis muscles. It usually happens because of wrist compression near the distal head of the radius by tight watches, handcuffs, bracelets, rubber gloves, tight casting, formation of scar tissue in patients with repetitive wrist movements at work, blunt injury to the radial aspect of the wrist, laceration during operations for De Quervain's tenosynovitis, vein cannulation, and rupture of synovial cysts. There is sensory loss in the cutaneous distribution of the nerve and painful dysethesias, occasionally associated with painful Tinel's points along the nerve.

Painful Thoracic Radiculopathies and Neuralgias

- *Chronic post-thoracotomy and post-chest tube insertion radiculopathy.* Pain is constant, worse with movement, associated with allodynia at the site of the previous scar and hypesthesia in the distribution of the nerve.
- *Postmastectomy pain.* Depending on the extent of axillary dissection, the intercostobrachial nerves may be severed, and the resulting pain is located in the axilla, anterior chest wall, and upper medial aspect of the arm. Pain may lead to development of frozen shoulder.
- *Acute compression of T5–spinal nerves* may mimic angina pectoris, and compression of the lower thoracic nerves may resemble various intra-abdominal processes. Pain is worse with palpation of the respective vertebral spinous process and relieved by intercostal nerve block. It may be triggered by heavy physical activity as a result of malposition of the costovertebral joints (slipping-rib syndrome). There is usually sensory deficit in the territory of the respective intercostal nerve.
- *Slipping rib syndrome.* Trauma causing thoracic disk herniations or degenerative disease of thoracic spine may cause pain in a radicular distribution, but they are rare in comparison with lumbar and cervical disk herniations (due to the immobility of the thoracic spine).
- *Pancoast syndrome (resulting from malignant invasion of the lung apex and stellate ganglion)* may be associated with pain and paresthesias in the axilla, upper arm (medial aspect), and chest wall (T1–2 distribution).
- *Postherpetic neuralgia* is described as continuous, deep, aching, or burning soreness with superimposed paroxysms of stabbing, shooting pain (which may or may not be triggered by touching the area, movement, or breathing). There is a marked allodynic component of the pain, and there may be an area of decreased sensitivity to light touch or pinprick surrounding the area of allodynia and spontaneous pain.
- *Chronic intercostal neuralgia* may be the expression of *herpes sine herpete* (postherpetic neuralgia not preceded by a herpetic eruption).
- *Vertebral metastases* may cause midline bony pain or pain in a unilateral or bilateral radicular distribution. Pain is reproduced by percussion of the affected vertebral spinous processes.

- *Diabetic thoracoabdominal neuropathy* is characterized by radicular, intercostal, or abdominal pain and paresthesias. The pathophysiology is assumed to be an infarction of the thoracic spinal nerves or their cutaneous branches. There is marked tactile allodynia. The sensory deficit may be patchy; on the anterior, lateral, or posterior aspect of the trunk; in the distribution of one or several adjacent spinal nerves; or at different levels on the two sides of the trunk. It usually resolves after 1 to 2 years but may recur or become chronic.
- *Lyme disease* may affect cervical and thoracic dermatomes and present as a polyradiculopathy.

Notalgia paresthetica is an idiopathic condition described as a burning, itchy, or "pins and needles" type of pain sometimes in the medial scapular region (T2–6 dermatomes) over an area equal to the size of the palm of a hand. The affected area may be hyperalgesic.

Painful Lumbosacral Radiculopathies

- L2 radiculopathy causes pain and sensory deficit in the anterior thigh and weakness in the hip flexors (ilipsoas), thigh adductors, and knee extensors (quadriceps). The bent-knee pulling test may be positive (the patient is positioned half prone on the table, and the examiner pulls the knee backward while putting forward pressure on the buttock and thus reproduces the patient's lumbar radicular pain).
- L3 radiculopathy causes pain and sensory deficit in the lower anterior thigh and medial aspect of the knee and weakness in the same muscle groups affected by L1 radiculopathy.
- L4 radiculopathy causes low back, buttock, anterolateral thigh, and anterior leg pain associated with sensory loss on the knee and medial calf, weakness of the knee extension (quadriceps) and foot dorsiflexion and inversion (tibialis anterior), and decreased knee jerk (L2–4).
- L5 radiculopathy causes low back, buttock, lateral thigh, and anterolateral calf pain and numbness, occasionally as low as the dorsum of the foot (medial aspect) and big toe. Weakness is in the gluteal muscles (medius and minimus), knee flexion (hamstrings), plantar flexion (tibialis posterior, flexor digitorum longus), foot inversion (tibialis posterior), foot eversion (peroneus longus and brevis), and extension of the big toe (extensor hallucis longus). Both knee and ankle jerks are spared with an L5 radiculopathy.
- S1 radiculopathy causes low back, buttock, lateral thigh, and lateral calf pain. There is numbness in the little toe, lateral foot, and most of the plantar surface (except for the plantar aspect of the big toe) and weakness in the gluteus maximus (hip extension), knee flexion (hamstrings), plantar flexion (gastrocnemius, soleus, flexor hallucis longus), toe flexion (flexor digitorum longus, flexor hallucis longus), and toe extension (extensor digitorum brevis). There is also decreased or absent ankle jerk (S1–2).
- S2–5 radiculopathy causes pain and numbness in the calf, posterior thigh, buttocks, and perianal region. There may be associated impairment of bowel and bladder control and absent anal wink.
- Neurogenic pseudoclaudication of the cauda equina results from congenital or acquired spinal stenosis and causes pain in one or both buttocks, thighs, or legs, sometimes associated with numbness, weakness (after exertion), and absent reflexes. Pain is relieved by rest, bending over, and sitting and is worsened by back extension and standing for a long period of time. Walking, coughing, and sneezing do not worsen the pain. The straight-leg-raising test may be negative (Fig. 9-23).

The Lumbosacral Plexus (Ventral Rami of L1–S4 Spinal Nerves) and Painful Neuropathies of the Lower Extremity

The *ventral rami of L4–S1* converge to form ventral and dorsal branches, then the *lumbar plexus* (within the psoas muscle), and then *peripheral nerves* (e.g., obturator, femoral, and lateral cutaneous nerve of the thigh; iliohypogastric, ilioinguinal, and genitofemoral nerves; and motor branches to the psoas and iliacus muscles). The lumbar part of the plexus gives rise to the femoral and obturator nerves, whereas the sacral part gives rise to the sciatic nerve. The *lumbosacral trunk (L4–5)* connects the lumbar and sacral plexuses and passes over the sacral ala adjacent to the sacroiliac joint.

The *sacral plexus* results from the convergence of the ventral rami of S1–4 on the posterolateral wall of the pelvis and exits it through the sciatic notch.

The *dorsal branches of the lumbosacral trunk* and the *dorsal branches of S1 and S2 ventral rami* form the lateral trunk of the sciatic nerve that continues as the common peroneal nerve. The *ventral branches of L4–S2 ventral rami* converge to form the medial trunk of the sciatic nerve (from which stems the tibial nerve). From the sacral plexus also arise the superior and inferior gluteal nerves, the pudendal nerves, the posterior cutaneous nerve of the thigh, and muscular branches to levator ani and external anal sphincter.

Malignant invasion of the lumbosacral plexus by cancer originating in the prostate, uterus, colon, urinary bladder, and lymph nodes or by metastasis presents with

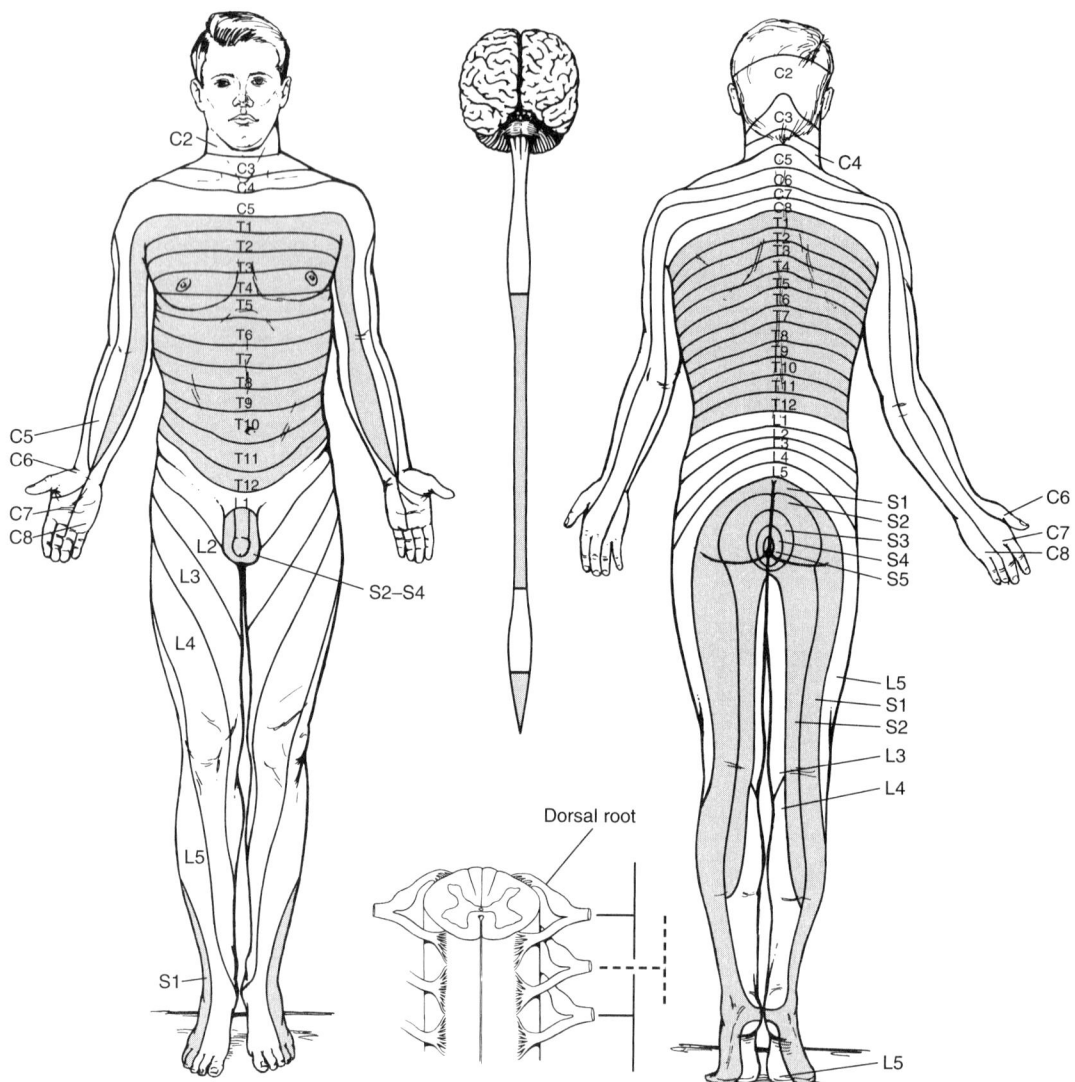

Figure 9-23. The spinal dermatomes and their nerve root correspondence. (*Used with permission from Martin JH:* Neuroanatomy: Text and Atlas, *3d ed. New York: McGraw-Hill, 2003, p. 116.*)

low back or pelvic pain radiating to the leg. It may affect the lumbar or sacral plexuses or both. There is progressive development of weakness and sensory deficit in the ipsilateral leg, as well as bowel and bladder symptoms (constipation and urinary retention) if there is bilateral involvement of the lumbosacral plexus or if there is involvement of the cauda equina. Sacral plexopathy needs to be differentiated from lesions affecting the conus or cauda equina by magnetic resonance imaging (MRI).

Traumatic causes of lumbosacral plexopathy include pelvic/sacral/sacroiliac joint fractures or dislocations and difficult labor. Radiation-induced lumbosacral plexopathy is less common than the brachial plexopathy due to the deep location of the lumbar plexus in the pelvis. It may occur several months to years after radiation and is seen most often after radiation for testicular, uterine, or ovarian malignancies. The patient presents with unilateral or bilateral leg weakness and sometimes with sensory deficit but usually no pain or much less pain than with malignant invasion.

Lumbosacral plexus compression may be caused by retroperitoneal fibrosis, hemorrhage, or abscess formation. Hemorrhages usually occur in patients with coagulopathies, after femoral artery catheterization, or after aortic aneurysm rupture. The patient reports sudden onset of pain radiating into the lower abdomen and groin or into the thigh and leg that is worse with hip extension (compression of the femoral, obturator, and femurocutaneous nerves). The neurological deficit may be confined to the quadriceps, iliopsoas, and hip adductors or may involve the whole leg (if the hematoma extends to the sacral plexus). Retroperitoneal abscess extends by contiguity from vertebral osteomyelitis, perinephric, pericolic or perirectal abscess; it usually involves the psoas muscle but may extend to the cauda equina or the lumbosacral nerve roots in the neural foramina.

On plexus lesions, electromyography (EMG) of the paraspinal muscles should be normal, but the sensory action potentials usually are abnormal (the opposite is true for radiculopathies). The diagnosis of plexopathy is made by showing clinically or by EMG that there is involvement of other nerves in the plexus (e.g., lateral cutaneous nerve of the thigh in the case of quadriceps weakness or superior and inferior gluteal nerves in the case of a sciatic neuropathy).

The *T12 and L1 ventral rami* join and then split and form a superior division (i.e., the ilioinguinal and iliohypogastric nerves) and an inferior division (which joins the L2 and forms the genitofemoral nerve). The *iliohypogastric nerve* comes from the retroperitoneal area from under the psoas muscle and passes through the transverse abdominis and under the internal oblique. It innervates the transverse abdominis and internal oblique. It gives rise to lateral cutaneous (sensory) branches to the posterolateral gluteal region and anterior cutaneous branches to the suprapubic region.

The *ilioinguinal nerve* originates in the retroperitoneal area, passes through the muscles of the abdominal wall medial to the anterior superior iliac spine and caudal to the iliohypogastric nerve, perforates the transversus abdominis and internal oblique, and enters the inguinal canal below the spermatic cord, which it exits through the superficial inguinal ring. It innervates the superomedial aspect of the thigh and the superior part of penis/scrotum or mons pubis/labia major. It may be entrapped spontaneously in the abdominal muscles or by lipomas, leiomyomas, and endometriosis. Various surgeries, such as herniorrhaphy, appendectomy, posterolateral abdominal wall incision for nephrectomy, Pfannenstiel horizontal suprapubic cesarean section, tubal ligation, bladder suspension for stress incontinence, and iliac crest bone grafting procedures may lacerate the nerve. Pain is located in the iliac fossa and inguinal region, radiating to the genitalia, and is made worse by walking and hip extension and better with hip flexion. There is tenderness in a point medial and inferior to the anterior superior iliac spine. Sensory abnormalities (e.g., hypesthesia, allodynia, or hyperalgesia) may appear in the distribution of pain.

The *genitofemoral nerve* descends retroperitoneally from the L3–4 level, passes through the psoas muscle inferomedially, behind the ureter, and divides at a variable level above the inguinal ligament into genital and femoral branches. The genital branch enters the inguinal canal via the deep inguinal ring and innervates the cremaster muscle/scrotal skin or mons pubis/labia major. The femoral branch descends lateral to the external iliac artery, passes behind the inguinal ligament, and enters the femoral sheath lateral to the femoral artery, innervating the skin of the upper part of the femoral triangle. Genitofemoral neuralgia occurs most commonly after inguinal herniorrhaphy but also after appendectomy, cesarean section, psoas abscess, psoas mass, or blunt groin trauma.

The *obturator nerve* (anterior divisions of the L2–4 ventral primary rami) originates in the psoas nerve and emerges from its medial border at the pelvic rim. It has an inferior and lateromedial course, exiting the pelvis through the obturator foramen, where it divides into muscular branches for the hip adductors. A sensory branch innervates the upper medial thigh. Injuries to the obturator nerve are rare because of its deep location. It may be injured by obturator hernias, pelvic fractures, hip surgery (by stretch or compression from a retractor), prostate or gynecologic sugery, pelvic lymph node dissection, and prolonged labor as a result of compression of the fetal head against the pelvic wall. It may be infiltrated by malignant processes in the pelvis. Newborns may have a transient obturator neuropathy due to prolonged abnormal in utero leg position. Patients with obturator neuropathy complain of leg weakness due to an inability to stabilize the hip. There is weakness of the hip adductors, sensory deficit, and paresthesias on the medial aspect of the upper thigh.

Meralgia paresthetica is caused by entrapment of the lateral cutaneous nerve of the thigh in the psoas muscle, pelvis, at the inguinal ligament, or in the thigh. Patients complain of burning pain associated with tingling and numbness over the middle lateral thigh. On examination, there may be allodynia, but patients may gain relief from rubbing the area; in addition, there are variable degrees of hyperesthesia, hypesthesia, and/or anesthesia and a Tinel's point on deep palpation over the inguinal ligament. Pain is worsened by standing and walking and improved by sitting down and flexing the hip. It may be idiopathic or associated with rapid weight loss in obese patients; may occur in persons wearing tight jeans, belts, or trusses; and peri- or postpartum. The nerve also may be entrapped by scar formation following lower abdominal surgery (e.g., gynecologic procedures, appendectomy, or laparoscopic herniorrhaphy), retroperitoneal tumors, or hematomas. It may be damaged by seatbelts during car accidents. In the thigh, it may be affected by injections, lacerations, or sport injuries. The differential diagnosis includes L2 radiculopathy and femoral neuropathy.

The *sciatic nerve* (L4–S3 ventral rami) has an inferolateral course from the lateral margin of the sacrum and inner wall of the pelvis, then exits the pelvis through the sciatic notch (greater sciatic foramen), passes through the buttock area, under or on top of the piriformis muscle, and then between the ischial tuberosity and the greater trochanter (near the hip joint), under the gluteus maximus, and then goes in the thigh (where it is deep). Similar to median nerve, the sciatic nerve is formed by a lateral trunk (continues as the peroneal nerve) and medial trunk (continues as the tibial nerve). The sciatic nerve may divide into trunks either from the origin or anywhere on its course above the popliteal fossa. It innervates the hamstrings and, in part, the adductor magnus. Compression in the gluteal region may be caused by trauma (motor vehicle accidents), fractures, and luxations of the hip joint, during hip arthroplasty, with entrapment of the nerve in callus, by sacroiliac joint dislocations, with compression by hematoma, during lithotomy positioning, with use of a tourniquet around the thigh (which may cause an ischemic sciatic neuropathy), by compartment syndrome of the posterior thigh muscles, in bedridden patients, by bicycling, with masses (e.g., endometriosis, tumors of the sciatic nerve, hematomas, and pseudoaneursyms), and by injections. Compression in the thigh may be caused by trauma, comminuted fractures of the femur, hematomas, compartment syndrome, fibrous bands, myositis ossificans of the hamstring muscles, nerve sheath tumors, nerve tumors, and popliteal fossa masses. Most sciatic nerve injuries are incomplete and may involve the medial or lateral trunks (the latter most commonly because of its more superficial location). The signs and symptoms are similar to those of peroneal or tibial neuropathies. Complete lesion of the sciatic nerve causes hamstring paralysis and paralysis of the muscles below the knee, as well as sensory deficit below the knee, with the exception of the medial calf (innervated by the saphenous branch of the femoral nerve). The pelvic and thigh masses may be appreciated during inspection and palpation along the course of the sciatic nerve and during rectal and/or vaginal examinations. Tenderness to palpation of the sciatic notch and abnormal straight-leg raising tests are nonspecific and may be caused by lumbosacral radiculopathies, plexopathies, sciatic neuropathy, and hip joint abnormalities. EMG and imaging studies are necessary for differentiation between these entities.

The *piriformis syndrome* is a frequent diagnosis and a nebulous entity. Patients present with signs, symptoms, and electrophysiologic evidence of proximal sciatic neuropathy but without evidence of lumbosacral neuropathy (i.e., normal paraspinal EMG) and without evidence of anatomic compression (i.e., no mass or disk compressing on the nerve roots, lumbosacral plexus, or nerve). The pain is in the buttock and may radiate in the distribution of the sciatic nerve but may involve the territory of the inferior gluteal nerve (which supplies the superior gluteal muscle). Most commonly, there are no anatomic abnormalities on surgical exploration. Very rarely, there are sciatic nerve abnormalities (e.g., anomalous course under the muscle or adhesions to the piriformis muscle) or piriformis muscle abnormalities (e.g., bifid muscle, fibrous bands constricting the nerve, or muscle hypertrophy). Occasionally, there is only a dynamic compression of the nerve during hip and leg movements. Tenderness on palpation in the gluteal area does not differentiate between the piriformis syndrome and lumbosacral radiculopathies, soft tissue injuries, or masses compressing the sciatic nerve. Some authorities say that pain relief after local anesthetic injections is also nonspecific and cannot be used for diagnostic purposes.

The *tibial nerve* (ventral divisions of L5–S1–2 ventral rami) is a continuation of the medial trunk of the sciatic nerve. It has its origin at the bifurcation of the sciatic nerve in the distal thigh and passes through the popliteal fossa between the two heads of the gastrocnemius muscle. In the calf it gives muscular branches to the gastrocnemius, soleus, tibialis posterior, flexor digitorum longus, and flexor hallucis longus. Above the ankle, the nerve becomes superficial (on the medial aspect of the Achilles tendon) and then passes under the flexor retinaculum in the tarsal tunnel. Here it divides into the medial and plantar nerves and into the calcaneal branch (innervating the sole of the heel). The medial and lateral plantar nerves divide into interdigital nerves that pass between the distal heads of the metatarsal bones, crossing the deep transverse metatarsal ligament. The plantar nerves supply all the muscles in the sole of the foot. Tibial neuropathy is much less common than peroneal neuropathy because the nerve is deep in its course at the hip, thigh, and calf. Sites of possible compression are

1. Hip joint trauma.
2. Knee.
3. Popliteal artery aneurysms or tumors.
4. At the flexor retinaculum, where it divides into lateral and medial plantar nerves (tarsal tunnel syndrome, similar to carpal tunnel). Tarsal tunnel syndrome is manifested by burning pain in the distribution of the medial and lateral plantar nerves (medial aspect of the foot and sole) associated with weakness and atrophy of the intrinsic foot muscles. There may be a positive Tinel's sign on tapping the nerve below the medial malleolus. The syndrome may be caused by a synovial cyst arising from the ankle joint.

Patients with proximal tibial nerve lesions develop atrophy of the calf muscles and weakness of the plantar flexors, foot invertors, long toe flexors, and intrinsic foot muscles. There is decreased sensation in the plantar aspect of the foot. The territory innervated by the sural nerve also may be involved. The differential diagnosis includes S1 and S2 radiculopathy, sacral plexopathy, and lesions of the medial trunk of sciatic nerve.

The *sural nerve* arises from the tibial nerve in the popliteal fossa and receives a communicating branch from the peroneal nerve; it descends (midline) in the calf and passes behind the lateral malleolus. It supplies the skin over the lateral aspect of the ankle and the foot to the base of the little toe.

Morton's metatarsalgia represents pain into the web and adjacent sides of the toes (particularly in the distribution of the interdigital nerve between the third and fourth metatarsals) and is worse on standing and walking. It is caused by compression of the interdigital nerves against the deep transverse metatarsal ligament and between the metatarsal heads. Occasionally, a neuroma may form and may be felt on palpation.

The *peroneal nerve* (posterior divisions of L4–5–S1–2 ventral rami) is the continuation of the lateral trunk of the sciatic nerve. It originates in the upper part of the popliteal fossa, passes behind the head of the fibula, winds along its neck, and lies against the periosteum of the fibula under the skin and subcutaneous tissue. It pierces the superficial head of the peroneus longus (fibular tunnel) and then exits the peroneus longus and divides into the superficial and deep peroneal nerves.

1. The *superficial peroneal nerve* descends along the lateral aspect of the fibula, close to the peroneus longus and brevis muscles. It gives muscular branches to the peroneus longus and brevis and sensory branches to the skin of the anterolateral aspect of the lower calf. The nerve then pierces the deep fascia above the lateral malleolus and divides into medial and lateral terminal branches, which pass on the dorsal aspect of the foot, which it innervates (except for a small portion of the skin between the first and second toes).
2. The *deep peroneal nerve* descends medial to the superficial peroneal nerve, in the anterior compartment of the leg, between the tibialis anterior and extensor hallucis longus, innervating the tibialis anterior, extensor hallucis longus, and extensor digitorum longus muscles. The nerve passes through the extensor retinaculum at the ankle and then divides into a lateral motor branch for the extensor digitorum brevis and a medial sensory branch (which is next to the dorsal pedal artery) for the skin between the first and second toes.

The peroneal nerve is frequently compressed at the fibular neck, where it is covered only by skin and fascia, after trauma or even after sitting with the legs crossed for a prolonged period of time, as well as with underlying medical conditions such as diabetes and connective tissue diseases. The condition may occur in the context of a mononeuritis multiplex, multiple compressive neuropathies, patients who have lost a lot of weight, and bedridden patients. Weakness is in a characteristic pattern: foot drop, variable degrees of weakness of dorsiflexion, inversion, and eversion of the foot. Sensory deficit usually is restricted to a small area over the dorsum of the foot.

Superior gluteal neuropathy may be caused by masses in the sciatic notch, hip arthroplasty, trauma with falls on the buttock, entrapment by the piriformis muscle, or buttock injections. The patient develops atrophy and weakness of the gluteus minimus and medius and tensor fasciae latae.

Inferior gluteal neuropathy can be caused by trauma, injections, and neoplastic infiltration of the nerve at the sciatic notch. It presents with numbness in the posterior thigh (if the posterior cutaneous nerve of the thigh is affected also), weakness of the gluteus maximus, and buttock pain. It may be associated with lesions of the sciatic and pudendal nerves and the posterior cutaneous nerve of the thigh.

Posterior cutaneous nerve of the thigh neuralgia presents with pain in the lower buttock and posterior aspect of the thigh or both. It may be positional only (while sitting). It may be caused by compressive lesions in the pelvis (e.g., tumors) or thigh (e.g., thigh injections, prolonged bicycling, and neuromas). It has to be differentiated from S2 neuropathy (the latter has, in addition, weakness in the gastrocnemius and decreased ankle jerk).

Pudendal neuropathy is rare because of the deep location of the nerve. It may occur after hip arthoplasties or internal fixations of the femur and after traction to the lower extremity during surgery. Entrapment of the pudendal nerve by sutures may occur following sacrospinal colpopexy. Compression may occur during bicycling and as a result of blunt perineal injuries, in the perineal canal, during labor and delivery, straining and defecation. Pudendal neuropathy may involve autonomic fibers (impotence, fecal incontinence). EMG shows prolongation of the anal reflex, pudendal nerve latencies, and denervation of the anal sphincter. *Proctalgia fugax* is a variety of pudendal neuralgia that has been associated with malignant infiltration of the pudendal nerves.

Conus medullaris lesions present with fecal and urinary retention because of loss of sensory input to rectum and bladder and motor output to pelvic muscles, followed by saddle anesthesia. Perineal, buttock, and thigh pain is a late occurrence.

▶ ANATOMY OF AUTONOMIC STRUCTURES AND VISCERAL PAIN

The *celiac plexus* is a sympathetic prevertebral plexus located in the epigastric region at the takeoff of the celiac and superior mesenteric arteries from the aorta. It has two components: sympathetic fibers (from the thoracic aortic plexus) and parasympathetic fibers (vagal fibers from the esophageal plexus). The *afferent sympathetic fibers* originate in the anterolateral horn of T5–12 spinal cord segments and then exit the spinal cord with the ventral roots, forming the white communicating rami; then they form the splanchnic nerves (greater, lesser, and least). The splanchnic nerves pass through the sympathetic chain but synapse only in the celiac ganglia. The *afferent parasympathetic fibers* originate in the vagus nerve. There are also somatic sensory afferents from the phrenic nerve. The number and diameter of celiac ganglia in the plexus are variable (one to five ganglia, 0.5 to 4.5 cm). They are located anterior and anterolateral to the aorta, anterior to diaphragmatic crura, at L1 level. There is a left and a right ganglion group. Both groups are located below the celiac artery, but the left group is lower than the right. The *celiac plexus efferents* innervate most of the abdominal viscera (i.e., distal esophagus, stomach, duodenum, small intestine, ascending and proximal transverse colon, adrenal glands, pancreas, spleen, liver, and biliary system). The efferents travel along the blood vessels originating in the aorta and form the phrenic, hepatic, splenic, superior gastric, superior and inferior mesenteric, abdominal aortic, suprarenal, renal, spermatic, and ovarian plexuses. The phrenic nerve afferents to the celiac plexus carry some of the nociceptive sensation, and this will be referred to the supraclavicular and scapular regions.

The *lumbar sympathetic chain* contains a variable number of ganglia, four to five on each side. L1 and L2 ganglia are common. A lesser contribution comes from ganglia that are aggregated at L3–4 and L4–5 disk levels. The size also varies between 5 and 15 mm. The sympathetic efferents (postganglionic fibers) join spinal nerves in the lumbar plexus and travel with the femoral, sciatic, and obturator nerves and give segmental branches to their corresponding blood vessels. Some of the preganglionic sympathetic fibers bypass the sympathetic chain and synapse with the postganglionic fibers in the somatic lumbosacral plexus and spinal nerves.

The *superior hypogastric plexus (presacral nerve)* is a sympathetic prevertebral (retroperitoneal) plexus located in front of the lumbosacral junction (lower third of the L5 vertebra and upper third of the S1 vertebra at the sacral promontorium). It is a paired structure. It lies medial and slightly inferior to the bifurcation of the common iliac arteries in a longitudinal plane. It receives sympathetic afferents from the aortic plexus (derived from the celiac and inferior mesenteric plexuses) and lumbar sympathetic chains. The parasympathetic afferents originate in the S2–4 cord segments via the pelvic (erigentes) nerves. The hypogastric plexus gives off efferents to the pelvic viscera and genitalia via rami traveling with the common and internal iliac (hypogastric) artery. Ultimately, these efferents form other plexuses (e.g., ureteric, testicular, ovarian, middle hemorrhoidal, vesical, prostatic, uterine, and vaginal).

The *inferior hypogastric plexus* is another paired structure lateral to the rectal wall and inferior part of the bladder, prostate/seminal vesicles, or uterine cervix/vaginal fornix. It is situated in a transverse plane, parallel to the pelvic floor.

BIBLIOGRAPHY

Bogousslavsky J, Caplan L (eds): *Stroke Syndromes,* 2d ed. Cambridge: Cambridge University Press, 2001.

Brazis PW, Masdeu JC, Biller J: *Localization in Clinical Neurology,* 3d ed. Boston: Little, Brown, 1996.

Carpenter MB: *Core Text of Neuroanatomy,* 2d ed. Baltimore: William & Wilkins, 1979.

Chusid JG: *Correlative Neuroanatomy and Functional Neurology,* 18th ed. Los Altos, CA: Lange Medical Publications, 1982.

Davidoff R: Cranial neuralgias and atypical facial pain, in Gilman S (ed), *MedLink Neurology.* San Diego: MedLink Corporation, 2003; available at *http://www.medlink.com.*

Kandel ER, Schwartz JH, Jessell TM: *Principles of Neural Science,* 4th ed. New York: McGraw-Hill, 2000.

Lanska DJ: Otalgia, in Gilman S (ed), *MedLink Neurology.* San Diego: MedLink Corporation, 2003; available at *http://www.medlink.com.*

Martin JH: *Neuroanatomy: Text and Atlas,* 3d ed. New York: McGraw-Hill, 2003.

Moore KL: Eye-related headache, in Gilman S (ed), *MedLink Neurology.* San Diego: MedLink Corporation, 2003; available at *http://www.medlink.com.*

Morgan GE, Mikhail MS, Murray MJ, Larson CP: *Clinical Anesthesiology,* 3d ed. New York: McGraw-Hill, 2002.

Patten J: *Neurological Differential Diagnosis,* 2d ed. Berlin: Springer-Verlag, 1996 (reprinted 2000).

Silberstein SD: Rhinosinus-related headache, in Gilman S (ed), *MedLink Neurology.* San Diego: MedLink Corporation, 2003; available at *http://www.medlink.com.*

Stewart JD: *Focal Peripheral Neuropathies,* 3d ed. Philadelphia: Lippincott, Williams & Wilkins, 2000.

Waldman SD: *Interventional Pain Management,* 2d ed. Philadelphia: W.B. Saunders, 2001.

CHAPTER 10

Brain Imaging of Pain

Karen D. Davis

Pain is the most common complaint brought to a physician's office. Despite the tremendous impact on quality of life, there is often a great difficulty in understanding, diagnosing, and treating chronic pain. Factors that may contribute to a patient's pain may not be readily apparent in standard radiologic, immunologic, or other pharmacologic tests. However, the emergence of new imaging techniques may provide significant information related to brain function and hence have an impact on the assessment of chronic pain and aid in monitoring treatment effects.

▶ OVERVIEW OF IMAGING TECHNOLOGIES

The two most widely used imaging technologies in brain research are positron-emission tomography (PET) and functional magnetic resonance imaging (fMRI), and so this chapter will focus on imaging pain with these technologies. Both PET and fMRI are considered indirect methods because they do not specifically measure neuronal activity but rather events that are coupled to neuronal activity.[1-4] fMRI is sensitive to hemodynamic changes (blood flow, oxygenation, perfusion, etc.), and PET can detect hemodynamic or metabolic (i.e., glucose consumption) events and neurotransmitter binding.[5,6]

There are three common types of PET imaging, all of which involve injection of a radiochemical to track particular types of events. One of the most popular PET imaging techniques is the oxygen-15 water method, which measures regional cerebral blood flow (rCBF). Another common technique is the 18-fluorodeoxyglucose (FDG) method, which tracks brain glucose metabolism. Finally, neurotransmitter receptor binding can be measured by injecting radiolabeled agonists or antagonists. Therefore, one of the main advantages of PET is that it can measure both resting-state activity and task- or stimulus-evoked changes. However, PET studies are relatively expensive and are limited by the need to inject radioactive agents (which limits the allowable number of scans per subject) and scan resolution.

There are both advantages and disadvantages to using fMRI compared with PET. The advantages include its noninvasiveness (i.e., it can acquire a large number of scans per subject), superior spatial and temporal resolution, and lower cost. However, a technical limitation is the need for special MR-compatible study equipment and the prohibition of subjects with internal ferromagnetic devices (e.g., pacemakers, stimulators, etc.). Furthermore, the popular fMRI blood oxygen level detection (BOLD) technique is limited to activation paradigms and cannot assess resting states. This is so because a "functional map" is constructed from a statistical comparison between MR signal intensities within brain areas measured during two series of scans. Thus it is imperative to appreciate that "brain activations" are always a reflection of the relative activity in that brain region during one state compared with another. A review of the PET technique, the related technique of single-photon-emission computed tomography (SPECT), and the fMRI technique and their use in studies of acute and chronic pain was published recently.[7]

▶ CAN WE USE IMAGING TO ASSESS INDIVIDUAL PATIENTS?

In order for functional brain imaging to have an impact on pain medicine, it must be reliable, sensitive, specific, and practical to implement for individual patient cases. Practicality has benefited by the development of faster and more routine imaging protocols. Enhanced technology in both PET (i.e., three-dimensional scanners) and MRI (i.e., high-field scanners) now allow for faster imaging with improved signal-to-noise ratios.

The issue of reliability is an active area of research but is not yet resolved for individual subject scans. There are some data concerning test-retest reliability,[8-10] but these data have not been examined carefully for pain tasks. A related topic is the issue of variability in pain responses between subjects. In imaging, this issue is complicated by the fact that brain activations can be related to processes triggered by the stimulus that may or may not reach consciousness. Whether one extracts stimulus-related brain activations or activations associated with specific perceptions is determined by the type of statistical analysis performed on the data. Studies of intersubject variability in noxious stimulus responses,[11-13] pain percept–specific responses,[14] and genetics[15] underscore the complexity that one must appreciate to develop imaging protocols for pain medicine.

The issue of sensitivity is complex and relates to statistical power and the choice of statistical thresholds. For individual-subject studies, it is important to acquire enough data during many task repetitions to ensure an adequate level of statistical power. However, even with adequate power, the results are affected by the choices made during the statistical analysis (e.g., smoothing, etc.).[16] Most critical in imaging data analysis is the choice of a method and level for statistical thresholding.[17-19] Most imaging studies control for the number of false-positive results (type 1 errors) by setting a p value deemed acceptable, say, 0.05. This is often coupled with methods to avoid the multiple-comparison problem (e.g., random field theory or Bonferroni corrections), or to control for a selected expected proportion of false-positive results (false discovery rate), or simply to choose a minimum cluster size threshold. However, one needs to appreciate that some of these methods are overly conservative and may result in an increase in the number of "misses," i.e., failure to identify true-positive results (type II error). Choosing a statistical threshold is always a fine balance between reducing type I and type II errors. Basic scientific discovery using imaging has tended to err on the side of minimizing the false-positive rate, sometimes at the expense of missing true-positive results. However, this approach may be too conservative for some clinical applications. For instance, fMRI has been used to identify eloquent cortex, which can help guide neurosurgical procedures such as tumor resection. In this type of clinical application, fMRI data are used to establish which cortical areas can be safely removed and which areas are essential for critical functions (i.e., language, sensory, motor, etc.). Therefore, it is important to consider controlling the number of false-negative results (i.e., misses, type II errors) in these types of clinical applications. This example highlights the need to carefully address the issue of statistical thresholding as we begin to adapt imaging for clinical applications.

▶ CHOOSING THE BEST IMAGING METHOD FOR IMAGING CHRONIC PAIN ABNORMALITIES

As noted earlier, the type of pain response can dictate the choice of imaging method. If one wishes to examine abnormalities associated with a chronic pain state (i.e., stimulus-independent pain), then PET is recommended rather than fMRI. PET is also the method of choice to examine endogenous opiate function via labeled opiate receptor ligands. However, if one wishes to identify abnormal stimulus-related response (e.g., allodynic or hyperalgesic responses), then either fMRI or PET can be used. In this case, the choice between fMRI and PET depends on whether temporal resolution is important (in which case fMRI is superior) or whether there are issues of MR compatibility (in which case PET is preferred). Longitudinal studies or those requiring multiple scan sessions (e.g., before and after treatments) can be done using PET or fMRI, although the total number of scans per year for each subject is limited in PET according to acceptable maximal levels of radioactive tracers.

▶ IMAGING CHRONIC PAIN

As noted earlier, imaging technologies available for identifying brain abnormalities in a chronic state are limited. However, some studies have developed protocols for such endeavors, mostly using PET or SPECT imaging.

Two different groups have studied low back pain patients. Derbyshire and colleagues[20] did not find any significant cortical abnormalities in these patients' responses to noxious thermal stimulation of the hand, suggesting that their back pain did not have an impact on the processing of noxious stimulation. Apkarian's group[21] developed a leg-raise task to augment patients' back pain to allow for fMRI assessment of leg-raise augmented pain. This group also has been studying the use of patient pain ratings of spontaneously fluctuating back pain to image cortical correlates of chronic back pain.

Forebrain involvement in neuropathic pain has been studied using two approaches. In one approach, PET scans were obtained in a small group of patients with chronic unilateral neuropathic pain.[22] The data revealed an abnormally low rCBF in the thalamus contralateral to the affected limb (internal control) as compared with the normal control subjects. In another approach, several groups assessed chronic pain abnormalities by comparing the ongoing pain state with a reduced pain state, with each patient serving as his or her own control. Hsieh and colleagues[23] used a regional nerve block to alleviate neuropathic pain and reported that the pain was associated with increased rCBF in the right anterior cingulate cortex (ACC), posterior parietal cortex, and lateral prefrontal cortex and decreased rCBF in the contralateral thalamus. Hsieh and colleagues[24] also reported an increased rCBF in the right ACC and the medial prefrontal cortex in patients with trigeminal neuropathic pain. Consistent with these findings were those of Apkarian and colleagues,[25] who used fMRI to examine sympathetically mediated chronic pain before and after sympathetic block for pain relief and found that the chronic pain was associated with increased prefrontal and ACC activity and decreased contralateral (to the pain) thalamic activity. Interestingly, a SPECT study of patients with chronic fibromyalgia pain also reported decreased rCBF in the thalamus and caudate.[26] Another confirmation of abnormally low thalamic blood flow associated with chronic pain comes from a study of pain sensitivity to cold in patients who suffer from cluster headache. Di Piero and colleagues[27] reported that during pain-free periods, cluster headache patients show abnormally low responses to noxious cold stimuli in the contralateral S1 and thalamus compared with normal control subjects despite similar pain ratings to the cold stimuli. However, Ness and colleagues[28] reported a case report of a patient with fluctuating spinal cord injury pain attenuated by gabapentin. In contrast to the other pain-related findings noted earlier, a comparison of SPECT scans obtained in this patient during high and low pain states indicated an increased rCBF in the ACC, somatosensory cortex, and thalamus (and decreased in the caudate).

Therefore, the most consistently reported abnormalities associated with chronic pain are a decreased blood flow in the thalamus (but see ref. 28) and increased blood flow in the ACC, prefrontal cortex, and possibly the somatosensory cortex.

A meta-analysis of data up to 1999 implicates the mid-ACC area 24 in acute pain and allodynia but a more extensive rostral extent in chronic pain and analgesia.[29]

▶ IMAGING HYPERALGESIA AND ALLODYNIA

Central mechanisms of hyperalgesia and allodynia have been studied with functional brain imaging in both normal subjects and patients with chronic pain. In one line of study, the capsaicin model of allodynia was used to examine allodynia and hyperalgesia in normal subjects. For these types of studies, capsaicin is applied to the skin to create a zone of allodynia (i.e., touch- or heat-evoked pain), and imaging is performed before and after the treatment to examine capsaicin-, mechanical-, and heat-evoked responses. All studies reported that many brain areas were commonly activated during capsaicin- and stimulus-evoked pain in a normal state and allodynia. However, some differences between "normal" and allodynic pain were noted in these studies. In particular, enhanced responses in the allodynic state compared with normal touch or capsaicin pain were noted in the prefrontal cortex,[30–32] parahippocampal gyrus,[30] and posterior parietal cortex.[32] Lorenz and colleagues[33] also demonstrated increased frontal activity via a medial thalamic pathway during heat allodynia compared with normal heat pain.

There is scant information from patient studies of allodynia. Peyron and colleagues[34] studied patients with central pain (Wallenberg's syndrome) following a lateral medullary infarct and found an increased rCBF in the contralateral thalamus and other lateral pain pathway cortical areas associated with the allodynic state. In contrast, there was an abnormally reduced pain-related ACC activation. Peyron and colleagues[35] also reported on a patient with chronic pain and allodynia who had a parietal infarct involving S1 and S2 and an ACC cortex lesion. Consistent with the previous findings, neither the chronic pain nor the allodynia was associated with activity in the residual ACC. In another study of allodynia in chronic pain (mononeuropathy) patients, Petrovic and colleagues[36] found many regions of increased activation in somatosensory and motor areas in the allodynic state versus nonpainful states.

Taken together, these studies in normal subjects and chronic pain patients demonstrate enhanced forebrain responses to pain-producing stimuli in allodynic skin compared with nonpainful stimuli in normal skin. The findings from the capsaicin model also suggest an increased emotional response to allodynic pain compared with normal acute pain stimuli.

There have been several studies of rectal-evoked pain in patients with irritable bowel syndrome (IBS). Since patients with IBS show visceral hypersensitivity, fMRI studies of rectal-evoked responses are possible. IBS patients show a greater ACC response to rectal stimulation compared with control subjects[37,38] or patients with inflammatory bowel disease[39] (but see ref. 40). Many other areas, including the prefrontal cortex, somatosensory cortex, insula, and thalamus, showed increased rectal-evoked responses in IBS patients compared with controls. However, it is not clear whether these abnormalities are primary to the disease or merely reflect the increased rectal-evoked pain experienced by IBS patients.[37] To control for this complication, recent studies by Davis and colleagues were designed to clamp the pain intensity equally in IBS patients and control subjects.

These ongoing studies also take into consideration the temporal pattern of the rectal-evoked perceived pain, which can vary significantly from the rectal stimulus in IBS patients but not in controls.[41–43]

▶ IMAGING FUNCTIONAL PAIN ABNORMALITIES

Apart from hyperalgesia and allodynia, some patients with chronic pain exhibit other sensory or cognitive abnormalities. One example is patients with chronic pain and unexplained nondermatomal widespread somatosensory deficits, sometimes considered *hysterical anesthesia*. An fMRI study of a small group of these patients[44] demonstrated normal cortical responses to touch and noxious stimuli applied to the normal-feeling limb. However, these same stimuli applied to the anesthetic limb failed to evoke responses in the anterior insula, caudal ACC, and prefrontal cortex (area 44/45), decreased activity in other areas of the prefrontal cortex, the posterior parietal cortex, S1, and S2 and recruited normally unresponsive areas of the perigenual ACC. These findings demonstrate that fMRI may be useful to detect cortical abnormalities in patients who exhibit unusual perceptual responses to somatosensory stimuli despite normal somatosensory evoked potentials.

▶ IMAGING ANALGESIA EFFECTS AND ENDOGENOUS PAIN CONTROL CAPACITY

Studies of the effects of analgesics and anesthetics suggest that functional brain imaging may provide an opportunity to assess pain control treatments. Functional brain imaging has been used to locate cortical opiate-binding sites and to examine the effects of exogenously delivered opiates and the endogenous release of opiates.

Foundation PET studies using [^{11}C]diprenorphine[45–47] reported opiate receptor binding in normal subjects in many somatosensory cortical areas but particularly concentrated in cortical projection sites of the medial pain system such as the ACC and prefrontal cortex. Another line of PET studies has been conducted using mu opiate agonists, such as fentanyl, remifentanil, and carfentanil. The mu agonists themselves activate the middle and perigenual ACC.[48,49] In addition, when fentanyl acted to attenuate acute pain, this action was associated with reduction in the normal pain-evoked response in the thalamus,[50] S1, and the middle ACC.[49] More recently, Petrovic and colleagues[51] implicated the rostral ACC in both placebo analgesia and opiate analgesia.

Zubieta and colleagues[52] have developed an acute muscle pain model to study endogenous opiate release in normal subjects. They found that the endogenous mu opioid system was indeed activated within the amygdala, thalamus, ACC, lateral prefrontal cortex, insula, and hypothalamus during painful stimulation of the masseter muscle. However, they also noted significant intersubject variability in mu opioid receptor binding during both the pain state and the baseline state. Interestingly, they found that across subjects, the amount of receptor activation in the dorsal ACC, thalamus, and nucleus accumbens was negatively correlated with their ratings of pain affect. In a more extensive study of individual differences in pain and endogenous mu opioids, Zubieta and colleagues[15] identified the catechol-O-methyltransferase (COMT) val^{158}met genetic polymorphism as a key factor influencing intersubject pain perception and endogenous opioid release.

▶ IMAGING CONNECTIVITY

In addition to the application of fMRI and PET data to brain function, there are two emerging imaging technologies that warrant mention here: diffusion tensor MR imaging (DTI) and network analysis using functional or effective connectivity. Both methods are designed to examine connectivity between brain structures.

The first method, DTI, is an anatomic technique developed to visualize white matter fiber tracts (for review, see ref. 53). Although still in its infancy, DTI holds promise for many clinical applications. For instance, since DTI can be used to assess changes in white matter fiber tracts over time, it could be developed to assess plasticity or alterations in connectivity between brain regions following injury, treatment, or rehabilitation.

The second method assesses brain connectivity using functional data and analytical and network modeling (for reviews and critical analysis, see refs. 54 and 55). Friston and colleagues defined *functional connectivity* as the "temporal correlation between spatially remote neurophysiologic events" and *effective connectivity* as the "influence that one neural system exerts over another either directly or indirectly."[56,57] In the future, these approaches could be useful clinically because they may be able to detect abnormal brain function (even in the absence of a stimulus or task) or treatment effects. For instance, a recent study demonstrated that the functional connectivity between the middle ACC and several cortical areas during a noxious stimulus could be modulated by an antinociceptive hypnotic state.[58] Another clinically relevant aspect of network analysis is its ability to identify intersubject variability.[59] This may have particular applicability to the study of chronic pain patients. Finally, of potential pertinence to chronic pain patients is the use of functional connectivity to assess the so-called default mode of brain function.[60–62]

▶ CONCLUSIONS: PAIN IMAGING AS A DIAGNOSTIC TOOL OR FOR ASSESSING TREATMENT EFFECTS

The studies noted earlier point to the potential for brain imaging in pain medicine. However, great care should be taken in applying these techniques to diagnose and/or assess treatment in an individual patient. At the moment, the technical limitations (in terms of statistical power, reliability, reproducibility, etc.) and the known intersubject variability in pain sensitivity and reactivity pose a risk to the proper use of imaging in pain medicine. However, this is an obstacle that can be overcome by careful study in individual subjects and by the knowledge of the range of "normal" pain responses. Like the other "vital signs," with a range of normal upper and lower limits, the brain's response to noxious stimuli must be defined by a range of potential activations and connections.

REFERENCES

1. Raichle ME: Functional brain imaging and human brain function. *J Neurosci* 2003;23:3959–3962.
2. Ugurbil K, Toth L, Kim DS: How accurate is magnetic resonance imaging of brain function? *Trends Neurosci* 2003;26:108–114.
3. Logothetis NK: The underpinnings of the BOLD functional magnetic resonance imaging signal. *J Neurosci* 2003; 23:3963–3971.
4. Lauritzen M, Gold L: Brain function and neurophysiological correlates of signals used in functional neuroimaging. *J Neurosci* 2003;23:3972–3980.
5. Magistretti PJ: Cellular bases of functional brain imaging: Insights from neuron-glia metabolic coupling. *Brain Res* 2000;886:108–112.
6. Berns GS: Functional neuroimaging. *Life Sci* 1999; 65:2531–2540.
7. Davis KD: Functional brain imaging in humans: Methodology and issues, in Kruger L (ed), *Methods in Pain Research*. Boca Raton, FL: CRC Press, 2001:225–240.
8. Maitra R, Roys SR, Gullapalli RP: Test-retest reliability estimation of functional MRI data. *Magn Reson Med* 2002;48:62–70.
9. Genovese CR, Noll DC, Eddy WF: Estimating test-retest reliability in functional MR imaging: 1. Statistical methodology. *Magn ResonMed* 1997;38:497–507.
10. Noll DC, Genovese CR, Nystrom LE, et al: Estimating test-retest reliability in functional MR imaging: 2. Application to motor and cognitive activation studies. *Magn Reson Med* 1997;38:508–517.
11. Kwan CL, Crawley AP, Mikulis DJ, et al: An fMRI study of the anterior cingulate cortex and surrounding medial wall activations evoked by noxious cutaneous heat and cold stimuli. *Pain* 2000;85:359–374.
12. Davis KD, Kwan CL, Crawley AP, et al: Functional MRI study of thalamic and cortical activations evoked by cutaneous heat, cold and tactile stimuli. *J Neurophysiol* 1998; 80:1533–1546.
13. Coghill RC, McHaffie JG, Yen YF: Neural correlates of interindividual differences in the subjective experience of pain. *Proc Natl Acad Sci USA* 2003;100:8538–8542.
14. Davis KD, Pope GE, Crawley AP, et al: Neural correlates of prickle sensation: A percept-related fMRI study. *Nature Neurosci* 2002;5:1121–1122.
15. Zubieta JK, Heitzeg MM, Smith YR, et al: COMT $val^{158}met$ genotype affects μ-opioid neurotransmitter responses to a pain stressor. *Science* 2003;299:1240–1243.
16. Brett M, Johnsrude IS, Owen AM: The problem of functional localization in the human brain. *Nature Rev Neurosci* 2002;3:243–249.
17. Loring DW, Meador KJ, Allison JD, et al: Now you see it, now you don't: Statistical and methodological considerations in fMRI. *Epilepsy Behav* 2002;3:539–547.
18. Genovese CR, Lazar NA, Nichols T: Thresholding of statistical maps in functional neuroimaging using the false discovery rate. *Neuroimage* 2002;15:870–878.
19. Machulda MM, Ward HA, Cha R, et al: Functional inferences vary with the method of analysis in fMRI. *Neuroimage* 2001; 14:1122–1127.
20. Derbyshire SW, Jones AK, Creed F, et al: Cerebral responses to noxious thermal stimulation in chronic low back pain patients and normal controls. *Neuroimage* 2002;16:158–168.
21. Apkarian AV, Krauss BR, Fredrickson BE, et al: Imaging the pain of low back pain: Functional magnetic resonance imaging in combination with monitoring subjective pain perception allows the study of clinical pain states. *Neurosci Lett* 2001;299:57–60.
22. Iadarola MJ, Max MB, Berman KF, et al: Unilateral decrease in thalamic activity observed with positron emission tomography in patients with chronic neuropathic pain. *Pain* 1995; 63:55–64.
23. Hsieh JC, Belfrage M, Stone-Elander S, et al: Central representation of chronic ongoing neuropathic pain studied by positron emission tomography. *Pain* 1995;63:225–236.
24. Hsieh JC, Meyerson BA, Ingvar M: PET study on central processing of pain in trigeminal neuropathy. *Eur J Pain* 1999; 3:51–65.
25. Apkarian AV, Thomas PS, Krauss BR, et al: Prefrontal cortical hyperactivity in patients with sympathetically-mediated chronic pain. *Neurosci Lett* 2001;311:193–197.
26. Mountz JM, Bradley LA, Modell JG, et al: Fibromyalgia in women. *Arthritis Rheum* 1995;38:926–938.
27. Di Piero V, Fiacco F, Tombari D, et al: Tonic pain: A SPECT study in normal subjects and cluster headache patients. *Pain* 1997;70:185–191.
28. Ness TJ, San Pedro EC, Richards JS, et al: A case of spinal cord injury–related pain with baseline rCBF brain SPECT imaging and beneficial response to gabapentin. *Pain* 1998; 78:139–143.
29. Peyron R, Laurent B, Garcia-Larrea L: Functional imaging of brain responses to pain: A review and meta-analysis. *Neurophysiol Clin* 2000;30:263–288.
30. Iadarola MJ, Berman KF, Zeffiro TA, et al: Neural activation during acute capsaicin-evoked pain and allodynia assessed with PET. *Brain* 1998;121:931–947.
31. Baron R, Baron Y, Disbrow E, et al: Brain processing of capsaicin-induced secondary hyperalgesia: A functional MRI study. *Neurology* 1999;53:548–557.

32. Witting N, Kupers RC, Svensson P, et al: Experimental brush-evoked allodynia activates posterior parietal cortex. *Neurology* 2001;57:1817–1824.
33. Lorenz J, Cross DJ, Minoshima S, et al: A unique representation of heat allodynia in the human brain. *Neuron* 2002;35:383–393.
34. Peyron R, García-Larrea L, Grégoire MC, et al: Allodynia after lateral-medullary (Wallenberg) infarct: A PET study. *Brain* 1998;121:345–356.
35. Peyron R, Garcia-Larrea L, Gregoire MC, et al: Parietal and cingulate processes in central pain: A combined positron emission tomography (PET) and functional magnetic resonance imaging (fMRI) study of an unusual case. *Pain* 2000;84:77–87.
36. Petrovic P, Ingvar M, Stone-Elander S, et al: A PET activation study of dynamic mechanical allodynia in patients with mononeuropathy. *Pain* 1999;83:459–470.
37. Mertz H, Morgan V, Tanner G, et al: Regional cerebral activation in irritable bowel syndrome and control subjects with painful and nonpainful rectal distention. *Gastroenterology* 2000;118:842–848.
38. Verne GN, Himes NC, Robinson ME, et al: Central representation of visceral and cutaneous hypersensitivity in the irritable bowel syndrome. *Pain* 2003;103:99–110.
39. Bernstein CN, Frankenstein UN, Rawsthorne P, et al: Cortical mapping of visceral pain in patients with GI disorders using functional magnetic resonance imaging. *Am J Gastroenterol* 2002;97:319–327.
40. Silverman DHS, Munakata JA, Ennes H, et al: Regional cerebral activity in normal and pathological perception of visceral pain. *Gastroenterology* 1997;112:64–72.
41. Davis KD, Bushnell MC, Strigo IA, et al: Imaging visceral sensations, in Dostrovsky JO, Carr DB, Koltzenburg M (eds), *Proceedings of the 10th World Congress on Pain*. Seattle: IASP Press, 2003:261–276.
42. Kwan CL, Mikula K, Diamant NE, et al: The relationship between rectal pain, unpleasantness, and urge to defecate in normal subjects. *Pain* 2002;97:53–63.
43. Kwan CL, Diamant NE, Mikulis DJ, et al: Percept-related fMRI of rectal-evoked sensations in irritable bowel syndrome. *Soc Neurosci Abst* 2002.
44. Mailis-Gagnon A, Giannoylis I, Downar J, et al: Altered central somatosensory processing in chronic pain patients with "hysterical" anesthesia. *Neurology* 2003;60:1501–1507.
45. Jones AKP, Qi LY, Fujirawa T, et al: In vivo distribution of opioid receptors in man in relation to the cortical projections of the medial and lateral pain systems measured with positron emission tomography. *Neurosci Lett* 1991;126:25–28.
46. Vogt BA, Watanabe H, Grootoonk S, Jones AKP: Topography of diprenorphine binding in human cingulate gyrus and adjacent cortex derived from coregistered PET and MR images. *Hum Brain Mapp* 1995;3:1–12.
47. Sadzot B, Price JC, Mayberg HS, et al: Quantification of human opiate receptor concentration and affinity using high and low specific activity [^{11}C]diprenorphine and positron-emission tomography. *J Cereb Blood Flow Metab* 1991;11:204–219.
48. Wagner KJ, Willoch F, Kochs EF, et al: Dose-dependent regional cerebral blood flow changes during remifentanil infusion in humans: A positron-emission tomography study. *Anesthesiology* 2001;94:732–739.
49. Casey KL, Svensson P, Morrow TJ, et al: Selective opiate modulation of nociceptive processing in the human brain. *J Neurophysiol* 2000;84:525–533.
50. Bencherif B, Fuchs PN, Sheth R, et al: Pain activation of human supraspinal opioid pathways as demonstrated by [^{11}C]-carfentanil and positron emission tomography (PET). *Pain* 2002;99:589–598.
51. Petrovic P, Kalso E, Petersson KM, et al: Placebo and opioid analgesia: Imaging a shared neuronal network. *Science* 2002;295:1737–1740.
52. Zubieta JK, Smith YR, Bueller JA, et al: Regional mu opioid receptor regulation of sensory and affective dimensions of pain. *Science* 2001;293:311–315.
53. Masutani Y, Aoki S, Abe O, et al: MR diffusion tensor imaging: Recent advance and new techniques for diffusion tensor visualization. *Eur J Radiol* 2003;46:53–66.
54. Horwitz B: The elusive concept of brain connectivity. *Neuroimage* 2003;19:466–470.
55. Lee L, Harrison LM, Mechelli A: A report of the functional connectivity workshop, Dusseldorf 2002. *Neuroimage* 2003;19:457–465.
56. Friston KJ, Frith CD, Liddle PF, et al: Functional connectivity: The principal-component analysis of large (PET) data sets. *J Cereb Blood Flow Metab* 1993;13:5–14.
57. Friston KJ, Frith CD, Frackowiak R: Time-dependent changes in connectivity measured with PET. *Hum Brain Mapp* 1993;1:69–79.
58. Faymonville ME, Roediger L, Del Fiore G, et al: Increased cerebral functional connectivity underlying the antinociceptive effects of hypnosis. *Brain Res Cogn Brain Res* 2003;17:255–262.
59. Mechelli A, Penny WD, Price CJ, et al: Effective connectivity and intersubject variability: Using a multisubject network to test differences and commonalities. *Neuroimage* 2002;17:1459–1469.
60. Gusnard DA, Raichle ME, Raichle ME: Searching for a baseline: Functional imaging and the resting human brain. *Nature Rev Neurosci* 2001;2:685–694.
61. Raichle ME, MacLeod AM, Snyder AZ, et al: A default mode of brain function. *Proc Natl Acad Sci USA* 2001;98:676–682.
62. Greicius MD, Krasnow B, Reiss AL, et al: Functional connectivity in the resting brain: A network analysis of the default mode hypothesis. *Proc Natl Acad Sci USA* 2003;100:253–258.

CHAPTER 11

Psychological Aspects of Pain: A Consciousness Studies Perspective

C. Richard Chapman

That pain, a subjective experience, should have psychological aspects is a curious notion. After all, it is entirely dependent on consciousness and is intrinsically aversive, and its expression is invariably emotional. Consequently, it is in the strictest sense entirely psychological, even though it normally occurs in response to tissue trauma or damage to nervous structures. Although pain does not reduce readily to physiologic explanation, it has well-defined neurophysiologic mechanisms such as nociception, sensitization, aberrant firing patterns, and deficient modulation. Do these mechanisms, which are the targets of medical intervention intended to prevent or relieve pain, constitute the essence of pain? Or are they necessary but insufficient in themselves for a scientific account of the complex, subjective psychological experience that patients call pain?

Scientific knowledge in other domains of human perception offers clues. The transduction of light at the retina and the transmission of neural signals to higher brain structures cannot account for why one person can perceive a work of art as engaging and beautiful, whereas another finds it boring and banal. And yet neither person can experience the art without the sensory processes that make visual perception possible. Is the experience of pain—or any other somatic awareness—fundamentally different? This question brings us into the classic problem of understanding the relationship between the body and the mind.

Classical thinking in science and philosophy, stemming from the work of Descartes[1] (1649), has long held that mind and body are separate domains. The body, Descartes held, operates mechanically and lends itself to mechanistic description. The mind, in contrast, is an inhabitant of the body. It processes the streams of sensory input from the various modalities in what Dennett[2] has described as the Cartesian theater. It makes meaning of salient bits of sensory input within the realm of the mind, which is essentially independent of the bodily mechanisms. Spinoza, a contemporary of Descartes, argued for the unity of the body and the mind but lost the contest.[3] Consequently, the dualism that Descartes instantiated in Western culture has so profoundly affected our view of human nature and our language that it is now extraordinarily difficult to discuss a problem such as pain without the implicit assumption of dualism. To approach the psychological aspect of pain without a Cartesian bias, we need a new frame of reference.

Contemporary consciousness studies, a multidisciplinary field, are beginning to make significant inroads into the problem of how a mechanistic nervous system can produce phenomenal reality. It is too early to describe this approach as monism, but it does break free of classic Cartesian assumptions. The mind appears to be an emergent property of brain activity and the result of complex patterns of preconscious processing. In this framework, pain is an emergent property of ongoing somatic monitoring, and it normally serves the adaptive function of calling attention to threat or damage to biologic integrity.[4]

The problem of understanding pain is essentially that of defining the relationship of the mind to the body. Line-labeling neurophysiologic explanations of nociception, while valuable scientifically, cannot account well for pain as patients experience and express it. Clinical pain rarely, if ever, reduces to sensory mechanisms, and skillful therapeutic interventions directed at such mechanisms can fail, particularly when the pain is chronic. Nor can current cognitive-behavioral models for pain explain it satisfactorily on the sole basis of cognitive processes, affective processes, and the patient's social environment. Physiologic mechanisms contribute to nearly all pain problems, albeit in ways that are not always straightforward. Pain appears to emerge from an interaction of neurophysiologic and psychological processes, and when it is chronic, psychological processes sometimes seem to dominate. From a clinical perspective, therefore, managing pain often extends beyond controlling nociception, especially chronic pain.

This chapter introduces the constructivist framework as a means of understanding how mechanistic bodily processes contribute to complex and highly individual pain experiences. This approach holds that pain, like all aspects of consciousness, is the end product of a complex process of construction and not simply the arrival of tissue injury signals at the threshold of consciousness. A clear distinction between nociception and pain is essential for this approach. A review of the literature on the complex impacts of nociception on the brain, including patterns of activity evident during functional brain imaging during pain, with the goal of showing that nociception initiates and sustains complex patterns of central processing that integrate nociceptive sensory input with sensory input from other modalities, memories, expectations, attentional biases, and preexisting affective states. The brain constructs from moment to moment a coherent, purposive model of reality that encompasses both the body in which it resides and the external environment that it must negotiate. Pain is a special, aversive, and compelling aspect of somatic awareness that has emotional and cognitive as well as sensory features. It may emerge as a consequence of immediate tissue trauma and subside with tissue healing, in which case it is acute pain. In some cases, chronic pain may exist due to chronic disease and tissue trauma. Alternatively, it may persist indefinitely with a weak relationship to tissue pathology, becoming a major psychological and social feature of the patient. This constructivist position has clinical implications for patient evaluation and therapeutic intervention.

▶ NOCICEPTION AND PAIN

Although the distinction is clear to anyone involved in pain studies, teachers and writers speaking in the vernacular continue to muddle nociception and pain, using the term *pain* as a synonym for *nociception*. Nociception refers to the detection of tissue trauma through the activation of nociceptors and the transmission of signals of tissue injury within the nervous system. It is never conscious, and nociception has important negative physiologic consequences because it elicits motor and sympathetic reflexes as well as a hypothalamically-mediated and noradrenergic stress response.[5] Pain, in contrast, is a complex, compelling, unpleasant bodily awareness normally associated with tissue trauma. It is a phenomenon of consciousness, and it has sensory, emotional, and cognitive features. Pain cannot exist apart from consciousness; it is in an approximate sense the psychological counterpart of nociception, and it normally informs consciousness of nociceptive traffic in the nervous system and the impacts of that traffic on the autonomic and neuroendocrine homeostatic equilibrium. The virtually ubiquitous terminology of pain receptor, pain pathway, and pain modulation within the neuraxis is unfortunate because it presumes a simplistic mind-body dichotomy in which the pain sensation is nociception realized. This Cartesian oversimplification can mislead well-intentioned physicians seeking to prevent or relieve pain.

The assumption that pain is simply nociception manifested in awareness is archaic and has no scientific basis. Nakamura and Chapman[6] proposed that pain, like other forms of perception, is the end product of dynamic construction. The brain does not passively receive and realize information about the external and bodily environments; instead, it constantly constructs and revises a model of the body and the world around it. Pain is a part of a larger ongoing construction and always has a place in the continuous self-organization that produces coherent momentary awareness. Individuals vary widely in their responses to a constant injurious stimulus because their individual learning histories, expectations, interpretations of the meaning of the injury, and general emotional status vary. Nociception, too, may vary across individuals as a function of age, overall health, and genetic differences, but pain varies to a far greater degree across persons because it is the end product rather than the predecessor of construction and is much more complex.

▶ PSYCHOLOGICAL CONSEQUENCES OF NOCICEPTION

The Multiple Impacts of Nociception

Older scientific explanations for pain tended to regard nociception as a purely sensory function that did little else besides make possible bodily awareness of tissue trauma by generating an unpleasant sensation that flows centrally and becomes somehow "realized" in the cortex. The main point of controlling nociception medically was to prevent or reduce the distressing bodily awareness that we call pain by interfering with the transduction and transmission of the noxious message. However, nociception has many medically-significant nonconscious consequences.[5] These consequences can create related symptom burdens such as fatigue and sleep deprivation that debilitate the patient and thereby contribute indirectly to the complex experience of pain. Reviewed briefly here to call attention to the medical relevance of nociception, they are examined in greater detail later in this chapter to demonstrate that nociception is not a simple message destined to break into awareness but rather a triggering mechanism for massive, parallel, distributed preconscious processing in the limbic brain and elsewhere.

Nociceptive input stimulates the hypothalamopituitary-adrenocortical (HPA) axis to produce a sympathomedullary stress response.[7] This response involves the release of cortisol and other humoral messenger substances into the circulation, generalized autonomic arousal, and noradrenergerically-mediated hypervigilance originating in limbic structures.[8] The resulting blood pressure elevation can activate mesencephalic structures that control descending, nociception-inhibiting pathways.[9] When sudden, life-threatening emergencies occur, the stress response can induce a short-term analgesic state that ensures that pain will not interfere with survival behaviors. In clinical settings, however, the stress response tends instead to accentuate the emotional dimension of pain because it mobilizes many of the central structures involved in the production of threat, and it creates sweating, tachycardia, hyperventilation, and other changes that the patient recognizes as bodily signs of distress.

Uncontrolled nociception contributes significantly to morbidity and mortality following surgery.[5] When the stress response persists, catabolism predominates. This leads to poor wound healing and poor energy storage, which produce hypersensitive muscle tissue and fatigue. Prolonged nociception creates an extended stress response that disturbs biorhythms so that patients experience poor, nonrestorative sleep, diminished appetite, and loss of libido. Moreover, trauma to the abdomen or chest can cause muscle splinting from nociception-induced motor reflexes, and these can compromise pulmonary function and increase the risk of adventitious pulmonary infection.

Psychologically, patients with uncontrolled, sustained nociception manifest impaired concentration and attention problems along with depression-like mood states. These problems demonstrate that uncontrolled nociceptive traffic within the nervous system has complex physiologic consequences that, in turn, alter and denigrate the cognitive-emotional state of the individual experiencing the pain. Put another way, nociception strongly and negatively influences psychological well-being. One might say that one of the psychological aspects of pain is nociception-driven degradation of psychological well-being. In the section that follows, some of the mechanisms by which this occurs are described.

Autonomic Consequences and Defense Response

The autonomic nervous system (ANS) plays an important role in regulating the constancy of the internal environment, and it does so in a feedback-regulated manner under the direction of the hypothalamus, the solitary nucleus, the amygdala, and other central nervous system (CNS) structures.[10,11] In general, it regulates activities that are not normally under voluntary control. The hypothalamus is the principal integrator of autonomic activity. Stimulation of the hypothalamus elicits highly integrated patterns of response that involve the limbic system and other structures.[12] Autonomic activity contributes to the pathophysiology of chronic pain.[13] It is also an important feature of the emotional aspect of pain.[14]

Many researchers hold that the ANS consists of three divisions, the sympathetic, the parasympathetic, and the enteric.[15,16] Others subsume the enteric under the other two divisions. Broadly, the sympathetic nervous system makes possible the arousal needed for "fight or flight" reactions, whereas the parasympathetic system governs basal heart rate, metabolism, and respiration. The enteric nervous system innervates the viscera via a complex network of interconnected plexuses.

The sympathetic and parasympathetic systems are largely mutual physiologic antagonists—if one system inhibits a function, the other typically augments it. There are, however, important exceptions to this rule that demonstrate complementary or integratory relationships. The mechanism most heavily involved in the affective response to tissue trauma is the sympathetic nervous system.

During emergency or injury to the body, the hypothalamus uses the sympathetic nervous system to increase cardiac output, respiration rate, and blood glucose level. It also regulates body temperature, causes piloerection, alters muscle tone, provides compensatory responses to hemorrhage, and dilates pupils. These responses are part of a coordinated, well-orchestrated response pattern called the *defense response*.[17–20] It resembles the better-known orienting response in some respects, but it only can occur following a strong stimulus that is noxious or frankly painful. It sets the stage for escape or confrontation, thus serving to protect the organism from danger. In a conscious cat, both electrical stimulation of the hypothalamus and infusion of norepinephrine into the hypothalamus elicit a rage reaction with hissing, snarling, and attack posture with claw exposure, and a pattern of sympathetic nervous system arousal accompanies this.[21–23] Circulating epinephrine produced by the adrenal medulla during activation of the hypothalamopituitary-adrenocortical axis accentuates the defense response, fear responses, and aversive emotional arousal in general.

Because the defense response and related changes are involuntary in nature, we generally perceive them as something that the environment does to us. We generally describe such physiologic changes not as the bodily responses that they are but rather as feelings. We might describe a threatening and physiologically arousing event by saying that "It upset me" or that "It made me really mad."

Phenomenologically, feelings seem to happen to us; we do not "do" them in the sense that we think thoughts or choose actions. They are not volitional. Emotions are who we are in a given circumstance rather than choices we make, and we commonly interpret events and circumstances in terms of the emotions that they elicit. ANS arousal, therefore, plays a major role in the complex psychological experience of injury and is a part of that experience.

Early views of the ANS followed the lead of Cannon[17] and held that emergency responses and all forms of intense aversive arousal are undifferentiated, diffuse patterns of sympathetic activation. While this is broadly true, research has shown that definable patterns characterize emotional arousal and that these are related to the emotion involved, the motor activity required, and perhaps the context.[10,11] An investigator attempting to understand how humans experience emotions must remember that the brain not only recognizes patterns of arousal but also creates them.

One of the primary mechanisms in the creation of emotion is feedback-dependent sympathetic efferent activation. The ANS has both afferent and efferent functions. The afferent mechanisms signal changes in the viscera and other organs, whereas efferent activity conveys commands to those organs. Consequently, the ANS can maintain feedback loops related to viscera, muscle, blood flow, and other responses. The visceral feedback system exemplifies this process. In addition, feedback can occur via the endocrine system, which under the control of the ANS releases neurohormones into the systemic circulation. Because feedback involves both autonomic afferents and endocrine responses, and because some feedback occurs at the level of unconscious homeostatic balance and other feedback involves awareness, the issue of how visceral change contributes to the creation of an emotional state is complex. The mechanisms are almost certainly pattern-dependent, dynamic, and at least partly specific to the emotion involved. Moreover, they occur in parallel with sensory information processing.

The feedback concept is central to emotion research: Awareness of physiologic changes elicited by a stimulus is a primary mechanism of emotion. The psychiatric patient presenting with panic attack, phobia, or anxiety is reporting a subjective state based on patterns of physiologic signals and not an existential crisis that exists somewhere in the domain of the mind, somehow apart from the body. Similarly, the medical patient expressing emotional distress during a painful procedure or during uncontrolled postoperative pain is experiencing the sensory features of that pain against the background of strong sympathetic arousal signals.

The concept of feedback underscores an essential point: Nociception stimulation does not have purely sensory effects. How nociceptive signaling undergoes parallel processing at affect-generating structures as well as at the somatosensory cortex is described below. When a neural signal involves threat to biologic integrity, it elicits strong patterns of sympathetic and neuroendocrine response. These, in turn, contribute to the awareness of the perceiver. Sensory processing provides information about the environment, but this information exists in awareness against a background of emotional arousal, either positive or negative, and that arousal may vary from mild to extreme.

Central Affective Consequences of Nociception

Evidence from several lines of inquiry makes it quite clear that nociceptive signaling serves multiple adaptive purposes and has multiple consequences in the CNS. Providing a message to the conscious mind is only one consequence. Physiologic arousal sufficient to mobilize flight or fight, the generating of threat as an emotional awareness, and cognitive hypervigilance appear to be the major consequences from an evolutionary standpoint. These responses assist adaptation and survival for acute injury in a primitive world, but chronic nociception in our contemporary world has only negative consequences such as circadian dysregulation and immune suppression.

Multiple pathways convey signals of tissue trauma to multiple destinations, where they provoke multiple complex, higher-order processes. Nociceptive centripetal transmission engages multiple pathways: spinoreticular, spinomesencephalic, spinolimbic, spinocervical, and spinothalamic tracts.[24,25] The spinoreticular tract contains somatosensory and viscerosensory afferent pathways that arrive at different levels of the brain stem. Spinoreticular axons possess receptive fields that resemble those of spinothalamic tract neurons projecting to the medial thalamus, and like their spinothalamic counterparts, they transmit tissue injury information.[26,27] Most spinoreticular neurons carry nociceptive signals, and many of them respond preferentially to noxious activity.[28,29] The spinomesencephalic tract consists of several projections that terminate in multiple midbrain nuclei, including the periaqueductal gray matter, the red nucleus, the nucleus cuniformis, and the Edinger-Westphal nucleus.[25] Spinolimbic tracts include the spinohypothalamic tract, which reaches both the lateral and the medial hypothalamus,[30,31] and the spinoamygdalar tract, which extends to the central nucleus of the amygdala.[32] The spinocervical tract, like the spinothalamic tract, conveys signals to the thalamus. All these tracts transmit tissue trauma signals rostrally.

Central processing of nociceptive signals to produce affect undoubtedly involves multiple neurotransmitter systems. Four extrathalamic afferent pathways project to neocortex: the noradrenergic medial forebrain bundle originating in the locus ceruleus (LC), the serotonergic fibers that arise in the dorsal and median raphe nuclei, the dopaminergic pathways of the ventral tegmental tract that arise from substantia nigra, and the acetylcholinergic neurons that arise principally from the nucleus basalis of the substantia innominata.[33] Of these, the noradrenergic and serotonergic pathways link most closely to negative emotional states.[8,34,35] The set of structures receiving projections from this complex and extensive network corresponds to the classic definition of the limbic brain.[36–38]

Although other processes governed predominantly by other neurotransmitters almost certainly play important roles in the complex experience of emotion during pain, the role of central noradrenergic processing and the medial forebrain bundle should be emphasized here. More extensive description and review of this subject appears elsewhere.[39] The *noradrenergic stress-response hypothesis* helps clarify the impact of nociception on the brain. Put simply, any stimulus that threatens the biologic, psychological, or psychosocial integrity of the individual increases the firing rate of the LC, and this, in turn, increases release and turnover of norepinephrine in the brain areas involved in noradrenergic innervation. The literature on the role of central noradrenergic pathways in anxiety, panic, stress, and post-traumatic stress disorder reveals that nociceptive signaling provokes deep and cortical limbic structures that in turn contribute to the construction of the experience of threat and hypervigilance,[8,40] as well as motor structures that can contribute to adaptive behaviors.[41] Although traditional thinking would have it otherwise, I suggest that it is arbitrary and pointless to insist that pain is purely sensory and that the emotional arousal produced concomitantly by nociceptive signaling is not pain or is secondary to pain. It is a part of pain, and it comprises the affective dimension of pain.

Although the contribution of central noradrenergic activation to chronic pain remains poorly defined, Bremner and colleagues[8] offer a perspective on how a chronic psychological disorder may affect central noradrenergic function. They postulated that chronic stress can affect regional norepinephrine turnover and thus contribute to the *response sensitization* evident in panic disorder and post-traumatic stress disorder. Chronic exposure to a stressor (including perseverating nociception) could create a situation in which noradrenergic synthesis cannot keep up with demand, thus depleting brain norepinephrine levels. Animals exposed to inescapable shock demonstrate greater LC responsiveness to an excitatory stimulus than animals that have experienced escapable shock.[42] In addition, such animals display "learned helplessness" behaviors; that is, they cease trying to adapt to, or cope with, the source of shock.[43] From an evolutionary perspective, this is a failure of the defense response as adaptation; it represents surrender to suffering. Extrapolating this and related observations to patients, Bremner and colleagues[8] suggested that persons who have once encountered overwhelming stress and suffered exhaustion of central noradrenergic resources may respond excessively to similar stressors that they encounter later.

Research on brain activation during the experience of pain supports the potential importance of limbic activation in pain. Studies involving positron-emission tomography (PET) of regional cerebral blood flow (rCBF) in volunteers experiencing pain and other studies of functional brain imaging in pain patients demonstrate that noxious stimulation activates limbic and motor structures as well as sensory structures. Changes in rCBF index reveal neuronal metabolic activity in specific brain regions. It is possible to demonstrate stimulus-induced responses in brain activation patterns as well as endogenously generated brain activity that makes cognition possible.

Several reviews have summarized and interpreted functional brain imaging studies designed to capture the complex central processing associated with pain.[44–49] Although not perfectly consistent, well over 100 studies demonstrate beyond any doubt that massive, parallel distributed processing occurs in the brain following tissue trauma or threat. Processing includes, but is not limited to, sensory structures. The most striking feature of this massive, parallel distributed processing is that a great deal of it occurs in limbic brain, which consists of structures that contribute heavily to emotional experiences such as threat. The basal ganglia and cerebellar vermis are often active as well, and this suggests that nociception readies the brain for escape or defense behaviors.[41]

This body of literature raises many intriguing questions. How a coherent awareness of injury in a specific part of the body emerges from specific patterns of activity within the brain remains a mystery. This quandary is a part of the larger question of how consciousness itself occurs. However, this literature demonstrates unequivocally that pain is not a simple sense modality involving the transmission of specific messages to specific brain areas. When patients are in a state of pain, the brain is engaging in massive processing that engages structures that contribute to emotion, sensation, and cognition.

Given the preceding considerations, this much is now clear: Nociceptive signals do not simply make their way through the neuraxis and then enter consciousness. Instead, such signals have multiple destinations within the brain, and they trigger complex, probably interdependent, higher-order processes that, following integration, create the conditions necessary for the emergence of pain as a feature in the dynamic pattern of phenomenal awareness. Pain is a property of bodily awareness that is contingent on certain conditions, normally tissue trauma. Bodily awareness, like other aspects of consciousness, is an emergent consequence of brain activity.

In a healthy nervous system and under normal conditions, nociception is necessary for acute pain but not sufficient to account for patient presentation. The individual in whom nociception is occurring must be conscious and must not have his or her consciousness fully engaged in something else if pain is to emerge in consciousness. Individuals differ markedly in the intensity of the pain that they experience to a common injurious event because each constructs a unique phenomenal reality and bodily awareness. Clinicians need to appreciate that, in humans, pain is always more complex than nociception because several factors contribute to it.

Although nociception can occur in an unconscious individual, pain cannot. Like other phenomena of consciousness, pain is an emergent product of complex, distributed activity within the brain. Put succinctly, pain is a complex, consciousness-dependent, unpleasant somatic experience with cognitive and emotional as well as sensory features. It is not a signal that "enters" consciousness but rather an unpleasant aspect of the moment-to-moment construction of consciousness, which consists of awareness of both the external and the internal, or somatic, environment. In the section that follows, the knotty problem of understanding how the process of construction might operate is addressed.

▶ CONSTRUCTIVISM

Basic Considerations

Many clinicians and some neurophysiologists still see pain as fundamentally sensory in nature, even though people in pain manifest strong negative emotions to a much greater degree than sensory descriptions. Those with a fundamentally sensory perspective presume that emotions can arise in one of two ways. One is that the "realization" (entry into consciousness) of nociceptive sensory signaling at the cortex elicits reflexive and higher-order responses that create both the arousal and the emotions inevitably occurring together with pain (primary pain-related emotions). The second is that emotions can occur in parallel to pain because an injurious stimulus has multimodal effects (secondary emotions). Primary pain emotions require that the conscious sensation of pain must occur before the emotional arousal can take place. In other words, pain is a bottom-up sensory experience, and the affective aspect of pain and related cognitions are consequent to the sensory experience. An overlay of secondary emotions normally occurs as well in humans because injurious events have many higher-order associations. For example, Damasio[50] (p. 71) asserted that "emotions can be caused by the same stimulus that causes pain, but they are a different result from that same cause."

Although this point of view advances understanding and accounts for the observations of functional brain imaging studies of pain, it suffers from three limitations: (1) It construes the brain as passive and purely reactive in its processing of nociception, (2) it offers no account for how pain emerges in consciousness and, when intense, dominates awareness, and (3) it cannot explain why the fit of tissue trauma to pain report is usually poor and, in the case of chronic pain, sometimes lacking altogether. Moreover, this perspective does not integrate naturally with ongoing, parallel research in the fields of emotion, cognition, and consciousness studies.

An alternative framework grounded in consciousness studies can better account for the complex nature of pain and other feelings, and it can bridge pain research more readily to other relevant fields of study. This position, termed *constructivism*, opposes the assumption that the brain maintains an accurate representation of the world based on incoming sensory information. Instead, constructivists argue, the brain constructs subjective experience from sensory input, memory, and expectations based on environmental context. Figure 11-1 illustrates a constructivist view of pain.

The Foundations of Constructivism

Pain is the end product of adaptive perceptual self-organization and not the product of central registration of nociceptive signaling. The brain actively and dynamically

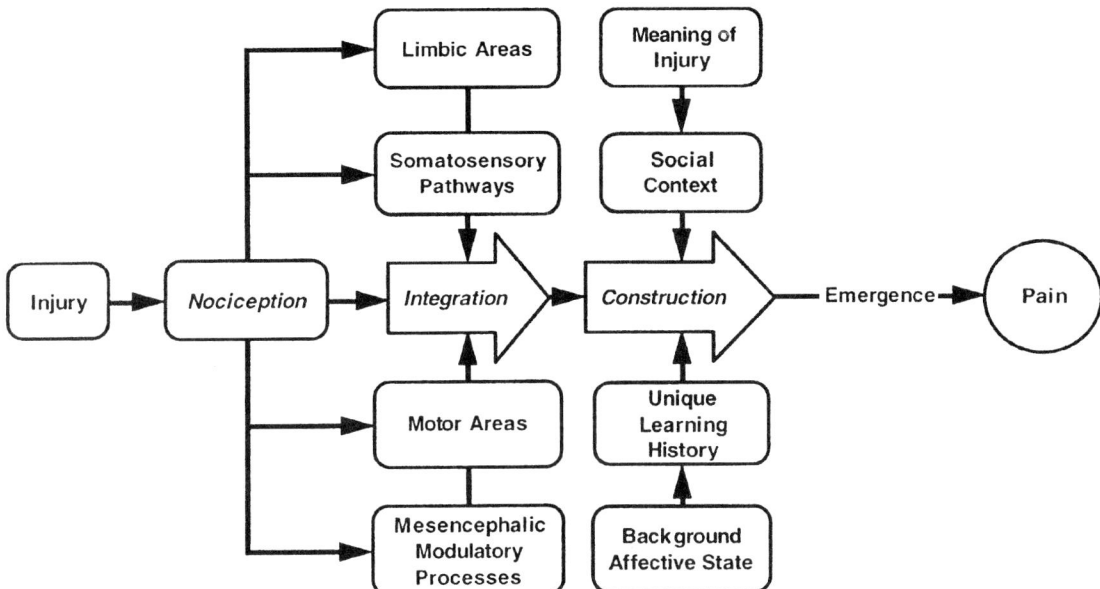

Figure 11-1. A constructivist model of the process by which injury leads to pain. The diagram indicates that nociception initiates multiple parallel processes that lead first to integration and then progress to construction of the pain experience. The process of construction is multimodal and multifaceted. The integrated products of nociception must fit into a dynamic model of the body and the world that is unique to the individual and the situation. Pain, like all aspects of consciousness, is an emergent feature of phenomenal awareness. Individuals differ markedly in the pain they experience to a common injury in small part because they differ neurophysiologically and in large part because the process of construction is unique to each individual.

constructs the stream of conscious experience from moment to moment to maintain a model of the body, the world around it, and the sense of self. Historically, the idea of "constructing reality" stems from Kantian epistemology and philosophy of science.[51,52] In contrast to the empiricists, who asserted that we passively receive knowledge and information from the external world, Kant proposed that we construct our experience on the basis of certain innate conceptual structures (such as space, time, and causality). This notion resurfaced in Piaget's work in genetic epistemology and developmental psychology.[53]

Constructivists reject many of the metatheoretical assumptions that have long characterized cognitive science and neuroscience. They refute, for example, the notion of strictly referential representations of the body and environment in the brain because it implies a correspondence between representations and stimulus properties. Constructivism also denies that we have direct access to the properties of an objective external world; instead, the brain intersubjectively constructs the experiential world by interacting with the body and the environment. At the same time, it avoids solipsism by recognizing the interdependent nature of the brain, the body, and the world. In this respect, constructivism incorporates the perspective of embodied cognition.[54]

The constructivist viewpoint asserts that the brain produces awareness from a complex array of massively parallel processes.[55,56] This applies readily to the question of emotion and pain. The constructivist assumes that the brain deals not with reality itself but rather with an internal, autonomous representation of reality that it builds and revises from moment to moment using sensory information and networks of association in memory. Subjective reality undergoes constant revision (self-organization) as it includes sensory information, emotion, ratiocination, and other aspects of cognition. Furthermore, constructed reality always has a point of view. The point of view is each individual's sense of self, which Dennett[2] described as a center of narrative gravity.

Constructivist orientations have started to appear in a few areas of contemporary cognitive science and neuroscience. Mountcastle[57] provided a timely survey of where neuroscience found itself at the end of the twentieth century. He listed some well-established generalizing principles, which are necessarily subject to the corrective effects of new discoveries. Two of these principles are important for constructivists: (1) Recalled memories are constructions, not replications, and (2) brain representations of external objects and events are constructions, not replications.

Mountcastle's bold generalizations have support from several contemporary neuroscientists. In the field of memory, Schacter and colleagues[58] proposed a notion of constructive memory that emphasizes the fabricated nature of human memory. Memory distortion phenomena such as false recognition, intrusions, and confabulations are consistent with the notion that the brain constructs memory. In visual perception, Hoffman[59] described visual intelligence as the power that we use to construct an experience of objects out of colors, lines, and motions. His book demonstrated the mysterious constructive powers of our eye-brain machines using simple drawings and diagrams to illustrate basic rules of the visual road. Taken together, these emerging signs of constructivist thinking in divergent domains support the generality and usefulness of the constructivist approach.

Basic Premises of Constructivism

Three premises define the constructivist approach. First, as Mountcastle[57] emphasized, the brain actively constructs conscious experience. Awareness is the consequence of self-organizing processes that evaluate incoming inputs against an internal model of the self and the world and then integrate the inputs into a coherent stable pattern with an extraordinary plasticity. Furthermore, this construction of consciousness proceeds in response to the intentional and situational imperatives of each person. This process involves integrating sensory signals, memory and prior experience, expectations, and immediate and long-term goals and plans. It weaves all these into the coherent, stable macro-emergent pattern that seems to underlie awareness of the world and one's self. Far from being a passive entity that merely registers information coming in from various sensory channels, the brain is an active, adaptive system that constantly "simulates" the world and the body within which it dwells.

Second, viewed this way, conscious experience resembles an attribution that each person actively constructs; it is not simply a veridical registration of information from the environment and body.[56] Conscious experience is subject to cultural influences because culture provides rules to guide social and other constructions. In this way, an individual's cultural meaning systems may determine the possible range of states of consciousness.

Third, consciousness plays a causal role in the course of information processing at higher levels of the CNS. The causal effects of consciousness typically manifest as "constraints on" or "selections for" subsequent cycles of information processing.[60] The brain is less open to information that is not congruent with the immediate model of constructed consciousness.

The constructivist approach escapes the narrow constraints imposed by the classical sensory neurophysiology of pain and thus makes it possible to conceptualize pain as a phenomenon of consciousness. For the constructivist, pain is a complex, emotionally negative bodily awareness characterized by sensory qualities that are normally consequent to tissue trauma. Individuals differ markedly in the pain they experience to a common type of tissue trauma because the many factors influencing construction vary greatly across persons.

The transduction, transmission, and modulation aspects of classical neurophysiologic thinking can serve as basic mechanisms for the constructivist framework. However, whereas the sensory neurophysiology framework focuses on the transmission of nociceptive impulses as specific sensory signals, the constructivist perspective emphasizes the central processing of such signals, rejecting the notion of simple sensory registration. The brain constructs pain from complex patterns of massive, parallel distributed processing as a focal element in the perceiver's model of the self and the world.

Schemata and Construction

The *schema*, a perceptual hypothesis, is a fundamental unit for the construction of feeling states. Historically, the term dates back to Kant, Piaget, and Bartlett, who all contributed to the notion.[61] Kant thought of schemata as structures of imagination that connect concepts with percepts and described them as procedures for constructing images. Later thinkers construed them as basic knowledge structures, narrative structures, or perceptual hypotheses. Some researchers see them as neural networks or patterns of physiologic processes that produce consistencies in subjective experience. Nakamura and Chapman understand schemata as fuzzy, preconscious, and dynamic, roughly related to dynamically stable patterns in neural networks.[6] Rumelhart and colleagues[60] usefully described schemata in this way: "Schemata are not single things. There is no representational object which is a schema. Rather schemata emerge at the moment they are needed from the interaction of large numbers of much simpler elements, all working in concert with one another. Schemata are not explicit entities, but rather are implicit in our knowledge and are created by the very environment that they are trying to interpret as it is interpreting them . . ." (p. 20).

Drawing on Rumelhart and colleagues,[60] Balkenius[62] summarized the main features of schemata:

- The brain uses schemata for representing objects, situations, and actions. This implies that schemata must have dynamic properties.
- Schemata have variables. This indicates that there are ways to change a schema to make it adapt to different situations.
- Schemata can embed. One schema can have another schema as a part or as an instantiation of a variable.

- Schemata support default assumptions about the environment. They can fill in missing information. Schemata are not passive data structures. They are active representations.
- Schemata are closely connected to memory.

These notions suggest that we do not experience bodily or environmental events directly as isolated incidents but rather as facets of one or another schema, which derive from past experiences, the immediate model of the situation, and expectations.

▶ CLINICAL IMPLICATIONS

Constructivist thinking suggests new ways of understanding many different clinical pain phenomena, as well as many curious social rituals in which people undergo tissue damage with apparent unconcern. It also helps us to approach the challenge of pain control from a new perspective. In addition to attenuating nociceptive traffic in the nervous system, clinicians can, in principle, intervene in the construction of the pain experience itself. A few of many possible applications of constructivism in the field of pain are discussed below.

First, constructivism helps clarify the nature of some chronic pain problems. For example, the schema notion helps to explain postamputation phantom feelings and phantom limb pain. Phantom limbs are kinesthetically vivid realizations of a body part that is physically absent. Sometimes a phantom limb will hurt, and no treatment directed at the stump or administered systemically can relieve the pain.[63] In a study of 68 patients with amputated limbs, Katz and Melzack[64] found that the patients acknowledged various pains, all localized to the missing limbs. The patients described these pains as immediate and real, quite unlike the recollection of a past pain state. In constructivist terms, the patients were reproducing or reconstructing the pain and not simply recalling a past event. Interestingly, some of the phantom experiences involved painless experience (e.g., feeling a shoe on a missing foot), and some had multimodal sensory qualities. Katz and Melzack speculated on the basis of neuromatrix theory that "a higher-order somatosensory memory component, once formed, can be activated when only some of its elements are present in the sensory input".[64p33]

The phantom limb story supports the fundamental notion that pain is a somatic construction. It is unique among pain problems in that deafferentation apparently has somehow frozen or locked in a schema that is very resistant to change. Ramachandran's work with phantom limb patients involves a virtual reality box and multimodal sensory feedback.[65] Patients with a painful, clenched phantom hand are able to feel what seems to be kinesthetic feedback and to open the clenched phantom hand, thus relieving the pain. This remarkable maneuver demonstrates that, in selected cases, it is possible to re-establish the construction of a particular aspect of frozen somatic awareness, in this case both the disquieting position and the pain of a phantom hand. This suggests that facilitating or altering construction may be a viable approach to chronic pain management.

This principle may hold for chronic pain in general. Pain is a form of somatic perception, and like all perception, it is the product of construction. If this is the case, then it is reasonable to expect that pain may become chronic if the mechanisms of construction can become self-sustaining and essentially independent of sensory input. Conceivably, some patients with a long history of experiencing noxious sensory input eventually will learn to inadvertently sustain the construction in the absence of that input. The schemata, operating like neural networks, become autonomous and persist despite alterations in transduction, transmission, or modulation of nociception. This is consistent with many puzzling clinical observations of pain, including, for example, the perseveration of pelvic pain following surgical removal of the offending organs.[66]

Second, the constructivist approach can help make sense of the vast differences in response to painful events that one sees in clinical settings, as well as the variation in other types of perception. Persons with similar types and degrees of tissue damage vary from having no pain to having disabling and severe pain. Many people suffer from disabling chronic pain even though they have no definable tissue damage. Moreover, some people, when examined for other medical reasons, have severe tissue damage but do not have pain. The constructivist approach submits that each person's psychological experience differs from that of others because he or she creates and lives with unique experiential realities. Two persons with identical lesions do not experience the same pain because they do not have identical experiential realities. In preventing or relieving pain, it is important to fit the intervention to the psychological uniqueness of the person. Constructivist thinking ultimately may provide a rationale for the development and refinement of new psychological interventions for pain.

Third, the constructivist approach can account for the effects of familiar psychological interventions such as hypnosis and distraction on pain.[67] These are interventions that recognize and address the complexity of human perceptual experience and the uniqueness of the individual. It has been proposed that effective hypnotic suggestion directed at removing pain alters the construction of the pain experience. If this is so, then the processes of competition and integration among multiple activated schemata are as viable targets for pain intervention as the neural structures and synapses that underlie nociception.

Finally, and scientifically, the constructivist approach allows us to make sense of parallel distributed processing as reflections of ongoing dynamic competition and integration of schemata that support the construction of consciousness. Thus far it is impossible to link activity in any single brain region to the qualities or scalable features of subjective pain experience or any other subjective experience. Complex brain activity patterns do not seem to relate in a linear fashion to the contents of consciousness. It is suggested that distributed processing may reflect the preconscious workings of the constructive process. Combining this insight with the knowledge that the temporal resolutions of current imaging technologies constrain what we can observe in imaged data, researchers can perhaps begin to design experiments that will shed light on how consciousness emerges out of nonconscious parallel distributed processes in the brain. Manipulation of attention, expectancy, as well as short- and long-term memory, may illuminate the construction of pain and other feeling states when combined with functional brain imaging.

This chapter began by addressing the essence of pain. Is it nociception? Does it reduce ultimately to transduction, transmission, and modulation of signals indicating tissue trauma? It has been asserted that pain is inherently psychological in nature; it is not a primitive sensory message of tissue trauma that somehow gets "realized" in consciousness when it reaches the somatosensory cortex. Rather, it is the end product of a complex construction of somatic awareness. It has been argued that the essence of pain is tied to the essence of consciousness itself. It is a multifaceted emergent feature of brain activity, and like all perception, pain is constructed. The reasoning of Newtonian mechanics will not help us understand our patients because they are all unique, and each constructs his or her pain from a complex and highly individual personal reality.

At the risk of overgeneralization, it would seem that the construction of acute pain depends heavily on sensory input, and it ceases or attenuates with diminished signaling from the periphery. Chronic pain, in contrast, sometimes can create enduring schemata through neuroplasticity and higher forms of learning that then become more or less autonomous. Interventions directed at nociceptive pathways may not eliminate pain because the brain is quite capable of sustaining the familiar construction in the absence of the original cause. The constructivist framework may provide a useful guide for the development of future research on chronic pain and its control.

REFERENCES

1. Descartes R: The passions of the soul, in Haldane ES, Ross GTR (trans), *The Philosophical Works of Descartes,* Vol 1. New York: Dover Press, 1649/1967:219–327.
2. Dennett DC: *Consciousness Explained.* Boston: Little, Brown, 1991.
3. Damasio A: *Looking for Spinoza: Joy, Sorrow, and the Feeling Brain.* Orlando, FL: Harcourt, 2003.
4. Chapman CR, Nakamura Y: Pain and consciousness: A constructivist approach. *Pain Forum* 1999;8:113–112.
5. Carr DB, Goudas LC: Acute pain. *Lancet* 1999; 353:2051–2058.
6. Nakamura Y, Chapman CR: Measuring pain: An introspective look at introspection. *Conscious Cogn* 2002;11:582–592.
7. Lilly MP, Gann DS: The hypothalamic-pituitary-adrenal-immune axis: A critical assessment. *Arch Surg* 1992; 127:1463–1474.
8. Bremner JD, Krystal JH, Southwick SM, Charney DS: Noradrenergic mechanisms in stress and anxiety: I. Preclinical studies. *Synapse* 1996;23:28–38.
9. Ghione S: Hypertension-associated hypalgesia: Evidence in experimental animals and humans, pathophysiological mechanisms, and potential clinical consequences. *Hypertension* 1996;28:494–504.
10. LeDoux JE: The neurobiology of emotion, in Ledoux JE, Hirst W (eds), *Mind and Brain: Dialogs in Cognitive Neuroscience.* Cambridge: Cambridge University Press, 1986:301–354.
11. LeDoux JE: *The Emotional Brain: The Mysterious Underpinnings of Emotional Life.* New York: Simon & Schuster, 1996:384.
12. Morgane PJ: Historical and modern concepts of hypothalamic organization and function, in Morgan PJ, Panksepp J (eds), *Handbook of the Hypothalamus,* Vol 1. New York: Marcel Dekker, 1981:1–64.
13. Benarroch EE: Pain-autonomic interactions: A selective review. *Clin Auton Res* 2001;11:343–349.
14. Chapman CR, Gavrin J: Suffering: The contributions of persisting pain. *Lancet* 1999;353:2233–2237.
15. Dodd J, Role LW: The anatomic nervous system, in Kandel ER, Schwartz JH, Jessell TM (eds), *Principles of Neural Science,* 3d ed. New York: Elsevier, 1991:761–775.
16. Burnstock G, Hoyle CHV (eds): *Autonomic Neuroeffector Mechanisms.* Philadelphia: Harwood Academic Publishers, 1992.
17. Cannon, WB: *Bodily Changes in Pain, Hunger, Fear and Rage.* New York: Appleton, 1929.
18. Sokolov EN: *Perception and the Conditioned Reflex.* Oxford: Pergamon Press, 1963.
19. Sokolov EN: The orienting response, and future directions of its development. *Pavlov J Biol Sci* 1990;25:142–150.
20. Donaldson GW, Chapman CR, Nakamura Y, et al: Pain and the defense response: Structural equation modeling reveals coordinated psychophysiological response to increasing painful stimulation. *Pain* 2003;102:97–108.
21. Hess WR: Hypothalamus und die Zantren des autonomen Nervensystems: Physiologie. *Arch Psychiatr Nervenkrankh* 1936;104:548–557.
22. Hilton SM: Hypothalamic regulation of the cardiovascular system. *Br Med Bull* 1966;22:243–248.
23. Barrett JA, Shaikh MB, Edinger H, Siegel A: The effects of intrahypothalamic injections of norepinephrine upon affective defense behavior in the cat. *Brain Res* 1987; 426:381–384.
24. Villanueva L, Bing Z, Bouhassira D, Le Bars D: Encoding of electrical, thermal, and mechanical noxious stimuli by

24. [continued] subnucleus reticularis dorsalis neurons in the rat medulla. *J Neurophysiol* 1989;61:391–402.
25. Willis WD, Westlund KN: Neuroanatomy of the pain system and of the pathways that modulate pain. *J Clin Neurophysiol* 1997;14:2–31.
26. Villanueva L, Cliffer KD, Sorkin LS, et al: Convergence of heterotopic nociceptive information onto neurons of caudal medullary reticular formation in monkey (*Macaca fascicularis*). *J Neurophysiol* 1990;63:1118–1127.
27. Craig AD: Spinal and trigeminal lamina I input to the locus coeruleus anterogradely labeled with *Phaseolus vulgaris* leukoagglutinin (PHA-L) in the cat and the monkey. *Brain Res* 1992;584:325–328.
28. Bing Z, Villanueva L, Le Bars D: Ascending pathways in the spinal cord involved in the activation of subnucleus reticularis dorsalis neurons in the medulla of the rat. *J Neurophysiol* 1990;63:424–438.
29. Bowsher D: Role of the reticular formation in responses to noxious stimulation. *Pain* 1976;2:361–378.
30. Burstein R, Cliffer KD, Giesler GJ: The spinohypothalamic and spinotelecephalic tracts: Direct nociceptive projections from the spinal cord to the hypothalamus and telencephalon, in Dubner R, Gebhart GF, Bond MR (eds), *Proceedings of the 5th World Congress on Pain*. New York: Elsevier, 1988:548–554.
31. Burstein R, Dado RJ, Cliffer KD, Giesler GJ: Physiological characterization of spinohypothalamic tract neurons in the lumbar enlargement of rats. *J Neurophysiol* 1991;66:261–284.
32. Bernard JF, Besson JM. The spino(trigemino)pontoamygdaloid pathway: Electrophysiological evidence for an involvement in pain processes. *J Neurophysiol* 1990;63:473–490.
33. Foote SL, Morrison JH: Extrathalamic modulation of cortical function. *Annu Rev Neurosci* 1987;10:67–95.
34. Gray JA: *The Neuropsychology of Anxiety: An Enquiry into the Functions of the Septohippocampal System*. New York: Oxford University Press, 1982.
35. Gray JA: *The Psychology of Fear and Stress*, 2d ed. Cambridge: Cambridge University Press, 1987.
36. Papez JW: A proposed mechanism of emotion. *Arch Neurol Psychiatry* 1937;38:725–743.
37. Isacson RL: *The Limbic System*, 2d ed. New York: Plenum Press, 1982.
38. MacLean PD: *The Triune Brain in Evolution: Role in Paleocerebral Functions*. New York: Plenum Press, 1990.
39. Chapman CR: Neuromatrix theory: Do we need it? (Commentary) *Pain Forum* 1996;5:139–142.
40. Charney DS, Deutch A: A functional neuroanatomy of anxiety and fear: Implications for the pathophysiology and treatment of anxiety disorders. *Crit Rev Neurobiol* 1996;10:419–446.
41. Wall PD: *Pain: The Science of Suffering*. New York: Columbia University Press, 2000.
42. Weiss JM, Simson PG: Depression in an animal model: Focus on the locus ceruleus. *Ciba Found Symp* 1986;123:191–215.
43. Seligman ME, Weiss J, Weinraub M, Schulman A: Coping behavior: Learned helplessness, physiological change and learned inactivity. *Behav Res Ther* 1980;18:459–512.
44. Casey KL: Concepts of pain mechanisms: The contribution of functional imaging of the human brain. *Prog Brain Res* 2000;129:277–287.
45. Ingvar M: Pain and functional imaging. *Philos Trans R Soc Lond B Biol Sci* 1999;354:1347–1358.
46. Davis KD: The neural circuitry of pain as explored with functional MRI. *Neurol Res* 2000;22:313–317.
47. Hudson AJ: Pain perception and response: Central nervous system mechanisms. *Can J Neurol Sci* 2000;27:2–16.
48. Bromm B: Brain images of pain. *News Physiol Sci* 2001;16:244–249.
49. Treede RD, Apkarian AV, Bromm B, et al: Cortical representation of pain: Functional characterization of nociceptive areas near the lateral sulcus. *Pain* 2000;87:113–119.
50. Damasio A: *The Feeling of What Happens: Body and Emotion in the Making of Consciousness*. New York: Harcourt Brace, 1999.
51. Beck LW: *Studies in the Philosophy of Kant*. Indianapolis: Bobbs-Merrill, 1965.
52. Hundert EM: *Lessons from an Optical Illusion: On Nature and Nurture, Knowledge and Values*. Cambridge, MA: Harvard University Press, 1995.
53. Piaget J: *The Mechanisms of Perception*. London: Rutledge and Kegan Paul, 1969.
54. Varela F, Thompson E, Rosch E: *The Embodied Mind*. Cambridge, MA: MIT Press, 1991.
55. Mandler G, Nakamura Y: Aspects of consciousness. *Personality Soc Psychol Bull* 1987;13:299–313.
56. Marcel A: Conscious and unconscious perception: Experiments on visual masking and word recognition. *Cogn Psychol* 1983;15:197–237.
57. Mountcastle VB: Brain science at the century's ebb. *Daedalus* 1998;127:1–36.
58. Schacter DL, Norman KA, Koutstaal W: The cognitive neuroscience of constructive memory. *Annu Rev Psychol* 1998;49:289–318.
59. Hoffman DD: *Visual Intelligence: How We Create What We See*. New York: Norton, 1998.
60. Rumelhart DE, Smolensky P, McClelland JL, Hinton GE: Schemata and sequential thought processes in PDP models, in McClelland JL, Rumelhart DE (eds), *Parallel Distributed Processing: Explorations in the Microstructure of Cognition*, Vol 2: *Psychological and Biological Models*. Cambridge, MA: MIT Press, 1986:7–57.
61. Martin B: The schema, in Gowan G, Pines D, Meltzer D (eds), *Complexity: Metaphors, Models and Reality*. Reading, MA: Addison-Wesley, 1994:263–285.
62. Balkenius C: Neural mechanisms for self-organization of emergent schemata, dynamical schema processing, and semantic constraint satisfaction. *Lund Univ Cogn Stud* 1992;14 (available at http://www.lucs.lu.se/Abstracts/LUCS_Studies/LUCS14.html).
63. Melzack R: The tragedy of needless pain. *Sci Am* 1990;262(2):27–33.
64. Katz J, Melzack R: Pain "memories" in phantom limbs: Review and clinical observations. *Pain* 1990;43:319–336.
65. Ramachandran VS, Rogers-Ramachandran DC: Synaesthesia in phantom limbs induced with mirrors. *Proc Roy Soc Lond Series B: Biol Sci* 1966;263(1369):377–386.
66. Baskin LS, Tonagho EA: Pelvic pain without pelvic organs. *J Urol* 1992;147:683–686.
67. Chapman CR, Nakamura Y: Hypnotic analgesia: A constructivist framework. *Int J Clin Exp Hypnosis* 1998;47:6–27.

PART II

The Pain Physician and the Patient with Pain

CHAPTER 12

Ethics

Carol R. Taylor

Seldom does ethics get the attention it deserves in health care. This may be linked to the assumption that since most individuals choose careers in the health professions because of a desire to "help others," they must be "good people," and thus ethics should take care of itself. Pain management is only one arena in which this assumption is not only false but actually proves to be dangerous. Even well-intended, good professionals can cause harm by failing to manage pain appropriately for multiple reasons.

The 5-year $25 million Study to Understand Prognoses and Preferences for Outcomes and Risks of Treatments (SUPPORT) was designed to identify and correct problems associated with end-of-life care. Among the more than 4000 SUPPORT patients who died in the postintervention phase, 40 percent had severe pain most of the time. (Support Principal Investigators 2002)

At the end of 2002, Last Acts *published the report, "Means to a Better End: A Report on Dying in America" (Last Acts, 2002). The 50 states and the District of Columbia were rated on eight key elements of end-of-life care (state advance directive policies, location of death, hospice use, hospital end-of-life services, care in intensive care units at the end of life, pain among nursing home residents, state pain policies, and palliative care-certified physicians and nurses). Most states earned C's, D's, and even E's on the majority of the criteria, which included the following:*

- Care in intensive care units at the end of life. *Nationwide, 28 percent of Medicare patients who die are treated in intensive care units (ICUs) in their last 6 months of life. The rate varies widely even within individual states. Patients in ICUs typically are subjected to heavy use of technology. This may be at the expense of attention to comfort or against expressed treatment preferences—often expressed as, "I don't want to die hooked up to machines."*
- Persistent pain among nursing home residents. *Nearly half of the 1.6 million Americans living in nursing homes have persistent pain that is not noticed and not treated adequately.*
- State pain management policies. *All states have laws addressing the use of controlled substances. Some are effective, but others create formidable barriers to good pain management.*

Even more worrisome are the media stories alleging that clinicians abused the trust of their patients in an intentional scheme to addict them in an effort to overbill insurance companies and sell painkillers, including Oxycontin (oxycodone). This chapter describes the ethical responsibilities of caregivers responsible for managing pain and highlights select ethical challenges.

▶ ETHICS: THE BIG PICTURE

Ethics is systematic inquiry into principles of right and wrong conduct, of virtue and vice, and of good and evil as they relate to conduct. It is important to distinguish ethics from religion, law, custom, and institutional practices. For example, the fact that an action is legal or customary does not in itself make the action ethically or morally right. Ethics poses questions about how we ought to act and how we should live. It is an inquiry into the justification of particular actions (are these actions right or wrong?), as well as a search for traits of moral character that promote human flourishing. One simple definition of ethics is that it is about how an individual or an institution (profession) *acts out its identity*. If pain relief and comfort are essential to our identity as healers, the ethics of pain management is about how we act out our identity as healers. Health care works best when *everyone* involved in health care design, implementation, administration, financing, and evaluation (1) has an understanding of self as a moral agent, (2) an understanding of the moral nature of health care administration, medicine, nursing, etc., and (3) an understanding of the possible ways in which one can explain and justify moral choices and decisions.

What Is Moral Agency and Why Does It Matter?

Moral agency is the capacity to act habitually in an ethical manner. It entails a certain set of competencies in matters ethical, as well as moral character and motivation.

1. *Moral sensibility.* Ability to recognize the "moral moment" when a moral challenge presents itself.
2. *Moral responsiveness.* Ability and willingness to respond to the moral challenge.
3. *Moral reasoning.* Knowledge of and ability to use sound theoretical and practical approaches to "think through" moral challenges; these approaches are used to *inform* as well as to *justify* moral behavior.
4. *Moral discernment.* Ability to select the best course of action in response to a *particular* challenge after weighing the pros and cons of alternative courses of action.
5. *Moral accountability.* Ability and willingness to accept responsibility for one's moral behavior and to learn from the experience of exercising moral agency.
6. *Moral character.* Cultivated dispositions that allow one to act as one believes one ought to act.
7. *Moral valuing.* Valuing in a conscious and critical way that which squares with good moral character and moral integrity.
8. *Transformative moral leadership.* Commitment and proven ability to create a culture that facilitates the exercise of moral agency, a culture in which people do the right thing because it is the right thing to do.

For instance, if I have mastered all the complexities of effective pain management but fail to use this expertise in my daily practice as I encounter individuals suffering with pain, there are two plausible explanations for my behavior. First, my moral agency may be deficient. For example, my character may be deficient in that I may lack the compassion or skill to recognize a patient's suffering, or I may not sufficiently value effective pain management to make it a priority. Or second, external variables at the level of the institution or society may impede my ability to effectively design and implement a pain regimen. Moral distress results from recognizing the ethically appropriate course of action but not taking it because of obstacles such as lack of time, supervisory reluctance, an inhibiting medical power structure, or institution policy or legal considerations.

External obstacles to effective pain management in the health care system include (1) control of pain is not considered a sufficient reason for admission to the hospital, and terminally ill patients cannot be admitted for this purpose alone, (2) there are inadequate community resources and expertise to manage patients with pain at home, (3) rationing pain management on a financial basis results in forcing some patients to consider death their only option, and (4) there remain significant legal restrictions placed on narcotic prescriptions. Often the failure to manage pain effectively is linked to both deficient moral agency *and* moral distress. In this case, we need to work with individual professional caregivers and, at the same time, target the institutional and societal structures that are constraining good pain management.

What Is Moral Integrity?

Moral integrity may be defined as that condition or state in which moral activity (valuing, choosing, acting) is intimately linked to a particular conception of the good, the good life. As professional healers, we are individuals with moral integrity if what we value in our clinical practice, how we make decisions, and how we behave square with what it means to be a good healer. To the extent that we meet the reasonable expectations of those who entrust themselves or who are entrusted to our care, we possess moral integrity. At issue here is whether it is reasonable for members of the public to expect their physical pain to be recognized and managed.

Moral Integrity and Pain Management

New advances in understanding and treating pain have focused on the real possibility of controlling most human pain. Misperceptions about the efficacy of analgesic regimens and a lack of commitment to achieving successful regimens are no longer justified or tolerated. The Joint Commission on Accreditation of Health Care Organizations (JCAHO) supports the patient's right to effective pain management and publishes standards for assessment and management of pain in hospitals, ambulatory care settings, and home care settings. The JCAHO recommendations include teaching all patients to use a pain-rating scale and determining a pain-rating goal with each patient.

Also, according to JCAHO guidelines, if a facility does not have the resources to treat a patient's pain adequately, the patient must be referred to a facility that does. The guidelines for acute pain and cancer pain developed by the Agency for Health Care Policy and Research (AHCPR) are just one example of clear standards for effective pain management. "Clinicians should reassure patients and their families that most pain can be relieved safely and effectively" (Clinical Practice Guideline No. 9, USDHHS, AHCPR Report, 1994, p. 7). *No longer can professional caregivers claim that it is unreasonable for patients to expect their pain to be managed effectively.*

Moral integrity matters because it is what allows patients to trust us. Since pain and human suffering create vulnerabilities in the patient that result in unequal relationships between professional caregivers and patients, fiduciary not contractual relationships are the norm. In a fiduciary relationship, the patient expects the professional to be trustworthy, not only to be competent—in this case in pain management—but also to be willing to use that competence to secure the interests (comfort) of the patient.

Are There Universal, Nonnegotiable Moral Obligations for Health Care Professonals?

While clinicians differ today in their beliefs about whether or not there are *any* moral obligations incumbent on all professional caregivers, most adhere to the Hippocratic injunction to act always so as to benefit patients, or at the very least, to do no harm, *primum non nocere*. A universal, nonnegotiable obligation to recognize and effectively manage pain falls easily within this injunction. It is difficult today to imagine a vigorous defense of the claim, "I am your doctor or nurse, but I am not responsible for recognizing and trying to alleviate your pain." While modern medicine's at times exclusive focus on cure has diminished its traditional role of comforting patients, the old French adage "to cure sometimes, to relieve often, to comfort always" remains the challenge.

The Anatomy of Moral Action

Four elements are analyzed when determining whether a particular human action is morally praiseworthy or blameworthy: the intent of the agent, the nature of the act itself, the consequences of the act, and the context in which the act takes place. Different systems of ethical justification are linked to each of these elements. Virtue theory highlights the role of the agent and the agent's character and intent. Deontologic theory highlights the nature of the act itself and holds that certain human actions are good or evil in themselves, independent of their consequences. Utilitarian or consequentialist theory highlights consequences and claims that of any two actions, the most ethical one is that which will produce the greatest balance of benefits over harms. Finally, situation ethics holds that unique circumstances ought to influence judgments about what is morally right or wrong in a particular situation. The best analysis of human action pays attention to all four elements; thus neither the act itself, its consequences, context, or the intent of the agent alone determines morality. Grounds exist to critique statements such as these: "*The end justifies the means;*" "*she is better off dead;*" "*So long as I don't intend* his death, I am OK;" and "It may not usually be right, *but in this case* it seems justified."

▶ ETHICS AND PAIN MANAGEMENT

There are clearly ethical challenges that present at the level of the profession, the institution, and society in light of their respective moral obligations:

- *Profession*—adequately preparing and then holding practitioners responsible for effective pain management of high quality
- *Institution*—developing and monitoring a culture in which effective pain management of high quality is the norm
- *Society*—designing, implementing, and evaluating a health care system in which effective pain management of high quality is guaranteed for all—regardless of ability to pay

This chapter, however, will focus on common ethical challenges encountered by individual clinicians "at the bedside," including the "virtual" bedside of the clinician whose practice is conducted solely on the Internet.

Figure 12-1 is a guide to ethical reasoning in health care decision making. Ethical issues include who makes decisions about pain management and what criteria and interventions should we use when there is conflict about how to manage pain.

Who Makes Decisions About Pain Management?

The rule of thumb is that patients with decision-making capacity (ability to understand the consequences of the decision at hand, ability to reason in accord with a relatively consistent set of values, and ability to communicate a preference) have the right to request, authorize, or refuse pain management interventions. Clinicians have an obligation not only to respect patient preferences and not interfere with a patient's autonomous requests, but also to provide the information and support patients need to make decisions consistent with their beliefs, values, and decisional history. The object is not simply that a patient expresses a preference regarding pain management that is accepted, but also that the patient reaches a

The ability to work up the ethical aspects of a case is an essential part of clinical reasoning. The emphasis in the ethics workup is on a sensible progression from the facts of the case to a morally sound decision. An ethics workup (this one or a similar version) may be used by a variety of health professionals, such as physicians, nurses, social workers, etc. With some adjustments, it also may be used by laypersons. Using the five principal steps of the ethics workup, health professionals holding a variety of philosophical and religious positions regarding ethics can share a basic framework for thinking about and discussing morally troubling cases.

1. **What are the facts?** It is vitally important to clarify the facts of the case in order to anchor the decision. These facts are both medical and social. For example, both an estimate of prognosis and an understanding of the patient's home situation are often relevant to an ethical decision.

   Persons involved (Who?)
   Diagnosis, prognosis, therapeutic options (What?)
   Patient preferences, beliefs, and values (What?)
   Chronology of events, time constraints on decision (When?)
   Medical setting (Where?)
   Reasons supporting claims, goals of current care (Why?)

   Nurses and social workers may be instrumental in ensuring that the patient/family and other nonmedical health professionals understand the medical facts and that the health care team understands pertinent nonmedical information about the patient and family.

2. **What is the issue?** It is necessary to identify the specific ethical issue in the case. The issue may not be ethical but rather a diagnostic problem or a simple miscommunication.

3. **Frame the issue.** Some health professionals will explore the issue using only one moral approach. Others will eclectically employ a variety of approaches. But no matter what one's underlying moral orientation, the ethical issue at stake in a given case can be framed in terms of several broad areas of concern representing aspects of the case that may be in ethical conflict. It is, therefore, useful, if somewhat artificial, to dissect the case apart along the lines of the following areas of concern:
   a. Identify the appropriate decision maker(s).
   b. Apply the criteria to be used in reaching clinical decisions.
      (1) **The specific biomedical good of the patient.** One should ask, What will advance the biomedical good of the patient? What are the medical options and likely outcomes?
      (2) **The broader goods and interests of the patient.** One should ask, What broader aspects of the patient's good, i.e., the patient's dignity, religious faith, other valued beliefs, relationships, and the particular good of the patient's choice, are pertinent to the decision at hand?
      (3) **The goods and interests of other parties.** Health professionals also must be attentive to the goods and interests of others, e.g., in the distribution of resources. One should ask, What are the concerns of other parties (family, health care professionals, health care institution, law, society, etc.) and What differences do they make, morally, in the decisions that need to be made about this case? In deciding about an individual case, however, these concerns generally should not be given as much importance as that afforded the good of the individual patient whom health professionals have pledged to serve. The physician explains the medical options to the patient/surrogates and, if indicated, makes a recommendation. The patient/surrogate makes an uncoerced, informed decision. Limits to patient/surrogate autonomy include the bounds of rational medicine/nursing/social work, the probability of direct harm to identifiable third parties, and violation of the consciences of involved health care professionals. In problematic cases, the interdisciplinary team may meet to ensure consistency in their recommendations to the patient/surrogate(s).
   c. Establish the health care professionals' moral/professional obligations. Each health care professional must decide what he or she owes the patient, herself/himself, the health care team, the health care institution, and other third parties. Conflicts may present.

4. **Decide.** In clinical ethics, as in all other aspects of clinical care, a decision must be made. There is no simple formula. The answer will require clinical judgment, practical wisdom, and moral argument. The health care professional must ask himself or herself, "What should I do? Where can I get help?" He or she must analyze the data, reflect on it morally, and draw a conclusion. He or she must be prepared to explain his or her decision and the moral reasons for it. Sources of justification include
   a. The nature of the health care professional–patient relationship; compatibility of recommended course of action with aims of profession (internal morality of profession).
   b. Approaches to ethical inquiry. Principle-based ethics, virtue-based ethics, casuistry, feminist/caring/existentialist ethics, theological ethics.
   c. Ethically-relevant considerations:
      (1) Balancing benefits and harms in the care of patients
      (2) Disclosure, informed consent, and shared decision making
      (3) The norms of family life
      (4) The relationships between clinicians and patients
      (5) The professional integrity of clinicians
      (6) Cost-effectiveness and allocation
      (7) Issues of cultural and religious variation
      (8) Considerations of power (Fletcher, Brody, Miller, and Spencer)
   d. Grounding and source of ethics. Philosophical (based in reason), theological (based in faith), sociocultural (based in custom).

5. **Critique.** It is important to be able to critique the decision that has been made by considering its major objections and then either responding adequately to them or changing one's decision. The health care professional also should seek his or her colleagues' input when time permits. Some cases can even be taken to an ethics committee for further reflection. Retrospective analysis is also useful in preparing "for the next time" such a situation is encountered.

Figure 12-1. The Ethics Workup: Georgetown University Center for Clinical Bioethics.

judgment about what is right for him or her in the light of who he or she is and what matters to him or her. Clinicians are never obligated, however, to sacrifice their integrity to satisfy patient preferences. Culture, religion, family experiences, and previous experiences with pain and suffering are just some of the variables that influence how different individuals respond to pain and suffering.

When a patient lacks decision-making capacity, a surrogate must make decisions about pain management. If the patient at one time had decision-making capacity, the surrogate's task is to use the *substituted judgment standard* to express the preference he or she believes the patient would express if asked. His or her role is merely to be a voice for a patient who can no longer speak for himself or herself. Ethical problems arise if clinicians have reason to believe that a surrogate is not communicating a preference that the patient himself or herself would have expressed. If the surrogate is speaking for a patient who never had decision-making capacity, an infant, child, or profoundly retarded adult, the *best interests standard* is used. The challenge for the surrogate is to decide, in concert with the clinicians, what is in the best interests of the patient absent knowledge of the patient's beliefs, values, or previous preferences. A common ethical challenge here is how to respond to a surrogate who is demanding something clinicians believe not to be in the best interests of the patient. Recent cases involved the mother of a 1-year-old infant demanding that her terminally ill and suffering child not be given any analgesics or sedatives because she wanted her to die fighting and to be awake during her visits, and the husband of a woman in the end stages of Parkinson's disease similarly objecting to his wife receiving analgesics or sedatives because he nurtured the hope that she could be successfully weaned from the ventilator—contrary to clinical opinion. In these cases, being an advocate for the patient is complicated by the need to respect the wishes of a morally valid surrogate. If the best efforts to reason with the surrogate are unsuccessful, an ethics consult may be indicated.

What Criteria and Interventions Should We Use When There Is Conflict About How to Manage Pain?

Helpful criteria when making clinical decisions about pain management are

- *The specific biomedical good of the patient.* One should ask, What will advance the biomedical good of the patient? What are the pain management options and likely outcomes?
- *The broader goods and interests of the patient.* One should ask, What broader aspects of the patient's good, i.e., the patient's dignity, religious faith, other valued beliefs, relationships, and the particular good of the patient's choice, are pertinent to the decision at hand? For example, if the patient believes in the redemptive power of suffering and autonomously chooses to experience physical pain and forego standard pain management, this ought to be respected if clinicians have ascertained that this choice is being made with full knowledge and without coercion. On the other hand, there are patients with low pain thresholds who are adverse to any experience of physical pain. Ethical challenges frequently result when clinicians don't believe the pain reports of patients and must balance a commitment to promoting the patient's physical health while simultaneously respecting the patient's preferences.
- *The goods and interests of other parties.* Some patients will choose to sacrifice some comfort in the interests of being fully alert to visit with family. Health professionals also must be attentive to the goods and interests of others. One should not, however, medicate a patient simply to make the professional or nonprofessional caregivers feel better or refuse to appropriately medicate a patient because of the interests of third parties. For example, if a relative is stealing a patient's analgesics, this is not sufficient justification to withhold needed medications. A consult might be in order to find a way to reduce the likelihood of theft while continuing to appropriately respond to and manage the patient's pain.

When the clinician, in concert with the patient and/or the patient's surrogate(s), fails to reach consensus about pain management, or when members of the professional caregiving team are in conflict about how to proceed, an ethics consult is indicated to help mediate the conflict. Ethics consultants can be helpful in clarifying the source of the conflict, identifying competing obligations, and evaluating alternative courses of action. Four common ethical challenges will now be addressed.

Ethical Challenge 1: "None or Not Enough"

As noted earlier, one of the most common ethical failures of professional caregivers is failure to adequately identify and manage pain. While holding clinicians *legally* responsible for undermanagement of pain is a relatively new phenomenon, health care ethicists have long been critical of clinicians who allow patients to suffer pain needlessly. There are multiple reasons for this failure, including (1) lack of knowledge, (2) the low priority clinicians historically have attached to pain management despite the rhetoric that suggested that comfort was a high priority, and (3) confusion about the addicting properties of analgesics. One of the most common

> Criteria that must be present to use the principle of double effect include
>
> - An action that is good itself (or at least indifferent) has two effects.
> - The good effect: an intended and otherwise not reasonably attained good effect.
> - The evil effect: a foreseen but merely permitted (unintended) side effect.
> - No relationship of causality: The evil effect is not the means to achieving the good effect.
> - Proportionate reason: The good to be realized must compensate for the evil that is permitted.

Figure 12-2. The principle of double effect.

reasons for failure to administer appropriate dosages of analgesics to persons with serious and life-threatening illness remains, however, the fear that the analgesics will compromise respiration and hasten death. More than one nurse has been falsely accused of trying to "kill" a patient when appropriately administering titrated doses of analgesics.

The American Nurses Association's position statement on pain management is quite clear: "Nurses should not hesitate to use full and effective doses of pain medication for the proper management of pain in the dying patient. The increasing titration of medication to achieve adequate symptom control, even at the expense of life, thus hastening death secondarily, is ethically justified" (Position Statement, ANA, 1991). This statement is firmly grounded in the *principle of double effect,* which holds that a good or morally neutral action (administration of analgesics) with two effects, the foreseen and intended good effect (analgesia) and the foreseen but unintended evil effect (hastens death), may be performed so long as both effects happen simultaneously and the evil effect is not the means to achieving the good effect (death is not the intended means to relief of pain and suffering), and there is a proportionate reason for risking the evil effect. According to this analysis, there is a marked difference between the administration of a large but titrated (in response to increasing physiologic tolerance) dose of analgesic (ethically justified) and the administration of a whopping dose of an analgesic (or potassium chloride) to cause the death of the patient and thereby relieve his or her pain (not ethically justified). Figure 12-2 specifies the criteria to be used with the principle of double effect. In this analysis it is clear that the intent of the agent, the nature of the act and its consequences, and circumstances are all relevant.

Ethical Challenge 2: "Too Much: The Lethal Dose"

The *Journal of the American Medical Association* in 1988 published a letter from a young gynecology resident who had injected a 20-year-old woman dying from ovarian cancer with a lethal dose of morphine. The resident had been paged during the night and told that the patient wanted to "get this over with." As a result of the letter's publication in a reputable medical journal, physician-assisted suicide and euthanasia emerged as hot issues in a contentious debate within both the medical community and the public forum. Is administering a lethal dose to those who have suffered long and hard enough the ultimate act of compassion or is this unjustified killing? Is the intent of the prescribing clinician to alleviate pain or to "deal with" a troublesome patient? Is a patient with full decision-making capacity the only party who can make this request, or can the legally and morally valid surrogate of a patient who lacks decision-making capacity make the request? In a similar vein, may a professional caregiver take it on himself or herself to make this determination for a patient who is suffering unbearably and unable to make the request for himself or herself? If it is morally justified to request this sort of "aid in dying," does it follow naturally that the physician or nurse should be the agent of the lethal dose, or is this incompatible with the tradition of medical and nursing ethics?

And finally, of what *ethical* significance is it that as this text goes to press, assisted suicide is only legal in one of the 50 states in the United States. It is important to note that even when assisted suicide is a legal option, there may be multiple involved parties, including attending physicians, residents, medical students, nurses, pharmacists, institutions, health care systems, and others who for ethical reasons of their own are unable to comply with such requests.

Dr. Edmund Pellegrino concisely summarized the chief arguments in favor of and against assisted suicide and direct voluntary euthanasia, and these follow.

Arguments in Favor of Assisted Suicide and Direct Voluntary Euthanasia

1. It is a beneficent and compassionate act.
2. It respects autonomy by preserving the patient's control of the manner, method, and timing of death.
3. It takes the matter outside the reach of "medical power" and scrupulosity.
4. It prevents the injustice that allows some patients to choose death by refusal of life-support measures while denying others the right to do so by active euthanasia.
5. In a pluralistic society, euthanasia must be accepted, whatever its intrinsic morality, because states have no moral justification for intruding into such private decisions.

Proponents of involuntary euthanasia argue that, in addition, it is irrational to afford rights of personhood to anencephalics or patients in a permanent vegetative state and unjust to deny them the right to die to which they are entitled. They argue that beneficence to patient, family, and health care professionals, as well as conservation of society's resources, makes involuntary euthanasia a positive moral duty of physicians.

Arguments Against Assisted Suicide and Direct Voluntary Euthanasia

1. It undermines the value of, and respect for, all human life.
2. Guidelines cannot avoid sliding down the slippery slope to involuntary euthanasia and selective devaluation of the lives of the most vulnerable among us.
3. Euthanasia should be unnecessary because the major reasons for requesting it—intolerable pain and fear of overtreatment—now can be handled by better palliative care, analgesia, and advance directives.
4. A focus on euthanasia will deviate attention from other valuable palliative techniques.
5. If euthanasia is legal, it is predicted that patients will feel a subtle pressure to conform so as to relieve the economic and emotional burdens they impose on family and friends.
6. Euthanasia is socially destructive; it undermines trust in physicians and the health care system, desensitizes society to killing, and imperils the ground already gained in legitimating passive euthanasia.
7. The religious belief of many Americans that human life is the gift of the creator and that humans are its stewards but not its absolute masters.

Since the object of all clinical decision making is action that promotes the health and well-being or good dying of the patient and which simultaneously respects the integrity of all participants in the decision-making process, it is critical for professional caregivers to know the patient well enough to understand his or her preferences and then to know themselves well enough to understand if they can honor the patient's preferences without sacrificing personal, professional, or institutional integrity. A clinician's obligation to respect a patient's capacity to be self-determining is not absolute. If what the patient wants violates the integrity of the clinician, profession, or institution, there is no obligation to satisfy the patient's preferences. If what is desired is legal, an obligation may exist to transfer the patient to another clinician or institution.

Ethical Challenge 3: "Too Much: The Addicting Dose"

A primary ethical obligation for all clinicians is to remain competent as research reveals new findings, such as Ballantyne and Mao's (2003) claim that two important concepts arising from their improved understanding of how opioids act are (1) that apparent opioid tolerance does not equal pharmacologic tolerance and (2) that prolonged, high-dose opioid therapy may have serious adverse consequences. As a result, they recommend for patients with chronic pain that is not associated with a terminal disease: "Whereas it was previously thought that opioid therapy in unlimited dose escalation was at least safe, evidence now suggests that prolonged, high-dose opioid therapy may be neither safe nor effective. It is therefore important that physicians make every effort to control indiscriminate prescribing, even when they are under pressure by patients to increase the dose of opioids." Interestingly, Ballantyne and Mao address the reality that busy practice settings are often a factor resulting in a physician's compromising and ordering very large doses of opioids, even in the absence of any real improvement in the patient's pain or level of functioning.

Most clinicians can quickly recall patients who appeared to be "gaming the system" in order to secure analgesic prescriptions to feed their chemical addictions. Clinicians can err by (1) falsely concluding that such patients are not experiencing pain and thus fail to appropriately identify and treat real pain (pity the substance abuser whose postoperative nurse decides to withhold appropriate analgesia because of the patient's history), (2) correctly concluding that there is no identifiable physiologic basis for the pain and make no prescription, thus failing to treat the underlying psychological or social problems, or (3) correctly concluding that there is no identifiable physiologic basis for the pain and then prescribe an analgesic regimen anyway because it is easier to accede to the patient's wishes than to treat the real problem.

Patients also can fall prey to clinicians whose self-serving intent is to capitalize on addictive personalities by prescribing more and more medications in an effort to secure the clinician's personal financial advantage. While it is certainly true that some clinicians who are appropriately prescribing analgesics have come under needless scrutiny, it is also true that a growing number of clinicians are creating financial enterprises based on the sale of painkillers to pseudopatients, a practice facilitated by Internet commerce. Since an altruistic commitment to secure the interest of the patient—even when this entails a certain degree of self-effacement—is the central characteristic of a moral profession, this practice is morally repugnant and cries for redress by appropriate legal and regulatory bodies.

Ethical Challenge 4: "Placebos: The Pseudodose"

In the event that patient requests for analgesics result from a general need for attention such that any professional intervention seems to result in a lessening of the alleged "pain," the suggestion is often made that placebos may be the "therapy" of choice. As a matter of fact, the placebo may indeed "work" effectively in relieving the reported symptoms, especially if accompanied by the professional suggestion that "this ought to do the trick!" Although thoughts on this vary, it is generally believed that the administration of placebos for analgesics is never justified because the therapeutic intervention is based on deception. And even in the event that the deception results in a short-term gain—the patient now reports feeling "fine"—the likelihood that the patient eventually will learn about the deception is great, and then the trust that is essential to the healing encounter in general is vitiated.

In conclusion, the identification and appropriate management of pain is a serious and binding moral obligation of all clinicians. Never have we been better positioned to effectively manage human pain thanks to countless researchers. The moral challenge facing us now is how to hold ourselves and one another accountable to use the science of pain management well in each clinical encounter. While JCAHO can sanction institutions who continue to place a low value on pain management, and medical and nursing boards and courts can sanction individual clinicians who ignore or under- or overmanage pain, the aim is for clinicians to value patient comfort sufficiently to master the science of pain management and then use this expertise to cure sometimes, relieve often, and comfort always.

BIBLIOGRAPHY

American Nurses Association: *Position Statement on Promotion of Comfort and Relief of Pain in Dying Patients*. Washington: ANA; 1991.

Ballantyne JC, Mao J: Opioid therapy for chronic pain. *New Engl J Med* 2003;349:1943–1953.

Pellegrino ED: Decisions to withdraw life-sustaining treatment. *JAMA* 2000;283:1065–1067.

Quill TE, Lee BC, Nunn S: Palliative treatments of last resort: Choosing the least harmful alternative. *Ann Intern Med* 2000;132:488–493.

Robert Wood Johnson Foundation: First State-by-State Report of Dying in American Today, 2002; *http://www.rwjf.org/news/releaseDetail.jsp?id=1037586523372*.

SUPPORT Principal Investigators: A controlled trial to improve care for seriously ill hospitalized patients: The study to understand prognoses and preferences for outcomes and risks of treatment (SUPPORT). *JAMA* 1995;274:1591–1598.

CHAPTER 13

The Epidemiology of Pain

Brenda Breuer

The dictionary definition of epidemiology is "the study of the distribution and determinants of health-related states or events in specified populations and the application of this study to control health problems." This chapter, which discusses the relationships of demographic characteristics (e.g., age, race/ethnicity, sex, and socioeconomic status) with the pain experience and pain management, will clarify this definition. An understanding of these relationships may help us to understand the etiology and characteristics of pain. For example, a condition characterized by a female predominance throughout the lifespan is more likely to be associated with gender roles or structural differences between the sexes than to hormonal factors that vary over time.[1] Demographic relationships may help to discriminate between genetic and environmental risk factors. A difference in the risk of developing a certain condition between Chinese individuals who live in China and those who live in the United States is evidence of the involvement of an environmental factor rather than a genetic one.

The following are additional definitions of epidemiologic terms used in this chapter, taken from A Dictionary of Epidemiology.[2]

Confounders *A variable that can cause or prevent the outcome of interest, is not an intermediate variable, and is associated with the factor under investigation. For example, the relationship between birth order and the risk of Down syndrome is confounded by maternal age.*

Cross-sectional studies *Studies that determine the relationships between diseases (or other health-related characteristics) and other variables of interest in a defined population at one particular time. Disease prevalence, rather than incidence, is normally determined in cross-sectional studies. The temporal sequence of cause and effect usually cannot be determined in a cross-sectional study. For example, if we find that on January 1, 2003, the prevalence of back pain is higher among teachers than among physicians, we cannot discriminate between the possibility (1) that people with back pain are more likely to become teachers and (2) that teachers are more likely to develop back pain.*

Gold standard *A method, procedure, or measurement that is widely accepted as being the best available, e.g., the accuracy of mammography findings may be compared with those of biopsy results, which may be considered the "gold standard."*

Incidence *The number of new events (e.g., cases of a disease) in a defined population within a specified period of time, e.g., the number of new cases of lung cancer during the year 2003 divided by the number of people at risk for developing lung cancer.*

Odds ratio *The ratio of the odds in favor of getting a disease, if exposed, to the odds in favor of getting the disease if not exposed, e.g., the odds of getting lung cancer for smokers divided by the odds of getting the disease for nonsmokers.*

Prevalence rate *The total number of individuals who have an attribute or disease at a particular time divided by the population at risk of having the particular attribute or disease at this time, e.g., the number of people with heart disease (regardless of when the disease began) on January 1, 2003, divided by the number of people at risk on that date.*

Reliability *The degree to which the results obtained by a measurement procedure can be replicated, e.g., paying attention to vocalizations and facial expressions may be a reliable measure of determining whether a noncommunicative patient is experiencing pain (i.e., there will be a high level of agreement among the observers); it is not a reliable way for determining the intensity of the pain (i.e., observers will disagree about the intensity).*

Validity *The degree that a measurement measures what it purports to measure. Consider the validity of a 0 to 10 numeric pain-rating scale for people who, because of their stoic nature, do not want to admit that they have pain. Note that in order for an instrument to be valid, it must be reliable. The reverse is not true.*

Pain is a significant national public health problem. It is the most frequent reason individuals seek medical care, and it accounts for millions of medical visits annually, costing the American public more than $100 billion each year in health care,[3] litigation, and compensation. Some studies indicate that more than a third of the American population suffers from a chronic pain condition at some point in their life. Pain-related disability presents a significant and costly liability to employers, workers, and society. In the workplace, about 14 percent of employees take time off from their jobs because of pain conditions. In hospitalized patients, pain is associated with increased recovery time, increased length of stay, and poorer patient outcomes, all of which have health care quality and cost implications.

Pain affects quality of life (i.e., sleep, appetite, affect, independence in activities of daily living, and length of hospital stays). While often the direction of the association is unclear (i.e., which is the cause and which is the effect), pain is associated with numerous comorbidities, such as depression. New pain standards from the Joint Commission on Accreditation of Healthcare Organizations (JCAHO) require that accredited facilities have procedures and policies in place to ensure that pain is managed and assessed appropriately.[4] Nevertheless, a majority of patients with serious illnesses suffer from pain despite clinical advances and efforts at reform.[5,6] In a regional cross-sectional survey conducted in Italian hospitals in 2002, on a 0 to 10 numeric rating scale, the median score of the worst and of the average pain during the last 24 hours was 7 and 5, respectively. The documented information about the status of pain management is not based solely on patient reports. One study found that 86 percent of physicians felt that the majority of their cancer patients were undertreated for pain.[7] It is likely that undertreatment is, in large part, attributable to the newness of the pain management field (pain medicine received accreditation as a subspecialty from the Accreditation Council for Graduate Medical Education in 1991), the shortage of pain/palliative care medicine specialists (Kimberly Kutska, American Academy of Pain Medicine, personal communication, May 2002), and the lack of appropriate pain/palliative care management training in medical schools and beyond.[8–10]

Pain is not limited to those who are hospitalized or have life-threatening diseases. In a study of the course of chronic pain in a community in the United Kingdom,[11] the overall prevalence of chronic pain (defined for this study as pain or discomfort present either all the time or on and off for 3 months or longer) increased from 45.5 to 53.8 percent after a 4-year follow-up. Of those with chronic pain at baseline, 79 percent still had it at follow-up. The average annual recovery rate was 5.4 percent, and the average annual incidence was 8.3 percent. In a multinational European survey of individuals between the ages of 15 and 100 years,[12] 7.6 percent of respondents reported headaches, 5.8 percent had pain in lower limbs, 3.2 percent had joint/articular disease, 3.1 percent had backache, and 1.5 percent had gastrointestinal disease. In contrast to this survey of the general population, in a representative sample of a working population in a county in Sweden,[13] 23 percent reported back pain.

While some states have mandated pain as the fifth vital sign,[14] unlike the other signs (i.e., respiratory rate, pulse, systolic blood pressure, and diastolic blood pressure), pain ratings reported by people with clinically painful conditions are subjective. Some aspects of pain may include its unpleasantness, emotional response, associated behaviors, and intensity.[15] The subjective nature of pain reports has led researchers to study pain ratings in humans who have been exposed to experimental, measurable pain.

This chapter is about the epidemiology of pain, i.e., the study of populations in order to determine the frequency and distribution of disease and measure risks. The associations of pain with socioeconomic status, sex/gender, race, ethnicity, culture, and age will be adderessed. Caution should be exercized in the inferences drawn from these findings. For example, if elderly people with impaired cognition report less pain, are they "hurting less" or are there other explanations, e.g., they cannot communicate their pain in the ways that others do. Keep in mind the following questions:

- *Are there clear, specific inferences that we can draw from associations, or can there be confounders that are responsible for reported associations? As mentioned earlier, elderly people with compromised cognition may report less pain because they are less communicative and not because they actually experience less pain.*
- *Are there clinical implications for the treatment of individual patients?*
- *Are there public health implications regarding health policy at a federal, state, or municipal level?*
- *Do the findings highlight gaps in knowledge that need to be researched?*

▶ RELATIONSHIP OF PAIN TO SOCIOECONOMIC STATUS (SES)

SES is related to health status. Possible reasons for this relationship include the association between SES and living in a toxic environment, poor housing, crowding, and noise; a lower level of awareness regarding what constitutes risky behavior, such as smoking; and a lower likelihood of avoiding such behaviors.[16] In the United States, hospitalizations occur more frequently among the less educated.[17] While this may suggest that differences in educational level in this country do not seem to be related to access to care, it also may indicate that lesser educated individuals use medical care less effectively.

Americans with less education not only die earlier than better-educated Americans, they also experience pain differently.[18] Education may be a marker for specific characteristics, such as acquisition of adaptive skills, intelligence, or awareness of health behaviors that pose risks.

Education is a good SES indicator, because it is unlikely to be affected by chronic diseases that start in adulthood, as would income and occupation. Consequently, in contrast to occupation, the use of educational level as an SES indicator can provide a clearer time sequence of cause and effect. As an example of the cause-and-effect ambiguity that may arise when using occupation as an SES, consider possible inferences regarding back pain and being a schoolteacher. There may be a higher prevalence of back pain among schoolteachers because the demands of the occupation can cause back pain. Alternatively, people with back pain may be more likely to become schoolteachers because they think that compared with the work required in other occupations, the work required for teaching is less likely to aggravate their pain.

Much of the literature on SES and pain deals with musculoskeletal pain. Evidence from a review of 64 articles published between 1966 and 2000, which documented the association of formal education with back pain,[19] suggests that the association of low education with longer duration and/or higher recurrence of back pain is stronger than its association with onset. The implication is that low education may be a risk factor for a less favorable course of back pain.

People with lower income and education may be less able to control their lives and more likely to have jobs with greater physical demands and that involve tasks with stress on the spine.[16] It is well documented that repetitive work or otherwise physically monotonous work is associated with an increase in musculoskeletal pain of the neck, shoulder, and low back. Recent studies also report an association between muscle pain syndromes and psychosocial factors. In addition, there are studies indicating that mental and physical demands may interact to further increase muscle tension and the risk of musculoskeletal disorders.

Musculoskeletal disorders differ from many other major health problems, such as cancer and cardiovascular disease, in that symptoms often appear at young ages and after a relatively short exposure to adverse environmental conditions. Pain syndromes often are reported after only 6 to 12 months on a job that involves repetitive work. Hence, people may suffer over very long time periods. The high prevalence of musculoskeletal disorders in psychologically stressful but light physical work, such as assembly work and data entry, suggests that mental stress may play an important role. Consistent with this, experimental studies show that mental stress, even in the absence of physical demands, increases muscle tension.

The disabling process associated with back pain is associated with premature retirement, which is a difficult burden for the individual and expensive for the society.[20] A prospective study of all employed men and women in Norway between the ages of 20 and 53 years in 1980 ($n = 1,333,556$) showed that back pain diagnoses accounted for approximately 15 percent of all individuals granted disability pensions from 1983 to 1993. The incidence was higher in women than in men and among those with fewer years of education and lower socioeconomic status. Compared with the highest social class, the incidence was 3- to 11-fold higher for unskilled workers. The persistence of the association between socioeconomic status and disability retirement from back pain even at the higher end of the socioeconomic scale leads one to conclude that the relationship cannot be attributed only to the manual/nonmanual nature of a job, and that the spine itself may not be the only focus of the medical treatment of back pain and its resulting disability. Rather, factors related to the occupational and social environment may play an important role, and appropriate management may require a psychosocial evaluation, which may uncover such problems as psychological workplace demands, job dissatisfaction, and poor coping skills. Further, since this study also demonstrated that individuals at high risk for future back pain disability pension consistently perceived their work as physically demanding and reported having angina pectoris and diabetes,[21] the prevention of disability may need an even broader interdisciplinary base. Finally, among low back pain patients, disability pensions and litigation claims are themselves associated with greater self-reported pain, depression, and disability before and after treatment.[22,23]

In a prospective study in Europe, Hemingway and colleagues[24] investigated the relation between employment grade and sickness absence because of back pain in office workers and found a strong inverse relation between employment grade and rate of absences. For low back pain specifically, a cross-sectional study found a relationship between education and work absenteeism,[17] with the relationship being stronger in men than in women.

Data from the 1989 National Health Interview Survey that covers both employed and nonemployed individuals in the United States[25] found relationships that were similar to those in Norway and Sweden. Among adults, the prevalence of a disabling back condition was higher among non-high-school graduates, the unemployed, and those with disabling nonback morbidities. Among workers, those in sales, clerical, technical, private household, service, precision production and repair, and transportation occupations were more likely to report disabling back conditions compared with workers in professional occupations. Being between 25 and 64 years of age is another risk factor. The high prevalence of back problems and related disability (although the magnitudes of the associations with risk factors are not large) may have dramatic public health implications.

The seriousness of a health problem is determined, in part, by the number of people who have the problem condition. The societal impact of a risk factor that is associated with a small increased incidence of a health problem depends on the prevalence of that problem. Thus, for example, if 2 percent of a population of 1 million people (i.e., 20,000 people) has a condition, then a risk factor associated with a 50 percent increase in the incidence would translate into an additional 10,000 people getting the condition. On the other hand, if the problem condition affects only 0.05 percent of that population (i.e., 500 people), a 50 percent increase would translate into only another 250 people getting the problem condition.

The relationships between pain and SES are not limited to back pain, nor are they limited to industrialized countries. A population-based cross-sectional survey conducted in Sweden found that both male and female blue-collar workers showed significantly more chest pain when excited than white-collar workers.[26] Overall, they also reported significantly worse self-rated health than the white-collar workers. Thus, physicians need to be especially vigilant in helping less educated people get the proper care and involving all the appropriate members of the interdisciplinary team, e.g., social workers, primary physicians, etc.

A study of Nigerian patients with sickle-cell disease further underscores the universality of these relationships. Patients of low SES had significantly more bone pain crises and a higher prevalence of leg ulcerations.[27] In addition, poorer patients were more likely to reside in neighborhoods with a greater exposure to infections, which might themselves lead to pain crises. The authors postulate that individuals with higher SES are more likely to have had a better education as well as motivation to comply with health-promoting habits.

▶ RELATIONSHIP OF PAIN TO GENDER

LeResche[1] reviewed the literature on the sex-prevalence ratios (prevalence of females/prevalence in males) for the following conditions: general headache, migraine, temporomandibular pain, burning mouth pain, neck pain, shoulder pain, back pain, knee pain, abdominal pain, and fibromyalgia. For each of these conditions, the prevalence was higher for females than for males. While the ratios for migraines and fibromyalgia were 2.5 and 4.3, respectively, those for the other conditions were 1.6 or less. Possible reasons for the higher prevalence of pain among females include their greater biologic sensitivity to pain.[28,29] Further, the additional route in women for internal trauma and infection via the vaginal canal puts them at a greater risk for developing hyperalgesia in multiple body regions. Additionally, the actions of sex hormones suggest pain-relevant differences in the operation of many neuroactive agents, opiate and nonopiate systems, nerve growth factor, and the sympathetic system.

The conditions for which women's health care differs from men's are not limited to sex-specific phenomena, such as childbirth and menopause. Women and men may have different symptoms for the same medical conditions and may react differently to the same medications. These differences, in combination with the historic lack of representation of women in clinical trials, suggest that the state of the art for treating women may lag behind that for treating men. Thus, treatment for women may be inferior to that for men because women have not been studied as much and because some clinicians may not be fully aware of the male/female differences that have been demonstrated.

As an example of sex-related pain manifestations, women are more likely to have painless progression of heart disease. Further, pain or discomfort in the stomach area may be dismissed mistakenly as heartburn or indigestion.[30,31] Women also may have nausea, fatigue, dizziness, pain in one or both arms, or shortness of breath as their heart-related symptoms.[30] While their heart-related pain may not be located on the left side of the chest, as it is frequently in men, among those with angiographically-confirmed coronary artery disease, women are more likely to report pain in the throat, neck, or jaw.[32,33] Philpott and colleagues[33] found that women with these pain patterns were less likely to be revascularized, even if they were deemed appropriate for revascularization. In many women, the first heart attack is fatal because previous symptoms and risk factors were ignored.[30] While, on balance, women and men with chronic stable angina appear to have more similarities than differences in chest pain characteristics, women report greater physical limitation related to anginal pain. It is unclear whether this greater limitation is related to the nature of the pain or perhaps to cultural/gender factors that affect the nature of the activities in which women across the social spectrum engage, such as housekeeping and caregiving. It is not known whether women with anginal pain view these activities as amenable to modification.

Responses to pain treatment also differ by sex. In a prospective survey of Chinese patients 20 to 70 years of age and receiving patient-controlled intravenous (IV) morphine subsequent to general anesthesia and surgery, sex was the strongest predictor for postoperative morphine requirements, with females consuming significantly less morphine via patient-controlled analgesia (PCA) in the first 3 postoperative days than men, even after adjusting for surgery site.[34] These findings are consistent with a U.S. study of post-dental surgery pain, where relief by opioids was greater among females than among males.[35] Similarly, a report from England indicates that after upper abdominal surgery in patients who receive morphine via PCA, female patients require significantly less morphine than male patients to achieve similar levels of pain relief.[36] Yet there is evidence that for many conditions women experience more pain than men. In another study of postoperative dental pain, men reported less pain than women regardless of ancestry.[37] Similarly, the reported intensity of chronic pain associated with immunodeficiency syndrome is greater for women than for men.[38] These reports of more intense pain among females can be reconciled with their need for less opioid analgesics if there are sex-associated differences in the pharmacokinetics as well as the pharmacodynamics of opioid treatment. Indeed, the finding of Gear and colleagues[35] of a greater kappa opioid (e.g., pentazocine and nalbuphine) analgesia among females than among males suggests a sex-related difference in kappa opioid-activated endogenous pain-modulating circuits.

Experimental studies provide other hypothetical causes for sex differences in reported pain experiences. For instance, the higher prevalence of fibromyalgia among women may be related to their deficiency of a second pain-inhibitory mechanism.[39] Single painful stimuli evoke two successive and qualitatively separate sensations referred to as *first* and *second* (windup) *pain* sensation. First pain alerts individuals to imminent danger and provides precise sensory information for an immediate withdrawal, whereas second pain attracts longer-lasting attention and prompts behavioral responses to restrict further injury and optimize recovery.[40] The relative lack of the second pain-inhibitory mechanism among women may be conducive to a transition from acute to chronic pain and, thus, to their higher prevalence of fibromyalgia.[39]

Paulson and colleagues[41] found that females reported greater pain intensity than males in response to a noxious heat stimulus. The difference in the reported intensity was associated with greater activation in the contralateral thalamus and anterior insula, as shown by positron-emission tomography (PET). These two structures have direct anatomic interconnections. It is possible, therefore, that the increased activation of these two forebrain structures in females reflected their higher ratings of heat pain intensity. Chesterton and colleagues[42] found that females have a lower pressure pain threshold in the first interosseous muscle. They cited the reports of others regarding sex-related differences in pressure pain threshold in other anatomic sites,[43–45] but underscored that sex differences depend on the anatomic site under investigation, and thus there is a need for consideration of gender differences in planning pain studies. It is important to bear in mind, however, that the relevance of experimental findings to clinical issues is unclear.[42] Some gender differences found in experimental pain may be related to the gender of the experimentee in relation to the experimenter and to gender-related levels of anxiety and fear.[46,47]

In addition to sex-related differences in pain symptomatology and in responses to pain treatment, *clinicians' treatment* may be related to the sex of the patient. Cleeland and colleagues[48] found that despite published guidelines for cancer pain management, female sex was predictive of inadequate pain management. Women's pain reports are taken less seriously than those of men, and they are treated less aggressively. One possible explanation is that the greater number of coping mechanisms that women have for dealing with pain may make it appear that they can endure pain more easily. Second, their pain is often discounted as "psychogenic" or "emotional" and, therefore, "not real." Third, individuals who are attractive often are perceived as healthy and not in pain, and women have been socialized to attend more to their physical appearance. Fourth, cultural influences make men feel obligated to display stoicism in response to pain,[28] and perhaps, therefore, more attention is paid to their pain reports. Fifth, men are more likely to delay seeking pain treatment and generally receive more aggressive treatment once they enter the health care system. Whatever the reasons, the difference in pain treatment may have resulted in some loss of confidence by women in the health care system and in greater use by women of alternative therapies.

Although women have a higher prevalence of chronic pain conditions and of multiple pain problems, the prevalence patterns differ by condition and by age. Conditions with more consistently reported patterns include joint pain, chronic widespread pain, fibromyalgia, and shoulder pain, all of which appear to increase with age until approximately age 65, and all are more prevalent among women.[1] In summary, as Berkeley points out, one should not assume that women report more pain intensity in all conditions but that patterns of sex-related pain differences change with such factors as age, location and type of pain, subject demographics, genetic profile, menopausal status, treatment utilization behavior, and analgesic.[29]

▶ RELATIONSHIP OF PAIN TO RACE, ETHNICITY, AND CULTURE

According to Edward and colleagues,[49] *race* distinguishes major groups of people by their ancestry or heredity (Mongoloid race, Caucasian race, etc.). *Ethnicity* not only includes race but also refers to characteristics that are cultural, social, psychological, and political in nature. Nowadays, however, federal forms use the word *ethnicity* to mean Hispanic/Latino or not Hispanic/Latino and the word *race* to mean American Indian or Alaska native, Asian, black or African-American, native Hawaiian or other Pacific Islander, or white. While different authors use the terms *ethnicity* and *race* differently, regardless of the precise use of the terms, data overwhelmingly indicate that people of different ethnic/racial origins have different health patterns. The issue of ethnic disparity regarding health and treatment is complex. In addition to genetics and clinician bias, cultural differences exist regarding the use of available health care.[50] For instance, while older black Americans are more likely than whites to rate their health as poor, they are less likely to use the services of formal health service agencies that are available to them.[51]

Pain management specialists need to be aware of cultural influences because of the large increase in the immigrant population in the United States, particularly immigrants from nontraditional regions such as Southeast Asia and Latin America. These new immigrants are more heterogeneous than their European predecessors with respect to educational and socioeconomic status and geographic background. In the course of developing a pain management plan, clinicians should conduct culturally-sensitive pain assessments that elicit information regarding the beliefs of the patient and the family members about the pain experience and approaches toward healing practices.[52]

Variation in the pain experience may be attributed, in part, to differences in treatment administered by clinicians to individuals of different ethnicity.[53,54] Language may be a barrier, as illustrated by a study in which two nurses, one of whom shared the mother tongue (Arabic) of the patients and one of whom did not, were asked to use a 10-point visual analogue scale to rate the patients' pain. Only nurses sharing a mother tongue with the patient provided pain ratings that correlated significantly with those of their patients.[55] Moore and colleagues[56] reported that preferred pain descriptors for birth, as well as for tooth drilling and injections, varied by ethnicity. A descriptor unique to Chinese individuals is *sourish (suan)*. This word has been used to describe bone, muscle, joint, tooth, and gingival pain.[56,57]

Cleeland and colleagues[58] developed a pain-assessment instrument, the Brief Pain Inventory (BPI), that was administered to patients with metastatic cancer who were from the United States, France, the Philippines, and China. This instrument was subsequently translated and validated in Vietnamese, French, Italian, Chinese, German, Japanese, and Spanish.[59–62] The BPI rates the intensity of pain, as well as the level of interference of pain in enjoyment of life, activity, walking, mood, sleep, work, and relations with others. Even with this instrument, however, in order to capture all the cultural nuances that may be relevant to pain management, one must talk to patients and listen to patients and not only administer a pain-assessment questionnaire.[63]

The effect of ethnicity on pain management is further illustrated by a study of postoperative pain treatment in a California hospital,[64] where a greater amount of narcotic medication via PCA was prescribed for whites and blacks as compared with Hispanics. In fact, the patient's ethnicity had a greater impact on the amount of narcotic prescribed than on the amount of narcotic self-administered by the patient, even after controlling for age, gender, pain site, preoperative use of narcotics, and insurance status. Cleeland and colleagues[48] found that patients treated at centers that predominantly cared for minorities were three times more likely than those treated elsewhere to have inadequate pain management.

However, the differences in pain levels cannot be explained entirely by treatment differences. Biologic and/or cultural parameters account for some of the difference. Ethnic factors may influence activity in higher nervous centers that alter pain through the activation of descending neural inhibitory controls. For example, African-American and Caucasian individuals with hypertension have different biologic reactions to intravenous catheterization: African-Americans display greater blood pressure and heart rate responses to stressors than do Caucasians.[49,65]

Ethnicity-related differences in the pain experience also may be associated with SES and its ramifications. Unfair treatment, discrimination, and racism put ethnic and racial minorities at a higher risk for chronic stress, which, in turn, can produce chronically high levels of physiologic exhaustion and sympathetic activation.[66] Another source for differences between African-Americans and white Americans may be an increased number of comorbidities among the former, which may be attributable to their high prevalence of hypertension.[67]

Attitudes regarding pain can be influenced by cultural factors. For example, Moore and colleagues[68] reported that in some cultures patients might prefer to suffer more pain at the end of life than to lessen social interaction because of medication-induced incoherent mental states. They also found that natives of India with chronic cancer pain experienced a spiritual dimension and a sense of higher good, whereas American patients saw pain as a symbol of their demise and as a punishment for actions that may have contributed to their deadly illness. In a comparison of Korean-American and white American obstetric patients,[69] Koreans rated their overall levels of pain intensity higher than whites. Although the Koreans tended to have a higher level of education, which has been reported to be associated with lower reported levels of pain, compared with the white women, they may have experienced more anxiety and embarrassment because of the little value placed on modesty during their care. This, in turn, may have caused their higher reported pain intensities.

Interestingly, in ranking the use of a local anesthetic as a remedy for childbirth labor pain, it was sixth among Americans, with first through fifth ranks being breathing deeply, being informed/prepared, being held or touched, relaxing/resting, and being with someone close. While use of a local anesthetic also was ranked sixth among Scandinavians, it was ranked fifteenth among Chinese.

While Susan Beck's ethnographic study of cancer pain management is about black Africans in South Africa,[70] its findings on the impact of culture on the pain experience have universal implications. Beck systematically recorded interviews of clinicians from diverse settings, as well as representatives of government, nongovernmental organizations, higher education, and pharmaceutical companies. Many of the comments ring true for her American readers. The following are some meaningful remarks:

- "I do not think white patients generally will suffer in silence to the same extent, and black people have been so used to being kind of victims, they unquestionably accept the status quo."
- "If you take the ... Zulu as a people, the instinct there is the 'warrior is the fighter,' so it is a strength of survival. So the tolerance levels of pain are incredibly high, so when the person is in pain I do not think it is expressed."
- Some African people are stoical, and so for that reason ..., *we are not mindful of their pain*" [italics added].

In South Africa, many black African patients are reluctant to use opioids because they think the need to use morphine is a prelude to death. The lack of knowledge is evident among professionals as well as patients, especially regarding the fears that the medical use of opioids can lead to addiction. Perhaps because of the newness of the pain management specialty in the

United States (pain medicine received accreditation as a subspecialty from the Accreditation Council for Graduate Medical Education in 1991), reports indicate similar problems here.[50,71] A 1993 survey of U.S. oncologists[7] revealed that (1) concerns about side-effect management and tolerance limited the prescribing of pain medications, (2) poor pain assessment was rated as the chief barrier to adequate pain management, and (3) other barriers included patient reluctance to report pain and to take analgesic medications.

Beck[70] found that a lack of resources poses another barrier to effective pain management, with a shortage of drugs in hospitals, clinics, and pharmacies. Shortages also have been documented in the United States. Morrison and colleagues[72] found that while 72 percent of pharmacies in predominantly white neighborhoods (those in which at least 80 percent of residents were white) had opioid supplies that were sufficient to treat patients with severe pain, only 25 percent of pharmacies in predominantly nonwhite neighborhoods (those in which less than 40 percent of residents were white) had an adequate supply. The shortage of pain doctors in South Africa is something we have in the United States as well (Kimberly Kutska, American Academy of Pain Medicine, personal communication, May 2002), although in all likelihood to a lesser degree. Similarly, both countries have geographic locations where access to appropriate care is more limited than other areas, i.e., rural areas in the United States and areas with poor roads and transportation in South Africa.

Finally, Beck found that doctors keep patients ignorant and do not tell them that if one medicine does not work, they could try another. Doctors and nurses often think that they can assess the patient's pain better than the patient. Once again, U.S. studies also have shown that clinicians often undertreat pain because they do not believe the patient, despite the fact that self-report of pain is the single most reliable indicator of pain intensity.[73]

▶ RELATIONSHIP OF PAIN TO AGE

Children

Pain is prevalent across all age groups. In a recent survey, parental reports indicated that in the last month of life, 89 percent of children who died of cancer experienced much pain, fatigue, or dyspnea.[74] However, pain is a matter of concern among healthy children as well. In a large Dutch study that focused on children aged 0 to 18 years,[75] and in which parents provided reports for the youngest children, 54 percent of participants reported a pain experience in the prior 3 months. The proportion of such reports ranged from 30 percent in boys younger than 4 years of age to 76 percent in girls aged 12 to 15 years (we should bear in mind, however, that pain is subjective and that pain reports can be influenced by gender.) The overall prevalence of chronic pain and acute pain were similar, i.e., approximately a quarter of the group in each case, with the intensity of chronic pain being significantly higher than nonchronic pain. The prevalence rates for headache, abdominal pain, and limb pain, which overall were reported most frequently, were 23, 22, and 22 percent, respectively. Although the prevalence of chronic pain increases with age until age 14 in girls and boys, both chronic pain and nonchronic pain decrease thereafter, perhaps because of exposure to higher levels of education. The association of chronic pain with emotional, social, and financial burden intensifies the impact of pain.[4,76–80] The specific likelihood of an association of low back pain in the growth period of childhood and adolescence with back pain in later life[81] means that attention to this particular condition is especially critical.

Other studies are consistent with the high prevalence of pain in children, specifically for headaches, abdominal pain, and spinal pain, with the spinal pain being most prevalent among older children.[75,82–85] While thoracic pain and lumbar pain are equally common in adolescence, thoracic pain is most common in childhood. Neck pain and pain in more than one area of the spine are uncommon in either group.[84]

Risk factors for spinal pain include parental history of treated low back pain, competitive sports activity, and time spent watching television. Since these data come from a *survey*,[83] the time sequence between watching television and spinal pain could not be ascertained, i.e., were children watching television because back pain limited their physical activity, or did they develop back pain because of factors related to watching television?

Recent attention has been paid to the development of guidelines for the conduct of clinical studies in children. They address aspects of consent, risk-benefit ratios, fair subject selection, and respect for subjects.[86] Grant applications to the National Institutes of Health now require inclusion of children or an explanation and justification for their exclusion. In addition, the Food and Drug Administration (FDA) has become proactive and now has the regulatory authority to force manufacturers to perform pediatric studies. Since these developments are relatively recent, however, the vast majority of available medications have not been labeled as safe and efficacious for pediatric use.[87] Thus, many medications for children who are in hospitals or at home are prescribed off label, i.e., they are used for an indication, dosage, age, or route of administration for which there is no FDA approval.[88]

Why is there a lack of adequate funding[87] from industry, government, and health care providers for clinical trials that include children? Pediatric studies may be perceived as difficult and expensive, especially in light of the relatively small market share that children represent. There are also ethical issues. The vulnerability of children to potential long-term adverse effects may help to explain why parents are reluctant to allow their children to be research subjects and why investigators may fear legal liability. Further, pain assessment in children younger than 8 years can be difficult.[89] In patients who were 3 to 7 years of age and had undergone surgery, self-reports of pain were poorly correlated with behavioral measures, such as crying, moaning, whimpering, grimacing, restlessness, tense torso or limb movements, and reaching for or touching the incisional site.[90] These findings suggest that while it may be necessary to depend on behavioral manifestations alone in the very young,[89] they may not discriminate between pain and other causes of distress. Therefore, whenever possible, pain assessments should be based on both behavioral manifestations and self-reports.[90]

Perhaps the possible reasons for the lack of attention to children in clinical trials also may explain, in part, what seems to be their inferior treatment in the hospital setting. A 1986 report found that, compared with adult hospital patients, children hospital patients were less likely to receive opioid medication for hernias, appendectomies, burns, and fractured femurs.[91] A 2002 article about 43,725 pediatric and 114,207 adult emergency department encounters strongly suggested that this trend persists despite the JCAHO requirement for safe and adequate pain management for children and adults.[92] It is possible, however, that clinicians feel that treatments deemed appropriate for adults may be unsafe for children.

What are some of the biologic and developmental considerations that might shed light on the pain experience of children? In the immature nervous system, invasive procedures can cause distress and delayed recovery.[93] First, the immaturity of sensory processing within the newborn spinal cord can lead to lower thresholds for sensitization and excitation, i.e., hyperalgesia, potentially maximizing the central effects of these tissue-damaging inputs. Hyperalgesia in the neonate can be explained, in part, by the substantial upregulation of neurotrophins in response to tissue damage and the resulting skin hyperinnervation. Injury-related upregulation of neurotrophic factors and the sprouting response are several times greater in the neonate than in the adult. Lower sensory potentials can spread outside the immediate area of injury. Further, because of the plasticity of both central and peripheral sensory connections in neonates, damage in early infancy can lead to prolonged structural and functional alterations in pain pathways that can last into adult life. Development of persistent pain in the infant may be related to the lack of descending inhibition in the neonatal dorsal horn. In adults, this descending inhibitory mechanism constitutes an endogenous analgesic system that can "dampen" persistent sensory inputs. In addition to the association between neurological development and the pain experience, the pharmacodynamics and pharmacokinetics of analgesics change with age,[89] and these changes are not necessarily linear. Thus, because of the incomplete maturation of their hepatic enzyme systems, neonates have reduced weight-normalized clearance of many drugs, but children 2 to 6 years of age have greater clearance than adults for many drugs. The quicker drug clearance in children than in adults may require more frequent drug dosing. For example, a sustained-release oral morphine drug administered twice daily in adults may require three daily doses in children.[89] Differences in body composition also dictate dosing. Thus, because neonates have a lower plasma protein concentration, they should receive lower doses of drugs with a high degree of protein binding so as to prevent an increased fraction of unbound drug and a concomitant increased toxicity.

Elderly

Increasing age is associated with an increased prevalence of chronic pain, a lower likelihood of recovering from chronic pain,[11] and poorer pain management.[48] The prevalence of pain among the elderly is twice as high as among younger individuals, with 25 to 50 percent of community-dwelling older people having important pain problems.[94,95] There are numerous possible explanations for the increasing pain prevalence with age. These include the development of pain-related conditions, such as arthritis, diabetic neuropathy or peripheral vascular disease; an accrual of cases of chronic pain (i.e., even with a stable incidence over the lifespan, with increasing age, more cases enter the pool of patients with incurable, non-life-threatening painful conditions); and an increased susceptibility to pain. While persistent pain can have numerous causes, it also can result in several consequences, e.g., anxiety, agitation, decreased socialization, depression, sleep disturbances, impaired ambulation, increased health care costs, and slow rehabilitation.[94]

Poorer pain management among the elderly may be related to the philosophy and knowledge of this group regarding pain and to how they view their clinicians. Elderly patients are more likely to have a fatalistic approach to their pain. For example, they are more likely to believe that pain is an inevitable part of cancer and that nothing can be done to alleviate it.[96] Further, they believe that they must be "good patients" and that "good patients" do not complain about pain because clinicians find such complaints to be annoying. Finally, some elderly patients do not complain about their pain because they think that it will distract their clinician from treating their cancer. Thus, the clinician must make it clear to such patients that they, i.e., the clinicians, see pain management as a vital part of patient care. Making it clear to the patient, however, may not be sufficient. Since many pain-assessment instruments have been validated in adult groups that did not include older individuals, one may wonder about the appropriateness of these instruments for frail elderly individuals with a high prevalence of visual, hearing, motor, and cognitive impairments.[97] This problem is especially critical because pain assessment is the first step in pain management.

Complicating the management of pain in the elderly is their risk of adverse events related to treatment with analgesic drugs. Among frail elderly individuals with multiple-system disease, an unacceptable risk of life-threatening gastrointestinal bleeding is associated with persistent use of nonselective nonsteroidal anti-inflammatory drugs (NSAIDs). Further, treatment with COX-2 inhibitors remains an active research area. Therefore, the chronic use of opioids for persistent pain in the elderly frequently may be the preferred treatment.[94] In fact, older people seem to experience greater and longer pain relief with opioids. Nevertheless, pain treatment is more likely to be complicated among the elderly because of polypharmacy, as well as physiologic changes. Such changes include slower absorption of medication and metabolic changes that favor or worsen opioid-related side effects, such as confusion, oversedation, urinary retention, and constipation.

Edwards and colleagues[98] present a possible biologic mechanism that could explain the higher pain susceptibility in the elderly. Their research suggests that there is an age-associated decrement in at least one form of endogenous analgesia, i.e., diffuse noxious inhibitory controls (DNICs). DNIC research involves measuring the response to a noxious test stimulus before, during, and after the application of a different noxious conditioning stimulus. DNICs lower the perception of pain produced by the test stimulus during and after the application of the conditioning stimulus. In the study of Edwards and colleagues,[98] each subject rated his or her response to a thermal test stimulus. The conditioning stimulus was a circulating cold-water bath into which participants immersed their hand. While younger subjects reported lower pain intensities in response to the thermal test stimulus, as would be expected with DNICs, older adults reported an increased pain response to the test stimulus.

In another experimental study of a clinically more relevant pain, ischemic pain was produced by exercising a hand as blood flow to the arm was occluded with a tourniquet.[99] Younger subjects demonstrated a higher pain threshold and tolerance. The authors cite the work of others[100–105] that indicates that naloxone, a selective opioid antagonist, increases ischemic pain but not thermal pain sensitivity. Taken together, the data point to age-related decrements in the inhibition of ischemic pain via endogenous opioids.

The Frail Elderly and Nursing Home Patients

Nursing homes are increasingly providing health care and support for older people who cannot function independently. With 45 to 84 percent of nursing home residents experiencing chronic pain, pain management in these facilities needs to improve. A 2002 survey of managers of 121 nursing homes[106] found that 75 percent of nursing homes did not use a standardized pain-assessment tool and that 69 percent did not have a written policy for pain management. Less than half the qualified staff had specialist knowledge regarding the management of pain in elderly people, and less than half the nursing homes provided educational or training sessions in pain management to their staff.

While clinicians need to be aware of the danger of prescribing too much opioid medication, they also need to consider the dangers of not prescribing enough. A 1990 lawsuit filed in North Carolina dealt with a patient who had been admitted to a nursing home with cancer of the prostate metastatic to the hip and spine and with a life expectancy of less than 6 months.[107] This lawsuit may have had a critical impact on the development of medical guidelines for pain management in nursing homes. It focused on the responsibilities of health care providers to ensure the proper administration of appropriate doses of pain medications. While the attending physician had prescribed morphine elixir to be given every 3 hours as needed, a nurse had deemed the patient to be "addicted to morphine." Therefore, without consulting the physician, the administration of morphine was withheld altogether, and at times only a minor tranquilizer was substituted. After 3.5 days of testimony, a jury took less than an hour to find the operator and owner of the nursing home negligent in failing to administer adequate pain medication. The jury awarded the patient's estate $7.5 million in compensatory and $7.5 million in punitive damages. Subsequently, a settlement was reached for an undisclosed amount. In the summary statement, the judge underscored the serious legal penalties health care providers risk if they negligently underuse appropriate pain medication.

Among elderly individuals residing in nursing homes, the reported prevalence rates of dementia and pain are higher than among community-dwelling elderly, i.e., the nursing home prevalence rates are from 40 to 78 percent for dementia[108,109] and, as noted earlier, from 45 to 84 percent for pain.[110–112] Krulewitch and colleagues[113] assessed the ability of community-dwelling elderly people with cognitive impairments to rate their pain via a number of pain-assessment tools: the Faces Pain Scale,[114] consisting of seven pictures of faces ranging from a smile reflecting "no pain" to a face portraying excruciating pain; the Nonverbal Analogue Scale,[115] a line labeled "no pain" at one end and "worst pain" at the other end; and the six-item Pain Intensity Scale, consisting of "not at all," "a little," "moderately," "quite a bit," and "extremely." The number of tools a participant was able to complete decreased with increased cognitive impairment. The instrument that could be used by the greatest number of cognitively impaired individuals was the Pain Intensity Scale. However, only 50 percent of the subjects with severe dementia (Minimental Status Examination score of 11 or below) were able to respond to the questions in this instrument. Thus, while the patient's report is the most accurate method for measuring pain,[116] a substantial percentage of patients with severe dementia cannot verbally report their experience of pain and, therefore, are at an increased risk for undetected and untreated pain.[117,118] Although the investigators found only fair agreement between subjects and their caregivers regarding the pain intensity of the subject, they concluded that because the subjects and their caregivers agreed in two-thirds of the cases, it is likely that caregiver assessment of pain is beneficial to demented patients whose pain often may be unrecognized.

Appropriate pain assessment by geriatricians also varies according to cognitive level. Cohen-Mansfield and Lipson[119] reported that geriatricians were able to evaluate pain successfully for persons with moderate cognitive impairment but not for those with severe cognitive impairment.

While pain assessment in nursing home residents with dementia has received quite a bit of attention recently,[112,120–123] this area remains distressingly complex. For example, one barrier for testing the validity of a pain-assessment instrument for noncommunicative individuals with advanced dementia is the lack of a "gold standard" for pain and the frequent intermittent nature of pain. Manfredi and colleagues[120] addressed this problem by interviewing communicative nursing home residents who could convey their pain experience. Each of the eight patients the investigators interviewed during the dressing change of an advanced decubitus ulcer reported that the procedure was painful. Thus, such dressing changes served as a "gold standard" for a procedure that produced acute pain. The investigators then videotaped the faces of noncommunicative, severely demented residents before and during such dressing changes and asked medical students and nurses to rate the pain they detected in the videotapes. The study showed that by paying attention to the vocalizations and facial expressions of noncommunicative patients with severe dementia, clinicians can reliably (i.e., they have a high degree of interrater agreement) determine whether a patient is experiencing pain. This means that clinicians need to *look for signs of pain* in patients who are not able to verbally relate their pain experience. Manfredi and colleagues, however, also showed that clinicians could not reliably rate the pain *intensity* of the patient, i.e., there was poor interrater agreement regarding intensity of pain. In a subsequent study, Manfredi and colleagues[124] demonstrated that opioid medication significantly reduced refractory agitation in severely demented nursing home residents 85 years of age and older, indicating that their agitation may have resulted from unrecognized pain.

What should clinicians look for?

- Changes in sleeping patterns, activity level of functioning
- Fidgeting, tense body language, wringing of hands, rubbing of body
- Frightened or sad facial expressions
- Vocalizations that can range from being mournful and groaning to being hushed
- Breathing that is audible or appears exaggerated or labored[125]

With the current state of knowledge regarding pain in elderly demented individuals, we cannot be absolutely certain whether lower pain reports are attributable to a dementia-related inability to report pain or to a truly lower rate of pain among those with dementia. McNamara and colleagues[126] reviewed what is currently known about aspects of neurology that are associated with both pain and cognitive impairment. For instance, the frontal cortex or its anatomic associations are critical for producing pain experiences. The frontal cortex is also the site for nueritic changes that develop with cognitive decline in the elderly.[127] Thus, lesions in the frontal cortex simultaneously may be associated with cognitive decline as well as attenuation of pain. Another hypothesis is that the chronic pain experience may be attributable, in part, to continuous activation of right frontal lobe networks and the concomitant perseverative retrieval of pain memories. Individuals with right frontal dysfunction and associated memory impairment may experience less distress associated with their pain. One needs to bear in mind, however, that there is no confirmatory evidence that people with dementia who have painful conditions experience less pain than their cognitively-intact counterparts. The current literature is overwhelmingly consistent in its appraisal that pain in elderly people with dementia is critically undertreated.

▶ CONCLUSION

Our epidemiologic review has highlighted many of the complex aspects of the pain experience and pain management. Clinicians must consider reported pain intensities, the differences in the progression of diseases according to demographic parameters, and the resulting effects on pain; demographic variability of the pharmacodynamics and pharmacokinetics of analgesic medications and of emotional responses to pain; cultural attitudes about pain; patient preferences regarding coherent thinking versus pain alleviation, especially at the end of life; and occupational demands and implications and their different relationships to chronic and acute pain.

While there is evidence that responses to pain management differ across demographic categories, historically, the elderly,[94] children, females, and minorities have been underrepresented in clinical trials, which are critical for progress in pain management. Thus it would seem that much of clinical medicine is based on, and generalized from, findings of trials involving middle-age males. Pain management is a relatively new field. While we have witnessed dramatic improvement in this field, there are still large gaps in our knowledge, many of which can be filled through a better understanding of the epidemiology of pain, including the epidemiology of responses to pain management. The current emphasis on inclusion in studies of minorities, women, and individuals of all ages should ensure great progress in this field in the future. Much improvement, however, is still needed in the training and educating of medical students, residents, and attending physicians.

REFERENCES

1. LeResche L: Epidemiologic perspectives on sex differences in pain, in Fillingim RB (ed), *Sex, Gender, and Pain.* Seattle: IASP Press, 2000:233–249.
2. Last JM: *A Dictionary of Epidemiology.* New York: Oxford University Press, 1995.
3. Cook L: The battle to relieve pain: Good news for women. *National Institutes of Health News & Features,* Fall 1997; Special Issue: *Research on Women's Health,* 2003.
4. Joint Commission on Accreditation of Healthcare Organizations (JCAHO), 2002.
5. Meier D: United States overview of cancer pain and palliative care. *J Pain Symptom Manage* 2002;24:265–269.
6. Costantini M, Viterbori P, Flego G: Prevalence of pain in Italian hospitals: Results of a regional cross-sectional survey. *J Pain Symptom Manage* 2002;23:221–230.
7. Von Roenn JH, Cleeland CS, Gonin R, et al: Physician attitudes and practice in cancer pain management: A survey from the Eastern Cooperative Oncology Group. *Ann Intern Med* 1993;119:121–126.
8. Benedetti C, Dickerson ED, Nichols LL: Medical education: A barrier to pain therapy and palliative care. *J Pain Symptom Manage* 2001;21:360–362.
9. Lebovits AH, Florence I, Bathina R, et al: Pain knowledge and attitudes of healthcare providers: Practice characteristic differences. *Clin J Pain* 1997;13:237–243.
10. Galer BS, Keran C, Frisinger M: Pain medicine education among American neurologists: A need for improvement. *Neurology* 1999;52:1710–1712.
11. Elliott AM, Smith BH, Hannaford PC, et al: The course of chronic pain in the community: Results of a 4-year follow-up study. *Pain* 2002;99:299–307.
12. Ohayon MM, Schatzberg AF: Using chronic pain to predict depressive morbidity in the general population. *Arch Gen Psychiatry* 2003;60:39–47.
13. Reigo T, Timpka T, Tropp H: The epidemiology of back pain in vocational age groups. *Scand J Prim Health Care* 1999;17:17–21.
14. Dahl JL: Working with regulators to improve the standard of care in pain management: The U.S. experience. *J Pain Symptom Manage* 2002;24:136–146.
15. Riley JL III, Wade JB, Myers CD, et al: Racial/ethnic differences in the experience of chronic pain. *Pain* 2002; 100:291–298.
16. Lundberg U: Stress responses in low-status jobs and their relationship to health risks: Musculoskeletal disorders. *Ann NY Acad Sci* 1999;896:162–172.
17. Deyo RA, Tsui-Wu YJ: Functional disability due to back pain: A population-based study indicating the importance of socioeconomic factors. *Arthritis Rheum* 1987; 30:1247–1253.
18. Wong MD, Shapiro MF, Boscardin WJ, Ettner SL: Contribution of major diseases to disparities in mortality. *New Engl J Med* 2002;347:1585–1592.
19. Dionne CE, Von Korff M, Koepsell TD, et al: Formal education and back pain: A review. *J Epidemiol Commun Health* 2001;55:455–468.
20. Hagen KB, Holte HH, Tambs K, Bjerkedal T: Socioeconomic factors and disability retirement from back pain: A 1983–1993 population-based prospective study in Norway. *Spine* 2000;25:2480–2487.
21. Hagen KB, Tambs K, Bjerkedal T: A prospective cohort study of risk factors for disability retirement because of back pain in the general working population. *Spine* 2002;27:1790–1796.
22. Vaccaro AR, Ring D, Scuderi G, et al: Predictors of outcome in patients with chronic back pain and low-grade spondylolisthesis. *Spine* 1997;22:2030–2034.
23. Rainville J, Sobel JB, Hartigan C, Wright A: The effect of compensation involvement on the reporting of pain and disability by patients referred for rehabilitation of chronic low back pain. *Spine* 1997;22:2016–2024.
24. Hemingway H, Shipley MJ, Stansfeld S, Marmot M: Sickness absence from back pain, psychosocial work characteristics and employment grade among office workers. *Scand J Work Environ Health* 1997;23:121–129.
25. Hurwitz EL, Morgenstern H: Correlates of back problems and back-related disability in the United States. *J Clin Epidemiol* 1997;50:669–681.
26. Baigi A, Marklund B, Fridlund B: The association between socio-economic status and chest pain, focusing on self-rated health in a primary health care area of Sweden. *Eur J Public Health* 2001;11:420–424.
27. Okany CC, Akinyanju OO: The influence of socio-economic status on the severity of sickle cell disease. *Afr J Med Sci* 1993;22:57–60.
28. Hoffmann DE, Tarzian AJ: The girl who cried pain: A bias against women in the treatment of pain. *J Law Med Ethics* 2001;29:13–27.

29. Berkley KJ: Female pain versus male pain? in Fillingim RB (ed), *Sex, Gender, and Pain*. Seattle: IASP Press, 2003:373–381.
30. Torpy JM, Lynm C, Glass RM: Men and women are different. *JAMA* 2003;289:510.
31. Kimble LP, McGuire DB, Dunbar SB, et al: Gender differences in pain characteristics of chronic stable angina and perceived physical limitation in patients with coronary artery disease. *Pain* 2003;101:45–53.
32. Torpy JM, Lynm C, Glass RM: JAMA patient page: Heart disease and women. *JAMA* 2002;288:3230.
33. Philpott S, Boynton PM, Feder G, Hemingway H: Gender differences in descriptions of angina symptoms and health problems immediately prior to angiography: The ACRE study. Appropriateness of Coronary Revascularisation Study. *Soc Sci Med* 2001;52:1565–1575.
34. Chia YY, Chow LH, Hung CC, et al: Gender and pain upon movement are associated with the requirements for postoperative patient-controlled IV analgesia: A prospective survey of 2298 Chinese patients [Les besoins d'analgesie IV auto-controlee sont lies au sexe du sujet et a la douleur au mouvement: une enquete aupres de 2298 Chinois]. *Can J Anesthesiol* 2002;49:249–255.
35. Gear RW, Miaskowski C, Gordon NC, et al: Kappa-opioids produce significantly greater analgesia in women than in men. *Nature Med* 1996;2:1248–1250.
36. Burns JW, Hodsman NB, McLintock TT, et al: The influence of patient characteristics on the requirements for postoperative analgesia: A reassessment using patient-controlled analgesia. *Anaesthesia* 1989;44:2–6.
37. Faucett J, Gordon N, Levine J: Differences in postoperative pain severity among four ethnic groups. *J Pain Symptom Manage* 1994;9:383–389.
38. Breitbart W, McDonald MV, Rosenfeld B, et al: Pain in ambulatory AIDS patients: I. Pain characteristics and medical correlates. *Pain* 1996;68:315–321.
39. Staud R, Robinson ME, Vierck CJ, Price DD: Diffuse noxious inhibitory controls (DNIC) attenuate temporal summation of second pain in normal males but not in normal females or fibromyalgia patients. *Pain* 2003;101:167–174.
40. Ploner M, Gross J, Timmermann L, Schnitzler A: Cortical representation of first and second pain sensation in humans. *Proc Natl Acad Sci USA* 2002;99:12444–12448.
41. Paulson PE, Minoshima S, Morrow TJ, Casey KL: Gender differences in pain perception and patterns of cerebral activation during noxious heat stimulation in humans. *Pain* 1998;76:223–229.
42. Chesterton LS, Barlas P, Foster NE, et al: Gender differences in pressure pain threshold in healthy humans. *Pain* 2003;101:259–266.
43. Jensen R, Rasmussen BK, Pedersen B, et al: Cephalic muscle tenderness and pressure pain threshold in a general population. *Pain* 1992;48:197–203.
44. Fischer AA: Pressure threshold meter: Its use for quantification of tender spots. *Arch Phys Med Rehabil* 1986;67:836–838.
45. Berkley KJ: Sex differences in pain. *Behav Brain Sci* 1997;20:371–380.
46. Levine FM, De Simone LL: The effects of experimenter gender on pain report in male and female subjects. *Pain* 1991;44:69–72.
47. Buchanan HM, Midgley JA: Evaluation of pain threshold using a simple pressure algometer. *Clin Rheumatol* 1987;6:510–517.
48. Cleeland CS, Gonin R, Hatfield AK, et al: Pain and its treatment in outpatients with metastatic cancer. *New Engl J Med* 1994;330:592–596.
49. Edwards CL, Fillingim RB, Keefe F: Race, ethnicity and pain. *Pain* 2001;94:133–137.
50. Lasch KE: Culture, pain, and culturally sensitive pain care. *Pain Manag Nurs* 2000;1:16–22.
51. Green CR, Baker TA, Smith EM, Sato Y: The effect of race on older adults presenting for chronic pain management: A comparative study of black and white Americans. *J Pain* 2003;4:82–90.
52. Lasch KE: Culture, pain, and culturally sensitive pain care. *Pain Manag Nurs* 2000;1:16–22.
53. Weisse CS, Sorum PC, Sanders KN, Syat BL: Do gender and race affect decisions about pain management? *J Gen Intern Med* 2001;16:211–217.
54. Todd KH, Samaroo N, Hoffman JR: Ethnicity as a risk factor for inadequate emergency department analgesia. *JAMA* 1993;269:1537–1539.
55. Harrison A, Busabir AA, al Kaabi AO, al Awadi HK: Does sharing a mother-tongue affect how closely patients and nurses agree when rating the patient's pain, worry and knowledge? *J Adv Nurs* 1996;24:229–235.
56. Moore R, Brodsgaard I, Mao TK, et al: Acute pain and use of local anesthesia: Tooth drilling and childbirth labor pain beliefs among Anglo-Americans, Chinese, and Scandinavians. *Anesth Prog* 1998;45:29–37.
57. Moore RA, Dworkin SF: Ethnographic methodologic assessment of pain perceptions by verbal description. *Pain* 1988;34:195–204.
58. Cleeland CS, Nakamura Y, Mendoza TR, et al: Dimensions of the impact of cancer pain in a four country sample: New information from multidimensional scaling. *Pain* 1996;67:267–273.
59. Ger LP, Ho ST, Sun WZ, et al: Validation of the Brief Pain Inventory in a Taiwanese population. *J Pain Symptom Manage* 1999;18:316–322.
60. Radbruch L, Loick G, Kiencke P, et al: Validation of the German version of the Brief Pain Inventory. *J Pain Symptom Manage* 1999;18:180–187.
61. Uki J, Mendoza T, Cleeland CS, et al: A brief cancer pain assessment tool in Japanese: The utility of the Japanese Brief Pain Inventory—BPI-J. *J Pain Symptom Manage* 1998;16:364–373.
62. Badia X, Muriel C, Gracia A, et al: Validation of the Spanish version of the Brief Pain Inventory in patients with oncological pain. *Med Clin* 2003;120:52–59.
63. Lasch KE: Culture and pain. *Pain: Clinical Updates* 2002;10:1–4.
64. Ng B, Dimsdale JE, Rollnik JD, Shapiro H: The effect of ethnicity on prescriptions for patient-controlled analgesia for post-operative pain. *Pain* 1996;66:9–12.
65. McNeilly M, Zeichner A: Neuropeptide and cardiovascular responses to intravenous catheterization in normotensive

and hypertensive blacks and whites. *Health Psychol* 1989;8:487–501.
66. Clark C, Robinson TM: Cultural diversity and transcultural nursing as they impact health care. *J Natl Black Nurs Assoc* 1999;10:46–53.
67. Green CT, Baker TA, Sato Y, et al: Race and chronic pain: A comparative study of young black and white Americans presenting for management. *J Pain* 2003;4:176–183.
68. Moore R, Brodsgaard I: Cross-cultural investigations of pain, in Crombie IK, Croft PR, Linton SJ, et al (eds), *Epidemiology of Pain*. Seattle: IASP, 1999:53–80.
69. Lee MC, Essoka G: Patient's perception of pain: Comparison between Korean-American and Euro-American obstetric patients. *J Cult Divers* 1998;5:29–37.
70. Beck SL: An ethnographic study of factors influencing cancer pain management in South Africa. *Cancer Nurs* 2000;23:91–99.
71. Adams NJ, Plane MB, Fleming MF, et al: Opioids and the treatment of chronic pain in a primary care sample. *J Pain Symptom Manage* 2001;22:791–796.
72. Morrison RS, Wallenstein S, Natale DK, et al: "We don't carry that"—failure of pharmacies in predominantly non-white neighborhoods to stock opioid analgesics. *New Engl J Med* 2000;342:1023–1026.
73. McCaffery M, Ferrell BR, Pasero C: Nurses' personal opinions about patients' pain and their effect on recorded assessments and titration of opioid doses. *Pain Manag Nurs* 2000;1:79–87.
74. Wolfe J, Grier HE, Klar N, et al: Symptoms and suffering at the end of life in children with cancer. *New Engl J Med* 2000;342:326–333.
75. Perquin CW, Hazebroek-Kampschreur AAJM, Hunfeld JAM, et al: Pain in children and adolescents: A common experience. *Pain* 2000;87:51–58.
76. Carlsson J, Larsson B, Mark A: Psychosocial functioning in schoolchildren with recurrent headaches. *Headache* 1996;36:77–82.
77. Larsson B: The role of psychological, health-behaviour and medical factors in adolescent headache. *Dev Med Child Neurol* 1988;30:616–625.
78. Smith MS, Martin-Herz SP, Womack WM, McMahon RJ: Recurrent headache in adolescents: Nonreferred versus clinic population. *Headache* 1999;39:616–624.
79. Scharff L: Recurrent abdominal pain in children: A review of psychological factors and treatment. *Clin Psychol Rev* 1997;17:145–166.
80. Huang RC, Palmer LJ, Forbes DA: Prevalence and pattern of childhood abdominal pain in an Australian general practice. *J Paediatr Child Health* 2000;36:349–353.
81. Harreby M, Neergaard K, Hesselsoe G, Kjer J: Are radiologic changes in the thoracic and lumbar spine of adolescents risk factors for low back pain in adults? A 25-year prospective cohort study of 640 school children. *Spine* 1995;20:2298–2302.
82. Dalsgaard-Nielsen T, Engberg-Pedersen H, Holm HE: Clinical and statistical investigations of the epidemiology of migraine. *Danish Med Bull* 1970;17:138–148.
83. Balague F, Nordin M, Skovron ML, et al: Non-specific low-back pain among schoolchildren: A field survey with analysis of some associated factors. *J Spinal Disord* 1994;7:374–379.
84. Wedderkopp N, Leboeuf-Yde C, Andersen LB, et al: Back pain reporting pattern in a Danish population-based sample of children and adolescents. *Spine* 2001;26:1879–1883.
85. Perquin CW, Hazebroek-Kampschreur AA, Hunfeld JA, et al: Pain in children and adolescents: A common experience. *Pain* 2000;87:51–58.
86. Shevell MI: Ethics of clinical research in children. *Semin Pediatr Neurol* 2002;9:46–52.
87. Cote CJ, Kauffman RE, Troendle GJ, Lambert GH: Is the "therapeutic orphan" about to be adopted? *Pediatrics* 1996;98:118–123.
88. Sutcliffe AG: Testing new pharmaceutical products in children. *Br Med J* 2003;326:64–65.
89. Berde CB, Sethna NF: Analgesics for the treatment of pain in children. *New Engl J Med* 2002;347:1094–1103.
90. Beyer JE, McGrath PJ, Berde CB: Discordance between self-report and behavioral pain measures in children aged 3–7 years after surgery. *J Pain Symptom Manage* 1990;5:350–356.
91. Schechter NL, Allen DA, Hanson K: Status of pediatric pain control: A comparison of hospital analgesic usage in children and adults. *Pediatrics* 1986;77:11–15.
92. Hostetler MA, Auinger P, Szilagyi PG: Parenteral analgesic and sedative use among ED patients in the United States: Combined results from the National Hospital Ambulatory Medical Care Survey (NHAMCS), 1992–1997. *Am J Emerg Med* 2002;20:83–87.
93. Fitzgerald M, Beggs S: The neurobiology of pain: Developmental aspects. *Neuroscientist* 2001;7:246–257.
94. AGS Panel on Persistent Pain in Older Persons: The management of persistent pain in older persons. *J Am Geriatr Soc* 2002;50:S205–S224.
95. Fulmer TT, Mion LC, Bottrell MM, NICHE Faculty: Pain management protocol: Inappropriate pain management leaves both the elder and the nurse feeling unfulfilled and unhappy with care. *Geriatr Nurs* 1996;17:222–226.
96. Gunnarsdottir S, Donovan HS, Serlin RC, et al: Patient-related barriers to pain management: the barriers questionnaire II (BQ-II). *Pain* 2002;99:385–396.
97. Kamel HK, Phlavan M, Malekgoudarzi B, et al: Utilizing pain assessment scales increases the frequency of diagnosing pain among elderly nursing home residents. *J Pain Symptom Manage* 2001;21:450–455.
98. Edwards RR, Fillingim RB, Ness TJ: Age-related differences in endogenous pain modulation: A comparison of diffuse noxious inhibitory controls in healthy older and younger adults. *Pain* 2003;101:155–165.
99. Edwards RR, Fillingim RB: Age-associated differences in responses to noxious stimuli. *J Gerontol [A] Biol Sci Med Sci* 2001;56:M180–M185.
100. Frid M, Singer G, Oei T, Rana C: Reactions to ischemic pain: Interactions between individual, situational and naloxone effects. *Psychopharmacology (Berl)* 1981;73:116–119.
101. Frid M, Singer G, Rana C: Interactions between personal expectations and naloxone: effects on tolerance to ischemic pain. *Psychopharmacology (Berl)* 1979;65:225–231.
102. Grevert P, Albert LH, Goldstein A: Partial antagonism of placebo analgesia by naloxone. *Pain* 1983;16:129–143.

103. Janal MN, Colt EW, Clark WC, Glusman M: Pain sensitivity, mood and plasma endocrine levels in man following long-distance running: Effects of naloxone. *Pain* 1984;19:13–25.
104. Lautenbacher S, Roscher S, Strian D, et al: Pain perception in depression: Relationships to symptomatology and naloxone-sensitive mechanisms. *Psychosom Med* 1994; 56:345–352.
105. Stacher G, Abatzi TA, Schulte F, et al: Naloxone does not alter the perception of pain induced by electrical and thermal stimulation of the skin in healthy humans. *Pain* 1988;34:271–276.
106. Allcock N, McGarry J, Elkan R: Management of pain in older people within the nursing home: A preliminary study. *Health Soc Care Commun* 2002;10:464–471.
107. Ferrell BA: Pain evaluation and management in the nursing home. *Ann Intern Med* 1995;123:681–687.
108. Rovner BW, Kafonek S, Filipp L, et al: Prevalence of mental illness in a community nursing home. *Am J Psychiatry* 1986;143:1446–1449.
109. Magaziner J, Zimmerman SI, German PS, et al: Ascertaining dementia by expert panel in epidemiologic studies of nursing home residents. *Ann Epidemiol* 1996;6:431–437.
110. Stein WM, Ferrell BA: Pain in the nursing home. *Clin Geriatr Med* 1996;12:601–613.
111. Parmelee PA: Pain in cognitively impaired older persons. *Clin Geriatr Med* 1996;12:473–487.
112. Fisher SE, Burgio LD, Thorn BE, et al: Pain assessment and management in cognitively impaired nursing home residents: Association of certified nursing assistant pain report, Minimum Data Set pain report, and analgesic medication use. *J Am Geriatr Soc* 2002;50:152–156.
113. Krulewitch H, London MR, Skakel VJ, et al: Assessment of pain in cognitively impaired older adults: A comparison of pain assessment tools and their use by nonprofessional caregivers. *J Am Geriatr Soc* 2000;48:1607–1611.
114. Herr KA, Mobily PR, Kohout FJ, Wagenaar D: Evaluation of the Faces Pain Scale for use with the elderly. *Clin J Pain* 1998;14:29–38.
115. Huskisson EC: Visual analog scales, in Melzack R (ed), *Pain Measurement and Assessment.* New York: Raven Press, 1983:33–37.
116. *Classification of Chronic Pain.* Seattle: IASP Press, 1994.
117. Feldt KS, Ryden MB, Miles S: Treatment of pain in cognitively impaired compared with cognitively intact older patients with hip fracture. *J Am Geriatr Soc* 1998; 46:1079–1085.
118. Morrison RS, Siu AL: A comparison of pain and its treatment in advanced dementia and cognitively intact patients with hip fracture. *J Pain Symptom Manage* 2000; 19:240–248.
119. Cohen-Mansfield J, Lipson S: Pain in cognitively impaired nursing home residents: How well are physicians diagnosing it? *J Am Geriatr Soc* 2002;50:1039–1044.
120. Manfredi PL, Breuer B, Meier DE, Libow L: Pain assessment in elderly patients with severe dementia. *J Pain Symptom Manage* 2003;25:48–52.
121. Villanueva MR, Smith TL, Erickson JS, et al: Pain Assessment for the Dementing Elderly (PADE): Reliability and validity of a new measure. *J Am Med Dir Assoc* 2003;4:1–8.
122. Warden V, Hurley AC, Volicer L: Development and Psychometric Evaluation of the Pain Assessment in Advanced Dementia (PAINAD) scale. *J Am Med Dir Assoc* 2003;4:9–15.
123. Cadogan MP: Assessing pain in cognitively impaired nursing home residents: The state of the science and the state we're in. *J Am Med Dir Assoc* 2003;4:50–51.
124. Manfredi P, Breuer B, Wallenstein S, et al: Opioid treatment for refractory agitation in patients with advanced dementia. *Int J Geriatr Psychiatry* 2003;18:700–705.
125. Brown University Center for Gerontology and Health Care Research: *Resource Guide to Use Toolkit of Instruments to Measure End-of-Life Care,* 2001:6–7.
126. McNamara P, Oscar-Berman M, Albert M: Frontal lobe function and pain in the elderly. *J Adult Dev* 2000;7:113–119.
127. Naslund J, Haroutunian V, Mohs R, et al: Correlation between elevated levels of amyloid beta-peptide in the brain and cognitive decline. *JAMA* 2000;283:1571–1577.

CHAPTER 14

Principles of Pain Assessment

Christine Miaskowski

Pain assessment is the cornerstone of effective pain management. The assessment of pain is a continuous and ongoing process that includes universal screening for pain, comprehensive pain assessments, and regular ongoing assessments of the patient's pain. Pain management standards published by the Joint Commission for the Accreditation of Healthcare Organizations (JCAHO) require that all health care organizations that are accredited by JCAHO implement policies and procedures that make pain assessment and effective management strategies a routine part of every patient's care.[1,2]

The two main purposes of the assessment (and reassessment) processes are to diagnose the cause of the pain and to evaluate the effectiveness of the pain management plan. In order to diagnose the cause of the pain, clinicians should perform a comprehensive pain assessment (i.e., the initial *pain assessment) and evaluate the multiple dimensions of the pain experience. Subsequent evaluations of the effectiveness of the pain management plan (i.e., the* ongoing *pain assessments) often focus on determining whether the intensity/severity of the pain has decreased as a result of pharmacologic and nonpharmacologic interventions. This chapter describes the multiple dimensions of the pain experience and their impact on pain assessment. In addition, approaches to conducting the initial pain assessment and ongoing pain assessments are described and evaluated for their clinical usefulness.*

▶ MULTIPLE DIMENSIONS OF THE PAIN EXPERIENCE

Pain is defined by the International Association for the Study of Pain (IASP) as "an unpleasant sensory and emotional experience associated with actual or potential tissue damage, or described in terms of such damage."[3] This definition suggests that pain has unique physiologic and psychological components that contribute to the pain experience. Pain is a uniquely *individual* experience. No two people, even those with similar etiologies for their pain, will experience pain in exactly the same way. The assessment of pain focuses on evaluating the individual patient's experience and determining the effectiveness of the pain management plan based on the responses of *that* individual.

Pain is conceptualized as a multidimensional experience. Ahles and colleagues[4] were the first to develop a five-dimensional model that they applied to cancer pain. The five dimensions of the pain experience are (1) physiologic (organic cause of the pain), (2) sensory (intensity, location, quality), (3) affective (anxiety, depression), (4) cognitive (meaning of the pain, thought processes, and views of oneself), and (5) behavioral (physical activity, medication intake). McGuire[5] added a sociocultural dimension and suggested that this multidimensional model could be applied to all patients who are experiencing pain. Table 14-1 depicts the major components of each of the various dimensions of

▶ **TABLE 14-1.** MULTIPLE DIMENSIONS OF THE PAIN EXPERIENCE

1. Physiologic dimension
   A. Etiology or cause of the pain
   B. Pain duration (acute or chronic)
2. Sensory dimension
   A. Location
   B. Intensity (severity)
   C. Quality
3. Affective dimension
   A. Emotional responses (depression, mood, anxiety, worry, helplessness, fear)
   B. Suffering
   C. Psychiatric disorders
4. Cognitive dimension
   A. Thought processes/views of self
   B. Meaning of pain
   C. Coping strategies
   D. Attitudes, beliefs, knowledge
   E. Level of cognition
5. Behavioral dimension
   A. Indicators of pain
   B. Pain control behaviors
   C. Communication of pain
   D. Associated symptoms (fatigue, sleep)
6. Sociocultural dimension
   A. Demographic variables
   B. Cultural background
   C. Personal, family, and work roles
   D. Family factors
   E. Caregiver perspectives

the pain experience. Each dimension is summarized in the next sections of this chapter.

Physiologic Dimension

One of the major aspects of the physiologic dimension is the etiology, or cause, of the patient's pain. Through a thorough assessment of all the various dimensions of the pain experience, the diagnosis or the cause of the patient's pain will be elucidated. In addition, the initial pain assessment should provide an impression of the impact of the pain on the patient's ability to function, as well as on the patient's mood and quality of life.

A second aspect of the physiologic dimension is the duration of the patient's pain; in other words, how long has the pain lasted? Pain often is classified as either acute (i.e., lasting less than 6 months) or chronic (i.e., persisting for greater than 3 months).[6] The most common characteristics of acute and chronic pain are listed in Table 14-2.

Sensory Dimension

The major aspects of the sensory dimension are location, severity or intensity, and quality of the pain. These aspects form the foundation of the initial pain assessment and often are used in the ongoing assessments of pain (particularly severity) to determine the effectiveness of the pain management plan.

The location of the pain often provides information about the etiology of the pain. Pain may occur in multiple locations. Sometimes having patients indicate on a body outline drawing (refer to Fig. 14-1) the location(s) of their pain will assist the clinician in determining the exact cause of the pain. Based on the location of the pain, one can determine the level of functional impairment that the patient may be experiencing. The location of the pain may affect both an individual's physical and affective responses to pain. For example, while the pain associated with a traumatic injury to the femur does limit mobility, the pain problem is relatively short-lived and does not have significant psychological sequelae. In contrast, multiple pathologic fractures associated with bone metastases may result in severe limitations in physical mobility, as well as have devastating psychological consequences, as a patient realizes that his or her cancer is progressing to the more terminal stages.

▶ **TABLE 14-2.** CHARACTERISTICS OF ACUTE AND CHRONIC PAIN

Acute Pain	Chronic Pain
• Time-limited	• Indefinite duration
• Well-defined etiology	• Etiology may be poorly defined
• Usually accompanied by activation of the sympathetic nervous system (e.g., tachycardia, hypertension)	• Adaptation of the sympathetic nervous system occurs
• Treatment has a high probability of success	• Treatment is not always effective and, in some cases, may contribute to iatrogenic complications (e.g., patient who undergoes multiple surgeries for back pain)
• Usually associated with minimal psychosocial consequences	• May be associated with significant psychosocial consequences (e.g., financial burden, mood disturbances)

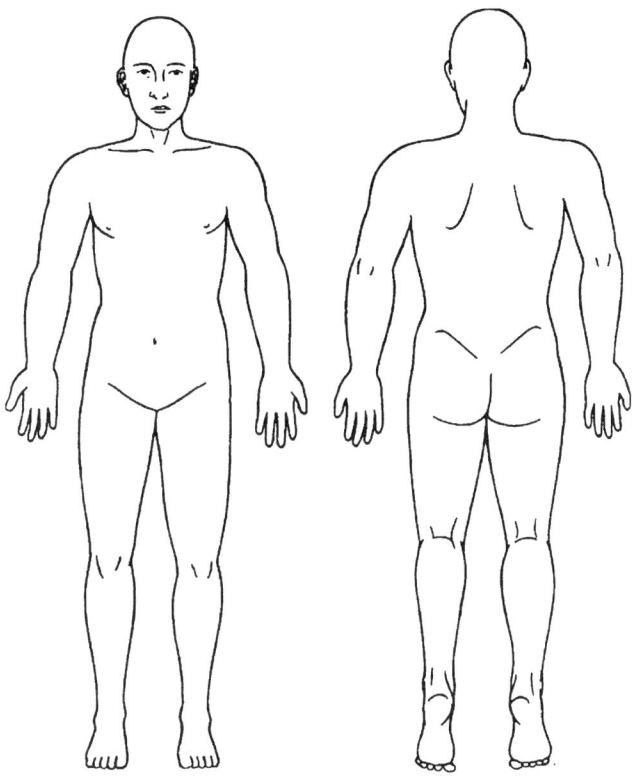

Figure 14-1. Example of a body outline drawing.

The intensity or severity of the pain is one of the most commonly assessed pain parameters. Numerous scales, discussed in subsequent sections of this chapter, are used to evaluate the intensity of pain. An evaluation of pain intensity is subjective in nature and influenced by a variety of physiologic, affective, cognitive, behavioral, and sociocultural factors. Two individuals may be experiencing the same type of pain yet report totally different levels of pain intensity. When using a pain intensity measure to evaluate the effectiveness of a pain management intervention, one needs to use the patient as his or her "own control." Clinicians should not compare pain intensity scores *across* different patients with similar pain problems. Pain is a subjective experience, and each person must be evaluated as an individual. As Ahles and Martin[7] point out, "an advantage of viewing pain from a multidimensional framework is that it forces one to recognize the multitude of factors that can potentially influence the answer to the seemingly simple question of 'How much does it hurt?'" (p. 29).

The third component of the sensory dimension is the quality of the pain or the descriptors that patients use to describe what their pain feels like. The quality of the pain often provides clues as to the etiology of the pain. For example, pain associated with a postoperative incision may be described as sharp, shooting, and tender. However, pain associated with diabetic neuropathy or postherpetic neuralgia may be described as burning, numbness, and tingling. The assessment of the quality of pain often is done using an adjective checklist such as the one found in the McGill Pain Questionnaire (MPQ).[8]

Affective Dimension

The affective dimension of the pain experience includes a variety of emotional responses to pain (e.g., depression, anxiety, fear), the concept of suffering, and the presence or absence of psychiatric disorders. Stated most simply, the affective dimension focuses on the feelings engendered by the experience of pain. Acute and chronic pain problems may result in different emotional responses to the pain experience. For example, acute pain associated with a dental problem may generate anxiety or fear in an individual, particularly if he or she has an aversion to going to the dentist. In contrast, cancer patients who are experiencing cancer-related pain may report higher levels of depression than cancer patients who are pain-free.[9]

The affective responses that are often associated with acute and chronic pain include depression, anxiety, and fear. Whether pain exacerbates some of these responses or is exacerbated by their presence, or both, is not clear at the present time. However, the interrelationships between the affective responses of an individual and his or her pain experience need to be acknowledged and evaluated by clinicians caring for these patients.

Cognitive Dimension

The cognitive dimension consists of the manner in which pain influences an individual's thought processes and the way in which he or she views himself or herself and the meaning of the pain to the individual. Several other components have been added to the cognitive dimension, including the choice of coping strategies used to manage pain; the attitudes, beliefs, and knowledge that an individual has about pain and pain management; and the patient's level of cognition that may influence his or her ability to report pain, react to a pain problem, or modify his or her behavior to adhere with the pain management plan.[10]

In most instances, the cognitive dimension of pain has been evaluated by asking patients to describe the meaning that they attribute to their pain problem. Particularly when patients have chronic pain clinicians should evaluate the meaning that patients attribute to their pain. Clinicians can clarify whether the meaning attributed to the pain is correct or inappropriate. By clarifying misconceptions (e.g., patients may erroneously think that increased pain indicates disease progression), patients may avoid needless anxiety, depression, or suffering.

Patients may use cognitive coping strategies to facilitate the management of acute and chronic pain. These strategies include self-talk, reinterpretation of pain as another sensation, selective inattention to the pain, and various forms of distraction.[11] A careful assessment of these strategies and their effectiveness may prove useful in the development of a pain management plan.

The attitudes, beliefs, and knowledge that patients have about pain and pain management can influence their perception of pain, as well as the success of the pain management plan. Little is known about the general public's attitudes, beliefs, and knowledge about pain and pain management. Most of the studies evaluating patients' knowledge and attitudes about pain and pain management have focused on oncology patients. In one study of the knowledge base of oncology outpatients about cancer pain management, Yeager and colleagues[12] found that over 60 percent of the patients who were experiencing a significant amount of cancer-related pain did not know that they should take their pain medicine on a regular schedule and that they should take a large enough dose of pain medication to relieve their pain.

Finally, the patient's level of cognitive functioning may influence his or her ability to report pain, react to a pain problem, or modify his or her behavior to adhere with the pain management plan. A discussion of the assessment and management of pain in cognitively impaired patients is found later in this chapter.

Another factor that can affect an individual's cognitive functioning is the ability of opioid analgesics to induce cognitive deficits. Opioids have the potential to produce psychomotor slowing; difficulties with memory, concentration, and attention; and confusion with regard to time, place, and person. However, two recent reviews[13,14] found moderate and consistent evidence that psychomotor skills and cognitive functioning were not impaired in patients who were taking opioids for chronic pain.

Behavioral Dimension

The behavioral aspects of pain are overt, observable behaviors that communicate to others in the environment that the patient is experiencing pain. Certain pain behaviors demonstrate the presence and perhaps the severity of pain (e.g., facial expressions such as grimacing, body postures such as muscle tension, and vocalizations such as crying or moaning). However, pain behaviors can vary depending on whether the pain is acute or chronic in nature. Since acute pain is often accompanied by activation of the sympathetic nervous system, behaviors associated with acute pain may be more demonstrative (e.g., moaning, grimacing, increased activity, and crying). In contrast, the sympathetic nervous system usually adapts in situations of chronic pain. Patients may not exhibit demonstrative pain behaviors but may exhibit withdrawal behaviors or a decrease in their activity levels. Methods for systematically assessing pain behaviors are reviewed in a subsequent section of this chapter. However, clinicians need to remember that each individual will exhibit unique pain behaviors. Therefore, one of the most useful questions a clinician can ask a patient when evaluating the behavioral dimension of pain is: "How do you behave when you are in pain?" These behaviors should be noted in the patient's medical record or on the nursing kardex and are particularly useful when patients require pain management in the inpatient setting.

Another component of the behavioral dimension is the identification of behaviors that patients use to decrease pain. In one study of oncology patients,[15] typical pain-control behaviors included analgesic use, applying pressure, manipulation, positioning, and distraction. In a subsequent study,[16] the pain-control behaviors used by patients with lung cancer included bracing to shift up and down, guarding the arm ipsilateral to the lung affected by the cancer, applying pressure to the painful area, shrugging or rotating the shoulders, and placing the hands on the thigh or the knee when shifting position. As with the identification of behaviors that communicate pain, the identification of pain-control behaviors may assist clinicians in the development of a more effective pain management plan. Pain-control behaviors can be reinforced or enhanced with the addition of other behavioral approaches (e.g., relaxation exercises or biofeedback to decrease muscle tension).

Sociocultural Dimension

The sociocultural dimension of the pain experience includes a variety of demographic variables (e.g., age, gender), as well as an individual's cultural background; personal, family, and work roles; and his or her family members' or caregiver's perspectives on the pain problem and various pain management approaches. Both the patient and his or her family members need to be assessed to determine how this component of the pain experience affects the patient, as well as his or her family members.

Summary

The various dimensions of the pain experience (refer to Table 14-1) cannot be viewed or assessed in isolation. All these dimensions interact to a greater or lesser degree to contribute to the human experience of pain. Clinicians need to consider and assess each of the six dimensions when caring for patients in acute and chronic pain.

The assessment of pain is a continuous and ongoing process that begins with universal screening, followed by an initial and comprehensive pain assessment and regular and ongoing reassessments of the patient's pain. Because of the high prevalence of unrelieved acute and chronic pain, all patients who enter the health care system should be screened for the presence of pain. Universal screening can be done by asking patients to indicate if they are experiencing any pain and to rate the intensity of the pain using a numeric rating scale (i.e., 0 = no pain to 10 = worst pain imaginable) or a verbal descriptor scale (e.g., none, mild, moderate, severe, very severe, intolerable). The initial and ongoing assessments of pain should be done in a systematic and organized fashion that provides clinicians with the data they need to diagnose the cause of the pain and develop an appropriate pain management plan.

▶ INITIAL PAIN ASSESSMENT

The initial pain assessment begins with a detailed history, followed by a physical examination and an appropriate diagnostic workup. Clinicians should develop and follow a routine approach to the assessment of pain. The initial pain assessment outlined in Table 14-3 should be done with each new report of pain. The purpose of the initial pain assessment is to identify the cause of the patient's

▶ **TABLE 14-3.** INITIAL PAIN ASSESSMENT

Pain History

A. *Persistent Pain*
- Onset: When did the pain start?
- Description: What does the pain feel like? What words would you use to describe your pain?
- Location/radiation: Show me where it hurts. Does the pain go anywhere else?
- Duration: How long does the pain last?
- Periodicity: Is the pain continuous or intermittent in nature?
- Severity/intensity: On a scale of 0 to 10, with 0 being no pain and 10 being the worst pain you can imagine, how much does it hurt—right now?, on average?, at its worst?, at its least?
- Aggravating factors: What makes the pain worse?
- Relieving factors: What makes the pain better?
- Pain behaviors: What do you do when you are in pain? What do you do to decrease your pain?
- Affective dimension: How does pain affect your mood or your ability to enjoy life?
- Cognitive dimension: What meaning do you give to your pain? How do you try to cope with your pain?
- Associated symptoms: Do you have any other symptoms that are associated with your pain?
- Sociocultural dimension: How do your spiritual or cultural beliefs influence your knowledge and attitudes about pain and pain management strategies?
- Previous treatment modalities and their effectiveness: What types of pain treatments are you currently using or have you used in the past? How effective are/were these pain treatments?

B. *Breakthrough Pain*
- Presence of breakthrough pain: Do you have sudden, severe flare-ups of pain that come and go?
- Frequency: How often do these episodes of breakthrough pain occur?
- Duration: How long do each of these episodes of breakthrough pain last?
- Intensity: On a scale of 0 to 10, with 0 being no pain and 10 being the worst pain you can imagine, how much does an episode of breakthrough pain hurt when it occurs?
- Occurrence of breakthrough pain: Does your breakthrough pain occur with activity or spontaneously?
- Previous and current treatments and their effectiveness: What types of treatments have you tried to relieve your breakthrough pain? Were they and are they effective?

Physical Examination

- Mental status
- Cranial nerve examination
- Motor function: Bulk and abnormal movements, tone, strength
- Sensory function: Pain, temperature, touch, vibration, proprioception (joint position)
- Coordination
- Gait
- Reflexes
- Autonomic function: Sweating, local swelling, trophic changes

pain. Appropriate treatment cannot be initiated until the specific cause of the patient's pain is identified.

History

The history should provide information on the various dimensions of the pain experience. As with any patient history, a review of the patient's past medical history, as well as a review of systems, should be done to provide the clinician with a complete health history of the patient.

The history of the pain complaint is outlined in Table 14-3. For patients with chronic pain, persistent pain needs to be distinguished from breakthrough pain. *Persistent pain* is constant pain that lasts for long periods. Breakthrough pain consists of sudden flare-ups of pain that come and go. These flare-ups are referred to as *breakthrough pain* because the pain "breaks through" the treatment for the persistent pain.

The history for persistent pain includes a description of the pain, its onset and temporal pattern, its location, and its intensity. In addition, information should be obtained about aggravating and relieving factors, as well as previous and current pain treatments and their effectiveness. The history regarding breakthrough pain evaluates the frequency and duration of each episode of breakthrough pain, its intensity, causes, and previous and current treatments and their effectiveness.[17,18]

Physical Examination

The physical examination for a patient in pain includes a general medical examination that is supplemented with a focused examination of the painful area or areas.[19] The focused examination should include a detailed neurological evaluation (refer to Table 14-3). If appropriate, functional testing of the painful area should be attempted and, if possible, should be done without severe exacerbation of the pain.

General Appearance and Physical Characteristics

Baseline information including temperature, heart rate, respiratory rate, blood pressure, height, and weight should be recorded. The patient's skin should be examined for skin lesions, scars, needle marks, and tattoos. A notation should be made of the patient's posture, and the spine should be examined for kyphosis, lordosis, scoliosis, and focal areas of tenderness.

Assessment of Mental Status

The mental status examination evaluates for major intellectual and psychiatric problems. The major areas of the mental status examination include orientation, calculations, memory, speech, and comprehension. An example of a "quick and dirty" mental status examination is found in Table 14-4.

▶ **TABLE 14-4.** THE "QUICK AND DIRTY" MENTAL STATUS EXAMINATION

Orientation	Ask the following questions: What is your full name? What is today's date? What is the year? Who is the president? Who is the vice-president?
Calculations	Ask the following questions: How many nickels are in a dollar? How many dollars do 60 nickels make?
Memory	Ask the following questions: What was your mother's maiden name? Who was president before Bill Clinton? Give the patient three items to remember (examples, a red ball, a blue telephone, and address 66 Hill Street). After several minutes of conversation, ask the patient to repeat the list.
Speech	Have the patient repeat two simple sentences, such as: Today is a lovely day. The weather this weekend is expected to be excellent. Have the patient name several objects in the room. Ask the patient to rhyme simple words, such as ball, pat, and can.
Comprehension	Ask the patient to: Put the right hand on the left hand. Point to the ceiling with the left index finger.

NOTE: This simple screening mental status examination uncovers many but not all cognitive defects. It can be performed in under 3 minutes and is useful in evaluating basic aspects of memory, language, and general intellectual capacity.
SOURCE: Donohoe, C. D. (1996). Targeted history and physical examination. In S. D. Waldman & A. P. Winnie (Eds.), *Interventional Pain Management* (pp. 73–84). Philadelphia: W. B. Saunders.

▶ TABLE 14-5. CLINICAL EVALUATIONS OF CRANIAL NERVE FUNCTION

Cranial Nerve(s)		
Number	Name	Evaluation Procedure(s)
I	Olfactory	Test ability to identify familiar aromatic odors, one naris at a time with eyes closed (not routinely tested)
II	Optic	Test vision with Snellen chart or Rosenbaum near vision chart
		Perform ophthalmoscopic examination of fundi
		Be able to recognize papilledema
		Test fields of vision using confrontation and double simultaneous stimulation
III, IV, and VI	Oculomotor, trochlear, and abducens	Inspect eyelids for drooping (ptosis)
		Inspect pupil size for equality (direct and consensual response)
		Check for nystagmus
		Assess basic fields of gaze
		Note asymmetric extraocular movements
V	Trigeminal	Palpate jaw muscles for tone and strength while patient clenches teeth
		Test superficial pain and touch sensation in each branch: V1, V2, V3
VII	Facial	Test corneal reflex
		Inspect symmetry of facial features
		Have patient smile, frown, puff cheeks, wrinkle forehead
		Watch for spasmodic, jerking movements of face
VIII	Acoustic	Test sense of hearing with watch or tuning fork
		Compare bone and air conduction of sound
IX	Glossopharyngeal	Test gag reflex and ability to swallow
X	Vagus	Inspect palate and uvula for symmetry with gag reflex
		Observe for swallowing difficulty
		Have patient take small sip of water
		Watch for nasal or hoarse quality of speech
IX	Spinal accessory	Test trapezius strength (have patient shrug shoulders against resistance)
		Test sternocleidomastoid muscle strength (have patient turn head to each side against resistance)
XII	Hypoglossal	Inspect tongue in mouth and while protruded for symmetry, fasciculations, and atrophy
		Test tongue strength with index fingers when tongue is pressed against cheek

SOURCE: Donohoe, C. D. (1996). Targeted history and physical examination. In S. D. Waldman & A. P. Winnie (Eds.), *Interventional Pain Management* (pp. 73–84). Philadelphia: W. B. Saunders.

Cranial Nerve Function

An evaluation of the 12 cranial nerves is an important part of pain assessment, particularly in the evaluation of headache and facial pain. Table 14-5 provides an efficient approach to the evaluation of cranial nerve functioning.[20]

Motor Examination

The motor examination begins with an inspection of muscle volume and contour. Any atrophy or hypertrophy of muscle and any abnormal muscle movements (e.g., fasciculations, jerking), contractures, or deformities should be evaluated. Strength is measured both proximally and distally in the upper and lower extremities. Muscle tone is best assessed by passive movement. Hypertonicity usually is seen in lesions rostral to the anterior horn cells. Hypotonicity is seen in diseases affecting the neuroaxis below this level.

Sensory Examination

The sensory examination should be kept simple and targeted based on the information obtained from the patient's history. Knowledge of the difference between the skin areas innervated by dermatomes (i.e., specific segments of the cord, roots, or dorsal root ganglion) and the corresponding peripheral nerve cutaneous sensory distributions, as well as changes in motor function and reflexes clinically, defines a nerve root lesion from a peripheral nerve abnormality.

Deep Tendon Reflexes

Deep tendon reflexes are muscle-stretch reflexes that are mediated through neuromuscular spindles. Unilateral absence of a deep tendon reflex suggests pathology at the level of a peripheral nerve or a nerve root. Diffuse reduction or absence of deep tendon reflexes suggests a more generalized process affecting peripheral nerves.

Gait

Testing a patient's gait evaluates the complex interaction between muscles, bones, tendons, joints, and the nervous system. Patients should be asked to walk both with their eyes opened and with their eyes closed.

▶ ONGOING PAIN ASSESSMENTS

Following the initial assessment and the determination of the cause of the pain, a pain management plan is developed in collaboration with the patient. The effectiveness of the pain management plan is evaluated through repeated and ongoing assessments of the patient's pain. Pain assessments should be done and documented at *regular* intervals following initiation of the treatment plan. Clinicians should develop a system for the ongoing assessment of pain within their health care setting.

In addition to an evaluation of pain intensity, usually done using unidimensional measures of pain intensity, clinicians may choose to evaluate the effectiveness of the pain management plan by evaluating the patient's level of physical functioning (e.g., activity level, ability to sleep, ability to return to work) and psychosocial functioning (e.g., improvement in mood, improvement in family interactions). The establishment of specific goals with the patient that can be used to judge the effectiveness of the pain management plan may be incorporated into all the ongoing pain assessments.

▶ PSYCHOMETRIC MEASUREMENT THEORY AND PAIN MEASUREMENT

Principles of Psychometric Measurement Theory

The basis for pain measurement scales is the assignment of numbers to objects or events to represent quantities for various pain attributes.[21] The measurement of an individual's self-report of pain may be qualitative or quantitative. Classification and scaling of attributes are common forms of qualitative and quantitative measurement.

Qualitative measures are associated with the classification of pain attributes with measurements made at the nominal level. Qualitative measures address the nature of the pain, including descriptions of the sensory, affective, evaluative, and/or temporal dimensions of the pain experience. The word lists on the MPQ[8] or the Adolescent Pediatric Pain Tool[22] are examples of pain measurement tools that contain lists of words that are classified in order to describe the quality of the patient's pain.

Scales are associated with the quantitative measurement of pain. Quantitative pain measures address the intensity of pain. The Visual Analog Scale (VAS), numeric rating scale, and word graphic rating scale are examples of pain measurement scales. Commonly, scales use hierarchical rules representing nominal, ordinal, interval, or ratio approaches to assign numbers to objects or persons that represent the amount or kind of attribute. Compared with the other three scales, the ratio scale offers the greatest range for statistical analysis and generalizability.

Precision and accuracy reflect a measurement instrument's psychometric rigor to control for random and for systematic errors in order to promote the generalizability of the instrument's findings. Both precision and accuracy are critical to the standardization of pain measures and the development of instruments. More specifically, *precision* refers to the measurement instrument's reliability or ability to generate reproducible data by controlling for error.[21,23] An instrument is accurate or valid when it measures what it purports to measure.[21]

Reliability

The psychometric rules for ensuring reliability include test-retest as well as internal and interrater estimates of reliability. Based on one's conceptualization of pain, researchers disagree about whether or not a pain measure can be tested for reliability.

Beyer and Knapp[24] and Erickson[25] argue that for self-report measures of pain, none of the standard methods for assessing reliability is appropriate. First, the view that pain is a state, not a trait, that varies over time makes test-retest reliability impossible. Although the reliability of a subject's memory or recall of past pain intensity and pain affect may be argued as a method to establish a pain measurement instrument's test-retest reliability, there is only minimal evidence supporting the ability of individuals to accurately recall past procedural pain intensity and affect.[26]

Second, Beyer and Knapp[24] and Erickson[25] question one's ability to test the precision of pain instruments using either internal consistency or interrater reliability rules. It is obviously inappropriate to consider reliability testing for internal consistency when measuring only one pain dimension such as pain intensity. In addition, if interrater reliability is viewed as reliability between two outside observers, then interrater reliability is only appropriate for behavioral measures of pain, not self-report measures.[24,25]

These psychometric measurement perspectives are not adhered to by all pain researchers. Other pain researchers[27,28] have reported reliability data for both qualitative and quantitative pain measures. For example, test-retest reliability has been reported for adolescents who demonstrated consistent sorting and resorting of pain words and for adolescents who were experiencing acute pain and who reported consistent ratings of pain intensity between five different pain-intensity measures mixed with one duplicated pain-intensity scale.[27,28] Although some researchers argue that determination of the reliability of pain measures is not possible, others have reported success through nontraditional reliability estimate approaches.

Validity

Unlike the controversy surrounding the establishment of reliability, testing a proposed pain measurement tool for validity is expected.[24] Although there are different ways to measure validity, three major categories provide evidence of validity: content validity, criterion or predictive validity, and construct-related validity. First, *content validity* refers to the conceptual or theoretical adequacy of an instrument as determined by experts.[29,30] The content validity of a pain measurement tool may be achieved by having experts participate in development of the instrument.

Second, an investigation of *criterion or predictive validity* requires that the external criterion against which the instrument is validated must correlate highly with the new instrument.[21] The merit of criterion-related validity depends on the quality of the external criteria chosen. One problem is to identify an external pain criterion that is more reliable and valid than the pain instrument that is being developed.

Concurrent *construct-related validity* of a pain measure involves comparisons between the individual's self-reports of pain and information provided by another informant. Construct-related validity may include comparing an individual's self-reports of pain intensity with health care providers' or family members' judgments of the individual's pain intensity or correlating an individual's self-reported pain intensity with a researcher's observations of the individual's distress behaviors.[31,32] However, due to the complex subjective nature of pain, the use of family members' and health care providers' ratings or ratings of behavioral distress as valid criteria for pain is questionable. Findings from several studies[31-33] suggest that behavioral distress does not necessarily correlate with perceived pain intensity and that inconsistencies in perceived pain intensities exist between an individual's self-report of pain and family members' and health care providers' projections of an individual's pain intensity.

▶ MULTIDIMENSIONAL PAIN ASSESSMENT MEASURES

Multidimensional pain assessment measures are used to assess two or more dimensions of the pain experience. Many of the tools listed in Table 14-6 were developed by psychologists, nurses, and physicians to assess various dimensions of the pain experience in a variety of patient populations. Most of these tools have undergone extensive testing and have excellent validity and reliability.

Adolescent Pediatric Pain Tool

Savedra and colleagues[22] developed the Adolescent Pediatric Pain Tool (APPT) for use with children ages 8 to 17 years. The APPT is modeled after the McGill Pain

▶ **TABLE 14-6.** EXAMPLES OF MULTIDIMENSIONAL PAIN ASSESSMENT MEASURES

Pain Assessment Tool	Clinical Population	References
Acute Pain Management—Initial Pain Assessment Tool	Acute pain	McCaffery & Beebe, 1989; Acute Pain Management Guideline Panel, 1992
Adolescent Pediatric Pain Tool	Acute pain in children and adolescents between 8 and 17 years of age	Savedra, Holzemer, Tesler, & Wilkie, 1993
Brief Pain Inventory	Cancer pain and pain caused by other diseases	Daut, Cleeland, & Flannery, 1983
McGill Pain Questionnaire—long and short forms	All types of pain	Melzack, 1975; Melzack, 1987
Memorial Pain Assessment Card	Cancer pain	Fishman, Pasternak, Wallenstein, Houde, Holland, & Foley, 1987

Questionnaire (MPQ) used with adults. The instrument includes a body outline to measure pain location, a word graphic rating scale to measure pain intensity, and a pain descriptor word list to describe the quality of the pain.

The three components of the tool were tested separately for reliability and validity, and the utility of the combined elements of the tool was tested in the measurement of postoperative pain in children and adolescents.[34,35] All components of the tool can detect changes in pain over time that are consistent with the course of postoperative recovery. Data from these studies support the use of the APPT in the measurement of postoperative pain in children between the ages of 8 and 17 years. However, use of a multidimensional tool to evaluate the intensity of pain on an ongoing basis may be too cumbersome and require too much time when obtaining multiple measures in close succession. The usefulness of the APPT in the initial assessment of chronic pain problems in children and adolescents remains to be elucidated.

Brief Pain Inventory

The Wisconsin Brief Pain Inventory (BPI)[36] is a multidimensional instrument that was modeled after the MPQ. The instrument was constructed to measure pain caused by cancer and other diseases (e.g., rheumatoid arthritis, chronic orthopedic problems). The BPI contains items for assessing pain history, etiology, intensity, location, quality, and interference with activities.

In a study of 1308 oncology outpatients who were treated in institutions that participated in the Eastern Cooperative Oncology Group[37] and completed the BPI, the data were subjected to psychometric analyses. The psychometric evaluation revealed two factors on the BPI—pain severity and pain interference. Internal consistency reliability of the severity and interference subscales of the BPI revealed alpha coefficients of 0.88 and 0.92, respectively.

Analysis of the BPI with several readability formulas indicates that its reading level ranges from 6.0 to 7.6.[38] The BPI is short, easy to use, designed for patients to complete, and easy to score. The BPI also comes in a short form that was recommended as a comprehensive cancer pain assessment tool in the Agency for Health Care Policy and Research (AHCPR) Clinical Practice Guideline: Management of Cancer Pain.[39]

McGill Pain Questionnaire: Long and Short Forms

Long Form of the McGill Pain Questionnaire

The long form of the MPQ was developed by Melzack and Torgerson in 1971.[8,40] The instrument consists of four parts: a body outline on which location(s) of pain are marked; 20 groups of word descriptors (each group contains two to six words) that are used to evaluate sensory, affective, cognitive, and miscellaneous aspects of the pain experience; pattern of pain (brief, transient, intermittent, continuous); and the present pain intensity index [PPI; ranging from 0 (no pain) to 5 (excruciating pain)].

The long form of the MPQ is scored to provide three pain indices: (1) Total Pain Rating Index (PRI-T), which is the sum of the rank value of the words chosen from the list, (2) the number of words chosen (NWC), and (3) the Present Pain Intensity (PPI). In addition, the 20 groups of words can be subdivided to yield subindexes of the PRI (i.e., PRI-sensory, groups 1–10; PRI-affective, groups 11–15; PRI-evaluative, group 16; and PRI-miscellaneous, groups 17–20).

The long form of the MPQ has been translated into Dutch,[41] German,[42] and Italian.[43] The validity and reliability of the long form of the MPQ were demonstrated in a variety of clinical pain problems (e.g., cancer pain, low back pain, dysmenorrhea, arthritis, and headache). The long form of the MPQ has the ability to discriminate between different types of acute and chronic pain, is useful in clinical practice and research. The major disadvantages of the long form of the MPQ are that it can take up to 30 minutes to complete, and some of the word descriptors may be difficult to understand.

Short Form of the McGill Pain Questionnaire

The short form of the MPQ was published in 1987[44] and consists of two sections. Section 1 contains a list of 11 sensory words (*throbbing, shooting, stabbing, sharp, cramping, gnawing, hot-burning, aching, heavy, tender, splitting*) and four affective words (*tiring/exhausting, sickening, fearful, punishing/cruel*). Three scores are derived from summing the intensity-rank values of all selected words (i.e., a sensory score, an affective score, and a total score). The second section of the short form of the MPQ contains two measures of pain intensity, namely, the Present Pain Intensity Index (PPI) from the original MPQ and a 10-cm visual analogue scale (VAS) with anchors of "no pain" and "worst possible pain."

The short form of the MPQ was tested in a variety of patient populations, including postoperative, obstetrical, and musculoskeletal patients;[44] patients with chronic pain problems;[45] and oncology patients.[46] The instrument takes approximately 2 to 5 minutes to complete. The short form of the MPQ has established validity and reliability and is a useful tool in both clinical and research settings.

Memorial Pain Assessment Card

The Memorial Pain Assessment Card (MPAC) was developed by Fishman and colleagues[47] to evaluate pain in an oncology patient population. The tool consists of an 8.5 × 11 inch card that is folded in the middle with four

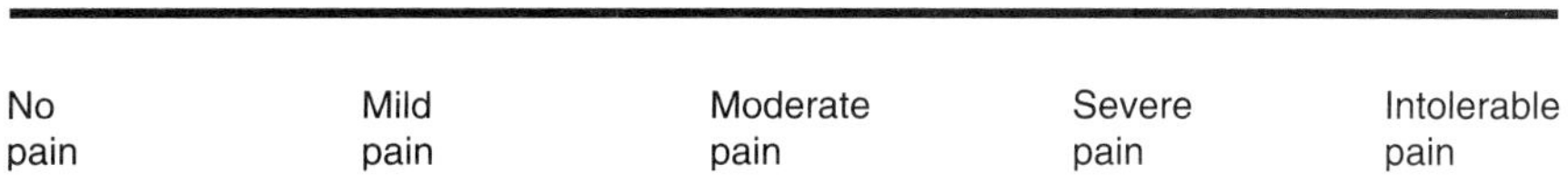

| No | Mild | Moderate | Severe | Intolerable |
| pain | pain | pain | pain | pain |

Figure 14-2. Word graphic rating scale.

measures (i.e., a VAS measuring pain intensity, a VAS measuring pain relief, a VAS measuring mood, and an adaptation of Tursky's pain adjective rating scale). The validity and reliability of the MPAC were established with 50 hospitalized oncology patients. It is a simple and easy tool to administer and can be used in a variety of clinical and research settings.

▶ UNIDIMENSIONAL PAIN ASSESSMENT MEASURES

Unidimensional pain assessment measures are used to assess a single dimension of the pain experience. Table 14-7 lists some examples of unidimensional pain assessment measures.

Behavioral Rating Scales

The best-developed and most extensively researched method for assessing pain behaviors in adults is the behavioral observation method developed by Keefe and Block[48] for the evaluation of patients with low back pain. The behavioral observation method consists of videotaping patients while they perform a standard set of tasks (lying down, sitting, standing, walking) that frequently effect pain behaviors. Five operationally-defined categories of pain behaviors are scored using a standardized coding system (i.e., guarding, bracing, rubbing, grimacing, and sighing). Keefe and Block demonstrated that the behavioral observation technique (1) can be scored reliably, (2) is sensitive to changes associated with treatment, (3) has construct validity, in that ratings of pain intensity made by untrained observers were highly correlated with the level of pain behaviors, and (4) has discriminative validity, in that pain patients demonstrated higher levels of pain behaviors than nonpain patients and depressed controls.[48] Similar systems, with data supporting their reliability and validity, were developed for patients with rheumatoid arthritis[49] and nonchronic back pain.[50] This approach for evaluating pain behaviors may be most useful in research settings because

▶ **TABLE 14-7.** EXAMPLES OF UNIDIMENSIONAL PAIN ASSESSMENT MEASURES

- Behavioral rating scales
- Graphic rating scales
- Verbal descriptor scales
- Visual analogue scales

of the need for extensive training in the scoring systems and the need for video equipment that may be cumbersome.

Graphic Rating Scales

The graphic rating scale (GRS) is composed of words presented on a continuum of increasing value[51] (Fig. 14-2). It is a version of a VAS. This measure is used to evaluate a single dimension of the pain experience, namely, one aspect of the sensory dimension (i.e., pain intensity). The GRS is relatively easy to use for research purposes. However, for clinical purposes, the GRS requires the use of paper and pencil to administer.

Verbal Descriptor Scales

Verbal descriptor scales (VDSs) are used to evaluate a major component of the sensory dimension of pain, namely, pain intensity. Typically, a VDS consists of a list of three to five numerically ranked descriptors: (0) none, (1) mild, (2) moderate, (3) severe, and (4) excruciating. The VDSs are brief, easy to administer, easy to score, and can be used with a variety of patient populations that are experiencing acute or chronic pain. Pain assessments done using the VDS appear to be valid and reliable. However, one of the major criticisms of the VDS is that the words chosen for a VDS may artificially categorize the intensity of pain and may not divide the perceptual continuum of pain into equal segments. Therefore, the VDS may not have sufficient sensitivity to measure pain for research purposes.[30,52]

Visual Analogue Scale

A visual analogue scale (VAS) measures pain intensity using a line that is typically 10 cm long. The line can be positioned horizontally or vertically and is anchored on each side with extreme descriptors (e.g., "no pain" and "pain as bad as it could possibly be") (Fig. 14-3). Patients are asked to place a mark through the line at the point that best describes how much pain they are experiencing at a given point in time.

A VAS is considered a more sensitive measure of pain intensity than a VDS because this scale causes individuals to conceptualize pain on a straight-line continuum rather than as categorical responses. The VAS has been used to measure pain intensity in a variety of patient populations, including patients with postoperative pain, cancer pain, labor pain, burn pain, and chronic pain. The VAS is used often in research studies to evaluate the effectiveness of pharmacologic agents.

No pain   Worst pain imaginable

Figure 14-3. Visual analogue scale.

▶ **TABLE 14-8.** DISADVANTAGES OF THE VISUAL ANALOGUE SCALE

- Patients sometimes have difficulty conceptualizing pain intensity along a continuum
- Patients may have difficulty making a mark on the line (e.g., patients with an intravenous line)
- It is unclear whether patients prefer horizontal or vertical formats for the visual analogue scale

For research purposes, the VAS is easy to administer and to score. However, several disadvantages of the VAS are listed in Table 14-8. Patients need to be taught how to use the tool prior to an evaluation of pain intensity. The VAS is not useful in clinical practice to assess pain intensity on a routine basis because it requires the use of paper and pencil, which is not convenient for either clinicians or patients.

▶ **SUMMARY**

While numerous unidimensional and multidimensional pain assessment tools are available for use in clinical practice and research, pain remains a subjective experience. Accurate assessments must rely on patients' self-reports of pain. A careful initial pain assessment must be done to determine the cause of the patient's pain. Following the development of a pain management plan, ongoing assessments are done to evaluate the efficacy of the treatment plan.

REFERENCES

1. Berry PH, Dahl JL: The new JCAHO pain standards: Implications for pain management nurses. *Pain Manag Nurs* 2000;1:3–12.
2. Berry PH: Getting ready for the JCAHO: Just meeting the standards or really improving pain management. *Clin J Oncol Nurs* 2001;5:110–112.
3. International Association for the Study of Pain: Pain terms: A current list with definitions and notes on usage. *Pain* 1986;3:S216–S221.
4. Ahles TA, Blanchard EB, Ruckdeschel JC: The multidimensional nature of cancer-related pain. *Pain* 1983;17:277–288.
5. McGuire DB: The multidimensional phenomenon of cancer pain, in McGuire DB, Yarbro CH (eds), *Cancer Pain Management*. Philadelphia: Saunders, 1987:1–20.
6. Merskey H: Classification of chronic pain: Description of chronic pain syndromes and definitions of pain terms. *Pain* 1986;3:1–226.
7. Ahles TA, Martin JB: Cancer pain: A multidimensional perspective, in Turk DC, Feldman CS (eds), *Noninvasive Approaches to Pain Management in the Terminally Ill*. New York: Hawthorn Press, 1992:25–48.
8. Melzack R: The McGill Pain Questionnaire: Major properties and scoring methods. *Pain* 1975;1:277–299.
9. Glover J, Miaskowski C, Dibble S, Dodd MJ: Mood states of oncology outpatients: Does pain make a difference? *J Pain Symptom Manage* 1995;10:120–128.
10. McGuire DB: The multiple dimensions of cancer pain: A framework for assessment and management, in McGuire DB, Yarbro CH, Ferrell BR (eds), *Cancer Pain Management*, 2d ed. Boston: Jones and Bartlett, 1995:1–17.
11. Devine EC: Meta-analysis of the effect of psychoeducational interventions on pain in adults with cancer. *Oncol Nurs Forum* 2003;30:75–89.
12. Yeager KA, Miaskowski C, Dibble SL, Wallhagen M: Differences in pain knowledge and perception of the pain experience between outpatients with cancer and their family caregivers. *Oncol Nurs Forum* 1995;22:1235–1241.
13. Fishbain DA, Cutler RB, Rosomoff HL, Rosomoff RS: Are opioid-dependent/tolerant patients impaired in driving-related skills? A structured evidence-based review. *J Pain Symptom Manage* 2003;25:557–577.
14. Tassain V, Attal N, Fletcher D, et al: Long-term effects of oral sustained release morphine on neuropsychological performance in patients with chronic non-cancer pain. *Pain* 2003;104(1–2):389–400.
15. Wilkie D, Lovejoy N, Dodd M, Tesler M: Cancer pain control behaviors: Description and correlation and pain intensity. *Oncol Nurs Forum* 1988;15:723–731.
16. Wilkie D, Keefe FJ, Dodd MJ, Copp LA: Behavior of patients with lung cancer: Description and associations with oncologic and pain variables. *Pain* 1992;51:221–240.
17. Hwang SS, Chang VT, Kasimis B: Cancer breakthrough pain characteristics and responses to treatment in a VA medical center. *Pain* 2003;101(1–2):55–64.
18. Mercadante S, Radbruch L, Caraceni A, et al: Episodic (breakthrough) pain: Consensus conference of an expert working group of the European Association for Palliative Care. *Cancer* 2002;94:832–839.
19. Portenoy RK: The physical examination in cancer pain assessment. *Semin Oncol Nurs* 1997;13:25–29.
20. Donohoe CD: Targeted history and physical examination, in Waldman SD, Winnie AP (eds), *Interventional Pain Management*. Philadelphia: Saunders, 1996:73–84.
21. Nunnally J, Bernstein I: *Psychometric Theory*, 3d ed. New York: McGraw-Hill, 1994.
22. Savedra MC, Holzemer WL, Tesler MD, Wilkie DJ: Assessment of postoperative pain in children and adolescents using the Adolescent Pediatric Pain Tool. *Nurs Res* 1993;42:5–9.

23. McLaughlin F, Marasculio L: *Advanced Nursing and Health Care Research: Quantitative Approaches*. Philadelphia: Saunders, 1990.
24. Beyer J, Knapp T: Methodological issues in the measurement of children's pain. *Child Health Care* 1986;14:233–241.
25. Erickson C: Pain measurement in children: Problems and directions. *Dev Behav Pediatr* 1990;11:135–137.
26. Lander J, Hodgins M, Fowler-Kerry S: Children's pain predictions and memories. *Behav Res Ther* 1992;30:117–124.
27. Tesler M, Savedra M, Holzemer W, et al: The word-graphic rating scale as a measure of children's and adolescent's pain intensity. *Res Nurs Health* 1991;14:361–371.
28. Wilkie D, Holzemer W, Tesler M, et al: Measuring pain quality: Validity and reliability of children's and adolescents' pain language. *Pain* 1990;41:151–159.
29. Lynn M: Determination and quantification of content validity. *Nurs Res* 1985;35:382–385.
30. Waltz C, Strickland O, Lenz E: *Measurement in Nursing Research*. Philadelphia: Davis, 1984.
31. Horgas AL, Dunn K: Pain in nursing home residents: Comparisons of residents' self-report and nursing assistants' perceptions. Incongruities exist in resident and caregiver reports of pain; therefore, pain management education is needed to prevent suffering. *J Gerontol Nurs* 2001;27:44–53.
32. Weiner D, Peterson B, Keefe F: Chronic pain-associated behaviors in the nursing home: Resident versus caregiver perceptions. *Pain* 1999;80:577–588.
33. Miaskowski C, Zimmer EF, Barrett KM, et al: Differences in patients' and family caregivers' perceptions of the pain experience influence patient and caregiver outcomes. *Pain* 1997;72:217–226.
34. Savedra MC, Tesler MD, Holzemer WL, et al: Testing a tool to assess postoperative pediatric and adolescent pain. *Adv Pain Res Ther* 1990;15:85–93.
35. Savedra MC, Tesler WL, Wilkie DJ, Ward JA: Pain location: Validity and reliability of body outline markings by hospitalized children and adolescents. *Res Nurs Health* 1989;12:307–314.
36. Daut RL, Cleeland CS, Flannery RC: Development of the Wisconsin Brief Pain Questionnaire to assess pain in cancer and other diseases. *Pain* 1983;17:197–210.
37. Cleeland CS, Gonin R, Hatfield AK, et al: Pain and its treatment in outpatients with metastatic cancer. *N Engl J Med* 1994;330:592–596.
38. McGuire DB: Measuring pain, in Frank-Stromborg M, Olsen SJ (eds), *Instruments for Clinical Health-Care Research*, 2d ed. Boston: Jones and Bartlett, 1997:528–564.
39. Jacox A, Carr DB, Payne R, et al: *Management of Cancer Pain*. Clinical Practice Guideline No. 9, AHCPR Publication No. 94-0592. Rockville, MD: Agency for Health Care Policy and Research, U.S. Department of Health and Human Services, Public Health Service, 1994.
40. Melzack R, Torgerson W: On the language of pain. *Anesthesiology* 1971;34:50–59.
41. Vanderiet K, Adriaensen H, Carton H, Vertommen H: The McGill Pain Questionnaire constructed for the Dutch language (MPQ-DV): Preliminary data concerning reliability and validity. *Pain* 1987;30:395–408.
42. Kiss I, Muller H, Abel M: The McGill Pain Questionnaire, German version. *Pain* 1987;29:195–207.
43. Maiani G, Sanavio E: Semantics of pain in Italy: The Italian version of the McGill Pain Questionnaire. *Pain* 1985;4:399–405.
44. Melzack R: The short-form McGill Pain Questionnaire. *Pain* 1987;30:191–197.
45. Swanston M, Abraham C, Nacrae WA, et al: Pain assessment with interactive computer animation. *Pain* 1993;53:347–351.
46. McGuire DB, Altomonte V, Peterson DE, et al: Patterns of mucositis and pain in patients receiving preparative chemotherapy and bone marrow transplantation. *Oncol Nur Forum* 1993;20:1493–1502.
47. Fishman B, Pasternak S, Wallenstein SL, et al: The Memorial Pain Assessment Card: A valid instrument for the evaluation of cancer pain. *Cancer* 1987;60:1151–1158.
48. Keefe FJ, Block AR: Development of an observation method for assessing pain behavior in chronic low back pain patients. *Behav Ther* 1982;13:363–375.
49. McDaniel LK, Andersen KO, Bradley LA, et al: Development of an observation method for assessing pain behavior in rheumatoid arthritis patients. *Pain* 1986;24:165–184.
50. Jensen IB, Bradley A, Linton SJ: Validation of an observation method of pain assessment in nonchronic back pain. *Pain* 1989;39:267–274.
51. Huskisson EC: Measurement of pain. *Lancet* 1974;2:1127–1131.
52. Heft MW, Parker SR: An experimental basis for revising the graphic rating scale for pain. *Pain* 1984;19:153–161.

CHAPTER 15

Comprehensive Evaluation of the Patient with Chronic Pain

E. Daniela Hord, Jennifer A. Haythornthwaite, and Srinivasa N. Raja

The International Association for the Study of Pain has defined pain as an unpleasant sensory and emotional experience associated with actual or potential tissue damage or described in terms of such damage. Pain can be described as acute or chronic based on the duration of the symptoms, and as nociceptive or neuropathic depending on the etiology.

Nociceptive pain results from stimuli that have the potential to cause tissue damage or may follow tissue injury and inflammation. Neuropathic pain is caused by a lesion, injury, or dysfunction of the peripheral or central nervous system. Unlike nociceptive pain, which is secondary to activation of nociceptors by an inflammatory, infectious, or infiltrative process, neuropathic pain occurs in the absence of such activation and is due to maladaptive changes of the pain pathways at the level of peripheral and/or central nervous systems. Neuropathic pain states are often chronic and refractory to treatment.

Neuropathic pain, which does not serve any known biologic function, is quite common, although precise estimates of the prevalence are not available.[1]

Chronic pain has a negative impact on many aspects of a patient's daily life,[2] including deterioration in quality of life and the development of psychological distress, psychiatric disorders, and strained interpersonal relationships. Health-related quality of life is the patient's subjective assessment of his or her current condition.[3] Alterations in quality of life may result from deterioration in physical function (impediments to leisure and daily activities) and sleep disturbances, and may be associated with a greater incidence of other somatic complaints. Psychological morbidity may manifest as pathologic anxiety, anger, depression, and loss of self-esteem. The higher the intensity and frequency of pain, the lower is the self-reported quality of life, especially psychological functioning, physical status, functional status, and family life.[4]

▶ MECHANISMS OF CHRONIC PAIN

Chronic pain is the result of plastic changes in the peripheral and/or central nervous systems induced by injury or dysfunction in the nervous system. A number of mechanisms are involved in the pathogenesis of pain, for example, peripheral and central sensitization.

Peripheral Sensitization

Tissue damage caused by injury, disease, or inflammation releases endogenous algogenic chemicals, including protons, K$^+$, serotonin, histamine, prostaglandins, bradykinin, and substance P. Specialized receptors, called *nociceptors*, in the peripheral nervous system detect and filter the intensity and type of noxious stimuli.

The primary afferent nociceptors can be activated either directly by a noxious stimulus or indirectly by the chemical soup released at sites of tissue injury and inflammation. Unlike other types of cutaneous receptors, the nociceptors respond to multiple stimulus modalities, including mechanical, heat, cold, and chemical stimuli. In peripheral sensitization, after prolonged and/or repeated noxious stimuli, the various sites of injury develop ectopic and evoked discharges along the length of the neuron, including the proximal stump of the injured nerve, focal areas of axonal demyelination, and the cell body in the dorsal root ganglion (DRG). Thus, the nociceptors exhibit lower thresholds to noxious stimuli, increased responses to suprathreshold stimuli, and spontaneous activity. There is also chemical sensitization of damaged as well as intact nerves following nerve injury, thus providing a possible explanation for the increased adrenergic sensitivity.

Central Sensitization

Central sensitization can occur at the cortical level or at the level of the spinal cord. The primary and secondary somatosensory cortex has a role in integration of somatic pain and in discriminatory aspect and localization of pain, whereas the limbic system and especially the cingulate cortex are involved in psychological expression of behavior, in emotion, and in integration of the affective component of pain. The anterior insula seems to have a role in the memory components of pain, and the supplemental motor area is thought to be involved in the integration of the motor response to pain. At the level of the spinal cord, central sensitization could occur with prolonged or repeated noxious stimuli, which can lower the threshold of dorsal horn neurons.[5] Sensitization of the dorsal horn neurons manifests as a reduction in the threshold, an increase in the responsiveness, and an expansion of the extent and recruitment of novel inputs to receptive fields,[6] and can be prevented or reversed by NMDA antagonists. There are also plastic changes in the central nervous system. In central sensitization, the release of L-glutamate and activation of glutamate receptors AMPA occur at low frequencies, with subsequent rapid activation of tetrodotoxin-dependent sodium currents, whereas NMDA receptors occur at high frequencies, with a subsequent slow increase in calcium currents. Sustained chronic pain is demonstrated to induce a special type of sodium current called tetrodotoxin-resistant sodium currents that fire independently, whereas NMDA agonists have been proven to induce excitotoxicity by overcoming the protective magnesium-induced blockade, thus allowing excessive calcium to enter the cell. Inactivation occurs by reuptake of glutamate at presynaptic terminals and by metabolization of the glutamate to γ-aminobutyric acid (GABA) with glutamic acid decarboxylase as required enzyme. GABA-A increases chloride conductance, and GABA-B increases potassium conductance, both leading to hyperpolarization of the neuron. New basic science theories suggest that one of the final pathways in the mechanism of central pain is an increase in the voltage-sensitive calcium currents. The pathophysiologic similarity between the well-demonstrated long-term potentiation of hippocampal neurons thought to be related to memory and nociceptive central sensitization suggests that their mechanisms may be similar.

▶ CLINICAL ASSESSMENT

Pain assessment is a comprehensive and complex process that is aimed primarily at establishing a specific etiology and pathophysiology for the pain and evaluating the patient's functional status such that a targeted therapy can be planned. In addition, the initial assessment should help to develop a treatment algorithm, establish objective criteria to follow the course of progression of the neuropathic pain state, and evaluate the success of the treatment strategy. The organization of the clinical evaluation of the patient with neuropathic pain should include a comprehensive medical and pain history, assessment of the multidimensional aspects of pain with use of questionnaires, a general clinical examination, and a focused neurological examination consistent with the pain problem. Finally, psychological, behavioral, and social factors should be assessed to identify comorbid conditions that may be associated with or contribute to the pain state.

History Taking

An important step in pain history taking is the clinician's acknowledgment that the patient experiences pain. Acceptance of the patient's pain reports as genuine is critical in developing a therapeutic doctor-patient relationship.

Similar to the assessment of all medical conditions, a thorough history is critical to evaluation of the patient suspected to have neuropathic pain. It should include all the elements of a complete review of the chief complaints and a review of systems. Constructing a questionnaire that can be sent to the patient prior to his or her clinic visit may facilitate the process of obtaining a thorough history. In addition to general demographic information, such as age, sex, birth date, address, and telephone numbers, information on the patient's education level, occupation, and current employment status should be obtained. To assess the patient's support system, information on marital status and relationships, ethnic origin, religious beliefs, and family environment should be obtained as well.

A key element in the history during evaluation of a patient with neuropathic pain is a detailed pain history.

Pain Characteristics

Pain characteristics obtained by a good history include the temporal occurrence, location, intensity, quality, and modifying factors of pain. *Temporal occurrence* implies the type of onset, duration, and fluctuation of pain characteristics since onset. It is important to find out if pain is acute, recurrent, or chronic. *Location* can be evaluated by asking the patient to mark appropriate parts of the body on a full-body drawing and by having the patient point to the painful area during the consultation. Radiation of pain always should be described. One has to remember that the value of localizing pain diminishes from the moment the pain ceases. *Intensity* can be assessed by using visual analogue or verbal numeric pain scales. Descriptors such as *mild, moderate, severe, very severe,* and *excruciating* also can be used. The McGill Pain Questionnaire (MPQ) is another modality commonly used to assess pain intensity with the use of anchor words. It is important to find out the degree of fluctuation by asking the patient to rate the average intensity of pain, as well as the worst and best ratings. Another way to determine the fluctuations in pain ratings is to have the patient maintain a pain diary for a week with ratings of pain several times during the day along with notes on activities performed during the day.

Quality of pain may suggest a superficial lesion if the pain is sharp, burning, and well localized or a deep lesion if the pain is dull, diffuse, and poorly localized. Burning pain is the most common quality associated with neuropathic pain; however, a number of descriptors, e.g., *aching, throbbing, stabbing,* and *sharp,* also are described in neuropathic pain syndromes. However, patients who have other pain syndromes, such as myofascial pain, at times also may use the same or similar descriptors for their pain.

Modifying factors could be relieving or exacerbating factors. In particular, the influence of day-to-day physical activities on the pain should be assessed. The presence of *associated symptoms,* such as bowel or bladder dysfunction, alterations in skin temperature or sweating in the affected region, color changes, edema, numbness, or paresthesias, also should be ascertained.

Pain Measurement

Pain measurement is an important step in the clinical evaluation of patients with pain, in research activity, and in medicolegal situations. The physician evaluating a patient with pain has to follow the results of pain treatment by quantifying pain. Pain is subjective, and patient self-report is the "gold standard" for assessment. Therefore, it is crucial to pay attention to the words used by patients to describe pain because those words can give clues to the underlying mechanism. It is difficult but important to differentiate between pain measurement and pain complaint. As a concept, one also has to distinguish between impairment and disability. Impairment is a medical concept, whereas disability is a legal or social concept, and disability may result from impairment.

There are limitations and biases in pain measurement because of several factors, such as the possible influence of cultural background, age, or gender on pain report; the influence of mood or affect on pain report; the potential influence of home environment versus clinic or hospital setting on pain perception; and patient, family, or physician expectations.

The basic approaches in measuring pain are the verbal rating scale (none, mild, moderate, severe, excruciating), the numeric rating scale (11-point numeric scale where 0 is no pain and 10 is the worst pain imaginable), and the visual analogue scale (VAS) (a 10-cm line anchored at one end by the descriptor "no pain" or "least possible pain" and anchored at the other end by "worst possible pain"). The intensity on a VAS should include worst, least, average pain, and pain now.[7] In the elderly, in children, and in neonates, there are specific issues and different methods for pain evaluation.

The specific scale used to measure pain intensity is less important than ensuring its consistent use over time.

Diagnostic and Treatment History

A list of previous laboratory, radiologic, and neurodiagnostic tests and their results should be obtained and reviewed. A complete history of prior treatments—pharmacologic, nonpharmacologic, complementary and alternative therapies, diagnostic and therapeutic nerve blocks, and surgeries—should be obtained to help in the formulation of a management strategy.

Substance Abuse History

A history of the use/abuse of alcohol, tobacco, and illicit drugs should be obtained. This is particularly important if the use of opioids is considered in the management of the patient.

Questionnaires

There are more complex tools in the multidimensional pain evaluation, such as the MPQ, the Brief Pain Inventory, and Neuropathic Pain Scale.

- The MPQ assesses a variety of pain qualities (78 descriptors classified into 20 categories). The questionnaire can be scored to obtain global measures of the sensory and affective aspects of pain. It is also useful for the assessment of pain descriptors.

- The Brief Pain Inventory quantifies subjective intensity of pain (pain worst, pain least, pain average, and pain right now) and the effect of pain on patient function (general activity, mood, ability to walk, normal work, socializing with others, enjoyment of life, and sleep). It is short and can be used for follow-up assessments.
- The Neuropathic Pain Scale is designed to assess distinct pain qualities associated with neuropathic pain.[8] The scale itself includes two items that assess the global dimensions of pain intensity and pain unpleasantness and includes eight items that assess eight specific qualities of neuropathic pain. This scale also can be used for follow-up assessments.
- The Leeds Assessment of Neuropathic Symptoms and Signs (LANSS) pain scale is based on analysis of sensory description and bedside examination of sensory dysfunction and provides immediate information in clinical settings[9] It was published recently and requires validation over time.

Clinical Examination

A traditional general examination should be performed in all patients evaluated for neuropathic pain. The most common etiologies of neuropathic pain are shown in Table 15-1. Any skin color changes, rashes, swelling, or temperature abnormalities should be documented. A particular evaluation of the musculoskeletal system should be done, including the status of joints, muscles, and ligaments, as well as the presence of swelling, laxity, and/or tenderness. Antalgic postures should be documented, if present.

A general neurological examination should be performed in all patients with chronic pain. The effect of pain on mood and affect must be part of the evaluation. Antalgic gait should be noted, if present. Depending on the etiology of the pain, careful cranial nerve examination and assessment of mental status and cognitive function may be important.

The targeted neurological examination should be governed by the anatomy of the structures in the painful region and the suspected underlying pathophysiology of the neuropathic pain. The main focus should be on sensory examination.

Sensory Examination

The traditional sensory examination, based on sharp-dull discrimination of pinprick, is not enough for the examination of patients with suspected neuropathic pain. Sensory signs can manifest as negative sensory phenomena.[10] This means that a stimulus such as light touch, pinprick, cold, warm, vibration, joint-position sensation, two-point discrimination, or sensory neglect is perceived as decreased or less. In contrast, positive sensory signs can be present, such as allodynia, hyperalgesia, and hyperpathia, where stimuli are perceived as increased in intensity or duration. The positive sensory signs are associated with positive phenomena such as summation and aftersensation. Since neuropathic pain is precipitated

▶ **TABLE 15-1.** COMMON ETIOLOGIES FOR PERIPHERAL NEUROPATHIC PAIN

Metabolic disorders	Diabetes
	Hypothyroidism
	Uremia
	Amyloidosis
Infectious	HIV infection
	Postherpetic neuralgia
Nutritional deficiencies	Alcoholic
	Thiamine deficiency
Toxins	Isoniazid (pyridoxine) deficiency
	Vincristine
	Nitrofurantoin
	Arsenic
	Thallium
	Dydeoxynosine
	Dydoxicitosyne
Malignancies	Paraneoplastic (small cell and other carcinoma, lymphoma)
	Paraproteinemia (multiple myeloma, Waldenstrom's)
Inherited	HSAN type 1
	Fabry's
Inflammatory	Acute inflammatory demyelinating polyneuropathy (Guillain-Barré)
Mechanical causes	Radiculopathy
Idiopathic	Cryptogenic sensory polyneuropathy
	Idiopathic trigeminal neuralgia

▶ **TABLE 15-2.** DEFINITION OF PAIN TERMINOLOGY

Pain Terminology	Definition
Analgesia	Absence of pain sensation
Anesthesia	Absence of all sensation
Allodynia	Abnormal perception of pain due to nonpainful stimulus
Dysesthesia	Unpleasant abnormal sensation
Hyperesthesia	Increased sensitivity to stimulus
Hypoesthesia	Decreased sensitivity to stimulus
Hyperalgesia	Exaggerated pain from a usually painful stimulus
Hypoalgesia	Decreased sensation to a painful stimulus
Hyperpathia	Exaggerated and increased pain reaction to repetitive nonpainful stimulus
Summation	Abnormally increasing painful sensation to a repeated stimulus although the actual stimulus remains constant
Aftersensation	Abnormal persistence of a sensory perception provoked by a stimulus even though the stimulus has ceased

by neuronal injury, there are some clinically distinctive pain characteristics, such as stimulation-induced paresthesias or dysesthesias, that may be associated with other neurological deficits or local autonomic dysregulation. The sensory examination should define not only the nature of the abnormality but also its pattern, particularly if it is in a dermatomal, peripheral nerve, or glove-and-stocking distribution. Such an examination will provide important clues to the possible etiology of the neuropathic pain.

An understanding of some commonly used terminology is helpful in evaluating the neuropathic pain patient (Table 15-2). *Analgesia* is absence of pain, whereas *anesthesia* is absence of all sensation. *Dysesthesia* is an unpleasant abnormal sensation. *Paresthesia* is an abnormal sensation, whereas *hyperesthesia* is increased sensitivity to stimulus. *Hypoesthesia* is decreased sensitivity to stimuli. *Allodynia* is abnormal perception of pain due to nonpainful stimulus and can be mechanical or thermal depending on the type of stimulus. Mechanical allodynia can be divided further as follows: static (gentle pressure on the skin evokes pain) and dynamic (light brush evokes pain). Thermal allodynia can be further divided into heat and cold allodynia. *Hyperalgesia* is exaggerated pain from a usually painful stimulus and can be mechanical (punctate hyperalgesia caused by punctate stimuli such as pinprick-evoked pain), thermal (cold or heat hyperalgesia), or chemical depending on the type of stimulus. *Hypoalgesia* is decreased sensation to a painful stimulus. *Hyperpathia* is an exaggerated and increased reaction to repetitive nonpainful stimulus.

Summation describes an abnormally increasing painful sensation to a repeated stimulus, although the actual stimulus remains constant. *Aftersensation* refers to the abnormal persistence of a sensory perception provoked by a stimulus even though the stimulus has ceased. Specific symptoms or signs of neuropathic pain seem to be associated with a specific mechanism.[11] A summary of symptoms and findings in neuropathic pain is outlined in Table 15-3.

Motor Examination

Motor examination should include a grading of the different muscle groups in the affected region using a 0 to 5 grading scheme:

5—Complete range of movement against gravity and with full resistance
4—Complete range of movement against gravity and with some resistance
3—Complete range of movement against gravity
2—Complete range of movement with effect of gravity eliminated
1—Evidence of mild muscle contraction but no joint movement
0—No evidence of muscle contraction

Reflexes

Deep tendon reflexes (DTRs) provide information on the integrity of specific spinal nerves. Grading is included in Table 15-4. The different nerve roots tested by the reflexes are indicated in Table 15-5.

Specific Tests

Straight-leg raising test, also known as the *Lasegue sign*, determines the mobility of the dura and dural sleeves from L4 to S2. The sensitivity of this test to diagnose lumbar disk herniation ranges between 0.6 and 0.97, with a specificity of 0.1 to 0.6. Tension on the sciatic nerve begins with 15 to 30 degrees of elevation in the supine position. This puts traction on the nerve roots from L4 to S2 and on the dura. The end of the range normally is restricted by the hamstring muscle at 60 to 120 degrees. More than 60 degrees of elevation causes movement in the sacroiliac joint and, therefore, may be painful in sacroiliac joint disorders.

▶ TABLE 15-3. SYMPTOMS AND FINDINGS IN NEUROPATHIC PAIN

Symptom/Finding	Stimulus	Clinical Presentation	Mechanism	Pharmacological Blockade
Static allodynia	Gentle mechanical pressure	In area of injury	Sensitized C nociceptors	Systemic and topical lidocaine, opioids
Dynamic allodynia	Light brush stimuli	In area of injury and outside (primary and secondary zone)	1. Central sensitization due to increased input 2. Central sensitization due to loss of input	Systemic NMDA antagonists; opioids?
Cold allodynia	Cool stimuli (acetone, alcohol)	Nerve injuries, neuropathies, and central pain	Central disinhibition because of loss of input	None?
Heat allodynia	Radiating heat	In area of injury (primary hyperalgesia)	Sensitized C nociceptors	Systemic and topical lidocaine, opioids
Punctate hyperalgesia	Pinprick stimuli	In area of injury and outside (primary and secondary zone)	Sensitized A-delta nociceptors and central sensitization	Systemic and topical lidocaine, opioids
Chemical hyperalgesia	Topical capsaicin or histamine	Evoked pain/itch or vasodilatation	Sensitized mechanosensitive VR1/histamine receptors	Topical lidocaine
Wind-up like pain	Light brush or pinprick >3Hz	Evoked pain by repetitive stimulation on and surrounding injury	Central sensitization due to increased input	Systemic NMDA antagonists and lidocaine systemically
Aftersensation	Any stimulus	Inside and outside injury zone	Central sensitization	?
Sympathetically maintained	Sympathetic stimulation or blockade	Present in nerve injuries	Sympathetic hyperactivity	Stimulation: norepinephrine Blockade: stellate block

ABBREVIATIONS: NMDA = N-methyl-D-aspartate; VR1 = vanilloid receptor subtype 1.
SOURCE: Adapted from Jensen et al, 2003.

▶ **TABLE 15-4.** GRADING OF DEEP TENDON REFLEXES

0	Absent, abnormal
1+	Diminished, may or may not be abnormal
2+	Normal
3+	Increased, may or may not be abnormal
4+	Markedly increased, abnormal. May be associated with clonus

Spurling's test requires the examiner to produce oblique extension of the neck on the affected side with axial compression to the head. By narrowing the affected foramen, the radicular pain may be reproduced in the upper extremity.

Tinel's test involves percussion over the course of a peripheral nerve. In carpal tunnel syndrome resulting from median nerve entrapment, a positive test elicits paresthesias in the first three fingers. This test also can be performed in other locations where nerve entrapment is possible.

Phalen's test involves passive flexion in the wrist for 1 minute, followed by sudden extension. If paresthesias are elicited, the test is positive for carpal tunnel syndrome.

Waddell's signs are used to assess the patient for possible symptom magnification, which sometimes may be seen in patients with compensation or litigation issues. To evaluate possible functional disorders in patients with low back pain and radiculopathy, a list of physical signs was developed by Waddell. Three or more positive signs of the five tests are considered to be significant clinically. The tests include (1) tenderness to superficial palpation that is diffuse and nonanatomic, (2) complaints of low back pain with axial loading on the head with the patient standing, (3) distraction tests with inconsistent results when the patient's attention is distracted, e.g., straight-leg raise in the supine and sitting positions, (4) nonanatomic weakness and unexplained giving way of muscle groups and nondermatomal sensory abnormalities, and (5) overreaction characterized by disproportionate verbalization, facial expressions, jumping, cringing, etc.

Observing the Patient

While conducting the clinical examination, the examiner should be observing the patient carefully for affect, pain behaviors, the emotional tone with which history is provided, and body language. In addition, marked discrepancies between reported and observed limitations of movement and ability to walk should be noted.

▶ ASSESSMENT OF PSYCHOLOGICAL, BEHAVIORAL, AND SOCIAL FACTORS

Anxiety and Depression in Chronic Pain

Pain is a significant physical stressor that may induce or exacerbate psychological distress. A significant proportion of patients who have chronic pain, regardless of its cause and origin, experience psychological symptoms in the course of their illness. In most chronic pain patients, psychological disturbances are not the primary cause of pain but are the consequence of unrelieved pain and its effects on the quality of life.[10] Psychological issues should be assessed properly and treated specifically, regardless of whether they are primary or secondary. The goals of psychological assessment include (1) determining the nature and extent of psychological disturbance prior to the initiation of therapy so that the treatment effectiveness can be judged, (2) assessing whether the patient may be appropriate for a particular treatment, and (3) helping to tailor the treatment to the patient's needs.

Some of the common psychological disturbances associated with chronic pain include depression, anxiety, sleep disturbances, and decreased sexual activity. These factors can lead to physical deconditioning and disability, and prolonged psychological distress often leads to pain behavior. Pain can be a strong negative factor leading to avoidance behavior and functional disability.

The pain-depression comorbidity is thought to be related to both dysphoric physical symptoms (including pain) and psychological symptoms (including depression) and a state of somatosensory amplification in which psychological distress amplifies dysphoric physical sensations, including pain.[12] In a recent study,[13] 43.4 percent of subjects with major depressive disorder reported having at least one chronic painful physical condition. This is four times more often than in patients without major depressive disorder. Conversely, the prevalence of major depressive disorder was 10 percent in subjects with a chronic painful physical condition compared with 2.7 percent in subjects without chronic painful physical condition.

Pain and anxiety often coexist. The latter can precede pain or can be exacerbated by pain. Pain also can lead to aggravation of social problems, such as family stress, work issues, legal issues, and financial concerns.

Assessment of Mood

A number of measures of mood are available in the chronic pain literature. Of these, the two most frequently used

▶ **TABLE 15-5.** NERVE ROOTS TESTED BY DEEP TENDON REFLEXES (DTR)

DTR	Nerve Root
Bicipital	C5-6
Brachioradialis	C5-6
Tricipital	C7
Knee jerk	L4
Ankle jerk	S1

and well-validated measures of depressive symptoms in chronic pain patients are the Beck Depression Inventory (BDI)[14] and the Center for Epidemiological Studies Depression Scale (CES-D).[15] The BDI is a 21-item self-report instrument. In each item, four statements reflect the increasing severity of a particular depressive symptom. Patients are to indicate which statement best describes how they have been feeling over the past week. The BDI total score represents a single, higher-order depression construct. This score consists of three well-correlated second-order factors: negative attitudes and feelings, performance impairment, and somatic disturbance.[16] While the BDI is a brief, easily administered and easily scored instrument, there is evidence that the total score may overestimate the degree of depression in chronic pain and other medical conditions. A number of items in the BDI assess somatic symptoms that may be related to the pain itself rather than to depression per se.[17]

The CES-D is a 20-item self-report instrument that is designed to measure depressive symptoms in the general population. Each item is rated by patients on a four-point scale according to how frequently they have experienced that symptom in the past week. The total score is a composite measure of the number and duration of symptoms endorsed. Like the BDI, the CES-D is a reliable and valid scale that is brief and easy to administer and score. Analysis of this measure suggests that four correlated factors may be assessed: depressive affect, lack of wellness, interpersonal complaints, and somatic complaints. Criticism of the CES-D is similar to that of the BDI, i.e., overestimating the degree of depression in persons with chronic pain because somatic items are included.

Assessment of Anxiety

In chronic pain patients, both general anxiety and pain-specific anxiety have been measured with different scales. Feelings of apprehension, nervousness, worry, and tension are assessed with the State-Trait Anxiety Inventory (STAI).[18] This reliable and valid measure is a 40-item self-report instrument in which patients indicate on four-point scales how they are feeling at the present time. The STAI has been used widely in medical and psychiatric patients to assess clinical anxiety.[19] Another 40-item self-report measure is the Pain Anxiety Symptoms Scale (PASS).[19] The four subscales within this instrument measure fear and anxiety behaviors related to pain: (1) cognitive anxiety symptoms, (2) escape and avoidance responses, (3) fearful appraisal of pain, and (4) physiologic anxiety symptoms related to pain. Patients indicate for each item on a six-point scale how often they engage in that thought or behavior when experiencing pain. The reliability and validity of PASS has been shown to be adequate in predicting pain disability and behavior in heterogeneous pain populations.[19–21]

The Symptom Checklist 90, Revised,[22] is a 90-item self-report inventory that measures multiple dimensions of mood, including depression and anxiety (somatization, obsessive-compulsive behaviors, interpersonal sensitivity, depression, anxiety, hostility, phobic anxiety, paranoid ideation, and psychoticism). Patients report on a five-point scale how often they have been bothered over the past week by each of the symptoms. This relatively brief scale, developed initially to measure psychopathology in psychiatric and medical outpatients, has been used frequently to assess psychological distress in chronic pain patients. However, it has been argued that the original scoring method may be inappropriate for use with chronic pain patients, and alternative scoring methods that allow for identification of distinct subgroups of chronic pain patients have been suggested.[23]

Disability and Physical Function

Painful neuropathies are associated with varying degrees of interference in many daily activities, such as work, recreation, socializing, and sleep.[24,25] Two measures—the Pain Disability Index[26] and the Multidimensional Pain Inventory[27]—are useful in assessing disability or interference in neuropathic pain. Another measure, the SF-36, is likely to be less useful because it lacks disease or pain specificity. The Pain Disability Index, a seven-question instrument, assesses disability related to pain in seven areas of function: family/home responsibilities, recreation, social activities, occupation, sexual behavior, self-care, and life support (eating, sleeping). Each item is rated on an 11-point scale (0 = no disability to 10 = total disability). The Multidimensional Pain Inventory includes 52 items and 12 scales. Two of these scales—pain-related interference and general activity level—are useful in the assessment of disability in neuropathic pain. The 9 items in the interference scale assess interference and change in satisfaction with vocational, social, and family functioning. Each item is rated on 7-point scales (0 = no interference/change to 6 = extreme interference/change).[28] The general activity scale is used to rate the frequency (0 = never to 6 = very often) of a list of 19 daily activities.

Pain Coping Strategies

Several studies have demonstrated a relationship between coping and adjustment in persons with chronic pain.[29] For example, Hill and colleagues[30] found that catastrophizing was a significant predictor of both physical and psychosocial function in a sample of amputees with phantom limb pain, whereas avoidance was related to increased pain and psychological distress.[31]

Several instruments have been used to assess coping strategies in chronic pain populations. The Coping

Strategies Questionnaire (CSQ)[32] and the Chronic Pain Coping Inventory (CPCI)[29] are two measures developed specifically for the study of pain coping strategies.

Social Support

Painful neuropathies, such as those associated with spinal cord injury or HIV infection, often are associated with interference in many daily activities, including social activities, sexual life, and family and social relationships.[24,25] Negative responses to pain of a significant other have been shown to be associated with higher levels of pain[33,34] and higher levels of guilt in individuals with pain after a spinal cord injury.

▶ INVESTIGATIONS AND DIAGNOSTIC PROCEDURES

Unfortunately, there are no specific tests to diagnose neuropathic pain.

Electrodiagnostic Tests

Electromyography (EMG) and nerve conduction studies are useful for diagnosing peripheral nerve injuries, large-fiber polyneuropathies, and radiculopathies. However, they are insensitive for lesions that affect only small unmyelinated and poorly myelinated axons that comprise pain fibers. Besides that, normal EMG and nerve conduction studies do not rule out a nerve injury.

Quantitative Sensory Testing

Quantitative sensory testing (QST) evaluates a variety of sensory modalities by stimulating with standardized stimuli, recording the results, and comparing them with population norms.[35] QST can be used to assess the thermal (cold and heat) sensory thresholds, as well as the thermal pain threshold. Thermal testing can provide information for differential diagnosis between pain of nociceptive origin and pain of neuropathic origin. In nociceptive pain, thermal (cold and heat) sensory thresholds are expected to be normal, whereas in neuropathic pain, they are expected to be elevated. Hyperalgesia may be present in either condition. Electrical stimulation offers a quantitative evaluation of sensory dysfunction of C fibers, Aδ nerve fibers, and Aβ nerve fibers.

MRI/CT Scans

Magnetic resonance imaging (MRI) and computed tomographic (CT) imaging are of great value in localizing a lesion producing central pain. They also may be helpful in diagnosing radiculopathies. However, one has to keep in mind that they do not provide functional assessment.

Nerve Blocks

Despite the preceding diagnostic tools, it is sometimes difficult to delineate the source of the nociceptive input in some patients. Diagnostic neural blockade can be used to define the source of the pain generator. Transcutaneous local anesthetic blocks of peripheral nerves or roots can help to localize peripheral lesions. Sympathetic blocks are useful for the diagnosis of sympathetically maintained pain, although there are concerns about their specificity because spillage of local anesthetic can affect the neighboring somatic fibers as well. A temporary block using local anesthetics may help as a prognostic tool in determining if longer-lasting relief with decompressive or neuroablative techniques is possible.

Local anesthetic blocks of adjacent nerves may be indicated as controls.[36,37] The success of the block should be confirmed with objective measures such as appropriate anesthesia in the innervation territory of the blocked nerve or cutaneous temperature changes after sympathetic blocks. In addition, it should be confirmed that other adjacent nerves that were not the targets of interest are not anesthetized inadvertently. Preblock and postblock pain scores should be obtained, and sensory examination for a change in intensity and/or an area of allodynia should be recorded.

Infusion Tests

Infusion of sodium channel blockers (e.g., lidocaine) can be effective in the treatment of neuropathic pain.[38] This helps to identify the class of oral medication that can be effective. Commonly used sodium channel blockers for patients responding to intravenous lidocaine are oxcarbazepine, carbamazepine, lamotrigine, and mexiletine.

Skin Biopsies

Recent experience suggests that punch skin biopsy can help to identify changes associated with neuropathic pain conditions. The intraepidermal density of small nerve fibers appears to be significantly reduced in patients with painful sensory neuropathies compared with age-matched subjects.[39,40]

▶ CONCLUSION

The management of pain is based on the thorough assessment of the patient's condition, and includes a detailed history of pain, a neurological examination with emphasis on the sensory examination, and a psychological evaluation. Social and behavioral factors also should be assessed. Several investigations and diagnostic procedures may be useful in supporting the clinical findings.

REFERENCES

1. Dworkin RH: An overview of neuropathic pain: Syndromes, symptoms, signs and several mechanisms. *Clin J Pain* 2002;18:343–349.
2. Gerstle DS, All AC, Wallace DC: Quality of life and chronic nonmalignant pain. *Pain Manag Nurs* 2001;2:98–109.
3. Bech P: Health-related quality of life measurements in the assessment of pain clinic results. *Acta Anaesthesiol Scand* 1999;43:893–896.
4. Hunfeld JA, Perquin CW, Duivenvoorden HJ, et al: Chronic pain and its impact on quality of life in adolescents and their families. *J Pediatr Psychol* 2001;26:145–153.
5. Woolf CJ, Thompson SWN: Segmental afferent fibre induced analgesia: Transcutaneous electrical nerve stimulation and vibration, in Wall PD, Melzack R (eds), *Textbook of Pain,* 3d ed.; Edinburgh:Churchill Livingstone, 1994:884–896.
6. Woolf CJ, King AE: Dynamic alterations in the cutaneous mechanoreceptor fields in the dorsal horn neurons in the rat spinal cord. *J Neurosci* 1990;10:2717–2726.
7. Jensen MP, Turner JM, Romano JM, et al: Comparative reliability and validity of chronic pain intensity measures. *Pain* 1999;83:157–162.
8. Galer BS, Jensen MP: Development and preliminary validation of a pain measure specific to neuropathic pain: The neuropathic pain scale. *Neurology* 1997;48:332–338.
9. Bennett M: The LANSS Pain Scale: The Leeds assessment of neuropathic symptoms and signs. *Pain* 2001;92:147–157.
10. Backonja M, Galer BS: Neuropathic pain syndromes. *Neurol Clin* 1998;16:775–789.
11. Jensen TS, Gottrup H: Assessment of neuropathic pain, in Jensen TS, Wilson PR, Rice ASC (eds): *Clinical Pain Management: Chronic Pain.* London: Arnold; 2003:113–124.
12. Korff M, Simon G: The relationship between pain and depression. *Br J Psychiatry* 1996;168:101–108.
13. Ohayon M, Schatzberg A: Using chronic pain to predict depressive morbidity in the general population. *Arch Gen Psychiatry* 2003;60:39–47.
14. Beck AT, Ward CH, Mendelson M, et al: An inventory for measuring depression. *Arch Gen Psychiatry* 1961;4:561–571.
15. Radloff LS: The CES-D scale: A self-report depression scale for research in the general population. *Appl Psychol Meas* 1977;1:385–401.
16. Novy DM, Nelson DV, Berry LA, et al: What does the Beck Depression Inventory measure in chronic pain? A reappraisal. *Pain* 1995;61:261–270.
17. Wesley AL, Gatchel RJ, Garofalo JP, et al: Toward more accurate use of the Beck Depression Inventory with chronic back pain patients. *Clin J Pain* 1999;15:117–121.
18. Spielberger CD, Gorsuch RL, Lushene R, et al: *Manual for the State Trait Anxiety Inventory.* Palo Alto, CA: Consulting Psychologists Press, 1983.
19. McCracken LM, Zayfert C, Gross RT: The Pain Anxiety Symptoms Scale: Development and validation of a scale to measure fear of pain. *Pain* 1992;50:67–73.
20. McCracken LM, Gross RT, Aikens J, et al: The assessment of anxiety and fear in persons with chronic pain: A comparison of instruments. *Behav Res Ther* 1996;34:927–933.
21. Burns JW, Mullen JT, Higdon LJ, et al: Validity of the Pain Anxiety Symptoms Scale (PASS): Prediction of physical capacity variables. *Pain* 2000;84:247–252.
22. Derogatis L: *The SCL-90R: Administration, Scoring, and Procedures Manual.* Towson, MD: Clinical Psychometric Research, 1983.
23. Williams DA, Urban B, Keefe FJ, et al: Cluster analyses of pain patients' responses to the SCL-90R. *Pain* 1995;61:81–91.
24. Galer BS, Gianas A, Jensen MP: Painful diabetic polyneuropathy: Epidemiology, pain description, and quality of life. *Diabetes Res Clin Pract* 2000;47:123–128.
25. Benbow SJ, Wallymahmed ME, MacFarlane IA: Diabetic peripheral neuropathy and quality of life. *Q J Med* 1998;91:733–737.
26. Tait RC, Chibnall JT, Krause S: The Pain Disability Index: Psychometric properties. *Pain* 1990;40:171–182.
27. Kerns R, Turk D, Rudy T: The West Haven–Yale Multidimensional Pain Inventory (WHYMPI). *Pain* 1985;23:345–356.
28. Kerns RD, Jacobs MC: Assessment of the psychosocial context of the experience of chronic pain, in Turk DC, Melzack R (eds), *Handbook of Pain Assessment.* New York: Guilford Press, 1992:235–253.
29. Jensen MP, Turner JA, Romano JM, et al: The chronic pain coping inventory: development and preliminary validation. *Pain* 1995;60:203–216.
30. Hill A, Niven CA, Knussen C: The role of coping in adjustment to phantom limb pain. *Pain* 1995;62:79–86.
31. Gallagher P, MacLachlan M: Psychological adjustment and coping in adults with prosthetic limbs. *Behav Med* 1999;25:117–124.
32. Rosenstiel AK, Keefe FJ: The use of coping strategies in chronic low back pain patients: Relationship to patient characteristics and current adjustment. *Pain* 1983;3:1–8.
33. Conant LL: Psychological variables associated with pain perceptions among individuals with chronic spinal cord injury pain. *J Clin Psychol Med Settings* 1998;5:71–90.
34. Summers JD, Rapoff MA, Varghese G, et al: Psychosocial factors in chronic spinal cord injury pain. *Pain* 1991;47:183–189.
35. Yarnitsky D: Quantitative sensory testing. *Muscle Nerve* 1997;20:198–204.
36. Hogan QH, Abram SE: Neural blockade for diagnosis and prognosis: A review. *Anesthesiology* 1997;86:216–241.
37. Raja SN: Nerve blocks in the evaluation of chronic pain. *Anesthesiology* 1997;86:4–6.
38. Mao J, Chen LL: Systemic lidocaine for neuropathic pain relief. *Pain* 2000;87:7–17.
39. Holland NR, Stocks A, Hauer P, et al: Intraepidermal nerve fiber density in patients with painful sensory neuropathy. *Neurology* 1997;48:708–711.
40. Oaklander AL: The density of remaining nerve endings in human skin with or without postherpetic neuralgia after shingles. *Pain* 2001;92:139–145.

CHAPTER 16

Psychological Evaluation of the Patient with Chronic Pain

Lynette A. Menefee

Chronic pain presents challenges to the person who experiences it. Patients with intractable pain are confronted with managing the sensations and emotions associated with pain. Often, individuals experience changes in psychological and social functioning, as well as changes in behavioral and daily activities. Therefore, psychological evaluation is an important part of the multidisciplinary treatment of chronic pain. This chapter presents an overview of the psychological evaluation of the patient with chronic pain.

▶ THE PURPOSE OF THE PSYCHOLOGICAL EVALUATION

The overarching purpose of the psychological evaluation of patients with chronic pain conditions is to assess the patient's psychological, social, and behavioral functioning with attention to how functioning is affected by pain, suffering, and disability.[1] Psychological evaluations are used for specific purposes, such as recommending treatment interventions that promote adaptive beliefs and patterns of coping and/or to assess for psychological difficulties prior to a medical intervention or treatment. Treatment that focuses on the patient's unique difficulties (e.g., pain-coping skills, psychosocial stress, depression) may be recommended as a result of these evaluations. Evaluations that occur prior to medical procedures or treatment are becoming more common, especially prior to interventional procedures (e.g., spinal cord stimulation, intrathecal pump placement) or the prescription of opioids.

The purpose of the psychological evaluation is *not* to determine whether the patient has "psychogenic pain" or to determine whether pain is "organic versus functional."[1] Given the current definition and science of pain, the coexistence of painful sensations and unpleasant emotional experience is presumed. The purpose of pretreatment or preinterventional evaluations is not to keep patients from procedures with the potential to reduce their pain but to optimize the outcome of such procedures. For example, patients who exhibit severe depression often are unable to judge whether the trial procedure was helpful and likely would benefit from treatment for depression prior to a surgical procedure for an implantable therapy.

Whether psychological evaluation is a standard part of pain treatment or whether the patient has been referred to a pain psychologist for a specific reason, some patients react negatively to being referred to a psychologist. Patients who react negatively to a psychological referral most often feel disbelieved or that the purpose of the evaluation is to determine the psychologic versus physical origin of their pain. Patients' fears generally can be allayed when the physician explains that patients with pain often experience changes in their lives and that a psychological evaluation will provide an opportunity to talk about these concerns. Additionally, patients can be informed that they can learn additional ways to cope with pain and some of the difficulties in their lives.

▶ COMPONENTS OF THE PSYCHOLOGICAL EVALUATION

A comprehensive psychological evaluation involves a clinical interview, psychological testing, and if possible, interviews with significant others in the patient's life. A full comprehensive evaluation is not always needed and depends on the referring question and the purpose of the evaluation.

Clinical Interview

The history of the patient's pain condition should be the initial focus of the clinical interview. This initial focus often puts the patient at ease, especially if the patient has never consulted with a psychologist. The psychologist should have received permission from the patient for pertinent treatment notes from the medical provider. However, the psychologist will want the patient to present his or her experiences from the onset of the pain condition. The patient's manner and emotional expression in relaying this history can provide invaluable information. The onset of the pain, subsequent treatments, responses to treatments, and the patient's reactions to his or her experiences prior to the psychological evaluation serve as a basis for further questioning and help to develop rapport with the psychologist. Specific questions about previous treatments should be asked to ascertain the patient's understanding of the treatments and to be sure that most previous treatments are included in the patient's recounting. Asking directly whether the patient has undergone certain treatments, such as biofeedback, opioid medication trials, and alternative medicine treatments (e.g., acupuncture, herbal treatments), can help the patient remember specific treatments that he or she may have forgotten to mention. The patient's perceptions of treatment and goals of current treatment should be discussed. During this discussion, patients often will volunteer their perceptions of past and current providers. The psychologist can ascertain the patient's attitudes about his or her previous treatment providers and note whether the patient has a history of dissatisfaction with multiple treatment providers.

The location, quality, and characteristics of pain should be discussed with the patient. Ratings of pain intensity on average, pain intensity at its least, variations in pain during different portions of the day, and exacerbating and relieving factors give the clinician an idea of the patient's current experience with pain. During the evaluation, the clinician observes pain behaviors, such as facial grimacing or rubbing the affected body part, that may occur as the patient talks about his or her pain.

An important area of questioning is how the patient's activities and functioning have changed since the onset of the pain condition. Patients generally will describe differences in work experiences, including changes in the ability to perform their previous occupational duties, family responsibilities, and social and recreational activities. Areas of primary importance will be highlighted by the patient, such as statements like, "I can't even play with my children anymore," and are often accompanied by emotional expression. Asking the patient how he or she copes with exacerbations of pain will suggest the types of coping strategies typically employed by the patient. Patients may define active ways to cope with pain (e.g., use it as a stimulus to perform a distracting activity), passive coping (e.g., wishing the pain would go away), and/or maladaptive patterns (e.g., the expression of anger toward others).

The clinical interview also should encompass questions related to changes the patient may have experienced in important relationships with family members and friends. Patterns of family relationships often are disrupted following the onset of a pain condition and may deteriorate further when pain persists. Communication between the spouse and/or significant others may change. The patient may experience reinforcement or punishment for communicating pain verbally or through pain behaviors, such as resting, taking medications, or groaning. Asking directly how the patient's primary relationship has changed can be more fruitful than asking a more general question about how the significant other has been coping with the situation. Patients often will describe the significant other's expectations that the patient return to his or her former level of functioning, changes in role responsibilities, and changes in their sexual relationship secondary to pain and/or a lack of desire for sex.

Current personal and family stressors should be detailed. Many patients report stressful circumstances, including changes in their financial situation, litigation, employment concerns, and fears about the future.

Changes in the patient's mood since the onset of pain should be covered in the clinical interview. Although depressed mood, major depressive disorder, and anxiety are common for patients with chronic pain, many patients are reluctant to admit such symptoms. Many patients who have never experienced mood disturbance prior to the onset of their pain condition believe that "solving" their pain complaint also will alleviate their mood disturbances. Although pain may have preceded difficulties with depressed mood and/or anxiety, patients need to learn that pain and mood disturbance are treated differently. Asking about the neurovegetative symptoms of depression, such as appetite and sleep disturbance, prior to asking about persistent sadness, irritability, and hopelessness, can reduce defensiveness because many patients experience these symptoms. Suicidal ideation or a passive death wish (i.e., "I wouldn't kill myself, but it would be okay if I died tomorrow.") are not uncommon expressions of hopelessness and helplessness felt by some patients with chronic pain conditions. An assessment of the lethality of these ideas (i.e., whether the patient has thought about a plan and has the means to enact it) should follow immediately. A history of psychiatric difficulties, including symptoms, previous psychotherapy, medication, and hospitalization, should be taken, with special attention to treatments that have and have not been effective for symptoms the patient exhibits or describes in the interview. Depending on the patient's previous history and presentation, a more thorough interview of psychiatric symptoms may need to ensue to rule out psychiatric disorders. Because of genetic predispositions to some psychiatric illnesses, the patient's family history of psychiatric illness should be ascertained.

The clinical interview also should contain the patient's description of his or her general medical history. This history is helpful in understanding the type and length of contact with medical professionals and the patient's general health. Medications and the dose and frequency taken by the patient, especially those medications prescribed for pain and psychiatric conditions, should be recorded. This information may be useful in framing the patient's complaints (e.g., excessive sedation).

The patient's past and present habit history, including use of alcohol, illegal drugs, and tobacco, should be included in the clinical interview. The extent and duration of use and whether the patient's consumption of substances has changed since the onset of the pain condition are important information, as is whether the patient has undergone treatment in the past for drug and/or alcohol abuse or dependence.

A personal and social history should be taken, but this personal history is generally less detailed than psychological evaluations performed for other purposes. Details about the patient's early life, parents, and siblings likely will be limited to description and the patient's overall feeling about his or her family of origin. Traumatic experiences, such as the presence of childhood physical and sexual abuse, should be recorded. Noting whether the patient met early childhood developmental milestones, such as walking and talking, and whether the patient was sick often as a child can be helpful. The patient's educational and work history is important both from a descriptive and a narrative point of view. A brief history of the patient's primary relationships gives the clinician a picture of the patient's stability in relationships. Finally, the patient should be asked to describe activities performed in a typical day.

Ideally, the psychologist also will obtain information from the patient's significant other (e.g., spouse, partner, family member) with the permission of the patient. Turner and Romano[1] suggest that the significant other be asked about his or her conceptualization of the pain problem and treatment recommendations, behavioral changes in the patient, the patient's medication, alcohol and substance use, recent life stress, and observations of the patient's psychological functioning. This information can give another perspective on the patient's reality and help to identify how the patient's pain condition has affected family members.

The clinician records behavioral observations, including the patient's appearance; gait; aids used in ambulation; the manner in which the patient approached the interview; the tone, rhythm, volume, and content of the patient's speech; the patient's affect and self-reported mood during the interview; and eye contact.

Psychological Testing

A comprehensive psychological evaluation will include standardized self-report measures. Domains the psychologist may want to measure with psychometric instruments are pain, perceptions of disability and functioning, depression and anxiety, coping, and social functioning. Pain can be measured with a variety of one-dimensional instruments (e.g., visual analogue scale, numerical rating scale) or multidimensional measures, such as the Multidimensional Pain Inventory (MPI)[2] or the McGill Pain Questionnaire (MPQ).[3,4] The MPI has subscales that relate to pain severity, affective distress, interference in life activities, general activity, and perceived social responses to pain from a significant other. The MPQ allows patients to describe their pain with classes of words that describe sensory qualities (e.g., temporal, spatial, pressure), affective qualities (e.g., tension, fear), and evaluative qualities (i.e., subjective intensity).

A number of assessment instruments can be helpful in the psychological evaluation of a patient with pain. A review of these instruments is beyond the scope of this chapter. Suggested instruments of physical and psychosocial disability, general psychopathology, depression, marital adjustment, gross screening of cognitive functioning, and alcohol abuse can be found in Turner and Romano.[1] An excellent comprehensive resource is the *Handbook of Pain Assessment* (2d ed) by Turk and Melzack.[5]

▶ PSYCHOLOGICAL EVALUATIONS PRIOR TO MEDICAL TREATMENT

Preinterventional treatment evaluations for spinal surgery and implantable therapies, such as spinal cord stimulation and intrathecal pump drug delivery, have been discussed in the literature.[6] More recently, psychological evaluation prior to the use of opioid medications also has been proposed. The purpose of these evaluations is to maximize surgical, interventional, and medical treatment outcomes. Generally, these evaluations contain a comprehensive clinical interview, psychological testing, and some determination about the psychological readiness of a patient to undergo a procedure or treatment.

Presurgical Psychological Evaluations

In a recent review of the literature on presurgical psychological evaluations, Epker and Block[7] identify several psychological factors predictive of suboptimal outcome of spinal surgery performed for the purpose of pain reduction. The factors include scale elevations on the Minnesota

Multiphasic Personality Inventory associated with pain sensitivity, depression, anger, and anxiety, as well as maladaptive coping strategies, litigation status, worker's compensation, and reinforcement of pain behaviors by a spouse. Historical variables that may impede optimal outcome, identified by Epker and Block,[7] are preexisting psychological difficulties, sexual and/or physical abuse, marital distress, and substance abuse. Psychological pre- or postsurgical follow-up may be recommended based on the evaluation. Treatment can be directed to improving a patient's compliance, increasing the patient's nonpharmacologic strategies for coping with pain (e.g., biofeedback, relaxation), providing guidance for smoking cessation or weight control, or recommending more intensive intervention.[7]

Preintervention Psychological Evaluation

Psychological evaluation prior to spinal cord stimulation or intrathecal therapies has increased over the past few years and is recommended by many manufacturers of devices. Like presurgical evaluations, preinterventional psychological evaluations are comprehensive and involve a clinical interview and psychological testing. Factors thought to be important in selection vary from the presence of severe psychiatric disorders (e.g., active psychosis, current suicidality or homicidality) and substance abuse to the lack of social support in the family.[8] Additionally, the patient's expectations and goals of treatment are important areas of discussion. Patients who have endured intractable pain for years may have unrealistic expectations for these interventions (i.e., perceiving them as a "cure" for pain).

Psychological Evaluations Prior to the Prescription of Opioids

The use of opioids for treatment of nonmalignant pain conditions has raised questions about which patients might be prone to misuse or abuse them. Complicating the question of whether a patient is abusing prescribed medications is some disagreement about the definition of addiction in the context of pain. The American Society of Addiction Medicine, the American Academy of Pain Medicine, and the American Pain Society recognize addictive behaviors as impaired control over drug use, compulsive use, continued use despite harm, and craving.[9] Some judgment must be made about whether a patient's behavior fits these categories, and these judgments sometimes can be difficult. Therefore, practitioners may want patients to undergo psychological evaluation prior to the initiation of opioids and/or at regular intervals during treatment. Although some predictive factors for problematic drug use have been proposed (e.g., multiple episodes of prescription "loss," seeking prescriptions from multiple sources),[10] few empirical studies have been devoted to this topic. Currently, there are no widely accepted instruments specifically developed for the purpose of predicting problematic drug use.[10] However, this is an area of active research, and there are likely to be instruments proposed in the future. Although there will never be a perfectly predictive instrument or interview, psychological evaluation can be helpful to the practitioner in determining the patient's previous compliance, his or her tendency toward impulsive behavior, attitudes about opioid use, and past history of substance use, abuse, or dependence.

Pretreatment evaluations culminate in a recommendation to the practitioner regarding the psychological and behavioral contraindications of proceeding with treatment. Psychologists vary in their recommendations, but the most common categories are no contraindications, contraindicated due to behavioral issues, or not recommended.[10] A "no contraindications" recommendation indicates that the patient is a good candidate for the intervention from a psychological perspective. A recommendation that the patient is "contraindicated from a psychological or behavioral perspective" means that the patient describes or demonstrates psychological difficulties (e.g., has psychological symptoms, such as depression or anxiety, cognitive impairment, or unrealistic expectations of the treatment). The number and severity of these symptoms generally determines the psychologist's recommendation for proceeding. A suggestion of concomitant psychological treatment and/or a psychiatric referral for a pharmacologic consultation may be indicated but would not significantly negatively affect proceeding with treatment. However, if the patient's symptoms are numerous and/or severe and the outcome of treatment likely will be affected by these symptoms, the recommendation may be to complete psychological treatment before proceeding with the medical intervention. Rarely, patients are judged to be inappropriate for the procedure from a psychological perspective. Patients in this category generally demonstrate intractable, severe psychiatric symptoms, relapsing substance abuse, or enduring characteristics or circumstances that pose a potential risk to compliance.[8]

▶ CONCLUSION

Psychological evaluations are an important part of multidisciplinary treatment of patients with chronic pain conditions. They also can be helpful in maximizing surgical, interventional, and medical treatment outcomes by providing a recommendation of the patient's psychological readiness for these treatments. Psychological evaluations are performed for the purpose of recommending treatment that will help the patient to reduce painful sensations, improve coping skills, address psychological and behavioral difficulties, and improve his or her quality of life.

REFERENCES

1. Turner JA, Romano JM: Psychological and psychosocial evaluation, in Loeser JD, Butler SH, Chapman CR, Turk DC (eds), *Bonica's Management of Pain,* 3d ed. Philadelphia: Lippincott, Williams & Wilkins, 2001.
2. Kerns RD, Turk DC, Rudy TE: The West Haven-Yale Multidimensional Pain Inventory (WHYMPI). *Pain* 1985; 23:345–356.
3. Melzack R: The McGill Pain Questionnaire: Major properties and scoring methods. *Pain* 1975;1:277–299.
4. Melzack R: The short-form McGill Pain Questionnaire. *Pain* 1987;30:191–197.
5. Turk DC, Melzack R (eds): *Handbook of Pain Assessment,* 2d ed. New York: Guilford Press, 2001.
6. Gatchel RJ: A biopsychosocial overview of pretreatment screening of patients with pain. *Clin J Pain* 2001;17:192–199.
7. Epkar J, Block AR: Presurgical psychological screening in back pain patients: A review. *Clin J Pain* 2001;17:200–205.
8. Prager J, Jacobs M: Evaluations of patients for implantable pain modalities: Medical and behavioral assessment. *Clin J Pain* 2001;17:206–214.
9. Savage S, Covington EC, Heit HA, et al: *Definitions related to the use of opioids for the treatment of pain:* A consensus document from the American Academy of Pain Medicine, the American Pain Society and the American Society of Addiction Medicine. American Academy of Pain Medicine: Glenview, IL; American Pain Society: Glenview, IL; American Society of Pain Medicine: Chevy Chase, MD, 2001.
10. Robinson RC, Gatchel RJ, Polatin P, et al: Screening for problematic prescription opioid use. *Clin J Pain* 2001; 17:220–228.

CHAPTER 17

Pain in Children

Stephen C. Brown, Patricia A. McGrath, and Kristina R. Krmpotic

As in adults, children's pain depends on complex neural interactions, where impulses generated by injury are modified by both ascending systems activated by innocuous stimuli and descending pain-suppressing systems activated by various situational and psychological factors. Children are not "little adults" with respect to how they perceive pain though, and their developing nociceptive systems respond differently to injury (i.e., increased excitability and sensitization) when compared with the mature adult system. Moreover, a child's pain appears to have a greater degree of plasticity when compared with that of adults—more influenced by cognitive, behavioral, and emotional factors.[1]

▶ UNIQUE FEATURES OF A CHILD'S PAIN

The Developing Nociceptive System

Until relatively recently, infants and children were regarded as having immature nervous systems that could not feel pain as intensely as adults. Consistent with this assumption, infants and children routinely did not receive adequate pain control following surgery.[2] Ethical concerns and increasing publicity about the lack of analgesia for children during surgery led to a dramatic upsurge in clinical research to document objectively how infants and children respond to surgical trauma and how analgesic administration affects postoperative outcome. At the same time, basic scientific research focused on the underlying nociceptive system of the child. Investigators traced the anatomic development of each component, starting from the primary sensory neurons innervating peripheral tissues to dorsal horn cells, spinal reflex connections, the pathways projecting to higher brain centers, and the descending inhibitory pathways from the brain. Physiologic studies detailed how the maturing pain system responded to different types of stimuli at different developmental periods.

Although the basic connections in pain pathways are formed prior to birth, they are immature, and there are considerable postnatal changes.[3] However, this does not mean that infants do not experience pain. The postnatal period is a time of considerable synaptic growth and reorganization in the dorsal horn of the spinal cord. The conduction velocity of afferent fibers, action potential shape, receptor transduction, firing frequencies, and receptive field properties change substantially over the postnatal period. (For a more comprehensive review of the basic science research, see refs. 4 and 5.)

The majority of research in developmental neurobiology has been conducted on rat pups because they have comparable developmental timetables with respect to the anatomy, chemistry, and physiology of maturing human pain pathways. At birth, some sensory neurons show altered responses to stimulation. High-threshold $A\delta$ mechanoreceptors (which respond maximally to noxious mechanical stimuli) and low-threshold $A\beta$ mechanoreceptors (which respond maximally to innocuous stimuli) respond with lower firing frequencies than those in the adult animal. The pattern of afferent innervation into the spinal cord also changes markedly in the postnatal period. $A\beta$ afferents extend dorsally into laminae II and I along with C fibers, rather than into only laminae III and IV, as in the adult animal. Activation of these $A\beta$ afferents evokes excitatory responses more typical of those evoked by $A\delta$ and C fibers in the adult animal.

The receptive fields of dorsal horn cells are larger in the newborn. These larger receptive fields and the dominant A-fiber input increase the likelihood of central cells being excited by peripheral sensory stimulation and act to increase the sensitivity of infant sensory reflexes.[4]

Neonatal somatosensory cortical cells also have larger receptive fields than those in adults. The development of these neural circuits, especially in the developing cortex, is influenced by sensory experiences during critical periods early in life.[6] Some inhibitory mechanisms in the dorsal horn are immature at birth; and thus, descending inhibition is delayed.[7,8] This lack of descending inhibition in the neonatal dorsal horn means that an important endogenous analgesic system that should attenuate noxious input as it enters the spinal cord is lacking, and the effects of this input may be more profound than in the adult.[4] Thus, considerable neuronal plasticity is evident in the developing pain system from the periphery to the brain.

Behavioral studies on human infants also have revealed findings of plasticity and increased excitability in the developing nervous system. In comparison with adults, young infants have exaggerated reflex responses (i.e., lower thresholds and longer-lasting muscle contractions) in response to certain types of trauma, such as needle insertion.[9] The threshold of the flexion withdrawal reflex in the neonate rises with increasing postconceptual age,[10] reflecting the change from a predominantly low-threshold A-fiber influence in the dorsal horn of the spinal cord early in development to increasingly Aδ- and C-fiber convergence in the superficial laminae with maturation.[11]

Repeated mechanical stimulation at strong (but not pain-inducing) intensities can cause sensitization in very young infants,[10] and preliminary studies have noted a striking hypersensitivity to touch as well as pain in infants after surgery.[12] Repeated painful procedures such as those required during intensive care can profoundly affect sensory processing in infants. Reflex thresholds are significantly less after tissue has been injured by repeated heel lances, and this altered sensitivity can persist for weeks.[10] Ultimately, we do not know how injuries such as these will affect the mature pain system or their influence on adult pain perception.

The Long-Term Impact of Early Pain Experiences

Much research attention is focused on the possible consequences of untreated pain, particularly in infants.[13,14] Circumcised infants displayed a stronger pain response to subsequent routine immunizations at 4 and 6 months than uncircumcised infants.[15] Preoperative treatment with lidocaine-prilocaine cream attenuated the pain response to subsequent vaccinations.[16] Studies of former premature infants who received multiple invasive procedures, including frequent heel sticks, venipunctures, and intubations, also have shown behavioral differences related to early pain experiences.[17] It is difficult, however, to differentiate the specific impact of pain from other physical, environmental, and family factors. Grunau[13] has emphasized the complexity of these effects and the need for careful study design and interpretation.

For many years, practitioners assumed that infants could not feel pain or would not remember it. Yet the results of anatomic, physiologic, and behavioral studies indicate increased responsivity rather than diminished sensitivity, as well as reduced inhibitory mechanisms. The key for practitioners is not to underestimate the invasiveness of the procedures they perform on infants.[14] If a procedure is noxious to adults, it also will be noxious to infants.[18]

The Plasticity of a Child's Pain

Throughout the last decade, we have gained an increasing appreciation for the plasticity of pain from biologic and psychological perspectives. As with adults, a child's pain is often initiated by tissue damage caused by noxious stimulation, but the consequent pain is neither simply nor directly related to the amount of tissue damage. The model shown in Fig. 17-1 provides a framework for assessing the factors that can modify a child's pain. Factors such as age, cognitive level, gender, temperament, previous pain experience, family learning, and cultural background (shown in the lower box in the figure) shape how children generally interpret and experience the various sensations caused by tissue damage. In contrast, situational factors such as cognition, behavior, and emotion (shown in the upper boxes) are fluid and represent a dynamic interaction between the child and the situation in which the pain is experienced.

Childhood Factors

There is a complex interrelationship among age, cognitive level, gender, temperament, previous pain experience, family learning, and culture in shaping a child's pain perception. A child's understanding of pain, the words they use to describe pain, and how they evaluate new pain experiences are influenced by all these factors. A child evaluates any new pains in comparison with what has already been experienced. The nature and diversity of a child's pain experience form the frame of reference for perceiving any new pain. Yet, unlike adults who generally have experienced a wide range of pains differing in intensity and quality, a child's frame of reference is more limited so that new injuries or invasive procedures may cause stronger and more distressing pains in comparison with adults.[19]

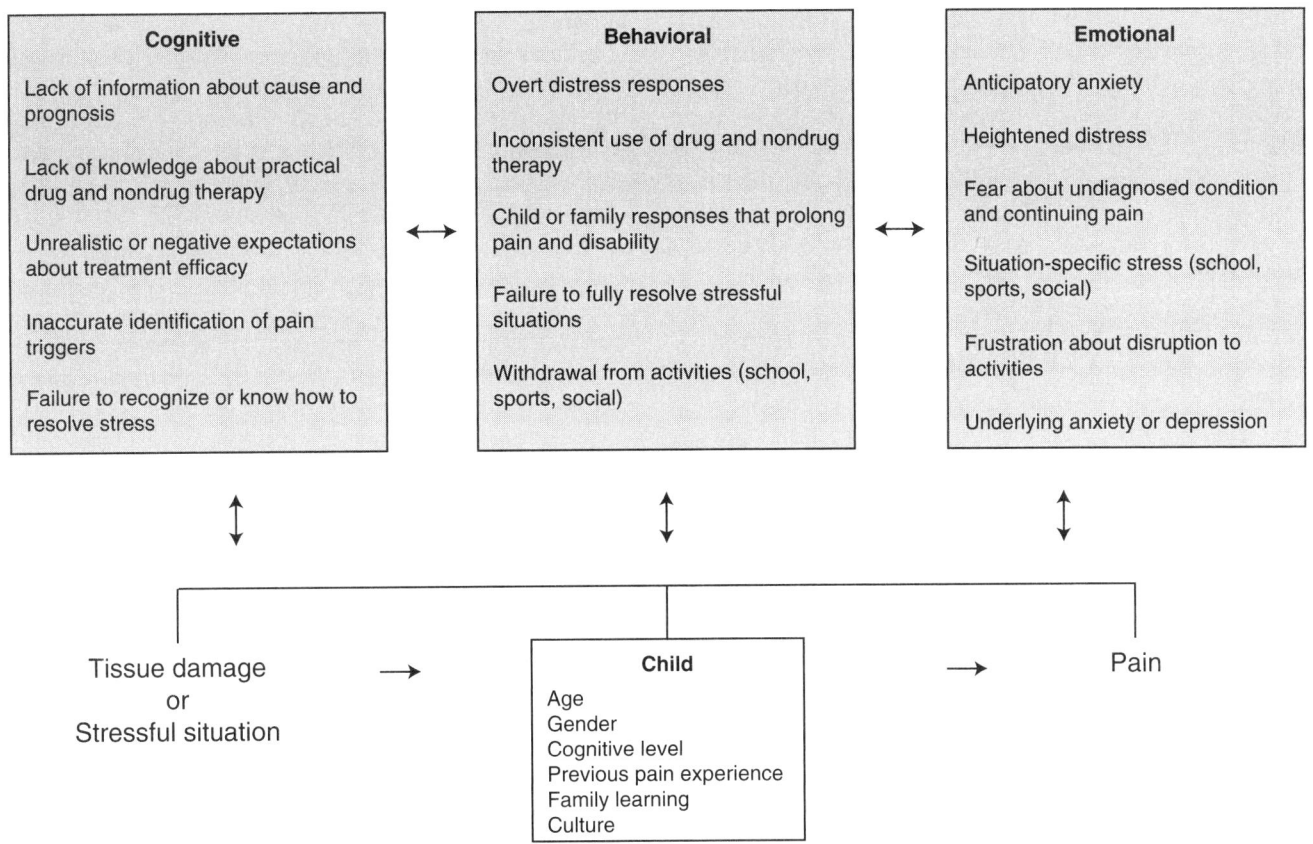

Figure 17-1. Child and situational factors that modify pain and disability. (*Adapted with permission from McGrath and Hillier.*[1])

Most studies of acute pain caused by invasive medical procedures (e.g., venipunctures) reveal age-related decreases in both children's pain ratings and their overt distress behaviors (e.g., crying, resisting) during these procedures (for review, see ref. 1). However, the effect of age probably varies depending on the type of pain and the nature of their previous pain.[20–22] Some studies of postoperative pain show increasing pain with age,[23] whereas others show age-related decreases[24] or no differences.[25] As children mature and experience a wider variety of pains varying in quality, intensity, location, and duration, they also learn new methods of coping to lessen pain in different situations. Thus, a young child depends greatly on parents and health care providers to understand and learn to cope with new pain experiences, whereas older children are more capable of independent coping strategies.

The results of studies evaluating sex- and gender-related trends in children's pain perception, pain expression, pain behaviors, and coping strategies yield equivocal results. Such inconsistent findings are explainable because studies differ widely with respect to the number of children sampled, their ages and health status, the type of pain studied, and the specific outcome measures used. The extent to which sex- and gender-related differences are present may vary according to the type of pain and to the particular pain features evaluated.

Although hereditary factors can contribute to the onset of certain painful diseases in childhood, which may account for a familial predisposition to some recurrent pains, the family's influence on a child's pain is not limited solely to genetic factors. Parents and siblings shape what a child learns about pain—how to express pain and how to cope with different types of pain, as readily evidenced when toddlers fall during play. Children quickly look toward their parents to see their reaction. Some parents remain calm, reassure children, and encourage them to get up and continue playing, whereas other parents are visibly distressed, are overly attentive, focus on the area hurt, and encourage them to be comforted and to stop their play. Parents may respond differently when children sustain mild injuries, depending on the age, sex, or birth order of the child. They may provide more attention to younger children but encourage older children to cope more independently. They may encourage boys to be stoic, suppress overt distress behaviors, and develop active behavioral responses. In contrast, parents may subtly reinforce girls to express their discomfort, exhibit more sick behaviors, and rely on passive pain-coping methods. Parents may be more anxious, overtly protective, and concerned about the routine bumps and scrapes of childhood for their first-born child than for later children.

The family and society, in addition to being the primary models for the development of childhood pain attitudes and behaviors, also may influence a child's actual pain experiences. Edwards and colleagues[26,27] demonstrated a positive relationship between the number of familial pain models and the frequency of an individual's pain complaints and the nature of his or her coping strategies. Schanberg and colleagues[28] also have demonstrated that parental pain history and family environment were related to the functional status and pain complaints of children with juvenile primary fibromyalgia. Parents who reported multiple chronic pain conditions were more likely to have children with higher impairment and disability.

Situational Factors

Situational factors can vary dynamically depending on the specific circumstance in which the child experiences pain. For example, a child receiving treatment for cancer may have repeated injections, central venous port access, and lumbar punctures—all of which can cause pain (depending on the analgesics, anesthetics, or sedatives used). Even though the tissue damage from these procedures is similar each time, the particular set of situational factors for each treatment is unique for a child. The expectations, behaviors, and emotional states of the child, parent, and health care provider all play a critical role. Although the causal relationship between injury and consequent pain seems direct and obvious, a child's pain experience is profoundly affected by what the child and parents understand and feel, as well as what they and their health care providers do (as listed in Fig. 17-1).

Cognitive factors include parents' and a child's understanding of the pain problem, beliefs about the source of the pain, perceptions about their ability to control the situation, their knowledge of effective pain control strategies, and their expectations regarding the current pain experience and for the future. In general, health care providers can lessen a child's pain when they provide accurate age-appropriate information about a medical procedure (e.g., emphasizing the specific sensations that children will experience, such as the stinging quality of an injection, rather than general "hurting" aspects), increase control and choices (e.g., which arm for an injection, whether to sit or lie down), and teach the child some independent pain-reducing strategies.[19]

Behavioral factors refer to a child's, parents', and health care providers' specific behaviors when a child is experiencing pain and also encompass parent's and children's wider behaviors in response to a chronic pain problem. Common behavioral factors include a child's distress or coping reactions (e.g., crying, using a pain control strategy, withdrawing from life) and parents' and health staff's subsequent reactions to them (e.g., displaying frustration, calmly providing encouragement for the child to use pain control strategies, engaging the child in conversation and activities). They also include the extent to which children are physically restrained during invasive or aversive treatments and the broader physical and social restrictions on a child and his or her family life. Distress behaviors and some altered family behaviors may initiate, exacerbate, or maintain a child's pain.

In general, a child's pain should lessen when staff and parents become more consistent so that invasive procedures are conducted with a clear and predictable routine and when staff assists children to use simple psychological pain control methods such as attention and distraction. Regrettably, staff often regard those health care providers who use psychological methods effectively with children as simply having positive personalities rather than having acquired a special skill set—a skill set that all staff could learn and apply routinely in clinical practice.

Emotional factors include a child's and parents' feelings about the painful episodes or conditions, the subsequent impact of pain and illness on the family, and the anxiety or depression that may underlie certain chronic conditions. Children's emotions affect their ability to understand what is happening, their ability to cope positively, their behaviors, and ultimately, their pain. A child's immediate emotional reactions to pain may vary from a relatively neutral acceptance to annoyance, anxiety, fear, frustration, anger, or sadness. The specific emotions depend on the nature of the pain and its impact on the child's life. In general, the more emotionally distressed children are, the stronger or more unpleasant is their pain.

When a child does not understand what is happening, lacks control, and does not know any independent pain control strategies, emotional distress increases, and pain intensifies. Similarly, when a child's behaviors are restricted, whether due to physical restraints applied during medical procedures or due to disruptions in usual athletic and social activities because of chronic pain, emotional distress and pain can intensify.

Certain *situational factors* also can intensify pain and distress, whereas others eventually can trigger pain episodes, prolong pain-related disability, or maintain the cycle of repeated pain episodes for children with a recurrent pain syndrome.[29] Thus, it is essential to recognize and evaluate the mitigating impact of these factors in order to relieve a child's pain. While parents and health care providers are unable to change the more stable childhood factors, they can dramatically improve a child's pain experience and minimize the child's disability by modifying situational factors.

▶ EVALUATING A CHILD'S PAIN

Pain assessment is an integral component of diagnosis and treatment for children. A thorough medical history, physical examination, and review of laboratory and radiologic results, and an assessment of pain characteristics and contributing factors are necessary to establish a correct clinical diagnosis. Subsequent assessments of pain intensity enable clinicians to determine when treatments are effective and to identify children for whom they are most effective. Health care providers need pain measures that are convenient to administer and whose resulting scores provide meaningful information about a child's pain experiences. An extensive array of pain measures has been developed and validated for use with infants, children, and adolescents.[30–32] However, no single pain measure is equally appropriate for all ages of children or for all types of pain.

Like adult pain measures, children's pain measures are classified as physiologic, behavioral, and psychological depending on what is monitored—physical parameters (e.g., heart rate, sweat index, blood pressure, cortisol level), distress behaviors (e.g., grimaces, cries, protective guarding gestures), or a child's own descriptions of what he or she is experiencing (e.g., words, drawings, numerical ratings). Physiologic and behavioral measures provide indirect estimates of pain because health care providers must infer the location and strength of a child's pain solely from the child's responses. In contrast, psychological measures can provide direct information about the location, strength, quality, affect, and duration of the pain.

The criteria for an accurate pain measure are similar to those required for any measuring instrument. A pain measure must be *valid,* in that it measures a specific aspect of pain so that changes in pain ratings reflect meaningful differences in a child's pain experiences. The measure must be *reliable,* in that it provides consistent and trustworthy pain ratings regardless of the time of testing, the clinical setting, or who is administering the measure. The measure must be *relatively free from bias,* in that children should be able to use it in a similar fashion regardless of differences in how they may wish to please adults. The pain measure should be *practical* and *versatile* for assessing different types of pain (e.g., disease-related, procedural pain) in many different children (according to age, cognitive level, cultural background) and for use in diverse clinical and home settings.

Behavioral and Physiological Pain Scales

Behavioral scales must be used when a child is unable to communicate directly about his or her pain experience. Most behavioral pain scales consist of checklists of different types of distress behaviors (e.g., crying, grimacing, guarding) that children exhibit when they experience a certain type of pain.[30,33,34] To develop these scales, trained clinicians carefully observe children when they are in pain (e.g., during invasive medical procedures, after surgery) and document any behaviors that seem caused by the pain. They then list these presumed "pain behaviors" on an itemized checklist. Parents complete the pain scale by checking which of the listed distress behaviors the child displays, often ranking the intensity of each behavior on a 0–2 or 0–4 scale. The intensity scores for each of the observed behaviors are added together to produce a composite pain score. This pain score provides an indirect estimate of pain severity. The vast majority of behavioral scales were developed for acute procedural or postoperative pain in otherwise healthy children.

More recently, increasing attention is focusing on the special problem of pain for children with developmental disabilities. Behavioral pain scales are being developed in consultation with parents and caregivers for children who are cognitively or physically impaired,[35–40] and practical guidelines for pain management have been detailed.[41] Health care providers and parents must work together to determine which behaviors provide the most relevant indices of pain for any particular child.

Although physiologic parameters can provide valuable information about a child's state of distress, more research is required to develop a sensitive system for interpreting how these parameters reflect pain strength. At present, there are no valid physiologic pain scales for children.

Self-Report Pain Scales

A child's understanding and descriptions of pain naturally depend on the child's age, cognitive level, and previous pain experience. Children begin to understand pain through their own hurting experiences. Most children can communicate meaningful information about their pain. Gradually, they develop an increasing ability to describe specific pain features—the quality (aching, burning, pounding, sharp), intensity (mild to severe), duration and frequency (a few seconds to years), location (from diffuse location on their skin to more precise internal localization), and unpleasantness (mild annoyance to intolerable discomfort). A child's understanding of pain and the language the child uses to describe pain come from the words and expressions used by the family and peers and from characters depicted in books, videos, and movies. (For a more extensive review of developmental factors in children's pain, see refs. 19, 32, 42, and 43.)

Physicians always should ask children directly about their pain. Pain onset, location, frequency (if recurring), quality, intensity, accompanying physical symptoms, and pain-related disability should be assessed as part of a child's clinical examination. Psychological or self-report pain measures potentially provide the most comprehensive information about a child's pain experience because they capture directly the subjective experience of pain. A broad spectrum of interviews, questionnaires, and qualitative descriptive scales and numerous quantitative pain intensity rating scales, each with some evidence of validity and reliability, are available to capture valuable diagnostic information about the causes for and contributing factors to a child's pain.[30,44] Most pain specialists use a combination of structured interviews, multidimensional pain questionnaires, and pain diaries to assess a child with a recurrent pain syndrome or a chronic pain condition.[31,45,46] These flexible measures identify social, physical, school, or family factors that contribute to a child's pain and disability.

Although clinical interviews are ideally suited for learning about the sensory characteristics of pain and contributing cognitive, behavioral, and emotional factors, clinicians also should use a simple rating scale to document a child's pain intensity. Pain intensity scales are easy to administer. A child selects a level on the scale, such as a number, a mark on a visual analogue scale, a face from a series of faces varying in emotional expression, or a particular word from adjective lists describing different intensities, to match the strength of his or her own pain. Health care providers must consider the age and cognitive ability of a child when selecting a pain scale. Most toddlers (approximately 2 years of age) first express the "hurting" aspect of pain using a few words learned from their parents to describe the sensations they feel when they hurt themselves. Gradually, children learn to differentiate and describe three levels of pain intensity—"a little," "some" or "medium," and "a lot." By age 5, most children can differentiate a wide range of pain intensities, and many children at this age can use simple quantitative scales to rate their pain intensity (Fig. 17-2). Visual and colored analogue scales are ideal for most children older than 5 years of age. In addition to excellent psychometric properties, these scales are versatile for use with acute, recurrent, and chronic pain and provide a convenient and flexible pain assessment tool for use in hospital and at home. They provide meaningful values that reflect a child's pain intensity and are ideal measures for evaluating treatment effectiveness.

In summary, assessing a child's pain requires an integrated approach. Physicians always should ask a child directly about his or her pain experience, noting its location, intensity, quality, and duration. They should regularly measure a child's pain intensity to monitor the effectiveness of their therapy. Pain specialists should also assess the relevant cognitive, behavioral, and emotional factors so that they may modify their potential pain-exacerbating impact.

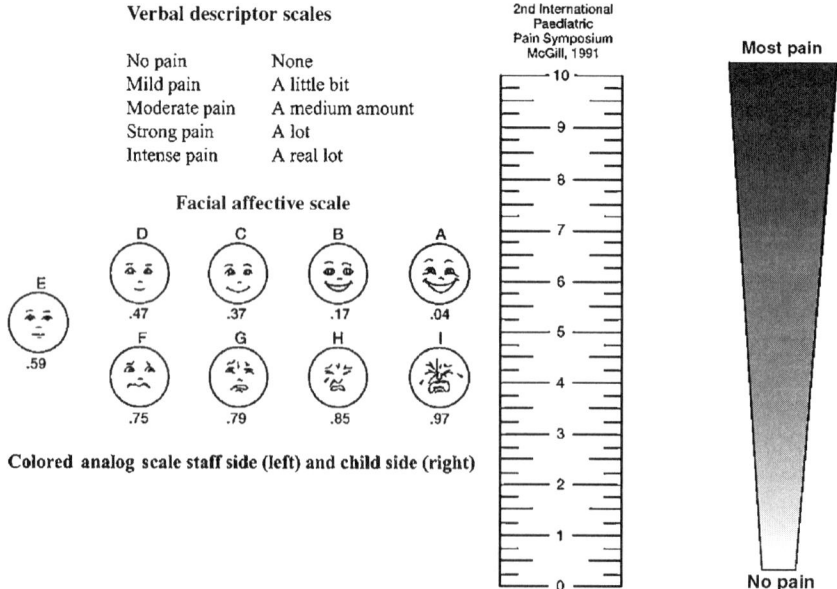

Figure 17-2. Pain scales used to assess pain intensity in children. (*Reprinted with permission from McGrath and Brown.*[71])

▶ TREATING A CHILD'S PAIN

Controlling a child's pain requires a child-centered approach that integrates both drug and nondrug therapies. Health care providers must carefully evaluate the varied causes and contributing factors that trigger, maintain, or exacerbate pain to select the most effective therapies for each child. Adequate analgesic prescriptions, administered at regular dosing intervals, must be complemented by a practical cognitive-behavioral approach to ensure optimal pain relief.

Drug Therapies

As reviewed in previous chapters, analgesics include acetaminophen, nonsteroidal anti-inflammatory drugs (NSAIDs), opioids, and adjuvant analgesics such as various anticonvulsants and tricyclic antidepressants. The use of adjuvant analgesics has become a cornerstone of pain control for children with chronic pain, especially when the pain has a neuropathic component.

Dosing Guidelines

Although specific drugs and dosages are determined by the needs of each child, general guidelines for drug therapies to control pain in children have been developed.[47–51] The drugs listed in this chapter are based on these sources and guidelines from our institution.[52] Recommended starting doses for analgesic medications are listed in Tables 17-1 and 17-2. Starting doses for adjuvant analgesic medications to control pain, drug-related side effects, and other symptoms are listed in Table 17-3. Neonates and infants require the same three categories of analgesic drugs as older children. However, premature and term newborns show reduced clearance of most opioids. The differences in pharmacokinetics and pharmacodynamics among neonates, preterm infants, and full-term infants warrant special dosing considerations for infants and close monitoring when they receive opioids.[53]

Children should receive analgesics at regular times—"by the clock"—to provide consistent pain relief and prevent breakthrough pain. The specific drug schedule (e.g., every 4 or 6 hours) is based on the drug's duration of action and the child's pain severity. Children with severe pain may require progressively higher and more frequent opioid doses due to drug tolerance, and they should receive the doses they need to relieve their pain.[51,54] The fear of opioid addiction in children has been greatly exaggerated.

Nondrug Therapies

An extensive array of nondrug therapies are available to treat a child's pain, including counseling, distraction, guided imagery, hypnosis, relaxation training, biofeedback, behavioral management, acupuncture, massage, homeopathic remedies, naturopathic approaches, and herbal medicines. Children are increasingly using complementary

▶ TABLE 17-1. NON-OPIOID DRUGS TO CONTROL CHILDREN'S PAIN

Drug	Dosage	Comments
Acetaminophen	10–15 mg/kg PO, every 4–6 h	Lacks gastrointestinal and hematological side-effects; lacks anti-inflammatory effects (may mask infection-associated fever) Dose limit of 65 mg/kg/day or 4 g/day, whichever is less
Ibuprofen	5–10 mg/kg PO, every 6–8 h	Anti-inflammatory activity Use with caution in patients with hepatic or renal impairment, compromised cardiac function or hypertension (may cause fluid retention, edema), history of GI bleeding or ulcers, may inhibit platelet aggregation Dose limit of 40 mg/kg/day; max dose of 2400 mg/day
Naproxen	10–20 mg/kg/day PO, divided every 12 h	Anti-inflammatory activity. Use with caution and monitor closely in patients with impaired renal function. Avoid in patients with severe renal impairment Dose limit of 1 g/day
Diclofenac	1 mg/kg PO, every 8–12 h	Anti-inflammatory activity. Similar GI, renal and hepatic precautions as noted above for ibuprofen and naproxen Dose limit of 50 mg/dose

NOTE: Increasing the dose of nonopioids beyond the recommended therapeutic level produces a "ceiling effect," in that there is no additional analgesia but there are major increases in toxicity and side effects.
ABBREVIATIONS: PO = by mouth; GI = gastrointestinal.
SOURCE: Modified with permission from McGrath and Brown, 2004.[71]

▶ **TABLE 17-2.** OPIOID ANALGESICS: USUAL STARTING DOSES FOR CHILDREN

Drug	Equianalgesic Dose (parenteral)	Starting Dose IV	IV:PO Ratio	Starting Dose PO/Transdermal	Duration of Action
Morphine	10 mg	Bolus dose = 0.05 mg/kg–0.1 mg/kg every 2–4 h Continuous infusion = 0.01–0.04 mg/kg/h	1:3	0.15–0.3 mg/kg/dose every 4 h	3–4 h
Hydromorphone	1.5 mg	0.015–0.02 mg/kg every 4 h	1:5	0.06 mg/kg every 3–4 h	2–4 h
Codeine	120 mg	Not recommended		1.0 mg/kg every 4 h (dose limit 1.5 mg/kg/dose)	3–4 h
Oxycodone	5–10 mg	Not recommended		0.1–0.2 mg/kg every 3–4 h	3–4 h
Meperidine[a]	75 mg	0.5–1.0 mg/kg every 3–4 h	1:4	1.0–2.0 mg/kg every 3–4 h (dose limit 150 mg)	1–3 h
Fentanyl[b]	100 μg	1–2 μg/kg/h as continuous infusion		25 μg patch	72 h (patch)
Controlled-release morphine[c,d]				0.6 mg/kg every 8 h or 0.9 mg/kg every 12 h	
Controlled-release hydromorphone[d]				0.18 mg/kg every 12 h	
Controlled-release codeine[d]				3 mg/kg every 12 h	
Controlled-release oxycodone[d]				0.3–0.6 mg/kg every 12 h	
Methadone	10 mg	0.1 mg/kg every 4–8 h	1:2	0.2 mg/kg every 4–8 h	12–50 h

NOTE: Doses are for opioid naïve patients. For infants under 6 months, start at one-quarter to one-third the suggested dose and titrate to effect.
Principles of opioid administration:
1. If inadequate pain relief and no toxicity at peak onset of opioid action, increase dose in 50% increments.
2. Avoid IM administration.
3. Whenever using continuous infusion, plan for hourly rescue doses with short onset opioids if needed. Rescue dose is usually 50–200% of continuous hourly dose. If greater than 6 rescues are necessary in 24 h period, increase daily infusion total by the total amount of rescues for previous 24 h ÷ 24. An alternative is to increase infusion by 50%.
4. To change opioids- because of incomplete cross-tolerance: if changing between opioids with short duration of action, start new opioid at 50% of equianalgesic dose. Titrate to effect. If changing between opioids from short to long duration of action (i.e., morphine to methadone), start at 25% of equianalgesic dose and titrate to effect.
5. To taper opioids- anyone on opioids over 1 week must be tapered to avoid withdrawal: taper by 50% for 2 days, and then decrease by 25% every 2 days. When dose is equianalgesic to an oral morphine dose of 0.6 mg/kg/day, it may be stopped. Some patients on opioids for prolonged periods may require much slower weaning.

[a]Avoid use in renal impairment. Metabolite may cause seizures.
[b]Potentially highly toxic. Not for use in acute pain control.
[c]Use may be hampered by child's difficulty in swallowing large tablets.
[d]The widely equianalgesic doses in adults are used as guidelines in pediatric practice but have not been substantiated in children.
ABBREVIATIONS: PO = by mouth; IV = intravenous.
SOURCE: Modified with permission from McGrath and Brown, 2004.[71]

and alternative therapies.[55–58] Although nondrug therapies generally are regarded as safe, with few contraindications for their use in otherwise healthy children, little is known about their safety and effectiveness for children. Pediatric research is just beginning on many of the therapies regarded as complementary to traditional medical approaches, such as acupuncture.[59] Thus the efficacy of complementary therapies for treating children's pain is unknown. In contrast, the evidence base supporting the efficacy of psychological therapies, particularly cognitive and behavioral approaches, is strong (for review, see refs. 60 and 61).

▶ **TABLE 17-3.** ADJUVANT ANALGESIC DRUGS FOR CHILDREN

Drug Category	Drug, Dosage	Indications	Comments
Antidepressants	Amitryptyline, 0.2–0.5 mg/kg PO. Titrate upward by 0.25mg/kg every 2–3 days. Maintenance: 0.2–3.0 mg/kg Alternatives: nortriptyline, doxepin, imipramine, venlafaxine.	Neuropathic pain (i.e., vincristine-induced, radiation plexopathy, tumor invasion, CRPS-1). Insomnia.	Usually improved sleep and pain relief within 3–5 days. Anticholinergic side effects are dose-limiting. Use with caution for children with increased risk for cardiac dysfunction.
Anticonvulsants	Gabapentin, 5 mg/kg/day PO. Titrate upward over 3–7 days. Maintenance: 15–50 mg/kg/day PO divided TID. Carbamazepine, Initial dosing: 10 mg/kg/day PO divided OD or BID. Maintenance: up to 20–30 mg/kg/day PO divided every 8h. Increase dose gradually over 2–4 weeks. Alternatives: phenytoin, clonazepam.	Neuropathic pain, especially shooting, stabbing pain.	Monitor for hematological, hepatic, and allergic reactions with carbamazepine. Side effects: gastrointestinal upset, ataxia, dizziness, disorientation, somnolence.
Sedatives, hypnotics, anxiolytics	Diazepam, 0.025–0.2 mg/kg PO every 6 h. Lorazepam, 0.05 mg/kg/dose SL. Midazolam, 0.5 mg/kg/dose PO administered 15–30 min prior to procedure; 0.05 mg/kg/dose IV for sedation.	Acute anxiety, muscle spasm. Premedication for painful procedures.	Sedative effect may limit opioid use. Other side effects include depression and dependence with prolonged use.
Antihistamines	Hydroxyzine, 0.5 mg/kg PO every 6 h. Diphenhydramine, 0.5–1.0 mg/kg PO/IV every 6 h.	Opioid-induced pruritus, anxiety, nausea.	Sedative side effects may be helpful.
Psychostimulants	Dextroamphetamine, Methylphenidate, 0.1–0.2 mg/kg BID. Escalate to 0.3–0.5 mg/kg as needed.	Opioid-induced somnolence. Potentiation of opioid analgesia.	Side effects include agitation, sleep disturbance, and anorexia. Administer second dose in afternoon to avoid sleep disturbances.
Corticosteroids	Prednisone, prednisolone, and dexamethasone dosage depends on clinical situation (i.e., dexamethasone initial dosing: 0.2 mg/kg IV. Dose limit 10 mg. Subsequent dose 0.3 mg/kg/day IV divided every 6 h).	Headache from increased intracranial pressure, spinal, or nerve compression; widespread metastases.	Side effects include edema, dyspeptic symptoms, and occasional gastrointestinal bleeding.

ABBREVIATIONS: CRPS-1 = Complex Regional Pain Syndrome, Type 1; PO = by mouth; IV = intravenous; SL = sublingual.
SOURCE: Modified with permission from McGrath and Brown, 2004.[71]

Psychological Therapies

Cognitive therapies are directed at a child's beliefs, expectations, and coping abilities. Accurate information about procedures and nonpainful sensations that may be experienced should improve a child's understanding, decrease his or her emotional distress, and reduce his or her pain and anxiety. Health care providers can teach children how to use simple independent pain control methods such as distraction. When a child's attention is fully absorbed in concentrating on an engaging topic or activity, distraction is a very active process that can reduce the neuronal responses to a noxious stimulus, lessening the intensity and unpleasantness of the pain. The choice of distraction is crucial and varies according to a child's age and interests.

Behavior therapy is often used in combination with cognitive therapy. The goals are to lessen the specific behaviors that may increase pain, distress, or disability while concomitantly increasing healthy behaviors that engage children in living as fully as possible. Children should learn that pain from some procedures is generally less when they are able to be involved with the procedure and when they are very relaxed. Relaxation training is also a common method used for children with chronic pain. Biofeedback can be a useful tool for teaching children to recognize when their bodies are relaxed but is not recommended for children whose pain problems are exacerbated by extremely high performance expectations because it may increase their stress and tension.

Children seem more adept than adults at using nondrug therapies, presumably because they are usually less biased than adults about the potential efficacy of these interventions. Health care providers should teach children a few basic attention and distraction methods to reduce pain and guide families to recognize the particular circumstances that exacerbate pain and distress. Together, health professionals and parents can relieve a child's pain not only by administering analgesic drugs but also by helping children to understand the situation and allowing them to make choices and to gain whatever control is possible within the setting.

▶ PRACTICAL GUIDELINES FOR TREATING COMMON PAIN PROBLEMS

Acute Pain from Medical Procedures

Preemptive analgesia is the key for managing acute pain from invasive medical procedures. Depending on the procedure, health care providers may choose psychological methods, anesthetic techniques, sedation, and analgesics. For example, a child requiring a diagnostic or therapeutic lumbar puncture and bone marrow aspiration may first receive a local anesthetic cream applied to the skin prior to the procedure to minimize pain from puncture of the skin. Additional local anesthetic also can be infiltrated into the periosteum prior to bone marrow aspiration, not only to minimize local pain, but also to minimize the total anesthetic required for the procedure. Acetaminophen may be administered rectally during the procedure or orally following the procedure. When appropriate, NSAIDs also provide analgesia and assist in pain control after the procedure. While these two procedures generally are performed under general anesthesia in the authors' institution, moderate sedation protocols combining appropriate sedation and analgesia can control acute pain effectively and may prevent long-term pain sequelae for those children who require repeated invasive procedures (for review, see refs. 51, 54, and 62).

During an invasive procedure, children are often uncertain about what to expect and may not know any simple pain-reducing strategies. Children who receive multiple invasive procedures throughout a prolonged time period are at particular risk for developing anxiety and fear. Inadvertently, health care providers can increase a child's pain and distress when they do not help a child to understand why treatments are needed, how to behave during treatments to increase control, and how to use a pain control method to gain mastery as an active participant rather than a passive victim. Moreover, unpredictable treatment schedules or inconsistency in how treatments are conducted can intensify a child's anxiety, distress, and treatment-related pain. Regrettably, the more procedures a child has, the more likely it is that the child will experience a difficult procedure—more pain, more nausea, etc. A child then becomes fearful and expects that another difficult procedure is inevitable, and a cycle of heightened anxiety, increased pain, and emotional distress ensues. In order to end this cycle, children should be referred to trained health care professionals who can assess and modify the specific factors affecting the child and family.

To minimize the possibility that this aversive cycle will develop, health care providers should prepare children adequately for what will happen; what to do to lessen anxiety, distress, and pain; and what they may feel. The simple provision of information about what is happening and why the procedure is conducted in a certain manner is helpful. What equipment will be used, who will be present, the nature of the child's sensations during different phases of the procedure, and some straightforward instructions about what the child can do to help lessen pain will reduce the pain caused by any invasive procedure dramatically. Since pain during procedures varies, with some children experiencing slight pain (assuming that appropriate analgesics or anesthetics are used), the most accurate preparation is, "It sometimes hurts a little, but not always." Emphasis should be placed on what sensations a child will feel—the warmth, coolness, or pressure associated with the procedure—rather than on the frightening label *pain*.

For example, before an injection, a child can be told: "I am not sure exactly what you will feel. Some children your age describe it as a tickling jab, a prick like a mosquito's bite, or like a sharp sting. It doesn't always hurt. We'll do things to make it hurt less! You can choose which finger [or arm], help to get the area ready, and then deeply rub the area to close your pain gate. Do you want to look, or do you want to look away?" Some children prefer to watch intently while others prefer to look away, watch something interesting, or use other forms of distraction. A practical form of distraction for children who choose to watch is asking them to describe the sensations they feel. The extent of information a child requires depends on his or her interest and needs.

Most children will state when they have heard enough or will ask more in-depth questions if needed. All children require some control, which can be provided by allowing them to have as much choice as possible during invasive treatments, by teaching them some simple pain control methods, and by motivating them to select and routinely use such methods during procedures. Procedures should be conducted as consistently as possible. In order that staff members do not inadvertently reinforce distress behaviors, they should agree on clear plans for children and encourage them to actively use independent pain control methods.

Even young children can easily learn to use several practical pain control methods, such as deep breathing, counting slowly and regularly, or progressive muscle relaxation.[1,19,63-65] Children seem to possess an enhanced ability to absorb themselves completely in a task, game, or imagined event and thus might be more able than adults to trigger endogenous pain-inhibitory mechanisms. For planned hospital admissions, children should attend relevant preparation programs to learn about what will happen during their stay, including the type of medical procedures they will receive. Most parents benefit from written guidelines for how they should respond to a child's questions, such as those summarized in Table 17-4 for parents of children with cancer.

Recurrent Pain Syndromes

Many children suffer from recurrent headache, abdominal pain, or limb pain. The medical care of children with recurrent pain varies greatly from the child with acute pain. While many acute pain or procedural pain services are run by individual health care providers with the short-term objective of acute pain relief and often quick discharge of children from hospital after medical procedure, the scope and difficulties of children with recurrent pain are vastly different. In this section we describe key aspects of managing a child's recurrent pain, using headache as an example.

Headache, particularly migraine and tension-type headache, is a major problem for many otherwise healthy children and adolescents.[66,67] Headache causes significant suffering and disability in children, creates anxiety and disruption for their families, and represents a substantial cost to parents and the health care system.[68,69] Recurrent headaches are not symptoms of an underlying disease that requires medical treatment;

▶ **TABLE 17-4.** GUIDELINES FOR PREPARING CHILDREN FOR PAINFUL PROCEDURES

Be honest with your child about what will happen.

Explain the procedure in age-appropriate language so he/she can understand the why's and what's and who's—with respect to the rationale for the procedure, the equipment that will be used, and who will be present. For young children, analogies with a pet are often beneficial. If your child has a pet and can understand that it needs injections and examinations, they may be less frightened when you explain what will happen as if their pet needed it. Ask them how they could make it easier for the pet. In this way, your child may focus more on actively coping than on being frightened.

Emphasize the qualitative sensations he/she may experience such as cold, tingling, or pressure, so that children focus on what they are feeling, not just on a hurting aspect.

If the procedure will cause some pain, describe the pain that the child may feel in familiar terms that he/she will understand. For example, use examples of pains that they have experienced already during play or pains that they may have observed other family members experience without distress.

Focus attention on what you and your child can do to make the procedure less distressing and less painful. Have a game plan of simple things that he/she can do such as taking slow deep breaths while you pace the rhythm, becoming immersed in a distracting image, or attending to the various stages of the procedure (depending on his/her preference for being involved or being distracted).

Choice, control, and predictability are very important for children receiving invasive treatments. Allow your child as much choice as possible such as which arm (for injections), whether to watch or look away, and which pain-reducing tool to use.

Remember that you and your child may have different preferences for coping with invasive procedures. You might prefer to be distracted while your child might prefer to be involved. It is important to follow your child's preferences rather than unintentionally requiring him/her to follow your preferences. You should know which is your child's preference from his/her past experiences—for example, does he/she watch or look away if you remove a sliver from his finger, does he/she like to remove bandages independently or prefer that you do it.

After the procedure is over, praise your child for coping and following the plan that you both chose. Even if your child showed some distress, praise him for trying and explain that you will work on the plan to make it better. Your child may need practice as with any other activity he/she has tried (e.g., riding a bike, roller-blading, skating). Be careful to not praise him for just getting through it—because you might be rewarding him/her for enduring it as a victim and inadvertently make him more frightened or distressed for future procedures.

SOURCE: *Reprinted with permission from McGrath and Hillier, 2003.*

instead, the recurrent headache syndrome is the disorder. Analgesic drugs can relieve the pain of individual headache attacks, but drug therapy alone does not alleviate the cycle of repeated attacks. Effective treatment of this syndrome requires a multimodal approach with cognitive-behavioral therapy in addition to abortive drug therapy. The treatment regimen is based on modifying the factors that trigger attacks and those which exacerbate pain and disability.

Although hereditary factors (especially positive maternal history) may predispose a child to develop the condition, other situational factors can trigger recurrent attacks, increase pain, prolong disability, or maintain the cycle of repeated attacks. The most common cause of headache attacks is stress—related to schoolwork, sports participation, or peer relationships. The relevant feature underlying different stressors is a child's inability to fully resolve the stressful situation rather than simply exposure to the stress.[29] Some children also may experience headache due to a major underlying emotional problems such as anxiety or depression.

Primary care providers should diagnose and explain the syndrome clearly to parents. In our clinical experience, parents are often confused by their child's initial headache diagnosis—accepting that the physician did not find anything wrong but often believing that something is wrong with their child that might be found with additional tests. Thus, the emphasis of the traditional diagnosis was somewhat negative—"a failure to find something" rather than a positive confirmation of an understandable and, therefore, treatable condition. We believe that it is essential to provide parents and children with a confident, positive diagnosis of the headache type—migraine, tension-type, mixed—with an emphasis on our improved understanding of this common recurrent-pain problem. Although some children require a specialized pain management program, many children (especially if they have had headaches for only a few months) may benefit from the brief educational intervention outlined in Table 17-5.[70]

Primary care providers know a child and family best. They can build on this trusting relationship to help explain the multifactorial etiology of headache and the need for a multimodal treatment approach, in direct contrast to the "single-cause and single-treatment" approach normally adequate for relieving acute pain.[71] Parents and children should understand that each causative factor must be treated and that the most effective treatments include education, basic drug and nondrug pain management techniques, and resolution of the stressful situations that provoke headache or intensify chronic pain. During the first consultation appointment, primary care providers can explain the common pain triggers and make explicit recommendations for managing pain and preventing disability.

▶ **TABLE 17-5.** COMPONENTS OF BRIEF TREATMENT FOR RECURRENT HEADACHE

Diagnostic Information
- Positive confirmation of recurrent migraine or tension-type headache
- Common pain problem for children and adolescents
- Multifactorial etiology, unlike acute pain

Treatment Information
- Unlike acute pain, a multimodal approach
- Targeted at all causes and contributing factors
- Drug or non-drug therapy effective for relief of headache pain
- Non-drug therapy effective for treating causes

Explicit Recommendations
- "Over-the-Counter" analgesics and non-drug methods to use during headache
- Guidelines on how to identify stressors through prospective monitoring
- Guidance on how to modify causes and contributing factors for child
- Guidance for parents on how to minimize disability behaviors

SOURCE: *Adapted with permission from McGrath and Hillier, 2001.*

Parents should leave the appointment with a concrete plan for what to do when their child develops a headache or when pain awakens a child, prevents school attendance, etc. In most instances, this plan should include an immediate pain intervention for the child and a subsequent discussion about what triggered the headache or exacerbated the pain and how that situation could be changed. This brief feedback intervention can lessen headache attacks significantly, decrease pain, and minimize pain-related disability. Some primary care providers may choose initially to review prospective records of pain activity to identify relevant contributing factors, subsequently evaluate a child's (and the family's) ability to manage the pain, and finally guide parents to identify and resolve those contributing factors. The common cognitive, behavioral, and emotional factors for children with recurrent headache are listed in Table 17-6.

Most children with long-term recurrent headache should be referred to a specialist for assessment (See Table 17-7). Even though children seemingly may have the same headache problem (i.e., migraine headaches at a frequency of one to two per month), the causative and contributing factors may be very different, and result in children requiring different treatment plans.[70] Parents often hold inaccurate beliefs about headache etiology, effective drug and nondrug therapies, and the role of environmental versus stress triggers for headache attacks or pain exacerbations. Parents usually try to understand recurrent headache from an acute pain perspective, where pain is often due to a single cause and requires a

▶ **TABLE 17-6.** COMMON SITUATIONAL FACTORS FOR CHILDREN WITH RECURRENT PAIN

Cognitive: Beliefs about etiology, pain control, and the role of stress
    Belief in single, as yet undiagnosed, cause
    Belief in "presumed" environmental triggers
    Inaccurate understanding of the primary and secondary causes, especially that stress can cause headache
    Poor knowledge about using and evaluating drug therapies
    Little knowledge of effective non-drug therapies
    Little knowledge about the situations that are continuing stressors for children
    Little understanding of the impact of a child's high expectations for achievement

Behavioral: Child and parent responses
    Inconsistent parental responses during headache attacks
    Ineffective use of analgesic drugs and independent non-drug therapies
    Children withdrawn from school, sports, or social activities
    Children relieved from routine family responsibilities
    Persistent diagnostic search for environmental triggers
    Failure to resolve continuing sources of stress

Emotional: Child's feelings
    Situation-specific stress
    Emotional suppression or denial
    Anxiety re: high expectations for achievement
    Fear re: an undiagnosed condition

single treatment. They do not understand that unlike most acute disease- or trauma-related pains their children have already experienced, the pain may have several interrelated causes. Parents expect that they eventually will find one treatment that will immediately stop recurrent attacks or relieve chronic pain complaints. Thus, they may reject potentially effective treatments after only one attempt, even though the treatment would address some of the causes and might help to lessen pain over time. The lack of understanding about the multifactorial etiology of pain leads some parents to continue to search for specific environmental triggers that children

▶ **TABLE 17-7.** SITUATIONAL FACTORS FOR CHILDREN WITH CHRONIC PAIN

Cognitive: Beliefs about etiology, pain control, and the role of stress
    Belief in single, as yet undiagnosed, cause
    Inaccurate understanding that chronic pain typically has multiple causes and contributing factors

Behavioral: Child and parent responses
    Ineffective use of analgesic drugs and independent non-drug therapies
    Children withdrawn from school, sports, or social activities
    Children relieved from routine family responsibilities
    Persistent diagnostic search for single etiology, with a primary focus on additional medical consultations and diagnostic investigations
    Parental responses that reinforce illness and disability
    Parental modeling of pain behaviors

Emotional: Child's feelings and family emotional expression
    Child's emotional suppression or denial of normal feelings associated with negative experiences[a]
    Child's anxiety re: failing to achieve unrealistically high expectations (social, academic, sports)[a]
    Child and family's fear re: an undiagnosed condition
    Child and family's anxiety re: life-threatening potential of condition
    Fear re: the likelihood of increasing pain and disability
    Frustration re: the unpredictability of painful episodes and disruption of life activities for child and family
    Anger towards health care providers for failing to cure the pain
    Underlying depression or anxiety

[a]Certain types of chronic pain where functional disability is prominent.

could avoid to prevent future attacks. Their children are then at risk for developing various learned pain triggers.[72]

A parent's and child's beliefs about therapies exert a powerful influence on children. Some parents, especially those whose own pain is relieved only by potent medication, may doubt the efficacy of any nondrug therapies. Even if they attempt to use them with their children, they often communicate their attitudes to their child. A child's negative expectations will counteract any potential benefits. Other parents may select only passive nondrug methods that increase a child's dependency and lack of control. Parents may not know how to assist children in using active and independent pain coping strategies that they could incorporate into their regular activities. These methods will improve a child's control, lessen pain, and minimize any maladaptive disability behaviors.

Parents' and children's beliefs guide what they do to relieve pain. Several behaviors are critical for the development of recurrent headache, as listed in Table 17-6. Typically, several of these behaviors are present for a child, with all contributing to the pain syndrome. Parents' uncertainty about how to stop the headaches can lead them to respond to a child's headache complaint in a manner that inadvertently increases distress, pain, and disability. Moreover, when parents seek additional diagnostic tests, attribute headaches solely to environmental triggers, and fail to identify and resolve relevant stress triggers, their behaviors directly maintain the cycle of repeat headache attacks.

Primary care providers can assist families to understand how their behaviors (as well as those of care providers, teachers, coaches, and instructors) shape a child's pain response. Parents may inadvertently reinforce pain-related disability when they allow a child to stay at home instead of attending school, to withdraw from potentially stressful sports or social situations, and to relieve them from routine family responsibilities—*without helping the child to address the underlying causes for pain*. When a child has headaches due in part to their high expectations for achievement, the child is relieved of some performance pressure (either self-imposed pressure or parent/coach's pressure) when they experience pain. Children whose headaches are triggered or exacerbated by their anxiety about coping in particular situations (e.g., gym class, social interactions at lunch, team games) begin to develop more frequent headaches in association with those situations. For example, initially they may miss a few classes or sports activities on an intermittent basis because of their headache attacks, which leads to avoiding these situations on a more consistent basis. Stress reduction, inadvertently reinforced by parents, is a major secondary gain that prolongs the headache problem.

Recurrent headaches are caused by cognitive, behavioral, and emotional factors. Effective treatment requires a multimodal approach with cognitive-behavioral therapy in addition to abortive drug therapy. The treatment approach is child-centered, based on modifying the factors that trigger headache attacks and those which exacerbate pain and disability.

Chronic Pain

Children can experience chronic pain from disease, injury, and psychological factors. Chronic pain causes significant suffering, disability, anxiety, and emotional distress for many children and adolescents.[68,73] Unlike acute and recurrent pain, chronic pain often has nociceptive and neuropathic components. In addition, environmental, family, and psychological factors influence a child's chronic pain, so a multimodal therapeutic regimen consisting of drug, physical, and psychological therapies usually is required. This section describes key aspects of managing a child's chronic pain and minimizing pain-related disability, using non-disease-related chronic pain as an example.

Multidisciplinary chronic pain care teams typically include a medical director with expertise in medications and anaesthetic block procedures, a physiotherapist, a psychiatrist and/or psychologist, and a nurse clinician. Children may be seen individually by team members or interviewed within the team setting. Following a complete history and physical examination, the varied team members share their findings to ascertain the child's pain diagnosis and to formulate a treatment plan. The medical director presents this information to the child and parents, concisely explaining the basis for the child's diagnosis and presenting a clear rationale for the recommended therapies. By the time children are referred to a specialized clinic, parents may have received some conflicting diagnostic or therapeutic information, so it is essential that the team allows sufficient time to answer the family's questions.

Drug therapies include the analgesics and adjuvant analgesics, listed in Tables 17-1 through 17-3. Most children with non-disease-related chronic pain (i.e., CRPS-I) can achieve pain control through an individualized regimen, using acetaminophen or NSAIDs for pain associated with tissue inflammation, and adjuvant analgesics for pain associated with nerve injury. Only rarely do children with non-disease-related pain require opioid analgesics for long-term management. Opioids are prescribed on occasion for children whose pain severity increases intermittently and is orthopedic in origin, pain that requires a period of bone healing after a fracture (i.e., osteogenesis imperfecta), or pain prior to joint replacement (i.e., avascular necrosis of femoral head after chemotherapy).

Frequently, children have already tried medications that should have been effective but either have used an insufficient dosing schedule or a trial duration. Medications are best prescribed on a sequential basis in that only one medication is prescribed at a time and for an adequate trial period. For example, the authors prescribe gabapentin to children whose pain has neuropathic features, such as allodynia and hyperesthesia. A standard dose of 300 mg is often increased over a 3-day period to a dose of 300 mg three times daily. If children experience appropriate pain relief, then the dose is maintained. However, if children experience only partial pain relief, the 900 mg dose can be adjusted gradually up to a maximum of 65 mg/kg per day. The dosage should be increased until the maximum positive effect is achieved with a minimum of side effects, which often limit this increase in dosage. A positive effect is often noticed within 2 to 3 weeks of commencing therapy. In our clinical experience, children with neuropathic pain whose sleep is severely disrupted may respond best to a trial of amitriptyline rather than gabapentin. Often the amount required remains a low therapeutic dose (i.e., 10 to 20 mg). Most of the pharmacologic management of neuropathic pain in children and adolescents is based on extrapolation from adult studies. While tricyclic antidepressants and gabapentin are well-established analgesics for these conditions in adults, evidence for efficacy in children is confined to case reports or very small series.[74,75]

While therapeutic blocks remain a standard therapy in adult pain clinics, in our experience they are employed only infrequently in children's pain clinics when all other medical therapeutic modalities have been exhausted. For example, a sympathetic block is required occasionally for the patient with CRPS-I when medication, physiotherapy and psychological interventions have failed to achieve the desired pain control.

Chronic pain adversely affects all aspects of a child's life. A child may endure a prolonged period of physical disability, continuing pain, and varied medical treatments. Parents are distressed by the pain itself, its implications for their child's future, its life-threatening potential (if any), and the prospect of continuing pain and progressive disability. The dynamics within the family (both for siblings and extended family members) inevitably change and often sadly deteriorate as chronic pain prevents children from pursuing their normal activities and as family schedules adjust accordingly to the health care needs of the child with pain.

Many families have not received concrete information about the pain, the environmental factors that intensify it, and effective nondrug therapies to complement the primary analgesic or anesthetic treatment regimen. This results in many children not knowing simple pain control methods that they can incorporate practically into their daily activities. Of paramount importance, most families do not receive any information about the plasticity and complexity of pain, so they continue to try to understand their child's chronic pain from an inappropriate acute pain model—wherein there is a single cause and a single treatment. Because a child with chronic pain typically has multiple sources of pain with nociceptive and neuropathic components affecting the peripheral and central nervous systems, parents and children become confused, anxious, and frustrated by the apparent failure of the health care system to find and treat the cause. Families become fearful that the child will continue to have pain or must rely on potent opioid analgesics to obtain even partial pain relief. A child's chronic pain is further complicated by decreased independence, reduced control, and an uncertain prognosis.

A child's usual behaviors are restricted by physical disability related to the condition or injury so that he or she gradually withdraws from social activities with peers and from most physical sports. As normal activities decrease, abnormal sensory input may increase, producing concomitant increases in pain. Parents, siblings, and health care providers can reward disability behaviors when they encourage a child to adopt a passive patient role, behave differently from other children, and depend primarily on others for pain control. These situational factors can intensify pain and distress and prolong disability. Parents who encourage a child to resume as many normal activities as possible create an environment where their child's pain will be minimized.

Physical therapy is essential for children who have gradually restricted their behaviors due to pain, particularly children who have heightened sensitivity to touch. These children need practical desensitization exercises and a flexible program of specific exercises to rehabilitate their bodies and resume normal activities. In addition to drug or physical therapies to address the primary pain sources, a child with chronic pain requires a cognitive-behavioral approach to address the secondary contributing factors and may require supportive counseling to ameliorate their emotional distress. (Note: Certain children with chronic pain also may have long-standing emotional problems suggestive of mood disorders, anxiety disorders, and somatoform disorders.) Childhood chronic pain must be viewed from a multidimensional perspective because multiple sensory, environmental, and psychological factors are responsible for the pain, no matter how seemingly clear-cut the etiology.

▶ SUMMARY

Optimal pain control for children with acute pain from medical procedures or children with recurrent and chronic pain problems requires an integrated treatment approach with both drug and nondrug therapies. However, the specific interventions must be selected after determination of the primary and secondary sources of noxious stimulation and after a thorough assessment of the unique situational, behavioral, emotional, and familial factors that affect a child's pain. It is impossible to relieve a child's pain adequately from a unidimensional perspective, in which pain is considered as synonymous with the nature and extent of tissue damage. Childhood pain must be viewed from a multidimensional perspective because multiple sensory, environmental, and emotional factors are responsible for the pain—no matter how seemingly clear-cut the etiology. Treatment begins with a thorough assessment of these multiple factors using structured interviews and standardized measures. Pharmacologic, physical, and psychological strategies must be incorporated into a flexible intervention program for children in which parents and siblings form an essential component of treatment.

Pain control should include regular pain assessments, appropriate analgesics administered at regular dosing intervals, adjunctive drug therapy for symptom and side-effect control, and nondrug interventions to modify the situational factors that can exacerbate pain and suffering. Children should learn some simple pain control strategies so they can reduce acute pain caused by invasive treatments and disease or therapy-related pain.

While certain aspects of a multimodal pain care plan may have been emphasized in this chapter, it is difficult to differentiate the importance of one aspect over another. As an analogy, if a child's chronic pain equated to that of a bicycle wheel and all the various methods used to treat that pain are represented by the spokes of the wheel, it is impossible to tell whether some spokes (such as drug therapies) versus other spokes (such as nondrug therapies) play a more important role in affecting a cure for the child's pain.

Effective pain control is possible when the goals are to reduce or block nociceptive activity by attenuating responses in peripheral afferents and central pathways, activating endogenous pain-inhibitory systems, and modifying situational factors that exacerbate pain. Thus, the choice for pain control is not merely "drug versus nondrug therapy" but rather a therapy that mitigates both the causative and contributing factors for pain. Different combinations of drug and nondrug therapies will be required at different times. Health professionals must continually assume as much responsibility for monitoring and relieving children's pain as for medically managing their diseases.

REFERENCES

1. McGrath PA, Hillier LM: Modifying the psychological factors that intensify children's pain and prolong disability, in Schechter NL, Berde CB, Yaster M (eds), *Pain in Infants, Children, and Adolescents,* 2d ed. Philadelphia: Lippincott, Williams & Wilkins, 2003:85–104.
2. Schechter NL, Berde CB, Yaster M: Pain in infants, children, and adolescents: An overview, in Schechter NL, Berde CB, Yaster M (eds), *Pain in Infants, Children, and Adolescents,* 2d ed. Philadelphia: Lippincott, Williams & Wilkins, 2003:3–18.
3. Fitzgerald M: Development of the peripheral and spinal pain system, in Anand KJS, Stevens BJ, McGrath PJ (eds), *Pain in Neonates*. Amsterdam: Elsevier, 2000:9–22.
4. Fitzgerald M, Howard RF: The neurobiologic basis of pediatric pain, in Schechter NL, Berde CB, Yaster M (eds), *Pain in Infants, Children, and Adolescents,* 2d ed. Philadelphia: Lippincott, Williams & Wilkins, 2003:19–42.
5. Andrews KA: The human developmental neurophysiology of pain, in Schechter NL, Berde CB, Yaster M (eds), *Pain in Infants, Children, and Adolescents,* 2d ed. Philadelphia: Lippincott, Williams & Wilkins, 2003:43–57.
6. Kaas JH, Catania KC: How do features of sensory representations develop? *Bioessays* 2002;24:334–343.
7. Fitzgerald M, Koltzenburg M: The functional development of descending inhibitory pathways in the dorsolateral funiculus of the newborn rat spinal cord. *Brain Res* 1986;389:261–270.
8. Boucher T, Jennings E, Fitzgerald M: The onset of diffuse noxious inhibitory controls in postnatal rat pups: A c-fos study. *Neurosci Lett* 1998;257:9–12.
9. Andrews K, Fitzgerald M: Cutaneous flexion reflex in human neonates: A quantitative study of threshold and stimulus-response characteristics after single and repeated stimuli. *Dev Med Child Neurol* 1999;41:696–703.
10. Fitzgerald M, Shaw A, MacIntosh N: Postnatal development of the cutaneous flexor reflex: Comparative study of preterm infants and newborn rat pups. *Dev Med Child Neurol* 1988;30:520–526.
11. Fitzgerald M, Jennings E: The postnatal development of spinal sensory processing. *Proc Natl Acad Sci USA* 1999;96:7719–7722.
12. Andrews K, Fitzgerald M: Wound sensitivity as a measure of analgesic effects following surgery in human neonates and infants. *Pain* 2002;99:185–195.
13. Grunau RE: Long-term consequences of pain in human neonates, in Anand KJS, Stevens BJ, McGrath PJ (eds), *Pain in Neonates*. Amsterdam: Elsevier, 2000:55–76.
14. Goldschneider KR, Anand KS: Long-term consequences of pain in neonates, in Schechter NL, Berde CB, Yaster M (eds), *Pain in Infants, Children, and Adolescents,* 2d ed. Philadelphia: Lippincott, Williams & Wilkins, 2003:58–70.
15. Taddio A, Goldbach M, Ipp M, et al: Effect of neonatal circumcision on pain responses during vaccination in boys. *Lancet* 1995;345:291–292.
16. Taddio A, Katz J, Ilersich AL, et al: Effect of neonatal circumcision on pain response during subsequent routine vaccination. *Lancet* 1997;349:599–603.
17. Grunau RE, Whitfield MF, Petrie JH: Children's judgements about pain at age 8–10 years: Do extremely low birth-weight (≤1000 g) children differ from full birthweight peers? *J Child Psychol Psychiatry* 1998;39:587–594.
18. Porter FL, Wolf CM, Miller JP: Procedural pain in newborn infants: The influence of intensity and development. *Pediatrics* 1994;104:e13.

19. McGrath PA: *Pain in Children: Nature, Assessment and Treatment*. New York: Guilford Press, 1990.
20. Bijttebier P, Vertommen H: The impact of previous experience on children's reactions to venipunctures. *J Health Psychol* 1998;3:39–46.
21. Dahlquist LM, Gil KM, Armstrong FD, et al: Preparing children for medical examinations: The importance of previous medical experience. *Health Psychol* 1986;5:249–259.
22. Thastum M, Zachariae R, Herlin T: Pain experience and pain coping strategies in children with juvenile idiopathic arthritis. *J Rheumatol* 2001;28:1091–1098.
23. Bennett-Branson SM, Craig KD: Postoperative pain in children: Developmental and family influences on spontaneous coping strategies. *Can J Behav Sci* 1993;25:355–383.
24. Palermo TM, Drotar D: Prediction of children's postoperative pain: The role of presurgical expectations and anticipatory emotions. *J Pediatr Psychol* 1996;21:683–698.
25. Gidron Y, McGrath PJ, Goodday R: The physical and psychosocial predictors of adolescents' recovery from oral surgery. *J Behav Med* 1995;18:385–399.
26. Edwards PW, O'Neill GW, Zeichner A, et al: Effects of familial pain models on pain complaints and coping strategies. *Percept Mot Skills* 1985;61:1053–1054.
27. Edwards PW, Zeichner A, Zuczmierczyk AR, et al: Familial pain models: The relationship between family history of pain and current pain experience. *Pain* 1985;21:379–384.
28. Schanberg LE, Keefe FJ, Lefebvre JC, et al: Social context of pain in children with juvenile primary fibromyalgia syndrome: Parental pain history and family environment. *Clin J Pain* 1998;14:107–115.
29. McGrath PA, Hillier LM: Recurrent headache: Triggers, causes, and contributing factors, in McGrath PA, Hillier LM (eds), *The Child with Headache: Diagnosis and Treatment*. Seattle: IASP Press, 2001:77–107.
30. McGrath PA, Gillespie JM: Pain assessment in children and adolescents, in Turk DC, Melzack R (eds), *Handbook of Pain Assessment*. New York: Guilford Press, 2001:97–118.
31. Finley GA, McGrath PJ (eds): *Measurement of Pain in Infants and Children*. Seattle: IASP Press, 1998.
32. Gaffney A, McGrath PJ, Dick B: Measuring pain in children: Developmental and instrument issues, in Schechter NL, Berde CB, Yaster M (eds), *Pain in Infants, Children, and Adolescents*, 2d ed. Philadelphia: Lippincott, Williams & Wilkins, 2003:128–141.
33. McGrath PJ: Behavioral measures of pain, in Finley GA, McGrath PJ (eds), *Measurement of Pain in Infants and Children*. Seattle: IASP Press, 1998:83–102.
34. Stevens B, Johnston C, Gibbins S: Pain assessment in neonates, in Anand KJS, Stevens BJ, McGrath PJ (eds), *Pain in Neonates*. Amsterdam: Elsevier, 2000:101–134.
35. Hunt AM, Goldman A, Mastroyannopoulou K, et al: Identification of pain cues of children with severe neurological impairment, in Devor M, Rowbotham MC, Wiesenfeld-Hallin Z (eds), *Proceedings of the 9th World Congress on Pain*. Seattle: IASP Press, 1999:abstract 84.
36. McGrath PJ, Rosmus C, Canfield C, et al: Behaviours caregivers use to determine pain in non-verbal, cognitively impaired individuals. *Dev Med Child Neurol* 1998;40:340–343.
37. Hadden KL, von Baeyer CL: Pain in children with cerebral palsy: Common triggers and expressive behaviors. *Pain* 2002;99:281–288.
38. Breau LM, McGrath PJ, Camfield CS, et al: Psychometric properties of the non-communicating children's pain checklist-revised. *Pain* 2002;99:349–357.
39. Stallard P, Williams L, Velleman R, et al: The development and evaluation of the pain indicator for communicatively impaired children (PICIC). *Pain* 2002;98:145–149.
40. Terstegen C, Koot HM, deBoer JB, et al: Measuring pain in children with cognitive impairment: pain response to surgical procedures. *Pain* 2003;103:187–198.
41. Oberlander TF, Craig KD: Pain and children with developmental disabilities, in Schechter NL, Berde CB, Yaster M (eds), *Pain in Infants, Children, and Adolescents*, 2d ed. Philadelphia: Lippincott, Williams & Wilkins, 2003:599–619.
42. Ross DM, Ross SA: *Childhood Pain: Current Issues, Research, and Management*. Baltimore: Urban & Schwarzenberg, 1988.
43. Bush JP, Harkins W (eds): *Children in Pain: Clinical and Research Issues from a Developmental Perspectives*. New York: Springer-Verlag, 1991.
44. Champion GD, Goodenough B, vonBaeyer CL, et al: Measurement of pain by self-report, in Finley GA, McGrath PJ (eds), *Measurement of Pain in Infants and Children*. Seattle: IASP Press, 1998:123–160.
45. McGrath PA, Koster AL: Headache measures for children: A practical approach, in McGrath PA, Hillier LM (eds), *The Child with Headache: Diagnosis and Treatment*. Seattle: IASP Press, 2001:29–56.
46. Varni JW, Thompson, KL, Hanson V: The Varni/Thompson pediatric pain questionnaire: I. Chronic musculoskeletal pain in juvenile rheumatoid arthritis. *Pain* 1987;28:27–38.
47. Acute Pain Management Guideline: *Clinical Practice Guideline: Acute Pain Management in Infants, Children, and Adolescents: Operative and Medical Procedures*. Rockville, MD: Agency for Health Care Policy and Research, 1992.
48. Finley GA: Pharmacological management of procedure pain, in Finley GA, McGrath PJ (eds), *Acute and Procedure Pain in Infants and Children*. Seattle: IASP Press, 2001:57–76.
49. Wolf AR: Local and regional analgesia, in Finley GA, McGrath PJ (eds), *Acute and Procedure Pain in Infants and Children*. Seattle: IASP Press, 2001:33–56.
50. Krane EJ, Leong MS, Golianu B, et al: Treatment of pediatric pain with nonconventional analgesics, in Schechter NL, Berde CB, Yaster M (eds), *Pain in Infants, Children, and Adolescents*, 2d ed. Philadelphia: Lippincott, Williams & Wilkins, 2003:225–240.
51. McGrath PA, Brown SC: Pain control in children, in Doyle D, Hanks GWC, MacDonald N (eds), *Oxford Textbook of Palliative Medicine*. Oxford: Oxford University Press, 2003:1–35.
52. Hospital for Sick Children: *Drug Formulary 2001–2002*. Toronto: Hospital for Sick Children, 2002.
53. Wong CM, McIntosh N, Gopi M, et al: The pain (and stress) in infants in a neonatal intensive care unit, in Schechter NL, Berde CB, Yaster M (eds), *Pain in Infants, Children, and Adolescents*, 2d ed. Philadelphia: Lippincott, Williams & Wilkins, 2003:669–692.

54. Collins JJ, Weisman SJ: Management of pain in childhood cancer, in Schechter NL, Berde CB, Yaster M (eds). *Pain in Infants, Children, and Adolescents,* 2d ed. Philadelphia: Lippincott, Williams & Wilkins, 2003:517–538.
55. Spigelblatt L, Laine-Ammara G, Pless IB, et al: The use of alternative medicine by children. *Pediatrics* 1994;94:811–814.
56. Ernst E: Homeopathic prophylaxis of headaches and migraine? A systematic review. *J Pain Symptom Manage* 1999;18:353–357.
57. Kemper KJ Gardiner P: Complementary and alternative medical therapies in pediatric pain treatment, in Schechter NL, Berde CB, Yaster M (eds), *Pain in Infants, Children, and Adolescents,* 2d ed. Philadelphia: Lippincott, Williams & Wilkins, 2003:449–461.
58. Lin YC: Acupuncture, in Schechter NL, Berde CB, Yaster M (eds), *Pain in Infants, Children, and Adolescents,* 2d ed. Philadelphia: Lippincott, Williams & Wilkins, 2003:462–470.
59. Zeltzer, LK, Tsao JC, Stelling C, et al: A phase I study on the feasibility and acceptability of an acupuncture/hypnosis intervention for chronic pediatric pain. *J Pain Symptom Manage* 2002;24:437–446.
60. Eccleston C, Morley S, Williams A, et al: Systematic review of randomised controlled trials of psychological therapy for chronic pain in children and adolescents, with a subset meta-analysis of pain relief. *Pain* 2002;99:157–165.
61. McGrath PA, Holahan AL: Psychological interventions with children and adolescents: Evidence for their effectiveness in treating chronic pain. *Semin Pain Med* 2003;1:99–109.
62. Brown SC, Roy WL: Anesthesia and sedation for satellite and remote locations, in Bissonnette B, Dalens B (eds), *Pediatric Anesthesia: Principles and Practice*. New York: McGraw-Hill, 2002:627–642.
63. Barrera M: Brief clinical report: Procedural pain and anxiety management with mother and sibling as co-therapists. *J Pediatr Psychol* 2000;25:117–121.
64. Kazak AE, Penati B, Brophy P, et al: Pharmacologic and psychologic interventions for procedural pain. *Pediatrics* 1998;102:59–66.
65. Powers SW, Blount RL, Bachanas PJ, et al: Helping preschool leukemia patients and their parents cope during injections. *J Pediatr Psychol* 1993;18:681–695.
66. McGrath PA: Headache in children: The nature of the problem, in McGrath PA, Hillier LM (eds), *The Child with Headache: Diagnosis and Treatment*. Seattle: IASP Press, 2001:1–27.
67. Rothner AD: Differential diagnosis of headaches in children and adolescents, in McGrath PA, Hillier LM (eds), *The Child with Headache: Diagnosis and Treatment*. Seattle: IASP Press, 2001:57–76.
68. Perquin CW, Hazebroek-Kampschreur AA, Hunfeld JA, et al: Pain in children and adolescents: A common experience. *Pain* 2000;87:51–58.
69. Merlijn VP, Hunfeld JA, van der Wouden JC, et al: Psychosocial factors associated with chronic pain in adolescents. *Pain* 2003;101:33–43.
70. McGrath PA, Hillier LM: Treating recurrent headache: An effective strategy for primary care providers, in McGrath PA, Hillier LM (eds), *The Child with Headache: Diagnosis and Treatment*. Seattle: IASP Press, 2001:159–182.
71. McGrath, PA, Brown SC: Special considerations in pediatric pain management, in Lipman AG (ed), *Pain Management for Primary Care Physicians*. Bethesda, MD. American Society of Health-System Pharmacists, 2004:199–217.
72. Hillier LM, McGrath PA: A cognitive-behavioral program for treating recurrent headache, in McGrath PA, Hillier LM (eds), *The Child with Headache: Diagnosis and Treatment*. Seattle: IASP Press, 2001:183–220.
73. McGrath PA: Chronic pain in children, in Crombie IK, Croft PR, Linton SJ, et al (eds), *Epidemiology of Pain*. Seattle: IASP Press, 1999:81–101.
74. McGraw T, Kosek P: Erythromelalgia pain managed with gabapentin. *Anesthesiology* 1997;86:988–990.
75. Rusy LM, Troshynski TJ, Weisman SJ: Gabapentin in phantom limb pain management in children and young adults: Report of seven cases. *J Pain Symptom Manage* 2001;21:78–82.

CHAPTER 18

Pain in the Elderly

Stephen W. Harkins

Senescence is not universal in nature, not even common. Most creatures out in the wild die off, or are killed off, at the first loss of physical or mental power. In a real sense, aging, real aging, the continuation of living through the whole long period of senescence, is a human invention, and perhaps a relative recent one at that.

Lewis Thomas[1] (pp. 24–25)

Some people consider that pain in the elderly is different from that in younger adults.[2] Certainly the pattern of chronic pain problems seems to change with age, and evidence for this is reviewed here. But does this mean that pain threshold or pain tolerance is different in older compared with younger individuals? Does this mean that pain-related suffering is decreased with aging or, as some think, increased?

While substantial advances in our basic understanding of the physiologic mechanisms of pain have been gained in the past 20 years, geriatric pain has not been researched systematically. It appears that many older adults, particularly the dependent and frail elderly, are at risk for undertreatment of pain problems.

Critchley,[2] in an influential and otherwise very accurate lecture on the neurology of old age, indicated that pain sensitivity is obtunded in the old, much as are the other senses. Critchley instructed his medical students and physician colleagues that the elderly tolerate minor surgical procedures and even dental extractions with little or no discomfort. This is consistent with the point of view that aging results in a systematic increase in pain threshold and pain tolerance, i.e., sensory losses, that parallels those seen in the aging of other senses. The evidence reviewed here does not support a consistent loss of pain sensitivity with aging in the later years of life.

Reviewed here are studies of acute pain evaluated under controlled conditions in the laboratory, pain as a symptom seen in primary and tertiary care pain settings, and preliminary findings from population-based epidemiologic surveys.

Laboratory studies have provided information concerning pain sensitivity in different age groups. It is unclear how aging in the later years of life influences presentation of pain as a symptom. Symptoms in the elderly, particularly the frail elderly, present with chronicity, multiplicity, and duplicity. Management of pain under these conditions is complex and challenging.

The laboratory studies on pain in different age groups are at best untidy, and for the present, it is wise to conclude that the effects of aging on pain sensitivity are minimal and likely are not clinically significant. This is in contrast to the known effects of age on other senses.

Several well-designed population-based surveys exist that, while not designed as studies of pain or aging, do contain information on pain, particularly musculoskeletal pain, in various age groups. From these, preliminary evidence indicates that pain in the general community-dwelling population is a national public health problem and is likely a risk factor for social isolation and disability.

▶ DEFINITIONS

Scientific studies of pain and of aging have shared definitional problems that have hampered systematic research on their interactions. This is due, in part, to the fact that there are no universally accepted measures or tests for either the degree or the rate of change of either pain or aging. They are often referred to as *subjective phenomena*. Certainly, many consider measures of pain. Regarding old age, we all consider old to be at least 10 years advanced of our own age.

Pain has been defined as being subjective, and it is, therefore, subject to social history.[3] *Pain* is defined by the International Association for the Study of Pain as being an "unpleasant sensory experience that is associated with actual or potential tissue damage."[3]

A commonly accepted definition of being *old* is attainment of the age of 65 years. This is artificial. The choice of 65 years of age for retirement and eligibility for Social Security is inherited from social policy developed in Prince Otto von Bismarck's Germany during the 1880s.[4] This age has been subject to change based on political and economic forces driven by demographics. This is reflected today in the increasing age for full Social Security retirement benefits for those born after 1940.

The first edition of the *Oxford English Dictionary*[5] (OED) defines *aging* as "any period in the life span without reference to a specific chronological period." Aging, in this definition, may be considered to begin at conception and continue across the life span. In contrast, the second edition of the OED[6] refers to aging in terms of "being old."

A more precise definition of aging in the later years of life is needed. Biogerontology uses the term *senescence* to indicate the physiologic processes that occur later in the life span. Senescence begins at some point following the onset of physiologic maturity, which in humans is during the third decade of life or earlier.

Senescence as a biologic term is defined as "species-specific, universal in multicellular animals, irreversible, and progressive."[7] In demographic research, the force of mortality or age-specific death rates defines senescence. It is unclear, but likely, that senescence is relatively independent of the selective process of evolution because organisms in nature seldom live long enough to senesce.

It is the very fact that the number of older individuals in the population is increasing that research on senescence has taken on new meaning. Many wish to delay or even reverse the processes of senescence, whereas others wish simply to improve their overall quality of life. The fact is that life expectancy has increased substantially from birth and that more and more people are living to the upper practical limit (about 85 years of age) of the human life span. This has resulted in a compression of mortality in the later years of life.

▶ COMPRESSION OF MORTALITY

Over the past century, life expectancy from birth in the United States has increased from approximately 48 to over 76 years. This increase in life expectance has resulted from reductions in mortality early in life and not to an addition of years later in life. This has resulted in a demographic shift over the past two centuries to an older population, with an increasing number of individuals surviving into their seventh, eighth, and even ninth decades of life. The increase in the population 85 years of age and older is of critical importance due to the increased levels of disability in those approaching the upper limit of the human life span (approximately 123 to 125 years). This has resulted in demands for a change in medical practice and health education due to the increased number of older individuals and the fact that illness in this group is chronic more than acute. Alzheimer's disease, for example, now accounts for about 60 percent of those in long-term care settings, and some estimates are that up to 40 percent of community-dwelling individuals over the age of 85 years meet diagnostic screening criteria for possible dementia of the Alzheimer's type.

Another way to view this demographic change is in terms of raw numbers. In 1900, there were approximately 3 million individuals aged 65 years and older in the United States (about 4.1 percent of the total population). By 1994, the number of those 65 years of age and older increased to over 33.2 million (about 12.5 percent of the total population). Between 2010 and 2030, the number of older adults inevitably will increase dramatically due to aging of the baby-boom generation (individuals born between 1946 and 1964 who will begin entering the over-65 generation in 2011).

By 2030, those 65 years of age and older will number somewhere between 70 and 78 million (based on different mortality, fertility, and migration assumptions) and will comprise between 19 and 21 percent of the total population. While in 1900 one in 25 people was 65 years of age or older, by 2030 the most likely demographic scenarios project that 1 in 5 persons will be 65 and older. This density most likely will characterize the population from 2030 to 2050. This may occur as early as 2025. There are substantial policy implications resulting from this broad shift in population structure, as well as a need for focusing clinical practice on diagnosis and treatment of pain in the elderly.

This compression of mortality into the practical upper limit of human life expectancy (from birth) is not just an academic issue for gerontologists and demographers. The U.S. Congressional Budget Office estimates that in 1998 Social Security and Medicare spending was approximately 6 percent of the national gross domestic product. This is projected to rise to approximately 14 percent by 2060, a time when the surviving baby-boomers will be entering the ranks of the oldest-old. A reduction from a current level of 13 percent of gross national product to about a level of 5 percent for defense, law enforcement, education, and other federal government spending to cover this increase in mandated expenses for Social Security and Medicare costs[8] will be required.

Numbers, however, do not tell the whole story. Increased between-individual variability is a major characteristic of the old. Growth of the elderly population, no matter how dramatic, misses the fact that as a group they are characterized by great diversity—greater social, psychological, physiologic, and genetic diversity—than younger adults. This increased variability with aging is complicated by the contributions of diverse ethnic backgrounds, race, and gender. For example, location of historic residence, rural versus urban, has a major impact on expectancies and demands on the health care system by the elderly. The increasing number and the increasing diversity of older individuals will affect the health care delivery system in ways that we cannot now imagine. If we are living healthier, as well as longer, then we would expect that morbidity would be compressed into the later years of life.

▶ COMPRESSION OF MORBIDITY

There is no question that the geriatric imperative, or "graying," of the more developed countries is due to a compression of mortality into the later years of the human life span. Simply put, people are dying nearer to the practical upper limit of life, or at about 85 years of age. A critical question concerns whether this longevity is accompanied by health. Basically, the question is, Are new cohorts of elders healthier than past cohorts? Are disability and chronic illness being compressed into a shorter and shorter period of time near death in old age? The possibility of increasing health until the upper limit of the human life span has been characterized as a *compression of morbidity*.[9] Certainly, improvements in quality of life and health will result in future savings in health care, long-term care, and social support costs.

An alternative view is there will be an *expansion of morbidity* as the population ages. The impact on the health care system of nonfatal but debilitating diseases that decrease ability to perform common and desired activities of daily living will increase according to this view. In this regard, in the geriatric community attention has focused on the impact of age-related conditions such as sensory impairment, incontinence, osteoporosis, Parkinson's disease, and Alzheimer's disease. Little specific attention has been focused on the impact of pain on successful aging.[10]

▶ PRESBYALGOS

If aging in the later years of life influences perceptions or responses to painful stimuli, these changes could be summarized by the term *presbyalgos* (*presby*—"old"; *algos*—"pain"). This would, of course, parallel the well-documented age-dependent changes associated with presbycusis and presbyopia. Presbycusis is associated with elevated thresholds for pure tones and decreased ability to discriminate speech sounds, particularly in an acoustically noisy environment. Hearing loss with age is greater in women than in men, is greater for higher frequencies, is progressive, and is irreversible. The pattern of hearing loss associated with presbycusis has been termed *sensorineural* hearing loss and is due primarily to damage to hair cells (cochlear) with secondary loss of spiral ganglion afferents. Presbyopia, in turn, is a decrease in accommodation with increasing loss of near-point vision due to loss of elasticity of the cornea and perhaps some weakening of ciliary muscles.

It is important to realize that the age-dependent changes in audition and vision are associated with properties of energy transduction and reception. It is likely that the more complex the anatomic and physiologic apparatus associated with a sensory process, the more likely it is that time (age) will act adversely in a consistent fashion on that sense.

▶ LABORATORY STUDIES OF PAIN

The view that aging dulls pain perception is supported by a number of early laboratory studies of age difference in pain sensation.[11-13] For the most part, these studies employed the psychophysical end points of threshold or tolerance and used radiant heat based on the Wolff-Hardy-Goodell dolorimeter.[14] In their own work, Wolff and colleagues indicated that they observed no effects of age on pain thresholds if the subjects were well instructed and practiced and if the psychophysical end point was specifically defined as the "pricking pain threshold."[14,15]

Figure 18-1 illustrates similarity of pain intensity ratings by young, middle-aged, and older adults.[16] In this study, pain intensity was rated using a validated magnitude-matching procedure.[17-19] Participants were instructed and trained to provide separate ratings for the perceived intensity and unpleasantness of the stimuli in separate sessions. Middle-aged individuals had the lowest sensory intensity ratings and the highest pain unpleasantness ratings. Although a significant age effect was present such that older adults tended to underrate low-level pain stimuli and overrate higher levels compared with young and middle-aged individuals, the magnitude of this was small.

Thermal stimuli delivered to hairy skin of the arms or legs can produce a sensation of two distinct pains. The first is a brief, well-localized, sharp or pricking pain sensation that seldom outlasts the stimulus. This sensation has been associated with activity in Aδ type II mechanoheat afferents. These fibers are small (1 to 5 μm), lightly myelinated, with a conduction velocity of between 10 and 30 m/s. Under appropriate conditions, the sensation of first pain is followed by a painless period of up to 1 s, after which a second pain may be perceived.

This second pain is characterized as a diffuse and poorly localized burning sensation that frequently outlasts the stimulus. This second sensation to thermal stimulation of hairy skin is associated with activation of slow nociceptive afferents or C fibers. This class of nociceptive afferents consists of small (0.05 to 2 μm) fibers with a conduction velocity of 0.5 to 2 m/s. The differences in the sensory qualities, including onset times, of these two classes of nociceptors allow evaluation of the selective effects of age on each.

Figure 18-2 shows pain intensity ratings of first and second pains in younger and older adults.[19] In this study, younger and older individuals rated pain intensity

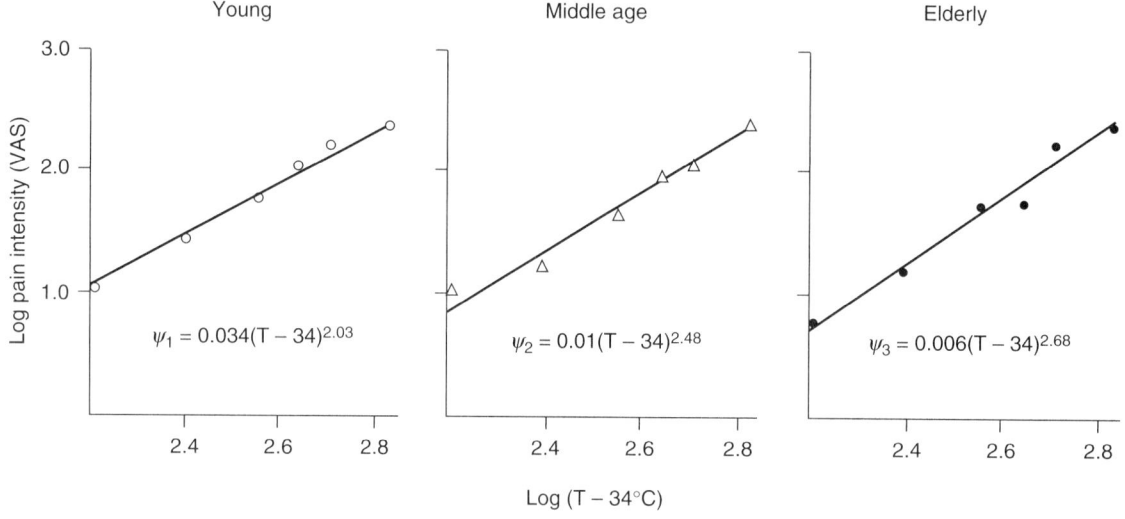

Figure 18-1. Pain psychophysical functions for younger, middle-aged, and older adults of contact thermal stimuli in the nociceptive range over a wide range of stimulus intensities. Stimuli were 5-s heat pulses delivered to unsensitized and unadapted skin on the volar surface of the forearm. Data are plotted in log-log coordinates in this figure. Pain intensity ratings were made by visual analogue scales (VAS). Older individuals had no difficulty using the VAS (younger individuals: $n = 21$, mean age 25.3 years, range 20–35 years; middle-aged: $n = 10$, mean age 53.4 years, range 45–60 years; older: $n = 13$, mean age 72.3 years, range 65–80 years). The five points plotted in each panel represent $\log_n$ of the stimulus intensity (43, 45, 47, 49, and 51°C) minus the adapting stimulus intensity (34°C). Note the lower intercept and higher slopes across the three age groups. Older individuals tended to rate lower-level stimuli (43 and 45°C) as less painful than younger individuals and rate higher-level stimuli (49 and 51°C) as more painful compared with younger individuals. These results may be due to slightly elevated pain thresholds with aging or birth-cohort differences in response bias for reporting low-level stimuli. The data indicate that the intensity growth curve grows more quickly in older groups. This effect was statistically significant, but its overall magnitude was slight. The similarity with age in the pain ratings is remarkable and suggests that aging does not markedly influence pain sensitivity.

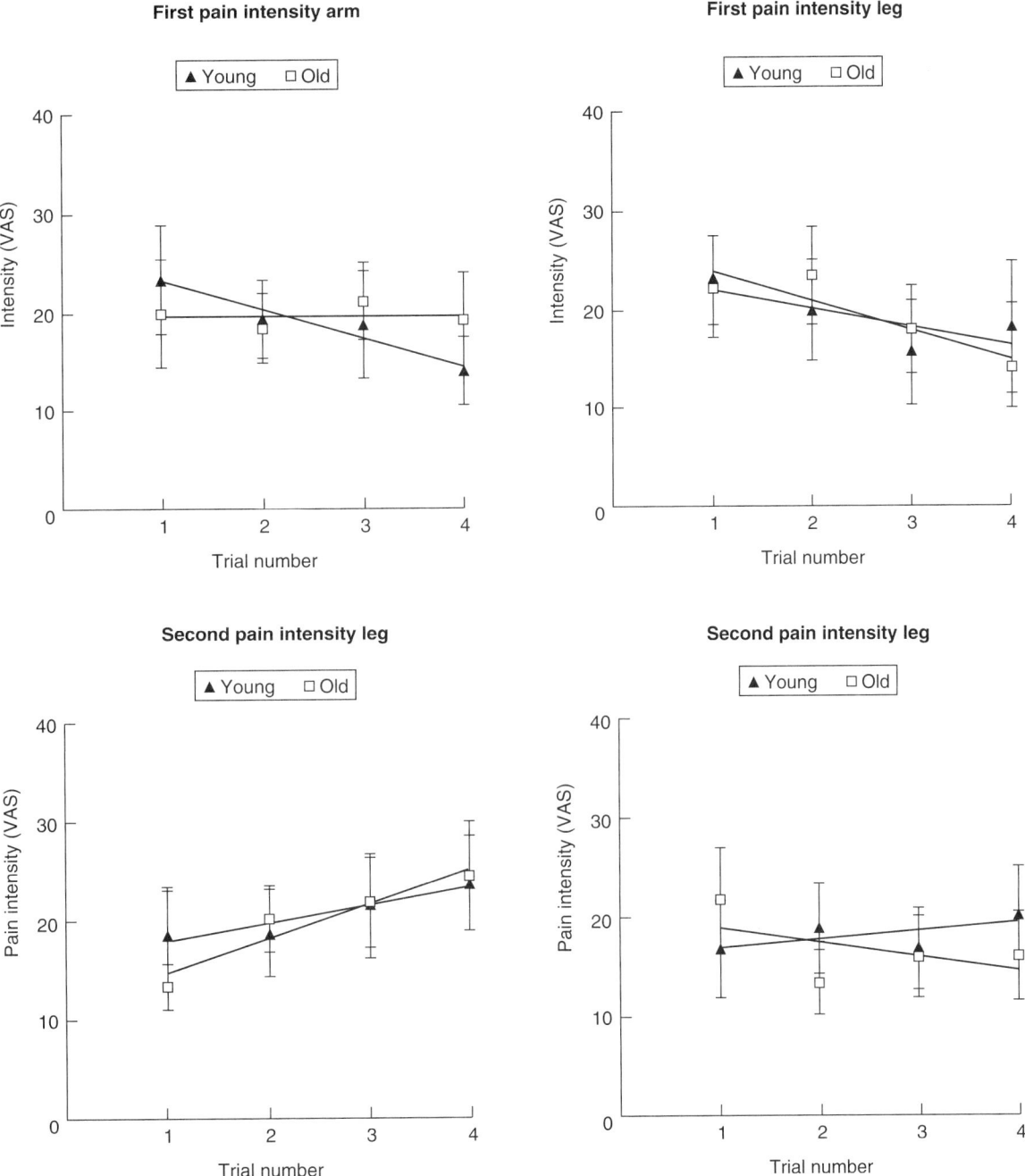

Figure 18-2. First and second pain. Pain intensity ratings of younger and older adults to brief (0.7-s) heat pulses from adapting temperature of 39°C to peak of 51°C; pulse rise time of 27°C. Series of one, two, three, and four pulses (labeled "Trial Number") were presented in different sessions to the arm and the leg. Pain ratings were made after the final pulse ("Trial") in each series (*Top two panels*). Visual analogue scale (VAS) ratings of pain intensity to repeated stimuli delivered under conditions producing suppression of first pain to arm (*left*) and leg (*right*). Across repeated stimuli, intensity of first pain decreased in both older and younger individuals. No age differences in ratings of intensity of first pain were present. (*Bottom two panels*) VAS ratings of pain intensity to repeated stimuli delivered under conditions producing slow temporal summation of second pain at arm (*bottom left*) and leg (*bottom right*). Intensity ratings of second pain increased across repeated stimuli from arm in both young and old groups. Results for leg were less clear, with the suggestion of the older group failing to evidence summation of second pain.

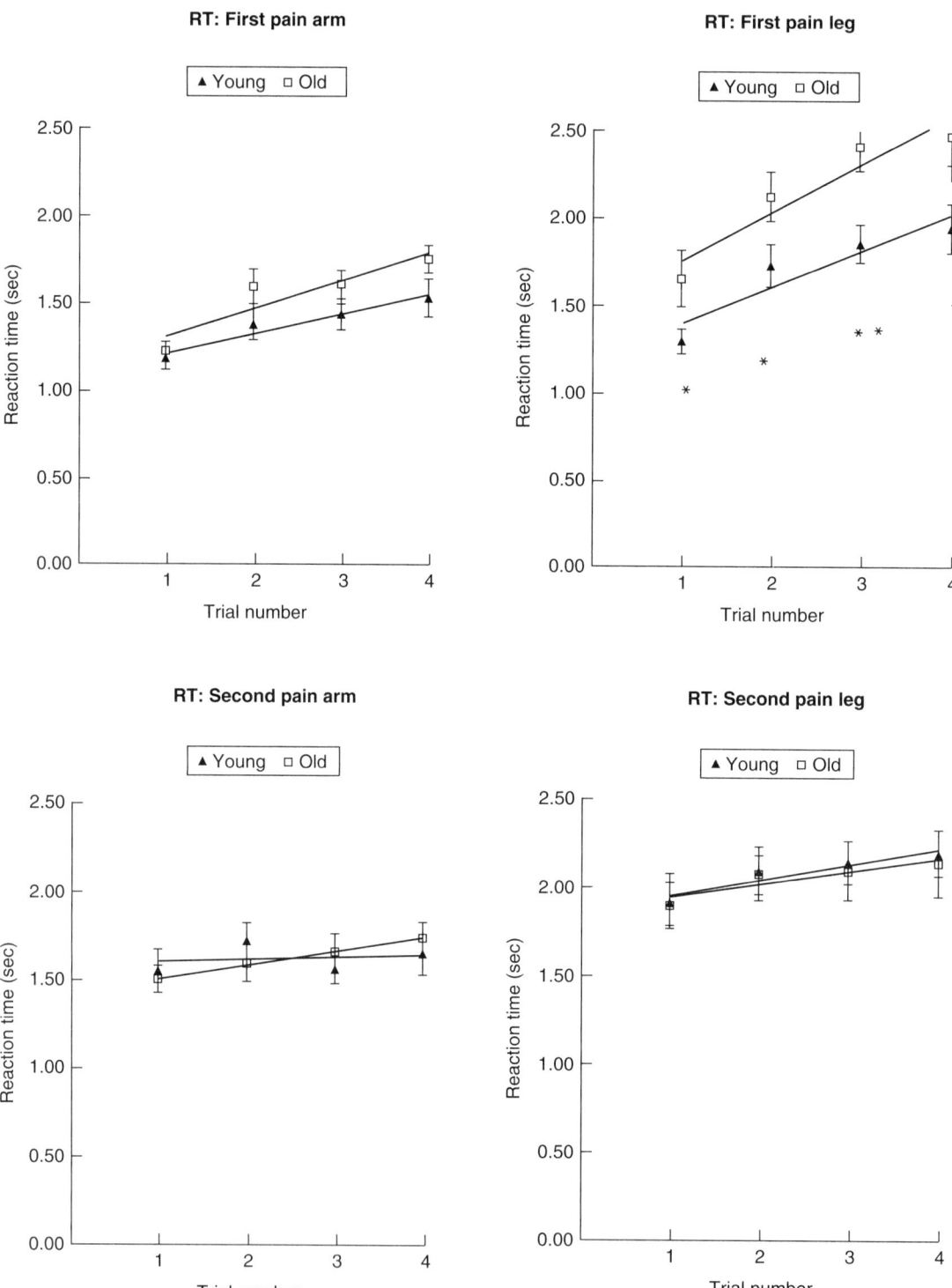

Figure 18-3. Pain onset response times for first and second pain. Reaction times (RTs) to repeated heat pulse (0.7-s duration) from 39°C adapting temperature to 51°C in young and older subjects as in Fig. 18-2. Note that RTs were identical in younger and older subjects to second but not first pain. Older subjects had longer first pain RTs compared with younger subjects, and this was statistically significant for responses to stimuli delivered to the leg. Auditory RTs were obtained in the same subjects and were substantially slower in older compared with younger individuals.

elicited by 700-ms heat pulses delivered from an adapting temperature of 39°C to a temperature of 51°C. Stimuli were delivered to arms and legs under conditions that allow separate assessment of first and second pain.

The two groups did not differ on intensity ratings of pain to the first stimuli in each series. This finding serves as a replication for the findings shown in Fig. 18-1 but selectively for first and second pains.

Use of brief thermal stimuli presented in series to the same location on the arm or the leg allows evaluation for adaptation (suppression) of first pain sensation and the temporal summation of second pain. In the first case, the process is due to peripheral adaptation of Aδ type II afferents with a decrease or loss of sensation of first pain. In contrast, repeated stimulation of the same location can result in an increase in intensity of second pain.

Older subjects failed to evidence slow temporal summation of second pain at the leg, suggesting a failure in mechanisms of central windup at the cord (see Fig. 18-2, lower right panel). This finding, if replicated,[20] has implications for atypical presentation of referred pain in the elderly because it is likely that the second-order mechanisms of slow temporal summation of second pain and summation of nociceptive-visceral afferent activity share similar mechanisms.

Age may influence conduction properties of Aδ but not C nociceptive afferents. First pain onset appears to be delayed in older adults, and sensory qualities (but not quantity, e.g., intensity) appear obtunded. Response slowing is one of the core findings with aging, except perhaps for painful stimuli. Figure 18-3 shows reaction times to first and second pains in younger and older adults across a series of stimuli delivered to the same location on the arm and leg. Older individuals had slightly longer reaction times to sensation onset of first pain from the arm compared with younger individuals. This effect of delayed response times was significant for stimulation of the leg. In contrast, there were no age group differences in onset of the sensation of second pain. These findings were interpreted to indicate an age-dependent change in conduction properties of Aδ type II nociceptive fibers consistent with an age-related small-fiber peripheral neuropathy. This is supported in part by the results reported by Gibson and colleagues.[20]

It is important to stress that the experimental evidence to date is that aging does not result in a marked reduction in the perception of the intensity of pain. Nociceptive processes do not seem to parallel the effects of aging on many other senses. This should not be surprising. The comparative simplicity of nociceptive afferents and their receptors suggests that "normal" aging will have minimal impact on pain sensation per se.

▶ RECURRENT AND CHRONIC PAIN IN THE ELDERLY

Specialty pain clinics historically have treated younger and middle-aged adults and not the geriatric population. Figure 18-4 shows the age composition of consecutive patients seen in two pain clinics over fixed periods of time.[21] Older individuals with chronic pain are underrepresented in these tertiary referral settings.

This may reflect the fact that there is substantial age-related change in presentation of pain complaints at the primary care level. With age, there is a decrease in visits for back pain and headache (Fig. 18-5) and an increase in visits for musculoskeletal pains[17,22] (Fig. 18-6). This reflects patterns of work-related injury and a

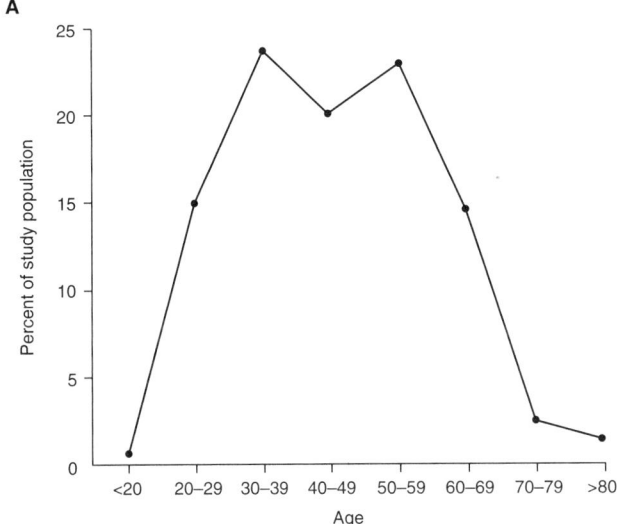

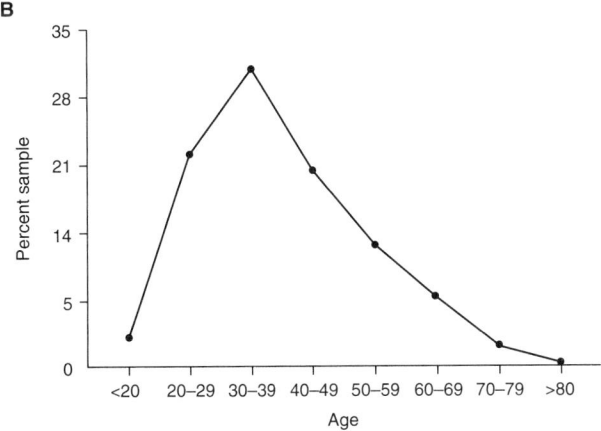

Figure 18-4. Age distributions of patients referred to two tertiary care university-based pain clinics. The top panel represents consecutive (*n* = 174) patients referred to a medical school clinic, and the bottom panel represents the age distribution of patients (*n* = 310) referred to an orofacial pain clinic. Older individuals are underrepresented in these clinics.

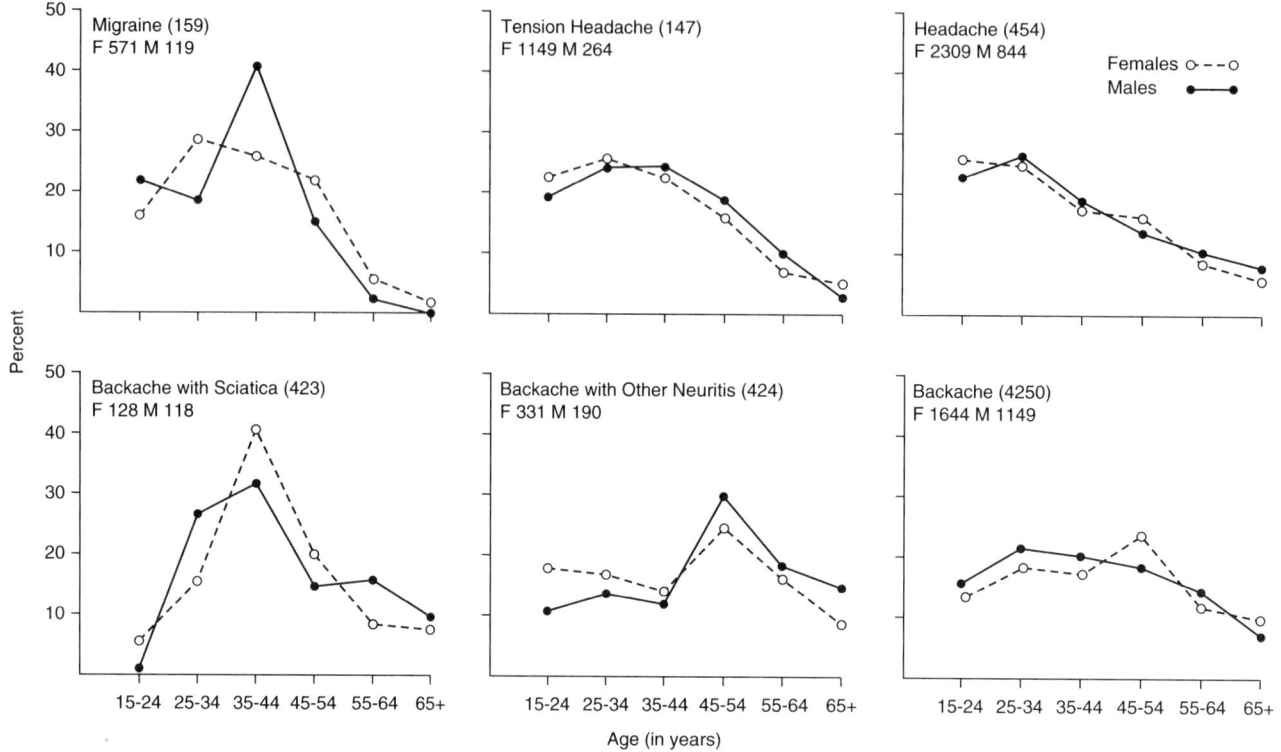

Figure 18-5. Age and sex distributions of selected reasons for visit of approximately 86,000 patients making over a half million visits to five family practice centers in the Commonwealth of Virginia over a 2-year period. Diagnosis was by problem-oriented adaptation of disease classification of the Royal College of General Practitioners (number in parentheses). Women, open circles; men, filled circles. This figure presents observed frequency of reasons for visit that often lead to referral to the pain clinic. This includes headaches and back pain. Note that these diagnoses as reasons for visit decrease with chronologic age equally in both men and women.

sociocultural expectancy that musculoskeletal pain associated with osteoarthritis is a "normal" part of the aging process.

Clinic samples are notoriously biased. National population-based surveys designed to assess pain and its consequences in the general population are lacking. Several well-constructed surveys have included questions concerning arthritis and thus allow estimation of prevalence and possible correlates of musculoskeletal pain in relation to aging.

Table 18-1 shows information from the National Health and Nutrition Follow-up Survey.[23] The data here represent the younger and older thirds of the sample, which was selected originally as a representative survey of ambulatory civilian Americans. The data presented in this table are not designed for estimates of prevalence of specific pains but rather allow contrast of the younger and older segments of the population with regard to pain intensity (visual analogue scale rating), depression scores, and activities of daily living. The results indicate (1) that musculoskeletal pain is common in older individuals, particularly knee pain and stiff, painful joints, (2) that older individuals report more intense pain compared with younger individuals, (3) that older individuals with chronic pain are only marginally more likely than younger adults to score high on a survey research instrument for depression,[24] and (4) that older individuals with pain are at a greater level of limitations in activities of daily living[25] compared with younger individuals with chronic pain. These results are cross-sectional and do not allow for causality by temporal relations of onset and resolution. The results do imply, however, that chronic pain is associated with a significant level of morbidity in terms of restriction in daily activities. Development of reliable and valid indicators of pain-specific limitations in activities of daily living as indicator variables of pain-related morbidity are needed and likely would establish that pain in the later years of life is a significant threat to public health.

Data from the Third National Health and Nutrition Examination Survey (NHANES III),[26] a nationally representative sample of community-dwelling older Americans

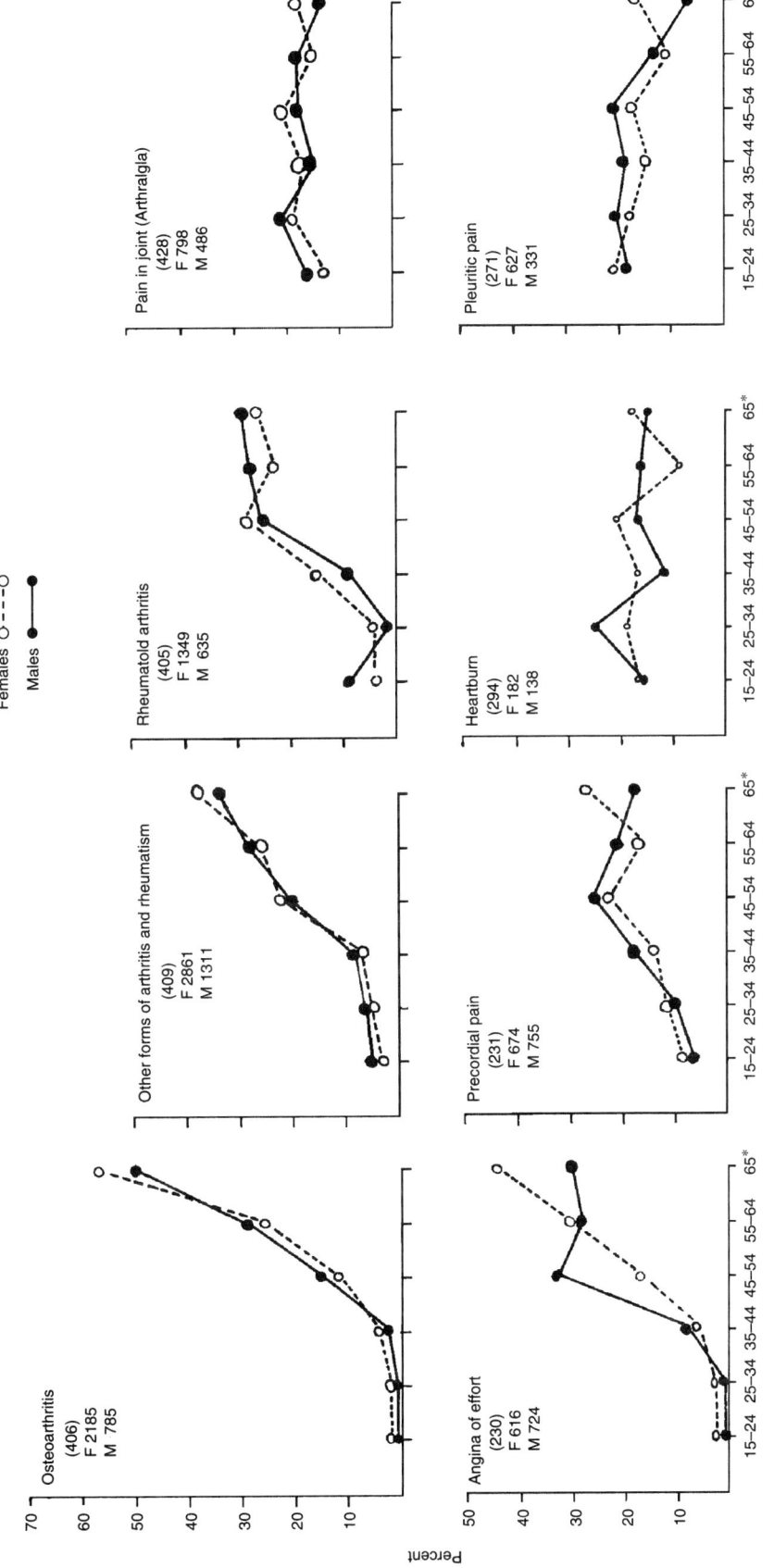

Figure 18-6. Age and sex distributions of selected reasons for visits as in Fig. 18-5 illustrating age-dependent increases in pain complaints and diagnoses associated with the musculoskeletal system and the cardiovascular system.

▶ **TABLE 18-1.** SUMMARY OF ORIGINAL TABULATIONS OF PREVALENCE OF PAIN LASTING AT LEAST ONE MONTH AND PRESENT DURING THE WEEK IMMEDIATELY PRIOR TO THE INTERVIEW IN 5,632 COMMUNITY-DWELLING INDIVIDUALS. PREVALENCE OF SIX SELECTED PAINS, PAIN INTENSITY, DEPRESSION AND ACTIVITIES OF DAILY LIVING. COMPILED FROM THE NATIONAL HEALTH AND NUTRITION FOLLOW-UP SURVEY.

		AGE	Neck	Back	Hip	Knee	Other Joint Pain	Stiff Joints
N	Mean	So	% (N)	% (N)	% (N)	% (N)	% (N)	% (N)
2,796	39.6	3.3	8.55 (239)	19.17 (536)	9.33 (261)	10.26 (287)	11.48 (321)	14.5 (406)
2,863	74.7	5.5	9.71 (278)	21.34 (611)	14.98 (429)	21.31 (610)	13.66 (391)	26.1 (748)
—[2]			N.S.	4.13*	42.2***	123.7****	6.09**	117.4****
VAS[1] Pain Intensity	Young		35.18 (30.2)	31.6 (30.8)	35.7 (32.3)	32.0 (31.0)	34.7 (31.9)	37.9 (28.9)
	Old		44.39 (30.6)	40.2 (32.0)	43.6 (32.0)	45.2 (31.3)	42.7 (30.5)	44.7 (29.6)
	F		11.7**	21.2****	9.8**	34.3****	11.4***	14.0***
	(df)		(1,513)	(1,1129)	(1,683)	(1,878)	(1,702)	(1,1137)
CESD	Young		10.8 (10.6)	10.6 (11.0)	10.2 (11.1)	10.9 (11.5)	10.67 (10.5)	12.3 (11.5)
	Old		12.9 (10.8)	11.9 (9.9)	12.2 (10.1)	12.2 (10.1)	11.7 (9.4)	12.9 (10.2)
	F		4.9*	4.2*	5.7*	2.62	1.85	0.74
	(df)		(1,515)	(1,1145)	(1,688)	(1,895)	(1,712)	(1,1152)
ADL Score	Young		28.2 (8.7))	27.4 (7.3)	28.0 (8.2)	28.0 (8.1)	27.5 (7.7)	28.1 (7.5)
	Old		35.7 (15.3)	34.6 (14.2)	36.8 (15.2)	36.0 (14.9)	35.5 (14.7)	36.3 (15.0)
	F		44.1****	112.8***	75.6****	72.6****	77.6***	107.0****
	(df)		(1,515)	(1,1145)	(1,688)	(1,895)	(1,710)	(1,1152)
Age	Young		40.2 (3.1)	39.8 (3.31)	40.0 (3.1)	39.5 (3.5)	40.1 (3.45)	40.0 (3.2)
	Old		75.0 (5.8)	74.6 (5.8)	75.7 (5.7)	75.1 (5.8)	74.8 (5.7)	74.7 (5.7)

*<0.05; **<0.01; ***<0.001; ****<0.0001.
ABBREVIATIONS: VAS = Visual Analogue Scale; CESD = Center For Epidemiology Studies Depression Inventory (84); ADL = Activities of Daily Living Inventory (85).
SOURCE: From Harkins, et al., 1994 (see Ref. 13).

(n = 6596; ages 65 to 90-plus) conducted between 1988 and 1994, indicate that approximately 50 percent of older Americans have at least one chronic musculoskeletal pain.[27] The prevalence of specific chronic musculoskeletal pain across the entire study population, in order of increasing frequency, was hip pain (14.1 percent, SE = 0.3), hand pain (18.2 percent, SE = 0.8), knee pain (21.2 percent, SE = 0.8), and back pain (29.2 percent, SE = 0.8). Logistic regression analyses employing these data showed a significant relationship between number of these pains and limitations in physical functioning after controlling for potential confounding variables (e.g., age, race, and health status). These findings are consistent with the suggestion earlier that pain is a significant source of morbidity in the elderly and an underrecognized factor in reduced quality of life for elders.[27]

▶ **SPECIAL POPULATIONS OF THE ELDERLY**

Old age is associated with a host of chronic medical conditions. There are increases in the prevalence of Alzheimer's disease, movement disorders, herpes zoster infection with risk of trigeminal neuralgia, incontinence, gait disorders with elevated risk of falls, and type II diabetes with aging. Brief inspection of the National Institute of Aging's Web pages will reveal significant effort toward identification of risk factors for these and other age-related chronic disorders. Pain, however, is not represented as a national issue in assessments of the health of the nation's old.[10]

Atypical Presentation of Pain in the Elderly

Taken together, the evidence concerning experimental studies of pain with age is untidy but seems to support limited loss of pain sensitivity with normal aging. This is in contrast to clinical observation and experimental findings for silent or painless acute myocardial infarction (MI) in older adults. Pain consequent to tissue ischemia develops when sufficient levels of afferent impulses are reached and when an appropriate activation of central ascending pathways has been established. In individuals presenting with a silent MI, such levels are apparently not achieved, perhaps due to insufficient stimulation of myocardium, decreased capacity for cephalad transmission, or some other unknown pathophysiologic reasons affecting peripheral transmission.[28-30]

Pain from skin (as used in experimental studies) and pain from deep structures, particularly referred pain, arise from different phenomena. It is inappropriate to apply the experimental findings to ischemia pain from deep tissues in the absence of further empirical data.

Pain and Parkinson's Disease

While pain has been associated with Parkinson's disease (PD) for over a century,[31] the natural history of pain in PD patients is poorly understood. Pain is reported as a major problem by up to 50 percent of persons with PD. Pain associated with motor symptoms may be relieved by treatment of the primary motor symptoms, suggesting that the pain is due primarily to these. Other evidence indicates a central origin of pain in some persons with PD.

PD is due to loss of striatal dopamine due to degeneration of dopaminergic substantia nigra neurons. Interestingly, chronologic age itself is associated with loss of nigra neurons, but clinical diagnosis of PD usually does not occur until sufficient depletion of dopamine occurs (striatal dopamine loss of approximately 80 percent).

▶ **TABLE 18-2.** CHARACTERISTICS OF PRESBYALGOS

Sensory Components

Characteristics:
  Determined by stimulus intensity, location, duration, type.
  Sensory qualities differ for types of pain (e.g., superficial versus deep pain).

POSSIBLE AGE EFFECTS:
- Increased pain thresholds. (Not likely)
- Increased pain tolerance. (Not likely)
- Reduced ability to discriminate between pain of various intensities. (Not likely)
- Reduced ability to discriminate among different pains. (Difficult to assess)
- Increased frequency of atypical pain as a symptom of disease processes. (Definitely)
- Increased frequency of chronic pain. (Definitely)

Primary Affective Components

Characteristics:
  Strongly related to pain intensity and ANS arousal.
  Related to appraisal of the present and short-term future.
  Mediated by meaning and cognitive appraisal.

POSSIBLE AGE EFFECTS:
- Reduced unpleasantness of pain, due to reduced sensory intensity of pain in general. (Not likely)
- Reduced unpleasantness of pain, due to decreased arousal, exteroceptive (sight, sound) and interoceptive (startle, autonomic) responses resulting in reduced segmental responses to painful injury. (No evidence exists for acute pain, may be true of chronic pain)
- Reduced general aversiveness of nociceptive stimuli. (Unlikely)
- Decreased perception of threat, distress, annoyance associated with the intensity of the painful sensation and its accompanying arousal. (Unlikely for acute pain)
- Differences or changes in cognitive appraisal. (Likely occur).

Secondary Affective Components of Pain

Characteristics:
  Related to past and long-term future.
  Cognitive appraisal.
  Related to or representative of suffering.
  Not measurable in experimental studies of pain.
  Pain affect and pain specific emotional distress can exist in the absence of nociceptive input. Pain-specific affect and distress share properties of emotional suffering in general. Suffering is defined here "as the state of severe distress associated with events that threaten the intactness of the person."[35] There is confusion between chronic pain and suffering due to the fact that disease models dominate thinking concerning pain.
  Unameliorated pain-related suffering (Stage II affect) requires different interventions than those traditionally used for control of the sensory intensity or the primary affective components of pain.
  No systematic studies exist concerning effects of age on this affective component of human pain.

Substantia nigra neurons are likely involved in processing nociceptive information. The evidence includes observation by neural imaging techniques (PET) that the basal ganglia are activated by painful stimuli. Other studies indicate that direct electrical stimulation of the substantia nigra (as well as caudate nucleus) reduces pain behavior (see the excellent review by Chulder and Dong[32]). Direct recording studies show that nigra neurons are sensitive to the intensity of painful stimuli but not the location (wide receptive field). The functional role of the substantia nigra with regard to nociceptive stimuli is likely that of somatosensory/motor integration involved in organization of protective responses to painful as well as other forms of somatosensory stimuli.[32]

Pain in Alzheimer's Disease (AD)

Pain sensitivity has been suggested to be decreased in Alzheimer's patients.[33,34] Several case reports exist of individuals with AD sustaining severe injuries without complaint. Two of these include patients sustaining first- and second-degree burns with indifference, according to the report of their primary caregiver. Care must be taken in generalizing these limited case reports. It is likely that AD is a multifactorial disease with a "common final pathway." Given the broad central effects and different etiologies of AD, it may well be clinical presentation of pain, and pain sensitivity may aid in the differential diagnosis of the type of dementia. Frontal type dementia is likely to be associated with pain indifference, and this would have implications for care in the long-term care setting.[33]

The perception that pain due to chronic disorders is part of the aging process, combined with the idea that aging is associated with decreased pain sensitivity, may lead to reduced efforts to assess pain in the older adult. This is particularly true for geriatric pain patients with severe cognitive impairments.

▶ SUMMARY

Table 18-2 presents a summary of adult age-related changes and differences in pain sensitivity, pain perception, pain-related emotions, and pain-related behaviors. In contrast to the effects of aging on vision (presbyopia) and audition (presbycusis), presbyalgos is not restricted to the description of losses in sensory acuity but incorporates sensory, cognitive, affective, and behavioral components. The definition of presbyalgos includes psychosocial history and thus birth-cohort effects that influence an individual's interpretation of pain and the definition of socially accepted behaviors used to express presence of pain and suffering.[35]

REFERENCES

1. Thomas L: Special report. *Discovery*, December 24–25, 1984.
2. Critchley M: The neurology of old age. *Lancet* 1931; 1:1221–1230.
3. Merskey H: Classification of chronic pain: Descriptions of chronic pain syndromes and definitions of pain terms. *Pain* 1986;3:S1–S222.
4. Dawson WH: *Bismarck and State Socialism: An Exposition of the Social and Economic Legislation of Germany Since 1870*. New York: H. Fertig, 1973.
5. *The Oxford English Dictionary*. Oxford, England: Clarendon Press, 1933.
6. *The Oxford English Dictionary*, 2d ed. Oxford, England: Clarendon Press, 1989.
7. Strehler BL: *Time, Cells and Aging*, 2d ed. New York: Academic Press, 1977.
8. Murray A: The outlook: Clinton plays to aging boomers. *Wall Street Journal* 233, p. A1, March 29, 1999.
9. Fries JF: Aging, natural death, and the compression of mortality. *N Engl J Med* 1980;303:130.
10. Harkins SW: Sans pain. *www.ampainsoc.org/pub/bulletin/mar97/qanda.htm*.
11. Gagliese L, Melzack R: Chronic pain in elderly people. *Pain* 1997;70:3–14.
12. Harkins S: Geriatric pain: Pain perception in the old. *Clin Geriatr Med* 1996;12:435–445.
13. Harkins SW, Price DD, Bush FM. et al: Geriatric pain, in Wall PD, Melzack R (eds), *Textbook of Pain*. Edinburgh: Churchill Livingstone, 1994:769–784.
14. Hardy JD, Wolff HG, Goodell H: The pain threshold in man. *Am J Psychiatry* 1943;99:744–751.
15. Schumacher GA, Goodell H, Hardy JD, Wolff HG: Uniformity of the pain threshold in man. *Science* 1940; 92:110–112.
16. Harkins SW, Price DD, Martelli M: Effects of age on pain perception: Thermonociception. *J Gerontol* 1996;41:58–63.
17. Harkins S: Geriatric pain, in Loeser J, et al (eds), *J J Bonica's Management of Pain in Clinical Practice*, 3d ed. Baltimore: Lippincott, Williams & Wilkins, 2001:809–819.
18. Harkins SW, Lagua B, Small R: Geriatric pain, in Roy R (ed), *Chronic Pain in Old Age: An Integrated Biopsychosocial Perspective*. Toronto, Canada: University of Toronto Press, 1995:127–163.
19. Harkins SW, Davis M, Bush F, et al: Suppression of first pain and slow temporal summation of second pain in reaction to age. *J Gerontol [A] Biol Sci Med Sc* 1996; 51:M260–M265.
20. Chakour MC, Gibson SJ, Bradbeer M, Helme RD: The effect of age on A-delta and C-fiber thermal pain perception. *Pain* 1996;64:143–152.
21. Harkins SW, Kwentus J, Price DD: Pain in the elderly, in Benedetti C, et al (eds), *Advances in Pain Research and Therapy*, Vol 14. New York: Raven Press, 1984.
22. Crook J, Rideout E, Browne G: The prevalence of pain complaints in a general population. *Pain* 1984;18:299–314.
23. National Center for Health Statistics: Plan and operation of the NHANES I epidemiologic follow-up study, 1982–1984,

Vital and Health Statistics, Series 1, No 22, DHHS Pub. No. (PHS)87-1324, 1987.
24. Radloff LS: The CES-D scale: A self-report depression scale for research in the general population. *Appl Psychol Measure* 1977;1:385–401.
25. Lawton MP, Brody EM: Assessment of older perople: Self-maintaining and instrumental activities of daily living. *Gerontologist* 1968;9:179–188.
26. U.S. Department of Health and Human Services (DHHS), National Center for Health Statistics: *Third National Health and Nutrition Examination Survey, 1988–1994,* NHANES III Laboratory Data File (CD-ROM). Public Use Data File Documentation Number 76200. Centers for Disease Control and Prevention, Hyattsville, MD, 1996. Available from National Technical Information Service (NTIS), Springfield, VA, Acrobat PDF format; includes access software: Adobe Systems, Inc., Acrobat Reader 2.1.
27. Harkins SW: Original tabulations based on the Third National Health and Nutrition Examination Survey.
28. Ambepitiya GB, Iyenger EN, Roberts ME: Review: Silent exertional myocardial ischaemia and perception of angina in elderly people. *Age Ageing* 1993;22:302–307.
29. Applegate WB, Graves S, Collins T, et al: Acute myocardial infarction in elderly patients. *South Med J* 1984;77:1127–1129.
30. Montague T, Wong R, Crowell R, et al: Acute myocardial infarction: Contemporary risk and management in older versus younger patients. *Can J Cardiol* 1990;6:241–246.
31. Charcot J-M: *Leçons sur les localisations dans les maladies du cerveau.* Paris: Progrés médical & V. A. Delahaye, 1876; English translation of Vol 2, New Sydenham Society.
32. Chudler EH, Dong WK: The role of the basal ganglia in nociception and pain. *Pain,* 1995;60:3–38.
33. Scherder EJ, Sergeant JA, Swaab DF: Pain processing in dementia and its relation to neuropathology. *Lancet Neurol* 2003;11:677–686.
34. Gabre P, Sjoquist K: Experience and assessment of pain in individuals with cognitive impairments. *Spec Care Dentist* 2002;5:174–180.
35. Cassell EJ: The importance of understanding suffering for clinical ethics. *J Clin Ethics* 1991;2:81–82.
36. Ferrell B: Overview of aging and pain, in Ferrell BR, Ferrell BA (eds), *Pain in the Elderly.* Seattle: IASP Press, 1996:1–10.

CHAPTER 19

Pain Medicine and Chemical Dependency

Seddon R. Savage

Concerns related to substance abuse may complicate the treatment of pain. Most often these concerns relate to one of three issues:

- *How to provide safe and effective treatment of pain in persons with concurrent substance-use disorders who may be at risk for medication abuse*
- *Whether the use of opioids may trigger substance-use problems in persons with no prior history of abuse or addiction*
- *How to identify and manage persons seeking drugs for purposes other than pain treatment, such as recreational use or diversion for sale*

Effective management of these concerns requires understanding of the neurobiology and phenomenology of substance abuse, clear conceptualization of the different ways in which persons use and misuse opioids, and an understanding of how various opioid formulations may differ in abuse potential.

▶ EPIDEMIOLOGY, NEUROBIOLOGY, AND PHENOMENOLOGY OF DRUG ABUSE

Prevalence of Drug Use

Substance-use disorders affect a significant portion of the population. The 2002 National Survey on Drug Use and Health (NSDUH) found a 9.4 percent prevalence of substance-use disorders in the population of the United States[1] 68 percent of these reflected alcohol disorders alone, 18 percent illicit drug use alone, and 14.5 percent combined alcohol and drug-use disorders. Relevant to pain treatment, opioid-use disorders have been rising. According to the Drug Abuse Warning Network, which collects data from sentinel emergency rooms around the country, prescription opioid abuse rose 123 percent and heroin abuse rose 47 percent between 1994 and 2001, with patients with prescription opioid abuse presenting to emergency departments more often than heroin-abuse patients in absolute numbers.[1] Of persons treated for substance-use disorders, NSDUH data for 2002 suggest that 6.3 percent received treatment for prescription opioid dependence, whereas 39 percent received treatment for alcohol, 17 percent for marijuana, 14 percent for cocaine, 4.8 percent for heroin, and 18 percent for other substances.

Reasons for Nonprescribed Drug Use

People who use drugs recreationally most often seek a "rush" and a "high" along with a transient increased sense of well-being. While different classes of drugs, such as alcohol, opioids, stimulants, and hallucinogens, have different pharmacologic actions, the sought-after high appears to occur due to a final common pathway of enhanced dopamine availability in synapses of the limbic reward centers, including the nucleus accumbens and the ventral tegmental areas.[2] Some drugs, such as opioids, when used in rapidly peaking bolus doses, stimulate interneurons to signal afferent neurons to increase dopamine release, whereas some other drugs, such as cocaine, interfere with dopamine reuptake by afferent neurons, allowing accumulation of dopamine in the synapse.³ Although reward is recognized as a key component in abuse and addiction to drugs, there are other reinforcing variables for use because reward does not appear to correlate completely with continued use.⁴

Etiology of Addiction

In some persons, repeated abuse of drugs may lead to addiction, whereas others may abuse drugs regularly without ever developing behaviors characteristic of addiction. Vulnerability to addiction appears to have an important biogenetic basis, although psychological, social, and environmental components contribute as well.[5,6] The neurobiologic changes that mediate transition from voluntary abuse to compulsive or addictive use are not known, but persistant neurophysiologic changes in brain function have been demonstrated in addicted persons.

While the evidence supporting an important genetic component of addictive disease is strongest for alcoholism, considerable evidence supports a genetic basis for other substance addictions as well.[7] A common vulnerability to the development of addiction to different substances that is mediated, in part, genetically has been appreciated, but there appears to be variance in the degree of individual susceptibility to specific drugs.[6] Cross-addiction or polysubstance dependence is common.[7,8] Individuals most often identify a drug or drug class of choice but may report secondary addictions or may recover from one addiction only to relapse to a different substance. Thus, an individual in recovery from alcoholism or other nonopioid drug addiction may have some increased risk of developing addiction to prescribed opioids over an individual with no history of any addiction, although this has not been demonstrated conclusively.

▶ TERMS AND CONCEPTS IN DRUG USE AND MISUSE

Confusing nomenclature related to the medical use of opioids often has led to misunderstandings and consequent mismanagement of pain and related problems.[9] In addition, not all aberrant behaviors related to prescribed opioids reflect addiction or voluntary abuse of medications.[10] It is important to discriminate addiction, physical dependence and tolerance, recreational abuse, undertreated pain, inability to comply with treatment due to psychic or cognitive problems, self-medication, and diversion because appropriate and effective approaches to management differ (Table 19-1).

Addiction

Addiction has been defined by the American Society of Addiction Medicine as "a primary, chronic, neurobiolog-

▶ **TABLE 19-1.** DIFFERENTIAL DIAGNOSIS OF BEHAVIORS SUGGESTIVE OF ADDICTION

- Inadequate pain management
  - Stable condition with sub-optimal analgesia
  - Progressive pathology
  - Opioid tolerance/opioid-induced hyperalgesia
- Inability to comply with treatment
  - Cognitive impairment
  - Psychiatric disorder
- Self-medication of
  - Mood
  - Sleep
  - Trauma flashbacks
  - Addictive disease (opioid maintenance therapy)
  - Other distresses
- Diversion by patient for
  - Analgesic use by others
  - Drug abuse/addiction
  - Profit
- Theft or diversion by others

SOURCE: *Savage, 2002.*

ic disease with genetic, psychosocial and environmental factors influencing its development and manifestations; it is characterized by one or more of the following behaviors: impaired control over use, compulsive use, continued use despite harm (adverse consequences), and craving."[11] Recovery and compliance rates are similar to other chronic conditions, such as diabetes, hypertension, and asthma.[12] Recovery from addiction usually requires an active patient role and a multidimensional approach that may include important psychospiritual, social, and pharmacologic components.[13] It is important to distinguish the condition of addiction from physical dependence and tolerance, which are common and expected forms of physiologic adaptation to opioids when they are used for pain treatment over time.

Physical Dependence

Physical dependence on opioids is signaled by an opioid-specific withdrawal syndrome that occurs when an opioid is discontinued abruptly, the dose is lowered precipitously, or an opioid antagonist administered. The withdrawal syndrome includes symptoms of central neurological excitation, including sleeplessnees, irritability, and psychomotor agitation that are thought to be mediated largely by a noradrenergic mechanism in the locus ceruleus, as well as peripheral signs of sympathetic arousal such as lacrimation, piloerection, mydriasis, and diarrhea.[14] Significant musculoskeletal pain and abdominal cramping are common during opioid withdrawal, and underlying pain may be increased when withdrawal is present.[15] Withdrawal may be avoided by gradual tapering of opioid. Opioid withdrawal symptoms may be attenuated somewhat by α-adrenergic blocking agents such as clonidine, by nonopioid analgesics, and by sedative hypnotics, among other medications. Some patients may continue to use opioids despite resolution of pain to avoid withdrawal; this does not in itself indicate addiction.

Tolerance

Tolerance is indicated by the need for increasing doses of a drug to induce the initial effects of the drug; tolerance occurs to both the analgesic effects of opioids and most opioid side effects. There are often multiple contributors to the evolution of tolerance; these may include, among other mechanisms, increased drug metabolism, reductions in cellular responses at the molecular level, and changes in synaptic processing of signal induction.[16] The occurrence of opioid tolerance is not uniform and is difficult to predict. Tolerance sometimes is addressed effectively by rotation from one opioid to another due to incomplete cross-tolerance, by coadministration of NMDA receptor antagonists, or by periods of nonuse of opioids.[17,18]

Neither physical dependence nor tolerance is a specific indicator of addiction, although they may coexist with addiction because individuals addicted to drugs often use the drugs regularly over time, and etiologic relationships between physical dependence, tolerance, and addiction have not been clearly demonstrated. Physical dependence occurs with many drugs that are not associated with addiction, such as corticosteroids, tricyclic medications, and many antihypertensive medications. Conversely, persons may experience addiction in the absence of physical dependence, as illustrated by the binge pattern of use often associated with cocaine addiction and the craving and drug-seeking behaviors of incarcerated heroin addicts who have no current use of opioids. Many older definitions and diagnostic criteria related to substance-use disorders, however, include physical dependence or tolerance as defining features of addiction, which may support overdiagnosis and trivialization of addiction when it does occur.[19,20]

Pseudoaddiction

Persons who experience severe pain that is not treated adequately may present in distress, with an apparent focus on obtaining opioids. Such behavior may be interpreted as reflecting addiction when in fact the underlying desire is not for opioids per se but for relief of pain. Such a scenario has been termed *pseudoaddiction*.[21] Pseudoaddiction usually can be distinguished from addiction by treating pain aggressively and observing that further behaviors that may reflect addiction, such as loss of control or compulsive use of medications and adverse consequences, do not occur but that pain and function improve.

Inability to Comply with Treatment

Patients with cognitive deficits related to developmental disorders, dementia, aging, cognitively impairing psychiatric illness, or other causes of intellectual or memory impairment may be unable to use medications reliably and appear to be overusing. Increased supervision and structure of care may improve both pain relief and safety.

Self-Medication

Patients may use opioids in the absence of pain in an attempt to treat a variety of symptoms. Some may continue to use opioids when pain is resolved because of persistent physical dependence and lack of knowledge regarding how to taper to avoid withdrawal or the inability to taper with the dosage forms available to them. Patients with anxiety, depression, post-traumatic stress disorder (PTSD), flashbacks, or other psychiatric symptoms may find transient relief with use of opioids and attempt to self-medicate their nonpain symptoms. Others may use opioids to assist in sleep induction. Careful questioning regarding an individual's actual medication use patterns may often reliably identify such misuse and permit implementation of more specific and appropriate treatment approaches.[22] It is not usually clinically helpful to simply lump all medication misuse together and label the patient as a "drug abuser."

Diversion for Profit

Diversion of opioids for financial profit may occur under a variety of circumstances. Types of diversions range from organized rings of entrenched criminals who purposefully mislead multiple doctors over large geographic areas for the sole purpose of financial gain, to persons with pain who exaggerate their pain to obtain excess opioids for sale, to impecunious elderly persons who sell a few of their opioid doses to others in pain in order to afford the medication. All diversion for profit poses a serious threat to public health and should be identified when possible and addressed firmly and appropriately.

▶ ABUSE POTENTIAL OF OPIOIDS

Variability of Individual Reward Responses

While all opioid medications appear to stimulate reward in some persons, a number of factors may modulate the degree of reward experienced. These include both host factors and drug characteristics.

Variability of Opioid Receptor Effects

The euphoria of limbic reward appears to be induced most prominently by stimulation of mu and delta opioid receptors with less or no reward effect through kappa receptor stimulation.[23] It is not surprising, then, that among different analgesics, the predominantly mu-opioid analgesics such as morphine, oxycodone, hydromorphone, and others appear to be associated most often with abuse. Not all persons using opioids experience euphoria with use of mu opioids; some experience dysphoria or no phoric or mood changes. In addition, some evidence suggests that the presence of pain may attenuate euphoria on administration of opioids.[24]

Abuse liability appears to be somewhat reduced with kappa-agonist/mu-antagonist analgesics, such as butorphanol, pentazocine, and nalbuphine, although some individuals do abuse these medications.[25,26] Similarly, partial mu agonists such as tramadol and buprenorphine appear to have lower abuse potential than full mu agonists.[27,28] The lower abuse potential of kappa-agonist and partial mu-agonist medications is reflected in lower Drug Enforcement Agency (DEA) scheduling than pure mu-agonist medications. However, there are limitations to the use of both kappa agonists and partial mu agonists in the treatment of pain. All have ceilings to their ability to effect analgesia or are limited in titration or use by side effects. The agonist/antagonist opioids may reverse mu analgesia and cause acute withdrawal in patients who are also using mu opioid analgesics. Because some drug users do experience reward with abuse of kappa agonists and partial mu agonists and may become addicted, the use of these medications must be monitored similarly to mu opioids.

Impact of Routes and Schedules of Administration

Variables in Opioid Administration

Several variables in the administration of mu opioids may affect their potential to produce reward. These principles, which have been understood through study of the pharmacologic treatment of opioid addiction, include route of administration, speed of drug release, and periodicity of effects.

A more rapid rise in brain blood level of a drug of abuse is associated with the greater likelihood that a reward—rush and high—is experienced with use.[29] Therefore, it is expected that for the same dose of opioid, bolus dose intravenous administration would be more likely to result in reward than slow intravenous administration, and subcutaneous or intramuscular administration would be expected to have greater reward than an equianalgesic dose given orally. Similarly, oral medications that have intrinsically slow onset, such as methadone, and those with slower release, such as sustained-release morphine or transdermal fentanyl (used as intended), appear less likely to produce reward than quick-onset or immediate-release opioids.

Opioid regimens that produce stable blood levels produce less reward than regimens that result in frequent peaks and falls in blood levels.[30] Therefore, continuous intravenous infusions or scheduled doses of long-acting or sustained-release medications are expected to produce less reward than intermittent intramuscular bolus doses or short-acting oral medications. It is important to note, however, that sustained-release oral opioid formulations and trancutaneous patches can be adulterated easily to provide quick-onset, high-bolus doses by those who wish to abuse them. Because they often are intended to provide sustained analgesia over time for opioid-tolerant individuals, such abuse may be life-threatening, especially in opioid-naïve persons.

Patient-controlled analgesia (PCA), which provides small self-administered increments of medication at timed intervals with limits of both doses and interval under control of the provider, does not usually result in significant reward. PCA has been documented to result in improved analgesia and faster recovery in the postoperative setting with less net opioid than other types of administration.[31] However, some persons in recovery may experience ambivalent feelings regarding self-injection of a potentially rewarding drug. In addition, relatively large quantities of injectable opioid are present in PCA machines and may be obtained through tampering by individuals seeking to abuse or divert medication. Thus, consideration of PCA must be individualized.

▶ RISK OF ADDICTION IN PAIN TREATMENT

Addiction Risk in Acute Pain Treatment

The risk of developing addiction to opioids in the course of pain treatment is not known. Two large, uncontrolled, retrospective surveys suggest that the risk of the development of addiction or serious drug abuse problems in medical patients who have no personal history of substance-use problems and who use opioids on a short-term basis is low, perhaps on the order of 1 in 10,000.[31,32] However, many patients do have addictive disorders and therefore, higher risk of abuse or addiction to therapeutically prescribed opioids, and many persons have persistent pain that may require longer-term opioid use. Therefore, these surveys do not accurately reflect risk in the general medical population.

Prevalence of Addictive Disorders in Medical Populations

Although the prevalence of addictive disorders in the general population is approximately 10 percent,[1] studies consistently show much higher rates of substance-use disorders in hospitalized populations, between 20 and

25 percent;[34,35] and for patients who sustain significant trauma, data indicate substance-use rates of 39 to 86 percent.[36] Thus, a relatively high rate of substance-use disorders is expected in persons requiring treatment for acute pain. Because some of these persons subsequently may develop chronic pain related to their injury or illness, a somewhat higher prevalence of addictive disorder is expected among persons with chronic pain, although actual studies suggest prevalence rates variously the same as or slightly greater than the general population.[37–39] Studies of the prevalence of addiction among cancer patients vary in their findings, but some suggest a somewhat higher rate than the general population.[40] Because alcoholism carries an increased risk of a number of malignancies and is highly associated with tobacco use, this is perhaps not surprising.

Similarly, a higher than expected prevalence of chronic pain in persons with addictive disorders has been noted. Some level of chronic pain has been reported in between 60 and 80 percent of persons in methadone clinics, and severe chronic pain has been reported in 37 percent.[41,42] In persons presenting for inpatient substance abuse treatment, one study showed some level of pain in 78 percent and severe pain in 24 percent.[41]

Individual Vulnerability to Addiction

Persons with a concurrent addictive disorder or in recovery from an addictive disorder appear to be at higher risk for developing abuse or addiction problems when opioids are used for pain than persons without such histories, but clearly, such problems are not inevitable. There is likely a spectrum of risk in terms of vulnerability to developing addiction to prescribed opioids. Based on an understanding of the biogenetic basis of addiction, persons with no family or personal history of addiction are likely at least risk. Persons with a family history but no personal history may have some elevated risk, although there is only extrapolated and observational evidence to support this view. Limited published evidence suggests that persons with a history of heroin or other opioid addiction have a higher risk of developing addictive behaviors related to prescribed opioids than individuals with a history of alcoholism.[43] Recent evidence suggests that the physiologic changes associated with stress may be an important factors in mediating the evolution of addiction in some settings.[44]

Medication-Host Interaction

It may be helpful to view risk of addiction to prescribed opioids as an interaction between host vulnerability and drug reward potential. Persons with low host vulnerability may be able to use any opioid by any regimen with little risk; persons with greater identifiable vulnerability may be managed best using opioid regimens with the least potential to cause drug reward. That said, clinical observation suggests that persons with past heroin addiction who are in active recovery and have intermittent pain sometimes do well with intermittent short-acting opioids and that persons with no identified risk can develop abusive patterns of use of scheduled long-acting opioids. Thus, all management decisions should be individualized, and all patients using opioids should be monitored carefully for safety and efficacy of treatment and treatment adapted accordingly. The adoption of universal precautions in opioid prescribing has been recommended[45] because, as with infectious diseases, risk cannot always be predicted, harm to both patient and doctor may occur when addiction is present and undetected, and early identification improves outcomes.

▶ ASSESSMENT FOR SUBSTANCE-USE PROBLEMS

Goals of Assessment

Ideally, all patients for whom opioids are contemplated as a component of pain treatment should be assessed for substance-use problems. While this does not always seem critical in the context of acute care, when subsequent problems of protracted pain complaints or aberrant opioid use develop, the value of routine assessment may be appreciated.

In assessing substance-use issues, the goals generally are to determine whether the individual has a personal or family history of drug or alcohol problems, what the individual's current patterns of drug or alcohol use are, and if the person is in recovery, how the individual is cultivating recovery. In addition, in an individual who acknowledges past or present heavy drug or alcohol use, it is important to learn if this more likely reflects or reflected voluntary or instrumental abuse associated with stressors or addiction, reflected in impaired control, craving, or compulsive use.

Substance-Use Interviewing

Persons with a past history of drug and alcohol problems usually are willing to discuss their past issues and often are proud of their recovery. Persons with current problems may be less willing to reveal them, at least in part because of what a former director of the National Institutes of Drug Abuse calls their "hijacked brain"[46] wanting to protect the ability to satisfy the craving. A nonjudgmental, supportive attitude and an explanation of the medical importance of the information are often helpful in encouraging an honest sharing of information. When substance-use problems are suspected or identified, clinicians can have an important role in promoting

► **TABLE 19-2.** CAGE QUESTIONS ADAPTED TO INCLUDE DRUGS (CAGE-AID)

1. Have you felt you ought to **C**ut down on your drinking or drug use?
2. Have people **A**nnoyed you by criticizing your drinking or drug use?
3. Have you felt bad or **G**uilty about your drinking or drug use?
4. Have you ever had a drink or used drugs first thing in the morning to steady your nerves or to get rid of a hangover (**E**ye-opener) ?

SOURCE: *Brown, 1995.*

► **TABLE 19-3.** CYR-WARTMAN SCREEN

Have you ever had a problem with alcohol [or drugs]? When was your last drink [or drugs]?

NOTE: A positive screen that roughly correlates with the CAGE in terms of specificity and sensitivity is "yes" and within 24 hours of the medical appointment.
SOURCE: *Cyr, 1984.*

recovery by sharing concerns with the patient, providing education on the medical consequences of the perceived problems, and providing resources for further assessment and care.[47] Such brief interventions are often successful over time in changing behavior.

Screening and Characterization of Use Patterns

A number of standard screens for addictive disease are useful. Probably the most frequently used is the brief CAGE screen, which asks four simple questions related to current use. Initially developed to screen for alcohol problems, it has been adapted as the CAGE-AID to screen for other drugs of abuse (see Table 19-2).[48] A shorter two-question screen that explores both current and past use and has been validated as equivalent to the CAGE in sensitivity and specificity is the Cyr-Wartman screen (Table 19-3).[49] This is often used clinically to screen for both alcohol and drugs, but it has not been validated for other drugs to my knowledge. It may be useful to screen using the Cyr-Wartman questions and then proceed to the CAGE if past problems are identified and there is any recent use. Other screens, such as the trauma screen[50] (which does not focus on substances), the AUDIT screen,[51] and the MAST screen,[52] may be useful in different clinical settings.

If a positive history of problems is obtained, it is important to identify drug classes abused and to determine whether the problematic use constituted abuse or addiction. For persons using drugs or alcohol currently, quantity and frequency of use should be described.

Objective Information

In addition to patient interview, physical examination and laboratory data sometimes may add useful information about substance use. Common physical findings in persons with drug or alcohol abuse or withdrawal include scent of alcohol or marijuana or tobacco, hepatic enlargement, hyper- or hyporeflexia, pupillary changes, plethoric facies, and/or intravenous injection sites in antecubital fossae or backs of hands or feet. In heavy alcohol use there may be an isolated elevation of GGT, an increase in MCV, and an increase in carbohydrate-deficient transferin (CDT).[53] Markers of infectious diseases that are often associated with intravenous drug use, such as HBAg, HCAg, or positive HIV screen, should prompt inquiry regarding drug use if another route of acquisition is not known (Table 19-4).

► TREATMENT OF PAIN IN PERSONS WITH ADDICTIVE DISORDERS

Barriers to Pain Treatment in Addiction

Pain is often undertreated in persons who have or are perceived to have addictive disorders.[54–56] Physicians tend to write for lower doses of medications at longer intervals, whereas nurses, when given choices, also administer the lowest dose at the longest intervals. Common contributing factors to this include not believing

► **TABLE 19-4.** COMMON PHYSICAL AND LABORATORY FINDINGS ASSOCIATED WITH ADDICTIVE DISORDERS

Note that none of the following are specific for addictive disorders, but should trigger further evaluation when they occur together with other suggestive information

Suggestive physical signs
  Hepatic enlargement
  IV injection scars, "tracks" (commonly antecubital fossae, dorsum of hand/feet)
  Alcohol withdrawal signs
  Flushing, hyperreflexia, elevated BP and pulse
  Opioid withdrawal signs
  Mydriasis, sweating, irritability, rhinorrhea
  Intoxication, nodding (alcohol or drugs)
  Constricted pupils (opioids)

Suggestive laboratory findings
  Positive urine, blood or breatholyzer screens
  Positive HIV (IV drug use)
  Positive HepBAg or Anti HepBAg (IV drug use)
  Positive HepCAg or AntiHep Cag (IV drug use)
  Elevated GGT (alcohol)
  Elevated MCV (alcohol)

the addicted person's reports of pain; fear of exacerbating addiction, or causing relapse in recovering persons; fear that patients will cause themselves harm through misuse of medications; and concerns about being manipulated. In addition, misunderstandings about pharmacologic issues related to opioids, such as tolerance or the belief that chronically administered opioids will provide analgesia for acute pain, may contribute. Persons with addictive disorders appear to have relatively high rates of pain,[41] and it is possible to address this pain safely and effectively in most circumstances.

Acute Pain Treatment

Plan Treatment When Possible
Persons with addictive disorders frequently require treatment of acute pain. When acute pain is predictable, such as that associated with elective surgery or in patients with recurrent sickle cell disease, care often is improved and stress on both patient and staff is reduced when a treatment plan is in place prior to the onset of pain. The patient's input, based on prior experience, is often an important part of planning. Information from family, primary care physician, addictions provider, and others also may be helpful.

Treat Pain Effectively
In the context of severe acute pain, such as that associated with trauma, perioperative pain, or acutely painful medical problems such as renal lithiasis or pancreatitis, the first concern should be management of the pain effectively. Often opioids are an important part of management. The choice of initial pain management technique is unlikely to have a significant impact on the course of an addictive disorder in this context, so effective use of opioids by the most expeditious route and schedule is appropriate. Once pain is under adequate control and the patient is stabilized, if modification of treatment seems necessary, a treatment plan that both effectively addresses pain and accommodates addiction issues can be implemented.

As with all patients in the acute setting, goals for pain treatment should be established, and the intensity of pain should be assessed at regular intervals to ensure that goals are being met. Unless there are compelling reasons to question the patient's report of pain (such as persistent intoxication and declining function with increasing doses of medication), treatment should be titrated to achieve satisfactory reported pain relief.

When mild to moderate acute pain is present, nonopioid approaches such as nonsteroidal anti-inflammatory drugs (NSAIDs), ice, transcutaneous electrical nerve stimulation (TENS), or other approaches may be effective. When these are not effective or pain is more severe, opioids are appropriate. In patients who are not tolerant to, or dependent on, opioids and have moderate pain, less reinforcing opioids such as kappa agonists or partial mu agonists may be effective. However, these cannot be titrated for more severe pain, and it is important to avoid agonist-antagonist medications in opioid-dependent patients because they may precipitate withdrawal. Schedules and routes that reduce reward are reasonable to use when they are effective, but concern regarding reward should be secondary to providing pain relief when significant acute pain is present.

Treatment of acute pain with epidural analgesia, plexus infusions, rib blocks, and other procedures is helpful in some patients with pain that is responsive to such approaches. It is important that the patient be amenable to such treatment and that the procedures be within the routine of the care team. Pain treatment plans that confuse or frustrate the patient and staff (for example, attempting to avoid parenteral opioids in a heroin addict by setting up epidural local anesthetic analgesia on a floor that has never managed an epidural) may be a recipe for failure and inhumane to the patient needing relief.

Address Withdrawal Symptoms
Drug and alcohol withdrawal should be treated when present. This will facilitate pain treatment and, in some cases, is critical for safety. The sympathetic arousal associated with acute alcohol withdrawal, including hypertension, tachycardia, and sweating, may be misinterpreted as reflecting undertreated pain. Alcohol withdrawal may exacerbate pain, and it is not attenuated with opioids. Untreated severe alcohol withdrawal may be associated with seizures, hallucinations, hyperpyrexia, and death, so it is critical to make this distinction. The most common, effective and safest treatment of alcohol withdrawal is treatment with judicious titration of benzodiazepines. Benzodiazepines have no direct analgesic effect; therefore, analgesic therapy must be provided as well. Close observation of the patient is always merited with parenteral use of benzodiazepines or opioids but is especially necessary when they are used together.

Opioid withdrawal is usually blocked by opioids titrated for acute pain, so specific other treatments for withdrawal are not required. However, it is important to recognize that the opioid-dependent persons may require significantly higher doses of opioids to achieve analgesia than nondependent persons.[15] This is so for several reasons: Accustomed doses of opioids do not provide analgesia for acute pain, some level of opioid tolerance is often present in opioid-dependent patients, and there is evidence that some patients using opioids chronically may experience hyperalgesia or increased pain sensitivity due to opioids.[57,58] Since a level of dependence—or opioid debt—and tolerance usually cannot be determined accurately, it is recommended frequently to begin with doses at the high end of the recommended starting dose range for a particular opioid and titrate as needed. Most pure mu opioids are interchangeable (if appropriate adjustment in dose is made to account for differences in potency and tolerance), with the exception of methadone. It has been noted that it is often difficult to achieve analgesia when methadone is stopped and another opioid initiated.[59]

Special Issues in Opioid Agonist Therapy of Addiction

Methadone Treatment

Effective treatment of acute pain requires that baseline doses of opioid must be provided in all patients who are physically dependent on opioids (whether due to addiction or therapeutic use) and additional treatment provided for analgesia. In patients receiving methadone treatment for addiction, the baseline dose of methadone is usually continued to meet chronic dosing requirements, and a different additional analgesic is provided for pain rather than discontinuing methadone and using another opioid for both pain control and agonist therapy of addiction. Use of methadone for addiction and a different opioid for pain allows the indications to be tracked separately, keeping clear what medication is for addiction and what is for pain. In addition, clinical observations of difficulty achieving pain control and/or withdrawal symptoms have been made in some patients whose methadone was discontinued abruptly, even with addition of an alternative opioid.[59,60]

Some clinicians elect to continue methadone for addiction and add additional methadone doses for pain. However, in addition to concerns regarding blurring of indications, methadone's relatively slow onset and long, unpredictable half-life limit its flexibility and efficacy in acute pain treatment, although with careful attention and skill it can be a safe and effective medication choice.

Methadone is usually given once daily for methadone maintenance but is needed most often at 6- to 8-hour intervals for pain. If the patient's reported methadone maintenance dose cannot be verified with a treatment program, it is important to give this in divided doses and observe the patient carefully for oversedation before giving the next dose because the daily doses used often in MMT programs may be fatal to nontolerant patients. Methadone can be given intravenously or intramuscularly at one-half the oral dose.

Buprenorphine Treatment

Buprenorphine was introduced in the United States for treatment of opioid addiction in 2002 and is approved for use in primary care practice by registered physicians.[61] While buprenorphine is itself an analgesic, there is little published experience on treating pain in individuals on buprenorphine maintenance therapy of addiction. Buprenorphine binds avidly to mu opioid receptors and may block or reverse mu opioid analgesia.[62] Patients who present with acute severe pain while on buprenorphine therapy for addiction may require aggressive titration of mu opioids to achieve analgesia. This must be done with careful monitoring. It has been suggested that opioids with high intrinsic efficacy, such as fentanyl, may be most effective in overcoming buprenorphine blockade and achieving analgesia, although this is not certain.

When pain is anticipated, it may be best to taper and stop buprenorphine a few days before the anticipated procedure, gradually substituting methadone to block addiction craving if needed.

Pain that Persists Beyond Apparent Healing

A relatively common clinical challenge occurs when the cause of acute pain appears to be healing, but a patient who has been using opioid analgesics for pain control complains of continued or increasing pain. When this occurs, especially in a person with a history of abuse or addiction problems, the immediate concern is often that the individual has become addicted to his or her medication. It is helpful to have a methodical way of addressing such issues (Fig. 19-1). While such an approach is suggested here as a series of steps, in practice, these steps most often are done simultaneously.

Search for Physiologic Causes

As with all patients, it is important to screen for an identifiable physical cause. Abscesses, infections, undetected tissue injury, or neuropathic mechanisms all may cause ongoing pain following apparent healing of the initially painful problem. Physical signs may be subtle and require meticulous evaluation to identify the cause. This is particularly true for neuropathic pain, in which signs may be absent or limited to subtle regionalized autonomic changes, such as thermal or sweating changes, or sensory changes, such as allodynia or hyperalgesia. In such cases, symptom description consistent with a neuropathic pain syndrome may be the only diagnostic clue.

Identify and Address Pain-Facilitating Factors

Whether or not a clear underlying physiologic mechanism of pain can be determined, screening for factors that may facilitate pain such as sleep disturbance, mood disorders, or major stressors may identify factors acting to perpetuate the experience of pain. The patient may be self-medicating nonpain symptoms with opioids. It is also important to ensure that the patient is not experiencing intermittent opioid withdrawal as opioids are tapered because this may cause exacerbated pain that

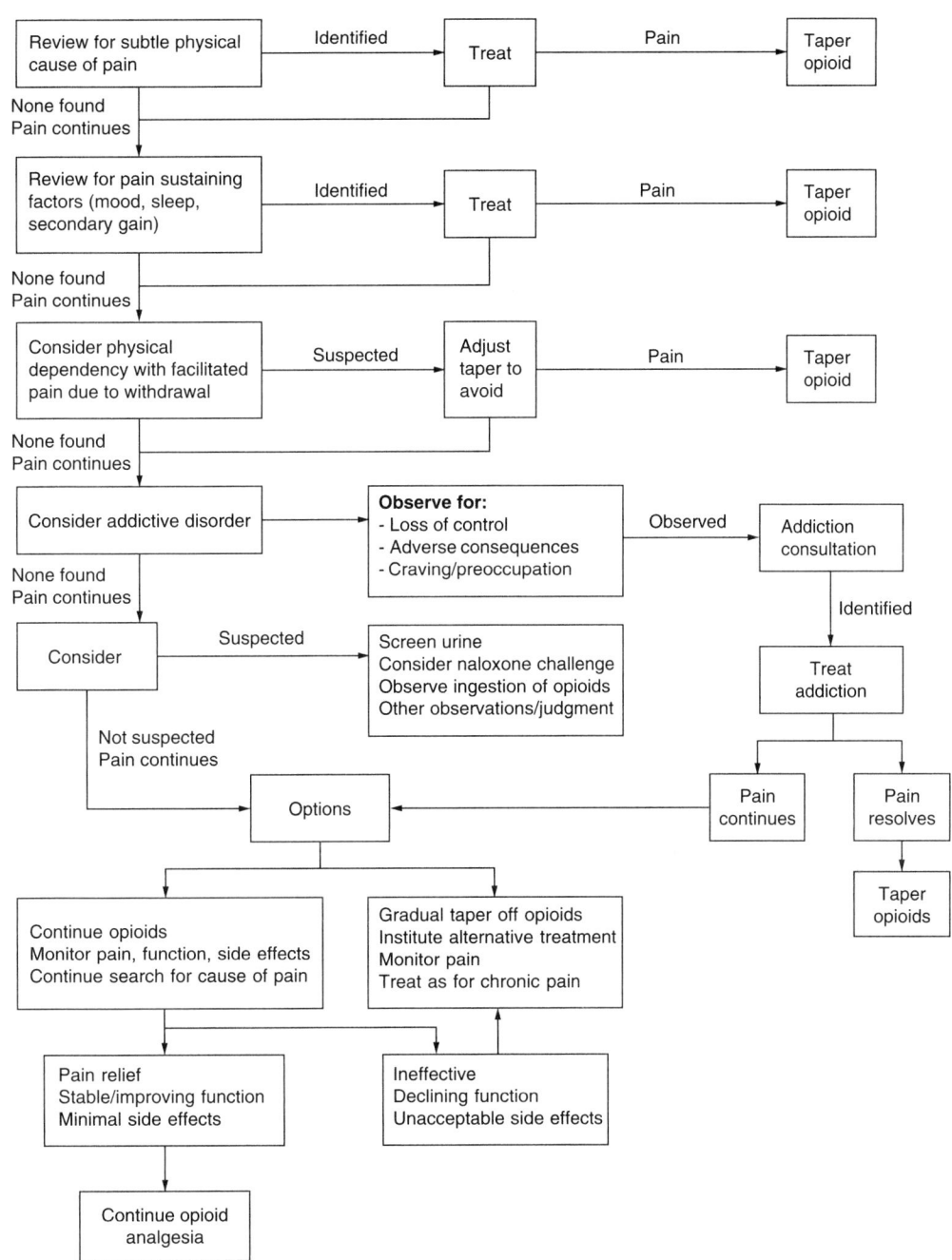

Figure 19-1. When pain persists after apparent healing.[20]

responds to continued opioids.[15] Dosage forms and intervals that allow gradual decrease in opioid blood levels without intermittent nadirs usually avoid withdrawal and associated rebound pain.

Assess for Abuse, Addiction, or Diversion

When no explanatory pain generators or facilitators are identified or those which are present fail to respond to treatment, it is reasonable to consider addiction, abuse, or diversion as possible reasons that the individual is seeking to continue opioids. Eliciting a clear history of how and for what symptoms the patient reports using the medications, a careful review of personal and family substance-use history, and urine or serum drug screens to document use and screen out use of nonprescribed substances are appropriate. In considering if addiction is present when opioids are used for pain, it is helpful to methodically review behaviors with respect to the four C's that relate to

▶ **TABLE 19-5.** ASSESSMENT FOR ADDICTION DURING OPIOID THERAPY OF PAIN: LOOKING FOR THE FOUR "C'S"

Pattern May Suggest Addiction*	Pattern Suggests Therapeutic Use
Adverse consequences/harm due to use	*Favorable therapeutic response to use*
Intoxicated/somnolent/sedated	No significantly altered consciousness
Declining activity	Stable or improving activity
Irritable/anxious/labile mood	Stable or improved mood
Increasing sleep disturbance	Stable or improved sleep
Increasing pain complaints	Stable or improving pain
Increasing relationship dysfunction	Improving relationships
Impaired control over use/Compulsive use	*Able to use as prescribed*
Reports lost or stolen scripts or meds	Rare or no medication incidents
Frequent early renewal requests	Uses meds as prescribed
Urgent calls or unscheduled visits	Doses discussed at clinic visits
Abusing other drugs or alcohol	No alcohol or drug abuse
Can not produce medications on request	Has expected amount of medication left
Withdrawal noted at clinic visits	No withdrawal signs
Observers report overuse or sporadic use	Observers report appropriate use
Preoccupation with use due to Craving	*Seeking pain relief not opioid reward*
Frequently misses appointments unless opioid renewal expected	Makes most appointments
Does not try non-opioid treatments	Shows up for recommended evaluations
Cannot tolerate most medications	Gives reasonable treatment recs a fair trial
Requests medications with high reward	Medication sensitivities and favorable responses not predictable by medication abuse liability
No relief with anything except opioids	Adopts self management strategies (Can demonstrate/discuss techniques)

*Any of these behaviors may occur from time to time in patients using opioids appropriately for pain relief or when pain is inadequately relieved. A pattern of these behaviors in the context of titrated pain therapy suggests the need for further evaluation.
SOURCE: Savage, 2002.

behaviors associated with addiction: consequences, control, compulsion, and craving (Table 19-5). If addiction is suspected, evaluation by a skilled addictions counselor or physician may be helpful. If diversion is suspected, observation of the patient taking the reported dose of medication may be enlightening, especially if this is a high dose, because the diverter who is not using the medication often will manifest sedation, intoxication, or respiratory depression to unaccustomed doses of opioid. An opioid antagonist should be readily available.

When pain persists and no clear causes are identified, a chronic pain model of management that addresses the pain itself, as well as psychological, social, spiritual, and functional issues, is often the most satisfactory approach to management

Pain Associated with Life-Threatening Illnesses

Persons with cancer experience coexisting addictive disorders at the same or possibly a higher rate than the general population.[40,63] In persons with HIV-infection-related pain, the prevalence of addictive disease is somewhat higher than in the general population because of the higher prevalence of HIV infection in the intravenous drug using population.[55,64] Pain is often undertreated in persons with HIV infection, and the World Health Organization (WHO) has recommended that the WHO guidelines for cancer pain treatment be applied to this population.[65]

Addictive behaviors often do not interfere with pain management in the setting of life-threatening disease, although behavioral problems can occur.[66] In addition to pain, persons with life-threatening illness often experience a variety of stressors in association with their illness. For persons who have often coped with life stressors with the use of drugs, it is natural to use available substances to help cope; thus, there is risk of use of opioid analgesics for this purpose. Close support and monitoring are helpful. People who have benefited from work with a recovery program may find significant psychological, social, and spiritual support from engaging closely with their recovery system. When opioid medications are needed to control significant pain associated with cancer or other life-threatening illness, they should never be withheld due to concerns related to addiction (unless alternatives satisfactory to the patient are available), but their use should be as tightly structured as needed to ensure safety and effective use for pain.

Figure 19-2. Synergy of chronic pain and addiction.

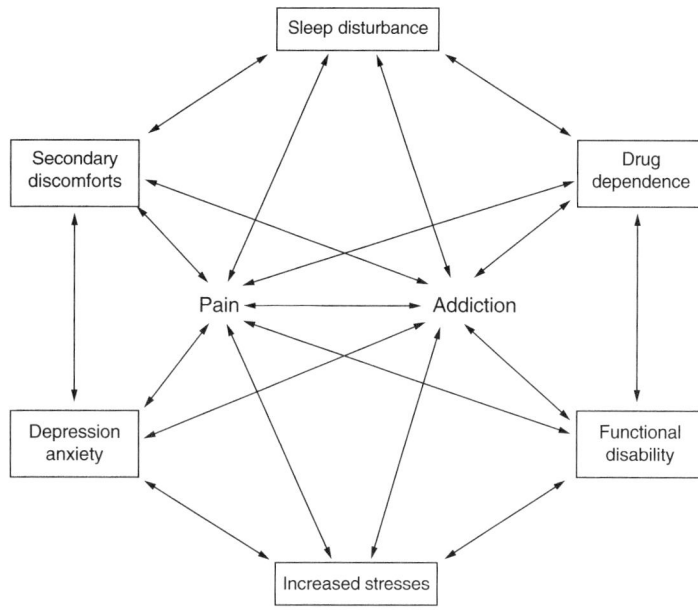

Management of Chronic, Non-Cancer-Related Pain

Synergy of Chronic Pain and Addiction

Chronic pain and active addiction often present as similar conditions in their impact on many dimensions of an individual's life. Each may cause adverse changes in mood, sleep, relationships, capacity to work and recreate, and the overall quality of the individual's life (Fig. 19-2). In addition, active addiction may make it difficult to comply with treatment recommendations. Frequent intoxication may mask pain and result in further injury- or activity-related exacerbation of pain. The physiologic stresses associated with intoxication and withdrawal may facilitate pain. When pain and addiction coexist, they may reinforce one another, making it important to address both issues.

Multidimensional Management

The management of chronic pain in individuals with addictive disorders is similar to chronic pain management in persons without addictive disease. Usually the approach is multidimensional, and the focus is on not only pain relief but also improvement in mood and sleep, relationships, and the ability to engage in meaningful work or avocational activities and recreation. While pain treatment is important, rehabilitation is often an equivalent goal. An active role of the patient usually is required in the treatment of chronic pain, as it is in recovery from addiction. Active involvement in an addiction recovery program may facilitate pain treatment. Twelve-step models, effective in addiction recovery, have been piloted successfully for the treatment of pain.[67] A wide variety of strategies and tools may be employed in chronic pain treatment, and they are discussed elsewhere. Generally, these tools fall into four categories: physical modalities, cognitive-behavioral approaches, invasive procedures, and medications, including both opioid and nonopioid medications.

Opioids for Chronic Pain in Recovery

As with all patients, consideration of the use of opioids in the treatment of chronic pain in persons with addictive disease requires a careful weighing of the potential risks and benefits of the use of opioids for that individual against the risks and benefits of a program of care that does not include opioids.[68] In an individual with an addictive disorder, an unquantifiable, but somewhat likely, increased risk of addiction must be considered. The beliefs and desires of the patient and his or her significant others must be considered. This is particularly important in individuals in recovery who may have strong beliefs regarding the impact of medications on their recovery and whose family and recovering associates may have similar or different perceptions. The feasibility and costs of various treatments also must be factored in. An active recovery program and strong social support have been associated with improved outcomes with opioid therapy of chronic pain in individuals with addictive disorders.[43] Unfortunately, there is no clear formula for treatment that best meets the needs of all patients and the capacities of all providers; treatment planning must be individualized.

Opioids for Chronic Pain in Active Addiction

Persons with active addiction, whether to alcohol, opioids, or other substances, are rarely, if ever, candidates for opioid therapy of chronic non-cancer-related pain because they cannot control their use of potentially rewarding substances reliably and risk the danger of overdose when opioids are available. Persons using illicit substances also have regular contact with the street culture and, therefore, the possibility to divert opioids. Sometimes the promise of aggressive analgesia can help an individual commit to recovery; in such cases, very tight monitoring and controls are necessary because the risk of relapse is highest in early recovery. In rare instances, opioids administered in a supervised manner have been used safely and effectively in persons with severe chronic pain and active addiction. However, such treatment requires a highly structured system to ensure patient safety and meticulous documentation of informed consent and agreement for treatment. Persons with a history of opioid addiction, particularly a recent history, who require opioids as a component of chronic pain treatment often need to be enrolled in a methadone treatment program for their addiction and receive additional, well-coordinated treatment of their pain.

Marijuana Use

There is debate within the pain community as to whether the occasional use of marijuana is a contraindication to the use of opioids for chronic pain. It is argued on one side that access to marijuana means contact with street markets and thus an increased risk of diversion, that marijuana use per se may have medical consequences, and that it is often associated with abuse of other drugs. On the other side, it is argued that cannabinoids have clearly documented analgesic effects, that the medical consequences of occasional marijuana use are minimal if activities requiring judgment and coordination are avoided while affected, and that occasional use is so common as to be normative behavior in some responsible segments of society. There is clearly some truth on both sides. Whatever a clinician's view of these issues, it is appropriate that persons found to be using marijuana have a thorough assessment of substance-use patterns and referral for expert evaluation if addiction or dangerous patterns of drug abuse are identified. In addition, it is important to consider that, whatever the medical truth, it is likely that physicians who prescribe opioids to patients who use marijuana likely have a greater risk of regulatory scrutiny and potential sanction than physicians who require that patients who elect to use opioids for pain control abstain from use of all illicit substances.

Structure and Monitoring of Opioid Therapy

When opioids are elected as a component of care for chronic pain in a person in recovery, they should be structured in a manner that permits close monitoring and early identification of relapse. A balance of safety and pain relief must be found. The patient should be reassessed regularly and opioids dispensed at a frequency that provides only that amount of opioids that the patient can control. For some patients this may require dispensing by a pharmacy or trusted other on a daily basis; other patients may handle a month's supply of medications without difficulty. Some persons have had signed and dated transdermal analgesic patches dispensed every 3 days in an emergency room; for others, even such dispensing is dangerous because of a risk of compulsive intravenous use. It is often helpful to engage significant others and co-care providers in problem solving. In some cases, trusted others may assist with monitoring or dispensing, although this can be a significant source of interpersonal stress in other cases.

Support Recovery

In most cases it is helpful for recovering persons to be engaged in an active recovery program and for the prescribing physician to communicate closely with that program. At regular visits, the clinician should assess pain-related issues as well as recovery issues. Urine drug screens may be a powerful support for recovery and help to identify relapse earlier than otherwise possible.

Opioid Regimens

When a patient has persistent around-the-clock pain, a medication regimen that provides stable blood levels is usually the best choice because it is least likely to cause reward and reduces focus on taking medication. Usually this requires use of long-acting opioids such as methadone or sustained-release medications. It is important to recall that individuals who wish to abuse sustained-release medications can do so without difficulty. For intermittent incident pain, an "as needed" regimen may be appropriate. When possible, it is less likely to reinforce pain and/or opioid use. The opioid analgesic is paired with an intended activity or a specific time of the day (or week) instead of pain. This avoids pairing the subjective experience of pain with a potentially reinforcing substance. For example, a person with severe knee DJD who has little resting pain but severe pain with ambulation might take an opioid ½ hour before an anticipated shopping trip. This both preempts the pain, providing better pain control, and avoids the patient having to decide "when the pain is bad enough," which can lead to both self-doubt in a recovering person and a gradual increase in indications for use.

Legal Issues in Pain Treatment and Addiction

Any schedule II–IV opioid, including methadone, may be used legally to treat pain in any patient, including patients with addictive disorders or histories of drug abuse.[69] Clinicians must be affiliated with a narcotic treatment program (NTP) and/or have a special waiver from the Center for Substance Abuse Treatment (CSAT), however, to use opioids to treat addiction in most circumstances.

There are two exceptional circumstances in which clinicians without special license or affiliation may use methadone to treat addiction.[69] If an opioid-addicted patient is hospitalized for a purpose other than addiction treatment, i.e., for trauma, medical condition, surgery, or a non-addiction-related psychiatric problem, methadone may be prescribed to prevent withdrawal or craving by any physician with a standard DEA controlled-substances license. This may be used throughout the hospitalization, whether or not the patient has been in an NTP outside the hospital. However, methadone prescribing must be stopped on discharge, or the patient must be transferred to a licensed NTP. Methadone also may be administered or dispensed (not prescribed) on a daily basis for 3 consecutive days to an opioid-addicted patient who is awaiting entry to an addiction treatment program. The 3-day treatment may not be repeated or extended.

Clinicians who have taken an approved 8-hour course and have received a waiver from CSAT may prescribe buprenorphine, a schedule III controlled substance, for the treatment of opioid addiction.[70] Buprenorphine may be used for the treatment of pain without special authorization other than a DEA controlled-substances license.

Many states have adopted guidelines or other policies related to the use of opioids for the treatment of pain that are intended to promote pain treatment and protect clinicians who use opioids appropriately to treat pain.[71,72] Interpretation and application of these guidelines by government attorneys differ significantly from jurisdiction to jurisdiction. Physicians who provide compassionate pain treatment for persons with histories of abuse and addiction sometimes may find their practices under scrutiny because of the behaviors of their patients. In order to treat pain safely and effectively in individuals with addictions and to protect one's ability to continue to provide it, it is critical to carefully address issues of substance abuse in the course of pain treatment.[73]

REFERENCES

1. *www.oas.samhsa.gov*.
2. Gardner E: Brain reward mechanisms, in Lowinson J, Ruiz P, Millman R (eds): *Substance Abuse: A Comprehensive Text*. Baltimore: Williams & Wilkins, 1997:51–85.
3. Gardner E: What we have learned about addiction from animal models of drug self-administration. *Am J Addict* 2000;9:285–313.
4. Lamb R, et al: The reinforcing and subjective effects of morphine in post-addicts: a dose response study. *J Pharmacol Exp Ther* 1991;259:1165–1173.
5. Van Ree JM, Gerrits MA, Vanderschuren LJ: Opioids, reward and addiction: An encounter of biology, psychology, and medicine. *Pharmacologic Rev* 1999;51:341–396.
6. Tsuang MT, BJ, Harley RM, Lyons MJ: The Harvard Twin Study of Substance Abuse: What we have learned. *Harvard Rev Psychiatry* 2001;9:267–279.
7. Enoch M, Goldman D: Genetics of alcoholism and substance abuse. *Psychiatr Clin North Am* 1999;22:289–299.
8. Kessler RC, Crum RM, Warner L, et al: Lifetime co-occurence of DSM-III-R alcohol abuse and dependence with other psychiatric disorders in the national comorbidity survey. *Arch Gen Psychiatry* 1997;54:313–321.
9. Zacny J, Bigelow G, Compton P, et al: College on Problems of Drug Dependence Taskforce on Prescription Opioid Non-medical Use and Abuse: Position statement. *Drug Alcohol Depend* 2003;69:215–232.
10. Passik Kirsh K, McDonald MV, et al: A pilot survey of aberrant drug-taking attitudes and behaviors in samples of cancer and AIDS patients. *J Pain Sympt Manage* 2000; 19:274–286.
11. ASAM: Public policy statement on definitions related to the use of opioids in pain treatment. *J Addict Dis* 1998; 17:129–133.
12. Mclellan AT, Lewis DC, O'Brien CP, Kleber HD: Drug dependence, a chronic medical illness: Implications for treatment, insurance, and outcomes evaluation. *JAMA* 2000; 284:1689–1695.
13. Blomqvist J: Paths to recovery from substance misuse: Change of lifestyle and the role of treatment. *Substance Use Misuse* 1996;31:1807–1852.
14. Melichar JK, Daglish MR, Nutt DJ: Addiction and withdrawal: Current views. *Curr Opin Pharmacol* 2001;1:84–90.
15. Brodner RA, Taub A: Chronic pain exacerbated by long-term narcotic use in patients with non-malignant disease: Clinical syndrome and treatment. *Mt Sinai J Med* 1978;45:233–237.
16. Foley K: Clinical tolerance to opioids, in Basbaum AI, Besson J-M (eds): *Towards a New Pharmacotherapy of Pain*. New York: Wiley, 1991:181–203.
17. Collett B: Opioid tolerance: The clinical perspective. *Br J Anaesth* 1998;81:58–68.
18. Mau J: NMDA and opioid receptors: Their interactions in antinociception, tolerance and neuroplasticity. *Brain Res Rev* 1999;30:289–304.
19. Heit H: Addiction, physical dependence, and tolerance: Precise definitions to help clinicians evaluate and treat chronic pain patients. *J Palliat Care Pharmacother* 2002:1–12.
20. Savage S, Joranson D, Covington E, et al: Definitions related to the medical use of opioids: Evolution towards universal agreement. *J Pain Sympt Manage* 2003;26:655–667.
21. Weissman DE, Haddock JD: Opioid pseudoaddiction: An iatrogenic syndrome. *Pain* 1989;36:363–366.
22. Kirsh K, Whitcomb LA, Donoghy K, and Passik SD: Abuse and addiction issues in medically ill patients with pain: Attempts at clarification of terms and empirical study. *Clin J Pain* 2002;18:S52–60.
23. Dworkin S, Porrino L, Smith J: Neurobiological substrates of opioid abuse, *The Neurobiology of Opiates*. Boca Raton, FL: CRC Press, 1993.
24. Zacny JP, McKay MA, Toledano AY, et al: The effects of cold-water immersion stressor on the reinforcing and subjective

effects of fentanyl in healthy volunteers. *Elseviers Drug Alcohol Depend* 1996;42:133–142.
25. Bidlack J, McLaughlin J, Wentland M: Partial opioids: Medications for the treatment of pain and drug abuse. *Ann NY Acad Sci* 2000;909:1–11.
26. Mamoon A, Barnes AM, Ho IK: Comparative rewarding properties of morphine and butorphanol. *Brain Res Bull* 1995; 38:507–511.
27. Cicero TJ, Adams E, Geller A, et al: A postmarketing surveillance program to monitor Ultram (tramadol hydrochloride) abuse in the United States. *Drug Alcohol Depend* 1999; 57:7–22.
28. Tzschentke T: Behavioral pharmacology of buprenorphine, with a focus on preclinical models of reward and addiction. *Psychopharmacology* 2002;16:1–16.
29. Kreek M: Methadone-related opioid agonist pharmacotherapy for heroin addiction: History, recent molecular and neurochemical research and future in mainstream medicine. *Ann NY Acad Sci* 2000;909:186–216.
30. Stimmel B, Kreek M: Neurobiology of addictive behaviors and its relationship to methadone maintenance. *Mt Sinai J Med* 2000;67:375–380.
31. Colwell C, Morris B: Patient-controlled analgesia compared with intramuscular injection of analgesics for the management of pain after an orthopaedic procedure. *J Bone Joint Surg* 1995;77:726–733.
32. Porter J, Jick H: Addiction rare in patients treated with narcotics. *N Engl J Med* 1980;302:123.
33. Perry S, Heindrich G: Management of pain during debridement: A survey of U.S. burn units. *Pain* 1982;13:12–14.
34. Moore R, Bone LR, Geller G: Prevalence, detection and treatment of alcoholism in hospitalized patients. *JAMA* 1989;261:934–958.
35. Graham AW: Screening for alcoholism by life-style risk assessment in a community hospital. *Arch Intern Med* 1991;151:958–964.
36. Madan A, Yu K, Beech D: Alcohol and drug use in victims of life-threatening trauma. *J Trauma Injury Infect Crit Care* 1999;47:568–571.
37. Chabal C, Erjavec MK, Jacobson L, et al: Prescription opiate abuse in chronic pain patients: Clinical criteria, incidence, and predictors. *Clin J Pain* 1997;13:150–155.
38. Hoffmann NG, Olefssen O, Salen B, Wickstrom L: Prevalence of abuse and dependency in chronic pain patients. *Int J Addict* 1995;30:919–927.
39. Fishbain D, Rosomoff H: Drug abuse, dependence and addiction in chronic pain patients. *Clin J Pain* 1992;8:77–85.
40. Bruera E, Moyano J, Seifert L, et al: The frequency of alcoholism among patients with pain due to terminal cancer. *J Pain Symptom Manage* 1995;10:599–603.
41. Jamison RN, Kauffman J, Katz NP: Characteristics of methadone maintenance patients with chronic pain. *J Pain Symptom Manage* 2000;19:53–62.
42. Rosenblum A, Joseph H, Fong C, Kipnis S, Cleland C, Portenoy RK: Prevalence and characteristics of chronic pain among chemically dependent patients in methadone maintenance and residential treatment facilities. *JAMA* 2003; 189:2370–2378.
43. Dunbar S, Katz N: Chronic opioid therapy for nonmalignant pain in patients with a history of substance abuse: Report of 20 cases. *J Pain Symptom Manage* 1996; 11:163–171.
44. Piazza P, Moal ML: Pathophysiological basis of vulnerability to drug abuse: Role of an interaction between stress, glucocorticoids and dopaminergic neurons. *Annu Rev Pharmacol Toxicol* 1996;36:359–378.
45. Gourlay D: "Universal precautions in pain management." Paper presented at: Conference on Common Threads in Pain and Addiction, April 22,2004; Washington, DC.
46. Leshner A: Why shouldn't society treat substance abusers? *Los Angeles Times,* 1999.
47. Miller W, Rallnick S: *Motivational Interviewing: Preparing People to Change Addictive Behaviors.* New York: Guilford Press,1991.
48. Brown R, Rounds L: Conjoint screening questionnaire for alcohol and drug abuse. *Wisc Med J* 1995;94:135–140.
49. Cyr M, Wartman A: The effectiveness of routine screening questions in the detection of alcoholism. *JAMA* 1984; 259:51–54.
50. Skinner H, Holt S, Schuller R, et al: Identification of alcohol abuse using laboratory tests and a history of trauma. *Ann Intern Med* 1984; 101:847–851.
51. Saunders J, Aasland OG, et al: AUDIT: Development of the alcohol use disorders identification test: Guidelines for use in primary health care. *Addiction* 1993;88:791–804.
52. Selzer ML, Vinokur A, et al: A self-administered Short Michigan Alcoholism Screening Test (SMAST). *J Stud Alcohol* 1975;36:117–126.
53. Lakshman M, Tsutsumi M: Alcohol biomarkers: Clinical significance and biochemical basis. *Alcohol Alcoholism* 2001; 25:171–172.
54. Portenoy RKJ, Lowinson H, Rice J, et al: Pain management and chemical dependency: Evolving perspectives. *JAMA* 1997;278:592–593.
55. Breitbart W, Rosenfeld BD, Passik SD, et al: The undertreatment of pain in ambulatory AIDS patients. *Pain* 1996;65:239–245.
56. Cleeland CS, Gonin R, Hatfield AK, et al: Pain and its treatment in outpatients with metastatic cancer. *N Engl J Med* 1994; 330:592–596.
57. Compton P, Charuvastra VC, Kintaudi K, Ling W: Pain responses in methadone-maintained opioid abusers. *J Pain Symptom Manage* 2000;20:237–245.
58. Doverty M, White JM, Somogyi AA, et al: Hyperalgesic responses in methadone maintenance patients. *Pain* 2000;90:90–96.
59. Moryl N, Santiago-Palma J, Kornick C, et al: Pitfalls of opioid rotation: Substituting another opioid for methadone in patients with cancer pain. *Pain* 2002;96:325–328.
60. Doverty M, Somogyi AA, White JM, et al: Methadone maintenance patients are cross-tolerant to the antinociceptive effects of morphine. *Pain* 2001;93:155–163.
61. Ling W, Smith D: Buprenorphine: Blending practice and research. *J Subst Abuse Treat* 2002;23:87–92.
62. Robinson S: Buprenorphine: An analgesic with an expanding role in the treatment of opioid addiction. *CNS Drug Rev* 2002;8:377–390.
63. Passik SD, Portenoy R, Ricketts PL: Substance abuse issues in cancer patients: 1. Prevalence and diagnosis. *Oncology (Hunt)* 1998;12:517–521, 524.

64. McCormack JP, Li R, Zarowny D, Singer J: Inadequate treatment of pain in ambulatory HIV patients. *Clin J Pain* 1993; 9:279–283.
65. World Health Organization: *Fact Sheet 8: HIV Palliative and Terminal Care.* Geneva: WHO, 1996.
66. Passik SD, Portenoy R, Ricketts PL: Substance abuse issues in cancer patients: 2. Evaluation and treatment. *Oncology (Hunt)* 1998;12:729–734;discussion 736, 741–742.
67. Kalt R: Twelve-step recovery from alcoholism, drug addiction and chronic physical pain. (unpublished manuscript, 1997).
68. Wesson D, Smith D: Prescription of opioids for treatment of pain in patients with addictive disease. *J Pain Symptom Manage* 1993;8:289–296.
69. *Code of Federal Regulations* 21-1306.07. *http://www.gpoaccess.gov/efr/index.html.*
70. Mitka M: Agency offers guidance on office-based drug treatment for opioid addiction. *JAMA* 2003;289:690.
71. Federation of State Medical Boards: Model guidelines for the use of opioids for the treatment of pain. *Bull Fed State Med Boards* 1998;85:84–87.
72. Joranson D, Gilson AM, Dahl JL: Pain management, controlled substances, and state medical board policy: A decade of change. *J Pain Symptom Manage* 2002;23:138–147.
73. Gilson A, Joranson D: U.S. policies relevant to the prescribing of opioid analgesics for the treatment of pain in patients with addictive disease. *Clin J Pain* 2002;18:S91–98.

CHAPTER 20

Management of the Patient with Chronic Pain: Aspects of Quality Assurance and Outcomes

Debra B. Gordon and Nathan J. Rudin

Quality is a nebulous concept, especially in the realm of health care. The definition of quality depends on both the purpose of measurement and the perspective of the measurer. For example, in any given instance of health care delivery, the patient, the provider, and the insurer may have very different perceptions of the quality of care. Although these differing points of view make quality difficult to define (and hence to measure), health care quality has become an increasingly important concept to patients, insurers, and care providers alike. Marketplace competition, the availability of online and other consumer health information, and the rising cost of health care have all helped to fuel this interest in quality. Current trends demand that health care quality be systematically measured, improved, and publicized, however ill defined quality itself may be.

Many factors affect the quality of health care, including practice patterns (assessment and treatment behaviors), decision making, policies, and patient outcomes. It is widely recognized that health care is laden with quality problems related to underuse, overuse, and misuse of services.[1] Fee-for-service payment systems promote resource overuse, whereas capitated systems encourage underuse. Unfortunately, no current payment system systematically rewards excellence in quality.[2]

When examining the burgeoning field of pain medicine, health care quality becomes even more difficult to define because pain itself is a highly individualized, subjective experience that does not lend itself easily to standardized measurement. Further, pain medicine itself is practiced in many different ways (e.g., pharmacologic, interventional, rehabilitative, interdisciplinary) and in many different settings. The discussion of pain treatment quality is deeply paradoxical. From a technical and scientific standpoint, our capabilities to manage pain are extraordinary, yet we know little about the effects of variations in pain management practices on patient outcomes.

Unrelieved pain can have a serious impact on physical, emotional, and social well-being. Dealing effectively with these diverse consequences of pain demands considerable effort, and pain treatment tends to be resource-intensive. In an increasingly complex and fractious fiscal environment, the survival of pain treatment programs is linked to our ability to clearly define services and quality, demonstrate cost-effectiveness, and show positive treatment outcomes.

Unfortunately for those involved in the process, quality management remains a moving target. Terminology, measurement, and improvement approaches continue to evolve and become ever more complex. As an aid to improving pain management quality in this challenging environment, this chapter provides definitions and a conceptual model of quality useful for the measurement and improvement of care in chronic noncancer pain. Strategies for improvement and quality measurement considerations also are addressed.

▶ HISTORICAL INFLUENCES ON DISCUSSION OF QUALITY

Concern about the quality of health care began more than 100 years ago when Ernest Codman first proposed standardized outcomes data collection to determine whether care was effective.[3] Although many viewed Codman's ideas as threatening, he was able to convince the American College of Surgeons to establish a Committee on the Standardization of Hospital Care. This ultimately led to formation of the Joint Commission on the Accreditation of Healthcare Organizations (JCAHO).[4]

The advent of Medicare and Medicaid in the 1960s was accompanied by steep inflation in health care costs. To keep costs down and maintain quality, both public and private health insurers turned to the development of standards of care as the basis for better, and presumably more cost-effective and defensible, coverage policy and quality measurement. Of particular interest was how care standards could be used as tools to shape policy, assess use of services, and define and identify inappropriate medical care.[5] To help address these needs and ensure delivery of health care in a manner consistent with available standards, accreditation programs such as JCAHO and the Commission on Accreditation of Rehabilitation Facilities (CARF) became prerequisites for participation in the Medicare program. Following the lead of these accreditation bodies, many third-party payors adopted quality assessment and management requirements linking strong economic incentives to the development of quality programs.

In recent years, quality measurement and improvement have changed significantly with new developments in clinical epidemiology, outcomes research, and information technology, and discussion of health care quality issues has accelerated.[6] Clinical epidemiologists were the first to identify wide variation in the practice and outcomes of care among patients treated for the same health care problems in different places and health care settings.[6] Awareness of these practice variations led to the need to better understand their effect on outcomes. The application of outcomes research has resulted in new measures of quality to help improve clinical practice, especially in the treatment of chronic illnesses such as back pain, where improved function is a primary objective.[7,8]

Outcomes research and public interest in health care quality have burgeoned with advances in information technology. New techniques have made it faster, cheaper, and easier to accumulate and analyze multiple types of data. Computerized medical records offer great capacity for data input throughout the course of a patient's care for subsequent use in quality assessment and reporting. The Internet, online health information, and the spread of computer technology to offices and homes have provided consumers, payors, and health care professionals with a wealth of information about available clinical choices and new opportunities to examine quality.

▶ DEFINING THE CONCEPT OF QUALITY

The quality of pain treatment is difficult to define or measure, in part, because of the unique nature of the outcomes involved. Pain is highly subjective and hence difficult to quantify reproducibly. The experience of pain is intertwined with its emotional impact (suffering), which is even more difficult to quantify and compare between individuals. Despite these difficulties, recent developments in quality-assessment theory and practice can provide a useful framework for the task.

Attributes of Quality

If quality is to be managed, it must be defined in terms of specific measurable attributes.[9] Health care quality generally is defined in terms of the attributes and outcomes of care provided by practitioners and received by patients.[10] Donabedian[11] has defined six important attributes of health care quality (Table 20-1). The degree of attention given to any given attribute will vary depending on the aspect of quality being assessed. A more specific definition emphasizes the technical aspects and characteristics of interactions between provider and patient.[12] Important attributes of the provider-patient interaction include communication, trust, empathy, sensitivity, and honesty. *Technical quality* consists of "doing the right thing right,"[10] i.e., performing the right tests or providing the right services to accomplish the desired result. *Outcomes* refer to the patient's health status (e.g., changes in pain, physical function, affect, social roles) after treatment concludes.[13]

Conceptualizing Quality

Donabedian[12] defined quality health care as "that kind of care which is expected to maximize an inclusive measure of patient welfare, after one [has] taken account of the balance of expected gains and losses that attend the

▶ **TABLE 20-1.** IMPORTANT ATTRIBUTES OF QUALITY IN HEALTH CARE

1. Effectiveness
The ability to attain the greatest improvements in health now achievable by the best care
2. Efficiency
The ability to lower the cost of care without diminishing attainable improvements in health
3. Optimality
The balancing of costs against the effects of care on health (or on the benefits of health care, meaning the monetary value of improvements in health) so as to attain the most advantageous balance
4. Acceptability
Conformity to the wishes, desires, and expectations of patients and responsible members of their families
5. Legitimacy
Conformity to social preferences as expressed in ethical principles, values, norms, laws, and regulations
6. Equity
Conformity to a principle that determines what is just or fair in the distribution of health care and of its benefits among the members of a population

SOURCE: *Reprinted with permission: Donabedian A. The role of outcomes in quality assessment and assurance. QRB, November: 356–360, 1992.*

process of care in all its parts." Several different formulations and approaches to the management of quality are possible and legitimate depending on the circumstances and purposes involved.[11] The structure-process-outcome model, first developed by Donabedian[14] and later adopted by many investigators,[15,16] is a useful framework for conceptualizing quality. *Structure* is defined as the physical and organizational properties of the setting in which care is provided; *process* is what is done for patients; and *outcome* is what is accomplished for patients.

According to this model, any examination of quality should include measurements in all three of these dimensions. Quality theory suggests that the delivery of high-quality services increases the likelihood of positive outcomes, but assessment of outcome alone may not adequately determine quality. Poor outcomes can occur despite the best health care; conversely, patients may do well despite poor care. By examining structure and process as well as outcome, we can explore not only the end result but also the factors that lead to it.

Structural Variables

Health care is delivered in a wide variety of settings, and the type of care delivered to the patient may vary considerably between venues. Since pain is such a ubiquitous complaint, patients may encounter pain management issues in virtually any setting, from the emergency department to the chronic pain clinic. Structural variables such as provider training, organizational resources, time allotted per patient, and staffing levels can provide insight into the factors that direct and shape health care quality.

Process Variables

Because outcomes are difficult to measure, process measures often are used as primary quality indicators. Processes include clinical activities (e.g., test ordering, drug prescribing, referral patterns), as well as interpersonal activities, such as teaching and other aspects of patient-provider communication.[4] Process measures are used widely by purchasers and Medicare peer-review organizations to evaluate quality, create practice guidelines, and stimulate quality-improvement activities. The American Medical Association (AMA) Council on Medical Service identified eight process components as essential elements of health care quality and stressed that processes should be judged only if they are clearly linked to outcome (Table 20-2).

Gordon and colleagues[17] defined several process variables that appear to be closely tied to quality pain management. These include appropriate assessment (screening for the presence of pain, completion of a comprehensive initial evaluation when pain is present, and frequent reassessments of patient response to treatment);

▶ **TABLE 20-2.** EIGHT DIMENSIONS OF QUALITY: AMERICAN MEDICAL ASSOCIATION (AMA) COUNCIL ON MEDICAL SERVICES REPORT, 1986

1. Produce the optimal improvement in health status possible
2. Emphasize health promotion and early detection
3. Be provided in a timely manner
4. Involve the informed participation of the patient
5. Be based upon accepted medical and scientific principles
6. Be provided with sensitivity to the stress and anxiety exacerbated by illness
7. Make efficient use of technology
8. Be well documented in the medical record

SOURCE: *Reprinted with permission: Katz JN, Sangha O. Assessment of the quality of care. Arthritis Care and Research 10(6):359–369, 1997.*

interdisciplinary, collaborative care planning that includes patient input; appropriate treatment that is efficacious, cost conscious, culturally and developmentally appropriate, and safe; and access to specialty care as needed.

Outcome Variables

While they are perhaps the most sought-after measures, outcome variables allow only an inference of the quality of care. In pain management, pertinent outcome variables may include changes in pain intensity, frequency, and location; changes in the disease process causing pain; changes in physical function and quality of life (e.g., mobility, activities of daily living, household and/or work tasks); and indicators of patient satisfaction.

Satisfaction is an increasingly frequent focus of quality measurement. The old adage, "Give the customers what they want," seems to be as applicable to health care as to commerce. Identifying and responding to consumer preferences and values is a critical aspect of quality in any setting.[18] The Institute of Medicine (IOM) defined quality health care in 1990 as "the degree to which health services for individuals and populations increase the likelihood of desired health outcomes and are consistent with current professional knowledge."[19] The IOM revisited this statement in 1998, redefining the phrase *desired health outcomes* to specifically include the health outcomes that *patients* desire.[2] Clearly, one must assess not only the medical outcome of treatment but also the patient's satisfaction with his or her care.

▶ APPROACHES TO IMPROVING QUALITY

Several approaches have been used to cultivate improvements in health care quality. For many years, accreditation (of programs) and certification (of individuals) were the primary mechanisms used to foster quality. Newer approaches have arisen in recent years and may have greater potential to improve health care quality. These approaches are applicable to pain medicine and may be useful in any clinical setting. They include quality assurance (QA), which has evolved into quality improvement (QI), evidence-based medicine (EBM) and clinical practice guidelines, and public accountability.[20] Given the complexity of quality, it is unlikely that any one approach will be successful if used alone.

Certification and Accreditation

Certification and accreditation are obtained through comprehensive peer-review processes that determine one's ability to meet standards of care developed by the particular field. Certification establishes that individual providers have the credentials and training necessary to deliver the care they provide. Accreditation produces summary information about a health system's processes, capabilities, and control over activities in the context of delivering high-quality care.[21] These review processes are thought to stimulate QI through education and the development of new or improved structures, processes, and assessments.

Relatively few accreditation and certification programs have focused on pain management. The most notable of these are the pain management accreditation standards developed by JCAHO in 1999. Since JCAHO plays such a prominent role in American health system accreditation, these standards have had a remarkable impact. Although the JCAHO standards have been plagued by controversy and are as yet unsupported by QI data, there is little question that they have helped to increase dramatically the visibility of and accountability for pain management in settings across the United States.

Accreditation and certification programs in pain management may be useful, but they should not be used alone to ensure the delivery of high-quality care. More detailed examinations of care quality are necessary.

Quality Assurance

The conventional QA paradigm popular in the 1970s and 1980s focused on identifying adverse outcomes and censuring responsible parties. QA is a broad-based form of audit and feedback in which outcomes such as percentages of satisfied patients, referral rates, and adverse events are used for comparison by administrative authorities. Standards of care are used as yardsticks, with a focus on reducing providers' errors in the care of individual patients. The difficulty with QA is that feedback comes too late in the process and does little to explain differences in practice or outcomes. Also, medical records, a primary source of QA review data, are notoriously incomplete and often are missing information important to evaluate the quality of pain management.[22]

In 1990, Mitchell Max, then chairman of the American Pain Society's (APS) Quality of Care Committee, suggested that the traditional formula of education and QA was insufficient to improve pain management and treatment outcomes.[23] Citing the failure of these efforts to change practice, Max suggested that a series of background factors must be addressed in the context of the clinical setting. In order to improve pain management, Max suggested the field must (1) increase the visibility of pain as a clinical problem, (2) give clinicians practical tools for change, and (3) increase clinician accountability. This proposition changed the focus from individual clinicians and disciplines to systems thinking, a perspective advocated by the growing QI movement. Max's ideas provided a framework for recommendations by the APS and the Agency for Health Care Policy and Research (AHCPR; now called the Agency for Health Care Research and Quality, or AHRQ) to develop a formal

▶ **TABLE 20-3.** KEY ELEMENTS OF THE AMERICAN PAIN SOCIETY'S QI GUIDELINES FOR THE TREATMENT OF ACUTE AND CANCER PAIN

1. Recognize and treat pain promptly; that is, pain intensity should be assessed and recorded in a way that makes it highly visible to the health care team in a manner that promptly draws attention to the problem
2. Make information about analgesics readily available to clinicians in a way that facilitates order writing and interpretation of orders
3. Promise patients attentive analgesic care in a way that makes it clear that the institution and health care professionals are accountable for pain relief
4. Define explicit policies for the use of advanced analgesic technologies that define the acceptable level of patient monitoring as well as appropriate roles, accountability, and limits of practice for all groups of health care providers involved
5. Examine the process and outcomes of pain management with the goal of continuous improvement

SOURCE: American Pain Society Quality of Care Committee. Quality improvement guidelines for the treatment of acute pain and cancer pain. JAMA 274(23):1874–1880, 1995.

means to monitor and improve the quality of pain management.[24–26]

Quality Improvement

Quality improvement is the broader and more comprehensive evolutionary descendant of QA. Used for years in the manufacturing industry, QI is a compilation of methods adapted from psychology, statistics, and operations research to avert predictable human errors, eliminate unnecessary and harmful variations in practice, and improve the production of goods and services.[27] There are numerous approaches to QI, but the general principles are similar. QI focuses on reducing variation in a production process, standardization, and continuous improvement in outcomes rather than on the identification and elimination of defects. QI involves enlisting an entire organization to work toward a goal of continuous improvement in quality by carefully studying the process one is attempting to improve and changing the responsibilities and power of frontline workers.[9]

Organizations are systems designed to serve customers (patients). A system is a group of related processes or a sequence of tasks necessary to achieve a particular outcome. Important processes for pain management include assessment and communication of pain, interdisciplinary communication and care planning, and goal-driven treatment plans that are readjusted based on patient outcomes. QI theory states that data can be collected to help understand a system's processes and uncover the root causes of any inconsistencies or variations that contribute to quality problems. Unlike QA, there is no predetermined or final yardstick of quality; instead, the goal is continuous improvement. Frontline workers are involved at all stages of QI.

To make care consistent and reduce statistical errors, data must be generated and compared against an established standard.[5] Thus, QI depends on the development and implementation of care standards in the health care environment and the generation of data for performance measurement. In fact, the use of data to track trends has been shown to be a critical factor in successful hospital-based QI initiatives.[28]

Recognizing the promise of QI efforts to improve the quality of pain management, the APS developed QI guidelines for the treatment of acute and cancer pain in 1995.[26] The QI guidelines emphasized five key elements to improve pain management as a starting point for institutional responsibility (Table 20-3). Each institution was encouraged to develop an interdisciplinary QI team to assess guideline implementation and to monitor outcomes, including assessment and prescribing practices (chart audits) and patient outcomes (patient surveys). The APS QI guidelines have been implemented by both hospitals and outpatient pain clinics, with mixed success.[29] Progress has been made in the areas of pain assessment, staff knowledge, and prescribing practices. Reductions in pain severity, however, are reported less frequently.[30–33]

Gordon and colleagues[17] performed a meta-analysis of 20 pain QI studies performed at eight large U.S. hospitals that used the APS QI guidelines. This review led to six revised quality indicators for hospital-based pain management (Table 20-4). These recommendations focus on acute inpatient pain management, and the authors point out the need to develop and test new measures that can evaluate the quality of multimodal chronic pain treatment programs.

A persistent paradoxical finding in hospital pain QI studies has been that despite high pain intensity ratings, the majority of patients surveyed report extremely high satisfaction ratings. Patient satisfaction has been particularly problematic in pain management. The IOM recommends including consumer satisfaction indices in any evaluation of quality, and Chapman[34] specifically recommends their use when evaluating chronic pain management programs. However, in the context of pain management, interpreting satisfaction is complex. Responses almost always are skewed toward the positive and may be more representative of the quality of the interpersonal relationship between

▶ **TABLE 20-4.** REVISED QUALITY INDICATORS AND SUGGESTED MEASURES FOR HOSPITAL PAIN MANAGEMENT QI INITIATIVES

Quality Indicator	Measure (tool)
Process (assessment and treatment)	
The intensity of pain is documented with a numeric (e.g., 0–10, 0–5) or descriptive (e.g., mild-moderate-severe) rating scale.	Is there any documentation of pain? __Yes __No In the charts where there is some documentation of pain, did the documentation include the use of either a numeric (e.g., 0–10, 0–5) or descriptive (e.g., mild-moderate–severe) pain intensity scale? __Yes __No
Pain intensity is documented at frequent intervals.	How many pain intensity ratings (either numeric or descriptive) were recorded during this (24-hour) period by the RNs? ___
Pain is treated by a route other than intramuscular (IM).	Percent of patients receiving intramuscular injections
Pain is treated with regularly administered analgesics, and when possible a multimodal approach is used (e.g., combinations of regional or local techniques, with nonopioid, opioid, adjuvant analgesics, and nonpharmacologic methods).	Percent of patients receiving nonopioid alone, opioid alone, regional techniques (e.g., neuraxial) and various combinations of nonopioid, opioid, and regional techniques Percent of patients receiving meperidine (patient question) Did you use any non-drug interventions in addition to analgesics to manage your pain? __Yes __No; if yes, please check all that apply: relaxation, meditation, heat, cold, deep breathing, walking, imagery or visualization, other (please describe)
Outcomes	
Pain is prevented and controlled to a degree that facilitates function and quality of life	On this scale (0–10), please indicate the worst pain you had in the first 24 hours. On this scale (0–10), please indicate the least pain you had in the first 24 hours. How often were you in moderate to severe pain in the first 24 hours? __always __almost always __often __almost never __never Circle the number that best described how, during the first 24 hours, pain interfered with your: Activity, mood, sleep *(may add other items for specific populations)* 0 = does not interfere, 10 = completely interferes
Patients are adequately informed and knowledgeable about pain management	Adequacy of information you received about pain and pain control options while in hospital: 1 = poor, 2 = fair, 3 = good, 4 = very good, 5 = excellent

SOURCE: *Reprinted with permission: Gordon, DB, Pellino TA, Miaskowski C, et al. A Ten Year Review of Quality Improvement Monitoring in Pain Management and Recommendations for Standardized Measures. Pain Management Nursing 3(4):116–130, 2002.*

patient and caregiver than the quality of the actual services or their outcomes. When evaluating the quality of pain management, questions about the adequacy of pain-related information provided to patients may provide a better measure than general patient satisfaction questions.

After a decade of experience with QI in all areas of health care, there are still no shining, organization-wide examples of revolutionary change and improvement like those seen in commercial business.[35] However, as a means to improve pain management, QI programs do seem a worthwhile endeavor. Although experience is still limited, and it may yet be too early to see the longer-term results of reduced pain, QI has been shown to affect the visibility of pain as a clinical priority, enhance interdisciplinary collaboration, and facilitate the implementation of clinical guidelines at the bedside.

Evidence-Based Medicine and Clinical Practice Guidelines

Guidelines are systematically developed statements designed to help practitioners and patients make appropriate health care decisions for specific clinical conditions.[24] The early 1990s saw a plethora of consensual and evidence-based clinical practice guidelines focused on the assessment and treatment of various pain conditions (Table 20-5). In a randomized, controlled trial of community-based oncology clinics, implementation of pain management practice guidelines has been shown to enhance patient outcomes.[36]

As a sequel to the APS QI guidelines, which focused on acute and cancer pain, Sanders and colleagues[37] published a set of guidelines to help pain treatment programs for chronic noncancer pain systematically evaluate the quality of their programs. This guideline was adopted by the American Academy of Physical Medicine and Rehabilitation (AAPM&R) and updated in 1999 based on additional new evidence for recommendations.[38] Echoing an earlier call to the field from the International Association for the Study of Pain,[39] the authors recommended a standardized reporting system for pain treatment facilities. Programs are encouraged to evaluate themselves based on eight goal-directed clinical and economic variables, including reduced medication misuse, health care service utilization, pain intensity, and cost of care and increased physical, social, and work productivity. The authors include specific recommendations for outcome instruments and the timing of measurements (Table 20-6). Instruments were selected based on their reliability and validity levels, availability, ease of use, and feasibility of application. Program measures should be applied to as many patients as possible, but meaningful estimates of treatment results can be obtained if at least 10 percent of patients treated over a 12-month period (or 25 patients) are evaluated.

The recommendations of Sanders and colleagues are supported by the medical literature. Routine, standardized

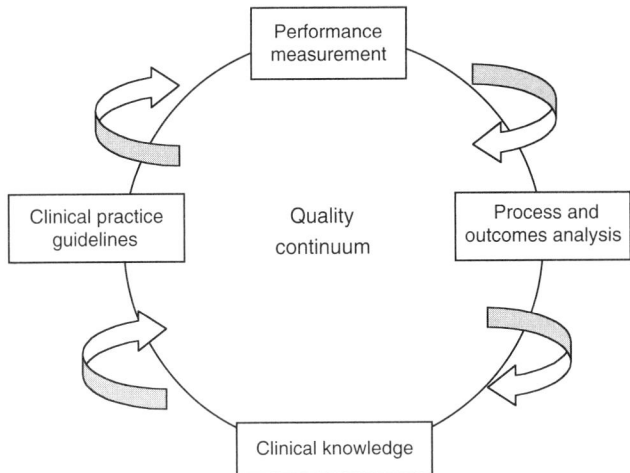

Figure 20-1. The quality continuum: interrelationships between clinical practice guidelines and measurement. (Adapted with permission from Kaegi L: AMA clinical quality improvement forum ties it all together: From guidelines to measurement and back to guidelines. Joint Comm J Qual Improv 1999; 25(2):95–106.)

patient evaluation has been shown to be useful in assessing the quality of acute and chronic pain treatment programs.[40] Additionally, multidisciplinary pain clinics produce superior results in all eight of the criteria of Sanders and colleagues when compared with conventional treatment.[41] Unfortunately, no additional published studies have evaluated the efficacy of pain guidelines or program evaluation standards, and there is as yet no movement to adopt a field-wide set of standardized program measures.

The trend toward clinical practice guidelines as a part of QI continues. In October of 1998, the AMA Department of Clinical Quality Improvement introduced the broader concept of a quality continuum linking clinical practice guidelines to performance measurement and, ultimately, quality improvement[42] (Fig. 20-1). Using new information technologies and other tools, better performance measures are being developed to assess guideline utilization, track long-term outcomes, and assess overall guideline impact. In the spirit of continuous QI, the knowledge gleaned from these efforts will be fed back into clinical practice in the form of revised guidelines, perpetuating the cycle of improvement.

Public Accountability (Performance Indicators)

Health care purchasers use accountability measures, or performance indicators, to compare services. Performance indicators should measure variations and improvements in performance and reflect the concerns of patients, providers, regulators, accreditation bodies, and other stakeholders.[43] These indicators often are rate-based and are reported as fractions or percentages of a total number of eligible events.

▶ **TABLE 20-5.** EVIDENCE-BASED CONSENSUS STATEMENTS AND PRACTICE GUIDELINES FOR PAIN MANAGEMENT

Agency for Healthcare Policy and Research (AHCPR). Acute Pain Management: Operative or Medical Procedures and Trauma: Clinical Practice Guideline No. 1. Rockville, MD, US Public Health Service, Agency for Health Care Policy and Research, 1992; AHCPR publication 92-0032.

Agency for Healthcare Policy and Research (AHCPR). Management of Cancer Pain: Adults: Clinical Practice Guideline No. 9. Rockville, MD, US Public Health Service, Agency for Health Care Policy and Research 1994; AHCPR publication 94-0593.

American Academy of Neurology. McCrory DC, Matchar DB, Gray RN, Rosenberg JH, Silberstein SD. Evidence-based guidelines for migraine headache: Overview of program description and methodology. The US Headache Consortium, American Academy of Neurology 1999; [Online]. Available: *www.aan.com*.

American Academy of Neurology. Matchar DB, Young WB, Rosenberg JH, et al. Evidence-based guidelines for migraine headache: Pharmacological management for acute attacks and prevention of migraine. The US Headache Consortium, American Academy of Neurology 1999; [Online]. Available: *www.aan.com*.

American Academy of Neurology. Campbell JK, Penzien DB, Wall EM. Evidence-based guidelines for migraine headache: behavioral and physical treatments. The US Headache Consortium, American Academy of Neurology 1999; [Online] Available: *www.aan.com*.

American Academy of Pain Medicine, American Pain Society, American Society of Addiction Medicine. Consensus Statement: Definitions Related to the Use of Opioids for the Treatment of Pain. In press

American Academy of Pediatrics and Canadian Paediatric Society. Policy Statement: Prevention and management of pain and stress in the neonate. *Pediatrics* 2000;105:454–461.

American Academy of Physical Medicine & Rehabilitation. Sanders, SH, Rucker KS, Anderson, KO, Harden RN, Jackson KW, Vicente PJ, Gallagher RM. Clinical practice guidelines for chronic non-malignant pain syndrome patients. *Journal of Back and Musculoskeletal Rehabilitation* 1996;5:115–120.

American Academy of Physical Medicine & Rehabilitation. Sanders SH, Rucker L. Clinical practice guidelines for chronic non-malignant pain management. *Journal of Back and Musculoskeletal Rehabilitation* 1996;7:19–25.

American Academy of Physical Medicine & Rehabilitation. Sanders SH, Harden RN, Benson SE, Vicente PJ. Clinical practice guidelines for chronic non-malignant pain syndrome patients II: An evidence-based approach. *Journal of Back and Musculoskeletal Rehabilitation* 1999;13:47–58.

American Geriatric Society Panel on Chronic Pain in Older Persons. The management of chronic pain in older persons. *Journal of the American Geriatrics Society* 1998;46:635–651.

American Medical Directors Association. Chronic Pain Management in the Long-Term Care Setting 1999; *www.amda.com*

American Pain Society Quality of Care Committee. Quality improvement guidelines for the treatment of acute pain and cancer pain. *Journal of the American Medical Association* 1995;23:1874–1880.

American Pain Society and the American Academy of Pain Medicine. Consensus Statement on the Use of Opioids in Chronic Pain 1996; *www.ampainsoc.org*.

American Pain Society. *Principles of Analgesic Use in the Treatment of Acute Pain and Cancer Pain* (4th Edition). 1999;Glenview, IL: Author.

American Pain Society. Guideline for the management of acute and chronic pain in sickle-cell disease. 1999; Glenview, IL: American Pain Society.

American Society of Addiction Medicine. Definitions Related to the Use of Opioids in Pain Treatment, a Public Policy Statement 1997; *www.asam.org*.

American Society of Anesthesiologists Task Force on Pain Management, Cancer Pain Section. Practice guidelines for cancer pain management. *Anesthesiology* 1996;84(5):1243–1257.

American Society of Anesthesiologists Task Force on Pain Management, Acute Pain Section. Practice guidelines for acute pain management in the perioperative setting. *Anesthesiology* 1995;82:1071–1081.

American Society of Anesthesiologists. Practice guidelines for chronic pain management. *Anesthesiology* 1997;87:995–1004.

American Society of Clinical Oncology. Cancer pain assessment and treatment curriculum guidelines. *Journal of Clinical Oncology* 1992;10:1976–1982.

Federation of State Medical Boards of the United States, Inc. Model Guidelines for the Use of Controlled Substances for the Treatment of Pain,1998; Euless, TX: author (817-868-4000).

Mersky H & Bogduk N. *Classification of Chronic Pain: Descriptions of Chronic Pain Syndromes and Definitions of Pain Terms* 1994; Seattle: IASP Press.

Society for Nuclear Medicine. Procedure guideline for bone pain treatment. Reston (VA): Society of Nuclear Medicine 1999; Version 2.0

Stanton-Hicks M, Baron R, Boas R, Gordh T, Harden RN, Hendler N, Koltzenburg M, Raj P, & Wilder R. Complex regional pain syndromes: Guidelines for therapy. *The Clinical Journal of Pain* 1998;14:115–116.

Wolfe F. The fibromyalgia syndrome: A consensus report on fibromyalgia and disability. *The Journal of Rheumatology* 1996;23(3):534–539.

SOURCE: *Reprinted with permission: Gordon DB, Pellino TA. Assessment and Management of Pain. In Maher AB, Salmnond SW, Pellino (eds), Orthopaedic Nursing, 3rd ed. Philadelphia: W.B. Saunders 2002; pp. 129–151.*

TABLE 20-6. EIGHT OUTCOME OBJECTIVES TO EVALUATE PROGRAMS THAT TREAT CHRONIC NON-CANCER PAIN SYNDROME PATIENTS

Objective	Measurement Instrument	Measurement Points in Time
Reduction in the misuse of medications (if present)	Medication Quantification Scale (MQS)	At least at start and end of initial treatment, if possible at 3, 6, & 12 month post-treatment
Increase in physical activity, capacity, and general function	Sickness Impact Profile (SIP) (or Medical Outcome Study (MOS) short form (SF) 36 or Multidimensional Pain Inventory (MPI))	At least at start and end of initial treatment, if possible at 3, 6, & 12 month post-treatment
Increase in productivity at home, socially, and/or at work	Sickness Impact Profile (SIP) (or alternative MOS-SF-36 or MPI) work status Estimation of actual "up-time" by patient self-monitoring diary	At least at start and end of initial treatment, if possible at 3, 6, & 12 month post-treatment
Improvement in general mood and well being	Beck Depression Inventory (BDI) long or short form	At least at start and end of initial treatment, if possible at 3, 6, & 12 month post-treatment
Reduction in patients' subjective pain intensity	0–10 Visual Analog Scale (VAS)	At every visit
Reduction in use of and expenditures for health care	Number and frequency of contacts with MDs for pain, including Emergency Room visits, and number and days of hospitalizations for pain in past 90 days	At least at start and end of initial treatment, if possible at 3, 6, & 12 month post-treatment
Where applicable, achievement of case settlement	Status of case	At least at end of treatment and 3 month post-treatment, 6 & 12 month post-treatment when possible
Minimizing of pain treatment program cost without compromising quality of care	Total cost for given chronic pain patient including service, materials, lodging, meals, and transportation	At least at end of treatment and 3 month post-treatment, 6 & 12 month post-treatment when possible

SOURCE: Sanders SH, Rucker KS, Anderson KO, et al. Guidelines for program evaluation in chronic non-malignant pain management. Journal of Back and Musculoskeletal Rehabilitation 7:19–25, 1996.

▶ **TABLE 20-7.** DESIRABLE ATTRIBUTES OF QUALITY MEASURES: DEVELOPED JOINTLY BY THE AMA, JCAHO, AND NCQA

Attribute	Definition
Importance of Topic Area Addressed by the Measure	
1A. High priority for maximizing the health of persons or populations	The measure addresses a process or outcome that is strategically important in maximizing the health of persons or populations. It addresses an important medical condition as defined by high prevalence, incidence, mortality, morbidity, or disability.
1B. Financially Important	The measure addresses a clinical condition or area of health care that requires high expenditures on inpatient or outpatient care. A condition may be financially important if it has either high per-person cost or if it affects a large number of people.
1C. Demonstrated Variation in Care and/or Potential for Improvement	The measure addresses an aspect of health care for which there is a reasonable expectation of wide variation in care and/or potential for improvement. If the purpose of the measure is internal quality improvement and professional accountability, then wide variation in care across physicians or hospitals is not necessary.
Usefulness in Improving Patient Outcomes	
2A. Based on Established Clinical Recommendations	For process measures, there is good evidence that the process improves health outcomes. For outcome measures, there is good evidence that there are processes or actions that providers can take to improve the outcome.
2B. Potentially Actionably by User	The measure addresses an area of health care that potentially is under the control of the physicians, health care organization or health care system that it assesses.
Measure Design	
3A. Well-Defined Specifications	The following aspects of the measure are to be defined: numerator, denominator, sampling methodology, data sources; allowable values, methods of measurement, and method of reporting.
3B. Documented Reliability	The measure will produce the same results when repeated in the same population and setting (low random error). Tests of reliability included (a) test-retest (reproducability): test-retest reliability is evaluated by repeating administration of the measure in a short time frame and calculating agreement among the repetitions; (b) inter-rater: agreement between raters is measured and reported using the kappa statistic; (c) data accuracy: data are audited for accuracy; and (d) internal consistency for multi-items measures; analyses are performed to ensure that items are internally consistent.
3C. Documented Validity	The measure has face validity—it should appear to a knowledgeable observer what is intended. The measure also should correlate well with other measures or the same aspects of care (construct validity) and capture meaningful aspects of care (content validity).
3D. Allowance for Risk	The degree to which data collected on the measure is risk adjusted or risk stratified on the purpose of the measure. If the purpose of the measure is for internal continuous quality improvement and professional accountability, then requirements for risk adjustment or risk stratification are not stringent.

(Continued)

▶ **TABLE 20-7.** (CONTINUED)

	Measure Design
	If the purpose of the measure is for comparison and accountability, then either the measure should not be appreciably affected by any variables that are beyond the user's control (covariates), or to the extent possible, any extraneous factors should be known and measurable. If case-mix and/or risk adjustment is required, there should be well-described methods for either controlling through risk stratification or for using validated models for calculating an adjusted result that corrects for the effects of covariates. (In some cases, risk stratification may be preferable to risk adjustment because it will identify quality issues of importance to different subgroups.)
3E. Proven Feasibility	The data required for the measure can be obtained by physicians, health care organizations, or health care system with reasonable efforts and within the period allowed for data collection.
	The cost of data collection and reporting is justified by the potential improvements in care and outcomes that result from the act of measurement.
	The measure should not be susceptible to cultural or other barriers that might make data collection infeasible.
3F. Confidentiality	The collection of data for the measures should not violate any accepted standards of confidentiality.
3G. Public Accountability	The measure's specifications are publicly available.

SOURCE: *Reprinted with permission American Medical Association, Joint Commission on Accreditation of Healthcare Organizations, and National Committee for Quality Assurance. Pain Management Performance Measurement Clinical Expert Panel: JCAHO, October 21, 2001, Chicago, Il.*

For example, of all health plan members carrying a diagnosis of arthritis, how many are screened on a regular basis for the presence of pain? Performance indicators are reported back to the health care purchaser/consumer and can be powerful drivers of health care choice.

The creation of public accountability through performance measurement is a relatively new feature of health care in the United States. The theory is that comparing performance among health care systems will create a market demand for the systems with the best performance scores. This creates competition in the marketplace, which ostensibly leads to improvements in health care quality and service. Leaders in the development of broad-based national performance data sets include the National Committee on Quality Assurance (NCQA), the Foundation for Accountability (FACCT), CARF, JCAHO, and the National Quality Forum (NQF). However, only the CARF and JCAHO measurement sets currently include pain management among their variables.

The need for nationally standardized performance measures related to pain management continues to grow. In response to these demands and to address challenges presented by emerging user needs in performance measurement, JCAHO, AMA, and NCQA convened a collaborative meeting in the spring of 2002. A joint clinical expert panel was formed to facilitate evidence-based justification, development, and field testing of new pain management performance measures applicable to cancer, back pain, and arthritis through the continuum of care. Thirty-four candidate measures initially developed by the panel were narrowed in the fall of 2002 to eight possible measures for field testing focused on screening, assessment, pain relief (reassessment), and treatment. Field testing of specific measures, including those for comprehensive assessment of pain in patients with arthritis and low back pain, and appropriateness of imaging studies for acute low back pain, was completed in the fall of 2003. The final measures will be presented to the National Quality Forum (*http://www.qualityforum.org*) for their inclusion into a developing national performance indicator data set.

An important by-product of the JCAHO/AMA/NCQA collaboration has been the development of a list of desirable attributes for quality measures (Table 20-7). Whether for purposes of QI, research, or accountability, measures must be constructed carefully so that they are accurate, reliable, and feasible. Developing quality measures that fulfill these requirements is a complex and difficult task and requires much further work, particularly in the area of chronic noncancer pain management.

▶ CONCLUSIONS

Economic and societal pressures to manage and improve quality continue to grow. In March 2001, the IOM released a report on the current state of health care quality and concluded that (1) quality can be defined and measured precisely, (2) serious quality problems exist and a large number of patients are harmed as a result, and (3) current efforts at improvement will not succeed unless we undertake a major, systematic effort to overhaul health care service delivery, educate and train clinicians, and measure quality.[2]

Examination of the literature clearly demonstrates that the classic models of continuing education and QA are insufficient to truly improve quality. Newer approaches, such as QI and performance indicators, have a better chance of success because of their systems-oriented perspective and emphasis on appropriate measurement tools. Integrating new knowledge and behaviors into day-to-day pain management practice is a complicated but necessary process when attempting to improve quality. Ultimately, our ability to modify the behavior of patients, purchasers, and health care providers depends on evidence-based standards of care and the collection of valid data about our performance; however, the process of developing these standards, measurements, and data sets for pain management is still in its infancy.

As our knowledge of pain management has grown, so too has the complexity of quality-measurement tools. A comprehensive evaluation of the quality of pain management involves measurement of both practice patterns and patient outcomes. Well-designed observational and experimental studies are needed to determine whether new approaches such as QI or patient-oriented empowerment will result in successful change. Much work is needed to develop and test valid and reliable measures of pain management quality and outcomes. Pain management clinicians must work in collaboration with accrediting bodies, quality researchers, and policymakers to better define and measure the quality and benefits of pain management. These tasks have barely begun, but the framework is in place and the direction is clear. With sufficient ambition, effort, and funding, we will gain a better understanding of how best to deliver care, measure results, justify and/or trim costs, and ultimately deliver the outcomes most desired by our patients—reduced pain and improved quality of life.

REFERENCES

1. Chassin MR: Assessing strategies for quality improvement. *Health Affairs* 1997;16:151–161.
2. Chassin MR, Galvin RW, and the National Roundtable on Health Care Quality: The urgent need to improve health care quality. *JAMA* 1998;280:1000–1005.
3. Codman E: *A Study in Hospital Efficiency*. Boston: Thomas Todd, 1920.
4. Katz JN, Sangha O: Assessment of the quality of care. *Arthritis Care Res* 1997;10:359–369.
5. Kinney ED: The brave new world of medical standards of care. *J Law Med Ethics* 2001;29:323–334.
6. Blumenthal D: The origins of the quality-of-care debate, part 4. *N Engl J Med* 1996;335:1146–1148.
7. Malmivaara A, Hakkinen U, Aro T, et al: The treatment of acute low back pain: Bed rest, exercises, or ordinary activity? *N Engl J Med* 1995;332:351–335.
8. Deyo R, Battie M, Beurskens AJ, et al: Outcome measures for low back pain research: A proposal for standardized use. *Spine* 1998;23:2002–2013.
9. Kritchevsky SB, Simmons BP: Continuous quality improvement: Concepts and applications for physician care. *JAMA* 1991;266:1817–1823.
10. Blumenthal D: Quality of health care: 1. What is it? *N Engl J Med* 1996; 335:891–893.
11. Donabedian A: The role of outcomes in quality assessment. *QRB* 1992;18:356–360.
12. Palmer RH: Considerations in defining quality of health care, in Palmer RH, Donabedian A, Povar GJ (eds), *Striving for Quality in Health Care: An Inquiry into Policy and Practice*. Ann Arbor: Health Administration Press, 1991:1–53.
13. Brook RH, McGlynn EA: Quality of health care: 2. Measuring quality of care. *N Engl J Med* 1996; 35:966–969.
14. Donabedian A: Promoting quality through evaluation of the process of patient care. *Med Care* 1968;6:181–202.
15. Tarlov A, Ware J, Greenfield S, et al: The medical outcomes study: An application of methods for monitoring the results of medical care. *JAMA* 1989;262:925–930.
16. Kelly KC, Huber DG, Johnson M, et al: The Medical Outcomes Study: A nursing perspective. *J Prof Nurs* 1994; 10:209–216.
17. Gordon, DB, Pellino TA, Miaskowski C, et al: A ten-year review of quality improvement monitoring in pain management and recommendations for standardized measures. *Pain Manag Nurs* 2002;3:116–130.
18. Mulley AG Jr: Industrial quality management science and outcomes research: responses to unwanted variation in health outcomes and decisions, in Blumenthal D, Scheck AC (eds), *Improving Clinical Practice: Total Quality Management and the Physician*. San Francisco: Jossey-Bass, 1995:73–105.
19. Lohr KN: *Medicare: A Strategy for Quality Assurance*. Washington: National Academy Press, 1990.
20. Grol R: Improving the quality of medical care: Building bridges among professional pride, payer profit, and patient satisfaction. *JAMA* 2001;286:2578–2585.
21. Sennett C: An introduction to the National Committee for Quality Assurance. *Pediatr Ann* 1998;27:210–214.
22. Solomon DH, Schaffer JF, Katz JN, et al: Can history and physical examination be used as markers of quality? An analysis of the initial visit note in musculoskeletal care. *Med Care* 2000;38:383–391.
23. Max MB: Improving outcomes of analgesic treatment: Is education enough? *Ann Intern Med* 1990;11:885–889.
24. Agency for Health Care Policy and Research: *Acute Pain Management: Operative or Medical Procedures and*

Trauma. Clinical Practice Guideline No. 1, AHCPR Publication 92-0032. Rockville, MD: US Public Health Service, Agency for Health Care Policy and Research, 1992.
25. Agency for Health Care Policy and Research: *Management of Cancer Pain.* Clinical Practice Guideline No. 9, AHCPR Publication 94-0592. Rockville, MD: US Public Health Service, Agency for Health Care Policy and Research, 1994.
26. American Pain Society Quality of Care Committee: Quality improvement guidelines for the treatment of acute pain and cancer pain. *JAMA* 1995;274:1874–1880.
27. Blumenthal D: Total quality management and physicians' clinical decisions. *JAMA* 1993;269:2775–2778.
28. Bradley EH, Holmboe ES, Mattera JA, et al: A qualitative study of increasing β-blocker use after myocardial infarction: Why do some hospitals succeed? *JAMA* 2001;285:2604–2611.
29. Ward S, Donovan M, Max MB: A survey of the nature and perceived impact of quality improvement activities in pain management. *J Pain Symptom Manage* 1998;15:365–373.
30. Dietrick-Gallagher M, Polomano R, Carrick L: Pain as a quality management initiative. *J Nurs Care Quality* 1994; 9:30–42.
31. Bookbinder M, Coyle N, Thaler H, et al: Implementing national standards for cancer pain management: program model and evaluation. *J Pain Symptom Manage* 1995;12:334–347.
32. Super A: Improving pain management practice. *Health Prog* 1996;July–August:50–54.
33. Gordon DB, Stewart JA, Dahl JL, et al: Institutionalizing pain management. *J Pharm Care Pain Sympt Control* 1999; 7:3–16.
34. Chapman SL: Patient satisfaction measures in the evaluation of chronic pain management programs. *APS Bull* 1992;2:6–7.
35. Blumenthal K, Kilo CM: A report card on continuous quality improvement. *Milbank Q* 1998;4:625–648.
36. Du Pen SL, Du Pen AR, Polissar N, et al: Implementing guidelines for cancer pain management: Results of a randomized, controlled trial. *J Clin Oncol* 1999;17:361–370.
37. Sanders SH, Rucker KS, Anderson KO, et al: Guidelines for program evaluation in chronic non-malignant pain management. *J Back Musculoskel Rehabil* 1996;7:19–25.
38. Sanders SH, Harden RN, Benson SE, et al: Clinical practice guidelines for chronic nonmalignant pain syndrome patients: II. An evidence-based approach. *J Back Musculoskel Rehabil* 1999;13:47–58.
39. International Association for the Study of Pain: *Desirable Characteristics for Pain Treatment Facilities.* Seattle: IASP Press, 1990.
40. Hadjistavropoulos HD, Clark J: Using outcome evaluations to assess interdisciplinary acute and chronic pain programs. *Joint Comm J Qual Improv* 2001;27:335–348.
41. Turk DC, Okifuji A: Treatment of chronic pain patients: Clinical outcomes, cost-effectiveness and cost-benefits of multidisciplinary pain centers. *Crit Rev Phys Rehabil Med* 1998;10:181–208.
42. Kaegi L: AMA clinical quality improvement forum ties it all together: From guidelines to measurement and back to guidelines. *Joint Comm J Qual Improv* 1999;25:95–106.
43. Joint Commission on Accreditation of Healthcare Organizations and National Pharmaceutical Council, Inc: *Improving the Quality of Pain Management Through Measurement and Action.* Oakbrook Terrace, IL: Joint Commission Resources, 2003.

PART III

Syndromes and Disorders in Pain Medicine

CHAPTER 21

Taxonomy of Pain Syndromes

Howard S. Smith

Taxonomy *refers to the principles and practice of classification. Taxonomy in the field of pain medicine is vitally important for communication, education, research efforts, and advancement of the field, as well as for providing optimal patient care and evaluation of treatment outcomes. It is paramount to providing quality clinical care, as well as performing quality research, that clinicians and investigators must "be on the same page." In research, agreed-on definitions and agreed-on criteria must be accepted before you can be certain that you are considering the same phenomena. Additionally, in applying principles of clinical treatment, clinicians need to be "speaking the same language." Appropriate accepted and used taxonomy is necessary to ensure that medical practitioners are talking about the same conditions and that there is consistency among comparisons. Pain taxonomy is dynamic and continues to evolve along with the field of pain medicine—advances continue to be made. Effective taxonomy needs to be concise, accurate, and practical. Medical classifications are often (1) dynamic and constantly evolving, (2) pragmatic but untidy (compared with the periodic table or divisions of flora and fauna), and (3) practical and clinically useful for practitioners. Effective medical classifications need to be useful clinically—they are a method of organizing what is in the clinic so that one may catalogue the case material of the clinic as one might arrange library books in a library. With the advent of electronic medical records and "billing codes" based on classifications, classifications have become essential to the coding and delivery of health care. Basic pain terminology should be understood before classifications are presented.*

▶ INTERNATIONAL ASSOCIATION FOR THE STUDY OF PAIN (IASP) PAIN TERMINOLOGY

Pain An unpleasant sensory and emotional experience associated with actual or potential tissue damage or described in terms of such damage.[1]

This definition is reasonable and has seemed to stand the test of time. However, it has been criticized. One issue regarding the IASP definition of pain that has been debated is the potentially suboptimal language for addressing noncommunicative patients (incapable of self-report).[2]

Unpleasantness is closely related to stimulus intensity.[3] Stimulus-bound unpleasantness has been referred to as *primary unpleasantness,* and *secondary unpleasantness* is reserved for the unpleasant experience that reflects a higher-level process having a more variable relationship to stimulus intensity (predominantly affected by memories and contextual features[3]). Fields[3] has indicated that other somatic sensations, such as dysesthesias, may have a high level of unpleasantness but may not be recognized as "painful" by many people. Fields proposed the term *algosity* for the quality that pain has—other than unpleasantness—that allows it to be uniquely identified by the patient as "pain."[3] Use of this term has not "caught on" in the pain literature or in popular clinician-used "pain jargon." However, according to the senior editor of the *IASP Classification of Chronic Pain,* dysesthesias (but not paresthesias) should be considered as pain (even though some patients may not consider them or describe them as such).

Allodynia Pain due to a stimulus that does not ordinarily provoke pain.[1]

The IASP Committee for Classification of Chronic Pain specifically avoided reference to the specific physical characteristics of the stimulation (e.g., pressure in kilopascals per square centimeter) in an effort to provide practical terms for clinical use.[1] All too often clinicians only document the presence of allodynia in an area of the body without any additional details. This is suboptimal even for clinical purposes. Documentation of allodynia should consist of at least four pieces of information:

1. *Stimulus type* (e.g., mechanical, thermal—including an actual description of stimuli, such as brush or finger)
2. *Location* (e.g., dorsum of hand)
3. *Size/extent* (e.g., 2 × 2 cm)
4. *Intensity*

Intensity can be documented by using a 0–10 numerical rating scale (NRS), with zero being no allodynia and 10 being the worst allodynia imaginable.

Alternatively, intensity may be documented by busy clinicians if simple terms are employed. The author proposes the use of *mild allodynia* (1 to 3 of 10), *moderate allodynia* (4 to 6 of 10), and *severe allodynia* (7 to 10 of 10) to describe allodynic intensity, which should be followed by stimulus type and stimulus intensity (e.g., minimal mechanical stimuli being allodynia due to air pressure but no direct contact or minimal light contact). This is barely touching, e.g., such as clothes or sheets at night, and it equals mechanical *s*timuli 0 to 1 [MS (0–1)]. In this system, more intense stimuli are referred to as *mild* [MS (1–2)] and *moderate* [MS (2–3)]. Therefore, severe pain provoked by bed sheets touching feet would be *severe allodynia from minimal mechanical stimuli* [SAMS (0–1)].

Hyperalgesia An increased response to a stimulus that normally is painful.[1]

Hypoalgesia Diminished pain in response to a normally painful stimulus.[1]

Hyperesthesia Increased sensitivity to stimulation, excluding the special senses.[1]

Dysesthesia An unpleasant abnormal sensation, whether spontaneous or evoked.[1]

Paresthesia An abnormal sensation, whether spontaneous or evoked.[1]

Analgesia Absence of pain in response to stimulation that normally would be painful.[1]

Hyperpathia A painful syndrome characterized by an abnormally painful reaction to a stimulus, especially a repetitive stimulus, as well as an increased threshold.[1]

Hypoesthesia Diminished sensitivity to stimulation, excluding the special senses.[1]

Neuralgia Pain in the distribution of nerve or nerves.[1]

Neuritis Inflammation of a nerve or nerves.[1]

Anesthesia dolorosa Pain in an area or region that is anesthetic.[1]

Neurogenic pain Pain initiated or caused by a primary lesion, dysfunction, or transitory perturbation in the peripheral or central nervous system.[1]

Neuropathic pain Pain initiated or caused by a primary lesion or dysfunction in the nervous system.[1]

Neuropathic pain presents some unique issues. Controversy exists regarding the definition of neuropathic pain and what it entails. Some clinicians have argued for removal of the words *or dysfunction* from the IASP definition and proposed that the definition for neuropathic pain be "pain initiated or caused by a primary lesion of the nervous system."[6] On the other hand, going back to a pure neuroanatomic description of neuropathic pain overlooks the plasticity of the nervous system and its continuous modulation of the response to pain and how this modulation may change after activation or injury.[7] Without the word *dysfunction* in the definition of neuropathic pain, the entity of trigeminal neuralgia may require two subcategories, one neuropathic with a definable lesion and one not.[8] Occasionally, there are instances where clinicians may have difficulty determining if a pain complaint represents neuropathic pain or not. One tool available as a helpful adjunct is the Leeds Assessment of Neuropathic Symptoms and Signs (LANSS) Pain Scale.[9] Additionally, the Neuropathic Pain Scale (NPS) is validated as a specific measurement tool for neuropathic pain, and NPS items appear to be sensitive to treatments known to have an impact on neuropathic pain.[10]

Neuropathy A disturbance of function or pathologic change in a nerve; in one nerve, mononeuropathy; in several nerves, mononeuropathy multiplex; if diffuse and bilateral, polyneuropathy.[1]

Central pain Pain initiated or caused by a primary lesion or dysfunction in the central nervous system.[1]

Noxious stimulus A stimulus that is damaging to normal tissues.[1]

Pain threshold The least experience of pain that a subject can recognize.[1]

Pain tolerance level The greatest level of pain that a subject is prepared to tolerate.[1]

Nociceptor A nerve fiber preferentially sensitive to a noxious stimulus or to a stimulus that would become noxious if prolonged.[1]

This is another term that has been subjected to criticism. An issue here is use of the word *preferential*—some nerve fibers may function in a "nociceptive" capacity under certain specific conditions or specific times and may not be preferentially sensitive to noxious stimuli (or stimuli that would become noxious if prolonged). Orstavik and colleagues,[4] in a systematic study of single C fibers in patients with chronic painful conditions, indicate an active contribution of mechanoinsensitive fibers to chronic pain.

Mechanoinsensitive C fibers (Cmi), also referred to as "silent" nociceptors, are insensitive to destructive mechanical stimuli (no matter how prolonged), but most Cmi nociceptors are responsive to chemical irritants (e.g., capsaicin), and various Cmi subgroups may be responsive to heat and histamine.[5]

Peripheral neurogenic pain Pain initiated or caused by a primary lesion or dysfunction or transitory perturbation in the peripheral nervous system.[1]

Peripheral neuropathic pain Pain initiated or caused by a primary lesion or dysfunction in the peripheral nervous system.[1]

▶ OTHER PAIN TERMINOLOGY

Sympathetically maintained pain (SMP) Pain that is maintained by sympathetic efferent innervation or by circulating catecholamines.[9,11]

Deafferentation Refers to conditions in which there is diminished or absent sensory function of a region. Deafferentation may lead to CNS nociceptive neurons with increased sensitization due to loss of primary afferent input.[11–13]

Dysafferentation Refers to conditions in which there is abnormal sensory function of a region without the loss of nerve supply. A variety of growth factors help to maintain the proper mix of skin nerve endings in a delicate equilibrium.[15] Neurological insult may disrupt this harmonious mix, resulting in an imbalance associated with pathologic pain in which some nerve endings are lost and others increase.[15] Such an imbalance could be referred to as *dysafferentation* as opposed to merely a "pure" loss of innervation referred to as *deafferentation*.[15]

The American Academy of Pain Medicine, the American Pain Society, and the American Society of Addiction Medicine recognize the following three definitions and recommend their use:

Addiction Addiction is a primary, chronic, neurobiologic disease with genetic, psychosocial, and environmental factors influencing its development and manifestations. It is characterized by behaviors that include one or more of the following: impaired control over drug use, compulsive use, continued use despite harm, and craving.

Physical dependence A state of adaptation that is manifested by a drug class-specific withdrawal syndrome that can be produced by abrupt cessation, rapid dose reduction, decreasing blood level of the drug, and/or administration of an antagonist.

Tolerance A state of adaptation in which exposure to a drug induces changes that result in a diminution of one or more of the drug's effects over time.

▶ PAIN SUBTYPES CRITERIA

There are eight major, commonly used, unidimensional subtypes for pain:

1. Pain subtypes based on pain *location* (e.g., low back pain, headache).
2. Pain subtypes based on pain *duration*. Chronic pain has been considered by some to be "pain that extends beyond the expected period of healing." However, "expected healing periods" are not extremely well studied for all types of insults and may be extremely variable depending on genetics, associated medical conditions (e.g., diabetes mellitus, renal failure), etc. Therefore, definitions of chronic pain based purely on duration have merit in that they are objective, straightforward, and concrete. The IASP definition of chronic pain addresses both duration and appropriateness.[16] Chronic pain (by IASP criteria) is pain without apparent biological value that has persisted beyond the normal tissue healing time (usually taken to be 3 months). Chronic pain (by the American College of Rheumatology [ACR] criteria) is widespread or regional pain for at least 3 months.[17] ACR criteria for chronic widespread pain include all the following: pain present for at least 3 months, pain in the left and right sides of the body, pain above and below the waist, and the presence of axial skeletal pain (cervical spine, anterior chest, thoracic spine, or low back).[17] Various studies have used IASP, ACR, and the

Diagnostic and Statistical Manual of Mental Disorders, 4th edition (DSM-IV) definitions of chronic pain. A 6-month criterion is still used in the DSM-IV and may be useful in certain research efforts (since after 6 months there is little question that this is *chronic* or persistent pain).[18] Pain duration shorter than 6 months but present beyond the "expected healing period"—up to 6 months—has been termed by some clinicians as *subacute pain.* Alder has argued that the term *chronic* with respect to the pain should be dropped.[19] The term *persistent pain* is now often used instead of chronic pain.

3. Pain subtypes based on pain *origin.* Portenoy[20] proposed three major pain categories: nociceptive (somatic and visceral pain), neuropathic, and psychogenic. Current opinion is that psychogenic pain should be used rarely. Psychogenic pain may have a place in conjunction with pain secondary to a conversion disorder, but this is exceedingly hard to "prove" and almost never should be diagnosed. The DSM IV Diagnostic criteria for pain disorder include
   - Pain in one or more anatomic sites is the predominant focus of the clinical presentation and is of sufficient severity to warrant clinical attention.
   - The pain causes clinically significant distress or impairment in social, occupational, or other important areas of functioning.
   - Psychological factors are judged to have an important role in the onset severity, exacerbation, or maintenance of the pain.
   - The symptom or deficit is not produced intentionally or feigned (as in factitious disorder or malingering).
   - The pain is not better accounted for by a mood, anxiety, or psychotic disorder and does not meet criteria for dyspareunia.[18]

4. Pain subtypes based on *pain condition/diagnosis etiology/association.* For example, cancer pain, sickle cell pain, and postherpetic neuralgia pain. [(*Note:* Two classifications may be combined (e.g., chronic noncancer pain).]

5. Pain subtypes based on *body system involved.* For example, myofascial, rheumatic, "causalgic," or "sympathetic-maintained," neurological, or vascular.[21]

6. Pain subtypes based on *pain severity.* For example, mild, moderate, or severe.

7. Pain subtypes based on *mechanism(s) involved.* For example, peripheral sensitization, disinhibition, or central sensitization.[22] Although a diagnostic-based approach to treatment constitutes a reasonable initial approach, some pain symptoms may not respond to traditional treatments. Conventional approaches may be supplemented by a mechanism-based approach in efforts to individually tailor targeted therapy.[22-24] A mechanism-based approach accounts for the observation that patients with one disease (e.g., diabetic neuropathy) may have different symptoms due to different mechanisms.[25] What precise pain mechanism is "at play" (e.g., ectopic discharges, disinhibition, peripheral, or central sensitization) is uncertain in any individual patient, but initial, educated guesses potentially may be useful. Physical examination should include documenting stimulus-independent pain versus evoked pain symptoms (e.g., mechanical such as brush-evoked, repetitive pressure, or pin prick; thermal; chemical; etc.), as well as the extent of zones of primary and secondary hyperalgesia, in addition to a careful neurological examination. Chemical challenges (e.g., norepinephrine, capsaicin) occasionally may be beneficial.[26] The data on the symptom/mechanism/treatment relationship are preliminary, and further investigation is necessary to substantiate existing observations.[27] However, current knowledge provides a framework with which to begin addressing specific mechanism-based targeted therapies.[27]

8. Pain subtypes based on *treatment responsiveness.* For example, opioid-responsive pain, opioid moderately responsive pain, and opioid poorly responsive pain.

▶ IASP CLASSIFICATION OF PAIN

In 1979, Dr. John Bonica noted that "taxonomy was vital to the field of pain, and the development and widespread adoption of universally accepted definitions of terms and a classification of pain syndromes are among the most important objectives and responsibilities of the IASP."[1,28]

Dr. Harold Merskey noted that it would be most convenient and helpful if there were some consensus on the technical meaning and usage of terminology and classification related to painful conditions.[1,29] Dr. Merskey was the senior editor of the first and revised second editions of the IASP classification of chronic pain, published through the efforts of the IASP Task Force on Taxonomy in 1994.[1] The *IASP Classification of Chronic Pain* is compatible with the *International Classification of Diseases* (ICD 9 and ICD 10) but reflects special provisions for more detailed identification of various chronic pain syndromes and major acute pain syndromes. The IASP classification of chronic pain, although a freestanding classification of pain syndromes, is not meant to be a rival to other classifications. It is intended to amplify rather than supplant other schemes/classifications.

The second edition of the IASP classification contained a number of changes from the first edition. The revised edition did not adopt the classification of the International Headache Society (IHS) entirely but overlapped with another IHS publication.[30] The IASP classification for fibromyalgia does follow the 1990 criteria of the American College of Rheumatology.[1,17] The 1994 IASP classification rejected the term *atypical facial pain* because it was felt that this term did not describe a definite syndrome but was used variously by numerous authors to refer to a variety of conditions.[1]

The IASP system for classification of chronic pain syndromes uses five axes that focus on physical manifestations of pain:

Axis I: Anatomic regions
Axis II: Organ systems
Axis III: Temporal characteristics, pattern of occurrence
Axis IV: Patient statement of intensity, time since onset of pain
Axis V: Etiology[1] (Table 21-1)

The second edition of the IASP classification establishes unique five-digit codes for each pain diagnosis. For example, the code for a patient with mild postherpetic neuralgia of T5 or T6 for 6 months' duration would be

300 = *Region:* Thoracic region
00 = *System:* The abnormal functioning is attributed to the nervous system
3 = *Temporal characteristics*
0.2 = *Intensity/duration:* Mild, 1 to 6 months
0.002 = *Etiology:* Infective, parasite
Axis I: 300
Axis II: 00
Axis III: 3
Axis IV: 0.2
Axis V: 0.02
Five-digit code: 303.22

Within the Task Force on Taxonomy, a subcommittee on back pain adopted categories for spinal pain (e.g., back pain and root pain [R] or if both present [C] for combined). (This was originally drawn up by Dr. Nikolai Bogduk.) It was suggested that the term *sciatica* be abandoned. Pain in the distribution of a spinal root is referred to as *radicular pain* or *radiculalgia* (a term not used in the IASP classification). *Radiculopathy* refers to objective loss of sensory and/or motor function as a result of conduction block in axons of a spinal nerve or its roots. Examples of lumbar spine pain from the IASP classification include lumbar spinal or radicular pain attributable to a fracture, infection, neoplasm, metabolic bone disease, arthritis, or congenital vertebral anomaly; pseudoarthrosis of a transitional vertebra, pain referred from abdominal viscera or vessels; lumbar spinal pain of unknown or uncertain origin; lumbar spinal or radicular pain after failed spinal surgery; lumbar discogenic pain; internal disk disruption, lumbar zygapophyseal joint pain; lumbar muscle sprain, spasm, trigger point syndrome; lumbar segmental dysfunction; lumbar ligament sprain, sprain annulus fibrosis; lumbar instability; interspinous pseudoarthrosis; lumbar spondylolysis; and prolapsed intervertebral disks.[1]

Perhaps one of the most difficult classifications that the taxonomy subcommittee of IASP wrestled with was the diagnosis of spinal pain.[1,31] The IASP taxonomy subcommittee apparently felt that it could not simply reliably and validly diagnose a specific etiology for low back pain in a particular patient and thought that the most intellectually and clinically honest and accurate diagnosis was "lumbar spine pain of unknown or uncertain origin."[1,31]

However, this term is long and cumbersome and conveys the sense that the doctor does not know what is going on.[31] Bogduk[31] argues for the development of a taxonomically correct terminology, which reassures patients that they can confidently resume normal activities.[31]

Although the IASP taxonomy still has problems, it is very reasonable and continues to evolve. Further evaluation should be sought regarding the consistency, reliability, validity, and utility of the IASP taxonomy. Preliminary data suggest that axis I coding demonstrated adequate interrater reliability but that axis V (etiology) coding did not.[32] "Overlap problems" present in the first edition of the IASP classification of pain[33] were addressed in the second edition.[1] The second edition also did not attempt to categorize pain that occurs after spinal cord injury.[34]

Taxonomy remains an evolving process. One area in which taxonomy has been criticized as being suboptimal is spinal cord injury (SCI) pain. Several classifications have been used by various authors,[35–37] but there is no universally accepted classification system. Members of the Task Force for Pain Following Spinal Cord Injury of the IASP felt that a lack of consistency in inclusion criteria and terminology has had negative implications for both SCI research and the evaluation of treatments.[38] As a result, a three-tier classification system was proposed as new taxonomy for SCI pain by members of the Task Force on Pain following SCI of the IASP[38] (Table 21-2).

The goal in this taxonomic approach is to be able to use tier three classifications, but if that is not achievable, then one would "fall back" to using a tier two classification or, rarely, a tier one classification.[38] This proposed taxonomy is not meant to be definitive, exclusive, or exhaustive, and there may be classifications that occur both above and below or with nociceptive and neuropathic components.[38]

Hicken and colleagues[39] argued for a classification scheme for pain following SCI that is psychometrically

TABLE 21-1. IASP: SCHEME FOR CODING CHRONIC PAIN DIAGNOSIS

Axis I Regions

Record main site first; record two important regions separately. If there is more than one site of pain, separate coding will be necessary. More than three major sites can be coded, optionally, as shown

Head, face, and mouth	000
Cervical region	100
Upper shoulder and upper limbs	200
Thoracic region	300
Abdominal region	400
Lower back, lumbar spine, sacrum, and coccyx	500
Lower limbs	600
Pelvic region	700
Anal, perineal, and genital region	800
More than three major sites	900

Axis II Systems

Nervous system (central, peripheral, and autonomic) and special senses; physical disturbances or dysfunction	00
Nervous system (psychological and social)*	10
Respiratory and cardiovascular systems	20
Musculoskeletal system and connective tissue	30
Cutaneous and subcutaneous and associated glands (breast, apocrine, etc.)	40
Gastrointestinal system	50
Genito-urinary system	60
Other organs or viscera (e.g., thyroid, lymphatic, hemopoietic)	70
More than one system	80
Unknown	90

Axis III Temporal Characteristics of Pain: Pattern of Occurrence

Not recorded, not applicable, or not known	0
Single episode. Limited duration (e.g., ruptured aneurysm, sprained ankle)	1
Continuous or nearly continuous, nonfluctuating (e.g., low back pain, some cases)	2
Continuous or nearly continuous, fluctuating severity (e.g., ruptured intervertebral disc)	3
Recurring irregularly (e.g., headache, mixed type)	4
Recurring regularly (e.g., premenstrual pain)	5
Paroxysmal (e.g., tic douloureux)	6
Sustained with superimposed paroxysms	7
Other combinations	8
None of the above	9

Axis IV: Patient's Statement of Intensity: Time Since Onset of Pain†

Not recorded, not applicable, or not known		.0
Mild	— 1 month or less	.1
	— 1 month to 6 months	.2
	— more than 6 months	.3
Medium	— 1 month or less	.4
	— 1 month to 6 months	.5
	— more than 6 months	.6
Severe	— 1 month or less	.7
	— 1 month to 6 months	.8
	— more than 6 months	.9

Axis V: Etiology

Genetic or congenital disorders (e.g., congenital dislocation)	.00
Trauma, operation, burns	.01
Infective, parasite	.02

(Continued)

TABLE 21-1. (Continued)

Inflammatory (no known infective agent) immune reactions	.03
Neoplasm	.04
Toxic, metabolic (e.g., alcoholic neuropathy, anoxia, vascular, nutritional, endocrine, radiation)	.05
Degenerative, mechanical	.06
Dysfunctional (including psychophysiological)	.07
Unknown or other	.08
Psychological origin (e.g., conversion hysteria, depressive hallucination)	.09

*To be coded for psychiatric illness without any relevant lesion
†Decide the time at which pain is recognized retrospectively as having started, even though the pain may occur intermittently. Grade for intensity in relation to the level of current pain problem.
NOTE:
1. The system is coded whose abnormal functioning produces the pain, e.g., claudication = vascular. Similarly, the nervous system is to be coded only when a pathological disturbance in it produces pain. Thus pain from a pancreatic carcinoma = gastrointestinal; pain from a metastatic deposit affecting bones = musculoskeletal.
2. No physical cause should be held to be present, not any pathophysiological mechanism.
SOURCE: *Reproduced with permission.*[1]

sound, universally accepted, and easily applied, similar to the American Spinal Injury Association's neurological standards. Hicken and colleagues stated that the psychometric properties of the proposed classification by Siddall and colleagues[38] or other classifications should be evaluated carefully.[39] Additionally, incorporation of psychosocial aspects may be beneficial.

The "revised" (second edition) IASP classification of pain did address the change in terminology from *reflex sympathetic dystrophy* to *complex regional pain syndrome*.[1] The older terms of *reflex sympathetic dystrophy* (RSD) and *causalgia* have been changed to *complex regional pain syndrome type I* and *complex regional pain syndrome type II*, respectively.[1] The Orlando consensus workshop was organized in 1993 to address the terminology of RSD in part because of inconsistencies with diagnostic criteria and testing, variability in clinical presentation, and overall suboptimal long-term treatment outcomes. The term *complex regional pain syndrome* (CRPS) is based on the multiaxial classification of chronic pain syndromes developed by IASP. The recommendations of the consensus workshop were accepted by the Committee for Classification of Chronic Pain of the IASP and published in the second edition of the *IASP Classification of Chronic Pain*.[1] The IASP diagnostic criteria for CRPS types I and II are presented in Tables 21-3 and 21-4.[1] The new taxonomy allows for patients who do not fulfill the criteria for either type I or

TABLE 21-2. PROPOSED TAXONOMY FOR SCI PAIN

Broad Type (Tier One)	Broad System (Tier Two)	Specific Structures and Pathology (Tier Three)
Nociceptive	Musculoskeletal	Bone, joint, muscle trauma or inflammation
		Mechanical instability
		Muscle spasm
		Secondary overuse syndromes
	Visceral	Renal calculus, bowel dysfunction, sphincter dysfunction, etc.
		Dysreflexic headache
Neuropathic	Above-level	Compressive mononeuropathies
		Complex regional pain syndromes
	At-level	Nerve root compression (including cauda equina)
		Syringomyelia
		Spinal cord trauma/ischemia (transitional zone, etc.)
		Dual-level cord and root trauma (double lesion syndrome)
	Below-level	Spinal cord trauma/ischemia (central dysesthesia syndrome, etc.)

SOURCE: *Reproduced with permission.*[33]

▶ **TABLE 21-3.** IASP DIAGNOSTIC CRITERIA FOR CRPS-I

1. The presence of an initiating noxious events, or a cause of immobilization.
2. Continuing pain, allodynia, or hyperalgesia with which the pain is disproportionate to any inciting event.
3. Evidence at some time of edema, changes in skin, blood flow, or abnormal sudomotor activity in the region of the pain.
4. The diagnosis is excluded by the existence of conditions that would otherwise account for the degree of pain and dysfunction.

NOTE: Criteria 2–4 must be satisfied.
SOURCE: Reproduced with permission.[1]

▶ **TABLE 21-4.** IASP DIAGNOSTIC CRITERIA FOR CRPS-II

1. The presence of continuing pain, allodynia, or hyperalgesia after a nerve injury, not necessarily limited to the distribution of the injured nerve.
2. Evidence at some time of edema, changes in skin blood flow, or abnormal sudomotor activity in the region of the pain.
3. This diagnosis is excluded by the existence of conditions that would otherwise account for the degree of pain and dysfunction.

NOTE: All three criteria must be satisfied.
SOURCE: Reproduced with permission.[1]

type II. These patients would be designated *CRPS III* or, in the International Classification of Diseases (ICD) terminology, *CRPS—not otherwise specified* (CRPS-NOS).[40] These criteria were meant to be modified over time as knowledge grows.

Internal and external validation research suggests that these IASP criteria are reasonably sensitive (e.g., "picking up most cases of CRPS") but not particularly specific, leading to problems with overdiagnosis.[41–43] The diminished specificity may result in part from (1) the combination of vasomotor changes and sudomotor changes/edema being grouped into the same criterion and (2) the exclusion of motor disturbances and trophic changes.[44] Reinders and colleagues[45] concluded that the 1994 IASP criteria for CRPS type I are not widely employed in clinical studies. Bruehl and colleagues have proposed modified research diagnostic criteria for CRPS type I that they report may decrease the rate of overdiagnosis significantly (i.e., improve the specificity) with only a modest drop in diagnostic sensitivity[42,44] (Table 21-5). The pros and cons of significantly increased specificity along with a modestly decreased sensitivity should be evaluated carefully in any future modifications of IASP criteria, but it would seem reasonable to use these proposed criteria for research purposes.

The term *chronic pain syndrome* (CPS) and other terms such as *chronic generalized pain syndrome* (CGPS) for conditions like fibromyalgia are not accepted by IASP.

More recently, some investigators have advanced "ultraspecialized classifications"[46] (Table 21-6). An initial attempt at distinguishing various types of chronic neuropathic pain after breast cancer surgery was made to stimulate research leading to a taxonomy of chronic pain following breast cancer surgery with reliable and valid diagnostic criteria and assessment methods.[46] Additionally, goals of such future research would include developing a taxonomy of chronic pain after breast cancer surgery that is valid with respect to differences in pain mechanisms and treatment response, leading to improved research efforts, education, communication, and patient outcomes.[46]

Turk and Rudy[47,48] identified three distinct subgroups of chronic pain patients based on cluster analyses

▶ **TABLE 21-5.** PROPOSED MODIFIED RESEARCH DIAGNOSTIC CRITERIA FOR CRPS-1

1. Continuing pain that is disproportionate to any inciting event
2. Must report at least one symptom in each of the four following categories:
    Sensory: Reports of hyperesthesia
    Vasomotor: Reports of temperature asymmetry and/or skin color changes and/or skin color asymmetry
    Sudomotor/Edema: Reports of edema and/or sweating changes and/or sweating asymmetry
    Motor/Trophic: Reports of decreased range of motion and/or motor dysfunction (weakness, tremor, dystonia) and/or trophic changes (hair, nails, skin)
3. Must display at least one sign in two or more of the following categories:
    Sensory: Evidence of hyperalgesia (to pinprick) and/or allodynia (to light touch)
    Vasomotor: Evidence of temperature asymmetry and/or skin color changes and/or skin color asymmetry
    Sudomotor/Edema: Evidence of edema and/or sweating changes and/or sweating asymmetry
    Motor/Trophic: Evidence of decreased range of motion and/or motor dysfunction (weakness, tremor, dystonia) and/or trophic changes (hair, nails, skin)

SOURCE: Reproduced with permission.[39]

▶ TABLE 21-6. CLASSIFICATION OF CHRONIC NEUROPATHIC PAIN SYNDROMES FOLLOWING BREAST CANCER SURGERY

Syndrome	Description
Phantom breast pain*	Sensory experience of removed breast that is still present and is painful
Intercostobrachial neuralgia (includes postmastectomy pain syndrome)	Pain, typically accompanied by sensory changes in the distribution of the intercostobrachial nerve following breast cancer surgery with or without axillary dissection
Neuroma pain (includes scar pain)	Pain in the region of a scar on the breast, chest, or arm that is provoked or exacerbated by percussion
Other nerve injury pain	Pain outside the distribution of the intercostobrachial nerve consistent with damage to other nerves during breast cancer surgery (e.g., medial and lateral pectoral, long thoracic, thoracodorsal, and other intercostal nerves)

*To be distinguished from nonpainful phantom breast sensations.
SOURCE: Reproduced with permission.[46]

of patient responses to the Multidimensional Pain Inventory (MPI).[49] The three subgroups include

1. *Dysfunctional (DYS)*. Very severe pain interfering with much of their lives and significantly restricting their activity.
2. *Interpersonally distressed (ID)*. Perception that significant others were not very supportive of their pain problems.
3. *Adaptive copers (AC)*. Mild to moderately severe pain with a lot of social support and without interfering significantly with their lives or restricting their activity.

This MPI-based classification has seemed to have potential for clinical utility.[50,51] Employing both the IASP and MPI-based classifications for individual patients may provide complementary profiles and a more comprehensive picture. However, this type of "combo classification" (med/psych) is not used widely. Commonly, people in distress require "extra" adaptation/coping skills. It is not uncommon that patients are given two diagnoses—one regarding their physical status and one concerning their mental status. Psychologic/psychiatric patient issues may include (1) preexisting conditions, (2) coincident conditions, (3) the impact of chronic pain on the patient's life, and (4) selection factors. *Selection bias* refers to the fact that most pain clinics work with highly "selected" patients. Patients with nonfatal chronic conditions generally are evaluated by successive layers of medical care, being seen by specialists/subspecialists of increasing standing. Patients who seek out multiple pain management evaluations at highly ranked academic medical schools tend to be more anxious, depressed, or hypochondriacal than a family practice patient population with the same conditions (more "advanced" clinics have a higher prevalence of selection bias).

▶ SUMMARY

Pain taxonomy remains a crucial link in the communication of addressing painful conditions. More universally accepted language is needed. Classifications of pain are hampered by a suboptimal knowledge base and understanding of various human painful conditions. Although some classifications may be useful for research purposes, if a classification scheme is too cumbersome or just not practical, it will not be universally adopted by clinicians. The utility of classification systems depends largely on their application. As more knowledge about various chronic pain syndromes is acquired, classifications would be expected to improve. Those who feel that classifications should reflect some sort of ultimate truth and universal consistency will never be fully satisfied with any medical classification.[1] Complete consistency is beyond the hopes of any medical system of classification.[1] The categories of an ideal classification system should be mutually exclusive and completely exhaustive with regard to data.[1] No classification in medicine has achieved such aims, nor can it be expected to do so.[1,52] The IASP classification of pain provides a reasonable framework for a common ground to discuss clinical pain syndromes "on the same page" that should be refined as knowledge expands. Persons/groups dissatisfied with current IASP classifications are encouraged to work "within the system" through/with the IASP Task Force on Taxonomy and convey their thoughts with rationales and evidence.

▶ ACKNOWLEDGMENT

I would like to acknowledge Dr. Harold Merskey for his help/comments in reviewing this chapter and Doris Eve Jensen for help in the preparation of this chapter.

REFERENCES

1. Merskey H, Bogduk N: *Classification of Chronic Pain*, 2d ed. Seattle: IASP Press, 1994.
2. Anand KJS, Craig KD: New perspective on the definition of pain (editorial). *Pain* 1996;67:3–6.
3. Fields HL: Pain: An unpleasant topic. *Pain* 1999; (suppl 6): S61–9.
4. Orstavik K, Weidner C, Schmidt R, et al: Pathological C-fibres in patients with a chronic painful condition. *Brain* 2003;126:567–578.
5. Koltzenburg M, Handwerker HO, Koerber HR: The differential effect of nociceptor subtypes for generating chronic pain, in Dostrovsky JO, Carr DB, Koltzenburg M (eds), *Proceedings of the 10th World Congress on Pain, Progress in Pain Research and Management*, Vol. 24. Seattle: IASP Press, 2003.
6. Max MB: Clarifying the definition of neuropathic pain. *Pain* 2002;96:406–407.
7. Jensen TS, Sindrup SH, Bach FW: Test the classification of pain: Reply to Mitchell Max. *Pain* 2002;96:407–408.
8. Merskey H: Clarifying definition of neuropathic pain. *Pain* 2002;96:408–409.
9. Bennett M: The LANSS Pain Scale: The Leeds assessment of neuropathic symptoms and signs. *Pain* 2001;92:147–157.
10. Galer BS, Jensen MP: Development and preliminary validation of a measure specific to neuropathic pain: The Neuropathic Pain Scale. *Neurology* 1997;48:332–338.
11. Stanton-Hicks M, Janig W, Hassenbusch S, et al: Reflex sympathetic dystrophy: changing concepts and taxonomy. *Pain* 1995;63:127–133.
12. Fields HL, Rowbotham M, Baron R: Post-herpetic neuralgia: Irritable nociceptors and deafferentation. *Neurobiol Dis* 1998;58:209–227.
13. Rowbotham MC, Yosipovitch G, Connolly MK, et al: Cutaneous innervation density in the allodynia form of post-herpetic neuralgia. *Neurobiol Dis* 1996;3:205–214.
14. Oaklander AL: The density of remaining nerve endings in human skin with and without post-herpetic neuralgia after shingles. *Pain* 2001;92:139–145.
15. Rice F, Pare M, Smith HS: A peripheral basis for neuropathic pain, in Smith HS (ed), *Drugs for Pain*. Philadelphia: Hanley and Belfus, 2003.
16. International Association for the Study of Pain: *Pain* 1986; 24:S1–S225.
17. Wolfe F, Smythe HA, Yunus MB, et al: The American College of Rheumatology 1990 criteria for the classification of fibromyalgia: Report of the Multicenter Criteria Committee. *Arthritis Rheum* 1990;33:160–172.
18. America Psychiatric Association (APA): *Diagnostic and Statistical Manual of Mental Disorders*, 4th ed. (DSM-IV). Washington: APA, 1994.
19. Alder RH: The term "chronic" with respect to pain should be dropped. *Clin J Pain* 2000;16:365.
20. Portenoy R: Mechanisms of clinical pain observations and speculations. *Neurol Clin North Am* 1989;7:205–230.
21. Friction J: Medical evaluation of patients with chronic pain, in Barber J, Adrian C (eds), *Psychological Approaches to the Management of Pain*. New York: Brunner Mazel, 1982:37–61.
22. Woolf CJ, Bennett GJ, Doherty M, et al: Towards a mechanism-based classification of pain? *Pain* 1998;77:227–229.
23. Woolf CJ, Decosterd I: Implications of recent advances in the understanding of pain pathophysiology for the assessment of pain in patients. *Pain* 1999;(suppl 6):S141–147.
24. Woolf CJ, Max MB: Mechanism-based pain diagnosis: Issues for analgesic drug development. *Anesthesiology* 2001;95:241–249.
25. CJ, Mannion RJ: Neuropathic pain: Aetiology, symptoms, mechanisms, and management. *Lancet* 1999;353:1959–1964.
26. Petersen KL, Fields HL, Brennum J, et al: Capsaicin evoked pain and allodynia in post-herpetic neuralgia. *Pain* 2000; 88:125–133.
27. Smith HS, Sang CN: The evolving nature of neuropathic pain: Individualizing treatment. *Eur J Pain* 2002;6(suppl B):13–18.
28. Bonica JJ: The need of a taxonomy (editorial). *Pain* 1979; 6:247–252.
29. Merskey H: Classification of chronic pain: descriptions of chronic pain syndromes and definitions. *Pain* 1986; (suppl 3):345–356.
30. Headache Classification Committee of the IHS (International Headache Society): Classification and diagnostic criteria for headache disorders, cranial neuralgias, and facial pain. *Cephalagia* 1988;8(suppl 7):1–96.
31. Bogduk N: What's in a name? The labeling of back pain. *Med J Aust* 2000;173:400–401.
32. Turk DC, Rudy TE: IASP taxonomy of chronic pain syndromes: Preliminary assessment of reliability. *Pain* 1987; 30:177–189.
33. Vervest A, Schimmer G: Taxonomy of pain of the IASP (letter). *Pain* 1988;34:318–321.
34. Sidall PJ, Taylor SA, Cousins MJ: Classification of pain following spinal cord injury. *Spinal Cord* 1997;35:69–75.
35. Frisbie JH, Aquilera EJ: Chronic pain after spinal cord injury: An expedient diagnostic approach. *Paraplegia* 1990;28:460–465.
36. Segatore M: Deafferentation pain after spinal cord injury: I. Theoretical aspects. *SCI Nurs* 1992;9:46–50.
37. Beric A: Post-spinal cord injury pain states. *Pain* 1997; 72:295–298.
38. Sidall PJ, Yezierski RP, Loeser JD: Taxonomy and epidemiology of spinal cord injury pain, in Yeziersi RP, Burchiel KJ (eds), *Spinal Cord Injury Pain: Assessment, Mechanisms, Management*, Vol 23: *Progress in Pain Research and Management*. Seattle: IASP Press, 2002.
39. Hicken BL, Putzke JD, Richards JS: Classification of spinal cord injury pain: literature review and future directions, in Yeziersi RP, Burchiel KJ (eds), *Spinal Cord Injury Pain: Assessment, Mechanisms, Management*, Vol 23: *Progress in Pain Research and Management*, Seattle: IASP Press, 2002.
40. Stanton-Hicks M: CRPS: Impact of the change in taxonomy, in Harder RN, Baron R, Janig WE (eds), *Pain Syndromes Complex Regional*, Vol. 22: *Progress in Pain Research and Management*. Seattle: IASP Press, 2002.
41. Galer BS, Bruehl S, Harden RN: IASP diagnostic criteria for complex regional pain syndrome: A preliminary empirical validation study. *Clin J Pain* 1998;14:48–54.
42. Bruehl S, Harden RN, Galer BS, et al: External validation of IASP diagnostic criteria for complex regional pain syndrome

and proposed research diagnostic criteria. *Pain* 1999; 81:147–154.
43. Harden RN, Bruehl S, Galer BS, et al: Complex regional pain syndrome: Are the IASP diagnostic criteria valid and sufficiently comprehensive? *Pain* 1999;83:211–219.
44. Bruehl S, Harden RN: An empirical approach to modifying IASP diagnostic criteria for CRPS, in Harden RN, Baron R, Janig WE (eds), *Complex Regional Pain Syndrome* Vol. 22: *Progress in Pain Research and Management*, Seattle: IASP Press, 2001.
45. Reinders MF, Geertzen JH, Dijkstra PU: Complex region pain syndrome type I: Use of the International Association for the Study of Pain diagnostic criteria defined in 1994. *Clin J Pain* 2002;18:207–215.
46. Jung BF, Ahrendt GM, Oaklander AL, et al: Neuropathic pain following breast cancer surgery: Proposed classification and research update. *Pain* 2003;104:1–13.
47. Turk DC, Rudy TE: Classification logic and strategies in chronic pain, in Turk DC, Melzack R (eds), *Handbook of Pain Assessment*. New York: Guilford Press, 1992.
48. Turk DC, Rudy TE: Toward an empirically derived taxonomy of chronic pain patients: Integration of psychological assessment data. *J Consult Clin Psychol* 1988; 56:233–238.
49. Kerns RD, Turk DC, Rudy TE: The West Haven–Yale Multidimensional Pain Inventory (WHYMPI). *Pain* 1985; 23:345.
50. Turk D, Okifuji A, Starz T, et al: Differential responses by psychosocial subgroups of fibromyalgia syndrome patients to an interdisciplinary treatment. *Arthritis Care Res* 1998; 11:397–404.
51. Turk D, Rudy T, Kubinski J, et al: Dysfunctional patients with temporomandibular disorders: Evaluating the efficiency of a tailored treatment protocol. *J Consult Clin Psychol* 1996;64:139–146.
52. Merskey H: Development of a universal language of pain syndromes, in Bonica JJ, Lindblom U, Iggo A (eds), *Proceedings of the Third World Congress on Pain,* Vol 5: *Advances in Pain Research and Therapy.* New York: Raven Press, 1983:37–52.

CHAPTER 22

Central Neuropathic Pain

Nadine Attal and Didier Bouhassira

For over 100 years,[1] it has been acknowledged that lesions in the brain or spinal cord may induce development of pain.[2,3] Although thalamic pain initially was the best known cause of central pain, particularly following the initial description by Dejerine and Roussy of their famous "syndrome thalamique" in 1906,[4] it was recognized early on that central pains also may occur after lesions of the cerebral cortex,[5] the brain stem,[6] and the spinal cord.[7,8]

Until recent times, general awareness of this condition was very limited because central pain was considered a very rare and even exceptional entity. However, recent epidemiologic studies have indicated that prevalence of central pain is much higher than thought previously. Such generally chronic pains represent a considerable burden for patients and greatly affect quality of life and activities of daily living.

At present, central pain is being studied increasingly by basic scientists and clinicians, with the development of animal models of spinal cord injury[9] and the publication of several epidemiologic, clinical, and pharmacologic studies. However, despite this growing interest, the mechanisms of central pain remain hypothetical, and there is still a lack of consensus regarding diagnostic criteria and therapeutic strategy for such conditions.

This chapter presents the definition, main etiologies, epidemiology, and symptomatology of central pains, along with the current theories of central pains and the pharmacologic and surgical treatment. Also some evidence showing that central pains and pains related to peripheral nervous lesions, both included in the category of neuropathic pains, may have more similarities than specificities in terms of clinical presentation, pathophysiology, and response to therapy, will be presented. These data tend to challenge the relevance of the classical distinction between these pain conditions.

▶ DEFINITION AND CLASSIFICATION

Central pain is a generic term referring to "pain initiated or caused by a primary lesion or dysfunction of the central nervous system," according to the definition proposed by the International Association for the Study of Pain (IASP).[10] Thus it includes pains associated with cerebral and spinal lesions. The term *central (neuropathic) pain* has replaced "deafferentation pain,"[2,11] which suggested a specific pathophysiologic mechanism. Central pains arising after a stroke (the most common cerebral cause of pains) are now called *central poststroke pains* instead of "thalamic pains"[12] because it is now recognized that painful strokes may involve many other brain regions apart from the thalamus. Central pains associated with spinal cord injury have been classified recently into two subgroups: *neuropathic pains below level* (referring to pains located diffusely at the lower limbs) and *neuropathic pains at level* (located in a segmental pattern at the level of injury).[13,14]

The concept of central pain is included as part of the broader category of *neuropathic pain*, which refers to "pain initiated or caused by a primary lesion or dysfunction in the nervous system."[10] The latter also

▶ TABLE 22-1. PREVALENCE OF CENTRAL PAINS IN PROSPECTIVE STUDIES OF CONSECUTIVE PATIENTS WITH STROKE AND SPINAL CORD INJURY

Etiology	Authors	N	Pain Site	Prevalence Rates (%)			
				1 mth	6 mths	1 year	5 years
Stroke	Andersen, 1995[44]	267	Upper/lower	4.8	6.5	8.4	
	Weimar, 2002[46]	119	Upper/lower			9.2	
	Gamble, 2002[45]	123/205	Shoulder		7		
Spinal cord trauma	Siddall, 1999, 2003[15-16]	100	At level	38	38		41
			Below level	14	19		34

ABBREVIATION: mth = month.

includes peripheral neuropathic pain, in which the primary lesion is initiated in the peripheral nervous system. It is well established that peripheral nerve lesions can induce secondary central changes, but the term *central pain* does not apply to the central consequences of peripheral nerve lesions.

▶ ETIOLOGIES AND EPIDEMIOLOGY

Which Lesions and Diseases Cause Central Pains?

Stroke (ischemic or hemorrhagic),[12] traumatic spinal cord injury,[15,16] and multiple sclerosis[17-19] represent by far the most common etiologies of central pains (over 80 percent of cases). Other cerebral causes include brain tumors, head injuries, epilepsy (1.2 to 3 percent of epilepsy patients),[20-25] and vascular malformations.[26] Other spinal causes include syringomyelia,[27,28] spinal cord infarction,[29,30] tumors, vascular malformations, and anterolateral cordotomy.[31] Parkinson's disease and multiple system atrophy are associated with several types of pains, some of which are considered to result from the degenerative processes and, therefore, may be considered central pains.[32-34]

Either rapidly developing or slowly progressing lesions may induce central pains.[35,36] Any lesion along the neuraxis located in the dorsal horn, the ascending pathways throughout the spinal cord and brain stem, the thalamus, the subcortical white matter, and the cerebral cortex can cause central pain. After a stroke, painful lesions involve the thalamus in only 50 to 60 percent of cases, and similar painful syndromes have been observed after lesions of the thalamocortical tract, the internal capsule,[29,36,37] the brain stem,[36,38] or the cortical parietal areas (primary and secondary somatosensory cortex).[39,40] It is not clearly established whether particular regions of the thalamus are more commonly associated with central pains because studies mainly have included painful patients (see, however, Bogousslavsky et al.[41]). According to several authors, the thalamic lesions preferentially may involve the ventroposterior nuclei[36,41,42] and may have a right-sided predominance[43] (see, however, Bowsher et al.[36]). There is no correlation between the intensity of central pain and the size of the primary brain lesion or the level or completeness of the injury after spinal cord trauma.[14]

Prevalence of Central Pains

The exact prevalence of central pains in the general population is unknown because most data are retrospective, and neuropathic pain has not always been distinguished from other types of pains. Few prospective studies have estimated the prevalence of central pains in specific etiologies (Table 22-1). One Scandinavian study of 191 patients with an ischemic or hemorrhagic stroke found a prevalence of central pain of 8.4 percent at one year, of which 5 percent was mild or moderate and 3 percent was severe,[44] and similar prevalence rates (7 to 9 percent) were reported recently at 6 to 12 months in two prospective studies of stroke patients.[45,46] The prevalence of central pain appears to be much higher after spinal cord injury. Thus, in the only prospective study performed so far in this condition, which included 100 patients with traumatic spinal cord injury, the prevalence of neuropathic pain at level (including pain classified as segmental and radicular) was 41 percent at 5 years, and that of pain below level (located diffusely below the level of the injury) was 19 percent at 6 months and increased to 34 percent at 5 years.[15,16] These rates are similar to or lower than those found in retrospective surveys of spinal cord injury patients (i.e., 30 to 60 percent).[2,14,47-52]

There are no prospective data regarding the prevalence of central pains in other etiologies. It has been estimated as 28 percent (ranging from 17 to 52 percent) in multiple sclerosis[3] and 30 to 57 percent in Parkinson's disease and multiple system atrophy,[32,34,53,54] although the latter estimation generally includes non-neuropathic pains, such as those due to painful dystonic spasms or pains related to fluctuations of motor symptoms.[32]

CLINICAL SYMPTOMATOLOGY

Onset of Central Pain

It has been reported generally that central pain occurs after a certain delay, up to several months or years after the initial injury, as sensory deficit and weakness improve.[5,15,35,55–57] However, in contrast with this classical notion, recent studies have suggested that in many cases pain appears after a short delay and may even start immediately after the lesion.[3,15,38,42–44] Thus, Andersen and colleagues reported that pain was present within 1 month after a stroke in 63 percent of patients.[44] Times of onset have been more difficult to determine after spinal cord injury because most investigators made no distinction between different types of pains. In their prospective study of patients with spinal cord trauma, Siddall and colleagues found that neuropathic pains at level occurred in 38 percent of patients within 2 weeks versus 14 percent for below-level pains, which were first experienced after 2 years in more than half the patients.[15,16] In multiple sclerosis, the prevalence of central pain appears to increase after 5 years rather than occurring early in the disease.[19]

Symptoms and Signs

Central pain syndromes include a variety of symptoms and signs: spontaneous ongoing/paroxysmal pains, evoked pains, paresthesias and dysesthesias, and sensory abnormalities in the painful area.[58–61]

Spontaneous Pains

Spontaneous pain refers to pain in the absence of any stimulus and may be ongoing (superficial or deep) or paroxysmal. Superficial burning pains are the most common symptoms, but pains also have been described commonly as nonburning (i.e., squeezing, aching, cold, etc.).[15,16,35,36,44,57,62–65] Paroxysmal pains may be described as shooting, electric shock-like, or stabbing. In many cases different pain types may coexist in one patient in the same region or in different parts of the body. Pain tends to be worsened by psychological factors, such as emotion, anxiety, and stress.[56]

Evoked Pains

Evoked pains, i.e., *allodynia* ("pain due to a stimulus that does not normally provoke pain") and *hyperalgesia* ("an increased response to a stimulus that is normally painful"),[10] are observed commonly after central lesions. These pains may be assessed using standard neurological examination, quantitative sensory testing,[66–68] or more simply, specific self-questionnaires, such as the recently validated Neuropathic Pain Symptom Inventory.[69] They are frequently evoked by brush (dynamic mechanical allodynia), punctate stimuli using pinprick or Von Frey filaments (punctate mechanical allodynia/hyperalgesia), or cold stimuli (cold allodynia/hyperalgesia) and less commonly by static large or heat stimuli.[35–37,44,47,55,56,64,65,70] A specific form of dynamic allodynia also has been reported and is referred to as *movement allodynia* (induced by muscle contraction).[56] In most cases, allodynia is associated with spontaneous pain, but dissociated allodynia in response to mechanical[71] or cold stimuli[72] also has been reported.

Evoked pains also may persist after the stimulation (aftersensation), appear with a prolonged latency after the stimulus (delayed sensation), spread outside the site of stimulation (radiation), and be increased or provoked by repetitive stimuli (temporal summation).[55] Many patients present with hyperpathia, i.e., "a painful syndrome characterized by an abnormally painful reaction to a stimulus, especially a repetitive stimulus, as well as an increased threshold."[10] Thus hyperpathia might be considered as a combination of allodynia, hyperalgesia, and temporal summation. Owing to its complex clinical characteristics, this term is often used inappropriately and tends to be employed less commonly.

Paresthesias and Dysesthesias

Abnormal sensations, whether spontaneous or evoked, are also common in central pains. They are referred to as *paresthesia* or *dysesthesia* (the latter referring to "abnormal unpleasant sensation") and often described as "tingling," "pins and needles," and sometimes "numbness."

Dysautonomia

Patients may present with signs of autonomic impairment on the painful side. The painful area may be cooler and vasoconstricted or warmer, and occasional changes in sweating have been reported.[56]

Sensory Deficit

Nearly all the patients with central pains present with a sensory deficit in their painful area, whereas other deficits, such as motor impairment and ataxia, differ according to the etiologic disease and may be absent. Sensory deficit may range from mild hypesthesia or hypalgesia (diminished pain in response to a normally painful stimulus) to complete anesthesia or analgesia. Studies using quantitative sensory testing in patients with central pains due to stroke, spinal cord injury, or multiple sclerosis all have indicated that the deficit largely predominates on temperature and pinprick, with proportions ranging from about 80 percent in multiple sclerosis[67] to nearly 100 percent of patients with poststroke pains[36–38,44,55,56,73,74] and spinal cord injury.[30,62–65,70,75–77] Thus, there are almost always increased detection thresholds to warm and cool stimuli and to a lesser extent, an increase in pain thresholds. In contrast, proprioceptive/tactile sensation is spared more commonly, particularly in Wallenberg syndrome, syringomyelia, and anterior spinal artery syndrome.[30,38,75]

Clinical Diagnosis of Central Pains

Central pain is a neuropathic pain and, as such, presents with several characteristics that may help differentiate this type of pain from possible coexisting nociceptive (in particular musculoskeletal) and visceral pains. Symptoms such as burning pain, electric shocks, tingling, itchiness, and numbness are those most discriminant between neuropathic and non-neuropathic pains.[78–83] The presence of a sensory deficit and of allodynia to brushing and cold stimuli may help to further differentiate both conditions.[80,82] An adequate algorithm for the positive diagnosis of neuropathic pains is still lacking. Although two pain questionnaires have been validated recently for this purpose, patients with central pains were not included,[82] and the etiologies of neuropathic pains were not mentioned.[83] A diagnostic scale based on symptoms and signs and including patients with peripheral and central neuropathic pains has been validated recently in France.[80]

Differences and Similarities Among Etiologies of Central Pain

The main differences between the multiple etiologies of central pains relate to the pain localization and coexisting painful symptoms, which vary according to the primary injury. After a stroke, pain generally involves one-half of the body with or without the ipsilateral or contralateral face,[36,37] but also may be more restricted and involve a quadrant of the body, one side of the face, the mouth and hand (cheiro-oral distribution), or even the ulnar or radial side of the hand, particularly after a lesion of the parietal cortex.[36,39] In Wallenberg syndrome, pain generally affects the contralateral side of the body,[38] but patients also may present with ipsilateral facial pain due to lesions of the lower trigeminal tract.[84] After spinal cord injury, patients may present with neuropathic pains below level located diffusely at the lower limbs and neuropathic pains at level located in a segmental pattern at the level of injury.[13,14] Both pain locations are common after spinal cord trauma or ischemia, such as in the anterior spinal artery syndrome,[29,30] whereas neuropathic pains at level only are more characteristic of syringomyelia and intraspinal tumors.[67,75]

Coexisting non-neuropathic painful symptoms and other causes of pains are particularly frequent in spinal cord injury pains. Thus, neuropathic pains are often associated with visceral pains (due to renal calculus, bowel and sphincter dysfunction, etc.) and musculoskeletal pains (due to bone, joint, or muscle trauma or inflammation, mechanical instability, muscle spasms, and secondary overuse syndrome).[13,14] Peripheral neuropathic pain due to nerve root entrapment at the level of spinal trauma or due to arachnoiditis is also common.[14] In multiple sclerosis, central ongoing pain often is associated with tonic spasms.[19] In poststroke pains, many patients also present with complex regional pain syndrome.[56]

In contrast, neither the type of lesion (i.e., vascular, tumoral, etc.) nor the site or etiology of the injury has a major impact on the clinical symptomatology.[3,56] Although some clinical differences are detectable, there are no pathognomonic features pertaining to some etiologies. Thus, in a recent clinical assessment of neuropathic pain patients using a specific neuropathic pain questionnaire, we found that similar proportions of patients with traumatic spinal cord injury, syringomyelia, and poststroke pains reported burning pain, deep squeezing/pressing pain, paroxysmal pain, and brush-induced and cold-evoked pains (Fig. 22-1). These patients also have comparable

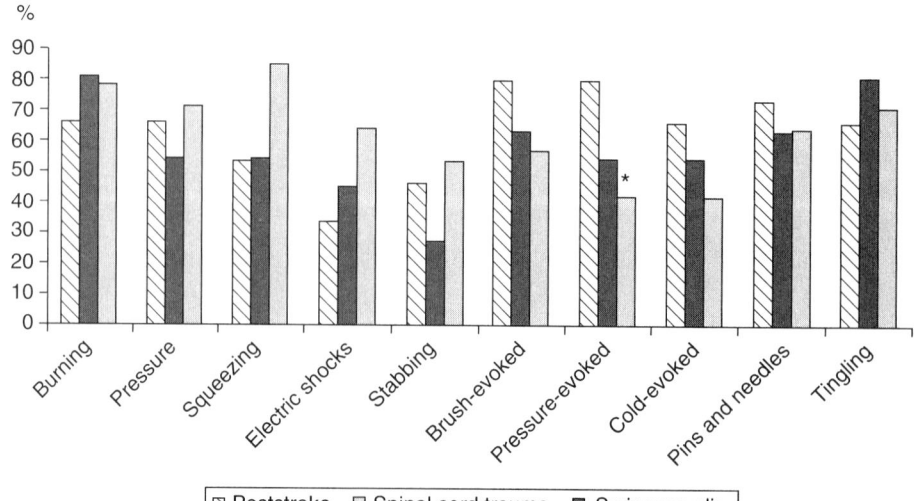

Figure 22-1. Comparison of the frequency of neuropathic symptoms (percent) between patients with central poststroke pains ($n = 15$), syringomyelia ($n = 11$), and traumatic spinal cord injury ($n = 14$). *$p < 0.05$ between poststroke pains and spinal cord trauma. (ref. 78)

thermal deficits and pains when assessed using quantitative sensory testing. Only one significant intergroup difference was identified: patients with spinal cord trauma reported less pressure-evoked pain than poststroke pain patients (see Fig. 22-1). This may relate to the fact that several of these patients had complete spinal cord injury with paraplegia.

▶ MECHANISMS OF CENTRAL PAINS

The mechanisms of central pains have been considered for over 100 years but remain largely unknown. Several hypotheses have been proposed that fall into two major categories: central disinhibition and/or imbalance and central sensitization leading to neuronal hyperactivity/hyperexcitability of spinal/supraspinal nociceptive neurons.[2,12,85,86] It also has been suggested that functional plasticity/reorganization of somatosensory circuits occurs in central (poststroke) pains on the basis of functional imaging studies[87] and thalamic microelectrode recordings.[88] Most theories were based initially on human studies, but the recent development of several animal models of at-level or below-level spinal cord injury due to excitotoxicity, contusion, section, or ischemia has allowed further investigatation of the molecular and neurochemical basis of neuronal hyperactivity.[9,52,89–91]

Neuronal Hyperactivity of Central Nociceptive Neurons

On the basis of psychophysical studies performed in patients with central poststroke and spinal cord injury pains, it has been proposed that hyperexcitability of central nociceptive neurons may be responsible for spontaneous pain and allodynia.[64,65,75–77] Direct evidence of abnormal excitation of spinal or supraspinal neurons comes mainly from experiments in animal models of chronic spinal cord injury[9,89,92–94] and a few microelectrode recordings performed in patients with central pains. Thus, abnormal spontaneous/evoked bursting activity within deafferented regions of the lateral and medial thalamic nuclei have been recorded in some patients with poststroke pains.[96–99] Similarly abnormal focal hyperactivity within the superficial dorsal horn of the injured cord has been evidenced in patients with pain due to spinal cord injury.[100–104]

The Notion of Central Imbalance as a Possible Cause of Central Pains

Psychophysical studies have shown repeatedly that central pain patients have abnormal temperature and pain sensibility but may present with normal touch and vibration sensation (see "Symptoms and Signs" above). These results indicate that the presence of a spinothalamic dysfunction is a necessary condition for the occurrence of central pains, whereas modalities subserved by the dorsal columns/medial lemniscus system are affected less commonly. This was further supported by studies showing abnormalities of laser-evoked potentials (reflecting the function of $A\delta$ nociceptive fibers) on the painful side of patients with poststroke pains or spinal cord injury,[105–109] whereas sensory evoked potentials, subserved by large-diameter fibers correlated with the decrease in touch and vibration sensibility, may be preserved.[62,110–112]

On the basis of these data, it was proposed initially that central pains and dysesthesias could be induced by an imbalance of integration between the spared dorsal column/medial lemniscus activity and the lesioned spinothalamic tract.[29,62] However, central pains also may occur following complete spinal transection or supraspinal lesions that affect all types of sensibilities.

Another form of imbalance between the lateral spinothalamic system (projecting via the lateral thalamic nuclei to the insular region) and the medial system (projecting via the medial thalamic nuclei to the anterior cingulate) also has been proposed to account for allodynia particularly after a stroke.[87,113] However, the mechanisms by which an imbalance of integration between these systems could induce central pains are currently unknown.

Disinhibition Theories

Since the original hypothesis of Head and Holmes,[114] the notion of a central disinhibition, particularly at the thalamic level, has been one of the most favored pathophysiologic theories of central (particularly poststroke) pains.[11,31,115] Head and Holmes[114] proposed that a lesion in the lateral thalamus, which is characteristic of thalamic stroke, could disinhibit activity in the medial thalamus and cause ongoing pain through disruption of an inhibitory pathway between the lateral and the medial thalamus. Some authors proposed an indirect route for such disinhibition via the thalamic reticular nuclei that contain inhibitory interneurons.[97,98,116] However, this hypothesis is not supported by direct anatomic evidence. More recently, Craig and colleagues suggested a new version of the thalamic disinhibition hypothesis of Head and Holmes.[85,117–120] Since thermosensory loss is the cardinal feature of nearly all central pain patients, these authors proposed that central poststroke pain (particularly burning pain and cold allodynia) might be due to a reduction of the physiologic inhibition of thermal (cold) systems on nociceptive neurons. Their "thermosensory disinhibition" theory aims to provide a general framework for understanding the homeostatic nature of pain.[85,120] A major limitation of this theory is that it is based mainly on studies performed in healthy volunteers and in animals. Interestingly, however, a psychophysical study comparing stroke patients with and without pain provided indirect support for such a theory, showing that painful patients with supratentorial lesions had larger cold deficits than pain-free controls.[36]

The neurochemical basis of central disinhibition[90,92,121] may involve the hypofunction of GABAergic, monoamine, and/or endogenous opioidergic systems. This would account for the beneficial effects of antidepressants and of GABAergic agonists (such as IV propofol) in central pains, as well as the reduced efficacy of opioids (see "Pharmacologic Treatment" below).

Central Sensitization

Another possibility could be that the hyperactivity of nociceptive neurons in central pain results from direct modifications of their electrophysiologic properties, i.e., central sensitization.[65] This could involve damage to central neurons from excitatory amino acids related to N-methyl-D-aspartic acid (NMDA) receptor activation[92] and possibly also from sodium channels.[122,123] Indirect evidence for the role of central sensitization in central pains is provided by the beneficial effects of NMDA antagonists and sodium channel blockers in animal models of spinal injury[92,90] and in patients with central pains (see "Pharmacologic Treatment" below).

What Is the Relationship Between the Mechanisms and Symptoms of Central Pains?

Animal studies alone cannot establish the precise relationship between painful symptoms and mechanisms, and in this respect, human studies are needed. Thus, it is important to determine whether the mechanisms observed in humans relate specifically to pain. Studies comparing patients with and without pains, therefore, are critical. Recent psychophysical studies have indicated that thermal deficits, which are considered necessary for the occurrence of central pains, were observed to a similar degree in nonpainful patients, showing that the spinothalamic lesion is not a sufficient condition to account for the development of central pains.[36,37,64,75,77] Similarly, the significance of the observed "bursting," at least in the thalamus, to account for central pain remains controversial because recent studies have identified a similar bursting pattern in patients without pain.[124] This indicates that bursting may result from the central deafferentation rather than from pain. It also should be noted that functional imaging studies have been performed exclusively in patients with neuropathic pains; however, their aim was not to compare patients with and without pains. Thus, a relative hypometabolism has been reported in the thalamus in functional imaging studies of patients with spontaneous pains after a stroke, but whether this finding is related to the pain or the deafferentation is still unclear.[96,125–127]

Furthermore, the mechanisms of the various symptoms of central pains may be distinct, as suggested by various psychophysical and pharmacologic studies. Preferential effects of some drugs on specific painful symptoms have been reported in patients with central pains (see "Pharmacologic Treatment" below).[128,129] In patients with neuropathic pains related to syringomyelia, two subgroups have been individualized on the basis of their symptoms (patients with allodynia and minimal deficits and patients with "pure" spontaneous pains and major thermal deficits), which suggests that distinct mechanisms may be at play in these cases.[75] In accordance with these findings, the antiepileptic lamotrigine has proved to be effective only in a subgroup of patients with incomplete spinal cord injury and mechanical allodynia but not in patients with "pure" spontaneous pains.[130] These data also may account for the apparently discrepant findings obtained in functional imaging studies, which included patients with various types of painful symptoms. They emphasize the complex relationship between symptoms and mechanisms of neuropathic pains.[59,60,78]

▶ TREATMENT

Like other chronic pain syndromes, central pain has a major impact on quality of life and affective state.[49,131,132] Pain may have more effect on quality of life than the extent of a spinal cord injury or a neurological dysfunction.[132] Catastrophizing is also associated with increased disability in these patients.[133] For these reasons, the treatment of central pain always should include management of the affective comorbid conditions and of disability through the use of rehabilitation programs, cognitive behavioral therapies, and drug treatments.[50] Transcutaneous electrical nerve stimulation (TENS) also has been proposed, despite the absence of a theoretical framework for use in central pain syndromes.[134] Therapy aiming to treat any nociceptive component related, for instance, to infections and musculoskeletal factors, including abnormal posture, should not be neglected.

Pharmacologic Treatment

The pharmacologic treatment of central pains still represents a challenge for the clinician because response to therapy generally is incomplete or insufficient, and many drugs used for this condition induce significant side effects.[135] Like other neuropathic pains, these pains generally are refractory to conventional analgesics, such as acetaminophen, aspirin, and nonsteroidal anti-inflammatory agents. Other pharmacologic classes shown to have some efficacy on central pains, on the basis of randomized, controlled trials (RCTs), include tricyclic antidepressants (TCAs), antiepileptics, opioids, local anesthetics, and NMDA receptor antagonists (Table 22-2). Intrathecal

Table 22-2. DOUBLE-BLIND, RANDOMIZED, CONTROLLED TRIALS OF SYSTEMIC DRUGS FOR CENTRAL PAINS: FULL PUBLICATIONS ARE PRESENTED IN WHICH DATA FOR AT LEAST 9 PATIENTS WITH CENTRAL PAINS WERE ANALYZED SPECIFICALLY.

Classes	Etiology	Authors	Active drugs	N	Daily doses Average (range)	Route	Methodology	Results
Antidepressants	Stroke	Leijon, 1989[138]	Amitriptyline	15	75 mg	po	cross over	amitriptyline > Pbo
		Davidoff, 1987[142]	Trazodone	19	150 mg	po	cross over	trazodone = Pbo
	Spinal cord	Cardenas, 2002[139]	Amitriptyline	84	50 mg (10-125)	po	parallel	ami = benztropine[1]
Antiepileptics	Stroke	Leijon, 1989[138]	Carbamazepine	15	800 mg	po	cross over	carbamazepine[5] = Pbo
		Vestegaard, 2001[151]	Lamotrigine	30	200 mg	po	cross over	lamotrigine > Pbo
	Spinal cord	Drewes, 1994[160]	Valproate	14	1800 mg	po	cross over	valproate = Pbo
		Finnerup, 2002[130]	Lamotrigine	30	400 mg	po	cross over	lamotrigine = Pbo[2]
Opioid agonists/ antagonists	Stroke	Bainton, 1992[242]	Naloxone	20	8 mg	iv	cross over	naloxone = Pbo
	Stroke/spinal	Eide, 1996[166]	Alfentanyl	9	0,6 mg/kg/mn	iv	cross over	alfentanyl=Pbo (sp. pain)[3]
		Attal, 2002[129]	Morphine	15	16 mg (9-30)	iv	cross over	morphine = Pbo (sp. pain)[3]
		Rowbotham, 2003[167]	Levorphanol	23/81	2.7 and 8.9 mg	po	parallel	high doses > low doses
Local anesthetics and derivatives	Stroke/spinal	Attal, 2000[170]	Lidocaine	16	5 mg/kg	iv	cross over	lidocaine > Pbo
	Stroke cord	Chiou-Tan, 1996[172]	Mexiletine	11	450 mg	po	cross over	mexiletine = Pbo
NMDA antagonists	Spinal cord	Eide, 1996[166]	Ketamine	9	0.15 mg/kg	iv	cross over	ketamine > Pbo
		McQuay, 1994[177]	Dextromethorphan	15	81 mg	po	cross over	dextrometorphan = Pbo
GABA agonists	Stroke	Canavero, 1994[188]	Propofol	16	0.3 mg/kg	iv	cross over	propofol > Pbo
Cannabinoids	Mixed (multiple sclerosis ++)	Wade, 2003[183]	Plant extract	18/20	2.5-120 mg	subl.	cross over	active drug> Pbo[4]

Abbreviations : Pbo = placebo; sp. pain = spontaneous pain; ami = amitriptyline; subl. = sublingual.
Comments
[1] The effects of amitriptyline were assessed on chronic pains including musculoskeletal ones.
[2] Lamotrigine was effective in a subgroup of patients with incomplete spinal cord injury and mechanical allodynia.
[3] Opioids were superior to placebo on mechanical allodynia, whereas they did not reduce thermally-evoked pains.
[4] In this pilot study, a subgroup of 12 patients had neuropathic pain, but its characteristics were not stated. The treatment consisted of whole plant extracts of delta-9-tetrahydrocannabinol and cannabidiol.

therapy may be proposed in refractory cases. However, despite the recent increased number of RCTs for central pains, their proportion does not exceed 10 percent of all neuropathic pain trials, and most of them have small sample sizes.

Antidepressants

It is largely acknowledged that antidepressants induce specific analgesic efficacy that is independent of their effects on depression. The mechanisms of their analgesic effects generally are considered to relate mainly to a central blockade of monoamine reuptake (serotonin and/or norepinephrine) resulting in enhancement of the descending monoaminergic inhibitory pathways.[136] However, other mechanisms have been proposed, such as NMDA antagonistic effects, action on endogenous opioid systems, and even sodium channel blockade.[137]

Among antidepressants, TCAs are still considered the most effective in the treatment of neuropathic pains and generally are recommended as first choice. Although they are largely used in these patients, the efficacy of TCAs in central pains is supported by only one small RCT comparing amitriptyline (75 mg/d), carbamazepine, and placebo in 15 patients with poststroke pains, and showed significant efficacy of amitriptyline.[138] A recent RCT failed to confirm the efficacy of the drug in a large cohort of spinal cord injury pains.[139] However, a possible moderate analgesic effect on neuropathic pain cannot be excluded in this study because most patients presented with several types of pain.

There is great interindividual variability as to the optimal dosage of TCAs for pain relief, and contradictory results have been reported concerning correlation with plasma levels.[128] However, in the study of amitriptyline in poststroke pain, greater pain relief was achieved at higher plasma concentrations.[138] Treatment guidelines recommend initiating them at low dosages (10 to 20 mg/d) and increasing titration weekly to intolerable side effects or efficacy.[58,140] The onset of efficacy is usually 4 to 5 days to 1 week after reaching optimal dosages. Currently, there are no data on the optimal duration of treatment. It is recommended to keep dosages stable for at least several months if pain has abated and then to attempt a reduction.[58]

The main limitations concerning the use of TCAs relate to their unfavorable side-effect profile. Thus, most patients do not reach the optimal dosage that would be effective for their pain,[141] and urinary retention is a noteworthy side effect, particularly in spinal cord injury patients.[139] More selective drugs generally have a better side-effect profile. However, the selective serotonin reuptake inhibitors, such as trazodone and citalopram, have not shown significant efficacy in central pain.[142,143]

Antiepileptics

Antiepileptics are the second largest pharmacologic class used in central pain patients. They are considered to act on these pains by reducing abnormal neuronal hyperexcitability through modulation of sodium/calcium channels and/or effects on excitatory amino acids and/or GABA-mediated inhibition.[137,144,145]

CARBAMAZEPINE AND OXCARBAZEPINE

In the comparative study of 15 patients with poststroke pains reported earlier, carbamazepine (800 mg/d) was not superior to placebo.[138] However, a third of the patients were considered responders to the active drug, which is similar to results obtained with most treatments of neuropathic pains.[146] This suggests that the lack of significance in this trial may be due to an insufficient number of patients. Carbamazepine may be particularly effective on paroxysmal (shooting) pains related to central nervous system (CNS) lesions, such as in multiple sclerosis,[147,148] but it is still unclear whether such effects are superior to the reduction of ongoing pains. The effects on evoked pains have not been assessed. The drug has a generally unfavorable side-effect profile (25 to 50 percent of patients with side effects in clinical trials, generally consisting of drowsiness, dizziness, and somnolence).[149] It caused more dose-limiting side effects than amitriptyline in the preceding comparative study. Titration should begin with a low initial dosage (100 mg/d) and be increased to efficacy or intolerable side effects. The average analgesic dosage is 800 mg, ranging from 600 to 1600 mg/d.

Oxcarbazepine is a keto analogue of carbamazepine with a distinct pharmacokinetic profile that causes less enzyme induction and, therefore, may be a possible substitute in patients intolerant to carbamazepine or with significant drug-drug interaction. This drug may have an analgesic profile similar to carbamazepine in the treatment of neuropathic pains, but published controlled trials are still lacking.[150]

LAMOTRIGINE

A recent RCT reported a significant efficacy for lamotrigine (200 mg/d) in patients with central poststroke pain, with a mean reduction of spontaneous pain of 30 percent.[151] In this trial, the drug also was effective on a measure of cold-evoked pain, but not on mechanical allodynia. In contrast, in patients with traumatic spinal cord injury, the same drug titrated to 400 mg/d had no overall efficacy but was superior to placebo in patients with incomplete lesion and mechanical allodynia, with a median reduction of 25 percent of spontaneous pain at or below level.[130] The initial titration must be very slow (increase of 50 mg every 2 weeks) in order to minimize the risk of serious complications, such as skin rashes. Adverse effects also include dizziness, nausea, headache, and fatigue.

GABAPENTIN

Several open-label studies have suggested the benefit of gabapentin in central pains due to spinal cord injury or stroke.[152-155] One placebo-controlled study used gabapentin, increased up to 1800 mg/d, in seven patients with central pain of spinal origin and reported a trend toward benefit on dysesthesia and burning pain.[156] However, the conclusions of this study are limited by the small sample size and the lack of individual titration. Another study including 307 patients with various peripheral and central neuropathic pains, 9 of whom were suffering from poststroke pains, reported a modest overall effect of gabapentin titrated to 2400 mg/d over 8 weeks.[157] However, in this study, patients previously unresponsive to gabapentin were not included, and it is not clear whether all had neuropathic pains. Gabapentin may be effective with regard to several pain components, including pain paroxysms and brush/cold-induced allodynia related to both peripheral or central nerve lesions.[153] The dosages used vary from 1200 to 3600 mg/d, with an optimal dose of 1800 mg/d.[158] Most side effects occur during titration and consist of dizziness and somnolence (about one-fourth of patients), but weight gain may be observed after long-term use. The exact mechanisms of the action of gabapentin in central pains are currently unknown, and it is unclear whether the drug preferentially targets calcium channels (particularly on the $\alpha_2\delta$ subunit) and/or GABAergic transmission.[144,159]

OTHER ANTIEPILEPTICS

There is currently no evidence showing the benefit of other antiepileptics in the treatment of central pains. Valproate appears to be ineffective, at least on the basis of an RCT of 14 patients with spinal cord injury pains.[160] Phenytoin has not been studied, and negative results have been reported with topiramate in an open study of 7 patients with central brain and spinal cord injury pain.[161]

Opioids

Although there is a large consensus concerning the effectiveness of strong opioids in nociceptive pain, their efficacy in neuropathic pain has been debatable until recent years.[162,163] On the basis of several controlled studies, it is now admitted that these drugs may relieve neuropathic pains, provided that sufficient doses are administered using individual titration.[162,164] The analgesic doses in neuropathic pains are at least twice as high as those required to relieve nociceptive pains.[165] Three double-blind RCTs have used opioids in patients with central pains.[129,166,167] In a trial of 15 patients with poststroke and spinal cord injury pains, titrated IV morphine showed no significant effect on ongoing pain in comparison with placebo.[129] However, 46 percent of the patients gained a significant effect from the active drug, and most were still relieved with oral sustained-release morphine after 1 month. Furthermore, morphine was effective on brush-induced allodynia,[129] as also observed with alfentanyl in spinal cord injury pains.[166] In a controlled trial of patients with peripheral and central pains comparing low and high doses of the mu agonist levorphanol (with no placebo group), a 39 percent pain relief was observed in patients with central pains receiving high dosages who completed the study.[167] Opioids also may be moderately effective in a subgroup of patients with multiple sclerosis, as suggested by a single-blind study using high dosages of IV morphine.[168]

These data suggest the benefit of opioid treatments at least in a subgroup of patients with central pains. However, the benefit of opioids after 1 to 2 years only concerns less than 20 percent of patients due to an unfavorable balance between efficacy and side effects.[129]

Local Anesthetics and Derivatives

Local anesthetics and derivatives (antiarrythmics) are considered to act mainly as sodium channel blockers at the periphery[145,169] but probably also at spinal or supraspinal sites.[139,123]

Intravenous lidocaine presents with moderate analgesic effects in central pains and relieves spontaneous pain and mechanical dynamic/static allodynia to the same extent, whereas thermally evoked pains are less affected.[170] The drug also has been shown to be particularly effective in patients with paroxysmal pains due to multiple sclerosis in a single-blind study.[171] Side effects consist of light-headedness, somnolence, nausea, and perioral numbness, but severe side effects such as bradycardia and convulsions are potential complications. Unfortunately, only a few patients (about 10 percent) may benefit from long-term efficacy with IV lidocaine.[170] For this reason, the drug essentially has been considered as a predictive test for the efficacy of oral congeners, such as the antiarrythmic mexiletine.

In contrast to IV lidocaine, the efficacy of *mexiletine*, a structural analogue that can be administered orally, has not been demonstrated in central pains. One placebo-controlled study of patients with spinal cord injury pain reported negative results with moderate dosages of the drug (450 mg/d),[172] also observed using individual titration (up to 900 mg/d) in an open-label trial subsequent to IV lidocaine test.[170] The poor therapeutic ratio of mexiletine may account for these disappointing results. Side effects generally consist of nausea, dizziness, headache, sleep disturbances, and fatigue. Although no serious cardiac effects have been reported in patients with neuropathic pain, transient tachycardia and palpitations have occurred, and the potential cardiotoxic effects warrant caution in elderly patients. Owing to its lack of confirmed efficacy and poor side-effect profile, mexiletine cannot be recommended for the treatment of central pains.

NMDA Receptor Antagonists
On the basis of the involvement of NMDA receptors in central sensitization, NMDA receptor antagonists have been studied for the treatment of neuropathic pains.[173] One double-blind, placebo-controlled trial reported a significant effect of IV *ketamine*, a traditionally used anesthetic that binds noncompetitively to the phencyclidine site of the NMDA receptor, on spontaneous pain and mechanical allodynia after spinal cord injury.[166] Use of this treatment is limited by psychomimetic effects, which may be prevented by haloperidol.[174] Clinical experience with oral ketamine in the treatment of central pain is limited,[174,175] and benefit may not exceed 10 percent of responders to the IV infusion.[176] Other weaker NMDA receptor antagonists, such as amantadine, memantine, and riluzole, have yielded inconsistent results in neuropathic pains, and negative results have been reported with moderate dosages of dextromethorphan in central pains.[177]

Thus, the current NMDA antagonists have a limited place in the treatment of central pains. More selective antagonists and drugs acting on non-NMDA receptors currently are under development and may be better tolerated.

Cannabinoids
Recently, the discovery of the cannabinoid receptors and of the endogenous cannabinoid system has led to renewed interest for the use of cannabis and its derivatives (cannabinoids) as potential analgesic agents.[178,179] Cannabinoids have been shown to reduce the hyperalgesia and allodynia-like symptoms associated with nerve injury in animals.[180,181] Two recent randomized, controlled trials have reported moderate effects of tetrahydrocannabinol (dronabinol) and sublingual cannabinoids on neuropathic pains related to multiple sclerosis,[182,183] but lack of efficacy and poor tolerability also were reported with long-term use of dronabinol for up to 6 months in a pilot study of eight patients with severe refractory neuropathic pains.[184] Large international studies are now in progress to investigate the safety and efficacy of cannabinoids in the treatment of central pains mainly related to multiple sclerosis.

Other Systemic Treatments
Several drugs have shown significant efficacy in peripheral neuropathic pains on the basis of RCTs but have not been studied specifically in central pains. These include tramadol, a centrally-acting analgesic drug that acts through monoaminergic and opioid mechanisms; venlafaxine, a mixed inhibitor of norepinephrine and serotonin reuptake; and to a lesser extent, phenytoin.[185-187] As discussed below, one cannot rule out a possible efficacy of these drugs in central pains.

The use of other systemic drugs is marginal in central pains. Sympatholytics have been proposed to relieve complex regional pain syndromes sometimes encountered in these patients, but this is not supported by RCTs, as for other neuropathic pains.[128] The GABA-A agonist propofol, used at subanesthetic dosages, proved effective in poststroke pains, particularly on mechanical allodynia on the basis of one double-blind trial,[188] but the clinical application of these data is limited, except possibly as a predictive test for the response to motor cortex stimulation (see "Surgical Treatment" below).

Intrathecal Therapy
In patients with refractory pain, drugs may be administered intrathecally via an implanted pump generally using continuous infusion. However, most trials have used an open-label design or have included a limited number of patients, which renders definite conclusions difficult.[189]

In spinal cord injury pains, several open or single-blind trials have suggested the benefit of epidural or intrathecal *clonidine* or *morphine*, but only one placebo-controlled study has been performed. This study has shown that the combination of both drugs during 6 days was significantly more effective than either treatment administered alone and superior to placebo after 4 hours in 15 patients.[190] The best results were obtained on at-level neuropathic pain. These results may relate to the synergistic action of intrathecal clonidine and morphine.[190] Sedation and low blood pressure are the most common side effects and may persist despite reduction of dosages, but tolerance to the combination develops more slowly than with morphine alone.

There is limited evidence for use of other intrathecal drugs in central pains. *Intrathecal baclofen*, a GABA-B agonist, has proved effective on spontaneous pain and allodynia after acute administration on the basis of a pilot placebo-controlled study of seven patients with neuropathic pain and spasticity mainly related to multiple sclerosis,[191] and its efficacy also has been reported in central poststroke and spinal cord injury pains in an open-label study.[192] However, long-term administration has been shown to enhance spinal cord injury pain in some patients.[193] *Spinally-administered ziconotide*, a neuron-specific N-type calcium channel blocker, has proved effective in refractory chronic pain on the basis of RCTs.[194] This drug is not associated with the development of tolerance after prolonged use. However, its efficacy has not been specifically investigated in neuropathic pains. Intrathecal *lidocaine* has been shown to reduce pain in a proportion of patients with spinal cord injury pain on the basis of one placebo-controlled trial,[195] but its effects seem to be temporary. Finally, intrathecal *midazolam* has been reported occasionally as beneficial in patients with central pains.[196]

Analgesic Combinations

Multiple analgesic combinations are used largely in clinical practice in neuropathic pain based on past experimentation and case reports, although no double-blind studies have compared such combinations with monotherapy. There is no need to prescribe multiple analgesics systematically as first-line therapy because of the risk of cumulative side effects and of drug-drug interactions. However, a combination of analgesics may be required to improve the balance between analgesia and adverse effects, notably in the case of additive or synergistic effects. Thus, NMDA receptor antagonists potentiate the effects of opioids in animals and humans and may reduce tolerance to morphine.[197] Opioids also may be combined with other drugs, as has been shown with clonidine after intrathecal administration (see "Intrathecal Therapy" above). Gabapentin has been shown to enhance the analgesic effect of morphine in acute experimental pain in healthy volunteers,[198] and this deserves specific studies in neuropathic pain. A combination of analgesics also may be advocated because of effects on multiple-pain symptoms or complementary-action mechanisms. Thus, antidepressants and antiepileptics often are used in combination and may have a broader spectrum of efficacy than each drug administered alone, although this has not been confirmed in systematic studies.

Surgical Treatment

Surgical treatment is only proposed for severe refractory central pains. The technical aspects of these techniques will not be detailed here.[196,199]

Stimulation Techniques

In the treatment of central pain, stimulation of the spinal cord and motor cortex is by far the most used technique, whereas deep brain stimulation (thalamus, periventricular gray matter, internal capsule, septal region, caudate) now has a very limited place.[200,201]

SPINAL CORD STIMULATION

Spinal cord stimulation involves the insertion of a stimulating lead in the spinal epidural space, followed by insertion of an implanted electrode after a successful screening trial. The effect of this technique on central pain is much less documented than in peripheral nerve lesions, particularly chronic radiculopathies.[196] On the basis of long-term experience with this technique, most neurosurgeons do not recommend it anymore in the treatment of central pains, except for particular cases of spinal cord injury.[202–205,57] The best results have been obtained in patients with incomplete lesions, suffering from steady pains, and with preservation of dorsal column function as assessed using somatosensory evoked potentials, whereas patients with diffuse below-level pains generally are unrelieved.[206,203,57,207] Complications include exceptional cases of epidural hematomas and abcesses.

MOTOR CORTEX STIMULATION

To date, the preferred surgical option for central pains is extradural stimulation of the motor cortex.[208–210] In this technique, the stimulating electrode is implanted extradurally on the motor cortex contralateral to the painful area, and the targeting of the motor cortex generally is performed using evoked potentials[210] and functional magnetic resonance imaging.[211] The physiologic basis of this technique has not been fully elucidated[212] but may involve a reinforcement of descending inhibitory controls at thalamic or brain stem sites.[213–214] Good results (i.e., mean pain reduction by at least 50 percent) have been reported in 52 percent of the 159 patients with poststroke pains published to date,[215–218] and 8 of 9 published patients with spinal cord injury pains.[210] Discrepant results also have been published.[219,220] Although motor cortex stimulation is the only neuromodulatory technique where controlled trials are possible, owing to a lack of paresthesia during therapeutic stimulation, such trials have not been published yet. Predictors of response mainly include a beneficial response to transcranial magnetic stimulation[221–223] and the absence of motor weakness and thermal impairment in the painful area.[213,224] A positive response to pharmacologic agents such as amytal,[225] propofol,[209] and ketamine[217] also may be predictive. Complications are rare: 2.2 percent epidural or subdural hematomas, 0.7 percent epileptic seizures, few system failures, wound dehiscences, and 2 percent noncerebral infections.[210] Trials are now in progress, notably in France, with various types of neuropathic pains in order to expand and better define the indications of these techniques.

Neurosurgical Ablative Procedures

Neurosurgical ablative procedures such as cordotomy, cordectomy, thalamotomy, and mesencephalic tractotomy now have a very limited place in the treatment of chronic noncancer pains.[14,199] To date, dorsal root entry zone (DREZ) lesions involving two to three spinal segments are the preferred technique.[205,226] The procedure aims at destroying the superficial part of the dorsal horn.[205] The best indications are represented by neuropathic pains at level after incomplete spinal cord injury, particularly paroxysmal shooting pains and evoked pains, whereas below-level neuropathic pains generally are refractory to such techniques.[57,227–230] Thus, in a recent prospective series of 44 patients, only the patients with segmental pain distributions had a significant effect for such procedure with good pain relief (pain relief exceeding 75 percent being achieved in 68 percent patients at 1 year).[229] However, owing to the possible risks of neurological complications (aggravation of sensory and/or motor deficit), cerebrospinal fluid leaks, infection, and hematoma, this technique should be reserved only for very refractory cases after failure of more conservative surgical techniques.

Therapeutic Strategy

There is currently no consensus concerning the optimal therapeutic strategy for central pain. Treatments generally are selected on the basis of evidence for efficacy and safety in randomized, placebo-controlled studies conducted in disease-based groups of patients (evidence-based medicine).[141,146,149] In this respect, TCAs and antiepileptics remain the first therapeutic option for central pains, followed by opioids, then other systemic treatments, and finally, more invasive procedures in refractory cases. Pharmacologic IV tests have been proposed to predict the effectiveness of long-term treatments, but few appear to be really helpful in clinical practice, with the exception of IV opioids, which may be predictive to the subsequent response to oral/transdermal opioids.[129,162] A rational strategy for selecting treatments targeted at mechanisms (mechanism-based approach) has been advocated.[60,231–233] However, this approach is difficult to apply for clinicians[233] because the exact relationship between the mechanisms and symptoms of central pains is still undetermined (see "Mechanisms of Central Pains" above). To date, an approach based on the detailed assessment of the pain symptoms and signs appears to be more realistic. It may be best achieved through the use of quantitative sensory tests but probably may be simplified owing to the development of specific self-questionnaires allowing assessment of the various dimensions of neuropathic pains.[69] This approach has proved to be helpful for revealing the differential antiallodynic or antihyperalgesic effects of drugs and for predicting the response to some treatments, such as sodium channel blockers (see "Pharmacologic Treatment" above).

▶ CENTRAL AND PERIPHERAL NEUROPATHIC PAINS: MORE SIMILARITIES THAN DIFFERENCES

Various clinical and experimental studies have underlined the similarities between central and peripheral neuropathic pains as regards symptoms, mechanisms, and response to therapy, thus challenging the clinical relevance of the classical distinction between central and peripheral nerve lesions.

Symptomatology

As described earlier (see "Symptoms and Signs"), the painful symptoms of central pains resemble those observed after peripheral nerve injury.[78,61] In a recent clinical study comparing the main neuropathic pain symptoms between patients with peripheral and central neuropathic pains using a specific neuropathic pain questionnaire, we detected very few clinical differences between central and peripheral neuropathic pains (Fig. 22-2). Slightly more patients with central pains presented with cold-evoked pains, whereas pressure-evoked pain was more common after peripheral nerve injury. These data suggest that the site of injury to the central or peripheral nervous system does not play a major role in the clinical symptomatology.

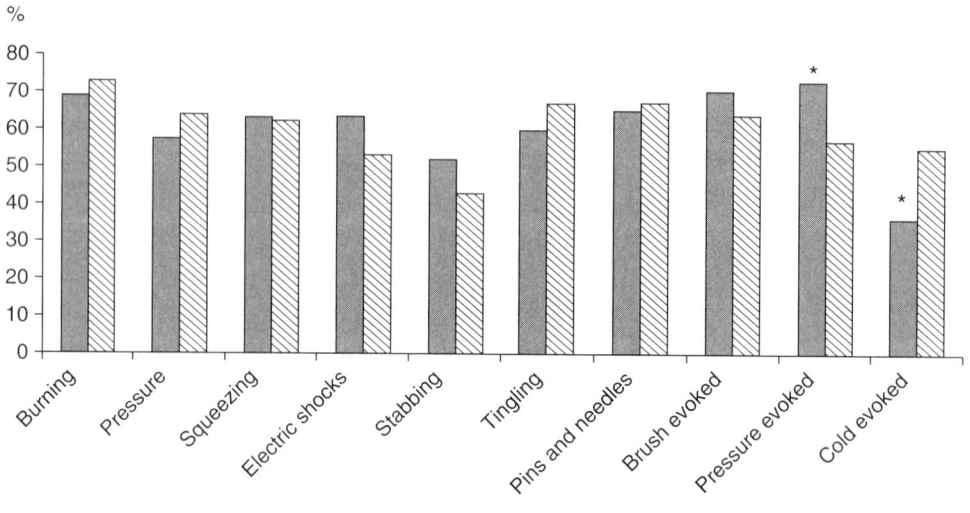

Figure 22-2. Comparison of the frequency of neuropathic symptoms (percent) between patients with peripheral nerve lesions ($n = 120$) and central lesions ($n = 56$). *$p < 0.05$. (ref. 78)

Pathophysiology

Several of the spinal and supraspinal mechanisms suggested as accounting for central pains also may apply to peripheral nerve lesions. Thus, the central neurophysiologic changes reported after central lesions, such as increased neuronal hyperactivity at the spinal cord level and in supraspinal structures (see "Mechanisms of Central Pains" above) present many similarities with those observed after peripheral nerve lesions.[97–99,234,235] Both central and peripheral nerve lesions may reduce inhibitory controls.[86,232] However, it remains to be determined whether the molecular basis of these various central mechanisms is similar or distinct in peripheral and central nervous system injuries. In any case, the major difference between these entities is the fact that the central changes reported in central pain are not maintained by abnormal peripheral input, in contrast to many cases of peripheral neuropathic pains.

Response to Systemic Treatments

With the exception of topical agents, most of the treatments used for neuropathic pains have several targets at spinal and/or supraspinal sites and, therefore, may be expected to induce analgesic effects in pains due to central lesions. Thus, data from animal models of peripheral neuropathy and spinal cord injury generally have shown a similar pattern of responses to opioids, NMDA receptor antagonists, antidepressants, and antiepileptics.[9,90,92,236] In humans, most of the drug classes that have proved effective on the basis of RCTs in central pains are also beneficial in peripheral neuropathic pains, such as postherpetic neuralgia and diabetic neuropathies, and their number needed to treat (NNT) for substantial pain relief ranges from 2.5 to 4, whatever the etiology, with rare exceptions.[146,237] Similarly, mexiletine, a weakly effective drug with an NNT of 10 or more in trials of peripheral nerve lesions,[146,237] is not useful in central pains. In the few trials that included patients with several types of neuropathic pains, the level of injury to the peripheral or central nervous system did not seem to influence the analgesic response. Thus, in the study by Rowbotham and colleagues[167] (see "Pharmacologic Treatment" above), the opioid levorphanol at high dosages had similar efficacy in patients with postherpetic neuralgia and spinal cord injury pains and seemed even more effective in multiple sclerosis. A similar pattern of analgesic response was observed using IV lidocaine in poststroke pains, spinal cord injury pains, and peripheral nerve lesions, although the effects of the drug on ongoing pain were longer lasting after peripheral nerve injury.[170,238] Opioids, NMDA antagonists, and sodium channel blockers have been shown to reduce mechanical allodynia due to peripheral or central lesions.[129,170,238,240–242] Finally, similar predictors of the response to some treatments also have been identified in peripheral and central pain patients. The sodium channel blocker IV lidocaine is significantly more active on spontaneous pains in patients with coexisting allodynia after peripheral nerve injury,[238] and a similar dissociated profile has been reported using lamotrigine, which also targets sodium channels, in patients with spinal cord injury pains.[130]

Although one cannot exclude that central pain patients are sometimes more refractory to certain drugs, particularly those with additional peripheral mechanisms of action, these data suggest that the level of the injury to the nervous system is not a major predictive factor for the analgesic response to centrally acting drugs.

▶ CONCLUSION

There is now increasing awareness that central pain, initially thought a very uncommon disorder, is in fact a prevalent neuropathic pain condition. Despite some recent improvements in the clinical assessment of patients with central pains and their therapeutic management, such pain continues to have a great impact on quality of life and still remains a very difficult condition to treat. The development of clinically relevent experimental models in animals has been a first step toward a better understanding of some pathophysiologic mechanisms and assessment of newer analgesic agents. However, progress in understanding the pathophysiology of central pains and their therapeutic management also requires the continued development of human studies. Such studies will necessitate a standardization of the diagnostic criteria for central pains; the use of advanced techniques, such as functional neuroimagery for the study of patients with various neuropathic symptoms; and a detailed evaluation of the effects of treatments in large cohorts of patients aiming to provide more rational therapeutic strategies.

REFERENCES

1. Greiff Zur Localisation des Hemichorea. *Arch Psychol Nervenkrantkh* 1883;14:598.
2. Bonica JJ: Introduction, in Nashold BS, Ovelmen-Levitt J (eds): *Advances in Pain Research and Therapy*. New York: Raven Press, 1991:1–19.
3. Boivie J: Central pain, in Walle PD and Melzeck (eds): *Textbook of Pain*. London: Churchill Livingstone, 1999: 879–913.
4. Dejerine J, Roussy G: Le syndrome thalamique. *Rev Neurol* 1906;12:521–532.
5. Edinger L: Giebt es central antstehender Schmerzen. *Deutche Z Nerrvenheilk* 1891;1:262–282.
6. Riddoch G: The clinical features of central pain. *Lancet* 1938;234:1093–1098, 1205–1209.
7. Holmes G: Pain of central origin, in Osler W (ed), *Contributions to Medical and Biological Research*. New York: Paul B. Hoeber, 1919:235–246.

8. Guillain G, Garcin R: Le syndrome de Brown Sequard d'origine traumatique. *Ann Med* 1931;29:361–385.
9. Vierck CJ Jr, Siddall P, Yezierski RP: Pain following spinal cord injury: Animal models and mechanistic studies. *Pain* 2000;89:1–5.
10. Merskey H, Bogduk N (eds): *Classification of Chronic Pain.* Seattle: IASP Press, 1994.
11. Casey KL (ed): *Pain and Central Nervous Disease: The Central Pain Syndromes.* New York: Raven Press, 1991.
12. Schott GD: From thalamic syndrome to central poststroke pain. *J Neurol Neurosurg Psychiatry* 1996;61:560–564.
13. Siddall PJ, Taylor DA, Cousins MJ: Classification of pain following spinal cord injury. *Spinal Cord* 1997;35:69–75.
14. Siddall PJ, Loeser JD: Pain following spinal cord injury. *Spinal Cord* 2001;39:63–73.
15. Siddall PJ, Taylor DA, McClelland JM, et al: Pain report and the relationship of pain to physical factors in the first 6 months following spinal cord injury. *Pain* 1999;81:187–197.
16. Siddall PJ, McClelland JM, Rutkowski SB, Cousins MJ: A longitudinal study of the prevalence and characteristics of pain in the first 5 years following spinal cord injury. *Pain* 2003;103:249–257.
17. Chatel M, Lanteri-Minet M, Lebrun-Frenay C: Pain in multiple sclerosis. *Rev Neurol (Paris)* 2001;157:1072–1078.
18. Clifford DB, Trotter JL: Pain in multiple sclerosis. *Arch Neurol* 1984;41:1270–1272.
19. Moulin DE: Pain in central and peripheral demyelinating disorders. *Neurol Clin* 1998;16:889–898.
20. Gates P, Nayernouri T, Sengupta RP: Epileptic pain: A temporal lobe focus. *J Neurol Neurosurg Psychiatry* 1984;47:319–320.
21. Mauguière F, Courjon J: Somatosensory epilepsy: A review of 127 cases. *Brain* 1978;101:307–332.
22. Potagas C, Avdelidis D, Singounas E, et al: Episodic pain associated with a tumor in the parietal operculum: A case report and literature review. *Pain* 1997;72:201–208.
23. Salanova V, Andermann F, Rasmussen T, et al: Parietal lobe epilepsy: Clinical manifestations and outcome in 82 patients treated surgically between 1929 and 1988. *Brain* 1995;118:607–627.
24. Young B, Blume WT: Painful epileptic seizures. *Brain* 1983;106:537–554.
25. Young B, Barr WK, Blume WT: Painful epileptic seizures involving the second sensory area. *Ann Neurol* 1986;19:412.
26. Stoodley MA, Warren JD, Oatey PE: Thalamic syndrome caused by unruptured cerebral aneurysm: Case report. *J Neurosurg* 1995;82:291–293.
27. Attal N, Parker F, Brasseur L, et al: Characterization of sensation disorders and neuropathic pain related to syringomyelia: A prospective study. *Neurochirurgie* 1999;45(suppl 1):84–94.
28. Milhorat TH, Kotzen RM, Mu HTM, et al: Dysesthetic pain in patients with syringomyelia. *Neurosurgery* 1996;38:940–947.
29. Beric A: Central pain: "New" syndromes and their evaluation. *Muscle Nerve* 1993;16:1017–1024.
30. Triggs WJ, Beric A: Sensory abnormalities and dysaesthesias in the anterior spinal artery syndrome. *Brain* 1992;115:189–198.
31. Pagni CA: Central pain due to spinal cord and brainstem damage, in Wall PD, Melzack R (eds), *Textbook of Pain.* Edinburgh: Churchill Livingstone, 1989:634–655.
32. Quinn NP, Koller WC, Lang AE, Marsden CD: Painful Parkinson's disease. *Lancet* 1986;1:1366–1369.
33. Schott GD: Pain in Parkinson's disease. *Pain* 1985;22:407–411.
34. Tison F, Wenning GK, Volonte MA, et al: Pain in multiple system atrophy. *J Neurol* 1996;243:153–156.
35. Leijon G, Boivie J, Johansson I: Central post-stroke pain-neurological symptoms and pain characteristics. *Pain* 1989;36:13–25.
36. Bowsher D, Leijon G, Thuomas KA: Central poststroke pain: Correlation of MRI with clinical pain characteristics and sensory abnormalities. *Neurology* 1998;51:1352–1358.
37. Vestergaard K, Nielsen J, Andersen G, et al: Sensory abnormalities in consecutive, unselected patients with central post-stroke pain. *Pain* 1995;61:177–186.
38. MacGowan DJ, Janal MN, Clark WC, et al: Central post-stroke pain and Wallenberg's lateral medullary infarction: Frequency, character, and determinants in 63 patients. *Neurology* 1997;49:120–125.
39. Michel D, Laurent B, Convers P, et al: Douleurs corticales: Etude clinique, électrophysiologique et topographique de 12 cas. *Rev Neurol (Paris)* 1990;146:405–414.
40. Schmahmann JD, Leifer D: Parietal pseudothalamic pain syndrome: Clinical features and anatomic correlates. *Arch Neurol* 1992;49:1032–1037.
41. Bogousslavsky J, Regli F, Uske A: Thalamic infarcts: Clinical syndromes, etiology, and prognosis. *Neurology* 1988;38:837–848.
42. Pacaroni M, Bogousslavsky J: Pure sensory syndromes in thalamic stroke. *Eur Neurol* 1998;39:211–217.
43. Nasreddine ZS, Saver JL: Pain after thalamic stroke: Right diencephalic predominance and clinical features in 180 patients. *Neurology* 1997;48:1196–1199.
44. Andersen G, Vestergaard K, Ingeman-Nielsen M, Jensen TS: Incidence of central post-stroke pain. *Pain* 1995;61:187–193.
45. Gamble GE, Barberan E, Laasch HU, et al: Poststroke shoulder pain: A prospective study of the association and risk factors in 152 patients from a consecutive cohort of 205 patients presenting with stroke. *Eur J Pain* 2002;6:467–474.
46. Weimar C, Kloke M, Schlott M, et al: Central poststroke pain in a consecutive cohort of stroke patients. *Cerebrovasc Dis* 2002;14:261–263.
47. Finnerup NB, Johannesen IL, Sindrup SH, et al: Pain and dysesthesia in patients with spinal cord injury: A postal survey. *Spinal Cord* 2001;39:256–262.
48. Ravenscroft A, Ahmed YS, Burnside IG: Chronic pain after SCI: A patient survey. *Spinal Cord* 2000;38:611–614.
49. Störmer S, Gerner HJ, Gruninger W, et al: Chronic pain/dysaesthesiae in spinal cord injury patients: Results of a multicentre study. *Spinal Cord* 1997;35:446–455.
50. Sjolund BH: Pain and rehabilitation after spinal cord injury: The case of sensory spasticity? *Brain Res Rev* 2002;40:250–256.
51. Turner JA, Cardenas DD, Warms CA, McClellan CB: Chronic pain associated with spinal cord injuries: a community survey. *Arch Phys Med Rehabil* 2001;82:501–509.
52. Yezierski RP: Pain following spinal cord injury: The clinical problem and experimental studies. *Pain* 1996;68:185–194.
53. Goetz CG, Tanner CM, Levy M, et al: Pain in Parkinson's disease. *Mov Disord* 1986;1:45–49.

54. Snider SR, Fahn S, Isgreen WP, Cote LJ: Primary sensory symptoms in parkinsonism. *Neurology* 1976;26:423–429.
55. Boivie J, Leijon G, Johansson I: Central post-stroke pain: A study of the mechanisms through analyses of the sensory abnormalities. *Pain* 1989;37:173–185.
56. Bowsher D: Central pain: Clinical and physiological characteristics. *J Neurol Neurosurg Psychiatry* 1996;61:62–69.
57. Tasker RR, Gervasio FRCS, DeCarvalho TC, Dolan EJ: Intractable pain of spinal cord origin: Clinical features and implications for surgery. *J Neurosurg* 1992;77:373–378.
58. Dworkin RH, Backonja M, Rowbotham MC, et al: Advances in neuropathic pain: diagnosis, mechanisms and treatment recommendations. *Arch Neurol* 2003;60:1524–1434.
59. Bouhassira D: Neuropathic pain: the clinical syndrome revisited. *Acta Neurol Belg* 2001;101:47–52.
60. Jensen TS, Baron R: Translation of symptoms and signs into mechanisms in neuropathic pain. *Pain* 2003;102:1–8.
61. Jensen TS, Gottrup H, Sindrup SH, Bach FW: The clinical picture of neuropathic pain. *Eur J Pharmacol* 2001;429:1–11.
62. Beric A, Dimitrijevic MR, Lindblom U: Central dysesthesia syndrome in spinal cord injury patients. *Pain* 1988;34:109–116.
63. Davidoff G, Roth E, Guarracini M, et al: Function-limiting dysesthetic pain syndrome among traumatic spinal cord injury patients: A cross-sectional study. *Pain* 1987b;29:39–48.
64. Defrin R, Ohry A, Blumen N, Urca G: Characterization of chronic pain and somatosensory function in spinal cord injury subjects. *Pain* 2001;89:253–263.
65. Eide PK, Jorum E, Stenehjem AE: Somatosensory findings in patients with spinal cord injury and central dysesthesia pain. *J Neurol Neurosurg Psychiatry* 1996;60:411–415.
66. Boivie J, Hansson P, Lindblom U (eds): *Touch, Temperature and Pain in Health and Disease: Mechanisms and Assessment*. Seattle: IASP Press, 1994.
67. Boivie J: Central pain and the role of quantitative sensory testing (QST) in research and diagnosis. *Eur J Pain* 2003;7:339–343.
68. Cruccu G, Anand P, Attal N, et al: EFNS guidelines on assessment of neuropathic pain and treatment. *Eur J Neurol* 2004;11:153–612.
69. Bouhassira D, Attal N, Fermanian J, et al: Development and validation of the Neuropathic Pain Symptom Inventory. *Pain* 2004;108:248–257.
70. Cohen MJ, Song ZK, Schandler SL, et al: Sensory detection and pain thresholds in spinal cord injury patients with and without dysesthetic pain, and in chronic low back pain patients. *Somatosens Mot Res* 1996;13:29–37.
71. Attal N, Brosseur L, Chauvin M, Bouhassira D: A case of "pure" dynamic mechanical allodynia due to a lesion of the spinal cord: pathophysiological considerations. *Pain* 1998;75:399–404.
72. Greenspan JD, Joy SE, McGillis SL, et al: A longitudinal study of somesthetic perceptual disorders in an individual with a unilateral thalamic lesion. *Pain* 1997;72:13–25.
73. Rousseaux M, Cassim F, Bayle B, Laureau E: Analysis of the perception of and reactivity to pain and heat in patients with Wallenberg syndrome and severe spinothalamic tract dysfunction. *Stroke* 1999;30:2223–2229.
74. Samuelsson M, Samuelsson L, Lindell D: Sensory symptoms and signs and results of quantitative sensory thermal testing in patients with lacunar infarct syndromes. *Stroke* 1994;25:2165–2170.
75. Bouhassira D, Attal N, Parker F, Brasseur L: Quantitative sensory evaluation of painful and painless patients with syringomyelia, in Devor M, Rowbotham MC, Wiesenfeld-Hallin (eds), *Proceedings of the IXth World Congress on Pain*. Seattle: IASP Press, 2000:401–411.
76. Finnerup NB, Johannesen IL, Bach FW, Jensen TS: Sensory function above lesion level in spinal cord injury patients with and without pain. *Somatosens Mot Res* 2003;20:71–76.
77. Finnerup NB, Johannesen IL, Fuglsang-Frederiksen A, et al: Sensory function in spinal cord injury patients with and without central pain. *Brain* 2003;126:57–70.
78. Bouhassira D, Attal N: Novel strategies for neuropathic pain, in Villanueva L, Dickenson AE, Ollat H (eds), *The Pain System in Normal and Pathological States, A Primer For Physicians*. Seattle: IASP Press, in press.
79. Backonja MM, Krause SJ: Neuropathic pain questionnaire-short form. *Clin J Pain* 2003;19:315–316.
80. Bouhassira D, Attal N, Boureau F, et al: Comparison of neuropathic and non-neuropathic pains and development of a new diagnostic questionnaire. *J Pain* 2004,5, (suppl. 1):115.
81. Boureau F, Doubrère JF, Luu M: Study of verbal description in neuropathic pain. *Pain* 1990;42:145–152.
82. Bennett M: The LANSS Pain Scale: The Leeds assessment of neuropathic symptoms and signs. *Pain* 2001;92:147–157.
83. Krause SJ, Backonja MM: Development of a neuropathic pain questionnaire. *Clin J Pain* 2003;19:306–314.
84. Fitzek S, Baumgartner U, Fitzek C, et al: Mechanisms and predictors of chronic facial pain in lateral medullary infarction. *Ann Neurol* 2001;49:493–500.
85. Craig AD: The functional anatomy of lamina I and its role in post-stroke central pain. *Prog Brain Res* 2000;129:137–151.
86. Nurmikko TJ: Mechanisms of central pain. *Clin J Pain* 2000;16:S21–25.
87. Peyron R, García-Larrea L, Grégoire MC, et al: Parietal and cingulate processings in central pain: A positron emission tomography (PET) study of one original case. *Pain* 1999;84:77–87.
88. Lenz FA, Lee JI, Garonzik IM, et al: Plasticity of pain-related neuronal activity in the human thalamus. *Prog Brain Res* 2000;129:259–273.
89. Christensen MD, Hulsebosch CE: Chronic central pain after spinal cord injury. *J Neurotrauma* 1997;14:517–537.
90. Wiesenfeld-Hallin Z, Aldskogius H, Grant G, et al: Central inhibitory dysfunctions: Mechanisms and clinical implications. *Behav Brain Sci* 1997;20:420–425.
91. Yezierski RP, Liu S, Ruenes GL, et al: Excitotoxic spinal cord injury: Behavioral and morphological characteristics of a central pain model. *Pain* 1998;75:141–155.
92. Eide PK: Pathophysiological mechanisms of central neuropathic pain after spinal cord injury. *Spinal Cord* 1998;36:601–612.
93. Yezierski RP, Park SH: The mechanosensitivity of spinal sensory neurons following intraspinal injections of quisqualic acid in the rat. *Neurosci Lett* 1993;157:115–119.
94. Weng HR, Lee JI, Lenz FA, et al: Functional plasticity in primate somatosensory thalamus following chronic lesion of the ventral lateral spinal cord. *Neuroscience* 2000;101:393–401.

95. Tasker RR: Microelectrode findings in the thalamus in chronic pain and other conditions. *Stereotact Funct Neurosurg* 2001;77:166–168.
96. Hirato M, Watanabe K, Takahashi A, et al: Pathophysiology of central (thalamic) pain: Combined change of sensory thalamus with cerebral cortex around central sulcus. *Stereotact Funct Neurosurg* 1994;62:300–303.
97. Jeanmonod D, Magnin M, Morel A: Thalamus and neurogenic pain: Physiological, anatomical and clinical data. *Neuroreport* 1993;4:475–478.
98. Jeanmonod D, Magnin M, Morel A: Low-threshold calcium spike bursts in the human thalamus: Common physiopathology for sensory, motor and limbic positive symptoms. *Brain* 1996;119:363–375.
99. Rinaldi PC, Young RF, Albe-Fessard D, et al: Spontaneous neuonal hyperactivity in the medial and intralaminar thalamic nuclei of patients with deafferentation pain. *J Neurosurg* 1991;74:415–421.
100. Lenz FA, Kwan HC, Martin R, et al: Characteristics of somatotopic organization and spontaneous neuronal activity in the region of the thalamic principal sensory nucleus in patients with spinal cord transection. *J Neurophysiol* 1994;72:1570–1587.
101. Edgar RE, Best LG, Quail PA, Obert AD: Computer-assisted DREZ microcoagulation: Posttraumatic spinal deafferentation pain. *J Spinal Disord* 1993;6:48–56.
102. Jeanmonod D, Sindou M, Magnin M, Baudet M: Intraoperative unit recordings in the human dorsal horn with a simplified floating microelectrode. *Electroencephalogr Clin Neurophysiol* 1989;72:450–454.
103. Lenz F, Kwan HC, Dostrovsky JO, Tasker RR: Characteristics of the bursting pattern of action potentials that occurs in the thalamus of patients with central pain. *Brain Res* 1989;496:357–360.
104. Loeser JD, Ward AA Jr, White LE Jr: Chronic deafferentation of human spinal cord neurons. *J Neurosurg* 1968;29:48–50.
105. Bromm B, Frieling A, Lankers J: Laser-evoked brain potentials in patients with dissociated loss of pain and temperature sensibility. *Electroencephalogr Clin Neurophysiol* 1991;80:284–291.
106. Casey KL, Beydoun A, Boivie J, et al: Laser-evoked cerebral potentials and sensory function in patients with central pain. *Pain* 1996;64:485–491.
107. Hansen HC, Treede RD, Lorenz J, et al: Recovery from brain-stem lesions involving the nociceptive pathways: Comparison of clinical findings with laser-evoked potentials. *J Clin Neurophysiol* 1996;13:330–338.
108. Garcia-Larrea L, Convers P, Magnin M, et al: Laser-evoked potential abnormalities in central pain patients: The influence of spontaneous and provoked pain. *Brain* 2002;125:2766–2781.
109. Treede RD, Lankers J, Frieling A, et al: Cerebral potentials evoked by painful, laser stimuli in patients with syringomyelia. *Brain* 1991;114:1595–1607.
110. Holmgren H, Leijon G, Boivie J, Johansson I, Ilievska L: Central post-stroke pain—somatosensory evoked potentials in relation to location of the lesion and sensory signs. *Pain* 1990;40:43–52.
111. Mauguiére F, Desmedt JE: Thalamic pain syndrome of Dejerine-Roussy: Differentiation of four subtypes assisted by somatosensory evoked potentials data. *Arch Neurol* 1988;45:1312–1320.
112. Wessel K, Vieregge P, Kessler C, Kompf D: Thalamic stroke: Correlation of clinical symptoms, somatosensory evoked potentials, and CT. *Acta Neurol Scand* 1994; 90:167–173.
113. Peyron R, García-Larrea L, Grégoire MC, et al: Allodynia after lateral-medullary (Wallenberg) infarct: A positron emission tomography (PET) study. *Brain* 1998;121:345–356.
114. Head H, Holmes G: Sensory disturbances from cerebral lesions. *Brain* 1911;34:102–254.
115. Cassinari V, Pagni CA: *Central Pain: A Neurological Survey*. Cambridge, MA: Harvard University Press, 1969.
116. Cesaro P, Mann MW, Moretti JL, et al: Central pain and thalamic hyperactivity: A single photon emission computerized tomographic study. *Pain* 1991;47:329–336.
117. Craig AD, Bushnell MC: The thermal grill illusion: Unmasking the burn of cold pain. *Science* 1994;265:252–255.
118. Craig AD, Chen K, Bandy D, Reiman EM: Thermosensory activation of insular cortex. *Nature Neurosci* 2000; 3:184–190.
119. Craig AD, Reiman EM, Evans A, Bushnell MC: Functional imaging of an illusion of pain. *Nature* 1996;384:258–260.
120. Craig AD: R A new view of pain as a homeostatic emotion. *Trends Neurosci* 2003;26:303–307.
121. Canavero S, Bonicalzi V: The neurochemistry of central pain: Evidence from clinical studies, hypothesis and therapeutic implications. *Pain* 1998;74:109–114.
122. Hains BC, Klein JP, Saab CY, et al: Upregulation of sodium channel Nav1.3 and functional involvement in neuronal hyperexcitability associated with central neuropathic pain after spinal cord injury. *J Neurosci* 2003;23:8881–8892.
123. Max MB, Hagen NA: Do changes in brain sodium channels cause central pain? *Neurology* 2000;54:544–545.
124. Radhakrishnan V, Tsoukatos J, Davis KD, et al: A comparison of the burst activity of lateral thalamic neurons in chronic pain and non-pain patients. *Pain* 1999;80:567–575.
125. Canavero S, Pagni CA, Castellano G, et al: The role of cortex in central pain syndromes: Preliminary results of a long-term technetium-99 hexamethylpropyleneamineoxime single photon emission computed tomography study. *Neurosurgery* 1993;32:185–191.
126. Hirato M, Kawashima Y, Shibazaki T, et al: Pathophysiology of central (thalamic) pain: A possible role of the intralaminar nuclei in superficial pain. *Acta Neurochir Suppl* 1991; 52:133–136.
127. Laterre EC, De Volder AG, Goffinet AM: Brain glucose metabolism in thalamic syndrome. *J Neurol Neurosurg Psychiatry* 1988;51:427–428.
128. Attal N: Chronic neuropathic pain: Mechanisms and treatment. *Clin J Pain* 2000;16:S118–S131.
129. Attal N, Guirimand F, Brasseur L, et al: Effects of IV morphine in central pain: A randomized, placebo-controlled study. *Neurology* 2002;58:554–563.
130. Finnerup NB, Sindrup SH, Bach FW, et al: Lamotrigine in spinal cord injury pain: A randomized, controlled trial. *Pain* 2002;96:375–383.

131. Nepomuceno C, Fine PR, Richards JS, et al: Pain in patients with spinal cord injury. *Arch Phys Med Rehabil* 1979;60:605–609.
132. Westgren N, Levi R: Quality of life and traumatic spinal cord injury. *Arch Phys Med Rehabil* 1998;79:1433–1439.
133. Turner JA, Jensen MP, Warms CA, Cardenas DD: Catastrophizing is associated with pain intensity, psychological distress, and pain-related disability among individuals with chronic pain after spinal cord injury. *Pain* 2002;98:127–134.
134. Stanton-Hicks M: Transcutaneous and peripheral nerve stimulation, in Simpson B (ed), *Electrical Stimulation and the Relief of Pain: Pain Research and Clinical Management*. New York: Elsevier, 2003;15:38–55.
135. Warms CA, Turner JA, Marshall HM, Cardenas DD: Treatments for chronic pain associated with spinal cord injuries: Many are tried, few are helpful. *Clin J Pain* 2002;18:154–163.
136. Max MB: Antidepressants as analgesics, in Fields HL and Liebeskind U (eds), *Progress in Pain Research and Therapy*, Vol 1. Seattle: IASP Press, 1994:229–246.
137. Brau ME, Dreimann M, Olschewski A, et al: Effects of drugs used for neuropathic pain management on tetrodotoxin-resistant Na+ currrents in rat sensory neurons. *Anesthesiology* 2001;94:137–144.
138. Leijon G, Boivie J: Central post-stroke pain: a controlled trial of amitriptyline and carbamazepine. *Pain* 1989; 36:27–36.
139. Cardenas DD, Warms CA, Turner JA, et al: Efficacy of amitriptyline for relief of pain in spinal cord injury: Results of a randomized controlled trial. *Pain* 2002;96:365–373.
140. Watson CPN: Antidepressant drugs as adjuvant analgesics *J Pain Symptom Manage* 1994;9:392–405.
141. McQuay HJ, Tramer M, Nye BA, et al: A systematic review of antidepressants in neuropathic pain. *Pain* 1996; 68:217–227
142. Davidoff G, Guarrancini M, Roth E, et al: Trazodone hydrochloride in the treatment of dysesthetic pain in traumatic myelopathy: A randomized, double-blind, placebo-controlled study. *Pain* 1987;29:151–161.
143. Vestegaard K, Andersen G, Jensen TS: Treatment of central post-stroke pain with a selective serotonin reuptake inhibitor. *Eur J Neurol* 1996;3(suppl 5):169.
144. Dickenson AH, Matthews EA, Suzuki R: Neurobiology of neuropathic pain: Mode of action of anticonvulsants. *Eur J Pain* 2002;6(suppl A):51–60.
145. Tanelian DL, Brose WG: Neuropathic pain can be relieved by drugs that are use-dependent sodium channel blockers: Lidocaine, carbamazepine, and mexiletine. *Anesthesiology* 1991;74:949–951.
146. Sindrup SH, Jensen TS: Efficacy of pharmacological treatment of neuropathic pain: An update and effect related to mechanism of drug action. *Pain*;1999;81:389–400.
147. Espir MLE, Millac P: Treatment of paroxysmal disorders in multiple sclerosis with carbamazepine (Tegretol). *J Neurol NeurosurgPsychiatry* 1970;33:528–531.
148. Swerdlow M, Cundill JG: Anticonvulsant drugs used in the treatment of lancinating pain: A comparison. *Anaesthesia* 1981;36:1129–1132.
149. McQuay H, Carroll D, Jadad AR, et al: Anticonvulsant drugs for management of pain: A systematic review. *Br Med J* 1995;311:1047–1052.
150. Carrazana E, Mikoshiba I: Rationale and evidence for the use of oxcarbazepine in neuropathic pain. *J Pain Symptom Manage* 2003;25:S31–S35.
151. Vestergaard K, Andersen G, Gottrup H, et al: Lamotrigine for central poststroke pain: A randomized controlled trial. *Neurology* 2001;56:184–190.
152. Ahn SH, Park HW, Lee BS, et al: Gabapentin effect on neuropathic pain compared among patients with spinal cord injury and different durations of symptoms. *Spine* 2003;28:341–346.
153. Attal N, Brasseur B, Parker F, et al: Effects of the anticonvulsant gabapentin on peripheral and central neuropathic pain: A pilot study. *Eur Neurol* 1998;40:191–200.
154. Putzke JD, Richards JS, Kezar L, et al: Long-term use of gabapentin for treatment of pain after traumatic spinal cord injury. *Clin J Pain* 2002;18:116–121.
155. To TP, Lim TC, Hill ST, et al: Gabapentin for neuropathic pain following spinal cord injury. *Spinal Cord* 2002;40:282–285.
156. Tai Q, Kirshblum S, Chen B, et al: Gabapentin in the treatment of neuropathic pain after spinal cord injury: A prospective, randomized, double-blind, crossover trial. *J Spinal Cord Med* 2002;25:100–105.
157. Serpell MG: Neuropathic pain study group. Gabapentin in neuropathic pain syndromes: A randomized, double-blind, placebo-controlled trial. *Pain* 2002;99:557–566.
158. Backonja MM, Glanzman RL: Gabapentin dosing for neuropathic pain: Evidence from randomized, placebo-controlled clinical trials. *Clin Ther* 2003;25:81–104.
159. Maneuf YP, Gonzalez MI, Sutton KS, et al: Cellular and molecular action of the putative GABA-mimetic, gabapentin. *Cell Mol Life Sci* 2003;60:742–750.
160. Drewes AM, Adreasen A, Poulsen LH: Valproate for treatment of chronic central pain after spinal cord injury: A double-blind cross-over study. *Paraplegia* 1994;32:565–569.
161. Canavero S, Bonicalzi V, Paolotti R: Lack of effect of topiramate for central pain. *Neurology* 2002b;58:831–832.
162. Dellemijn P: Are opioids effective in relieving neuropathic pain? *Pain* 1999;80:453–462.
163. Foley KM: Opioids and chronic neuropathic pain. *N Engl J Med* 2003;348:1279–1281.
164. Jensen TS, Sindrup SH: Opioids: A way to control central pain? *Neurology* 2002;58:517–518.
165. Benedetti F, Vighetti S, Amanzio M, et al: Dose-response relationship of opioids in nociceptive and neuropathic postoperative pain. *Pain* 1998;74:205–211.
166. Eide PK, Stubhaug A, Stenehjem AE: Central dysesthesia pain after traumatic spinal cord injury is dependent on N-methyl-d-aspartate receptor activation. *Neurosurgery* 1995;37:1080–1087
167. Rowbotham MC, Twilling L, Davies PS, et al: Oral opioid therapy for chronic peripheral and central neuropathic pain. *N Engl J Med* 2003;348:1223–1232.
168. Kalman S, Osterberg A, Sorensen J, et al: Morphine responsiveness in a group of well-defined multiple sclerosis patients: A study with IV morphine. *Eur J Pain* 2002; 6:69–80.

169. Mao J, Chen LL: Systemic lidocaine for neuropathic pain. *Pain* 2000;87:7–17.
170. Attal N, Gaude V, Dupuy M, et al: Intravenous lidocaine in central pain: A double-blind placebo-controlled psychophysical study. *Neurology* 2000;544:564–574.
171. Sakurai M, Kanazawa I: Positive symptoms in multiple sclerosis: Their treatment with sodium channel blockers, lidocaine and mexiletine.channel blockers, lidocaine and mexiletine. *Neurol Sci* 1999;162:162–168.
172. Chiou-Tan FY, Tuel SM, Johnson JC, et al: Effect of mexiletine on spinal cord injury dysesthetic pain. *Am J Phys Med Rehabil* 1996;75:84–87.
173. Fisher K, Coderre TJ, Hagen NA: Targeting the N-methyl-d-aspartate receptor for chronic pain management: Preclinical animal studies, recent clinical experience and future research directions. *J Pain Symptom Manage* 2000;20:358–373.
174. Fisher K, Hagen NA: Analgesic effect of oral ketamine in chronic neuropathic pain of spinal origin: a case report. *J Pain Symptom Manage* 1999;18:61–66.
175. Vick PG, Lamer TJ: Treatment of central post-stroke pain with oral ketamine. *Pain* 2001;92:311–313.
176. Haines DR, Gaines SP: N of 1 randomised controlled trials of oral ketamine in patients with chronic pain. *Pain* 1999;83:283–287
177. McQuay HJ, Carroll D, Jadad AR, et al: Dextromethorphan for the treatment of neuropathic pain: A double-blind randomised crossover trial with integral n-of-1 design. *Pain* 1994;59:127–133.
178. Croxford JL: Therapeutic potential of cannabinoids in CNS disease. *CNS Drugs* 2003;17:179–202.
179. Iversen L: Cannabis and the brain. *Brain* 2003;126:1252–1270.
180. Bridges D, Ahmad K, Rice AS: The synthetic cannabinoid WIN55,212-2 attenuates hyperalgesia and allodynia in a rat model of neuropathic pain. *Br J Pharmacol* 2001;133:586–594.
181. Fox A, Kesingland A, Gentry C, et al: The role of central and peripheral cannabinoid$_1$ receptors in the antihyperalgesic activity of cannabinoids in a model of neuropathic pain. *Pain* 2001;92:91–100.
182. Svendsen KB, Jensen TS, Bach FW: Dronabinol (delta-9-tetrahydrocannabinol) alleviates pain in multiple sclerosis. Fourth Congress of EFIC, Prague, September 2–6, 2003, abstract 523.
183. Wade DT, Robson P, House H, et al: A preliminary controlled study to determine whether whole-plant cannabis extracts can improve intractable neurogenic symptoms. *Clin Rehabil* 2003;17:21–29.
184. Attal N, Brasseur L, Guirimand D, Clemont-Gnamien S, Atianie S, Bauhassira D: Are oral cannabinoids safe and effective in refractory neuropathic pain? *Eur J Pain* 2004;8:173–177.
185. Sindrup SH, Andersen G. Madsen C, Smith T, Brosen K, Jensen TJ: Tramadol relieves pain and allodynia in polyneuropathy: a randomised, double blind, controlled trial. *Pain* 1999;83:85–90.
186. Sindrup SH, Bach FW, Madsen C, et al: Venlafaxine versus imipramine in painful polyneuropathy: A randomized, controlled trial. *Neurology* 2003;60:1284–1289.
187. McCleane GJ: Intravenous infusion of phenytoin relieves neuropathic pain: A randomized, double-blinded, placebo-controlled, crossover study. *Anesth Analg* 1999;89:985–998.
188. Canavero S, Bonicalzi V, Pagni CA, et al: Propofol analgesia in central pain: Preliminary clinical observations. *J Neurol* 1995;242:561–567.
189. Bennett G, Serafini M, Burchiel K, et al: Evidence-based review of the literature on intrathecal delivery of pain medication. *J Pain Symptom Manage* 2000;20:S12–36.
190. Siddall PJ, Molloy AR, Walker S, et al: The efficacy of intrathecal morphine and clonidine in the treatment of pain after spinal cord injury. *Anesth Analg* 2000;91:1493–1498.
191. Herman RM, D'Luzansky SC, Ippolito R: Intrathecal baclofen suppresses central pain in patients with spinal lesions: A pilot study. *Clin J Pain* 1992;8:338–345.
192. Taira T, Kawamura H, Tanikawa T, et al: A new approach to the control of central deafferentation pain-spinal intrathecal baclofen. *Acta Neurochir Suppl (Wien)* 1995;64:136–138.
193. Loubser PG, Akman NM: Effects of intrathecal baclofen on chronic spinal cord injury pain. *J Pain Symptom Manage* 1996;12:241–247.
194. Jain KK: An evaluation of intrathecal ziconotide for the treatment of chronic pain. *Exp Opin Invest Drugs* 2000;9:2403–2310.
195. Loubser PG, Donovan WH: Diagnostic spinal anaesthesia in chronic spinal cord injury pain. *Paraplegia* 1991;29:25–36.
196. Canavero S, Bonicalzi V: Neuromodulation for central pain. *Exp Rev Neurother* 2003;3:591–607.
197. Bossard AE, Guirimand F, Fletcher D, et al: Interaction of a combination of morphine and ketamine on the nociceptive flexion reflex in human volunteers. *Pain* 2002;98:47–57.
198. Eckhardt K, Ammon S, Hofmann U, et al: Gabapentin enhances the analgesic effect of morphine in healthy volunteers. *Anesth Analg* 2000;91:185–191.
199. Meyerson BA: Neurosurgical approaches to pain treatment. *Acta Anaesthesiol Scand* 2001;45:1108–1113.
200. Tronnier VM: Deep brain stimulation, in Simpson B (ed), *Electrical Stimulation and the Relief of Pain: Pain Research and Clinical Management*. New York: Elsevier, 2003;15:211–236.
201. Nandi D, Aziz T, Carter H, Stein J: Thalamic field potentials in chronic central pain treated by periventricular gray stimulation: A series of eight cases. *Pain* 2003;101:97–107.
202. Gybels JM, Sweet WH: Neurosurgical treatment of persistent pain, in Gildenberg PL (ed), *Pain and Headache*. Basel: Karger, 1989:1–442.
203. Richardson RR, Meyer PR, Cerullo LJ: Neurostimulation in the modulation of intractable paraplegic and traumatic neuroma pains. *Pain* 1980;8:75–84.
204. Siegfried J: Therapeutic neurostimulation: Indications reconsidered. *Acta Neurochir Suppl (Wien)* 1991;52:112–117.
205. Sindou M, Mertens P: Neurosurgical management of neuropathic pain. *Stereotact Funct Neurosurg* 2000;75:76–80.
206. Lazorthes Y, Siegfried J, Verdie JC, Casaux J: Chronic spinal cord stimulation in the treatment of neurogenic pain: Cooperative and retrospective study on 20 years of follow-up. *Neurochirurgie* 1995;41:73–86.
207. Sindou MP, Mertens P, Bendavid U, et al: Predictive value of somatosensory evoked potentials for long-lasting pain

207. relief after spinal cord stimulation: Practical use for patient selection. *Neurosurgery* 2003;52:1374–1383;
208. Brown JA, Barbaro NM: Motor cortex stimulation for central and neuropathic pain: Current status. *Pain* 2003; 104:431–435.
209. Canavero S, Bonicalzi V: Therapeutic extradural cortical stimulation for central and neuropathic pain: A review. *Clin J Pain* 2002;18:48–55.
210. N'Guyen JP, Lefaucheur JP, Keravel Y: Motor cortex stimulation, in Simpson B (ed), *Electrical Stimulation and the Relief of Pain: Pain Research and Clinical Management*, Elsevier, 2003;15:197–209.
211. Roux FE, Ibarrola D, Tremoulet M, et al: Methodological and technical issues for integrating functional magnetic resonance imaging data in a neuronavigation system. *Neurosurgery* 2001;49:1149–1157.
212. Drouot X, Nguyen JP, Peschanski M, Lefaucheur JP: The antalgic efficacy of chronic motor cortex stimulation is related to sensory changes in the painful zone. *Brain* 2002;125:1660–1664.
213. Garcia-Larrea L, Peyron R, Mertens P, et al: Electrical stimulation of motor cortex for pain control: A combined PET-scan and electrophysiological study. *Pain* 1999;83:259–273.
214. Peyron R, Garcia-Larrea L, Deiber MP, et al: Electrical stimulation of precentral cortical area in the treatment of central pain: Electrophysiological and PET study. *Pain* 1995;62:275–286.
215. Tsubokawa T, Katayama Y, Yamamoto T, et al: Chronic motor cortex stimulation for the treatment of central pain. *Acta Neurochir* 1991;52(suppl):137–139.
216. Tsubokawa T, Katayama Y, Yamamoto T, et al: Chronic motor cortex stimulation in patients with thalamic pain. *J Neurosurg* 1993;78:393–401.
217. Saitoh Y, Kato A, Ninomiya H, et al: Primary motor cortex stimulation within the central sulcus for treating deafferentation pain. *Acta Neurochir Suppl* 2003;87:149–152.
218. N'Guyen JP, Lefaucheur JP, Decq P, et al: Chronic motor cortex stimulation in the treatment of central and neuropathic pain: Correlations between clinical, electrophysiological and anatomical data. *Pain* 1999;82:245–251.
219. Meyerson BA, Lindblom U, Linderoth B, et al: Motor cortex stimulation as treatment of trigeminal neuropathic pain. *Acta Neurochir (Wien)* 1993;58:150–153.
220. Herregodts P, Stadnik T, De Ridder F, D'Haens J: Cortical stimulation for central neuropathic pain: 3D surface MRI for easy determination of the motor cortex. *Acta Neurochir Suppl (Wien)* 1995;64:132–135.
221. Canavero S, Bonicalzi V, Dotta M, et al: Transcranial magnetic cortical stimulation relieves central pain. *Stereotact Funct Neurosurg* 2002;78:192–196.
222. Lefaucheur JP, Drouot X, Keravel Y, Nguyen JP: Pain relief induced by repetitive transcranial magnetic stimulation of precentral cortex. *Neuroreport* 2001;12:2963–2965.
223. Migita K, Uozumi T, Arita K, Monden S: Transcranial magnetic coil stimulation of motor cortex in patients with central pain. *Neurosurgery* 1995;36:1037–1039.
224. Katayama Y, Fukaya C, Yamamoto T: Poststroke pain control by chronic motor cortex stimulation: Neurological characteristics predicting a favorable response. *J Neurosurg* 1998;89:585–591.
225. Yamamoto T, Katayama Y, Hirayama T, Tsubokawa T: Pharmacological classification of central post-stroke pain: Comparison with the results of chronic motor cortex stimulation therapy. *Pain* 1997;72:5–12.
226. Denkers MR, Biagi HL, Ann O'Brien M, et al: Dorsal root entry zone lesioning used to treat central neuropathic pain in patients with traumatic spinal cord injury: A systematic review. *Spine* 2002;27:177–184.
227. Friedman AH, Nashold BS Jr: DREZ lesions for relief of pain related to spinal cord injury. *J Neurosurg* 1986;65:465–469.
228. Sampson JH, Cashman RE, Nashold BS Jr, Friedman AH: Dorsal root entry zone lesions for intractable pain after trauma to the conus medullaris and cauda equina. *J Neurosurg* 1995;82:28–34.
229. Sindou M, Mertens P, Wael M: Microsurgical DREZotomy for pain due to spinal cord and/or cauda equina injuries: Long-term results in a series of 44 patients. *Pain* 2001;92:159–171.
230. Spaic M, Markovic N, Tadic R: Microsurgical DREZotomy for pain of spinal cord and cauda equina injury origin: Clinical characteristics of pain and implications for surgery in a series of 26 patients. *Acta Neurochir (Wien)* 2002;144:453–462.
231. Woolf CJ, Decosterd I: Implications of recent advances in the understanding of pain pathophysiology for the assessment of pain in patients. *Pain* 1999;6(suppl):S141–147.
232. Woolf CJ, Mannion RJ: Neuropathic pain: Aetiology, symptoms, mechanisms and management. *Lancet* 1999; 35:1959–1964.
233. Woolf CJ, Max MB: Mechanism-based pain diagnosis: Issues for analgesic drug development. *Anesthesiology* 2001;95:241–249.
234. Lenz FA, Garonzik IM, Zirh TA, Dougherty PM: Neuronal activity in the region of the thalamic principal sensory nucleus (ventralis caudalis) in patients with pain following amputations. *Neuroscience* 1998;86:1065–1081.
235. Lenz FA, Gracely RH, Baker FH, et al: Reorganization of sensory modalities evoked by microstimulation in region of the thalamic principal sensory nucleus in patients with pain due to nervous system injury. *J Comp Neurol* 1998;399:125–138.
236. Bridges D, Thompson SW, Rice AS: Mechanisms of neuropathic pain. *Br J Anaesth* 2001;87:12–26.
237. Sindrup SH, Jensen TS: Pharmacologic treatment of pain in polyneuropathy. *Neurology* 2000;55:915–920.
238. Attal N, Rouaud J, Brasseur L, et al: Systemic lidocaine in pain due to peripheral nerve injury and predictors of response. *Neurology* 2004;62:218–255.
239. Eide PK, Jorum E, Stubhaub A, et al: Relief of post-herpetic neuralgia with the *N*-methyl-d-aspartic acid receptor antagonist ketamine: A double-blind, cross-over comparison with morphine and placebo. *Pain* 1994;58:347–354.
240. Eide PK, Strubhaug A, Oye I, Breivik H: Continuous subcutaneous administration of the *N*-methyl-d-aspartic acid (NMDA) receptor antagonist ketamine in the treatment of post-herpetic neuralgia. *Pain* 1995a;61:221–228.
241. Felsby S, Nielsen J, Arendt Nielsen L, Jensen TS: NMDA receptor blockade in chronic neuropathic pain: A comparison of ketamine and magnesium chloride. *Pain* 1996;64:283–291.
242. Bainton T, Fox M, Bowsher D, Wells C: A double-blind trial of naloxone in central post-stroke pain. *Pain* 1992;48:159–162.

CHAPTER 23

Peripheral Neuropathic Pain

Marco Pappagallo

Pain may arise from a pathologic process or dysfunction occurring along and within the nervous system pain pathways. In this instance, pain can be diagnosed or classified as neuropathic.[1,2] *It is thought to originate ectopically and spontaneously from abnormal nociceptors.*

Pain also may arise as a response to an inflammatory process occurring within tissues. In this instance, pain has been diagnosed or classified as nociceptive. *It is thought to originate from "healthy" tissue nociceptors sensitized by local algogenic molecules (e.g., protons, prostaglandins, bradykinin). Nociceptive pain has been related to tissue inflammation. However, in the context of some inflammatory or posttraumatic conditions (e.g., cancer, rheumatologic disorders, complex regional pain syndrome), a number of diffusible factors can be involved in the development of neuropathic pain. Specific proinflammatory cytokines (e.g., tumor necrosis factor α, interleukin 1β) in combination with tissue-released growth factors (e.g., nerve growth factor) are known not only to sensitize uninjured tissue nociceptors but also possibly mediate upregulation or expression of channels and membrane changes sufficient to generate ectopic and spontaneous activity in otherwise "healthy" nociceptors. In this instance, neuropathic pain would be classified more appropriately as* inflammatory neuropathic pain.[3-6]

There is a complex interplay between pain mechanisms of peripheral nervous system (PNS) and central nervous system (CNS) origin. This interplay is exemplified in Fig. 23-1. The chapter will focus on the painful pathologies or disorders affecting the PNS.

▶ CLINICAL ASSESSMENT OF THE PATIENT WITH PERIPHERAL NEUROPATHIC PAIN

Medical History

The clinical interview should focus on questions about onset, duration, progression, and nature of complaints suggestive of *neurological deficits* (e.g., persistent numbness in a body area or limb weakness, such as tripping episodes, the progressive inability to open jars, etc.), as well as complaints suggestive of *sensory dysfunction* (e.g., touch-evoked pain, intermittent abnormal sensations, spontaneous burning and shooting pains).[7]

A number of painful neuropathies are related to systemic diseases. Complaints such as urinary frequency, weight loss, fatigue, edema, somnolence, skin discoloration (e.g., icterus), fever, persistent cough, dry eyes, joint swelling, skin rash, tremors, gait unsteadiness, nail changes, and hair loss should suggest underlying systemic medical conditions. These conditions can include diabetes mellitus, hypothyroidism, chronic renal failure, malabsorption, malignancy, connective tissue diseases, chronic infections, intravenous use of illicit drugs, alcoholism, abnormal dietary habits, or chronic intoxication (Table 23-1).[2]

This chapter has been updated, revised, and adapted from Pappagallo M: Neuropathic pain in peripheral neuropathies, in Tollison CD, Satterthwaite JR, Tollison JW (eds), *Practical Pain Management*. Philadelphia: Lippincott, Williams & Wilkins, 2002: 431–448.

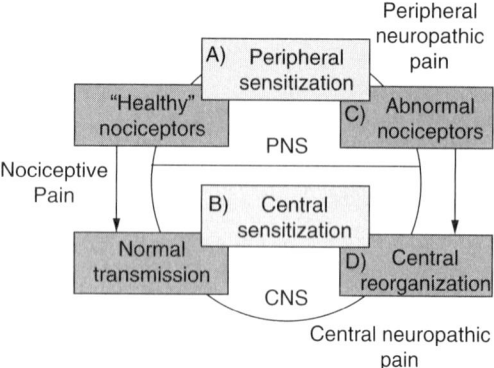

Figure 23-1. The interplay of physiologic and pathologic pain processes and how they may interact during (*A*) peripheral sensitization of nociceptors (may occur in both nociceptive and neuropathic states), (*B*) central sensitization of dorsal horn pain-signaling neurons (may occur in both nociceptive and neuropathic states), (*C*) dysfunction or pathologic changes of the peripheral nociceptive pathways, and (*D*) dysfunction or pathological changes of the CNS pain pathways. (*Revised and adapted with permission*).[2]

Physical Examination

Examination of the symptomatic region is essential. For example, complex regional pain syndrome (CRPS) usually involves a single distal extremity, and at the time of the evaluation, the affected limb may show the classic CRPS-associated features, such as skin color and temperature changes, edema, and sudomotor abnormalities. Light tapping and palpation along the course of peripheral nerves are important steps of the examination. A positive *Tinel's sign* is evidence of mechanosensitivity of the peripheral nerve and suggests the presence of a neuroma or nerve entrapment. Hypertrophy of a nerve trunk suggests either a nerve tumor (e.g., neurofibroma) or a localized hypertrophic neuropathy. Multifocal nerve hypertrophies are found in some peripheral nerve disorders, including leprosy, neurofibromatosis, or Charcot-Marie-Tooth (CMT) disease.

In addition, some diagnostic clues may derive from the general examination.[7] Assessment of the hair can show alopecia, which is seen in thallium poisoning. Wide transverse nail bands called *Mees' lines* are observed in arsenic and thallium intoxication. Telangiectasias over the abdomen and inguinal and gluteal regions are found in Fabry's disease. Purpuric skin eruptions are observed in cryoglobulinemia and in some vasculitidies. Inspection of the extremities can show pes cavus and hammer toes, which are suggestive of CMT disease.

Abnormal Sensory Findings Associated with Neuropathic Pain

Patients with neuropathic pain states usually present with a spontaneous ongoing or intermittent pain typically characterized by a burning and/or shooting, electric-shock quality.[8] They also may present with some or all of the following abnormal sensory symptoms and signs:

▶ **TABLE 23-1.** ETIOLOGIES OF PAINFUL PERIPHERAL NEUROPATHIES

METABOLIC AND ENDOCRINOLOGIC DISORDERS
- Diabetic neuropathies
- Hepatic disease
- Renal disease and hemodialysis
- Hypothyroidism

INFECTIONS
- Human immunodeficiency virus
- Varicella zoster virus
- Hepatitis B and C virus
- Human T-cell lymphotrophic virus
- Leprosy
- Lyme disease

DEMYELINATING INFLAMMATORY DISORDERS
- Guillain-Barré syndrome

MALIGNANCIES

ENTRAPMENT NEUROPATHIES

CONNECTIVE TISSUE DISEASES, GRANULOMA-RELATED DISORDERS, AND VASCULITIDIES
- Sjögren syndrome
- Systemic lupus erythematosus
- Rheumatoid arthritis
- Sarcoidosis
- Polyarteritis nodosa
- Churg-Strauss vasculitis
- Wegener's granulomatosis
- Giant-cell arteritis or temporal arteritis

IMMUNOGLOBULINEMIAS
- Monoclonal (M) proteins
- Primary and secondary amyloidosis
- Cryoglobulinemia

DIETARY OR ABSORPTION ABNORMALITIES
- Alcoholic neuropathy
- Celiac disease
- B_{12}, thiamine and other vitamin deficiencies
- Strachan's syndrome

TOXIC NEUROPATHIES
- Heavy metals
- Chemotherapeutic agents

HEREDITARY NEUROPATHIES
- Charcot-Marie-Tooth disease
- Fabry's disease
- Familial amyloid polyneuropathy
- Porphyric neuropathy

CRYPTOGENIC PAINFUL NEUROPATHIES
- Idiopathic polyneuropathy
- Complex regional pain syndromes
- Essential trigeminal and glossopharyngeal neuralgias
- Cryptogenic brachial plexus neuropathy

SOURCE: *Revised and adapted.*[2]

- *Paresthesias*—spontaneous, intermittent, painless, but often annoying, sensations.
- *Dysesthesias*—spontaneous or evoked unpleasant distressing sensations, triggered, for example, by cold stimuli or pinprick testing.
- *Allodynia*—pain elicited by non-noxious stimuli (i.e., clothing, air movement, tactile stimuli) when applied to the symptomatic cutaneous area. Allodynia may have mechanical static (e.g., induced by a light pressure) or dynamic (e.g., induced by moving a soft brush) features, as well as thermal (e.g., induced by a nonpainful cold or warm stimulus) characteristics.
- *Hyperalgesia*—an exaggerated pain response to a mildly noxious (mechanical or thermal) stimulus applied to the symptomatic area.
- *Hyperpathia*—a delayed and explosive pain response to a noxious stimulus applied to the symptomatic area.

Mechanical allodynia can be assessed by stroking the painful skin with a cotton swab or a soft brush, and *mechanical hyperalgesia* can be assessed by using a pin or slightly pinching the skin. Cold allodynia can be assessed by placing a cold stimulus, such as a cold tuning fork, on the painful region for a few seconds.

Mechanical and cold allodynia can be present in the cutaneous area contiguous to a nociceptive focus (e.g., a burn injury). These sensory symptoms define the area of *secondary hyperalgesia* and indicate the occurrence of CNS (dorsal horn) sensitization. In contrast, the absence of cold allodynia along with the presence of heat and mechanical allodynia/hyperalgesia characterizes the area of *primary hyperalgesia*, i.e., the zone corresponding to the burn injury.[9]

Cold and mechanical allodynia/hyperalgesia are observed in neuropathic pain states. In these instances, the sensory findings are chronic and not necessarily linked to any local ongoing inflammatory process. For example, patients affected by sympathetically-maintained pain (SMP; see below) typically present with cold and mechanical allodynia/hyperalgesia.[10]

Diagnostic Workup

A methodical diagnostic workup of patients with neuropathic pain can effectively clarify the etiology. Laboratory studies, analysis of tissue samples from skin and nerve, anatomic imaging, and functional neurophysiologic studies all have a role to play in the evaluation of patients with chronic neuropathic pain.

Laboratory Assessment

A comprehensive diagnostic laboratory workup[2] (Table 23-2) includes all the following: complete cell blood count (CBC), sedimentation rate, general chemistry profile, thyroid function tests, vitamin B_{12} and folate serum levels, fasting blood sugar and glycosylated hemoglobin, serum protein electrophoresis with immunofixation, Lyme disease antibody titers, hepatitis virus B and C screening titers, antinuclear antibodies (ANA), rheumatoid factor (RF), Sjögren titers (SS-A, SS-B), antigliadin and antitransglutaminase titers cryoglobulins, antisulfatide IgM antibody and anti-HU titers, heavy metal serum and urine screening, HIV testing, and cerebrospinal fluid (CSF) study for ruling out demyelinating neuropathies (Guillain-Barré syndrome) and polyradiculopathies related to meningeal carcinomatosis.[7,11]

It is important to screen patients with a chronic neuropathy of undetermined etiology, particularly individuals older than age 60, for monoclonal proteins because a portion of these patients may be affected by monoclonal gammopathies. Several autoantibodies with reactivity to various components of the peripheral nerve have been associated with peripheral neuropathies. However, despite all diagnostic efforts, chronic neuropathies of idiopathic origin affect up to 30 percent of the patients with peripheral neuropathies.

▶ **TABLE 23-2.** COMPREHENSIVE LABORATORY WORK-UP

- Cell blood count
- Sedimentation rate
- C reactive protein
- Comprehensive metabolic panel
- Thyroid function tests
- Vitamin B_{12} and folate serum levels
- Fasting blood sugar and glycosylated hemoglobin
- Sero-protein electrophoresis with immunofixation
- Lyme disease antibody titer
- Hepatitis virus B and C screening titers
- Antinuclear antibodies
- Rheumatoid factor
- Sjögren titers (SS-A, SS-B)
- Anti-gliadin, Anti-transglutaminase titers
- Cryoglobulins
- Angiotensin-converting enzyme
- Anti-sulfatide IgM antibody titer
- Anti-HU titers
- Heavy metal serum and urine screening
- HIV testing
- Cerebrospinal fluid study (for demyelinating neuropathies and meningeal carcinomatosis)

SOURCE: *Revised and adapted.*[3]

Nerve and Skin Biopsies

The sural nerve is usually selected for biopsy because the sensory deficit following the procedure is limited to a small area along the dorsolateral aspect of the ankle and foot. The biopsy is useful for the diagnosis of vasculitis, amyloidosis, sarcoidosis, IgM monoclonal gammopathies, CMT disease, chronic inflammatory demyelinating polyneuropathies, and small-fiber neuropathies.

Skin biopsy[4,13] is a technique to evaluate the density of unmyelinated fibers within the dermis and epidermis. Immunostaining with the panaxonal marker PGP 9.5 has been used to demonstrate the intraepidermal network of C fibers. Patients with diabetic sensory neuropathies, idiopathic neuropathies, and HIV-associated neuropathies have been found to have a significantly diminished intraepidermal density of small fibers.

Imaging Studies

Magnetic resonance imaging (MRI) studies are an essential tool in the diagnosis of structural pathologies causing neuropathic pain states. MRI can help with the anatomic evaluation of CNS and PNS structures, including spinal cord, nerve roots, plexuses, and major peripheral nerves.

Electrodiagnostic Studies

Electromyography (EMG) and nerve conduction velocity (NCV) studies help to localize the lesion and can indicate an axonal versus a focal segmental demyelinating process. For example, the EMG-NCV study may identify a radiculopathy or mononeuropathy, i.e., a focal process affecting a specific root or nerve. A traumatic injury, compression or nerve entrapment, ischemia, and cancers are common causes of mononeuropathy. The study also may reveal a *mononeuropathy multiplex,* a pathology of multiple but noncontiguous nerves. Mononeuropathy multiplex is often the result of vasculitis and microangiopathy causing axonal disease in multiple noncontiguous nerves. EMG-NCV studies may help to detect diffuse abnormalities, indicating a polyneuropathy. Of note, diabetes mellitus, hypothyroidism, and hereditary neuropathy with liability to pressure palsy may cause neuropathies that predispose affected patients to develop superimposed entrapment neuropathies.

▶ ETIOLOGIES

As shown in Table 23-1, neuropathic pain may have numerous etiologies, including metabolic or endocrinologic disorders, infections, cancer, autoimmune disorders, immunoglobulinemias, dietary or absorption abnormalities, entrapment due to anatomic abnormalities, or exposure to toxins or drugs. In addition, neuropathic pain may arise as a result of hereditary abnormalities.[7,11]

Peripheral Neuropathic Pain Associated with Metabolic and Endocrinologic Disorders

Diabetes Mellitus

The neuropathies associated with diabetes mellitus, which include symmetric or asymmetric neuropathies, mononeuropathies due to infarction or entrapment, truncal neuropathy, and femoral amyotrophy, are the most prevalent peripheral neuropathic disorders. The diagnosis of diabetic neuropathy is based on evidence of hyperglycemia, clinical symptoms, and objective neurological signs. Fasting plasma glucose levels greater than 125 mg/dL on more than one occasion, any plasma glucose levels higher than 200 mg/dL, or a sustained plasma glucose level of more than 200 mg/dL at 2 hours after ingesting 75 g of glucose confirms the diagnosis of diabetes mellitus. Glycosylated hemoglobin (hemoglobin A1c) is a useful indicator of long-term control of blood sugar levels. EMG-NCV studies can confirm the examination findings.[14]

Many mechanisms have been proposed for the pathogenesis of diabetic neuropathy (e.g., toxic-metabolic factors, microvascular abnormalities, specific neurotropic factors, and neuroinflammatory/immunological events). Most individuals with diabetic polyneuropathy have pathologic involvement of both large- and small-diameter sensory fibers, as well as autonomic fibers. However, a distal neuropathy primarily affecting small-diameter axons, i.e., Aδ and C fibers, can occur, and patients affected by this condition often present with the complaint of "burning feet."

Asymmetric neuropathies include single or multiple cranial mononeuropathies, single or multiple limb or truncal mononeuropathies, and lumbosacral plexopathies. The most common diabetic cranial neuropathy is the painful, pupil-sparing third nerve palsy. Involvement of single nerves in diabetes mellitus may occur from either ischemic events or entrapment. *Truncal neuropathy* usually affects intercostal nerves and can cause pain and dysesthesias in the chest and abdomen. The term *diabetic femoral amyotrophy* implies a severe and acute painful state affecting a proximal lower extremity. Patients complain of pain in the anterior thigh. Proximal leg weakness associated with reduction in or loss of the patellar reflex develops rapidly within a few weeks. Weight loss occurs in more than 50 percent of patients. Pain usually subsides spontaneously over months, followed by improvement in strength. However, a comprehensive evaluation to rule out any concurrent and perhaps treatable pathology affecting the lumbar plexus or femoral nerve must be performed.

Hepatic Disease

Neuropathic syndromes have been associated with acute and chronic liver diseases.[7,11] Hepatitis B has been linked with demyelinating neuropathies and polyarteritis nodosa (see below). Most patients with neuropathy-associated cryoglobulinemia have hepatitis C infection. Patients with primary biliary cirrhosis may develop a sensory polyneuropathy or dorsal root ganglionopathy.

Renal Failure and Hemodialysis

Peripheral neuropathy develops in a large majority of patients with renal failure who require chronic dialysis.[7]

The pathogenesis of uremic neuropathy is likely secondary to the accumulation of toxic factors, although no specific causes have been identified. The clinical features are those of a slowly progressive sensory polyneuropathy characterized by a constellation of symptoms, including burning feet, muscular cramps, dysesthesias, and restless leg syndrome. Accumulation of β_2-microglobulin amyloid in the carpal tunnel may cause a painful median nerve entrapment neuropathy.

Hypothyroidism

Patients suffering from hypothyroidism may develop a polyneuropathy characterized by paresthesias and muscle cramps. This polyneuropathy has reportedly been observed in approximately 30 to 40 percent of patients with hypothyroidism. Thyroid hormone replacement may improve the neuropathy-associated symptoms. Patients with hypothyroidism are also predisposed to develop a carpal tunnel syndrome and other entrapment neuropathies.[11]

Peripheral Neuropathic Pain Associated with Infections

Painful peripheral neuropathies occur in a number of infectious diseases, including viral, bacterial, and parasitic infections.[7,11]

Human Immunodeficiency Virus

Human immunodeficiency virus (HIV) infection is associated with a variety of peripheral neuropathy syndromes, including axonal neuropathy, inflammatory demyelinating neuropathy, mononeuropathies, and polyradiculopathies. HIV-related neuropathies also may be caused by nutritional deficiencies (see below) or the use of antiviral or antibacterial drugs (e.g., didioxynucleosides, dapsone, isoniazide, etc.).[15] The most common neuropathy related to HIV infection is a distal symmetric polyneuropathy. Patients present with the complaint of burning feet, distal sensory loss, weakness, and autonomic dysfunction. It is advisable to test for HIV antibodies in patients who present with a neuropathy and are known to have HIV risk factors or positive hepatitis B serology, polyclonal hypergammaglobulinemia, and CSF pleocytosis. Painful mononeuropathy and multiple mononeuropathies may be secondary to superinfections, including herpes zoster, hepatitis C, and cytomegalovirus (CMV). Multiple mononeuropathies are likely to occur in immunocompromised AIDS patients with a CD4+ T-lymphocyte count below 200 cells/µL. A subset of HIV-positive patients with polyneuropathy has been found to have necrotizing vasculitis. Acute lumbosacral polyradiculopathy is often associated with CMV infection and is one of the most devastating neurological complications in AIDS. In CMV-related lumbosacral polyradiculopathy, CSF examination shows pleocytosis with more than 40 percent polymorphonuclear cells, a high level of proteins, and a low concentration of glucose. An MRI with gadolinium may demonstrate enhancement of the cauda equina nerve roots.

Varicella-Zoster Virus

Varicella-zoster virus (VZV) may cause a variety of neuropathic conditions. One of the most common is postherpetic neuralgia (PHN), the pain that persists beyond the time of normal healing of the acute skin rash (herpes zoster). PHN occurs in approximately 50 percent of patients aged 60 years and older following herpes zoster (or shingles). The VZV manifestation can cause extensive inflammatory, hemorrhagic, and necrotic changes in the dorsal root ganglia, as well as in the corresponding dorsal horns of the spinal cord.[16]

Hepatitis B Virus and Hepatitis C Virus

Guillain-Barré syndrome, multiple painful neuropathies secondary to vasculitis, and cryoglobulinemia infections have been reported in association with hepatitis B virus (HBV) and hepatitis C virus (HCV). The combination of HCV infection and cryoglobulinemia has recently been recognized as a cause of painful neuropathy.[17]

Human T-Cell Lymphotrophic Virus Type 1

Infection with human T-cell lymphotrophic virus type 1 (HTLV-1) may cause a painful neuropathy. Although most patients affected by HTLV-1 present with spastic paraparesis, some patients may develop a painful neuropathy.[18] Because HTLV-1 is much more common in tropical regions, the term *tropical spastic paraparesis* has also been used.

Leprosy

Mycobacterium leprae may cause a painful neuropathy.[19] Leprous neuritis is very prevalent worldwide, especially in tropical and subtropical regions. *M. leprae* is likely transmitted through the upper respiratory tract. Once the nasal mucosa is colonized, then the mycobacterium begins to spread throughout the body. The skin and superficial nerve trunks are particularly vulnerable. In all forms of leprous neuritis (i.e., tuberculoid, dimorphous, and lepromatous), the primary neurological manifestation is sensory loss. However, neuropathic pain may occur and be quite excruciating. The complication is thought to be due to a vasculitis affecting nerve trunks. Clinically, there is nodular inflammation of the skin and painful, swollen nerve trunks.

Lyme Disease

The spirochete of Lyme borreliosis[20] can be responsible for a variety of neurological conditions. Infection by the tick-borne spirochete *Borrelia burgdorferi* can induce a painful polyradiculopathy and neuropathy. This may present with distal sensory symptoms (i.e., burning feet) or radicular pains. EMG-NCV studies may reveal a multifocal axonal neuropathy. A sural nerve biopsy should show axonal loss as well as perivascular and perineural mononuclear inflammation. Positive serology for Lyme disease confirms previous exposure. High CSF titers of anti-*B. burgdorferi* antibodies may help to establish a causal relationship between the spirochete and the neuropathy or other neurological disturbances.

Peripheral Neuropathic Pain Associated with Demyelinating Inflammatory Neuropathies

Guillain-Barré Syndrome and Chronic Inflammatory Demyelinating Polyneuropathy

Guillain-Barré syndrome (GBS) is an acute, inflammatory, multifocal, demyelinating disease affecting spinal roots and peripheral nerves supposedly caused by an autoimmune disorder initiated by T-lymphocytes and maintained by antibodies against peripheral nerve antigens.[19] During the acute stage of GBS, approximately 50 percent of patients experience back pain and body aches.[21] Pain usually subsides over time. In contrast, body pains are more rarely reported by patients who are diagnosed with chronic inflammatory demyelinating polyneuropathy (CIDP). CSF examination and electrophysiologic studies are essential to confirm the diagnosis of both GBS and CIDP. In GBS and CIDP, CSF study shows albuminocytologic dissociation (i.e., high levels of proteins and normal cell counts). Electrophysiologic workup reveals a pattern of nerve conduction "blocks" and other abnormalities supporting the diagnosis of multifocal demyelination.

Peripheral Neuropathic Pain Associated with Malignancies and Cancer Therapy

Neuropathy may result from one or more cancer-related mechanisms (e.g., compression, mechanical traction, inflammation, or infiltration of nerve trunks or plexi caused by the primary cancer or metastatic disease).[22] Head and neck cancer and skull-based tumors can cause painful cranial neuropathies.[23] Salivary gland cancers can cause painful facial neuropathies. Breast or lung cancers can infiltrate the brachial plexus and cause painful plexitis. Pelvic or retroperitoneal cancers can invade the lumbosacral plexus. If the meninges are affected (meningeal carcinomatosis), the involvement of adjacent roots, spinal nerves, and plexi can occur. Metastatic disease or lymphomas can cause meningeal carcinomatosis and affect multiple spinal roots. Peripheral neuropathies with pain and dysesthesia are also observed in the presence of lymphomas. Of note, inflammatory demyelinating polyneuropathy of the GBS type may occur with lymphomas, particularly Hodgkin's disease.

Antineoplastic therapeutic agents such as cisplatinum, taxoids, and vincristine may cause painful neuropathies. Postradiation plexopathies may arise when more than 60 Gy (6000 rad) of irradiation is given to the patient. Surgical resection of cancers may result in traumatic injuries to peripheral nerves with development of painful neuromas. For example, post-thoracotomy pain can be caused by injury to the intercostal nerves, and postmastectomy pain may arise through injury to the intercostal brachial nerve.

Compression or entrapment neuropathies occur in the presence of cachexia. For example, cancer patients who have lost substantial fat and muscle body weight are prone to develop peroneal neuropathies.

Paraneoplastic autoimmune syndromes due to antineuronal antibodies may present as painful neuropathies. Patients who complain of burning dysesthesias in their feet, hands, and face (in the setting of diagnosed or undiagnosed carcinoma) may have antineuronal nuclear antibodies type I (ANNA-1), also known as anti-HU. Patients who present with sensory neuronopathy and small cell carcinoma of the lung may have significantly elevated titers of anti-HU. All patients with burning dysesthesias of face, hands, and legs and positive titers for anti-HU should undergo a computed tomographic (CT) scan or MRI of the chest. In fact, small cell carcinoma of the lung may remain undetected by plain chest x-ray. In any case, anti-HU positivity should prompt a careful search for malignancy, especially for a small cell carcinoma of the lung. Painful dysesthesias develop first in one limb and then progress to involve other limbs, face, scalp, and trunk over weeks or months. In these patients, deep tendon reflexes are reduced or absent, and muscle strength is preserved. Patients may be disabled in their ambulation because of the sensory ataxia that is often associated with their painful symptoms.

Peripheral Neuropathic Pain Associated with Autoimmune Connective Tissue Diseases, Granuloma-Related Disorders, and Vasculitidies

Autoimmune and connective tissue diseases[23] can affect peripheral nerves, including cranial nerves. The pathogenesis of peripheral neuropathy in the course of such disorders may involve neuro-immunological events and ischemic lesions.

Sjögren Syndrome

Sjögren syndrome (SS) is an autoimmune disease primarily affecting exocrine glands. It is characterized by reduced lacrimal and salivary gland secretions, resulting in dry eyes and dry mouth (sicca complex). The sicca complex is due to lymphocyte infiltration and dysfunction of lacrimal and salivary glands. SS primarily occurs in women. Positive laboratory tests for autoantibodies to extractable nuclear antigens (ENAs, also called SS-A and SS-B), elevated sedimentation rate, rheumatoid factor (RF), and hypergammaglobulinemia may all be present. A sensory neuropathy occurs in almost one-third of patients with SS.[24] Painful symptoms may be the major and only complaint in SS-affected patients. If the clinical suspicion is high and laboratory tests remain unrevealing, a biopsy of the salivary glands of the lower lip is necessary to confirm the diagnosis. This should show lymphocytic gland infiltrations. A nerve biopsy also can demonstrate perivascular lymphocytic infiltration.

Systemic Lupus Erythematosus

Systemic lupus erythematosus (SLE) is an autoimmune connective tissue disease. PNS involvement develops in about 30 percent of SLE-affected patients. Painful cranial nerve neuropathies, painful plexopathy, GBS, and CIDP have all been observed in association with SLE. Neuroimmunological and immune-mediated vasculitic events have been proposed as pathogenetic mechanisms for the SLE-related PNS involvement.

Rheumatoid Arthritis

Rheumatoid arthritis (RA) is a chronic disease characterized by severe joint inflammation. The exact cause is unknown. There is, however, a genetic predisposition to RA. An abnormal immunologic response characterized by activated T-lymphocytes and macrophage infiltrations of the synovium with subsequent release of proinflammatory cytokines (e.g., tumor necrosis factor α, interleukin 1, etc.) occurs. Rheumatoid vasculitis can complicate the overall clinical picture. Painful polyneuropathy, painful mononeuropathies, or mononeuropathy multiplex may occur. Entrapment neuropathies secondary to nerve compression due to joint inflammation may develop. The sedimentation rate is elevated. The rheumatoid factor (RF) titer (i.e., corresponding to a serum titer of antibodies to abnormal gammaglobulins) is often present and elevated. The C4 complement serum level is often low. The synovial fluid is cloudy (rich in neutrophils) but sterile. Cutaneous nerve or muscle biopsies may demonstrate a necrotizing vasculitis.

Sarcoidosis

Sarcoidosis is a granulomatous disease affecting multiple organs, including peripheral nerves. Sarcoidosis can cause cranial neuropathies, truncal sensory mononeuropathies, a GBS-like polyradiculopathy, cauda equina syndrome, or mononeuropathy multiplex. Although rare, painful symptoms may accompany the more common motor and sensory deficits. Nerve damage is probably related to granulomas affecting the vasa nervorum. An elevated serum level of angiotensin-converting enzyme (ACE), and positive biopsies of lymph nodes, muscles, and sural nerves, confirms the diagnosis.

Vasculitidies

A number of immune-based vascular conditions may affect the peripheral nervous system.[25]

Polyarteritis nodosa (PAN) is the most common vasculitis, characterized by necrotizing inflammation of medium-sized and small arteries. Hepatitis B antigens are found in one-third of the PAN patients. Painful mononeuritis multiplex is characteristically associated with PAN.

Churg-Strauss vasculitis (CGV) typically presents with asthma, pulmonary infiltrates, and eosinophilia and may be associated with a painful neuropathy. A high titer of anti-neutrophilic cytoplasm antibodies (ANCAs) is relevant to the diagnosis of CGV.

Wegener's granulomatosis (WG) is a necrotizing granulomatous vasculitis that affects the upper and lower respiratory tract and may be accompanied by glomerulonephritis. The PNS is involved in up to a third of WG patients. Patients may present with severe facial pain secondary to cranial nerve neuropathy. Of note, ANCAs are almost constantly present in WG.

Giant cell arteritis or *temporal arteritis* (TA) affects large and medium-sized arteries. The vasculitis often involves arteries of cranial peripheral nerves, causing ischemic neuropathies or mononeuritis multiplex. The sedimentation rate is usually markedly elevated (>100 mm/h), and a normochromic, normocytic anemia is often present. Of note, the vast majority of the patients with TA are older than 50 years of age. They may present to the neurologist with severe daily headaches, scalp tenderness, and pain on mastication. They also may present with fatigue, joint pain, carpal tunnel syndrome, and weight loss. The diagnosis of TA is confirmed by a temporal artery biopsy. The appropriate treatment (e.g., corticosteroid therapy) should be started immediately if the clinical diagnosis of TA is entertained. In fact, this likely will prevent the patient from developing blindness from ischemic optic neuropathy or other serious clinical events related to ischemia.

Peripheral Neuropathic Pain Associated with Monoclonal Proteins and Amyloidosis

Monoclonal Proteins

Patients undergoing evaluation for chronic peripheral neuropathies of unknown cause should be screened for the presence of monoclonal immunoglobulins (IgG, IgM, IgA, IgD, IgE), also called *M proteins*.[26] Of note, a subset of patients presenting with painful peripheral neuropathy of cryptogenic or idiopathic origin may be affected by monoclonal gammopathy. Free monoclonal immunoglobulin light chains (also called *Bence-Jones proteins*) are excreted in the urine and detected by testing a 24-h urine collection. Bence-Jones protein is associated with multiple myeloma and amyloidosis. Routine serum/urine protein electrophoresis may lack the sensitivity required to detect small amounts of M protein. Serum immunoelectrophoresis (or immunofixation) is then required to confirm the monoclonal nature.

The pathophysiologic relationship between M proteins and neuropathy is unclear, but some specific M proteins may have pathogenetic effects on components of myelin or axolemma. Some M proteins (IgM) have been implicated in the pathogenesis of painful neuropathies.[27,28] For example, antisulfatide and anti-chondroitin sulfate M proteins (IgM) have been associated with a sensory painful neuropathy. Finding an M protein in patients with a neuropathy may lead to the discovery of underlying disorders such as primary systemic amyloidosis, multiple or osteosclerotic myeloma, macroglobulinemia, cryoglobulinemia, lymphoma, or a malignant proliferative disease. For example, Waldenström's macroglobulinemia (WM) is characterized by a malignant plasma cell dyscrasia producing monoclonal IgM proteins; it typically affects elderly men and may cause amyloidosis and a sensory neuropathy. However, in more than 50 percent of patients with an M protein, no detectable underlying disease is found. These patients are described as having a monoclonal gammopathy of undetermined significance (MGUS). The underlying mechanism of nerve fiber damage in MGUS neuropathy remains unknown, although a neuro-immunological mechanism is likely. Of these patients, a subset may be diagnosed as having WM at long-term follow-up.

Cryoglobulinemia

Cryoglobulins are immunoglobulins that precipitate at cold temperatures. Cryoglobulinemia may occur as a primary condition without any apparent underlying disease (*essential cryoglobulinemia*). Alternatively, cryoglobulinemia may be secondary to autoimmune diseases, infections (e.g., hepatitis C), and lymphoproliferative disorders. Cryoglobulins are classified into the following types: type 1—M proteins associated with myeloma or other lymphoproliferative disorders; type 2—M proteins with anti-RF activity; and type 3—polyclonal IgM and IgG immunoglobulins. Hepatitis C viral infection appears to be the most common cause of type 2 cryoglobulinemia. Cryoglobulinemia is often associated with painful vasculitis and mononeuritis multiplex.[29,30] In most cases, nerve biopsy shows loss of myelinated fibers, axonal degeneration, and necrotizing vasculitis.

Amyloidosis

Primary amyloidosis is a systemic disorder characterized by tissue deposition of abnormal proteins.[31] In primary systemic amyloidosis, a clone of plasma cells synthesizes light-chain peptides that accumulate in the extracellular spaces of tissues as amyloid. In both primary amyloidosis and secondary amyloidosis complicating multiple myeloma or WM, the extracellular amyloid is made of fragments of immunoglobulin light chains. Primary amyloidosis usually occurs in individuals of age 60 and older. The neuropathy begins with painful dysesthesias in the feet and follows a slowly progressive course. Most patients develop autonomic dysfunction, such as postural hypotension, gastrointestinal disturbances, impaired sweating, and loss of bladder control. Some patients develop a superimposed carpal tunnel syndrome caused by amyloid infiltration. The combination of painful dysesthesias, autonomic dysfunction, and a history of carpal tunnel syndrome should alert the pain physician to the possible diagnosis of amyloidosis. An M protein (light chains) can be found in the vast majority of the patients by means of serum or urine immunoelectrophoresis. In patients with suspected amyloid neuropathy, a nerve biopsy represents the most sensitive diagnostic technique.

Peripheral Neuropathic Pain Associated with Dietary or Absorption Abnormalities

Alcoholic Neuropathy and Vitamin Deficiencies

Neuropathy usually begins insidiously in patients affected by alcoholism and associated nutritional deficiencies. Patients complain of painful burning feet and numbness. Leg cramps and gait difficulty are common.[7,11] *Thiamine deficiency,* which is due to inadequate dietary intake, impaired absorption, or perhaps to a higher metabolic demand in alcoholic patients, is another possible cause of painful neuropathy. The clinical constellation of painful neuropathy, orogenital dermatitis, and amblyopia has been named *Strachan syndrome*. This syndrome originally was referred to as "Jamaican neuritis" and appears to occur in malnourished individuals and occasionally in alcoholic patients.[32] The treatment is reportedly based on the prescription of high doses of thiamine and other vitamins (e.g., vitamin B_{12}, folate riboflavin, niacin). A painful sensory neuropathy has been found to be associated with celiac disease (celiac neuropathy).[131] *Pyridoxine (vitamin B_6) deficiency* can cause a painful neuropathy. Deficiency of pyridoxine, however, may occur during treatment with isoniazide (INH), hydralazine, or penicillamine. These drugs resemble vitamin B_6 structurally and interfere with pyridoxine coenzyme activity. When using INH, supplementary pyridoxine (100 mg daily) is recommended. Paradoxically, megadoses (e.g., more than 200 to 300 mg/d) of pyridoxine have also been found to cause a painful sensory neuropathy. *Vitamin B_{12} and folate deficiencies* can cause a sensory polyneuropathy, as well as spasticity secondary to myelopathy. Acquired malabsorption of vitamin B_{12} and folate may occur following gastric and/or ileum surgical resections. Strict vegetarians and individuals affected by intestinal parasites (e.g., tapeworms) also may develop B_{12} and folate deficiencies. Low serum levels of vitamin B_{12} and/or folate confirm the diagnosis of vitamin deficiency. Individuals who abuse or are often exposed to nitrous oxide (which inactivates some vitamin B_{12}–dependent enzymes, perhaps in the setting of low serum levels of vitamin B_{12}) may develop signs of neuropathy and myelopathy. Pernicious anemia is an autoimmune process directed against the parietal cells of the gastric mucosa that cause malabsorption of vitamin B_{12}. The full-blown clinical picture of vitamin B_{12} deficiency consists of megaloblastic anemia, glossitis, and neurological signs such as painful neuropathy and myelopathy.[33]

Peripheral Neuropathic Pain Associated with Entrapment Neuropathies

Entrapment neuropathies are caused by a focal chronic compression of a peripheral nerve. The compression of entrapment neuropathies, therefore, is caused by anatomic structures (e.g., carpal tunnel, cubital groove, etc.) surrounding the course of a nerve. Patients with entrapment neuropathies may complain of sharp shooting pains often radiating along the distribution of the affected nerve. Motor and sensory findings arise over time. A positive Tinel's sign or tenderness at the site of nerve compression can be elicited during the neurological examination. Electrophysiologic testing and diagnostic peripheral nerve blocks are useful in confirming the clinical diagnosis of entrapment neuropathy.[34,35]

Peripheral Neuropathic Pain Associated with Toxic Neuropathies

A variety of toxic substances are known to induce a peripheral neuropathy.[7,36] The following are examples of painful neuropathies caused by heavy metals and drugs.

Heavy Metals
Thallium and arsenic intoxications consistently induce a painful neuropathy, as well as abnormal lines in nails (Mees' lines). Alopecia occurs with thallium intoxication. Twenty-four-hour urine collection, serum levels, and hair testing confirm the diagnosis of heavy metal–induced neuropathy.

Medications
Examples of drug-induced painful neuropathies are

1. *Taxoid-induced painful neuropathy.* Taxoids are alkaloids used in the treatment of ovarian cancer. These agents disrupt microtubule assembly and axonal transport. A painful sensory neuropathy develops with doses above 200 mg/m².
2. *Cisplatinum-induced painful neuropathy.* Cisplatinum is an agent used in the treatment of ovarian, bladder, and testicular malignancies. The neuropathy may develop several months after discontinuation of the drug.
3. *Vincristine-induced painful neuropathy.* Vincristine is a vinca alkaloid used in chemotherapeutic regimens. Vincristine, when used at the conventional doses, invariably causes a mild polyneuropathy. Of note, paresthesias and dysesthesias often start in fingers and hands.
4. *Amiodarone-induced painful neuropathy.* Amiodarone is a class III antiarrhythmic agent used in the management of refractory ventricular arrhythmias. A painful polyneuropathy may develop in some patients receiving long-term amiodarone therapy.
5. *Metronidazole-induced painful neuropathy.* Metronidazole is an antibiotic used in the treatment of protozoa and anaerobic bacterial infections. It may cause a painful polyneuropathy.
6. *Pyridoxine-induced painful neuropathy.* Vitamin B_6 can cause a severe painful sensory neuropathy when taken at megadoses (higher than 200 to 300 mg/d).

Peripheral Neuropathic Pain Associated with Hereditary Neuropathies

Eliciting a history of long-standing complaints of numbness and limb weakness; obtaining a detailed family history; and examining family members, looking for musculoskeletal abnormalities such as hammer toes, high arches in the feet, or spinal scoliosis, are some clinical ways of identifying hereditary neuropathies.[7,37]

Charcot-Marie-Tooth Disease
Charcot-Marie-Tooth disease (CMT) is the most common inherited neuropathy, estimated to affect 1 of every 2500 people in the United States.[38] Painful symptoms of CMT disease[39] have been reported in the hypertrophic or demyelinating form (CMT-1). Neuropathic pain appears to be a relevant and underrecognized problem in this disorder, with patients complaining of shooting, sharp, burning pains in their toes, feet, ankles, and knees. Patients with CMT-1 develop their initial complaints, such as difficulties

in walking or running, during the first or second decade. Adult patients have absent ankle jerks. Enlarged hypertrophic peripheral nerves can be found on examination. The characteristic "onion bulb" finding, which is due to the progressive process of demyelinization and subsequent remyelinazation of the large fibers, is found on nerve biopsy. In this form, there is a marked decline in nerve conduction velocities (NCVs). In the axonal form (CMT-2), NCVs are minimally impaired, and nerve biopsy findings show axonal loss without "onion bulb" formation. Both CMT-1 and CMT-2 have an autosomal dominant inheritance. Of note, CMT disease is also known to have a third clinical form (or Dejerine-Sottas disease), a severe demyelinating neuropathy, with onset in early childhood.

Fabry's Disease

Fabry's disease (FD) is an X-linked recessive disorder in which a deficiency of the lysosomal enzyme α-galactosidase causes accumulation of the glycolipid ceramidetrihexoside in the endothelial cells of blood vessels.[7] Patients develop cutaneous angiokeratomas (telangiectasias) mainly over the lower part of the trunk, glutei, and scrotum. A painful small-fiber neuropathy develops in childhood or adolescence. Fever and hot environments may intensify distal dysesthesias and lancinating pains. Autonomic dysfunction occurs and includes anhidrosis, impaired lacrimation, and salivation, as well as decreased intestinal motility. Deposits of the glycolipid in vascular structures lead to ischemic disease affecting heart, kidneys, and brain. Leukocytes or skin fibroblasts are used to test the activity of the lysosomal enzyme α-galactosidase and to determine the diagnosis of FD.

Porphyric Neuropathy

High serum levels of porphyrins have toxic effects on the autonomic, peripheral, and central nervous systems.[7] Porphyrias are dominant inherited disorders and include acute intermittent porphyria (AIP), variegate porphyria (VP), and hereditary coproporphyria (HCP). Hereditary enzyme defects affecting the synthesis of heme cause porphyrins to accumulate. These partial enzyme defects remain latent until precipitating factors trigger acute attacks, often characterized by neuropsychiatric manifestations. Precipitating factors include certain drugs, alcohol, and unbalanced dieting during prolonged fasting. Porphyric attacks usually occur during the second and third decades of life and are more common and severe in women. All clinical symptoms can be explained by dysfunction of the autonomic, peripheral, and central nervous systems. Patients with AIP may present with attacks of abdominal pain, urinary retention nausea, and vomiting, tachycardia and difficulty with urination. A subacute neuropathy characterized by paresthesias, dysesthesias, sharp pains, and cramps may affect extremities or the trunk. Patients with VP and HCP also may develop cutaneous photosensitivity during adult life. High concentrations of aminolevulinic acid (ALA) and porphobilinogen (PBG) in blood and urine confirm the diagnosis of porphyria.

Familial Amyloid Polyneuropathy

Familial amyloid polyneuropathy (FAP) is a group of autosomal dominant disorders characterized by the extracellular deposition of amyloid along and within peripheral nerves and visceral organs.[11] Most patients with FAP have mutations of the plasma protein transthyretin. This is a transport protein for thyroxin and is synthesized in the liver. The neuropathy begins insidiously in the third and fourth decades of life, often presenting with lancinating pains and paresthesias. Patients also develop signs of autonomic dysfunction, such as postural hypotension, bladder dysfunction, and anhidrosis. A limited form of FAP, called *FAP type 2,* presents with carpal tunnel syndrome in the fourth or fifth decade. Autonomic manifestations are absent. Surgical decompression of the carpal tunnel may provide not only pain relief but also the diagnosis. In fact, the biopsy of the flexor retinaculum taken at the time of the surgery demonstrates amyloid infiltration. In order to differentiate FAP from the nonfamilial forms of amyloidosis, serum and urine immunoelectrophoresis for an M protein (light chains) should be performed. If no evidence of M proteins is found, serum electrophoresis should be focused on transthyretin. The finding of a variant transthyretin should prompt genetic testing.

Peripheral Neuropathic Pain Associated with Cryptogenic Disorders

Idiopathic Polyneuropathy

Patients with idiopathic painful axonal neuropathy commonly complain of burning feet. Of note, idiopathic neuropathy more commonly afflicts patients older than 60 years of age. Patients have unrevealing diagnostic tests. In some patients, elevated sensory thresholds to cold, heat, and pinprick, as well as impaired distal sweating, can be found on examination. A skin biopsy may show reduced density of small nerve fibers in the epidermis. This finding provides evidence of distal small-fiber neuropathy.

Complex Regional Pain Syndromes

In 1994, the International Association for the Study of Pain (IASP) renamed the disorder formerly called "reflex sympathetic dystrophy" (RSD) as *complex regional pain syndrome (CRPS) type 1* and the disorder previously called "causalgia" as *CRPS type 2.*[40] Following a distal limb traumatic injury and subsequent limb immobilization (often an important predisposing factor in the medical history of this disorder), some patients develop CRPS. CRPS may be secondary to an abnormal neuroimmunologic inflammatory process. An outgrowth of this concept is the number of observations addressing the role of macrophages and macrophage-released cytokines, as well as tissue-released nerve growth factor (NGF) (from mast cells, fibroblasts, keratinocytes, etc.) in the genesis of hyperalgesia. Specific inflammatory cytokines (e.g., tumor necrosis factor α, interleukin 1β, etc.), in combination with tissue-released growth factors (e.g., NGF), not only are known to bring on hyperalgesia by sensitizing tissue nociceptors but also, possibly, set off altered gene expression in primary sensory pain neurons (e.g., changes in sodium channels, expression of altered receptors and channels, including *adrenoreceptor excitability*). The end result is a chronic neuropathic pain condition.[2,5,6,41,42]

Neurotrophic factors, including NGF, can induce miniature axon sprouts not only from the endings of afferent fibers but also from the sympathetic efferent nerves, perhaps favoring a coupling between somatic and sympathetic nerves. In this context, norepinephrine released locally from the sympathetic efferents would be the molecule that activates the altered nociceptors expressing *adrenoreceptor excitability*.[43]

α-Adrenoreceptor excitability represents the pathophysiologic basis of what has been called *sympathetically maintained pain* (SMP). SMP is a mechanism (and not a clinical entity) that can complicate the pathogenesis of CRPS. The SMP mechanism may well be just a reversible phenomenon in the course of CRPS. Of interest is that in several instances CRPS appears to be maintained by plastic mechanisms (e.g., the SMP mechanism) that can be switched off by early and aggressive therapeutic interventions. SMP can complicate CRPS as well as other painful disorders, including shingles, neuralgias, and metabolic or autoimmune neuropathies.[3]

Pathologic studies in patients affected by longstanding intractable CRPS type 1 have shown not only the occurence of small fiber axonal sprouts, but also the presence of a microangiopathic process.[44] Lastly, there may be a genetic predisposition to the development of CRPS type 1, with a higher incidence in certain HLA types, e.g., HLA DQ1.[45]

In summary, while the multiple pathogeneses of CRPS have not been defined, it appears that a traumatic injury to the distal extremity of predisposed individuals may initiate a process in the affected limb that results in a regional neuropathic pain syndrome.

Essential Trigeminal and Glossopharyngeal Neuralgias

Idiopathic trigeminal neuralgia or *tic douloureux* is a neuropathic painful state characterized by severe, lancinating, brief (<60 s), unilateral facial pains. The paroxysmal symptoms follow the distribution of one or two branches of the trigeminal nerve. Facial trigger points are commonly present, and chewing or talking can induce the attacks. The onset is after age 50 in the vast majority of patients. The etiology of the condition may be related to a vascular structure, e.g., a tortuous artery compressing the preganglionic segment of the trigeminal nerve within the posterior cranial fossa. Secondary trigeminal neuralgia should be suspected in young patients. Etiologies include multiple sclerosis, meningioma, or another posterior fossa tumor. Diagnostic evaluation includes MRI of the posterior fossa with gadolinium.

Idiopathic glossopharyngeal neuralgia appears to have a similar vascular mechanical pathogenesis as tic douloureux. However, the clinical features of the attacks are more variable, including more prolonged pains that may radiate to the tongue, throat, ear, or along the lateral and anterior aspects of the neck. The pain may be accompanied by autonomic symptoms such as salivation or syncope. Trigger points may be located in the tonsil or posterior third of the tongue, and swallowing may induce the painful attacks. Oropharyngeal carcinoma, retropharyngeal or paratonsillar abscess, or elongated styloid process all can cause secondary glossopharyngeal neuralgia. Diagnostic evaluation includes imaging studies of the posterior fossa, skull base, and neck.

Cryptogenic Painful Brachial Plexopathy

Cryptogenic painful brachial plexopathy also has been known as *Parsonage-Turner syndrome* or *acute brachial plexus neuropathy*. The pain is extremely severe and abrupt in onset. It radiates to the scapula, arm, and neck. Weakness and atrophy of the muscles innervated by the brachial plexus, such as the serratus anterior, deltoid, biceps, and triceps, occur. Sensory abnormalities include paresthesias, dysesthesias, and sensory loss. The condition can be bilateral and, in this case, affects the limbs asymmetrically. Severe pain usually lasts a few weeks. A chronic, more diffuse, but less intense pain can last several months. Events that have been reported to precede the painful plexopathy include immunizations, surgical or traumatic injuries, and upper respiratory infections. The pathogenesis is unknown, but it is thought to be secondary to a neuroimmunological vasculitic process.[46]

▶ THERAPEUTIC INTERVENTIONS

Pharmacology of Neuropathic Pain

The pain signal is transmitted from the peripheral nociceptors, through the dorsal horn of the spinal cord, through the thalamus, and up to the cortex. In the periphery, nociceptors can be activated by the chemical products of tissue damage and inflammation, which include prostanoids, serotonin, bradykinin, cytokines, adenosine, adenosine-5′-triphosphate (ATP), histamine, protons, free radicals, and neurotrophins. These molecules can directly activate afferent fibers or sensitize them to a range of mechanical, thermal, and chemical stimuli. A proportion of the afferent fibers that are normally unresponsive to noxious stimuli ("silent" or "sleeping" nociceptors) can be "awakened" by inflammatory chemicals and contribute to pain and hyperalgesia. In some instances, specific factors, such as some proinflammatory cytokines and growth factors can be internalized and undergo retrograde axonal transport to the dorsal root ganglion and initiate altered gene expression. These intracellular transcriptional changes will lead to the expression of altered channels (or changes in channel density) and receptors in the injured as well uninjured nociceptors.

Some recently identified cellular components that might be relevant to the development of pathologic peripheral pain (and therefore potential targets for the treatment of neuropathic pain) include[42,47,48]

- *Voltage-gated sodium channels.* These include the tetrodotoxin-sensitive $Na_v1.3$ channels and the tetrodotoxin-resistant $Na_v1.8$ channels. Of note, the tetrodotoxin-resistant $Na_v1.8$ channel is expressed exclusively in primary nociceptive sensory neurons.[49–51] The sodium channel blockers include the traditional local anesthetics, such as bupivacaine and lidocaine; the oral antiarrhythmic agent mexiletine; several antiepileptic drugs, such as carbamazepine, oxcarbazepine, and lamotrigine; and the tricyclic antidepressants, such as amitriptyline, doxepin, and imipramine.
- *Voltage-gated calcium channels.* There are several lines of evidence indicating that gabapentin and pregabalin, both antiepileptic agents effective in the treatment of postherpetic neuralgia and painful diabetic neuropathy produce their analgesic effect by modulating N- and P/Q-type voltage-gated calcium channels. Both gabapentin and pregabalin bind with high affinity to the $\alpha_2\delta$ subunit of the calcium channels and produce a decrease in intracellulor calcium influx.[50] Ziconotide, a peptide analgesic derived from the venom of the predatory marine snail *Conus magellaris*, is a neuron-specific (N-type) calcium channel blocker that may become available for clinical use.[52–54]
- *Vanilloid receptors.* These receptors (e.g., the VR-1; now termed TRPV1 or transient receptor potential vanilloid-1) are activated by noxious heat and low pH.
- *Acid-sensing ion channels (ASICs).* These are amiloride-sensitive channels activated by low pH. The ASIC3 channels are possibly involved in the development of chronic muscle pain.[55]
- *Tyrosine kinase (TrkA) receptor for nerve growth factor (NGF).* The TrkA-NGF complex is internalized and transported retrogradely to the dorsal root ganglion cell body, where it initiates gene transcription. This may produce upregulation of receptors, ion channels, and neuropeptides involved in pain transmission.[56] However, a recent study[57] indicates that another growth factor, the glial cell line–derived neurotrophic factor (GDNF), may have potent analgesic effects in neuropathic pain models. It has been reported that following a nerve injury GDNF can prevent and abolish ectopic discharges from afferent fibers by blocking the expression of certain subtypes of sodium channels involved in pain transmission.
- *Interleukin 1 (IL-1) receptor for IL-1β*. This promotes gene transcription for fiber excitability.[58]
- *Adrenergic receptors in SMP.*[43]
- *The purinergic receptors P2X2/3.* These receptors are localized on peripheral sensory afferents. Activation of these channels by ATP causes nociception and contributes to the development of hyperalgesia and mechanical allodynia after chronic nerve constriction injury in animal models of neuropathic pain. P2X3 receptors are found in a subset of small-fiber afferents that are also sensitive to capsaicin.[59,60]
- *Adenosine A_1 and A_3 receptors.* Adenosine is a nucleoside that acts as an autocoid and activates G protein–coupled membrane receptors, known as A_1, A_2a, A_2b, and A_3. In humans, the peripheral administration of adenosine produces pain that is similar in quality to ischemic pain. However, spinally injected adenosine induces antinociception in animal models of neuropathic pain, and intrathecal adenosine appears to induce analgesia.[61,62] The activation of adenosine A_3 produces nociceptive behaviors in animals. Of note, ATP may activate not only the P2X but also adenosine receptors.[63–65]
- *Glutamate receptors.* Peripheral glutamate receptors have been identified recently and linked to peripheral nociception.[66] Neurotransmitters of nociception within the dorsal horn of the spinal cord include the excitatory amino acids (EAAs) glutamate and aspartate, as well as a number of neuropeptides (e.g., substance P, calcitonin gene-related peptide, cholecystokinin, and neurokinin A). EAAs can induce central sensitization. They act on several receptors, including *N*-methyl-D-aspartate (NMDA), α-amino-3-hydroxy-5-methyl-4-isoxazole-propionic acid (AMPA), kainate, and the metabotropic receptors. The NMDA receptors seem to play a relevant role in pain modulation.

Some recently identified receptors that might be relevant to the modulation of peripheral neuropathic pain include:

- *Cannabinoid receptors (CB1 receptors primarily located in the CNS and CB2 receptors in the peripheral tissues) and the endogenous cannabinoid system.*[67,68] Recent studies suggest a potential role of cannabinoids not only as anti-inflammatory and antihyperalgesic agents but also in the potentiation of opioid analgesia[69].

Lastly, factors that may be relevant to the mechanisms of pathological peripheral pain include:

- *Tumor Necrosis Factor α (TNF-α) and other* proinflammatory cytokines (e.g., IL-1β, TNF-α, IL-6, leukemia inhibitory factor). After nerve injury, TNF is known to be released from macrophages and upregulated. Evidence suggests that following a nerve injury, TNF may be essential for the development and maintenance of neuropathic pain.[70–72] Emerging peripheral analgesics for neuropathic inflammatory states include agents with anti-inflammatory cytokine properties (e.g., TNF-α antagonists, or drugs blocking the synthesis of TNF-α) and drugs that block the recruitment of macrophages at the site of a nerve injury.
- *NF-Kappa B (nuclear factor Kappa B)* is a transcription factor that once activated and transported in to the nucleus can induce pro-inflammatroy cytokine gene expression and iNDS (inducible nitric oxide synthase) expression in macrophages. Inhibition of NF-Kappa B activation has shown to reduce hyperalgesia in neuropathic and inflammatory animal models.[132]

Classes of Pharmacologic Agents with Established Efficacy for Neuropathic Pain

Antiepileptic Drugs

Antiepileptic drugs (AEDs) are becoming the most promising agents for the management of neuropathic pain. In May 2002, gabapentin, an anticonvulsant approved for the treatment of partial seizures, gained U.S. Food and Drug Administration (FDA) approval for postherpetic neuralgia (PHN) (a neuropathic state characterized by allodynia and spontaneous burning pain). In recent controlled trials, efficacy of gabapentin has been shown not only for PHN but also for painful diabetic neuropathy (PDN) (a state predominantly characterized by spontaneous burning pain).[73–75] The higher potency gabapentin analog, pregabalin, has shown to have efficacy for a number of painful states, including PDN and fibromyalgia.[133] Pregabalin has a much higher bioavailability than gabapentin. Recent evidence indicates that gabapentin and pregabalin modulate voltage-gated calcium channels (by binding to the $\alpha_2\delta$ subunit of the channel) and decrease intracellular calcium influx.[52] Trigeminal neuralgia (a neuropathic condition characterized by brief excruciating, lancinating pains) responds extremely well to carbamazepine, whereas another AED, lamotrigine, has shown some efficacy for carbamazepine-resistant trigeminal neuralgia.[76] Topiramate has been used anecdotally in the treatment of CRPS type 1.[77] Several new AEDs (i.e., levetiracetam, zonisamide, oxcarbazepine, and tiagabine) have become available for medical use in the United States, and these, along with topiramate, may have analgesic effect in primary headaches (e.g. migraine) and neuropathic pain.[78–80]

Opioids

Opioids are currently the most potent and effective analgesics used to treat acute and chronic pain, and as such, they have been prescribed to patients suffering from intractable pain. Morphine, a mu agonist, represents the mainstay of treatment for moderate to severe nociceptive cancer pain.[81]

Long considered ineffective for neuropathic pain, opioids have demonstrated efficacy in several recent clinical trials.[82–87]

The analgesic action of the pure opioid agonists (e.g., morphine, methadone, fentanyl, oxycodone, hydromorphone, etc.) is well known. The mu, kappa, and delta opioid receptors are located not only in the CNS (primarily in the dorsal horn) but also peripherally on the nociceptors.[1] Opioids may have a relevant peripheral analgesic effect during inflammatory states.[88,89]

Opioid agonists (e.g., morphine, fentanyl, methadone, levorphanol, oxycodone, hydromorphone, and hydrocodone) are the mainstay for the treatment of severe disabling pain. The treatment of chronic pain may rely on the use of long-acting agents (e.g., methadone and levorphanol) or controlled-release preparations of morphine, fentanyl, and oxycodone. Among the pure opioid agonists, methadone has peculiar properties. It has an intrinsic N-methyl-D-aspartate (NMDA) receptor antagonistic effect that may add adjuvant analgesia in neuropathic pain (see below). Of interest, recent animal studies indicate that the addition of an extremely low dose of an opioid receptor antagonist (e.g., naltrexone) to morphine in a ratio of 1:1000 may enhance the analgesic efficacy of the opioid agonists.[90] Tramadol is an analgesic agent with a weak mu opioid agonistic effect. Its potency is comparable with that of a codeine-acetaminophen preparation. In controlled trials, tramadol has shown efficacy for the treatment of neuropathic pain.[91,92]

Unlike anti-inflammatory drugs, opioid agonists have no true "ceiling dose" for analgesia and do not cause direct organ damage. Except for constipation, some degree of tolerance occurs for most of the opioid-related side effects (e.g., nausea, vomiting, respiratory depression, and drowsiness). *Opioid titration* and *opioid rotation* are essential concepts in the management of neuropathic pain. In order to determine adequate opioid responsiveness, a careful titration of the opioid dose is necessary.[1] However, tolerance to opioid side effects, degree of analgesia, and development of analgesic tolerance are extremely variable among patients receiving these medications. If severe pain persists or side effects become intolerable during the initial drug trial, trials of different opioids (i.e., opioid rotation) are recommended. Studies indicate that patients on a stable opioid regimen do not report significant impairment in their driving ability, attention, mood, and general cognitive functioning.[93]

Antidepressants

Antidepressants also play an important role in the treatment of chronic pain. TCAs such as amitriptyline, nortriptyline, and desipramine[94] have established efficacy for neuropathic pain. They have been used successfully for painful diabetic neuropathy and postherpetic neuralgia and provide pain relief in nondepressed patients affected by neuropathic pain. Of note, TCAs such as amitriptyline, doxepin, and imipramine have been found to have potent local anesthetic properties. Amitriptyline appears to be more potent than bupivacaine as a Na^+ channel blocker.[95] TCAs frequently cause poorly tolerated adverse effects, including cardiotoxicity, confusion, urinary retention, orthostatic hypotension, nightmares, weight gain, drowsiness, dry mouth, and constipation.

Venlafaxine is a newer antidepressant that lacks the anticholinergic and antihistamine effects of the TCAs. Venlafaxine appears to possess an analgesic mechanism of action, with similar TCA-like beneficial properties but fewer side effects.[96,97] More recently, a slow-release preparation of bupropion, an atypical antidepressant, at the dose of 150 mg twice a day, was found to be effective and well tolerated for the treatment of neuropathic pain.[98]

Selective serotonin reuptake inhibitors (SSRIs) such as paroxetine and fluoxetine are effective antidepressants but ineffective analgesics. While being used for the management of comorbidities such as anxiety, depression, and insomnia that frequently affect patients with chronic neuropathic pain, SSRIs have not shown the same efficacy as the TCAs in the treatment of neuropathic pain.[94]

Local Anesthetics

The FDA has approved transdermal lidocaine for postherpetic pain.[99] In a controlled clinical trial, the transdermal form of 5% lidocaine relieved pain associated with PHN without significant adverse effects.[100] There is also early evidence to suggest that the patch provides benefit for other neuropathic pain states,[101] including diabetic neuropathy,[102] CRPS, postmastectomy pain, and HIV-related neuropathy.[103]

Intravenous lidocaine and oral mexiletine also have been used in patients with neuropathic pain.[54] The antiarhythmic local anesthetic mexiletine is a sodium channel blocker with analgesic properties for neuropathic pain similar to those of some AEDs (e.g., lamotrigine and carbamazepine). Mexiletine is contraindicated in the presence of second- and third-degree atrioventricular conduction blocks. Also, the incidence of gastrointestinal side effects (e.g., diarrhea and nausea) is quite high in patients taking mexiletine.

Adjuvants Analgesics for Neuropathic Pain

In addition to the classes of agents discussed earlier, many drugs from a variety of pharmacologic classes can be classified as adjuvant analgesics and can be used "off label" in the management of patients with chronic intractable pain. In many cases, the mechanisms supporting this analgesic enhancement are still unknown. At present, the evidence that adjuvants and emerging analgesics may possess analgesic properties for neuropathic pain derives mostly from preclinical investigations and preliminary observations.

α_2-Adrenergic Agonists

Drugs acting on the α_2-adrenergic spinal receptors (e.g., clonidine and tizanidine) have been recognized clinically as analgesics.[1,104] α_2-Adrenergic agonists are known to have a spinal antinociceptive effect. Controlled trials have shown the effectiveness of intraspinal clonidine for pain control.[104,105] Clonidine has been found to potentiate intrathecal opioid analgesia. Moreover, transdermal clonidine has a local antiallodynic effect in patients with SMP.[106] Topical clonidine, an α_2-adrenergic agonist, is analgesic in SMP. Clonidine causes local inhibition of norepinephrine release by acting on the adrenergic α_2 autoreceptors of the sympathetic endings.[106]

Tizanidine is a relatively short-acting oral α_2-adrenergic agonist with a much lower hypotensive effect than clonidine. Tizanidine has been used for the management of spasticity. However, animal studies and clinical experience indicate the usefulness of tizanidine for a variety of painful states.[107-109] The most common side effects of the α_2-adrenergic agonists are somnolence and dizziness (to which tolerance usually develops).

Capsaicin

Capsaicin is the natural substance present in hot chili peppers. Capsaicin activates the recently cloned vanilloid neuronal membrane receptor.[110] After an initial depolarization, a single administration of a large dose of capsaicin appears to produce a prolonged deactivation of capsaicin-sensitive nociceptors. The analgesic effect is dose-dependent and may last for several weeks. Capsaicin must be compounded topically at high concentrations (>1%) and administered under local or regional anesthesia.[111] Over-the-counter creams must be applied several times a day for many weeks. Controlled studies at low concentrations (0.075% or less) show mixed results, possibly due to noncompliance.

NMDA Antagonists and Cannabinoids

Evidence gleaned from animal experiments shows that NMDA receptors play an important role in the central mechanisms of hyperalgesia and chronic pain.[112,113] Dextromethorphan, memantine, and ketamine are NMDA antagonists that may be used as adjuvants in the management of hyperalgesic neuropathic states poorly responsive to opioid analgesics. Ketamine and dextromethorphan may be used in conjunction with opioids (1) in the prevention and treatment of analgesic tolerance and (2) in the management of allodynia and hyperalgesia. However, these agents (in particular ketamine) have a very narrow therapeutic window. Parenteral ketamine can cause intolerable side effects, such as hallucinations and memory impairment.

The opioid agent methadone is a racemic mixture of the isomers D- and L-methadone. D-Methadone, while reportedly lacking the opioid agonistic effect, has been shown to possess NMDA receptor antagonist activity.[114] Methadone's role in the treatment of neuropathic pain[114] may be limited by its long and unpredictable half-life, interindividual variations in pharmacokinetics, and lack of knowledge regarding appropriate use.

Of interest is the possibility that NMDA antagonists may prevent or counteract opioid analgesic tolerance.[115]

Evidence from preclinical studies and clinical observations indicate that cannabinoids have analgesic properties.[116] Interestingly, the addition of inactive doses of cannabinoids to low doses of opioid mu agonists appears to potentiate opioid antinociception. Moreover, cannabinoids appear to have a predominant antihyperalgesic effect.[68,69]

Anti-Inflammatory Drugs and Bisphosphonates

Steroid therapy may be considered for severe inflammatory pain due to cancer infiltrating structures such as the brachial or lumbosacral plexi, roots, or nerve trunks. Nonsteroidal anti-inflammatory drugs (NSAIDs), e.g., COX type 1 and 2 inhibitors and acetaminophen, have been of little benefit in the treatment of neuropathic pain. Several lines of evidence indicate that TNF-α, as well as other proinflammatory interleukins, play a key role in the mechanism of inflammatory neuropathic pain. Neutralizing antibodies to TNF-α and interleukin 1 receptor may become an important therapeutic approach for severe inflammatory pain resistant to NSAIDs, as well as for forms of neuropathic inflammatory pain.[117] Of note, thalidomide has been shown to prevent hyperalgesia caused by nerve constriction injury in rats,[118,119] and thalidomide is known to inhibit TNF-α production. TNF-α antagonists or newly developed thalidomide analogues with a better safety profile might be effective in the prevention and treatment of otherwise intractable painful disorders.[120]

High doses of intravenous bisphosphonates were reported to be efficacious in the treatment not only of bone pain secondary to metastatic disease but also of CRPS.[121,122] Their analgesic effect may be related to the inhibition of activated osteoclasts and macrophages. This leads to a decreased release of proinflammatory cytokines in the area of inflammation. In animal models of neuropathic pain (sciatic nerve ligature), bisphosphonates reduced the number of activated macrophages infiltrating the injured nerve, reduced Wallerian nerve fiber degeneration, and decreased experimental hyperalgesia.[123]

GABA Agonists

Baclofen is an analogue of the inhibitory neurotransmitter GABA and has a specific action on the GABA-B receptors. It has been used for many years as an effective spasmolytic agent. Baclofen also has shown effectiveness in the treatment of trigeminal neuralgia.[1] Clinical experience supports the use of low-dose baclofen to potentiate the antineuralgic effect of carbamazepine for trigeminal neuralgia. Baclofen also has been used intrathecally to relieve intractable spasticity, and it may have a role as an adjuvant when added to spinal opioids for the treatment of intractable neuropathic pain and spasticity. The most common side effects of baclofen are drowsiness, weakness, hypotension, and confusion. It is important to note that discontinuation of baclofen always requires a slow tapering in order to avoid the occurrence of seizures and other severe neurological manifestations.

Benzodiazepines (e.g., alprazolam, lorazepam, and diazepam) are GABA-A agonists. Their clinical use in patients with chronic pain is controversial.[1,124] In a controlled trial, patients with postherpetic neuralgia did worse on lorazepam than on placebo or amitriptyline.[125] Benzodiazepine-related side effects include depression and disruption of physiologic sleep. In combination with opioids, benzodiazepines cause significant cognitive impairment, whereas opioid analgesics alone do not.[93]

Invasive Treatment Interventions

Implantable devices, such as intrathecal pumps (IPs), spinal cord stimulators (SCSs), and motor cortex stimulators,[126] have become available recently for the treatment of neuropathic pain that responds poorly to standard pharmacologic and conservative therapeutic modalities. Among the most commonly used implantable devices are SCSs and IPs. SCSs have been used successfully in patients with severe limb pain that does not respond to conventional methods. A recent randomized trial[127] showed the efficacy of spinal cord stimulation in CRPS. Patients who received spinal cord stimulation in combination with physical therapy demonstrated a significant decrease in pain intensity compared with those receiving physical therapy alone. However, while significant pain relief was obtained in the SCS-treated patients, no significant improvement in functional status was observed.

IPs are used to deliver analgesics, such as opioids, clonidine, local anesthetics, and baclofen, into the CSF.[133] They likely will be used in the near future for other emerging analgesic agents (e.g., ziconotide and adenosine). Clinical experience and several reports indicate that clonidine and/or local anesthetics intrathecally can potentiate opioid analgesia for neuropathic pain.[128,133] Intrathecal morphine is currently the most commonly used analgesic administered by pump. However, prior to implantation of an intrathecal morphine pump, opioid trials must performed to show that the patient's pain is somewhat opioid-responsive. Pumps can be implanted permanently once trials are successful.

For some specific intractable neuropathic pain disorders, neuroablative procedures may be considered. For example, the dorsal root entry zone (DREZ) lesion has been recommended for the treatment of intractable pain from brachial plexus avulsions.[129,130] The decision to perform neuroablative surgery should come only after a thorough, comprehensive assessment has been carried out by a multidisciplinary team of pain medicine specialists and conservative management has failed to produce improvement in the patient's quality of life.

Analgesic Stepladder for Neuropathic Pain

Management of neuropathic pain can be a challenge, and combination therapies employing agents with established efficacy sequential and classes represent the contemporary standard approach.[34,134] The number and variety of options for pain control can be confusing and daunting, even for physicians specializing in the treatment of pain. First line medications for neuropathic pain include: gabapentin (and pregabalin), lidocaine 5% patch, opioids (and tramadol), and tricyclic antidepressants. Specific agents can be employed in an escalating regimen (Table 23-3) that

▶ **TABLE 23-3.** ANALGESIC STEPLADDER FOR NEUROPATHIC PAIN

Pain, Predominant Feature	Proposed Steps of Polyanalgesic Intervention
STEP 1: Mild Functional Impairment/Mild pain **Pain with a score of <4/10 on the Brief Pain Inventory**	
Allodynia, hyperalgesia[1]	Topical therapies +/− gabapentin or tramadol or antidepressant
Spontaneous, burning, constant[2]	Gabapentin or antidepressant or tramadol +/− topical therapies[2]
Spontaneous, lancinating, intermittent[3]	Gabapentin or lamotrigine or antidepressant
STEP 2: Moderate to Severe Functional Impairment/Moderate to Severe Pain **Pain with a score of >4/10 on the Brief Pain Inventory** **Pain Intractable to step 1 treatments**	
Allodynia, hyperalgesia[1]	Topical therapies + opioid and/or gabapentin
Spontaneous, burning, constant[2]	Gabapentin + opioid and/or antidepressant
Spontaneous, lancinating, intermittent[3]	Gabapentin + opioid and/or lamotrigine
STEP 3: Severe Functional Impairment/Severe pain[4] **Pain intractable to step 2 treatments**	
Allodynia, hyperalgesia[1]	Opioid rotation (e.g., methadone) + new AED[5] or gabapentin +/− topical therapies
Spontaneous, burning, constant[2]	Opioid rotation (e.g., methadone) + new AED[5] or gabapentin +/− antidepressant
Spontaneous, lancinating, intermittent[3]	Opioid rotation (e.g., methadone) + new AED[5] or gabapentin +/− mexiletine
STEP Procedure: Pain Poorly Responsive to Steps 1–3 **Pain treatment non amenable to conventional drug routes**	
Neurostimulatory procedure (e.g., spinal cord stimulation, motor cortex stimulation)	
Intrathecal pump for opioid (e.g., morphine, hydromorphone) +/− bupivicaine +/− clonidine[4]	

[1]If SMP, consider topical clonidine and sympatholytic interventions.
[2]If clinically feasible, trials of topical therapies, e.g., lidocaine 5% patch, may be considered for a variety of neuropathic pain states and features.
[3]For essential trigeminal/glossopharyngeal neuralgias, consider trials of carbamazepine and/or lamotrigine +/− baclofen.
[4]On a compassionate basis, according to the patient's clinical condition and pain mechanism, the physician may want to consider for intractable pain an empirical trial of one or more of the emergent topical, oral or parenteral/intrathecal therapies,[132] as discussed in the text.
[5]New AEDs: lamotrigine, oxcarbazepine, levetiracetam, topiramate, zonisamide, tiagabine.

matches the degree of pain-related impairment. Dose titration is an important principle to recognize when using analgesics, in particular AEDs, antidepressants, and opioids. Physicians must know how to titrate the dose appropriately while assessing the pain and recognizing and managing drug-related side effects. Patients suffering from complex pain states need to have treatment plans tailored to their individual problems. A trial of a single medication may be considered initially if the patient presents with an overall high level of function and mild pain or discomfort. However, as the patient becomes less functional and incapacitated by pain, a more aggressive intervention based on opioid titration and combination therapy may be necessary. Table 23-3 proposes a treatment algorithm for selection and implementation of different analgesic regimens. The treating physician needs to balance the efficacy, safety, and tolerability of a variety of drugs used for neuropathic pain. Moreover, the physician who wishes to use the neuropathic analgesic stepladder should (1) know how to assess the patient's pain and function, (2) perform a differential diagnosis and determine if possible the etiology underlying the neuropathic pain, (3) have a good fund of knowledge of the pharmacology of the existing analgesics for neuropathic pain, and (4) know how to recognize and manage side effects of each drug because multiple agents will be used concurrently.

▶ CONCLUSIONS

Peripheral neuropathic pain states are challenging disorders to treat. Advances are being made in comprehension of the various mechanisms and etiologies underlying neuropathic pain syndromes. Patients suffering from these disorders need to have treatment plans tailored to their individual problems. As indicated in Table 23-3, a trial of a single medication with established efficacy for neuropathic pain, is considered initially if the patient presents with an overall high level of function. However, as the patient becomes less functional or presents with incapacitating pain, a more aggressive intervention is necessary. The employment of agents from a variety of pharmacologic classes, including opioid analgesics, represents a contemporary standard approach to neuropathic pain management. Treatment for severe disabling pain may call for a rational pharmacologic therapy that includes a balanced combination of two or more analgesic medications and adjuvants. While progress is being made in the field of pain medicine, the key goals of care in the management of these patients are:

- A comprehensive diagnostic evaluation of the suffering patient
- Determining the cause of the pain syndrome, if necessary by obtaining appropriate consultations
- Longitudinal assessment of the clinical, functional, and psychosocial status of the patient
- Striving for the maximal amount of pain relief that hopefully will allow the patient to have as functional a lifestyle as possible.

REFERENCES

1. Pappagallo M: Aggressive pharmacologic treatment of pain, in Pisetsky DS, Bradley L (eds), *Pain Management in the Rheumatic Diseases: Rheumatic Disease Clinics of North America*. Philadelphia: Saunders, 1999:193–213.
2. Pappagallo M: Neuropathic pain in peripheral neuropathies, in Tollison CD, Satterthwaite JR, Tollison JW (eds), *Practical Pain Management*, 3d ed. Philadelphia: Lippincott Williams & Wilkins, 2002:431–448.
3. Pappagallo M: Complex regional pain syndromes, in Galer B (ed), *A Supplement to Neurology Reviews: Clinical Trends and News in Neurology*. 2000:25–29.
4. Griffin et al: Painful peripheral neuropathies and C-fiber nociceptors, in Dostrovsky JO, Carr DB, Koltzenburg M (eds), *Proceedings of the 10th World Congress on Pain*. Seattle: IASP Press, 2003:155–172.
5. Campbell JN: Nerve lesions and the generation of pain. *Muscle Nerve* 2001; 24:1261–1273.
6. Bennett GJ: Scientific basis for the evaluation and treatment of RSD/CRPS syndromes: Laboratory studies in animals and man, in Max M (ed), *Pain 1999—An Updated Review*. Seattle: IASP Press, 1999:331–337.
7. Asbury AK: Diseases of the peripheral nervous system, in Isselbacher KJ, et al (eds), *Harrison's Principles of Internal Medicine*, 14th ed. New York, McGraw-Hill, 2000:2457–2469.
8. Backonja MM, Galer BS: Pain assessment and evaluation of patients who have neuropathic pain. *Neurol Clin* 1998; 16(4):775–790.
9. Raja SN, Campbell JN, Meyer RA: Evidence for different mechanisms of primary and secondary hyperalgesia following heat injury to the glabrous skin. *Brain* 1984; 107(4):1179–1188.
10. Campbell JN: Complex regional pain syndrome and the sympathetic nervous system, in Campbell JN (ed), *Pain 1996—An Updated Review*. Seattle: IASP Press, 1996:89–96.
11. Bosch EP, Smith BE: Disorders of peripheral nerves, in Bradley WG, Daroff RB, Fenichel GM, Marsden CD (eds): *Neurology in Clinical Practice*, 3d ed. Boston, Butterworth-Heinemann, 2000:2045–2130.
12. Wolfe GI, Barohn RJ: Cryptogenic sensory and sensorimotor polyneuropathies. *Semin Neurol* 1998;8:105–111.
13. Holland NR, Stocks A, Hauer P, et al: Intra-epidermal nerve fiber density in patients with painful sensory neuropathy. *Neurology* 1997;48:708–711.
14. Waldman SD: Diabetic neuropathy: Diagnosis and treatment for the pain management specialist. *Curr Rev Pain* 2000;4:383–387.
15. Sadler M, Nelson M: Peripheral neuropathy in HIV. *Int J STD AIDS* 1997;8:16–21.

16. Pappagallo M, Haldey EJ: Pharmacological management of postherpetic neuralgia. *CNS Drugs* 2003;17:771–780.
17. Tembl JI, Ferrer JM, Sevilla MT, et al: Neurologic complications associated with hepatitis C virus infection. *Neurology* 1999;53:861–864.
18. Douen AG, Pringle CE, Guberman A: Human T-cell lymphotropic virus type 1 myositis, peripheral neuropathy, and cerebral white matter lesions in the absence of spastic paraparesis. *Arch Neurol* 1997;54:896–900.
19. Asbury AK: Pain in generalized neuropathies, in Fields HL (ed), *Pain Syndromes in Neurology*. Butterworths Neurology Medical Reviews, Vol 10. London: Butterworths, 1990;131–144.
20. Logigian EL: Peripheral nervous system Lyme borreliosis. *Semin Neurol* 1997;17:25–30.
21. Moulin DE, Hagen N, Feasby TE, et al: Pain in Guillain-Barré syndrome. *Neurology* 1997;48:328–331.
22. Amato AA, Collins MP: Neuropathies associated with malignancy. *Semin Neurol* 1998;18:125–144.
23. Olney RK: Neuropathies associated with connective tissue disease. *Semin Neurol* 1998;18:63–72.
24. Grant IA, Hunder GG, Homburger HA, et al: Peripheral neuropathy associated with sicca complex. *Neurology* 1997;48:855–862.
25. Moore PM: Vasculitic neuropathies. *J Neurol Neurosurg Psychiatry* 2000;68:271–274.
26. Kissel JT, Mendell JR: Neuropathies associated with monoclonal gammopathies. *Neuromuscul Disord* 1996;6:3–18.
27. Latov N: Pathogenesis and therapy of neuropathies associated with monoclonal gammopathies. *Ann Neurol* 1995;37:S32–542.
28. Dabby R, Weimer LH, Hays AP, et al: Antisulfatide antibodies in neuropathy: Clinical and electrophysiologic correlates. *Neurology* 2000;54:1448–1452.
29. Caniatti LM, Tugnoli V, Eleopra R, et al: Cryoglobulinemic neuropathy related to hepatitis C virus infection: Clinical, laboratory and neurophysiological study. *J Peripher Nerv Syst* 1996;1:131–138.
30. Authier FJ, Pawlotsky JM, Viard JP, et al: High incidence of hepatitis C virus infection in patients with cryoglobulinemic neuropathy. *Ann Neurol* 1993;34:749–750.
31. Falk RH, Comenzo RL, Skinner M: The systemic amyloidosis. *N Engl Med J* 1997;337:898–909.
32. Roman GC: An epidemic in Cuba of optic neuropathy, sensorineural deafness, peripheral sensory neuropathy and dorsolateral myeloneuropathy. *J Neurol Sci* 1994;127:11–28.
33. Toh BH, van Driel IR, Gleeson PA: Pernicious anemia. *N Engl J Med* 1997;337(20):1441–1448.
34. Cox JM, Pappagallo M: Contemporary and emergent pharmacological therapies for chronic pain: nonopioid analgesia. *Exp Rev Neurother* 2001;1:81–91.
35. Williams VB, Pappagallo M: Entrapment neuropathies, in Benton HT, et al (eds), *Essentials of Pain Medicine and Regional Anesthesia*. New York: Churchill-Livingston, 1999:295–298.
36. Schaumburg HH, Kaplan JG: Toxic peripheral neuropathies, in Asbury AK, Thomas PK (eds), *Peripheral Nerve Disorders 2*. Oxford, England: Butterworth-Heinemann, 1995:238–261.
37. Dyck PJ, Chance PF, Lebo RV, et al: Hereditary motor and sensory neuropathies, in Dyck PJ, Thomas PK, Griffin JW, et al (eds), *Peripheral Neuropathy*, 3d ed. Philadelphia: Saunders, 1993:1094–1136.
38. Martyn CN, Hughes RA: Epidemiology of peripheral neuropathy. *J Neurol Neurosurg Psychiatry* 1997;62:310–318.
39. Carter GT, Jensen MP, Galer BS, et al: Neuropathic pain in Charcot-Marie-Tooth disease. *Arch Phys Med Rehabil* 1998;79:1560–1564.
40. Merskey H, Bogduk N: *International Association for the Study of Pain (IASP) Classification of Chronic Pain*, 2d ed. Seattle: IASP Press, 1994.
41. Bennett GJ: Does a neuroimmune interaction contribute to the genesis of painful peripheral neuropathies? *Proc Natl Acad Sci USA* 1999;96:7737–7738.
42. Baron R: Peripheral neuropathic pain: from mechanisms to symptoms. *Clin J Pain* 2000;16:S12–520.
43. Ali Z, Raja SN, Wesselmann U, et al: Intradermal injection of norepinephrine evokes pain in patients with sympathetically maintained pain. *Pain* 2000;88:161–168.
44. Van der Laan L, ter Laak HJ, Gabreels-Festen A, et al: Complex regional pain syndrome type I (RSD): Pathology of skeletal muscle and peripheral nerve. *Neurology* 1998;51:20–25.
45. Kemler MA, van de Vusse AC, van den Berg-Loonen EM, et al: HLA-DQ1 associated for reflex sympathetic dystrophy. *Neurology* 1999;53:1350–1351.
46. Stewart JD: The brachial plexus, in Steward JD (ed), *Focal Peripheral Neuropathies*, 2d ed. New York: Raven Press, 1993:111–140.
47. Costigan M, Woolf CJ: Pain: Molecular mechanisms. *J Pain* 2000;1:35–44.
48. Hill RG: Peripheral analgesic pharmacology: An update, in Max M (ed), *Pain 1999—An Updated Review*. Seattle: IASP Press, 1999:391–395.
49. Lai J, Hunter JC, Porreca F: The role of voltage-gated sodium channels in neuropathic pain. *Curr Opin Neurobiol* 2003;13:291–297.
50. Hains BC, Klein JP, Saab CY, et al: Upregulation of sodium channel $Na_v1.3$ and functional involvement in neuronal hyperexcitability associated with central neuropathic pain after spinal cord injury. *J Neurosci* 2003;23:8881–8892.
51. Roza C, Laird JM, Souslova V, et al: The tetrodotoxin-resistant Na^+ channel $Na_v1.8$ is essential for the expression of spontaneous activity in damaged sensory axons of mice. *J Physiol* 2003;550:921–926.
52. Matthews EA, Dickenson AH: Effects of spinally delivered N- and P-type voltage-dependent calcium channel antagonists on dorsal horn neuronal responses in a rat model of neuropathy. *Pain* 2001;92:235–246.
53. Wang YX, Pettus M, Gao D, et al: Effects of intrathecal administration of ziconotide, a selective neuronal N-type calcium channel blocker, on mechanical allodynia and heat hyperalgesia in a rat model of postoperative pain. *Pain* 2000;84:151–158.
54. Wallace MS: Calcium and sodium channel antagonists for the treatment of pain. *Clin J Pain* 2000;16:S80–S85.
55. Sluka K, Price MP, Wemmie JA, Welsh MJ: ASIC 3 but not ASIC1 channels are involved in the development of

55. chronic muscle pain, in Dostrosvky JO, Carr DB, Koltzenburg M (eds), *Proceedings of the 10th World Congress on Pain*. Seattle: IASP Press, 2003:71–79.
56. Mamet J, Lazdunski M, Voilley N: How nerve growth factor drives physiological and inflammatory expressions of acid-sensing ion channel 3 in sensory neurons. *J Biol Chem* 2003;278:48907–48913.
57. Boucher TJ, Okuse K, Bennett DL, et al: Potent analgesic effects of GDNF in neuropathic pain states. *Science* 2000; 290:124–127.
58. Shamash S, Reichert F, Rotshenker S: The cytokine network of Wallerian degeneration: Tumor necrosis factor-α, interleukin-1α, and interleukin-1β. *J Neurosci* 2002; 22:3052–3060.
59. Jarvis MF, Burgard EC, McGaraughty S, et al: A-317491, a novel potent and selective non-nucleotide antagonist of P2X3 and P2X2/3 receptors, reduces chronic inflammatory and neuropathic pain in the rat. *Proc Natl Acad Sci USA* 2002;99:17179–17184.
60. McGaraughty S, Wismer CT, Zhu CZ, et al: Effects of A-317491, a novel and selective P2X3/P2X2/3 receptor antagonist, on neuropathic, inflammatory and chemogenic nociception following intrathecal and intraplantar administration. *Br J Pharmacol* 2004; 1009;223–227.
61. Khandwala H, Zhang Z, Loomis CW: Inhibition of strychnine-allodynia is mediated by spinal adenosine A1- but not A2-receptors in the rat. *Brain Res* 1998;808:106–109.
62. Gomes JA, Li X, Pan HL, et al: Intrathecal adenosine interacts with a spinal noradrenergic system to produce antinociception in nerve-injured rats. *Anesthesiology* 1999;91:1072–1079.
63. Sawynok J: Adenosine receptor activation and nociception. *Eur J Pharmacol* 1998;347:1–11.
64. Pappagallo M, Gaspardone A, Tomai F, et al: Analgesic effect of bamiphylline on pain induced by intradermal injection of adenosine. *Pain* 1993;53:199–204.
65. Zhu CZ, Mikusa J, Chu KL, et al: A-134974: A novel adenosine kinase inhibitor, relieves tactile allodynia via spinal sites of action in peripheral nerve injured rats. *Brain Res* 2001;29;905:104–110.
66. Carlton SM, McNearney TR, and Cairns BE: Peripheral glutamate receptors: novel targets for analgesics, in Dostrosvky JO, Carr DB, Koltzenburg M (eds), *Proceedings of the 10th World Congress on Pain*. Seattle: IASP Press, 2003:125–139.
67. Rice AS, Farquhar-Smith WP, Nagy I: Endocannabinoids and pain: Spinal and peripheral analgesia in inflammation and neuropathy. *Prostaglandins Leukot Essent Fatty Acids* 2002;66:243–256.
68. Karst M, Salim K, Burstein S, et al: Analgesic effect of the synthetic cannabinoid CT-3 on chronic neuropathic pain: A randomized, controlled trial. *JAMA* 2003; 290:1757–1762.
69. Richardson JD, Aanonsen L, Hargreaves KM: Antihyperalgesic effects of spinal cannabinoids. *Eur J Pharmacol* 1998;345:145–153.
70. Schafers M, Lee DH, Brors D, et al: Increased sensitivity of injured and adjacent uninjured rat primary sensory neurons to exogenous tumor necrosis factor-α after spinal nerve ligation. *J Neurosci* 2003;23:3028–3038.
71. Lindenlaub T, Sommer C: Cytokines in sural nerve biopsies from inflammatory and non-inflammatory neuropathies. *Acta Neuropathol (Berl)* 2003;105:593–602.
72. Empl M, Renaud S, Erne B, et al: TNF-α expression in painful and nonpainful neuropathies. *Neurology* 2001; 56:1371–1377.
73. Backonja M, Beydoun A, Edwards KR, et al: Gabapentin for the symptomatic treatment of painful neuropathy in patients with diabetes mellitus: A randomized, controlled trial. *JAMA* 1998;280:1831–1836.
74. Rowbotham M, Harden N, Stacey B, et al, for the Gabapentin Postherpetic Neuralgia Study Group: Gabapentin for the treatment of postherpetic neuralgia: a randomized controlled trial. *JAMA* 1998;280:1837–1842.
75. Rice AS, Maton S: Gabapentin in postherpetic neuralgia: A randomized, double-blind, placebo controlled study. *Pain* 2001;94:215–224.
76. Zakrzewska JM, Chaudhry Z, Nurmikko TJ, et al: Lamotrigine in refractory trigeminal neuralgia: Results from a double-blind, placebo-controlled crossover trial. *Pain* 1997;73:223–230.
77. Pappagallo M: Preliminary experience with topiramate in the treatment of chronic pain syndromes. Poster presented at the 17th Annual Meeting, American Pain Society, San Diego, CA, 1998.
78. Laughlin TM, Tram KV, Wilcox GL, et al: Comparison of anti-epileptic drugs tiagabine, lamotrigine, and gabapentin in mouse models of acute, prolonged, and chronic nociception. *J Pharmacol Exp Ther* 2002; 302:1168–1175.
79. Tremont-Lukats IW, Megeff C, Backonja M-M: Anticonvulsants for neuropathic pain syndromes: Mechanisms of action and place in therapy. *Drugs* 2000; 60:1029–1052.
80. McQuay H, Carroll D, Jadad AR, et al: Anticonvulsant drugs for management of pain: A systematic review. *Br Med J* 1995;311:1047–1052.
81. Portenoy RK: Opioid therapy for chronic nonmalignant pain: A review of the critical issues. *J Pain Symptom Manage* 1996;11:203.
82. Suzuki R, Chapman V, Dickenson AH: The effectiveness of spinal and systemic morphine on rat dorsal horn neuronal responses in the spinal nerve ligation model of neuropathic pain. *Pain* 1999;80:215–228.
83. Watson CPN, Babul N: Efficacy of oxycodone in neuropathic pain: A randomized trial in postherpetic neuralgia. *Neurology* 1998;50:1837–1841.
84. Dellimijn P, Vanneste J: Randomized double-blind active-placebo-controlled crossover trial of intravenous fentanyl in neuropathic pain. *Lancet* 1997;349:753–758.
85. Raja S, Haythornthwaite J, Pappagallo M, et al: Opioids versus antidepressants in postherpetic neuralgia: A randomized, placebo-controlled trial. *Neurology* 2002; 59:1015–1021.
86. Rowbotham MC, Twilling L, Davies PS, et al: Oral opioid therapy for chronic peripheral and central neuropathic pain. *N Engl J Med* 2003;348:1223–1232.
87. Gimbel JS, Richards P, Portenoy RK: Controlled-release oxycodone for pain in diabetic neuropathy: A randomized, controlled trial. *Neurology* 2003;60:927–934.

88. Maekawa K, Minami M, Masuda T, et al: Expression of mu- and kappa-, but not delta-, opioid receptor mRNAs is enhanced in the spinal dorsal horn of the arthritic rats. *Pain* 1996;64:365–371.
89. Zhang Q, Schaffer M, Elde R, Stein C: Effects of neurotoxins and hindpaw inflammation on opioid receptor immunoreactivities in dorsal root ganglia. *Neuroscience* 1998;85:281–291.
90. Crain SM, Shen KF: Antagonists of excitatory opioid receptor functions enhance morphine's analgesic potency and attenuate opioid tolerance/dependence liability. *Pain* 2000;84:121–131.
91. Harati Y, Gooch C, Swenson M, et al: Double-blind, randomized trial of tramadol for the treatment of the pain of diabetic neuropathy. *Neurology* 1998;50:1842–1846.
92. Sindrup SH, Madsen C, Brosen K, et al: The effect of tramadol in painful polyneuropathy in relation to serum drug and metabolite levels. *Clin Pharmacol Ther* 1999;66:636–641.
93. Haythornthwaite JA, Menefee LA, Quatrano-Piacentini AL, et al: Outcome of chronic opioid therapy for non-cancer pain. *J Pain Symptom Manage* 1998;15:185–194.
94. Max MB, Lynch SA, Muir J, et al: Effects of desipramine, amitriptyline and fluoxetine on pain in diabetic neuropathy. *N Engl J Med* 1992;326:1250–1256.
95. Sudoh Y, Cahoon EE, Gerner P, Wang GK: Tricyclic antidepressants as long-acting local anesthetics. *Pain* 2003;103:49–55.
96. Lang E, Hord AH, Denson D: Venlafaxine hydrochloride (Effexor) relieves thermal hyperalgesia in rats with an experimental mononeuropathy. *Pain* 1996;68:151–155.
97. Schreiber S, Backer MM, Pick CG: The antinociceptive effect of venlafaxine in mice is mediated through opioid and adrenergic mechanisms. *Neurosci Lett* 1999;273:85–88.
98. Semenchuk MR, Sherman S, Davis B: Double-blind, randomized trial of bupropion SR for the treatment of neuropathic pain. *Neurology* 2001;57:1583–1588.
99. Galer BS, Rowbotham MC, Perander J, et al: Topical lidocaine patch relieves postherpetic neuralgia more effectively than a vehicle topical patch: results of an enriched enrollment study. *Pain* 1999;80:533–538.
100. Rowbotham MC, Davies PS, Verkempinck C, et al: Lidocaine patch: Double-blind controlled study of a new treatment method for post-herpetic neuralgia. *Pain* 1996;65:39–44.
101. Devers A, Galer BS: Topical lidocaine patch relieves a variety of neuropathic pain conditions: an open-label study. *Clin J Pain* 2000;16:205–208.
102. Hart-Gouleau S, Gammaitoni A, Galer B, et al: Open-label study of the effectiveness and safety of lidocaine patch 5% (Lidoderm) in patients with painful diabetic neuropathy (abstract 173), in *Program and Abstracts of the IASP 10th World Congress on Pain*, San Diego, CA, August 17–22, 2002:169.
103. Berman SM, Justis JC, Ho M, et al: Lidocaine patch 5% (Lidoderm) significantly improves quality of life (QOL) in HIV-associated painful peripheral neuropathy (abstract 177), in *Program and Abstracts of the IASP 10th World Congress on Pain*, San Diego, CA, August 17–22, 2002:173.
104. Khan ZP, Ferguson CN, Jones RM: Alpha-2 and imidazoline receptor agonists: Their pharmacology and therapeutic role. *Anaesthesia* 1999;54:146–165.
105. Eisenach JC, DuPen S, Dubois M, et al: Epidural clonidine analgesia for intractable cancer pain. The Epidural Clonidine Study Group. *Pain* 1995;61:391–399.
106. Davis KD, Treede RD, Raja SN, et al: Topical application of clonidine relieves hyperalgesia in patients with sympathetically-maintained pain. *Pain* 1991;47:309–317.
107. Fromm GH, Aumentado D, Terrence CF: A clinical and experimental investigation of the effects of tizanidine in trigeminal neuralgia. *Pain* 1993;53:265–271.
108. McCarthy RJ, Kroin JS, Lubenow TR, et al: Effect of intrathecal tizanidine on antinociception and blood pressure in the rat. *Pain* 1990;40:333–338.
109. Fogelholm R, Murros K: Tizanidine in chronic tension-type headache: A placebo controlled, double-blind, crossover study. *Headache* 1992;32:509–513.
110. Caterina MJ, Schumacher MA, Tominaga M, et al: The capsaicin receptor: A heat-activated ion channel in the pain pathway. *Nature* 1997;389:816–824.
111. Robbins WR, Staats PS, Levine J, et al: Treatment of intractable pain with topical large-dose of capsaicin: Preliminary report. *Anesth Analg* 1998;86:579–583.
112. Bennett GJ: Update on the neurophysiology of pain transmission and modulation: Focus on the NMDA-receptor. *J Pain Sympt Manage* 2000;19:S2–S6.
113. Bennett G, Deer T, Du Pen S, et al: Future directions in the management of pain by intraspinal drug delivery. *J Pain Sympt Manage* 2000;20:S44–S50.
114. Davis AM, Inturrisi CE: D-Methadone blocks morphine tolerance and N-methyl-D-aspartate-induced hyperalgesia. *J Pharmacol Exp Ther* 1999;289:1048–1053.
115. Price DD, Mayer DJ, Mao J, Caruso FS: NMDA-receptor antagonists and opioid receptor interactions as related to analgesia and tolerance. *J Pain Sympt Manage* 2000;19:S7–S11.
116. Richardson JD: Cannabinoids modulate pain by multiple mechanisms of action. *J Pain* 2000;1:2–14.
117. Schafers M, Brinkhoff J, Neukirchen S, et al: Combined epineurial therapy with neutralizing antibodies to tumor necrosis factor-α and interleukin-1 receptor has an additive effect in reducing neuropathic pain in mice. *Neurosci Lett* 2001;310:113–116.
118. Sommer C, Marziniak M, Myers RR: The effect of thalidomide treatment on vascular pathology and hyperalgesia caused by chronic constriction injury of rat nerve. *Pain* 1998;74:83–91.
119. Ribeiro RA, Vale ML, Ferreira SH, Cunha FQ: Analgesic effect of thalidomide on inflammatory pain. *Eur J Pharmacol* 2000;391:97–103.
120. George A, Marziniak M, Schafers M, et al: Thalidomide treatment in chronic constrictive neuropathy decreases endoneurial tumor necrosis factor-α, increases interleukin-10 and has long-term effects on spinal cord dorsal horn met-enkephalin. *Pain* 2000;88:267–275.
121. Varenna M, Zucchi F, Ghiringhelli D, et al: Intravenous clodronate in the treatment of reflex sympathetic dystrophy syndrome: A randomized, double-blind, placebo-controlled study. *J Rheumatol* 2000;27:1477–1483.

122. Cortet B, Flipo RM, Coquerelle P, et al: Treatment of severe, recalcitrant reflex sympathetic dystrophy: Assessment of efficacy and safety of the second generation bisphosphonate pamidronate. *Clin Rheumatol* 1997;16:51–56.
123. Liu T, van Rooijen N, Tracey DJ: Depletion of macrophages reduces axonal degeneration and hyperalgesia following nerve injury. *Pain* 2000;86:25–32.
124. Dellemijn PL, Fields HL: Do benzodiazepines have a role in chronic pain management? *Pain* 1994;57:137–152.
125. Max MB, Schafer SC, Culnane M, et al: Amitriptyline, but not lorazepam, relieves postherpetic neuralgia. *Neurology* 1988;38:1427–1432.
126. Garcia-Larrea L, Peyron R, Mertens P, et al: Electrical stimulation of motor cortex for pain control: A combined PET-scan and electrophysiological study. *Pain* 1999;83:259–273.
127. Kemler MA, Barendse GA, van Kleef M, et al: Spinal cord stimulation in patients with chronic reflex sympathetic dystrophy. *N Engl J Med* 2000;343:618–624.
128. Katz N: Neuropathic pain in cancer and AIDS. *Clin J Pain* 2000;16:S41–S48.
129. Campbell JN, Solomon CT, James CS: The Hopkins experience with lesions of the dorsal horn (Nashold's operation) for pain from avulsion of the brachial plexus. *Appl Neurophysiol* 1988;51:170–174.
130. Thomas DG, Kitchen ND: Long-term follow up of dorsal root entry zone lesions in brachial plexus avulsion. *J Neurol Neurosurg Psychiatry* 1994;57:737–738.
131. Chin RL, Sander HW, Brannaganth, et al: Celiac neuropathy. *Neurology* 2003, May 27;60:1581–1585.
132. Tegeder I, Niederberger E, Schmidt R, et al: Specific Inhibition of I kappa B kinase reduces hyperalgesia in inflammatory and neuropathic pain models in rats. *J. Neurosci* 2004, Feb 18:24:1637–1645.
133. Hassenbusch SJ, Portenoy RK, Cousins M, et al: Polyanalgesic consensus conference 2003: An update on the management of pain by intra spinal drug delivery. *J Pain Symptom Manage* 2004;27:540–563.
134. Dworkin RH, Backonja M, Rowbotham MC, et al: Advances in neuropathic pain, diagnosis, mechanisms and treatment recommendatations, *Arch Neuro* Vol 60. Nov 2003; 1524–1534.
135. Stahl, SM: Anticonvulsants and the relief of chronic pain pregabalin and gabapentin as $\alpha_2\delta$ ligands at voltage-gated calcium channels. *J Clin Psychiatry* 2004;65:596–597.

CHAPTER 24

Pain in Neurological Disorders

E. Daniela Hord and Anca Popescu

Pain is a common symptom in many neurological disorders, but the exact prevalence is not known due to a lack of large studies. Both peripheral and central processes contribute to many chronic neuropathic pain syndromes. The various combinations of these mechanisms may explain the qualitatively different symptoms and signs that patients experience.[1]

Central pain most commonly develops after a delay of weeks or months, is associated with sensory change involving the spinothalamic pathways, and has a poor prognosis for spontaneous recovery. It is seen in patients with stroke, multiple sclerosis, syringomyelia, and spinal cord injury. Hypotheses to explain the pathogenesis and varied clinical manifestations can be divided into two categories: those stressing aberrant neural activity in the deafferented circuits and those focusing on the postlesion imbalance between facilitory and inhibitory neural pathways.[2]

▶ PAIN IN STROKE

Acute central pain is a very rare symptom in acute stroke. It has been described in association with hemiballismus in a patient with an anterior parietal artery stroke. In this particular case, disconnection of the parietal lobe from deeper structures may have played a role in the genesis of pain.[3]

Acute pain in stroke also may be related to the pathogenetic mechanism. Neck pain is common in patients with extracranial arterial dissections. Patients with dissection of the vertebral artery complain more often of neck pain than patients with carotid artery dissection.[4]

Chronic poststroke neuropathic pain (central poststroke pain, CPSP) affects 2 to 8 percent of all stroke patients[5,6] but accounts for about 90 percent of all supraspinal central pain.[7] Classically, the lesion causing CPSP is in the thalamus or the neospinothalamocortical tract, but other lesions (subcortical, capsular, or lower brain stem) may lead to development of CPSP.[7] It affects the part of the body contralateral to the lesion.

The mechanisms of CPSP are complex and poorly understood. Lesions in the neospinothalamic pathway have been shown to cause denervation hypersensitivity in the paleoreticulothalamic connections.[7] Other authors believe that there is a significant contribution from the thalamic and cortical neurons that have lost their normal input and have become hyperexcitable and discharge spontaneously.[8] Intriguingly, pain may disappear with a new lesion (stroke or hemorrhage) ipsilateral or contralateral to the original event.[9,10] In patients who have residual hemisensory symptoms following strokes in various locations (thalamus, thalamic–internal capsular area, putamen, lateral medulla), sensory symptoms may worsen and become painful following another stroke on the contralateral side of the brain or brainstem.[11]

The natural history of CPSP may include spontaneous improvement, stabilization, or sometimes worsening.[12] In the latter situations, painful symptoms may spread to the contralateral side to the CPSP in a mirror image.[12]

CPSP usually has delayed onset after the stroke and develops in the area of the initial sensory impairment.[13] Pain is described as numb, cold, burning, aching, swollen, achy, pricking, lancinating, and squeezing in various combinations.[13,14] It is associated with dysesthesias, allodynia, and sometimes hyperpathia. It is constant, with paroxysms that occur spontaneously or are triggered by various factors (cold, heat, mechanical stimuli, psychological stress, and fatigue).[13] Some nociceptive stimuli may evoke pain, whereas others may not.[15]

CPSP occurs in about 25 percent of patients with Wallenberg syndrome.[7] In these patients, pain affects the ipsilateral cheek alone or in combination with the contralateral limbs. This pain syndrome has a significant correlation with the extent of clinical sensory deficits to pinprick and temperature but not with the size of the infarction on magnetic resonance imaging (MRI).[7] An autopsy performed in a patient with Wallenberg syndrome revealed a cavity in the previously infarcted area of the dorsolateral medulla with aberrant nerve fibers and neuroma-like formations. These originated on the surface of the medulla and coursed along blood vessels toward the infarcted area.[16]

CPSP occurs in 13 to 17 percent of thalamic infarctions;[17,18] lesions that cause pain often involve the nucleus ventrocaudalis or nucleus ventrooralis intermedius[17] (in the paramedian or anterolateral thalamus).[19] Pain onset in the syndrome is variable. A meta-analysis of patients with thalamic vascular lesions (infarct and hemorrhage)[20] found that in 36 percent of cases pain occurred in the first week after the stroke. Thalamic pain was more common in geniculothalamic artery territory stroke and more often in patients with right-sided (nondominant) lesions. Wessel[19] divided patients with post-thalamic stroke CPSP into two categories: those with coexistent somatosensory deficits (classic thalamic pain syndrome) and those with normal sense perception (pure algetic thalamic syndrome). Phenomenologically, a category of these patients develops a syndrome of paresthesia-hyperpathia.[21] Thalamic lesions have differential effects on the various pain modalities (mechanical, heat, and cold); thermal allodynia may develop without mechanical allodynia; improvement in sensory modalities may occur before the resolution of spontaneous paresthesias and pathologic pain.[22]

CPSP may occur after lenticulocapsular hemorrhages involving the dorsal part of the posterior limb of the internal capsule, probably damaging the thalamocortical sensory pathway.[14]

Various painful syndromes, paroxysmal or continuous, sometimes indistinguishable from CPSP, may develop in patients with brain tumors involving the thalamus or parietal lobe.[23–26]

Shoulder pain affects about 40 percent of stroke patients in the first year after the stroke.[27] Early but careful mobilization may reduce the incidence and intensity of this pain syndrome. Spontaneous reductions in shoulder subluxation may occur only when there is significant motor recovery of the affected upper limb.[28]

Reflex sympathetic dystrophy (CRPS) of the paretic arm occurs in about 15 percent of patients in the first year after the stroke.[27] Early and aggressive treatment may decrease the incidence and the degree of disability.[29]

▶ PAIN IN SPINAL CORD INJURY (SCI)

SCI can have traumatic, inflammatory, ischemic, or degenerative causes. Irrespective of etiology, pain syndromes that develop are similar.

Chronic pain is a major problem following SCI, affecting about 65 percent of these patients, with 30 percent of them rating their pain as severe.[30] The mechanism of this pain is not completely understood, and it seems to involve functional and structural plastic changes in the central nervous system (CNS), with changes in receptor function and loss of normal inhibition resulting in increased neuronal activity.[30] There are similarities to the phantom pain syndrome.

Siddall and coworkers[31,30] divided pain following SCI into two categories (as subsequently did Sjolund)[32]

- Nociceptive pain
  - Musculoskeletal—from bone, joint, muscle trauma or inflammation, mechanical instability, muscle spasm, secondary overuse syndromes
  - Visceral—renal calculi, bowel, sphincter dysfunction, and dysreflexic headache
- Neuropathic pain
  - Above-level pain—compressive mononeuropathies, CRPS
  - At-level pain (segmental pattern)—nerve root compression (including cauda equina), syringomyelia, spinal cord trauma/ischemia (transitional zone), dual-level cord and root trauma (double lesion syndrome)
  - Below-level pain (diffuse)—spinal cord trauma/ischemia (e.g., central dysthesia syndrome)

In a study by Siddall and colleagues[33] on 100 patients with recently diagnosed SCI followed for 6 months, 40 percent of patients had musculoskeletal pain, 36 percent had at-level neuropathic pain, and 19 percent had below-level neuropathic pain. Musculoskeletal pain and at-level pain were more prominent early after SCI, whereas below-level pain was more prominent 3 to 6 months afterwards. With time elapsed after SCI, the incidence of below-level pain increases. Also, the authors of the study observed that at-level neuropathic pain is more common in hemicord injuries and not seen in anterior cord injuries.

Vogel and colleagues[34] studied the prevalence of musculoskeletal and neurological complications in adults with pediatric-onset SCI (<18 years). The 216 individuals who were interviewed had mean age at injury of 14 years and a mean age at follow-up of 29 years. Most common complications were pain at any site (69 percent), spasticity (57 percent), shoulder pain (48 percent), scoliosis (40 percent), hip contractures (23 percent), and back pain (22 percent). Whites were more likely than nonwhites to experience pain. Younger age at injury was significantly associated with scoliosis and hip subluxation, and older age at injury was associated with ankle pain and spasticity. Ankle pain, elbow contractures, and spasticity were more common in those with quadriplegia, and hip contractures were associated with paraplegia.

At-injury neuropathic pain may be of two types. There may be monoradicular pain affecting the dermatome at the level of injury. This is described as lancinating and stabbing and is believed to result from entrapment of the nerve root. It may benefit from decompression. The second type is a polyradicular pain affecting several dermatomes, in band-type distribution, uni- or bilateral, at the transitional zone that encompasses the injury level. It may be due to segmental deafferentation, it is stimulus-independent and associated with allodynia and hyperalgesia.[32]

Below-injury level neuropathic pain has a continuous, stimulus-independent, and dysesthetic quality. It is common in incomplete SCI.[32] The imbalance between activation of the spinothalamic versus the dorsal column pathway (in cases in which the latter pathway is spared) and the degree of involvement of the descending nociceptive inhibitory system probably plays a role in its pathogenesis.

Traumatic SCI may lead to the development of syringomyelia. The highest incidence is found in patients with complete thoracic lesions.[35] Post-traumatic syrinxes may extend many cord segments rostral to the level of the SCI and significantly dilate the spinal cord.[36] Development of a syrinx may be heralded by new or changing clinical features, such as increasing myelopathy, ascending neurological level, pain, or increasing muscle spasms.[37] A study from Denmark[38] followed patients with SCI over a 30-year period. This occurred in 20 patients. The mean time from trauma to diagnosis of syringomyelia was 12 years; the mean time from the first symptom to diagnosis was 3 years. Thirteen patients were operated on with syringoperitoneal drain, myelotomy, or decompression of the spinal cord and dural reconstruction. The best effect of the operations was on pain; little or no improvement was found in activities of daily living, motor function, spasticity, sensory deficit, and bladder or bowel function. Therefore, the authors recommended that patients with SCI have an MRI performed 3 months after the injury independent of recovery. If a cyst or syrinx is present or relevant clinical symptoms emerge, MRI should be performed every 3 months. If no cyst or syrinx or change in neurological status develop, MRI should be performed every 6 months. If no changes are seen in 2 years, MRI may be performed every 2 or more years.

Lee[39] studied retrospectively 45 patients with syringomyelia following SCI. The mean follow-up was 23.9 months for the patients available for follow-up (range 12 to 102 months). Pain was the most frequent manifestation, followed by motor deterioration and spasticity. Postoperative improvements in more than 50 percent of the patients were noted in those presenting with worsening motor function or spasticity.

Frisbie and colleagues[40] surveyed 55 patients with SCI (median duration 10 years, range 3 weeks to 42 years). They found that burning, stabbing, pins and needles, numbness, and location at or distal to the level of paralysis suggest central pain. Musculoskeletal pain was aching in quality located at or distal to the level of paralysis; it was associated with degenerative joint disease, scoliosis, shoulder dislocation, contractures, fractures, and soft tissue calcium deposits. Syringomyelic pain was suggested solely on the basis of pain location above the level of paralysis in 11 patients and confirmed by MRI in 8 patients.

Oral or intrathecal baclofen, used for treatment of spasticity and painful spasms after SCI, also has been shown to decrease the neuropathic pain.[41]

▶ SYRINGOMYELIA

Preservation of proprioception (conveyed by the dorsal columns) in the face of severe pain and temperature deficit is characteristic for the syringomyelic sensory dissociation, but this pattern, although classic, is present in only about 49 percent of patients.[42]

In the course of enlargement of the syringomyelic cavity, dorsal columns and corticospinal tracts become affected. Again, the clinical picture may not correlate with the MRI and pathologic examination. A patient described by Goldstein[36] developed a large syrinx after an SCI with T4 level of neurological deficit. The cavity extended from T1 to C6 and tapered to a small unilateral syrinx at C2. Light microscopy of sections from T1 to C2 showed massive loss of intermediate to intermediolateral gray neurons and moderate reduction of motor neurons at the T1 to C6 levels. Despite these findings, manual muscle testing results remained normal for wrist and elbow extensors, and the patient continued to perform independent sliding-board transfers.

Milhorat and colleagues[43] studied 137 patients with syringomyelia. At the time of clinical presentation, 37 percent had dysesthesias, with burning pain, hyperesthesia, and trophic changes. MRI showed extension of the syrinx into the dorsolateral quadrant of the spinal cord on the same side and at the level of pain in 84 percent. Surgical treatment of syringomyelia resulted in relief or improvement of dysesthetic pain in 59 percent, but 41 percent reported no improvement or an intensification of pain despite collapse of the syrinx. Postoperative dysesthetic pain often was a disabling complaint that responded poorly to medical therapy, including analgesics, sedatives, antiepileptics, antispasmodics, and anti-inflammatory agents. In most cases, there was a gradual improvement of symptoms, although 6 patients continued to complain of pain 24 to 74 months postoperatively. Prompt but transient relief was achieved in 2 of 2 patients with regional sympathetic blocks, and prolonged relief was achieved in 1 patient by stellate ganglionectomy. The authors suggested that the painful dysesthesias were caused by a disturbance of pain-modulating centers in the dorsolateral quadrant of the spinal cord. Their causalgia-like features responded in an unpredictable way to surgical collapse of the syrinx.

Attal and colleagues[44] performed quantitative sensory testing (QST) in patients with painful or painless syringomyelia before and after surgical treatment of their syrinx (at 3 and 9 months). They used Von Frey hairs, a vibrameter, and a thermotest device to determine the mechanical, vibratory, and thermal detection thresholds and the mechanical and thermal pain thresholds. Of the 18 patients, 12 had central neuropathic pain, and 17 had deficits in temperature and pain sensitivity; in 11 patients there also were deficits in vibration and touch sensitivity. MRI demonstrated good correlation between paramedian extension of the syrinx and the laterality of thermal deficits. Somatosensory evoked potentials (11 patients) were abnormal in 9 patients at level and showed good correlation with deficits in vibration. The magnitudes of the thermal and tactile deficits were similar between areas of spontaneous pain and adjacent nonpainful areas. Surgery induced a significant decrease in tactile deficits and, to a lesser extent, in thermal deficits. Effects on neuropathic pain were positive in 3 patients (total disappearance of pain) and negligible or negative in 3 patients despite collapse of the syrinx (in 2 patients).

The natural history and prognosis of syringomyelia are variable. Mariani and colleagues[45] followed 50 patients; 36 of them had surgery. In 30 percent of the nonsurgically treated patients, the clinical course was benign. In 26 of the surgically treated patients, an improvement was noted at the short-term assessment both for spasticity and for pain, but in most of them it was not maintained. Boiardi and colleagues[46] followed 69 syringomyelic patients. Half the patients deteriorated whether they were operated on or not; only 1 in 5 improved, and the rest remained stable. The authors concluded that for surgical treatment to be successful, the disease must be in rapid evolution but without definite paraparesis.

▶ NEUROPATHIC OSTEOARTHROPATHY (CHARCOT JOINT)

Syringomyelia, SCI, meningomyelocele, diabetes mellitus, congenital insensitivity to pain, steroid injections, syphilis, and leprosy may lead to joint destruction, disorganization, and effusion with osseous debris. The pathogenesis is unclear, but neurovascular and neurotraumatic factors play an important role. An important contribution to the pathogenesis is played by an abnormal vasomotor reflex, analogous to reflex sympathetic dystrophy, on a background of severe peripheral neuropathy.[47] Radiographic findings include resorption of the ends of tubular bones and fractures.[48] The result is significant bone deformities.

Neuropathic osteoarthropathy is a common complication of diabetes (occurs in up to 35 percent of patients with diabetic neuropathy). Charcot's joint should be considered in patients with a unilateral, warm, erythematous, swollen foot without other systemic symptoms. Early recognition of a Charcot's joint is important in ultimate outcome. Immobilization of the joint, patient education, and proper foot care and footware are essential in preventing further complications, including ulceration and amputation.[49]

▶ PAIN IN MULTIPLE SCLEROSIS (MS)

Pain is common at some point in the course of MS. Studies have reported variable prevalence estimates—up to 50 percent of patients,[50] or from 30 to 90 percent.[51-58] In a mail survey in Denmark, the incidence of pain in MS patients did not appear to differ significantly from the background population (79.4 versus 74.7 percent), but pain intensity, daily analgesic requirements, and impact on daily life were higher in patients with MS than in the general population.[59]

Pain may be the presenting symptom in 8 percent of patients with MS.[58] Pain in MS does not correlate with age, sex, or duration of disease.[58] The association of pain with disease severity, although found by some authors,[58] has been disproved by others.[50] Pain seems to occur more often in patients with spinal cord involvement.[58]

For didactic purposes, pain syndromes associated with MS have been divided into acute or chronic,[54] but this classification does not have any etiopathogenetic underpinning. Among the acute pain syndromes, the most characteristic for MS are trigeminal, glossopharyngeal, and occipital neuralgia, paroxysmal dysesthetic and acute radicular pain without radicular compression,[60] Lhermitte's sign (26 percent of patients),[58] optic neuritis, and painful tonic spasms (19 percent).[58] Paroxysmal symptoms may be the first manifestation of multiple sclerosis.[55]

Chronic pain syndromes associated with MS are dysesthesic limb pain (45 percent),[58] joint pain, spasms, musculoskeletal low back and shoulder pain (34 percent)[58] (due to spasticity and deconditioning, overuse of muscle groups, alteration of body mechanics, and decubitus ulcers), headaches (8 percent migraine headaches and 26 percent tension headaches), and abdominal pain.[53,61] Complications of disuse (osteoporosis, frozen shoulder, aseptic hip necrosis, compression fractures) also may cause pain.[62]

Painful cranial neuropathies are frequent in MS. Their spectrum is wide. Ocular, periorbital, or retroorbital pain, worsened by eye movements, is characteristic for the optic neuritis associated with MS. The risk of developing MS in patients who present with an episode of optic neuritis depends on the presence of demyelinating lesions on MRI of the brain. Those with no demyelinating lesions at the presentation of optic neuritis have a 22 percent risk of developing MS at 10 years, whereas those who have one or more plaques have a 56 percent risk.[63] Treatment with steroids leads to faster return of visual acuity and resolution of pain. Painful pupil-involving third nerve palsy has been reported in a patient with a demyelinating lesion in the vicinity of the third nerve nucleus in the midbrain, and it recovered with administration of intravenous steroids.[64] About 1 percent of patients with MS will have trigeminal neuralgia during the course of the disease,[65] in these patients, it may be bilateral. The mechanism of trigeminal neuralgia in MS has been hypothesized to include ephaptic transmission and demyelination at the junction of peripheral and central myelin (dorsal root entry zone of the trigeminal nerve, proximal part of the nerve root, with associated gliosis, inflammation, and juxtaposition of axons).[66] Of note, trigeminal neuralgia associated with MS has good response to the administration of prostaglandins[67,68] and gamma knife radiosurgery,[69] but these patients may have a higher incidence of sensory and motor dysfunction[70,69] and poor response to microvascular decompression.[70] Central phonophobia (painful paresthesias evoked by noise) has been reported in 3 patients with MS and demyelinating lesions in the central auditory pathways.[71] Glossopharyngeal neuralgia in MS is much less common than trigeminal neuralgia.

About 50 percent of MS patients have chronic headaches, most frequently of the vascular or tension type.[72] In patients treated with interferon-β agents but not glatiramer acetate, headache frequency and intensity may increase.[73] MS has been reported to cause acute headache in the context of a meningeal syndrome, with nuchal pain, fever, and malaise in a patient with multiple brain, brain stem, and cerebellar plaques,[74] possibly reflecting a high degree of autoimmune response such as occurs in acute demyelinating encephalomyelitis (ADEM).

Painful leg spasms are common in patients with MS and demyelinating lesions in the spinal cord. They are treated by antispasticity agents (GABA-A and B agonists, dantrolene, tizanidine, vigabatrin, and botulinum toxin A and B).[75] Botulinum toxin administration into the spastic muscles of the lower extremities (rectus femoris, vastus lateralis, hamstrings, adductor longus, gastrocnemius) decreases the frequency of painful spasms and increases the distance that the patient can walk before significant pain ensues.[76] Some authors advocate performance of percutaneous neurolytic procedures as soon as spasticity becomes painful and disabling.[77] Cannabinoids (such as nabilone) have been shown to improve the neuropathic pain and painful spasms in MS but may have intolerable side effects.[78,79] However, there is an accumulating body of data pointing to a significant loss of cannabinoid receptors in neurodegenerative diseases, so synthetic cannabinoid derivatives with little psychodynamic effect may have a future not only in the symptomatic but also in the pathogenetic treatment of MS and other neurodegenerative disorders.[80]

Central pain is one of the most challenging from the therapeutic perspective. It occurs in up to 28 percent of patients.[81] Demyelinating lesions located in eloquent areas may cause dissociation in perception of thermal and pain modalities (such as in a patient who developed central pain in the arm and hand contralateral to an upper cervical cord plaque; the patient had heat hyperalgia and cold allodynia).[82] Central pain in MS is refractory to high doses of opioids; there are anecdotal reports of resolution of painful lancinating dysthesias in the lower extremities associated with demyelinating lesions in the spinal cord with administration of intrathecal baclofen even in the absence of spasticity.[83]

Demyelinating plaque in the spinal cord may progress into a syrinx; a patient with a cervical syrinx following a cervical demyelinating lesion developed, 1 year later, a syndrome similar to complex regional pain syndrome (CRPS) in the weak hand, with erythema, edema, and severe pain of the affected extremity.[84]

Despite the novel pharmacologic, surgical, psychological, and rehabilitative approaches, unresolved pain, associated distress, and affective distress are common in patients with MS. Efforts to manage MS and its associated symptoms may cause osteoporosis and may exacerbate pain and pain-related disability.[61] Treatment with immunosuppressant agents (cyclophosphamide) may cause painful peripheral neuropathies and hemorrhagic cystitis.[60] Painful indurations may appear at the sites of injections with interferon β that may have significant impact on quality of life.[85] Treatment modalities aimed at reduction of pain may worsen MS-associated fatigue.[61]

▶ PAIN IN AMYOTROPHIC LATERAL SCLEROSIS (ALS)

Cramps are present in early stages of ALS and may be very painful, especially at night or during strong muscle contraction. Baclofen, quinine, tizanidine, or verapamil may be beneficial. The cramps need to be differentiated from periodic leg movements in sleep (PLMS) and fragmentary myoclonus.[86] Later on, cramps disappear as weakness progresses; musculoskeletal pain becomes more prominent. Between 40 and 70 percent of patients have pain in later stages. Painful sites are the shoulders (adhesive capsulitis or subluxation of the shoulder joint), pressure points (on the low back, buttocks), traction on the ligaments and joints during passive mobilization of the patient, or pressure sores.[87] Pain is managed following the principles of palliative care, including the use of opioids.

▶ PAIN IN DISORDERS OF THE PLEXUSES AND PERIPHERAL NERVES

Idiopathic brachial plexitis (neuralgic amyotrophy, Parsonage-Turner syndrome) usually presents with prominent periscapular and shoulder pain described as deep, aching, boring, or throbbing that is worsened by arm movements. The most commonly affected group is 20 to 50 years.[88,89] Pain leads to an antalgic position, with the shoulder adducted and internally rotated and the elbow flexed (the flexion-adduction sign).[90] The pain may wake the patient up at night. It usually lasts for 2 to 3 weeks. There may be sensory deficit, but this is usually confined to the sensory distribution of the axillary nerve and rarely down the radial aspect of the forearm. Pain may be concomitant with or precede the motor deficit, usually affecting the upper trunk of the brachial plexus.

Rare cases of brachial plexopathy without pain have been reported, and these seem to be more common in children than in adults. Children with brachial plexitis have a poorer prognosis than adults.[91]

Guillain-Barré syndrome (GBS) often has prominent painful symptomatology during its course. Pain either is caused by the direct effects of nerve injury or is the result of paralysis and prolonged immobilization.[92] In addition, some authors believe that a deafferentation pain syndrome is common in the early stages of recovery.[93]

Moderate to severe pain is a common and early symptom of GBS; it may precede the development of weakness by several days.[94] On admission, approximately 80 percent of patients with GBS will have pain,[95] and 100 percent of the recovering GBS ICU patients will have severe pain requiring opioid treatment.[96] About 90 percent of GBS patients describe pain during the course of their illness.[97]

Younger children with GBS may present with acute, severe pain.[48] Pain is present on admission in 79 percent of children; it is often the most important symptom and may lead to misdiagnosis. The most common pain syndrome is back and lower limb pain, present in 83 percent of patients.[95] In those with severe back leg pain, and bladder dysfunction, MRI with contrast may show enhancement of the cauda equina.[98] In an adult series, 60 percent of patients with prominent enhancement had severe back or leg pain; only 10 percent of patients with mild or no enhancement had back or leg pain.[99] Severe musculoskeletal pain in cases of severe and rapidly progressive GBS may herald the development of myositis ossificans.[100]

Several pain syndromes associated with GBS have been described: deep aching back and leg pain, dysesthesias, paresthesias, axillar and radicular pain, myalgia, joint pain, visceral discomfort, and meningismus.[101] The most common pain syndromes observed are deep aching back and leg pain and dysesthetic extremity pain. Pain intensity on admission correlates poorly with initial neurological disability and is not a predictor of poor prognosis.[97]

A recently described entity, *acute small-fiber sensory neuropathy* (ASFSN), is probably a variant of GBS. It presents with acute onset of numbness and sometimes burning dysesthesias. The dysesthesias disappear after 4 months, whereas the sensory deficit persists longer.[102]

Rarely, GBS following *Campylobacter jejuni* enteritis is accompanied by severe cramping pain and a marked increase in serum creatine kinase (CK) level. It is hypothesized that the axonal degeneration causes muscle hyperexcitability.[103]

Back and leg pain usually resolves over the first 8 to 12 weeks, but dysesthetic extremity pain may persist longer in 5 to 10 percent of patients.[97,92] Musculoskeletal pain due to joint contractures, hypercalcemia of immobilization, and decubitus ulcers may develop in the early stages of recovery and interfere with the rehabilitation program.[93]

Chronic demyelinating inflammatory neuropathy (CIDP) may have pain in 42 percent of cases.[104]

Painful peripheral neuropathies include but are not limited to the categories listed in Table 24-1. Pain represents

▶ TABLE 24-1. COMMON CAUSES OF PAINFUL PERIPHERAL NEUROPATHIES

Toxic-metabolic	Endocrine (diabetic, hypothyroid); medication-related neuropathies (isoniazid, metronidazole, chloramphenicol, misonidazole); nutritional causes (beri-beri, alcoholic, pellagra); heavy metal intoxication (thallium, arsenic); organophosphorus intoxication
Compressive	Nerve entrapment syndromes
Autoimmune	Vasculitic, paraneoplastic, and parainfectious etiologies
Infectious	Viral (HIV, herpes zoster), spirochetal, bacterial, Lyme disease
Hereditary	Fabry's disease, amyloid neuropathy, hereditary sensory neuropathy
Unknown	Idiopathic sensory neuropathy

one of the "positive" phenomena. It is described as a burning, stinging, or prickling sensation or as paresthesias, and dysesthesias. Pain usually affects the extremities symmetrically. Nevertheless, any part of the body can be involved. Other "positive" phenomena include hyperalgesia, mechanical allodynia, thermal cold and/or warm allodynia, summation, and after-sensation.[105] Cold allodynia seems to be a result of central plasticity changes,[106] whereas warm allodynia is related to C-nociceptor sensitization.[107] Summation and after-sensation may be secondary to the wind-up phenomenon.[108] "Negative" phenomena are represented by subjective numbness and objective sensory deficit. Frequently, positive and negative phenomena coexist.

Diabetic neuropathy encompasses several characteristic syndromes:

- *Cranial neuropathies* (third, sixth, and seventh nerve palsies).
- *Acute painful peripheral neuropathy.* This is rare. It may present soon after the diagnosis of diabetes mellitus type 1. Nerve biopsy revealed acute axonal degeneration with some inflammatory features in two patients; the pain resolved after a few months.[109]
- *Distal sensorimotor polyneuropathy (DSP).* This may be acute, subacute, or chronic. It is the most common microvascular complication of both type 1 and type 2 diabetes. There is evidence of selective sympathetic denervation in the painful feet of patients with diabetic neuropathy.[110] Early dysesthesia associated with painful neuropathy is a marker for the development of the "at risk" foot.[111] The Semmes-Weinstein (10-g monofilament) examination is a simple clinical modality for the prediction of early DSP, foot ulceration, and amputation and, in turn, a predictor of mortality in patients with diabetes.[112] Intensive glucose control is the only therapy proven to prevent or slow the progression of diabetic polyneuropathy.[113]
- *Thoracoabdominal neuropathy.* This occurs with subacute onset of painful paresthesias in unilateral and bilateral patches on the trunk.[114]
- *Mononeuropathies.* These are due to a vascular insult and resolve spontaneously. They are best treated by supportive therapy.[115] Diabetic amyotrophy (Bruns-Garland syndrome, proximal diabetic neuropathy) is associated with weight loss and poor metabolic control in both types of diabetes. Pain is located most commonly in the thigh but also may occur in the hip and is associated with proximal leg weakness. Pain resolves before the resolution of weakness. Diabetic amyotrophy is believed to have an autoimmune pathophysiology with inflammatory infiltrates in the nerves and vasa nervora.[116]
- *Entrapment syndromes.* These are treated by nerve decompression.

Neuropathic pain in patients with diabetes is more severe than pain experienced by patients with other types of painful peripheral neuropathies.[117] The pathogenesis of diabetic neuropathy is not completely understood. Polyol pathway abnormalities, anomalous nerve regeneration, dysfunctional sodium and calcium channels, neurotrophic and vascular endothelial growth factors have been shown to play a potential role.[118]

It is unclear why some diabetic patients develop painful neuropathy, whereas in others the neuropathy is painless. Recent studies suggest that patients with painful neuropathy have greater fluctuations in serum glucose levels and poorer diabetes control (higher mean glucose) compared with patients with painless neuropathy.[119]

Pain due to diabetic neuropathy is described as burning, electric, sharp, and dull ache, frequently worse at night and facilitated by fatigue and stress. It markedly interferes with sleep and enjoyment of life, recreational activities, normal work, mobility, general activity, social activities, and mood. More than 50 percent of these patients have a family member with painful diabetic neuropathy.[120] In an observational study, the most common pain phenomena were deep aching (88 percent) and pressure pain (69 percent), followed by pain paroxysms (59 percent) and, less frequently, pain on light touch (31 percent). Patients with increased cold and heat detection thresholds were more likely to have pain paroxysms. Patients with summation on repetitive mechanical stimulation more often had touch-evoked pain than patients without these phenomena. Pain mechanisms seemed to be increased small-fiber response in 7.4 percent of patients, deafferentation in 50.6 percent, and unclassified in 42 percent. The authors of the study inferred that the association between pain paroxysms with decreased small-fiber function and touch-evoked pain with abnormal pain summation on mechanical stimulation indicates the participation of the CNS, that sensitized small fibers as the single mechanism of pain are rare, and that symptomatology could not predict pain mechanisms as being mainly deafferentiation or sensitized small fibers.[121]

Studies on cohorts of patients with idiopathic polyneuropathy showed that about 35 percent of patients with neuropathic pain have *impaired glucose tolerance* (IGT), defined as abnormal 2-hour glucose tolerance test results with normal fasting glucose or hemoglobin A1c. This is more than twice the prevalence found in large, unselected population studies.[122]

TABLE 24-2. COMMON PAIN SYNDROMES ASSOCIATED WITH AIDS

Acute neurological pain syndromes	Acute demyelinating neuropathies
	Mononeuritis multiplex
	Myositis
	Headache
	Meningitis (the most common being toxoplasmosis and cryptococcal), HIV encephalitis, atypical aseptic meningitis or sinusitis, cerebral toxoplasmosis, or other opportunistic infections
Acute non-neurological pain syndromes	Necrotizing gingivitis
	Dental abscesses
	Oral ulcerations
	Esophageal ulcerations secondary to infections or AZT
	Arthritis (septic, reactive, aseptic)
	Acute obstruction of bowel (infection or invasion by lymphoma or Kaposi syndrome)
	Perforation related (infection or invasion by lymphoma or Kaposi syndrome)
	Abdominal pain related to cholangitis, cholecystitis or biliary tract obstruction (caused by opportunistic infections or HIV/AIDS therapy), toxic hepatitis
	Pancreatitis
Chronic neurological pain syndromes	Headache (CNS neoplasms, AZT-induced headache, meningitis of opportunistic infections, tension headache, migraine headache)
	Predominantly sensory neuropathy of AIDS
	Other neuropathies (infectious, toxic and nutritional, chronic demyelinating)
	Myopathy
Chronic non-neurological pain syndromes	Dysphagia often caused by candidiasis
	Arthropathies
	Anorectal pain
	Chest pain (besides the usual causes it could be related to opportunistic infections or local invasion by opportunistic cancers like lymphoma or Kaposi sarcoma)
	Abdominal pain caused by opportunistic infections or by invasion of opportunistic cancers

HIV/AIDS neuropathy. The most common pain syndromes associated with AIDS are listed in Table 24-2. They can be divided in neurological and nonneurological, acute and chronic. In a survey of 438 ambulatory AIDS patients in New York City, approximately 60 percent experienced significant pain.[123] Other surveys reported a prevalence of pain between 25 and 80 percent,[124–126] increasing with advanced disease.[127] In HIV/AIDS patients, the pain could be related to the disease or to therapy. The most common painful illnesses reported were HIV-related headaches, herpes simplex, painful peripheral neuropathy, back pain, and zoster, AZT-induced headaches, throat pain, and arthralgia. There is an association between the frequency, number of pain sites, increased disability on the Karnofsky scale, and higher depression scores on the Brief Symptom Inventory (BSI).[127]

There are several forms of neuropathy associated with HIV:

- *Distal sensory polyneuropathy (DSP).* This is seen in 15 to 50 percent of patients with AIDS.[128] It occurs in later stages of HIV-1 infection and has a slow and protracted clinical course.[129] The most common complaint is pain in the soles, with paresthesias involving the entire foot and absent or reduced ankle reflexes.[130] Its pathogenesis includes treatment with dideoxynucleosides (toxic to the mitochondria) and immune-mediated mechanisms (triggered by HIV infection). There is a strong correlation between plasma HIV-1 RNA levels, severity of pain, and QST results in HIV-associated DSP.[131] Intraepidermal nerve fiber density in skin biopsies is used in the detection and monitoring of this form of neuropathy.[132] On nerve biopsy, both the HIV-mediated and toxic antiretroviral forms of DSP have "dying back" distal axonal degeneration, loss of unmyelinated fibers, and variable degrees of macrophage infiltration in peripheral nerves and dorsal root ganglia.[133] Patients with painful DSP have a more frequently and significantly impaired warmth perception threshold than patients with painless DSP; also, the former have increased pain sensitivity to cold.[128] Also, patients with painful DSP have a static mechanical allodynia/hyperalgesia suggestive of a selective alteration in the processing of mechanoreceptive signals.[134] We do not know why some patients have asymptomatic and other patients symptomatic DSP. It has been shown that use of dideoxynucleosides is not a risk factor for the symptomatic form.[135]

- *Acute and chronic inflammatory demyelinating polyradiculoneuropathies (AIDP and CIDP)*. These cause global limb weakness. AIDP may occur at seroconversion, and it can be the initial manifestation of HIV-1 infection. CIDP generally occurs in the middle to late stages of HIV-1 infection.[129]
- *Mononeuropathy multiplex (MM)*. This infection is immune-mediated in the early stages of HIV-1 infection and presents as self-limited sensory and motor deficits in the distribution of individual peripheral nerves. In advanced AIDS, it is most commonly associated with cytomegalovirus (CMV) infection and affects multiple nerves in two or more extremities and even cranial nerves.[136,129]
- *Progressive polyradiculopathy (PP)*. This occurs in patients with advanced immunodeficiency and generally is caused by CMV infection. It may cause rapidly progressive flaccid paraparesis, radiating pain and paresthesias, areflexia, and sphincter dysfunction. Rapid diagnosis and treatment with anti-CMV therapy are necessary to prevent irreversible neurological deficits resulting from nerve root necrosis.[136]
- *Diffuse infiltrative lymphocytosis syndrome (DILS)*. This presents as a Sjögren's-like disorder with CD8+ T-cell infiltration of multiple organs. It may respond to antiretroviral therapy and steroids.[136]
- *Vasculitic neuropathy in HIV*. This is a rare entity. It may present as a subacute painful neuropathy (axonal sensorimotor on nerve conduction studies). Nerve biopsy reveals a necrotizing vasculitis. Pain resolves rapidly with steroids.[137]
- *Painful legs and moving toes syndrome*. This has been reported to be associated with neuropathy in HIV-infected patients.[138]
- *Tropical spastic paraparesis (HTLV-I-associated myelopathy)*. This also may occur in patients with HIV. Distal paresthesias are in the context of a myelopathic syndrome (with spastic paraparesis, sensory deficit to pinprick with a thoracic level, back pain, impotence, and sphincter disturbances). It may not respond to antiretrovirals but may improve spontaneously.[139]
- *Coexisting metabolic neuropathies*. These may be due to alcohol abuse or vitamin B_{12} or E deficiency, and must be identified and treated.

▶ PAIN IN MYOPATHIES

There are scant data on pain related to *muscular dystrophies*, although pain is present in many of these patients. Myalgia and cramps often are described either focally or involving the extremities, sometimes being present before muscle strength is affected.[140] Epigastric pain secondary to acute gastric dilatation or abdominal pseudo-obstruction in patients with Duchenne muscular dystrophy was described in several anecdotal reports,[141-143] and so were the fractures of the extremities[144] secondary to frequent falls. Steroids, which are used in an attempt to prolong the duration of ambulation, also have been reported to be helpful for myalgias.[145]

Idiopathic inflammatory myopathies (polymyositis, dermatomyositis, and overlap syndrome) may have a monophasic, relapsing-remitting, or chronic progressive course. The latter has a higher intensity of pain than the other forms.[146]

Polymyositis also has been reported as a complication of graft-versus-host disease. The disease has similar features to the idiopathic polymyositis (muscle atrophy, proximal muscle weakness, pain, elevated CK, aldolase, and SGPT). Likewise, treatment with immunosuppressants leads to resolution of symptoms.[147]

Infectious myositides (bacterial, mycoplasma, viral, fungal, parasitic) have prominent pain in the clinical picture. Of note, cases of myositis due to influenza virus with rhabdomyolysis have been reported.[148] Cases of polymyositis in patients with AIDS are still seen despite new anti-HIV agents.[149] In patients on antiretroviral medications, myalgias on pain and pressure with rhabdomyolysis may be caused by antiretroviral medications.[150] HTLV-1 myositis has an insidious presentation, more protracted course, and poorer response to steroid treatment than myositis in seronegative patients.[151]

Skeletal muscle vasculitis presents most commonly with myalgias (approximatively 22 percent of patients); other symptoms are paresthesias and weight loss. Most of these cases arising outside the spectrum of inflammatory myopathy are necrotizing; they are frequently associated with neurogenic atrophy in the setting of concomitant involvement of the peripheral nervous system.[152]

Eosinophilic myositis frequently presents with myalgias associated with swelling, tenderness, and cramps, most commonly affecting the lower extremities.[153] Differential diagnosis includes eosinophilia-myalgia, eosinophilic myositis, connective tissue diseases, parasitic infections, and Lyme disease. It usually responds to steroids.[153]

Rarely, *metabolic myopathies* (such as McArdle's disease) may mimic myositis, with muscle pain and weakness, increased erythrocyte sedimentation rate (ESR), creatinine phosphokinase (CPK), and biochemical autoimmune markers. A similar syndrome, with myalgias, stiffness, weakness, and increased muscle enzyme levels, has been described in hypothyroidism.[154] Lack of response of a myositic syndrome to steroids should make one suspect an underlying metabolic myopathy.[155]

Fibromyalgia-like symptoms with muscle pain are common in Sjögren's syndrome (44 percent pain, 27 percent fulfilling the criteria for fibromyalgia). Most of these patients (72 percent) also had inflammatory changes on muscle biopsy.[156]

Orbital myositis (orbital pseudotumor) is a nonspecific orbital inflammation confined to one or more extraocular muscles. Patients typically present with orbital pain exacerbated by eye movements, diplopia, and periorbital edema. There is enlargement of one or more extraocular muscles. These symptoms and signs usually resolve with administration of steroids.[157,158]

Statin-induced myopathy is characterized by myalgias, normal or mildly elevated CK level, muscle weakness, muscle cramps, and persistent elevation of CK and myalgias after withdrawal of these medications. Muscle pain and weakness probably affect 1 to 5 percent of patients. Rarely, myositis or rhabdomyolysis may occur. The most common statin causing myopathy is Cerivastatin. The mechanism of toxicity appears to involve inhibition of production of small regulatory proteins that are important for myocyte maintenance.[159]

Treatment of rheumatoid arthritis with D-*penicillamine* may induce a myositis with proximal muscle pain and weakness.[160]

▶ PAIN IN MOVEMENT DISORDERS

Parkinson's disease (PD) may be associated with disabling pain symptoms. Pain is estimated to occur in approximately 40 to 50 percent of patients with PD.[161,162] In a minority of individuals, pain becomes severe enough to overshadow the motor symptoms of the disease.[161,163] One of the six most frequent causes of emergency room visits in PD patients is pain.[164]

The dopaminergic system appears to play a major role in the processing of nociceptive information in the striatum and limbic areas. Neurodegenerative diseases affecting the midbrain dopaminergic system have been reported to produce spontaneous pains as in PD.[165]

Pain in patients with PD can be primary (central or peripheral neuropathic pain) or secondary (painful dystonia, muscle cramps and spasms, akathitic discomfort, autonomic dysregulation, edema, vascular disorders).[161] The various pain syndromes experienced by patients with PD suggest that there are many etiopathogenetic mechanisms.

Central pain occasionally precedes the onset of PD.[166] A cerebral or spinal cord origin is suggested by the pseudoradicular or thalamic pain pattern.[162] Pain is frequently more severe on the side affected the most by PD.[166] Sensory symptoms include paresthesias, burning dysthesias, coldness, numbness, and deep aching in a nerve or root distribution and affecting the legs more than the arms (and face the least).[162] An interesting variety of central pain, the "burning mouth" syndrome, is reported to occur in 24 percent of PD sufferers, which is five times more common than that of the general population.[168]

Pain in the abdomen, genitalia, and lower limbs may be associated with the "off" state.[169] Off-state pain responds to adjustment of antiparkinsonian agents. Abdominal pain also may be a side effect of dopaminergic medications.[170]

In general, pain is one of the dopa-resistant nonmotor symptoms, along with autonomic dysfunction, mood and cognitive impairment, and sleep problems.[171] In patients with advanced PD who underwent pallidal deep brain stimulation (DBS), there was a significant improvement in unilateral surgical pain, off pain, dystonia, cramps, and dysesthesias.[172] Pallidotomy also was found to be effective in reducing pain related to PD.[173]

Dystonia-related pain probably arises by abnormal firing in afferent nerve fibers in the dystonic muscles;[162] this is suggested by the decrease in pain with treatment of dystonia (muscle relaxants, botulinum toxin).[174] Subcutaneous apomorphine administration also may help intractable pain in PD.[189]

Arthritic pain or pain caused by shoulder bursitis may be early or presenting features of PD. In these situations, pain responds to dopaminergic medications.[162] Total shoulder arthroplasty in patients with PD is successful in terms of pain relief, but the functional results are poor, especially in patients older than 65 years of age, in whom complications are more frequent.[175]

Pain in the legs, cramping and aching, paresthesias and dysesthesias, may be a symptom of restless legs syndrome (see below). It usually responds to dopaminergic agents.

Pain in the "off" state often is compounded by hypomobility and depression. Therefore, pain may not respond completely to optimization of the dose of dopaminergic agents and may require analgesics and muscle relaxants.[176]

Restless legs syndrome (RLS) is characterized by a desire to move the extremities, often associated with sensory phenomena. These are described as aching, pulling, pricking, numbness and tingling, prickling, crawling, and creeping sensations in the calves and feet. Dysesthesias occur at rest and are most bothersome in the evening or at night, on attempting to fall asleep.[177] Patients may get relief by rubbing or moving the legs, walking, turning in bed, body rocking, or marching in place.[178] Primary RLS may be familial or idiopathic. Secondary RLS has been associated with peripheral neuropathies, iron-deficiency anemia and low serum ferritin, uremia, diabetes mellitus, rheumatoid arthritis, peripheral vascular disease,[177] and PD.[179] Symptoms usually improve with dopaminergic medications, opioids, or benzodiazepines.

Idiopathic cervical dystonia (ICD or torticollis) may be associated with pain in two-thirds of patients. Clonic torticollis is less frequently associated with pain than the tonic variety.[180]

Pain in ICD may be continuous or intermittent, diffuse over the neck and shoulders, radiating most often on the side toward which the head is twisted. The intensity of pain is not proportional to the severity of motor signs. Based on this discrepancy, it was hypothesized that there are other mechanisms involved in the genesis of pain apart from muscle spasm.[181]

Painful trauma may be involved in the pathogenesis of idiopathic torticollis. Nociception is processed by the basal ganglia and may lead to the abnormal physiology underlying dystonia.[182] Post-traumatic ICD is associated with pain onset immediately after the trauma.[183]

The threshold and discrimination of pain perception are decreased in patients with ICD compared with control subjects at the same levels of muscle contraction. Pain ipsilateral to the side of neck deviation is less intense and unpleasant than pain in patients for whom the painful side is opposite to the deviated one.[184]

Pain responds to botulinum toxin administration, paralleling the improvement in the severity of torticollis, disability, and degree of head turning.[185] There is a better response with multiple injection points per muscle than with single injections per muscle.[186] Pain also improves significantly after surgery (intradural or extradural denervation and myotomy).[187] Resolution of pain occurs before the clinical improvement in dystonia.[188]

▶ STIFF PERSON SYNDROME (SPS)

Described by Moersch and Woltman (1956),[190] its etiology and pathogenesis are still a matter of debate. It probably has both spinal and brainstem origin and an autoimmune mechanism, related to the intrathecal synthesis of antibodies to glutamic acid decarboxylase (anti-GAD).[191]

Symptoms start with slowly progressive hypertonia (board-like stiffness) of the axial and proximal limb muscles. This leads to hyperlordosis of the lumbar spine and low back pain. The back pain initially has a mechanical time pattern, but later pain becomes continuous; it is worsened by hyperextension of the lumbar spine and improved with positions that decrease the lumbar curvature.

Apart from stiffness, there are contractions of the thoraco-lumbar muscles. These are episodic spasms precipitated by noise, sensory stimulation, emotion or sudden movement. They spread to the cervical and abdominal muscles and to the upper extremities. Ambulation becomes difficult due to both the stiffness and the cramps of the lower back and legs. As a consequence, patients walk with a rigid and straight back. Muscle spasms may become severe enough to fracture the neck of the femur. Symptoms of SPS improve with sleep, anesthesia, myoneural or nerve block, intravenous administration of diazepam and intrathecal baclofen. Paraspinal muscle administration of botulinum toxin A reduces the tone of paraspinal and thigh muscles, relieves pain and improves ambulation.[192]

Also, significant improvements in the VAS score, SF-36 subscores for pain, and social functioning, general mental health, and energy-vitality have been reported with intravenous immunoglobulin (IVIG) treatment.[193]

Physical therapy with stretching of posterior musculature, postural awareness exercises, and analgesic pool therapy with water jets to induce muscle relaxation may render patients pain-free and decrease the lumbar lordosis.[194]

REFERENCES

1. Dworkin RH: An overview of neuropathic pain: Syndromes, symptoms, signs, and several mechanisms. *Clin J Pain* 2002;18:343–349.
2. Nurmikko TJ: Mechanisms of central pain. *Clin J Pain* 2000;16:S21–25.
3. Rossetti AO, Ghika JA, Vingerhoets F, et al: Neurogenic pain and abnormal movements contralateral to an anterior parietal artery stroke. *Arch Neurol* 2003;60:1004–1006.
4. Dziewas R, Konrad C, Drager B, et al: Cervical artery dissection: Clinical features, risk factors, therapy and outcome in 126 patients. *J Neurol* 2003;250:1179–1184.
5. Bowsher D: The management of central post-stroke pain. *Postgrad Med J* 1995;71:598–604.
6. Andersen G, Vestergaard K, Ingeman-Nielsen M, et al: Incidence of central post-stroke pain. *Pain* 1995;1:187–193.
7. MacGowen DJ, Janal MN, Clark WC, et al: Central post-stroke pain and Wallenberg's lateral medullary infarction: Frequency, character, and determinants in 63 patients. *Neurology* 1997;49:120–125.
8. Vestergaard K, Nielsen J, Anderson G, et al: Sensory abnormalities in consecutive, unselected patients with central post-stroke pain. *Pain* 1995;61:177–186.
9. Helmchen C, Lindig M, Petersen D, Tronnier V: Disappearance of central thalamic pain syndrome after contralateral parietal lobe lesion: Implications for therapeutic brain stimulation. *Pain* 2002;98:325–330.
10. Daniele O, Fierro B, Brighina F, et al: Disappearance of haemorrhagic stroke-induced thalamic (central) pain following a further (contralateral ischaemic) stroke. *Funct Neurol* 2003;18:95–96.
11. Kim JS: Aggravation of poststroke sensory symptoms after a second stroke on the opposite side. *Eur Neurol* 1999;42:200–204.
12. Kim JS: Delayed-onset ipsilateral sensory symptoms in patients with central poststroke pain. *Eur Neurol* 1998;40:201–206.

13. Bowsher D: Central pain: Clinical and physiologic characteristics. *J Neurol Neurosurg Psychiatry* 1996;61:62–69.
14. Kim JS: Central post-stroke pain or paresthesia in lenticulocapsular hemorrhages. *Neurology* 2003;61:679–682.
15. Mailis A, Bennett GJ: Dissociation between cutaneous and deep sensibility in central post-stroke pain (CPSP). *Pain* 2002;98:331–334.
16. Moffie D, Hamburger HL: Pain and neuroma formation in Wallenberg's lateral medullary syndrome. *Clin Neurol Neurosurg* 1986;88:217–221.
17. Paciaroni M, Bogousslavsky J: Pure sensory syndromes in thalamic stroke. *Eur Neurol* 1998;39:211–217.
18. Bogousslavsky J, Regli F, Uske A: Thalamic infarcts: Clinical syndromes, etiology, prognosis. *Neurology* 1988;38:837–848.
19. Wessel K, Vieregge P, Kessler C, Kompf D: Thalamic stroke: Correlation of clinical symptoms, somatosensory evoked potentials, and CT findings. *Acta Neurol Scand* 1994;90:167–173.
20. Nasreddine ZS, Saver JL: Pain after thalamic stroke: Right diencephalic predominance and clinical features in 180 patients. *Neurology* 1997;48:1196–1199.
21. Schott B, Laurent B, Mauguiere F: Thalamic pain: Critical study of 43 cases. *Rev Neurol (Paris)* 1986;142:308–315.
22. Greenspan JD, Joy SE, McGillis SL, et al: A longitudinal study of somesthetic perceptual disorders in an individual with a unilateral thalamic lesion. *Pain* 1997;72:13–25.
23. Retif J: Central pain and suprathalamic lesion: Temporoparietal meningioma presenting as a paroxysmal pain syndrome of pseudoradicular character of the contralateral lower extremity *Acta Neurol Belg* 1963;63:955–969.
24. Weisberg LA, Dunn D: Thalamic gliomas: Clinical and computed tomographic correlations. *Comput Radiol* 1983;7:229–235.
25. Potagas C, Avdelidis D, Singounas E, et al: Episodic pain associated with a tumor in the parietal operculum: A case report and literature review. *Pain* 1997;72:201–208.
26. Amancio EJ, Peluso CM, Santos AC, et al: Central pain due to parietal cortex compression by cerebral tumor: Report of 2 cases. *Arq Neuropsiquiatr* 2002;60:487–489.
27. Pinedo S, de la Villa FM: Complications in the hemiplegic patient in the first year after the stroke. *Rev Neurol* 2001;32:206–209.
28. Zorowitz RD: Recovery patterns of shoulder subluxation after stroke: A six-month follow-up study. *Top Stroke Rehabil* 2001;8:1–9.
29. Petchkrua W, Weiss DJ, Patel RR: Reassessment of the incidence of complex regional pain syndrome type 1 following stroke. *J Neurol Rehabil* 2000;14:59–63.
30. Siddall PJ, Loeser JD: Pain following spinal cord injury. *Spinal Cord* 2001;39:63–73.
31. Siddall PJ, Yezierski RP, Loeser JD: Pain following spinal cord injury, clinical features, prevalence and taxonomy. *IASP Newsletter* 2000;3:3–7.
32. Sjolund BH: Pain and rehabilitation after spinal cord injury: The case of sensory spasticity? *Brain Res Rev* 2002;40:250–256.
33. Siddall PJ, Taylor MJ, McClelland SB, et al: Pain report and the relationship of pain to physical factors in the first 6 months following spinal cord injury. *Pain* 1999;81:187–197.
34. Vogel LC, Krajci KA, Anderson CJ: Adults with pediatric-onset spinal cord injury: 2. Musculoskeletal and neurological complications. *J Spinal Cord Med* 2002;25:117–123.
35. Perrouin-Verbe B, Robert R, Lefort M, et al: Post-traumatic syringomyelia. *Neurochirurgie* 1999;45:58–66.
36. Goldstein B, Hammond MC, Stiens SA, Little JW: Post-traumatic syringomyelia: Profound neuronal loss, yet preserved function. *Arch Phys Med Rehabil* 1998;79:107–112.
37. Potter K, Saifuddin A: Pictorial review: MRI of chronic spinal cord injury. *Br J Radiol* 2003;76:347–352.
38. Nielsen OA, Biering-Sorensen F, Mosdal C: Post-traumatic syringomyelia. *Ugeskr Laeger* 2003;165:2877–2882.
39. Lee TT, Alameda GJ, Camilo E, Green BA: Surgical treatment of post-traumatic myelopathy associated with syringomyelia. *Spine* 2001;26:S119–127.
40. Frisbie JH, Aguilera EJ: Chronic pain after spinal cord injury: An expedient diagnostic approach. *Paraplegia* 1990;28:460–465.
41. Finnerup NB, Yezirski RP, Sang CN, et al: Treatment of spinal cord injury pain. *Pain Clin Updates* 2001;9:1–6.
42. Honan WP, Williams B: Sensory loss in syringomyelia: Not necessarily dissociated. *J R Soc Med* 1993;86:519–520.
43. Milhorat TH, Kotzen RM, Mu HT, et al: Dysesthetic pain in patients with syringomyelia. *Neurosurgery* 1996;38:940–946; discussion 946–947.
44. Attal N, Brasseur L, Parker F, et al: Characterization of sensation disorders and neuropathic pain related to syringomyelia: A prospective study. *Neurochirurgie* 1999;45:84–94.
45. Mariani C, Cislaghi MG, Barbieri S, et al: The natural history and results of surgery in 50 cases of syringomyelia. *J Neurol* 1991;238:433–438.
46. Boiardi A, Munari L, Silvani A, et al: Natural history and postsurgical outcome of syringomyelia. *Ital J Neurol Sci* 1991;12:575–579.
47. Jeffcoate W, Lima J, Nobrega L: The Charcot foot. *Diabetes Med* 2000;17:253–258.
48. Jones EA, Manaster BJ, May DA, Disler DG: Neuropathic osteoarthropathy: Diagnostic dilemmas and differential diagnosis. *Radiographics* 2000;20:S279–293.
49. Shah MK, Hugghins SY: Charcot's joint: An overlooked diagnosis. *J La State Med Soc* 2002;154:246–250;discussion 250.
50. Solaro C, Lunardi GL, Mancardi GL: Pain and MS. *Int MS J* 2003;10:14–19.
51. Archibald CJ, McGrath PJ, Rivo PG, et al: Pain prevalence, severity and impact in a clinic sample of multiple sclerosis patients. *Pain* 1994;58:89–93.
52. Clifford DB, Trotter JL: Pain in multiple sclerosis. *Arch Neurol* 1984;41:1270–1272.
53. Moulin DE, Foley KM, Ebers GC: Pain syndromes in multiple sclerosis. *Neurology* 1988;38:1830–1834.
54. Stenager E, Knudsen L, Jensen K: Acute and chronic pain syndromes in multiple sclerosis. *Acta Neurol Scand* 1991;84:197–200.
55. Twomey JA, Espir ML: Paroxysmal symptoms as the first manifestation of multiple sclerosis. *J Neurol Neurosurg Psychiatry* 1980;43:296–304.
56. Vermote R, Ketelaer P, Carton H: Pain in multiple sclerosis patients: A prospective study using the McGill Pain Questionnaire. *Clin Neurol Neurosurg* 1986;88:87–93.

57. Warnell P: The pain experience of a multiple sclerosis population: A descriptive study. *Axone* 1991;13:26–28.
58. Fryze W, Zaborski J, Czlonkowska A: Pain in the course of multiple sclerosis. *Neurol Neurochir Pol* 2002;36:275–284.
59. Svendsen KB, Jensen TS, Overvad K, et al: Pain in multiple sclerosis: a population based study. *Arch Neurol* 2003;60:1089–1094.
60. Ramirez-Lassepas M, Tulloch JW, et al: Acute radicular pain as a presenting symptom in multiple sclerosis. *Arch Neurol* 1992;49:255–258.
61. Kerns RD, Kassirer M, Otis J: Pain in multiple sclerosis: A biopsychosocial perspective. *J Rehabil Res Dev* 2002;39:225–232.
62. Krupp LB, Rizvi SA: Symptomatic treatment for underrecognized manifestations of multiple sclerosis. *Neurology* 2002;58:S32–39.
63. Beck RW and Optic Neuritis Study Group: High- and low-risk profiles for the development of multiple sclerosis within 10 years after optic neuritis: Experience of the optic neuritis treatment trial. *Arch Ophthalmol* 2003;121:944–949.
64. Bentley PI, Kimber T, Schapira AHV: Painful third nerve palsy in MS. *Neurology* 2002;58:1532.
65. Brett DC, Ferguson GG, Ebers GC, et al: Percutaneous trigeminal rhizotomy: Treatment of trigeminal neuralgia secondary to multiple sclerosis. *Arch Neurol* 1982;39:219–221.
66. Love S, Gradidge T, Coakham HB: Trigeminal neuralgia due to multiple sclerosis: Ultrastructural findings in trigeminal rhizotomy specimens. *Neuropathol Appl Neurobiol* 2001;27:238–244.
67. Anthony TR, Arnason B: Trigeminal neuralgia in multple sclerosis relieved by prostaglandin E analogue. *Neurology* 1995;5:1097–1100.
68. DMKG Study Group: Misoprostol in the treatment of trigeminal neuralgia associated with multiple sclerosis. *J Neurol* 2003;250:542–545.
69. Huang E, Teh BS, Zeck O, et al: Gamma knife radiosurgery for treatment of trigeminal neuralgia in multiple sclerosis patients. *Stereotact Funct Neurosurg* 2002;79:44–50.
70. Tacconi L, Miles JB: Bilateral trigeminal neuralgia: A therapeutic dilemma. *Br J Neurosurg* 2000;14:33–39.
71. Weber H, Pfadenhauer K, Stohr M, Rosler A: Central hyperacusis with phonophobia in multiple sclerosis. *Multiple Sclerosis* 2002;8:505–509.
72. Rolak LA, Brown S: Headaches and multiple sclerosis: A clinical study and review of the literature. *J Neurol* 1990;237:300–302.
73. Pollmann W, Erasmus LP, Feneberg W, et al: Interferon beta but not glatiramer acetate therapy aggravates headaches in MS. *Neurology* 2002;59:636–639.
74. Yoshihara A, Yamanoi T, Takahashi S, Yamamoto T: A case of childhood multiple sclerosis presenting with meningitis and multiple silent plaques. *Rinsho Shinkeigaku* 2003;43:98–101.
75. Shakespeare D, Boggild M, Young C: Anti-spasticity agents for multiple sclerosis. *Cochrane Database Syst Rev* 2003;4:CD001332.
76. Grazko MA, Polo KB, Jabbari B: Botulinum toxin for spasticity, muscle spasms and rigidity *Neurology* 1995;45:712–717.
77. Viel E, Pelissier J, Pellas F, et al: Alcohol neurolytic blocks for pain and muscle spacificity. *Neurochirurgie* 2003;49:256–262.
78. Notcutt W, Price M, Blossfeld P, Chapman G: Clinical experience of the synthetic cannabinoid nabilone for chronic pain, in Nahas GG, Sutin KM, Harvey D, Augurell S (eds), *Marijuana and Medicine*. Totowa: Humana Press, 1999:567–572.
79. Notcutt W, Price M, Sansom C, et al: Medical cannabis extract in chronic pain: Overall results of 29 "N of 1" studies (CBME-1), in *Symposium on the Cannabinoids; http:www.cannabinoidsociety.org/progab2.pdf*, April 1, 2003:55.
80. Baker D, Pryce G, Giovannoni G, Thompson AJ: The therapeutic potentials of cannabis. *Lancet Neurol* 2003;2:291–298.
81. Kalman S, Osterberg A, Sorensen J, et al: Morphine responsiveness in a group of well-defined multiple sclerosis patients: A study with IV morphine. *Eur J Pain* 2002;6:69–80.
82. Morin C, Bushnell MC, Luskin MB, Craig AD: Disruption of thermal perception in a multiple sclerosis patient with central pain. *Clin J Pain* 2002;18:191–195.
83. Becker R, Uhle EI, Alberti O, Bertalanffy H: Continuous intrathecal baclofen infusion in the management of central deafferentation pain. *J Pain Symptom Manage* 2000;20:313–315.
84. Das A, Puvanendran K: Syringomyelia and complex regional pain syndrome as complications of multiple sclerosis. *Arch Neurol* 1999;56:1021–1024.
85. Heinzerling L, Dummer R, Burg G, Schmid-Grendelmeier P: Panniculitis after subcutaneous injection of interferon beta in a multiple sclerosis patient. *Eur J Dermatol* 2002;12:194–197.
86. Sufit RL, Miller RG: Symptomatic management: Amyotrophic lateral sclerosis. *Continuum* 2002;8:82–109.
87. Bradley WG: Palliative care in the terminal stages of amyotrophic lateral sclerosis. *Continuum* 2002;8:125–139.
88. Parsonage MJ, Turner JW: Neuralgic amyotrophy, the shoulder girdle syndrome. *Lancet* 1948;1:973–978.
89. Berghi E, Kurland LT, Mulder DW, Nicolosi A: Brachial plexus neuropathy in the population of Rochester, Minnesota, 1970–1981. *Ann Neurol* 1985;18:320–323.
90. Aymond JK, Goldener JL, Hardaker WT Jr: Neuralgic amyotrophy. *Orthop Rev* 1989;18:1275–1279.
91. Van Alfen N, Schuuring J, van Engelen BGM, et al: Idiopathic neuralgic amyotrophy in children: A distinct phenotype compared to adult form. *Neuropediatrics* 2000;31:328–322.
92. Moulin DE: Pain in central and peripheral demyelinating disorders. *Neurol Clin* 1998;16:889–898.
93. Meythaler JM: Rehabilitation of Guillain-Barré syndrome. *Arch Phys Med Rehabil* 1997;78:872–879.
94. Yamauchi Y, Abe S, Sudo S, et al: A case of Guillain-Barré syndrome with severe sciatica preceding motor paralysis in the lower extremities. *Masui* 1999;48:198–200.
95. Nguyen DK, Agenarioti-Belanger S, Vanasse M: Pain in the Guillain-Barré syndrome in children under 6 years old. *J Pediatr* 1999;134:773–776.
96. Tripathi M, Kaushik S: Carbamazepine for pain management in Guillain-Barré syndrome patients in the intensive care unit. *Crit Care Med* 2000;28:655–658.

97. Moulin DE, Hagen N, Feasby TE, et al: Pain in Guillain-Barré syndrome. *Neurology* 1997;48:328–331.
98. Wilmshurst JM, Thomas NH, Robinson RO, et al: Lower limb and back pain in Guillain-Barré syndrome and associated contrast enhancement in MRI of the cauda equina. *Acta Paediatr* 2001;90:691–694.
99. Gorson KC, Ropper AH, Muriello MA, Blair R: Prospective evaluation of MRI lumbosacral nerve root enhancement in acute Guillain-Barré syndrome. *Neurology* 1996;47:813–817.
100. Hung JC, Appleton RE, Abernethy L: Myositis ossificans complicating severe Guillain-Barré syndrome. *Dev Med Child Neurol* 1997;39:775–776.
101. Pentland B, Donald SM: Pain in the Guillain-Barré syndrome: A clinical review. *Pain* 1994;59:159–164.
102. Seneviratne U, Gunasekera S: Acute small fibre sensory neuropathy: Another variant of Guillain-Barré syndrome? *J Neurol Neurosurg Psychiatry* 2002;72:540–542.
103. Satoh J, Okada K, Kishi T, et al: Cramping pain and prolonged elevation of serum creatine kinase levels in a patient with Guillain-Barré syndrome following *Campylobacter jejuni* enteritis. *Eur J Neurol* 2000;7:107–109.
104. Gorson KC, Allam G, Ropper AH: Chronic inflammatory demyelinating polyneuropathy: Clinical features and response to treatment in 67 consecutive patients with and without a monoclonal gammopathy. *Neurology* 1997; 48:321–328.
105. Backonja M: Painful neuropathies, in Loeser J (ed), *Bonica's Management of Pain,* 3d ed. Philadelphia: Lippincott, Williams & Wilkins, 2001:371–388.
106. Koltzenburg M: Afferent mechanisms mediating pain and hyperalgesia in neuralgia, in Janig W, Stanton-Hicks M (eds), *Reflex Sympathetic Dystrophy: A Reappraisal.* Seattle: IASP Press, 1996:123–150.
107. Cline MA, Ochoa J, Torebjork HE: Chronic hyperalgesia and skin warming caused by sensitized C nociceptors. *Brain* 1989;112:621–647.
108. Mendell LM: Physiological properties of unmyelinated fiber projection to the spinal cord. *Exp Neurol* 1996;16:316–322.
109. Vital C, Vital A, Dupon M, et al: Acute painful diabetic neuropathy: Two patients with recent insulin-dependent diabetes mellitus. *J Peripher Nerv Syst* 1997;2:151–154.
110. Tack CJ, van Gurp PJ, Holmes C, Goldstein DS: Local sympathetic denervation in painful diabetic neuropathy. *Diabetes* 2002;51:3545–3553.
111. Spruce MC, Potter J, Coppini DV: The pathogenesis and management of painful diabetic neuropathy: A review. *Diabetes Med* 2003;20:88–98.
112. Perkins BA, Bril V: Diagnosis and management of diabetic neuropathy. *Curr Diabetes Rep* 2002;2:495–500.
113. Poncelet AN: Diabetic polyneuropathy: Risk factors, patterns of presentation, diagnosis, and treatment. *Geriatrics* 2003;58:16–18, 24–25, 30.
114. Stewart JD: Diabetic truncal neuropathy: Topography of the sensory deficit. *Ann Neurol* 1989;25:233–238.
115. Vinik AI: Diabetic neuropathy: Pathogenesis and therapy. *Am J Med* 1999;107:S17–S26.
116. Stewart JD: *Focal Peripheral Neuropathies,* 3d ed. Philadelphia: Lippincott, Williams & Wilkins, 2000.
117. Vrethem M, Boivie J, Arnqvist H, et al: Painful polyneuropathy in patients with and without diabetes: Clinical, neurophysiologic, and quantitative sensory characteristics. *Clin J Pain* 2002;18:122–127.
118. Simmons Z, Feldman EL: Update on diabetic neuropathy. *Curr Opin Neurol* 2002;15:595–603.
119. Oyibo SO, Prasad YD, Jackson NJ, et al: The relationship between blood glucose excursions and painful diabetic peripheral neuropathy: A pilot study. *Diabetes Med* 2002;19:870–873.
120. Galer BS, Gianas A, Jensen MP: Painful diabetic polyneuropathy: Epidemiology, pain description, and quality of life. *Diabetes Res Clin Pract* 2000;47:123–128.
121. Otto M, Bak S, Bach FW, et al: Pain phenomena and possible mechanisms in patients with painful polyneuropathy. *Pain* 2003;101:187–192.
122. Singleton JR, Smith AG, Bromberg MB: Painful sensory polyneuropathy associated with impaired glucose tolerance. *Muscle Nerve* 2001;24:1225–1228.
123. Lebovits AH, Lefkowitz M, McCarthy D, et al: The prevalence and management of pain in patients with AIDS: A review of 134 cases. *Clin J Pain* 1989;5:245–248.
124. McCormack JP, Li R, Zarowny D, et al: Inadequate treatment of pain in ambulatory HIV patients. *Clin J Pain* 1993;9:279–283.
125. O'Neill WM, Sherrard JS: Pain in human immunodeficiency virus disease: A review. *Pain* 1993;54:3–14.
126. Breitbart W, McDonald MV, McCarthy D, et al: Pain in ambulatory AIDS patients: Pain characteristics and medical correlates. *Pain* 1996;68:315–321.
127. Singer EJ, Zorilla C, Fahy-Chandon B, et al: Painful symptoms reported by ambulatory HIV-infected men in a longitudinal study. *Pain* 1993;54:15–19.
128. Martin C, Solders G, Sonnerborg A, Hansson P: Painful and non-painful neuropathy in HIV-infected patients: An analysis of somatosensory nerve function. *Eur J Pain* 2003;7:23–31.
129. Verma A: Epidemiology and clinical features of HIV-1 associated neuropathies. *J Peripher Nerv Syst* 2001;6:8–13.
130. Cornblath DR, McArthur JC: Predominantly sensory neuropathy in patients with AIDS and AIDS-related complex. *Neurology* 1988;38:794–796.
131. Simpson DM, Haidich AB, Schifitto G, et al: ACTG 291 study team: Severity of HIV-associated neuropathy is associated with plasma HIV-1 RNA levels. *AIDS* 2002;16:407–412.
132. Luciano CA, Pardo CA, McArthur JC: Recent developments in the HIV neuropathies. *Curr Opin Neurol* 2003; 16:403–409.
133. Pardo CA, McArthur JC, Griffin JW: HIV neuropathy: Insights in the pathology of HIV peripheral nerve disease. *J Peripher Nerv Syst* 2001;6:21–27.
134. Bouhassira D, Attal N, Willer JC, Brasseur L: Painful and painless peripheral sensory neuropathies due to HIV infection: A comparison using quantitative sensory evaluation. *Pain* 1999;80(1–2): 265–272.
135. Schifitto G, McDermott MP, McArthur JC, et al: Dana Consortium on the Therapy of HIV Dementia and Related Cognitive Disorders: Incidence of and risk factors for HIV-associated distal sensory polyneuropathy. *Neurology* 2002;58:1764–1768.
136. Wulff EA, Wang AK, Simpson DM: HIV-associated peripheral neuropathy: Epidemiology, pathophysiology and treatment. *Drugs* 2000;59:1251–1260.

137. Bradley WG, Verma A: Painful vasculitic neuropathy in HIV-1 infection: Relief of pain with prednisone therapy. *Neurology* 1996;47:1446–1451.
138. Pitagoras de Mattos J, Oliveira M, Andre C: Painful legs and moving toes associated with neuropathy in HIV-infected patients. *Mov Disord* 1999;14:1053–1054.
139. Berger JR, Svenningsson A, Raffanti S, Resnick L: Tropical spastic paraparesis-like illness occurring in a patient dually infected with HIV-1 and HTLV-II. *Neurology* 1991;41:85–87.
140. Samaha FJ, Quinlan JG: Myalgia and cramps: Dystrophinopathy with wide-ranging laboratory findings. *J Child Neurol* 1996;11:21–24.
141. Bensen ES, Jaffe KM, Tarr PI: Acute gastric dilatation in Duchenne muscular dystrophy: A case report and review of the literature. *Arch Phys Med Rehabil* 1996;77:512–514.
142. Camelo AL, Awad RA, Madrazo A, et al: Esophageal motility disorders in Mexican patients with Duchenne's muscular dystrophy. *Acta Gastroenterol Latinoam* 1997;27:119–122.
143. Lunshof L, Schweizer JJ: Acute gastric dilatation in Duchenne's muscular dystrophy. *Ned Tijdschr Geneeskd* 2000;144:2214–2217.
144. Hsu JD: Extremity fractures in children with neuromuscular disease. *Johns Hopkins Med J* 1979;145:89–93.
145. Higuchi I, Nakamura K, Nakagawa M, et al: Steroid-responsive myalgia in a patient with Becker muscular dystrophy. *J Neurol Sci* 1993;115:219–222.
146. Sultan SM, Ioannou Y, Moss K, Isenberg DA: Outcome in patients with idiopathic inflammatory myositis: Morbidity and mortality. *Rheumatology (Oxf)* 2002;41:22–26.
147. Couriel DR, Beguelin GZ, Giralt S, et al: Chronic graft-versus-host disease manifesting as polymyositis: An uncommon presentation. *Bone Marrow Transplant* 2002;30:543–546.
148. Ozawa H, Noma S, Nonaka I: Myositis and rhabdomyolysis with influenza infection. *Nippon Rinsho* 2000;58:2276–2281.
149. Reveille JD: The changing spectrum of rheumatic disease in human immunodeficiency virus infection. *Semin Arthritis Rheum* 2000;30:147–166.
150. Mendila M, Walter GF, Stoll M, Schmidt RE: Rhabdomyolysis in antiretroviral therapy with Lamivudin. *Dtsch Med Wochenschr* 1997;122:1003–1006.
151. Gilbert DT, Morgan O, Smikle MF, et al: HTLV-1-associated polymyositis in Jamaica. *Acta Neurol Scand* 2001;104:101–104.
152. Prayson RA: Skeletal muscle vasculitis exclusive of inflammatory myopathic conditions: A clinicopathologic study of 40 patients. *Hum Pathol* 2002;33:989–995.
153. Kaufman LD, Kephart GM, Seidman RJ, et al: The spectrum of eosinophilic myositis: Clinical and immunopathogenic studies of three patients, and review of the literature (review). *Arthritis Rheum* 1993;36:1014–1024.
154. Ciompi ML, Zuccotti M, Bazzichi L, Puccetti L: Polymyositis-like syndrome in hypothyroidism: Report of two cases. *Thyroidology* 1994;6:33–36.
155. Horneff G, Paetzke I, Neuen-Jacob E: Glycogenosis type V (McArdle's disease) mimicking atypical myositis. *Clin Rheumatol* 2001;20:57–60.
156. Lindvall B, Bengtsson A, Ernerudh J, Eriksson P: Subclinical myositis is common in primary Sjögren's syndrome. *J Rheumatol* 2002;29:717–725.
157. Le Gal G, Ansart S, Boumediene A, et al: Idiopathic orbital myositis. *Rev Med Interne* 2001;22:189–193.
158. Attarian S, Fernandez C, Azulay JP, et al: Clinical and radiological features and clinical course of orbital myositis. *Rev Neurol (Paris)* 2003;159:307–312.
159. Thompson PD, Clarkson P, Karas RH: Statin-induced myopathy. *JAMA* 2003;289:1681–1690.
160. Chappel R, Willems J: d-Penicillamine-induced myositis in rheumatoid arthritis. *Clin Rheumatol* 1996;15:86–87.
161. Ford B: Pain in Parkinson's disease. *Clin Neurosci* 1998;5:63–72.
162. Olanow CW, Watts RL, Koller WC: An algorithm (decision tree) for the management of Parkinson's disease. *Neurology* 2001;56:S1–88.
163. Goetz CG, Tanner CM, Levy M, et al: Pain in Parkinson's disease. *Mov Disord* 1986;1:45–49.
164. Factor SA, Molho ES: Emergency department presentations of patients with Parkinson's disease. *Am J Emerg Med* 2000;18:209–215.
165. Saade NE, Atweh SF, Bahuth NB, Jabbur SJ: Augmentation of nociceptive reflexes and chronic deafferentation pain by chemical lesions of either dopaminergic terminals or midbrain dopaminergic neurons. *Brain Res* 1997;751:1–12.
166. Schott GD: Pain in Parkinson's disease. *Pain* 1985;22:407–411.
167. Quinn NP, Koller WC, Lang AE, Marsden CD: Painful Parkinson's disease. *Lancet* 1986;1:1366–1369.
168. Clifford TJ, Warsi MJ, Burnett CA, et al: Burning mouth in Parkinson's disease sufferers. *Gerodontology* 1998;15:73–78.
169. Raudino F: Non motor off in Parkinson's disease. *Acta Neurol Scand* 2001;104:312–315.
170. Pogarell O, Gasser T, van Hilten JJ, et al: Pramipexole in patients with Parkinson's disease and marked drug resistant tremor: A randomised, double-blind, placebo-controlled multicenter study. *J Neurol Neurosurg Psychiatry* 2002;72:713–720.
171. Rascol O, Payoux P, Ory F, et al: Limitations of current Parkinson's disease therapy. *Ann Neurol* 2003;53(suppl 3):S3–12;discussion S12–15.
172. Loher JJ, Burgunder JM, Weber S, et al: Effect of chronic pallidal deep brain stimulation on "off" period dystonia and sensory symptoms in advanced Parkinson's disease. *J Neurol Neurosurg Psychiatry* 2002:73;395–399.
173. Honey CR, Stoessi AJ, Tsui JK, et al: Unilateral pallidotomy for reduction of parkinsonian pain. *J Neurosurg* 1999;91:198–201.
174. Cordivari C, Misra VP, Catania S, Lees AJ: Treatment of dystonic clenched fist with botulinum toxin. *Mov Disord* 2001;16:907–913.
175. Koch LD, Cofield RH, Ahlskog JE: Total shoulder arthroplasty in patients with Parkinson's disease. *J Shoulder Elbow Surg* 1997;6:24–28.
176. Stein WM, Read S: Chronic pain in the setting of Parkinson's disease and depression. *J Pain Symptom Manage* 1997;14:255–258.
177. Gorman CA, Dyck PJ, Pearson JS: Symptom of restless legs. *Arch Intern Med* 1965;115:155–160.
178. Walters AS, Henning W, Chokroverty S: Frequent occurrence of myoclonus while awake and at rest, body rocking

and marching in a subpopulation of patients with restless legs syndrome. *Acta Neurol Scand* 1998;77:418–421.
179. Strang RR: The symptom of restless legs. *Med J Aust* 1967;1:1211–3.
180. Astarloa R, Morales B, Penafiel N, et al: Craniocervical dystonia and facial hemispasm: Clinical and pharmacological characteristics of 52 patients. *Rev Clin Esp* 1991;189:320–324.
181. Kutvonen O, Dastidar P, Nurmikko T: Pain in spasmodic torticollis. *Pain* 1997;69:279–286.
182. Dauer WT, Burke RE, Greene P, Fahn S: Current concepts on the clinical features, aetiology and management of idiopathic cervical dystonia. *Brain* 1998;121:547–560.
183. Truong DD, Dubinsky R, Hermanowicz N, et al: Post-traumatic torticollis. *Arch Neurol* 1991;48:221–223.
184. Lobbezoo F, Tanguay R, Thon MT, Lavigne GJ: Pain perception in idiopathic cervical dystonia (spasmodic torticollis). *Pain* 1996;67:483–491.
185. Greene P, Kang U, Fahn S, et al: Double-blind, placebo-controlled trial of botulinum toxin injections for the treatment of spasmodic torticollis. *Neurology* 1990;40:1213–1218.
186. Borodic GE, Pearce LB, Smith K, Joseph M: Botulinum a toxin for spasmodic torticollis: Multiple vs single injection points per muscle. *Head Neck* 1992;14:33–37.
187. Krauss JK, Toups EG, Jankovic J, Grossman RG: Symptomatic and functional outcome of surgical treatment of cervical dystonia. *J Neurol Neurosurg Psychiatry* 1997;63:642–648.
188. Domzal TM: Botulinum toxin in the treatment of pain. *Neurol Neurochir Pol* 1998;32:57–60.
189. Factor SA, Brown DL, Molho ES: Subcutaneous apomorphine injections as a treatment for intractable pain in Parkinson's disease. *Mov Disord* 2000;15:167–169.
190. Moersch FP, Woltman HW: Progressive fluctuating muscular rigidity and spasm ("stiff man" syndrome): report of a case and some observations in 13 others cases. *Proc Staff Meet Mayo Clin* 1956;31:421–427.
191. Dalakas MC, Li M, Fujii M, Jacobowitz DM: Stiff person syndrome: quantification, specificity, and intrathecal synthesis of GAD65 antibodies. *Neurology* 2001;11;57:780–784.
192. Davis D, Jabbari B: Significant improvement of stiff-person syndrome after paraspinal injection of botulinum toxin A. *Mov Disord* 1993;8:371–373.
193. Gerschlager W, Brown P: Effect of treatment with intravenous immunoglobulin on quality of life in patients with stiff-person syndrome. *Mov Disord* 2002;17:590–593.
194. Gallien P, Durufle A, Petrilli S, Verin M, Brissot R, Robineau S: Atypical low back pain: stiff-person syndrome. *Joint Bone Spine* 2002;69:218–221.

CHAPTER 25

Complex Regional Pain Syndromes

Ralf Baron and Andreas Binder

▶ DEFINITION

The term complex regional pain syndrome describes "a variety of painful conditions following injury which appears regionally having a distal predominance of abnormal findings, exceeding in both magnitude and duration the expected clinical course of the inciting event [and] often resulting in significant impairment of motor function, and showing variable progression over time" (IASP classification).[1] These chronic pain syndromes show different clinical features, including spontaneous pain, allodynia, hyperalgesia, edema, autonomic abnormalities, and trophic signs.

▶ OVERVIEW

After the American Civil War in 1872, Weir Mitchell described a pain syndrome accompanied by sensory and trophic symptoms that he named causalgia. It followed a traumatic partial nerve injury with a distribution of symptoms and signs beyond the territory of the injured peripheral nerve. In the beginning of the twentieth century, the German surgeon Paul Sudeck described a similar pain syndrome that was initiated by distal bone fractures without any overt nerve lesion. Because of a remarkable pain relief in patients with this syndrome owing to blockade of the sympathetic nerves supplying the injured extremity, in 1946 Evans introduced the term reflex sympathetic dystrophy (RSD). Different classifications and nomenclatures were used in the past. The most common classification consisted of two different terms. RSD described a pain syndrome with sensory, motor, and trophic symptoms and signs without a detectable nerve lesion, and causalgia was used in patients showing an accompanying nerve lesion. Research stressed the importance of the sympathetic nerve system in patients with this disease. Hence, showing that only a subgroup of patients is identified as having sympathetically maintained pain (SMP) and implementing SMP as a nonobligatory sign for the diagnosis, a new classification was introduced in 1995. The International Association for the Study of Pain (IASP) grouped the disorder in the second edition of IASP Classification of Chronic Pain Syndromes under the term complex regional pain syndromes (CRPS), type 1 corresponding to RSD and type 2 corresponding to causalgia. The term sympathetically maintained pain (SMP) was "considered to be [a] variable phenomenon associated with a variety of disorders, including CRPS types 1 and 2" only.[2]

▶ **TABLE 25-1.** SYNOPSIS OF POSSIBLE INCITING EVENTS OF CRPS

Peripheral nerve and dorsal root: peripheral nerve trauma, brachial plexus lesions, postherpetic neuralgia, root lesions
Central nervous system: spinal cord lesions, head injury, cerebral infarction, cerebral tumor
Viscera: abdominal disease, myocardial infarction
Peripheral tissues: fractures and dislocations, soft-tissue injury, fasciitis, tendonitis, bursitis, ligamentous strain, arthritis, mastectomy, deep vein thrombosis, immobilization
Idiopathic: CRPS often follows minor trauma that could not be remembered by the patient. Also some patients negate any inciting event.

SOURCE: *Used with permission.*[109]

▶ BASICS

Inciting Events

The inciting events of CRPS are widespread. The most common events are trauma (~65 percent), especially fractures, postoperative conditions, traumatic nerve injuries, contusions, and strains or sprains. Less common are inciting injuries following venipuncture, lacerations, spinal cord injuries, cerebrovascular accidents, and cardiac ischemia.[3,4] A synopsis of possible inciting events is given in Table 25-1.

Incidence

A population-based study on CRPS type 1 calculated an incidence of about 5.5 per 100,000 person-years at risk and a prevalence of about 21 per 100,000. An incidence of 0.8 per 100,000 person-years at risk and a prevalence of about 4 per 100,000 was reported for CRPS type 2.[5] Thus, CRPS type 1 develops more often than CRPS type 2. Estimations suggest an incidence of CRPS type 1 of 1 to 2 percent after fractures,[6] 1 to 5 percent of CRPS type 2 after peripheral nerve injury,[4] 12 percent after brain lesions,[7] and 5 percent after myocardial infarction.[8]

Sex

Females are affected more often than males, with a ratio of 2:1 to 4:1.[3–5,9,10]

Age

CRPS shows a distribution over all ages, with a mean age peak at 37 to 50 years.[3–5,9–11] No gender differences regarding the age of onset have been described.[5]

Socioeconomics

Retrospective studies in two different populations showed a mean time before pain center attendance after onset of CRPS of from 405 days or 30 months. In 113 patients with CRPS type 1, the mean number of different physicians consulted before referral to a U.S. pain center was 4.8, and the most common physicians referring the patients were surgeons, anesthesists, and neurologists. Only 7 percent of patients were referred by their primary care physician. Fifty-four percent were receiving workers' compensation related to CRPS, and 17 percent were involved in a related lawsuit. On average, 5.2 different treatments had been prescribed before, ranging from physiotherapy and medication to invasive treatments, such as spinal cord stimulation.[3,4]

Predisposition

No clear predisposing factors have been identified so far. Genetic studies indicate phenotypic differences of CRPS patients and healthy volunteers. Also, differences in HLA class I and II molecules have been encoded (see "Genetics" below). The role of these findings in relation to the occurrence of CRPS is not known.

▶ PATHOPHYSIOLOGY

Only a few animal models have been used to explain the clinical phenomena of CRPS.[12–14] The investigation of possible underlying mechanisms relies primarily on neuropathic pain models in humans and animals; and it has focused, so far, on the role of the sympathetic nervous system and inflammatory processes, as well as possible genetic predisposition and changes in the central nervous system (CNS).

Autonomic Disturbances

Various signs of autonomic dysfunction such as alterations in skin blood flow, temperature, and sweating suggest involvement of the sympathetic nervous system in CRPS. Physiologically, the sympathetic preganglionic neurons are under a strong central control, leading to characteristic patterns of ongoing and reflex discharges in each efferent pathway, e.g., skin and muscle vasoconstrictor activity and sudomotor activity.[15,16]

It has been assumed that abnormalities in blood flow and sweating in CRPS are caused by a functional (CRPS type 1) or anatomic (CRPS type 2) impairment of efferent sympathetic neurons.

Anatomic Impairment

This hypothesis was deduced from an animal experimental model using skin nerve injuries. Nerve lesion led to an immediate vasodilatation within the denervated area, followed by a vasoconstriction due to a supersensitivity of the vessels to circulating catecholamines.[17-19] Thus, the skin on the lesioned side was abnormally warm at first and then changed to a chronically cold status, mimicking the course of many patients with CRPS. Accordingly, measurements of the skin temperature and cutaneous blood flow demonstrated changes over the duration of the CRPS, with patients in whom the initially warm extremity showed a decrease of skin temperature and cutaneous blood flow after 6 months.[20]

An increased density of cutaneous α-adrenoreceptors in CRPS patients has been reported.[21,22] This may explain why increased unilateral vasoconstriction and low skin temperature can occur while sympathetic postganglionic nerve activity is reduced or normal; the physiologic or reduced release of norepinephrine leads to an increased vasoconstriction due to a pathologic upregulation of α-adrenergic receptors.[23,24]

Functional Impairment

CRPS type 1 does not show any overt nerve lesion, and in patients with CRPS type 2, the symptoms spread beyond the denervated territory, indicating that sympathetic denervation and denervation supersensitivity cannot explain the vasomotor and sudomotor abnormalities exclusively. Hence the autonomic symptoms and signs of CRPS also indicate a dysfunction in the CNS.[25]

Wasner and colleagues[20,26] showed that the warmth of the affected extremity in patients with CRPS type 1 is due to a functional inhibition of central cutaneous vasoconstrictor activity in the acute stage of the disease (<6 months) that leads to cutaneous vasodilatation. Accordingly, venous norepinephrine levels are lowered in the affected extremity.[27,28] However, a functional inhibition of vasoconstrictor activity may lead with time to an adrenergic supersensitivity of the cutaneous blood vessels, although a structural impairment of sympathetic nerve fibers is absent.[15,17,29,30] Disturbed sympathetic reflex patterns in the affected extremity were thought to be the underlying mechanism.[31,32] Consistent findings in animal experiments show excited reflex patterns in cutaneous vasoconstrictor neurons of impaired peripheral nerves, whereas the pathophysiologic changes are thought to be located within the CNS.[33,34]

Furthermore, patients with CRPS show unilateral sweating abnormalities.[35] The sweat production at the affected extremity is increased in acute as well as chronic stages, demonstrating unilateral enhanced sympathetic sudomotor activity. It is well known that sweat glands do not develop any denervation supersensitivity,[36,37] which rules out this phenomenon as being responsible for this hyperhidrosis.

It is to be concluded that the function of the CNS, i.e., controlling sudomotor activity and cutaneous sympathetic vasoconstrictor activity, is impaired in CRPS patients, resulting in a unilateral autonomic impairment. Predominantly spinal circuits are thought to be involved.

Sympathetically Maintained Pain (SMP)

Historically, the term *reflex sympathetic dystrophy* (RSD) was introduced by the observation that many patients with CRPS experienced pain relief as a result of sympatholytic procedures. Since then, SMP, RSD, and causalgia have been used as a mix of terms signifying pain entities and underlying mechanisms. However, the new diagnostic criteria of CRPS are based on the patient's history, symptoms, and signs only. Furthermore, SMP is not specific for CRPS because, on the one hand, not all patients with CRPS respond positively to sympatholytic procedures and, on the other hand, several different neuropathic pain syndromes also may show an SMP component, e.g., acute herpes zoster, phantom limb pain, and traumatic neuralgias. Therefore, the IASP renewed the definition of *sympathetically maintained pain* (SMP) as "pain maintained by sympathetic efferent innervation or by circulating catecholamines," whereby "SMP may be a feature of several types of painful conditions and is not an essential requirement of any one condition."[2] Thus, a positive response to those interventions is not mandatory for the diagnosis of CRPS. In patients with CRPS, subgroups can be identified by the negative (sympathetically independent pain, SIP) or positive (SMP) effect of selective sympatholytic procedures or α-adrenoreceptor antagonism.[38,39] The incidence of SMP in CRPS patients remains unclear and may change with time. However, only a correctly applied sympatholytic intervention can provide a diagnosis of SMP.[2]

Under normal circumstances, there is no interaction between sympathetic and peripheral afferent nociceptive neurons[33,40,41]; thus an intense sympathetic excitation (forced mental arithmetic) does not induce any change in the activity of afferent neurons physiologically.[42]

However, it is hypothesized that the pathophysiology of SMP results from an abnormal coupling between sympathetic efferent and nociceptive afferent neurons.[43] Neurophysiologic and neuroanatomic animal experiments found evidence for this pathomechanism following a mechanically induced peripheral nerve lesion. It may be an interaction of sympathetic efferent and intact or regenerating peripheral nociceptive C-fiber neurons and/or between sympathetic vasoconstrictor neurons and afferent somata within the dorsal root ganglion (DRG).[44] This coupling is mediated by norepinephrine, released from sympathetic terminals, and adrenoreceptors that are newly expressed on afferent nociceptive neurons. Thus, increased mRNA for α_{2A}-adrenoreceptors has been demonstrated in DRG neurons following nerve lesion.[45]

Clinical studies in humans support the hypothesis of a sympathetic-afferent coupling. In patients with limb amputation, the injection of epinephrine next to a stump neuroma increased pain.[46] In patients with CRPS type 2, the intraoperative sympathetic trunk stimulation led to an increase in spontaneous pain.[47] Furthermore, in patients with CRPS type 2 and post-traumatic neuralgia, the intracutaneous application of norepinephrine within the affected territory evoked pain that had been abolished by sympathetic blockade before. This gave further support to the hypothesis that human nociceptive neurons acquire noradrenergic sensitivity following partial nerve lesion.[48] Intradermally injected norepinephrine within the affected area in physiologically comparable dosages caused more intense pain in patients with SMP than in the contralateral nonaffected area and in healthy volunteers.[49] Concordantly, the intensity of spontaneous pain and the area of dynamic and punctate hyperalgesia were enhanced experimentally in patients with CRPS type 1 by alteration of sympathetic vasoconstrictor neuron activity within a physiologic range by controlled thermoregulation and sympathetic arousal.[50,51]

One can argue that the sympathetic cutaneous vasoconstrictor activity is inhibited functionally and unilaterally in the beginning of CRPS (see above), and the relevance of possible SMP may be questioned. However, this is not really contradictory, because it is known that loss of integrity of nociceptive fibers, as well as the activity of those neurons, forms the basis for the initiation and maintenance of painful processes.[52] Thus, a reduced sympathetic activity may be able to maintain pain when targeting a functionally upregulated number of α-adrenoreceptors on lesioned nociceptive afferent fibers.

Sensory Dysfunction

Basically, peripheral sensitization and central sensitization are involved in the pathophysiology of spontaneous pain and hyperalgesia.[53] Clinical and quantitative sensory testing of patients with CRPS demonstrated quadrant or hemisensory impairment and hyperalgesia that spreads beyond the area of spontaneous pain. These patterns indicate altered central afferent processing, and this has been shown in imaging studies.[54–58] Prefrontal cortical networks are involved in SMP in CRPS patients, and in one CRPS patient, a traumatic cerebral contusion of the left temporal lobe resolved the symptoms.[59]

Motor Dysfunction

An impairment of the muscle strength that affects the complete distal extremity without being caused by pain, edema, disuse, trophic changes, or a nerve lesion is likely to involve abnormalities in central motor processing. The phenomenon that passive movements of the joints are less impaired than those performed actively by the patient supports this idea. Kinematic analysis of reach and grip movements in patients with CRPS type 1 of the upper extremity revealed motor deficits that probably are due to an impaired integration of the visual and sensory afferent input within the parietal cortex.[60] About 50 percent of patients under investigation showed an increased physiologic tremor owing to changes in the CNS.[61] Further on, some patients complain of a neglect-like syndrome for motor tasks.[62,63] Additionally, magnetoencephalography demonstrated in CRPS patients a distinct disinhibition of the motor cortex.[57]

Thus, in addition to autonomic abnormalities, sensory and motor symptoms and signs of CRPS type 1 support an involvement of the CNS. If all these data are taken together, it has to be assumed that the CNS plays an important role in the pathophysiology of CRPS. It is not known whether these findings are primary or secondary abnormalities, e.g., induced by a constant painful afferent input.

Peripheral Inflammatory Processes

Earlier, Paul Sudeck postulated an exaggerated inflammatory reaction underlying CRPS.[64] Clinical signs, especially in the early phase of the disease, such as redness and higher temperature of the skin at the affected side, could be explained by an inflammatory process.[65,66] This hypothesis is supported clinically by a partial relief of signs following therapy with steroids.[67]

Increasing scientific evidence corroborates that a neurogenic inflammation is involved in the pathogenesis of edema, vasodilatation, and sweating in CRPS. In acute CRPS type 1, extensive plasma extravasation was shown using scintigraphy with radiolabeled immunoglobulins, indicating a high vascular permeability as a characteristic sign of inflammation.[68] Synovial biopsies and the analysis of joint fluid in CRPS patients have shown an infiltration with neutrophil granulocytes, synovial hypervascularity, and high protein concentrations.[69,70] Synovial effusion was found to be enhanced in patients in CRPS type 1 using magnetic resonance imaging (MRI).[71] Weber and colleagues[72] elicited a neurogenic vasodilatation in acute untreated CRPS type 1 patients and healthy controls by intense transcutaneous electrical stimulation via intradermal electrodes. Simultaneous measurement of protein extravasation using microdialysis showed a continuous protein extravasation in CRPS patients only that was accompanied by an increased vasodilatation. Importantly, the time course of protein extravasation can be approximated by exogenous substance P application, pointing toward a distinct involvement of neuropeptide release in the pathophysiology of CRPS. Accordingly, significantly increased serum calcitonin gene (CGRP) levels were detected in CRPS patients that were correlated with the incidence of a nerve lesion and hyperhidrosis. After 9 months of therapy, the CGRP levels declined to control levels with a reduction in inflammatory signs but without pain relief.[73] Furthermore, increased skin lactate levels suggest a chronic tissue hypoxia, possibly due to chronic inflammation.[74] Increased proinflammatory cytokine serum levels of interleukine 6 (IL-6) and tumor necrosis α (TNF-α) at the affected extremity and an increased production of nitric oxide in peripheral blood monocytes after stimulation with interferon-γ indicate an inflammatory process.[75,76]

A significantly higher seroprevalence of parovirus B19 was observed in CRPS type 1 patients.[77] The importance of antecedent infections in the pathophysiolgy of CRPS, e.g., in the generation of a facilitated chronic inflammation, is as yet unknown.

Taking all these results together, it must be acknowledged that inflammatory processes are involved in the pathophysiology of CRPS. However, the mechanisms of generation and maintenance of inflammation are only incompletely understood. One of the still unsolved questions is whether this inflammatory component is acting autonomously or is controlled by the sympathetic nervous system. Several animal studies have shown a modulating role of the sympathetic activity on the intensity of inflammation,[78,79] and sympatholytic procedures are able to reduce pain and inflammation in humans.[78] This concept of sympathetic-inflammatory interaction has still not been proven in CRPS.

Genetics

The clinical importance of genetic factors in CRPS is not clear. A Mendelian law does not seem to have an impact on the incidence and prevalence. However, limited evidence on HLA associations with different phenotypes has been obtained so far. Miailis and colleagues[80] found an increase in A3, B7, and DR(2) major histocompatibility complex (MHC) antigens in a small group of CRPS patients in whom resistance to treatment was associated with positivity of DR(2). HLA-DQ1 was increased significantly in CRPS patients, and HLA-DR13 is associated with a progression toward multifocal or generalized dystonia in CRPS.[81,82] Recently, a new HLA I locus was detected to predict a spontaneous onset of CRPS.[83] The clinical relevance of these results is unknown. It could be speculated that a genetically predisposed neuroimmune facilitation induces CRPS in distinct phenotypes. However, the innate cytokine profile was demonstrated to be normal in CRPS patients.[84]

Pathophysiologic Concepts in CRPS Following Stroke and Spinal Cord Injury

CRPS also may develop after lesions of the CNS.[7] The pathophysiology relies on peripheral traumatic injuries secondary to the preceding stroke deficits. Typical features of stroke are paresis, neglect and somatosensory deficits, and decreased skin temperature of the paretic extremity, possibly due to disuse.[85-87] However, unilateral sweating abnormalities, burning pain, and allodynia, e.g., after thalamic infarction, led to the hypothesis that autonomic disturbances and pain in stroke and CRPS may share a common pathophysiology within the CNS. Furthermore, visual deficits, neglect, paresis of the shoulder girdle, and somatosensory deficits are risk factors for a recurrent initiating event, e.g., trauma of the affected extremity, that self-perpetuate a vicious cycle of a CRPS.[88] Accordingly, affected extremities after brain injury are at higher risk of developing CRPS than unaffected extremities.[7]

CRPS following spinal cord injury is relatively rare, ranging from 5 to 12 percent in selected cohorts.[89-91] It develops within a few months, more often unilaterally at the upper extremity in tetraplegic patients. Bilateral affection of the lower extremities has been reported.[92,93] CRPS-like symptoms occur without the complete picture of CRPS being present.[94] Medullar gunshot wounds seem to predict the development of CRPS.[92] Similar to stroke patients, the association of paresis and limb trauma may initiate a vicious cycle in the pathophysiology (see above). Additionally, CRPS may contribute to contractures in the course of spinal cord injury.[89]

▶ CLINICAL FEATURES

Spatial Distribution

CRPS predominantly occurs in an extremity. Retrospective studies in large cohorts showed a similar distribution in the upper (44 to 59 percent) and lower (41 to 63 percent) extremities. In 113 retrospectively reviewed patients, the symptoms occurred in 47 percent on the right, in 51 percent on the left side, and in 2 percent bilaterally. In another survey of 134 patients, symptoms in multiple extremities were reported in 7 percent.[3,4,10] Sandroni and colleagues[5] reported symptoms in the upper extremities twice as often as in the lower extremities.

CRPS Type 1

CRPS type 1 develops following a noxious event mostly in one extremity. The clinical picture is characterized by pain and sensory, autonomic, trophic, and motor abnormalities and inflammatory symptoms. In contrast to CRPS type 2, a peripheral nerve lesion is not verifiable. A prospective study of 829 patients with CRPS gave a quantitative overview of possible signs and symptoms in relation to the disease duration (Table 25-2).[4]

Pain and Sensory Disturbances

Spontaneous pain, mostly of burning quality and felt deep inside the limb, is present in nearly every patient in the course of the disease. The quality may vary in each patient over time. Typically, the pain shows a distribution

TABLE 25-2. SIGNS AND SYMPTOMS OF CRPS

Signs and Symptoms	Duration of CRPS		
	2–6 Months	>12 Months	Total from 0–>12 Months
Inflammatory			
Pain	88%	97%	93%
Color difference	96%	84%	92%
Edema	80%	55%	69%
Temperature difference	91%	91%	92%
Limited movement	90%	83%	88%
Increase of complaints after exercise	95%	97%	96%
Neurological			
Hyperesthesia	75%	85%	76%
Hyperpathy	79%	81%	79%
Incoordination	47%	61%	54%
Tremor	44%	50%	49%
Involuntary movements	24%	47%	36%
Muscle spasm	13%	42%	25%
Paresis	93%	97%	95%
Pseudoparalysis	7%	26%	16%
Atrophy			
Skin	37%	44%	40%
Nails	23%	36%	27%
Muscle	50%	67%	55%
Bone (diffuse/spotty osteoporosis on x-ray)	41%	52%	38%
Sympathetic			
Hyperhidrosis	56%	40%	47%
Changed growth hair	71%	35%	55%
Changed growth nails	60%	52%	60%

SOURCE: *Used with permission.*[4]

in the distal part of the affected extremity, and its intensity is disproportionate regarding severity, duration, and distribution to the inciting event. Often an increase in the pain intensity is reported if the affected extremity is in a loose-hanging position, and relief is gained by keeping it up. Additionally, pain attacks can occur. *Evoked pain* is one of the striking signs. It consists of mechanical and thermal allodynia and/or hyperalgesia. Thus, any contact with the skin causes pain and leads to a protective behavior that might possibly contribute additionally to loss of motor function. Also, "deep somatic allodynia" can be evoked by any movements of the affected extremity and pressure to the joints, although they may be remote to the affected area. *Somatosensory deficits* have been demonstrated in up to 73 percent of patients diagnosed with CRPS type 1. These include a decreased perception of touch, pinprick stimuli, temperature, and thermal pain. The distribution could be a hemisensory impairment, deficits in one quadrant, or limited to the affected limb. Interestingly, patients with a generalized sensory deficit (hemisensory deficit, quadrant impairment) are identified as having a significantly longer illness duration and higher percentage of allodynia/hyperalgesia.[55,56]

Autonomic Abnormalities

Autonomic signs are variable over time and are important for the diagnosis but not necessarily present at the time of physical examination. Thus, taking the patient's history is critical to assess these signs. Common autonomic abnormalities are edema, changes in sweating, skin blood flow alterations, skin color changes, and skin temperature differences.[20,95] The edema commonly is restricted to the distal part of the affected extremity and often is present in the early stage of the disease and can resolve later on. The swelling depends on external aggravating stimuli, especially movement and stockings, and can be diminished by sympathetic blocks in some patients[96] (Fig. 25-1).

Changes in sweating are present in about half of patients. Either hypohidrosis or, more frequently, hyperhidrosis is present.[32,37,95] After rest at normal room temperature, significant skin temperature side differences (mean

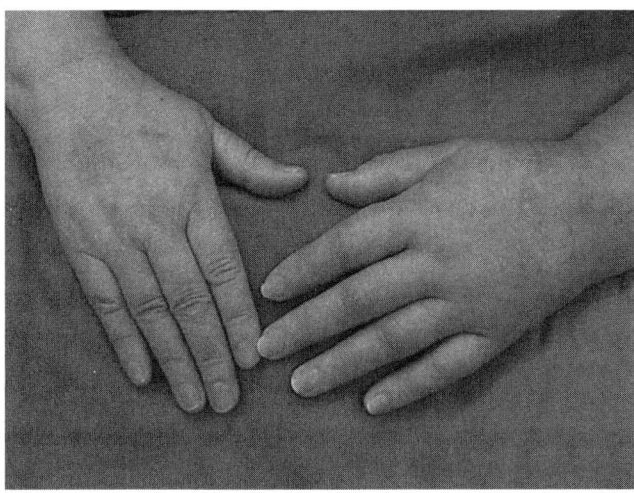

Figure 25-1. Clinical picture of a patient with CRPS type 1 of the upper left extremity following distortion trauma of the left wrist.

2.1°C) are detectable. During controlled whole-body warming and cooling, the temperature changes dynamically, and side differences become most prominent (mean 4.5°C) at medium to high sympathetic vasoconstrictor activity (cold environment). Only minor differences were found in patients with limb pain of other origin and healthy controls.[97] Accordingly, the skin color changes between paleness and redness, and skin blood flow is decreased or increased.

Trophic Abnormalities

Trophic changes are typical features of CRPS but also may result from prolonged disuse. Hair loss or hypertrichosis; decreased or increased nail growth or structure (e.g., hourglass nails; striation); thinning or thickening of the skin; and shiny or brown pigmented skin are common. Frequently, osteoporosis is diagnosed, predominantly in chronic stages, as well as atrophy of the subcutaneous tissues and muscles. Less commonly, palmar or plantar nodular fasciitis and clubbing of fingers or toes are present.[4]

Motor Dysfunction

Weakness of the muscles at the affected extremity is present in most patients. The initiation of movements and the performance of precise small motions are impaired without signs of apraxis.[98] In about 50 percent of patients, a postural and/or action tremor with a high amplitude and a mean frequency of 7 Hz is obvious clinically and is thought to be an enhanced physiologic tremor.[61] Dystonia of the affected limb, especially in the hand or foot, is reported in about 10 percent of patients.[99,100] In patients with long disease duration, restrictions of active and passive movements occur, e.g., contractures. Motor impairment, including contractures, paresis, tremor, and impairment of movement initiation, is associated with distributed sensory deficits and a higher incidence of allodynia/hyperalgesia.[55] Hand involvement often is accompanied by complaints of the shoulder, such as frozen shoulder and tendinitis of the biceps.[4] Myofascial dysfunction, diagnosed by the presence of trigger points, tenderness in the taut muscle, and exacerbation of pain caused by palpitation of the trigger point at the remote muscles, was shown in 61 percent of 41 CRPS patients. Also, in 65 percent of these, a neglect-like syndrome of the affected extremity was present.[101]

CRPS Type 2

The symptoms of CRPS type 2 are similar to those of CRPS type 1. The only exception is that a lesion of nerve structures and subsequently focal deficits are mandatory for the diagnosis. The symptoms and signs usually are not limited to the territory of a single nerve, but a restriction is not in conflict with the current definition.[1] Thus, a complete patient history and physical examination are needed to differentiate CRPS from other neuropathic pain conditions, e.g., post-traumatic neuralgia.

CRPS in Children

CRPS types 1 and 2 also occur in children. The diagnosis seems to be delayed more often than in adults.[102,103] The incidence increases with puberty, and females are affected more frequently (with a ratio of 4:1), as is the lower limb (5.3:1). The mean age of onset was 12.5 years in a cohort of 396 children.[103] Small differences are reported in the value of diagnostic tests and treatment strategies (see below).

▶ DIAGNOSTIC CRITERIA

The diagnosis of CRPS types 1 and 2 follows the IASP clinical criteria from 1995[2] that have to be matched. When two clinical signs are joined by *or*, if *either* sign is present, or if both are present, the condition of the statement is satisfied.

CRPS Type 1 (Formerly RSD)

1. Type 1 is a syndrome that develops after an initiating noxious event.
2. Spontaneous pain and allodynia/hyperalgesia occur, are not limited to the territory of a single peripheral nerve, and are disproportionate to the inciting event.
3. There is or has been evidence of edema, skin blood flow abnormality, or abnormal sudomotor activity in the region of the pain since the inciting event.
4. This diagnosis is excluded by the existence of conditions that otherwise would account for the degree of pain and dysfunction.

CRPS Type 2 (Formerly Causalgia)

1. Type 2 is a syndrome that develops after nerve injury. Spontaneous pain and allodynia/hyperalgesia occur and are not necessarily limited to the territory of the injured nerve.
2. There is or has been evidence of edema, skin blood flow abnormality, or abnormal sudomotor activity in the region of the pain since the inciting event.
3. This diagnosis is excluded by the existence of conditions that otherwise would account for the degree of pain and dysfunction.

Pain is essential for the diagnosis, whereby *spontaneous* indicates pain without external cause. Motor symptoms and findings are not included in this classification, although they are common and can include tremor, dystonia, and weakness.[61,98,99,104]

Sensitivity and Specifity of the Diagnostic Criteria

An external validation of the IASP criteria in 117 patients with CRPS and 43 patients with neuropathic pain without CRPS demonstrated a high sensitivity (0.98) but low specifity (0.36). Thus, it could be concluded that a positive diagnosis of CRPS is likely to be correct in about 40 percent only and bears the risk of overdiagnosis.[105,106] These results foster a modification of the criteria by additional and prompt clinical validation. New revised preliminary diagnostic criteria consisting of at least two sign and four symptom categories that additionally comprise motor dysfunction may improve the accuracy of diagnosis up to 85 percent.[105]

Standardized diagnostic tests help to confirm the clinical diagnosis of both CRPS types (see below). Possibly due to the low specific criteria of CRPS, the new terminology did not replace the former denominations immediately. Alvarez-Lario and colleagues[107] demonstrated that in 1995 to 1999, 576 Med-Line–listed publications used the term RSD but only 24 used CRPS type 1.

Stages

A sequential progression of untreated CRPS has been described in the literature.[98,108–110] Accordingly, each stage (usually three stages are proposed) differs in patterns of signs and symptoms. This concept has come into question in the last years. Bruehl and colleagues[10] tested 113 patients with CRPS for the clinical validity of this concept of disease course. Using a cluster analysis, three subgroups could be identified, differentiated by their symptoms and signs, but each showed no differences in disease duration. These results argue against the sequential concept and are in line with pervious reports.[111–113] However, the sequential concept relies on the course of untreated CRPS. All studies performed so far to prove its clinical validity investigated patients who had already been under treatment. Furthermore, vascular disturbances and skin temperature measurements indicate some different stages in CRPS (see above). Therefore, a sequential course cannot be determined.

In conclusion, the concept of sequential progression of CRPS should be used carefully and has not been included in the current taxonomy.

▶ DIAGNOSTIC TESTS

The current criteria for CRPS types 1 and 2 are based mainly on the patient's history and a careful physical examination. Some further diagnostic tests could add valuable information to confirm the diagnosis, although the absence of abnormal results does not argue against the diagnosis of CRPS.

Patient History

A detailed exploration of the patient's history and symptoms is compulsory. Special attention should be paid to a possible inciting trauma, symptom onset in relation to the trauma, time course of the disease, distribution and characteristics of pain, and all sensory, autonomic, and motor symptoms. Especially, autonomic and trophic changes, such as edema, skin blood flow abnormality, or abnormal sudomotor activity in the region of the pain since the inciting event, according to the current definition, have to be checked. Psychological stressors also should be explored carefully (see below).

Physical Examination

A neurological examination is necessary. At first, an inspection is useful to assess autonomic (e.g., hyper-/hypohidrosis, edema, skin color, and skin temperature) and trophic changes (e.g., skin structure, nail appearance, hyper-/hypotrichosis). The investigator must become aware of the extent and distribution of the sensory symptoms (e.g., sensory deficits, hyperalgesia, mechanical and/or thermal allodynia, and deep hyperalgesia caused by joint movement or pressure to the joints) and motor symptoms (e.g., muscle strength, dystonia, tremor, exaggerated tendon reflexes, and myoclonic jerks) at the affected limb. Interestingly, passive movements of the joints are less impaired than those performed actively by the patient. In a detailed examination, hemisensory or quadrant deficits of touch and warm, cold, and heat perception ipsilateral to the affected side may be detectable.[56] A neglect-like syndrome will become obvious with neurological motor neglect testing.[63]

The examination has to be carried out in comparison with the status of the contralateral limb and has to take the involvement of multiple limbs into account.

Autonomic Function Tests

These tests consist of infrared thermometry, infrared thermography, quantitative sudomotor axon reflex test (QSART), thermoregulatory sweat test (TST), and laser Doppler flowmetry.

Skin temperature differences can be assessed easily by infrared thermometry or thermography.[114] These depend on environmental conditions, can change dynamically within minutes, and are most prominent in cold to warm environments. Skin temperature side differences at rest (22 to 24°C room temperature, supine position for 30 minutes) of less than 2°C show a poor sensitivity of 32 percent but a specificity of 100 percent. Under controlled thermoregulation temperature, side differences of more than 2.2°C achieve a sensitivity of 76 percent and a specificity of 100 percent. Thus, to improve the sensitivity, i.e., the attempt to assess the maximum asymmetry, in clinical practice, repetitive measurements should be carried out at the beginning, in the middle, and at the end of the patient's visit. If detected, differences of more than 2.2°C are specific and sensitive for the diagnosis of CRPS.[26,31,97]

In a study in 21 patients with CRPS, the QSART and the TST showed an enhanced sudomotor output compared with the contralateral limb within a mean disease duration of 5 weeks. At a mean duration of 94 weeks, the TST remained pathologic, whereas the QSART showed no side differences.[32]

Wasner and colleagues[20] used the laser Doppler flowmetry to assess the vascular reflex response in 25 patients with CRPS type 1. By controlled whole-body cooling and warming, the sympathetic vasoconstrictor activity was altered, and the cutaneous blood flow in the upper or lower extremities was monitored simultaneously. Based on the results obtained, the authors emphasized three different vascular regulation patterns in CRPS type 1. In short-lasting CRPS, the so-called acute phase, with a mean disease duration of 4 months, the affected limb showed higher skin perfusion values. In patients with a history of 15 months (mean), the intermediate phase, the affected limb showed either higher or lower skin perfusion. If the duration was longer, with a mean of 28 months, the affected limb showed lower perfusion of the skin on the affected side. Subsequently, the skin temperature was altered the same way.[20]

Bone Scintigraphy

Osseous changes are common in CRPS. Thus a three-phase bone scintigraphy can provide valuable information.[115] A homogeneous unilateral hyperperfusion in the perfusion (30 s after injection) and blood pool phases (2 min after injection) is characteristic and will help to exclude differential diagnosis, e.g., osteoporosis due to inactivity. Three hours after injection, the mineralization phase will show increased unilateral periarticular tracer uptake (Fig. 25-2). A pathologic uptake in the

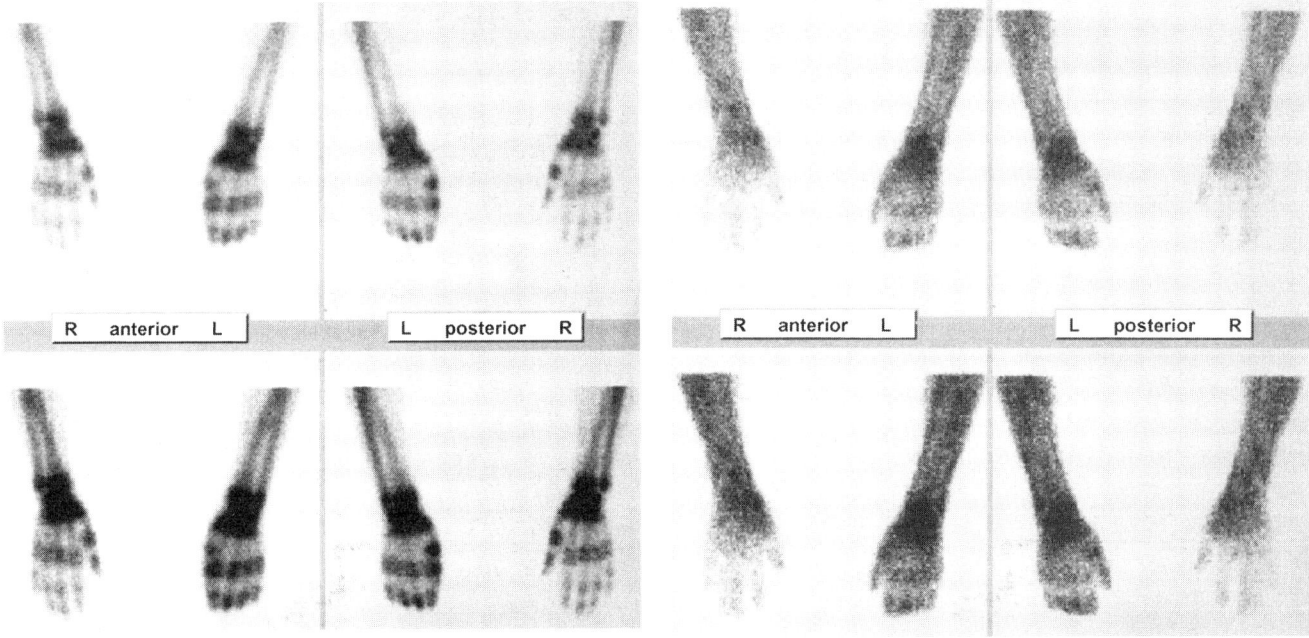

Figure 25-2. Three-phase-bone scintigraphy in a patient with CRPS type 1 of upper left extremity. Characteristic unilateral hyperperfusion in the blood pool phase (*right*) and increased unilateral periarticular tracer uptake (*left*).

metacarpophalangeal or metacarpal bones is thought to be highly sensitive and specific for CRPS.[116,117] It should be noted that bone scintigraphy only shows significant changes in the first year of the disease. However, a "gold standard" to compare with is not yet known, but the test is useful to rule out pain syndromes of other origin.[7]

The value of this test in children seems to be less than in adults, showing a higher variability and, interestingly, often decreased uptake. Therefore, the test should be performed in children mainly to rule out other etiologies.[103]

Plain Radiographs and X-Ray Bone Densitometry

Endostal and intracortical excavation, subperiostal and trabecular bone resorption, spotty and localized bone demineralization, and osteoporosis have been thought to be specific signs of CRPS. However, a comparison of radiography and three-phase scintigraphy in early postfracture CRPS, although performed in a small cohort, showed a lower sensitivity (73 versus 97 percent) and specifity of the radiography (57 versus 86 percent).[117] MRI is thought to be more reliable than radiographic examination and scintigraphy but has to prove its value in further studies.[118] X-ray bone densitometry was performed in a small sample of 12 patients with CRPS and showed a high sensitivity and specifity for the diagnosis in comparison with healthy controls.[119]

Quantitative Sensory Testing (QST)

Bedside testing should be part of the physical examination, as outlined earlier, to confirm allodynia and hyperalgesia. Additionally, standardized psychophysical tests of the thermal, thermal pain, and vibratory thresholds to assess the function of large and small myelinated and unmyelinated afferent fibers are useful. Impairment of warm and cold sensation, as well as heat pain, has been demonstrated in patients with CRPS.[55,56] Further detailed sensory testing, including static and dynamic allodynia, pinprick allodynia, heat and mechanical hyperalgesia, and temporal summation, has shown abnormal results.[120–123] However, no characteristic sensory pattern of CRPS has been identified so far, but the testing is still useful to determine and quantify the individual signs of each patient and to document successful response to treatment.

Psychological Investigations

In 1996, Covington[124] drew several conclusions about the psychological factors in CRPS: (1) No evidence was found to support that CRPS is a psychogenic condition; (2) because anxiety, stress, and chemical dependency increase nociception, relaxation and antidepressive treatment is helpful; (3) the pain in CRPS is the cause of psychiatric problems and not the converse; (4) maladaptive behavior by patients, such as volitional or inadvertent actions, are mostly due to fear, regression, or misinformation and do not indicate psychopathology; and (5) some patients with conversion disorders and factious diseases have been diagnosed incorrectly with CRPS. Their poor response to treatment sometimes leads the pain specialist to the opinion that CRPS is a psychiatric condition. In summary, the widely proposed[125,126] "RSD personality" is clearly unsubstantiated. Accordingly, an even distribution of childhood trauma, pain intensity, and psychological distress was confirmed by Ciccone and colleagues[127] in patients with CRPS in comparison with patients with other neuropathic pain and chronic back pain. Further studies demonstrated a high psychiatric comorbidity, especially depression, anxiety, and personality disorders, in CRPS patients. These findings are also present in other chronic pain patients and are more likely a result of the long and severe pain disease.[128] In comparison with patients with low back pain, CRPS patients showed a higher tendency to somatization but did not show any other psychological differences.[129] In 145 patients, "42 percent reported of stressful life events in close relationship to the onset of CRPS, and 41 percent had a history of chronic pain before."[35] Thus, stressful life events could be risk factors for the development of CRPS.[130]

Significant emotional dysfunction was demonstrated in a small number of children with CRPS.[102] Wilder and colleagues[103] hypothesized a possible relation of intensive parental-forced sports and leisure activities to the occurrence of trauma leading to CRPS as a sign of escape from parents' excessive demands.

▶ DIFFERENTIAL DIAGNOSIS

Owing to the lack of a "gold standard" in the diagnosis of CRPS, overdiagnosis has to be taken into account. To differentiate CRPS from other neuropathic and pain syndromes, a detailed history and physical examination based on the specifications outlined earlier are mandatory.

Neuralgia is defined according to IASP criteria as a pain syndrome that develops after a lesion of a distinct nerve and that is restricted to its innervation territory. *Post-traumatic neuralgia* may present with pain that is largely located within the territory of the injured nerve.[131] Characteristic symptoms are spontaneous pain, hyperalgesia, and mechanical and cold allodynia. The symptoms also may spread a little beyond the innervation territory. Thus, the symptoms and signs and their distribution can lead to confusion with regard to the diagnosis. However, the signs in most of the cases are confined to the innervation territory. The pain is felt superficially and is not dependent on the position of the extremity. No signs of edema, trophic bone changes, and distal progressive generalized distribution are seen.

Neuropathies, such as diabetic polyneuropathy, also may present with spontaneous pain, skin color changes, and motor deficits but are distinguished by the symmetric distribution and the patient's history. Furthermore, *inflammation, infections* (e.g., arthritis, rheumatism, and ulcerations), and *vascular occlusive diseases* can cause unilateral pain and vascular abnormalities and have to be excluded. In addition, *malingering* or *factitious disorders* may mimic CRPS symptoms and signs.

▶ THERAPY

The pathophysiologic mechanisms of CRPS are incompletely understood at present. Additionally, the pain specialist has to cope with the lack of objective diagnostic criteria and sensitive tests. Thus, only a few reliable trials for different treatment options have been conducted so far. Three meta-analyses of all studies presented limited consistent evidence regarding effective pharmacologic, interventional, and physical therapy in patients with CRPS.[132–134] Moreover, the methodology is often of low quality within the only 30 studies available.[134]

The following treatment options and proposed algorithm should be taken as a recommendation that is based on the clinical experience and on the results taken from other trials in patients with different neuropathic pain syndromes.

General Rules

In the treatment of CRPS, a multidisciplinary approach is necessary. Neurologists, anesthesists, orthopedic specialists, physiotherapists, psychologists, and the general practitioner should be involved in the diagnosis and the treatment regimen. At first, the diagnosis and effective therapy are to be initiated as soon as possible. The therapeutic concept is aimed at the patient's full recovery. This can be achieved only by an individualized pharmacologic, physiotherapeutical, and psychological treatment that has to be performed for a sufficient period of time to assess a successful or unsuccessful response. Surgical interventions on the peripheral or central nervous system should be avoided to protect the patient from further deafferentation and the additional risk of pain increase and prolongation.

Pharmacologic Treatment

Nonsteroidal Anti-Inflammatory Drugs
Nonsteroidal anti-inflammatory drugs (NSAIDs) have not been investigated in the treatment of CRPS. However, from clinical experience, mild to moderate pain can be controlled. Trials in acute and chronic pain are necessary.

Opioids
The effect of opioids in CRPS is not known. In controlled trials, tramadol and oxycodone have been shown to be effective in other neuropathic pain conditions, such as painful polyneuropathy (number of patients needed to treat [NNT] to achieve 50 percent pain reduction, NNT = 2.5) and postherpetic neuralgia (NNT = 3.4).[135–137] A recent study in a heterogeneous population with neuropathic pain that was refractory to previous treatment demonstrated a significant response to the mu agonist levorphanol.[138] Thus, without profound scientific evidence, opioids have become part of the treatment regimen in CRPS. Because opioids are potent analgesics in other neuropathic pain syndromes, they should be prescribed immediately if other agents do not achieve sufficient analgesia.

Antidepressants
Tricyclic antidepressants (TCAs) have been studied intensely in different neuropathic pain conditions but not in CRPS. Serotonin and norepinephrine reuptake blockers, such as amitriptyline, and selective norepinephrine blockers, such as desipramine, are active in painful diabetic neuropathy (NNT = 2.0–2.3) and postherpetic neuralgia (NNT = 1.4–2.0), and amitriptyline additionally is active in central pain (NNT = 1.7) and painful post-traumatic neuropathy (NNT = 2.5) (for review, see ref. 139). The mean analgesic dosage is lower than necessary for antidepressant effects (e.g., amitriptyline 75 to 100 mg/d), with onset of pain relief in about 2 weeks and peaking at 4 to 6 weeks.[140] The effectiveness of selective serotonin reuptake inhibitors in neuropathic pain states is still being discussed. Only one of four studies performed, so far, gave evidence for a significant pain reduction in painful diabetic neuropathy (NNT = 6.7). None has been performed on CRPS patients.

Sodium Channel Blocking Agents
Lidocaine administered intravenously is effective in CRPS types 1 and 2 regarding spontaneous and evoked pain.[141] Mexiletine is not active in central pain and of poor efficacy in painful diabetic neuropathy (NNT > 10).[142,143] Contraindications and side effects, e.g., cardiac conduction abnormalities, reduced left ventricular function, and coronary heart disease, limit use of these compounds.

GABA Agonists
Intrathecally administered baclofen is effective in the treatment of dystonia in CRPS.[144] Oral baclofen has been effective in the treatment of trigeminal neuralgia.[145] No further trials in CRPS are available, and there is no evidence for an analgesic effect of baclofen, valproic acid, vigabatrine, or benzodiazepines in CRPS or other neuropathic pain conditions.

Gabapentin and Carbamazepine
Promising limited evidence was revealed by two studies on patients with CRPS who showed an analgesic response to gabapentin.[146–148] Gabapentin is effective in painful diabetic neuropathy (NNT = 3.2) and postherpetic neuralgia (NNT = 3.7).[149,150] Carbamazepine has not been tested in CRPS but is effective in postherpetic neuralgia (NNT = 3.3) and central pain (NNT = 2.6).[151,152]

Steroids

Orally administered prednisone, 10 mg three times daily, clearly has demonstrated its efficacy in improvement of up to 75 percent of the entire clinical status in CRPS patients.[67] No evidence has been obtained about other immune-modulating therapies, such as intravenous immunglobulines or immunosuppressive drugs.

NMDA Receptor Blockers

Dextromethorphan is effective in the treatment of painful diabetic neuropathy and not effective in postherpetic neuralgia and central pain.[153,154] Memantine is not effective in postherpetic neuralgia.[155] Intravenous administration of ketamine has been effective in postherpetic neuralgia and phantom limb pain.[156,157] None of these drugs has been administered to CRPS patients in controlled trials.

Calcium-Regulating Drugs

Calcitonin administered intranasally three times daily demonstrated significant pain reduction in CRPS patients.[158] Clodronate 300 mg daily intravenously and alendronate 7.5 mg daily intravenously showed a significant improvement in pain, swelling, and movement range.[159,160] The mode of action of these compounds in CRPS is unknown.

Free-Radical Scavengers

Recently, Perez and colleagues[161] conducted a placebo-controlled trial using the free-radical scavengers dimethylsulfoxide 50% topically (DMSO) or N-acetylcysteine (NAC) orally for the treatment of CRPS type 1. Both drugs were found to be equally effective, but DMSO seemed more favorable for "warm" and NAC for "cold" CRPS type 1. The results were negatively influenced by longer disease duration. A previous trial with DMSO failed to show a positive result in CRPS,[162] whereas DMSO has been more effective than regional blocks with guanethidine in a small population of CRPS patients.[163]

Interventions at the Sympathetic Nervous System

Blockage of the sympathetic trunk or a sympathectomy is used to treat CRPS. Currently, two techniques of sympathetic blocks are used: (1) injections of local anesthetics next to the sympathetic paravertebral ganglia projecting to the affected extremity (sympathetic ganglion blocks), and (2) regional intravenous application of phentolamine, guanethidine, bretylium, or reserpine (norepinephrine-depleting drugs) into the affected extremity blocked with a tourniquet (intravenous regional sympatholysis, IVRS or RIS).

The efficacy of these procedures is still under discussion and has been questioned in the past.[125,132,164–166] Indeed, the specifity, long-term results, and techniques used have rarely been evaluated adequately. One placebo-controlled trial in CRPS patients failed to show any difference of sympathetic blockade in the immediate effect of pain reduction compared with saline application. However, patients who received local anesthetics experienced a much better pain relief after 24 hours than controls.[167] Furthermore, perioperative blocks of the ganglion stellatum reduce the recurrence rate of CRPS in patients with a CRPS history.[168] Neither phentolamine nor guanethidine used for sympathetic blocks (RIS) was effective in CRPS.[165] No improvement was found with reserpine given together with guanethidine or guanethidine only (RIS).[169,170] However, stellate blocks with bupivacaine and regional blocks with guanethidine (RIS) demonstrated significant improvement in pain compared with baseline, but no differences between these two drugs.[171] A crossover study performed by Hord[172] showed significant pain reduction from sympathetic blockade with lidocaine and bretylium. No differences were obtained between sympathetic blocks with guanethidine and lidocaine (RIS), or guanethidine and reserpine (RIS).[173] No effect was obtained by droperidol (RIS).[174] Hanna and colleagues[175] demonstrated significant improvement in pain from a single bolus (RIS) of ketanserin. Bounameaux and colleagues[176] failed to show any significant effect with the same procedure.

There is only limited evidence regarding the efficacy of thoracoscopic or surgical sympathectomy. Four studies reported partial long-lasting benefits in CRPS types 1 and 2.[177–180] The most important independent factor in determining the outcome of sympathectomy is the time between injury and intervention.[180] A procedure within 12 months is favorable.[179] Baron and colleagues[31] investigated skin blood flow, sympathetic vasoconstrictor reflexes, and pain after surgical sympathectomy. Postoperatively, no vasoconstriction due to deep inspiration (vasoconstrictor reflex) could be elicited at the affected extremity, indicating complete sympathetic denervation. Additionally, the skin temperature at the affected hand increased. After 4 weeks, skin temperature decreased without signs of reinnervation. This denervation supersensitivity was associated with the recurrence of pain and is thought to rely on a vascular supersensitivity to cold and circulating catecholamines. Only 2 of 12 patients experienced long-term pain relief. Videoscopic lumbar sympathectomy is as effective as open surgical intervention.[181] Interestingly, alterations in the three-phase bone scan in acute CRPS are similar to those resulting from sympathectomy without being related to the success of the intervention.[182]

Sympatholytic procedures can achieve marked and reliable pain relief. As outlined earlier, there is empirical and scientific evidence that an adequate sympatholytic procedure could be effective in some CRPS patients. Thus, if SMP can be identified by sympatholytic blocks, these could contribute to a successful treatment regimen for an SMP component. Irreversible sympathectomy may be recommended in selected patients because of the risk of adaptive supersensitivity and subsequent pain increase and prolongation.

Stimulation Techniques and Spinal Drug Application

Epidural spinal cord stimulation (SCS) is effective in selected patients with chronic CRPS who have been refractory to previous sympathectomy.[183,184] The pain relief of SCS does not seem to rely on a change in the microcirculation of the skin, i.e., modulation of the cutaneous sympathetic vasoconstrictor activity, suggesting that central inhibition processes are involved.[185] Sensory detection thresholds are not altered.[186] SCS is more effective if it is combined with physiotherapy.[187] Case reports of successful peripheral nerve stimulation and deep brain stimulation (stimulation of the thalamus and medial lemniscus) have been published.[188,189]

The epidural application of clonidine showed significant pain reduction in higher dosages (700 μg) than in lower dosages (300 μg).[190] Ketamine induced analgesia when administered epidurally.[191] Both drugs were associated with marked side effects, such as sedation and hypotension. Intrathecally administered baclofen is effective in the treatment of dystonia in CRPS.[144]

Transcutaneous electrical nerve stimulation (TENS) is effective in patients with neuropathic pain but has not been evaluated in CRPS patients.

Acupuncture and Qigong

Two studies examined the efficacy of acupuncture in patients with CRPS.[192,193] In both, an improvement was achieved compared with placebo without reaching statistical significance. Another trial failed to show any improvement.[194] Qigong exercises twice a week led to a significant improvement in CRPS type 1 compared with sham exercises.[195]

Physical and Occupational Therapy

Standardized physiotherapy has shown long-term relief of pain and physical dysfunction in children.[196] Oerlemans and colleagues[197,198] demonstrated that physical therapy and, to a lesser extent, occupational therapy are able to reduce pain and improve active mobility in CRPS type 1. Lymph drainage provides no benefit when applied with physiotherapy in comparison with physiotherapy only.[199] Patients with less pain and less affected motor function initially are predicted to benefit to a greater extent than others.[186] Physical therapy for CRPS is more effective and less costly than occupational therapy or less-effective control treatments.[200]

Psychological Therapy

Although there is evidence of a psychological impact on CRPS patients, only one study has addressed the efficacy of psychological treatment. A prospective, randomized, single-blind trial of cognitive-behavioral treatment versus physical therapy in children and adults with CRPS was conducted and showed a similar long-lasting reduction with therapy in all symptoms in all groups.[201]

Proposed Therapy Algorithm

The severity of the disease determines the therapeutic regimen. The general rules outlined earlier must be followed throughout treatment. Reduction in pain is the precondition with which all other interventions must comply. Pain treatment should be initiated immediately. First-line analgesics and coanalgesics, based on scientific evidence, are opioids, tricyclic antidepressants, gabapentin, and carbamazepine. Additionally, corticosteroids should be considered. Calcium-regulating agents will be used in cases of refractory pain. Sympatholytic procedures for SMP testing, preferably sympathetic ganglion blocks, are used if no, or insufficient, pain relief is achieved and should be perpetuated in case of efficacy. If a relief of rest pain is obtained, physiotherapy and occupational therapy up to the pain threshold have to be started and continued until complete motor function returns. Psychological treatment has to flank the regimen to strengthen coping strategies and discover contributing factors.

In the case of refractory therapy, spinal cord stimulation could be considered. If refractory dystonia develops, intrathecal baclofen application is worth considering.

Treatment of Children

Limited attention has been paid to differences in the therapeutic response of children. Conservative strategies such as TENS and cognitive-behavioral pain management have been successful in children.[103] Physical therapy seems to be effective to treat childhood CRPS and to prevent symptom prolongation.[196,201] No further information has been obtained so far. Thus, with the knowledge outlined earlier, children should receive an adapted treatment like that for adults.

Prevention Studies

Only two reliable randomized, placebo-controlled prevention studies have been conducted so far. Zollinger and colleagues[202] proved a significantly lower incidence of CRPS following Colle's fracture with vitamin C (500 mg/d) treatment. The recurrence rate of CRPS in patients with a history of CRPS undergoing surgery of the formerly affected extremity was reduced significantly by a perioperative stellate ganglion block.[168] Guanethidine

(20 mg, RIS) administered preoperatively did not prevent CRPS in patients undergoing faciectomy for Duputryen's disease.[203]

PROGNOSIS

Duration

The disease duration is variable and may persist over decades.[4] In rare cases, therapy directed at the underlying causes of the CRPS, e.g., the decompression of a compressed nerve, may lead to complete recovery.[204,205] A 5.5-year follow-up study showed that 62 percent of the patients were still limited in their activities of daily living, pain and motor impairments being the most important factors.[130] The impairments may change to chronic stages, and in more than 60 percent of patients with CRPS type 2, the complaints remained unchanged even after 1 year of intensive therapy,[206] and further studies demonstrate long-lasting impairments.[130,205,207] In contrast, a retrospective population-based study reported a resolution of symptoms in 74 percent of patients with CRPS type 1.[5] The severity more than the etiology seems to determine disease course,[109,208] and age, sex, and affected side are not associated with outcome.[5] Fractures may have a higher resolution rate (91 percent) than sprains (78 percent) and other inciting events (55 percent).[5]

In 1183 patients, the incidence of recurrence was 1.8 percent per year. Patients with a recurrent CRPS were significantly younger but did not differ in gender or primary localization. The symptoms and signs were few in case of recurrence, were in half the patients of spontaneous origin, and affected the symmetric limb if a second limb was involved. A low skin temperature at the onset of disease may predict an unfavorable course and outcome.[4] This study hypothesis is weakened by a possible discrepancy between patient history and a missing objective assessment of skin temperature.[198]

Between 50 and 88 percent of the children with CRPS type 1 showed good long-term resolution of all symptoms.[103,196]

Complications

A retrospective analysis of 1006 CRPS patients showed an incidence of severe complications in about 7 percent. These consisted of infection, ulceration, chronic edema, dystonia, and/or myoclonus. Mostly younger female patients with CRPS of the lower limb were affected.[209]

REFERENCES

1. Merskey H, Bogduk N: *Classification of Chronic Pain: Descriptions of Chronic Pain Syndromes and Definition of Terms.* Seattle: IASP Press, 1995.
2. Stanton-Hicks M, Jänig W, Hassenbusch S, et al: Reflex sympathetic dystrophy: Changing concepts and taxonomy. *Pain* 1995;63:127–133.
3. Allen G, Galer BS, Schwartz L: Epidemiology of complex regional pain syndrome: A retrospective chart review of 134 patients. *Pain* 1999;80:539–544.
4. Veldman PH, Reynen HM, Arntz IE, et al: Signs and symptoms of reflex sympathetic dystrophy: Prospective study of 829 patients. *Lancet* 1993;342:1012–1016.
5. Sandroni P, Benrud-Larson LM, Mcclelland RL, et al: Complex regional pain syndrome type 1: Incidence and prevalence in Olmsted County, a population-based study. *Pain* 2003;103:199–207.
6. Bohm E: Das Sudecksche Syndrom. *Hefte Unfallheilkunde* 1985;174:241–250.
7. Gellman H, Keenan MA, Stone L, et al: Reflex sympathetic dystrophy in brain-injured patients. *Pain* 1992;51:307–311.
8. Rosen P, Graham W: The shoulder-hand syndrome: Historical review with observations on 73 patients. *Can Med Assoc J* 1957;77:86–91.
9. Low P, Wilson PR, Sandroni P, et al: Clinical characteristics of patients with reflex sympathetic dystrophy (sympathically maintained pain) in the United States, in Jänig W, Stanton-Hicks M (eds), *Progress in Pain Research and Management: Reflex Sympathetic Dystrophy: A Reappraisal.* Seattle: IASP Press, 1996:49–66.
10. Bruehl S, Harden RN, Galer BS, et al: Complex regional pain syndrome: Are there distinct subtypes and sequential stages of the syndrome? *Pain* 2002;95:119–124.
11. Baron R, Blumberg, H, Jänig, W: Clinical characteristics of patients with complex regional pain syndrome in Germany, with special emphasis on vasomotor function, in Jänig W, Stanton-Hicks M (eds), *Progress in Pain Research and Management: Reflex Sympathetic Dystrophy: A Reappraisal.* Seattle: IASP Press, 1996:25–48.
12. Kim SH, Chung JM: An experimental model for peripheral neuropathy produced by segmental spinal nerve ligation in the rat. *Pain* 1992;50:355–363.
13. Choi Y, Yoon YW, Na HS, et al: Behavioral signs of ongoing pain and cold allodynia in a rat model of neuropathic pain. *Pain* 1994;59:369–376.
14. Vatine JJ, Argov R, Seltzer Z: Brief electrical stimulation of C-fibers in rats produces thermal hyperalgesia lasting weeks. *Neurosci Lett* 1998;246:125–128.
15. Jänig W, McLachlan EM: Neurobiology of the autonomic nervous system, in Mathias CJ (ed), *Autonomic Failure: A Textbook of Clinical Disorders of the Autonomic Nervous System,* 4th ed. Oxford: Oxford University Press, 1999:3–15.
16. Jänig W, Häbler HJ: Organization of the autonomic nervous system: Structure and function, in Vinken PJ (ed), *Handbook of Clinical Neurology: The Autonomic Nervous System,* Part I: *Normal Functions.* Amsterdam: Elsevier Science, 1999:1–52.
17. Fleming W, Westfall D: Adaptive supersensitivity, in Trendelenburg U, Weiner N (eds), *Handbook of Experimental Pharmacology.* New York: Springer-Verlag, 1988:509–559.
18. Kurvers HA, Tangelder GJ, De Mey JG, et al: Skin blood flow abnormalities in a rat model of neuropathic pain: Result of decreased sympathetic vasoconstrictor outflow? *J Auton Nerv Syst* 1997;63:19–29.

19. Wakisaka S, Kajander KC, Bennett GJ: Abnormal skin temperature and abnormal sympathetic vasomotor innervation in an experimental painful peripheral neuropathy. *Pain* 1991;46:299–313.
20. Wasner G, Schattschneider J, Heckmann K, et al: Vascular abnormalities in reflex sympathetic dystrophy (CRPS I): Mechanisms and diagnostic value. *Brain* 2001;124:587–599.
21. Drummond PD, Skipworth S, Finch PM: α_1-Adrenoceptors in normal and hyperalgesic human skin. *Clin Sci (Lond)* 1996;91:73–77.
22. Arnold JM, Teasell RW, Macleod AP, et al: Increased venous α-adrenoceptor responsiveness in patients with reflex sympathetic dystrophy. *Ann Intern Med* 1993;118:619–621.
23. Casale R, Elam M: Normal sympathetic nerve activity in a reflex sympathetic dystrophy with marked skin vasoconstriction. *J Auton Nerv Syst* 1992;41:215–219.
24. Torebjörk E: Clinical and neurophysiological observations relating to pathophysiological mechanisms of reflex sympathetic dystrophy, in Stanton-Hicks M, Jänig W, Boas RA (eds), *Reflex Sympathetic Dystrophy*. Boston: Kluwer, 1989:71–80.
25. Jänig W, Baron R: Complex regional pain syndrome is a disease of the central nervous system. *Clin Auton Res* 2002;12:150–164.
26. Wasner G, Heckmann K, Maier C, et al: Vascular abnormalities in acute reflex sympathetic dystrophy (CRPS I): Complete inhibition of sympathetic nerve activity with recovery. *Arch Neurol* 1999;56:613–620.
27. Drummond PD, Finch PM, Smythe GA: Reflex sympathetic dystrophy: The significance of differing plasma catecholamine concentrations in affected and unaffected limbs. *Brain* 1991;114:2025–2036.
28. Harden RN, Duc TA, Williams TR, et al: Norepinephrine and epinephrine levels in affected versus unaffected limbs in sympathetically maintained pain. *Clin J Pain* 1994;10:324–330.
29. Goldstein DS, Tack C, Li ST: Sympathetic innervation and function in reflex sympathetic dystrophy. *Ann Neurol* 2000;48:49–59.
30. Bossut DF, Shea VK, Perl ER: Sympathectomy induces adrenergic excitability of cutaneous C-fiber nociceptors. *J Neurophysiol* 1996;75:514–517.
31. Baron R, Maier C: Reflex sympathetic dystrophy: Skin blood flow, sympathetic vasoconstrictor reflexes and pain before and after surgical sympathectomy. *Pain* 1996;67:317–326.
32. Birklein F, Riedl B, Neundorfer B, et al: Sympathetic vasoconstrictor reflex pattern in patients with complex regional pain syndrome. *Pain* 1998;75:93–100.
33. Jänig W, Koltzenburg M: Sympathetic reflex activity and neuroeffector transmission change after chronic nerve lesions, in Bond MR, Charlton EJ, Woolf CJ (eds), *Proceedings of the VIth World Congress on Pain*. Amsterdam: Elsevier Science, 1991:365–371.
34. Blumberg H, Jänig W: Reflex patterns in postganglionic vasoconstrictor neurons following chronic nerve lesions. *J Auton Nerv Syst* 1985;14:157–180.
35. Birklein F, Riedl B, Sieweke N, et al: Neurological findings in complex regional pain syndromes: Analysis of 145 cases. *Acta Neurol Scand* 2000;101:262–269.
36. Birklein F, Spitzer A, Riedl B: The assessment of sudomotor function for diagnosis of autonomic diseases: Principles and methods. *Fortschr Neurol Psychiatr* 1999;67:287–295.
37. Low PA, Amadio PC, Wilson PR, et al: Laboratory findings in reflex sympathetic dystrophy: A preliminary report. *Clin J Pain* 1994;10:235–239.
38. Arner S: Intravenous phentolamine test: Diagnostic and prognostic use in reflex sympathetic dystrophy. *Pain* 1991;46:17–22.
39. Raja SN, Treede RD, Davis KD, et al: Systemic alpha-adrenergic blockade with phentolamine: A diagnostic test for sympathetically maintained pain. *Anesthesiology* 1991;74:691–698.
40. Baron R, Wasner G, Borgstedt R, et al: Effect of sympathetic activity on capsaicin-evoked pain, hyperalgesia, and vasodilatation. *Neurology* 1999;52:923–932.
41. Wasner G, Binder A, Kopper F, et al: No effect of sympathetic sudomotor activity on capsaicin-evoked ongoing pain and hyperalgesia. *Pain* 2000;84:331–338.
42. Elam M, Olausson B, Skarphedinsson JO, et al: Does sympathetic nerve discharge affect the firing of polymodal C-fiber afferents in humans? *Brain* 1999;122:2237–2244.
43. Janig W, Levine JD, Michaelis M: Interactions of sympathetic and primary afferent neurons following nerve injury and tissue trauma. *Prog Brain Res* 1996;113:161–184.
44. Mclachlan EM, Jänig W, Devor M, et al: Peripheral nerve injury triggers noradrenergic sprouting within dorsal root ganglia. *Nature* 1993;363:543–546.
45. Shi TS, Winzer-Serhan U, Leslie F, et al: Distribution and regulation of alpha(2)-adrenoceptors in rat dorsal root ganglia. *Pain* 2000;84:319–330.
46. Chabal C, Jacobson L, Russell LC, et al: Pain response to perineuromal injection of normal saline, epinephrine, and lidocaine in humans. *Pain* 1992;49:9–12.
47. Walker A, Nulsen F: Electrical stimulation of the upper thoracic portion of the sympathetic chain in man. *Arch Neurol Psychiatr* 1948;59:559–560.
48. Torebjork E, Wahren L, Wallin G, et al: Noradrenaline-evoked pain in neuralgia. *Pain* 1995;63:11–20.
49. Ali Z, Raja SN, Wesselmann U, et al: Intradermal injection of norepinephrine evokes pain in patients with sympathetically maintained pain. *Pain* 2000;88:161–168.
50. Baron R, Schattschneider J, Binder A, et al: Relation between sympathetic vasoconstrictor activity and pain and hyperalgesia in complex regional pain syndromes: A case-control study. *Lancet* 2002;359:1655–1660.
51. Drummond PD, Finch PM, Skipworth S, et al: Pain increases during sympathetic arousal in patients with complex regional pain syndrome. *Neurology* 2001;57:1296–1303.
52. Green PG, Jänig W, Levine JD: Negative feedback neuroendocrine control of inflammatory response in the rat is dependent on the sympathetic postganglionic neuron. *J Neurosci* 1997;173234–3238.
53. Woolf CJ, Mannion RJ: Neuropathic pain: Aetiology, symptoms, mechanisms, and management. *Lancet* 1999;353:1959–1964.
54. Sieweke N, Birklein F, Riedl B, et al: Patterns of hyperalgesia in complex regional pain syndrome. *Pain* 1999;80:171–177.
55. Rommel O, Gehling M, Dertwinkel R, et al: Hemisensory impairment in patients with complex regional pain syndrome. *Pain* 1999;80:95–101.

56. Rommel O, Malin JP, Zenz M, et al: Quantitative sensory testing, neurophysiological and psychological examination in patients with complex regional pain syndrome and hemisensory deficits. *Pain* 2001;93:279–293.
57. Juottonen K, Gockel M, Silen T, et al: Altered central sensorimotor processing in patients with complex regional pain syndrome. *Pain* 2002;98:315–323.
58. Fukumoto M, Ushida T, Zinchuk VS, et al: Contralateral thalamic perfusion in patients with reflex sympathetic dystrophy syndrome. *Lancet* 1999;354:1790–1791.
59. Apkarian AV, Thomas PS, Krauss BR, et al: Prefrontal cortical hyperactivity in patients with sympathetically mediated chronic pain. *Neurosci Lett* 2001;311:193–197.
60. Schattschneider J, Wenzelburger R, Deuschl G, et al: Kinematic analysis of the upper extremity in CRPS, in Harden RN, Baron R, Jänig W (eds), *Progress in Pain Research and Management: Complex Regional Pain Syndrome*. Seattle: IASP Press, 1999:119–128.
61. Deuschl G, Blumberg H, Lucking CH: Tremor in reflex sympathetic dystrophy. *Arch Neurol* 1991;48:1247–1252.
62. Galer BS, Butler S, Jensen MP: Case reports and hypothesis: A neglect-like syndrome may be responsible for the motor disturbance in reflex sympathetic dystrophy (complex regional pain syndrome-1). *J Pain Symptom Manage* 1995;10:385–391.
63. Galer BS, Jensen M: Neglect-like symptoms in complex regional pain syndrome: Results of a self-administered survey. *J Pain Symptom Manage* 1999;18:213–217.
64. Sudeck P: Über die akute (trophoneurotische) Knochenatrophie nach Entzündungen und Traumen der Extremitüten. *Dtsch Med Wochenschr* 1902;28:336–342.
65. Van Der Laan L, Goris RJ: Reflex sympathetic dystrophy: An exaggerated regional inflammatory response? *Hand Clin* 1997;13:373–385.
66. Leitha T, Korpan M, Staudenherz A, et al: Five-phase bone scintigraphy supports the pathophysiological concept of a subclinical inflammatory process in reflex sympathetic dystrophy. *Q J Nucl Med* 1996;40:188–193.
67. Christensen K, Jensen EM, Noer I: The reflex dystrophy syndrome response to treatment with systemic corticosteroids. *Acta Chir Scand* 1982;148:653–655.
68. Oyen WJ, Arntz IE, Claessens RM, et al: Reflex sympathetic dystrophy of the hand: An excessive inflammatory response? *Pain* 1993;55:151–157.
69. Kozin F, Mccarty DJ, Sims J, et al: The reflex sympathetic dystrophy syndrome: I. Clinical and histologic studies: evidence for bilaterality, response to corticosteroids and articular involvement. *Am J Med* 1976;60:321–331.
70. Renier JC, Arlet J, Bregeon C, et al: The joint in algodystrophy: Joint fluid, synovium, cartilage. *Rev Rhum Mal Osteoartic* 1983;50:255–260.
71. Graif M, Schweitzer ME, Marks B, et al: Synovial effusion in reflex sympathetic dystrophy: an additional sign for diagnosis and staging. *Skeletal Radiol* 1998;27:262–265.
72. Weber M, Birklein F, Neundorfer B, et al: Facilitated neurogenic inflammation in complex regional pain syndrome. *Pain* 2001;91:251–257.
73. Birklein F, Schmelz M, Schifter S, et al: The important role of neuropeptides in complex regional pain syndrome. *Neurology* 2001;57:2179–2184.
74. Birklein F, Weber M, Ernst M, et al: Experimental tissue acidosis leads to increased pain in complex regional pain syndrome (CRPS). *Pain* 2000;87:227–234.
75. Huygen FJ, De Bruijn AG, De Bruin MT, et al: Evidence for local inflammation in complex regional pain syndrome type 1. *Mediators Inflamm* 2002;11:47–51.
76. Hartrick CT: Increased production of nitric oxide stimulated by interferon-gamma from peripheral blood monocytes in patients with complex regional pain syndrome. *Neurosci Lett* 2002;323:75–77.
77. Van De Vusse AC, Goossens VJ, Kemler MA, et al: Screening of patients with complex regional pain syndrome for antecedent infections. *Clin J Pain* 2001;17:110–114.
78. Levine JD, Taiwo YO, Collins SD, et al: Noradrenaline hyperalgesia is mediated through interaction with sympathetic postganglionic neurone terminals rather than activation of primary afferent nociceptors. *Nature* 1986;323:158–160.
79. Perl ER: Cutaneous polymodal receptors: Characteristics and plasticity. *Prog Brain Res* 1996;113:21–37.
80. Mailis A, Wade J: Profile of Caucasian women with possible genetic predisposition to reflex sympathetic dystrophy: A pilot study. *Clin J Pain* 1994;10:210–217.
81. Kemler MA, Van De Vusse AC, Van Den Berg-Loonen EM, et al: HLA-DQ1 associated with reflex sympathetic dystrophy. *Neurology* 1999;53:1350–1351.
82. Van Hilten JJ, Van De Beek WJ, Roep BO: Multifocal or generalized tonic dystonia of complex regional pain syndrome: A distinct clinical entity associated with HLA-DR13. *Ann Neurol* 2000;48:113–116.
83. Van De Beek WJ, Roep BO, Van Der Slik AR, et al: Susceptibility loci for complex regional pain syndrome. *Pain* 2003;103:93–97.
84. Van De Beek WJ, Remarque EJ, Westendorp RG, et al: Innate cytokine profile in patients with complex regional pain syndrome is normal. *Pain* 2001;91:259–261.
85. Wanklyn P, Forster A, Young J, et al: Prevalence and associated features of the cold hemiplegic arm. *Stroke* 1995;26:1867–1870.
86. Korpelainen JT, Sotaniemi KA, Myllyla VV: Asymmetrical skin temperature in ischemic stroke. *Stroke* 1995;26:1543–1547.
87. Butler S: Disuse and CRPS, in Harden RN, Baron R, Jänig W (eds), *Progress in Pain Research and Management*. Seattle: IASP Press, 2001:141–150.
88. Braus DF, Krauss JK, Strobel J: The shoulder-hand syndrome after stroke: A prospective clinical trial. *Ann Neurol* 1994;36:728–733.
89. Dalyan M, Sherman A, Cardenas DD: Factors associated with contractures in acute spinal cord injury. *Spinal Cord* 1998;36:405–408.
90. Gellman H, Eckert RR, Botte MJ, et al: Reflex sympathetic dystrophy in cervical spinal cord injury patients. *Clin Orthop* 1988;233:126–131.
91. Subbarao J, Stillwell GK: Reflex sympathetic dystrophy syndrome of the upper extremity: Analysis of total outcome of management of 125 cases. *Arch Phys Med Rehabil* 1981;62:549–554.
92. Gallien P, Nicolas B, Robineau S, et al: The reflex sympathetic dystrophy syndrome in patients who have had a spinal cord injury. *Paraplegia* 1995;33:715–720.

93. Cremer SA, Maynard F, Davidoff G: The reflex sympathetic dystrophy syndrome associated with traumatic myelopathy: Report of 5 cases. *Pain* 1989;37:187–192.
94. Aisen PS, Aisen ML: Shoulder-hand syndrome in cervical spinal cord injury. *Paraplegia* 1994;32:588–592.
95. Chelimsky TC, Low PA, Naessens JM, et al: Value of autonomic testing in reflex sympathetic dystrophy. *Mayo Clin Proc* 1995;70:1029–1040.
96. Blumberg H, Jänig W: Clinical manifestation of reflex sympathetic dystrophy and sympathetically maintained pain, in Melzack R, Wall PD (eds), *Textbook of Pain*. Edinburgh: Churchill Livingstone, 1994:685–697.
97. Wasner G, Schattschneider J, Baron R: Skin temperature side differences: A diagnostic tool for CRPS? *Pain* 2002;98:19–26.
98. Schwartzman RJ, Kerrigan J: The movement disorder of reflex sympathetic dystrophy. *Neurology* 1990;40(1):57–61.
99. Bhatia KP, Bhatt MH, Marsden CD: The causalgia-dystonia syndrome. *Brain* 1993;116:843–851.
100. Marsden C: Muscle spasms associated with Sudeck's atrophy after injury. *Br Med J (Clin Res Ed)* 1984;288:173–176.
101. Rashiq S, Galer BS: Proximal myofascial dysfunction in complex regional pain syndrome: A retrospective prevalence study. *Clin J Pain* 1999;15:151–153.
102. Barbier O, Allington N, Rombouts JJ: Reflex sympathetic dystrophy in children: Review of a clinical series and description of the particularities in children. *Acta Orthop Belg* 1999;65:91–97.
103. Wilder R: Reflex sympathetic dystrophy in children and adolescents: Differences from adults, in Jänig W, Stanton-Hicks M (eds), *Progress in Pain Research and Management: Reflex Sympathetic Dystrophy: A Reappraisal*. Seattle: IASP Press, 1996:67–78.
104. Jankovic J, Van Der Linden C: Dystonia and tremor induced by peripheral trauma: Predisposing factors. *J Neurol Neurosurg Psychiatry* 1988;51:1512–1519.
105. Bruehl S, Harden RN, Galer BS, et al: External validation of IASP diagnostic criteria for complex regional pain syndrome and proposed research diagnostic criteria. International Association for the Study of Pain. *Pain* 1999;81:147–154.
106. Galer BS, Bruehl S, Harden RN: IASP diagnostic criteria for complex regional pain syndrome: A preliminary empirical validation study. International Association for the Study of Pain. *Clin J Pain* 1998;14:48–54.
107. Alvarez-Lario B, Aretxabala-Alcibar I, Alegre-Lopez J, et al: Acceptance of the different denominations for reflex sympathetic dystrophy. *Ann Rheum Dis* 2001;60:77–79.
108. Schwartzman RJ, Mclellan TL: Reflex sympathetic dystrophy: A review. *Arch Neurol* 1987;44:555–561.
109. Bonica J: Causalgia and other reflex sympathetic dystrophy, in Bonica J (ed), *The Managment of Pain*, 2d ed. Philadelphia: Lea & Febiger, 1990:220–243.
110. Gibbons JJ, Wilson PR: RSD score: Criteria for the diagnosis of reflex sympathetic dystrophy and causalgia. *Clin J Pain* 1992;8:260–263.
111. Bickerstaff DR, Kanis JA: Algodystrophy: An underrecognized complication of minor trauma. *Br J Rheumatol* 1994;33:240–248.
112. Zyluk A: The natural history of post-traumatic reflex sympathetic dystrophy. *J Hand Surg [Br]* 1998;23:20–23.
113. Galer BS, Henderson J, Perander J, et al: Course of symptoms and quality of life measurement in complex regional pain syndrome: A pilot survey. *J Pain Symptom Manage* 2000;20:286–292.
114. Gulevich SJ, Conwell TD, Lane J, et al: Stress infrared telethermography is useful in the diagnosis of complex regional pain syndrome, type I (formerly reflex sympathetic dystrophy). *Clin J Pain* 1997;13:50–59.
115. Kozin F, Soin JS, Ryan LM, et al: Bone scintigraphy in the reflex sympathetic dystrophy syndrome. *Radiology* 1981;138:437–443.
116. Zyluk A: The usefulness of quantitative evaluation of three-phase scintigraphy in the diagnosis of post-traumatic reflex sympathetic dystrophy. *J Hand Surg [Br]* 1999;24:16–21.
117. Todorovic-Tirnanic M, Obradovic V, Han R, et al: Diagnostic approach to reflex sympathetic dystrophy after fracture: Radiography or bone scintigraphy? *Eur J Nucl Med* 1995;22:1187–1193.
118. Sintzoff S, Sintzoff S Jr, Stallenberg B, et al: Imaging in reflex sympathetic dystrophy. *Hand Clin* 1997;13:431–442.
119. Arriagada M, Arinoviche R: X-ray bone densitometry in the diagnosis and followup of reflex sympathetic dystrophy syndrome. *J Rheumatol* 1994;21:498–500.
120. Price DD, Bennett GJ, Rafii A: Psychophysical observations on patients with neuropathic pain relieved by a sympathetic block. *Pain* 1989;36:273–288.
121. Price DD, Long S, Huitt C: Sensory testing of pathophysiological mechanisms of pain in patients with reflex sympathetic dystrophy. *Pain* 1992;49:163–173.
122. Wahren LK, Torebjork E, Nystrom B: Quantitative sensory testing before and after regional guanethidine block in patients with neuralgia in the hand. *Pain* 1991;46:23–30.
123. Wahren LK, Torebjork E: Quantitative sensory tests in patients with neuralgia 11 to 25 years after injury. *Pain* 1992;48:237–244.
124. Covington E: Psychological issues in reflex sympathetic dystrophy, in Jänig W, Stanton-Hicks M (eds), *Progress in Pain Research and Management*. Seattle: IASP Press, 1996:191–216.
125. Ochoa JL: Truths, errors, and lies around "reflex sympathetic dystrophy" and "complex regional pain syndrome." *J Neurol* 1999;246:875–879.
126. Verdugo RJ, Ochoa JL: Abnormal movements in complex regional pain syndrome: Assessment of their nature. *Muscle Nerve* 2000;23:198–205.
127. Ciccone DD, Bandilla EB, Wu W: Psychological dysfunction in patients with reflex sympathetic dystrophy. *Pain* 1997;71:323–333.
128. Monti DA, Herring CL, Schwartzman RJ, et al: Personality assessment of patients with complex regional pain syndrome type I. *Clin J Pain* 1998;14:295–302.
129. Bruehl S, Husfeldt B, Lubenow TR, et al: Psychological differences between reflex sympathetic dystrophy and non-RSD chronic pain patients. *Pain* 1996;67:107–114.
130. Geertzen JH, De Bruijn-Kofman AT, De Bruijn HP, et al: Stressful life events and psychological dysfunction in complex regional pain syndrome type I. *Clin J Pain* 1998;14:143–147.
131. Baron R, Jänig W: Pain syndromes with causal participation of the sympathetic nervous system. *Anaesthetist* 1998;47:4–23.

132. Kingery WS: A critical review of controlled clinical trials for peripheral neuropathic pain and complex regional pain syndromes. *Pain* 1997;73:123–139.
133. Perez RS, Kwakkel G, Zuurmond WW, et al: Treatment of reflex sympathetic dystrophy (CRPS type 1): A research synthesis of 21 randomized clinical trials. *J Pain Symptom Manage* 2001;21:511–526.
134. Forouzanfar T, Koke AJ, Van Kleef M, et al: Treatment of complex regional pain syndrome type I. *Eur J Pain* 2002;6:105–122.
135. Harati Y, Gooch C, Swenson M, et al: Double-blind randomized trial of tramadol for the treatment of the pain of diabetic neuropathy. *Neurology* 1998;50:1842–1846.
136. Sindrup SH, Andersen G, Madsen C, et al: Tramadol relieves pain and allodynia in polyneuropathy: A randomized, double-blind, controlled trial. *Pain* 1999;83:85–90.
137. Watson CP, Babul N: Efficacy of oxycodone in neuropathic pain: A randomized trial in postherpetic neuralgia. *Neurology* 1998;50:1837–1841.
138. Rowbotham MC, Twilling L, Davies PS, et al: Oral opioid therapy for chronic peripheral and central neuropathic pain. *N Engl J Med* 2003;348:1223–1232.
139. Sindrup SH, Jensen TS: Efficacy of pharmacological treatments of neuropathic pain: An update and effect related to mechanism of drug action. *Pain* 1999;83:389–400.
140. Max MB: Treatment of post-herpetic neuralgia: Antidepressants. *Ann Neurol* 1994;35:S50–53.
141. Wallace MS, Ridgeway BM, Leung AY, et al: Concentration-effect relationship of intravenous lidocaine on the allodynia of complex regional pain syndrome types I and II. *Anesthesiology* 2000;92:75–83.
142. Stracke H, Meyer UE, Schumacher HE, et al: Mexiletine in the treatment of diabetic neuropathy. *Diabetes Care* 1992;15:1550–1555.
143. Wright JM, Oki JC, Graves L: Mexiletine in the symptomatic treatment of diabetic peripheral neuropathy. *Ann Pharmacother* 1997;31:29–34.
144. Van Hilten BJ, Van De Beek WJ, Hoff JI, et al: Intrathecal baclofen for the treatment of dystonia in patients with reflex sympathetic dystrophy. *N Engl J Med* 2000;343:625–630.
145. Fromm GH, Terrence CF, Chattha AS: Baclofen in the treatment of trigeminal neuralgia: Double-blind study and long-term follow-up. *Ann Neurol* 1984;15:240–244.
146. Mellick GA, Mellick LB: Gabapentin in the management of reflex sympathetic dystrophy. *J Pain Symptom Manage* 1995;10:265–266.
147. Serpell MG: Gabapentin in neuropathic pain syndromes: A randomized, double-blind, placebo-controlled trial. *Pain* 2002;99:557–566.
148. Mellick LB, Mellick GA: Successful treatment of reflex sympathetic dystrophy with gabapentin. *Am J Emerg Med* 1995;13:96.
149. Backonja M, Beydoun A, Edwards KR, et al: Gabapentin for the symptomatic treatment of painful neuropathy in patients with diabetes mellitus: A randomized, controlled trial. *JAMA* 1998;280:1831–1836.
150. Rowbotham M, Harden N, Stacey B, et al: Gabapentin for the treatment of postherpetic neuralgia: A randomized controlled trial. *JAMA* 1998;280:1837–1842.
151. Rull JA, Quibrera R, Gonzalez-Millan H, et al: Symptomatic treatment of peripheral diabetic neuropathy with carbamazepine (Tegretol): Double-blind crossover trial. *Diabetologia* 1969;5:215–218.
152. Leijon G, Boivie J: Central post-stroke pain: A controlled trial of amitriptyline and carbamazepine. *Pain* 1989;36:27–36.
153. Nelson KA, Park KM, Robinovitz E, et al: High-dose oral dextromethorphan versus placebo in painful diabetic neuropathy and postherpetic neuralgia. *Neurology* 1997;48:1212–1218.
154. Mcquay HJ, Carroll D, Jadad AR, et al: Dextromethorphan for the treatment of neuropathic pain: A double-blind, randomized, controlled crossover trial with integral n-of-1 design. *Pain* 1994;59:127–133.
155. Eisenberg E, Kleiser A, Dortort A, et al: The NMDA (N-methyl-D-aspartate) receptor antagonist memantine in the treatment of postherpetic neuralgia: A double-blind, placebo-controlled study. *Eur J Pain* 1998;2:321–327.
156. Eide PK, Jorum E, Stubhaug A, et al: Relief of post-herpetic neuralgia with the N-methyl-D-aspartic acid receptor antagonist ketamine: A double-blind, crossover comparison with morphine and placebo. *Pain* 1994;58:347–354.
157. Nikolajsen L, Hansen CL, Nielsen J, et al: The effect of ketamine on phantom pain: A central neuropathic disorder maintained by peripheral input. *Pain* 1996;67:69–77.
158. Gobelet C, Waldburger M, Meier JL: The effect of adding calcitonin to physical treatment on reflex sympathetic dystrophy. *Pain* 1992;48:171–175.
159. Varenna M, Zucchi F, Ghiringhelli D, et al: Intravenous clodronate in the treatment of reflex sympathetic dystrophy syndrome: A randomized, double-blind, placebo-controlled study. *J Rheumatol* 2000;27:1477–1483.
160. Adami S, Fossaluzza V, Gatti D, et al: Bisphosphonate therapy of reflex sympathetic dystrophy syndrome. *Ann Rheum Dis* 1997;56:201–204.
161. Perez RS, Zuurmond WW, Bezemer PD, et al: The treatment of complex regional pain syndrome type I with free radical scavengers: A randomized, controlled study. *Pain* 2003;102:297–307.
162. Zuurmond WW, Langendijk PN, Bezemer PD, et al: Treatment of acute reflex sympathetic dystrophy with DMSO 50% in a fatty cream. *Acta Anaesthesiol Scand* 1996;40:364–367.
163. Geertzen JH, De Bruijn H, De Bruijn-Kofman AT, et al: Reflex sympathetic dystrophy: Early treatment and psychological aspects. *Arch Phys Med Rehabil* 1994;75:442–446.
164. Schott GD: Interrupting the sympathetic outflow in causalgia and reflex sympathetic dystrophy. *Br Med J* 1998;316:792–793.
165. Verdugo RJ, Ochoa JL: Sympathetically maintained pain: I. Phentolamine block questions the concept. *Neurology* 1994;44:1003–1010.
166. Verdugo RJ, Campero M, Ochoa JL: Phentolamine sympathetic block in painful polyneuropathies: II. Further questioning of the concept of "sympathetically maintained pain." *Neurology* 1994;44:1010–1014.
167. Price DD, Long S, Wilsey B, et al: Analysis of peak magnitude and duration of analgesia produced by local anesthetics injected into sympathetic ganglia of complex regional pain syndrome patients. *Clin J Pain* 1998;14:216–226.

168. Reuben SS, Rosenthal EA, Steinberg RB: Surgery on the affected upper extremity of patients with a history of complex regional pain syndrome: A retrospective study of 100 patients. *J Hand Surg [Am]* 2000;25:1147–1151.
169. Blanchard J, Ramamurthy S, Walsh N, et al: Intravenous regional sympatholysis: A double-blind comparison of guanethidine, reserpine, and normal saline. *J Pain Symptom Manage* 1990;5:357–361.
170. Jadad AR, Carroll D, Glynn CJ, et al: Intravenous regional sympathetic blockade for pain relief in reflex sympathetic dystrophy: A systematic review and a randomized, double-blind crossover study. *J Pain Symptom Manage* 1995;10:13–20.
171. Bonelli S, Conoscente F, Movilia PG, et al: Regional intravenous guanethidine vs stellate ganglion block in reflex sympathetic dystrophies: A randomized trial. *Pain* 1983;16:297–307.
172. Hord AH, Rooks MD, Stephens BO, et al: Intravenous regional bretylium and lidocaine for treatment of reflex sympathetic dystrophy: A randomized, double-blind study. *Anesth Analg* 1992;74:818–821.
173. Ramamurthy S, Hoffman J: Intravenous regional guanethidine in the treatment of reflex sympathetic dystrophy/causalgia: A randomized, double-blind study. Guanethidine Study Group. *Anesth Analg* 1995;81:718–723.
174. Kettler RE, Abram SE: Intravenous regional droperidol in the management of reflex sympathetic dystrophy: A double-blind, placebo-controlled, crossover study. *Anesthesiology* 1988;69:933–936.
175. Hanna MH, Peat SJ: Ketanserin in reflex sympathetic dystrophy: A double-blind, placebo-controlled cross-over trial. *Pain* 1989;38:145–150.
176. Bounameaux HM, Hellemans H, Verhaeghe R: Ketanserin in chronic sympathetic dystrophy: An acute controlled trial. *Clin Rheumatol* 1984;3:556–557.
177. Singh B, Moodley J, Shaik AS, et al: Sympathectomy for complex regional pain syndrome. *J Vasc Surg* 2003;37:508–511.
178. Bandyk DF, Johnson BL, Kirkpatrick AF, et al: Surgical sympathectomy for reflex sympathetic dystrophy syndromes. *J Vasc Surg* 2002;35:269–277.
179. Schwartzman RJ, Liu JE, Smullens SN, et al: Long-term outcome following sympathectomy for complex regional pain syndrome type 1 (RSD). *J Neurol Sci* 1997;150:149–152.
180. Aburahma AF, Robinson PA, Powell M, et al: Sympathectomy for reflex sympathetic dystrophy: Factors affecting outcome. *Ann Vasc Surg* 1994;8:372–379.
181. Lacroix H, Vander Velpen G, Penninckx F, et al: Technique and early results of videoscopic lumbar sympathectomy. *Acta Chir Belg* 1996;96:11–14.
182. Mailis A, Meindok H, Papagapiou M, et al: Alterations of the three-phase bone scan after sympathectomy. *Clin J Pain* 1994;10:146–155.
183. Kumar K, Nath RK, Toth C: Spinal cord stimulation is effective in the management of reflex sympathetic dystrophy. *Neurosurgery* 1997;40:503–508;discussion 508–509.
184. Kemler MA, Barendse GA, Van Kleef M, et al: Spinal cord stimulation in patients with chronic reflex sympathetic dystrophy. *N Engl J Med* 2000;343:618–624.
185. Kemler MA, Barendse GA, Van Kleef M, et al: Pain relief in complex regional pain syndrome due to spinal cord stimulation does not depend on vasodilation. *Anesthesiology* 2000;92:1653–1660.
186. Kemler MA, Rijks CP, De Vet HC: Which patients with chronic reflex sympathetic dystrophy are most likely to benefit from physical therapy? *J Manip Physiol Ther* 2001;24:272–278.
187. Kemler MA, Furnee CA: Economic evaluation of spinal cord stimulation for chronic reflex sympathetic dystrophy. *Neurology* 2002;59:1203–1209.
188. Hassenbusch SJ, Stanton-Hicks M, Schoppa D, et al: Long-term results of peripheral nerve stimulation for reflex sympathetic dystrophy. *J Neurosurg* 1996;84:415–423.
189. Kumar K, Toth C, Nath RK: Deep brain stimulation for intractable pain: A 15-year experience. *Neurosurgery* 1997;40:736–746;discussion 746–737.
190. Rauck RL, Eisenach JC, Jackson K, et al: Epidural clonidine treatment for refractory reflex sympathetic dystrophy. *Anesthesiology* 1993;79:1163–1169;discussion 1127A.
191. Takahashi H, Miyazaki M, Nanbu T, et al: The NMDA-receptor antagonist ketamine abolishes neuropathic pain after epidural administration in a clinical case. *Pain* 1998;75:391–394.
192. Fialka V, Resch Kl, Ritter-Dietrich D, et al: Acupuncture for reflex sympathetic dystrophy. *Arch Intern Med* 1993;153:661–665.
193. Kho H: The impact of acupuncture on pain in patients with reflex sympathetic dystrophy. *Pain Clin* 1995;8:59–61.
194. Korpan MI, Dezu Y, Schneider B, et al: Acupuncture in the treatment of posttraumatic pain syndrome. *Acta Orthop Belg* 1999;65:197–201.
195. Wu WH, Bandilla E, Ciccone DS, et al: Effects of qigong on late-stage complex regional pain syndrome. *Altern Ther Health Med* 1999;5:45–54.
196. Sherry DD, Wallace CA, Kelley C, et al: Short- and long-term outcomes of children with complex regional pain syndrome type I treated with exercise therapy. *Clin J Pain* 1999;15:218–223.
197. Oerlemans HM, Oostendorp RA, De Boo T, et al: Adjuvant physical therapy versus occupational therapy in patients with reflex sympathetic dystrophy/complex regional pain syndrome type I. *Arch Phys Med Rehabil* 2000;81:49–56.
198. Oerlemans HM, Oostendorp RA, De Boo T, et al: Pain and reduced mobility in complex regional pain syndrome I: Outcome of a prospective, randomized, controlled clinical trial of adjuvant physical therapy versus occupational therapy. *Pain* 1999;83:77–83.
199. Uher EM, Vacariu G, Schneider B, et al: Comparison of manual lymph drainage with physical therapy in complex regional pain syndrome type I: A comparative randomized controlled therapy study. *Wien Klin Wochenschr* 2000;112:133–137.
200. Severens JL, Oerlemans HM, Weegels AJ, et al: Cost-effectiveness analysis of adjuvant physical or occupational therapy for patients with reflex sympathetic dystrophy. *Arch Phys Med Rehabil* 1999;80:1038–1043.
201. Lee BH, Scharff L, Sethna NF, et al: Physical therapy and cognitive-behavioral treatment for complex regional pain syndromes. *J Pediatr* 2002;141:135–140.

202. Zollinger PE, Tuinebreijer WE, Kreis RW, et al: Effect of vitamin C on frequency of reflex sympathetic dystrophy in wrist fractures: A randomised trial. *Lancet* 1999; 354:2025–2028.
203. Gschwind C, Fricker R, Lacher G, et al: Does peri-operative guanethidine prevent reflex sympathetic dystrophy? *J Hand Surg [Br]* 1995;20:773–775.
204. Wilhelm A, Suden R: Proximal radial nerve compression syndrome: Treatment and results. *Handchir Mikrochir Plast Chir* 1985;17:219–224.
205. Maier C: Sympathische Reflexdystrophie-M. Sudeck, in Dienr H, Maier C (eds), *Das Schmerz-Therapie Handbuch*. München: Urban und Schwarzenberg, 1996:170–180.
206. Karstetter K, Sherman RA: Use of thermography for initial detection of early reflex sympathetic dystrophy. *J Am Podiatr Med Assoc* 1991;81:437–443.
207. Atkins RM, Duckworth T, Kanis JA: Features of algodystrophy after Colles' fracture. *J Bone Joint Surg [Br]* 1990;72:105–110.
208. Gold B, Brickner D, Sukenik S: Reflex sympathetic dystrophy syndrome following minor trauma. *Isr J Med Sci* 1989;25:107–109.
209. Van Der Laan L, Veldman PH, Goris RJ: Severe complications of reflex sympathetic dystrophy: Infection, ulcers, chronic edema, dystonia, and myoclonus. *Arch Phys Med Rehabil* 1998;79:424–429.

CHAPTER 26

Chronic Daily Headaches

Anan Srikiatkhachorn

In early 1980, Saper and colleagues described a series of patients who experienced a progressive form of migraine. These patients had developed episodic headache of migraine features in early life. Headache became more frequent and eventually occurred on an almost daily basis. The authors also documented the prevalence of ergotamine overuse and comorbid depression in these patients. They called this syndrome the chronic headache complex.[1] This frequent-headache syndrome gained much more attention after the publication by Mathew et al. in 1982. In their landmark paper, Mathew and colleagues described an evolutional pattern of migraine from episodic to daily or near-daily headache.[2] Later, in 1987, they introduced the term transformed or evolutive migraine to describe the migraine syndrome with daily or near-daily occurrence. The authors were impressed by the high prevalence of this condition, which constituted 77 percent of daily headache patients seen at The Houston Headache Clinic.[3]

In addition to migraine, episodic tension-type headache also can develop into a more frequent form. The term chronic daily headache (CDH) was introduced subsequently to cover all forms of primary headache that occur on a daily or near-daily basis. Since diagnosis of CDH is based entirely on the frequency of headache, it should be regarded as a syndrome that covers a broad spectrum of headache disorders, rather than a specific headache entity. Although there are no diagnostic criteria for CDH, it is generally accepted that this designation does not include paroxysmal headaches that have a duration of less than 4 hours, e.g., cluster headache, paroxysmal hemicrania.

▶ EPIDEMIOLOGY

CDH is a widespread clinical problem. The prevalence is high in all age groups and all geographic distributions. Prevalence in the general population is around 5 percent.[4–6] The prevalence of CDH is much higher in the specialty headache clinic population, where it contributes to approximately 30 to 40 percent of patients visiting headache clinics.[3,7]

Besides its high prevalence, CDH is also a disabling headache condition that is refractory to treatment. The situation is worsened by the fact that at least one-third of patients with CDH are overusing analgesics, either prescribed medications or over-the-counter drugs. Overconsumption of analgesics can lead to the development of an intractable headache that is called medication-overuse headache.

▶ NOSOLOGY AND CLASSIFICATION

Nosology of CDH is one of the most controversial issues in headache classification. Despite the magnitude of this problem, no widely accepted classification for this condition exists. When the Headache Classification Committee of the International Headache Society (IHS) defined the various headache disorders in 1988, the issue of very frequent primary headaches was not addressed.[8] As a consequence, this classification system does not work well when it is applied to this condition. Several reports showed that at least one-third of CDH patients could not be categorized using IHS criteria.[7,9] The reason underlying this deficiency is that this classification system aims at diagnosing individual headache attack and does not take into enough consideration the temporal profile or natural history of headache. Therefore, when this classification is applied to CDH, the condition of which headache can vary widely, more than one diagnosis has to be given to each patient in order to describe all forms of headache adequately.

Silberstein and colleagues[10] proposed a classification scheme for CDH. They defined CDH as any primary headache that occurs more than 15 days per month and each attack lasts more than 4 hours per day untreated. Headaches of organic origin and paroxysmal chronic headache disorders, i.e., daily headache with a duration of less than 4 hours, such as cluster headache, or disorders not included in the IHS classification, such as hypnic headache and episodic paroxysmal hemicrania, were not included in this classification. Under this operating criterion, Silberstein and colleagues classified CDH into transformed migraine (TM), chronic tension-type headache (CTTH), new persistent daily headache (NPDH), and hemicrania continua.[10] Each of these disorders was further categorized into those with or without analgesic overuse.

Transformed Migraine

Transformed migraine (TM) is considered the most common headache among CDHs. Mathew and colleagues reported that TM accounted for 77 percent of CDH patients seen at The Houston Headache Clinic. This number is comparable with that reported by Manzoni and colleagues,[11] who demonstrated that 70 percent of CDH patients transformed their headache from episodic pattern to daily or near-daily headache. A recent population survey in France showed that two-thirds of the CDH subjects presented migraine features.[12] CDH with migraine features also accounts for a majority of children with daily continuous headache.[13]

Typically, this condition starts as a distinct occasional episodic migraine (with or without aura). The frequency is increased gradually and eventually evolves to daily headache. Several factors, such as medication overuse, underlying psychiatric disorders, and chronobiologic factors, are thought to account for transforming episodic migraine to CDH. Two distinct headache features are present in TM patients. The first is persistent low-grade headache, resembling tension-type headache. This headache is usually diffuse and pressure-like, nonthrobbing in character. Neck tightness is common. In addition to this background headache, TM patients also have superimposed intermittent headache of moderate to severe intensity with some migraine characteristics. This headache is more throbbing, is associated with nausea, with or without vomiting, and can be worsened by head motion. Full-blown migraine attacks occur more frequently in females and sometimes are related to menstrual cycle. Photophobia is more common in female patients, whereas sleep and emotional disturbances are more prevalent in male patients.[14] It should be noted that this combined headache does not indicate the coexistence of migraine and chronic tension-type headache. Recent analysis of CDH in adolescents and children showed that rather than having two coexisting headache types, CDH patients had a single syndrome that paroxysmally worsened and gathered migraine features.[15]

Manzoni and colleagues proposed another classification system for TM based on the evolving pattern.[16] They suggested that TM should be subdivided into (1) migraine with interparoxysmal headache; and (2) chronic migraine. The former diagnosis refers to TM in which typical migraine attacks are maintained and the interparoxysmal headache either fulfills or does not fulfill criteria for tension-type headache. The latter refers to TM that loses the distinct migraine and develops a continuous headache that exhibits most of the features of migraine (e.g., nausea, throbbing pain, unilaterality, etc.), except the temporal profile. In 1998, Spierings and colleagues classified the temporal pattern of CDH development into those with abrupt and gradual onsets.[17] They showed that besides the difference in temporal pattern, these two forms also differ in age of onset of their initial and daily headaches. Patients with abrupt-onset CDH tend to be younger and have less familial history. The onset of CDH may relate to some circumstance such as head, neck, or back injury; flulike illness; and other medical illnesses or surgical procedures. Patients with gradual-onset CDH, accounting for four-fifths of CDH cases, are usually older, and occurrence of headache in parents is more common.[17,18]

Chronic migraine will be included in the section on complications of migraine in the upcoming revision of the IHS classification system. In this revision, *chronic migraine* is defined as migraine headache that occurs more than 15 days per month for a period of not less than 3 months, and with no drug overuse. If drug overuse is present, i.e., acute migraine drugs and/or analgesics for more than 10 days per month, chronicity is most likely caused by medication overuse. Therefore, such patients should be diagnosed according to their primary migraine diagnosis and as medication-overuse headache. If chronicity persists after drug withdrawal, chronic migraine then should replace medication-overuse headache.

Chronic Tension-Type Headache (CTTH)

Similar to migraine, episodic tension-type headache (ETTH) can evolve to a more chronic and persistent form. CTTH is to be diagnosed if the headache frequency exceeds 15 days per month and persists more than 3 months. These headaches more often are diffuse or bilateral, frequently involving the posterior aspect of the head and neck. Duration of each headache attack can be varied, ranging from minutes to days. The pain typically is pressing, tightening in quality, and of mild to moderate intensity. Unlike migraine headache, CTTH does not worsen with routine physical activity. There is no nausea or photophobia, but phonophobia may be present. In CTTH, prior episodic migraine and most features of migraine are absent.

According to the IHS classification system, CTTH is further divided into those associated and not associated with increased pericranial muscle tenderness. Pure CTTH most often is not associated with drug overuse. Analgesic overconsumption can modify ETTH and foster the development of CTTH. According to the general principles of this classification, patients who get a new type of headache during drug overuse or whose headache is made worse should have both the diagnosis of the pre-existing headache and the diagnosis medication-overuse headache. Therefore, most patients with tension-type headache that has been exaggerated by drug overuse should receive both diagnosis of their primary headache and diagnosis of drug-overuse headache.

Silberstein and colleagues used a different approach to classify CTTH.[10] Emphasizing the clinical importance of concomitant medication overuse, they divided CTTHs into those with and without medication overuse. It should be noted that according to this classification, and unlike the IHS system, the presence of mild nausea or mild photophobia and phonophobia does not preclude the diagnosis of CTTH.

New Persistent Daily Headache (NPDH)

NPDH is a new subtype of CDH first described by Varnast.[19] In the first description, NPDH was used to describe daily headache that occurred in patients with no prior history of primary headaches. Later, this nomenclature was used to cover abrupt-onset daily headache occurring in patients with migraine or tension-type headache. Therefore, NPDH can be diagnosed in these patients only if their primary headache disorders do not increase in frequency to give rise to NPDH.

In this condition, daily headaches occur with abrupt onset (over less than 3 days).[10] More than 80 percent of patients can recall the exact date their headache started. Usually, the headache starts without precipitants such as trauma or psychological tension. NPDH affects females more than males, with female-to-male ratio of 1.5 to 2.5:1.[19,20] Patients with NPDH generally are younger than those with TM, especially when female patients are concerned. The peak age of onset is the second and third decades in females and the fifth decade in males.

The character, location, and duration of pain in NPDH are similar to those of CTTH. Approximately 80 percent of patients have a continuous headache on all days without a pain-free period. Headache is moderately severe and can be either pulsating or pressure-like. Diffuse headache with occipitonuchal predominance is common. Unilateral headache occurs in about a third of patients. Headache can be accompanied by migraine features. Nausea, photophobia, or phonophobia occurs in about 60 percent of patients. Light-headedness is also common.

One-third of patients developed NPDH with a flu-like illness. This finding raised the possibility of infection as an etiology of NPDH. Increased frequency of Epstein-Barr virus excretion has been reported in these patients.[21] In a study of 32 NPDH patients, almost 85 percent of patients were found to have an active Epstein-Barr virus infection. Activation of a latent Epstein-Barr virus has been proposed to be the trigger for development of daily headache.

Hemicrania Continua

Hemicrania continua constitutes a rare indomethacin-responsive headache disorder. Less than 100 cases have been reported in the literature. This headache syndrome is characterized by a strictly unilateral continuous headache with absolute responsiveness to indomethacin. Cases with alternating hemicrania and bilateral involvement have been reported, but they should be considered exceptions. Hemicrania continua affects females more frequently than males. The ratio of females to males is approximately 2.4:1.[22] This disorder may be subdivided into continuous and remitting forms. The continuous variety, which accounts for most cases, may evolve from a previous remitting form or continuous unremitting headache from the onset. In the remitting form, periods of daily headache alternate with pain-free remissions. Most patients with the remitting course in the beginning usually develop continuous headache. Only a few cases have the persistent feature of a remitting course. It should be noted that the idea of subdivision of hemicrania continua is not universally accepted. No subdivision is considered in the revised version of the IHS classification system.

Headache in hemicrania continua is usually moderately severe. Migrainous features, i.e., nausea, vomiting, photophobia, or phonophobia, are common, particularly in the exacerbation period. Headache is triggered by neck movements, but tender spots may be present. Sharp pain of brief duration (<1 minute), so-called jabs and jolts syndrome or benign stabbing headache, can occur in one-third of patients. This brief superficial headache responds well to indomethacin. It is not specific to hemicrania continua. Other conditions in which jab and jolt headache can occur include migraine, tension-type headache, and cluster headache or even headache-free individuals.

Associated autonomic dysfunctions are important clinical features. At least one of the autonomic features must be present in association with exacerbations of pain on the affected side to fulfill the diagnosis of hemicrania continua. Autonomic symptoms consist of conjunctival injection, tearing, ptosis, meiosis, rhinorrhea, eyelid edema, nasal stuffiness, and forehead sweating. Transient neurological deficits resembling migraine aura, such as hemiparesis and unilateral paresthesia, can occur but are not usual features.

▶ COMORBIDITY

Problems that usually coexist with CDH are medication overuse and psychiatric disorders. Several groups of drugs are overused by patients with CDH. These include nonopioid and opioid analgesics, ergot-containing preparations, decongestants, and anxiolytics. In recent years, triptans have been used and misused widely by migrainous patients. Besides other adverse effects, medication overuse can worsen headache and foster transforming episodic headache to daily headache. The paradoxical effect of analgesics in increasing headache susceptibility may be restricted to the patients with primary headaches. It is recognized that medication-overuse headaches do not develop in nonheadache patients who daily consume analgesics for other medical conditions. A recent population survey revealed that CDH patients who overuse analgesics had a lower quality of life compared with those who do not.[23] Analgesic overuse also conveys a poor prognosis.[6]

Patients with CDH have high frequencies of psychiatric comorbidity or psychological distress in clinic-based studies. Anxiety and mood disorders (ranging from dysthymia to major depression) are the most prevalent psychiatric conditions coexisting with CDH. The presence of psychological distress contributes to poor quality of life in patients with CDH. Comorbid major depression is a poor outcome predictor for CDH, whereas the presence of psychological distress does not predict prognosis.[24]

▶ PATHOPHYSIOLOGY AND PATHOGENESIS

The biologic mechanism underlying the pathogenesis of CDH is not well defined. This intractable headache disorder can be seen as a form of neuropathic pain. Spontaneous pain, perception of pain in response to non-noxious stimulation (e.g., scalp allodynia), and poor opioid responsiveness to opioids are characteristics of neuropathic pain.

Some clinical evidence implies that transformation of an initial episodic headache, either migraine or tension-type headache, to CDH may share common mechanisms. For instance, Spierings and colleagues compared clinical features of CDH between those developed from migraine and those from tension-type headache. They found that daily headaches in both groups did not differ in their features. They also found that when daily headaches became intermittent again, they reassumed the features of the initial headache.[25] The similarities in the characteristics of daily headache, regardless of the type of initial headache, imply that the process of transformation finally results in the common pathophysiologic state.

Peripheral Mechanism: Nociceptor Sensitization

Sensitization of neurons in the cranial nociceptive pathway is the main hypothesis for CDH development. Sensitization can be classified into peripheral and central forms. *Peripheral sensitization* refers to the state in which peripheral nociceptors increase their responses to suprathreshold stimuli and become responsive to subthreshold stimuli. Therefore, low-intensity stimuli such as touch, vascular pulsation, and so on, can result in the painful sensation. Sensitization also turns on the "silent" nociceptors. Several mediators released during tissue injury and inflammation (e.g., prostaglandins and bradykinin) can sensitize nociceptors. Sensitization of nociceptors in the pericranial vessels and pericranial myofascial tissue has been proposed to cause throbbing headache during migraine attack and pressure-like headache in ETTH, respectively.[26,27]

Repeated episodes of trigeminal nerve activation may sensitize nociceptors chronically and contribute to the development of CDH. Although the hypothesis of nociceptor sensitization is plausible, the evidence of peripheral sensitization in CDH patients is still lacking. The only evidence in favor of the peripheral mechanism hypothesis is the increase in activity of the surface electromyogram (EMG) over EMG of the temporal and frontal areas, which was observed in CTTH patients.[28] On the other hand, a finding of normal interstitial levels of inflammatory mediators and metabolites in tender trapezius muscle in patients with CTTH indicates that persistent, ongoing inflammation is unlikely to be the cause of pain in this condition.[29]

Central Mechanism: Central Sensitization

Central axons of the nociceptive trigeminal and cervical dorsal ganglion neurons terminate by synapse on neurons in laminae I and V of the trigeminal nucleus caudalis (TNC). Patterns of activation by these primary afferents are important in shaping the response of TNC neurons. After repetitive stimulation, N-methyl-D-aspartate (NMDA) receptors are recruited via the removal of voltage-dependent blockade by magnesium ions. Activation of the NMDA receptor results in a transient increase in intracellular calcium in the postsynaptic neurons. The rise in intracellular calcium activates calcium-calmodulin-dependent protein kinase II, protein kinase A, and protein kinase C. Neuronal proteins that are substrates for these kinase enzymes include neurotransmitter receptors, ion channels, transcription proteins, and so on. Phosphorylation of these proteins leads to various postsynaptic changes ranging from increasing membrane excitability to changes in genetic transcription. Calcium also activates neuronal nitric oxide synthase (nNOS), hence increasing nitric oxide (NO) production. This gaseous molecule can diffuse back to the presynaptic terminals and stimulates transmitter release. The facilitating effect of NO on TNC function was demonstrated in animals. Jones and colleagues showed that the responses of TNC neurons to cutaneous and visceral stimulations were enhanced after pretreatment with an NO-donating agent.

The preceding molecular mechanism underlies the development of central sensitization, a condition wherein sensitivity of central nociceptive neurons is enhanced. In this condition, TNC become responsive to low-intensity, normally subthreshold stimuli, and increase their responses to suprathreshold stimuli. The process of *central sensitization* also enlarges the receptive fields of these neurons, thus expanding the pain area. Central sensitization may contribute to several features of CDH. Allodynia in the form of scalp tenderness reflects the decrease in nociceptive threshold, which is the key feature of sensitization. Expansion of the headache area may reflect the recruitment of nociceptive neurons of higher level (i.e., thalamus, etc.) in the process of pain transmission. Some clinical evidence implies that sensitization of the trigeminal nociceptive system may develop more easily in patients with functional headaches. For instance, Katsarava and colleagues demonstrated facilitation of nociception-specific blink reflex responses predominantly on the headache side in migraine patients but not in patients with sinusitis.[30]

Derangement of Endogenous Pain Control System

The excitability of TNC neurons is under the strong influence of supraspinal controls. Several brain stem nuclei exert their nociceptive modulation effect via downward projection to TNC. Derangement of this endogenous pain control system can alter susceptibility to pain. Attenuation of endogenous antinociceptive activity can decrease the pain threshold, resulting in increasing pain susceptibility. In addition, experiments in animals showed that the development of long-term potentiation of dorsal horn synapses, a condition comparable to central sensitization, is facilitated when this endogenous antinociceptive system is impaired.[31]

Several lines of evidence indicate a possible relationship between periaqueductal gray (PAG) matter dysfunction and headache pathogenesis. PAG is the center of a powerful descending antinociceptive network that has a strong influence on TNC. Migraine-like headache sometimes occurs in patients with PAG electrode implantation for pain relief. The role of PAG in the pathogenesis of primary headache is confirmed by functional imaging studies. In 1995, Weiller and colleagues, using position-emission tomography (PET), demonstrated an increase in mesencephalic blood flow during spontaneous attacks of migraine without aura.[32] Such findings reflect possible involvement of PAG, dorsal raphe nucleus, and locus ceruleus in the pathogenesis of migraine. Involvement of PAG in the pathogenesis of CDH has been reported in a study using functional magnetic resonance imaging (fMRI). In 2001, Welch and colleagues demonstrated that the *R2'* value, a measure for nonheme iron, was increased in migraine and CDH patients compared with control subjects.[33] They also demonstrated that this *R2'* value correlated positively with duration of illness. An increased *R2'* value also could be viewed as an indicator of hyperoxic neuronal dysfunction. The authors concluded that the iron homeostasis in the PAG was impaired in migraine patients. Repeated migraine attacks may further deteriorate this homeostasis. In this regard, increased frequency of migraine attacks can alter PAG function. Dysfunction of PAG may lead to headache through interruption of its normal antinociceptive function.

The hypothesis that an endogenous antinociceptive system in patients with CDH is altered and defective is supported by several physiologic and biologic observations. Higher levels of nerve growth factor and substance P were evident in the cerebrospinal fluid (CSF) of patients with CDH compared with nonheadache controls.[34] In patients with CDH, there may be derangement of various steps required for information processing in the central nervous system (CNS). Fusco and colleagues demonstrated that the temporal summation of second pain, the psychological correlate of the excitatory pain circuits, was greater in patients with CDH than in patients with episodic migraine or tension-type headache.[35] De Tommaso and colleagues showed that cognitive tasks failed to distract cutaneous pain induced in laser-thermal stimulation applied on facial and hand areas in chronic migraine patients.[36] These findings suggest an abnormal cortical processing of nociceptive input in chronic migraine patients.

Analgesic Overuse, Dysfunction of Endogenous Pain Control System, and CDH

The mechanism by which analgesics foster the development of CDH is still unclear. Accumulating evidence shows that medication overuse may compromise the endogenous antinociceptive system.

Recent studies have indicated the close relationship between nonnarcotic analgesics and changes in the endogenous serotonin (5-HT)–dependent antinociceptive system. Administration of nonnarcotic analgesics (i.e., acetaminophen, acetylsalicylic acid) led to a rapid increase in 5-HT level in the cerebral cortex, hypothalamus, striatum, hippocampus, brain stem, and pons.[37,38] Acetaminophen administration also led to a reduction of dihydroxyphenylacetic acid (a dopamine metabolite) in the striatum and elevation of norepinephrine in the posterior cortex. Elevation of platelet 5-HT after acetaminophen administration also was reported.[39] The mechanism by which analgesics induce an increase in 5-HT level is still unclear. Besides altering the 5-HT level, exposure to analgesics also changes the expression of 5-HT receptors in the CNS. Acute administration of aspirin or acetaminophen downregulates the expression of 5-HT$_{2A}$ receptors in rat cortex tissue.[39–41]

The pattern of analgesic-induced plasticity within the central 5-HT system changes over time. In rats, a 15-day course of acetaminophen led to downregulation of the 5-HT$_{2A}$ receptor and upregulation of 5-HT transporter in the frontal cortex. Receptor downregulation and transporter upregulation became less evident following more prolonged administration of the drug, and these changes coincided with a decrease in the analgesic efficacy of acetaminophen.[42] These findings suggest that chronic analgesic use can alter the central 5-HT system.

Alterations in the 5-HT system have been demonstrated in patients with medication-induced CDH. Compared with patients with migraine, patients with CDH have a lower level of platelet 5-HT and a greater density of 5-HT$_{2A}$ receptors.[43–45] These changes can be reversed after drug withdrawal, and normalization of platelet 5-HT and its receptor correlates with improvement in clinical headache.[44] Morphologic study of platelets from patients with analgesic-induced CDH has demonstrated excessive enlargement of the intraplatelet canaliculi system, reflecting the abnormality in release mechanism.[46] An increase in platelet NO and cyclic guanosine monophosphate (GMP) production, as well as increased intracytosolic calcium after stimulation with collagen, have been reported in migraine patients who overused analgesics.[45] Increased NOS activity, decreased 5-HT content, and increased glutamate level have been demonstrated in platelets taken from patients with CTTH.[47] These findings may reflect an analogous central upregulation of NOS activity in the trigeminal pathway and supraspinal structures involved in the modulation of nociceptive input contributing to central sensitization.

The preceding experimental and clinical evidence suggests that the central, mainly 5-HT-dependent antinociceptive system is impaired in patients with CDH. Analgesic overuse may further derange this system by inducing a low level of 5-HT. The relative depletion of 5-HT subsequently leads to upregulation of the pronociceptive 5-HT$_{2A}$ receptor and changes in intracellular signaling. Reduction of nociceptive inhibitory control may facilitate the process of central sensitization, activate the nociceptive facilitating system, or promote kindling. Since hydrolysis of phosphoinositol is a transduction cascade of the 5-HT$_{2A}$ receptor, occupying these receptors will activate the release of calcium from its intracellular store and increase the cytoplasmic calcium concentration. Elevation of intracellular calcium is an important step in the development of long-term potentiation and central sensitization. Activation of 5-HT$_{2A}$ receptors also activates nNOS, thereby increasing NO production.[48] Thus, derangement in the central antinociceptive system with upregulation of the 5-HT$_{2A}$ pro-nociceptive receptor as a result of chronic medication use may increase sensitivity to pain perception and foster or reinforce CDH (see Figure 26-1).

▶ DIFFERENTIAL DIAGNOSIS

Secondary headache disorders must be excluded in patients with CDH. Patients with recent onset of headaches (<3 months to 6 months) require more attention. Differential diagnoses include mass lesions such as subdural hematoma or brain tumor causing intracranial hypertension. A syndrome of increased intracranial pressure with or without papilledema (pseudotumor cerebri) often has clinical features similar to chronic migraine. Occlusion of the cerebral venous system can cause chronic headache with variable features and should be excluded, especially in female patients with a history of taking contraceptive pills. Intracranial hypotension or low-CSF-volume headache and post-traumatic headache are important differential diagnoses in patients with suspected NPDH. Orthostatic headache is a key feature of low-CSF-volume headache. The presence of a history of an index event, i.e., lumbar puncture or epidural injection, or a vigorous Valsalva maneuver helps in the diagnosis. Absence of such a history does not preclude diagnosis because spontaneous CSF leakage can occur. In suspected cases, MRI with gadolinium enhancement is the investigation of choice. Diffuse pachymeningeal enhancement is the characteristic feature of this condition. Low-grade viral, tuberculous, or cryptococcal meningitis, as well as chronic pachymeningitis, should be excluded especially when neck stiffness is present. A toxic metabolic disturbance, such as anemia or hypothyroidism, or a medication-induced metabolic disturbance can cause daily headaches, but this is uncommon. Disorders of pericranial structures, e.g., glaucoma, frontal, or sphenoid sinusitis or cervical spine disorders, should be sought.

▶ DIAGNOSTIC WORKUP

As with other primary headaches, no diagnostic test is available for confirming the diagnosis of CDH. All diagnostic tests are conducted to exclude possible structural causes.

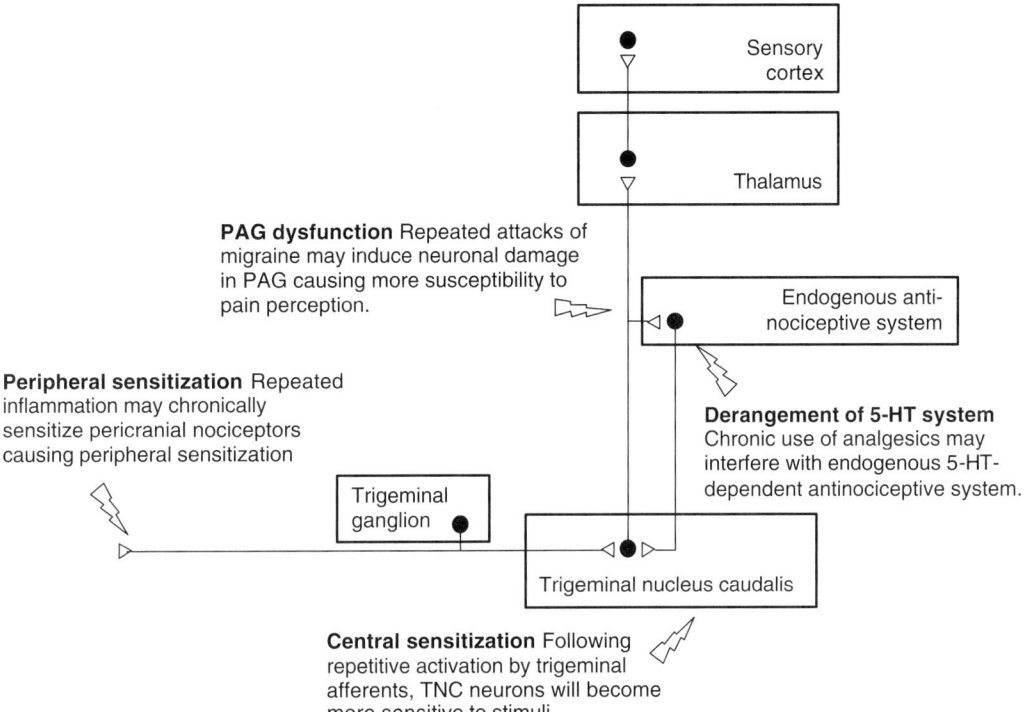

Figure 26-1. Proposed mechanisms underlying the development of CDH. Pericranial nociceptors can be sensitized after being stimulated repeatedly. An increase in activity of trigeminal afferents will lead to sensitization of central nociceptive neurons in the TNC (central sensitization). The function of neurons in the endogenous antinociceptive system may be altered by repeated migraine attacks (e.g., PAG neurons) or chronic medication overuse (e.g., 5-HT neurons). Attenuation of the central pain control system and sensitization of TNC neurons will increase pain susceptibility and may underlie the development of CDH.

A general panel of blood tests, including a complete blood count, electrolytes, kidney function tests, liver function tests, and possibly an erythrocyte sedimentation rate, is often obtained as a baseline and is used to exclude secondary causes of headache. Imaging studies such as MRI are required in patients with suspected intracranial pathology or in those who do not respond to treatment. The imaging study should not be repeated unless there is a significant change in the neurological examination or in the headache characteristics. Adverse effects of chronic analgesic usage, i.e., nephropathy, gastrointestinal ulcer, etc., should be looked for. An electrocardiogram is advised for patients who may overuse vasoconstrictors (e.g., triptans or dihydroergotamine).

▶ TREATMENT

General Considerations

Compared with other forms of primary headache, CDH is quite refractory to treatment. Patient education and motivation are crucial for successful treatment, especially for those with medication overuse. Secondary headache disorders must be excluded by careful clinical examination and proper investigation. The diagnosis of CDH should be made as comprehensive as possible to specify from which subtype patients suffer. This may be significant in forecasting the treatment response because patients with TM usually are more responsive to treatment than those with CTTH. Comorbidity, medical and psychiatric conditions, and analgesic overuse are important factors determining the prognosis of CDH and must be identified. The present and prior use of prescribed or over-the-counter drugs should be recorded. Besides analgesics, some patients also abuse other substances, such as tranquilizers, opioids, decongestants, etc.

Behavioral and psychological aspects of CDH need to be considered when treating CDH patients. A combination of pharmacologic and nonpharmacologic interventions, such as cognitive behavioral therapy, is essential for satisfactory results. Generally, the treatment for CDH, especially that complicated by medication-induced headache, is divided into two phases: acute detoxification, and long-term prophylaxis.

Acute Detoxification Therapy

Discontinuation of abused medication is the first step in the treatment of CDH with medication overuse. This step is extremely important because patients usually do not respond to preventive medications while overusing drugs. Moreover, it has been shown recently that analgesic withdrawal alone can result in reduction of headache. In 2000, Linton-Dahlöf and colleagues reported that 56 percent of their patients were improved significantly (defined as at least 50 percent reduction of headache days) after abrupt drug withdrawal therapy with no concurrent use of prophylactic medication.[49]

Analgesic discontinuation can be performed either in an inpatient or outpatient setting. The type and amount of medication being overused will determine the setting in which detoxification can take place. The German Migraine Society recommends outpatient withdrawal for patients who do not take barbiturates or tranquilizers with their analgesics and are highly motivated. Inpatient treatment is recommended for those who take tranquilizers, barbiturates, or opioids, who failed to withdraw the drugs as outpatients, or who have coexisting depression.[50]

Withdrawal symptoms include withdrawal headache, nausea, vomiting, agitation, arterial hypertension, tachycardia, sleep disturbance, restlessness, and nervousness. The withdrawal symptoms last for 2 to 10 days and are shorter in patients abusing only triptans. Hallucination and seizures are uncommon features even in patients abusing barbiturate-containing migraine drugs. However, it is recommended to replace short-acting barbiturates and benzodiazepines with long-acting medications that are tapered gradually to avoid a serious withdrawal syndrome.

Outpatient Treatment

Outpatient treatment is recommended for patients who take simple analgesics not containing barbiturates or opioids and whose durations of abuse are less than 5 years.[51] There are two approaches for analgesic discontinuation, namely, abrupt or gradual. Ergots, triptans, and nonopioids can be withdrawn abruptly. Opioids and barbiturates should be withdrawn more gradually, depending on dose and duration of intake. Abrupt analgesic withdrawal can be performed successfully on an outpatient basis by patient education and rescue medication for relief of headache. Besides the reduction in headache frequency, analgesic withdrawal also led to an improvement in sleep, as measured by polysomnography.[52]

Another outpatient approach is gradual reduction of overused medication. "Transitional medication," such as long-acting NSAIDs or naratriptan, can be helpful during the withdrawal period. Breakthrough headache should be treated with specific antimigraine drugs such as triptans. The patient can use this medication daily for approximately 1 week and then limit its use. A short course of corticosteroids such as oral prednisone may be helpful in detoxification treatment.[53] A neuroleptic or other rescue medication may be necessary during outpatient detoxification.

Inpatient Treatment

Detoxification can be difficult and sometimes requires hospitalization. Inpatient treatment is recommended for CDH patients whose duration of abuse is more than 5 years; who fail outpatient treatment; who consume barbiturates, opioids, or tranquilizers; or who have significant psychological and behavioral comorbidities.

Intravenous dihydroergotamine (DHE) at a dosage of 1 to 2 mg every 8 hours coadministered with 10 mg metoclopramide for 48 to 72 hours is an effective treatment for breaking a daily headache cycle. This regimen is recommended for CDH patients of migraine subtypes who have not abused ergotamine, DHE, or triptans. DHE also can be administered via continuous intravenous infusion. Approximately 60 to 70 percent of patients respond to DHE. In addition to its short-term efficacy, DHE also can decrease headache frequency, severity, medication-use headache, and absence from work at 3 months after initial treatment.[54] Diarrhea, paresthesia, dizziness, chest discomfort, and leg cramps are adverse effects in some patients. Owing to its vasoconstrictive effect, DHE is contraindicated in coronary and peripheral vascular diseases.

Alternative treatments for CDH patients in whom DHE is contraindicated or ineffective include intravenous valproate, lignocaine, corticosteroids, and propofol. Intravenous valproate sodium (initial dose of 15 mg/kg, followed by 5 mg/kg every 8 hours) should be considered a second-line drug.[55] The possible beneficial effect of intravenous lignocaine (100 mg bolus, followed by a 2 mg/min drip) was reported in a small open study.[56] Parenteral corticosteroids such as dexamethasone (4 mg/d) intramuscularly for 2 weeks are also alternatives.[57]

Opioid withdrawal symptoms can be controlled with clonidine (0.1 to 0.3 mg twice or three times daily). The dosage can be tritrated up and down depending on withdrawal symptoms such as palpitation, tremor, etc. A short course (no longer than a week) of anxiolytic medication may be necessary in some patients. If the patient has been using significant quantities of butalbital, a decreasing dosage of phenobarbital should be used to prevent serious withdrawal symptoms.[58]

Long-Term Prophylactic Therapy

Initiation of prophylactic medication is recommended for those who have considerable frequency of headache after drug withdrawal. The exact frequency that requires prophylactic medication is not known. More than three headache attacks per month is widely accepted. Choice of medication depends on the type of CDH. The preventive medication should be continued for 4 to 6 months or until the headache frequency is less than two to three attacks per month. Combined pharmacologic and nonpharmacologic treatments are the mainstay of prophylaxis for CDH.

Pharmacologic Treatment

Prophylactic pharmacotherapy for CDH includes antidepressants, anticonvulsants, muscle relaxants, 5-HT_1 agonists, 5-HT_2 antagonists, anxiolytics, and miscellaneous other drugs. It should be noted that several recommendations regarding pharmacologic treatment of CDH are based on retrospective case reviews or are even anecdotal. Only a few studies are randomized, placebo-controlled, double-blind clinical trials (RCTs).

ANTIDEPRESSANTS

Tricyclic antidepressants (TCAs) are recommended widely for CDH. TCAs are effective in the treatment of several subtypes of CDH, including medication-overuse headache. Besides preventing headache attacks, these agents also can control comorbid depression, which is quite common in patients with CDH. TCAs can be used alone or in conjunction with corticosteroids or NSAIDs. Amitriptyline (10 to 75 mg/d) is most widely used. Its efficacy in terms of reduction in headache frequency (more than 50 percent reduction) or duration is shown in several RCTs. Nortriptyline (25 to 75 mg/d), imipramine (30 to 75 mg/d), and doxepin (100 mg/d) are alternatives. Common side effects include drowsiness, dry mouth, and weight gain. Blurred vision, constipation, and orthostatic hypotension are less frequent. The use of TCAs is contraindicated in patients taking monoamine oxidase inhibitors. Special cautions should be paid if a TCA is to be used in a patient with urinary retention, glaucoma, a history of seizures, or hepatic, renal, cardiovascular, or thyroid disease.

The selective serotonin reuptake inhibitors (SSRIs) have fewer anticholinergic effects than TCAs and can be used with greater safety in the elderly and in patients with heart diseases, glaucoma, or urinary retention. RCTs showed that fluoxetine (20 to 40 mg/d), paroxetine (20 to 30 mg/d), and fluvoxamine (50 to 100 mg/d) are clinically effective in reducing headache frequency (more than 50 percent reduction in headache frequency) and severity.[59] It also should be noted that the efficacies of SSRIs and TCAs are not different. A combination of amitriptyline and fluoxetine does not yield more efficacy than amitriptyline alone.[60]

ANTICONVULSANTS

Anticonvulsants act primarily to stabilize the neuronal membrane, hence reducing its sensitivity. This action is theoretically suitable for treatment of CDH, where neurons in the trigeminal nociceptive pathway are supersensitive. Anticonvulsants are particularly useful in patients with comorbid epilepsy and manic-depressive illness and, possibly, anxiety disorders. Four anticonvulsants, including valproate (including divalproex), gabapentin, lamotrigine, and topiramate, have been tested for their efficacy in CDH. A study on divalproex sodium in the long-term treatment of chronic daily headache showed that 75 percent of 642 patients had at least a 50 percent reduction in headache frequency.[61] The dose of valproate used in control of CDH is lower than that used for the treatment of epilepsy. Doses between 500 and 1000 g/d are usually enough. Topiramate (maximum of 200 mg/d) reduced headache frequency by more than 50 percent in a retrospective trial of 96 patients. The common side effects include paresthesia, cognitive effect, weight loss, and dizziness.[62] Gabapentin (400 to 1200 mg/d) and lamotrigine are also effective.

MUSCLE RELAXANTS

Tizanidine, a central α_2 agonist, has antinociceptive effects that are independent of the endogenous opioid system. This drug has been proven effective in the treatment of CDH. Saper and colleagues conducted an RCT to investigate the effectiveness of tizanidine.[63] They found that patients taking tizanidine (mean dose 18 mg/d, range 2 to 24 mg/d) had a significant reduction in headache frequency, intensity, and duration. Botulinum toxin A is an effective treatment, especially for patients with CTTH. The toxin should be injected topically into the frontalis, occipitalis, temporalis, and trapezius muscles. Local muscle weakness and local swelling are side effects. Baclofen (15 to 50 mg/d), a γ-aminobutyric acid (GABA) analogue, is an effective alternative.

5-HT_2 ANTAGONISTS

Antagonists to 5-HT_2 receptors are effective preventive medications for migraine. Ritanserine (10 mg/d) has comparable effects to amitriptyline (50 mg/d) in the treatment of CTTH. It was noted that mianserine, a tetracyclic antidepressant that also blocks 5-HT_2 receptors, decreased headache frequency and intensity.

MISCELLANEOUS DRUGS

Other drugs that may be effective in the treatment of CDH include 5-HT_1 agonists, e.g., sumatriptan (2 to 4 mg/d); anxiolytics, e.g., buspirone (30 mg/d) or alprazolam (0.75 mg/d); and antipsychotics, e.g., olanzapine (2.5 to 35 mg/d). Less evident medications include methadone, clonidine, calcitonin nasal spray, ketamine, and chlorpromazine.

Nonpharmacologic Treatment

Nonpharmacologic interventions, including biofeedback therapy, individual cognitive therapy, physical exercise, and dietary instruction, are important for successful treatment of patients with CDH. Cognitive-behavioral therapies focus on preventing mild pain from becoming disabling pain; improving headache-related disability, affective distress, and quality of life; and reducing overreliance on medication. Adequate instruction about the nature of the patient's disorder and adverse effects of analgesics must be emphasized. The patients must understand that several factors, e.g., stress, anxiety, depression, and medication intake, have a major impact on the disease process, as well as the treatment outcome. Patients with medication overuse should accept that the beneficial effect of pharmacologic treatment will take place not earlier than 3 to 6 weeks after drug withdrawal. Psychiatric consultation is required for patients with severe psychiatric comorbidities, especially severe depression.

▶ PROGNOSIS

The long-term outcome of CDH cannot be defined as satisfactory. Despite a variety of initial treatments, their long-term outcomes are comparable. A 4-year follow-up study showed that only one-third of patients treated initially for CDH and medication overuse were successful in refraining from chronic overuse.[64] The relapse rate of mediation overuse after 1 and 2 years of follow-up were 30 and 35 percent, respectively.[65,66] Medication overuse is a major factor conveying poor prognosis. Mathew and colleagues[67] compared patients with successful treatment with those whose headaches were persistent and intractable. They found that comorbid psychiatric disorders, abnormal personalities, and alcohol consumption were more prevalent in patients with persistent, intractable headache. NPDH is also refractory to treatment.

▶ PREVENTION

To prevent the development of medication-induced headache, frequent migraine and tension-type headache should be treated aggressively with preventive medications. Abortive medication should be kept to the minimum. Patient education about the paradoxical effect of analgesics is helpful. Nonpharmacologic intervention for headache control, e.g., biofeedback, should be advocated to minimize analgesic use.

REFERENCES

1. Saper JR, Winter M: Chronic "mixed" headaches. Profile and analysis of 100 consecutive patients experiencing daily headache [abstract]. *Headache* 1982;22:145–146.
2. Mathew NT, Stubits E, Nigam M: Transformation of migraine into daily headache: Analysis of factors. *Headache* 1982;22:66–68.
3. Mathew NT, Reuvani U, Perez F: Transformed or evolutive migraine. *Headache* 1987;27:102–106.
4. Scher AI, Stewart WF, Liberman J, et al: Prevalence of frequent headache in the population sample. *Headache* 1998;38:497–506.
5. Castello J, Munoz P, Guitera V, et al: Epidemiology of chronic daily headache in the general population. *Headache* 1999;39:190–196.
6. Wang SJ, Fuh JL, Lu SR, et al: Chronic daily headache in Chinese elderly: Prevalence, risk factors, and biannual follow-up. *Neurology* 2000;54:314–319.
7. Srikiatkhachorn A, Phanthumchinda K: Prevalence and clinical features of chronic daily headache in a headache clinic. *Headache* 1997;37:277–280.
8. Headache Classification Committee of the International Headache Society: Classification and diagnostic criteria for headache disorders, cranial neuralgias and facial pain. *Cephalalgia* 1988;8 [suppl 7]:1–96.
9. Solomon S, Lipton RB, Newman LC: Evaluation of chronic daily headache: Comparison to criteria for chronic tension-type headache. *Cephalgia* 1992;12:365–368.
10. Silberstein SD, Lipton RB, Sliwinski M: Classification of daily and near-daily headaches: Field trial of revised IHS criteria. *Neurology* 1996;47:871–875.
11. Manzoni GC, Micieli G, Granella F, et al: Daily chronic headache: Classification and clinical features—Observation on 250 patients. *Cephalalgia* 1987;7 [Suppl 6]:169–170.
12. Lanteri-Minet M, Auray JP, El Hasnaoui A, et al: Prevalence and description of chronic daily headache in the general population in France. *Pain* 2003;102:143–149.
13. Hershey AD, Powers SW, Bentti AL, et al: Characterization of chronic daily headaches in children in a multidisciplinary headache center. *Neurology* 2001;24:1032–1037.
14. Krymchantowski AV, Moreira PF. Clinical presentation of transformed migraine: Possible differences among male and female patients. *Cephalalgia* 2001;21:558–566.
15. Koenig MA, Gladstein J, McCarter RJ, et al: Chronic daily headache in children and adolescents presenting to tertiary headache clinics. *Headache* 2002;42:491–500.
16. Manzoni GC, Granella F, Sandrini G, et al: Classification of chronic daily headache by International Headache Society criteria: Limits and new proposals. *Cephalalgia* 1995;15:37–43.
17. Spierings ELH, Schroevers M, Honkoop PC, et al: Development of chronic daily headache: A clinical study. *Headache* 1998;38:529–533.
18. Spierings ELH: Chronic daily headache: A review. *Headache Q* 2000;11:181–196.
19. Vanast WJ: New daily persistent headaches: Definition of a benign syndrome. *Headache* 1986;26:318.
20. Li D, Rozen TD. The clinical characteristics of new daily persistent headache. *Cephalalgia* 2002;22:66–69.
21. Diaz-Mitoma F, Vanast WJ, Tyrrell DL: Increased frequency of Epstein-Barr virus excretion in patients with new daily persistent headaches. *Lancet* 1987;1:411–415.
22. Peres MF, Silberstein SD, Nahmias S, et al: Hemicrania continua is not that rare. *Neurology* 2001;57:948–951.

23. Guitera V, Munoz P, Castillo J, et al: Quality of life in chronic daily headache: A study in a general population. *Neurology* 2002;58:1062–1065.
24. Wang SJ, Juang KD: Psychiatric comorbidity of chronic daily headache: Impact, treatment, outcome, and future studies. *Curr Pain Headache Rep* 2002;6:505–510.
25. Spierings ELH, Ranke AH, Schroevers M, et al: Chronic daily headache: A time perspective. *Headache* 2000;40:306–310.
26. Burstein R, Cutrer MF, Yarnitsky D: The development of cutaneous allodynia during a migraine attack: clinical evidence for sequential recruitment of spinal and supraspinal nociceptive neurons in migraine. *Brain* 2000;123:1703–1709.
27. Jansen R: Mechanism of tension-type headache. *Cephalalgia*. 2001;21:786–789.
28. Sakai F, Ebihara S, Akiyama M, et al: Pericranial muscle hardness in tension-type headache. A non-invasive measurement method and its clinical application. *Brain* 1995;118:523–531.
29. Ashina M, Stallknecht B, Bendtsen L, et al: Tender points are not sites of ongoing inflammation: In vivo evidence in patients with chronic tension type headache. *Cephalalgia* 2003;23:109–116.
30. Katsarava Z, Lehnerdt G, Duda B, et al: Sensitization of trigeminal nociception specific for migraine but not pain of sinusitis. *Neurology* 2002;59:1450–1453.
31. Sandkühler J, Liu X: Induction of long-term potentiation at the spinal synapse by noxious stimulation or nerve injury. *Eur J Neurosci* 1998;10:2476–2480.
32. Weiller C, May A, Limmroth V, et al: Brainstem activation in spontaneous human migraine attacks. *Nature Med* 1995;1:658–660.
33. Welch KMA, Nagesh V, Aurora SK, et al: Periaquiductal gray matter dysfunction in migraine: Cause or burden of illness? *Headache* 2001;41:629–637.
34. Sarchielli P, Alberti A, Floridi A, et al: Levels of nerve growth factor in cerebrospinal fluid of chronic daily headache patients. *Neurology* 2001;57:132–134.
35. Fusco BM, Colantoni O, Giacovazzo M: Alteration of central excitation circuits in chronic headache and analgesic misuse. *Headache* 1997;37:486–491.
36. de Tommaso M, Valeriani M, Guido M, et al: Abnormal brain processing of cutaneous pain in patients with chronic migraine. *Pain* 2003;101:25–32.
37. Pini LA, Sandrini M, Vitale G: The antinociceptive action of paracetamol is associated with changes in the serotonergic system in the rat brain. *Eur J Pharmacol* 1996;308:31–40.
38. Courade JP, Caussade F, Martin K, et al: Effects of acetaminophen on monoaminergic systems in the rat central nervous system. *Naunyn Schmiedebergs Arch Pharmacol* 2001;364:534–537.
39. Srikiatkhachorn A, Tarasub N, Govitrapong P: Acetaminophen-induced antinociception via central 5-HT2A receptors. *Neurochem Int* 1999;34:491–498.
40. Pini LA, Vitale G, Sandrini M. Serotonin and opiate involvement in the antinociceptive effect of acetylsalicylic acid. *Pharmacology* 1997;54:84–91.
41. Vitale G, Pini LA, Ottani A, et al: Effect of acetylsalicylic acid on formalin test and on serotonin system in the rat brain. *Gen Pharmacol* 1998;31:753–758.
42. Srikiatkhachorn A, Tarasub N, Govitrapong P: Effect of chronic analgesic exposure on the central serotonin system: A possible mechanism of analgesic abuse headache. *Headache* 2000;40:343–350.
43. Srikiatkhachorn A, Anthony M: Platelet serotonin in patients with analgesic-induced headache. *Cephalalgia* 1996;16:423–426.
44. Srikiatkhachorn A, Puangniyom S, Govitrapong P: Plasticity of 5-HT$_{2A}$ serotonin receptor in patients with analgesic-induced transformed migraine. *Headache* 1998;38:534–539.
45. Sarchielli P, Alberti A, Russo S, et al: Nitric oxide pathway, Ca_{2+}, and serotonin content in platelets from patients suffering from chronic daily headache. *Cephalalgia* 1999;19:810–816.
46. Srikiatkhachorn A, Maneesri S, Govitrapong P, et al: Derangement of the serotonin system in migraine patients with analgesics abuse headache: Clues from platelets. *Headache* 1998;38:43–49.
47. Sarchielli P, Alberti A, Floridi A, et al: L-arginine/nitric oxide pathway in chronic tension-type headache: relation with serotonin content and secretion and glutamate content. *J Neurol Sci* 2002;198:9–15.
48. Srikiatkhachorn A, Suwattanasophon C, Reungpattanatawee U, et al: 5-HT$_{2A}$ receptor activation and nitric oxide synthesis: A possible factor determining migraine attacks. *Headache* 2002;42:566–574.
49. Linton-Dahlöf P, Linde M, Dahlöf C: Withdrawal therapy improves chronic daily headache associated with long-term misuse of headache medication: A retrospective study. *Cephalalgia* 2000;20:658–662.
50. Haag G, Baar H, Grotemeyer KH, et al: Prophylaxis and treatment of drug-induced persistent headache. Therapy recommendation of the German Society for Migraine and Headache. *Schmerz* 1999;13:52–57.
51. Diener HC, Katasarva Z: Analgesic/abortive overuse and misuse in chronic daily headache. *Curr Pain Headache Rep* 2001;5:545–550.
52. Hering-Hanit R, Yavetz A, Dagan Y: Effect of withdrawal of misused medication on sleep disturbances in migraine sufferers with chronic daily headache. *Headache* 2000;40:809–812.
53. Krymchantowski AV, Barbosa JS: Prednisone as initial treatment of analgesic-induced daily headache. *Cephalalgia* 2000;20:107–113.
54. Pringsheim T, Howse D: In-patient treatment of chronic daily headache using dihydroergotamine: a long-term follow-up study. *Can J Neurol Sci* 1998;25:146–150.
55. Schwartz TH, Karpitskiy VV, Sohn RS: Intravenous valproate sodium in the treatment of daily headache. *Headache* 2002;42:519–522.
56. Hand PJ, Stark RJ: Intravenous lignocaine infusions for severe chronic daily headache. *Med J Aust* 2000;172:157–159.
57. Bonuccelli U, Nuti A, Lucetti C, et al: Amitriptyline and dexamethasone combined treatment in drug-induced headache. *Cephalalgia* 1996;16:197–200
58. Loder E, Biondi D: Oral phenobarbital loading: a safe and effective method of withdrawing patients with headache from butalbital compounds. *Headache* 2003;43:904–909.

59. Redillas C, Solomon S: Prophylactic pharmacological treatment of chronic daily headache. *Headache* 2000;40:83.
60. Krymchantowski AV, Silva MT, Barbosa JS, et al: Amitriptyline versus amitriptyline combined with fluoxetine in the preventative treatment of transformed migraine: a double-blind study. *Headache* 2002;42:510–514.
61. Freitag FG, Diamond S, Diamond ML, et al: Divalproex in the long-term treatment of chronic daily headache. *Headache* 2001;41:271–278.
62. Mathew NT, Kailasam J, Meadors L: Prophylaxis of migraine, transformed migraine, and cluster headache with topiramate. *Headache* 2002;42:796–803.
63. Saper JR, Lake AE III, Cantrell D, et al: Chronic daily headache prophylaxis with tizanidine: A double-blind, placebo-controlled, multicenter outcome study. *Headache* 2000;42:470–482.
64. Pini LA, Cicero AF, Sandrini M: Long-term follow-up of patients treated for chronic daily headache with analgesic overuse. *Cephalalgia* 2001;21:878–883.
65. Bigal ME, Rapoport AM, Sheftell FD, et al: Long-term follow-up of patients treated for chronic daily headache with analgesics overuse. *Cephalalgia* 2002;22:327–328.
66. Katsavara Z, Fritsche G, Muessig M, et al: Clinical features of withdrawal headache following overuse of triptans and other headache drugs. *Neurology* 2001;57:1694–1698
67. Mathew NT, Kurman R, Peres F: Intractable chronic daily headache. A persistent neurobiobehavioral disorder. *Cephalalgia* 1989;9:180.

CHAPTER 27

Migraine and Cluster Headaches

Alexander Mauskop

It is estimated that up to 30 million people in the United States suffer from migraine headaches and about 1 to 2 million have cluster headaches. Approximately 4 to 5% of the population suffers from daily headaches. Half of migraine sufferers have not had a diagnosis made by a physician.[1] There are several possible reasons for this, including the misconceptions that the problem is too trivial to consult a doctor, or that no treatment is available, or that the treatment may cause serious side effects. Some patients who do consult a physician are often not given optimal treatment because physicians may lack knowledge about the diagnosis and treatment of migraine and cluster headaches. Assessment of headache patients must include assessment of their disability status. Knowing the precise impact of migraine and other headaches on patients' quality of life helps design a proper treatment plan.

▶ PATHOPHYSIOLOGY

Migraine

The first modern explanation of the phenomenon of migraine as a vascular disorder was based on patients' reports of throbbing pain, change of facial coloring, and other observations. Research conducted in the past two decades indicates that vascular changes are not primary, but most likely occur as a result of neuronal perturbations. The onset of vasodilation that follows a period of cerebral vasoconstriction does not precisely coincide with the onset of migraine pain. PET and fMRI studies revealed that the earliest brain activation occurs in a circumscribed part of the midbrain, which is located in the proximity of the 5th nerve nucleus.[2] The first ophthalmic branch of the 5th nerve supplies innervation of the entire intracranial cavity. Activation of the trigeminal nerve leads to neurogenic inflammation of the meningeal vessels with the release of substance P, neurokinin A, calcitonin gene-related peptide and other neurotransmitters. This inflammation may induce sensitization of the *nervi vasorum* and this sensitization appears to be more responsible for the throbbing nature of pain than the modest vasodilation. In addition, as migraine progresses it leads to central sensitization, which manifests clinically by an extensive area of cutaneous allodynia and by poor relief of pain by triptans.[3] Successful treatment, notably complete pain relief, is much more likely before central sensitization occurs. This is an argument for the early use of triptans, which may result in much higher rates of pain-free responses. It is also possible that frequent episodes of untreated or inadequately treated migraine attacks, at least in some patients, leads to the development of chronic migraines. The initial activation of the brainstem migraine center is thought to occur due to various inputs, such as excessive photic, olfactory, or auditory stimuli; alcohol and other dietary substances; change in sleep patterns; and stress. Almost anyone can have an attack of migraine given enough of these stimuli, but those with frequent attacks seem to have a lower threshold and require fewer, or only one, trigger.

Those who suffer from migraines regularly tend to have a genetic predisposition to migraine. This is clear both from twin studies and from recent genetic research. A single gene was found to be responsible for the rare familial hemiplegic migraine.[4] This gene was found on chromosome 19 in one family and chromosome 1 in another. The gene codes for calcium channel, which leads some to suggest that migraine belongs to channelopathies, a group of diseases caused by abnormal ion channels. Common migraine appears to be polygenetic in origin, which explains the wide variety of its clinical manifestations. Studies showing successful prevention of migraines by riboflavin[5] and co-enzyme Q_{10}[6] suggest a possible role of mitochondrial dysfunction, at least in some patients with migraines.

NMR spectroscopy established that patients with migraines have lower brain magnesium levels than normal controls.[7] It is estimated that up to 50% of patients during an acute migraine attack have low magnesium levels. Magnesium regulates a variety of receptor and neurotransmitter activities, including nitric oxide and substance P, serotonin, and NMDA receptors.[8] Multiple studies have shown systemic magnesium deficiency as measured by magnesium loading test[9] measurements of serum ionized[10,11] and intracellular[12-15] levels. Potential causes of magnesium deficiency include stress, alcohol, caffeine, gastrointestinal disorders, genetic factors, chronic illness, and low dietary intake.

Cluster Headache

Pathogenesis of cluster headaches is less clear. Patients with cluster headaches do share many clinical features with patients with migraines and historically have been lumped together into the "vascular headache" category. However, along with many similarities, there are clear differences between these two headache types. The most conspicuous difference is in the time patterns of cluster headaches. Hypothalamic dysfunction has been shown to be present in patients with cluster headaches.[16] Prominence of autonomic symptoms, such as lacrimation, nasal congestion, and Horner's syndrome is another major difference. Similarities include unilateral pain and excellent response to injectable sumatriptan, which indicate involvement of the same serotonin receptors and trigeminal nerve system.

▶ DIAGNOSIS

History

Most of the information leading to the diagnosis of the headache type is obtained from the patient's history.

Frequency, Duration, and Time Patterns

Increasing frequency or duration of headaches indicates the need for a reevaluation of the patient, including an imaging study, such as an MRI scan. Migraines can occur anywhere from several times a year to daily. A typical intermittent migraine can last anywhere from 4 hours up to 3 days. A migraine that lasts more than 3 days is defined as *status migraine*. Daily headaches with migraine features are given the diagnosis of *chronic migraine* (a recent addition to the classification of headaches). Cluster headaches usually occur once or several times a day, often awakening the patient at the same time of night and lasting a relatively short time (30 to 90 minutes). Cluster periods last from 2 weeks to 3 to 4 months, while in some patients, cluster headaches never remit and become chronic. Very brief (5 to 15 minutes), but intense and very frequent (5 to 20 times a day) unilateral headaches, which occur mostly in women (and can be accompanied by nasal congestion and lacrimation), suggest the diagnosis of *chronic paroxysmal hemicrania*. This condition almost always responds to indomethacin and rarely to any other treatment. Hemicrania continua is a rare disorder with unilateral pain with ipsilateral nasal congestion and lacrimation that is present daily and continuously. It is also highly responsive to indomethacin. Rebound headaches are continuous or very frequent and can be indistinguishable from chronic migraines, except for the fact that they usually improve with discontinuation of the offending drug or dietary caffeine. These headaches are resistant to treatment, tend to be worse upon awakening, are not due to an organic cause, and are often the result of excessive caffeine intake. Both patients and physicians may be unaware of the problem since the patient may be consuming only modest amounts of coffee, soda, and over-the-counter medications for migraine that contain caffeine. However, one or two cups of coffee, two cans of soda with caffeine, and two tablets of Excedrin can result in a high daily dose of caffeine, more than sufficient to induce rebound headaches. A double-blind study established that as little as 2.5 cups of coffee a day was sufficient to induce a withdrawal headache even in people who normally do not suffer from headaches.[17]

The most common time of onset of migraines is 6 to 7 A.M. Headaches that occur mostly on weekends are usually due to "let-down" after stress is over, due to sleeping longer than usual, or due to caffeine withdrawal because of the delay in the intake of the first dose of caffeine.

Character and Location of Pain

Burning occipital pain suggests a focal neuropathy of the occipital nerve, often seen in older patients. Sharp, boring, and very intense type of pain is characteristic of cluster headaches. Unilateral and pulsatile pain is most common in migraine and cluster headaches. Location of pain does not necessarily correlate with the location of pathology. For example, patients with occipital posterior fossa tumors can present with a headache, which is felt in the front of the head. Cervicogenic headaches can also present with frontal pain. Perinasal pain is often mistaken for sinus headaches, but most commonly this is due to migraines, unless a purulent discharge is present. Very intense and very brief stabbing pains in various areas of the head, or so-called "ice pick" headaches, tend to occur in patients who also have migraines.

Precipitating Factors

Alteration of sleep patterns, tyramine-rich foods, alcohol, chocolate, food preservatives, additives, aspartame, and many other foods can provoke a migraine attack. Overexertion and emotional stress is one of the most common precipitating factors for migraine headaches. Strong sensory stimuli such as loud noise, strong odors, or bright and flashing lights can induce a headache in a susceptible individual. Changes in barometric pressure such as with weather changes, flying, or climbing a mountain can provoke a headache. During a cluster period, the smallest amount of alcohol can trigger an attack.

Preceding and Accompanying Symptoms

Migraine headaches are preceded by a visual or other type of aura in about 15–20% of patients. Nausea, sensitivity to light, noise, touch, and head movement are typical, but not necessary accompaniments of migraine headaches. Some patients with migraines report a prodrome that consists of elation, depression, yawning, and other symptoms that precede an attack of migraine by up to 24 hours. Postdrome is a more common phenomenon and is typically a sense of exhaustion. Agitation, unilateral nasal congestion, and tearing frequently occur with an attack of cluster headache. Dizziness can occur with migraine and cervicogenic headaches.

Disability Assessment

MIDAS (Migraine Disability Assessment Scale) is a validated and easy-to-use tool for assessment of disability in migraine patients.[18] MIDAS improves physician-patient communication and identifies patients with high treatment needs. For example, patients with higher MIDAS scores may need triptans rather than analgesics as the first step in their treatment. Using such stratified care improves patient outcomes as compared with step care, where every patient is first given a mild analgesic. A MIDAS questionnaire is included in the appendix to this chapter. Unfortunately, no such assessment tool is available for cluster or other headache types. Even though MIDAS was not validated in patients with other types of headaches, it can be used for those patients as well.

Physical Examination

A general medical examination is necessary to detect many of the systemic conditions that can lead to headaches. After a detailed history, a neurological examination is the most important diagnostic step. This examination should be normal in patients with migraine and cluster headaches with few exceptions. Migraine aura can result in transient visual field deficit, hemiparesis, or sensory deficits, as well as transient aphasia and confusion. Patients with cluster headaches often have Horner's syndrome, that can transiently persist for some time after the attack. Patients with occipital neuralgia and other types of cervicogenic headaches frequently have tenderness of the occipital nerve and can have decreased sensation in the occipital area; however, muscle spasm of suboccipital and trapezius muscles is frequently present in migraineurs as well.

Ancillary Tests

Imaging Studies

If the history or physical examination raises any doubt about the etiology of headaches, an imaging procedure such as CT scan or, preferably, MRI scan should be performed. Concern over a possible brain tumor or another serious condition often makes the headache worse. A negative CT or MRI scan reassures the patient and can sometimes reduce headaches. A new onset of headaches in an individual over 40 years of age requires an imaging study because of a possible malignancy. In the elderly, CT or MRI scan of the brain is routinely performed to exclude a subdural hematoma, which may develop from a trivial head injury suffered many weeks or months earlier. Up to 40 percent of elderly patients with a chronic subdural hematoma give no history of a head injury. Conditions such as metastatic brain tumor and cerebrovascular disease are also more common in the elderly than in younger people. MRI of the intracranial vessels is indicated in patients with attacks of severe headaches and family history of aneurysms and strokes. Young women developing severe headaches postpartum or during pregnancy may require an MRI to rule out venous sinus thrombosis.

Laboratory Tests

Patients with recent onset or recent worsening of headaches should have a complete blood count, thyroid function tests, and a metabolic panel. These tests may detect anemia, systemic infections, renal insufficiency, hypothyroidism, and other conditions that can cause headaches. An erythrocyte sedimentation rate (ESR) and C-reactive protein (CRP) must be obtained in a patient over 60 years of age with a recent onset of headaches. If these tests are abnormal, a temporal artery biopsy is necessary to confirm the diagnosis of giant cell arteritis. Magnesium deficiency has been reported to be present in up to 50% of migraine sufferers. The published reports, however, indicate that total serum magnesium levels do not reflect true magnesium stores, since 99% of the body's magnesium is intracellular. Serum magnesium levels in the lower half of normal range may be treated as potentially abnormally low.

Lumbar puncture

Lumbar puncture is indicated in patients with chronic headaches who are suspected of having benign intracranial hypertension. Presence of papilledema formerly was thought to be present in all patients, but recent reports indicate that this is not the case. A patient with daily headache not responding to a variety of therapies may also require a lumbar puncture. In addition to increased pressure, chronic or sub-acute infections, such as tuberculosis, and syphilitic and cryptococcal meningitis, are sometimes found in chronic headache patients.

▶ TREATMENT OF MIGRAINES

Nonpharmacological Treatment

Since migraine is a genetic disorder, no cure is possible. However, genetic predisposition is usually modified by a variety of triggering factors. Therefore, treatment begins with an attempt to avoid potential triggers, such as stress, lack of sleep, not eating for prolonged periods of time, dehydration, and food triggers. The best antidote for stress is regular aerobic exercise. For many patients, 20 to 30 minutes of vigorous exercise is sufficient, if done 4 to 5 times a week. Dietary changes can occasionally completely stop migraine headaches, but in many patients they only reduce the frequency of attacks. Some foods that can provoke migraine headaches include chocolate, yogurt, bananas, dried fruit, beans, aged cheese, citrus fruit or juice, pickled and marinated foods, and buttermilk. Monosodium glutamate and aspartame should be avoided. Among alcoholic beverages, red wine and beer are more likely to induce a migraine headache than vodka.

Mind–Body Techniques

Biofeedback, meditation, and other relaxation techniques are effective for the prevention of migraine headaches.[19-21] Biofeedback is a computerized relaxation technique that can be learned in 6 to 8 weekly sessions. Children are particularly adept at biofeedback and can master it in as few as 3 sessions.

Nutritional Supplements and Herbal Remedies

Magnesium

Magnesium can be effective for the prevention of migraine headaches in up to 50% of patients. An intravenous infusion of 1 gram of magnesium sulfate was given to 40 consecutive patients with acute migraine.[11] Of these 40 patients, 21 had very good and sustained relief of their headaches. Of the responders, 86% had low serum ionized magnesium levels, whereas of the nonresponders, only 16% had low values. An intravenous administration of magnesium in the treatment of cluster headaches suggests a 40% response rate. A correlation between clinical response and serum ionized magnesium levels was also found. Oral magnesium supplementation was used as prophylactic therapy for migraines in four double-blind trials. Two of the three trials in adults[12,22] had positive findings, and one was negative. The negative study[23] used what turned out to be a poorly absorbed salt of magnesium, since almost half of the patients in the active group developed diarrhea. A fourth study was performed with children,[24] and it showed a strong positive trend, but not statistical significance for the primary outcome measure. Magnesium oxide and magnesium diglycinate, in a dose of 400–600 mg are inexpensive and can be recommended routinely to all migraine sufferers. If these preparations cause diarrhea or other gastrointestinal side effects, slow-release magnesium chloride may be tried.

Riboflavin

Riboflavin, or vitamin B-2, has been reported to relieve migraine headaches better than placebo in a double-blind study.[5] The maximum effect was achieved after three months of daily intake of 400 mg riboflavin.

Co-Enzyme Q_{10}

An anecdotal report[25] and a double-blind study[6] suggest that this co-enzyme can prevent migraine headaches at a daily dose of 300 mg. Recent positive double-blind studies of its efficacy in Parkinson's disease (at 1200 mg a day) suggest it may indeed have a beneficial effect on at least some aspects of brain function.

Feverfew

Feverfew is the only herbal remedy that was submitted to multiple well-designed double-blind trials.[26-28] These trials showed that feverfew, when taken daily as a prophylactic therapy for migraines, tended to be better than placebo. Because feverfew is fairly safe and it may help some patients, it is the herb to recommend to patients interested in herbal remedies.

Butterbur Root

Two double-blind studies of the extract of butterbur root have shown benefit in migraine patients.[29,30] This plant and its root contain carcinogens; therefore, homemade preparations can be dangerous. Only a commercial product made in Germany, called *Petadolex*, is safe to use.

Combination Products

Several products combining several supplements and herbal remedies have become available. A combination of magnesium, riboflavin, and Feverfew (MigreLief, MigraHealth) is a convenient and expeditious way to try three safe and effective nonpharmacological products. It is possible, although not proven, that these three agents have synergistic effects, offering another advantage.

Acupuncture

The currently available data indicate a biological effect, specifically analgesic action, which is mediated through two possible mechanisms. Based on animal studies, one mechanism is presumed to be endorphin-mediated and is naloxone-reversible, while the second is probably serotonin-mediated and is not naloxone-reversible.[31] Review of 14 trials comparing true and sham acupuncture showed a clear trend in favor of acupuncture.[32]

Pharmacologic Treatment

Abortive Therapy

Nonsteroidal anti-inflammatory agents (NSAIDs) can be effective for milder migraine headaches. Rapid onset of action can be achieved by using an effervescent form of aspirin (Alka-Seltzer) or solubilized ibuprofen (Advil Migraine, Advil Liquigel).

Triptans

Sumatriptan (Imitrex) and other triptans are "designer" drugs specifically developed to bind to $5HT_{1B}$ and $5HT_{1D}$ serotonin receptor subtypes, which are operational in the pathogenesis of migraine headaches. The $5HT_{1B}$ receptors are present in the cerebral blood vessels (with very few of these receptors in the coronary blood vessels), while $5HT_{1D}$ receptors are found in neurons. Triptans relieve the pain, photophobia, phonophobia, and the nausea and allow the patient to return to normal functioning. Sumatriptan is available in an injection, which can be easily self-administered by the patient, tablets, and nasal spray. Side effects are more common with injection and include a flushed sensation, chest pressure (noncardiac in origin), paresthesias, and injection site pain. Rizatriptan (Maxalt), almotriptan (Axert), zolmitriptan (Zomig), naratriptan (Amerge), eletriptan (Relpax), and frovatriptan (Frova) are other triptans, that may have different efficacy, pharmacokinetics and side-effect profiles in each individual patient.[33] For example, rizatriptan may have good efficacy with a faster onset of action, while naratriptan and frovatriptan offer longer duration of action, but slow onset. Almotriptan may be least likely to cause chest pressure sensation, while offering good efficacy, and zolmitriptan may have a more consistent response. Sumatriptan nasal spray, particularly, has a very poor consistency of response from attack to attack. These characterizations apply to most patients taking these drugs, but individual variations are often present. Patients may respond very well to one, but not another, triptan; or the slower acting one, such as naratriptan, may be quickest acting for a particular patient. Side effects also vary greatly among patients. Therefore, patients should be given trials of various triptans in the hope of finding one with rapid onset, complete relief, no side effects, and no recurrence. Zolmitriptan and rizatriptan are available in tablet formulation and in orally disintegrating tablets, with the latter offering the convenience of not having to have water to take the medication. However, they contain a very small amount of aspartame, which, in a rare patient, can worsen the headache. Zolmitriptan is also available in nasal spray, which offers good efficacy with rapid onset of action and better consistency of response than sumatriptan nasal spray. Sumatriptan and other triptans are contraindicated in patients with uncontrolled hypertension, ischemic heart disease, multiple risk factors for coronary artery disease, and complicated migraines (migraines that are accompanied by a transient neurological deficit). Complicated migraine, which is a rare condition, needs to be differentiated from migraine with sensory or motor aura. In the latter, neurological symptoms precede the headache and triptans are not contraindicated, while in complicated migraines, neurological symptoms persist into the headache phase. This distinction may not be as important as the triptan package inserts suggest. Reports of safe use of triptans in patients with complicated migraine are not surprising, since the cause of these symptoms is more likely to be due to neuronal inhibition, rather than vasoconstriction, which follows neuronal inhibition. Various triptans are not supposed to be given within 24 hours of each other, although no good theoretical or experiential explanation for this prohibition exists either.

Combination Medications

The combination of acetaminophen or aspirin, caffeine, and a short-acting barbiturate (butalbital) (Fiorinal, Fioricet, Esgic, Medigesic) is very popular with many patients and physicians; however, excessive intake can lead to refractory rebound headaches due to caffeine, and addiction to butalbital. The safe limit for these drugs is about 20 tablets a month. Isometheptene, a sympathomimetic amine with vasoconstrictive properties, is available in combination with dichloralphenazone, a mild sedative, and acetaminophen (Midrin, Isocom, Duradrin). This combination can be effective in many patients who do not respond to other drugs. Drowsiness is a potential side effect. A similar limit of 20 tablets a month is placed on this drug to avoid rebound headaches. If a patient takes more than that amount, the medication may begin to worsen the headache through a rebound mechanism. Addition of codeine to some of the combinations (Fiorinal with codeine) improves their efficacy for severe headaches.

Ergots
This is the first group of drugs developed specifically for the treatment of migraines. Ergotamine alone (Ergostat, sublingual) or with caffeine (Cafergot, tablets and suppositories; Wigraine, tablets) can be quite effective. However, these drugs can sometimes worsen or cause nausea and other side effects. Reducing the dose, particularly of Cafergot suppositories, to one-quarter or one-half of a suppository can avoid nausea and provide effective and rapid relief. Ergots are contraindicated in patients with cardiac or peripheral ischemia, and in pregnant women. Dihydroergotamine (DHE-45) is effective for abortive treatment of migraines. This ergot derivative is available only in a parenteral form and can be given subcutaneously, intramuscularly, intravenously or intranasally. An injected dose of 1 mg is sufficient for most patients, but some may require 2 or 3 mg. The starting dose should be 0.5 mg repeated in 45 minutes if necessary. Once a total effective dose is established for a patient, that amount is given for future attacks. A nasal spray preparation of dihydroergotamine (Migranal) is more convenient to take, but it is less effective.

Antiemetics
If the headache is accompanied by nausea, an injection, tablet, or suppository of prochlorperazine (Compazine), or a tablet or injection of metoclopramide (Reglan) are usually effective. Other drugs in this category can be tried since they all may have different efficacy and side effects in individual patients. These include Phenergan (promethazine), Tigan (trimethobenzamide), and Zofran (ondansetron).

Corticosteroids
A prolonged or refractory headache that does not respond to triptans and other drugs may respond to a single large dose of steroids, such as dexamethasone (Decadron), 6 to 8 mg given orally or intravenously. A Medrol (methylprednisolone) dose pack is another option.

Opioid Drugs
These medications seem to be less effective for migraine headaches than for many other pain syndromes, mostly because side effects such as drowsiness and nausea appear to precede or accompany pain relief. Long-term opioid maintenance for patients with chronic migraines can be attempted following the standard rules of opioid maintenance described elsewhere in this book.

Treatment of an Intractable or Prolonged Attack
A very severe or prolonged attack can be treated with injectable medications much more efficiently in the office than in an emergency room. Intravenous magnesium (1 gram of magnesium sulfate in 10 cc of normal saline, given by slow push over five minutes) will provide prompt and complete relief in up to half of the patients. If ineffective within 10 to 15 minutes, it can be followed by injectable sumatriptan, then IV push of 30 to 60 mg of ketorolac (Toradol), followed in 15 minutes by IV push of 6 to 8 mg of dexamethasone and then by IV push of 500-1000 mg of valproic acid (Depakene). Unlike opioid drugs, which are commonly given in an emergency room, none of these drugs are sedating, are much more effective, and allow the patient to go home on his or her own. If nausea is not relieved by these measures, intravenous metoclopramide 10 mg, or prochlorperazine (Compazine) 10 mg, may have some sedative effect.

Prophylactic Therapy
Tricyclic (TCA) and other antidepressants can be as effective for migraine headaches, but their potential to cause weight gain deters many young women who are typical migraine sufferers. Severe persistent headaches may respond well to nortriptyline (Pamelor) or another TCA. Among the TCAs, amitriptyline (Elavil) has been studied most extensively; but nortriptyline (Pamelor), imipramine (Tofranil), protriptyline (Vivactil), and desipramine (Norpramine) are effective, as well, and may have fewer anticholinergic side effects. If one TCA is ineffective or produces unacceptable side effects, another one should be tried. The starting dose for most TCAs is 25 mg in a young or middle-aged individual and 10 mg in an elderly person. The average effective dose, however, is 50 to 75 mg taken once every 24 hours in the evening. The patients must be told that these medications are anti-depressants, but they are also used for chronic painful conditions even if there is no associated depression. When patients discover from other sources that these are antidepressant medications, they often become angry and noncompliant. They may think that their complaints were interpreted as depressive symptoms and not as real pain. Patients should be told that, besides an antidepressant effect, these drugs have analgesic properties. Warning patients about possible side effects, such as dryness of the mouth, drowsiness, and constipation, tends to improve their compliance. It also allows them to take early preventive measures against constipation. Some of the contraindications for the use of TCAs include concomitant use of monoamine oxidase inhibitors, recent myocardial infarction, cardiac arrhythmias, glaucoma, and urinary retention. An electrocardiogram should

be obtained before the initiation of treatment in all elderly patients or those with heart disease.

Antidepressants, which are selective serotonin reuptake inhibitors (SSRIs), such as fluoxetine (Prozac), sertraline (Zoloft), and paroxetine (Paxil), have fewer side effects, but they lack the analgesic effect of TCAs. They can sometimes help indirectly, by reducing anxiety and depression. SSRIs do not cause weight gain, but often cause sexual dysfunction—most commonly, loss of libido. Effexor (venlafaxine) is a serotonin and norepinephrine reuptake inhibitor and appears to have better efficacy for migraine headaches than SSRIs. Its potential side effects include insomnia and gastrointestinal symptoms.

Propranolol, nadolol, and other beta-blockers are good prophylactic drugs. The effective dose for propranolol can be as low as 40 mg daily but is usually 80 to 240 mg. Contraindications for the use of beta-blockers include bronchial asthma, sinus bradycardia, greater than first degree block, congestive heart failure, and diabetes. The angiotensin II receptor antagonist, candesartan (Atacand), was shown to prevent migraines in a double-blind trial. Calcium channel blockers are less effective than beta-blockers for the treatment of migraines.

Divalproex sodium (Depakote, Depakote ER) can relieve migraine headaches in patients who do not respond to beta-blockers or antidepressants. The starting dose is usually 250 mg a day with a gradual increase up to 2000 mg. Potential side effects include nausea, drowsiness, alopecia, and weight gain. Other anticonvulsants that can be useful in the prophylaxis of migraines are topiramate (Topamax) and gabapentin (Neurontin). Divalproex sodium is teratogenic and women must observe strict contraceptive measures.

Botulinum toxin injections into pericranial and neck muscles can be effective in prevention of migraines for periods of about three months.[34–42] The mechanism of action remains unclear. The observation of relief of headache but not of visual aura in patients with migraines with aura suggests a peripheral mechanism. Reduction of afferent input from pericranial muscles, which normally contract during a migraine attack, is a possible explanation. The two available preparations of botulinum toxin are botulinum toxin type A (Botox) and type B (Myobloc). The type B preparation is more painful to inject due to its low pH, while type A has a neutral pH. The effect of type A may last somewhat longer, although time of onset is shorter with type B. This treatment is easy to administer and carries no risk of serious complications. In addition to avoiding side effects of systemic prophylactic medications, botulinum toxin offers better compliance and high acceptance by patients. Despite high up-front cost, preliminary research suggests that this treatment is cost-effective.[43]

▶ TREATMENT OF CLUSTER HEADACHE

Abortive Treatment

Treatment of cluster headaches begins with measures designed to reduce pain of each attack while prophylactic drugs take effect. The most benign and frequently effective treatment is inhalation of oxygen. It is done through a mask (not nasal prongs) using 100% oxygen at 8 to 10 liters per minute. It should be used for patients who get most of their attacks at home. If headaches occur during the day, patients can store another oxygen tank at work.

Sumatriptan (Imitrex) injection is very effective in most patients and has few side effects. It can also be administered by the patient using an auto-injector. Zolmitriptan nasal spray can be an effective alternative, while, of the oral drugs, 10 mg rizatriptan (Maxalt), or another rapid-acting triptan may provide some relief.

Ergotamine (Cafergot, Wigraine, Ergostat) given in a suppository or sublingually can abort a cluster headache. Dihydroergotamine (DHE-45) can be self-injected by the patient if sumatriptan is ineffective or does not provide lasting relief.

Prophylactic Treatment

A short course of prednisone will often abort the entire cluster period. Starting dose is 60 to 80 mg daily with tapering over a period of 2 weeks. Long-term use of prednisone is fraught with potential complications; and other approaches listed below should be explored.

Calcium channel blockers are usually effective in cluster headaches. Verapamil (Calan SR, Isoptin) is used in a daily dose of 240 to 960 mg and higher, if tolerated.

Divalproex sodium (Depakote ER) 500 to 2000 mg daily, in divided doses, can provide relief for some patients; as can topiramate (Topamax), in doses of up to 800 mg a day. Lithium carbonate, 300 mg taken two to four times a day, is often effective. It can work for both episodic and chronic forms of cluster headaches, sometimes transforming chronic into episodic.

Ergotamine in a dose of 2 mg can provide good relief if taken at bedtime before the expected nocturnal attack. Regular intake of 1 to 2 mg of ergotamine three times a day has been reported to be effective in some patients.

Melatonin, an over-the-counter supplement, 9 to 10 mg nightly, has been reported to help a few patients with clusters. Intranasal capsaicin has been subjected to controlled trials and has been shown to prevent clusters. It is applied into the nostril on the side of the headache. Intense burning upon application limits its usefulness. A nonirritating version (civamide) is currently being tested in a double-blind trial.

Patients not responding to a single agent can be given a combination of two or more of the above drugs.

APPENDIX 27

▶ MIDAS QUESTIONNAIRE

Instructions

Please answer the following questions about ALL the headaches you have had over the last 3 months. Write your answer in the box next to each question. Write zero if the activity does not apply in the last 3 months:

1. On how many days in the last 3 months did you miss work or school because of your headaches? __days

2. How many days in the last 3 months was your productivity at work or school reduced by half or more because of your headaches? (Do not include days you counted in question 1 where you missed work or school) __days

3. On how many days in the last 3 months did you not do household work because of your headaches? __days

4. How many days in the last 3 months was your productivity in household work reduced by half or more because of your headaches? (Do not include days you counted in question 3 where you did not do household work) __days

5. On how many days in the last 3 months did you miss family, social, or leisure activities because of your headaches? __days

Total __**days**

On how many days in the last 3 months did you have a headache? (If a headache lasted more than one day, count each day) __days

On the scale 0 to 10, on average how painful were these headaches? (Where zero equals no pain at all, and 10 equals pain as bad as it can be) __

Once you have filled in the questionnaire, add up the total number of days from questions 1 to 5

Grading system for the MIDAS questionnaire

Grade	Definition	Score
I	Little or no disability	0 to 5
II	Mild disability	6 to 10
III	Moderate disability	11 to 20
IV	Severe disability	21+

REFERENCES

1. Lipton RB, Diamond SD, Reed M, Diamond ML, Stewart WF: Migraine diagnosis and treatment: Results from the American migraine study II. *Headache* 2001;41:638–45.
2. Weiller C, May A, Limmroth V, et al: Brain stem activation in spontaneous human migraine attacks. *Nature Med.* 1995; 1:658–660.
3. Burstein R, Yarnitzky D, Goor-Aryeh I, et al: An association between migraine and cutaneous allodynia. *Annals of Neurology* 2000;47:614–624.
4. Ducros A, Denier C, Joutel A, et al: The clinical spectrum of familial hemiplegic migraine associated with mutations in a neuronal calcium channel. *N Engl J Med* 2001; 345:17–24.
5. Schoenen J, Jacquy J, Lenaerts M: Effectiveness of high-dose riboflavin in migraine prophylaxis. *Neurology* 1998; 50:466–470.
6. Sandor PS, Di Clemente L, Coppola G, et al: Coenzyme Q10 for migraine prophylaxis: A randomized trial. *Cephalalgia* 2003;23:577.
7. Ramadan NM, Halvorson H, Vande-Linde A, Levine SR, Helpern JA, Welch KM: Low brain magnesium in migraine. *Headache* 1989;29:590–593.
8. Mauskop A, Altura BM: Role of magnesium in the pathogenesis and treatment of migraines. *Clinical Neuroscience* 1998;5:24–28.
9. Trauninger A, Pfund Z, Koszegi T, et al: Oral magnesium load test in patients with migraine. *Headache* 2002; 42:114–119.
10. Mauskop A, Altura BT, Cracco RQ, Altura BM: Deficiency in serum ionized Mg but not total Mg in patients with migraine. Possible role of ICa^{2+}/IMg^{2+} ratio. *Headache* 1993;33:135–138.
11. Mauskop A, Altura BT, Cracco RQ, Altura BM: Intravenous magnesium sulfate relieves migraine attacks in patients with low serum ionized magnesium levels: A pilot study. *Clin Science* 1995;89:633–636.
12. Facchinetti F, Sances G, Borella P, Genazzani A.R, Nappi G: Magnesium prophylaxis of menstrual migraine: effects on intracellular magnesium. *Headache* 1991;31:298–301.
13. Sarchielli P, Coata G, Firenze C, Morucci P, Abbritti G, Gallai V: Serum and salivary magnesium levels in migraine and tension-type headache. Results in a group of adult patients. *Cephalalgia* 1992;12:21–27.
14. Schoenen J, Sianard-Gainko J, Lenaerts M: Blood magnesium levels in migraine. *Cephalalgia* 1991;11:97–99.
15. Soriani S, Arnaldi C, De Carlo L, et al: Serum and red blood cell magnesium levels in juvenile migraine patients. *Headache* 1995;35:14–16.
16. May A, Goadsby PJ: Hypothalamic involvement and activation in cluster headache. *Curr Pain Headache Rep*: 2001; 5:60–66.
17. Silverman K, Evans SM, Strain EC, et al: Withdrawal syndrome after the double-blind cessation of caffeine consumption. *N Engl J Med* 1992;327:1109–1114.
18. Lipton RB, Stewart WF, Sawyer J, et al: Clinical utility of an instrument assessing migraine disability: The migraine disability assessment (MIDAS) questionnaire. *Headache* 41:854–861.

19. Chapman SL: A review and clinical perspective on the use of EMG and thermal biofeedback for chronic headaches. *Pain* 1986;27:1–43.
20. Blanchard EB: Long-term effects of behavioral treatment on chronic headache. *L Ther* 1987;8:375–385.
21. Gauthier JG, Carrier S: Long-term effects of biofeedback on migraine headache: A prospective follow-up study. *Headache* 1991;31:605–612.
22. Peikert A, Wilimzig C, Kohne-Volland R: Prophylaxis of migraine with oral magnesium: Results from a prospective, multi-center, placebo-controlled and double-blind randomized study. *Cephalalgia* 1996;16:257–263.
23. Pfaffenrath V, Wessely P, Meyer C, et al: Magnesium in the prophylaxis of migraine: A double-blind, placebo-controlled study. *Cephalalgia* 1996;16:436–440.
24. Wang F, Van den Eeden SK, Ackerson LM, Salk SE, Reince RH, and Elin RJ: Oral magnesium oxide prophylaxis of frequent migrainous headache in children: A randomized, double-blind, placebo-controlled trial. *Headache* 2003;43:601–610.
25. Rozen TD, Oshinsky ML, Gebeline CA, et al: Open label trial of coenzyme Q10 as a migraine preventive. *Cephalalgia* 2002;22:137.
26. Vogler BK, Pittler MH, Ernst E: Feverfew as a preventive treatment for migraine: A systematic review. *Cephalalgia* 1998;18:704–708.
27. Murphy J, Heptinsall S, Mitchell JRA: Randomized double-blind placebo-controlled trial of feverfew in migraine prevention. *Lancet* 1988;2:189–192.
28. Pfaffenrath V, Diener HC, Fischer M, Friede M, Henneicke-von Zepelin HH: The efficacy and safety of Tanacetum parthenium (feverfew) in migraine prophylaxis—a double-blind, multicentre, randomized placebo–controlled dose-response study. *Cephalalgia.* 2002;22:523–532.
29. Grossman M, Schmidramsl H: An extract of *Petasites hybridus* is effective in the prophylaxis of migraine. *Int J Clin Pharmacol Ther* 2000;38:430-435.
30. Lipton RB, Gobel H, Wilks K, Mauskop A: Efficacy of Petasites (an extract from Petasites Rhizone) 50 and 75 mg for prophylaxis of migraine: Results of a randomized, double-blind, placebo-controlled study. *Neurology* 2002;58 A472.
31. Han JS, Terenius L: Neurochemical basis of acupuncture analgesia. *Ann Rev of Pharm and Toxico* 1982;22:193–220.
32. Melchart D, Linde K, Fischer P, et al: Acupuncture for recurrent headaches: A systematic review of randomized controlled trials. *Cephalalgia* 1999;19:779–786.
33. Ferrari MD, Roon KI, Lipton RB, et al: Oral triptans in acute migraine treatment: A meta-analysis of 53 trials. *Lancet* 2001;358:1668–1675.
34. Binder WJ, Brin MF, Blitzer A, Schoenrock LD, Pogoda JM: Botulinum toxin type A (BOTOX) for treatment of migraine headaches: An open-label study. *Otolaryngol Head Neck Surg* 2000;123:669–676.
35. Smuts JA, Barnard PWA: Botulinum toxin type A in the treatment of headache syndromes: A clinical report of 79 patients. *Cephalalgia* 2000;20:332.
36. Mauskop A, Basdeo R: Botulinum toxin A is an effective prophylactic therapy of migraines. *Cephalalgia* 2000;20:422.
37. Brin MF, Swope DM, O'Brian C, Abbasi S, Pogoda JM: Botox for migraine: double-bind, placebo-controlled region-specific evaluation. *Cephalalgia* 2000;20:421–422.
38. Silberstein S, Mathew N, Saper J, et al: Botulinum toxin type A as a migraine preventive treatment. *Headache* 2000; 40:445–450.
39. Gobel H, Heinze A, Heinze-Kuhn K, Jost WH: Evidence-based medicine: botulinum toxin A in migraine and tension-type headache. *J Neurol* 2001;248:1:34–38.
40. Mauskop A: The use of botulinum toxin in the treatment of headaches. *Curr Pain Headache Rep* 2002; 6:320–323.
41. Blumenfeld A: Botulinum toxin type A as an effective prophylactic treatment in primary headache disorders. *Headache* 2003;43:853–860.
42. Mathew N, Kallasam K, Meadors L: Disease modification in chronic migraine with botulinum toxin type A—long-term experience. *Headache* 2002;42:389–463.
43. Blumenfeld A: Decrease in headache medication use and cost after botulinum toxin type A (Botox) treatment in a high triptan use population. *Headache* 2002;42:420.

CHAPTER 28

Trigeminal Neuralgia and Orofacial Pains

Giorgio Cruccu and Andrea Truini

▶ ANATOMIC-FUNCTIONAL ORGANIZATION OF THE TRIGEMINAL SYSTEM

Peripheral Pathways

The trigeminal nerve is the fifth cranial nerve. It is the largest cranial nerve and is named for its three major sensory branches: the ophthalmic (V1), maxillary (V2), and mandibular nerves (V3), which convey information about touch, temperature, pain, and proprioception from the mouth, face, and scalp to the brain stem (Fig. 28-1). The trigeminal nerve originates in the posterior cranial fossa, emerging from the pons, with a small motor root (portio minor, 7500 myelinated fibers) and a large sensory root (portio major, 170,000 myelinated fibers).[1]

The fibers of the sensory root arise from the pseudounipolar cells of the *semilunar* ganglion (gasserian ganglion), which is located in a dura mater cavity (Meckel's cave) in the middle cranial fossa. The three divisions depart from the convex border of the gasserian ganglion, heading toward their exits from the skull.

The ophthalmic nerve (25,000 fibers), the smallest of the three trigeminal divisions, is a purely sensory nerve. It runs along the lateral wall of the cavernous sinus, below the oculomotor and trochlear nerves. Just before entering the orbit through the superior orbital fissure, it divides into three branches, the lachrymal, frontal, and nasociliary nerves. These branches supply the cornea, nasal cavity, skin of the upper eyelid, dorsum of the nose, forehead, and scalp as far back as the border between the anterior two-thirds and the posterior one-third of the scalp (which is innervated by the great occipital nerve).

The maxillary nerve (50,000 fibers), again purely sensory, has an intermediate position and pathway between the ophthalmic and mandibular nerves. It originates at the middle of the semilunar ganglion and, running horizontally forward, exits the skull through the foramen rotundum, enters the orbit through the inferior orbital fissure and then the infraorbital canal on the floor of the orbit, and finally reaches the facial skin through the infraorbital foramen. The main terminal branches of the maxillary nerve convey sensory information from the lower eyelid, zygoma, nose, medial cheek, and upper lip. While running in the maxillary bone, the nerve also gives off a series of tiny branches to innervate the nasal and oral cavity, including the upper teeth.

The mandibular nerve (78,000 fibers) is the largest of the trigeminal divisions and a mixed nerve, made up of a large sensory root and a small motor root, that passes beneath the ganglion and unites with the sensory root immediately after exit from the skull through the foramen ovale. In the infratemporal fossa, below the skull base, the nerve divides in several motor and sensory branches. Motor nerves innervate the jaw closers (masseter, temporalis, and medial and lateral pterygoid muscles) and jaw openers (mylohyoid and anterior belly of the digastric muscles), as well as the tensor veli palatini and the tensor tympani muscles. The main sensory nerves (buccal, lingual, inferior alveolar, mylohyoid, mental, and auriculotemporal nerves) innervate the mandibular portion of the oral cavity, including the anterior two-thirds of the tongue, teeth, and periodontium; the skin anterior to the ear; the tympanic membrane; the temporomandibular joint; the skin overlying the mandible; and the lower lip.

Among the peculiarities of the trigeminal system, it is worth recalling that the corneal mucosa and the dental pulp have a special innervation probably because the only sensory function is protective. They are exclusively

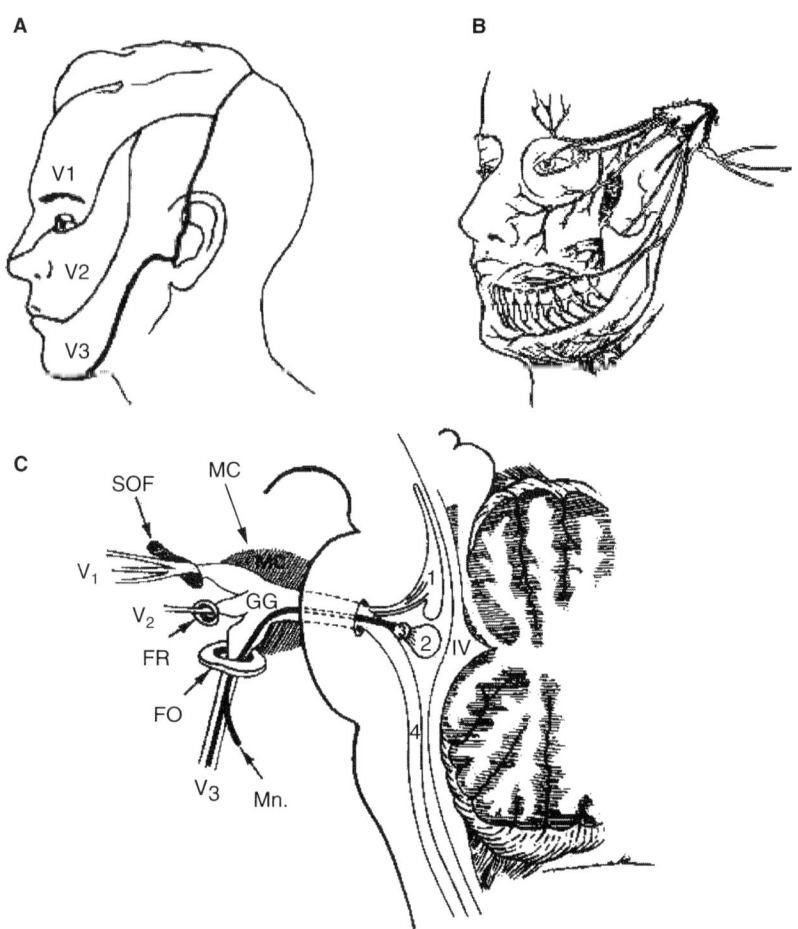

Figure 28-1. Anatomy of the trigeminal nerve. *A.* Cutaneous territories of the first (V1), second (V2), and third (V3) trigeminal divisions. *B.* Peripheral distribution of the trigeminal nerves. *C.* Brain stem trigeminal nuclei and trigeminal divisions arising from the gasserian ganglion. MC = Meckel's cave; Mn = motor nerve; SOF = superior orbital fissure; FR = foramen rotondum; FO = foramen ovale; GG = gasserian ganglion; 1 = mesencephalic nucleus; 2 = main sensory nucleus; 3 = motor nucleus; 4 = spinal trigeminal nucleus. (*Modified from Hutchins et al:* Am J Neuroradiology 1989, 10. 1031–1038)

provided with Aδ and C free nerve endings. Corneal sensations are mostly unpleasant, if not frankly painful. Indeed, the corneal nerve endings are dense with substance P (SP) and calcitonin gene–related peptide (CGRP).

The proximal axons of the gasserian ganglion cells form the sensory root, which enters the midventrolateral pons together with the motor root (which keeps a ventromedial position). The transition from Schwann cell myelination to oligodendroglial myelination begins a few millimeters from the root entry zone (Redlich-Obersteiner zone). The pathway of the sensory root, from the middle posterior fossa to the cerebellopontine angle, explains why trigeminal pain is often the presenting symptom of tumors of the skull base.

The intra-axial fibers of the primary neurons head toward the various nuclei that constitute the trigeminal brainstem complex, which extends from midbrain to the C2 segment of the spinal cord.

Motor Nucleus

The trigeminal motor nucleus is located in the dorsolateral pontine tegmentum, ventromedial to the trigeminal main sensory nucleus. Trigeminal motoneurons have been studied less extensively than spinal motoneurons and differ from them in several ways. They receive a strong inhibitory input from mechanoreceptors and free nerve endings.[2] The powerful inhibition exerted by cutaneous and intraoral A-beta mechanoreceptors and A-delta nociceptors probably compensates for the unusual organization of the jaw-closing motoneurons. In fact—unlike spinal motoneurons—jaw-closing motoneurons undergo neither reciprocal inhibition nor Renshaw inhibition; the jaw openers are devoid of muscle spindles; and all trigeminal motoneurons lack recurrent axons. This organization, in particular, the inhibition arising from peri- and intraoral receptors, contributes to speech control, exerts a defensive action during mastication, and has been thought to play a role in masticatory myofascial pains.

Mesencephalic Sensory Nucleus

The mesencephalic nucleus of the trigeminal sensory complex is a thin column that extends in the dorsomedial tegmentum from the level of the trigeminal motor nucleus in the pons to the rostral midbrain. Unique in the nervous system, this sensory nucleus, rather than the ganglion, contains the cell bodies of primary sensory neurons; these convey information from proprioceptors of the oculomotor and masticatory systems. The Ia axons of mesencephalic neurons that innervate the muscle spindles of jaw-closing muscles at a short distance from their cell bodies give off short collaterals that connect monosynaptically with jaw-closing motoneurons in the pons and mediate the jaw jerk (or mandibular stretch reflex).

Principal Sensory Nucleus

Compared with the other trigeminal sensory nuclei, this is a small gray mass that lies in the dorsolateral pontine tegmentum, close to the motor nucleus. Although small in size, it receives a most important input from the Aβ myelinated afferents that convey tactile information from capsulated mechanoreceptors in all trigeminal territories.[3]

Trigeminal Spinal Complex

This consists of the trigeminal descending tract (intra-axial primary afferents) and spinal nucleus (second-order neurons). The primary afferents descend caudally, always keeping lateral to the nucleus, down to the C2 spinal segment. The tract contains Aβ large-myelinated, Aδ small-myelinated, and unmyelinated C afferents conveying tactile, thermal, and nociceptive inputs from all trigeminal territories, as well as primary afferents of other cranial nerves (VII, IX, X), conveying sensory input from the ear, pharynx, and larynx. All these afferents connect with the second-order neurons in the spinal nucleus, which extends from midpons to C2 and is rostrocaudally divided in three subnuclei: oralis, interpolaris, and caudalis.[4] The nucleus oralis, the most rostral, merges in the pons with the main sensory nucleus. The nucleus interpolaris, located in between the other two nuclei, extends from the rostral pole of the hypoglossal nucleus to the obex.

The nucleus caudalis merges with the spinal dorsal horn and—having a similar laminar organization—is also known as the *medullary dorsal horn*. Gobel and colleagues,[5,6] who included the adjacent medullary reticular formation within the nucleus caudalis, divided this medullary dorsal horn in layers: Lamina I corresponds to the marginal layer and includes the interstitial nucleus of Cajal; lamina II corresponds to the substantia gelatinosa and contains a larger number of small-diameter axons than laminae III and IV, which correspond to the magnocellular layer and mostly contain large myelinated axons; lamina V corresponds to the subnucleus reticularis dorsalis; and lamina VI corresponds to the subnucleus reticularis ventralis of the lateral reticular formation.

The trigeminal spinal nucleus has a multiple anatomic-functional organization, two dealing with somatotopy and one with sensory modality. It is traditionally accepted that the nucleus caudalis has an "onion skin" organization, with oral and perioral areas rostrally (rostrum) and then all the other lateral territories disposed progressively more caudally, with forehead and scalp reaching the first spinal segments. It is also known, however, that in the trigeminal spinal nucleus the somatotopic representation of the ophthalmic, maxillary, and mandibular regions follows a ventrolateral to dorsomedial disposition, with the ophthalmic region ventrolateral, the mandibular region dorsomedial, and the maxillary region in between.[4] Finally, it is generally agreed that the rostral subnuclei of the trigeminal spinal nucleus contribute with the main sensory nucleus to relay and modulate the orofacial touch sense; they mostly contain low-threshold mechanoreceptive (LTM) neurons that provide the higher brain levels with detailed information on tactile sensations.[4] In contrast, the nucleus caudalis is universally considered as the main brain stem nucleus serving orofacial nociception.

In the nucleus caudalis, nociceptive terminals project to laminae I, II, V, and VI. Most neurons of the outer layers (I–II) are nociceptive-specific (NS); those in lamina V are wide dynamic range (WDR). The former respond selectively to noxious stimuli conveyed by small (Aδ and C) afferents. The latter, excited both by noxious and non-noxious stimuli, receive both large-fiber (Aβ) and small-fiber terminals. WDR neurons can encode and project different types of sensory information, nociceptive and nonnociceptive, varying their firing rate (higher for noxious and lower for non-noxious stimuli). NS neurons have a fairly localized receptive field and probably play an important role in spatial detection of nociceptive stimuli. In contrast, since WDR neurons have a large receptive field and a stimulus-response function (the higher the stimulus intensity, the higher is the firing rate of their output), their main function consists of detection and discrimination of intensity of the noxious stimuli.[7,8]

Although the nucleus caudalis is the most important trigeminal relay for pain transmission, it also contains LTM neurons in laminae III and IV, mostly receiving tactile afferents from the ophthalmic division. Conversely, the nucleus oralis and nucleus interpolaris also receive small-diameter afferents and contain NS and WDR neurons. Hence the whole trigeminal spinal complex probably contributes to orofacial nociception and may play a role in pathophysiology of orofacial pains.

Central Projections

Most axons of the trigeminal second-order neurons cross the midline and convey their signals through the trigeminothalamic pathways. The fibers arising from the main sensory nucleus and the rostral part of the spinal nucleus form the trigeminal lemniscus, which ascends with the medial lemniscus and projects to the ventral postero-medial nucleus (VPM) of the thalamus.[4] Those arising from the nucleus caudalis merge with the spinothalamic tract, which conveys thermal-pain information, and project both to the lateral thalamus (VPM) and to the medial and intralaminar thalamic nuclei.

Similar to the spinal system, the trigeminal system finally projects to the somatosensory primary and secondary cortices (SI and SII) in the parietal lobe, and thermal-nociceptive inputs also reach the insula and the anterior cingulated gyrus. In contrast with the classic description of the sensory homunculus, whereas the tactile input from the lower (V3) and upper (V1) facial territories is largely overlapped in SI, the lower territories are represented mostly contralaterally and the upper territories bilaterally in both SI and SII.[9] Regarding the cortical processing of thermal-pain sensation, the trigeminal system does not appear to differ from the spinal system: sensory-discriminative aspects probably are processed in the lateral parietal-insular cortices and affective-motivational aspects in several areas, including the cingulate gyrus, both for the Aδ and C inputs.[10]

▶ TRIGEMINAL NEURALGIA

Trigeminal neuralgia (TN) or tic douloureux is universally considered the facial pain most and best known in medical practice. The International Association for the Study of Pain (IASP) defined TN as "a sudden, usually unilateral, brief, stabbing recurrent pain in the distribution of one or more branches of the fifth cranial nerve." TN is infrequent, with an estimated prevalence of 0.1 percent.[11] The disorder is more common in women than in men (female-to-male ratio 2:3). Occasionally, trigeminal neuralgia may occur in more than one member of the same family.[12]

TN may have no apparent cause (idiopathic, primary, essential, or classic TN) or be secondary to multiple sclerosis or benign compressions in the posterior fossa (symptomatic *or* secondary TN). Because the debate about the possible causes of the so-called idiopathic TN is still open, we believe that the labels and definitions proposed by the International Headache Society (IHS) are the most appropriate: classic TN (with no apparent cause other than vascular compressions) and symptomatic TN (pain indistinguishable from that of classic TN but caused by a demonstrable structural lesion other than vascular compressions).

Symptoms

TN symptoms are unmistakable, and usually TN is diagnosed on the basis of the patient's symptoms.

Although the McGill Pain Questionnaire has been shown to be useful in the differential diagnosis of facial pains such as TN,[13] taking an accurate patient history remains the most reliable tool. Pain distribution is unilateral (bilateral TN sometimes may occur in multiple sclerosis) and follows the sensory distribution of the trigeminal divisions, typically radiating to the maxillary (V2) or mandibular (V3) territories; because it is most uncommon in classic TN, onset in the ophthalmic (V1) area is usually indicative of symptomatic TN and must promote all the relevant investigations. Even more indicative of a symptomatic form is pain affecting the tongue, which is never affected in classic TN. The right side of the face is involved more frequently than the left. Pain, usually referred to as stabbing or electric shock-like, is brief and paroxysmal, lasting a few seconds, with no pain in between paroxysms. Paroxysms may occur several times a day. The mechanical stimulation of specific cutaneous or mucosal areas of trigeminal innervation (trigger zones) will set off the painful paroxysms. For example, gently touching the face, washing or shaving, talking, brushing the teeth, chewing, swallowing, or even a slight breeze, but never thermal or painful stimuli, can trigger the paroxysms. Adjunctive signs may occur during paroxysms. Pain provokes brief muscle spasms of the facial muscles, thus producing the tic. Lacrimation or rhinorrhea is very rare.

Classic TN typically occurs in the sixth or seventh decade of life. If the onset occurs before age 50, TN is very likely to be symptomatic.

Etiology

Symptomatic TN can be related to slowly growing tumors, such as cholesteatomas, meningiomas, or neurinomas of the VIIIth nerve, that compress the trigeminal nerve root near the dorsal root entry zone. Tumors affecting the gasserian ganglion and neurinomas of the Vth nerve rarely are associated with typical TN. Rather, they cause sensory deficits, and if present, pain is constant.[14] Multiple sclerosis (MS) typically is associated with TN (2 to 4 percent of patients with TN).[15] Neurophysiologic, neuroimaging, and pathologic studies indicate that the demyelinating plaque that provokes TN affects the intrapontine presynaptic primary afferents near the root entry zone (Fig. 28-2).[16,17]

As already anticipated, many investigators refute the term *idiopathic TN* because they support the view that when no lesion affecting the trigeminal system can be demonstrated, TN is constantly related to a vascular compression of the trigeminal nerve root by tortuous or aberrant vessels (Fig. 28-3). Microsurgical interventions in the posterior fossa have shown that the compressing vessel is most often the superior cerebellar artery (about 75% of cases). A vein may contribute to compression and is the only compressing vessel in about 10% of patients.[18,19] Further support for this view comes from magnetic resonance imaging (MRI) studies reporting a

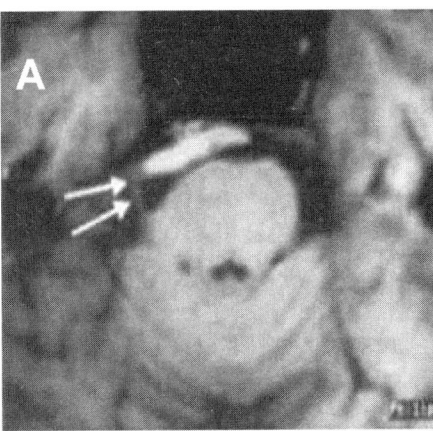
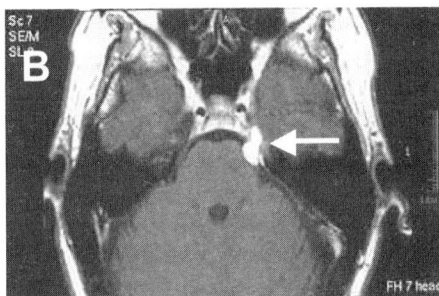
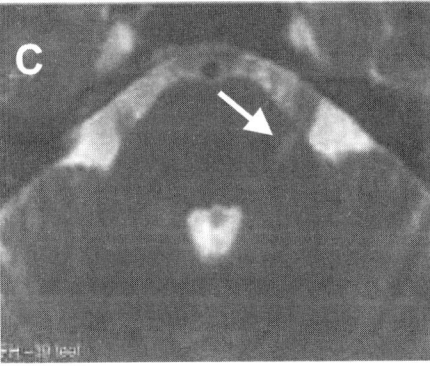

Figure 28-2. Causes of symptomatic TN. MRI scans in patients with symptomatic trigeminal neuralgia. *A.* Gradient-echo MRI showing basilar artery ectasia compressing the right trigeminal root. White arrows indicate the trigeminal root. *B.* Axial gadolinium-enhanced T_1-weighted MRI showing a small benign tumor compressing the left trigeminal root. *C.* Axial T_2-weighted MRI showing a demyelinating lesion just proximal to the root entry into the pons.

frequent contact between vessels[20] and the trigeminal root. Consistently, microvascular decompression relieves TN pain.[18] Furthermore, observations during posterior fossa surgery in patients with tumors and TN demonstrated a vessel compressing the nerve at the root entry zone in almost all patients.[21]

Nevertheless, other investigators do not support this view because autopsy findings seem to demonstrate that a vascular compression cannot be the main factor. Vascular compression of the trigeminal nerve root is found often during standard autopsies of patients with no history of TN, and the neurovascular contact often occurs precisely near the root entry zone.[22-24]

Classic TN must have a cause. Although vascular compression remains the most studied and supported hypothesis, it cannot be regarded as the universal cause of classic TN. Possibly, several factors may contribute to the development of TN. Indeed, many patients do not experience complete and persistent pain relief after microvascular decompression.[25] It is also worth reporting here that many neurosurgeons found that manipulation alone or even just exposure of the trigeminal nerve causes a temporary remission of TN.[26,27]

Pathophysiology

Since the symptoms of classic and symptomatic TN are identical and the latter is always secondary to a peripheral lesion, i.e., extra-axial or near the root entry zone, the primary site of damage in classic TN is also thought to be peripheral, near the root entry zone. Of note, nerve fibers change their myelination (from Schwann cells to oligodendroglia) at this site.

Demyelination of the primary afferents, whether produced by MS or chronic compression exerted by a blood vessel or a benign tumor, increases the susceptibility of the nerve fibers to ectopic excitation, ephaptic transmission, and high-frequency discharges.[28] Ephaptic transmission between large myelinated, non-nociceptive afferents and nociceptive afferents may explain how innocuous stimuli can trigger the painful paroxysms.[29] The large number of afferents innervating the perioral region would explain why trigger zones are most frequently located in that region. The perioral location of the trigger point would also be explained by the large representation of the perioral region in the spinal trigeminal nucleus, which makes it the most likely source of paroxysmal activity.[30] Neurophysiologic studies using trigeminal reflexes or evoked potentials have shown impairment of large Aβ fibers in classical TN. Using laser stimuli, however, recent studies have also disclosed an important dysfunction of the Aδ fibers.[31] In summary, the primary cause of TN is thought to originate from a dysfunction of the peripheral afferents, but the pathophysiological mechanism may secondarily involve the brain stem neurons.

Pharmacological Treatment

Patients with TN do not respond to conventional analgesic drugs. The traditional first-choice medical treatment of TN is carbamazepine (CBZ).[32] Almost all patients respond to CBZ. The starting dose should be 200 mg/d, to be increased by 200 mg every second day until the patient reaches satisfactory pain relief or encounters noteworthy

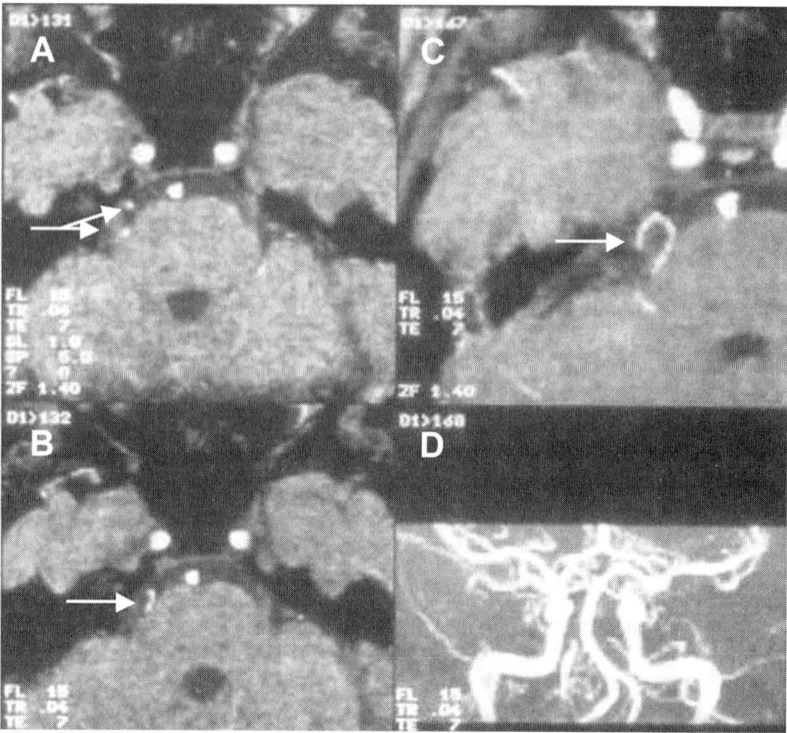

Figure 28-3. Neurovascular conflict in classic TN. MRIs from a patient with classic TN on the right V2–V3 divisions showing neurovascular conflict between right trigeminal nerve root and superior cerebellar artery. *A, B.* Thin-slice axial scans. White arrows indicate a tiny vessel (superior cerebellar artery, SCA), which engages with the trigeminal root. *C.* Reconstruction showing the anomalous loop of the SCA. *D.* General view of the angio-MRI.

side effects. In patients with classic TN, carbamazepine has a quick and intense analgesic effect. Usually 400 to 800 mg is sufficient. Although in epilepsy CBZ is administered at up to 2400 mg/d, its functional dose/response is nonlinear in TN. The increments in its analgesic effect reach a plateau at about 1200 mg. Higher doses of CBZ do not induce further pain relief.[33] Unresponsiveness to CBZ suggests the patient may not have classical TN and should promote further investigations.

The most common side effects of CBZ involve the central nervous system (CNS): drowsiness, dizziness, diplopia, gait and unsteadiness. These side effects usually subside within days or weeks. In a few elderly patients, however, the CNS disturbances may cause discontinuation of CBZ. Starting with a lower dose may reduce the occurrence of these side effects. Cutaneous reactions have been reported in about 6 percent of patients; in most cases these reactions are of minor importance, but rare cases of Stevens-Johnson syndrome have also been reported.

Although very rare (about 1 in 200,000 patients), the complication the physicians most worry about is aplastic anemia. Mild and transient leukopenia and thrombocytopenia occur in about 10% of patients. A complete blood count should be performed before the first administration of CBZ and every 2 weeks for the first 2 months of treatment. Afterwards, two blood counts a year are sufficient as a monitoring measure.

In order to avoid the potential CBZ-related side effects, other drugs can be considered for the management of TN.

Oxcarbazepine is a recent modification of carbamazepine, conceived to minimize the side effects on the CNS. However, oxcarbazepine may induce a depletion of Na+ ions.[34] Several observations suggest that oxcarbazepine may be as effective as carbamazepine in TN treatment.[35,36] Usually, the starting dose is 600 mg/d, increasing 150 to 300 mg every few days based on clinical response. The effective dose ranges widely between 600 and 1800 mg/d.

Baclofen was shown to be effective in the treatment of TN. It has the advantage of a synergistic action with carbamazepine.[37] However, probably because of the frequent side effects of this $GABA_B$ agonist, baclofen is prescribed rarely.

The new generation of antiepileptic drugs has brought new hopes for the treatment of neuropathic pain. We are aware of a number of observations reporting the effectiveness of gabapentin, lamotrigine, and topiramate.[32,38-42] However, randomized controlled trials of these agents conducted in large populations for TN are lacking.

Oral phenytoin (i.e., historically the first antiepileptic drug used for the treatment of TN) is effective in only 25 percent of patients. Its chronic administration has serious adverse effects.[33,43] However, phenytoin can be administrated intravenously. Thus, it may become a useful agent in an emergency setting, when extremely frequent TN paroxysms preclude oral therapy.

Surgical Treatment

Surgical therapy is indicated in three situations: when medical therapy does not succeed in providing sufficient pain relief, when contraindications or excessive side effects limit the use of other effective drugs, and when there is a removable cause.

Besides cerebellopontine angle tumors, which can be totally or partially removed, major interventions (i.e., necessitating a craniotomy) are aimed at freeing the trigeminal root from compression or stretching by anomalous vessels.

Microvascular decompression is a relatively safe and effective treatment for TN. Although craniotomy and general anesthesia are safer in younger patients, there is no fixed age limit for microvascular decompression. After a retromastoid craniotomy, the cerebellopontine angle is exposed. Using an operative microscope, the vessel in contact with the trigeminal root is disengaged, and a small Teflon implant is placed between the vessel and the nerve. In a study by Barker and colleagues,[18] major complications included deaths after the operation (0.2 percent), brain stem infarction, and ipsilateral hearing loss (1 percent). Most patients have complete and persistent resolution of their pain after microvascular decompression. Ten years after surgery, 64% of the operated patients have remained pain free.[18] In the unsuccessful cases, TN usually resurfaces within the first 2 years after the operation.

Alternatives for poor surgical candidates are percutaneous rhizotomies. This procedure is technically simpler than microvascular decompression. A needle is inserted between the maxilla and the mandible and directed through the foramen ovale into the Meckel's cave, where the trigeminal ganglion is located. This route is common to several techniques, all aiming at damaging the ganglion cells or their proximal axons mechanically (balloon compression), chemically (glycerol gangliolysis), or thermically (radiofrequency thermocoagulation).

Glycerol injection and balloon compression preferentially damage large-size fibers, thus sparing the thin Aδ and C fibers directed to the cornea. The severity of the lesion can be graded (varying, for example, the glycerol concentration, balloon degree of pressure, or duration of the application), and pain relief is achieved without major axonal degeneration. Facial numbness and a transient masticatory muscle weakness are the usual complications. Major complications such as meningitis, subarachnoid hemorrhage, and even death are extremely rare.

Radiofrequency (RF) thermocoagulation[47] is based on the concept that heat mainly affects small-myelinated and unmyelinated afferents, thus producing analgesia with a negligible sensory loss. The tip of the RF needle can be guided by paresthesias elicited in response to low-intensity electrical stimuli or by stimulating the peripheral territories and recording the nerve potentials from the intracranial operating needle.[48] In a 25-year follow-up of 1600 patients,[49] immediate pain relief was obtained in 98 percent; after 5 years 92 percent of patients with a single procedure or with multiple procedures still had complete pain relief; after 20 years, the pain relief persisted in 41 percent of the patients who underwent a single procedure and 100 percent of those who had multiple procedures. Complications included decreased corneal reflex in 6 percent of patients, masseter weakness in 4 percent, dysesthesia in 1 percent, anesthesia dolorosa in 0.8 percent, keratitis in 0.6 percent, and transient paralysis of oculomotor nerves in 0.8 percent. Patients who have the involvement of the first (VI) division may not undergo radiofrequency thermocoagulation, due to the inevitability of keratitis following the procedure.

Lesions by alcohol injection are no longer performed, due to the frequent complications of anesthesia dolorosa and keratitis. Anesthesia dolorosa is a constant neuropathic pain due to massive ganglion cell degeneration. Keratitis follows degeneration of the thin fibers that innervate the cornea, and may lead to blindness.

Neurophysiological recordings of trigeminal responses have shown that balloon compression produces transient damage to large-size fibers, whereas radiofrequency thermocoagulation provokes a far longer-lasting damage not only to small- but also to large-sized fibers, including motor fibers.[50]

With percutaneous rhizotomies, the experience has taught us that the more intense the lesion, the longer the pain-free period before recurrence will last.

▶ OTHER CRANIAL NEURALGIAS

Glossopharyngeal Neuralgia

Glossopharyngeal neuralgia is a severe transient stabbing pain experienced in the ear, base of the tongue, tonsillar fossa, or beneath the angle of the jaw. The pain is felt in the distribution of the auricular and pharyngeal branches of the glossopharyngeal, as well as vagus nerve. It is commonly provoked by swallowing, talking, or coughing. The neuralgia may remit and relapse over time.

Glossopharyngeal neuralgia is a rare pain condition that occurs in younger patients than those affected by classical trigeminal neuralgia (40 percent of patients are under 50 years of age). It is more common in females (67 percent) than males (33 percent).[51] Although glossopharyngeal neuralgia usually occurs without any evident lesion affecting the glossopharyngeal nerve, most authors suggest that it is related to a vascular compression of the glossopharyngeal nerve at the root entry zone. According to MRI findings and observations during posterior fossa surgery, the posterior inferior cerebellar artery is the most frequent vessel compressing the glossopharyngeal nerve.[52] The symptomatic forms are often secondary to intra- or extracranial compressions near the jugular foramen. An elongated or fractured styloid process, a calcified stylohyoid ligament (Eagle's syndrome), cerebellopontine angle tumors, parapharyngeal space lesions, carcinoma of the parapharyngeal space, carcinoma of the pharynx, nasopharyngeal carcinoma, posterior fossa arteriovenous malformation, and multiple sclerosis are causes of symptomatic glossopharyngeal neuralgia.[53,54]

It is worth emphasizing that pain attacks may lead to asystolia and syncope. The glossopharyngeal nerve innervates the carotid sinus; hyperactivity of the glossopharyngeal afferents can give rise to activation of the dorsal motor nucleus of the vagus nerve, resulting in a parasympathetic vagal efferent response causing severe bradycardia and eventually asystolia.[53,54]

The traditional first-line medical treatment for glossopharyngeal neuralgia is carbamazepine. The alternative pharmacological treatments described for TN can also be considered for glossopharyngeal neuralgia. Surgical approaches for glossopharyngeal neuralgia vary from nerve sectioning to microvascular decompression. The nerve section approach can be via posterior fossa, neck, or transtonsillary. Currently, microvascular decompression is the most common surgical treatment of glossopharyngeal neuralgia.[51]

Nervous Intermedius Neuralgia

This is a rare condition characterized by brief paroxysms of deep pain in the ear. Pain paroxysms are intermittent, last for seconds or minutes, and may be triggered by touching the posterior wall of the auditory canal. Pain sometimes is accompanied by a disorders of lacrimation, salivation, or taste. There is a common association with herpes zoster.

Superior Laryngeal Neuralgia

This is another rare disorder characterized by severe pain affecting the lateral aspect of the throat, submandibular region, and underneath the ear precipitated by swallowing, shouting, or turning the head. A trigger point is present on the lateral aspect of the throat overlying the hypothyroid membrane. The condition is relieved by local anesthetic block and cured by section of the superior laryngeal nerve.

▶ TRIGEMINAL SENSORY NEUROPATHY

Peripheral neuropathies often cause trigeminal nerve dysfunction, which is not always clinically relevant and usually nonpainful. An isolated trigeminal sensory neuropathy (TSN) (i.e., not associated with systemic neuropathy) is an infrequent clinical condition.

Following the first report of idiopathic trigeminal neuropathy,[55] several studies demonstrated an association between chronic trigeminal sensory neuropathy and connective tissue diseases (CTDs). Although systemic sclerosis is the most frequent CTD associated with trigeminal neuropathy, systemic lupus erythematosus, mixed connective tissue disease, and Sjögren syndrome also have been related to TSN. Rarely, TSN precedes the diagnosis of a CTD.

The few available supraorbital nerve biopsy studies have shown severe axonal degeneration of myelinated fibers and almost complete sparing of unmyelinated fibers; blood vessels appeared normal, and signs of inflammation were absent.[10,56] When associated with CTD, TSN may be provoked by vasculitis involving the vasa nervorum. Fibrosis of perineurium and epineurium and the subsequent increase in endoneurial pressure may participate in myelinated fiber damage.

Patients with TSN gradually develop sensory loss; first, it is usually localized to the perioral region on one side and then extends, over a period of months or a few years, to the other side, to the intraoral mucosa, the tongue, and to the whole face, sometimes sparing the first division. Impairment or loss of taste may occur. The motor trigeminal fibers are consistently spared. Pain is a common symptom. In the initial stage it may mimic TN, being paroxysmal, pricking, or electric shock-like, but it is never as severe as that of TN, and—most of all—it is never elicited by stimulation of trigger zones. Progressively, as the axonal loss proceeds, pain becomes constant, usually described as burning or aching. Allodynia also may occur.

Pain in TSN is related to degeneration of myelinated primary afferents. Ectopic discharges arising from suffering axons or regenerating sprouts may be responsible for spontaneous paroxysmal pain, which frequently is present in the initial stage of TSN.

▶ OPHTHALMIC POSTHERPETIC NEURALGIA

Herpes zoster (HZ) is a localized infection caused by the varicella-zoster virus. After remaining dormant in the sensory ganglia since the primary infection, the virus reactivates and spreads along the nerve fibers to the skin, causing a dermatomally distributed painful rash. In the ganglion, the virus causes neuronal death, followed by degeneration of spinal and peripheral axons. The main complication of HZ is postherpetic neuralgia (PHN), i.e., a neuropathic pain persisting more than 3 months after skin eruption.[57] Of all patients with HZ, about 10 percent will develop PHN, with a higher frequency in elderly patients and diabetic patients. Most often, PHN involves thoracic dermatomes (about 50 percent of patients), but the ophthalmic division of the trigeminal nerve also is a very common distribution (22 to 25 percent of patients).[58,59]

Ophthalmic PHN involves the supraorbital region and the eye. In ophthalmic PHN, as well as in nontrigeminal PHN, the areas involved show skin changes such as hyperpigmentation and scarring. The sensory disturbances consist of hypesthesia, involving all sensory modalities, and pain. Pain is both constant (burning, aching, dull) and paroxysmal (stabbing, electric shock–like); most PHN patients also have allodynia, usually of the dynamic mechanical type.[60] Neurophysiologic studies confirm damage to both large and small trigeminal afferents.[61]

This trigeminal localization of PHN presents the problem of eye involvement, which strongly limits the use of the various topical agents currently employed in PHN treatment. Hence medical treatment is the first choice. Although several RCTs have been carried out in PHN, none was dedicated to trigeminal PHN alone. We are unaware, however, of differences in responsiveness to drugs. Amitriptyline is the standard therapy for PHN, but nortriptyline also has been found to be effective (good pain relief in 67 percent of patients).[62] Unfortunately, since PHN affects elderly patients, often amitriptyline is discontinued because of its adverse effects (dry mouth, constipation, urinary retention) or cannot be initiated at all because of its contraindications (glaucoma, prostatic hypertrophy, conduction abnormalities, myocardial infarction). More recently, an RCT on over 300 PHN patients showed that gabapentin (1800 or 2400 mg/d) was significantly effective in inducing pain relief in over 50 percent of patients.[63]

Although several studies have reported the effectiveness of many drugs (various antidepressants, antiepileptics, and even opiates) in PHN, in our experience, few patients with ophthalmic PHN respond to treatment; for most, the pain is intractable. Luckily, also in the trigeminal territory, PHN may subside spontaneously.

▶ WALLENBERG SYNDROME

Wallenberg syndrome is caused by an infarction of the lateral tegmentum of the medulla owing to thrombosis of the posterior inferior cerebellar artery or one of its branches supplying the lower portion of the brainstem.[64] Lateral medulla infarction affects the spinal trigeminal complex and spinothalamic tract, thus producing thermal-pain hypesthesia on the ipsilateral face and contralateral body. In addition to these sensory pathways, the infarction often involves the descending pupillodilator fibers, nucleus ambiguous (nerves IX and X), and spinocerebellar and olivocerebellar tracts, thus provoking ipsilateral Bernard-Horner syndrome, dysphagia, dysphonia, dysarthria, and ipsi-, contra-, or bilateral ataxia.

Wallenberg syndrome has an overall good prognosis; motor, vegetative, and cerebellar functions recover promptly. However, residual sensory disturbances are frequent. They include thermal-pain hypesthesia and pain on the face and body. Pain is both spontaneous (burning, aching, dull) and provoked (generally allodynia to light mechanical stimuli). As in thalamic syndrome, in Wallenberg syndrome, overreactions with hyperalgesia and hyperpathia may occur. Pain correlates with the degree of clinical sensory loss but not with the size of infarction seen on MRI. Pain pathophysiology in Wallenberg syndrome, as well as in the other central pain syndromes, is still unclear. Naturally, nociceptive pathway (trigeminal and spinal) dysfunction should play a central role. Spinothalamic deafferentation alters thalamic activity, thus inducing pain.[65] Components of the spinothalamic pathway may serve to control the activity of nociceptive-specific neurons of spinothalamic tract. Lesion of the spinothalamic tract could remove this normal suppressing mechanism.[66] The spinothalamic tract and the adjacent spinoreticulothalamic tract are interrelated in such a way that degeneration of the former renders the normally nonexcitable reticulothalamic system abnormally hyperactive, thus provoking pain.[65,67] Support for this hypothesis comes from evidence that lesions extending to the reticular formation rarely produce pain.[65,67]

Some light on the role of the spinal trigeminal complex comes from a recent study correlating clinical, MRI, and neurophysiologic findings in patients with central pain following lateral medullary infarctions.[68] Although facial pain was correlated with lesions of the spinal trigeminal complex, none of the lesions involved the subnucleus caudalis, which contains most nociceptive neurons. These findings suggest that facial pain after medullary infarction is due to lesions of the lower spinal trigeminal tract (axons of primary afferent neurons), leading to deafferentation of spinal trigeminal nucleus neurons.

There are no studies dedicated to the treatment of neuropathic pains arising from Wallenberg syndrome or trigeminal sensory neuropathy. In our experience, the treatment of these painful conditions is similar to that for neuropathic pain in general. Carbamazepine, reducing the focal hyperactivity of injured nerve fibers, is sometimes effective in the initial stage of TSN, when paroxysmal, shooting pain occurs. Amitriptyline and gabapentin, with mechanisms (only partly understood) that may include a partial suppression of abnormal activity in secondary sensory neurons and strengthening of descending inhibitory control pathways, are useful in the treatment of allodynia and deafferentation pain.

▶ BURNING MOUTH SYNDROME

The IASP definition of burning mouth syndrome (BMS) is "glossodynia and sore mouth, also known as burning tongue or oral dysesthesiae, as burning pain in the tongue or other oral mucous membranes." It mainly affects women over 50 years of age.

Burning intraoral pain also may be a symptom of various dental or medical diseases; thus the term *BMS* should be reserved for patients without evidence of local or systemic diseases that may cause burning intraoral pain, in particular idiopathic trigeminal neuropathy and Sjögren neuronopathy, which may present with symptoms very similar to BMS.

BMS has been related to psychogenic factors. Patients with BMS have high scores on scales of somatization, obsession-compulsion, or psychoticism, and these psychiatric abnormalities may be present prior to symptom onset.[69] Other studies, however, using quantitative sensory testing and electrophysiologic methods, have demonstrated abnormalities of both nociceptive and nonnociceptive trigeminal pathway function, thus indicating a possible neuropathic etiology of BMS.[70] Furthermore, positron-emission tomographic (PET) studies in patients with BMS demonstrated a decrease in striatal dopamine level, which probably plays an important role in pain modulation.[71] As a further hypothesis, it also has been proposed that damage to the cranial nerves that mediate taste function decreases the inhibition of trigeminal nociceptive fibers, thus leading to oral burning symptoms.[72] In summary, BMS pathophysiology is still unknown.

Although BMS symptoms usually are unmistakable, a careful examination of the oral cavity is necessary to exclude other causes. Some patients report a spontaneous onset of pain with no identifiable precipitating factor; other patients link the onset of symptoms to a dental procedure, recent illness, or some medication (including antibiotic therapy). The burning sensation usually is constant and bilateral, with the anterior two-thirds of the tongue, the anterior hard palate, and the lower lip mucosa involved most frequently. Although absent during the night in many patients, pain may disturb sleep, thus worsening preexisting mood disturbances. Several patients also report changes in taste and smell, oral dryness, and difficulty swallowing. Scarce information is available on the natural course of BMS. Usually pain persists for several years, and no clinical factors predicting recovery have been noted.

BMS treatment is often unsatisfactory, and there is no definitive cure.[73] Reassurance and counseling have a powerful therapeutic action. Most patients ask for medical treatment anyway. Some observations suggest that patients may obtain partial benefit from amisulpiride, selective serotonin reuptake inhibitors (SSRIs; e.g., paroxetine, sertraline), clonazepam and cognitive-behavioral therapy.[73,75] Experience with hormone-replacement therapy and vitamins did not yield any significant improvement in symptoms.[74]

▶ ATYPICAL FACIAL PAIN

Atypical facial pain (AFP) is a diagnosis of exclusion. It describes a facial chronic pain that does not have the characteristics of cranial neuralgias and is not associated with identified lesions affecting the trigeminal system or the facial tissues. This diagnosis is not accepted by the IASP. Since there is no agreement on diagnostic criteria, it is impossible to gain information about AFP epidemiology. The first pathophysiologic theory proposed a vascular origin, similar to that of migraine. Psychogenic factors have been reported to precede the onset of pain.[69] Electrophysiologic studies have indicated trigeminal pathway abnormalities. PET studies have suggested a decrease of striatal dopamine in AFP patients.[76,78] Based on these findings, AFP might be a neuropathic pain disorder.

Often the onset of pain is preceded by medical or dental treatment or trauma. Usually, the pain is diffuse, constant, aching, dull, and fluctuates in intensity. It is poorly localized and does not follow known anatomic distributions. It is felt deep in the bones, in the nose, around and deep in the eyes, or in the upper and lower jaws. It is never associated with sensory loss or other physical signs; standard laboratory tests do not disclose noteworthy abnormalities.

Medical treatment of AFP is usually unsatisfactory. Simple analgesics are ineffective. Tricyclic antidepressants are the most effective drugs in AFP treatment.[71,77] However, many patients do not report a complete and persistent pain relief.[69]

▶ TEMPOROMANDIBULAR DISORDERS

Temporomandibular disorders (TMDs) are an interrelated set of clinical conditions that affect the jaw muscles and the temporomandibular joint, manifesting with facial pain, clicking sounds in the joint, and limited movement of the jaw. TMDs are seen very frequently. However, because of the lack of widely agreed diagnostic criteria, few studies investigating the epidemiology of TMDs have been performed. The prevalence of severe symptoms and signs ranges from 4 to 12 percent. Symptoms and signs of TMDs decrease with age. There is a strong female prevalence in reported symptoms, which also are more severe in women than in men.[78]

TMDs may be related to a temporomandibular joint dysfunction caused by a dislocated jaw or displaced disk, injury to the condyle, or degenerative joint disease, such as osteoarthritis or rheumatoid arthritis in the jaw joint.

However, in most patients, TMDs are not associated with an anatomic dysfunction of the temporomandibular joint and should be considered myofascial pain involving the jaw muscles and the face.[75]

Anxiety, and sleep disturbances are commonly associated with TMDs.[69,75]

Malocclusion has been proposed as the main cause of TMDs.[79]

Several studies demonstrated dysfunction of the trigeminal nociceptive pathways in TMD patients. It is not possible to know whether this dysfunction has a pathophysiologic role in TMDs or is a consequence of chronic pain.[80]

In summary, the pathophysiology of TMDs remains unclear.

A number of symptoms may be linked to TMDs. Pain, particularly in the chewing muscles and/or jaw joint, is the most common symptom. Limited movement of the jaw; radiating pain in the face, neck, or shoulders; and painful clicking, popping, or grating sounds in the jaw joint when opening or closing the mouth occur frequently. Often symptoms such as headaches and dizziness may be related to TMDs.

The patient's history and clinical examination usually provide useful information for diagnosing TMDs. Clinical examination includes palpating the jaw joints and chewing muscles for pain or tenderness, listening for clicking, and examining for limited motion or locking of the jaw while opening or closing the mouth.

Antidepressants, such as amitriptyline, are the most widely used drugs in orofacial pain treatment. They are effective analgesics in both depressed and nondepressed patients.[75,81]

Injections of steroids into painful muscle sites, physical treatments such as bite appliance, occlusal rehabilitation, and psychological therapies have all been studied, with controversial findings.[69,75]

▶ NEUROPHYSIOLOGICAL DIAGNOSTIC TESTS

Trigeminal Reflexes

According to the recommendations of the International Federation of Clinical Neurophysiology[82] and the European Federation of Neurological Societies,[83] the neurophysiologic recording of trigeminal reflexes represents the most useful and reliable test in the laboratory diagnosis of trigeminal pains (Fig. 28-4). The trigeminal reflexes consist of a series of reflex responses (R1 and R2 components of the blink reflex after electrical stimulation of the ophthalmic division, SP1 and SP2 components of the masseter inhibitory reflex after electrical stimulation of the maxillary or mandibular division, and the jaw jerk to chin taps) that assess function of the trigeminal afferents from all trigeminal territories, as well as the trigeminal central circuits in the midbrain, pons, and medulla.

In patients reporting pain in the trigeminal territory, trigeminal reflexes offer the clinician useful information. Abnormalities often are disclosed in divisions that appear unaffected clinically. An objective demonstration of dysfunction is provided in all patients with pain secondary to a documented disease, such as symptomatic trigeminal neuralgia, postherpetic neuralgia, vascular malformations, benign tumors of the cerebellopontine angle, and multiple sclerosis. Reflex responses are affected more extensively and more markedly in patients with constant pain than in those with paroxysmal pain. Although, like others, we occasionally have seen patients with mild reflex abnormalities, in the majority of patients with classic TN, all reflexes are normal. A diagnostic protocol for patients with trigeminal pain should rely primarily on trigeminal reflexes. The technique is easier and less invasive than that for evoked potentials, and the finding of any abnormality implies an underlying structural lesion. The most commonly reported causes of "symptomatic" neuralgia are benign tumors of the cerebellopontine angle, vascular anomalies in the posterior fossa impinging on the proximal portion of the trigeminal root, and multiple sclerosis with a plaque in the root entry zone.

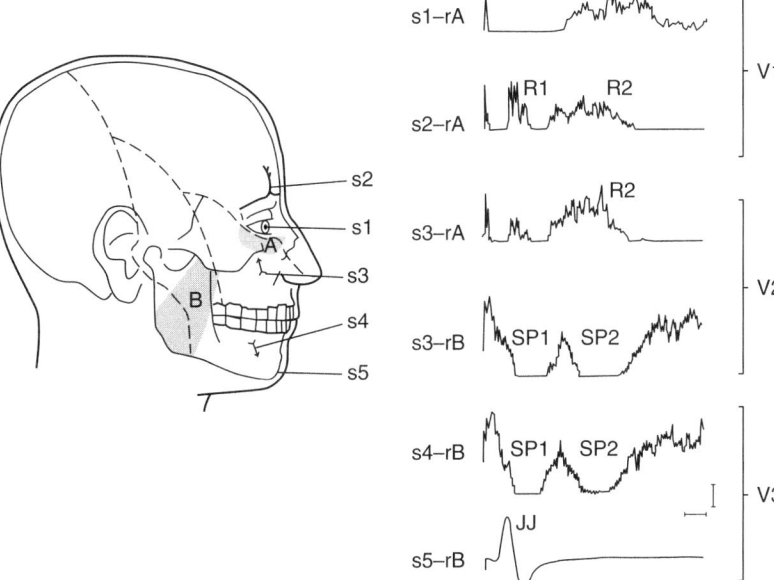

Figure 28-4. Trigeminal reflexes. (*Left*) Schematic drawing showing stimulation and recording sites. s1 = cornea; s2 = supraorbital nerve; s3 = infraorbital nerve; s4 = mental nerve; s5 = taps to the chin; A = orbicularis oculi muscle; B = masseter muscle. (*Right*) Trigeminal reflexes in one normal subject. From top to bottom: Corneal reflex (CR), early (R1) and late (R2) blink reflex, first (SP1) and second (SP2) masseter silent period, jaw jerk (JJ). Rectified and averaged signals. Calibration 200 mV, 10 ms. The signal of the bottom trace s5–rB is not rectified, and calibration is 500 mV, 10 ms. *Used with permission from Cruccu, et al.*[50]

In these cases, the trigeminal reflexes constantly show abnormalities of the short-latency responses, i.e., R1, SP1, and jaw jerk.

Whereas in symptomatic TN all trigeminal reflex abnormalities are unilateral, in TSN the responses to perioral stimuli (in particular, SP1) usually are abnormal bilaterally. In Wallenberg syndrome, only the long-latency R2 and SP2 responses (because they are mediated by medullary circuits) are affected. In patients with BMS and AFP, all trigeminal reflexes are normal.

Several investigators studied the masseter inhibitory reflex and jaw jerk in patients with TMDs, but their findings are inconsistent, and there is no general agreement. Trigeminal reflex testing in TMDs has an uncertain diagnostic value and may yield false-positive diagnoses. Although in clinical practice the most common abnormality in patients with TMDs consists of side asymmetries in jaw-jerk amplitude, the finding of a jaw-jerk asymmetry in a given patient should by no means lead to a diagnosis of TMD. In contrast, a diagnostic point to bear in mind is that an isolated jaw-jerk abnormality in a patient with no other reflex abnormalities does not necessarily imply nerve fiber or brainstem damage, and should warrant stomatognathic investigation.[84]

Trigeminal Evoked Potentials

Over the last few years, several studies have investigated trigeminal evoked potentials elicited by surface stimulation of the lips or gums. Although many investigators have discussed the clinical applicability of these responses, their neural origin has never been proved. Indeed, electrical stimulation of the facial skin unavoidably evokes facial muscle responses, which contaminate or hide the genuine neural signals. A definitive demonstration of this came from a study in curarized subjects showing that the scalp potentials elicited by surface stimulations consist of myogenic artifacts only.[85]

The only certainly genuine and reliable evoked potentials are the very early waves of the scalp potentials. Leandri and associates,[48,85] inserted two fine needles into the infraorbital foramen (thus avoiding direct stimulation of motor nerve fibers) and recorded the far fields generated by trigeminal primary afferents from the scalp (thus, before any reflex could appear). Scalp recording at the vertex with a noncephalic reference is employed to record the neural signal, which consists of three main components: W1, W2, and W3. According to the intraoperative recordings, W1 originates from the proximal part of the maxillary nerve, W2 originates from the retrogasserian root, and W3 from the intrapontine portion of trigeminal afferents directed toward the brainstem nuclei. This method, undeniably invasive and technically difficult, is excellent to assist thermal rhizotomies; the position of the intracranial operating needle can be located and the severity of the lesion monitored without awaking the patient from anesthesia.[48]

To assess nociceptive pathway function, lasers are the best tool. Laser-generated radiant heat pulses selectively excite free nerve endings in the superficial skin layers, activate Aδ and C mechanothermal nociceptors, and evoke scalp potentials generated by the operculoinsular cortex and cingulate gyrus.[86] By varying the area of the irradiated spot and stimulus intensity, it is possible to excite preferentially Aδ (evoking pinprick sensations) or C receptors (evoking warmth or burning sensations).

Because of the short conduction distance and high receptor density, the trigeminal territory is particularly advantageous for laser-evoked-potential (LEP) recording. Trigeminal LEPs are of higher amplitude and are recorded more easily than LEPs after limb stimulation. Trigeminal LEPs have been studied recently in classic and symptomatic TN, TSN, PHN, Wallenberg syndrome, TMDs, and migraine.[61,87]

So far, trigeminal LEPs have provided useful pathophysiologic information in facial pain syndromes.

REFERENCES

1. Pennisi E, Cruccu G, Manfredi M, et al: Histometric study of myelinated fibers in the human trigeminal nerve. *J Neurol Sci* 1991;105:22–28.
2. Nakamura Y: Brain stem neuronal mechanisms controlling the trigeminal motoneuron activity, in Desmesdt JE (ed), *Spinal and Supraspinal Mechanisms of Voluntary Motor Control and Locomotion*, Vol 8. Basel: Karger, 1980:181–202.
3. Darian Smith I: Neural mechanism of facial sensation. *Int Rev Neurobiol* 1966;9:301–395.
4. Johnson LR, Westrum LE, Henry MA: Anatomic organization of the trigeminal system and the effect of deafferentation, in Fromm GH, Sessle BJ (eds), *Trigeminal Neuralgia: Current Concepts Regarding Pathogenesis and Treatment*. Boston: Butterworth-Heinemann, 1991:27–71.
5. Gobel S, Purvis MB: Anatomical studies of the organization of the spinal V nucleus: The deep bundles and the spinal V tract. *Brain Res* 1972;48:27–48.
6. Gobel S, Falls WM, Hockfield S: The division of the dorsal and ventral horns of the mammalian caudal medulla into eight layers using anatomical criteria, in Anderson DJ, Matthews B (eds), *Pain in the Trigeminal Region*. Amsterdam: Elsevier North Holland Biomedical Press, 1977:443–453.
7. Dubner R: Recent advances in our understanding of pain, in Klineberg, Sessle B, (eds), *Oro-Facial Pain and Neuromuscular Dysfunction: Mechanism and Clinical Corelates*. Oxford, England: Pergamon Press, 1985:3–17.
8. Sessle BJ: Physiology of the trigeminal system, in Fromm GH, Sessle BJ (eds), *Trigeminal Neuralgia: Current Concepts Regarding Pathogenesis and Treatment*. Boston: Butterworth-Heinemann, 1991:71–105.
9. Iannetti GD, Porro CA, Pantano P, et al: Representation of different trigeminal divisions within the primary and

10. Cruccu G, Pennisi E, Truini A, et al: Unmyelinated trigeminal pathways as assessed by laser stimuli in humans. *Brain* 2003;126:2246–2256.
11. Munoz M, Dumas M, Boutros-Toni F, et al: A neuropidemiologic survey in a Limousin town. *Rev Neurol (Paris)* 1988;144:266–271.
12. Smyth P, Greenough G, Stommel E: Familial trigeminal neuralgia: Case reports and review of the literature. *Headache* 2003;43:910–915.
13. Melzack R, Terrence C, Fromm G, et al: Trigeminal neuralgia and atypical facial pain: use of the McGill Pain Questionnaire for discrimination and diagnosis. *Pain* 1986;27:297–302.
14. Bullit E, Tew JM, Boyd J: Intracranial tumors in patients with facial pain. *J Neurosurg* 1986;64:865–871.
15. Jensen TS, Rasmussen P, Reske-Nielsen E: Association of trigeminal neuralgia with multiple sclerosis: Clinical and pathological features. *Acta Neurol Scand* 1982;65:182–189.
16. Cruccu G, Leandri M, Feliciani M, et al: Idiopathic and symptomatic trigeminal pain. *J Neurol Neurosurg Psychiatry* 1990;53:1034–1042.
17. Gass A, Kitchen N, MacManus DG, et al: Trigeminal neuralgia in patients with multiple sclerosis: lesion localization with magnetic resonance imaging. *Neurology* 1997;49:1142–1144.
18. Barker FG II, Jannetta PJ, Bissonette DJ, et al: The long-term outcome of microvascular decompression for trigeminal neuralgia. *N Engl J Med* 1996;334:1077–1083.
19. Jannetta PJ: Arterial compression of the trigeminal nerve at the pons in patients with trigeminal neuralgia. *J Neurosurg* 1967;26:159–162.
20. Meaney JF, Eldridge PR, Dunn LT, et al: Demonstration of neurovascular compression in trigeminal neuralgia with magnetic resonance imaging: Comparison with surgical findings in 52 consecutive operative cases. *J Neurosurg* 1995;83:799–805.
21. Barker FG II, Jannetta PJ, Babu RP, et al: Long-term outcome after operation for trigeminal neuralgia in patients with posterior fossa tumors. *J Neurosurg* 1996;84:818–825.
22. Keller JT, Van Loveren H: Pathophysiology of the pain of trigeminal neuralgia and atypical facial pain: A neuroanatomical perspective. *Clin Neurosurg* 1985;32:275–293.
23. Adams CB: Microvascular compression: an alternative view and hypothesis. *J Neurosurg* 1989;70:1–12.
24. Adams CB: Trigeminal neuralgia: pathogenesis and treatment. *Br J Neurosurg* 1997;11:493–495.
25. Burchiel KJ, Steege TD, Howe JF, et al: Comparison of percutaneous radiofrequency gangliolysis and microvascular decompression for the surgical management of tic douloureux. *Neurosurgery* 1981;9:111–119.
26. Shelden CH, Pudenz RH, Freshwater DB, et al: Compression rather than decompression for trigeminal neuralgia. *J Neurosurg* 1955;12:123–126.
27. Fields WS, Lemak NA: Trigeminal neuralgia: historical background, etiology and treatment. *BNI Q* 1987;3:47–56.
28. Burchiel KJ: Abnormal impulse generation in focally demyelinated trigeminal roots. *J Neurosurg* 1980;53:674–683.
29. Calvin WH, Devor M, Howe JF: Can neuralgias arise from minor demyelination? Spontaneous firing, mechanosensitivity, and afterdischarge from conducting axons. *Exp Neurol* 1982;75:755–763.
30. Fromm G: Pathophysiology of trigeminal neuralgia, in Fromm GH, Sessle BJ (eds), *Trigeminal Neuralgia: Current Concepts Regarding Pathogenesis and Treatment.* Boston: Butterworth-Heinemann, 1991:105–131.
31. Cruccu G, Leandri M, Iannetti GD, et al: Small-fiber dysfunction in trigeminal neuralgia: Carbamazepine effect on laser-evoked potentials. *Neurology* 2001;56:1722–1726.
32. Backonja MM: Use of anticonvulsants for treatment of neuropathic pain. *Neurology* 2002;59:S14–S17.
33. Fromm G: Medical treatment of patients with trigeminal neuralgia, in Fromm GH, Sessle BJ (eds) *Trigeminal Neuralgia: Current Concepts Regarding Pathogenesis and Treatment.* Boston: Butterworth-Heinemann, 1991:131–145.
34. Zakrzewska JM, Patsalos PN: Long-term cohort study comparing medical (oxcarbazepine) and surgical management of intractable trigeminal neuralgia. *Pain* 2002;95:259–266.
35. Beydoun A, Kutluay E: Oxcarbazepine. *Exp Opin Pharmacother* 2002;3:59–71.
36. Carrazana E, Mikoshiba I: Rationale and evidence for the use of oxcarbazepine in neuropathic pain. *J Pain Symptom Manage* 2003;25:S31–S35.
37. Fromm G, Terrence CF, Chatta AS: Baclofen in the treatment of trigeminal neuralgia: Double-blind study and long term follow up. *Ann Neurol* 1984;15:240–244.
38. Khan OA: Gabapentin relieves trigeminal neuralgia in multiple sclerosis patients. *Neurology* 1998;51:611–614.
39. Solaro C, Lunardi GL, Capello E, et al: An open-label trial of gabapentin treatment of paroxysmal symptoms in multiple sclerosis patients. *Neurology* 1998;51:609–611.
40. Cheshire WP: Defining the role for gabapentin in the treatment of trigeminal neuralgia: A retrospective study. *J Pain* 2002;3:137–142.
41. Gilron I, Booher SL, Rowan JS, et al: Topiramate in trigeminal neuralgia: A randomized, placebo-controlled multiple crossover pilot study. *Clin Neuropharmacol* 2001;24:109–112.
42. Jensen TS: Anticonvulsants in neuropathic pain: Rationale and clinical evidence. *Pain* 2002;6:61–68.
43. Loeser JD: Tic douloureux and atypical face pain, in Wall PD, Melzack R (eds), *Texbook of Pain.* Edinburgh: Churchill Livingstone, 1984:426.
44. Fujimaki T, Fukushima T, Miyazaki S: Percutaneous retrogasserian glycerol injection in the management of trigeminal neuralgia: Long-term follow-up results. *J Neurosurg* 1990;73:212–216.
45. Saini SS: Retrogasserian anhydrous glycerol injection therapy in trigeminal neuralgia: Observations in 552 patients. *J Neurol Neurosurg Psychiatry* 1987;50:1536–1538.
46. Skirving DJ, Dan NG: A 20-year review of percutaneous balloon compression of the trigeminal ganglion. *J Neurosurg* 2001;94:913–917.
47. Sweet WH, Wepsic JG: Controlled thermocoagulation of trigeminal ganglion and rootlets for differential destruction of pain fibers: 1. Trigeminal neuralgia. *J Neurosurg* 1974;40:143–156.
48. Leandri M, Gottlieb A: Trigeminal evoked potential-monitored thermorhizotomy: A novel approach for relief of trigeminal pain. *J Neurosurg* 1996;84:929–939.

49. Kanpolat Y, Savas A, Bekar A, et al: Percutaneous controlled radiofrequency trigeminal rhizotomy for the treatment of idiopathic trigeminal neuralgia: 25-year experience with 1600 patients. *Neurosurgery* 2001;48:524–532.
50. Cruccu G, Inghilleri M, Fraioli B, et al: Neurophysiologic assessment of trigeminal function after surgery for trigeminal neuralgia. *Neurology* 1987;37:631–638.
51. Patel A, Kassam A, Horowitz M, et al: Microvascular decompression in the management of glossopharyngeal neuralgia: Analysis of 217 cases. *Neurosurgery* 2002;50:705–710.
52. Fischbach F, Lehmann TN, Ricke J, et al: Vascular compression in glossopharyngeal neuralgia: Demonstration by high-resolution MRI at 3 tesla. *Neuroradiology* 2003;45:810–811.
53. Rushton JG, Stevens JC, Miller RH: Glossopharyngeal (vagoglossopharyngeal) neuralgia: A study of 217 cases. *Arch Neurol* 1981;38:201–205.
54. Soh KB: The glossopharyngeal nerve, glossopharyngeal neuralgia and the Eagle's syndrome: Current concepts and management. *Singapore Med J* 1999;40:659–665.
55. Spillane JD, Wells CE: Isolated trigeminal neuropathy: A report of 16 cases. *Brain* 1959;82:391–416.
56. Lecky BR, Hughes RA, Murray NM: Trigeminal sensory neuropathy: A study of 22 cases. *Brain* 1987;110:1463–1485.
57. Dworkin RH, Portenoy RK: Pain and its persistance in herpes zoster. *Pain* 1996;33:241–251.
58. Loeser JD: Herpes zoster and postherpetic neuralgia. *Pain* 1986;25:149–164.
59. Watson CP, Evans RJ, Watt VR, et al: Post-herpetic neuralgia: 208 cases. *Pain* 1988;35:289–297.
60. Rowbotham MC, Fields HL: The relationship of pain, allodynia and thermal sensation in post-herpetic neuralgia. *Brain* 1996;119:347–354.
61. Truini A, Haanpaa M, Zucchi R, et al: Laser-evoked potentials in post-herpetic neuralgia. *Clin Neurophysiol* 2003;114:702–709.
62. Watson CP, Evans RJ, Reed K, et al: Amitriptyline versus placebo in postherpetic neuralgia. *Neurology* 1982;32:671–673.
63. Rice AS, Maton S and the Postherpetic Neuralgia Study Group: Gabapentin in postherpetic neuralgia: A randomized, double-blind, placebo-controlled study. *Pain* 2001;94:215–224.
64. Kim JS, Choi-Kwon S: Sensory sequelae of medullary infarction: Differences between lateral and medial medullary syndrome. *Stroke* 1999;30:2697–2703.
65. Peyron R, Garcia-Larrea L, Gregoire MC, et al: Allodynia after lateral-medullary (Wallenberg) infarct: A PET study. *Brain* 1998;121:345–356.
66. Craig AD, Bushnell MC: The thermal grill illusion: Unmasking the burn of cold pain. *Science* 1994;265:252–255.
67. MacGowan DJ, Janal MN, Clark WC, et al: Central post-stroke pain and Wallenberg's lateral medullary infarction: Frequency, character, and determinants in 63 patients. *Neurology* 1997;49:120–125.
68. Fitzek S, Baumgartner U, Fitzek C, et al: Mechanisms and predictors of chronic facial pain in lateral medullary infarction. *Ann Neurol* 2001;49:493–500.
69. Madland G, Feinmann C: Chronic facial pain: A multidisciplinary problem. *J Neurol Neurosurg Psychiatry* 2001;71:716–719.
70. Forssell H, Jaaskelainen S, Tenovuo O, et al: Sensory dysfunction in burning mouth syndrome. *Pain* 2002;99:41–44.
71. Hagelberg N, Forssell H, Rinne JO, et al: Striatal dopamine D1 and D2 receptors in burning mouth syndrome. *Pain* 2003;101:149–154.
72. Grushka M, Epstein JB, Gorsky M: Burning mouth syndrome and other oral sensory disorders: A unifying hypothesis. *Pain Res Manag* 2003;8:133–141.
73. Zakrzewska JM, Glenny AM, Forssell H: Interventions for the treatment of burning mouth syndrome. *Cochrane Database Syst Rev* 2001;3:CD002779.
74. Maina G, Vitalucci A, Gandolfo S, et al: Comparative efficacy of SSRIs and amisulpride in burning mouth syndrome: A single-blind study. *J Clin Psychiatry* 2002;63:38–43.
75. Zakrzewska JM, Harrison SD: Facial pain, in Jensen TS, Wilson PR, Rice (eds), *Clinical Pain Management: Chronic Pain*. London: Arnold, 2003:481–505.
76. Jaaskelainen SK, Forssell H, Tenovuo O: Electrophysiological testing of the trigeminofacial system: Aid in the diagnosis of atypical facial pain. *Pain* 1999;80:191–200.
77. Hagelberg N, Forssell H, Aalto S, et al: Altered dopamine D2 receptor binding in atypical facial pain. *Pain* 2003;106:43–48.
78. Carlsson EG, LeResche L: Epidemiology of temporomandibular disorders, in Sessle B, Bryant P, Dionne R (eds), *Temporomandibular Disorders and Related Pain Conditions: Progress in Pain Research and Management*. Seattle: IASP Press, 1995:211–227.
79. Magnusson T, Carlsson GE: Occlusal adjustment in patients with residual or recurrent signs of mandibular dysfunction. *J Prosthet Dent* 1983;49:706–710.
80. Romaniello A, Cruccu G, Frisardi G, et al: Assessment of nociceptive trigeminal pathways by laser-evoked potentials and laser silent periods in patients with painful temporomandibular disorders. *Pain* 2003;103:31–39.
81. Dionne RA: Pharmacological treatment for temporomandibular disorders, in Sessle B, Bryant P, Dionne R (eds), *Temporomandibular Disorders and Related Pain Conditions: Progress in Pain Research and Management*. Seattle: IASP Press, 1995:363–375.
82. Cruccu G, Ongerboer de Visser BW: The jaw reflexes, in Deuschl G, Eisen A (eds), *Recommendations for the Practice of Clinical Neurophysiology: Guidelines of the International Federation of Clinical Neurophysiology*. Elsevier, Amsterdam 1999:243–249.
83. Cruccu G, Anand P, Attal N, et al: EFNS guidelines on assessment of neuropathic pain and treatment. *Eur J Neurol* 2004 Mar;11:153–162.
84. Cruccu G, Frisardi G, Pauletti G, et al: Excitability of the central masticatory pathways in patients with painful temporomandibular disorders. *Pain* 1997;73:447–454.
85. Leandri M, Parodi CI, Zattoni J, et al: Subcortical and cortical responses following infraorbital nerve stimulation in man. *Electroencephalogr Clin Neurophysiol* 1987;66:253–262.
86. Treede RD, Lorenz J, Baumgartner U: Clinical usefulness of laser-evoked potentials. *Neurophysiol Clin (Paris)* 2003;33:303–314.
87. Romaniello A, Iannetti GD, Truini A, et al: Reponses a la stimulation trigeminale par laser. *Neurophysiol Clin (Paris)* 2003;33:315–324.

CHAPTER 29

Cervicogenic Headaches

Robert A. Duarte and S. Farhan Zaidi

Cervicogenic headache *has been defined as a recurrent, continuous, unilateral headache without side shift.*[1] *The reported prevalence for cervicogenic headache in the general population is said to be 17.8 percent. In the headache population, the range is from 13.8 to 35.4 percent.*[2] *It must be understood, however, that there continues to remain a controversy as to the definitive diagnostic criteria for cervicogenic headache (CEH). A main issue surrounding this controversy is the fact that there is a considerable overlap between symptoms of CEH and other headache syndromes, such as migraine and tension-type headache.*

▶ EPIDEMIOLOGY AND CONTROVERSY

The origins of CEH began in 1926 when Barré[3] hypothesized that cervical osteophytes irritated the sympathetic nerves leading to a syndrome including headache. In 1946, Bartschi-Rochaix[4] wrote about a compression of the vertebral artery by osteophytes leading to symptoms of head and neck pain. Despite these interesting hypotheses, there is no substantial evidence to support them. In 1983, Sjasstad and colleagues[5] resurrected the concept of CEH.

Headache was found to be a major complaint (40 percent) in patients with cervical disease such as cervical spondylosis.[6] These headaches typically are described as unilateral, occipital in location, sometimes spreading to the anterior head region, and often aggravated by head or neck movements. Since patients with cervical spondylosis often are asymptomatic, an absolute connection between headaches and cervical disease cannot be made.[7,8] Other structural possibilities for CEH include craniovertebral anomalies such as Arnold-Chiari malformation, atlantoaxial dislocation in rheumatoid arthritis, traumatic subluxation, Paget's disease of the skull, and primary or secondary tumors.[9]

▶ NEUROANATOMY AND PHYSIOLOGY

The posterior branch of the C2 root makes up the greater occipital nerve. In addition, the C2 root innervates the C1–2 joint.[10] Clinically, the use of local anesthetic blockade of the greater occipital nerve promotes pain relief in the neck and frontal region of the head.

The C3 root constitutes the lesser occipital nerve. This nerve innervates the C2–3 joint and supplies the semispinalis capitus muscle, providing cutaneous innervation to the posterior scalp.[11] Along its path, the lesser occipital nerve crosses the C2–3 zygapophyseal (facet) joint, to which it supplies sensory innervation. Dysfunction at this joint is frequently an important source of neck pain. Maigne and colleagues[12] wrote about intervertebral dysfunction of the "mobile segment," which includes the facet joint, the intervertebral disk, and the ligament. Mechanical dysfunction of this segment potentially leads to referred pain with possible skin changes to the specific dermatome and associated myotome.

There is an intimate relationship between the sensory fibers from the upper cervical systems and the trigeminal systems as they converge on the brain stem, and this ultimately leads to activation of the trigeminal nucleus caudalis at the level of C2–4.[13] The fibers from the ophthalmic division of the trigeminal nerve have the greatest degree of overlap with the upper cervical roots. This may explain the frontal head pain often seen with symptomatic upper cervical disease. Activation of the trigeminal nucleus caudalis will result in pain referred to the posterior head and neck region.

The motor fibers to the trapezius muscle originate in the spinal accessory nerve nucleus located in the ipsilateral ventral horn of C1–5. A study by Bansevicius and Sjaastad[14] described the electromyographic (EMG) amplitudes from the trapezius muscle on the symptomatic side to be higher before and during the test compared with the asymptomatic side. The examiner often palpates spasm of the trapezius muscle in patients with CEH.

▶ ARE THERE CRITERIA TO HELP THE CLINICIAN DIAGNOSE CEH?

Sjaastad and colleagues[15] proposed criteria to help clinicians diagnose CEH. Patients are predominantly female, usually with a prior history of head or neck trauma.

In contrast to CEH without injury, some authors suggest that postwhiplash CEH can be provoked from the asymptomatic cervical region as well as the symptomatic side.[16]

The criteria describe the pain as a unilateral, mild-to-moderate-intensity, nonthrobbing headache triggered by head movement or sustained awkward head positioning. The dominant pain is side-locked unilaterally, extending to the oculofrontotemporal region and possibly migrating to the ipsilateral shoulder and arm. In addition, there is a reduction in the cervical range of motion. Transient but complete pain relief can be achieved by an anesthetic blockade of the greater occipital nerve or the C2 region. Less important criteria include nausea, vomiting, dizziness, phonophobia, photophobia, blurred vision and difficulty swallowing.

▶ HOW DOES THE CLINICIAN DIFFERENTIATE CEH FROM OTHER HEADACHE CONDITIONS?

There is considerable overlap between CEH and other headache disorders, such as migraine and tension-type headaches, making a pinpoint diagnosis of cervicogenic headaches based on a detailed history frustrating and difficult.[17] Sjasstad and colleagues[18] reported that 20 percent of patients with cervicogenic headache fulfill criteria for migraine headache.

For example, migraineurs usually present with unilateral headache, photophobia, phonopobia, blurred vision, dizziness, nausea, or vomiting. It has been well documented that migraine and tension-type headache sufferers also potentially have trigger points on the ipsilateral cervical region. However, the head pain in CEH can be induced by neck movement more frequently than in migraines.[19] A retrospective analysis by Kaniecki and colleagues[20] looked at the prevalence and clinical characteristics of neck pain in migraine. Of the 144 patients who met International Headache Society (IHS) criteria for migraine, 75 percent reported neck pain. Forty-three percent described the neck pain as bilateral and 57 percent as unilateral. The neck pain occurred during all the phases of migraine attack, including the prodrome, headache, and postheadache phases.

In migraines, the headaches frequently begin anteriorly. Tension-type headaches tend to be diffuse. In contrast, CEHs always begin in the occipital nuchal region, and then possibly refer to the anterior portion of the head.[21,22]

In general, the examiner should perform a detailed neurological and musculoskeletal examination, including range of motion (ROM) of the cervical spine and identifying tender points along the cervical area, temporal area, temporomandibular joint, and occipital nerve region.

There have been techniques described to help differentiate CEH from other headache disorders. Maigne[23] explained the pinch and roll test, where the skin is first rolled over the jaw and into the C2 and C3 dermatomal territory at the angle of the jaw. In patients with painful intervertebral dysfunction of the upper cervical spine, he described an abnormal texture of the skin known as *cellulargia*. Pain on palpation of the C2 facet joint in CEH patients also can be seen in other benign headache conditions. Jaeger[24] discussed possible techniques to evaluate head posture, position, and tenderness of the transverse process of the atlas, certain neck movements, and trigger points in patients with CEH. These tests have not been validated for this condition and, therefore, should not be considered diagnostic.

Imaging studies, such as plain radiographs of the cervical spine, have not proven to be helpful unless there is a history of head/neck trauma to rule out fracture; or there is a recent past medical history or suggestion of malignancy. Coskun and colleagues[25] evaluated magnetic resonance imaging (MRI) of the cervical spine in 22 patients with CEH compared with control patients. They concluded that MRI of the cervical spine in patients with CEH compared with controls did not demonstrate significant differences and, therefore, should not be considered an adequate method to detect pathologic findings. Despite a paucity of literature, MRI of the cervical spine should be considered when first evaluating a patient with CEH to document structural disease or dysfunction.

▶ CERVICOGENIC HEADACHE DIAGNOSTIC CRITERIA

Localization of Pain

Primarily unilateral pain is seen that is localized to the neck and occipital region and projects to the forehead, orbital region, temples, vertex, or ears.

Examination Findings

On examination, there should be resistance to, or limitation of, passive neck movements; changes in neck muscle contour, texture, or tone; and tenderness to palpation at the C2–3 facet joint.

Imaging Studies

Imaging must be performed to rule out arthritis or other distinct pathology, such as an anomaly of the craniovertebral junction, as in Arnold-Chiari malformation or malignancy. If imaging does not reveal other known pathologic causes for the head pain, an occipital nerve blockade with a local anesthetic should demonstrate transient but total relief.

Other Criteria

In order to make a definitive diagnosis of CEH, the patient should not fulfill IHS criteria for another benign headache disorder.[8]

▶ CONSERVATIVE THERAPEUTIC INTERVENTIONS

Modalities

In clinical practice, patients with CEH initially receive conservative interventions. Local heat, transcutaneous electrical nerve stimulation (TENS), acupuncture, and manipulation are often considered therapies helpful for CEH headache. Unfortunately, studies have been sparse and not of good quality.

TENS of the nerve or muscle can help to increase ROM, strength, and circulation and can provide relief from muscle spasm or pain.[26] The notion of "cracking my neck" is often frightening to some patients. Obviously, the recommendation is to have the patient evaluated by a physician to rule out structural pathology. Once this has been completed, manipulative therapy, by a competent clinician, can be quite helpful. One must realize that manipulative therapy can, although rarely, lead to stroke from carotid artery or vertebral dissection.[27]

Two systematic reviews evaluating the role of manipulative therapy showed possible short-term benefit.[28,29]

A randomized, controlled trial of exercise and manipulative therapy, done by Jull and colleagues[30] showed some promise for these techniques in patients with CEH. A limitation of the study was the inability to blind the patient or the therapist. For a period of 6 weeks, the patient either received manipulative therapy with both low-velocity techniques for cervical mobilization and high-velocity techniques for cervical joint disorders, or therapeutic exercise in which cervicocranial flexion exercises were performed. A third group of patients was assigned to a combination of manipulative and exercise therapy. Results showed significant improvement across all three interventional groups as compared with controls, with the best results in the combined manipulative and exercise therapy group. This trial showed a significant reduction in headache frequency and intensity, as well as neck pain, when manipulative therapy was performed. In a 12-month follow-up, patients continued to report significant relief of symptoms.

Acupuncture is one of the oldest forms of recorded medical therapy, with documented cases going back more than 4000 years. Puncturing the skin with disposable needles along certain channels or meridians produces an increase in endorphin release. However, the underlying mechanisms continue to be controversial. Nabeta and Kawakita[31] compared acupuncture with sham acupuncture on tender points in volunteers with chronic pain in the neck and shoulder. They were treated once a week for 3 weeks. The results showed short-term effectiveness in the true acupuncture arm but did not show long-term superiority over sham acupuncture. One criticism of the study is the short duration of treatment. Most of the studies on acupuncture suffer from significant methodologic flaws, and the benefits are modest at best.

Pharmacologic Therapies

There is a paucity of clinical pharmacotherapeutic trials for CEH. Only one study by Bovim and Sjaastad reported 13 patients with CEH not responsive to ergots.[32,33] More studies are needed to determine if CEH potentially could respond to antimigraine agents because migraines have a similar clinical presentation and share some underlying mechanisms. Nonsteroidal anti-inflammatory drugs (NSAIDs) have been used on the basis of inflammation. Medications such as muscle relaxants have been employed when a muscle is in apparent spasm. If the symptoms are associated with a sleep disturbance, tricyclic antidepressants may be considered. Other agents include anticonvulsants and opioids. The usual recommendation is to provide such pharmacotherapies for a 2-week trial and then to reevaluate.

Injection Therapies

Travell and Simons[34] popularized the concept of a myofascial trigger point. A *trigger point* is defined as a focus of hyperirritability in a tissue that, when palpated, is locally tender and, if sufficiently hypersensitive, gives rise to referred pain and tenderness. On EMG testing, these active trigger points show spontaneous bursts of activity. The cervical and shoulder muscles commonly responsible for myofascial pain include the suboccipital, trapezius, levator scapuli, rhomboid, and supra- and infraspinatus muscles. A noninjection technique called "spray and stretch" can be performed, whereby the patients' skin is sprayed with vapor coolants at the tender points followed by gentle stretching and relaxation exercises. Trigger point injections usually combine local anesthetics with or without corticosteroids. Depending on the number of muscles and trigger points involved, varying numbers of injections can be administered in multiple sessions.[35]

Blockade of the greater occipital nerve with a local anesthetic such as lidocaine or bupivacaine is a simple technique that can be performed at the bedside. The headache should be resolved completely within minutes. Pollmann and colleagues[36] reported that about 66 percent of patients with CEH had relief following occipital block. Anthony and colleagues[37] demonstrated complete relief in 169 of 180 patients with CEH. Gawal and colleagues[38] demonstrated greater pain relief using occipital blockade in patients with CEH compared with patients with other headache disorders such as migraine. However, paroxysmal hemicrania and hemicrania continua usually do not respond to such a blockade.

Epidural steroid injections often are considered when there is a lack of response to occipital nerve or facet joint blockade. Despite some pain relief in the short term, results over the long term failed to demonstrate statistical significance in the CEH population.[39,40]

Bovim and colleagues[32] reported that if an occipital blockade provided relief, a C2–3 block usually was not necessary. However, if the occipital block is not helpful, then a C2–3 block should be considered. One study revealed equal benefit from greater occipital nerve and C2–3 nerve blockade in CEH. Slipman and colleagues[42] demonstrated relief from a single diagnostic C2–3 block in patients with postwhiplash injury.

Radiofrequency neurotomy of the superficial medial branch of the C3 sinuvertebral nerve or ablation of the C3–4 medial branches have shown some promise in patients with intractable CEH. Blume[43] found relief following radiofrequency of the sinuvertebral nerves to the upper cervical disks in patients with CEH.

Botulinum toxin is a neurotoxin that inhibits the release of acetylcholine from presynaptic nerve terminals, resulting in muscle relaxation. Our recent understanding of its mechanism of pain relief is by decreasing the external stimulation (muscle contraction) by reversible chemical denervation. This leads to a diminished input to the upper cervical sensory afferents, inhibiting the potential convergence onto the trigeminal nucleus caudalis.[44]

▶ CONCLUSION

In general, clinicians must think of CEH as a syndrome rather than a single disease entity.[5,10] There continues to be some controversy as to the specific criteria for diagnosing CEH owing to overlap with other headache disorders, such as migraine. A better understanding of the pathophysiology of CEH will allow clinicians to move toward better diagnostic testing and improved therapeutic options for their patients.

REFERENCES

1. D'Amico D, Leone M, Bossone G: Side-locked unilaterality and pain localization of long-lasting headaches: Migraine, tension-type headache and cervicogenic headache. *Headache* 1994;34:526–530.
2. Nilsson N: The prevalence of cervicogenic headache in a random population sample of 20–59-year-olds. *Spine* 1995; 20:1884–1888.
3. Barré J: Sur une syndrome sympathique cervical posterieur et sa cause frequente. *Rev Neurol* 1926;33:1246–1248.
4. Bartschi-Rochaix W: *Migraine cervicale: Das Encephale Syndrom nach Halwirbelttrauma*. Bern: Medizinischer Verlag Hans Huber, 1949.
5. Sjaastad O, Saunte H, Hovdal, et al: Cervicogenic headache: A hypothesis. *Cephalalgia* 1983;3:249–256.
6. Peterson D, Austin G, Dayes L: Headache due to discogenic disease of the cervical spine. *Bull La Neurol Soc* 1975;40:96–100.
7. Brain W, Northfield D, Wilkinson W: The neurological manifestations of cervical spondylosis. *Brain* 1952;75:187–225.
8. Edmonds JG: Disorders of the neck: Cervicogenic headache, in *Wolff's Headache*. Oxford University Press: Oxford: 2001;447–458.
9. McCrae DL: Bony abnormalities at the craniospinal junction. *Clin Neurosurg* 1969;16:356–375.
10. Sjaastad O, Fredricksen T, Pfaffenrath V: Cervicogenic headache: Diagnostic criteria. *Headache* 1998;38:442–345.
11. Drottning M: Cervicogenic headache after whiplash injury. *Curr Headache Rep* 2003;
12. Chou LH, Lenrow DA: Cervicogenic headache. *Pain Phys* 2002;5:215–225.
13. Sjaastad O, Fredrickson T, Pfaffenrath V: Cervicogenic headache: Diagnostic criteria. *Headache* 1990;30:725–726.
14. Mark BM: Cervicogenic headache differential diagnosis and clinical management: Literature review. *Cranio* 1990; 8:332–338.
15. Kaneicki et al: Poster presentation at 10th IHS Meeting, 2001.
16. Leone M, D'Amico D, Mosschiano F, et al: Possible identification of cervicogenic headache among patients with

migraine: An analysis of 374 headaches. *Headache* 1995;35:461–464.
17. Leone M, D'Amico D, Grazzi L, et al: Cervicogenic headache: A critical review of the current diagnostic criteria. *Pain* 1998;78:1–5.
18. Maigne R: Signes cliniquesdes cephalees cervicales: leur traitment. *Med Hygiene* 1981;39:1174–1195.
19. Jaeger B: Are cervicogenic headaches due to myofascial pain and cervical spine dysfunction? *Cephalalgia* 1989;9:157–164.
20. Coskun O, et al: Magnetic resonance imaging of patients with cervicogenic headache. *Cephalalgia* 2003;8:842–845.
21. Anthony M: Headache and the greater occipital nerve. *Clin Neurol Neurosurg* 1992;94:297–301.
22. Bogduk N: The anatomical basis for cervicogenic headache. *J Manip Physiol Ther* 1992;15:67–70.
23. Meloche JP, et al: Headache: Painful intervertebral dysfunction. Robert Maigne's Original Contribution to Headache of Cervical Region. *Headache* 1993;33(6):328–334.
24. Kerr FWL: Structural relation of the trigeminal tract to the upper cervical roots and the solitary nucleus of the cat. *Exp Neurol* 1961;4:134–148.
25. Bansevicius D, Sjaastad O: Cervicogenic headache: The influence of mental load on pain level and EMG of shoulder-neck and facial muscles. *Cephalalgia* 2003;23:842.
26. Tan JC (ed): *Practical Manual of Physical Medicine and Rehabilitation.* St Louis: Mosby, 1998.
27. Fabio RP: Manipulation of the cervical spine: Risks and benefits. *Phy Ther* 1999;79:50–65.
28. Vernon HT: Spinal manipulation and headaches of cervical origin. *J Manip Physiol Ther* 1989;12:455–468.
29. Hurwitz EL, et al: Manipulation and mobilization of the cervical spine: A systematic review of the literature. *Spine* 1996;21:1746–1759.
30. Jull G, et al: A randomized, controlled trial of exercise and manipulative therapy in cervicogenic headache. *Spine* 2002;17:1835–1843.
31. Nabeta T, Kawakita K: Relief of chronic neck and shoulder pain by manual acupuncture to tender points: A sham-controlled, randomized trial. *Complement Ther Med* 2003; 10:217–222.
32. Bovim G, Sjaastad O: Cervicogenic headache: Responses to nitrogylcerin, oxygen, ergotamine and morphine. *Headache* 1999;33:249–252.
33. Fishbain DA, et al: Do the proposed cervicogenic headache diagnostic criteria demonstrate specificity in terms of separating cervicogenic headache from migraine? *Curr Headache Rep* 2003;2:165–172.
34. Travel JG, Simons DG: *Myofascial Pain and Dysfunction: The Trigger Point Manual.* Baltimore: Williams & Wilkins, 1992.
35. Grabois, et al. (eds): *Physical Medicine and Rehabilitation: The Complete Approach.* Malden, MA: Blackwell Science, 2000.
36. Pollman W, Keidel M, Pfaffenrath S: Headache and the cervical spine: A critical review. *Cephalalgia* 1997;17:801–816.
37. Anthony M: Cervicogenic headaches: Prevalence and response to local steroid therapy. *Clin Exp Rheumatol* 2000;18S:59–64.
38. Gawal MJ, Rothbart PJ: Occipital nerve block in the management of headache and cervical pain. *Cephalalgia* 1992;12:9–13.
39. Marteletti P, Di Sabato F, Granata M, et al: Epidural corticosteroid blockade in cervicogenic headache. *Eur Rev Med Pharmacol Sci* 1998;2:31–36.
40. Marteletti P, DiSabato F, Granata M, et al: Failure of long-term results of epidural steroid injections in cervicogenic headache. *Eur Rev Med Pharmacol Sci* 1998;2:10.
41. Inan N, et al: C2/C3 nerve blocks and greater occipital nerve block in cervicogenic headache treatment. *Funct Neurol* 2001;3:239–243.
42. Slipman CW, Lipetz JS, Plastaras CT, et al: Therapeutic zygapophyseal joint injections for headache emanating from the C2–3 joint. *Am J Phys Med Rehabil* 2001;80:182–188.
43. Blume HG: Treatment of cervicogenic headaches: Radiofrequency neurotomy to the sinovertebral nerves, to the upper cervical disc and to the outer layer of the C3 nerve root or C4 nerve root, respectively. *Funct Neurol* 1998;13:83–84.
44. Hobson DE, Gladish DF: Botulinum toxin injection for cervicogenic headache. *Headache* 1997;37(4):253–255.

CHAPTER 30

Spinal Pain: Pathogenesis, Evolutionary Mechanisms, and Management

Anthony H. Wheeler and Daniel B. Murrey

Like a modern skyscraper, the human spine defies gravity and defines us as vertical bipeds. It forms the infrastructure of a biologic machine that anchors the kinetic chain and transfers biomechanical forces into coordinated functional activities. The spine acts as a conduit for precious neural structures and possesses the physiologic capacity to act as a crane for lifting and a crankshaft for walking. Subjected to aging, it adjusts to the wear and tear of gravity and biomechanical loading through structural and neurochemical changes, some of which can be maladaptive and cause pain, functional disability, and altered neurophysiologic circuitry. Some reactions are compensatory, but others are destructive and interfere with the organism's capacity to function and cope. Spinal pain is multifaceted, involving structural, biomechanical, biochemical, medical, and psychosocial influences that result in dilemmas of such complexity that treatment application is often difficult or ineffective.

▶ EPIDEMIOLOGY

Low back pain (LBP) is the "most expensive, benign condition in industrialized countries."[1] Experts have estimated that as many as 80 percent of Americans will experience LBP during their lifetime.[1-4] Early studies suggested rapid recovery of LBP in up to 50–90% of cases in 6 weeks.[1-7] However, more recent evaluations of primary care patients showed that one-third to one-fourth of patients will still be having problems after one year.[4,7] It is the most common cause of disability in persons under age 45.[2,3] Each year, 3 to 4 percent of the population is temporarily disabled, and 1 percent of the working-age population is totally and permanently disabled.[1,4,5] LBP has been cited as second to the common cold as a cause of lost work time,[6] the fifth most frequent cause for hospitalization, and the third most common reason for surgical procedures.[4]

Large population-based surveys of LBP estimate a point prevalence ranging between 15 and 30 percent, a 1-month prevalence between 19 and 43 percent, and a lifetime prevalence between 60 and 70 percent.[7] Between 1971 and 1981, the number of Americans disabled by LBP grew at a rate 14 times the population growth rate.[8]

Four large population surveys performed outside the United States cited a point prevalence for neck pain ranging between 11.5 and 22.2 percent.[7] A random sampling of 10,000 Norwegian adults revealed that more than 34.4 percent of the responders had experienced neck pain during the previous year.[9] The 1-year prevalence of neck pain in a Finnish rural population was 18 percent in women and 16 percent in men.[10] A Swedish study reported a similar 1-year prevalence: 18 percent of the population with neck-shoulder problems (20 percent in women and 16 percent in men) and 12 percent with neck pain alone.[11] Spinal pain and disability have reached endemic proportions in Western culture with enormous socioeconomic consequences.

▶ THE SPINAL PAIN DILEMMA

Most commonly, diagnoses of acute spinal conditions are nonspecific, such as neck or back strain, although injury may affect any pain-sensitive structures, including the disk, facet joints, spinal musculature, and ligamentous support.[12,13] The origin of chronic pain often is attributed to degenerative conditions of the spine; however, controlled studies have indicated little correlation between clinical symptoms and radiologic signs of degeneration.[3,8,13–16] Inflammatory arthropathy, metabolic bone conditions, and fibromyalgia are cited as a cause in some cases of chronic spine-related pain.[12,13] Although disk herniation has been popularized as a cause of spinal and radicular pain, asymptomatic disk herniations on computed tomography (CT), and magnetic resonance imaging (MRI) are common.[15–18] Furthermore, there is no clear relationship between the extent of disk protrusion and the degree of clinical symptoms.[19] Degenerative change, whether related to acute injury or to chronic recurrent trauma, produces a syndrome of varying combinations of axial and limb pain.[13] Activity-related spinal pain is aggravated most often by static loading of the spine (prolonged sitting or standing), by long lever activities (vacuuming or working with the arms elevated and away from the body), or by levered postures (forward bending of cervical or lumbar spine).[13] Pain is reduced when the spine is balanced by multidirectional forces (walking or constantly changing positions) or when the spine is unloaded (reclining); however, a strictly mechanical or pathoanatomic explanation for spinal and radicular pain syndromes is inadequate, and therefore, the role of biochemical and inflammatory factors remains under investigation.[13,19] In fact, it is this failure of the pathology model to predict spine pain that often leads to an ironic predicament. If diagnostic studies are unrevealing for a structural cause, physicians and patients alike often call into question whether the pain has a physical or psychological cause.[13]

The transition from acute tissue damage to a chronic pain state is influenced by both endogenous and exogenous factors, which alter function in the individual far beyond the initiating pathologic process. Chronic spine-related pain may result from impaired tissue healing or persistent pathoanatomic instability. When combined with nonphysical factors, a complex milieu of interwoven physiologic, psychological, and social factors may result. Identification of contributing physical and nonphysical factors enables the treating physician to enact a comprehensive approach with the best chance for success.[13]

▶ ANATOMY

Functional Anatomy

The lumbar spine forms the caudal flexible portion of an axial structure that supports the head, upper extremities, and internal organs over a bipedal stance. The sacrum forms the foundation of the spine through which it articulates with the sacroiliac joints to the pelvis. The cervical and lumbar portions of the spine are capable of supporting heavy loads in relationship to their cross-sectional area.[1] They maintain lordosis in neutral posture and resist an anterior gravitational moment. Unlike the thoracic spine, which is stabilized circumferentially by the ribs, the cervical and lumbar portions of the spine are unsupported laterally and display considerable mobility in both sagittal and coronal planes. The bony vertebrae act as specialized structures for transmitting loads through the spine.[1] Parallel lamellae of highly vascularized cancellous bone form trabeculae, which are oriented along lines of biomechanical stress and encapsulated in a cortical shell. Vertebral bodies become progressively larger in cross-sectional area as gravitational loads increase from the cervical to the lumbar spine.[1] Projections from the lumbar vertebra, the transverse and spinous processes, maintain ligamentous and muscular connections to the segments above and below. Cervical and lumbar extension cause the posterior articulations to be juxtaposed in a close-pack position, which allows optimal load sharing between the intervertebral disk and facet joints and permits minimal rotation. When cervical and lumbar portions of the spine are flexed, however, the posterior articulations are disengaged, with transfer of the load to the anterior column, especially the intervertebral disk. This open-pack position is the most vulnerable for injury.[1,13]

The cervical spine consists of seven cervical vertebrae with multiple articulations. The first cervical vertebra (atlas) articulates with the occiput and permits the head to bob in a vertical plane. The articulation of the atlas and axis allows rotation of the head through approximately 50 percent of its total range, with the remaining rotation occurring between C2 and C7.[1,13] The posterior articulations of each motion segment are the facets. The superior facet at each level faces upward and posteriorly, whereas the inferior facet faces downward and anteriorly. Flexion, extension, and side bending occur with sliding or gliding of these posterior articulations on each other.[1,13] It is rotational movements, however, that play the primary role in cervical segmental, specifically disk, degeneration because the axis of rotation occurs primarily over the disk.[20] The disk absorbs these biomechanical forces, which, in turn, cause cumulative wear and tear. Rotational movements and torsional injury are the usual cause of degenerative change in the disk, which usually succumbs before the posterior articulations and, in the cervical spine, before the uncinate process.[20]

The spinal cord extends from the foramen magnum to the approximate L1 spinal segmental level, where it terminates as the *conus medullaris*. Paired lumbosacral nerves passing distally within the dural sac compose the cauda equina. The posterior portion of the spinal nerve conveys sensory information to the dorsal nerve root through its cell bodies in the dorsal root ganglion, whereas the ventral portion of the nerve root consists primarily of motor efferent nerve fibers. After passage through the intervertebral foramen, the spinal nerve divides into the anterior and posterior rami. The posterior portion of the annulus fibrosis, facet joints, posterior longitudinal ligaments, erector spinae muscles, dura matter, and overlying skin are innervated by the posterior primary ramus, which is principally responsible for transmission of spinal nociception. The ventral ramus innervates muscle and skin that are anterior or lateral to the axis of the vertebral column, as well as the lateral aspect of the diskal annulus fibrosis.[1]

In the cervical spine, facet joints below C2–3 are supplied by medial branches of the cervical dorsal rami above and below the joint, which also innervate deep paramedian muscles. The C2–3 facet joint is supplied by the third occipital nerve.[21] Innervation of the atlantooccipital and atlantoaxial joints is derived from the C1 and C2 roots, respectively.[22,23] Thoracic and lumbar facet joints are innervated by at least two branches of the posterior rami of the spinal nerves, the intervertebral foramen at the same level and an ascending branch from the level below. In some cases, a third branch descends from the level above. In these cases, complete denervation of the segment would require the destruction of all three branches.[24]

Structures within the three-joint complex (composed of the disk and two posteriorly situated facet joints) that are pain-sensitive include nerve roots, dura, posterior and longitudinal ligaments, external annular fibers of the disk, facet joints, joint capsules, and cancellous bone.[25–27] Experiments in which hypertonic saline was injected into pain-sensitive paraspinal muscles and other structures revealed consistent segmental pain referral, as well as local pain patterns.[28,29] Interspinal structures without proven pain innervation include the ligamentum flavum, inner annulus, and nucleus pulposus.[25–27]

Upper cervical spine dysfunction and pain often are associated with craniofacial pain syndromes, including tension-type headache, migraine, occipital neuralgia, and temporomandibular joint dysfunction owing to kinetic chain and anatomic relationships and because of the cephalad distribution of C2 and C3 sensory innervation. Furthermore, widespread, nonspecific patterns of pain referral may result from dural irritation because the cervical spine receives innervation from multiple cervical segmental levels. The diameter of the central canal is the smallest in the thoracic spine; however, patency of the canal is protected by the stability of articulating ribs.[13] Risk of spinal cord ischemia is highest here because of the presence of an avascular watershed area between the radicular arteries and the artery of Adamkiewicz. Consequently, the thoracic spinal cord is most susceptible to neurological injury as the result of degenerative, diskogenic, ischemic, or traumatic causes.[30]

Pathoanatomy: The Degenerative Cascade

The intervertebral disk is composed of the outer annulus fibrosis and the inner nucleus pulposus. The outer annulus inserts into the vertebral body by Sharpies fibers and accommodates nociceptors and proprioceptive nerve endings.[31] The outer annulus provides strength to the disk for bending, whereas the inner annulus encapsulates the nucleus, providing the disk with extra strength during compression.[31] Greater than 70 percent of a compressive load on a healthy disk is borne by the nucleus. The nucleus pulposus constitutes two-thirds of the surface area of a disk and is composed of proteoglycan megamolecules, which are the largest molecules in the human body and capable of imbibing water to a capacity approximately 250 percent of their weight.[31] Until the third decade of life, the nucleus pulposus is composed of approximately 90 percent water; however, over the next four decades, the water content diminishes gradually to approximately 65 percent.[31] Only the outer third of the annulus receives a blood supply from the epidural space. Therefore, nutrition to the disk depends on diffusion of small-molecular substances across the vertebral endplates.[31] Repeated eccentric and torsional loading and recurrent microtrauma result in circumferential and then radial tears in the annular fibers. Tears may be accompanied by endplate separation with subsequent loss of nuclear nutrition. Coalescence of tears may result in radial tears through which nuclear material may migrate or herniate into the epidural space and cause nerve root compression or irritation.[31]

Initially, 80 to 90 percent of the weight of the tri-joint complex is borne across the posterior third of the disk; however, as degeneration causes disk height to decline, the center of biomechanical loading shifts posteriorly, causing a greater percentage of the weight distribution to be borne by the facet joints. Greater stress is accommodated by bone growth (osteophytes) to stabilize the trijoint complex. Hypertrophy of the facets and bony overgrowth of the vertebral endplates results in progressive foraminal and central canal stenosis.[31]

Kirkaldy-Willis described the evolution of spinal degeneration as proceeding from stability to instability and then back to stability. Often, compensatory mechanisms cannot adequately prevent the instability that is created during the early stages of disk degeneration. During this "instability phase," mechanical cervical and lumbar pain and radicular syndromes reach peak incidence.[20,31] Later in the stabilization phase, when bony arthrosis occurs, reduced segmental mobility and increased segmental stability are seen. However, osteophytic hypertrophy, spondylosis, and arthrosis eventually cause progressive reduction of the anteroposterior canal diameter and foraminal patency, which may lead to neural compression (i.e., radiculopathy and myelopathy) and vascular syndromes (i.e., pseudoclaudication and spinal cord ischemia).[31]

▶ ALGOGENIC MECHANISMS

Diskogenic Pain

Multiple studies have been performed that demonstrate the capacity of the vertebral disk and other various structures of the spinal motion segment to cause pain. Kuslich and colleagues[26] used progressive regional anesthesia in 193 patients who were about to undergo lumbar decompressive surgery for disk herniation or spinal stenosis. Pain was reported by 30 percent of patients who had stimulation of the paracentral annulus and by 15 percent who had stimulation of the central annulus with blunt surgical instruments or through an electric current of low voltage.[26]

Mechanical pain syndromes commonly become chronic in the absence of nonphysical or operant influences. This can be explained by identifying a tissue component of the spine processing unique pathophysiologic processes such that it obeys different rules than other connective tissues found elsewhere in the body.[8] Six weeks to 2 months is usually sufficient time for healing to occur in most soft tissue or joint injuries; however, 10 percent of LBP injuries do not resolve in 2 months. This divergent behavior is best explained by the intervertebral disk, which is composed of unique, large, water-imbibing proteoglycan molecules. During adulthood, these large molecules begin to break up into smaller molecules that bind less water. Repair by proteoglycan synthesis is very slow.[8] Fissuring and disruption of annular lamellae further exacerbate the molecular breakdown and dehydration of the disk. The meager blood supply to the peripheral one-third of the outer annulus provides no barrier to subsequent internal degeneration,[31] and the inner annulars and muscles are only able to receive nutrition by diffusion through adjacent vertebral endplates.

Although this sluggish healing may explain chronicity of the spinal lesion, we have already disclaimed a strict relationship between degeneration and spinal pain. Recent elucidation of biochemical behaviors of the disk and other pain-sensitive tissues may account for this discrepancy. Painful disks have a lower pH than nonpainful disks in humans.[32] Diskography on canine disks that are deformed normally and experimentally reveals an increase in concentrations of neuropeptides, i.e., substance P (SP) and vasoactive intestinal peptide (VIP), in the dorsal root ganglion.[33] Inflammatory factors may be responsible for pain in some cases for which epidural steroid injections provide relief. Neuropeptides such as SP, VIP, and calcitonin gene-related peptide (CGRP) exist in capsular and joint nerve fibers of the facets.[34]

CGRP and SP have been purported to be mediators in various stages of arthritis.[34] SP levels correlate with the severity of joint arthritis. SP infusion into joints with milder disease has been reported to accelerate the degenerative process.[35] Furthermore, these chemicals and inflammatory mediators have been linked to proteolytic and collagenolytic enzymes that cause cartilage matrix degradation and osteoarthritis.[34] The release of neuropeptides such as SP, VIP, and CGRP may occur in response to noxious biochemical forces and environmental factors (i.e., biomechanical stress, microtrauma, vibrations), and to the release of inflammatory agents (i.e., cytokines, prostaglandin E_2) and degradative enzymes (i.e., proteases, collagenase).[36]

Phospholipase A_2 (PLA_2), which plays a role in numerous models of inflammation, is elevated in surgically extracted samples of human herniated disks.[37,38] It has been hypothesized that PLA_2 also may play a dual role, inciting disk degeneration and sensitizing annular nerve fibers.[38] A nociceptive role for nitric oxide (NO) in diskogenic pain syndromes is under investigation.[13,39] NO is elevated with human disk herniations and when hydrostatic pressure of the disk is increased due to biomechanical stressors.[39] NO inhibits proteoglycan synthesis in cells within the nucleus pulposus, which leads to proteoglycan loss, reduced water content, and disk degeneration.[39]

It is unclear whether these biochemical changes that occur with disk degeneration are the consequence or cause of these painful conditions. However, chemical and inflammatory factors may create the environmental substrata through which biochemical stress and forces cause variable degrees in the character of axial or limb pain.[13]

Radicular Pain

The pathophysiology of spinal radicular pain is a subject of ongoing research. Proposed etiologies include neural compression with dysfunction, vascular compromise, inflammation, and biochemical influences. Spinal nerve roots have unique properties that may explain their proclivity to produce symptoms.[8] Unlike peripheral nerves, spinal nerve roots lack a well-developed intraneural blood-nerve barrier, which probably makes them more susceptible to symptomatic compression injury than peripheral nerves and more vulnerable to endoneural edema formation.[8,40,41] Endoneural edema can be induced by increased vascular permeability caused by mechanical nerve root compression.[40] Furthermore, elevated endoneural fluid pressure, caused by intraneural edema, can impede capillary blood flow and may cause intraneural fibrosis.[35] Also, spinal nerve roots receive approximately 58 percent of their nutrition from surrounding cerebral spinal fluid (CSF).[8,40,41] Perineural fibrosis, which interferes with CSF-mediated nutrition, has been shown to render nerve roots hyperesthetic and more sensitive to compressive forces.[8,40,41]

Research has elucidated multiple vascular mechanisms that are capable of producing nerve root dysfunction. Experimental nerve root compression shows that venous blood flow can be stopped at low pressures, i.e., 5 to 10 mm Hg, in some.[40] The occlusion pressure for radicular arterioles is significantly higher, approximating the mean arterial blood pressure and showing correlation with systolic blood pressure, creating a greater potential for venous stasis.[40,41] Some investigators postulate that venous and then capillary stasis causes congestion that may contribute to symptomatic nerve root syndromes.[40,41] Nerve root ischemia or venous stasis also may generate pathologic biochemical changes that cause pain, as opposed to the usual progressive sensory and then motor dysfunction typically seen with peripheral nerve compression.[40] Experimentally induced ischemia by nerve root compression has shown that compensatory nutrition from CSF diffusion during low-pressure radicular compression is probably inadequate when epidural inflammation or fibrosis is present.[8,40] Rapid onset of neural and vascular compromise is more likely to produce symptomatic radiculopathy than slow or gradual mechanical deformity.[41–44]

Research efforts have revealed other possible causative mechanisms for symptomatic radiculopathy. A 1987 animal study showed that autologous nucleus pulposus placed in the epidural space of dogs produced a marked epidural inflammatory reaction that did not occur with comparison to saline injections.[45] Similar studies have shown myelin and axonal injury, as well as reduced nerve conduction velocities, in nerve roots exposed to autologous nucleus pulposus.[41,46,47] A recent study suggests that experimental radicular exposure to degenerative nucleus pulposus and annulus fibrosis does not produce similar changes in nerve root function.[48,49] Research to find the cause of diskal irritation has been directed primarily at the proteoglycan components of the nucleus pulposus; however, cells of the nucleus pulposus are capable of inducing local neural dysfunction[41,48] and generating algogenic agents such as metalloproteases, i.e., collagenase and gelatinase, as well as interleukin 6 and prostaglandin-E_2.[50] Other biochemical substances, including tumor necrosis factor (TNF), have been implicated as causative.[50] TNF increases vascular permeability and appears capable of inducing neuropathic pain.[41,51–53] When injected into nerve fascicles, TNF produces changes that are characteristic of those seen when nerve roots are exposed to the nucleus pulposus.[41,52–55] In addition, scientists have long questioned whether an autoimmune response occurs when nucleus pulposus is exposed to the systemic circulation because it is sequestered by the annulus fibrosis and may not be recognized as normal by the immune system.[41] Indeed, research to date suggests that the cause of symptomatic radiculopathy is multifactorial and more complex than just neural dysfunction from structural impingement.

Facet Joint Pain

The superior and inferior articular processes of adjacent vertebral laminae join to form the facet or zygapophyseal joints, which are paired diarthrodial synovial articulations that share compressive loads and other biomechanical forces with the intervertebral disk.[56,57] Like other synovial joints, the facets react to trauma and inflammation by manifesting pain, stiffness, and dysfunction and secondary muscle spasm leading to joint stiffness and degeneration. This process is borne out, as described, through the degenerative cascade of the three-joint complex. Indeed, numerous radiologic and histologic studies have shown that diskal and facet degeneration are mutually linked and that over time, degeneration of the segment leads to osteoarthritis of the facets.[56,58–63]

Studies using provocative intra-articular injection techniques have demonstrated local and referred pain into the head and upper extremities from cervical facets;[64–67] local and referred pain into the upper midback and chest wall from thoracic facets;[68] and local, plus referred, pain into the lower extremity from the lumbar facets.[24,69–74] The fibrous capsule of the facet joint contains encapsulated, unencapsulated, and free nerve endings. Immunohistochemical studies have demonstrated nerve fibers containing neuropeptides that mediate nociception (e.g., SP, CGRP, and VIP).[75,76] SP nerve fibers have been found in subchondral bone and degenerative lumbar facets subjected to aging and to cumulative biomechanical loading. Evidence of nociceptive afferents and the presence of the neuropeptides SP and CGRP in facets and periarticular tissues support a role for these structures as spinal pain generators.[34] Clinical research studies have demonstrated facet pain in 54 to 67 percent of neck pain patients;[77–80] 48 percent of thoracic pain patients;[81] and 15–45 percent of LBP patients.[82–88]

Sacroiliac Joint Pain

The sacroiliac joint is a diarthrodial synovial joint that receives its primary innervation from the dorsal rami of the first four sacral nerves.[89–95] Arthrography or injection of irritant solutions into the sacroiliac joint causes pain provocation with variable local and referred pain patterns into regions of the buttock, lower lumbar area, lower extremity, and groin.[96–98] Using a variety of blocking techniques, reports of the prevalence of sacroiliac pain have been widely variable, with estimations as low as 2 percent and as high as 30 percent in patients evaluated for chronic LBP.[56,82,98,99]

Muscle Pain

Pain receptors in muscle are sensitive to a variety of mechanical stimuli, including pressure, pinching, cutting, and stretching.[100] Pain and injury occur when the musculotendinous contractual unit is exposed to single or recurrent episodes of biomechanical overloading. Injured muscles usually are shortened abnormally with increased tone and tension due to spasm or overcontraction. Injured muscles often meet diagnostic criteria for "a myofascial pain syndrome," a condition originally described by Drs. Janet Travell and David Simon.[101] Myofascial pain (MP) is characterized by muscles that are in a shortened or contracted state with increased tone and stiffness and containing trigger points (TrPs).[100,101] TrPs are tender, firm nodules 3 to 6 mm in size that are identified by palpatory examination of the muscles.[100,101] TrP palpation provokes radiating, aching-type pain into localized reference zones. Mechanical stimulation of the *taut band*, a hyperirritable spot within the TrP, by needling or rapid transverse pressure, often will elicit a localized muscle twitch. Sometimes TrP palpation can elicit a "jump sign," an involuntary reflex or flinching of the patient that is disproportionate to the palpatory pressure applied.[100,101] MP may become symptomatic through direct or indirect trauma, exposure to cumulative and repetitive strain, postural dysfunction, and physical deconditioning.[100] MP can occur at the site of tissue damage, but MP due to neuropathic disorders occurs at sites where pain is referred. Muscles affected by neuropathic pain may be injured due to prolonged spasm, mechanical overload, or metabolic and nutritional shortfalls. The pathogenesis of MP and TrPs remains unproven.

Spontaneous activity and spike discharges have been observed in TrPs when examined by needle or electromyography (EMG).[102,103] A recent review of the neurophysiology of MP cited anecdotal EMG findings supporting the idea that TrPs may represent a microscopic area of focal dystonia.[104] Animal studies have demonstrated that TrPs can be abolished by transsection of afferent motor nerves or infusion of lidocaine; however, spinal transsection above the level of segmental innervation of the TrP-containing muscle does not alter the TrP response.[104] Therefore, research suggests that myofasical dysfunction with characteristic TrPs is a spinal segmental reflex disorder.[104]

Most recently, Simon and colleagues[101] have postulated that abnormally increased production and excessive release of acetylcholine at the neuromuscular junction cause sustained depolarization of the postjunctional muscle cell membrane under resting conditions with persistent shortening of sarcomeres.[101] Continued contractile unit shortening and spasm can cause distortion and damage of involved tissues, which may precipitate the synthesis and release of endogenous algogenic biochemical and inflammatory substances that enhance nociception.[100]

▶ NEUROPHYSIOLOGY

Nociception Due to Spinal Injury

Nociception is the neurochemical process whereby specific nociceptors convey pain signals through peripheral neural pathways to the central nervous system (CNS).[105] In the setting of spinal injury, acute tissue damage to the spinal motion segment and associated soft tissues activates these pathways. Transduction is the process whereby noxious afferent stimuli are converted from chemical to electrical neural messages within the spinal cord that communicate cephalad to the brain stem, thalamus, and cerebral cortex.[105] Mechanical, thermal, and chemical stimuli activate peripheral nociceptors that transmit the pain message through lightly myelinated Aδ fibers and unmyelinated C fibers.[105,106] Nociceptors are present in the nucleus fibrosis facet capsule, posterior longitudinal ligament, associated muscles, and other structures of the spinal motion segment[25–27,106] Algogenic substances that typically are involved in tissue damage and capable of inducing transduction peripherally are several, and include low Ph, potassium, bradykinin, histamine, prostaglandins, leukotrienes, and serotonin, pro-inflammatory cytokines.[106] Transduction leads to transmission, which is the afferent conduction of pain signals to the dorsal root ganglion and dorsal horn of the spinal cord. The dorsal root ganglion contains cell bodies of primary afferent nociceptors and is a rich source of nociceptive chemicals including SP, VIP, and CGRP.[106] The dorsal root ganglion is mechanically sensitive and capable of independent pain transduction, transmission, and modulation. The dorsal horn is the initial major site for nociceptive modulation that occurs in Rexed laminae I, II, and V of the substantia gelatinosa.[105,106] Pain reception at this level is subject to modification by other afferent stimuli arriving at the dorsal horn via Aβ fibers and/or by the descending endogenous opioid system. Subsequent to spinal cord influences, the resulting pain message travels cephalad by several routes to the brain, where further modulation may occur. Pain perception is strongly influenced by psychological and environmental factors, as well as by excitatory and inhibitory neurophysiologic factors. These influences determine the final perception of pain, making each individual's pain experience unique and complex.

Chronic Nociception: Neural Processing and Biochemical Influences

When the peripheral source of pain persists, it is influenced by intrinsic mechanisms that reinforce nociception. The nervous system is capable of enhancing a pain stimulus generated by tissue damage to a level far greater than it signifies as a threat to the human organism, a common clinical scenario in cases of mechanical spinal pain.[107] Peripheral tissue changes produce sensitization of nociceptors so that they respond to mild or normal sensory stimuli, such as light touch or temperature change, in an exaggerated adverse manner (allodynia).[34,108]

Persistent pain may result when nociceptive afferents that project to internuncial neurons in the spinal cord set up abnormal, self-sustaining, reverberating activity within closed neuronal loops.[107-112]

CNS sensitization and chronicity are fostered through afferent processing by second-order nociceptive-specific neurons and wide dynamic range (WDR) neurons in the spinal cord. WDR-type neurons usually contribute to greater sensitivity than nociceptor-specific neurons because nociceptive and non-nociceptive afferents converge to synapse on a single WDR neuron. WDR neurons respond with equal intensity regardless of whether or not the neural signal is noxious (hyperalgesia).[112] Hyperalgesia and allodynia initially develop at the injury site; however, when central sensitization occurs through WDR neural activity, the area of pain expands beyond the initial region of tissue pathology.[112]

Finally, a phenomenon termed *windup* results from repetitive activation of C fibers sufficient to recruit second-order neurons that respond with progressively greater magnitude[111,113-115] and can be blocked by NMDA receptor antagonists.[111] Windup probably also contributes to CNS sensitization, including hyperalgesia, allodynia, and persistent pain.[112]

▶ FACTORS THAT INFLUENCE PAIN PERCEPTION AND CHRONICITY

Psychological Factors

Behavioral, cognitive-affective, and psychophysiologic reactions occur in individuals experiencing pain. For example, pain behaviors exhibited with an acute exacerbation of LBP often include guarded movement, grasping the painful area, verbal expression of pain, reduced activity levels, and body movements such as constant shifting of position or reclining. These behaviors are normal during the acute pain experience but become aberrant if they persist and extinguish normal well-behavior patterns. Furthermore, these abnormal verbal and nonverbal pain behaviors are prone to reinforcement by nonphysical and environmental factors, sometimes resulting in heightened or exaggerated behaviors that appear dissociated from any change in the intensity or character of the pain stimulus.[116]

Preexisting psychological factors frequently influence an individual's risk for injury occurrence and the development of chronic spinal pain.[116-118] Depressed patients are more prone to develop chronic pain or pain as a symptom.[116-123] Symptoms of depression, including psychomotor retardation and cognitive dysfunction, make these patients more prone to injury and more likely to respond poorly to treatment.[116] Somatizing and hypochondriacal patients are more likely to develop pain as a symptom and less likely to respond to treatments aimed at a presumed organic cause.[116,124]

Commonly, personality disorders or traits may influence the character or chronicity of pain. For example, borderline personalities may acquire pain as a way of structuring an otherwise empty existence. Narcissistic patients may acquire pain to seek medical attention as a way of preventing more serious illness or disability. Antisocial personality types are exploitative, adapt easily to game-playing roles, and are prone to complications.[116-118,124-126]

Previous learning and role models can influence treatment outcome. Prognosis may be affected by coping and attribution styles, such as the patient's tendency to catastrophize, overgeneralize, personalize, or selectively attend to negative aspects of the experience.[116,127-131]

Post-traumatic depression usually occurs in cases of prolonged or chronic spinal pain due to loss of physical function, self-esteem, employment, social relationships, and financial security. Anger or hostility directed at the workplace or perceived ineffective medical care may hinder communication with employer, physician, family, and friends.[116] As the length of time from injury increases, the aggregate of preexisting and post-traumatic psychosocial influences is complicated further by avoidance learning and deactivation.[116] As nonphysical elements of the illness accumulate, the probability of a poor prognosis grows (Table 30-1).

Physical Factors

Medical and surgical factors often complicate treatment of the chronic spinal pain patient. Examples include failed spinal surgery or medical problems such as diabetes or heart disease that make the patient a poor rehabilitation or surgical candidate.[116] Other physical factors include the patient's inherent capacity to exercise. In addition, patients may develop "deconditioning syndrome" due to reduced activity that follows injury.[1,14,116] This gradual

▶ TABLE 30-1. FACTORS THAT INFLUENCE PAIN PERCEPTION AND CHRONICITY

- **Psychological influences**
  - **Premorbid factors**
    - Depression
    - Cognitive-affective dysfunction
    - Anxiety/panic disorder
    - Substance abuse
    - At risk personality disorders
    - Predisposed to somatoform disorders
    - Low intelligence
    - Attribution/coping style
  - **Post-traumatic factors**
    - Persistent aberrant pain behaviors
    - Exaggerated or prolonged psychophysiologic reaction and symptoms
    - Depression
    - Panic/anxiety
    - Post-traumatic stress disorder
    - Somatoform pain syndrome
    - Symptom magnification
    - Anger/hostility
    - Loss of control/abnormal dependence
    - Brain injury
  - **Physical factors**
    - Medical illness that interferes with treatment
    - Failed surgery
    - Poor physical capacity for rehabilitation
    - Comorbid neurological or musculoskeletal disorders
    - Deconditioning syndrome
  - **Psychosocial factors**
    - Compensated unemployment
    - Out of work, disabled or seeking disability
    - Job dissatisfaction or conflicts
    - Family or spousal dynamics
    - Legal or adverse insurance influences
    - Age-related factors
    - Environmental stressors
    - Limited education or vocational potential

reduction in muscle strength, joint mobility, and cardiovascular fitness increases over time and may become a self-sustaining and independent component of the patient's musculoskeletal condition[1,14,116] (see Table 30-1).

Psychosocial Factors

The complex psychosocial milieu following a spinal injury is the most refractory and difficult treatment dilemma facing the medical provider in most cases of chronic pain, especially involving disability. Job dissatisfaction or conflict has been well documented as the most important nonphysical risk factor for predicting chronic LBP disability.[132] Furthermore, compensated unemployment may encourage chronicity. The injured patient may be unable to return to a previous strenuous job activity and may face limited alternatives owing to education and vocational potential.[116] Conflict can be created in older individuals with a decreased capacity for work as unemployment and loss of compensation become overriding issues. Family, financial, and legal issues also may contribute to chronicity.[116] The development of chronic spinal pain is a complex mixture of interwoven physiologic, psychological, and social factors that often creates a disease state unique to each individual (see Table 30-1).

▶ DIAGNOSTIC PRINCIPLES

Overview

The natural history of spine-related pain is usually benign. However, prompt physician evaluation, including appropriate radiographic, laboratory, and electrophysiologic testing, is indicated in cases of persistent severe neurological deficit, intractable limb pain, suspicion of a systemic illness, or a change in bowel or bladder control.[133]

In complex spinal disorders with chronic intractable pain and disability, provocative interventional techniques (e.g., diskography and facet injections) can help to identify spinal pain generators through the use of needles or radiologically guided catheters directed at selective spinal structures or neural elements using irritant solutions (e.g., radiopaque dyes) to provoke a characteristic pain pattern, followed by a local anesthetic to ablate it.[56,108,133] When the cause of neck and back pain remains inconclusive after a comprehensive diagnostic evaluation including interventional methods, a multidisciplinary evaluation combining surgical, medical, physical medicine, and psychological subspecialists may help to better define peripheral and central pain generators.[13,108,133]

The Spinal Lesion

The spinal lesion occurs when injury or disease adversely affects the capacity of the vertebral motion segment and its components, both contractile and noncontractile, to maintain normal function and manage normal biomechanical forces owing to impending or actual tissue damage. The magnitude of actual or threatened tissue damage elicits an appropriate or parallel nociceptive stimulus that provokes the organism to behave or react protectively. Injuries typically are caused by singular or recurrent episodes of biomechanical overloading, particularly in cases of body movements performed when spinal and extremity muscles are fatigued, when muscle power is reduced, when truncal protective mechanisms are at fault, when errors occur due to poor neuromuscular coordination or cognitive dysfunction, when tasks have poor ergonomic design, or when movements are nonphysiologic, such as simultaneous bent twisting and lifting. Details of a specific event or trauma allow the practitioner to develop a thorough understanding of the biomechanics of the injury, and therefore, enable him or her to determine the likelihood of a mechanical pathogenesis. Identifying spinal pain and associated neurological symptoms as mechanical is the first most important diagnostic objective.

The Patient History

When interviewing a patient with spinal pain, it is important to establish the portion of the pain that is axial relative to its distribution in the ipsilateral extremity. Cumulative microtrauma that occurs over time and injury-induced macrotrauma to spinal structures produce spinal and extremity pain in various combinations but with similar characteristics. This process of aging and recurrent trauma is thought to cause progressive degeneration of spinal motion segments. Nevertheless, the presence of degenerative changes typically is seen in patients who complain of nociceptive or mechanical pain syndromes. For example, patients with mechanical LBP typically report that their pain is aggravated by static loading of the spine (e.g., prolonged sitting or standing), long lever activities (e.g., vacuuming), or levered postures (e.g., bending forward) and is eased when the spine is balanced by multidirectional forces (e.g., walking) or is unloaded (e.g., reclining). Similarly, neck pain is aggravated when our arms, which are long levers, are supported away from the body with repetitive lifting or work activities, especially when a third-levered weight, the head, is added when the cervical spine is flexed. Mechanical conditions of the spine, including disk disease, spondylosis, spinal stenosis, and fractures, account for up to 98 percent of cases of spinal pain, with the remaining cases due to systemic and visceral disease[133,134] (Table 30-2).

Neuropathic pain often occurs with limb involvement and is characterized as thermal, jabbing, shooting, or like pins and needles, whereas nociceptive pain usually is described as dull and aching and is aggravated by the biomechanical loading (Table 30-3). Although neuropathic pain may increase with exposure to increased biomechanical forces, it is usually worse at night or at rest and is associated with other neurological symptoms, such as reduced sensation or weakness. Determining the character of pain as neuropathic is helpful clinically but does not differentiate among various neuropathic conditions that may be causative.[135] The presence and location of neuropathic pain within the peripheral or central nervous system can be determined by neurological abnormalities found on physical examination. Furthermore, allodynia and hyperalgesia are considered cardinal signs of neuropathic pain.[135,136] Allodynia is pain that results from a stimulus that does not normally evoke pain, whereas hyperalgesia is an exaggerated response to a stimulus that normally is painful.

Following injury, there is a critical period of time when historic recall is optimal and reports of the details of an injury are most explicit. These details should be well documented to establish biomechanical influences, especially in cases of personal or work injury. It is important to request disclosure of any previous similar problems, compensation-related or not. Physician evaluation of chronic or previously treated disorders requires examination of accumulated historical, diagnostic, and treatment data.

After establishing the characteristics and behavior of the pain, any associated symptoms are determined through questioning by review of systems. Any systemic, neurological, or visceral dysfunction and any weight loss, fever, or other suspicious symptoms that would mitigate a mechanical cause should be scrutinized and investigated thoroughly (Table 30-4).

Physical Examination

The purpose of the physical examination is to confirm or determine a working differential diagnosis of the anatomy

▶ **TABLE 30-2.** MECHANICAL OR ACTIVITY-RELATED CAUSES OF SPINAL PAIN

- Motion segment and diskal degeneration
- Myofascial and muscular pain disorders
- Diskogenic pain with/without radicular symptoms
- Radiculopathy due to structural impingement
- Axial or radicular pain due to biochemical or inflammatory reaction to spinal injury
- Motion segment/vertebral osseous fracture
- Spondylosis with/without central or lateral canal stenosis
- Macro- or micro-instability of the spine with/without radiographic hypermobility or evidence of subluxation.

▶ **TABLE 30-3.** SENSORY DESCRIPTIONS OF NEUROPATHIC PAIN AND ASSOCIATED SYMPTOMS

- Paresthesia
  - Tingling, painful (dysesthesiae)
  - Pins and needles
  - Limb is asleep
  - "Bugs crawling" (formication)
- Thermal pain
  - Burning, on fire
  - Icy cold
  - Frost bite
- Paroxysmal
  - Shooting
  - Electric
  - Lancinating
  - Jabbing
- Sharp
  - Raw skin
  - Broken glass—knife-like
- Dull
  - Deep ache
  - Bone pain
  - Toothache pain

▶ **TABLE 30-4.** DISORDERS THAT MAY BE ASSOCIATED WITH NONMECHANICAL SPINAL PAIN

Neurologic syndromes
- Myelopathy myelitis from intrinsic or extrinsic structural or vascular process
- Lumbosacral plexopathy (e.g., diabetes, vasculitis, malignancy)
- Brachial plexopathy (e.g., trauma, autoimmune, postradiation)
- Polyneuropathy, acute, subacute or chronic (e.g. chronic inflammatory demyelinating polyneuropathy, Guillain-Barré Syndrome, diabetes)
- Mononeuropathy, including causalgia (e.g., trauma, diabetes)
- Myopathy, including myositis and various metabolic causes
- Dystonia, cervical, truncal, or generalized

Systemic disorders
- Primary or metastatic neoplasms
- Osseous, diskal or epidural infection
- Inflammatory spondyloarthropathy
- Metabolic bone disease, including osteoporosis
- Vascular disorders such as atherosclerosis or vasculitis

Referred pain
- Esophageal and gastrointestinal disorders (e.g., pancreatitis, pancreatic cancer, cholecystitis)
- Cardiorespiratory disorders (e.g., pericarditis, pleuritis, pneumonia)
- Disorders of the ribs or sternum
- Genitourinary disorders (e.g., nephrolithiasis, prostatitis, pyelonephritis)
- Gynecologic disorders (e.g., ectopic pregnancy, pelvic inflammatory disease)
- Thoracic or abdominal aortic aneurysm
- Shoulder and hip disorders (e.g., injury, inflammation or end stage degeneration of joint and associated soft tissues—tendons, bursae, ligaments)

and etiology of the patient's spinal pain and related symptoms. Observations of verbal and nonverbal pain behaviors suggesting "symptom magnification," inconsistencies of requested tasks during the examination, nonphysiologic findings, and other "red flags" that may suggest contributing psychological factors should be documented. During the examination, it is essential to have the patient disrobed to view the spine. Open-backed gowns leave the physician with only one view of the spine, and often swimming attire is more appropriate for a complete 360-degree inspection of the trunk. Leg-length discrepancy and pelvic obliquity, scoliosis, postural dysfunction with forward head and shoulders, and/or accentuated kyphosis should be noted. Physician preferences vary with regard to the importance that is attributed to testing range of motion; however, just asking the patient to move into forward bending is often the most worthwhile observation.

Inspection of the cervical spine begins with evaluation of head and neck alignment. Head turning or tilt, especially in combination with focal or diffuse muscle hypertrophy, may account for abnormal head posture and indicate the presence of cervical injury with spasm or dystonia. In the absence of pain and abnormal muscular tone, neurological disorders or dysfunction of the craniocervical junction should be considered. Neurological and other disorders can cause craniocervical malalignment (Table 30-5). When testing cervical spine rotation in patients with cervical dystonia or spasmodic torticollis, the head turns with less resistance and with greater range forward in the direction of chin rotation, whereas patients with deformity caused by dysfunction of cervical motion segments due to joint and/or diskal injury or degeneration

▶ **TABLE 30-5.** CRANIOCERVICAL MALALIGNMENT WITH PAIN

- **Orthopedic or musculoskeletal**
  - Acute disk herniation
  - Advanced cervical spondylosis
  - Trauma with fracture or segmental subluxation
  - Asymmetric muscle spasm/myofascial disorder
  - Osteomyelitis, metabolic disorder or other causes of "bone softening"
  - Soft tissue disorder (e.g., hemorrhage, postradiation fibrosis)
- **Infections (e.g., pharyngitis or painful lymphadenopathy)**
- **Neurologic disorders**
  - Primary cervical dystonia
  - Secondary cervical dystonia (e.g., head trauma, encephalitis, metabolic, post-surgical or toxic/drug etiology)
  - Structural disorder of posterior fossa or cerebellum including neoplasm
  - Vestibular disorder
  - Spinal cord tumor or syrinx
  - Focal seizure disorder
- **Psychiatric disorder (conversion)**

often demonstrate marked stiffness and reduced rotation in either direction accompanied by substantial pain, muscle spasm, and guarding.

Observations with the patient standing should include body habitus, posture, stance, pelvic obliquity, spinal curvature, and a general overview of the contour, strength, and symmetry of truncal musculature. Next, the patient is asked to drop his or her head and shoulders forward and then to move slowly into forward bending. Normal forward bending is revealed when the patient recruits from each cephalad segment to the level below and so on, progressing from the cervical spine through the thoracic to the lumbar region, where flexion of the hips completes the excursion into full flexion. Patients with significant mechanical back pain or lumbar segmental instability usually stop cephalad to caudal recruitment on reaching the thoracolumbar junction or, in some cases, the involved lumbar segmental level. In order to continue forward bending, they then brace the lumbar muscles to protect the mechanically compromised segment en masse and then complete forward bending through hip flexion. To rise back to an erect posture, recruitment occurs in the same manner but reversed, beginning with motion through the hips, followed by bracing the lumbar spine, and then continuing cephalad from the cephalad segmental level above the muscularly braced lumbar region. Movement may stop or become "shuddering" while surrounding muscles contract to brace the painful segment(s). When the area is sensed as secure by spinal mechanisms, cephalad segmental recruitment again commences, completing the spine's excursion to the erect posture.[134]

In cases of severe mechanical thoracolumbar pain and segmental instability with regional muscular spasm, the patient may not be able to demonstrate any flexion below the symptomatic spinal level. Any soft tissue abnormalities and tenderness to palpation should be recorded. Palpation of lumbar paraspinal, buttock, and other regional muscles should be performed early in the examination. The examiner should palpate pertinent truncal and limb-girdle muscles and make note of areas with superficial and deep muscle spasm and TrPs, as well as any characteristic pattern(s) of referred pain.

Disassociation of physical findings from physiologic or anatomic principles is key in those patients in whom psychological factors are suspected to be influential. Examples of this phenomenon include nondermatomal patterns of sensory loss, nonphysiologic demonstration of weakness (give-way weakness when not caused by pain or ratchety weakness related to simultaneous agonist and antagonist muscular contraction), and disassociation between lumbar spine movements during the history or counseling sessions and during the examination.

Waddell signs have been popularized as a physical examination technique to identify patients who have "nonorganic" or psychogenic embellishment of their LBP syndrome. Examination techniques proposed by Waddell consist of a series of maneuvers that would not normally cause pain, including simulated rotation of the hips en masse with the lumbar spine without allowing spinal rotation, light pressure on the head, gentle effleurage of superficial tissues, disassociation between sitting and supine straight-leg raising, and demonstration of nonphysiologic weakness or sensory patterns by the patient.[137]

Straight-leg raising (SLR) is used commonly to determine the presence of sciatic nerve tension or irritability when patients experience lower extremity radicular symptoms. SLR is tested with the patient supine and should produce ipsilateral leg pain between 10 and 60 degrees to be declared positive. SLR that produces pain in the opposite leg signifies a high probability of disk herniation, and CT or MRI investigation should be considered, especially if neurological evidence for radiculopathy is present. Nonspecific complaints, overtly excessive pain behavior, patient contraction of antagonist muscles that limit the examiner's testing, and tightness of buttock and hamstring muscles are commonly mistaken examples of reported "positive" SLR. Reverse SLR, tested with the patient prone, may elicit symptoms of pain by inducing neural tension on irritated or compressed nerve roots in the middle to upper lumbar region. Additionally, this maneuver helps the clinician identify the presence of iliopsoas and/or quadriceps muscular tightness, which can contribute to chronic lumbar pain.

Neurological evaluation is performed to determine the presence or absence and level(s) of radiculopathy or myelopathy. Anatomic localization is derived through muscle and reflex testing combined with historical information obtained during the interview and coupled with the absence of neurological symptoms or signs that would implicate cerebral or brainstem involvement. Consistent myotomal weakness and sensory findings that at least "seem to coincide" with segmental radiculopathy or polyradiculopathies should not be ignored.

Lower motor neuron versus upper motor neuron syndromes and the level of spinal dysfunction should be identified by the spinal specialist. For example, hyperreflexia in caudal spinal levels with reduced or absent reflexes in the upper extremities suggests a cervical cord level of dysfunction and implicates both radicular and spinal cord involvement. Rectal examination is indicated in patients in whom myelopathy, especially cauda equina syndrome, is of diagnostic concern. Tone of the anal sphincter and the presence or absence of anal wink should be correlated with motor, sensory, and reflex findings in these patients. In all spinal examinations, a general overview of the patient's health must be confirmed by examination. Extremities affected by chronic pain may demonstrate abnormal skin with a rough, leathery texture or shiny skin with trophic changes, including hair loss, edema, abnormal temperature, and discoloration (bluish, reddish, or brownish hues). These changes may confer the presence of chronic pain, sympathetic nervous system involvement, or vascular insufficiency. Knowledge of cardiovascular and peripheral vascular status obtained by examination is pivotal in cases of claudication or reduced exercise tolerance for determining a diagnostic and treatment plan.

Pain Measurement

Because pain is a sensory experience, an objective means to quantify the patient's perception of pain intensity and its impact is a necessary component of the evaluation to determine diagnostic and treatment strategies and to measure treatment outcome. Several measures of pain and disability have been validated in the medical literature and have been advocated by spine specialists for use in the evaluation and treatment of spinal disorders.[138] These instruments include pain drawings,[139,140] the McGill Pain Questionnaire,[141] the Visual Analogue Scale,[142,143] the Verbal Analog Scale,[142,143] the Ten Point Numerical Rating Scale,[142,143] the Oswestry Disability Questionnaire,[144] the Pain Disability Index,[145] and the Rowland-Morris Disability Scale.[146] These psychometric instruments are helpful for measuring functional disability and various aspects of the generalized or LBP experiences.

Questionnaires have been developed for more specific measurement of cervical pain and dysfunction, including the Neck Disability Index (NDI), the Copenhagen Neck Functional Disability Index, the Northwick Park Scale, the Neck Pain and Disability Scale (NPDS), and the Patient-Specific Function Scale (PSFS).[147] The NPDS has been validated in the medical literature and shows high internal consistency with four underlying dimensions, including dysfunctional or disabling neck problems, pain intensity, an emotional or affective dimension, and interference in life activities.[148,149] A recent review of these measures that did not include final validation of the NPDS suggested the NDI as most frequently cited and best validated for prospective studies until further studies comparing all instruments (head to head) can be performed.[147] On the other hand, Pietrovon and colleagues[147] suggested that the PSFS was most responsive to even small but important changes between patients and, therefore, might best be suited for clinical decision making among individuals.

Pain questionnaires used for initial patient evaluations are helpful for making individual decisions, especially when dealing with chronic spinal pain. We also advocate use of the Beck Depression Inventory and the Dallas Pain Questionnaire as effective and easy screening devices for detecting the presence and degree of expected psychopathology usually seen with chronic back pain.[150] Furthermore, some of these psychometric instruments should be considered as requisite for clinical spinal research studies, including the Medical Outcome Study 36-Item Short-Form Health Survey (SF-36) scale, which is designed to measure overall quality of life (Table 30-6).

▶ **TABLE 30-6.** PSYCHOMETRICS FOR SPINAL PAIN

Low Back Pain	Clinical Management	Research
Pain Drawing	X	
Visual Analogue or Numeric Scale	X	X
Oswestry Disability Questionnaire	X	X
Pain Disability Index	X	
Dallas Pain Questionnaire	X	
McGill Pain Questionnaire	X	
Rowland-Morris Disability Index	X	X
Beck Depression Inventory	X	X
Global Assessment Scale[1]	X	X
SF-36[2]		X
Neck Pain		
Neck Disability Index	X	
Copenhagen Neck Functional Disability Scale	X	
Northwick Park Scale	X	
Patient-specific Functional Scale	X	
Neck Pain and Disability Index	X	X
Global Assessment Scale[1]	X	X
SF-36[2]		X

[1]Global Assessment Scale asks patient to estimate percent of improvement or worsening since treatment.
[2]Medical Outcome Study 36-Item Short Form Health Survey.

Diagnostic Strategies

When mechanical spine pain syndromes persist for 6 to 12 weeks despite adequate treatment, appropriate consultation and diagnostic imaging are indicated. Referral to a physician with expertise in spinal disorders should be considered before embarking on an expensive diagnostic workup. In such cases, when an underlying serious etiology is suspected or progressive neurological deficit is present, appropriate consultation and diagnostic workup are in order. CT scanning is usually most effective when spinal or neurological level is clear and when bony pathology is suspected.[151] MRI is more useful when the exact spinal or neurological level is unclear, when a pathologic condition of the spinal cord or soft tissues is suspected, when an underlying infectious or neoplastic process is possible, or when an acute osseous process affecting bone is suspected.[108] Contrast-enhanced MRI is more accurate for delineating recurrent postoperative diskal herniation.[108] Myelography is useful to clarify nerve root pathology, particularly in patients with previous spine surgery, degenerative scoliosis, dynamic neural compression, or a metal fixation device in place.[108] CT-myelography may be more informative for studying patients who have had multiple spinal operations or who have spinal stenosis.[108,151]

When limb pain predominates and imaging studies provide ambiguous information, clarification may be gained by electromyography (EMG), somatosensory evoked potentials (SSEPs), or selective nerve root blocks.[108] When axial and limb pain become chronic, a multidisciplinary evaluation may provide prognostic insight by uncovering unrecognized physical and psychosocial influences that may be contributing to prolonged pain and disability.[108] Refractory spinal pain may be investigated using provocative interventional radiologic techniques to identify which tissues are pain generators. Diskography has remained controversial as a diagnostic procedure but is still advocated by many to confirm a suspected diskogenic source of pain.[152] Other commonly accepted indications include assessment for surgical fusion, failed spine surgery, or percutaneous disk treatments.[152]

▶ NONOPERATIVE TREATMENT

Overview

The rationale for nonoperative treatment has been supported by clinical and autopsy studies that demonstrate that resorption of protruded and extruded disk material can occur over time.[153,154] Other studies have correlated MRI or CT scan improvement with successful nonoperative treatment in patients who have lumbar disk herniations and clinical radiculopathy.[154–156] The greatest reduction in size typically occurred in patients who had the largest herniations. Recent uncontrolled studies have shown that patients with definite herniated disks and radiculopathy, satisfying criteria for surgical intervention, were treated successfully with aggressive rehabilitation and medical therapy. Good to excellent results were achieved in 83 percent of cervical and 90 percent of lumbar patients.[157,158] An important longitudinal study was performed by the German neurologist Henrik Weber,[159] who randomly divided patients who had sciatica and confirmed disk herniations into operative and nonoperative treatment groups. He found significantly greater improvement in the surgically treated group at 1-year follow-up; however, the two groups showed no statistically significant difference in improvement at 4 to 10 years.[159]

In general, nonoperative treatment can be divided into three phases based on the duration of symptoms. Primary nonoperative care consists of passively applied physical therapy during the acute phase of soft tissue healing. Secondary treatment includes spine care education and active exercise programs, bridging the acute or postoperative phase with return to work at previous levels of function. Tertiary treatment focuses on interdisciplinary care to address low back disability, as well as physical and psychological deconditioning that developed as a result of chronic dysfunction.[160]

When spinal pain persists into the chronic phase, therapeutic interventions shift from rest and applied therapies to exercise and physical restoration. This shift is primarily a behavioral evolution, with the responsibility of care passed from physician and therapist to the patient.[14,108] Bed rest, therapeutic injections, manual therapy, and other externally applied therapies should be used sparingly for chronic spinal pain to treat a severe exacerbation of symptoms. When spinal pain is chronic or recurrent, traction or modalities such as heat and ice can be self-administered by patients for flare-ups to provide temporary relief.[14,108] Rational physical, medical, and surgical therapies can be selected by determining the relevant pathoanatomy and causal pain generators. Acute spinal injuries are first managed by elimination of biomechanical stressors, including short-term rest and supplemented by physical and pharmacologic therapies aimed directly at the nociceptive or neuropathic lesion(s).

The paradigm that best represents elimination of activity or causative biomechanical loading is bed rest. Bed rest usually is considered an appropriate treatment for acute back pain. However, 2 days of bed rest for acute LBP has been demonstrated to be as effective as 7 days and resulted in less time lost from work.[161] Furthermore, prolonged bed rest can have deleterious physiologic effects, leading to progressive hypomobility of joints, shortened soft tissues, reduced muscle strength, reduced cardiopulmonary endurance, and loss of mineral content from bone.[1,13,14] For these reasons, and because inactivity may reinforce abnormal illness behavior, bed rest usually is avoided when treating chronic spinal conditions.[1,13,14]

Oral Pharmacology

Overview

Rational pharmacology for the treatment of spinal pain is aimed at causative peripheral and central pain generators and is determined by the types of pain under therapeutic scrutiny (e.g., neuropathic and/or nociceptive) and modified additionally to deal with the evolving neurochemical and psychological influences that occur with chronicity. In general, published research for evaluation of medication efficacy to treat neck and back pain has demonstrated faulty methodology and inadequate patient/subject description.[162] However, medications continue to be used as adjuncts to other measures because of anecdotal reports, perceived standards of care, and some supportive clinical research.

During the acute period following biomechanical injury to the spine, excluding fracture, subluxation, other serious osseous lesions, or significant neurological sequelae, opioid analgesics may assist patients in minimizing inactivity and safely maximizing gradual progression of activity, including prescribed therapeutic exercises. Nonsteroidal anti-inflammatory drugs (NSAIDs) and muscle spasmolytics used during the day or at bedtime also may provide benefit.[13,108,133,134]

Nonsteroidal Anti-Inflammatory Drugs (NSAIDs)

NSAIDs have both analgesic and anti-inflammatory properties and, therefore, may affect mediators of the pathophysiologic process. Clinical trials have demonstrated NSAIDs to be useful as a treatment for pain; however, long-term use of NSAIDs should be discouraged owing to the frequent occurrence of adverse renal and gastrointestinal side effects.[13,162] Furthermore, the effects of these medications in the management of chronic musculoskeletal pain remains unclear, and no studies have demonstrated clear superiority over aspirin.[162] Additionally, no research has supported any specific agent over others, but sometimes switching chemical families through sequential trials may result in benefit.[13]

Spasmolytics

Spasmolytics or muscle "relaxants" traditionally are used to treat painful musculoskeletal disorders,[163] but because of their frequent sedative side effects, they often are prescribed only for bedtime use.

Tizanidine is a central γ_2-adrenoreceptor agonist that was developed for the management of spasticity due to cerebral or spinal cord injury but also has demonstrated efficacy when compared with other muscle spasmolytics. The muscle spasmolytic effects of tizanidine are thought to relate primarily to centrally acting α_2-adrenergic activity at both the spinal cord and supraspinal levels. Several clinical trials have demonstrated clinical efficacy of tizanidine for the treatment of acute neck and back pain.[164–167] Although not approved by the Food and Drug Administration (FDA) for pain treatment, tizanidine has been used for this purpose in Europe and Japan for approximately 12 years. Tizanidine exerts no significant effect on normal muscle tone; therefore, patients report muscle weakness less often as a side effect when compared with diazepam or other muscle relaxants.[168] Onset of action of tizanidine is rapid, with peak plasma concentrations occurring 1 to 2 hours following oral administration.[168]

Neuropathic Analgesics (see chapter on Peripheral Neuropathic Pain)

Conventional treatments for neuropathic pain, including anticonvulsants, can be considered for trial therapies in patients in whom nervous system structures are symptomatic, as well as for myofascial pain, which also may be a spinal-mediated disorder. A popularly prescribed anticonvulsant for chronic pain is gabapentin, which was shown to be effective for the treatment of neuropathic pain.[169]

Antidepressants (see chapter on Antidepressants)

Tricyclic antidepressants (TCAs) are used commonly for chronic pain treatment to alleviate insomnia, enhance endogenous pain suppression, reduce painful dysesthesia, and eliminate other painful disorders such as headaches.[170–172] The presumed mechanism of action is related to the capacity of TCAs to block serotonergic uptake, resulting in a potentiation of noradrenergic synaptic activity in the CNS–brainstem–dorsal horn nociceptive-modulating system. Also, recent studies in animals suggest that TCAs may act as local anesthetics through sodium channel blockade where ectopic discharges are generated.[173,174] There appears to be little evidence to support the use of selective serotonin reuptake inhibitors (SSRIs) to attenuate pain intensity, and recent studies have suggested that these agents are inconsistently effective for neuropathic pain at best.[135] Venlafaxine is a structurally novel antidepressant shown to have strong uptake inhibition of both serotonin and norepinephrine and anesthetic properties similar to the TCAs. An uncontrolled case series reported that venlafaxine provided pain relief in a variety of neuropathic pain disorders.[175]

The TCAs are limited, particularly in geriatric populations, owing to cardiovascular effects, such as tachycardia; anticholinergic side effects (including dry mouth, increased intraocular pressure, and constipation); oversedation; and dizziness (including orthostatic hypotension). SSRIs should be considered for symptoms that are commonly associated with chronic pain, including reduced coping, depression, anxiety, and fatigue. Overall, SSRIs have fewer adverse side effects than TCAs. Side effects associated with SSRIs include anxiety, nervousness, and insomnia; drowsiness and fatigue; tremor; increased sweating; appetite and gastrointestinal dysfunction; and male sexual dysfunction. Many pain specialists consider TCAs as first-line pain medications for the treatment of persistent neuropathic pain, especially as an adjunct to peripheral therapies and to manage the adverse influences of chronic illness.

Opioid Analgesics (see chapter on Opioid Pharmacology and Opioid Therapy)

Opioid medications may be useful for the treatment of spinal pain disorders.[13] Opioid analgesics may be used to assist a person in acute exacerbation of chronic pain.

Over the past decade, physicians have adopted a greater willingness to prescribe opioid analgesics for the treatment of refractory spinal pain and radiculopathy. The greater proportion of patients reclaim "what life they can." Side-effect profiles among long-acting opioids are similar, but cost is variable among current pharmaceutical offerings. Several principles apply to prescribing long-acting opioids for chronic pain. Medication should be taken in a time-contingent rather than pain-contingent manner and only provided by one prescribing physician and pharmacy. An agreement regarding the need and purpose of opioids should be signed by both the patient and the physician and placed in the medical record. Achievement of vocational, recreational, and social goals is a better measure of medication efficacy than subjective estimates of pain relief.[13,108]

Physical Therapy

Overview: Active versus Passive Therapy

Physical therapy for the spine may be divided into passive and active therapies. Passive therapies include modalities such as ultrasound, electrical stimulation, traction, heat and ice, and manual therapy. Passive modalities are most appropriate when used short term with acute injury or an exacerbation of a chronic problem. An often-advocated principle in chronic care is that such modalities should be self-administered by the patient.[14]

Corsets and braces have long been used as adjuncts to treatment, although their efficacy has not been demonstrated by studies without methodologic flaws.[13] As many as 89 percent of brace users have reported benefit from this prescription.[177] Also, a rigid orthosis has been demonstrated to be more effective than a simple support aid.[178] The primary mechanisms of action are unclear and probably differ based on such variables as the type of brace, patient morphology, pathoanatomy, and spinal activities.[13] Mechanisms of action likely are related to abdominal compression, as well as direct or indirect unloading of trunk muscles or structures, and to gross rather than intersegmental motion restriction.[13,179,180]

Traction also has long been used to treat conditions of the spine. Acute pain or an exacerbation of chronic pain is the recommended indication. In the lumbar spine, at least 60 percent of the body weight must be applied to produce dimensional changes in the lumbar disk; however, no evidence exists that this reduces a disk herniation.[12]

Manual therapy includes passive stretching, soft tissue mobilization, myofascial release, manual traction, muscle energy techniques, joint mobilization, and manipulation. Joint mobilization is a low-velocity passive stretch applied to a joint within or at the limit of its range. Manipulation uses a high-velocity thrust maneuver beyond a joint's restrictive range of motion.[181] More controlled trials have been carried out to evaluate manipulation and other conservative measures[182]; however, it is difficult to interpret these studies because of a variety of methodologic issues. Manipulative therapy may vary owing to the skill levels and techniques of different practitioners, e.g., physiotherapists, osteopaths, physicians, and chiropractors. In general, joint manipulation and mobilization are compared with other treatment groups rather than true placebo, making statistical interpretation difficult. Spinal manipulation probably is most beneficial for the treatment of acute axial spine pain without radiculopathy or neurological impairment.[183] The most serious complications of manipulation have resulted from cervical manipulation in which damage has resulted to vertebrobasilar blood vessels of the spinal cord.

Although neck and back schools to educate and train patients have been popular internationally, they have been ineffective as a preventative measure.[13,86] With variations in class size and emphasis, these schools usually provide information to patients about anatomy, pathophysiology, ergonomics (i.e., correct postures and body mechanics), exercise programs, self-management techniques, psychology, and activities of daily living. Back schools have received high grades for patient satisfaction (94 to 96 percent),[13,184] and in a prospective, randomized, controlled trial comparing back school education with exercise to exercise alone, the back school group showed significantly greater improvements in pain and disability.[185] Furthermore, at 16 weeks, the exercise-only group had reverted to their original level of disability, whereas the back school group showed continued improvement.[185] Other studies have shown that back schools prompt LBP patients to return to work earlier,[186] to seek less follow-up medical attention, and to suffer less frequent attacks of pain.[187] Overall, spine care education appears to be a valid, cost-effective treatment approach for acute and chronic spine pain patients.[13]

Therapeutic Exercise

Most physicians support the use of active exercise for chronic spinal pain, and there is growing support for this treatment in the medical literature.[1,189,190] Several studies incorporating exercise treatment have reported improved return to work,[1,189–192] functional gains,[1,189–192] and reduced pain.[192] Koes and colleagues[193] reviewed 23 random controlled studies of exercise therapy for back pain and concluded that poor methodology prevented any definite conclusion about the benefit of exercise therapy versus other conservative treatments for back pain or whether a specific type of exercise is most effective. Cervicothoracic and lumbar stabilization exercises are performed by teaching patients to obtain and sustain the spine in a posture with the least pain and potential risk for injury. In neutral spine posture, the motion segment shares biomechanical forces across the three-joint complex, with the degree of lordosis determined, in part, by unloading anterior versus posterior column pain generators, load sharing, and "close-packed" or engaged status of the facets that prevents nonphysiologic movements that may cause further injury to the disk. The patient is taught to maintain this position while surrounding muscles brace the spine isometrically. Extremity movements are performed while maintaining neutral spine postures in varied positions from supine to standing and eventually using weight machines, free weights, or no resistance other than the weight of the arms and legs.[188] The goal of treatment is to maintain a neutral spine posture while performing general strengthening and flexibility exercise with the least amount of pain and injury potential while continuing to advance to increasingly complex daily or work-related activities.[13,108] Furthermore, recent studies have shown that stabilization exercises are effective in the nonoperative treatment of patients with confirmed lumbar and cervical radiculopathy[157,158] and that spine stabilization exercise may reduce LBP recurrence.[194] A review of randomized, controlled studies in the medical literature supports long-term use of back exercises as a preventive intervention.

Functional Restoration

To address deconditioning, Mayer and colleagues and others[1,121,161,189] have advocated functional restoration (FR), which uses a sports medicine approach to address industrial back injuries through a program of physical training to restore normal flexibility, strength, and endurance and emphasizes the multifactorial nature of chronic back pain. FR programs, as advocated, must be highly structured, interdisciplinary, and intensive. Patients who are cleared surgically and found to be competent medically undergo a comprehensive entrance evaluation to identify and quantify physical, psychological, and socioeconomic factors that must be monitored and addressed during the treatment. The program then consists of daily and intensive physical, psychological, and behavioral reconditioning. With measured physical and functional improvement, patients participate in increasing levels of task-oriented rehabilitation and work simulation. This physical training is coupled with cognitive-behavioral support, including didactic sessions and disability management, and it culminates in an exit evaluation that again measures physical and functional parameters correlated with consistency of effort in the form of a work-capacity assessment. This evaluation can be used for return to work.

In 1987, Mayer and colleagues[195] published a prospective 2-year study of FR in industrial low back injury. Although patients were neither truly randomized into treatment and comparison groups nor representative of the general population, 87 percent of the treatment group who could be contacted were working, in contrast to 41 percent of the comparison group and 25 percent of the drop-outs after 1 year. In addition, the authors demonstrated a reduction in additional surgical and medical care in the treatment group compared with the other two groups. Although an FR program by Hazard and colleagues,[189] modeled after Mayer's project, demonstrated similar data, other programs have had difficulty replicating these results.[161] A multicenter FR study that involved 11 treatment centers in 7 states and that emphasized work hardening but excluded psychosocial programs showed statistically higher returns to work rate than the comparison group at the 6- (66 percent) and 12-month (77 percent) follow-ups ($p < 0.0001$).[196] Postoperative patients showed the same return to work rate as nonsurgical patients. FR treatment for spine-related pain, especially chronic LBP and disability, appears appropriate for selected patients; however, the ability to predict which patients will respond is not yet possible. Although studies suggest that these intensive programs may save money, treatment is still costly by most standards.[13]

Therapeutic Injections

Therapeutic injections of local anesthetics, or other substances may be administered directly into painful soft tissues, facet joints, nerve roots, the epidural space, or intrathecally.[13] Therapeutic injections have been advocated to alleviate acute pain or an exacerbation of chronic pain, to help patients maintain an ambulatory outpatient status, to participate in a rehabilitation program, to decrease the need for analgesics, and to avoid surgery.[13] Although injections into myofascial TrPs are widely advocated, a double-blind study evaluating local anesthetic versus saline[197] and a prospective, randomized, double-blind study evaluating dry needling versus groups in which lidocaine, corticosteroids, or vaporized coolants sprayed with acupressure were used showed no statistically significant differences in benefit between treatment groups.[198] Two double-blind studies evaluating paravertebral injections of botulinum toxin A as a treatment for chronic LBP demonstrated improvement in pain and function in the treatment groups compared with a group given saline injections. Improvements in both studies were statistically significant for up to 8 weeks.[199,200]

Recent clinical research has suggested a potential benefit from botulinum toxin A (BTXA) when applied as a treatment for muscular or myofascial neck pain.[201–206] Porta[206] showed that improvement was superior in BTXA-treated patients compared with patients treated with methylprednisolone and bupivacaine injections for myofascial pain at 2 months, when initial improvement in the comparison group waned. Porta stressed the importance of combining BTXA injections with physiotherapy. A prospective, double-blind, randomized, controlled study revealed no statistically significant benefit from BTXA for the treatment of cervical myofascial pain when using doses similar to those used to treat cervical dystonia without adjunctive physiotherapy.[205] Randomized, double-blind, controlled studies with injectable agents such as BTXA are necessary before any evidence-based practice guidelines are forthcoming.

▶ THERAPEUTIC SPINAL INTERVENTIONAL TECHNIQUES

Intra-articular Blocks

Four studies of intraarticular corticosteroid injections in lumbar spine facet joints[208–211] and one study in cervical spine joints[212] were performed using comparison groups that were similar demographically to the treatment group but received another treatment. Two trials were randomized, one by Carette and colleagues involving lumbar facet injections[208] and another by Barnsley and colleagues involving cervical facet injections.[212] Nonrandomized trials and observational studies have shown results of long-term pain relief (≥6 months) of 28 percent,[214] 38 percent,[215] and 54 percent.[216] Retrospective evaluations included a paper by Lippitt,[217] who reported initial relief at 50 percent that declined to 14 percent at 6 months and 8 percent at 12 months, and a paper by Lau and colleagues,[218] who reported initial relief in 56 percent that declined to 44 percent at 3 months and 35 percent after 6 months. *Although some physicians advocate*[209,216] *facet injections as a treatment method, a large prospective study*[219] *showed no long-term benefit*. In summary, in isolation, intra-articular facet injections, which are costly and invasive, have dubious therapeutic value[220] based on diagnostic and procedural variables that have not yet been identified. However, intra-articular facet injections generally are supported for use as a diagnostic tool.

Medial Branch Blocks

Medial branch blocks traditionally have been used for both diagnostic and prognostic purposes, with limited use therapeutically. The therapeutic role of medial branch blocks has been evaluated in three randomized clinical trials[221–223] and three nonrandomized clinical trials.[224–226] Only one randomized trial used adequate criteria to diagnose facet joint pain and showed adequate long-term follow-up of outcome data.[221] Combined evidence by Manchikanti[213] and colleagues[221] suggests strong support for short-term pain relief and moderate support for long-term pain relief of facet joint origin from this procedure.

Radiofrequency Medial Branch Neurotomy

Manchikanti and colleagues[227] also cited strong evidence that radiofrequency (RF) denervation provides short-term relief and moderate evidence for long-term relief of chronic cervical, thoracic, and lumbar spine pain of facet origin. A randomized trial by Lord and colleagues[228] compared 12 patients receiving RF lesions to the medial branches of the cervical dorsal ramus with the same number of patients receiving a sham procedure. Seven patients in the treatment group and one in the control group remained free of pain. Overall, patients receiving medial branch neurotomies had a long-term success rate of 75 percent. In another randomized trial, 47 percent of the treatment group showed sustained improvement following RF denervation at 12 months. Improvement measures included pain reduction, functional disability, and physical impairment. These and other studies show strong support for both a short- and long-term benefit to RF medial branch neurotomy for the treatment of lumbar facet syndrome in chronic LBP patients.[56] A nonrandomized, prospective trial of RF medial branch neurotomies for cervical pain revealed similar long-term benefit.[56] Potential side effects of RF denervation include painful cutaneous dysesthesia or hyperesthesia, pneumothorax, and deafferentation pain.[229]

Epidural Injections

Epidural injections have been used widely with direct placement near the involved nerve root or with midline presentation, as well as a caudal presentation, combining corticosteroid and local anesthetic of varying volumes. An intralaminar entry is directed more closely to the site of assumed pathology and requires less volume of injectate than a caudal route. However, the caudal entry usually is considered a safe approach with a small risk of inadvertent dural or neural structure puncture. Transforaminal corticosteroid injections are more target-specific and require the least volume of injectate to reach the presumed pathoanatomic site of pain through the ventrolateral epidural space.

When considering an epidural injection, each approach has advantages and disadvantages. The caudal approach requires a large fluid volume and thus greater dilution of the active ingredient within the injectate.

Because the needle cannula is threaded initially on a plane relatively parallel to the spinal canal, there is greater risk of intravascular, subcutaneous, subperiosteal, or interosseous needle puncture. Disadvantages of the intralaminar approach include dilution of the injectate, extraepidural placement of the needle, intravascular placement of the needle, preferential cranial and posterior flow of the solution, and dural puncture. In addition, the intralaminar approach is more difficult in postoperative patients and below the L4–5 innerspace.[56] The transforaminal approach is difficult in the presence of postoperative/ osseous fusion or when hardware is present. Other risks include intraneural/intravascular injection and spinal cord trauma. The use of fluoroscopy to direct needle placement and observe contrast flow should be considered necessary to reduce potential adverse events.[56]

Evidence synthesis by Manchikanti and colleagues[56] was achieved by reviewing eight randomized or double-blind trials, five of which supported short-term relief[226-239] (defined as less than 3 months) and five of which supported long-term relief (defined as 3 months or more) when caudal injections were performed.[226,228-231] In addition, three prospective trials[232-234] and four retrospective trials[235-238] showed support for short- and long-term pain relief with a series of injections. In the same evidence synthesis, Manchikanti and colleagues[56] found that 7 of 10 randomized trials were positive for short-term relief and 3 were positive for long-term relief for intralaminar epidural injections.[56] Numerous nonrandomized trials showed benefit for patients receiving cervical or lumbar intralaminar epidural steroid injections.[56] At present, there is strong literature support for the use of intralaminar corticosteroid epidural injections to provide short-term pain relief when treating cervical or lumbar radicular syndromes, even chronic cases, but this treatment is best reserved for use as an adjunctive therapy or during a flare-up of symptoms.[56]

Transforaminal epidural injections have been demonstrated to show positive short- and long-term results in multiple randomized trials.[56] Their effectiveness in post-lumbar laminectomy syndrome and disk extrusions is unclear.[56]

There continues to be considerable debate within the literature as to whether there is any benefit to be gained by offering patients epidural steroid injections. Reviews by Koes and colleagues[239] in 1995 and 1999 supported the usefulness of lumbar and caudal epidural injections for LBP and sciatica. Meta-analyses in 1995 by Watts and Silagy[240] and van Tulder and colleagues[241] reported conflicting evidence and inconsistent findings regarding the effectiveness of epidural steroids. A 1998 review of the literature[246] concluded that epidural corticosteroid injections are effective for back pain and sciatica, and a 2000 review by Vroomen and colleagues[243] cited epidural steroids as beneficial for some patients with nerve root compression and sciatica.

In summary, the spinal interventionalist should use clinical judgment as to the rationale and safety of such injections, combining clinical experience with the chosen procedure. There is no evidence that these procedures will provide long-term pain relief or any benefit. Epidural injections may be useful as a method of pain control and may provide benefit that is adjunctive to other therapies.

Epidural Adhesiolysis

Percutaneous adhesiolysis with or without spinal endoscopy is another interventional technique used to manage chronic refractory chronic LBP.[244-254] This procedure is performed to disrupt presumed epidural adhesions that may affect nerves or other pain-sensitive tissues. Percutaneous lysis of epidural adhesions allows improved delivery of injected drugs to targeted painful structures. Epidurolysis of adhesions with direct deposition of corticosteroids in the spinal canal can be achieved by a three-dimensional view provided by a spinal or epidural endoscope. In an evaluation of the clinical effectiveness of percutaneous epidural adhesiolysis using a spring-guided catheter with or without hypertonic saline neurolysis, Manchikanti and colleagues reported short-term relief as lasting less than 3 months and long-term relief as lasting longer than 3 months. Two randomized trials[255,256] demonstrated both positive short- and long-term relief. One retrospective study showed both short- and long-term relief,[257] whereas two others showed only short-term improvement.[258,259]

The effectiveness of spinal endoscopic adhesiolysis was evaluated by reviewing two prospective[260,261] and two retrospective studies[257,261] for short-term relief (≤6 months) and for long-term relief (>6 months). All studies showed support for short-term improvement but did not demonstrate any evidence to support long-term benefit.

Complications of adhesiolysis with spinal endoscopy include dural puncture, spinal cord compression, catheter sheering, infection, injury from the endoscope, and administration of high volumes of fluid.[262] Epidural infusion of high volumes of fluid, especially hypertonic saline, potentially can cause excessive epidural hydrostatic pressure, resulting in spinal cord compression, excessive interspinal or intracranial pressures, epidural hematoma, bleeding, infection, increased intraocular pressures with resulting visual deficiencies including blindness, and dural rupture.[262] Unintended subarachnoid or subdural puncture with injection of local anesthetic or hypertonic saline also can occur with resulting neural catastrophe.[262] Hypertonic saline injection into the subarachnoid space has been reported to cause cardiac arrhythmias and myelopathy.[253,262,263] Arachnoiditis and sheering of the catheter with retention also have been reported following epidural adhesiolysis with hypertonic saline.[264–266] In summary, these procedures should be performed only under fluoroscopic control by well-trained interventionalists.

Intradiskal Therapies

A prospective intradiskal therapies review of numerous procedures has been directed at the disk, presumed causative for many painful spinal and radicular syndromes. Over the years these have included chymopapain injections to achieve nucleolysis, percutaneous manual nucleotomy with a nucleotome, thermal vaporization with a laser, and percutaneous decompression with nucleotomy using coblation technology (nucleoplasty). Intradiskal electrothermal therapy (IDET) is a minimally invasive technique in which the annulus is subjected to thermomodualation.[56] A prospective, randomized, double-blind study of interdiskal injections into diskography-confirmed painful disks showed no statistically significant benefit between corticosteroids and local anesthetics.[267]

Recently, the use of diskography as a diagnostic procedure has been combined with therapeutic percutaneous intradiskal procedures in patients who demonstrate a concordant pain response. These include intradiskal electrothermal annuloplasty (IDET), percutaneous laser disk decompression (PLDD), percutaneous RF annular neurolysis, and nucleoplasty.[268] These procedures are postulated to shrink collagen fibers and coagulate neural tissues, thereby alleviating neurochemical nociceptive responses produced by mechanical loading on a painful disk. IDET is performed using radiographic placement of a 17-gauge introducer needle through a posterior annular wall into the nucleus of a symptomatic disk, as determined by diskography. A navigable catheter with a temperature-control thermal-resistant coil is passed through the needle so that it curls along the posterior inner annulus. Catheter temperatures are raised slowly to 90°C, causing thermocoagulation of intradiskal and annular collagen, as well as associated nociceptors. Reduction in pain symptoms may result from denervation or shrinking and remodeling of the diskal structure or both.[268] Karasek and Bogduk[269] introduced a flexible electrode into a "diskography symptomatic" disk with internal disk disruption. The electrode was passed circumferentially along the inner annulus to heat and coagulate annular collagen and nociceptive nerve fibers. Of the 35 treated patients, 23 percent achieved complete pain relief, and 60 percent improved. Improvements were sustained at 6 and 12 months. The 17 patients comprising a parallel comparison group did not benefit from a physical rehabilitation program alone, except 1 patient who had dramatic pain reduction.[269] At 2-year follow-up, 54 percent of patients had achieved at least 50 percent pain relief with concomitant functional improvement.[270] A randomized, double-blind study by Pauza and colleagues[271] in 2002 showed IDET to be effective, and it was reviewed by Wetzel and colleagues.[272] However, a number of prospective and retrospective studies have been published.[269–278] A study by Saal and Saal[272] of 62 patients showed improvement in multiple parameters in more than 70 percent of the patients at 12 months or more. Wetzel and colleagues[271] reported that all studies shared common designs using prospective cohorts with historical or noninterventional groups as controls. They concluded that pain due to lumbar disk disease may be diminished by IDET, with all studies suggesting an overall positive effect.

However, Manchikanti and colleagues,[56] in their comprehensive 2003 review to determine evidence-based practice guidelines for interventional techniques, reported that IDET only meets criteria for moderate evidence to support short-term relief but limited evidence to support long-term relief. Complications from IDET have included catheter breakage and cauda equina syndrome.

Percutaneous disk decompression (PLDD) with nucleoplasty is performed using RF energy to dissolve nuclear material through molecular disassociation. RF coagulation is thought to denature proteogylcans, to change the internal environment of the nucleus pulposus by debulking and modifying the contour of a symptomatic disk, and to reduce the overall volume of disk material, thereby reducing its nociceptive capacity.[268–279] The effectiveness of PLDD was reported in two prospective and two retrospective trials.[280–282] Both prospective trials showed statistically significant improvement over the short- (<6 months) and long- term (>6 months) without significant complications for the treatment of axial diskogenic back pain. PLDD users have suggested treatment as beneficial in approximately 80 percent of patients for painful internal disk disruptions or annular tears with protrusions or herniations.[268] However, some investigators warn that intradiskal laser energy is potentially hazardous to bone and nerve tissue because of inadequate temperature control and the extent to which tissues are heated.[266]

These RF procedures are advocated by spinal interventionalists to denervate the annular nerve for treatment of proven diskogenic pain.

▶ OPERATIVE TREATMENT OF SPINAL PAIN

Overview

Axial and radicular spinal pain have numerous causes that can result from systemic conditions or any of a number of localized traumatic, degenerative, or neurocompressive pathologies. The vast majority can be treated with nonoperative means, but a thorough diagnostic workup is necessary to identify patients who may benefit from surgical management or require urgent surgical treatment.

The causes of axial spinal pain include serious systemic conditions such as tumors, infection, systemic arthritides, and spondyloarthropathies. Traumatic conditions such as fracture or ligamentous disruption also can cause localized pain, and numerous degenerative conditions, including degenerative disk disease, facet arthropathy, and arthritis-related instability, can cause localized pain. Stenosis or other forms of neurocompression typically cause more radicular complaints. Occasionally, though, axial pain complaints can predominate or be associated with radicular numbness and weakness without pain.

The diagnostic workup of these patients should begin with a thorough history and physical examination. Plain x-ray films, including anteroposterior (AP) and lateral views of the affected region, typically are reviewed. Oblique studies of the lumbar spine occasionally can show fractures of the pars interarticularis. Cervical oblique films can show foraminal stenosis. However, the clinical utility of these studies have been called into question. Flexion and extension views of the cervical or lumbar spine are necessary occasionally to evaluate for instability, particularly in the setting of trauma or unexplained back or extremity pain.

In addition to plain films, advanced imaging such as MRI, CT, with IV contrast or myelography can be performed to rule out the rare tumor or infection. In the presence of long tract signs, an MRI is appropriate to rule out spinal cord compression or intrinsic cord pathology. Laboratory tests and MRI can be helpful in patients in whom systemic or serious illnesses are suspected. MRI is also used in ruling out neurological compression or certain types of trauma. Occasionally, interventional techniques such as facet blocks, nerve root blocks, and diskograms are also used in the more advanced evaluation of axial or radicular spinal pain.

Axial Neck Pain

Axial neck pain rarely comes to surgical management. The most common causes are degenerative and mechanical, and they rarely respond well to operative treatment. The role of degenerative disk disease in the etiology of axial neck pain is not well understood. The presence of degenerative disk disease alone typically is not considered adequate justification for surgical fusion. Limited research on cervical diskograms to identify pain-producing cervical disk has not shown reproducible results. Although research on cervical disk replacement is ongoing, the ultimate utility of cervical fusion or disk replacement for degenerative disk disease without neurocompression remains unclear. Still, axial neck pain can be a part of a symptom complex that includes radicular or myelopathic complaints. Typically, axial neck pain associated with radiculopathy that is treated with surgery does resolve along with the radicular complaints. Even in this case, however, reduction of axial pain with surgical treatment is less predictable than the reduction of radicular pain. Finally, cervical instability can produce significant neck pain and place the patient at risk for spinal cord injury. Patients with evidence of instability (>3.5 mm of subluxation or 11 degrees of angulation on "bending" films)[283] are candidates for surgical fusion at the unstable segments. These patients typically do experience a reduction in axial neck pain after surgery.

The preoperative evaluation typically includes plain x-rays, which frequently show loss of disk height, as well as endplate osteophyte formation and occasionally facet arthropathy. Alignment can become kyphotic, particularly in the setting of degenerative disk disease. Some authors speculate that cervical malalignment or loss of sagittal balance in the cervical spine may create mechanical neck pain in much the same way that sagittal imbalance in the thoracolumbar spine creates pain. However, this relationship has not been studied extensively, nor has surgery for it been validated (Fig. 30-1).

Cervical Radiculopathy

Cervical radiculopathy caused by foraminal stenosis, uncovertebral osteophytes, or cervical disk herniation responds quite well to surgical treatment. Typically, patients considered for surgery have undergone a series of nonoperative treatments, including physical therapy and traction, anti-inflammatory medications, oral steroids, and a period of observation of at least 4 to 6 weeks. Those who fail this trial typically undergo an MRI or CT/myelogram of the cervical spine. Flexion/extension views can reveal occult instability. For patients in whom the imaging findings match the pain complaints and physical examination findings, anterior cervical diskectomy and fusion or cervical foraminotomy with diskectomy provides reliable results. Typically, patients with greater degeneration do better with fusion surgery in addition to diskectomy. Those with soft disk herniations that are easily accessible from the posterior spine can be treated with diskectomy alone without fusion. Recent advances with endoscopic techniques for posterior cervical surgery may allow for adequate neurological diskectomy with less postoperative axial neck pain than is seen typically after traditional foraminotomy. However, these results await long-term studies for validation (Fig. 30-2).

Axial Thoracic Pain

Axial thoracic pain rarely requires surgical treatment. In younger patients this is primarily mechanical or muscular. In older patients it is seen most often with osteoporotic compression fractures. These compression fractures can be treated with injections of bone cement if conservative treatment with pain medication and bracing fails. These bone cement injections can be performed either directly

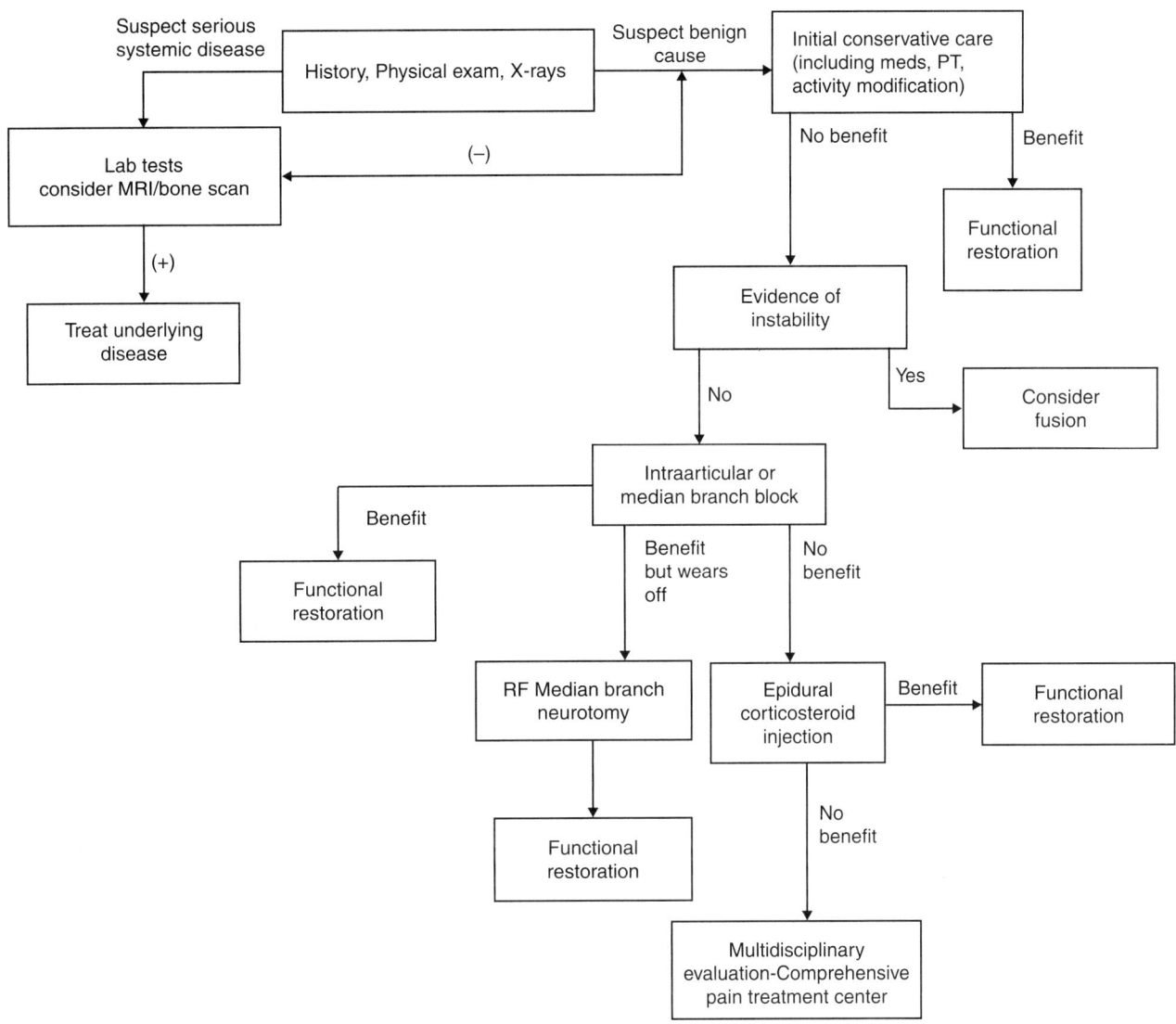

Figure 30-1. Axial neck pain.

through a needle or after balloon inflation within the vertebral body to restore alignment.[284] In most patients, these injection procedures provide immediate and substantial pain relief in patients with compression fractures of less then 3 months' duration. Other, more rare causes of axial thoracic pain include infection and tumor, particularly in older or immunosuppressed patients. Surgical treatment for thoracic degenerative disk disease has not been described or studied extensively.

Thoracic Radiculopathy

Thoracic radiculopathy is caused most commonly by disk herniation. However, thoracic disk herniations are rare, affecting less than 0.3 percent of the population and accounting for less then 1 percent of disk operations performed.[285–289] The onset of symptoms of a thoracic disk herniation may be gradual, occurring over weeks or months, and rarely is associated with trauma. Symptoms vary widely, with most people complaining of poorly localized thoracic back pain and nonspecific leg complaints. If the disk herniation is central and compressing the spinal cord, causing myelopathy, urgent surgical treatment is indicated. More lateral disk herniations compressing thoracic nerve roots without cord compression or myelopathy have less reliable results with surgical treatment. Typically, these are managed with physical therapy and, occasionally, injection therapy followed by functional restoration. In rare instances, when radicular pain complaints are persistent despite conservative treatment, diskectomy may be considered. Results of surgical treatment of thoracic diskectomy have been highly variable but generally not as good as surgical treatment for cervical or lumbar disk herniation.

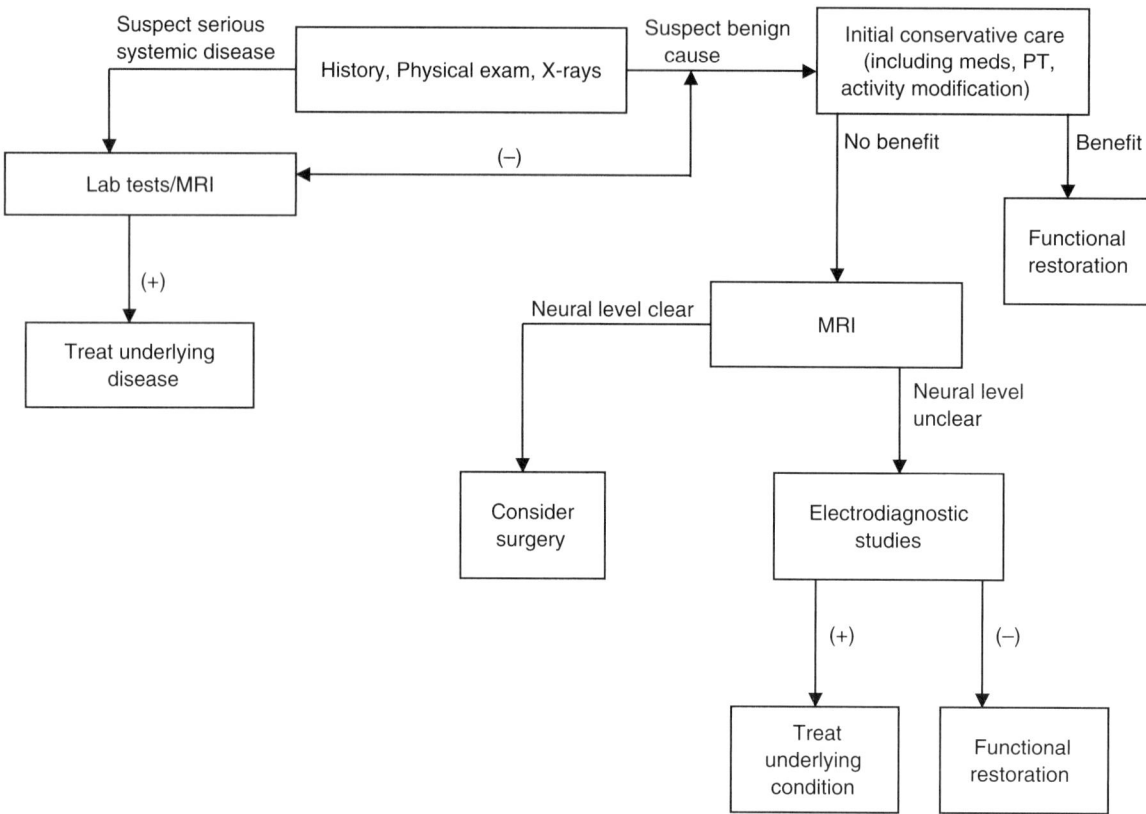

Figure 30-2. Cervical redicular pain.

Axial Lumbar Pain

Surgical treatment for lumbar back pain related to tumor, infection, and instability is reliable. Initial evaluation of these patients should include plain films, bending films, and MRI to identify these processes. Once identified, these patients should be referred directly for surgical management.

The surgical treatment of axial or mechanical LBP related to degenerative disk disease is controversial. The vast majority of patients with mechanical LBP are not candidates for surgical treatment. Part of the difficulty in these patients is identifying the pain generator, particularly when the cause is multifactorial. In recent years, interest has grown in identifying patients with disk-related LBP who might be candidates for surgical fusion or disk replacement.

Most often before a patient is considered for surgical treatment, he or she has failed extensive physical therapy and anti-inflammatory and pain medication trials and has a long-standing history of activity-related back pain. Epidural steroids and facet blocks also typically have been tried and failed before the degenerative disk is considered.

Diagnostic evaluation of these patients typically includes plain films, which should show some element of degenerative disk disease. This may simply be loss of disk height or sclerosis of the endplates with or without endplate osteophyte formation. Flexion and extension views usually do not show evidence of instability, and MRI shows darkening of the disk and frequently reactive changes within the bony endplates. An annular tear or high-intensity zone may be present, generally in the central posterior disk. Frequently, neurological compression is not present, although it may be. A diskogram also may be performed. This involves injection of dye under pressure into both the diseased and adjacent normal disks. The patient with a positive diskogram should note his or her typical and concordant pain on administration of pressure within the disk and should note unfamiliar pain or discomfort at other levels. Although the diskogram is used extensively as a diagnostic tool, considerable debate about its reliability remains.[290] Most authors recommend correlating the combination of patient's symptoms, MRI findings, and diskogram findings before considering surgical management. In patients for whom all three correlate, surgical fusion frequently yields good results for single-level disease. Patients in whom multilevel disease is present, even when multiple levels are concordant on diskogram, generally do more poorly with surgical fusion. For this reason, fusion of more than two disks for degenerative disk disease is rarely, if ever, indicated owing to the usually poor results.

Disk replacement is an area of extensive research. Numerous manufacturers are developing and testing intervertebral implants. Most implants fall into two types, either an artificial nucleus or a complete artificial disk. Both types are currently in trial and are available only through IDE studies. Early reports from these studies indicate that the intervertebral disk may show some promise, particularly in patients who do not have extensive facet arthropathy. Further research is necessary before disk replacement can be widely recommended (Fig. 30-3).

Lumbar Radiculopathy

There are numerous causes of lumbar radiculopathy, and many are not amenable to surgery. Only those in which documented neurocompression has occurred are truly candidates for surgical treatment. The most common causes of lumbar radiculopathy are disk herniation and stenosis. Disk herniation is more common in the

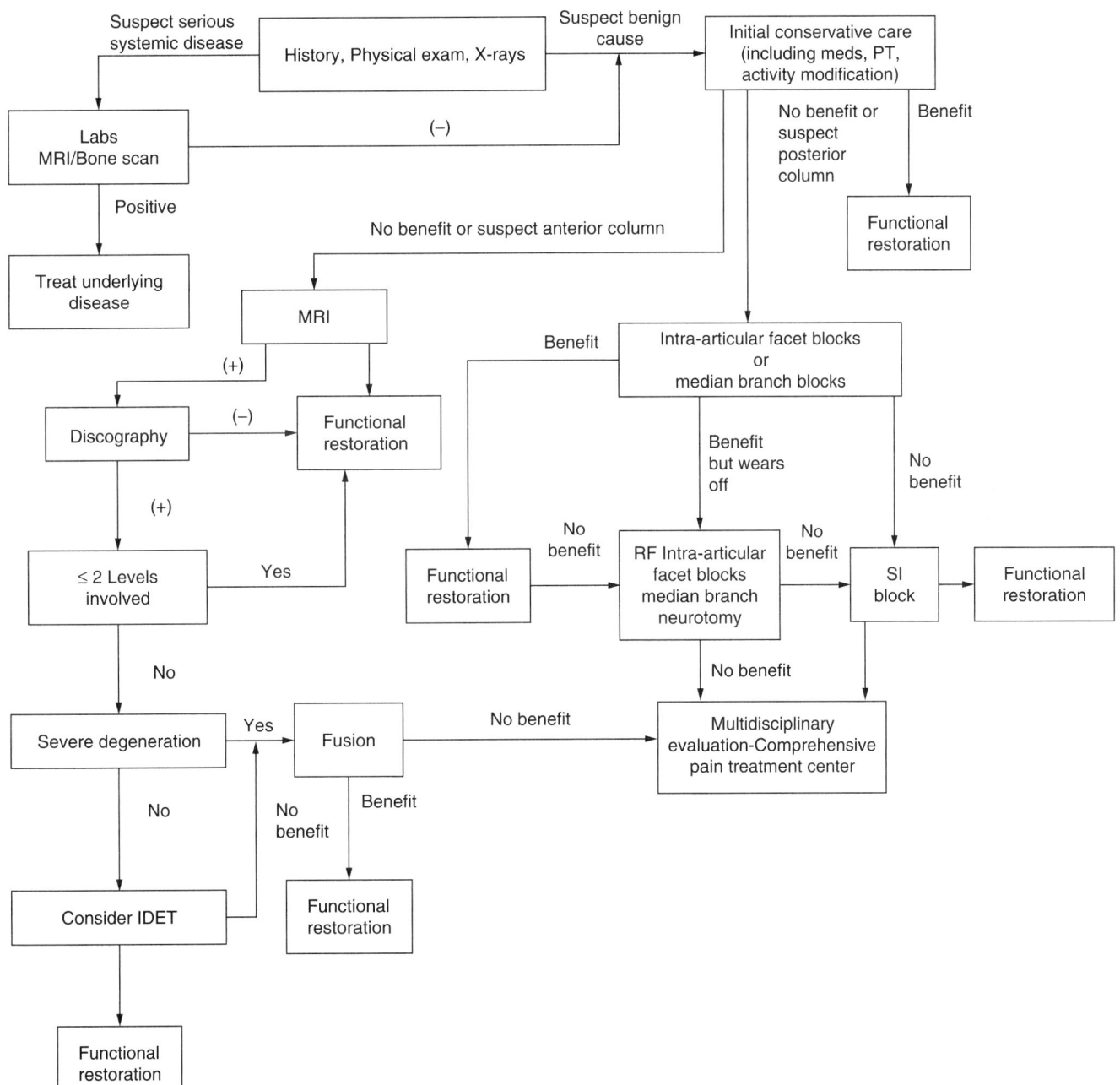

Figure 30-3. Axial lumbar pain.

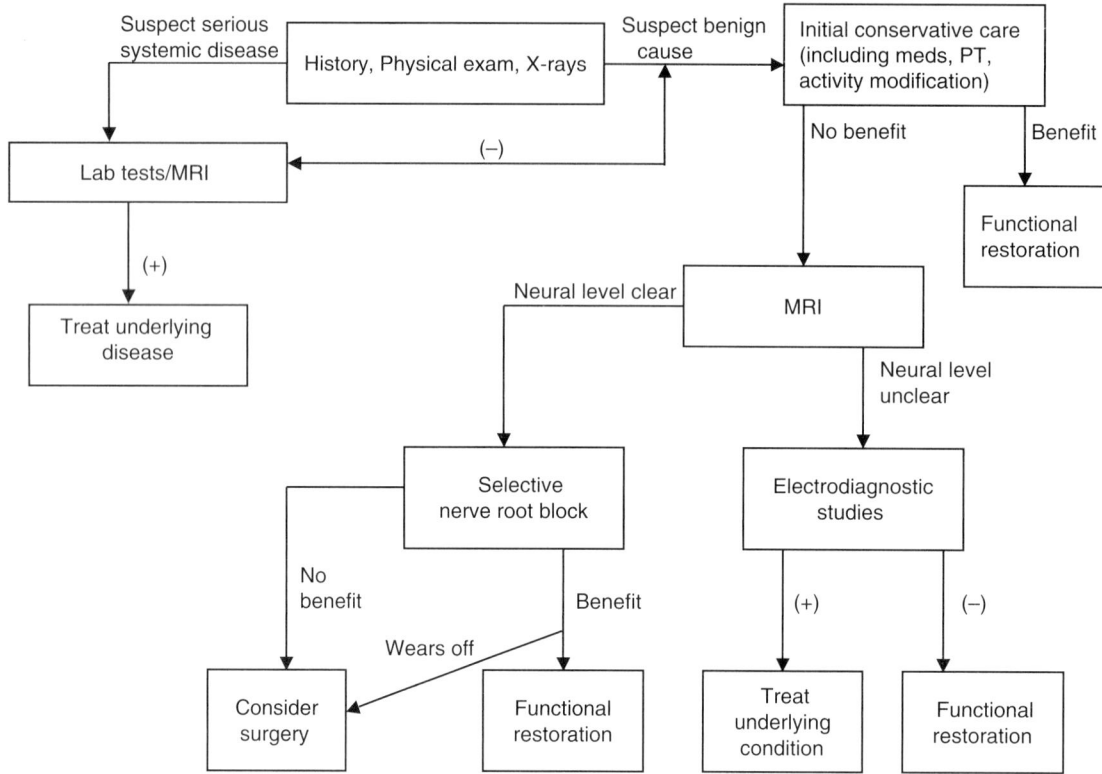

Figure 30-4. Lumbar radicular pain.

20- to 50-year-old age group, whereas stenosis is more common in older patients (Fig. 30-4).

Disk Herniation

Four pertinent criteria must be present in order to diagnose acute radiculopathy owing to a lumbar disk herniation.[291] Leg pain, not back pain, must be the dominant symptom. Second, pain distribution should be specific and attributable to a particular dermatome. Third, nerve tension signs should be present, including an abnormal straight-leg raise, contralateral straight-leg, or bowstring test. A positive test should reproduce the radicular pain. The fourth and final criterion is that neurological signs such as weakness, sensory loss, atrophy, or reflex alteration should correlate with the level.

Long-term studies comparing operative and nonoperative treatment of patients with herniated disks have shown no statistical differences in long-term outcome; however, patients treated with surgery do tend to have more rapid relief of symptoms. At 4-year follow-up, Weber[159] found no differences between the two groups. The surgical group had better results after 1 year, but this difference diminished with time. Hakelius[292] also found that the surgical group had better initial results from surgery; however, by 6 months, there were no significant differences between the two groups. In his study, at 7-year follow-up, the nonoperative group had more back pain than did patients in the surgical group. In contrast, Weber found no clinical difference with reference to back pain between groups at 10-year follow-up.[292]

Because most patients will improve with time, surgery should be used selectively. Traditionally, five clinical criteria have been used to select surgical candidates: (1) loss of bowel and bladder function, (2) gross motor weakness, (3) evidence of increasing impairment of nerve root function, (4) severe sciatic pain persisting or increasing despite 4 to 6 weeks of conservative treatment, and (5) recurrent incapacitating episodes of sciatica pain. In general, these indications should be considered relative; only cauda equina syndrome should be considered an emergent indication for surgery. Typically, a lumbar microdiskectomy can be performed through a small incision as an outpatient and provide a 90 to 95 percent success rate, alleviating or dramatically reducing radicular complaints.

Stenosis

In cases of spinal stenosis, the clinical picture is more variable. Patients may complain of predominantly back pain or leg pain. They frequently have activity-related fatigue symptoms in the lower extremities that worsen with extended standing or walking and generally are alleviated by rest. Nerve tension signs on examination are rare, and in fact, physical examination often is unremarkable in these patients.

In contrast to patients with lumbar disk herniations, patients with stenosis frequently have a long history of waxing and waning symptoms. Often stenosis symptoms have been present for up to a decade prior to presentation and have worsened gradually, progressively limiting the patient's activities of daily living. Diminished standing or walking tolerance is the most common clinical complaint, along with leg pain, fatigue, and weakness. Frequently, epidural steroid injections can provide temporary relief and ultimately may allow patients to forego operative treatment. However, as the disease progresses, epidural steroids no longer may be effective, and when they fail, lumbar decompression can provide reduction or resolution of symptoms in 80 to 90 percent of patients. Most often lumbar fusion is unnecessary unless evidence of instability or spondylolisthesis is present.

▶ FUTURE DIRECTION

Subspecialists trained in neurology, orthopedic medicine, and psychology will be necessary to work with spine surgeons to identify the complex physical and nonphysical ingredients that formulate the difficult chronic spinal pain patient. Fellowship training is common for operative spinal specialists, and similar training should be offered to neurologists or physiatrists to manage the complex behavorial and biochemical aspects of chronic spinal pain, combined with establishing competence in using electrodiagnostic skills for diagnosis and performing spinal interventional treatments with spinal injections.

As is the case of all human diseases, clinical practice should be based more firmly on the results of rigorous outcomes research and less on theory, emphasizing results of therapy in terms of functional status and return to work rather than symptoms. Only when prospective, double-blinded, controlled, randomized studies compare treatment groups with control groups that better represent the natural history of the disorder can the essential components of various therapies be assessed. Back care education and fitness programs beginning as early as high school may instill better health behaviors and prevent injury, whereas employer counseling and employee relations programs may reduce the incidence of disability.

REFERENCES

1. Mayer TG, Gatchel RJ: *Functional Restoration for Spinal Disorders: The Sports Medicine Approach*. Philadelphia: Lea & Febiger, 1988.
2. Kelsey JL, White AA: Epidemiology of low back pain. *Spine* 1980;6:133–142.
3. Waddell G: A new clinical model for the treatment of low back pain. *Spine* 1987;12:632–644.
4. Anderssen GBJ: The epidemiology of spinal disorders, in Frymoyer JW (ed), *The Adult Spine: Principles and Practice*. New York: Raven Press, 1997:93–141.
5. Cunningham LS, Kelsey JL: Epidemiology of musculoskeletal impairments and associated disability. *Am J Public Health* 1984;74:574–579.
6. Frymoyer JW, Mooney V: Current concepts review: Occupational orthopedics. *J Bone Surg* 1986;68A:469–473.
7. Nachemson A, Waddell G, Norlund AI: Epidemiology of neck and low back pain, in Nachemson A, Jonsson E (eds): *Neck and Back Pain: The Scientific Evidence of Causes, Diagnosis and Treatment*. Philadelphia: Lippincott Williams & Wilkins, 2000:165–187.
8. Mooney V: Where is the pain coming from? *Spine* 1987;12:754–759.
9. Bovim G, Schrader H, Sand T: Neck pain in the general population. *Spine* 1994;19:1307–1309.
10. Takala J, Sievers K, Klaukka T: Rheumatic symptoms in the middle-aged population in southwestern Finland. *Scand J Rheumatol Suppl* 1982;47:15–29.
11. Westerling D, Jonsson BG: Pain from the neck-shoulder region and sick leave. *Scand J Soc Med* 1980;8:131–136.
12. Frymoyer JW: Back pain and sciatica. *N Engl J Med* 1988;318:291–300.
13. Argoff CE, Wheeler AH: Spinal and radicular pain syndromes, in Backonja M-M (ed), *Neurologic Clinics*. Philadelphia: Saunders, 1998:833–845.
14. Wheeler AH, Hanley EN: Nonoperative treatment of low back pain: Rest to restoration. *Spine* 1995;20:375–378.
15. Modic MT, Brant-Zawadzki MN, Obuchowski N, et al: Magnetic resonance imaging of the lumbar spine in people without back pain. *N Engl J Med* 1994;331:69–73.
16. Powell MC, Szypryt P, Wilson M, et al: Prevalence of lumbar disk degeneration observed by magnetic resonance in symptomless women. *Lancet* 1986;2:1366–1367.
17. Weinreb JC, Wolbrsht LB, Cohen JM, et al: Prevalence of lumbosacral intervertebral disk abnormalities on MR images in pregnant and asymptomatic nonpregnant women. *Radiology* 1989;170:125–128.
18. Wiesel SW, Tsourmas N, Feffer HL, et al: A study of computer-assisted tomography: I. The incidence of positive CAT scans in an asymptomatic group of patients. *Spine* 1984;9:549–551.
19. Haldeman S: Presidential Address, North American Spine Society: Failure of the pathology model to model to predict back pain. *Spine* 1990;15:718–724.
20. Handal JA, Knapp JA, Poletti SC: The structural degenerative cascade: The cervical spine, in White AH, Schofferman JA (eds), *Spine Care: Diagnosis and Conservative Treatment*. St Louis: Mosby, 1995:16–23.
21. Bogduk N: The clinical anatomy of the cervical dorsal rami. *Spine* 1982;7:35–45.
22. Barnsley L, Bogduk N: Medial branch blocks are specific for the diagnosis of cervical zygapophyseal joint pain. *Reg Anaesth* 1993;18:343–350.
23. Dreyfuss P, Michaelsen M, Fletcher D: Atlanto-occipital and lateral atlanto-axial joint pain patterns. *Spine* 1994;19:1125–1131.
24. Mooney V: Facet syndrome, in Weinstein JN, Wiesel SW (eds), *The Lumbar Spine*. Philadelphia: Saunders, 1990:422.

25. Cavanaugh JM, Weinstein JN: Low back pain: Epidemiology, anatomy and neurophysiology, in Wall PD, Melzack R (eds), *Textbook of Pain,* 3d ed. London: Churchill Livingstone, 1994:441–455.
26. Kuslich SD, Ulstrom CL, Michael CJ: The tissue origin of low back pain and sciatica: A report of pain response to tissue stimulation during operation on the lumbar spine using local anesthesia. *Orthop Clin North Am* 1991; 22:181–187.
27. Murphy P: Sources and patterns of pain in disk disease. *Clin Neurosurg* 1968;15:343–350.
28. Inman VT, Saunders JB: Referred pain from skeletal structures. *J Nerv Mont Dis* 1944;99:660.
29. Kellgren JH: The anatomical source of back pain. *Rheumatol Rehabil* 1977;16:3–12.
30. Katz N: Neck and arm pain, in Samuels M, Feske S (eds), *Office Practice of Neurology.* New York: Churchill Livingston, 1996:1193–1206.
31. Selby DK: The structural degenerative cascade: The lumbar spine, in White AH, Schofferman JA (eds), *Spine Care: Diagnosis and Conservative Treatment.* St Louis: Mosby, 1995:8–16.
32. Nachemson AL: Intradiskal measurements of pH in patients. *Acta Orthop Scand* 1969;40:23–41.
33. Weinstein JN, Claverie W, Gibson S: The pain of diskography. *Spine* 1988;13:1344–1348.
34. Kawakami, Chatenia K, Weinstein JN: Anatomy, biochemistry, and physiology of low back pain, in White AH, Schofferman JA (ed), *Spine Care: Diagnosis and Conservative Treatment.* St Louis: Mosby, 1995:84–103.
35. Payan DG, McGillis JP, Goetzl ET: Neuroimmunology. *Adv Immunol* 1986;39:299.
36. Weinstein JN: The role of neurogenic and non-neurogenic mediators as they related to pain in the development of osteoarthritis (a clinical view). *Spine* 1992;17:S356.
37. Saal JA: The role of inflammation in lumbar pain: Physical medicine and rehabilitation. *State of the Art Rev* 1990;4:191–199.
38. Saal JS, Franson R, Dobrow J, et al: High levels of inflammatory phospholipase A_2 activity in lumbar disk herniations. *Spine* 1990;15:674–678.
39. Lui G, Ishihara H, Ryusuke O, et al: Nitric oxide mediates the change of proteoglycan synthesis in the human lumbar intervertebral disk in response to hydrostatic pressure. *Spine* 2001;26:134–141.
40. Rydevik BL: The effects of compression on the physiology of nerve roots. *J Manip Physiol Ther* 1992;1:62–66.
41. Olmarker K, Rydevik B: Pathophysiology of spinal nerve roots as related to sciatica and disk herniation, in Herkowitz HN, Garfin SR, Balderston RA, et al (eds), *Rothman-Simeone Studies: The Spine.* Philadelphia: Saunders, 1999:159–172.
42. Olmarker K, Rydevik B, Holm S: Edema formation in spinal nerve roots induced by experimental, graded compression: An experimental study on the pig cauda equina with special reference to differences in effects between rapid and slow onset of compress. *Spine* 1989;14:579–583.
43. Olmarker K, Holm S, Rydevik B: Importance of compression onset rate for the degree of impairment of impulse propagation in experimental compression injury of the porcine cauda equina. *Spine* 1990;15:416–419.
44. Olmarker K, Holm S, Rosenqvist A-L, et al: Experimental nerve root compression: Presentation of a model for acute, graded compression of the porcine cauda equina, with analysis of neural and vascular anatomy. *Spine* 1991;16:61–69.
45. McCarron RF, Wimpee MW, Hudkins PG, et al: The inflammatory effect of nucleus pulposus: A possible element in the pathogenesis of low-back pain. *Spine* 1987;12:760–764.
46. Olmarker K, Rydevik B, Nordborg C: Autologous nucleus pulposus induces neurophysiological and histologic changes in porcine cauda equina nerve roots. *Spine* 1993; 18:1425–1432.
47. Olmarker K, Rydevik B, Nordborg C: Ultrastructural changes in spinal nerve roots induced by autologous nucleus pulposus. *Spine* 1996;21:411–414.
48. Olmarker K, Brisby H, Yabuki S, et al: The effects of normal, frozen and hyaluronidase digested nucleus pulposus on nerve root structure and function. *Spine* 1997;24:471–475.
49. Iwabuchi M, Rydevik B, Kikychi S, et al: Effects of annulus fibrosis and experimentally degenerated nucleus pulposus on nerve root conduction velocity. *Spine* 2001;26:1651–1655.
50. Kang JD, Georgescu HI, MacIntyre-Larkin L, et al: Herniated lumbar intervertebral disks spontaneously produce matrix metalloproteinases, nitric oxide, interleukin-6 and prostaglandin E_2. *Spine* 1996;21:271–277.
51. Myers RR: The pathogenesis of neuropathic pain. *Reg Anaesth* 1962;31:91–114.
52. Sorkin LS, Xiao WH, Wagner R, et al: TNF-alpha applied to the sciatic nerve trunk elicits background firing in nociceptive primary afferents fibers, in 8th World Congress for Pain, IASP Abstracts. Seattle: IASP Press, 1996:354.
53. Wagner R, Myers RR: Endoneurial injection of TNF-alpha procedures nociceptive pain behaviors. *Neuroreport* 1996;7:2897–2901.
54. Probert L, Akassoglou K, Kassiotos G, et al: TNF-alpha transgenic and knockout models of CNS inflammation and degeneration. *J Neuroimmunonol* 1997;72:137–141.
55. Redford EJ, Hall SM, Smith KJ: Vascular changes and demyelination induced by the intraneural injection of tumor necrosis factor. *Brain* 1995;118:869–878.
56. Manchikanti L, Staats PS, Singh VJ, et al: Evidence-based guidelines for interventional techniques in the management of chronic spinal pain. *Pain Phys* 2003;6:3–81.
57. Bogduk N: The zygapophyseal joints, in *Clinical Anatomy of the Lumbar Spine and Sacrum,* 3d ed. New York: Churchill Livingstone, 1997:34–41.
58. Fujiwara A, Tamai K, Yamato M, et al: The relationship between facet joint osteoarthritis and disk degeneration of the lumbar spine: An MRI study. *Eur J Spine* 1999;8:396–401.
59. Fujiwara A, Tamai K, An HS: The relationship between disk degeneration, facet joint osteoarthritis, and stability of the degenerative lumbar spine. *J Spinal Disord* 2000;13:444–450.
60. Thompson RE, Pearcy MJ, Downing KJW: Disk lesions and the mechanics of the intervertebral joint complex. *Spine* 2000;25:3026–3035.
61. Fujiwara A, Lim T, An H, et al: The effect of disk degeneration and facet joint osteoarthritis on the segmental flexibility of the lumbar spine. *Spine* 2000;25:3036.
62. Moore RJ, Crotti TN, Osti OL, et al: Osteoarthrosis of the facet joints resulting from annular rim lesions in sheep lumbar disks. *Spine* 1999;24:519–525.

63. Indehl A, Kaigle AM, Reikeras O, et al: Interaction between the porcine intervertebral disk, zygapophyseal joints, and paraspinal muscles. *Spine* 1997;22:2834–2840.
64. Fukui S, Ohseto K, Shiotani M, et al: Referred pain distribution of the cervical zygapophyseal joints and cervical dorsal rami. *Pain* 1996;68:79–83.
65. Dwyer A, Aprill C, Bogduk N: Cervical zygapophyseal joint pain patterns: A study in normal volunteers. *Spine* 1990;6:453–457.
66. Aprill C, Dwyer A, Bogduk: The prevalence of cervical zygapophyseal joint pain patterns: II. A clinical evaluation. *Spine* 1990;6:458–461.
67. Pawl RP: Headache, cervical spondylosis, and anterior cervical fusion. *Surg Ann* 1977;9:391–498.
68. Dreyfuss P, Tibiletti C, Dreyer SJ: Thoracic zygapophyseal joint pain patterns: A study in normal volunteers. *Spine* 1994;19:807–811.
69. Mooney V, Robertson J: The facet syndrome. *Clin Orthop* 1976;115:149–156.
70. McCall IW, Park WM, O'Brien JP: Induced pain referral from posterior elements in normal subjects. *Spine* 1979;4:441–446.
71. Marks R: Distribution of pain provoked from lumbar facet joints and related structures during diagnostic spinal infiltration. *Pain* 1989;39:37–40.
72. Fukui S, Ohseto K, Shiotani M, et al: Distribution of referral pain from the lumbar zygapophyseal joints and dorsal rami. *Clin J Pain* 1997;13:303–307.
73. Hirsch C, Ingelmark BE, Miller M: The anatomical basis for low back pain. *Acta Orthop Scand* 1963;33:1–17.
74. Windsor RE, King FJ, Roman SJ, et al: Electrical stimulation induced lumbar medial branch referral patterns. *Pain Phys* 2002;5:405–418.
75. Ashton IK, Ashton BA, Gibson SJ, et al: Morphological basis for back pain: The demonstration of nerve fibers and neuropeptides in the lumbar facet joint capsule and not in libamentum flavum. *J Orthop Res* 1992;10:72.
76. Giles LG, Harcy AR: Immunohistochemical demonstration of nociceptors in the capsule and synovial folds of human zygapophyseal joints. *Br J Rheumatol* 1987;26:362.
77. Barnsley L, Lord SM, Wallis BJ, et al: The prevalence of chronic cervical zygapophyseal joint pain after whiplash. *Spine* 1995;20:20–26.
78. Lord SM, Barnsley L, Wallis BJ, et al: Chronic cervical zygapophyseal joint with whiplash: A placebo-controlled prevalence study. *Spine* 1996;21:1737–1745.
79. Manchikanti L, Singh V, Rivera J, et al: Prevalence of cervical facet joint pain in chronic neck pain. *Pain Phys* 2002;5:243–249.
80. Manchikanti L, Singh V, Pampati S, et al: Is there correlation of facet joint pain in lumbar and cervical spine? *Pain Phys* 2002;5:365–371.
81. Manchikanti L, Singh V, Pampati S, et al: Evaluation of the prevalence of facet joint pain in chronic thoracic pain. *Pain Phys* 2002;5:354–359.
82. Manchikanti L, Singh V, Pampati S, et al: Evaluation of the relative contributions of various structures in chronic low back pain. *Pain Phys* 2002;4:308–316.
83. Schwarzer AC, Aprill CN, Derby R, et al: Clinical features of patients with pain stemming from the lumbar zygapophyseal joints: Is the lumbar facet syndrome a clinical entity? *Spine* 1994;19:1132–1137.
84. Schwarzer AC, Aprill CN, Derby R, et al: The relative contributions of the disk and zygapophyseal joint in chronic low back pain. *Spine* 1994;19:801–806.
85. Schwarzer AC, Wang SC, Bogduk, et al: Prevalence and clinical features of lumbar zygapophyseal joint pain: A study in an Australian population with chronic low back pain. *Am Rheum Dis* 1995;54:100–106.
86. Manchikanti L. Pampati V, Fellows B, et al: Prevalence of lumbar facet joint pain in chronic low back pain. *Pain Phys* 1999;2:59–64.
87. Manchikanti L, Pampati V, Fellows B, et al: The diagnostic validity and therapeutic value of medial branch blocks with or without adjuvants. *Curr Rev Pain* 2000;4:337–344.
88. Manchikanti L, Pampati V, Fellows B, et al: The inability of the clinical picture characterize pain from facet joints. *Pain Phys* 2000;3:158–166.
89. Slipman CW, Huston CW: Diagnostic sacroiliac joint injections, in Manchikanti L, Slipman CW, Fellows B (eds), *Interventional Pain Management: Low Back Pain—Diagnosis and Treatment.* Paducah, KY: ASIPP Publishing, 2002:269–274.
90. Fortin JD, Kissling RO, O'Connor BL, et al: Sacroiliac joint innervation and pain. *Am J Orthop* 1999;28:687–690.
91. Grob Kr, Neuhuber WL, Kissling RO: Innervation of the sacroiliac joint of the human. *Z Rheumatol* 1995;54:117–122.
92. Vilensky JA, O'Connor BL, Fortin JD, et al: Histologic analysis of neural elements in the human sacroiliac joint. *Spine* 2002;27:1202–1207.
93. Sakamoto N, Yamashita T, Takebayashi K, et al: An electrophysiologic study of mechanoreceptors in the sacroiliac joint and adjacent tissues. *Spine* 2001;26:E468–E471.
94. Murata Y, Takahashi K, Yamagata M, et al: Origin and pathway of sensory nerve fibers to the ventral and dorsal sides of the sacroiliac joint in rates. *J Orthop Res* 2001;19:379–383.
95. Fortin JD, Dwyer AP, West S, et al: Sacroiliac joint: Pain referral maps upon applying a new injection/arthrography technique: I. Asymptomatic volunteers. *Spine* 1994;19:1475–1482.
96. Fortin JD, Aprill CN, Ponthieux B, et al: Sacroiliac joints: Pain referral maps upon applying a new injection/arthrography technique: II. Clinical evaluation. *Spine* 1994;19:1483–1489.
97. Slipman CW, Jackson HB, Lipetz JK, et al: Sacroiliac joint pain referral zones. *Arch Phys Med Rehabil* 2000;20:31–37.
98. Schwarzer AC, Aprill CN, Bogduk M: The sacroiliac joint in chronic low back pain. *Spine* 1995;20:31–37.
99. Maigne JY, Aivakiklis A, Pfefer F: Results of sacroiliac joint double block and value of sacroiliac pain provocation test in 54 patients with low back pain. *Spine* 1996;21:1889–1892.
100. Wheeler AH, Aaron GW: Muscle pain due to injury. *Curr Pain Headache Rep* 2001;5:441–446.
101. Simons DG, Travell JG, Simons LS: *Myofascial Pain and Dysfunction: The Trigger Point Manual,* 2d ed. Baltimore: Williams & Wilkins, 1999.
102. Hubbard DR, Berkoff GM: Myofascial trigger points show spontaneous needle EMG activity. *Spine* 1993;18:1803–1807.
103. Hubbard DR: Chronic and recurrent muscle pain: Pathophysiology and treatment, and review of pharmacologic studies. *J Musculoskel Pain* 1996;4:123–143.

104. Rivner MH: The neurophysiology of myofascial pain syndrome. *Curr Pain Headache Rep* 2000;5:432–440.
105. Fields HL: *Pain.* New York: McGraw-Hill, 1987.
106. Schofferman JA: Applied neurophysiology of pain, in White AH, Schofferman JA (eds), *Spine Care: Diagnosis and Conservative Treatment.* St Louis: Mosby, 1995:23–26.
107. Russell IJ: Neurochemical pathogenesis of fibromyalgia syndrome. *J Musculoskel Pain* 1996;4:61–92.
108. Wheeler AH, Murrey DB: Chronic lumbar spine and radicular pain: Pathophysiology and treatment. *Curr Pain Headache Rep* 2001;6:97–105.
109. Mense S: Biochemical pathogenesis of myofascial pain. *J Musculoskel Pain* 1996;4:145–162.
110. Pillemer SE, Bradley LA, Crofford LJ, et al: The neuroscience and endocrinology of fibromyalgia. *Arthritis Rheum* 1997;40:1928–1939.
111. Carlton SM, Zhou S, Coggeshal RE: Evidence for the interaction of glutamate and NK1 receptors in the periphery. *Brain Res* 1998;790:160–169.
112. Bennett RM: Emerging concepts in the neurobiology of chronic pain: Evidence of abnormal sensory processing in fibromyalgia. *Mayo Clin Proc* 1999;74:385–398.
113. Wall PD, Woolf CJ: Muscle but not cutaneous C-afferent input produces prolonged increases in the excitability of the flexion reflex in the rat. *J Physiol* 1984;356:443–458.
114. Mendell LM, Wall PD: Responses of single dorsal cord cells to peripheral cutaneous unmyelinated fibers. *Nature* 1965;206:97–99.
115. Roberts WJ: A hypothesis on the physiological basis for causalgia and related pains. *Pain* 1986;24:297–311.
116. Wheeler AH: Evolutionary mechanisms in chronic low back pain and rationale for treatment. *Am J Pain Manage* 1995;5:62–66.
117. Killian LE: Psychological barriers to recovery, in Isernhagen SJ (ed), *Work Injury Management and Prevention.* Gaithersburg, MD: Aspen, 1988:247–257.
118. Polatin PB, Kinny RK, Gatchel RJ, et al: Psychiatric illness in chronic low back pain: The mind and the spine, which goes first? *Spine* 1993;18:66–71.
119. Krishnan KRR, France RD, Pelton S, et al: Chronic pain and depression: I. Classification of depression in chronic low back pain patients. *Pain* 1985;22:279–287.
120. Krishnan KRR, France RD, Pelton S, et al: Chronic pain and depression: II. Symptoms of anxiety in chronic low back pain patients and their relationship to subtypes of depression. *Pain* 1985;22:289–294.
121. France RD, Krishnan KRR, Trainor M: Chronic pain and depression: III. Family history study of depression and alcoholism in chronic low back pain patients. *Pain* 1986;24:185–190.
122. Dersh J, Gatchel RJ, Polatin P: Chronic spinal disorders and psychopathology: Research findings and theoretical considerations. *Spine J* 2001;1:88–94.
123. Blumer D, Heilbronn M: Chronic pain as a variant of depression: The pain prone disorder. *J Nerv Ment Dis* 1982;170:381–406.
124. Thompson TL II, Byyny RL: Pain problems in primary care medial practice, in Tollison CD (ed), *Handbook of Chronic Pain Management.* Baltimore: Williams & Wilkins, 1989:532–559.
125. Gross RJ, Doerr H, Caldiorola D, et al: Borderline syndrome and incest in chronic pelvic pain patients. *Int J Psychiatr Med* 1981;10:79–96.
126. Blazer DG: Narcissism and development of chronic pain. *Int J Psychiatr Med* 1981;10:69–71.
127. Ingram RE, Atkinson JH, Slater MA, et al: Negative and positive cognition in depressed and nondepressed chronic pain patients. *Health Psychol* 1990;9:300–314.
128. Kaplan HI, Sadock BJ: *Synopsis of Psychiatry: Behavior Sciences, Clinical Psychiatry,* 6th ed. Baltimore: Williams & Wilkins, 1991.
129. Keefe FJ, Bedcham JC, Fillingim: The psychology of chronic back pain, in Frymoyer JW (ed), *The Adult Spine: Principles and Practice.* New York: Raven Press, 1991:185–197.
130. Lee PWH, Chow FL, Chan KC, Wong S: Psychosocial factors in influencing outcome in patients with low back pain. *Spine* 1989;14:838–843.
131. Cheatle MD, Brady JP, Ruland T: Chronic low back pain: Depression and attribution styles. *Clin J Pain* 1990;6:114.
132. Bigos S, Battie MC, Spengler DM, et al: A prospective study of work perceptions and psychosocial factors affecting the reports of back injuries. *Spine* 1991;16:1–6.
133. Wheeler AH: Diagnosis and management of low back pain and sciatica. *Am Fam Phys* 1995;552:133–141.
134. Wheeler AH, Stubbart J, Hicks B: Pathophysiology of chronic low back pain. *EMedicine* 2000 (online).
135. Galer B: Neuropathic pain of peripheral origin: Advances in pharmacologic treatment. *Neurology* 1995;45:S17–S25.
136. Backonja M-M, Galer BS: Pain assessment and evaluation of patients who have neuropathic pain. *Neurol Clin* 1998;16:775–789.
137. Waddell G, McCulloch JA, Kummel E, Venner RM: Nonorganic physical signs in low-back pain. *Spine* 1980;5:117–124.
138. Schofferman JA: Lumbar spine disorders: Taking and interpreting the history, in White AH, Schofferman JA (eds), *Spine Care: Diagnosis and Conservative Treatment.* St Louis: Mosby, 1995:52–70.
139. Margolis RB, Tait RC, Krause SJ: A rating system for use with patient pain drawings. *Pain* 1986;24:57–65.
140. Ransford AO, Cairns D, Mooney V: The pain drawing as an aid to the psychological evaluation of patients with low-back pain. *Spine* 1976;1:127–134.
141. Melzack R: The McGill Pain Questionnaire: Major properties and scoring methods. *Pain* 1986;24:57–65.
142. Murphy DF, et al: Measurement of pain: A comparison of the visual analogue with a nonvisual analogue scale. *Clin J Pain* 1988;3:197–199.
143. Duncan GH, Buchnell C, Lavigne GJ: Comparison of verbal and visual analogue scales for measuring the intensity and unpleasantness of experimental pain. *Pain* 1989;37:295–303.
144. Fairbank JC, Cooper J, Davies JB, O'Brien JP: The Oswestry low back pain disability questionnaire. *Physiotherapy* 1980;66:271–273.
145. Chibnall JT, Tait RC: The pain disability index: Factor structure and normative data. *Arch Phys Med Rehabil* 1994;75:1082–1086.

146. Roland M, Morris R: Study of natural history of back pain: I. Development of a reliable and sensitive measure of disability in low-back pain. *Spine* 1983;8:141.
147. Pietrobon R, Coeytaux RR, Carey TS, et al: Standard scales for measurement of functional outcome for cervical pain of dysfunction: a systematic review. *Spine* 2002; 27:515–522.
148. Wheeler AH, Goolkasian P, Baird AC, et al: Development of the neck pain and disability scale: Item analysis, face and criterion related validity. *Spine* 1999;24:1290–1294.
149. Goolkasian P, Wheeler AH, Gretz SS: The neck pain and disability scale: Test-retest reliability and construct validity. *Clin J Pain* 2002;18:–250.
150. Lawlis GF, Cuencas R, Selby D, McCoy CE: The development of the Dallas Pain Questionnaire: An assessment of the impact of spinal pain on behavior. *Spine* 1989;14:511–516.
151. The North American Spine Society's Ad Hoc Committee on Diagnostic and Therapeutic Procedures: Common diagnostic and therapeutic procedures of the lumbosacral spine. *Spine* 1991;16:1161–1167.
152. Anderson SR, Flanagan B: Diskography. *Cur Rev Pain* 2000;4:345–352.
153. Lindblom K, Hultqvist G: Absorption of protruded disk tissue. *J Bone Joint Surg* 1950;32A:557–560.
154. Saal JA, Saal JS, Herzog RJ: The natural history of lumbar intervertebral disk extrusions treated nonoperatively. *Spine* 1990;15:683–686.
155. Maigne JY, Rime B, Deligne B: Computed tomographic follow-up study of forty-eight cases of non-operatively treated lumbar intervertebral disk herniation. *Spine* 1992;17:1071–1074.
156. Komori H, Okawa A, Hirataka H, et al: Contrast-enhanced magnetic resonance imagining in conservative management of lumbar disk herniation. *Spine* 1998;23:67–62.
157. Saal JA, Saal JS: Nonoperative treatment of herniated lumbar intervertebral disk with radiculopathy: An outcome study. *Spine* 1989;14:431–437.
158. Saal JA, Saal JS, Yurth EF: Nonoperative management of herniated cervical intervertebral disk with radiculopathy. *Spine* 1996;21:1877–1883.
159. Weber H: Lumbar disk herniation: A controlled, prospective study with ten years of observation. *Spine* 1983;8:131–140.
160. Gatchel RJ, Mayer TG, Hazard RG, et al: Editorial: Functional restoration: Pitfalls in evaluating efficacy. *Spine* 1992;17:988–995.
161. Deyo RA, Diehl AK, Rosenthal M: How many days of bed rest for low back pain? A randomized clinical trial. *N Engl J Med* 1986;315:1064–1070.
162. Deyo RA: Nonoperative treatment of low back disorders: Differentiated useful from useless therapy, in Frymoyer JW, Ducker TB, Hadler NM, et al (eds), *The Adult Spine: Principles and Practice*. Philadelphia: Lippincott-Raven, 1997:1777–1793.
163. Harkens S, Linford J, Cohen J, et al: Administration of clonazepam in the treatment of TMD and associated myofascial pain: a double-blind pilot study. *J Craniomandib Disor* 1991;179–186.
164. Waldman SD: Recent advances in analgesic therapy—Tizanidine. *Pain Digest* 1999;9:40–43.
165. Berry H, Hutchinson DR: A multicenter placebo-controlled study in general practice to evaluate the safety and efficacy of tizanidine in acute low back pain. *J Int Med Res* 1988;16:75–82.
166. Berry H, Hutchinson DR: Tizanidine and ibuprofen in acute low back pain: Results of a multicenter double-blind study in general practice. *J Int Med Res* 1988;16:83–91.
167. Fryda-Kaurimsky Z, Muller-Fassbender H: Tizanidine in the treatment of acute paravertebral spasms: A controlled trial comparing tizanidine with diazepam. *J Int Med Res* 1981;9:501–505.
168. Tse FLS, Jaffe JM, Bhuta S: Pharmacokinetics of tizanidine in health volunteers. *Fundam Clin Pharmacol* 1987; 1:479–488.
169. Rosenberg JM, Harrell C, Rishi H, et al: The effect of gabapentin on neuropathic pain. *Clin J Pain* 1997;13:251–255.
170. Deyo RA: Drug therapy for back pain: Which drugs help which patients? *Spine* 1996;21:2840–2850.
171. Sindrup SH, Jensen TS: Efficacy of pharmacological treatments of neuropathic pain: An update and effect related to mechanism of drug action. *Pain* 1999;83:389–400.
172. Watson CP: The treatment of neuropathic pain: antidepressants and opioids. *Clin J Pain* 2000;16:S49–S55.
173. Pancrazio JJ, Kamatchi GL, Roscoe AK, et al: Inhibition of neuronal Na^+ channels by antidepressant drugs. *J Pharmacol Exp Ther* 1998;284:208–214.
174. Jacobson LO, Bley K, Hunter JC, et al: Anti-thermal hyperalgesic properties of antidepressants in a rat model of neuropathic pain. Abstract presented at the American Pain Society Annual Meeting, Los Angeles, CA, 1995.
175. Taylor K, Rowbotham MC: Venlafaxine for chronic pain. Abstract presented at the American Pain Society Annual Meeting, Los Angeles, CA, 1995.
176. Tunali D, Jefferson JW, Geist JH: *Depression and Antidepressants: A Guide*. Madison, WI: Information Centers, Madison Institute of Medicine, 1999.
177. Ahlgren SA, Hansen T: The use of lumbosacral corsets prescribed for low back pain. *Prosthet Orthot Int* 1978;2:101–104.
178. Million R, Nilsen H, Jayson MIV, et al: Evaluation of low back and assessment of low back pain and assessment of lumbar corsets with and without back supports. *Ann Rheum Dis* 1981;40:449–454.
179. Nachemson A, Schultz A, Andersson G: Mechanical effectiveness studies of lumbar spine orthoses. *Scand J Rehabil Med Suppl* 1983;15:139–149.
180. Axelsson P, Johnsson R, Stromqvist B: Effect of lumbar orthosis on intervertebral mobility: A roentgen stereophotogrammertric analysis. *Spine* 1992;17:678–681.
181. Ottenbacher K, Difabio RP: Efficacy of spinal manipulation/mobilization therapy: A meta-analysis. *Spine* 1985; 10:833–837.
182. Triano J: Standards of care: Manipulative procedures, in White AH, Anderson R (eds), *Conservative Care of Lower Back Pain*. Baltimore: Williams & Wilkins, 1991:159–168.
183. Shekelle PG: Spinal manipulation. *Spine* 1996;19:858–861.
184. Hall H: *Back School and Education: Non-Operative Care of Lumbar Pain Syndromes*. Boston: North American Spine Society and Seton Medical Center, 1992:69–76.

185. Moffett JA, Chase SM, Portek I, et al: A controlled, prospective study to evaluate the effectiveness of a back school in the relief of chronic low back pain. *Spine* 1986;11:120–122.
186. Bergquist-Ullman M, Larsson U: Acute low back pain in industry. *Acta Orthop Scand Suppl* 1977;17:1–150.
187. Hall H, Iceton J: Back school: An overview with specific reference to the Canadian back education units. *Clin Orthop* 1983;179:10–17.
188. Robison R: Low back school and stabilization: Aggressive conservative care, in White AG, Schofferman JA (eds), *Spine Care: Diagnosis and Conservative Treatment*. St. Louis: Mosby, 1995:394–412.
189. Hazard R, Fenwick J, Kalisch S, et al: Functional restoration with behavioral support: A one-year prospective study of chronic low back pain. *Spine* 1989;14:157–161.
190. Lindstrom I, Ohlund C, Eck C, et al: Mobility, strength and fitness after a graded activity program for patients with subacute low back pain. *Spine* 1992;17:641–652.
191. Manniche C, Asmussen K, Lauritsen B, et al: Intensive dynamic exercises for chronic low back pain: A clinical trial. *Pain* 1991;47:53–63.
192. Power RA, Taylor GJ, Fyfe IS: Lumbar epidural injection of steroid in acute prolapsed intervertebral disks. *Spine* 1992;17:453–455.
193. van Tulder MW, Goossens M, Waddell G, Nachemson A. Conservative treatment of chronic low back pain, in Nachemson A, Jonsson E (eds): Neck and back pain. The sceintific evidence of causes, diagnosis and treatment. Philadelphia Lippincott Williams & Wilkins, 2000: 271–303.
194. Hides JA, Tull GA, Richardson CA: Long-term effects of specific stabilization exercises for first-episode low back pain. *Spine* 2001;26:E23–E248.
195. Mayer T, Gatchel R, Mayer H, et al: A prospective two-year study of functional restoration in industrial low back injury: An objective assessment procedure. *JAMA* 1987;258:1763–1767.
196. Burke SA, Harms C, Aden PS: The impact of physical and non-physical factors on return to work within a functional restoration program: A multi-center, prospective study with comparison group. *Spine* 1994;19:1880–1885.
197. Frost FA, Jessen B, Siggaard-Anderson J: A controlled, double-blind comparison of mepivacaine injection versus saline injection for myofascial pain. *Lancet* 1980; 1:499–501.
198. Garvey TA, Marks MR, Wiesel SW: A prospective, randomized, double-blind evaluation of trigger-point injection therapy for low back pain. *Spine* 1989;14:962–964.
199. Knusel B, DeGryse R, Grant M, et al: Intramuscular injection of botulinum toxin type A (Botox) in chronic low back pain associated with muscle spasm (poster abstract), at American Pain Society Annual Meeting, San Diego, Nov. 5–8, 1998.
200. Foster L, Clapp L, Erickson M, et al: Botulinum toxin A and chronic low back pain: A randomized double-blind study. *Neurology* 2001;56:1920–1923.
201. Wheeler AH, Goolkasian P: Open label assessment of botulinum toxin A for pain treatment in a private outpatient setting. *J Musculoskel Pain* 2001;9:67–82.
202. Cheshire WP, Abashjan SW, Mann JD: Botulinum toxin in the treatment of myofascial pain syndrome. *Pain* 1994;59:65–69.
203. Yue SK: Initial experience in the use of botulinum toxin A for the treatment of myofascial related muscle dysfunction. *J Musculoskel Pain* 1995;3:22.
204. Wheeler AH, Goolkasian P, Gretz SS: A randomized, double-blind, prospective pilot study of botulinum toxin injection for refractory, unilateral, cervicothoracic paraspinal, myofascial pain syndrome. *Spine* 1998; 23:1662–1667.
205. Wheeler AH, Goolkasian P, Gretz SS: Botulinum toxin A for the treatment of chronic neck pain. *Pain* 2001;94:255–260.
206. Porta M: A comparative trial of botulinum toxin type A and methylprednisolone for the treatment of myofascial pain syndrome and pain from chronic muscle spasm. *Pain* 2000;85:101–105.
207. Freund BJ, Schwartz M: Treatment of whiplash associated with neck pain with botulinum toxin-A: A pilot study. *J Rheumatol* 2000;27:481–484.
208. Carette S, Marcoux S, Truchon R, et al: A controlled trial of corticosteroid injections into facet joints for chronic low back pain. *N Engl J Med* 1991;325:1002–1007.
209. Marks RC, Houston T, Thulbourne T: Facet joint injection and facet nerve block: A randomized comparison in 86 patients with chronic low back pain. *Pain* 1992;49:325–328.
210. Nash TP: Facet joints. Intra-articular steroids or nerve blocks? *Pain Clin* 1990;3:77–82.
211. Lilius G, Laasonen EM, Myllynen P, et al: Lumbar facet joint syndrome: A randomized clinical trial. *J Bone Joint Surg* 1989;71B:681–684.
212. Barnsley L, Lord SM, Wallis BJ, et al: Lack of effect of intra-articular corticosteroids for chronic pain in the cervical zygapophyseal joints. *N Engl J Med* 1994;330:1047–1050.
213. Mironer YE, Somerville JJ: Protocol for diagnosis and treatment of facet joint pain syndrome: A modified three-step approach. *Pain Digest* 1999;9:188–190.
214. Desoutet JM, Gilula LA, Murphy WA, et al: Lumbar facet joint injection: Indication, technique, clinical correlation, and preliminary results. *Radiology* 1982;145:321–325.
215. Murtagh FR: Computed tomography and fluoroscopy guided anesthesia and steroid injection in facet syndrome. *Spine* 1988;13:686–689.
216. Lippitt AB: The facet joint and its role in spine pain: Management with facet joint injections. *Spine* 1984; 9:746–750.
217. Lau LS, Littlejohn GO, Miller MH: Clinical evaluation of intra-articular injections for lumbar facet joint pain. *Med J Aust* 1985;143:563–565.
218. Jackson RP, Jacobs RR, Montesano PX: Facet joint injection in low-back pain. *Spine* 1988;13:966–971.
219. Jackson RP: The facet syndrome: Myth or reality? *Clin Orthop* 1992;279:110–121.
220. Manchikanti L, Pampati V, Bakhit CE, et al: Effectiveness of lumbar facet joint nerve blocks in chronic low back pain: A randomized clinical trial. *Pain Phys* 2001;4:101–117.
221. Barnsley L, Bogduk N: Medial branch blocks are specific for the diagnosis of cervical zygapophyseal joint pain. *Reg Anaesth* 1993;18:343–350.

222. North RB, Han M, Zahurak M, et al: Radiofrequency lumbar facet denervation: Analysis of prognostic factors. *Pain* 1994;57:77–83.
223. Manchikanti K, Singh V, Vilims B, et al: Medial branch neurotomy in management of chronic spinal pain: Systematic review of the evidence. *Pain Phys* 2002;5:405–418.
224. Lord SM, Barnsley L, Bogduk N: Percutaneous radiofrequency neurotomy in the treatment of cervical zygapophyseal joint pain: A caution. *Neurosurgery* 1995;35:732–739.
225. Hammer M, Meneese W: Principles and practice of radiofrequency neurolysis. *Cur Rev Pain* 1998;2:267–278.
226. Breivik H, Hesla PE, Molnar I, et al: Treatment of chronic low back pain and sciatica: Comparison of caudal epidural injections of bupivacaine and methylprednisolone with bupivacaine followed by saline, in Bonica JJ, Albe-Fessard D (eds), *Advances in Pain Research and Therapy,* Vol 1. New York: Raven Press, 1976:927–932.
227. Bush K, Hillier S: A controlled study of caudal epidural injections of triamcinolone plus procaine for the management of intractable sciatica. *Spine* 1991;16:572–575.
228. Manchikanti L, Pampati V, Rivera JJ, et al: Caudal epidural injections for treatment of chronic low back pain and sciatica. *Pain Phys* 2001;4:322–335.
229. Helsa PE, Breivik H: Epidural analgesia and epidural steroid injection for treatment of chronic low back pain and sciatica. *Tidsskr Nor Laegeforen* 1979;99:936–939.
230. Revel M, Auleley GR, Alaoui S, et al: Forceful epidural injections for the treatment of lumbosciatic pain with post-operative lumbar spinal fibrosis. *Rev Rhum Engl Ed* 1996;63:270–277.
231. Matthews JA, Mills SB, Jenkins VM, et al: Back pain and sciatica: Controlled trials of manipulation, traction, sclerosant and epidural injections. *Br J Rheumatol* 1987;26:416–423.
232. Manchikanti L, Singh V, Rivera J, et al: Effectiveness of caudal epidural injections in diskogram positive and negative chronic low back pain. *Pain Phys* 2002;5:18–29.
233. Yates DW: A comparison of the types of epidural injection commonly used in the treatment of low back pain and sciatica. *Rheum Rehabil* 1978;17:181–186.
234. Waldman SD: The caudal epidural administration of steroids in combination with local anesthetics in the palliation of pain secondary to radiographically documented lumbar herniated disk: A prospective outcome study with 6-months follow-up. *Pain Clin* 1998;11:43–49.
235. Hauswirth R, Michot F: Caudal epidural injection in the treatment of low back pain. *Ischw Med Wochenschr* 1982;112:222–225.
236. Manchikanti L, Pakanati RR, Pampati V: Comparison of three routes of epidural steroid injections in low back pain. *Pain Digest* 1999;9:277–285.
237. Goebert HW, Jallo SJ, Gardner WJ, et al: Painful radiculopathy treated with epidural injections of procaine and hydrocortisone acetate: Results in 113 patients. *Anesth Analg* 1961;140:130–134.
238. Clocon JO, Galindo-Clocon D, Amarnath L, et al: Caudal epidural blocks for elderly patients with lumbar canal stenosis. *J Am Geriatr Soc* 1994;42:593–596.
239. Koes BW, Scholten RJPM, Mens JMA, et al: Efficacy of epidural steroid injections for low back pain and sciatica: A systematic review of randomized clinical trials. *Pain* 1995;63:279–288.
240. Watts RW, Silagy CA: A meta-analysis on the efficacy of epidural corticosteroids in the treatment of sciatica. *Anaesth Intensive Care* 1995;23:565–569.
241. van Tulder MWV, Koes BW, Bouter LM: Conservative treatment of acute and chronic nonspecific low back pain: A systematic review of randomized controlled trials of the most common interventions. *Spine* 1997;22:2128–2156.
242. McQuay HJ, Moore RA: Epidural corticosteroids for sciatica, in *An Evidence-Based Resource for Pain Relief*. Oxford, England: Oxford University Press, 1998:216–218.
243. Vroomen PC, De Krom MC, Slofstra PD, et al: Conservative treatment of sciatica: A systematic review. *J Spin Disord* 2000;13:463–469.
244. Racz GB, Holubec JT: Lysis of adhesions in the epidural space, in Racz GB (ed), *Techniques of Neurolysis*. Boston: Kluwer Academic, 1989:57–72.
245. Manchikanti L, Singh V: Epidural lysis of adhesions and myeloscopy. *Curr Pain Headache Rep* 2002;6:427–435.
246. Anderson SR, Racz GB, Heavner J: Evolution of epidural lysis of adhesions. *Pain Phys* 2000;3:262–270.
247. Racz GB, Sabonghy M, Gintautas J, et al: Intractable pain therapy using a new epidural catheter. *JAMA* 1982;248:579–581.
248. Manchikanti L, Saini B, Singh V: Lumbar epidural adhesiolysis, in Manchikanti L, Slipman CW, Fellows B (eds), *Interventional Pain Management: Low Back Pain—Diagnosis and Treatment*. Paducah, KY: ASIPP Publishing, 2002:353–390.
249. Manchikanti L, Saini B, Singh V: Spinal endoscopy and lysis of epidural adhesions in the management of chronic low back pain. *Pain Phys* 2001;4:240–265.
250. Lewandowski EM: The efficacy of solutions used in caudal neuroplasty. *Pain Digest* 1997;7:323–330.
251. Saberski KR, Kitahata L: Review of the clinical basis and protocol for epidural endoscopy. *Conn Med* 1995;50:71–73.
252. Heavner JE, Chokhavatia S, Kizelshteyn G: Percutaneous evaluation of the epidural and subarachnoid space with the flexible fiberscope. *Reg Anaesth* 1991;15S1:85.
253. Saberski LR: Spinal endoscopy: current concepts, in Waldman SD (ed), *Interventional Pain Management,* 2d ed. Philadelphia: Saunders, 2000:143–161.
254. Heavner JE, Racz GB, Raj P: Percutaneous epidural neuroplasty: Prospective evaluation of 0.9% NaCl versus 10% NaCl with or without hyaluronidase. *Reg Anaesth Pain Med* 1999;24:202–207.
255. Manchikanti L, Pampati V, Fellows B, et al: Role of one day epidural adhesiolysis in management of chronic low back pain: A randomized clinical trial. *Pain Phys* 2001;4:153–166.
256. Manchikanti L, Pampati V, Bakhit CE, et al: Non-endoscopic and endoscopic adhesiolysis in post lumbar laminectomy syndrome: A one-year outcome study and cost effective analysis. *Pain Phys* 1999;2:52–58.
257. Racz GB, Holubec JT: Lysis of adhesions in the epidural space, in Racz GB (ed), *Techniques of Neurolysis*. Boston: Kluwer Academic, 1989:57–72.
258. Manchikanti L, Pakanati R, Bakhit CE, et al: Role of adhesiolysis and hypertonic saline neurolysis in management

of low back pain: Evaluation of modification of Racz protocol. *Pain Digest* 1999;9:91–96.
259. Geurts JW, Kaliewaard JW, Richardson J, et al: Targeted methylprednisoione acetate/hyaluronidase/clonidine injection after diagnostic epiduroscopy for chronic sciatica: A prospective, 3-year follow-up study. *Reg Anaesth Pain Med* 2002;27:343–352.
260. Richardson J, McGurgan P, Cheema S, et al: Spinal endoscopy in chronic low back pain with radiculopathy: A prospective case series. *Anaesthesia* 2001;56:454–460.
261. Manchikanti L: The value and safety of epidural endoscopic adhesiolysis. *Am J Anesthesiol* 2000;275–278.
262. Kim RC, Porter RW, Choi BH, et al: Myelopathy after intrathecal administration of hypertonic saline. *Neurosurgery* 1988;22:942–944.
263. Aldrete JA, Zapata JC, Ghaly R: Arachnoiditis following epidural adhesiolysis with hypertonic saline report of two cases. *Pain Digest* 1996;6:368–370.
264. Lou L, Racz G, Heavner J: Percutaneous epidural neuroplasty, in Waldman SD (ed), *Interventional Pain Management,* 2d ed. Philadelphia: Saunders, 2000:434–445.
265. Manchikanti L, Bakhit CE: Removal of torn Racz catheter from lumbar epidural space. *Reg Anaesth* 1997;22:579–581.
266. Simmons JW, McMillin JN, Emery SF, et al: Intradiskal steroids: A prospective, double-blind clinical trial. *Spine* 1992;17S:172–175.
267. Pinzon EG: Treating lumbar back pain. *Pract Pain Manag* 2001;April–May:14–20.
268. Karasek M, Bogduk N: Twelve-month follow-up of a controlled trial of intradiskal thermal anuloplasty for back pain due to internal disk disruption. *Spine* 2000; 25:2601–2607.
269. Bogduk N, Karasek M: Two-year follow-up of a controlled trial of intradiskal electrothermal anuloplasty for chronic low back pain resulting from internal disk disruption. *Spine J* 2002;2:343–350.
270. Pauza K, Howell S, Dreyfuss P, et al: A randomized, double-blind, placebo-controlled trial evaluating the efficacy of intradiskal electrothermal anuloplasty (IDET) for the treatment of chronic diskogenic low back pain: 6-month outcomes, in *Proceedings of the International Spinal Injection Society,* Austin, September 7, 2002.
271. Wetzel FT, McNally TA, Phillips FM: Intradiskal electrothermal therapy used to manage chronic diskogenic low back pain. *Spine* 2002;27:2621–2626.
272. Saal JA, Saal JS: Intradiskal electrothermal treatment for chronic diskogenic low back pain: A prospective outcome study with minimum 1-year follow-up. *Spine* 2000; 25:2622–2627.
273. Derby R, Eek B, Chen Y, et al: Intradiskal electrothermal anuloplasty (IDET): A novel approach for treating chronic diskogenic back pain. *Neuromodulation* 2000;3:82–88.
274. Singh V: Intradiskal electrothermal therapy: A preliminary report. *Pain Phys* 2000;3:367–373.
275. Endres SM, Fielder GA, Larson KL: Effectiveness of intradiskal electrothermal therapy in increasing function and reducing chronic low back pain in selected patients. *Wis Med J* 2002;101:31–34.
276. Saal JA, Saal JS: Intradiskal electrothermal treatment for chronic diskogenic low back pain: Prospective outcome study with a minimum 2-year follow-up. *Spine* 2002;27:966–974.
277. Gerszten PC, Welch WC, McGrath PM, et al: A prospective outcomes study of patients undergoing intradiskal electrothermy (IDET) for chronic low back pain. *Pain Phys* 2002;5:360–364.
278. Chen YC, Lee SH: Intradiskal pressure study with nucleoplasty in human cadaver, in *Proceedings of the International Spinal Injection Society (ISIS) 9th Annual Meeting,* Boston, September 2001.
279. Singh V, Piryani C, Liao K, et al: Percutaneous disk decompression, using Coblation (nucleoplasty) in the treatment of diskogenic pain. *Pain Phys* 2002;5:250–259.
280. Sharps LS, Isaac Z: Percutaneous disk compression using nucleoplasty. *Pain Phys* 2002;5:121–126.
281. Chen YC, Lee SH, Date ES, et al: Nucleoplasty (volumetric tissue ablation and coagulation of the nucleus) for chronic diskogenic back pain and/or radiculopathy: A preliminary 6-month follow-up study, in *Proceedings of the International Spinal Injection Society (ISIS) 9th Annual Meeting,* Boston, September 2001.
282. Slipman CW, Sharps L, Isaac Z, et al: Preliminary outcomes of percutaneous nucleoplasty for treatment of axial low back pain: A comparison of patients with versus without an associated central focal protrusion, in *Proceedings of the International Spinal Injection Society (ISIS) 10th Annual Meeting,* Austin, September 2002.
283. White AA, Southwick WO, Panjabi MM: Clinical instability in the lower cervical spine: A review of past and current concepts. *Spine* 1976;1:15.
284. Garfin SR, Yuan H, Reilly MA: New technologies in spine: Kyphoplasty and vertebroplasty for the treatment of painful vertebral compression fractures. *Spine* 2001;1:1311–1315.
285. Acre CA, Dohrmann GJ: Herniated thoracic disks. *Neurol Clin* 1985;3:383–392.
286. Benjamin V: Diagnosis and management of thoracic disk disease. *Clin Neurosurg* 1983;30:577–605.
287. Bigos S, Bowyer O, Braen G, et al: Acute low back problems in adults, in *Clinical Practice Guidelines No. 14,* AHCPR Publication No. 95-0642. Rockville, MD: Agency for Health Care Policy and Research, Public Health Service, U.S. Department of Health and Human Services, 1994.
288. Logue V: Thoracic intervetebral disk prolapse with spinal cord compression. *J Neurol Psychiatry* 1952;15:217–224.
289. Bohlman HH, Zdeblick TA: Anterior excision of herniated thoracic disks. *J Bone Joint Surg* 1988;70A:1038–1047.
290. Murrey DB, Hanley E: Drawback of diskograms, in Zdeblick et al (eds), *Controversies in Spine Surgery: Surgical Techniques and Medical Treatment,* Vol 2. Quality Medical Publishing, St. Louis 2001:234–239.
291. Boden SD, Davis DO, Dina TS, Patronas NJ, and Wiesel SW: Abnormal magnetic-resonance scans of the lumbar spine in asymptomatic subjects. *J Bone Joint Surg* 1990; 72A:403–408.
292. Hakelius A: Prognosis in sciatica: A clinical follow-up of surgical and non-surgical treatment. *Acta Orthop Scand* 1970;129:1–76.

CHAPTER 31

Chronic Pelvic Pain

Beverly J. Collett, Christine J. Cordle, Charles R. Stewart, and Kathleen Vits

Chronic pelvic pain (CPP) is nonmalignant pain perceived in the structures related to the pelvis of either men or women. CPP is a common, debilitating, and complex condition whose etiology remains poorly understood. Pain can be described as chronic if it has been present continuously or intermittently in the lower abdomen or pelvis for at least 6 months or when a nociceptive focus has healed and yet pain continues. For most clinicians, it is a frustating and complex problem to treat. Patients are distressed by their continuing symptoms, by extensive and repeated investigations, and often by the inability of the medical profession to diagnose and treat them effectively.

Many patients describe frustration at their inability to have their pain taken seriously or by suggestions that the pain may be due to psychological causes. Although rarely life threatening, significant morbidity can be associated with this pain, with loss of physical and sexual functioning. Treatment has focused on identifying pathology and using medical and surgical interventions to alleviate CPP. However, the traditional medical and surgical models have failed many patients with pelvic pain. Surgical and medical interventions give rise to their own disabilities. Some patients undergo multiple courses of medication and numerous operations of ever-increasing complexity with little or no improvement in symptoms or quality of life. As medical understanding of the complexity of pelvic pain has advanced to incorporate the psychosocial aspects of pain, the consensus has shifted to employing a multidisciplinary approach to the management of CPP.

An excellent review of urogenital and rectal pain syndromes has been published.[1] This chapter will focus on CPP in women, possible causes, and suggestions for treatment.

▶ NEUROBIOLOGY

The innervation of the pelvis is well described.[1] However, what is more complex is the neurobiology of this region. It is known that somatic and visceral pains are different in terms of both mechanisms and clinical features. Visceral pain is an indistinct, poorly defined sensation that is always perceived at the same site—usually the midline of the thorax or abdomen. It is accompanied by marked autonomic signs and emotional reactions. Subsequently, the symptom is referred to parietal somatic structures (skin, subcutaneous tissues, and muscle), usually in the same metameric field as the affected organ, where it may or may not be accompanied by secondary hyperalgesia of superficial or deep body wall tissues. At this stage, pain of visceral origin becomes sharper, better localized, and no longer accompanied by marked autonomic signs, and thus may be difficult to differentiate from pain arising primarily in somatic structures.

Much research effort has been devoted to the phenomenon of hyperalgesia because it has become evident in patients that painful visceral pathology frequently triggers hypersensitivity to painful stimuli, and this complicates the clinical expression of visceral pain.

Giamberardino[2] has described three types of visceral hyperalgesia that may be relevant to the clinical management of patients with pelvic pain:

1. *Visceral hyperalgesia.* Hyperalgesia of a viscus from inflammation and/or excess stimulation of the same viscera, e.g., irritable bowel syndrome.
2. *Referred hyperalgesia from viscera.* Hyperalgesia of somatic tissues in the area of referred pain from viscera, e.g., trigger points in body wall tissues identified by local compression in patients with pelvic pain.
3. *Viscerovisceral hyperalgesia.* Hyperalgesia of a viscus rendered clinically manifest by a painful condition of another viscera, e.g., exacerbation of urinary colic pain in patients with urinary calculus plus dysmenorrhea.

Clinical evidence and controlled studies in patients and in human volunteers have shown clearly that the phenomenon of hyperalgesia from internal organs exists and has clinical relevance. The challenge is to be able to explain these phenomena to patients in easily understandable language.

▶ EPIDEMIOLOGY

The prevalence of CPP in the community appears to be high. In the United States, a 3-month period prevalence of 15 percent was found in women aged 18 to 50 years.[3] In the United Kingdom, the annual prevalence in primary care was 38 per 1000, a rate similar to that reported for asthma or back pain.[4] Women complaining of CPP symptoms contribute to 15 to 20 percent of all consultations in the general gynecologic clinic and up to 10 percent of all female attendances in general practice.[5] CPP is the indication for 10 to 15 percent of hysterectomies performed in the United States. Women with CPP were found to have undergone almost five times more surgeries and to have sought treatment for four times as many somatic conditions unrelated to CPP compared with pain-free age-matched controls.

▶ EVALUATION OF WOMEN WITH PELVIC PAIN

Consultation

The assessment of women with CPP requires a systematic and comprehensive approach. A standardized medical and pain questionnaire can be useful in gathering information and can be used for audit. Psychological assessment tools and scales for somatization also can be administered. The assessment is a prime opportunity to establish rapport with the patient. Allowing time and opportunity for the patient to tell her story without interruption and to express her fears and concerns initiates the concept that the clinician and the patient are working together to manage symptoms.

A history of the nature of the pain, its cyclicity, and its exacerbating and relieving factors is important. Previous diagnoses, the effects of administered treatments, and the patient's understanding of these are important. The impact of symptoms on the patient's everyday life and family and sexual relationships is important. Questions regarding sexual and physical abuse must be asked in a sensitive way, and the clinician must gauge when it is appropriate to broach this subject. The patient may only disclose this information when she develops trust and confidence in her medical advisor. The consultation should be concluded by asking the patient for her expectations of the consultation (Table 31-1).

▶ **TABLE 31-1.** PELVIC PAIN ASSESSMENT

Location
  dermatomal distribution
  superficial or deep
  focal or diffuse
  radiation
Duration
  onset
  constant/intermittent
  frequency
  cyclicity
Associated events
  menses
  intercourse
  aggravating/relieving factors
  effect of activity
  urination/defecation
Character
Past medical and gynecological history including response to previous treatments
Somatic symptoms
Affect of pain on patient's life
Patient's views, concerns, expectations

Examination

General observation of the patient, especially her posture, is important. This is discussed in greater detail later. Scars can be a source of pain. These should be assessed for allodynia, anesthesia, and tenderness. Abdominal wall trigger points are identified by palpation. Evaluation of the abdominal wall with the patient's head raised off the table and with the rectus muscle tensed distinguishes abdominal wall from intra-abdominal pathology. The tense rectus muscles protect the peritoneum, and tenderness that continues or is exacerbated by rectus muscle tensing is likely to originate in the abdominal wall.

Vaginal examination may be performed as part of the evaluation of a patient with CPP. However, many women presenting to pain management services previously have undergone full gynecologic, urologic, and gastrointestinal investigation, and repeating this may not be appropriate. Vaginal examination does give the opportunity to assess pelvic floor musculature.

Investigation

Transvaginal ultrasound scanning is often the initial imaging modality to investigate women with CPP. Magnetic resonance imaging (MRI) is emerging as an important diagnostic tool in the evaluation of women with CPP. It lacks ionizing radiation and can be useful for characterizing an adnexal mass (e.g., endometrioma, dermoid) and in diagnosing uterine abnormalities (e.g., adenomyosis, fibroids).

Role of Laparoscopy

More than 40 percent of laparoscopies are performed for the diagnosis of CPP.[6] This investigation is expensive and not without morbidity. The decision to perform a laparoscopy should be based on history, examination, and the findings of noninvasive tests. Combining the results of published studies of laparoscopies for CPP shows that no visible pathology is detected in 35 percent (range 3 to 92 percent) of patients, endometriosis is diagnosed in 33 percent (range 2 to 80 percent), and adhesive disease is diagnosed in 24 percent (range 0 to 52 percent).[6] Thus, the predominant role of laparoscopy is to rule out endometriosis and adhesions. It is important that both physicians and patients understand this prior to laparoscopy.

Laparoscopic Conscious Pain Mapping

Laparoscopic conscious pain mapping is a diagnostic laparoscopy under local anesthesia, with or without conscious sedation, performed with the goal of identifying sources of pain in women with CPP. Currently, the utility and limitations of this technique are being evaluated.

Effective management depends on identification of the underlying cause. However, the symptoms of CPP in a female patient are complicated by the normal physiologic changes occurring in a cyclic manner and the presence, either confirmed or suspected, of gynecologic pathologies. The physician's difficulty lies in being able to attribute the complaint with certainty to observed pathology. Making an accurate diagnosis obviously is essential, but the negative aspects of repeated investigations, especially laparoscopy, should not be underestimated. Investigations, both invasive and noninvasive, compound the patient's and the physician's expectations that a cause or previously missed diagnosis will be found and encourage further medical or surgical intervention.

There are obvious pathologies associated with the complaint of CPP. However, many of these pathologies are to be found in women with no complaint of pain. Conversely, there are women who present with similar symptoms in whom no pathology is evident. These observations stress the importance to the physician of being open in discussions with the patient. It is also important to stress the relevance of the first assessment. This must create an expectation for the patient that the initial diagnosis is not absolute. Recognition at an early stage in the care of the patient that not all symptoms are due to a disease process and that not all pathologies are diseases assists the physician in avoiding the patient being precipitated into the spiral of increasing expectations. This also allows the patient to understand the role of physiologic processes in symptom development.

▶ GYNECOLOGIC DIAGNOSES

Tumor

The initial history and examination will identify the likelihood of tumor, either benign or malignant, being responsible for the symptoms of CPP. Should initial investigation with transvaginal ultrasound be inconclusive and there remains concern that tumor needs to be excluded, laparoscopy, which is essential within the overall primary assessment, will be definitive.

Treatment of a patient with a confirmed tumor obviously is directed at the tumor.

Infection

Proven and active infection within the uterus, fallopian tubes, and parametrium is an obvious source of chronic pain, but it is essential that the diagnosis is made with appropriate bacteriologic confirmation. The knowledge of an episode of pelvic inflammatory disease, either confirmed or suspected, in the past does not necessarily imply that persisting symptoms are a continuation of the infective process.

Once a chronic infective source of the symptoms is identified, treatment with appropriate antibiotics is indicated. Surgery may be indicated if there is abscess formation.

Endometriosis

Endometriosis is defined as "the presence of ectopic tissue that possesses the histological structure and function of the uterine mucosa." Endometriosis is common at laparoscopy and may even be a normal physiologic state. Some women with laparoscopic evidence of endometriosis have no pain, and no correlation has been found between pain complaints and the severity of disease.

The pain of endometriosis varies with the menstrual cycle, typically building up toward menstruation and then declining as menses commences. In addition, dysmenorrhea and dyspareunia are cardinal symptoms. If a patient has all three, she is 3.1 times more likely to have endometriosis at laparoscopy than a woman with no symptoms.[7]

Early superficial inflammatory endometriosis probably releases mediators such as prostaglandins and bradykinins and is more painful than the "burnt out" forms. Extensive nodular disease in the rectovaginal space may produce pain by traction on tissues or by infiltration or constriction of the nerves themselves. The formation of blood-filled cysts (endometriomata) will be evident on ultrasound imaging or on laparoscopy. Active endometriosis without the formation of endometriomata should be managed by appropriate nonsteroidal anti-inflammatory drugs (NSAIDs), analgesics, or suppression of gonadotrophin release. Significant distortion of the anatomy with endometrial cysts visible on ultrasound and/or laparoscopic examination is unlikely to respond to medical treatment. However, caution must be exercised when contemplating surgery. Immediate and long-term implications for fertility must be considered, with a preference being exercised for consecutive forms of surgery and use of minimal access techniques.

Adhesions

The presence of pelvic adhesions is not a reliable predictor of pelvic pain. Rapkin[8] observed no difference in the prevalence of pelvic adhesions in patients with chronic pain when compared with pain-free infertile controls, nor was she able to correlate the location or severity of adhesions with pain. Peters and colleagues[9] have reported the results of a prospective, randomized trial in which they observed no benefit from endoscopic lysis of adhesions for treatment of CPP, except in a small subset of women with dense adhesions involving bowel.

It is in the management of these patients with known pathologies, but pathologies of limited relevance, that the practicing gynecologist is most at risk of causing a spiral of increasing dependence and intervention. The most valuable information will be obtained from the history and the association between changes in symptoms and previous therapies. As soon as it becomes evident that one form of management, whether medical or surgical, has failed to influence the complaint significantly, then the association between the symptoms and the diagnosis must be questioned.

The symptoms of CPP frequently arise in women in whom either a coincidental pathology is identified or in whom an earlier "diagnosis" has not been confirmed. This is the group of patients most at risk for repeated therapies and interventions. It is essential that the history, relevance of clinical findings, and association of results of investigations are related. An apparent abnormality found on the left of the pelvis is unlikely to contribute to a symptom that is predominantly on the right.

Equally, surgical adhesions resulting from earlier surgery performed in an attempt to treat the complaint are irrelevant to the persisting symptoms.

▶ NONGYNECOLOGIC DIAGNOSES

Bowel-Related Pain

Irritable bowel syndrome (IBS) is a common gastrointestinal disorder affecting 10 to 20 percent of the population, with a female preponderance. It is a functional disease, meaning that pain and changes in bowel habit arise from abnormal behavior of bowel or from abnormal perceptions of physiologic events rather than structural abnormality. It is defined on the basis of the Rome criteria as the presence for at least 12 weeks (not necessarily consecutive) in the preceding 12 months of abdominal discomfort or pain in the absence of a structural or biochemical explanation that has at least two of the following three features: Pain is relieved with defecation, its onset is associated with a change in the frequency of bowel movements (diarrhea or constipation), or its onset is associated with a change in the form of the stool (loose, watery, or pellet-like).[10] IBS is common among women with CPP and is associated with a disproportionately high prevalence of abdominal and pelvic surgery.[11] The menstrual cycle affects rectal visceral sensitivity in patients,[12] and 50 percent of women with IBS have perimenstrual increase in symptoms.[13]

Bladder-Related Pain

The principal urologic causes of pelvic pain are interstitial cystitis and urethral syndrome. Interstitial cystitis is an inflammatory condition of unknown etiology characterized by pain and bladder symptoms such as urgency and frequency. It is a heterogeneous condition, but the presence of submucosal hemorrhages or Hunner's ulcer confirms the diagnosis. Dyspareunia may occur in 60 percent of women with this condition. Pain typically increases as the bladder fills and is relieved by passing urine, and may be recreated by pressure over the bladder base on vaginal examination.

Urethral syndrome represents a less well-defined entity. Positive diagnostic tests are urethral tenderness or pain on palpation and a slightly inflamed urethral mucosa found during endoscopy. Clinically, the diagnosis is commonly given to patients with dysuria (with or without frequency), nocturia, urgency, and urge incontinence in the absence of urinary infection. Women with pelvic floor dysfunction and postmenopausal women with estrogen deficiency may report these symptoms. Etiology is unknown, although it is possible that it is an earlier, milder form of interstitial cystitis.

Nerve-Related Pain

Nerves may be trapped at the edge of the rectus muscle and in scar tissue or fascia. Pain tends to be sharp and stabbing in nature, or may be dull and aching. Pain is usually highly localized and may be aggravated by particular movements. In a study of patients who had one Pfannensteil incision, the incidence of nerve entrapment (defined as pain persisting longer than 5 weeks after surgery or occurring after a pain-free interval) was 3.7 percent.[14] The injection of local anesthetic at the site of maximal tenderness should relieve symptoms at least temporarily. This can help the patient understand her condition and occasionally results in long-term relief. If there is documented nerve damage with anesthesia and paresthesia, drugs used for neuropathic pain, such as tricyclic antidepressants and anticonvulsants, may be helpful. However, further randomized, controlled trials of these medications in CPP need to be done.

▶ MUSCULOSKELETAL CAUSES AND THE ROLE OF THE PHYSIOTHERAPIST

Several authors have advocated physiotherapists as one of the multidisciplinary pelvic pain team members.[15–17] However, there is limited research to date to determine which patients with CPP should be considered for physiotherapy, and the most advantageous time to refer them for treatment. Research into the most effective treatment regimens is also very limited. King-Baker and colleagues[17] advocated individual musculoskeletal screening followed by individual treatment, which in their study was found to be effective, with 20 percent reporting complete relief of symptoms. A further 50 percent reported significant relief. In Albert's paper,[18] a group approach was used, resulting in statistically significant reduction in pain scores and reduced usage of analgesics. Further research is required to evaluate the most beneficial physiotherapeutic interventions for women with CPP.

The involvement of physiotherapists in the treatment of CPP is not surprising. Physiotherapists are recognized experts in musculoskeletal dysfunction and are widely involved in chronic pain management for other groups of patients suffering with chronic pain syndromes. It is perhaps more surprising that physiotherapy often has been a last resort for patients with CPP. King-Baker et al.'s study consisted of patients who had been treated unsuccessfully by gynecologists and psychologists.[17] Previously, patients would only be referred to physiotherapists when exhaustive medical testing found "nothing wrong." Since medical teams involved in treating women with CPP now advocate a multidisciplinary team approach from the outset of treatment, physiotherapy may begin to be developed further in this field.

To understand how physiotherapy may be effective in the treatment of CPP, it is important to understand the anatomy, physiology, and neurology involved, as well as the psychological aspects.

The musculoskeletal system is the largest system in the human body. It consists of the skeletal bony structure, which provides the framework with which the muscular system works to provide locomotion, and it also has joints, nerves, voluntary skeletal muscle, fascia, and connective tissue. Therefore, it is not surprising that it would be influenced by chronic pain, or indeed may be the primary site for the original pain.

Anatomy

The bony pelvis consists of two innominate bones joined anteriorly to form the symphysis pubis joint, and posteriorly with the sacrum to form the sacroiliac joint. Other joints associated with the bony pelvis are the hip joint, where the bony acetabulum of the innominate articulates with the femoral head. The superior base articulates with the fifth lumbar vertebra and the apex of the sacrum with the first coccygeal segment.

Neurology

The segmental innervation of the muscles, joints, and pelvic viscera is shared. This means that any referred pain from one of these structures may be confused. Table 31-2 summarizes this shared innervation.

Physiology

All skeletal muscles have the potential to adapt in response to changes in environment, work load, etc. There are two main reasons why these changes may occur:

- Factors beyond the control of the individual, e.g., aging, metabolic conditions, pathologic conditions
- Choice of the individual, e.g., taking up a new sport, reducing activity due to pain

In women with CPP, it is not always easy to identify whether the problems being dealt with are primary, i.e., the original cause of the pain, or secondary. It may be that a musculoskeletal problem is the primary cause of the pelvic pain.[17] However, Selfe and colleagues[19] have suggested that the reduced levels of activity adopted in

▶ TABLE 31-2. SEGMENTAL INNERVATION

Pelvic viscera	Innervation	Muscles and joints	Innervation
Ovaries	T10-T11	Abdominal muscles	T5-L1
Uterus	T10-L1	Quadratus lumborum	T12-L3
Fallopian tubes	T10-L1	Lumbar ligaments, facets and discs	T12-S1
Perineum	S2-S4	Hip	T12-S1
External genitalia	L1-L2, S3-S4	Iliopsoas	L1-L4
Kidney	T10-L1	Piriformis	L5-S3
Bladder	T11-L2, S2-S4	Obturator internal/external	L3-S2
		Pubococcygeus	S1-S4
		Sacroiliac joint	L4-S3

response to the pain may be the secondary cause of the adaptive musculoskeletal dysfunction. Poor outcome could be predicted by pain causing reduction in physical activity levels (secondary problem).

The relevant literature on reduced activity falls broadly within three categories[20]:

- Effects of detraining on athletes
- Neuromuscular pathology or trauma
- Effects of weightlessness caused by lack of gravity during space travel and training programs

Research investigating muscular changes due to enforced immobility in a plaster cast found that this immobility leads to absorption of sarcomeres owing to their being held in a shortened position.[21] To date, there is no consensus in the literature whether inactivity in humans produces preferential atrophy of one fiber type over another, i.e., type 1 or type 2 fibers. However, there is agreement that if muscle activity is reduced, muscle atrophy will occur with associated loss of peak force and power.[22] Even a few hours of muscle inactivity leads to changes within muscle, with the rate of protein synthesis—which is the main characteristic of muscle atrophy—starting to decrease.[23] A decrease in contractile protein may lead to a reduction in the number of active crossbridges per volume of muscle, and hence reduced electromedical efficiency.[24] The amount of muscle power lost varies from study to study. There is agreement that the greatest loss occurs within the first few days to 2 weeks.

The following factors also influence the degree of atrophy:

- *The initial muscle fiber type.* Postural muscles atrophy quicker than muscles with a higher proportion of fast-twitch fibers.
- *The position of the muscle.* Most atrophy occurs when the muscle is fixed in the shortened position.
- *The anatomic location of the muscle.* Muscles of the lower limb atrophy faster than upper limb muscles.
- *The length of time during which the muscle is inactive.*[21]

▶ PHYSIOTHERAPY INTERVENTIONS IN CPP

Physiotherapy Assessment

Owing to the complexity of CPP, it is important that all women receive an in-depth initial assessment. Although the physiotherapist's role is to identify and treat existing musculoskeletal dysfunction, he or she also has a health education role in the prevention of future secondary musculoskeletal dysfunction. Physiotherapists should work closely with other members of the multidisciplinary team.

Assessment will include the following:

Postural Assessment

The importance of postural assessment when dealing with women with CPP was examined in a study by King-Baker and colleagues[17] that identified a common posture in 75 percent of subjects with CPP. Unfortunately, the study failed to report on whether the faulty posture described was improved by the physiotherapeutic interventions used. The sole reported outcome measure was a subjective one, measuring reduction of pain. We, therefore, are unable to ascertain which aspects of the faulty posture may be crucial to improve a woman's pain. What is not disputed is that faulty alignment of posture leads to musculoskeletal changes, with muscles held in a shortened position tending to become stronger and muscles that are elongated becoming weaker, leading to muscular imbalance.

Thus, a thorough assessment of posture and any musculoskeletal dysfunction caused by habitual poor posture should be identified and therapeutic exercises and interventions used to help improve them.

Joint Assessment

The range of movement of all the joints in the pelvis, lower extremity, and spine should be assessed and any abnormalities noted.

Muscular Assessment

A thorough assessment of the musculature of the trunk and lower extremities should be performed, and any problems

▶ **TABLE 31-3.** GLAZER PROTOCOL

One minute rest, prebaseline
Five rapid contractions (flicks) with 10-second rest between (phasic)
Five 10-second contractions with a 10-second rest between (tonic)
One single endurance contraction of 60 seconds (endurance)
One minute rest, postbaseline
This will allow the pelvic floor musculature to be evaluated for the instability that is present
Start with two 20-minute exercise regimens daily with home machine
60 repetitions of 10-second contractions with 10 seconds rest
Seen once every two weeks, then monthly; treatment may take between 3 to 10 months

▶ **TABLE 31-4.** TREATMENT MODALITIES

Heat/cold therapy	Relaxation techniques
Ultrasound	Breathing techniques
Electrical therapy	Massage
Biofeedback	Alternative therapies, e.g., craniosacral therapy, reflexology, acupuncture
Joint mobilization	Soft tissue mobilization
Myofascial release	Therapeutic release

noted. Muscles should be checked for tone, length, strength, and trigger points. Particular care should to be taken with assessment of the musculature of the pelvic floor. Alterations in tone and contractility have been implicated in several studies in women with CPP, including work by Glazer and colleagues[25] and by Weiss.[26] Patients with vulvodynia have been helped by use of a protocol of pelvic floor musculature rehabilitation.[25] Glazer and colleagues postulated that it is the instability of the pelvic floor musculature that is a main component in the pain involved in vulvodynia. Once biofeedback has determined the particular type of pelvic floor musculature instability, women are instructed in a detailed "exercise prescription." This regimen is summarized in Table 31-3.

This protocol is very demanding for women, especially since many of them experience extreme pain on inserting the probe required for the home biofeedback machine. Many women suffering with vulvodynia abstain from penetrative intercourse. Studies are required to determine if the home machine is essential for successful completion of this treatment. It may be possible to treat these patients with simple pelvic floor regimens, concentrating on relaxation of the musculature of the pelvic floor, which would be less invasive and less time-consuming.

Palpation

When performed by a skilled practitioner, palpation of the joints and muscles involved will reveal musculoskeletal dysfunction.

Treatment Modalities

The range of treatment modalities (Table 31-4) available to the physiotherapist is wide and will vary from patient to patient depending on the dysfunctions discovered during assessment. A list of possible modalities is not meant to be exclusive, and the range of modalities used will depend on both the problems encountered and the skill of the therapist involved in care of the woman.

Heat and Cold Therapy

Moist heat in the form of compresses, warm baths, or hot water bottles, or hot/cold therapy using commercial packs, can help to alleviate pain in the short term by reducing muscle spasm and trigger points. Heat also will enhance the effects of scar mobilization, muscle stretching, and myofascial techniques.

Relaxation and Breathing Techniques

Relaxation and breathing techniques can be used to help women with CPP to reduce the body's natural "fight or flight" response to stress. Methods of both physical and mental relaxation should be taught. Breathing exercises will help patients to understand the importance of deep, slow breathing to help normalize faulty breathing patterns caused by stress.

Ultrasound

Ultrasound can be used to reduce pain by reducing swelling. It is also used to treat scar tissue, adhesions, muscle spasms, and joint pain and to help increase healing in soft tissues.

Electrical Therapies

A variety of electric currents can be used to alleviate pain, stimulate muscle contraction, and increase circulation, allowing women to establish contraction in weakened muscles. The choice of current used will depend on the response required.

Biofeedback

Biofeedback can be used to retrain muscular activity, whether it is needed to improve muscular stability, decrease muscular activity, or increase muscular strength. More research is required into the use of biofeedback in CPP of etiology other than vulvodynia. Increased pelvic floor muscle tone has been implicated in patients with interstitial cystitis and in other women with CPP.

Soft Tissue Myofascial Release, Joint and Soft Tissue Mobilization

These manual techniques may be used to reduce pain and improve function where musculoskeletal dysfunction has been found at initial assessment.

Following initial assessment, the physiotherapist will select which of the preceding manual therapies are indicated for the individual women seeking treatment.

Massage

Massage can be used to alleviate pain in musculoskeletal dysfunction. It has the advantage of allowing the physiotherapist to further evaluate musculature. Patients usually enjoy the sensation and the individual nature of the treatment. Aromatherapy oils can be used if the physiotherapist is skilled in this area. Therapeutic massage to stretch soft tissue also may be used. Massage has been directed at the levator ani musculature, using firm downward strokes from origin to insertion. Patients must be informed first that although this is termed massage, it is not comfortable, and in some patients is extremely painful.

Alternative Therapies

Acupuncture and reflexology have been used to treat pelvic pain, and anecdotal reports chronicle success.

Therapeutic Exercise

The fact that the pain is chronic means that members of the multidisciplinary team involved in the care of these women have an important health education role to fulfill. Education into the complexity of pain, a discussion of the difference between acute and chronic pain, and a clear explanation of how the physiotherapist hopes to help the woman deal with her pain are paramount. When people suffer from pain, their instinctive reaction is to rest and restrict their activity to allow healing of the injury/disease that is causing their pain. Unfortunately, this is counterproductive in chronic pain because it leads to patients becoming generally deconditioned. These women need to understand why this is not helpful and, in a safe environment with professional help, begin to experience the benefits of gentle exercise. This can be achieved either on an individual basis or in a group setting; and further research is required to determine the most beneficial environment for these women.

▶ PSYCHOLOGICAL FACTORS AND THE ROLE OF THE PSYCHOLOGIST

The role of psychological factors in the etiology of CPP has been debated for many years. Studies have not revealed any consistent psychological profile for patients with CPP without obvious pathology. In their review of the literature, Savidge and Slade[27] concluded that women with CPP have higher rates of distress and psychopathology than women without pain. However, they point out that these findings are common in people suffering from other kinds of chronic pain. The literature examining psychological factors in CPP has been reviewed in detail elsewhere.[1,28,29] There are a number of methodologic limitations in much of this research, including small sample sizes, selected samples, and lack of appropriate control groups.

It is commonly accepted that psychological factors may use both contribute to the experience of pain, and be a consequence of that pain. As with other chronic pain conditions, it is essential that a psychological assessment be part of the evaluation of CPP. Ideally, a clinical psychologist with a special interest in pain management would be an integral member of the clinical team and part of the initial assessment process. The importance of including a clinical psychologist as a member of the CPP management team has been well documented in the literature.[30–32] Many of the women will be suspicious about seeing a psychologist and may feel that their pain is viewed as "all in their mind." Therefore, one of the challenges facing the team is getting the patient to agree that psychological factors may be important in understanding her pain condition and its management. A dualistic approach is not helpful. Women need to be helped to understand how the mind and body work together.

Women with CPP repeatedly search for explanations for their pain.[29] A lack of any identifiable physical cause for the pain is not always reassuring. It may not be enough simply to tell a woman that she has no serious pathology. Some explanation is needed as to what might be causing her pain, e.g., normal structures in her pelvis such as muscles. Savidge and colleagues[33] found that women who had had a negative laparoscopy often felt that they were not believed. A key area of dissatisfaction reported by women is the extent to which they feel that their problem has been understood and acknowledged. Therefore, it is very important that a therapeutic relationship is created where the patient feels heard and understood; and that questions are asked specifically about any concerns and beliefs the woman may have about what is causing her pain. More idiographic exploration of women's experiences of CPP would help to expand our information regarding women's appraisals and beliefs, as well as their processes of adjustment.[27] Cognitive-behavioral interventions can be effective in modifying illness beliefs.

Many CPP patients describe a range of negative, pessimistic beliefs that often lead to catastrophizing and an increase in pain. Women can be taught how to challenge and reduce negative and unhelpful beliefs and thoughts about their pain. Thoughts such as, "I can't cope," and, "I should be able to cope," can be destructive, put pressure on the individual, and give rise to feelings of hopelessness and helplessness. These can be replaced with more positive thoughts, such as, "I can cope if I plan what I do." This will lead to positive action and feelings of achievement.

Studies have shown that women presenting with CPP often have high levels of anxiety and depression.[35]

The woman's current mood state should be assessed, together with any previous psychiatric history. Standardized questionnaires can be used, such as the Hospital Anxiety and Depression Scale[36] and the Modified Zung Depression Inventory.[37] Fry and colleagues[38] found that a high percentage of women with CPP reported significant fear regarding the origin of their pain and the prospect of a bleak prognosis. In another study, 50 percent of women with CPP feared that they had a serious disease as yet undiagnosed.[39] A recent study investigating CPP from the general practitioner's perspective found that general practitioners were very resistant to sharing with a woman their belief that her symptoms might turn out to be nonorganic until very late in a referral process. By this time, the woman may have convinced herself that she has something seriously wrong with her.[40] In addition to worrying about what is causing her pain, a patient with CPP may be very anxious about the impact the pain is having on her life and her relationships. She may have financial worries because of her inability to work; and a younger woman may have significant concerns about her fertility if she has been given a diagnosis of endometriosis or pelvic inflammatory disease. Pregnancy and infertility issues associated with diagnoses and treatment approaches are often overlooked with CPP patients.

Depression may be a consequence of CPP, or may predate symptoms of pain, as was found to be the case in 75 percent of women in one study.[33] Depression may increase isolation and perception of pain, as well as motivation and ability to adopt a self-management approach. If present, depression needs to be treated. Antidepressants, particularly tricyclic antidepressants, have been used to treat a variety of chronic pain syndromes, although there has been little study of the use of antidepressants in women with CPP. In one small study, 14 women with CPP were given nortriptyline. Seven of the women dropped out of the study because of adverse side effects; 6 of the 7 remaining women in the trial were pain-free at 1 year after the doses were increased to 100 mg/d.[41] More research is needed in this area.

Some women experience significant guilt because they are not able to be a "proper" wife and mother and have to depend on help with housework and shopping. They also may experience shame and embarrassment because of the location of their pain. Attitudes of disbelief, anger, and resentment are common and may signal a poor prognosis.[34] The woman's anger may be fueled by her often fruitless search for an explanation for her pain, together with the failure of conventional treatments to cure her pain and possibly to have made the pain worse. The woman's anger may be directed toward the professionals trying to help her. Such women need to be helped to recognize and manage the common antecedents of anger, such as disappointment and sadness.

A psychosocial history will help to understand the environment and context of a woman's pain. It is important to help the patient become aware of factors in her environment that affect her pain, e.g., family arguments, conflict at work, etc. Many women are aware that their pain is exacerbated by worry or tension. Self-monitoring can help to identify current stressors, and advice on stress management may be beneficial for some women. This would begin with an explanation of the psychophysiology of stress. Management techniques might include training in muscle relaxation, cognitive restructuring, and assertiveness. Many CPP patients admit to always being on the go and constantly responding to the demands of others. These women need to be encouraged to take time out to rest and to learn to say no. In addition to advice on stress management, patients also may benefit from advice on activity management and goal setting. CPP patients, in common with other pain patients, often base what they do and how much they do on how they feel. When they are having a better day, they try to "catch up" on tasks and often end up overdoing it. The pain is then worse the following day, and they are able to do very little. This is called *activity cycling,* and to break the cycle, women are encouraged to pace themselves, i.e., to keep to a regular amount of activity each day and base what they do on a plan and not on how they feel. Women also typically avoid doing things that they think will make the pain worse. They need to understand that hurt does not necessarily mean harm. Avoiding is not helpful in the long term, can make life increasingly negative, and can undermine self-confidence. Goal setting is the process that helps the CPP patient identify long- and short-term goals and break these down into achievable steps employing the principle of pacing.

Stress management and pain management strategies can be taught individually or in a group setting. It is well recognized that patients with a range of physical and mental health problems benefit from meeting others who are facing similar difficulties. Group therapy has been shown to be effective for chronic pain patients. It has been suggested that such groups can provide group acceptance, mutual comparison, and mutual support, which are valuable in bringing about change.[42] One of the most important factors appears to be the feeling that the patient is not alone, as well as the opportunity to benefit from observing how other people have coped, and the instillation of hope and opportunities for altruism.[43] Group therapy using a combination of cognitive-behavioral techniques and physical exercises has been shown to be effective in treating chronic pain patients. The effectiveness of group treatment for women with CPP involving a combination of psychological and physical interventions is reported in one study.[18]

Very few studies have reported attempts to include current partners of CPP in assessment.[44] Yet, women with

CPP often report relationship distress. The partners of CPP patients often become less supportive over time and may resent the lack of sex. Sexual dysfunctions, most notably dyspareunia and decreased sexual desire, are common in CPP patients.[32,45,46] A careful sexual history, therefore, is important, preferably involving the partner. Patients may benefit from specific interventions such as Kegel's exercises, use of graded dilators, and advice on lubrication and sensate focus exercises. Advice also can be offered on positions that limit depth of penetration and give the woman a greater sense of control (e.g., female astride). In addition to psychosexual therapy, couples may benefit from psychological therapy aimed at improving their communication and general relationship. Partners often can benefit from having a professional explain the physical and psychological aspects of CPP. This may help to reduce feelings of animosity, distrust, and hurt.

A number of controlled studies have shown that women with chronic pelvic pain have a higher incidence of previous sexual and physical abuse.[32,45,47,48] If women have a previous history of sexual abuse, they may have a significantly higher risk for having a current diagnosis of major depression and somatoform pain disorder compared with those with no abuse, or less severe abuse.[49] Questions should be asked in a supportive and open environment about any previous history of unwanted or unpleasant sexual experiences. If a history of sexual or physical abuse is affecting current functioning, psychotherapy may be appropriate. The mechanism linking a history of sexual or physical abuse with pelvic pain is unclear. In our clinic, we have seen a small percentage of women whose pelvic pain developed following a traumatic sexual experience such as rape. These women responded well to psychotherapy, and their pain symptoms improved significantly. We also have seen women with CPP in abusive relationships in whom the pain improves after the women have been helped to leave the relationships.

The main aim of psychological approaches to pain management is to reduce distress and help the patient regain some sense of control over her pain and her life and to maximize the quality of her life. Reports of controlled trials of psychological interventions specifically for CPP are very limited but do suggest they can be very beneficial.[50,51] More controlled studies with good follow-up data are needed. Psychological treatment should be provided early on. However, in practice, it is often only as a last resort that psychological and subjective dimensions of these women's experiences of pain are acknowledged, and a psychological consultation is arranged. A clear dichotomy still exists as to whether CPP is physical or psychological, and further education is needed to stress the importance of addressing physical and psychological factors simultaneously with these women.

▶ TREATMENT

Initial Consultation

The importance of the initial consultation with a woman with CPP cannot be overemphasized. It is vital to establish rapport and trust, and engage in a positive therapeutic relationship. The relevance of potential or previous diagnoses to the overall pain complaint needs to be discussed. The limitations of laparoscopy in the diagnosis of CPP and the fact that not all symptoms can be attributed to disease entities may need to be explained. Patients, especially if their previous treatment has been heavily based on the medical model, may find it difficult to comprehend that our understanding of CPP is still extremely limited. Reassurance regarding the absence, the presence, or the extent of pathology may be necessary. An explanation for the pain should be given, if possible. Discussion regarding the contribution of associated physical or psychosocial factors should be constructive. A treatment plan that includes reassurance and discharge then can be formulated between the health care professional(s) and the patient. Several treatment modalities may be offered simultaneously to try to address the various pathophysiologic mechanisms that may be contributing. This is preferable to offering treatments consecutively.

Multidisciplinary Approach

The use and limitations of gynecologic treatments in women with CPP have been highlighted. The importance of physiotherapy and psychology assessments and treatments in the care of these women cannot be overemphasized. The pain management physician can help with explaining the complexity of pain; and can advise on medication and nerve blocks, or trigger point injections, in selected patients. Sacral neuromodulation has been performed in an extremely small number of patients with refractory pelvic floor dysfunction and pelvic pain, but much larger studies are needed before this therapy can be recommended.[52]

Oral Analgesics

Optimization of oral analgesic therapy is an obvious initial treatment for chronic pelvic pain. NSAIDs can be helpful, especially for primary dysmenorrhea and for endometriosis. It is more effective to take analgesics on a regular, rather than an "as necessary" basis. It is more effective to take regular analgesics for mild-to-moderate pain. In addition, "prn" prescribing may increase the pain because it encourages the patient to focus on her symptoms and to monitor pain severity. Weak opioids are commonly taken for CPP. They may or may not be helpful. Our advice is to stop any medication that is not producing tangible relief.

The use of strong opioid analgesics in chronic nonmalignant pain is currently under debate. There have been no studies so far that have systematically investigated the use of opioids in patients with CPP.

Antidepressants and Anticonvulsants

TCAs have been used to treat many chronic pain syndromes. They have an action on pain independent of their antidepressant action. There is a correlation between dose and analgesic action, and the dose should be reviewed regularly and increased, if necessary. There is only one paper on their use in CPP, and further research in this area is needed.[41]

Anticonvulsant drugs may help if the pain has features of neuropathic pain. These are reviewed elsewhere in this book.

Hormonal Drugs

A careful history will identify the association between symptoms and the menstrual cycle, and may lead to a suspicion of a hormonal cause for the complaint. Moderation of the natural menstrual cycle by either continuous combined hormonal preparations, or by downregulation of the pituitary gland with gonadotropin-releasing hormone analogues, provides effective treatment and confirmation of the correct diagnosis.

Trigger Point Injections

Trigger points in the lower back, abdominal wall, and pelvis can contribute to CPP. Blockade of these with local anesthetic can be used as a diagnostic, and sometimes, therapeutic procedure.

Neural Blockade

Ilioinguinal and genitofemoral neuralgias can cause CPP, especially in patients with a history of prior abdominal or pelvic surgery. Injections of local anesthetic alone or combined with steroid can be useful. They also may assist in demonstrating to patients the source of their pain. If these are of only temporary benefit, then radiofrequency denervation may be considered.

The superior hypogastric plexus receives visceral input from the upper vagina, body of the uterus, medial fallopian tubes, broad ligament, upper bladder, cecum, appendix, and terminal large bowel. A superior hypogastric plexus block can be performed percutaneously under x-ray screening. This block was described originally as a neurolytic procedure for pelvic pain of malignant origin, and efficacy is documented for this indication. The role of this block as a diagnostic tool prior to presacral neurectomy has not been established. Neither the safety nor the efficacy of neurolytic superior hypogastric plexus block has been established for chronic nonmalignant pain.

Joint Injections

Skeletal structures of the spine, such as zygapophyseal or facet joints, and the pelvis can produce patterns of referred pain that can mimic visceral pain. Physiotherapy should be the first treatment considered. Occasionally, therapeutic facet joint injections with local anesthetic and steroid or radiofrequency denervation of the appropriate facet nerves may be considered.

Self-Help Groups

Many patients with CPP from a variety of etiologies have experienced prolonged pain relief, but total relief of their symptoms may not be a realistic option. Self-help or support groups, organized internationally, nationally, or locally, can be a valuable resource for the patient. These groups often benefit from links with or input by health care professionals.

▶ CONCLUSION

CPP can be a diagnostic and therapeutic dilemma, and it has proved difficult to manage some patients using a uniprofessional model. Women with CPP in whom there is no clear diagnosis or in whom diagnoses overlap need to be given clear explanations that do not undermine the legitimacy of their own experience of pain or convey a message of dismissal.[53] A multidisciplinary team can undertake assessment and treatment using a broader psychosocial concept of the pain experience. Integrated approaches using somatic and behavioral therapies have been shown to reduce pain and other symptoms and to improve functional ability.[51,54] This approach will not be needed and may not be practical for all patients. However, early identification of patients who may benefit from such an approach may be helpful and could provide the opportunity to work simultaneously with the many contributing factors that are often present. Further research is needed to identify the essential elements of interdisciplinary working in order to secure optimal outcomes for individual patients.

REFERENCES

1. Wesselmann U, Burnett AL, Heinberg LJ: The urogenital and rectal pain syndromes. *Pain* 1997;73:269–294.
2. Giamberardino MA: Visceral hyperalgesia, in Devor M, Rowbotham MC, Wiesenfeld-Hallin Z (eds), *Proceedings of the 9th World Congress on Pain, Progress in Pain Research and Management*. Seattle: IASP Press, 2000:523–550.

3. Matthias SD, Kupperman M, Liberman RF, et al: Chronic pelvic pain: Prevalence, health-related quality of life, and economic correlates. *Obstet Gynecol* 1996;87:321–327.
4. Zondervan KT, Yudkin PL, Vessey MP, et al: The prevalence of chronic pelvic pain in women in the UK: A systematic review. *Br J Obstet Gynaecol* 1998;105:93–99.
5. Jamieson DJ, Steege JF: The prevalence of dysmenorrhoea, dyspareunia, pelvic pain and irritable bowel syndrome in primary care practices. *Obstet Gynecol* 1996;87:55–59.
6. Howards FM: The role of laparoscopy in chronic pelvic pain: Promise and pitfalls. *Obstet Gynecol Surv* 1993; 48:357–387.
7. Fedel L, Bianchi S, Bocciolone L, et al: Pain symptoms associated with endometriosis. *Obstet Gynecol* 1992; 79:767–769.
8. Rapkin A: Adhesions and pelvic pain: A retrospective study. *Obstet Gynecol* 1986;68:13–15.
9. Peter AAW, Trimbos-Kemper GCM, Admiraal C, et al: A randomized clinical trial on the benefit of adhesion lysis in patients with intraperitoneal adhesions and chronic pelvic pain. *Br J Obstet Gynaecol* 1992;99:59–62.
10. Drossman DA, Corazziari E, Talley NJ, et al: Rome II: A multinatioanl consensus document on functional gastrointestinal disorders. *Gut* 1999;45:1–15.
11. Hasler WL, Schoenfeld P: Abdominal and pelvic surgery in patients with irritable bowel syndrome. *Aliment Pharmacol Ther* 2003;17:997–1005.
12. Jackson NA, Houghton LA, Whorwell PJ: The menstrual cycle affects rectal visceral sensitivity in patients with irritable bowel syndrome (IBS) but not healthy volunteers. *Gut* 1997;40:52A.
13. Moore J, Barlow DH, Jewell D, et al: Do gastrointestinal symptoms vary with the menstrual cycle? *Br J Obstet Gynaecol* 1998;105:1322.
14. Luijendijk RW, Jeekel J, Storm RK, et al: The low transverse Pfannensteil incision and the prevalence of incisional hernia and nerve entrapment. *Ann Surg* 1997;225:365.
15. Collett BJ, Cordle C, Stewart C: Setting up a multidisciplinary clinic. *Best Pract Res Clin Obstet Gynaecol* 2000;14:541–556.
16. Gambone JC, Reiter RC: Nonsurgical management of chronic pelvic pain: A multidisciplinary approach. *Clin Obstet Gynecol* 1990;33:205–211.
17. King-Baker PM, Myers CA, Ling FW, et al: Musculoskeletal factors in chronic pelvic pain. *J Psychosom Obstet Gynaecol* 1991;12:87–98.
18. Albert H: Psychosomatic group treatment helps women with chronic pelvic pain. *J Psychosom Obstet Gynaecol* 1999;40:216–225.
19. Selfe SA, Matthews Z, Stones RW: Factors influencing outcome in consultations for chronic pelvic pain. *J Womens Health* 1998;7:1041–1048.
20. Talmadge RJ: Myosin heavy chain isoform expression following reduced neuromuscular activity: Potential regulatory mechanisms. *Muscle Nerve* 2000;23:661–679.
21. St. Pierre D, Gardiner PF: The effect of immobilisation and exercise on muscle function: A review. *Physiother Can* 1987;39:24–35.
22. Fitts RH, Riley DR, Widrick JJ: Microgravity and skeletal muscle. *J Appl Physiol* 2000;89:823–839.
23. Rennie MJ, Edwards RHT, Emery PW, et al: Depressed protein synthesis is the dominant characteristic of muscle wasting and cachexia. *Clin Physiol* 1983;3:387–398.
24. Berg HE, Larsson L, Tesch PA: Lower limb skeletal muscle function after six weeks of bed rest. *J Appl Physiol* 1997; 82:182–188.
25. Glazer HI, Rodke G, Swencionis C, et al: Treatment of vulvar vestibulitis syndrome with electromyographic biofeedback of pelvic floor musculature. *J Reprod Med* 1995;40:283–290.
26. Weiss J: Pelvic floor myofascial trigger points: Manual therapy for interstitial cystitis and the urinary-frequency syndrome. *J Urol* 2001;166:2226–2231.
27. Savidge C, Slade P: Psychological aspects of chronic pelvic pain. *J Psychosom Res* 1997;42:433–444.
28. Fry RPW, Crisp AH, Beard RW: Sociopsychological factors in chronic pelvic pain: A review. *J Psychosomc Res* 1997; 42:1–15.
29. Grace VM: Mind/body dualism in medicine: The case of chronic pelvic pain without organic pathology. *Int J Health Serv* 1998;28:127–151.
30. Rapkin AJ, Kames LD: The pain management approach to chronic pelvic pain. *J Repro Med* 1987;32:323–327.
31. Gambone JC, Reiter RC: Non-surgical management of chronic pelvic pain: A multidisciplinary approach. *Clin Obstet Gynecol* 1990;33:205–211.
32. Price JR, Blake F: Chronic pelvic pain: The assessment as therapy. *J Psychosom Res* 1999;1:7–14.
33. Savidge CJ, Slade P, Stewart P, et al: Women's perspectives on their experiences of chronic pelvic pain and medical care. *J Health Psychol* 1998;3:103–116.
34. Bass C (ed): *Somatization: Physical Symptoms and Psychological Illness.* Oxford: Blackwell Scientific, 2000.
35. Walker EA, Katon W, Harrop-Griffiths J, et al: Relationship of chronic pelvic pain to psychiatric diagnoses and childhood sexual abuse. *Am J Psychiatry* 1998;145:75–80.
36. Zigmond AS, Snaith RP: The hospital anxiety and depression scale. *Acta Psychiatr Scand* 1983;67:361–370.
37. Main CJ, Waddell G: The detection of psychological abnormality using four simple scales. *Curr Concepts Pain* 1984; 2:10–16.
38. Fry RP, Crisp AH, Beard RW: Patients' illness models in chronic pelvic pain. *Psychother Psychosom* 1991;55:158–163.
39. Kellner R, Slocumb JC, Rosenfeld RC, Pathak D: Fears and beliefs in patients with the pelvic pain syndrome. *J Psychosom Res* 1988;32:303–310.
40. McGowan L, Pitts M, Clark CD: Chronic pelvic pain: The general practitioner's perspective. *Psychol Health Med* 1999;4:303–317.
41. Walker EA, Sullivan MD, Stenchever MA: Use of antidepressants in the management of women with chronic pelvic pain. *J Psychosom Res* 1998;44:203–207.
42. Nichols K, Jenkinson J: *Leading a Support Group.* London: Chapman & Hall, 1991.
43. Yalom I: *The Theory and Practice of Group Psychotherapy.* New York: Basic Books 1985.
44. Benson RC, Hanson KH, Matarazzo JD: Atypical pelvic pain in women: Gynecologic-psychiatric considerations. *Am J Obstet Gynecol* 1959;77:806–825.

45. Collett BJ, Cordle CJ, Stewart CR, Jagger C: A comparative study of women with chronic pelvic pain, chronic non-pelvic pain and those with no history of pain attending general practitioners. *Br J Obstet Gynaecol* 1998;105:87–92.
46. Stout AL, Steege JF: Psychosocial and behavioral self-reports of chronic pelvic pain patients. Paper presented at the meeting of the American Society for Psychosomatic Obstetrics and Gynecology, Houston, Texas, March 1991.
47. Rapkin AJ, Kames LD, Darke LL, et al: History of physical and sexual abuse in women with chronic pelvic pain. *Obstet Gynecol* 1990;76:92–95.
48. Walling MK, Reiter RC, O'Hara MW, et al: Abuse history and chronic pelvic pain in women: Prevalences of sexual abuse and physical abuse. *Obstet Gynecol* 1994;84:193–199.
49. Walker EA, Katon WJ, Hansom J, et al: Medical and psychiatric symptoms in women with childhood sexual abuse. *Psychosom Med* 1992;54:658–664.
50. Pearce S, Knight C, Beard RW: Pelvic pain: A common gynaecological problem. *J Psychosom Obstet Gynaecol* 1982;1:12–17.
51. Kames LD, Rapkin AJ, Naliboff BD, Afifi S: Effectiveness of an interdisciplinary pain management program for the treatment of chronic pelvic pain. *Pain* 1990;41:41–46.
52. Siegal S, Paszkiewicz E, Kirkpatrick C, et al: Sacral nerve stimulation in patients with chronic intractable pelvic pain. *J Urol* 2001;166:1742–1745.
53. Selfe SA, van Vugt M, Stones RW: Chronic gynecological pain: An exploration of medical attitudes. *Pain* 1998;77:215–225.
54. Peters AAW, van Dorst E, Jellis B, et al. A randomized clinical trial to compare two different treatment approaches in women with chronic pelvic pain. *Obstet Gynecol* 1990;77:740–744.

CHAPTER 32

Cancer Pain: Essentials of Pharmacological Management

Paolo Manfredi, Ling Hsiu-Hsien, and Gil Schreier

Thirty percent of patients with cancer have pain at diagnosis, and 85 percent have pain when the disease is advanced. It now has been more than two decades since the World Health Organization (WHO) established the primary role of opioid analgesia for the relief of cancer pain in 1982. The "three-step analgesic ladder"[1,2] advocates the use of opioids, nonopioid analgesics, and adjuvant analgesics alone or in combination and titrated to the needs of the individual patient. The clinical pharmacology of the drugs used commonly to treat pain has been well outlined and divulgated.[3] Still, many patients with cancer pain suffer needlessly. The main reasons for undertreatment of pain in patients with cancer are lack of assessment (patients are not asked about pain) and misconceptions about tolerance and addiction.

▶ PATHOPHYSIOLOGY OF PAIN

Pain is an unpleasant sensory and emotional experience associated with actual or potential tissue damage or described in terms of such damage.[4] This definition recognizes the fact that ongoing tissue injury may be absent in the face of severe pain (e.g., postherpetic neuralgia) and holds a window open for psychological contributors.

Noxious stimuli activate free nerve terminals (nociceptors). Nociceptive signals travel via primary afferent fibers (Aδ for thermal and mechanical nociceptors and C fibers for polymodal nociceptors) and form ascending nociceptive tracts (spinothalamic and spinohypothalamic tracts). Pain signals then reach the thalamus and cortex. From higher centers, descending pathways modulate transmission of pain signals at the level of the dorsal horn.

Neuropathic pain defines a condition where signals are not originating at the free nerve terminals but at different levels of an abnormally functioning nervous system, with diverse postulated mechanisms. The prevalence of neural tissue injury in patients with cancer ranges from 34 to 50 percent. In a study of 187 hospitalized patients with cancer pain referred to a neurology-based pain service, 103 patients had pain caused by injury to the nervous system.[5] Injury at all levels of the nervous system can cause pain; the most frequent localizations are the spinal cord, spinal roots, lumbosacral and brachial plexuses, and peripheral nerves. Injury to the brain as a cause of cancer pain is uncommon but has been described.[6]

Personality, previous experiences, and psychiatric disorders will modify the interpretation of nociceptive signals as more or less painful and the evaluation of pain as creating or contributing to suffering. Furthermore, the painful event itself can be loaded with powerful emotional forces that influence the amount of suffering, as shown by Beecher in men wounded in battle.[7] Similar mechanisms are likely to play a role in pain from cancer; in cancer patients, recurrence and/or worsening of pain often signals tumor progression and carries a heavy psychosocial burden that can benefit from supportive psychotherapy and spiritual care.[8]

▶ ROLE OF THE CONSULTING NEUROLOGIST

The highly subjective nature of suffering, the importance of psychosocial factors in determining both suffering and adequate pain treatment, the difficulties in quantifying pain and diagnosing specific somatic and psychologic etiologies, and the scarcity of controlled studies are all obstacles to a standardized approach that will be effective for all patients with cancer pain.

The hospital consulting neurologist confronted with patients with cancer pain should assess the pain complaint thoroughly, review pertinent tests, formulate an educated pathophysiologic diagnosis (including psychiatric aspects), administer the appropriate pharmacologic treatment, and when necessary, suggest additional diagnostic tests and consultations from ancillary services.

The ability to access help promptly from physical medicine and rehabilitation, psychiatry, anesthesia, and neurosurgery, as well as psychological and spiritual support, will add considerable efficacy to the therapeutic intervention.

Although patients are likely to obtain maximal benefit from a structured interdisciplinary approach, even in the first few hours after evaluation and initial treatment, most can achieve a substantial pain reduction from relatively straightforward pharmacologic modifications.[9]

An additional task that the consulting neurologist may face, especially with some forms of neuropathic pain where tissue injury is not evident, is legitimizing pain complaints that are viewed as exaggerated by the health care staff or the family.

▶ ASSESSMENT

A comprehensive medical history must include a detailed cancer history, including past cancer treatment, prognosis, and plan of care. The pain history should focus on temporal features, location, severity, quality, and provoking and relieving factors; past pains, treatments, and psychosocial issues also should be explored in detail. Understanding the pathophysiology of the pain complaint is the basis for all diagnostic and therapeutic steps (Table 32-1).

Temporal Features

Acute pain follows a straightforward course with a well-defined cause, a typical behavior (grimacing, sobbing, splinting), sympathetic nervous system hyperactivity (tachycardia, tachypnea, diaphoresis), and anxiety. While awaiting etiologic treatment and/or healing, symptomatic management relies on opioids and nonsteroidal anti-inflammatory drugs (NSAIDs). Although guidelines for the treatment of postoperative pain are adequate in providing relief for most patients undergoing surgery, it is crucial to keep in mind that patients already on opioids before surgery will require much higher opioid doses than opioid-naive patients after similar operations.

Spinal cord compression, compressive radiculopathy, compressive or inflammatory plexopathies, and herpes zoster are some of the neurological processes that can present as pain crises in patients with cancer. An individualized blend of opioids, steroids or NSAIDs, "neuropathic pain medications," and nerve blocks (e.g., sympathetic blocks in acute herpes zoster) often is needed to control the acute pain that accompanies neural tissue injury.

▶ **TABLE 32-1. DIAGNOSIS AND TREATMENT OF CANCER PAIN**

History, with emphasis on cancer history, prognosis, plan of care, present pain, previous pains and treatments, and psychosocial issues.
Physical examination, with emphasis on neurological examination and findings suggestive of neuropathic pain (neurological deficits, allodynia, hyperalgesia, hyperpathia).
Additional assessment: radiologic studies, electrophysiologic studies, diagnostic nerve blocks, and psychiatric evaluation. It is important to personally review old and new pertinent diagnostic studies.
Pain diagnosis: pathophysiology of the pain complaint, including localization of the inciting injury, cause of the pain (cancer, cancer treatment, unrelated) and classification in the main two subtypes:
    Nociceptive Pain
    Neuropathic Pain
Treatment approaches (integrated with ongoing cancer care):
    Pharmacological (analgesics, adjuvant analgesics, including drugs to treat side effects)
    Interventional (nerve blocks, intraspinal therapy, other)
    Palliative radiotherapy and radiopharmaceuticals
    Palliative chemotherapy
    Palliative surgery
    Psychological/behavioral/spiritual interventions
    Physical therapy and rehabilitation
    Integrative medicine approaches (massage, hypnosis, acupuncture)

Chronic pain defines pain that persists beyond the usual course for the inciting event, or pain caused by a chronic, often progressive illness such as cancer. Chronic pain is associated with vegetative symptoms (poor sleep, appetite, libido) and depression. The pathophysiology is often less clear-cut compared with acute pain, and psychosocial factors play a stronger role and may be independent, sometimes primary, treatment targets.

Continuous pain from ongoing nociception often requires around-the-clock treatment with opioids. If an improving course is expected (e.g., postoperative pain), switching from parenteral to oral medications, with a tapering plan as the pain diminishes, is routinely effective. Patients with cancer tend to have a more prolonged course of postoperative pain, especially if pain was present before surgery, even if the surgery is considered curative.

Pain from progressive cancer requires around-the-clock opioids and NSAIDs for continuous pain, and additional doses of opioids for daily fluctuations of pain (breakthrough pain). Medications to treat neuropathic pain (damaged nervous system from tumor progression, chemotherapy, surgery, radiation therapy) may be useful, although opioids, at adequate doses, do relieve neuropathic pain.[10,11]

Continuous dysesthetic pains from peripheral neuropathy or postherpetic neuralgia may be helped by a stable dose of tricyclic antidepressants or an anticonvulsant.

Intermittent pains of neuropathic origin, often lancinating in quality, also will require around-the-clock medication (e.g., anticonvulsants), usually for prolonged periods (stable course). Intermittent movement-related pain (e.g., metastatic bone cancer) may need relatively high doses of short-acting opioids before planned active or passive movements.

Site of Pain and Localization of Injury

The first distinction is among focal (e.g., postherpetic neuralgia), multifocal (e.g., multiple bone metastasis), and generalized pain (e.g., chemotherapy-related joint and muscle pain). Multifocal pains can originate from a single lesion, e.g., metastatic epidural cancer at L5–S1 causing back pain, posterior thigh pain, and pain at the fifth toe. The consulting neurologist usually is aware of patterns of referred pain (e.g., radiculopathy) but may be less familiar with the distribution of referred pain from less-commonly involved nerves such as the intercostobrachial nerve. This nerve supplies the cutaneous innervation of axilla, inner arm, and anterolateral chest wall. It is not infrequently injured after radical mastectomy[12] and by apical lung tumors.

Back pain from pancreatic cancer and right shoulder pain caused by liver capsular distension, or a process involving the pleura at the base of the lung, are commonly seen patterns of visceral referred pain. Bone, joints, muscles, and visceral organs all can refer pain to distant sites.[13]

Intensity, Quality, and Affective Component

Pain out of proportion to the nociceptive inciting event or an inconsistent response to treatment may suggest a neuropathic component or overlooked psychosocial issues. Intense pain, especially if acute in onset, should be considered a medical emergency, and assessment and treatment should be prompt. Intravenous opioids represent the route and medication of choice. Unless there are obvious contraindications, an opioid-naive patient should be given 8 to 10 mg of intravenous morphine or 1.5 mg of intravenous hydromorphone, followed by the same dose every 10 to 15 minutes until the pain is controlled. If the pain is expected to persist beyond the duration of action of the opioid (3 to 4 hours), an infusion or an oral regimen should be started (see Table 32-1).

Dysesthetic and lancinating pains suggest a neuropathic origin, and may be best relieved by tricyclic antidepressants (dysesthetic), anticonvulsants (lancinating), and local anesthetics, although opioids are also effective.[10,11]

Reactive anxiety usually accompanies acute pain, and reactive depression follows chronic pain. These psychological states will influence the intensity of perceived pain and the evaluation of pain as suffering. In specific instances, reactive anxiety and depression, which usually are best managed with supportive psychotherapy, may become independent pharmacologic targets.

Measurement of Pain

Pain is a subjective feeling, and therefore, its assessment is based on the patient's report. Patients may not report pain for a number of reasons: previous experiences with lack of response to their pain complaints, perceived sense of uneasiness coming from heath care staff and family, desire to be stoic and to be viewed as a "good patient," or even cognitive dysfunction. It is, therefore, imperative to ask patients repeatedly about their pain. A few major institutions have added to the hospital chart vital signs' sheet a 0–10 verbal numerical scale, where 0 is no pain and 10 is the worst pain. Other validated pain scales are the Visual Analogue Scale, a simple categorical scale (none, mild, moderate, severe),[14] the Memorial Pain Assessment Card,[15] and the Brief Pain Inventory.[16]

▶ DIAGNOSIS: NOCICEPTIVE AND NEUROPATHIC PAIN

Nociceptive and neuropathic are broad pathophysiologic categories of pain with important diagnostic and therapeutic implications (see Table 32-1).

Nociceptive pain is caused by ongoing noxious stimuli activating nociceptors (free nerve terminals) in somatic or visceral structures. The cause is usually apparent, and the pain is described as aching or pressure-like and, for hollow visceral organs, cramping. It is often readily relieved by opioids and/or NSAIDs.

The pain from acute compression or inflammation of nerve roots or trunks (e.g., by progressive cancer) has a neuropathic quality (electric, burning), but the mechanism can be nociceptive—activation of nociceptors innervating the nerve trunk (nervi nervorum). Therefore it also may respond best to opioids and steroids or NSAIDs. Cramping pain from obstruction of the intestinal tract may respond to surgery and gastric decompression with a nasogastric tube or a percutaneous gastrostomy. Parenteral octreotide also may help by decreasing painful peristalsis and gastric secretions.

Neuropathic pain originates at different levels of an abnormally functioning, damaged nervous system, with diverse postulated mechanisms. It is important to distinguish between peripheral and central neuropathic pain, although the medications used are similar for both categories. The importance of making a distinction lies in the fact that peripheral neuropathic pain is more responsive to pharmacologic treatment, may respond to procedures (neuroma resection-injection, surgery, or radiation for decompression of neural structures), and carries a better overall prognosis.

With peripheral nerve injury, aside from activation of the free nerve terminals of nervi nervorum, which temporally (acute process), therapeutically (response to anti-inflammatory and opioids), and pathophysiologically best falls in the nociceptive category, pain can be caused by (1) increased sensitization of nociceptors to mechanical, thermal, inflammatory, or adrenergic stimuli, (2) neurogenic inflammation, i.e., release of inflammatory mediators from nerve terminals, and (3) ectopic activation of afferent fibers within the injured nerve.

With central nervous system (CNS) injury, pain can be caused by (1) loss of central inhibitory effects of myelinated afferent fibers, (2) reorganization of spinal cord connections with non-nociceptive fibers connecting to pain transmission neurons, (3) spontaneous activity and/or prolonged excitation after activation of spinal or thalamic neurons, and (4) abnormal function of descending modulatory pathways.

Although a central pathophysiology for generation of pain signals may be present after peripheral nervous system lesions, the term *central pain* usually refers to pain originating after a lesion at any level of the CNS, including the spinal cord, brain stem, and thalamus. Central pain may be delayed in onset by months or years after spinal or supraspinal injury, is more resistant to treatment compared with peripheral pain, and although much less frequent than peripheral neuropathic pain, has been described in patients with cancer.[6]

In addition to the above-mentioned peripheral and central mechanisms of neuropathic pain, another postulated mechanism primarily involves the autonomic nervous system, which is in charge of nociceptive transmission from autonomic organs but also has been involved in maintenance of superficial pain states following cancer-related injury (causalgia or complex regional pain syndrome type 2) or herpes zoster. In patients with cancer, complex regional pain syndrome type 1, where no detectable nerve injury is present, is seen infrequently.

Allodynia, hyperesthesia, and hyperpathia are all phenomena typical of neuropathic pain. Allodynia is pain perception with stimuli normally not painful, e.g., light touch causing burning pain. Hyperalgesia is an increased pain perception with normally painful stimuli, e.g., pinprick perceived as shooting, electric shock-like pain. Hyperpathia is an abnormally painful response to a stimulus, especially a repetitive stimulus, in an area where the threshold for nonpainful stimuli is increased, e.g., hypesthesia to pinprick and severe pain on repetitive pinpricking. All three phenomena can be present in the acute and chronic stages of herpes zoster.

In the best interest of the patient, a comprehensive approach should be carried out whenever the pain cannot be explained on the basis of current biomedical knowledge.[17]

▶ PHARMACOLOGIC PAIN MANAGEMENT

General Principles

The results of numerous studies strongly suggest that the "three-step analgesic ladder" advocated by the WHO is effective in relieving cancer pain in the great majority of patients. The pathophysiologic basis for the pain (nociceptive, neuropathic, and psychogenic), however, should remain the guide to the choice of treatment approach most likely to help. While we will focus here on the pharmacology of analgesics, it must be kept in mind that the pain in patients with cancer is associated most often with progression of the disease, and the role of the pain consultant, therefore, may extend to the recommendation of specific anticancer treatments that may provide symptomatic relief, including radiation treatment, chemotherapy, and surgery.

Even within one pathophysiologic entity, cancer pain can be produced with a multiplicity of mechanisms, and each one can be a target for treatment. Brachial plexopathy from different etiologies can cause intermittent lancinating pains that may respond to anticonvulsants; continuous dysesthetic pains that may respond to antidepressants; a continuous deep ache relieved by opioids; and superficial burning pain relieved by sympathetic blockade.

When brachial plexopathy is caused by cancer infiltration, steroids and/or radiation therapy also may provide relief and help to preserve neural function.

With neuropathic pain, interindividual variability in medication response and side-effect susceptibility occurs often. An initially positive response just may reflect a placebo effect that will soon vanish. The patient with chronic neuropathic pain must be informed that multiple trials of different medications, gradually titrated to pain relief or side effects, often are undertaken before even a partial response is achieved. The lowest dose should be started, with gradual increases. Side effects should be treated with reassurance when mild, with medication when mild or moderate (symptomatic management of constipation, nausea, and sedation), and by discontinuation of the drug when moderate or severe. If possible, only one drug should be titrated at a given time. Treatment goals should be discussed with the patient and, when possible, should include measures of improved function. Medications that do not work should be tapered before going to the next agent. Medications that result in a partial response (e.g., by ameliorating one component of the pain), if well tolerated, can be continued; frequently, a combination of medications from different classes offers advantages over a single agent. (See Table 32-2).

The need for psychotherapy and/or antidepressant therapy should be assessed early on in the treatment course. Depression often coexists with cancer pain and, if not addressed, may render less effective even the most carefully thought-out medication trials.

Opioids

Opioids are the mainstay of cancer pain treatment[1-3] (Table 32-2). They decrease the intensity of nociception and alter the unpleasant experience associated with nociception by mimicking the action of naturally occurring opioid peptides (enkephalins, dynorphins, β-endorphin, α-neoendorphin) through recognition of common receptors.[18] Mu receptors, subdivided into mu_1 and mu_2, are the binding sites for the main opioids used in the treatment of cancer pain, including morphine, oxycodone, hydromorphone, fentanyl, and methadone. Mu_1 receptors predominate at the supraspinal level. In the dorsal horns of the spinal cord, morphine analgesia is instead mediated by mu_2 receptors. In addition to supraspinal and spinal morphine analgesia, peripheral mechanisms of opioid analgesia are suggested by the report of analgesia following intra-articular injection of morphine[19] and by experimental studies.[20]

Drugs that bind to opioid receptors are classified as agonists (e.g., morphine) if they produce analgesia, antagonists (e.g., naloxone) if they block the action of an agonist without other effects, and agonist-antagonist (e.g., pentazocine) if they produce analgesia by interacting with a specific receptor (e.g., kappa) but also have antagonistic activity at other receptors (e.g., mu). Partial agonists (e.g., buprenorphine) bind to receptors and produce analgesia, but unlike morphine, they exhibit a ceiling effect, i.e., increases in doses do not parallel increases in analgesia or respiratory depression. The clinical use of mixed agonist-antagonists is limited by their ability to produce dysphoria and hallucinations; these effects are mediated by kappa receptors and nonopioid sigma receptors.[21] Mixed agonist-antagonists and partial agonists both will cause a withdrawal syndrome when administered to patients taking opioids chronically.

Opioids are, among all medications, the closest agents we have to the ideal analgesic. Opioids have no ceiling effect; therefore, progressive increases in dosage result in more profound analgesia. Opioids are the most effective tool for relief of any type of acute pain and cancer-related pain because of their predictable, dose-dependent response. The most common side effect, constipation, is routinely treated preemptively by the administration of a stool softener and bowel stimulant. Opioids have no significant long-term organ toxicity and can be used for years. Addiction is a quite rare phenomenon when opioids are used in the context of medical care.[21] Tolerance to a drug implies the need to increase the dose to obtain an effect. Tolerance to the different opioid effects develops at different degrees. There is no tolerance to constipation (therefore, it is necessary to institute a bowel regimen at the start of opioid therapy and to continue it for the duration of opioid therapy). There is little tolerance to the analgesic effect; therefore, the need for dose escalation reflects either disease progression[23] or ineffectiveness of opioid therapy (e.g., pain unresponsive to opioids). Tolerance to the sedative effects of opioids usually develops in 3 or 4 days; when sedation persists, it can be counteracted by methylphenidate or dextroamphetamine, which also may provide additive analgesic effects. There is rapid tolerance to respiratory depression, and therefore, clinically significant hypoventilation is a rare complication in the setting of careful opioid titration. Methadone, with its long half-life, requires extra caution; signs of overdose can be delayed 3 to 7 days after a dose increase, catching both patient and physician by surprise. There is usually rapid tolerance to the emetic effects of opioids, especially if constipation is addressed properly.

Physical dependence is an altered physiologic state produced by repeated administration of a drug. This condition necessitates the continued administration of the drug to prevent withdrawal symptoms. When opioids are discontinued abruptly, the withdrawal symptoms consist of lacrimation, rhinorrhea, restlessness, tremors, diarrhea with abdominal cramping, and nausea and vomiting. These symptoms are usually mild and do not have the dramatic appearance seen in patients with opioid addiction. Even patients who receive opioids

► **TABLE 32-2.** OPIOID TREATMENT GUIDELINES: ORAL AND PARENTERAL OPIOID ANALGESIC EQUIVALENCES (SINGLE DOSES), RELATIVE POTENCY COMPARED TO MORPHINE AND ANALGESIC DURATION

Opioid Agonists	Parenteral (mg)	Oral (mg)	Conversion IV to PO	Duration
Morphine	10	30	3	3–4 hours
Long-acting morphine		30		12 hours
Oxycodone controlled-release		20		3-4 hours
Oxycodone		20		12 hours
Hydromorphone	1.5	7.5	5	3 hours
Oxymorphone	1	6	6	3–6 hours
Methadone[1]	10	20	2	4–24 hours (8 hours)
Levorphanol[1]	2	4	2	3–8 hours (6 hours)
Meperidine[2]	75	300	4	2–3 hours
Fentanyl[3]	0.1			1 hour
Codeine	130	200	1.5	3–4 hours

Starting an Opioid Infusion

Divide the amount of parenteral medication that effectively controls the pain without side effects by the duration of action of that opioid (use longest duration). If PCA is available, half of that amount is the hourly rate (e.g., morphine 40 mg IV effectively controls the pain of a pathologic fracture: infusion rate = 40 mg : 4 = 10 mg : 2 = 5 mg /hour). The PCA dose (in addition to the infusion of 5 mg hour), should be 2.5 mg of morphine (50% of the hourly dose), with a lock-out of 10 minutes. If PCA is not available, the basal infusion rate can be 25% higher (in the case outlined above, 7.5mg / hour).

Converting to Transdermal Fentanyl

Determine the 24-hour parenteral morphine equivalent requirement using the table above. For dosages of transdermal fentanyl over 100 μg/hr multiple patches can be used. Patch duration = 48–72 hours. It takes 12–24 hours before achieving full analgesic effect after the first patch is applied. The patch can be titrated every 24 hours. Do not use PRN (use short-acting opioids to treat acute pain).

Parenteral Morphine Equivalent (mg/24 hours)	Transdermal Fentanyl Equivalent (μg/hr)
8–22	25
23–37	50
38–52	75
53–67	100
68–82	125
83–97	150

Bowel Regimen Protocol

Virtually all patients on opioid therapy need an individualized bowel regimen.
Docusate 100 mg bid plus Senna 5–10–15 mg bid.
Docusate 100 mg bid, Senna 20 mg bid plus Lactulose 15 cc bid
Docusate 100 mg bid, Senna 20 mg bid plus Lactulose 30 cc bid
Docusate 100 mg bid, Senna 20 mg bid plus Lactulose 30 cc qid.
If there has been no bowel movement in 4 days, the nurse will administer mineral oil or soap suds enema or high colonic tap water enema. Be aware of the possibility of bowel obstruction or fecal impaction.

[1]Long half-life: when converting to around-the-clock chronic dosing with methadone or levorphanol, use much lower doses (see methadone section and Tables 32-1 and 32-3).
[2]Meperidine should not be used for chronic pain or in high doses because of CNS toxic metabolites (see meperidine section).
[3]Short half-life: when converting to around-the-clock chronic dosing, the fentanyl to morphine conversion is morphine 1 mg = fentanyl 25 μg.
- Short-acting strong opioids (morphine, hydromorphone, oxycodone) should be used to control moderate and severe pain. Long-acting preparations (MS Contin, Oxycontin, Transdermal Fentanyl), should be started after the pain is controlled on short-acting opioids. With the exception of methadone and levorphanol, long-acting opioids should not be used for acute pain.
- Titrate the opioid dose at least every 24 hours when the pain is moderate and at least every 2 hours when the pain is severe. Increase the dose by at least 30–50%. For equivalent dose purposes: rectal = oral route. SQ = IM = IV route.
- Manage breakthrough pain (acute pain in patients with otherwise controlled pain) with short-acting opioids using 1/3 of the single dose amount (e.g., patient on long-acting morphine 90 mg q 12 h, breakthrough morphine dose = 30 mg q 3 h PRN).
- Manage opioid side effects. Constipation must be treated prophylactically (see bowel regimen above).

(Continued)

▶ **TABLE 32-2.** (CONTINUED)

- Patient controlled analgesia (PCA) is a safe and effective modality for delivery of opioids for pain that is expected to resolve (post-operative pain) and for acute exacerbations of chronic pain (pathologic fracture in a patient with chronic pain from metastatic bone cancer): the patient self-delivers fixed doses of opioid by pressing a button. Overdose is very infrequent because the patient must be alert in order to press the button. Safe PCA starting doses for opioid-naive patients are: Morphine 1 mg every 10 minutes or hydromorphone 0.25 mg every 10 minutes. A continuous rate of opioid infusion is needed for patients who suffer from pain not expected to resolve shortly.
- Mixed agonists-antagonists (pentazocine, nalbuphine, butorphanol) should not be used because of their high incidence of cognitive effects and because they cause withdrawal in patients on chronic therapy with opioid agonists. Partial agonists (buprenorphine) may also cause withdrawal in patients on chronic opioids.
- Naloxone should be used only for life-threatening opioid-induced respiratory depression, an exceedingly rare occurrence in patients on chronic stable opioid doses. In order to minimize agitation, fever, emesis, and pain, when naloxone is needed, dilute 1 vial (0.4 mg) in 10 cc of NS and administer 1 cc every minute PRN. This careful titration will reverse respiratory depression without causing withdrawal symptoms. The half-life of naloxone (1 hour) is shorter than the half-life of opioid agonists; therefore, additional doses of naloxone might be needed.

over a period of a few weeks often will have only mild symptoms when the drug is discontinued abruptly. It is, however, proper practice to taper the medication gradually by approximately 25 percent of the previous dose every day or every other day. Symptoms of withdrawal are severe when an opioid antagonist (e.g., naloxone) is administered to a patient on chronic opioid therapy. In patients with cancer on chronic opioid therapy, naloxone is often administered inappropriately and should be reserved for life-threatening respiratory depression.[24]

Addiction, or psychological dependence, is defined by compulsive drug-seeking behavior and overwhelming involvement in drug procurement and use. It should not be confused with the behavior of the patient with pain who is given insufficient medication (pseudoaddiction). It has been shown that addiction does not depend on the characteristics of the opioid medication but on the individual's genetics and psychosocial milieu. This is corroborated by the rarity of this phenomenon when opioids are used in the medical setting.[22]

Opioid rotation defines a therapeutic strategy well known to clinicians treating cancer pain that involves switching from one opioid to another when the first opioid is ineffective at doses that cause side effects[25] (Tables 32-2, 32-3, and 32-4). Different reasons may explain why a patient responds to one opioid and not to another. There are interindividual differences in opioid receptor genes and their expression in different tissues. Improved knowledge of pharmacogenetics may help to predict beforehand which patients may respond to which opioids. The main scrutiny by investigators at this time is on the mu opioid receptor gene. Other genetic reasons for determining individual responses to one or another opioid involve genes that code for enzymes involved in opioid metabolism, e.g., approximately 10 percent of individuals cannot metabolize codeine to morphine and, therefore, are unresponsive to the analgesic effects of codeine. The buildup of toxic opioid metabolites is another factor that may help explain why patients respond initially to one opioid, and then require opioid rotation. Stopping the first opioid is, in this case, therapeutic because the buildup of toxic metabolites ceases and a gradual washout ensues. Finally, actions on other receptors may be responsible for the superior efficacy of the second opioid; when the second opioid is methadone, actions at the *N*-methyl-D-aspartic acid (NMDA) receptor or even on the catecholamine system may determine analgesia (see section on methadone below).

"Weak" Opioids and Oxycodone

For practical purposes, opioids have been categorized as weak or strong depending on their potency relative to morphine and their use in fixed combination with aspirin or acetaminophen. Weak opioids are used for less severe pain; their efficacy is limited by an increased incidence of side effects at higher doses (e.g., nausea and constipation with codeine, CNS excitation with propoxyphene,

▶ **TABLE 32-3.** OPIOID ROTATION FROM ORAL MORPHINE TO ORAL METHADONE: INDICATIVE SAFE AND EFFECTIVE STARTING DOSES

The ratio depends on the dose of previous opioid
- morphine 30 mg to 90 mg PO use ratio of 4:1
  (e.g., morphine 30 mg is approximately equivalent to 7 mg of methadone)
- morphine 90 mg to 300 mg PO use ratio of 8:1
  (e.g., morphine 300 mg PO is approximately equivalent to 35 mg methadone PO)
- morphine >300 mg PO use ratio of 12:1
  (e.g., morphine 400 mg PO is approximately equivalent to 35 mg methadone PO)

If previous morphine dose is *much* higher than 300 mg, the dose ratio will be higher than 12:1.

▶ **TABLE 32-4.** OPIOID ROTATION DURING CHRONIC OPIOID THERAPY WITH IV PATIENT-CONTROLLED ANALGESIA (PCA): INDICATIVE SAFE AND EFFECTIVE STARTING DOSES WHEN ROTATING FROM OTHER OPIOIDS TO METHADONE

Initial Opioid	Basal	New Opioid	Basal	Demand	CAB
Morphine	10	Methadone	1	1	5
Hydromorphone	1.5	Methadone	0.3	0.3	5
Fentanyl (µg)	250 (µg)	Methadone	1.25	1.25	5

NOTE:
1. Basal: continuous hourly infusion. Demand: dose available every 15 minutes by pressing the PCA button. CAB (Clinician Activated Bolus): dose administered by the nurse upon request if the pain persists despite the use of demand doses. All doses are in mg except for fentanyl (µg).
2. Decrease the initial dose by 25–50% for high doses (50 mg/hour of morphine) and increase the dose by 25–50% for low doses (5 mg/hour of morphine).

and aspirin or acetaminophen intoxication with fixed combinations).

Although weak opioids have a role in the management of mild to moderate cancer pain, they also can be substituted effectively with small doses of strong opioids.

Codeine, an alkaloid of opium, is the prototype of the weak analgesics. It is generally given by mouth, often in a fixed mixture with a nonopioid analgesic. A 200 mg dose is approximately equipotent to 30 mg morphine. The affinity of codeine for mu receptors is several thousand-fold less than morphine. The half-life of codeine is 2.5 to 3 hours; approximately 10 percent of orally-administered codeine is demethylated to morphine; free and conjugated morphine can be found in the urine. Most of the analgesic action is likely due to its conversion to morphine. Up to 10 percent of Caucasians and over 30 percent of Asians cannot metabolize codeine to morphine and, therefore, obtain limited analgesia from codeine.

Hydrocodone is a codeine derivative available in the United States only in combination with acetaminophen, aspirin, or ibuprofen in doses of 2.5, 5, 7.5, and 10 mg. It is thought to be more potent than codeine but less potent than oxycodone, although good data are lacking.

Propoxyphene is a synthetic analgesic structurally related to methadone. It is approximately equipotent to codeine as an analgesic but lacks its antitussive properties; it binds to mu receptors. Its analgesic activity lasts 3 to 5 hours. The half-life is 6 to 12 hours, and its major metabolite is norpropoxyphene, which has a half-life of 30 to 36 hours and may be responsible for some of the observed toxicity. Norpropoxyphene has local anesthetic effects similar to lidocaine, and high doses may cause arrhythmias. Seizures occur more often with norpropoxyphene intoxication than with other opioids.

Meperidine, a synthetic phenylpiperidine derivative mu agonist with anticholinergic properties, remains a commonly prescribed analgesic for acute pain, and it is still used widely for chronic pain. The reasons for this enthusiasm are unclear and not supported by scientific data. The CNS excitatory effects that appear after chronic use are well substantiated; the accumulation of its metabolite normeperidine causes multifocal myoclonus and grand mal seizures[26] not reversed by naloxone. In this context, dilated pupils and hyperactive reflexes are characteristic. The half-life of meperidine is 3 hours; it is, in part, demethylated to normeperidine, the only active metabolite, which has a half-life of 15 to 30 hours. Normeperdine accumulates only after chronic treatment, particularly in patients with renal dysfunction. When meperidine is given to patients being treated with monoamine oxidase (MAO) inhibitors, two different patterns of toxicity have been observed: severe respiratory depression and excitation, delirium, hyperpyrexia, and convulsions. There is no rationale for its use in a context where alternatives are available.

Oxycodone is a semisynthetic derivative of thebaine, an opium alkaloid. Because of its high bioavailability (>50 percent), it is suitable for oral administration, and by this route it is more potent than morphine. It has a half-life of 2 to 3 hours and a duration of action of 4 to 5 hours. Oxycodone often has been included among the weak opioids because of its use in fixed combinations with acetaminophen and aspirin. When oxycodone is used alone, it is a strong mu opioid agonist (see Table 32-2) that can be used to treat pain of any intensity that requires an opioid analgesic. Controlled-release preparations, which permit a twice- to three-times-a-day regimen, are effective, although there have been recent concerns about a rise in overdoses caused by diversion and grinding of the tablets.

"Strong" Opioids

Morphine is a phenanthrene derivative and is the prototypical opioid agonist. All others are compared with morphine in determining their relative analgesic potency (see Table 32-2). It is the drug of choice for the treatment of severe pain associated with cancer.[1–3] As with other strong opioids, there is no ceiling to the analgesic effect, although side effects, particularly sedation and confusion, may intervene before optimal analgesia. The duration of analgesia is 4 hours. Slow-release preparations, which permit a twice-a-day regimen, are effective and should be used after dose titration with morphine sulfate. Morphine is metabolized in the liver, where it undergoes glucuronidation at the 3 and 6 positions. Morphine-3-glucuronide (M3G) and morphine-6-glucuronide (M6G) accumulate with chronic morphine administration. M6G binds to mu receptors with affinity similar to morphine. With single-dose morphine studies, the relative parenteral-oral potency ratio is 1:6, but after chronic use, the ratio changes to 1:3; this is likely due to the accumulation of active metabolites. M3G has negligible affinity for opioid receptors and does not produce analgesia but has excitatory effects on neurons and can cause myoclonus and, possibly, a hyperalgesic state. Morphine metabolites are eliminated by glomerular filtration; they accumulate in renal failure, leading to an increased incidence of side effects. Morphine should be used with caution in renal failure, increasing the time interval between doses. The use of an alternative opioid (e.g., methadone) may be preferable.

Hydromorphone is a semisynthetic phenanthrene derivative opioid agonist available commercially as a highly water-soluble salt. When administered parenterally, 1.3 mg hydromorphone is equipotent to 10 mg morphine; it is somewhat shorter acting but has a higher peak effect. Its oral bioavailability is low, with an oral-parenteral ratio of 5:1. It has a half-life of 1.5 to 2 hours, and its chronic use also can cause accumulation of active and neurotoxic metabolites. It is the opioid of choice for continuous subcutaneous infusions with similar potency to the intravenous route.

Fentanyl is a synthetic phenylpiperidine-derivative opioid agonist that interacts primarily with mu receptors. Parenterally, it is eighty times more potent than morphine and highly lipophilic; these properties render it a suitable candidate when a transdermal route for opioid analgesia is sought. Since the first clinical study in patients with cancer pain,[27] transdermal fentanyl has become a useful adjunct in cancer pain management. It is most useful for patients with chronic pain who are unable to take drugs by mouth and do not require a rapid titration. The transdermal fentanyl therapeutic system delivers drug continuously to the systemic circulation for up to 72 hours. After application of the transdermal patch, the systemic absorption is very low in the 0- to 4-hour period, increases during the 4- to 8-hour period, and remains relatively constant with a coefficient of variation of 28 percent from 8 to 24 hours. The reason for the initial delay is the time required to establish a reservoir of fentanyl in the stratum corneum. Following removal of the transdermal patch, the serum fentanyl concentration falls about 50 percent in approximately 16 hours. This long half-life is likely due to the slow washout of the cutaneous reservoir. These considerations translate clinically in a several-hour delay in the onset of analgesia after an initial application and a persistence of analgesia and other effects for several hours after its removal. Oral transmucosal fentanyl is used for expedient treatment of breakthrough pain and may have advantages over oral morphine with initial pain relief starting within a few minutes, with maximum effect in 20 to 30 minutes.[28] Fentanyl also has an important role in epidural and intrathecal routes. In addition, since patients with cancer pain have frequent exacerbations that require parenteral titration of opioids, protocols for going from IV fentanyl to transdermal fentanyl and vice versa have been described.[29,30] Although when given in single doses fentanyl is approximately 80 to 100 times more potent than morphine, in the pragmatic clinical context of opioid rotation, the appropriate conversion is IV *morphine 1 mg = fentanyl 25 μg*. This lower dose of fentanyl can be explained by its shorter half-life compared with morphine.

Levorphanol is a synthetic opioid agonist structurally related to the phenanthrene-derivative opioids. It is a potent mu agonist but also binds delta and kappa receptors. When administered parenterally, 2 mg levorphanol is equianalgesic to 10 mg morphine. The drug also has good oral efficacy, with an intramuscular-oral ratio of 0.5. It has a half-life of 12 to 30 hours and a duration of analgesia of 4 to 6 hours; therefore, repeated administration is associated with accumulation. A dose reduction may be required 2 to 4 days after starting the drug in order to avoid side effects from overdosage. For the same reason, it is best to avoid it in patients with impaired renal function or encephalopathy. It is sometimes useful as a second-line drug in patients who cannot tolerate morphine because of inadequate analgesia and intolerable side effects. In the last few years, its availability has been limited.

Methadone is a synthetic diphenylheptane-derivative opioid mu agonist. Methadone has unique pharmacodynamic properties, with activity at the NMDA receptor and inhibition of catecholamine uptake.[31,32] Both these actions may render it particularly effective against neuropathic pain, although controlled studies are lacking. Numerous papers have addressed methadone's rediscovery as an effective and safe analgesic, with emphasis on its NMDA antagonistic activity and improved dosing guidelines.[33,34] Its widespread use remains limited by the need for a careful individualized dose and interval titration. Methadone is rapidly absorbed from the gastrointestinal tract, with measurable plasma concentrations within 30 minutes of oral administration. Its oral bioavailability is around 90 percent, and its half-life is long and variable (13 to 58 hours). Although single parenteral doses of methadone are equipotent to morphine, with repeated dosing the potency of methadone is much higher than expected from the original equianalgesic studies. Since the frequency of opioid rotation is increasing, there is a need for suggestions on safe and effective starting doses when rotating from other opioids to methadone[33] (see Tables 32-2 and 32-3). The higher than expected potency of methadone noted with chronic therapy is probably secondary to its long half-life with increasing serum drug levels, although its activity at receptors other than the opioid receptor may account for some cases of efficacy at doses much lower than expected.[33] Because of its high hepatic clearance and lack of toxic metabolites, methadone should be considered the primary opioid for patients with renal impairment.

▶ NONOPIOID ANALGESICS AND ADJUVANT ANALGESICS

Nonsteroidal anti-inflammatory drugs (NSAIDs) are the best known and most used family of nonopioid analgesics. They have a ceiling effect, which limits their use to mild and moderate pain, and well-known side effects on gastric mucosa and kidneys, which can occur with short- and long-term use, occur more frequently at higher doses, and are especially dangerous for the elderly. The mechanism of action of NSAIDs involves inhibition of the enzyme cyclooxygenase, with reduction of inflammatory mediators that are known to sensitize and activate nociceptors. A central effect is also postulated due to the poor correlation between anti-inflammatory activity and analgesic effects of different NSAIDs. They can be used alone for mild pain or pain particularly responsive to this class of medications, or concomitantly with opioids for the treatment of pain associated with cancer progression and any time inflammation is important in the pathophysiology of the pain syndrome. Their efficacy in relieving neuropathic pain is minimal. Different NSAIDs can offer specific advantages: Choline magnesium salicylate has less effect on platelet aggregation, ibuprofen is relatively inexpensive, piroxicam is given once a day only, and ketorolac can be administered parenterally. Sequential trials with different NSAIDs, until the most effective one for the individual patient is found, are advocated by some experts. Selective Cox-2 inhibitors produce anti-inflammatory and analgesic actions with substantially fewer effects on the gastric mucosa and without antiplatelet effects. The selective Cox-2 inhibitors do, however, retain kidney toxicity and may increase the risk of ischemic cardiovascular events in some patients. All NSAIDs, therefore, should be used with great caution in the elderly and in patients at increased cardiovascular risk.

Acetaminophen is a useful analgesic for mild pain and has very little, if any, anti-inflammatory action. It does not have adverse effects on gastric mucosa and platelet aggregation. Caution should be used when the combination opioid-acetaminophen is prescribed because patients in severe pain may exceed the 4-g-daily limit on acetaminophen, beyond which the risk of hepatic toxicity is substantial.

Tramadol is an analgesic with weak activity on opioid receptors and TCA-like effects on neurotransmitters. It has analgesic potency similar to mild opioids (e.g., acetaminophen with codeine preparations) but a longer duration of action (6 hours). Nausea and vomiting are possible side effects. Its usefulness is limited to the treatment of mild pain without clear advantages over mild opioids except for the lack of physical dependence.

Tricyclic antidepressants (TCAs) have been shown to relieve neuropathic pain independent of their effects on mood.[35] The postulated mechanism of action involves pain-modulating systems converging on dorsal horn neurons. Pain associated with peripheral neuropathy and postherpetic neuralgia is relieved effectively by amitriptyline and desipramine. Nortriptyline, imipramine and doxepin also appear to be useful analgesics. Selective serotonin reuptake inhibitors (SSRIs) do not seem to be as effective in treating neuropathic pain. In patients with cancer pain, antidepressants are used commonly to treat neuropathic pain, particularly of a continuous, burning or dysesthetic character, but they also may be used for lancinating pains.[36] TCAs should be started at very low doses at bed time (5 to 10 mg), with slow titration (10 mg every 3 days) up to pain relief or side effects, keeping blood levels in the safe range. TCAs are the mainstay of treatment of neuropathic pain, although a number of elderly patients (e.g., patients with postherpetic neuralgia) may get more relief from, and tolerate better, opioids at low doses.[11] The main side effects are sedation, orthostatic hypotension, urinary hesitancy and retention, xerostomia, constipation, cardiac conduction abnormalities, and worsening of closed-angle glaucoma. Elderly patients are particularly vulnerable to all these side effects. Following newly-instituted therapy or a dose increase, the acute onset of urinary retention, syncopal episodes, or delirium is not uncommon. Desipramine has the least anticholinergic and sedating properties among TCAs. Newer antidepressants with mixed reuptake properties appear to be promising, with fewer side effects compared with TCAs.[37]

Anticonvulsants are useful in treating paroxysms of lancinating pain that can occur with peripheral nerve injury but are also effective for the treatment of continuous dysesthetic pain and central pain. Carbamazepine is the most time-honored agent, but it should be used with caution in patients with cancer pain because of its dose-dependent bone marrow suppressive action. Over the past few years, gabapentin has become the most commonly used anticonvulsant for patients with cancer pain.

Its analgesic effect and safety were proven in randomized, double-blind, placebo-controlled trials in patients with peripheral neuropathy and postherpetic neuralgia.[38,39] As with antidepressants, there is a high individual variability in the dose required to achieve analgesic effect. The starting dose of gabapentin is usually low, 300 mg at night or divided into three equal daily doses, followed by a gradual increase until benefit or side effects are reported. The most common side effects are daytime sedation and ataxia. Evidence for efficacy exists also for phenytoin (which can be useful for more urgent situations that require a loading dose, e.g., tic douloureux), valproic acid, and other anticonvulsants (lamotrigine, topiramate, oxcarbazepine, zonisamide, felbamate).

CNS stimulants counteract sedation from opioids, can improve mood, and may have additive or synergistic analgesic activity when used with opioids. Caffeine can be used as a first step, either as a caffeinated beverage or in pill form. Methylphenidate is the most frequently used stimulant in the management of cancer pain, but newer agents such as modafinil may have fewer side effects. When sedation does not respond to the preceding agents, dextroamphetamine can be used.

Baclofen has been found to be a useful adjunct for lancinating pains refractory to anticonvulsants alone. Some patients are particularly vulnerable to its cognitive side effects.

Intravenous *lidocaine* has been reported to relieve neuropathic pain, at times for prolonged periods after a single injection. Contraindications include second-degree atrioventricular (AV) block and intraventricular conduction defects. The slow (5 to 10 minutes) IV injection of 100 mg lidocaine also can indicate if the patient will respond to *mexiletine;* if the pain is significantly diminished by lidocaine, mexiletine is started (150 mg/d with increases by 150 mg every 3 days to a maximum of 1200 mg/d in three divided doses). Lidocaine infusions (1 to 3 mg/min) can be useful for neuropathic pain refractory to opioids, antidepressants, and anticonvulsants.

Clonidine, an α_2-receptor agonist, has some value in the treatment of neuropathic pain, especially if it is maintained sympathetically, and pain associated with spasticity. It also has been used intrathecally for patients with pain refractory to opioids.

Benzodiazepines have anecdotal value in the treatment of acute muscle spasm (e.g., diazepam) and neuropathic pain (e.g., clonazepam).

Antihistamine drugs have mild analgesic properties. The speculated mechanism of action involves the action of histamine in modulation of pain. They also have antianxiety and antiemetic properties and, therefore, can be useful adjuncts for selected patients (e.g., hydroxyzine 50 mg every 6 hours).

Neuroleptics do not have analgesic properties, with the exception of methotrimeprazine, which has been found to have analgesic potency similar to morphine, although use of the IV formulation is limited by sedation and orthostatic hypotension in ambulatory patients. Haloperidol is useful to treat delirium of different etiologies in patients with cancer. Newer antipsychotic agents, such as olanzapine, also can help delirium and improve pain assessment.

Corticosteroids have strong anti-inflammatory action, and their analgesic properties are most evident when edema is causing compression of neural structures (lesions in proximity to the spinal cord or brachial plexus). Because of well-known side effects, their chronic use for analgesia is limited to patients with advanced cancer. In this setting, they also can improve appetite and mood. Corticosteroids can be used more liberally for analgesia when preservation of neural function is one of the objectives.

Dextromethorphan and *ketamine* have been used alone or in conjunction with opioids for refractory neuropathic pain. The postulated site of action is the NMDA receptor. Their narrow therapeutic window limits clinical use.

Bisphosphonates, such as pamidronate, have analgesic properties, activity inhibit osteoclastic and reduce the incidence of fractures in patients with bone metastases and multiple myeloma. Intravenous pamidronate has been used to treat bone pain.

Calcitonin may have some utility in pain associated with bone metastases and in phantom limb pain.

Topical treatments with local anesthetics, capsaicin, and even NSAIDs, antidepressants, ketamine, and opioids, have been claimed to offer relief in the treatment of pain related to herpes zoster, especially for elderly patients with poor tolerance to systemic drugs. Formulations of topical local anesthetics designed to penetrate the skin are available (EMLA cream). A recent method for treatment of neuropathic pain is the topical 5% lidocaine patch. In patients with postherpetic neuralgia (PHN), a double-blind, controlled study showed that application of the patch to allodynic skin caused analgesia and was safe without systemic adverse effects.[40]

Radiopharmaceuticals, such as strontium 89 or samarium 153, can be added to opioids and/or to nonopioid analgesics for the treatment of bone pain, especially when the pain is diffuse and has not responded to external radiation therapy, chemotherapy, or hormonal therapy. The main side effect is bone marrow suppression with thrombocythemia.

▶ INTERVENTIONAL APPROACHES

The number of procedures for the treatment of cancer pain has declined steadily as opioids and cancer pain guidelines have become more available (Table 32-5). The most effective procedure remains the injection of alcohol around the celiac plexus. This procedure can be

> **TABLE 32-5. INTERVENTIONAL APPROACHES TO TREAT CANCER PAIN**

Intraspinal (epidural and intrathecal) opioids, local anesthetics, clonidine

- *Celiac Plexus Block:* This is the only nerve block that should be thought of early in the treatment of pain from retroperitoneal cancer, e.g., pancreatic cancer; effective in more than 80% of patients.
- *Hypogastric Plexus Block: Pelvic pain.* Not as well documented as celiac plexus block.
- *Cordotomy, cervical:* Percutaneous procedure, radiofrequency. Effective for unilateral pain below the nipple line.
- *Commissural myelotomy:* requires a laminectomy. Bilateral pain below the waist.
- *Rhizotomy (chemical, radiofrequency, surgical):* Chemical rhizotomy can be accomplished with injection of alcohol in the subarachnoid space and proper positioning: alcohol is hypobaric compared to CSF, therefore it floats reaching the uppermost areas of the subarachnoid space.
- *Dorsal Root Entry Zone* (DREZ) lesion: Requires a laminectomy; indicated for pain from brachial plexus injury.
- Neurectomy (chemical, radiofrequency, surgical).

dramatically effective and should be considered early in the treatment of progressive inoperable pancreatic cancer. Sometimes, less risky procedures may avoid more invasive treatments (e.g., epidural opioids to avoid cordotomy). Intraspinal delivery of opioids generally is indicated when adequate trials of oral and parenteral opioids provide analgesia but side effects (e.g., sedation), although properly managed, prevent their use. The epidural route permits the delivery of local anesthetics and an easier titration of medications and combinations of medications (opioids and local anesthetics). There is, however, a significant risk of epidural fibrosis that will interfere with medication flow. The subarachnoid route requires a completely internalized device. Therefore, changing the solution ingredients and their concentrations is more cumbersome. However, the system is more durable and better tolerated and probably more beneficial for patients with a longer life expectancy. A recent multicenter trial showed that intrathecal opioids were more effective than comprehensive medical treatment for the management of cancer pain.[41] Although there were several methodologic points open for discussion, including a restrictive definition of comprehensive medical treatment, this study shows that some cancer patients benefit from early intraspinal opioid delivery.

REFERENCES

1. World Health Organization: *Cancer Pain Relief.* Geneva: WHO, 1986.
2. World Health Organization: *Cancer Pain Relief and Palliative Care: Report of a WHO Expert Committee.* Geneva: WHO, 1990.
3. Jacox A, Carr DB, Payne R, et al: Management of cancer pain. *Clinical Practice Guidelines No. 9,* AHCPR Publication No. 94-0592. Rockville, MD: Agency for Health Care Policy and Research, U.S. Department of Health and Human Services, Public Health Service, 1994.
4. Merskey H, Bogduk N (eds): *Classification of Chronic Pain: Descriptions of Chronic Pain Syndromes and Definitions of Pain Terms.* Seattle: IASP Press, 1994:210.
5. Manfredi PL, Gonzales GR, Ribeiro S, et al: Neuropathic pain in patients with cancer. *J Pall Care* 2003;19:115–118.
6. Gonzales GR, Tuttle S, Thaler HT, Manfredi PL: Central pain in patients with cancer. *J Pain* 2003;4:351–354.
7. Beecher HK: Pain in men wounded in battle. *Ann Surg* 1946;96:104–105.
8. Foley KM: The treatment of cancer pain. *N Engl J Med* 1985;313(2):84–95.
9. Manfredi PL, Chandler SJ, Pigazzi A, Payne R: Outcome of cancer pain consultations. *Cancer* 2000;920–924.
10. Gimbel JS, Richards P, Portenoy RK: Controlled-release oxycodone for pain in diabetic neuropathy: A randomized controlled trial. *Neurology* 2003;60:927–934.
11. Raja SN, Haythornthwaite JA, Pappagallo M, et al: Opioids versus antidepressants in postherpetic neuralgia: A randomized, placebo-controlled trial. *Neurology* 2002;59:1015–1021.
12. Assa J: The intercostobrachial nerve in radical mastectomy. *J Surg Oncol* 1974;6:123–126.
13. Kellegren JG: On distribution of pain arising from deep somatic structures with charts of segmental pain areas. *Clin Sci* 1939;4:35–46.
14. Jensen MP, Karoly P, Braver S: The measurement of clinical pain intensity: A comparison of six methods. *Pain* 1986;27:117–126.
15. Fishman B, Pasternak S, Wallenstein SL, et al: The Memorial Pain Assessment Card: A valid instrument for the assessment of cancer pain. *Cancer* 1987;60:1152–1158.
16. Daut RL, Cleeland CS, Flanery RC: Development of the Wisconsin Brief Pain Questionnaire to assess pain in cancer and in other diseases. *Pain* 1983;17:197–210.
17. Eisendrath SJ: Algorithm for assessing the psychiatric component of chronic pain: Psychiatric aspects of chronic pain. *Neurology* 1995;45:S26–S34.

18. Gutstein HB, Akil H: Opioid analgesics, in Hardman JG, Limbird LE (eds), *Goodman & Gilman's The Pharmacological Basis of Therapeutics*. New York: McGraw-Hill, 2001:569–619.
19. Stein C, Comisel K, Haimerl E, et al: Analgesic effect of intraarticular morphine after arthroscopic knee surgery. *N Engl J Med* 1991;325:1123–1126.
20. Kolesnikov YA, Jain S, Wilson R, Pasternak GW: Peripheral morphine analgesia: Synergy with central sites and a target of morphine tolerance. *J Pharmacol Exp Ther* 1999; 279:502–506.
21. Quirion R, Chicheportiche R, Contreras PC, et al: Classification and nomenclature of phencyclidine and sigma receptor sites. *Trends Neurosci* 1987;10:444–446.
22. Porter J, Hershel J: Addiction rare in patients treated with narcotics. *N Engl J Med* 1980;302:123.
23. Galer BS, Coyle N, Pasternak GW, Portenoy RK: Individual variability in the response to different opioids: Report of five cases. *Pain* 1992;49:87–91.
24. Manfredi PL, Ribeiro S, Chandler SW, Payne R: Inappropriate use of naloxone in cancer patients. *J Pain Symptom Manage* 1996;11:131–133.
25. Indelicato RA, Portenoy RK: Opioid rotation in the management of refractory cancer pain. *J Clin Oncol* 2002; 20:348–352.
26. Kaiko RF, Foley KM, Grabinski PY, et al: Central nervous system excitatory effects of meperidine in cancer patients. *Ann Neurol* 1983;13:180–185.
27. Miser AW, Narang PK, Dothage JA, et al: Transdermal fentanyl for pain control in patients with cancer. *Pain* 1989; 37:15–21.
28. Coluzzi PH, Schwartzberg L, Conroy JD, et al: Breakthrough cancer pain: A randomized trial comparing oral transmucosal fentanyl citrate (OTFC) and morphine sulfate immediate release (MSIR). *Pain* 2001;91:123–130.
29. Kornick C, Santiago-Palma J, Schulman G, et al: A safe and effective method for converting patients from transdermal to intravenous fentanyl for the treatment of acute cancer-related pain. *Cancer* 2003;97:3121–3124.
30. Kornick C, Santiago-Palma J, Khojainova N, et al: A safe and effective method of converting cancer patients from intravenous to transdermal fentanyl. *Cancer* 2001; 92:3056–3061.
31. Gorman AL, Elliott KJ, Inturrisi CE: The d- and l-isomers of methadone bind to the non-competitive site on the *N*-methyl-d-aspartate (NMDA) receptor in rat forebrain and spinal cord. *Nerurosci Lett* 1997;223:5–8.
32. Codd E, Shank R, Schupsky J, Raffia R: Serotonin and norepinephrine uptake inhibiting activity of centrally acting analgesics: Structural determinants and role in antinociception. *J Pharmacol Exp Ther* 1995;274:1263–1270.
33. Manfredi PL, Houde RW: Prescribing methadone: A unique analgesic. *J Supp Oncol* 2003;1:216–219
34. Bruera E, Sweeney C: Methadone use in cancer patients with pain: A review. *J Pall Med* 2002;5:127–138.
35. Max MB, Culnane M, Schatga SC, et al: Amitriptyline relieves diabetic neuropathy pain in patients with normal and depressed mood. *Neurology* 1987;37:589–594.
36. McQuay HJ, Tramer N, Nye BA, et al: A systematic review of antidepressants in neuropathic pain. *Pain* 1996; 68:217–227.
37. Galer BS: Neuropathic pain of peripheral origin: Advances in pharmacologic treatment. *Neurology* 1995;45:S17–S25.
38. Backonja M, Beydoun A, Edwards KR, et al for the Gabapentin Diabetic Neuropathy Study Group: Gabapentin for the symptomatic treatment of painful neuropathy in patients with diabetes mellitus: A randomized, controlled trial. *JAMA* 1998;280:1831–1836.
39. Rowbotham M, Harden N, Stacey B, et al for the Gabapentin Postherpetic Neuralgia Study Group: Gabapentin for the treatment of postherpetic neuralgia: A randomized, controlled trial. *JAMA* 1998;280:1837–1842.
40. Rowbotham MC, Davies PS, Verkempinck C, et al: Lidocaine patch: Double-blind, controlled study of a new treatment method for post-herpetic neuralgia. *Pain* 1996; 65:39–44.
41. Smith TJ, Staats PS, Deer T, et al : Randomized clinical trial of an implantable drug delivery system compared with comprehensive medical management for refractory cancer pain: Impact on pain, drug-related toxicity, and survival. *J Clin Oncol* 2002;20(19):4040–4049.

CHAPTER 33

Internal Medicine Aspects of Pain

Marco Maresca

Pain is frequently present in internal medicine. In many diseases concerning internal medicine, pain is the prominent symptom and must be evaluated carefully in the diagnostic review.

Internal medicine aspects of pain include
1. Pain due to diseases affecting deep somatic tissues (bone, joints, ligaments, and muscles)
2. Pain due to visceral diseases

▶ PAIN DUE TO DISEASES AFFECTING DEEP SOMATIC TISSUES

Different pain mechanisms may be active in diseases of deep somatic tissues. In most cases, pain is due to mechanical stimuli or to chemical stimuli (pain-producing substances released in inflammation).

Pain has the clinical characteristics of deep somatic pain;[1] it has a dull, aching quality and is rather well localized but tends to radiate (it is, therefore, different from well-localized cutaneous pain and from diffuse, poorly localized true visceral pain). The typical phenomenon of referred pain may be present; pain is felt in a site that is different from the site in which pain-producing conditions are localized. The intensity of pain depends on the characteristics of the pathologic process and on the pain sensitivity of the structures involved. Among the different deep somatic structures, the periosteum is the most sensitive, followed by ligaments, joints, tendons, fasciae, and muscles.

Pain in Bone Diseases

Distension of the periosteum is an important cause of pain in diseases of bones. Other bone-related pain-producing mechanisms also may be active. Chemical factors are prominent when inflammation or necrosis provokes the release of pain-producing substances.

Pain in Osteomyelitis

Infection of bone may be caused by different kinds of bacteria and by other microorganisms (fungi, viruses) that reach the bone by the hematogenous route or from a contiguous focus of infection. Among chronic infections of bones, infection due to *Mycobacterium tuberculosis* is especially important. Osteomyelitis, i.e., the inflammatory process due to infection of bone, may induce bone necrosis, often followed by the formation of a large fragment of necrotic bone (sequestrum) that maintains inflammation. Subperiosteal and soft tissue abscesses may develop. Sinus tracts may be present; they drain purulent material and fragments of necrotic bone. In children, hematogeneous osteomyelitis usually involves long bones, with a typical localization in the metaphysis, which is well perfused. In adults, the spine is infected more frequently. In diabetic patients, contiguous-focus osteomyelitis often involves the small bones of the feet following lesions due to peripheral vascular disease and diabetic neuropathy.

The pain of osteomyelitis may be very intense and usually is exacerbated by movement. In chronic osteomyelitis, however, pain may be absent. In acute osteomyelitis, pain usually is associated with fever. Children with osteomyelitis of the long bones may avoid walking or moving the affected limb. Tenderness over the metaphysis of the involved bone is found frequently upon clinical examination. In adults, vertebral osteomyelitis usually is characterized by persistent back pain with tenderness to percussion and palpation over the affected vertebrae and reflex spasm of paraspinal muscles. Pain is localized more frequently to the lumbar spine in pyogenic infections and to the thoracic spine in tuberculous spondylitis (*Pott's disease*). In some cases, pain due to vertebral osteomyelitis is felt in the chest, in the abdomen, or in the limbs; in these cases, pain is provoked by irritation of nerve roots. Bone destruction, with radiopaque sequestra, is observed on radiographic examination. Bone biopsy may be necessary for diagnosis, especially in cases of vertebral osteomyelitis. In the treatment of osteomyelitis, antibiotics are a fundamental therapeutic tool; antibiotics must be chosen based on the results of culture of blood or bone specimens. Surgical procedures are necessary to remove necrotic bone and abscesses.

Pain in Osteoporosis

The reduction in mass of bone per unit volume that occurs in osteoporosis provokes an abnormal bone fragility. Pain is mainly due to fractures, which more frequently involve vertebrae, wrist, hip, humerus, and tibia. Pain due to vertebral body fractures is characterized by an acute onset, often following movement of the spine, such as bending. Pain is localized in the lumbar or dorsal region, at the level of the affected vertebra, and may radiate to the flanks and abdomen. Pain intensity decreases after some days but may be exacerbated by movement of the spine. Tenderness frequently is detected by applying pressure on the spinous process of the affected vertebra. In some patients, several episodes of vertebral fracture occur, involving different vertebrae. Collapse of vertebral bodies induces loss in height; anterior collapse of dorsal vertebral bodies provokes a typical dorsal kyphosis. Adequate treatment of osteoporosis may reduce the risk of fracture. Calcium supplements, vitamin D, and bisphosphonates currently are used to reduce the phenomenon of bone resorption. It must be noted that an analgesic effect of bisphosphonates has been observed in conditions not related to bone demineralization.[2] It may be deduced that these drugs also may have a direct analgesic action.

Pain in Paget's Disease of Bone

Excessive resorption of bone and formation of irregular and very vascular new bone are the typical pathologic phenomena that occur in Paget's disease of bone (osteitis deformans). This disease is characterized by a decreased resistance of bones to mechanical stimuli and, consequently, by the onset of typical deformities, such as enlargement of the skull, accentuated dorsal kyphosis, and anterior bowing of the tibia. Pain is a frequent and important symptom and may be provoked by different mechanisms: stimulation of the periosteum owing to the decreased resistance of bone, compression on nerves with the onset of neuropathic pain, secondary osteoarthritis following bone deformities, and the occurrence of pathologic fractures. Back pain is frequent; in most cases, it is localized in the lumbar region. Pain frequently is felt in the legs and in the hip joint. Headache and facial pain also are common. An increase in pain may signal the onset of a severe complication, e.g., the development of an osteosarcoma. A marked increase in serum levels of alkaline phosphatase is a typical laboratory finding of Paget's disease of bone. Characteristic changes are observed by radiographic examination: lytic areas and zones of increased density due to new bone formation with bone enlargement and irregularly widened cortex. Treatment of the disease with bisphosphonates or calcitonin, both of which inhibit bone resorption, usually is accompanied by a decrease in pain.

Pain in Bone Tumors

Both primary and metastatic bone tumors may give rise to severe pain, which is mainly related to mechanical factors. The normal bone structure is replaced by neoplastic tissue, which stimulates periosteum and other pain-sensitive structures; the resistance of bone to mechanical stimuli is greatly decreased, and pathologic fractures may occur.

Pain in Joint Diseases

In many diseases of the joints, pain is provoked by inflammation. The genesis of pain is mainly due to chemical factors consequent to the release of pain-producing substances. An important role is played by the phenomenon of the so-called neurogenic inflammation, in which active neuropeptides are released by sensory nerve endings.[3] Mechanical factors may be active; when joint effusion is produced, pain is also due to high intra-articular pressure. Joint inflammation may be provoked by different agents: microorganisms, as in septic arthritis; the action of different types of crystals in the synovial fluid, as in gout and other crystal-induced arthritides; and immune reactions, as in rheumatoid arthritis, rheumatic fever, and reactive arthritis. Joint inflammation may be acute, as in rheumatic fever and acute gouty attack, or chronic, as in rheumatoid arthritis, psoriatic arthritis, and ankylosing spondylitis.

In acute arthritis, joint pain is accompanied by the other typical signs of inflammation, such as swelling (due to joint effusion and edema of periarticular tissues), warmth, redness, and functional impairment. Joint tenderness is well evident; pain is exacerbated by joint movements and by pressure on the joints. In the skin, allodynia is often present and is especially evident in some conditions, such as acute gouty arthritis. Pain may be elicited even by slight tactile stimuli. In chronic arthritis, some signs of inflammation, such as warmth and redness, are less evident and may be absent. Swelling may be especially evident when joint effusion is produced.

Inflammatory joint pain is often very intense during rest. Joint stiffness is an important symptom in all arthritides; it has been considered to be a consequence of the accumulation of edema fluid within inflamed tissues during sleep. Pain and stiffness are worse in the early morning and improve with activity. The severity of joint stiffness is evaluated by the length of time that is necessary to obtain maximal improvement. Morning stiffness lasting at least 1 hour before maximal improvement is considered a typical symptom of rheumatoid arthritis and has been included in classification criteria.[4] The different arthritides are characterized by different clinical features: acute monarticular involvement in most cases of septic arthritis, acute involvement of the large joints with the typical migrating character in rheumatic fever, recurrent monarticular involvement of small or large joints in gout (with typical involvement of the first metatarsophalangeal joint in most cases), symmetric involvement of small joints of the hands and feet in most cases of rheumatoid arthritis, and sacroiliac and vertebral involvement in ankylosing spondylitis.

Pain is also an important symptom of osteoarthritis, which is considered a typical degenerative disease of joints. Many hypotheses have been proposed to explain the genesis of pain in osteoarthritis. Destruction of articular cartilage and remodeling of subchondral bone, which are the characteristic phenomena of osteoarthritis, may induce pain by several mechanisms.[5] Bony changes that occur in osteoarthritis may provoke pain by elevation and stretching of the periosteum. Synovial inflammation may occur in osteoarthritis and provoke the onset of pain. It has been supposed that inflammation is activated by cartilage-derived macromolecules and calcium-containing crystals. Moreover, pain-producing conditions may be present in soft tissues: joint capsule, ligaments, bursae, tendons, and muscles. Stretching of the capsule and ligaments and onset of muscular tender spots in particular have been considered. Pain of osteoarthritis increases after use of the affected joint; it is usually worse in the evening and is relieved by rest. In some patients, however, pain also may be present at rest, even during the night.

Muscle Pain Syndromes

Pain in skeletal muscle may be induced by mechanical factors, such as stretching or increased muscle tension, and by chemical factors, such as pain-producing substances released by inflammation, ischemia, or necrosis. Some muscle pain syndromes are the consequence of vascular alterations that induce muscular ischemia or necrosis. In some cases, muscle pain is due to necrosis induced by toxic agents. Two frequently occurring muscle pain syndromes, i.e., myofascial pain syndrome and fibromyalgia, are characterized by persistent muscle pain that does not appear to be due to inflammation, ischemia, or necrosis. These syndromes are considered in other chapters of this book.

Pain in Inflammatory Myopathies

Pain is often present in inflammatory myopathies. It has typical features in the main idiopathic inflammatory myopathies polymyositis and dermatomyositis. It is associated with the most important symptom of these diseases, i.e., muscle weakness, which is more evident in the proximal limb muscles. Patients usually experience a deep, aching, continuous pain in the buttocks, thighs, or calves. Muscle tenderness is often present. Dermatomyositis is characterized by associated skin changes: lilac discoloration of the eyelids and periorbital area, localized or diffuse erythema, and maculopapular eruptions. Increased serum levels of muscular enzymes, such as creatine kinase, aldolase, and aspartate and alanine aminotransferase, are typical laboratory findings. Elevation of erythrocyte sedimentation rate and positive tests for circulating rheumatoid factor and antinuclear antibodies are observed frequently. Muscle biopsy demonstrates inflammatory and degenerative changes. Glucocorticoids, often associated with immunosuppressive drugs, usually are administered for the treatment of polymyositis and dermatomyositis. Improvement of muscle weakness induced by these drugs usually is accompanied by relief of pain.

Pain in Polymyalgia Rheumatica

Polymyalgia rheumatica usually occurs in persons older than 55 years of age, often with an abrupt onset. Pain is localized in the muscles of the shoulder girdle, pelvic girdle, and neck. In most cases, it is bilateral. Pain is more often intense during the night, and severe stiffness, especially morning stiffness, is also present. A marked increase in the erythrocyte sedimentation rate is a characteristic laboratory finding. The frequent association with giant cell arteritis is a typical feature. Polymyalgia rheumatica accompanies, precedes, or follows giant cell arteritis in 40 to 60 percent of cases.[6] The relationship between these two diseases has been the subject of many investigations. It has been suggested that polymyalgia rheumatica and giant cell arteritis are variants of the same disease entity. Pain and other symptoms of polymyalgia rheumatica are relieved promptly by the administration of glucocorticoids. The administration of NSAIDs alone is usually ineffective.

Pain Due to Muscular Ischemia

Intermittent claudication is a typical pain syndrome provoked by transient muscular ischemia, occurring during exercise in patients with occlusive arterial disease of the lower limbs. Ischemia may be due to different causes; it is usually provoked by atherosclerosis or arteritis. Pain is induced by a disproportion between muscular blood flow, which is reduced by arterial obstruction, and oxygen demand of the muscles of the lower limbs, which increases during exercise. Intermittent claudication is characterized by the onset of pain in the lower limbs during walking. The patient, after walking for a given distance, begins to feel pain in the lower limbs, usually in the calf muscles; pain becomes progressively stronger so that the patient is obliged to stop. During rest, pain decreases progressively and then disappears. The severity of the syndrome may be evaluated by considering the intensity of effort that provokes the onset of pain, i.e., the distance after which pain appears and the patient must stop walking. When arterial obstruction becomes more severe, causing a further decrease in blood flow, the onset of pain occurs after walking for a shorter distance. In Leriche's syndrome, which is caused by aortoiliac obstruction, intermittent claudication is characterized by pain localized in low back, buttocks, thigh, and calf muscles and is associated with impotence, atrophy of the lower limbs, and pallor of the skin of the feet and legs.

▶ PAIN DUE TO VISCERAL DISEASES

Many diseases affecting visceral organs may provoke pain-producing conditions. The onset of visceral pain depends on the characteristics of the pathologic process and the sensitivity of the structures involved. The stimuli that induce pain in viscera are different from those which induce pain in somatic structures. The main pain-producing conditions for viscera are (1) abnormal distension and contraction of the muscle wall of hollow viscera, (2) rapid stretching of the capsule of such solid visceral organs as the liver, spleen, and pancreas, (3) abrupt anoxemia of visceral muscle, (4) formation and accumulation of pain-producing substances, (5) direct action of chemical stimuli, (6) traction or compression of ligaments and vessels, (7) inflammation, and (8) necrosis.[7,8] In some cases, different pain-producing conditions are present simultaneously and interact. Pain sensitivity is different in different visceral structures. Serosal membranes are the most sensitive, followed by the wall of hollow viscera and by parenchymatous organs.

True visceral pain is deep and poorly localized. It may be difficult to define the characteristics of this type of pain and to distinguish it from other unpleasant visceral sensations. True visceral pain often is accompanied by a sense of malaise, and by reflex phenomena (nausea, vomiting, diffuse sweating, and changes in heart rate and arterial pressure). In experimental investigations comparing the psychophysical characteristics of visceral and cutaneous pain, it has been observed that visceral pain is associated with greater unpleasantness than is cutaneous pain of equal perceived intensity, a higher level of anxiety is associated with visceral pain compared with that associated with cutaneous pain of similar intensity, and visceral pain produces less defined sensations than cutaneous pain of similar intensity and is perceived as more diffuse than cutaneous pain.[9] True visceral pain usually is experienced at the beginning of a painful visceral disease.

When pain-producing visceral conditions persist or recur frequently, the typical phenomenon of referred pain may occur. Pain becomes better localized and is felt in more superficial structures, sometimes far from the site of origin. Referred pain usually is localized to the areas of segmental innervation of the skin (dermatomes) and of deep somatic tissues (myotomes and sclerotomes) corresponding to the segmental innervation of the affected organ. The phenomenon of referred pain is especially important in chronic pain syndromes due to visceral diseases, which are characterized by long-lasting pain or by repeated pain attacks.

Referred pain may be accompanied by typical sensory changes in the skin and deep somatic tissues.[10] Allodynia and hyperalgesia may be detected on the skin, and deep tenderness is often present in the muscles or in other deep somatic tissues of the abdominal or thoracic wall. In some cases, infiltration of referred pain areas with local anesthetic may cause the pain to disappear. Deep tenderness often is localized to small areas of the abdominal and thoracic wall. Different painful points, named after different authors, have been described on the abdominal and thoracic wall and are considered specific signs of painful diseases of different organs. In most cases, these points are due to the reference of visceral pain to somatic structures.

Different hypotheses have been proposed to explain the pathogenetic mechanisms of referred pain. Two pathophysiologic phenomena in particular have been considered: (1) convergence of visceral and somatic afferent impulses in the central nervous system (CNS); and (2) onset of somatic and autonomic reflexes. Convergence of visceral and somatic impulses in the CNS may induce the onset of referred pain. Painful stimuli arising from internal organs are interpreted as originating from the surface of the body, according to the hypothesis proposed by Head.[11] A similar mechanism was considered in Ruch's "convergence-projection theory."[12] Central convergence of impulses also may provoke referred pain by facilitation of impulses arising from somatic tissues, according to Ruch's theory.[12] Reflex activation of somatic and sympathetic efferent fibers, first considered by MacKenzie,[13] may hypothetically play a role in the origin of somatic pain.

Pain in Diseases of the Cardiovascular System

Heart Pain
The main pain-producing conditions in the heart are myocardial ischemia, necrosis, inflammation, and distension of the atrial or ventricular wall. Different pain-producing

conditions may be present in different diseases of the heart.

Cardiac pain may have different clinical characteristics.[14] True visceral pain caused by heart diseases is deep and poorly localized. It is usually retrosternal or epigastric, and it also may be interscapular. Pain often is accompanied by nausea, vomiting, diffuse sweating, and an intense alarm reaction. Allodynia, cutaneous hyperalgesia, and deep tenderness are not detected. Cardiac pain-producing conditions may induce chest discomfort that cannot be defined as frank pain by patients and is described as heaviness, pressure, smothering, tightness, choking, or strangling. Patients may experience other unpleasant sensations, such as a sense of gastric fullness, instead of pain.

Heart pain often has the characteristics of referred pain. This type of pain is more localized; it is frequently described as pressing or constricting and may be associated with allodynia, cutaneous hyperalgesia, or deep tenderness. Pain frequently is localized behind the sternum or in the precordial area and may radiate to both sides of the anterior chest wall, to the left upper limb, to both upper limbs, or rarely, only to the right upper limb. In some cases, pain radiates to the neck, jaw, and temporomandibular joints on both sides. Referred heart pain also may be localized to the left interscapulovertebral region. A particular type of referred heart pain has the characteristics of cutaneous pain; it is well localized within the dermatomes C8–T1 (i.e., in the ulnar side of the arm and forearm) and frequently is accompanied by cutaneous hyperalgesia or allodynia.

Pain Due to Ischemic Heart Disease

Ischemic heart pain has great clinical importance because ischemic heart disease may be characterized by severe complications and may provoke death. A careful evaluation of pain is useful for a correct diagnosis and correct treatment. Atherosclerosis of the coronary arteries is the most frequent cause of myocardial ischemia; arteritis may be another cause.

Pain of Myocardial Infarction

Pain is a typical symptom of myocardial infarction. However, pain intensity may be different in different patients. Many cases are characterized by a strong, unbearable pain that is relieved only by opioids, but in other cases pain is less intense and sometimes absent. It has been observed that the incidence of painless myocardial infarction is greater in patients with diabetes mellitus and that it increases with age.

Pain experienced by patients in early phases of myocardial infarction is usually deep, felt on the midline, and poorly localized, with the characteristics of true visceral pain. It is often felt behind the lower part of the sternum and less frequently in the epigastrium or in both these sites. Back pain may be associated; in some cases, only back pain is present. Pain may arise during or after muscular exercise or after a large meal but also during rest or sleep. Instead of frank pain, chest discomfort or an unpleasant sensation of gastric fullness may be present. Pain or other unpleasant sensations often are accompanied by nausea and vomiting. Psychic alarm reactions may be very strong; in some cases, patients experience a dreadful sense of impending death. This type of pain may last from a few minutes to some hours.

In a following phase, a different type of pain is experienced. This pain, which has the typical characteristics of referred pain, may be the first type of pain perceived by the patient. It is more superficial and well localized in the typical sites of cardiac referred pain. In patients not treated with drugs, this type of pain lasts from half an hour to 12 or more hours. Allodynia, cutaneous hyperalgesia, and deep tenderness may be present on the chest wall.

The diagnosis of myocardial infarction is confirmed by typical electrocardiographic changes (ST-segment elevation and development of abnormal Q waves) and specific laboratory findings (increase of serum levels of MB isoenzyme of creatine kinase and of cardiac-specific troponin T and I). Pain usually is relieved by the administration of morphine. In the first hours after the onset of symptoms, myocardial revascularization may be obtained by intracoronary or intravenous thrombolysis with fibrinolytic drugs, by coronary angioplasty, or by coronary artery bypass grafting.

Persistent or Recurrent Pain after Myocardial Infarction

Chest pain after acute myocardial infarction may be due to different pathologic conditions, such as extension of infarction, onset of angina pectoris, or pericarditis. In a few cases, postinfarction chest pain is due to *Dressler's syndrome*, characterized by pleuropericardial pain with fever, which may begin 1 to 6 weeks after myocardial infarction.[15] It has been supposed that *Dressler's syndrome* is caused by pericarditis and pleuritis due to autoimmune mechanisms.

Persistent pain in the left shoulder after myocardial infarction may be due to a scapulohumeral periarthritis; pain is associated with stiffness and reduced range of movement with the typical clinical picture of frozen shoulder. In some cases, the "*shoulder-hand syndrome*" may develop, with the clinical features of a complex regional pain syndrome Type 1: Pain progressively involves the whole upper limb and is accompanied by dystrophy of the skin (glossy skin, loss of hair) and of deep tissues (muscular atrophy, bone demineralization).

Pain of Angina Pectoris

Angina pectoris is characterized by repeated pain attacks. In some cases, anginal attacks are always provoked by effort (effort angina); in other cases, they also may occur at rest (rest angina). According to the classification of angina pectoris currently used, two types of angina are distinguished: stable angina and unstable angina. Stable angina is a typical effort angina in which pain attacks have the same characteristics. Unstable angina includes (1) new-onset (<2 months) angina that is severe and frequent (≥3 episodes per day); (2) accelerating (crescendo) angina, i.e., stable chronic angina in which pain attacks become more frequent, severe, and prolonged or are precipitated by less exertion than previously; and (3) angina at rest.[16] This classification is especially useful for its prognostic value because unstable angina is associated with a high risk of myocardial infarction.

Anginal pain is due to myocardial ischemia, i.e., a disproportion between myocardial oxygen demand and oxygen supply to the myocardium. Consequently, the onset of anginal attacks may be precipitated by two pathophysiologic conditions: (1) increased myocardial oxygen demand; and (2) decreased oxygen supply. Unstable angina has been considered to be provoked by several pathophysiologic factors, including progression of atherosclerosis, platelet aggregation, and thrombus formation.

Pain of anginal attacks may be vague, central, and poorly localized, with the characteristics of true visceral pain; instead of frank pain, chest discomfort may be experienced. A sensation of strangling may be present; this sensation was considered to be an important characteristic of the syndrome by Heberden, who used the Latin term *angina*, which means "strangling." Anginal pain often has the characteristics of referred pain; it may be localized to different areas of the chest or of the upper limbs and may be accompanied by cutaneous hyperalgesia, allodynia, or deep tenderness. Anginal attacks usually last from 1 to 5 minutes. In effort angina, pain is relieved promptly by rest. In all types of angina, pain is relieved by sublingual nitroglycerin.

Prinzmetal's variant angina is a particular form of angina characterized by attacks that occur at rest and are associated with ST-segment elevation in the electrocardiogram.[17] This type of angina is due to transient spasm of an epicardial coronary artery. In most patients with variant angina, a fixed obstruction is also present near the site of spasm. In *microvascular angina*, or syndrome X, typical anginal attacks are present in patients with normal coronary arteriography. In this type of angina, transient myocardial ischemia is provoked by episodes of abnormal vasoconstriction of the coronary microcirculation.

Specific drugs are used for the treatment of pain in angina pectoris. Nitroglycerin and amyl nitrite promptly relieve the pain of anginal attacks. Other drugs, such as long-acting organic nitrates, calcium antagonists, and β-blocking agents, are administered to prevent the onset of anginal attacks. The administration of aspirin is useful to induce an antithrombotic effect by inhibiting platelet aggregation. Coronary angioplasty and coronary artery bypass grafting, because they improve coronary blood flow, usually result in the disappearance of pain attacks.

PAIN IN VALVULAR HEART DISEASES

A typical anginal pain may be present in *aortic stenosis;* it usually has the characteristics of effort angina. The main pathogenetic factors that provoke angina are increased oxygen demand due to myocardial hypertrophy and reduced coronary flow due to increased pressure compressing the coronary arteries; moreover, coronary atherosclerosis may be associated.

Angina pectoris also may occur in *aortic regurgitation*, but less frequently than in aortic stenosis. Angina is a consequence of increased oxygen demand due to left ventricular dilatation and increased left ventricular systolic tension associated with reduced coronary blood supply due to low diastolic arterial pressure.

Pain with the same characteristics as that of angina pectoris also may be present in *mitral stenosis*. Some patients experience another type of pain, which is localized to the left interscapulovertebral region, where an area of muscular tenderness may be evident; this type of pain has been considered to be due to left atrial distension.

Chest pain may be reported by patients with *mitral valve prolapse*. In some cases, in which mitral valve prolapse is associated with coronary artery disease, a typical anginal pain is present. In other cases, pain has different characteristics. It is sharp, stabbing, and it is not precipitated by effort and is not relieved by nitroglycerin. It is difficult to explain the genesis of this type of pain. It has been thought to be provoked by an abnormal tension of the papillary muscles. Moreover, in some patients with mitral valve prolapse, chest pain may be a typical myofascial pain involving the muscles of the chest.

PAIN IN HYPERTROPHIC CARDIOMYOPATHY

A typical anginal pain also may be present in hypertrophic cardiomyopathy. Different factors may be active in the genesis of pain: narrowing of small intramyocardial coronary arteries, increased left ventricular wall stress due to high systolic pressure, and reduced coronary flow due to high diastolic pressure.

PAIN IN PERICARDITIS

Pain of pericarditis is caused mainly by the release of pain-producing substances as a result of inflammation. Pain of acute pericarditis usually is accompanied by fever. It lasts from a few hours to days. Recurrent episodes of pain may occur. Pain is often deep and retrosternal, usually corresponding to the upper two-thirds of the sternum; less frequently, pain is localized to the precordial region. Pain is exacerbated by deep inspiration, cough, and lateral movements of the chest; it is reduced by sitting up and leaning forward. Pain may be referred to the chest wall in the sternal or precordial region and may radiate to the left shoulder, scapula, and arm or to the neck and epigastrium. Allodynia and hyperalgesia may be observed. Deep tenderness often is present in the subclavicular fossa, on the superior ridge of the trapezius muscle, on the coracoid process, and rarely, on the left chest base.

Pain frequently is absent in chronic pericarditis. When pain is present, a deep, dull, slight pain sensation usually is felt. Some patients report other unpleasant sensations, such as sensations of heaviness or fullness in the chest.

PAIN IN COR PULMONALE

In acute cor pulmonale owing to pulmonary hypertension caused by massive pulmonary embolism, a severe deep retrosternal pain, similar to the pain of myocardial infarction, may be present. It must be noted that the pain of acute cor pulmonale does not radiate to the arms or jaw and rarely radiates to the back. Different mechanisms may be active in the genesis of pain—sudden distension of the pulmonary artery wall or of the right ventricle and ischemia of the right ventricle resulting from a disproportion between myocardial oxygen supply and increased oxygen demand provoked by the sudden increase of pulmonary artery pressure. In pulmonary embolism, another type of chest pain may be felt; this type of pain has the characteristics of pleuritic pain and is due to pleural inflammation caused by pulmonary infarction.

Patients also may report chest pain with chronic cor pulmonale due to pulmonary hypertension provoked by chronic pulmonary vascular or parenchymal diseases. Some characteristics of the pain, such as localization, radiation, quality, and intensity, are similar to those of angina. Other characteristics are different. In chronic cor pulmonale, pain attacks, which may be provoked by effort or occur at rest, are accompanied by an increase of cyanosis, whereas in pain attacks of ischemic heart disease patients are pale. Pain attacks in chronic cor pulmonale often last longer than those in ischemic heart disease and even may last several hours. In chronic cor pulmonale, pain may worsen during inspiration; this phenomenon is not observed in ischemic heart disease. Pain in chronic cor pulmonale may be caused by dilatation of the root of the pulmonary artery and by right ventricular ischemia; the increase of pain during inspiration probably is due to an increase in pulmonary pressure.

Pain Due to Diseases of the Vessels

In arterial diseases, pain may be provoked mainly by the following mechanisms: (1) arterial obstruction due to atherosclerosis or inflammatory changes, which induces ischemia or necrosis; (2) inflammation of the arterial wall, with local release of pain-producing substances and excitement of sensory nerve endings in the arterial wall and in adjacent tissues; and (3) enlargement of the affected arteries, which may provoke direct excitation of sensory nerve endings or compression of adjacent structures; this mechanism is evident mostly in some diseases of large-diameter vessels. Different mechanisms are associated in some conditions, such as aortic dissection.

PAIN DUE TO ATHEROSCLEROSIS

Atherosclerosis is an important cause of vascular lesions that may give rise to pain. Some phenomena occurring in atherosclerotic lesions, e.g., necrosis and thrombosis, are especially important in the genesis of pain-producing conditions. Two main pathogenetic mechanisms may be present: (1) a decrease in the resistance of the arterial wall to blood pressure with consequent enlargement of the arteries and the formation of aneurysms; and (2) the occurrence of arterial obstruction with consequent ischemia of the tissues that receive blood supply through the involved arteries. Typical syndromes are caused by atherosclerotic obstruction of different arteries. Intermittent claudication is usually the first symptom of atherosclerotic obstruction of the arteries of the lower limbs; it is due to a disproportion between blood flow and oxygen demand of limb muscles during exercise. Progression of the disease induces a further decrease in blood flow, which becomes inadequate for the oxygen demand of tissues even when the limbs are at rest. This condition provokes pain at rest and may be the cause of other pain-producing lesions, such as skin ulcers, necrosis of the extremities, and ischemic neuropathy. Obstruction of the coronary arteries is the cause of both angina pectoris and myocardial infarction. Atherosclerosis of mesenteric arteries induces the typical syndrome of angina abdominis, which is characterized by intermittent abdominal pain occurring after eating, i.e., when demand for splanchnic blood flow is increased.

PAIN DUE TO INFLAMMATORY ARTERIAL DISEASES

Two main pain-producing conditions may be active in vasculitis involving arterial vessels: (1) inflammation of the arterial wall with local release of pain-producing substances and excitement of sensory nerve endings in the arterial wall and in adjacent tissues, and (2) arterial obstruction with consequent ischemia of the tissues that receive blood supply from the affected arteries. Different pain syndromes are present in the different diseases characterized by involvement of different arteries.

Giant cell arteritis, also called *temporal arteritis* and *cranial arteritis,* occurs almost exclusively in subjects older than 55 years of age and more frequently in women than in men. It is characterized by inflammatory lesions of large- and medium-sized arteries, usually involving branches of the carotid artery. The temporal artery is affected in most cases. The presence of giant cells in the inflammatory lesions is a characteristic pathologic finding. When the temporal artery is involved, headache is a typical symptom. Pain usually is localized to the temporal region but also may be felt in other parts of the head. Pain is associated with induration of the affected vessel. Tenderness is present along the artery and in the surrounding area. Ischemic optic neuropathy may occur and may produce severe symptoms, even blindness. Weakness of the masticatory muscles is observed frequently. Intermittent claudication, myocardial infarction, and infarctions of other organs may be provoked by involvement of other arteries. Fever usually is present. A high erythrocyte sedimentation rate and anemia are typical laboratory findings. Giant cell arteritis and polymyalgia rheumatica, which are frequently associated, have been considered to be variants of the same disease entity.[6]

Takayasu's arteritis (aortic arc syndrome) is characterized by the involvement of large- and medium-sized arteries of the upper part of the body arising from the aortic arc. Giant cells are also observed in the inflammatory lesions of this disease. Young women are affected in most cases. Malaise, fever, low-grade stiffness, and arthralgias are present in a prodromic phase. In a following phase, pain is localized to the affected arteries and is associated with symptoms due to ischemia provoked by arterial obstruction. Headache, transient visual disturbances, and motor disorders may be due to cerebral ischemia. Weakness or pain in the masticatory muscles also may be present. Loss or weakness of pulses in the upper extremities is a typical sign of ischemia provoked by vascular obstruction and has given rise to the term *pulseless disease*. Chest pain due to myocardial ischemia also may occur but is rare. A high erythrocyte sedimentation rate, mild anemia, and elevated immunoglobulin levels are characteristic laboratory findings.

Polyarteritis nodosa is characterized by segmental necrotizing vasculitis of medium-sized and small arteries. Different tissues and visceral organs may be affected. Renal involvement is especially important because it may result in renal failure. Pain may be due to ischemia, caused by intimal proliferation and thrombosis, or to hemorrhage, caused by degeneration and necrosis of the vessel wall. The involvement of some visceral organs may give rise to severe pain, e.g., chest pain provoked by myocardial infarction or abdominal pain provoked by intestinal infarction or perforation.

In *thromboangiitis obliterans (Burger's disease)*, segmental inflammatory and proliferative phenomena involve medium-sized and small arteries of the limbs, causing arterial obstruction. The typical symptoms of occlusive arterial disease are present; intermittent claudication is followed by pain in the lower limbs at rest and by pain due to ulceration and gangrene. Inflammation also involves veins. Consequently, other pain-producing mechanisms result from thrombophlebitis and venous insufficiency.

PAIN DUE TO ANEURYSMS

Stretching of an arterial wall and compression on adjacent structures are the main pain-producing mechanisms in patients with aneurysms. Aortic aneurysms may give rise to thoracic or abdominal pain syndromes with typical characteristics. Thoracic aortic aneurysms may provoke pain by stimulation of nerve endings in the adventitia of the aorta, by pressure on adjacent nerves, or by erosion of bone. Patients may experience anterior chest pain, epigastric pain, or back pain. Pain often is persistent and may be accompanied by other symptoms of compression on other adjacent structures, such as the esophagus or airways. Abdominal aortic aneurysms usually remain asymptomatic for a long time. Pain appears after progressive enlargement of the aneurysm, when there is a high risk of rupture. Pain may be provoked by stimulation of nerve endings in the adventitia, by pressure on retroperitoneal nerves, or by erosion. It is usually localized to the midabdomen or the lumbar region. It may be referred to the groin, scrotum, or lower limbs. Rupture of an abdominal aneurysm is characterized by the sudden onset of severe back pain or of abdominal pain radiating to the back. Pain also may radiate to the flank, groin, scrotum, or thigh and is accompanied by symptoms of shock.

PAIN DUE TO AORTIC DISSECTION

Very severe, often unbearable pain is the common presenting symptom of aortic dissection. Most patients with dissection of the ascending aorta experience anterior chest pain that may be associated with back pain. Pain is almost always localized to the back when the dissection involves the descending aorta. Chest pain is similar to that of myocardial infarction, but some features are specific to aortic dissection. The quality of the pain is often described as "tearing," "ripping," or "stabbing." Pain usually is maximal at its onset and tends to migrate from the point of origin, following the progressive extension of the dissecting hematoma. In early phases of dissections involving the ascending part of the aorta, pain is stronger in the anterior part of the chest; when dissection progressively involves the descending part of the aorta, pain becomes stronger in the back. Pain frequently is accompanied by nausea, vomiting, diffuse sweating, and symptoms due to ischemia of other organs caused by obstruction of the orifices of aortic branches provoked by the dissecting hematoma. When ischemia of the mesenteric arteries is produced, abdominal pain due to intestinal infarction may be concomitant with pain due to the lesion of the aortic wall.

Pain Due to Vasculitis Involving Small Vessels and to Functional Disorders of the Peripheral Circulation

In vasculitis of small vessels, such as Henoch-Schönlein purpura, and in functional disorders of peripheral circulation, such as Raynaud's phenomenon and erythromelalgia, different pathophysiologic mechanisms may be active: (1) ischemia provoked by vasculitic lesions or abnormal arteriolar constriction, (2) action of pain-producing substances released by inflammation, which induce changes in the microenvironment and sensitization or direct excitation of sensory nerve endings, and (3) neurogenic inflammation, with antidromic release from afferent C fibers of substance P, calcitonin gene–related peptide, neurokinin A, neurokinin B, and other substances. Different pain-producing mechanisms are prominent in different diseases.

Pain is an important symptom of *Henoch-Schönlein purpura*, which is a syndrome, usually occurring in children, provoked by a systemic vasculitis involving small vessels. Polyarthralgias or frank arthritis, due to joint involvement, and abdominal pain, due to intestinal involvement, are associated with cutaneous hemorrhagic lesions and glomerulonephritis. Immune complex deposition is the presumptive pathogenetic mechanism of this syndrome.

Raynaud's phenomenon, due to cold-induced episodic vasospasm of digital arteries, is a typical functional disorder of the peripheral circulation. It is characterized by the sequential development of blanching, cyanosis, and rubor of the fingers or toes provoked by cold exposure and subsequent rewarming. A sensation of throbbing pain is often associated with the rubor, which is due to reactive hyperemia induced by rewarming. Pain caused by skin ulcers and necrosis of peripheral tissues may be present in Raynaud's disease (primary or idiopathic Raynaud's phenomenon) and, more frequently, in secondary Raynaud's phenomenon due to connective tissue diseases such as scleroderma.

Erythromelalgia (from the Greek words: *erythros,* meaning "red," *melos,* meaning "limb," and *algos,* meaning "pain") is a syndrome characterized by pain localized to the extremities and associated with vasodilatation, which provokes red discoloration and increased skin temperature. It is also called *erythermalgia;* this term includes the Greek names of the three fundamental symptoms: redness (*erythros*), warm (*thermos*), and pain (*algos*). Pain attacks are induced by increased temperature and last from a few minutes to many hours. The temperature that induces the onset of pain is called the *critical point;* it is different in different patients and usually is between 32 and 36°C. Pain and vasodilatation are localized to the distal part of the limbs. The lower limbs are affected more frequently. Pain has a typical burning quality and is aggravated by a dependent position and relieved by exposure of the affected extremities to cool air or water or by elevation. Erythromelalgia may be a primary disorder, or it may be secondary to different diseases, namely myeloproliferative diseases, such as polycythemia vera or essential thrombocytosis. Many hypotheses have been proposed to explain the genesis of erythromelalgia. It has been supposed that pain and other symptoms may be provoked by conditions that induce the release of vasodilating and pain-producing substances, by a disturbance of receptors for warm and cold, or by a dysfunction of the sympathetic nervous system. As a matter of fact, aspirin may relieve symptoms, especially in patients with erythromelalgia secondary to myeloproliferative disorders.

Venous Pain

Pain originating from pathologic conditions involving veins is present in thrombophlebitis. Three factors initially identified by Virchow play a fundamental role in the genesis of thrombophlebitis: stasis, vascular damage, and hypercoagulability. Consequently, an increased risk for the onset of thrombophlebitis is present in a series of clinical conditions, including surgical procedures, tumors, trauma, immobilization, pregnancy, use of oral contraceptives, and hypercoagulable states. Two main factors are active in the genesis of pain: a mechanical factor, i.e., venous stasis, and a chemical factor, i.e., the release of pain-producing substances as a result of inflammation of the vessel wall. Tenderness is often evident. In superficial thrombophlebitis, the affected vein may appear as a red, tender cord. Deep vein thrombophlebitis is asymptomatic in many cases. Thrombophlebitis of the iliac, femoral, and popliteal veins produces unilateral leg swelling, which in some patients is associated with cyanosis due to the high concentration of deoxygenated hemoglobin in the blood. This condition is called *phlegmasia coerulea dolens*. In other patients with marked edema, the affected leg appears pale because the interstitial tissue pressure exceeds the capillary perfusion pressure. The term *phlegmasia alba dolens*, therefore, is used to indicate this condition.

Thrombophlebitis of the veins of the calf may give rise to calf pain. The swelling of the limb may be modest when only one of multiple veins is involved so that venous return is not greatly reduced. Pain induced by dorsiflexion of the foot (Homan's sign) is an important diagnostic sign of this condition. Pulmonary embolism is the most severe complication of thrombophlebitis. Anticoagulant therapy may prevent thrombus propagation and allow the endogenous lytic system to operate. Heparin is administered in a first phase of treatment; it is replaced in a following phase by warfarin or other oral anticoagulants.

Pain may be present in chronic venous insufficiency due to valvular incompetence or to occlusion of deep veins following thrombophlebitis. Patients often experience a dull ache in the leg that increases with prolonged standing and is relieved by elevation of the limb. Edema and superficial varicose veins usually are evident. Dystrophic changes of the skin develop in the distal part of the affected limb, and ulcers may appear near the medial or lateral malleolus. In some cases, persistent pain and dystrophy after thrombophlebitis are not associated with venous insufficiency or other conditions that may account for these symptoms. This is due to the development of complex regional pain syndrome type 1.[18]

Pain in Diseases of the Respiratory System

Pleura has a high pain sensitivity and may be involved in different diseases of the lungs, such as pneumonia, pulmonary embolism, and lung tumors. Pain is an important symptom of pleurisy. Mechanical and chemical factors are active in the genesis of pain in pleurisy. Friction between the two opposite pleural surfaces and stretching of pleura during inspiration are the main mechanical factors. The action of chemical factors is due to pain-producing substances released by inflammation and contained in the pleural fluid. Pain of pleurisy often is localized to typical areas of reference, according to the location and extent of inflammation. In effusive pleurisy, pain frequently is localized to the submammary area. In diaphragmatic pleurisy, pain is often referred to the trapezius ridge and to the base of the affected side of the chest. In apical pleurisy, the typical area of pain reference is the interscapulovertebral region. Pain of effusive pleurisy usually is exacerbated by deep inspiration, cough, movements of the chest, and by lying on the affected side. These stimuli, in addition, usually do not increase the shoulder pain of diaphragmatic pleurisy and the interscapulovertebral pain of apical pleurisy. Allodynia, hyperalgesia, and deep tenderness may be present in the areas of referred pain. Mechanical factors are prominent in the genesis of pain in other pleural pathologic conditions, such as acute pneumothorax.

In lung tumors, pain may be due to the involvement of pleura. In some cases, pain is caused by compression or infiltration of peripheral nerves, which provokes the onset of neuropathic pain. *Pancoast's syndrome* is a typical example of pain due to the involvement of peripheral nerve trunks. Pain often is experienced even in early phases of the disease, when the tumor has not yet involved the pleura. In these cases, pain may be referred to the anterior or posterior surface of the chest. A good correlation is observed between the area of reference and the site of the tumor.

Pain in Diseases of the Digestive System

Pain in Esophageal Diseases

Pain due to esophageal diseases has great clinical importance. Some diseases of the esophagus are characterized by chest pain, which must be distinguished from heart pain. Moreover, pain may be a symptom of esophageal cancer; a careful evaluation of pain is useful for early diagnosis. Both mechanical and chemical factors may be active in the genesis of esophageal pain. Mechanical factors, such as distension or strong contraction of the esophageal muscle wall, are prominent in esophageal motor disorders and in all diseases characterized by esophageal stenosis. Chemical factors are prominent in esophagitis and in gastroesophageal reflux.

When esophageal pain has the typical characteristics of true visceral pain, it is felt behind the sternum and is deep and poorly localized. Esophageal referred pain is better localized to superficial structures. When pain-producing conditions involve the upper third of the esophagus, pain is referred to the upper part of the sternum; when the lower third of the esophagus is involved, pain is referred to the xyphoid process and to the epigastric area and also may be felt posteriorly in the midline at the level of the sixth and seventh thoracic vertebrae. Pain-producing conditions involving the middle third of the esophagus may give rise to pain referred either to the upper anterior area of reference or to the lower and posterior areas.

Esophageal pain frequently is associated with dysphagia. Heartburn is another symptom that may be present in painful esophageal diseases. In many cases, however, patients do not report either dysphagia or heartburn, and pain is similar to pain from diseases of other thoracic viscera. Differential diagnosis may be very difficult, especially when pain is similar to the ischemic heart pain of angina pectoris. Manometric evaluation is useful for the diagnosis of esophageal motor disorders, such as achalasia, diffuse esophageal spasm, and nutcracker esophagus (symptomatic peristalsis). Esophagitis and pathologic conditions that may provoke gastroesophageal reflux, such as hiatal hernia, may be detected by endoscopic examination. An increased sensitivity of the esophageal mucosa to chemical stimuli is demonstrated by a positive Bernstein test (pain induced by esophageal infusion of 0.1 N hydrochloric acid). With regard to the effects of drugs, esophageal pain may be relieved by drugs that are also effective in relieving anginal pain, such as nitroglycerin and calcium antagonists, but unlike angina, it also may be relieved by antacids.

Gastric Pain

The pain-producing action of chemical stimuli is prominent in the stomach. The action of HCl and of irritant foods has been demonstrated clearly by experimental and clinical observations. The analgesic effect of antacids and drugs inhibiting HCl secretion is an indirect proof of the pain-producing action of HCl in the stomach. Experimental and endoscopic findings demonstrate that inflammation plays an important role in the genesis of gastric pain. Stimuli that are ineffective in producing pain when applied to normal gastric mucosa are painful when mucosal inflammation is evident; the pain of gastric ulcer frequently is associated with endoscopic evidence of inflammation. In a series of well-known investigations, Wolf and Wolff[19] studied gastric pain-producing conditions in a subject with a large gastric stoma. They applied different kinds of stimuli, e.g., mechanical, electrical, and chemical, to the healthy mucosa of the fundus of the stomach. When the wall of the mucosa was taken between the blades of a forceps, no pain was induced. Even electrical stimulation intense enough to induce pain when applied to the tongue and chemical stimulation (50% and 95% alcohol, 1.0 N HCl, 0.1 N NaOH, and a 1:30 mustard suspension) did not provoke pain when applied to normal mucosa. When, instead, the gastric mucosa was inflamed, congested, and edematous, all these stimuli evoked intense pain. Mechanical factors, such as altered motility or distension of the gastric muscle wall, also may play a role in the genesis of pain in several gastric diseases.

Gastric pain usually is felt in the epigastrium; in some patients, it is referred to the left upper quadrant of the abdomen. Tenderness may be found by palpation of the epigastrium. Some clinical features have been considered typical of pain due to gastric ulcer—the onset of pain some hours after eating, when the stomach is empty, and the occurrence of pain relief when food is introduced. These characteristics are thought to be related to the pain-producing action of HCl, which is buffered by food.

Pain of Gastric Ulcer

In most cases, gastric ulcer is provoked either by infection with *Helicobacter pylori* or by NSAIDs.[20] Gastric acid secretion usually does not play a primary role in the genesis of peptic ulcer, but it contributes to mucosal injury. Increased gastric acid secretion is the main pathogenetic factor for the gastric ulcers observed in *Zollinger-Ellison syndrome,* due to a gastrinoma giving origin to hypergastrinemia. Gastric ulcer is an important cause of gastric pain, although some patients may be asymptomatic. To explain the genesis of pain in patients with gastric ulcers, two main factors have been considered: (1) a chemical factor, i.e., HCl contained in gastric juice, that directly stimulates sensory nerves in the ulcer base, and (2) a mechanical factor, i.e., gastric contraction in the area of ulceration. Pain of gastric ulcer usually is felt in the epigastric region, localized to the midline. It frequently occurs 90 minutes to 3 hours after eating and is relieved by eating food or by the administration of antacids. Nocturnal pain also is present in some patients but is not frequent. Instead of pain, some patients experience ill-defined unpleasant epigastric sensations. Epigastric tenderness may be evident. It must be noted that the clinical features of gastric ulcer are not specific, because similar symptoms may be present in patients with duodenal ulcer and in patients with so-called nonulcer dyspepsia, also called *functional dyspepsia* or *essential dyspepsia*. These terms are used to indicate a condition characterized by upper abdominal pain, which often is similar to the pain of gastric or duodenal ulcer but without the presence of an ulcer. In *nonulcer dyspepsia*, a pathogenetic role of infection with *H. pylori* has been suggested. Pain of gastric ulcer may be associated with other symptoms. Nausea and vomiting owing to gastric outlet obstruction may occur in patients who have a gastric ulcer near the pylorus. Vomiting frequently occurs several hours after eating. Changes in the characteristics of pain must be evaluated carefully because they may be provoked by the onset of some complications. If pancreatic penetration occurs, pain may become better localized, with radiation to the back, and is no longer relieved by food or antacids. Perforation provokes the sudden onset of severe diffuse abdominal pain, rapidly followed by signs of peritoneal inflammation, with abdominal tenderness and spasm of the abdominal musculature. Other severe complications, such as bleeding from the ulcer with hematemesis or melena, may occur after a long period in which the disease has been characterized only by persistent epigastric pain. Occult blood loss may provoke anemia.

Radiographic examination with a barium study of the proximal gastrointestinal tract may be used as a first diagnostic test, but endoscopy is the most sensitive procedure for the diagnosis of gastric ulcer. Moreover, since gastric ulcer may be malignant, direct examination of the ulcer and biopsy performed during endoscopy may rule out malignancy. When infection with *H. pylori* is diagnosed by the analysis of gastric biopsy specimen or by noninvasive methods, such as a urea breath test or serologic testing, the infection must be eradicated with specific antibacterial therapy. Cytoprotective agents, such as sucralfate, bismuth-containing preparations, and misoprostol, often are associated with antibacterial agents. The combination of bismuth, metronidazole, and tetracycline administered for a period of 14 days has been used widely with good results. Antacids or drugs inhibiting acid secretion, such as H_2 receptor antagonists and proton pump inhibitors, frequently are associated with antibacterial and cytoprotective agents and are especially useful for relief of pain. Surgical interventions may be necessary for treatment of medically refractory ulcers or for treatment of complications. Zollinger-Ellison syndrome must be suspected when multiple ulcers are present, when medical therapy is ineffective, and when other typical clinical features of the syndrome, such as erosive esophagitis or diarrhea, are associated. The diagnosis is confirmed by measurement of fasting serum gastrin levels and of gastric acid output; by gastrin provocative tests; and by imaging techniques (ultrasound examination, CT scan, or MRI) by which the gastrinoma is identified. Proton pump inhibitors are considered to be the drugs of choice for the treatment of ulcers in Zollinger-Ellison syndrome. Surgery may provide a definitive cure by removal of the gastrinoma when the location of the tumor is known and metastatic tumors are not detected.

PAIN IN GASTRIC TUMORS

Epigastric pain may be present in early gastric cancer and is experienced frequently by patients with advanced cancer. The characteristics of pain may be similar to those of gastric ulcer; in some cases, instead of pain, a vague upper abdominal discomfort or a sense of fullness is present. Pain usually is associated with anorexia and weight loss. Bleeding may occur, resulting in hematemesis or melena. Other symptoms may be related to the location of the cancer (e.g., dysphagia when the cancer is in the cardia or vomiting when it is in the antrum). Epigastric pain may be present in other tumors of the stomach, such as gastric lymphoma, polyps, and carcinoid.

Duodenal Pain

Inflammation (duodenitis) and peptic ulcer are the main causes of duodenal pain, which usually is localized to the epigastrium and may be referred to the right upper quadrant of the abdomen. It may be accompanied by tenderness in the epigastrium.

PAIN IN DUODENAL ULCER

Local injury provoked by NSAIDs and gastric infection with *H. pylori* also cause most cases of duodenal ulcer. Other factors, such as increased gastric acid secretion or decreased bicarbonate secretion, may play a pathogenetic role. Increased gastric acid secretion is the main pathogenetic factor of the duodenal ulcers observed in Zollinger-Ellison syndrome, due to hypergastrinemia provoked by a gastrinoma. Malignant duodenal ulcers are extremely rare. Symptoms of duodenal ulcer are similar to those of gastric ulcer. The pain of a duodenal ulcer usually is localized to the epigastrium. Instead of pain, patients may experience epigastric, ill-defined, unpleasant sensations. Pain or other unpleasant sensations occur 90 minutes to 3 hours after meals and disappear or are greatly reduced after ingestion of food or administration of antacids. Nocturnal pain that awakens the patients from sleep (between midnight and 3 A.M.) is a typical symptom that is present in two-thirds of patients with duodenal ulcer. However, it should be noted that this symptom is also present in one-third of patients with nonulcer dyspepsia.[20] Some complications of duodenal ulcer may induce the same changes in the characteristics of pain induced by complications of gastric ulcer—pain better localized, with radiation to the back, and no longer relieved by food or antacids is provoked by pancreatic penetration; and severe diffuse abdominal pain arising suddenly and followed by signs of peritoneal inflammation is provoked by perforation. Other complications may give rise to typical symptoms—bleeding may provoke hematemesis and melena, occult blood loss may cause anemia, and gastric outlet obstruction may induce vomiting. The diagnostic evaluation of duodenal ulcer requires the same procedures used for gastric ulcer evaluation (radiographic examination with barium study, endoscopy). It is important to diagnose gastric infection with *H. pylori* because eradication of the infection leads to healing of the ulcer. Zollinger-Ellison syndrome is suspected when endoscopy reveals an unusual location of the ulcer (e.g., in the second part of the duodenum and not in the duodenal bulb); when the ulcer is associated with other clinical features, such as erosive esophagitis or diarrhea; or when the ulcer is refractory to medical therapy. Increased gastric acid production due to hypergastrinemia may be demonstrated by measurement of gastric acid output, measurement of fasting serum gastrin levels, and gastrin provocative tests. The gastrinoma may be identified by imaging techniques. Duodenal ulcers associated with gastric infection with *H. pylori* are treated by the same combinations of antibacterial and cytoprotective drugs used for treatment of gastric ulcers. Antacids and drugs inhibiting gastric acid secretion also may be administered; it must be considered that these drugs also have a specific analgesic action. Surgical interventions are used when medical therapy is ineffective and when the urgent treatment of a complication is necessary. In Zollinger-Ellison syndrome, gastric acid output is reduced by the administration of proton pump inhibitors; the gastrinoma may be removed by surgical intervention.

Pain in Intestinal Diseases

As in other parts of the alimentary tract, two types of factors are active in the genesis of pain in small and large intestine: mechanical factors and chemical factors. In some pathologic conditions, such as irritable bowel syndrome and intestinal obstruction and pseudo-obstruction, pain is due mainly to mechanical factors, i.e., distension and contraction of intestinal muscle wall. Chemical factors are the main pain-producing factors in other conditions characterized by inflammation, ischemia, or necrosis, such as inflammatory bowel disease, chronic intestinal ischemia, and intestinal infarction.

True visceral pain caused by diseases of the small and large intestine is deep and poorly localized; it is often accompanied by nausea, vomiting, diffuse sweating, and a psychic alarm reaction. Frequently, it is felt in the right part of the abdomen in painful diseases of the cecum and ascending colon and in the left part of the abdomen in painful diseases of the descending and sigmoid colon; in other cases, a generalized abdominal pain is experienced or pain is felt around the midline (in the epigastrium, in the periumbilical region, or in the lower midabdomen). Referred pain is better localized and often accompanied by allodynia, hyperalgesia, or deep tenderness in the abdominal muscle wall. Pain due to diseases of the small intestine may be localized to the periumbilical region and the right lower abdominal quadrant. Pain provoked by diseases of the appendix, cecum, and ascending colon may be felt in the same areas. Painful diseases involving the hepatic flexure of the colon provoke pain localized to the right hypochondriac region. When the splenic flexure of the colon is affected, pain is localized to the left hypochondriac region. Pain provoked by diseases of the descending or sigmoid colon is localized to the left lower abdominal quadrant. Intestinal pain often is described as a typical "cramping" pain.

Pain in Inflammatory Bowel Disease

The term *inflammatory bowel disease* includes two main clinical entities: Crohn's disease and ulcerative colitis. An altered regulation of the mucosal immune system is thought to be the common pathogenetic mechanism that provokes a chronic inflammatory response in the bowel, giving origin to the typical pathologic and clinical features of Crohn's disease and ulcerative colitis.[21]

The main pathologic characteristic of *Crohn's disease* is a segmental transmural inflammatory process that may involve any part of the gastrointestinal tract. The presence of noncaseating granulomas is a typical histologic feature. Focal inflammation may give rise to local abscesses and fistulous tracts that resolve by fibrosis, causing thickening of the bowel wall and narrowing of the lumen. Abdominal pain is a typical symptom of Crohn's disease and is experienced by most patients. The characteristics of the pain depend on the location of the disease. When the inflammatory process is confined to the ileocecal area, typical onset of the disease is characterized by pain in the right lower quadrant of the abdomen associated with diarrhea and low-grade fever. Tenderness and a palpable mass usually are present in the area in which pain is felt. Recurrent painful episodes may occur in the first phases of the disease. Steady pain in the right lower abdominal quadrant may be associated with intermittent periumbilical pain. When the colon is involved, cramping pain may be present in one or both lower abdominal quadrants. Severe abdominal pain may be provoked by episodes of partial or complete intestinal obstruction. Somatic pain may be provoked by some extraintestinal manifestations of the disease, such as peripheral polyarthritis, involving larger joints, or ankylosing spondylitis. The disease usually is associated with some laboratory findings, such as an increased erythrocyte sedimentation rate, increased serum levels of C-reactive protein, and in more severe cases, anemia, leukocytosis, and hypoalbuminemia. Radiographic examination with a barium study and endoscopy are the most important diagnostic procedures. Typical changes may be observed: thickened intestinal folds and aphtous ulcerations in early phases of the disease and inflammatory masses, strictures, abscesses, and fistulas in advanced phases. Endoscopy is also useful for biopsy of lesions. For the treatment of Crohn's disease, sulfasalazine and 5-aminosalicylic acid are used in patients with mild to moderate disease. In more severe cases, glucocorticoids must be administered to induce clinical remission. Immunosuppressive agents, such as azathioprine, 6-mercaptopurine, methotrexate, and cyclosporine, also are used for maintenance therapy in severe cases. Surgery may be necessary for the treatment of abscesses, fistulas, and strictures causing intestinal obstruction. Adequate treatment of the disease induces relief of pain, together with improvements in other symptoms. However, analgesic therapy may be necessary to relieve abdominal pain. Analgesic anti-inflammatory drugs should not be administered because they may exacerbate symptoms. Anticholinergic agents may be used. Some opioids, such as diphenoxylate and loperamide, which are administered as antidiarrheal agents, also may induce an analgesic effect. It must be noted that the administration of high doses of agents that inhibit intestinal motility may increase the risk for the onset of toxic megacolon.

In *ulcerative colitis,* mucosal inflammation involving the rectum and all or a part of the colon and giving rise to hemorrhagic lesions and ulcers is the characteristic pathologic finding. Cryptitis, with distortion of the crypt architecture and plasma cell and lymphoid aggregates, is observed by histologic examination. Cramping abdominal pain associated with diarrhea, rectal bleeding, and fever usually is present. This pain frequently is localized to the left lower abdominal quadrant. It may awaken the patient at night and often is relieved by defecation. The abdomen is tender and distended. Pain having origin in deep somatic structures may be due to the rheumatologic manifestations of ulcerative colitis, i.e., peripheral polyarthritis and ankylosing spondylitis. Increases in the erythrocyte sedimentation rate and the serum levels of C-reactive protein are observed in acute phases of the disease; anemia frequently is present. Radiologic examination of the colon with a single- or double-contrast barium enema and endoscopy are the main diagnostic procedures. Edematous and thickened haustral folds associated with ulcers are typical radiographic findings. Mucosal inflammation and ulcers are seen directly by endoscopic examination. In the treatment of ulcerative colitis, good results are obtained with the same drugs used in Crohn's disease: sulfasalazine and 5-aminosalicylic acid for mild to moderate disease, glucocorticoids to induce clinical remission, and immunosuppressive agents for maintenance therapy in severe cases. As in Crohn's disease, abdominal pain may be relieved by anticholinergic drugs and by some opioids, such as diphenoxylate and loperamide, which are administered for the control of diarrhea. High doses of agents inhibiting intestinal motility should be avoided, especially in severe disease, because they may precipitate toxic megacolon. Analgesic anti-inflammatory drugs also should be avoided because they may induce worsening of the disease. Surgery is necessary when the disease is refractory to medical therapy. Some severe complications, such as toxic megacolon, colonic perforation, massive colonic hemorrhage, colonic obstruction, and cancer of the colon, also may require surgery. In many patients, a total proctocolectomy must be performed.

Pain in Diverticular Disease of the Colon

In most cases, uncomplicated diverticulosis is asymptomatic. When symptoms occur, pain is the most common symptom, and it is localized to the lower abdominal quadrants, more often in the left, and persists with variable intensity over a period of a few hours to several days. Pain often increases after eating and may be relieved by defecation or passage of flatus. Pain is believed to be provoked by increased tension in the colonic wall, with increased intraluminal pressure. Tenderness may be evident in the sites in which spontaneous pain is felt.

Diverticulitis, i.e., inflammation of a diverticulum, is the most frequent complication of diverticulosis. The inflammatory process of the diverticular wall spreads to the adjacent bowel wall and surrounding tissues, giving rise to a peridiverticulitis. In the genesis of pain, chemical factors, owing to inflammation, are associated with the mechanical factors that provoke pain in diverticulosis. In most cases, diverticulitis occurs in diverticula of the sigmoid colon. Pain usually is localized to the left lower abdominal quadrant and often radiates to the back. When part of the sigmoid colon is displaced to the right or inflammation involves a diverticulum of the right colon, pain is suprapubic or in the right lower abdominal quadrant. Pain usually is accompanied by fever. Tenderness is evident in most cases. A tender, sausage-like mass may be palpable. Diverticular inflammation may provoke necrosis of the bowel wall, with microperforation and the development of an intra-abdomnal abscess, which may rupture into adherent structures, giving rise to fistulas. Free perforation of a diverticulum into the peritoneal cavity also may occur; it is characterized by the onset of diffuse abdominal pain due to widespread peritonitis. Pain due to intestinal obstruction may result from severe peridiverticular inflammation of the colonic wall followed by fibrotic reactions causing segmental narrowing of the bowel.

Pain in Intestinal Tumors

Pain is present frequently in small bowel carcinomas. The localization of pain depends on the intestinal segment involved. Duodenal carcinomas may produce epigastric pain, cramping midabdominal pain due to bowel obstruction may be caused by carcinomas of the jejunum, and cramping lower abdominal pain may be provoked by ileal carcinomas. Other clinical features, such as weakness, weight loss, and iron-deficiency anemia due to chronic occult blood loss, may be associated with pain and are very important for a correct diagnosis. In some patients, an abdominal mass may be found by palpation. Pain also may be present in other tumors of the small intestine. Diagnosis is confirmed by barium study of the small bowel demonstrating a large mass or a constrictive lesion.

In patients with tumors of the large bowel, pain may be due to intestinal obstruction, which is provoked more frequently by cancer of the left colon because this segment has a narrower lumen than the proximal colon, and the tumor often involves the bowel circumferentially. Episodes of abdominal pain, frequently occurring after meals, may be accompanied by a change in bowel habits with constipation or increased frequency of defecation owing to the passage of small amounts of stool through the obstruction. Rectal bleeding also may be present. Pain due to intestinal obstruction instead is produced rarely by tumors of the right colon, which has a wide lumen; symptoms usually appear in an advanced phase of tumor development. The presenting clinical picture often is characterized by symptoms due to anemia, by the presence of a palpable mass, or by vague abdominal discomfort. Colonscopy is the most accurate diagnostic technique for detecting tumors of the large bowel.

Pain in Irritable Bowel Syndrome

Abdominal pain is a prominent symptom of the condition currently known as *irritable bowel syndrome*. Pain has different characteristics in different patients. It may be diffuse or localized, and it is felt more frequently in the left lower abdominal quadrant. Rectal pain and tenesmus may be present. Severe pain may be associated with back pain. Pain may be constant or intermittent, and episodes of severe, sharp pain may occur. The onset of pain often follows food ingestion, and relief frequently is induced by defecation.

Abdominal pain is associated with other typical features of irritable bowel syndrome, such as altered bowel habits and abdominal distension. Constipation or diarrhea may be present. In some patients, periods of constipation are interrupted by episodes of diarrhea. Persistent constipation may be associated with increased severity of pain. Defecation often induces pain relief, but in some cases it is followed by persistent pain or by an unpleasant sense of incomplete evacuation. Moreover, evacuation is painful in some patients. When diarrhea is present, evacuation frequently is preceded by abdominal pain that is relieved by defecation, but sometimes only for a brief period. Abdominal distension and increased belching or flatulence are reported by patients with irritable bowel syndrome, but intraluminal gas actually is increased in only a few cases. It has been suggested that most patients have a decreased tolerance to bowel distension. A variable amount of mucus also is observed in the stool. Other gastrointestinal symptoms, such as dyspepsia, heartburn, nausea, and vomiting, may be concomitant. Moreover, irritable bowel syndrome may be associated with dysmenorrhea, urinary bladder dysfunction, headache, and fibromyalgia.

No pathologic alterations or biochemical changes are detected by diagnostic tests in patients with irritable bowel syndrome. This condition is thought to be caused by a primary motor disorder associated with an increased sensitivity to bowel distension and to other types of painful visceral stimuli. The visceral hyperalgesia is accompanied by cutaneous hyperalgesia.[22] The onset of irritable bowel syndrome frequently occurs during or after periods of emotional stress, and symptoms may be exacerbated by stressful events. Abnormal psychological features have been observed in patients with this syndrome and have been considered to be important pathogenetic factors. Antispasmodic, usually anticholinergic, drugs are used to induce pain relief. Tranquilizers and antidepressants and psychological support may be useful for the treatment of anxiety or depression, which may exacerbate pain and other symptoms.

Pain in Diseases of the Liver

The liver, like other parenchymatous organs, has a low pain sensitivity. Many diseases of the liver, such as hepatic cirrhosis, are severe and long lasting but usually are not accompanied by pain. In diseases of the liver, the main pain-producing mechanism is the rapid stretching of Glisson's capsule. True visceral pain owing to diseases of the liver is poorly localized to the right hypochondriac region or the epigastrium. Referred pain is better localized and may be felt in the right hypochondriac region, in the epigastrium, in the right scapular region, and in the right shoulder.

Pain may be present in acute viral hepatitis because hepatic enlargement provokes capsular stretching. Pain is reported frequently by patients with hepatic abscesses and hepatic tumors. Persistent pain in the right hypochondriac or epigastric region may be the first symptom of hepatocellular carcinoma. Metastatic tumors also are accompanied frequently by pain.

Pain in Diseases of the Biliary Tract

Mechanical factors are prominent in the genesis of pain in the biliary tract. Distension and contraction of the muscle wall of the bile ducts is the main cause of the strong pain that is characteristic of an acute visceral pain syndrome, e.g., the biliary colic consequent to obstruction of a duct provoked by a calculus. Biochemical factors also may be active. Pain-producing substances released as a result of inflammation and the bile itself, which has an irritant action, may play a role in the genesis of pain in biliary colic and other conditions, such as cholecystitis, in which inflammatory phenomena are evident.

True visceral pain owing to diseases of the biliary tract is felt in the right hypochondriac and epigastric regions; it is deep, vague, and poorly localized. When pain is referred, it is better localized to the right hypochondriac region or the epigastric region; the right scapular region and the right shoulder are other typical areas of reference. Allodynia, hyperalgesia, and deep tenderness may be localized to the right upper quadrant of the abdomen or the right scapular region.

PAIN IN BILIARY COLIC

Transient obstruction of the biliary tract by a stone is the cause of severe abdominal pain that is called *biliary colic*. Biliary calculosis may be asymptomatic, and the frequency and severity of biliary colic are not related to pathologic changes induced by stones in the gallbladder. The pain of biliary colic usually is localized to the right upper abdominal quadrant or the epigastrium and may radiate to other parts of the abdomen or to the back; it frequently radiates to the right scapular region or the right shoulder. Colic may be precipitated by eating, but in most cases, no precipitating factor can be identified. Typical painful episodes are characterized by a first phase, lasting from 15 minutes to 1 hour, during which pain increases progressively; during the second phase, pain intensity remains constant for 1 hour or more; and then pain intensity diminishes slowly during the third phase. Pain frequently is accompanied by vomiting and diffuse sweating. During the attack, patients usually are restless. Tenderness may be present in the right upper abdominal quadrant or in the right scapular region, and it often is localized to the cystic point, corresponding to the external border of the right rectus abdominis muscle, under the costal border. The cystic point is the typical site of pain referred from the gallbladder. In patients with biliary calculosis, the characteristics of muscle hyperalgesia were investigated by threshold measurement performed at the level of the cystic point.[10,23] It was observed that hyperalgesia was not present in subjects affected with biliary stones who had never suffered from colic; it was evident instead in patients with painful dysfunction of the gallbladder but with no organic visceral pathology. The extent of hyperalgesia appeared to be a function of the pain experienced, and hyperalgesia persisted beyond the duration of spontaneous pain. It could be deduced that the occurrence, extent, and persistence of hyperalgesia depend on the perception of spontaneous visceral pain.

The occurrence and severity of further pain episodes after the first biliary colic are not predictable.

Ultrasonography is the main imaging technique used for diagnosis of cholelithiasis. The pain of biliary colic commonly is treated with antispasmodic drugs. In the treatment of cholelithiasis, medical therapy with ursodeoxycholic acid may be effective in inducing dissolution of cholesterol stones. Good therapeutic results may be obtained by lithotripsy with extracorporeal shock waves. Cholecystectomy is the definitive treatment when other therapies are contraindicated or have been ineffective.

PAIN IN ACUTE CHOLECYSTITIS

Most cases of acute cholecystitis are due to cholestasis, which usually results from cystic duct obstruction by a gallstone. Acute cholecystitis frequently occurs in patients who previously suffered from biliary colic. Pathologic changes in the gallbladder wall are characterized by the typical phenomena of inflammation, hyperemia, edema, and cellular infiltration. The gallbladder is distended by bile and exudate. Bacterial infection is present in most cases. Suppurative inflammation may develop, giving rise to empyema. Necrosis of the gallbladder wall may cause perforation. The first phase of acute cholecystitis is characterized by visceral pain felt in the right upper abdominal quadrant or in the epigastric region, and poorly localized. In a following phase, referred pain is present, and it is localized to the somatic structures of the right upper abdominal quadrant and may radiate to the back and right shoulder. Fever usually is present, and vomiting occurs often. Referred pain frequently is accompanied by tenderness, which is especially evident at the cystic point. Murphy's sign may be present; palpation of the right subcostal region provokes pain and inspiratory arrest when the patient takes a deep breath. This phenomenon is thought to result from direct mechanical stimulation of the inflamed gallbladder. Diagnosis usually is confirmed by ultrasonography, which may demonstrate gallstones, enlargement of the gallbladder, or thickening of the wall. Antibiotics usually are administered to treat infection. Cholecystectomy may be performed after complete recovery; it is performed during the acute attack if symptoms are severe and persistent. Urgent surgical treatment is necessary when perforation occurs.

PAIN IN CHOLANGITIS

Bacterial infection of bile in the bile duct is the cause of cholangitis, which occurs in different conditions leading to bile duct obstruction. Choledocolithiasis, biliary strictures, and tumors of the bile duct are the main pathologic conditions giving rise to cholangitis. Pain in the right upper quadrant of the abdomen is a typical symptom; it usually is associated with chills and fever. In most cases, jaundice is evident. Tenderness is present in the right upper abdominal quadrant. Common laboratory findings include leukocytosis, increased serum bilirubin levels, and an elevated alkaline phosphatase concentration. Bacterial species that cause infection may be identified by blood culture: *Escherichia coli, Klebsiella, Pseudomonas,* enterococci, and Proteus are cultured in most cases. Sepsis may develop, along with septic shock. The administration of antibiotics usually induces rapid improvement. Antibiotics should be chosen based on the results of blood culture and sensitivity tests. Surgery may be necessary to remove the cause of bile duct obstruction.

Pain in Diseases of the Pancreas

Among the factors that may be active in producing pain in pancreatic diseases, chemicals are especially important. Pain-producing substances released by inflammation and necrosis play a fundamental role in the genesis of the severe pain of acute pancreatitis but also are active in other conditions, such as chronic pancreatitis and pancreatic tumors. Mechanical factors, such as sudden distension of pancreatic ducts or the capsule, also may be important in the genesis of pain.

True visceral pain owing to diseases of the pancreas usually is felt in the upper midabdomen. Referred pain is localized to the epigastric region and the back in an area corresponding to the lower thoracic spine.

PAIN IN ACUTE PANCREATITIS

There are many causes of acute pancreatitis, including biliary tract diseases, abdominal trauma, pancreatic ischemia, alcoholism, hypertriglyceridemia, viral infections, and drugs. It is commonly believed that autodigestion of the pancreas is the fundamental pathogenetic mechanism of this disease. Autodigestion results from pancreatic enzymes that are activated in the pancreas rather than in the intestinal lumen. Different pathologic changes may be present, giving rise to different clinical pictures. In mild cases, pathologic changes are characterized by interstitial edema (edematous pancreatitis). Necrosis of glandular cells and surrounding fatty tissue causes severe clinical manifestations (necrotizing pancreatitis). Hemorrhages may occur, leading to collections of blood in the pancreas or retroperitoneal spaces (hemorrhagic pancreatitis). Confluent areas of necrosis containing tissue debris, pancreatic juice, blood, and fat droplets are indicated with the term *pseudocysts*. Secondary infection of necrotic tissue by enteric bacteria may give rise to pancreatic and peripancreatic abscesses. Acute pancreatitis is characterized by poorly localized pain that is felt in the epigastrium or left upper abdominal quadrant and often radiates to the back in the lower thoracic vertebral region. Pain increases progressively, and within 15 minutes to 1 hour, it becomes extremely severe. It is usually more intense in the supine position. Pain often is accompanied by fever, nausea, and vomiting. Intense epigastric tenderness is present, but signs of peritoneal irritation, such as abdominal wall rigidity, are not evident in the first phase of the disease. Peritoneal irritation is not induced in this phase because the pancreas is located in the retroperitoneal space. Hypotension and shock are present frequently. The clinical picture may be characterized by symptoms of severe complications, such as intra-abdominal hemorrhage, adult respiratory distress syndrome, and renal failure. Increases in the serum amylase and lipase concentrations and hyperglycemia are characteristic laboratory findings. Ultrasonography and CT scan demonstrate diffuse pancreatic enlargement that may be associated with peripancreatic edema. General supportive care of patients with acute pancreatitis includes adequate treatment of pain by the administration of opioids; meperidine is preferred to morphine because it provokes less spasm of the sphincter of Oddi. Intravascular volume, reduced by fluid loss owing to peripancreatic exudate, hemorrhage, and vomiting, must be restored and maintained. Vital signs and physical and laboratory findings are controlled to detect the first signs of life-threatening complications, which must be treated promptly.

Pain in Chronic Pancreatitis

Chronic inflammation of the pancreas is caused by the same mechanisms that cause acute pancreatitis, but the most frequent is chronic alcoholism. Many pathogenetic hypotheses have been suggested to explain the development of chronic pancreatic inflammation. Ductal obstruction, which may be due to precipitation of protein, has been considered to be especially important because it may cause the typical pathologic changes observed in chronic pancreatitis—dilatation of the ducts, atrophy of acinar cells, and fibrosis. Pain is an important symptom of chronic pancreatitis and is absent in only a few cases. Pancreatic juice outflow obstruction, with subsequent increase in ductal and parenchymal pressure within the pancreas, has been considered to be a fundamental pain-producing mechanism. It has been suggested that functional alterations of the nervous system and neurogenic inflammation may play an important role in the genesis of pain.[24] In most cases, pain is localized to the epigastric region, with radiation to the back. Pain may be continuous or intermittent and may be very severe. It is often exacerbated by the ingestion of alcohol or heavy meals (especially when meals are rich in fat) and is not relieved by antacids. Tenderness may be present in the epigastric region. Pain is associated with symptoms and signs of pancreatic exocrine insufficiency, such as weight loss and steatorrhea. Reduced glucose tolerance is often present; fasting blood glucose levels are also increased in some patients. Diffuse pancreatic calcifications are typical radiographic findings. The diagnosis of chronic pancreatitis is confirmed by ultrasonography, CT scan, and endoscopic retrograde colangiopancreatography (ERCP), which show dilated pancreatic ducts. Treatment of pain is an important therapeutic goal. Severe pain often requires the administration of opioids.

Pain in Pancreatic Tumors

Ductal adenocarcinoma is the most frequent malignant pancreatic tumor. In most cases, the tumor is located in the pancreatic head. Pain is a frequent and important symptom; it is usually a deep, poorly localized visceral pain that is felt in the epigastrium and may radiate to the back. A typical characteristic is a decrease in pain when patients bend forward. Pain is considered to be mainly due to infiltration of the splanchnic nerves, which is a result of retroperitoneal invasion. In some cases, pain is transient and accompanied by hyperamylasemia; it is considered to be due to acute pancreatitis resulting from ductal obstruction. Pain usually is associated with weight loss. Jaundice frequently is present when the tumor is located in the pancreatic head; the gallbladder usually is enlarged and may be palpable (Courvoisier's sign). Reduced glucose tolerance may be observed. The diagnosis is confirmed by increased serum levels of tumor-associated antigens, such as carcinoembryonic antigen (CEA) and CA19-9, and by imaging techniques such as ultrasonography and CT scan.

Pain in Diseases of the Kidney and Urinary Tract

Many important acute and chronic diseases of the kidney, such as glomerular nephropathies, may be very severe and may lead to renal insufficiency but are not usually accompanied by pain. In some painful renal diseases, such as renal infarction, the genesis of pain is related mainly to chemical factors, e.g., the release of pain-producing substances owing to ischemia, necrosis, or inflammation. Pain occurring in other renal diseases is produced mainly by mechanical factors, such as distension of the renal capsule and traction on the pedicle of the kidney.

However, pain is a frequent and important symptom in diseases of the urinary tract. Distension and contraction of the muscle wall of the different sections of the urinary tract are considered fundamental pain-producing mechanisms; inflammation is another mechanism that in some cases may be very important The ureteral colic provoked by an obstructing calculus is a typical example of strong pain having origin in the urinary tract. Distension and contraction of the ureteral muscle wall usually are associated with inflammation of the wall by the traumatic action of the calculus.

True visceral pain owing to diseases of the kidneys or ureters is poorly localized to the lumbar and iliac regions. Referred pain is localized more precisely to the region of the costovertebral angle, the flank, the iliac region, the suprapubic region, and the inguinal and scrotal or labial region. Allodynia, hyperalgesia, and deep tenderness often are present in the areas of referred pain.

Pain in Acute Pyelonephritis

Pain frequently is present in acute pyelonephritis. It is caused by inflammation from urinary tract and interstitial kidney tissue infection, which is caused in most cases by gram-negative bacteria. Pain is localized to the lumbar region and usually is accompanied by high fever. Deep tenderness may be evident in one or both costovertebral angles. Leukocytosis and leukocyturia are typical laboratory findings. Urine culture and antimicrobial susceptibility testing may be useful for the choice of treatment.

Pain in Renal Infarction

Renal infarction is caused by occlusion of the major renal arteries or their branches. Occlusion may be due to thrombosis induced by pathologic changes of the arterial wall (caused by inflammation or trauma) or to emboli, usually originating in the left side of the heart (from bacterial endocarditis or aseptic vegetations). The clinical presentation is characterized by the sudden onset of severe flank pain. Local muscle tenderness may be observed. Fever, hematuria, leukocytosis, and increased serum levels of renal enzymes (aspartate aminotransferase, lactic dehydrogenase, and alkaline phosphatase) may be present. If a large renal mass is involved, acute renal failure may occur.

Pain in Renal Vein Thrombosis

Thrombosis of renal veins may be caused by trauma, extrinsic compression, or hypercoagulable states (due to pregnancy, oral contraceptive use, nephrotic syndrome, or dehydration). It may be characterized by acute flank pain due to renal swelling with consequent capsular distension. Hematuria and proteinuria may be observed.

Pain in Polycystic Kidney Disease

Polycystic kidney may give rise to chronic flank pain or acute pain episodes. Chronic pain is caused by the enlargement of the kidney, which causes distension of the capsule and traction on the pedicle. Acute pain is due to infection, urinary tract obstruction by clot or stone, or sudden hemorrhage into a cyst. The association of pain with other clinical characteristics of this disease, such as hypertension, hematuria, and signs of impaired renal function, may be useful for a correct diagnosis, which is confirmed by ultrasound examination.

Pain in Ureteral Colic

Typical episodes of severe pain, i.e., ureteral colic, are caused by the passage of stones in the ureters. Mechanical factors, i.e., distension and strong contraction of the ureteral muscle wall, have been considered to be especially important in the genesis of pain. A chemical factor also must be considered; inflammation, provoked by the stone in the ureteral wall, may induce the release of pain-producing substances.

The intensity of pain in ureteral colic increases gradually and may become so severe that even the administration of opioids may be ineffective in bringing relief. Patients often try to find a comfortable position by moving continuously. In the first phase of ureteral colic, a poorly localized flank pain usually is present, with the characteristics of true visceral pain. In a following phase, a more localized referred pain is experienced. The localization of referred pain depends on the position of the stone in the urinary tract. If the stone is in the kidney pelvis, referred pain is felt in the costovertebral angle; if the stone is in the upper ureter, pain is referred to the flank; stones in the middle part of the ureter give rise to pain referred to the anterosuperior iliac crest and the inguinal region; stones in the lower part of the ureter refer pain to the suprapubic region and the scrotal or labial skin.[25] Allodynia, hyperalgesia, and deep tenderness may be evident in the somatic structures corresponding to the site to which pain is localized. MacLellan and Goodell[26] performed an interesting study on referred pain originating from the urinary tract. After direct electrical stimulation or local distension of the ureter and kidney pelvis, they observed that the muscles of the abdominal wall on the stimulated side remained contracted, and after about half an hour, pain appeared. Pain was localized to the muscles of the stimulated side and became progressively stronger, lasting about 6 hours; the following day the side was still tender. In a series of clinical investigations, Vecchiet and colleagues[27,28] studied muscular hyperalgesia in areas of referred pain in patients suffering from unilateral colic due to calculosis of the upper urinary tract. Muscular hyperalgesia was evaluated by pain threshold measurement to electrical and pressure stimulation of the obliquus externus muscle at the lumbar level. It was observed that muscular hyperalgesia was (1) mainly ipsilateral to the affected urinary tract, within the metameric field; (2) already detectable after a few painful episodes; (3) accentuated by repetition of the colics; and (4) usually persisted for a long time, even after the calculus had been eliminated. It must be noted that not all individuals who pass ureteral stones suffer colic episodes, referred pain, or referred hyperalgesia. It has been suggested that the onset of pain, at least in part, may be related to the phenomenon of *viscerovisceral hyperalgesia,* in which increased afferent input from another visceral organ to common spinal segments facilitates the central effect of input from the urinary tract.[29] Antispasmodics and NSAIDs are administered commonly to relieve the pain of ureteral colic. Hydration may be useful to induce spontaneous passage of the stone. If the ureteral obstruction persists, the stone must be removed. Extracorporeal shock wave lithotripsy is effective in many cases. If lithotripsy fails, it is necessary to use other procedures, such as extraction during cystoscopy, percutaneous nephrolithotomy, or open surgery.

The passage of clots may give rise to ureteral colic that is similar to the colic caused by nephrolithiasis. Clots usually are caused by renal bleeding due to trauma, tumors, or hemophilia.

Pain in Renal and Ureteral Tumors

Flank pain frequently is present in patients with renal tumors. It usually results from stretching of the renal capsule caused by the expanding tumor mass or by bleeding. Pain also may be due to the passage of clots in the ureters, which causes a typical ureteral colic. Pain may be associated with other clinical features of renal tumors, such as hematuria or palpable flank mass. In patients with ureteral tumors, pain results mainly from ureteral obstruction by the tumor mass.

Pain in Hemopathies

Pain may be present in many hematologic diseases, and in some cases it may be the prominent symptom. Different pathogenetic mechanisms may cause pain, including vascular obstruction and resulting ischemia, infiltration of different tissues by proliferating cells, and release of pain-producing substances.

Pain in Hemolytic Anemias

In hemolytic anemias characterized by hemolytic crises, abdominal or back pain frequently accompanies acute episodes. Pain is especially important in sickle cell disease, in which pain attacks may be the cause of severe suffering and disability.

PAIN IN SICKLE CELL DISEASE

Acute ischemic pain is a characteristic clinical feature of sickle cell disease and is due to vascular occlusion caused by the phenomenon of sickling, i.e., the change in red blood cell shape that resembles a sickle. The phenomenon of sickling is accompanied by a change in the physical characteristics of red cells, which become more rigid and traverse small vessels with great difficulty, so that vascular occlusion may occur. The phenomenon of sickling is due to hemoglobin S, which differs from normal hemoglobin in the replacement of normal glutamic acid with valine at position 6 in the β chain of globin. Hemoglobin S is the typical characteristic of sickle cell syndromes and is inherited in an autosomal codominant manner. Sickle cell anemia is the homozygous state, in which the sickle gene is inherited from both parents; other sickle cell syndromes are due to inheritance of the gene of hemoglobin S from one parent and the gene of another hemoglobinopathy, such as β-thalassemia, from the other parent; the heterozygous state (sickle cell trait) is usually asymptomatic. The high frequency of sickle cell disease in some African populations or persons of African ancestry is thought to be due to a biologic advantage conferred by hemoglobin S to heterozygotes through partial protection against *Plasmodium falciparum,* which causes severe malaria.[30] The phenomenon of sickling is caused by decreased solubility of the deoxygenated form of hemoglobin S and depends on the degree of oxygenation of hemoglobin, the concentration of hemoglobin S in red cells, the pH, and temperature; sickling is due to the formation of hemoglobin polymers that cause the typical change in red cell shape.

Acute pain owing to ischemia or infarction of different organs frequently is the first symptom of sickle cell disease. Precipitating factors, such as cold, infection, fever, dehydration, menses, alcohol consumption, and exposure to low oxygen tension, may be evident, but in most cases no precipitating factor is identified. Acute abdominal pain may be similar to the pain of other diseases of abdominal organs, such as biliary colic, appendicitis, or perforation of a hollow viscus. Acute chest pain may be accompanied by fever, dyspnea, leukocytosis, and radiographic evidence of pulmonary infiltrates; in this case, vascular occlusion is associated with infection. Pain localized to the back and extremities usually is caused by muscle and bone infarction. Joint pain, accompanied by swelling due to inflammatory joint effusion, may be the result of periarticular bone infarction. The occurrence of pain episodes is associated with other clinical features of the disease, such as hemolytic anemia with evidence of sickled red cells on peripheral blood smears; this may be accompanied by the clinical manifestations of other complications, such as hematuria due to renal papillary necrosis or stroke due to cerebral infarction. Frequent and severe pain attacks often cause anxiety and depression, which may give origin to psychosocial problems.

In the management of painful crises of sickle cell disease, the administration of NSAIDs is common; when pain is severe, mild or strong opioids are added. Antibiotics also are administered when infection is present. The administration of hydroxyurea induces a decrease in the frequency of pain attacks and an improvement in other clinical features of sickle cell disease. The therapeutic effect of hydroxyurea results mainly from interference with normal erythropoiesis, which causes an increased production of γ-globin chains and, hence, of fetal hemoglobin; the increased levels of fetal hemoglobin inhibit sickling phenomena. Psychosocial problems resulting from the disease often require administration of antidepressant drugs and provision of psychological support. Allogeneic bone marrow transplantation is a curative

Pain in Myeloproliferative Disorders

Different types of pain may be present in myeloproliferative disorders. Myeloid leukemias may cause neuropathic pain with various clinical features. In essential thrombocytosis and other myeloproliferative disorders, such as myeloid leukemias and polycythemia vera, the typical syndrome of erythromelalgia may be present.

Pain in Multiple Myeloma

Bone pain is a frequent symptom of multiple myeloma. It is due to bone destruction by proliferating plasma cells derived from a single clone. Pain is localized more frequently to the chest and the back. A typical clinical characteristic is the increase in pain produced by movement. Persistent localized pain usually is caused by a pathologic fracture. Vertebral collapse may provoke spinal cord compression with the onset of radicular pain and loss of bowel and bladder control. Neuropathic pain also may be due to mono- or polyneuropathies caused by amyloid infiltrating peripheral nerves.

Pain in Hemorrhagic Disorders

Pain due to hemorrhage into joints, muscles, and soft tissues may be present in many hemorrhagic disorders. It is especially important in hemophilia.

PAIN IN HEMOPHILIA

Both hemophilia A, due to hereditary X-linked deficiency of plasma clotting factor VIII, and hemophilia B, due to hereditary X-linked deficiency of clotting factor IX, cause severe pain from joint, muscle, and soft tissue hemorrhage. Joint hemorrhage, involving more often the knees, ankles, elbows, and shoulders, causes a specific hemophiliac arthropathy characterized by various clinical features.[31] Recurrent traumatic or spontaneous hemarthroses are accompanied by joint swelling and load pain; tenderness usually is present. Septic arthritis is an important complication that occurs more frequently in HIV-positive hemophiliac patients and usually is characterized by fever, leukocytosis, and signs of joint inflammation, such as severe pain, tenderness, warmth, and swelling. Subacute or chronic arthritis also may occur; synovitis is accompanied by intra-articular effusion, pain on awakening in the morning, and stiffness lasting up to 2 to 3 hours. In end-stage hemophiliac arthropathy, pathologic changes are similar to those of rheumatoid arthritis. Joint laxity, subluxation, and malalignment may be observed. Osteophytic bone overgrowth may provoke an enlargement of joints, which appear "knobby." Pain is continuous and is exacerbated by small movements. The range of motion may be greatly reduced, and fibrous ankylosis may occur. Muscle hemorrhage often involves the ileopsoas muscle, giving rise to acute groin pain provoked by flexion or extension of the hip. When hemorrhage occurs in volar forearm muscles, if the pressure remains high for several hours, a compartment syndrome may result; local pain increases progressively, and muscle and nerve function is impaired. Muscle infarction and ischemic nerve lesions may induce Volkmann's ischemic contracture, accompanied by continuous severe pain on the volar surface of the hand. Hemorrhages due to hemophilia are treated by infusions of factor VIII in hemophilia A and factor IX in hemophilia B. With this treatment, it is possible to stop the bleeding. NSAIDs must not be used to relieve pain because they may increase bleeding.

REFERENCES

1. Coda BA, Bonica JJ: General considerations of acute pain, in Loeser JD (ed), *Bonica's Management of Pain,* 3d ed. Philadelphia: Lippincott, Williams & Wilkins, 2001:222–240.
2. Bonabello A, Galmozzi MR, Bruzzese T, et al: Analgesic effect of bisphosphonates in mice. *Pain* 2001;91:269–275.
3. Zoppi M, Beneforti E: Joint pain. *Curr Rev Pain* 1999;3:121–129.
4. Arnett FC, Edworthy SM, Bloch DA, et al: The American Rheumatism Association 1987 revised criteria for the classification of rheumatoid arthritis. *Arthritis Rheum* 1988;31:315–328.
5. Creamer P: Osteoarthritis, in Wall PD, Melzack R (eds), *Textbook of Pain,* 4th ed. Edinburgh: Churchill Livingstone, 1999:493–504.
6. Weyand CM, Goronzy JJ: Polymyalgia rheumatica and giant cell arteritis, in Koopman WJ (ed), *Arthritis and Allied Conditions,* 14th ed. Philadelphia: Lippincott, Williams & Wilkins, 2001:1784–1798.
7. Ayala M: Douleur sympathique et douleur viscérale. *Rev Neurol* 1937;68:222–242.
8. Procacci P, Zoppi M, Maresca M: Clinical approach to visceral sensation, in Cervero F, Morrison JFB (eds), *Visceral Sensation.* Amsterdam: Elsevier, 1986:21–28.
9. Strigo IA, Bushnell MC, Boivin M, et al: Psychophysical analysis of visceral and cutaneous pain in human subjects. *Pain* 2002;97:235–246.
10. Giamberardino MA: Visceral hyperalgesia, in Devor M, Rowbotham MC, Wiesenfeld-Hallin Z (eds), *Proceedings of the 9th World Congress on Pain.* Seattle: IASP Press, 2000:523–550.
11. Head H: On disturbances of sensation with especial reference to the pain of visceral disease. *Brain* 1893;16:1–133.
12. Ruch TC: Visceral sensation and referred pain, in Fulton JF (ed), *Textbook of Physiology.* Philadelphia: Saunders, 1946:385–401.
13. MacKenzie J: *Symptoms and Their Interpretation.* London: Shaw and Sons, 1909.
14. Procacci P, Zoppi M, Maresca M: Heart, vascular and haemopathic pain, in Wall PD, Melzack R (eds), *Textbook of Pain,* 4th ed. Edinburgh: Churchill Livingstone, 1999:621–639.

15. Dressler W: Post-myocardial infarction syndrome: Preliminary report of a complication resembling idiopathic, recurrent benign pericarditis. *JAMA* 1956;160:1379–1383.
16. Selwyn AP, Braunwald E: Ischemic heart disease, in Braunwald E, Fauci AS, Kasper DL, et al (eds), *Harrison's Principles of Internal Medicine,* 15th ed. New York: McGraw-Hill, 2001:1399–1410.
17. Prinzmetal M, Kennamer R, Merliss R, et al: A variant form of angina pectoris. *Am J Med* 1959;27:375–388.
18. Scadding JW: Complex regional pain syndrome, in Wall PD, Melzack R (eds), *Textbook of Pain,* 4th ed. Edinburgh: Churchill Livingstone, 1999:835–849.
19. Wolf S, Wolff HG: *Human Gastric Function: An Experimental Study of a Man and His Stomach.* New York: Oxford University Press, 1947.
20. Del Valle J: Peptic ulcer disease and related disorders, in Braunwald E, Fauci AS, Kasper DL, et al (eds), *Harrison's Principles of Internal Medicine,* 15th ed. New York: McGraw-Hill, 2001:1649–1655.
21. Friedman S, Blumberg RS: Inflammatory bowel disease, in Braunwald E, Fauci AS, Kasper DL, et al (eds), *Harrison's Principles of Internal Medicine,* 15th ed. New York: McGraw-Hill, 2001:1679–1692.
22. Verne GN, Robinson ME, Price DD: Hypersensitivity to visceral and cutaneous pain in the irritable bowel syndrome. *Pain* 2001;93:7–14.
23. Giamberardino MA, Affaitati G, Iezzi S, et al: Referred muscle pain and hyperalgesia from viscera. *J Musculoskel Pain* 1999;7:61–69.
24. Shrikhande SV, Friess H, di Mola FF, et al: NK-1 receptor gene expression is related to pain in chronic pancreatitis. *Pain* 2001;91:209–217.
25. Vasavada SP, Comiter CV, Raz S: Painful diseases of the kidney and ureter, in Loeser JD (ed), *Bonica's Management of Pain,* 3d ed. Philadelphia: Lippincott, Williams & Wilkins, 2001:1309–1325.
26. MacLellan AM, Goodell H: Pain from the bladder, ureter and kidney pelvis. *Proc Assoc Res Nerv Ment Dis* 1943;23:252–262.
27. Vecchiet L, Giamberardino MA, Dragani L, et al: Pain from renal/ureteral calculosis: Evaluation of sensory thresholds in the lumbar area. *Pain* 1989;36:289–295.
28. Giamberardino MA, Valente R, Vecchiet L: Muscular hyperalgesia of renal/ureteral origin, in Vecchiet L, Albe-Fessard D, Lindblom U, et al (eds), *New Trends in Referred Pain and Hyperalgesia.* Amsterdam: Elsevier, 1993:149–160.
29. Giamberardino MA, Berkley KJ, Affaitati G, et al: Influence of endometriosis on pain behaviors and muscle hyperalgesia induced by a ureteral calculosis in female rats. *Pain* 2002;95:247–257.
30. Ballas SK: *Sickle Cell Pain.* Seattle: IASP Press, 1998.
31. Heck LW Jr: Arthritis associated with hematologic disorders, storage diseases, disorders of lipid metabolism, and dysproteinemias, in Koopman WJ (ed), *Arthritis and Allied Conditions,* 14th ed. Philadelphia: Lippincott, Williams & Wilkins, 2001:1903–1924.

CHAPTER 34

The Myalgic Syndromes

Robert D. Gerwin

Muscular pain may be one of the most common reasons for visits to physicians when one includes complaints associated with low back pain, neck and shoulder pain, arthritis, and tension headache, in addition to primary myalgias such as fibromyalgia. The prevalence of localized muscle pain is reported to be 20 percent and that of widespread muscle pain as high as 10 percent.[1] In this chapter, the two major types of myalgic syndromes, myofascial pain syndrome (MPS) and fibromyalgia syndrome (FMS), will be discussed. The diagnostic features of each syndrome will be outlined, and current therapeutic options will be presented. The importance of including MPS in a discussion of fibromyalgia comes from the observation that many cases diagnosed as FMS are really MPS that has not been properly examined for myofascial trigger points.

Muscle pain differs from cutaneous pain in that it is diffuse and poorly localized. Muscle pain is described as an aching or cramping type of pain rather than sharp or stabbing pain, which is characteristic of cutaneous pain. Muscle pain is caused by activation of muscle peripheral nociceptors by such endogenous substances as bradykinin, 5-hydroxytryptamin (5-HT, serotonin), and potassium ions.[2] Ischemia is a particular activator of some muscle nociceptors and is of special interest because muscle contraction that exceeds a certain force becomes ischemic as local capillaries are compressed. Two receptors, vanilloid and purinergic, are especially important in the genesis of muscle pain.[2] The release of adenosine triphosphate (ATP) by damaged or traumatized muscle fibers activates purinergic receptors and thus induces muscle pain. Ischemia also lowers the pH within muscle at the site of trauma or injury, which increases the H^+ (proton) concentration, which, in turn, activates the vanilloid VR-1 receptor. Muscle injury also releases K^+ ions, which activate muscle nociceptors. The muscle nociceptive nerve endings release calcitonin gene-related peptide (CGRP), substance P (SP), and somatostatin, each of which modifies the local extracellular environment. SP causes local edema and an increase in local blood flow and vascular permeability. The actions of prostaglandin E_2 (PGE_2) and 5-HT potentiate the effect of bradykinin sensitizing and then activating the nociceptors in muscle.[2] The sensitized muscle nociceptor best explains the peripheral mechanism of tenderness seen in myalgic conditions.

Nociceptor activity also causes central sensitization that leads to allodynia and referred pain. The changes in the central nervous system (CNS) take place in the dorsal horn of the spinal cord, in the trigeminal nucleus caudalis in the brain stem, and in supraspinal centers. Expansion of the activated dorsal horn neuronal population begins within hours of input from muscle nociceptors and is related to the release of glutamate and SP, which activates N-methyl-D-aspartic acid (NMDA) receptors. Second-messenger activation by Ca^{2+} ions leads to further activation of cell surface ion channels. The end result of these and related processes is an expanded neuronal population that is hyperexcitable. Moreover, there is spread of afferent input within the spinal cord.

Nociceptive afferent input spreads upward and downward as much as 7 to 10 segments within the spinal cord, allowing for expansion of dorsal horn neuronal activation over a wider area. The consequence of these changes is the development of allodynia and referred pain in an enlarged area of the body than primarily involved by the myalgic process but still within regional body segments. Thus, pain from a myalgic process in the head and neck may be referred to cervical segments (shoulder and arm) but less to upper thoracic segments and is unlikely to be spread to lumbar segments. Chronic pain can develop as a result of the persistence of acute pain that lasts long enough to induce neuroplastic changes in the CNS. Finally, morphologic changes occur in the dorsal horn with sprouting of new afferent nerve fibers that enlarge the population of excited neurons.[2]

The question arises as to how muscle develops pain in the course of ordinary activity. As stated previously, ischemic muscle releases substances that activate peripheral nociceptors. Concentric muscle exercise can produce pain from impaired blood flow resembling ischemic muscle pain. Eccentric muscle exercise can produce delayed-onset muscle pain with soreness maximal 1 to 2 days after exercise. Both conditions may result in peripheral and central sensitization, as described earlier.[3]

Experimental studies of induced referred pain have shown enlargement of the zones of referred pain in patients with fibromyalgia, osteoarthritis, whiplash injuries, and temporomandibular myofascial pain syndromes.[3] These findings are consistent with the observation that patients with fibromyalgia have a generalized hypersensitivity to a variety of noxious and benign stimulation.[4-6] Myofascial pain syndrome has been considered by some to be a regional pain syndrome, but in 45 percent of chronic cases it was generalized to three or four quadrants of the body,[7] resembling fibromyalgia. Generalized hypersensitivity has not been studied in myofascial pain. There is a resemblance between widespread myofascial pain and fibromyalgia; however, palpation of muscle can differentiate myofascial trigger points from fibromyalgia tender points. There are reports in the literature that claim that fibromyalgia as widespread muscle tenderness can start as regional myofascial pain and then spread, but none of these reports credibly examine for myofascial trigger points and, therefore, do not exclude the more likely possibility that regional MPS becomes generalized.

▶ FIBROMYALGIA

FMS is a chronic, widespread muscular pain syndrome. It is the second most common disorder seen by rheumatologists. FMS is not considered a disease per se, but rather a collection of symptoms and signs that can have multiple etiologies. The etiology appears related to neuroplastic changes of the CNS that result in general hypersensitivity to all types of stimuli through the unmasking of dormant synapses and reduction of central inhibition[8] as described earlier. FMS patients have decreased serum serotonin levels and impaired central serotonin metabolism, as well as elevated levels of cerebrospinal fluid (CSF) SP.[9,10] They have low levels of insulin-like growth factor 1,[11] and they have decreased function of the hypothalamic-pituitary-adrenal axis with low serum androgen levels that correlate with poor physical functioning and with pain.[12]

The hallmark of FMS is widespread muscular pain of 3 or more months' duration. Widespread pain means that pain is present in three or four quadrants of the body (i.e., upper and lower and right and left sides). Pain need not be present in every area at each examination, but over repeated examinations, it should be generalized. Pain often is accompanied by unusual fatigue and disturbed sleep. FMS occurs in about 3.5 percent of women and 0.5 percent of men and is more common with advancing age, peaking at about age 70 years. It occurs in children as well as adults. It is comorbid with many other conditions, such as rheumatoid arthritis, systemic lupus erythematosus, Lyme disease, hepatitis, Sjögren's syndrome, and MPS. In such circumstances, it is sometimes termed *secondary fibromyalgia,* as contrasted with *primary fibromyalgia,* a term used to describe fibromyalgia occurring alone.

Clinical Presentation

Patients with FMS complain of widespread muscle pain that interferes with activity. They are unusually fatigued out of proportion to any sleep disorder, but they also have impaired sleep and awaken feeling tired and unrested. They often have associated problems of depression, headache, joint pain, morning stiffness,

▶ **TABLE 34-1.** CLINICAL FEATURES OF FIBROMYALGIA

Primary

1. Widespread muscular pain
2. Chronicity: 3 months or more
3. Unusual fatigue
4. Nonrefreshing or nonrestorative sleep

Associative

a. Headaches
b. Joint pains
c. Raynaud's phenomenon
d. Irritable bowel syndrome
e. Irritable bladder syndrome/interstitial cystitis
f. Dyspareunia
g. Vulvodynia
h. Morning stiffness

Raynaud's phenomenon, bladder irritability, irritable bowel syndrome, and painful intercourse (Table 34-1). Many of these associated phenomena are the result of myofascial trigger point syndromes that coexist with FMS (e.g., muscular headaches, morning stiffness, pelvic floor or viscerosomatic pain syndromes, and dyspareunia). Patients complain of difficulty concentrating and of short-term memory impairment. Symptoms can be so severe as to cause many to seek disability retirement. In fact, the number of persons with FMS seeking disability has become so great that it has been characterized as a disability epidemic.

Diagnosis

The diagnosis is based on (1) the history of chronic, widespread pain; and (2) the presence of muscle tenderness (tender points, or TePs) that is bilateral and affects the upper and lower halves of the body (Table 34-2). The American College of Rheumatology (ACR) established criteria for the diagnosis of FMS to aid in the development of clinical studies.[13] These criteria have been adopted for clinical use as well and have been responsible for an explosion of FMS studies that has greatly increased the understanding of muscle pain in general and FMS in particular.

▶ **TABLE 34-2.** DIAGNOSTIC FEATURES OF FIBROMYALGIA

1. History of chronic muscle pain
2. Tenderness of 11 or more of 18 preselected examination sites*
3. Exclusion of other causes of widespread muscle pain
4. Other signs and symptoms are suggestive but not required

*11 tender points on average; there may be more or less on any given examination.

▶ **TABLE 34-3.** TENDER POINT SITES FOR FIBROMYALGIA EXAMINATION

1. Suboccipital
2. Lower anterior cervical
3. Upper trapezius
4. Supraspinatus
5. Parasternal at 2nd rib
6. Lateral epicondylar region
7. Anterior gluteal fold
8. Greater trochantor of the hip
9. Medial fat pad above the knee (vastus medialis)

The presence of tenderness at 11 or more of 18 preselected sites for examination had a diagnostic sensitivity of 88 percent and a specificity of 81 percent in the ACR study. There need not be 11 or more TePs at any given examination in clinical practice, but over time, muscle tenderness must be widespread, which the criteria ensure.

The TeP examination is conducted by palpating muscle with a force sufficient to blanche the fingernail (~4 kg). The standardized sites for examination are listed in Table 34-3. The sites are meant to be examined bilaterally. In clinical practice, tenderness is not confined to the sites designated in the ACR criteria but is distributed widely in muscles throughout the body. Care should be taken when examining for TePs to distinguish them from tender, taut muscle bands of myofascial trigger points so as not to confuse fibromyalgia TePs with myofascial pain trigger points. The associated problems of joint stiffness, Raynaud's phenomenon, interstitial cystitis, and so on mentioned above do not need to be present in order to diagnosis FMS. Other causes of widespread, chronic muscle pain should be excluded because a number of medical conditions can present with myalgia (Table 34-4). Among these is orthostatic neurogenic hypotension, which is seen in many patients with FMS. To diagnosis this, the blood pressure and pulse should be taken supine and immediately and again 2 minutes after standing.[14]

▶ **TABLE 34-4.** DIFFERENTIAL DIAGNOSIS OF FIBROMYALGIA

1. Myofascial pain syndrome
2. Drug induced: statin cholesterol-lowering drugs
3. Hypothyroidism
4. Iron deficiency
5. Vitamin D and B_{12} deficiency
6. Infections
   candidiasis
   parasitic
   bacterial (mycoplasma)
7. Sleep apnea, restless legs syndrome
8. Myoadenylate deaminase deficiency
9. Sjögren's syndrome
10. Lyme disease

A tilt-table test is commonly used to evaluate patients for this problem. The degree of orthostatic hypotension can be so profound as to prevent the individual from being upright long enough to be functional. Orthostatic neurogenic tachycardia also has been described as an autonomic dysfunction when orthostatic hypotension is not present. The pulse should rise at least 20 beats per minute when going from supine to standing.

Numerical rating scales for pain (0 = no pain, 10 = the worst possible pain) are useful for assessing the degree of pain that a patient is experiencing. The Fibromyalgia Impact Questionnaire[15] assesses the impact of pain on the patient's life and, therefore, is a quality-of-life measurement. A polysomnogram or a sleep disorder consultation may be necessary to diagnosis treatable sleep disturbances, such as sleep apnea and restless leg syndrome, which can affect muscle pain.

Treatment

Treatment consists of pharmacologic, physical, nutritional, hormonal, and psychological (including cognitive and educational) therapies.

Pharmacologic therapies often include antidepressant drugs (Table 34-5). Tricyclic antidepressants (TCAs) have been used for many years. They inhibit the reuptake of serotonin and norepinephrine and provide short-term improvement. Amitriptyline, at doses of 25 to 50 mg/d at bedtime, has been the most thoroughly studied, but it has not been studied for long-term improvement beyond 6 months. Fluoxetine, at 20 mg/d, and sertraline, at 50 mg/d, also have been shown to be effective. Antidepressants improve sleep, fatigue, and pain but not the tender point count.

Opioid analgesics may be used to treat pain when necessary (see point 6d under "Summary" below). Tramadol may cause seizures when used with a variety of different antidepressant drugs, and either should be avoided or should be used with caution in association with such common drugs used in fibromyalgia treatment as amitriptyline.

▶ **TABLE 34-5.** TREATMENT OF FIBROMYALGIA: PHARMACOLOGIC

1. OTC analgesics
2. Nonsteroidal anti-inflammatory drugs
3. Muscle relaxants and antispasticity drugs; cyclobenzaprine 10 mg tid
4. Antidepressants; amitriptyline 25–50 mg qhs
5. Anticonvulsants; gabapentin, pregabalin
6. Antispasticity drugs; tizanidine starting at 1–2 mg qhs, slowly titrate up to effectiveness
7. Opioids
8. Nutritional supplements; SAMe

Anticonvulsants will reduce neuropathic pain and can be used nonspecifically in FMS. Pregabalin has been shown to be effective in FMS, although there was a dropout rate of 19 percent because of side effects.

A number of other treatments have been tried for FMS, some with more success and some with less or no proven benefit. Among these is the use of growth hormone titrated to an insulin-like growth factor 1 (IGF-1) level of 250 mg/mL, which improves FMS after about 6 months of treatment. Treatment costs about $1000 per month and is not recommended for routine management. Symptoms recur after treatment is stopped.

Thyroid supplementation may result in the resolution of FMS symptoms when hypothyroidism coexists with FMS. There are no data to support the use of other kinds of hormonal therapy. S-adenosyl-L-methionine (SAMe), at 200 mg/d, improves pain, fatigue, mood, and morning stiffness in some studies. Acupuncture is an effective complementary therapy. Biofeedback, especially combined with exercise, is effective in improving function and reducing tender points. Hypnosis and meditation-based stress reduction both improve pain ratings and function and can reduce the number of tender points.

The combination of behavioral modification, education, and physical training is effective (Table 34-6).[16] Exercise is the most effective physical form of therapy both in the short term and for long-term outcome as well (up to 4 years). Graded exercise therapy, as moderately intense aerobic exercise two to three times per week, with minimal eccentric muscle activity results in an improvement in physical functioning, cardiovascular fitness, and self-efficacy, but pain levels, tender point counts, mood and depression, and sleep and fatigue may not improve.[17] Physical therapy is useful otherwise primarily to identify and treat coexisting problems such as MPS and postural dysfunctions.

Treatment of sleep disturbance is important. Treatment should include attention to sleep hygiene (darkened room and use the bed only for sleep and sex, not for reading, watching television, working, or eating) and medications to promote sleep such as melatonin 3 mg before bedtime, trazodone 50 to 150 mg at bedtime, or zolpidem 5 to 10 mg or zaleplon 10 mg at bedtime. Amitriptyline also improves sleep.

There is no evidence to support the use of magnesium supplements, DHEAS, guanefesine, corticosteroids, sex hormones, or herbal supplements.

▶ **TABLE 34-6.** TREATMENT OF FIBROMYALGIA: PHYSICAL MODALITIES

1. Physical therapy
2. Exercise
3. Electrical stimulation
4. Acupuncture

▶ **TABLE 34-7.** TREATMENT OF FIBROMYALGIA: PSYCHOLOGICAL

Education
Cognitive-behavioral therapy
Hypnosis
Psychotherapy

▶ **TABLE 34-8.** CLINICAL FEATURES OF MYOFASCIAL PAIN SYNDROME

Motor Features

1. Taut (hardened) band of muscle that runs the length of the muscle
2. Twitch or local contraction of muscle band upon mechanical stimulation
3. Restricted range of motion
4. Weakness

Sensory Features

1. Tenderness (allodynia, hypersensitivity) of the taut band (known as the myofascial trigger point or zone)
2. Referred pain

Autonomic Features

1. Skin temperature changes
2. Lacrimation
3. Piloerection (goose bumps)

The associative disorders of headache, interstitial cystitis, or irritable bowel syndrome and irritable bladder syndrome and dyspareunia should be treated symptomatically. The muscular (myofascial) bases for these should be assessed for muscular causes, e.g., a muscular basis for headache or muscular (myofascial) pelvic floor syndromes that can be treated specifically.

Trigger-point injections (see MPS below) are used to treat coexisting MPS, but injection of tender points with local anesthetics or corticosteroids has no proven benefit.

Psychological stress always should be evaluated and treated when it is thought to be contributing to fibromyalgia symptomatology (Table 34-7). Depression has been found in at least 30 percent of persons with fibromyalgia but has been said to be no more common than in the general population or in other chronic conditions. It also has been reported as being much more frequent than 30 percent. The general consensus is that depression is not a major factor in the development of fibromyalgia but may complicate the condition, as it does other chronic disorders. Other stress factors, such as anxiety, job or marital dissatisfaction, or parental conflicts (even in the older patient when there is a living parent), aggravate the pain of fibromyalgia. Anger from perceived mistreatment by the workers' compensation system, by employers, by fellow employees, or by family members or friends can smolder and perpetuate pain. Acute stress also can result in biomechanical stress, affecting low back pain in patients with fibromyalgia or without.

▶ MYOFASCIAL PAIN SYNDROME

Myofascial pain may be acute or chronic, regional or widespread, but in every case it is associated with tenderness or pain that is localized to a linear or nodular hardening in the muscle that is called a *myofascial trigger point* (Table 34-8). Referred pain, or pain that is felt at a distance from the point of stimulation, is characteristic of myofascial pain. MPS occurs as the result of a muscle overload, either acute, as in an injury such as whiplash or a sports injury, or chronic overuse, as in repetitive-strain syndromes seen in computer operators or musicians. Muscle that is maximally loaded through concentric exercise to the point of fatigue and muscle that is eccentrically loaded are susceptible to injury[18] and to ischemia. The mechanism that leads overloaded muscle to develop taut bands that extend from one myotendinous junction to the other is unknown but is thought to involve abnormal motor end plate function.[19] It is known that muscle that is injured or inflamed rapidly leads to central sensitization, lowering the threshold to nociceptive and nonnociceptive stimulation and producing hypersensitivity, allodynia (the phenonemon of nonpainful stimulation being perceived as painful), and referred pain.[20] MPS can persist long after the initial injury has resolved and may be forgotten. Myofascial trigger zones also develop in the referred pain zone, as well as in muscles that are agonists or antagonists of the muscle(s) that was injured initially.

Clinical Presentation

The individual with MPS complains of pain that generally occurs with activity but in more severe cases can be present at rest and interfere with sleep. Pain often is felt in an area of referred pain with or without pain felt in the primary area of the muscle trigger zone. Range of motion may be limited because of shortened taut muscle bands or because of pain. Weakness is found in affected muscles; the mechanism may involve central fatigue from persistently contracted taut bands within the muscle.[21] Mechanical stimulation of the primary trigger zone will reproduce the patient's pain, including referred pain. Referred pain syndromes can resemble other painful conditions and may be mistaken for them.

Some common referred pain syndromes are those which mimic cervical and lumbar radiculopathy and viscerosomatic syndromes, such as the pelvic floor pain syndromes of interstitial cystitis and irritable bowel syndrome (Table 34-9). Referred pain is both a spinal cord and a thalamic phenomenon.

▶ **TABLE 34-9.** SOME TYPICAL MYOFASCIAL PAIN SYNDROMES

1. Piriformis syndrome (entrapment of the sciatic nerve)
2. Interscalene compartment syndrome (entrapment of the brachial plexus)
3. Thoracic outlet-like syndrome (entrapment of the brachial plexus between the clavicle and 1st rib)
4. Hyperabduction syndrome
5. Viscerosomatic syndromes (cardiac, gastrointestinal, hepatic, genito-urinary)
6. Headaches (chronic tension type, with or without migraine)
7. Temporomandibular joint syndrome
8. Frozen shoulder

Taut (hardened) bands in affected muscle result in muscle shortening and increasing muscle diameter or cross-sectional bulk. This can result in nerve entrapment syndromes such as the piriformis syndrome of the sciatic nerve, brachial plexus compressions in the interscalene compartment or between the first rib and the clavicle, and the hyperabducation syndrome of the pectoralis minor muscle compressing the distal brachial plexus under the coracoid process. In these syndromes of intermittent compression, electromyography and nerve conduction studies usually are normal unless the nerve compression is severe and advanced. Disputed or non-neurogenic thoracic outlet syndrome often is caused by myofascial trigger points either as a direct result of nerve entrapment, as described earlier, or as a manifestation of trigger point referred pain.

Diagnosis

The diagnosis of MPS is based on identification of the primary trigger zone that reproduces the patient's pain complaint (Table 34-10). The diagnosis is made by physical examination. Limited range of motion is an important finding but may appear normal in persons who are hypermobile. Hypermobility, a condition affecting females more than males, is itself a risk factor for myofascial pain. The diagnosis of MPS is made by reproducing pain through palpation of the trigger point. A painful region in a linear or nodular hardness in the muscle is identified. It may elicit referred pain symptoms, if present, but should reproduce a part or all of the patient's pain complaint. In the case of deep muscles, such as the multifidi, needling the muscle may be necessary to elicit the symptoms. Laboratory tests are useful for identifying conditions that either coexist with the MPS or aggravate or perpetuate it, but there are no laboratory tests suitable for general clinical use that will aid in the diagnosis of MPS. The characteristic electromyographic abnormalities of the trigger points themselves are useful for research studies of MPS but are not practical for routine clinical use. Other techniques, such as the determination of skin impedance changes, have not come into general usage. Consequently, palpation remains the most practical diagnostic technique. It has proven to be reliable in the identification of the trigger point.[22,23]

Treatment

Treatment of MPS is both specific (treating the trigger point directly), general (treating pain and sleeplessness), and corrective (identifying and correcting the predisposing and perpetuating factors that may lead to and aggravate MPS). The first step is to identify and inactivate the myofascial trigger point in order to relieve pain and restore normal function. Second, obvious conditions that create and maintain trigger points, such as significant foot pronation, leg-length inequality, and ergonomic stresses, must be identified and corrected or eliminated.

Inactivation of the trigger point is achieved manually or by needling the trigger point. A large number of physical approaches have been used and found effective (Table 34-11). Most of these have in common compression and stretching of the trigger point, followed by muscle reeducation and restoration of normal muscle

▶ **TABLE 34-10.** DIAGNOSTIC FEATURES OF MYOFASCIAL PAIN SYNDROME

1. Restricted range of movement
2. Palpable taut band of muscle
3. Tenderness on the taut band
4. Reproduction of usual pain by palpation of the tender zone

▶ **TABLE 34-11.** TREATMENT OF MPS, PHYSICAL MODALITIES

Primary
1. Local trigger point compression
2. Local trigger point stretch
3. Myofascial release
4. Muscle play
5. Therapeutic stretch
6. Self-stretch
7. Muscle reeducation
Adjunctive
a. Intermittent cold
b. Postisometric relaxation
c. Strain-counterstrain
d. Dry needling or injection (local anesthetic or botulinum toxin)
e. Massage
f. Ultrasound
g. Electrical stimulation
h. Acupuncture

▶ **TABLE 34-12.** NEEDLING OR INJECTION OF THE TRIGGER POINT

A. Purpose
  1. Diagnostic
  2. Rapid relief of pain
  3. Facilitation of manual (physical) therapy
B. Medications
  1. Short-acting local anesthetics without epinephrine
  2. "Dry needling" with no drug (mechanical stimulation of the trigger point alone)
  3. Botulinum toxin

▶ **TABLE 34-13.** TREATMENT OF MPS: PHARMACOLOGICAL

1. OTC drugs
2. Nonsteroidal anti-inflammatory drugs
3. Antidepressant drugs (those that inhibit reuptake of both serotonin and norepinephrine like the tricyclic antidepressants and venlafaxine)
4. Muscle relaxants
5. Antispasticity drugs (tizanidine)
6. Anticonvulsants
7. Opioid analgesics (preferably long-acting, slow-release)
8. Botulinum toxin

sequencing in movement. Postisometric relaxation, reciprocal inhibition, and contract-relax techniques are all useful in stretching muscle. Therapeutic stretching and self-stretching are relatively contraindicated in persons who are hypermobile. Treatment in these individuals is directed locally to the trigger point. Ultrasound can aid in inactivation of trigger points, although no controlled study has been done to confirm this. Electrical stimulation can reduce pain and allow manual treatment to proceed more comfortably. Acupuncture can be used to reduce the pain of myofascial trigger points. Superficial dry needling, in which an acupuncture needle is inserted 2 to 4 mm under the skin over the point of trigger point pain, also can reduce trigger point local and referred pain.[24] Strengthening to maintain improvement should be reserved until after there has been a significant reduction in muscle pain.

Trigger point needling or injection (Table 34-12) is effective in providing relief of pain, although it has not been compared with placebo in any controlled trial. It also can confirm a suspected diagnosis if the individual's pain is relieved rapidly by needling or by injection of local anesthetic. Needling and injection therapy are precise; the needle should enter the trigger zone and elicit a twitch response for best results. There is no difference in outcome whatever the material injected or if only dry needling (no material injected) is done.[25] The effect of needling or injection is often temporary, lasting days, and should be combined with manual (physical) therapy (see Table 34-11). Lidocaine 0.25% is recommended for injection. This concentration causes less postinjection pain than 1% or 2% lidocaine. Use 0.1 or 0.2 mL to inactivate the trigger point, injecting the anesthetic when a twitch response is elicited.

Botulinum toxin injections into the trigger zone act as longer-lasting trigger point injections when compared with either needling or injection of local anesthetics. Thus, 25 units of botulinum toxin type A or 1250 units of botulinum toxin type B is injected into the trigger zone, except in head and neck muscles, where smaller amounts are injected (e.g., 5 to 10 units of botulinum toxin type A). Likewise, the muscles of the forearm also need a lesser amount of botulinum toxin. Since the purpose of injecting botulinum toxin is to reduce trigger point activity and not to paralyze the muscle, a smaller amount of toxin than that used to treat dystonia or spasticity suffices. A number of open-label studies have been done that showed promise in treating MPS with botulinum toxin, but one small randomized, double-blind, placebo-controlled trial showed only a trend toward improvement, and then only in patients injected a second time 6 weeks after the initial injection. However, a recent randomized, controlled study of the use of botulinum toxin type A in low back pain showed significantly greater efficacy of the toxin compared with saline.[26] Myofascial trigger points most certainly were injected in that study.

Pharmacologic treatment is directed toward pain relief and sleep improvement (Table 34-13). Acetaminophen or aspirin can be used to provide short-term pain relief in mild pain states. Nonsteroidal anti-inflammatory drugs (NSAIDs) are not used for any anti-inflammatory purpose in muscle but for their analgesic activity, including postinjection pain. If NSAIDs are used for a long period of time, the selective COX-2 inhibitors should be used because they have less likelihood of causing gastrointestinal bleeding, even though they are no more effective in pain relief than mixed COX-1–COX-2 inhibitors. Some anticonvulsant drugs (e.g., carbamazepine, gabapentin, and pregabalin) have been shown to be effective in the treatment of neuropathic pain. Other of the newer anticonvulsants (e.g., topirimate, lamotrigine) have had mixed results in trials for treatment of neuropathic pain. Nevertheless, anticonvulsants also may reduce muscle pain and can be tried in MPS for pain relief. Tizanidine, an α_2-adrenergic agonist, has been shown to have efficacy in the treatment of MPS. The major adverse effects are daytime drowsiness, dry mouth, and the possibility of an initial drop in blood pressure. The drug is started at a low dose of 1 to 2 mg at night and then is titrated upward on a twice or thrice daily schedule to effectiveness.

Cyclobenzaprine, related to the TCAs, has mixed reports concerning efficacy in MPS and is not recommended except as adjunctive treatment. Carisoprodol

▶ **TABLE 34-14.** TREATMENT OF MPS: PSYCHOLOGICAL

Education
Cognitive-behavioral therapy
Psychotherapy

and benzodiazepines have abuse potential and cause sedation. Clonazepam, a benzodiazepine derivative, was shown in one study to be effective in treating MPS. Start with 0.5 mg/d at bedtime and titrate upward. Sleep disruption can be treated with drugs such as zolpidem, 5 to 10 mg/d at bedtime, or with antidepressants such as trazodone, 50 to 150 mg/d at bedtime.

Opioids are used strictly for pain relief. They can be liberating for someone who is disabled by pain, but they have abuse potential and create constipation. If considered for long-term use, only long-acting, slow-release forms are recommended. Slow-release, long-acting preparations are available for morphine sulfate, oxycodone, methadone, and fentanyl. These drugs should be used only by physicians experienced in their management. Tramadol used alone at 50 to 100 mg every 4 to 6 hours or with acetaminophen up to a maximum of 800 mg/d can be used for pain of moderate intensity. However, tramadol in combination with certain antidepressant drugs can lower the seizure threshold.

Psychological treatment is indicated when the clinician identifies depression or other emotional stresses that might underlie and aggravate the MPS (Table 34-14). Psychological stress is often considered a predisposing factor for the development of muscular pain. Although there is no evidence that such stress can create trigger points, there is evidence that stress will greatly increase the abnormal electromyographic activity at the trigger point. Screening by a social worker, psychologist, or psychiatrist is appropriate in selected patients and is recommended for many, if not most, persons with a long history of myofascial pain. A skilled physician will treat with antidepressants when indicated.

▶ **SUMMARY**

For diagnosis and treatment of fibromyalgia:

1. Diagnosis is positive in terms of identifying widespread muscle tenderness in a person with a history of chronic, widespread pain and negative in terms of excluding other causes of widespread myalgia or muscle pain.
2. Use an antidepressant drug that inhibits both serotonin and norepinephrine uptake, such as amitriptyline 25 to 50 mg/d at bedtime, to relieve pain and improve sleep.
3. Improve sleep with trazodone 50 to 150 mg/d at bedtime, melatonin 3 mg/d at bedtime, or a drug such as zolpidem 5 to 10 mg/d at bedtime. Get a sleep consultation if there is a serious sleep disorder or significant daytime hypersomnia.
4. Develop a moderately intensive aerobic exercise program that is graded and that avoids excessive eccentric resistive exercise.
5. Treat pain with a suitable analgesic if necessary.
6. Use a combination of education and cognitive-behavioral therapy.
7. Treat coexisting conditions such as hypothyroidism, iron deficiency, or vitamin B_{12} deficiency.
8. Consider complementary/alternative methods of treatment such as supplementation with SAMe or the use of acupuncture.

For the treatment of myofascial pain syndrome:

1. Treat MPS patients with manual inactivation of the trigger point (physical therapy, including trigger point compression, massage, local and therapeutic stretching, and self-stretching). *Warning:* Stretching should be done cautiously or not at all in hypermobile individuals.
2. If trigger points do not release manually, they are needled or injected with local anesthetic. Botulinum toxin is used as a long-lasting trigger point injection in cases where needling or local anesthetic combined with physical therapy does not give long-term relief. Inject 25 units of botulinum toxin type A into the trigger point, except in the head and neck, where only 5 to 10 units are used.
3. Identify and correct postural, ergonomic, mechanical (structural), hormonal, nutritional, and other medical precipitating and perpetuating factors.
4. Perform strengthening exercises, including lumbar stabilization, when pain levels are reduced enough to allow the patient to perform resistive and stabilizing exercises without an undue increase in pain.
5. Use counseling and cognitive-behavioral therapy where warranted.
6. Use the following medications to treat symptoms when necessary:
   a. Sleep disturbance: trazodone 50 to 150 mg/d at bedtime, amitriptyline 25 to 50 mg/d at bedtime, or zolpidem 5 to 10 mg/d at bedtime.
   b. Antispasticity drugs: tizanidine starting at 1 to 2 mg/d at bedtime and titrating to effectiveness twice or thrice daily; up to 8 mg three times daily can reduce pain significantly.
   c. Muscle relaxants: cyclobenzaprine 10 mg three times daily or methylcarbamol 500 to 750

mg three times daily for short-term use. Avoid carisoprodol.

d. Opioid analgesics for short-term use, the type depending on the severity of the pain. For acute, severe pain, use rapid-onset, short-duration opioids such as oxycodone/APAP 5/325 or hydrocodone/APAP 5/500 every 4 hours as needed. For pain that is subacute or chronic severe pain, use slow-release, long-acting opioids such as slow-release morphine sulfate or oxycodone, methadone, or transdermal fentanyl starting at low doses and titrating upward to efficacy. For lesser pain, tramadol starting at 50 mg two to four times daily and titrating to effectiveness (50 to 100 mg every 4 to 6 hours, maximum dose of 800 mg/d), or NSAIDs can be used. For mild pain, over-the-counter preparations are satisfactory.

REFERENCES

1. Kissel J, Miller R: Muscle pain and fatigue, in *Muscle Diseases*. Woburn, MA: Butterworth-Heinemann, 1999:33–58.
2. Mense S: The pathogenesis of muscle pain. *Curr Pain Headache Rep* 2003;7:419–425.
3. Graven-Nielsen T, Arendt-Nielsen L: Induction and assessment of muscle pain, referred pain, and muscular hyperalgesia. *Curr Pain Headache Rep* 2003;7:443–451.
4. Granges G, Littlejohn G: Pressure pain thresholds in pain-free subjects, in patients with chronic regional pain syndromes, and in patients with fibromyalgia syndrome. *Arthritis Rheum* 1993;36:642–646.
5. Lautenbacher S, Rollman GB, McCain GA: Multi-method assessment of experimental and clinical pain in patients with fibromyalgia. *Pain* 1994;59:45–53.
6. Sorensen J, Graven-Nielsen T, Henriksson KG, et al: Hyperexcitability in fibromyalgia. *J Rheumatol* 1998;25:152–155.
7. Gerwin R: A study of 96 subjects examined both for fibromyalgia and myofascial pain. *J Musculoskel Pain* 1995;3:121.
8. Coderre T, Katz J, Vacvcarino A, Melzack R: Contribution of central neuroplasticity to pathological pain: Review of clinical and experimental evidence. *Pain* 1993;52:259–285.
9. Russell IJ: Neurochemical pathogenesis of fibromyalgia syndrome. *J Musculoskel Pain* 1996;4:61–92.
10. Pillemer S, Bradley L, Crofford L, et al: The neuroscience and endocrinology of fibromyalgia. *Arthritis Rheum* 1997;40:1928–1939.
11. Bennett R, Cook D, Clark S, et al: Hypothalamic-pituitary-insulin-like growth factor axis dysfunction in patients with fibromyalgia. *J Rheumatol* 1997;24:1384–1389.
12. Dessein PH, Shipton EA, Joffe BI, et al: Hyposecretion of adrenal androgens and the relation of serum adrenal steroids, serotonin and insulin-like growth factor-1 to clinical features in women with fibromyalgia. *Pain* 1999;83:313–319.
13. Wolfe F, Smythe HA, Yunus MB, et al: The American College of Rheumatolgy criteria for the classification of fibromyalgia. *Arthritis Rheum* 1990;33:160–172.
14. Rowe PC, et al: Neurally mediated hypotension and chronic fatigue syndrome. *Am J Med* 1998;105:15S–21S.
15. Burckhardt C, Clark S, Bennett R: The Fibromyalgia Impact Questionnaire: Development and validation. *J Rheumatol* 1991;18:728–233.
16. Burckhardt C, Mannerkorpi K, Hedenberg L, Bjelle A: A randomized, controlled clinical trial of education and physical training for women with fibromyalgia. *J Rheumatol* 1994;21:714–720.
17. Clark S, Jones K, Burckhardt C, Bennett R: Exercise for patients with fibromyalgia: Risks versus benefits. *Curr Rheumatol Rep* 2001;3:135–146.
18. Newham D, McPhail G, Mills K, Edwards R: Ultrastructural changes after concentric and eccentric contractions of human muscle. *J Neurol Sci* 1983;61:109–122.
19. Simons DG, Travell JG, Simons LS: *Myofascial Pain and Dysfunction: The Trigger Point Manual*. Baltimore: Williams & Wilkins, 1999.
20. Mense S, Simons DG: *Muscle Pain*. Philadelphia: Lippincott, Williams & Wilkins, 2001.
21. Loscher W, Nordlund M: Central fatigue and motor excitability during repeated shortening and lengthening actions. *Muscle Nerve* 2002;25:864–872.
22. Gerwin RD, Shannon S, Hong C-Z, et al: Interrater reliability in myofascial trigger point examination. *Pain* 1997;69:65–73.
23. Sciotti VM, Mittak VL, DiMarco L, et al: Clinical precision of myofascial trigger point location in the trapezius muscle. *Pain* 2001;93:259–266.
24. Baldry P: *Myofascial Pain and Fibromyalgia Syndromes*. Edinburgh: Churchill Livingstone, 2001.
25. Cummings T, White A: Needling therapies in the management of myofascial trigger point pain: A systematic review. *Arch Phys Med Rehabil* 2001;82:986–992.
26. Foster L, Clapp L, Erickson M, Jabbari B: Botulinum toxin A and chronic low back pain: A randomized double-blind study. *Neurology* 2001;56:1290–1293.

CHAPTER 35

Assessment and Treatment of the Myofascial Trigger Point

Charles Argoff

The term myofascial pain syndrome *(MPS) is used to describe a common pain disorder associated with the presence of active myofascial trigger points. The purpose of this chapter is to describe the historical background, epidemiology, pathophysiology, diagnosis (including examination and symptoms and signs), and treatment of this syndrome. It is acknowledged from the outset that the discussion of MPS in a neurology textbook is not without potential controversy because many neurologists deny the very existence of this disorder.*[1,2] *While many challenges exist regarding taxonomy, the lack of a specific diagnostic tool, and the chronic nature of the disorder, MPS is, in fact, one of the most common pain states that a neurologist will see. It is vital that the neurologist recognizes and treats MPS to the fullest extent possible.*

It might be best at this point to provide definitions for some of the more frequently used terms in MPS. The myofascial trigger point *is a localized area within tissue that may be tender and may trigger pain, not only at the site of palpation but also at a site distant from it (referred pain). Autonomic phenomena are frequently present as well. A myofascial trigger point can be* active, *or it can be* latent. *An* active myofascial trigger point *is painful at rest and with motion of the affected muscle. The active myofascial trigger point is always painful, and this pain limits full lengthening of the muscle fibers. In contrast, a* latent myofascial trigger point *is a focus of hypersensitivity and irritability within the muscle or fascia, and it is painful only when palpated. The term* jump sign *is used literally to describe the response of a patient who may cry out and wince when pressure is applied to a myofascial trigger point. The* local twitch response *is a brief contraction of muscle fibers within a trigger point in response to some type of stimulation. Most commonly, the stimulation is either the insertion of a needle (for diagnostic or therapeutic purposes) or "snap" palpation, in which the fingertip is placed on a tight band of muscle perpendicular to the direction of the muscle fibers and the finger is drawn back and rolled against the muscle. A* palpable band, *also known as a* taut band, *of muscle fibers is found within the trigger point and can be identified by manual examination of the muscle.* Muscle spasm *refers to increased muscle tension due to involuntary motor activity. It may or may not be associated with shortening of muscle fibers. The* zone of reference *is an area at a distance from the myofascial trigger point where various effects of the trigger point, including sensory, motor, and autonomic responses, may occur.*[3] *The term* fibromyalgia *often has been used erroneously as a synonym for MPS. In contrast to MPS, which is a regionalized or local pain disorder, the term* fibromyalgia *is used to describe a chronic painful disorder characterized by bilateral, widespread musculoskeletal pain associated with typical physical examination findings of painful tenderness at 11 or more* of 18 anatomically-defined soft tissue tender points. It is not appropriate to describe a painful state as fibromyalgia if it involves only one region of the body. *The fact that the pain of fibromyalgia is so widespread and the observation that patients with fibromyalgia have been noted to experience low pain thresholds have led to the notion that a neurochemical or neuromodulatory abnormality underlies this disorder.*[4]

▶ HISTORICAL BACKGROUND

There has been interest in "muscle-related" pain for centuries. One of the most vexing problems has been the difficulty in correlating the palpable abnormalities of muscle with any definable and consistent histopathologic and/or neurophysiologic changes. In addition, as in many other neurological disorders, a notable obstacle to improved understanding of MPS has been the use of multiple, perhaps overlapping terms in describing the disorder. For example, *muscular rheumatism, nonarticular rheumatism, idiopathic myalgia, muscular sciatica, fibrositis,* and *myalgia spots* are among the many terms that have been used historically to describe MPS.[3] Again, this is not dissimilar to the classification of headache, which for years suffered as well from a plethora of overlapping definitions for similar syndromes. Although the use of MPS is currently widely accepted, the reader is advised that older literature may still be confusing because of the use of fairly synonymous terms in the description of this disorder.

▶ EPIDEMIOLOGY

The first point that needs to be made regarding the epidemiology of this disorder is that no large studies have been completed to date that address the incidence and prevalence of MPS. The most widespread data is from the Nuprin Report, a publication designed in 1985 to address the prevalence of pain. The data from this report suggested that 53 percent of Americans experienced muscle pain, but it did not formally identify these individuals as having MPS. Sola and colleagues,[5] in their examination of 200 otherwise healthy young adults, found latent trigger points in the shoulder girdle muscles of 54 percent of the females studied and in 45 percent of the males studied. In another study performed by the same investigators, a survey of 1000 outpatients demonstrated that 32 percent of those surveyed met criteria for MPS, with a prevalence of 36 percent among the 598 women and of 26 percent among the 402 men involved in this study.[6] In another study by Gerwin[7] in 1998, 98 percent of patients with the diagnosis of chronic tension-type headache were found to have myofascial trigger points that reproduced all or part of their headache complaints.

▶ PATHOPHYSIOLOGY

In the past, the development of MPS and myofascial trigger points has been viewed as a local muscle disturbance. Histologic studies of myofascial trigger points have been completed, but the results do not demonstrate clearly why their presence is associated with pain. In one study, the observation, on review of biopsy samples from active myofascial trigger points, that adenosine triphosphate, phosphocreatine, and glycogen concentrations were reduced and that lactate levels were normal led investigators to hypothesize that the development of a myofascial trigger point could be related to a primary muscular disturbance or the result of excessive muscle tension.[8]

Others, including most notably Travell and Simons,[3] have maintained that myofascial trigger points are the result of repeated microtrauma and muscle overload. Taking this concept one step further, Calliet[9] has proposed that myofascial trigger points result from blood and other materials that are not completely degraded and resorbed following soft tissue damage. Perhaps one of the most interesting concepts regarding the pathophysiology of this syndrome is one proposed by Gunn.[10] He postulated that the etiology of the hypersensitivity and pain associated with myofascial trigger points is the result of a neuropathy affecting the neural input of the affected muscle. Mense has studied and reviewed the mechanisms of nociception in skeletal muscle extensively, and his work has demonstrated that peripheral nociception in muscle can lead rapidly to central sensitization. In one experiment, a microelectrode was first placed into the dorsal horn of the spinal cord. After mapping out the receptive field of the muscle, bradykinin was injected into it. Following the bradykinin injection, the dorsal horn neuron was seen to respond to a pinch of normal strength, e.g., the pain threshold for activation of the neuron remained normal. However, within 15 minutes of bradykinin injection, not only was the receptive field expanded, but the neuron now responded to a weaker pinch consistent with its being sensitized. This process is similar to that seen in sensory neuropathic phenomenon in general.[11,12]

Recently, the potential relationship between intramuscular hypoperfusion, altered sympathetic function, and chronic muscle pain has been reviewed. The results are not uniform, and hence there are not sufficient data to support this general hypothesis at this time.[13,14] Others have noted previously and suggested similar hypotheses regarding the pathogenesis of MPS. Zimmerman has postulated that because the muscle fibers that "form" the myofascial trigger point are contracted for an extended period of time, muscle fatigue and ischemic changes within the muscle may occur. Hypothetically, a change in the extracellular milieu of the affected fibers might result in the release of neuropeptides, such as histamine, prostaglandins, and kinins, that can promote nociception. Zimmerman also emphasized the role of the central nervous system in facilitating this process.[15]

In an attempt to develop a human model of myofascial pain, Mork and colleagues[16] infused numerous endogenous substances, such as bradykinin, serotonin, histamine, prostaglandin E, adenosine triphosphate, and combinations, into the trapezius muscle of normal humans to determine if any of these reproduce myofascial pain. They found that a combination of bradykinin, serotonin, histamine, and prostaglandin E in humans produced prolonged, moderately severe pain and muscle tenderness similar to that seen in MPS.[16] Further study of this model is clearly required to determine its suitability as a human model of MPS. The same group used this model to see if prolonged muscle tenderness could be induced in patients with tension-type headache compared with controls. There was a trend toward muscle tenderness being increased in patients with tension-type headache, but the difference was not statistically significant. The authors concluded that increased excitability of peripheral muscle afferents may be an important pathophysiologic mechanism of tension-type headache.

Mense and colleagues[18] have proposed recently that a dysfunctional muscle end plate may lead to increased acetylcholine release, which, in turn, may lead to abnormal muscle contraction and provide the basis for the formation of myofascial trigger points. They have tested this hypothesis in rats, lesioning rat skeletal muscle after small amounts of an acetylcholinesterase inhibitor is injected into the muscle and neuromuscular stimulation completed for approximately 60 minutes. Only the fibers that were injected with the acetylcholinesterase inhibitor demonstrated abnormalities of contraction and damage.

Audette and colleagues[19] measured electromyographic activity within an affected myofascial trigger point, as well as in the contralateral unaffected muscle, in patients with MPS, as well as age-matched controls, following placement of an acupuncture needle into the taut band of an active myofascial trigger point (dry needling). They observed that during the dry needling of the myofascial trigger point, the muscles contralateral to the muscle that was being treated also reacted. The same observations were not seen in controls. These observations led these investigators to conclude that the perpetuation of pain associated with an active myofascial trigger point may be due, in fact, not only to local factors but also to changes within the central nervous system, particularly within the dorsal horn of the spinal cord. These findings together support the role of both the peripheral and central nervous systems in the development and maintenance of some of the features of MPS.

▶ DIAGNOSIS AND EXAMINATION

While there are no universally agreed-on diagnostic guidelines for MPS, data from a survey of pain management practitioners who see patients with MPS indicated that there are generally agreed-on points regarding the signs and symptoms of this disorder.[20] Myofascial pain may appear acutely following obvious trauma or may develop in a more insidious manner over a longer period. For some patients with acute MPS, the pain will dissipate with time, e.g., most adults have experienced waking up with a "stiff neck" with painful cervical muscles (and myofascial trigger points) only to have the stiffness and pain dissipate over several days with little or no treatment. Regrettably, for some the acute myofascial pain becomes more chronic and may persist indefinitely. The diagnosis of MPS is made by taking a history and by conducting a thorough general physical, neurological, and musculoskeletal examination. The distribution of the pain complaint, local and referred, helps to provide clues to the origin of the pain. Is the pain related to a primary process, or is it secondary to another systemic or local process? For example, MPS often is seen in patients with spinal degenerative disorders, postlaminectomy states, and neuropathic pain states. In these settings, the myofascial pain would be considered secondary.[21]

Clearly, the key sign of MPS is the presence of a myofascial trigger point. Although pain is the main presenting symptom of MPS, other important symptoms that patients complain of include reduced movement and stiffness. The pain caused by a myofascial trigger point is often increased by using the affected muscle, by passively stretching the muscle, by placing direct pressure on the trigger point within the muscle, and by sustained and repeated contraction of the affected muscle. In addition, patients often will complain of increased pain in affected trigger points during a viral infection, during periods of increased emotional stress and anxiety, or when exposed to cold and damp weather conditions. The pain caused by a myofascial trigger point is often decreased by a period of rest or relaxation for that muscle, by slow and consistent stretching of the affected muscles, by the application of heat, by short periods of light activity, and through other medical/nonmedical approaches that will be discussed in the next section. Reduced range of motion and stiffness tend to be increased after periods of inactivity, such as in the morning. Patients with MPS may complain of muscle weakness that on manual muscle testing frequently appears to be related to pain avoidance as opposed to frank loss of strength. Abnormalities of autonomic function, including changes in skin temperature, hyperhidrosis in an affected extremity, and rhinorrhea, have been reported in patients with MPS.[22]

Obviously, the distribution of the pain gives the examiner clues as to the origin of the pain. Clinicians must be aware of the fact that the pain caused by a myofascial trigger point may be experienced at the site of the trigger point itself or at a distal or referred site. For example, myofascial pain that is experienced in the area inferior and medial to the scapula can be the result of myofascial dysfunction in the inferior aspect of the trapezius muscle, the latissimus dorsi muscle, the infraspinatus muscles, or the scalene muscles. The viscera also may refer pain to muscle and, therefore, may be the source of the myofascial abnormalities. In evaluating the patient with MPS, one must be careful to exclude a primary structural or systemic etiology to the disorder. Numerous structural (significant degenerative joint or disk disease), metabolic (electrolyte disturbance, hypothyroidism, connective tissue disease, vitamin D deficiency), and infectious etiologies exist that either may result in an MPS-like presentation or may exacerbate the pain for someone with known MPS.

The purpose of the physical examination in a patient with MPS is to identify the presence of myofascial trigger points, as well as to evaluate for the presence of any general, musculoskeletal, or neurological findings that may give the examiner a clue regarding the etiology of the MPS. Evaluation of a patient for MPS involves several components, including:

1. A search for taut bands. First, the muscle should be placed into an intermediate position in which the muscle is neither overly lengthened nor overly shortened. Recognizing that palpation of the muscle is always perpendicular to the muscle fiber direction, using a pincer or flat palpation, the examiner should feel the presence or absence of a taut band.
2. The range of motion of the affected areas should be determined.
3. The examiner should question the patient regarding any abnormal sensations that are experienced during direct muscle palpation. These abnormal sensations may be painful or nonpainful.
4. The local muscle twitch response should be elicited, if possible.
5. The patient should be examined for musculoskeletal and neurological abnormalities that can lead to biomechanical stress, including scoliosis, leg-length or other limb length discrepancies, and hemiparesis.

▶ TREATMENT

The purpose of treatment of MPS is to reduce pain sufficiently so that function can be restored. Fortunately, there are numerous approaches to consider; however, given the facts that there are few controlled trials of any type of treatment in MPS, that there is no specific test for MPS, and that there is a lack of uniform agreement regarding the diagnosis of MPS, comments regarding the treatment of this disorder need to be interpreted in light of such. Possible treatments of MPS will be divided into medical and nonmedical, but by no means is this section designed to provide the reader with every possible treatment approach that has been considered. All treatments are designed to interrupt the pain cycle caused by the trigger point.

Nonmedical Therapies

Nonmedical therapies for the management of MPS include the *stretch and spray technique*. This term emphasizes the importance of muscle stretching in this technique. After the patient is placed in a comfortable position, the involved muscles are placed into a position so that they are relaxed. A bottle of flourimethane or ethylcholoride spray is held 18 inches away from the patient, and the spray applied at an angle of 30 degrees. The spray is applied slowly in one direction and not in a haphazard manner. Ideally, the spray direction should be from the trigger point toward the direction of the referred pain. The involved muscle can then be passively or actively stretched depending on the severity of the pain. The patient is likely to feel better following this treatment but needs to be reminded that overloading the muscle after treatment is likely to reinjure it.[23] Pressure and massage have been shown to help some patients with MPS; however, other patients have reported exacerbations of their pain with such treatment. Ice massage has been shown to be effective for chronic low back pain, but formal studies for the pain associated specifically with MPS are not available. In some studies, ice massage for pain relief has been shown to be comparable with transcutaneous electrical nerve stimulation (TENS) and acupuncture for patients with chronic low back pain.[24]

Other nonmedical approaches that have been used for the treatment of MPS include acupuncture, biofeedback, ultrasound, cognitive-behavioral training, and lasers. Chiropractic care is used often by some patients for symptom relief. Despite the fact that these treatments are used commonly, and despite the fact that the literature suggests a trend in favor of their benefit, most studies do not have proper control groups, studies are often unblinded, and the sample sizes are often too small.[25–27]

Pharmacological Therapies

While numerous options exist for the pharmacotherapeutic management of chronic pain in general, few medications have been studied specifically for the treatment of MPS. Nevertheless, medications should be considered part of a comprehensive treatment program designed for patients with MPS to decrease pain and to facilitate function as much as possible. In choosing a particular agent, the clinician must balance the evidence for effectiveness of that agent for MPS, if any, with its ease of use and cost. Examples of this would include the potential use of benzodiazepines for MPS. Although benzodiazepines are known to be helpful for a number of conditions associated with muscle spasm, use of these agents is associated with sedation, as well as physical and psychological dependence.

Nonsteroidal anti-inflammatory agents and nonselective or Cox-2-specific drugs are known to have multiple actions within the peripheral and central nervous systems. Neither group of medications has been found to be specifically helpful to patients with MPS, but they are prescribed often.[28] Nonopioid analgesics such as acetaminophen are also used often, but caution must be exercised in using acetaminophen because long-term use of even recommended doses of this agent can be associated with renal and hepatic toxicities.[29] Topical analgesics such as the 5% lidocaine patch have been studied in the treatment of MPS with results that suggest that it may be helpful to some patients.[30] While opioid analgesics are used often in the short-term management of a variety of acute painful states, including those associated with myofascial dysfunction, no study has systematically addressed the role of long-term use of opioids in the management of MPS. Regardless, this class of medications can be used in the treatment of chronic pain of multiple origins, including MPS, if more conservative measures have failed. Tricyclic antidepressants and other antidepressant agents also are used often for the treatment of chronic musculoskeletal pain, but again, no methodologically sound formal studies have confirmed their benefit for MPS. The α-adrenergic agent tizanidine appears to have a positive effect on the treatment of MPS, according to some small studies.[31]

Despite their being prescribed widely, few formal studies have been completed regarding the use of "muscle relaxants" such as cyclobenzaprine or metaxolone in the treatment of MPS. In one recent randomized study, the combination of cyclobenzaprine with ibuprofen was compared with ibuprofen alone in the treatment of acute myofascial pain. No significant differences in analgesia were found.[32] Although studied in neuropathic pain, no formal investigations have been conducted to address the potential benefit of N-methyl-D-aspartic acid (NMDA) receptor antagonists in the treatment of chronic MPS.

Injection Therapies

Different injection therapies have been employed frequently in the treatment of MPS. The most commonly used injection therapy is the myofascial trigger point injection. These injections are designed to reduce the activity of the trigger point, thus reducing pain and facilitating improved function. Several techniques have been employed, including performing these without any medication (dry needling) and performing them with medication (most often a local anesthetic). Frequently, multiple injections are necessary, and the local twitch response should be elicited during the injection. Although commonly performed with local anesthetic, e.g., lidocaine, prilocaine, and bupivacaine, dry needling is just as likely to provide a similar duration of pain relief as those trigger point injections performed with local anesthetic. One might consider the use of local anesthetic in these injections in order to reduce the degree of post-treatment soreness.[33] Other injection therapies include the use of botulinum toxin, the use of which has grown considerably over the past decade for a number of chronic pain disorders. Published data have been limited to open-label and case-study approaches, yet there is a suggestion that this medication may be useful in the treatment of patients with MPS.[34–37]

▶ SUMMARY

MPS affects numerous individuals and must be recognized by the neurologist. Although the precise pathophysiology remains uncertain, there is increasing evidence for the role of the peripheral and central nervous systems in the development and maintenance of MPS. Early and appropriate diagnosis of MPS is necessary so that treatment can be completed in a timely fashion. The major goals of treatment include pain reduction and functional restoration.

REFERENCES

1. Bohr T: Problems with myofascial pain syndrome and fibromyalgia syndrome. *Neurology* 1996;46:593–597.
2. Bohr TW: Fibromyalgia syndrome and myofascial pain syndrome: Do they exist? *Neurol Clin* 1995;13:365–384.
3. Travell JG, Simons DG: *Myofascial Pain and Dysfunction: The Trigger Point Manual,* Vol 1. Baltimore: Williams & Wilkins, 1983:1–4.
4. Russell IJ. Fibromyalgia syndrome, in Loeser JD (ed), *Bonica's Management of Pain.* Philadelphia: Lippincott, Williams & Wilkins, 2001:543–556.
5. Sola AE, Rodenberger ML, Gettys BB: Incidence of hypersensitive areas in posterior shoulder muscles. *Am J Phys Med* 1955;34:585–590.

6. Sola AE, Bonica JJ: Myofascial pain syndromes, in Loeser JD (ed), *Bonica's Management of Pain*. Philadelphia: Lippincott, Williams & Wilkins, 2001:530–542.
7. Gerwin RD: Unpublished data (personal communication), 2001.
8. Henriksson KG, Bengtsson A, Larsson J, et al: Muscle biopsy findings of possible diagnostic importance in primary fibromyalgia (fibrositis, myofascial syndrome) (letter). *Lancet* 1982;2:1395.
9. Calliet R: *Soft Tissue Pain and Disability*. Philadelphia: FA Davis, 1977.
10. Gunn CC: Prespondylosis and some pain syndromes following denervation suprasensitivity. *Spine* 1980;5:185–192.
11. Mense S: Nociception from skeletal muscle in relation to clinical muscle pain. *Pain* 1993;54:241–289.
12. Simons DG, Mense S: Understanding and measurement of muscle tone as related to clinical muscle pain. *Pain* 1998;75:1–17.
13. Maekawa K, Clark GT, Kuboki T: Intramuscularly hypoperfusion, adrenergic receptors, and chronic muscle pain. *J Pain* 2002;3:251–260.
14. Graven-Nielsen, Arendt-Nielsen L: Is there a relation between intramuscular hypoperfusion and chronic muscle pain? *J Pain* 2002;3:261–263.
15. Zimmerman M: Peripheral and central nervous system mechanisms of nociception, pain and pain therapy: Facts and hypotheses, in Bonica JJ, Liebskind JC, Albe-Fessard DG (eds), *Advances in Pain Research and Therapy*, Vol 3. New York: Raven Press, 1979:3–32.
16. Mork H, Ashina M, Bendtsen L, et al: Experimental muscle pain and tenderness following infusion of endogenous substances in humans. *Eur J Pain* 2003;7:142–153.
17. Mork H, Ashina M, Bendtsen L, et al: Induction of prolonged tenderness in patients with tension-type headache by means of a new experimental model of myofascial pain. *Eur J Neurol* 2003;10:249–256.
18. Mense S, Simons DG, Hoheisel U, et al: Lesions of rat skeletal muscle after local block of acetylcholinesterase and neuromuscular stimulation. *J Appl Physiol* 2003;94:2494–2501.
19. Audette J, Wang F, Smith H: The electrophysiological characteristics of myofascial pain: characteristics of the local twitch response in subjects with active myofascial pain of the neck compared to a control group with latent trigger points. *NEPA J* 2002;7:10–14.
20. Harden RN, Bruehl SP, Fass S, et al: Signs and symptoms of the myofascial pain syndrome: A national survey of pain management providers. *Clin J Pain* 2000;16:64–72.
21. Rashiq S, Galer BS: Proximal myofascial dysfunction in complex regional pain syndrome: A retrospective prevalence study. *Clin J Pain* 1999;15:151–153.
22. Travell JG, Simons DG: *Myofascial Pain and Dysfunction: The Trigger Point Manual*, Vol 1. Baltimore: Williams & Wilkins, 1983:53–55.
23. Kraus H: *Clinical Treatment of Back and Neck Pain*. New York: McGraw-Hill, 1970.
24. Melzack R, Jeans ME, Stratford JG, et al: Ice massage and transcutaneous electrical stimulation: Comparison of treatment for low back pain. *Pain* 1980;9:209–217.
25. Sherman JJ, Turk DC: Nonpharmacological approaches to the management of myofascial temporomandibular disorders. *Curr Pain Headache Rep* 2001;5:421–431.
26. Audette JF, Blinder RA: Acupuncture in the management of myofascial pain and headache. *Curr Pain Headache Rep* 2003;7:395–401.
27. Harris RE, Clauw J: The use of complementary medical therapies in the management of myofascial pain disorders. *Curr Pain Headache Rep* 2002;6:370–374.
28. Miyoshi HR: Systemic nonopioid analgesics, in Loeser JL (ed), *Bonica's Management of Pain*, 3d ed. Philadelphia: Lippincott, Williams & Wilkins, 2001:1667–1681.
29. Clissold SP: Paracetamol and phenacetin. *Drugs* 1986;32:46–59.
30. Argoff CE: Targeted topical peripheral analgesics in the management of pain. *Curr Pain Headache Rep* 2003;7:34–38.
31. Argoff CE (ed): The role of alpha-adrenergic agonists in the pain management, in *Abstract Review: Current Research and Expert Commentary*. New York: Medical Education Network, 2002.
32. Turturro MA, Frater CR, D'Amico FJ: Cyclobenzaprine with ibuprofen alone in acute myofascial strain: A randomized, double-blind clinical trial. *Ann Emerg Med* 2003;41:818–826.
33. Hong CZ: Lidocaine injection versus dry needling to myofascial trigger point: The importance of the local twitch response. *Am J Phys Med Rehabil* 1994;73:256–263.
34. Argoff CE: The use of botulinum toxins or chronic pain and headaches. *Curr Treat Options Neurol* 2003;5:483–492.
35. DeAndres J, Cerda-Olmedo G, Valia JC, et al: Use of botulinum toxin in the treatment of chronic myofascial pain. *Clin J Pain* 2003;19:269–275.
36. Lang AM: A preliminary comparison of the efficacy and tolerability of botulinum serotypes A and B in the treatment of myofascial pain syndrome: A retrospective, open label chart review. *Clin Ther* 2003;25:2268–2278.
37. Smith HS, Audette J, Royal MA: Botulinum toxin in pain management of soft tissue syndromes. *Clin J Pain* 2002;18(6 suppl):S147–S154.

CHAPTER 36

Acute Pain Management

Christopher G. Gharibo, Sheng Ping Zou, and Andrew Rosenberg

The latest emphasis on the undertreatment of acute pain and the recognition of postoperative morbidity attributable to poorly treated pain, coupled with introduction of the Joint Commission on Accreditation of Healthcare Organizations (JCAHO) Pain Management Standards in 2001, have led to a new drive to use the more advanced acute pain management modalities such as intravenous and epidural patient-controlled analgesia and regional anesthesia techniques.[1,2]

▶ PERIPHERAL NERVE BLOCKS IN ACUTE PAIN MANAGEMENT

The perioperative period usually is associated with significant pain. Peripheral nerve blocks can be used to decrease the pain, as well as the stress response, associated with surgery, thus allowing patients to rest comfortably or perform important physical rehabilitation exercises. Peripheral nerve blocks can be used in the preoperative period for patients suffering pain from fractures, intraoperatively for surgery, and postoperatively for postoperative pain relief. Peripheral nerve blocks performed both intraoperatively or postoperatively provide excellent postoperative pain relief. For upper extremity surgery, an interscalene or an infraclavicular nerve block can be used for pain relief. After lower extremity surgery, two blocks that can help provide pain relief are the femoral nerve or fascia iliaca blocks. A long-acting sciatic nerve block performed prior to surgery also can be used for postoperative pain relief after lower extremity surgery.

It is important that patients be educated to appropriately protect a numb extremity by not letting anything too cold or hot near that extremity and not walking on a numb leg.

Most nerve blocks are performed with the use of a nerve stimulator. An electric current is used to place a needle in as close proximity to the nerve as possible without actually touching the nerve. Then local anesthetic is deposited around the nerve to anesthetize it. The electric current that is emitted by the needle causes a motor stimulus (twitch) of the muscles innervated by the nerve being stimulated. The end point that is sought is the best twitch that can be obtained in the distribution of the nerves to be blocked at the lowest milliamperage (mA) possible, usually 0.2 to 0.3 mA. Only by obtaining a twitch at such a low milliamperage can one have confidence that the needle is on the same side of the fascia sheath as the nerves being blocked and that the twitch is not being obtained by electrical stimulation across a fascial sheath. Since motor fibers are stimulated by less current than sensory or pain fibers, the procedure does not have to be uncomfortable for the patient. While a patient can receive some sedation in order to be comfortable while a block is being performed, peripheral nerve blocks using the nerve stimulator technique should not be performed while the patient is under general anesthesia. Searching for a twitch response while a patient is under general anesthesia can result in the patient not being able to inform you of the significant pain accompanying nerve impalement or injection of local anesthetic into a nerve. Reports of permanent nerve damage have been seen in the literature with the use of nerve stimulators to block nerves while adult patients are under general anesthesia.

Proper Equipment for Performing Peripheral Nerve Blocks

It is important to use proper equipment when performing peripheral nerve blocks because it will help to improve the success rate. A nerve stimulator with an adjustable dial that has a readout in 0.1-mA increments from 0 to 10 mA and provides impulses of at least one per second is important. In conjunction with a nerve stimulator, an insulated needle should be used. Insulated needles isolate the source of electric current being emitted from the needle to the tip of the needle. This is important so that the nerve is not stimulated from the side of the needle while local anesthetic is injected from the tip of the needle. With the electric current being emitted from the tip of the needle, the current and the local anesthetic are coming from the same point.

Technique for Performing Peripheral Nerve Blocks

- The landmarks for the block are determined and marked.
- The area to be blocked is prepped with antiseptic solution.
- The positive lead of the nerve stimulator is connected to an electrocardiographic (ECG) pad and placed on the contralateral side of the extremity to be blocked. (This is the ground electrode.)
- If the block to be performed is not a superficial block, local anesthetic can be used to produce a skin wheal. However, local anesthetic can interfere with superficial blocks.
- The nerve stimulator is set at 1.0 to 1.5 mA, and the negative lead of the nerve stimulator is attached to the needle.
- The needle is advanced through the skin as twitches are sought in the distribution of the nerves being blocked.
- Once a twitch is obtained, the needle is positioned so that a maximal twitch is generated at 0.2 to 0.3 mA. This is done by obtaining a twitch at the initial stimulating current and then dialing the nerve stimulator down to the desired current in small increments so as to maximize the twitch response at each level. If the twitch is lost, the nerve stimulator current is then increased until a twitch is obtained, and the direction or angle of the needle is changed as the needle is advanced or withdrawn as twitches are maximized.
- It is easier to locate the optimal position if the needle is not moved at the same time as the current is being changed.
- When the desired needle location is obtained, the needle is stabilized, and after aspiration to ensure that the needle is not intravascular, 2 mL of local anesthetic is injected. The twitches should be abolished by the 2 mL of local anesthetic, confirming close proximity of the needle to the nerve.
- The remainder of the local anesthetic is injected, with aspiration performed every 5 mL to make certain that the needle is not intravascular.

Interscalene Block

The interscalene block is an excellent block to perform when it is necessary to anesthetize an area from the shoulder to the midshaft of the humerus. It is not a block to be performed when the medial aspect of the forearm and hand need to be pain-free. The interscalene block can be used to provide both intraoperative and postoperative pain relief. Long-acting anesthetic agents can be used for an operative procedure in the shoulder region, and this can provide pain relief that can last long into the postoperative period. Alternatively, if the patient has surgery performed under general anesthesia, an interscalene block can be performed in the postoperative care unit after the patient is awake. Blocks can be performed on subsequent days, or an indwelling catheter can be placed to provide continuous pain relief via an infusion. It has been our experience that intermittent blocks are more successful than a continuous infusion because the neck catheter can be dislodged easily.

It is important to note that the phrenic nerve runs in close proximity to the site where an interscalene nerve block is performed and is blocked 100 percent of the time by an interscalene block. This results in diaphragmatic hemiparalysis for the length of time that the block is in effect. This can cause potential problems in patients with respiratory compromise, such as patients with chronic obstructive pulmonary disease (COPD) or asthma. A chest x-ray will reveal an elevated hemidiaphragm on the side of the block.

Performing the Block

- The head is turned away from the side of the neck to be blocked.
- The cricoid cartilage is noted, and a line is drawn laterally from the cricoid cartilage that is perpendicular to the long axis of the neck. The block is performed at this level in the neck.
- The posterior border of the sternocleidomastoid muscle is palpated, and the fingers are moved laterally from this to the border of the anterior scalene muscle. The fingers are then rolled further laterally to the area between the anterior and middle scalene muscles. This is the interscalene groove. The index and middle fingers are spread apart, and the needle is advanced through the prepped skin in a direction that is toward the sternal notch.
- The needle is advanced as twitches are sought in the distribution of the nerves in the upper arm and shoulder.

- Once a twitch is obtained, the needle is positioned so that a maximal twitch is generated at 0.2 to 0.3 mA. This is done by obtaining a twitch at the initial stimulating current and then dialing the nerve stimulator down to the desired current in small increments so as to maximize the twitch response at each level. If the twitch is lost, the nerve stimulator current is then increased until a twitch is obtained, and the direction or angle of the needle is changed as the needle is advanced or withdrawn as twitches are maximized.
- If a twitch of the diaphragm is obtained, then the needle is too anterior, becauses the phrenic nerve courses along the anterior scalene muscle.
- If a twitch of the posterior shoulder is obtained, the needle is a little too posterior.
- When the desired needle location is obtained, at 0.2 to 0.3 mA, the needle is stabilized, and after aspiration to ensure that the needle is not intravascular, 2 mL of local anesthetic is injected. The twitches should be abolished by the 2 mL of local anesthetic, confirming close proximity of the needle to the nerve.
- The remainder of the local anesthetic is injected, with aspiration performed every 5 mL to make certain that the needle is not intravascular.

Infraclavicular Nerve Block

The infraclavicular nerve block is an excellent block to provide perioperative pain relief to the area of the midshaft of the humerus distally to include the forearm, hands, and fingers. This can be obtained by a single injection or the use of an indwelling catheter.

A block of the brachial plexus in the infraclavicular region incorporates a block of the musculocutaneous nerve at the same time as the median, ulnar, and radial nerves are blocked. The advantage of this block is that a separate block of the musculocutaneous nerve does not have to be performed in the coracobrachialis muscle. The caveat, however, is that the musculocutaneous nerve is a very superficial nerve in the brachial plexus at this level, and it should not be used as a final end point for nerve stimulation to identify the brachial plexus because it can be stimulated across the fascial sheath or directly after it has exited from the brachial plexus distal to the coracoid process.

Performing the Block
- The patient is placed supine on the operating room table with the head away from the side to be blocked.
- The arm is abducted 90 degrees.
- The midpoint of the clavicle is identified, and a 1-in line is drawn from this point that is perpendicular to the long axis of the clavicle. This is the needle insertion point. Do not move the arm after identifying the landmarks.
- The axillary artery is palpated as high in the axilla as possible. Alternatively, the brachial plexus, the nerves surrounding the axillary artery, can be identified as high in the axilla as possible by using the negative lead of the nerve stimulator at 5 mA and placing it in the axilla until twitches are noted in the distribution of the nerves of the brachial plexus.
- After prepping the area, an insulated needle is inserted through the spot identified 1 in below the midpoint of the clavicle and perpendicular to the clavicle to the point identified as high in the axilla as possible.
- The needle is advanced at a 45-degree angle as twitches are sought.
- Once a twitch is obtained, the needle is positioned so that a maximal twitch is generated at 0.2 to 0.3 mA. This is done by obtaining a twitch at the initial stimulating current and then dialing the nerve stimulator down to the desired current in small increments so as to maximize the twitch response at each level. If the twitch is lost, the nerve stimulator current is then increased until a twitch is obtained, and the direction or angle of the needle is changed as the needle is advanced or withdrawn as twitches are maximized.
- If a biceps twitch is obtained, this is a musculocutaneous twitch and is not accepted as an end point but as an identifier that you are in the region of the brachial plexus and need to move the needle slightly more caudad and advance it at a slightly steeper angle.
- Twitches are sought in the forearm or hand at 0.2 to 0.3 mA. Twitches of the median, ulnar, or radial nerve are acceptable.
- When the desired needle location is obtained, at 0.2 to 0.3 mA, the needle is stabilized, and after aspiration to ensure that the needle is not intravascular, 2 mL of local anesthetic is injected. The twitches should be abolished by the 2 mL of local anesthetic, confirming close proximity of the needle to the nerve.
- The remainder of the local anesthetic is injected, with aspiration performed every 5 mL to make certain that the needle is not intravascular.

Femoral Nerve Block

The femoral nerve innervates the anterior thigh, a major portion of the knee, and the medial aspect of the skin of the leg. A nerve block provides excellent pain relief after knee surgery. It also can be used preoperatively to provide pain relief so that a patient can be moved onto the operating room table without significant pain or as part of a femoral-sciatic nerve block for intraoperative surgical anesthesia. A nerve stimulator can be used to perform this block.

Performing the Block
- The patient is positioned supine on the operating room table.
- The femoral nerve is palpated just below the midpoint of the inguinal ligament.
- After a sterile prep, an insulated needle is passed perpendicular to the skin, and stimulation is started at 0.5 to 1 mA. Since the femoral nerve is located approximately 1 cm below the skin, the skin twitches will be noted very soon after needle insertion.
- Twitches of the thigh are maximized at 0.2 mA.
- After aspiration, 2 mL of local anesthetic is injected, and twitches should be abolished.
- The remainder of the local anesthetic is injected, with aspiration performed every 5 mL to make certain that the needle is not intravascular.

Fascia Iliaca Block

The fascia iliaca block does not require use of the nerve stimulator. When performing this block, the needle is placed in an area between the femoral nerve and the lateral femoral cutaneous nerve away from the femoral artery. The advantage of this is that the local anesthetic can course medial to block the femoral nerve and lateral to anesthetize the lateral femoral cutaneous nerve while the chance of nerve damage or arterial or venous puncture is minimized. The block is performed by advancing a spinal needle caudad at a 30-degree angle at a point that is 1 in below the junction of the middle and lateral thirds of a line drawn from the anterosuperior iliac spine to the pubic tubercle.

Two distinct pops are felt as the needle is advanced—first the fascia lata is pierced, and then a more subtle pop is heard as the fascia iliaca is pierced. Deep to the fascia iliaca is the area where the nerves are located. This block is very helpful in providing pain relief after knee surgery, including total knee replacement and anterior cruciate ligament repair.

▶ PATIENT-CONTROLLED EPIDURAL ANALGESIA

Infusion of analgesics such as opioids and local anesthetics into the epidural space is one of the most popular and effective methods of pain control. Patients with postoperative pain ranging from routine cesarean sections to upper abdominal and thoracic surgeries make suitable candidates for use of patient-controlled epidural analgesia (PCEA).

The first description of epidural analgesia dates back to 1949, when bolus doses of local anesthetic were administered via an epidural catheter in the postoperative period.[3] Owing to the intermittent bolus delivery of high-concentration local anesthetics, this initial effort was effective but complicated by varying levels of pain relief and significant hypotension that required the use of vasopressors.

Physiologic Basis and Behavior of Local Anesthetics and Opioids as Epidural Analgesics

The physiologic basis for using epidural analgesics is similar to the basis for using the epidural space for operative anesthesia. The extent of the neural sensorimotor blockade is a function of the concentration and volume of local anesthetic delivered epidurally. Although high concentrations, such as 0.25 percent or 0.5 percent bupivacaine, are necessary for operative anesthesia, much more dilute local anesthetic concentrations are required in the epidural space.

The goal in an acute pain management setting such as the postoperative period is effective pain management without hemodynamic or neurological effects that can impede recovery. In fact, often any extent of sensorimotor blockade is undesirable because it tends to interfere with ambulation, an important parameter of postinjury recovery. Therefore, the concentrations of local anesthetics for epidural analgesia are two to four times more dilute than those used in the intraoperative setting.

The spinal level of epidural catheter placement is also of significance given that the analgesics and anesthetics administered through the catheter will spread in cephalad and caudal directions in the spinal canal to produce a segmental analgesia, i.e., bandlike analgesia that corresponds to the dermatomes of the spinal segments bathed by the analgesic solution. Therefore, catheter placement needs to be at a spinal level that corresponds to the spinal roots or dermatomes that will transmit the acute pain.

An epidural catheter infusing local anesthetics at the level of the nerve roots that are transmitting the pain directly blocks these nerves and provides superior static and dynamic pain relief. The dynamic pain relief afforded by epidural segmental analgesia is particularly advantageous in upper abdominal and thoracic surgery, where other pain management modalities frequently are inadequate. Furthermore, the sparing of the spinal segments above and below the surgery that usually affect the lower extremities and the upper thoracic muscles allows earlier ambulation and improved ability to cough and take deep breaths.

The discovery of the opioid receptors in the spinal cord in the 1970s popularized the use of opioids in the epidural and intrathecal space for operative anesthesia and acute pain management. Morphine sulfate was the first opioid to receive Food and Drug Administration approval for epidural and intrathecal use.

Unfortunately, similar to initial dosing efforts for local anesthetics, where excessive concentrations of local anesthetics were bolused intermittently, the initially suggested dosing for epidural morphine was one of bolus doses as high as 10 to 15 mg for postoperative analgesia. As can be expected today, these bolus deliveries of high doses of epidural morphine resulted in not just intense but intolerable side effects such as refractory pruritus and nausea but also more serious complications such as profound sedation, respiratory depression, and death.[4]

Furthermore, it also was observed that identical doses of morphine and fentanyl, whether given epidurally or intravenously, produced similar serum concentration and elimination curves. However, the duration of analgesia with intravenous morphine was about 10 to 15 minutes, whereas the same doses of epidural morphine produced variable-quality analgesia with a duration of action ranging from 4 to 24 hours depending on the type of surgery, placement of the epidural catheter, and the patient's age and overall medical status.[5]

Identical doses of fentanyl, when given epidurally or intravenously, also produced similar serum concentration and elimination curves. However, by contrast, the rapid onset and short duration of analgesia with intravenous and epidural fentanyl were similar between epidural and intravenous administration.

Adverse Effects

Certain adverse effects of epidurally administered analgesics are often extensions of their therapeutic effects. For example, the neural blockade induced by local anesthetics can cause undesirable numbness and weakness, as well as hypotension and bradycardia due to chemical spinal sympathectomy.

Hydrophilic epidural opioids, especially morphine, when given as a bolus, can produce latent respiratory depression up to 8 to 12 hours after administration. More common side effects of epidural opioids are similar to the side effects observed during its systemic administration, such as pruritus, nausea, sedation, and urinary retention.

Current State of Epidural Analgesia in Acute Pain Management

Continued understanding of the behavior of epidural opioids, such as morphine and fentanyl, and epidural local anesthetics that preferentially block sensory fibers over motor fibers at low concentrations, such as bupivacaine, further changed clinical practice. In the 1980s and 1990s, *combination analgesia*, achieved by infusing mixtures of different classes of analgesics into the epidural space using a patient-controlled analgesia pump, gradually developed into the current standard of care. The goal was to target acute pain pathways at multiple points through the use of different analgesic agents. This approach consisted of multidrug infusions (e.g., local anesthetics, opioids, clonidine, and epinephrine) in low concentrations, thereby making best use of each agent's analgesic properties while minimizing each agent's dose so as to minimize the associated side effects.[6]

The intermittent and continuous dosing of epidural analgesics have been replaced by delivery of these agents by patient-controlled analgesic pumps.

"All Epidurals Are Not Created Equal"

When we critically analyze the literature for efficacy of epidural analgesia in the acute setting and compare and contrast the modalities' benefits and the disadvantages with other pain management options, the evidence clearly favors epidural analgesia over other methods. For example, a recent meta-analysis published by Block and colleagues[7] showed that epidural analgesia provided greater reduction in the VAS scores (9.40 versus 29.40 mm) and better pain relief at each postoperative day.

Although differences in individual studies and meta-analyses do exist, there is a growing body of literature to support the idea that segmental epidural blockade provides static and dynamic analgesia that can improve postoperative outcomes without increasing overall mortality. There is decreased incidence of cardiopulmonary complications through reduced incidence of cardiac ischemia, pneumonia, atelectasis, and hypoxemia. There is a reduced incidence of vascular graft occlusion in lower extremity revascularization surgery secondary to the vasodilatory effects of epidural local anesthetics. And furthermore, there is earlier return of gastrointestinal function because neural blockade diminishes sympathetic flow to the gut.[8–12]

Indeed, the use of epidural analgesia and intermittent dosing of the epidural space continues to this day and is more popular than ever. The modality has undergone a steady evolution that has improved its efficacy, safety, and reliability, but with important differences. The statement, "All epidurals are not created equal," rings true for this modality more than ever owing to its inherent technical complexity. The main parameters that the physician needs to optimize fall into the following categories:

1. The site of epidural catheter placement needs to be at the center of dermatomes that are the subject of tissue injury.
2. The options in the mode of delivery include intermittent dosing, continuous infusion, and

patient-controlled epidural analgesia (PCEA). The current standard of care is PCEA that combines a basal dose with a demand delivery that provides the best analgesia and side-effect profile. A PCEA pump allows patients to autoregulate their analgesic needs with least inconvenience to themselves and the health care staff while optimizing the safety risks inherent in infusing local anesthetics and opioids into the epidural space.

3. Optimal analgesia is achieved by combining analgesics with different mechanisms of actions, such as local anesthetic and opioids, that improves the quality of the analgesia while minimizing the incidence of hemodynamic effects, profound sensory blockade, motor blockade, and other opioid-related side effects.

In conclusion, combination analgesia achieved through infusion of opioids and local anesthetics into the epidural space, while technically demanding and resource-intensive, is one of the most effective methods of pain control. Epidural analgesia through a PCEA pump can be used for a variety of surgeries from cesarean sections to abdominal and thoracic surgeries to improve the quality of analgesia, patient satisfaction, and postoperative morbidity and mortality.

▶ INTRAVENOUS PATIENT-CONTROLLED ANALGESIA

Intravenous patient-controlled analgesia (IV PCA) is provided by a programmable electronic infusion pump attached to an intravenous line. The PCA device allows the patient to self-administer a preset dose of analgesic medication when the patient experiences pain. Because of advances in computer technology, the machine has a programmable lockout period after each dose of analgesic administration. During each lockout time, the device will not deliver another dose of medication even when activated by the patient. This lockout time interval provides a safety feature for the patient and greatly reduces the probability of analgesic overdose.[13]

IV PCA therapy provides better pain control than intermittent intramuscular opioid administration in patients undergoing a variety of surgical procedures. Patients usually require smaller opioids doses via the IV PCA route when compared with the as-needed intramuscular route. Thus, side effects or potential life-threatening complications, such as respiratory depression or oversedation, are less with use of IV PCA than those with patients receiving intramuscular opioid injections.[14] Patients who have undergone coronary artery bypass grafting and have received IV PCA experience less postoperative pulmonary atelectasis.[15,16] Patient satisfaction with pain control is also higher with IV PCA than with intramuscular opioid injections, presumably owing to more prompt pain control and a greater sense of control afforded by IV PCA.[17-19]

Patient selection for IV PCA therapy is based on the patient's ability to understand the concept of IV PCA therapy and appropriately interact with the device. IV PCA also has been used widely in the pediatric population. IV PCA provides effective pain relief and attenuates the hormonal stress response without life-threatening complications among pediatric patients undergoing major abdominal and genitourinary surgery.[18,20,21] "Nursing IV PCA" also can be an efficient and safe method in the treatment of the postoperative pain in infants and toddlers.[21]

IV PCA is useful for elderly patients, however, caution is required among those with respiratory, renal, or hepatic insufficiency. Appropriate patient selection is necessary to exclude those who may be incapable of using the device, such as those with evidence of cognitive dysfunction or physical disability.[22]

The most popular opioids delivered through the IV PCA pump are morphine, hydromorphone, and fentanyl.[21] Meperidine also has been used in IV PCA pumps in certain conditions. However, meperidine is associated with the development of a normeperidine-induced central nervous system toxicity.[23]

Among breast-feeding parturients after cesarean delivery, IV PCA with meperidine is associated with significantly more neonatal neurobehavioral depression than IV PCA with morphine. Therefore, morphine is the IV PCA opioid of choice for postcesarean analgesia among breast-feeding parturients.[24]

The data gathered during IV PCA use also can facilitate conversion to oral opioids during the postoperative period, as well as in chronic pain patients.[25] The use of a basal infusion with IV PCA is controversial. Several studies suggest that use of a basal infusion does not appear to improve the analgesia and actually may increase the incidence of side effects of opioids.[26-28]

Although IV PCA therapy is considered to be safe, the potential for serious adverse outcomes exists. Life-threatening respiratory depression associated with the use of IV PCA, with basal dose, has been reported. Human error is also a significant contributor to poor outcomes, such as when mistakes are made during PCA pump programming or analgesic preparation.

REFERENCES

1. JCAHO Pain Management Standards, effective 2001; published at *www.jcaho.org*.
2. Liu S, Carpenter RL, Neal JM: Epidural anesthesia and analgesia: Their role in postoperative outcome. *Anesthesiology* 1995;82:1474–1506.

3. Cleland JG: Continous peridural caudal analgesia in surgery and early ambulation. *Northwest Med* 1949;48:26.
4. Bromage PR, Camporesi E, Durant P, et al: Nonrespiratory side effects of epidural morphine. *Anesth Analg* 1982;61:490–495.
5. Bromage PR, Camporesi E, Chestnut D: Epidural narcotics for post-operative analgesia. *Anesth Analg* 1980;59:472–480.
6. Chestnut DH, Owen CL, Bates JN, et al: Continous infusion epidural analgesia during labor: A randomized, double-blind comparison of 0.0625% bupivacaine/0.0002% fentanyl versus 0.125% bupivacaine. *Anesthesiology* 1988;68:754–759.
7. Block BM, Liu SS, Rowlingson AJ, et al: Efficacy of postoperative epidural analgesia: A meta-analysis. *JAMA* 2003; 290:2455–2463.
8. Yeager MP, Glass DD, Neff RK, et al: Epidural anesthesia and analgesia in high-risk surgical patients. *Anesthesiology* 1987;66:729–736.
9. Tuman KJ, McCathy RJ, March RJ, et al: Effects of epidural anesthesia and analgesia on coagulation and outcome after major vascular surgery. *Anesth Analg* 1991;66:696–704.
10. Reiz S, Balfors E, Sorenson M, et al: Coronary hemodynamic effects of general anesthesia and surgery: Modification by epidural analgesia in patients with ischemic heart disease.
11. Shulman M, Sandler AN, Bradley JW, et al: Post-thoracotomy pain and pulmonary function following epidural and systemic morphine. *Anesthesiology* 1984;61:569–575.
12. Carpenter RL: Gastrointestinal benefits of regional anesthesia/analgesia. *Reg Anaesth Pain Med* 1996; 21:13–17.
13. Badner NH, Doyle JA, Smith MH, Herrick IA: Effect of varying intravenous patient-controlled analgesia dose and lockout interval while maintaining a constant hourly maximum dose. *J Clin Anesth* 1996;85:382–385.
14. Ashburn M, Smith K: The management of postoperative pain. *Surg Rounds* 1991;14:129–134.
15. Gust R, Pecher S, Gust A, et al: Effect of patient-controlled analgesia on pulmonary complications after coronary artery bypass grafting. *Crit Care Med*. 1999;27:2218–2223.
16. Petterson PH, Lindakog EA, Owall A: Patient-controlled versus nurse-controlled pain treatment after coronary artery bypass surgery. *Acta Anaesthesiol Scand* 2000; 44:43–47.
17. Snell CC, Fothergill-Bourbonais F, Durocher-Hendriks S: Patient-controlled analgesia and intramuscular injections: A comparison of patient's pain experiences and postoperative outcomes. *J Adv Nurs* 1997;25:681–690.
18. Shin D, Kim S, Kim CS, Kim HS: Postoperative pain management using intravenous patient-controlled analgesia for pediatric patients. *J Craniofac Surg* 2001;12:129–133.
19. Breme K, Altmeppen J, Taeger K: Patient-controlled analgesia: Psychological predictors of pain experience, analgesic consumption and satisfaction. *Schmerz* 2000;14:137–145.
20. Bozkurt P: The analgesic efficacy and neuroendocrine response in paediatric patients treated with two analgesic techniques using morphine-epidural and patient-controlled analgesia. *Paediatr Anaesth* 2002;12:248–254.
21. Aguirre Corcoles E, Durn Gonzalez ME, Zambudio GA, et al: Post-surgical paediatric pain: Nursing-PCA vs continuous IV infusion of tramadol. *Cir Pediatr* 2003;16:30–33.
22. Mann C, Pouzeratte Y, Eledjam JJ: Postoperative patient-controlled analgesia in the elderly: Risks and benefits of epidural versus intravenous administration. *Drugs Aging* 2003;20:337–345.
23. Simopoulous TT, Smith HS, Peeters-Asdourian C, Stevens DS: Use of meperidine in patient-controlled analgesia and the development of a normeperidine toxic reaction. *Arch Surg* 2002;137:84–88.
24. Wittels B, Glosten B, Faure EA, et al: Postcesaren analgesia with both epidural morphine and intravenous patient-controlled analgesia: neurobehavioral outcomes among nursing neonates. *Anesth Analg* 1997;85:600–606.
25. Ginsberg B, Sinatra RS, Adler LJ, et al: Conversion to oral controlled-release oxycodone from intravenous opioid analgesic in the postoperative setting. *Pain Med* 2003;4:31–38.
26. Smythe MA, MB Zak, O'Donnell MP, et al: Patient-controlled analgesia versus patient-controlled analgesia plus continuous infusion after hip replacement surgery. *Ann Pharmcother* 1996;30:224–227.
27. Peters JW, Bandell Hoekstra IE, Haijer Abu-saad H, et al: Patient-controlled analgesia in children and adolescents: A randomized, controlled trial. *Paediatr Anaesth* 1999; 9:235–241.
28. Dal D, Kanbak M, Calgar M, Aypar U: A background infusion of morphine does not enhance postoperative analgesia after cardiac surgery. *Can J Anesth* 2003;50:476–479.

CHAPTER 37

Psychiatric, Somatic, and Behavioral/Psychological Comorbidities in Chronic Pain Patients: Diagnostic and Treatment Approaches

David A. Fishbain

Comorbidity *is defined as "any distinct clinical entity that has existed or that may occur during the patient's clinical course that has the index disease under study."*[1] *The presence of comorbid disease can affect the clinical course of the index disease dramatically by complicating, interfering with, or making treatment of the index disease more difficult, and its prognosis worse.*[2] *Thus, because of its impact on treatment of the index disease, comorbid disease can lead to spurious medical outcomes, especially if the comorbid disease is not classified, not analyzed, and its effect not controlled for.*[2]

Psychiatric comorbidity diagnosable according to DSM-IV[3] *is frequently found in chronic pain patients (CPPs).*[4] *For example, in an early study, Fishbain and colleagues*[4] *reported that in a large group of CPPs treated at a pain facility, only 5.2 percent had no DSM-IV diagnosis on Axis I (where diagnoses such as depression, anxiety, and so on are to be coded). In all, 34.9 percent had one diagnosis on Axis I, and 59.9 percent had more than one diagnosis.*[4] *Therefore, the vast majority of CPPs had some diagnosable (by DSM-IV criteria) psychiatric* comorbidity *on Axis I. In addition, CPPs are prone to develop some psychological issues (comorbidities) that are not diagnosable by DSM-IV criteria but often interfere with treatment outcome. These psychological comorbidities develop because CPPs consider themselves to suffer from a physical illness for which physicians cannot seem to develop a cure.*[5] *This physical illness is associated with significant impairment and disability that has tremendous impact on CPPs' lives.*[5] *In addition, CPPs often are placed on psychoactive substances that have depen-dence and addiction potential.*[6] *Finally, because often no apparent tissue damage can be found to explain the cause of the chronic benign nonmalignant pain, physicians frequently attribute the CPPs' pain to underlying psychiatric illness or to faking.*[6] *This last issue often leads to conflicts with the medical and insurance systems and to litigation.*[6] *This issue, in turn, leads to increasing environmental stress, which, in turn, leads to either psychiatric or psychological comorbidity. Finally, most CPPs have what can be characterized as* somatic comorbidities,[7] *such as headaches, sleep problems, and so on. These somatic comorbidities interfere with treatment and, therefore, have an impact on treatment outcome.*

This chapter will (1) outline the most common psychiatric, psychological, and somatic comorbidities encountered within CPPs, (2) delineate diagnostic and measurement approaches for these comorbidities, and (3) describe specific treatment approaches for each of these comorbidities.

▶ PSYCHIATRIC COMORBIDITIES DIAGNOSABLE BY DSM-IV CRITERIA

Affective Disorders

Of the psychiatric comorbidities presented in Table 37-1, *affective disorders* (mood disorders, depression) are the most common group of psychiatric comorbid disorders found within CPPs. The prevalence for depression has been reported variously to be from 10 to 100 percent[8,9] and for major depression from 1.5 to 54 percent.[9] In a recent review, Fishbain and colleagues[9] concluded that depression is found more commonly in CPPs than in non-CPPs, and it may be more common in some subtypes of CPPs versus others, e.g., those with chronic low back pain versus those with cancer-associated pain. In addition, Fishbain and colleagues[9] also concluded that that the evidence indicated that depression found in CPPs was, in most cases, the consequence of the pain and resulting disability and not the reason for the pain. In addition, the evidence indicated that vulnerability to depression could predispose to the development of the depression after pain onset.[9]

As noted earlier, the prevalence of depression in CPPs reported by various authors falls into a wide range. At issue is, why is this so? These discrepancies between authors relate to the following issues:[9] (1) There are differences in pain center CPPi selection criteria;[4] (2) a number of the criteria in the DSM-IV diagnosis of major depression relate to somatic problems, e.g., sleep, weight, etc., that are significantly affected by the pain itself. Because of lack of operationally-specified instructions for determining whether the somatic criteria should or should not be counted in determining if the CPP meets the diagnosis of major depression, it is likely that a number of authors have either under- or overcounted CPPs with major depression; and (3) it is clear that pain affects mood. As such, this could lead to actual over- or underestimates as chronic pain fluctuates, thereby leading to interpretation of the mood state according to the current pain level.

Approaches to the diagnosis and treatment of depression in CPPs are presented in Table 37-1. Although major depression can be diagnosed only by DSM-IV criteria, these criteria are not operationalized to measure improvements in treatment and, therefore, treatment outcome. Improvements in treatment outcome are best measured by depression rating scales such as the Beck Depression Inventory[10,11] and the Hamilton Depression Scale.[10,11] The Beck is a patient-completed inventory, whereas the Hamilton is a clinician-rated tool. Of the two, the Beck has been used in research more extensively in CPPs. Its advantage is that it has a significant proportion of cognitive symptoms of depression[45] and, therefore, does not give greater weight to somatic symptoms. Its use is recommended at present to measure treatment effects.

The adequate treatment of depression is extremely important in CPPs. This is so because it appears that depression predicts disability in CPPs.[46] Conversely, the treatment of pain in depressed CPPs is as important or more important because recent evidence indicates that the greater the severity of bodily pain, the less likely is the depression to remit.[47] Thus, any treatment plan for depression should include adequate treatment of pain, as shown in Algorithm A. In reference to specific antidepressant (AD) treatment, there is significant evidence that the dual-action ADs (nonadrenergic/serotonergic) may have greater efficacy for pain than the nonadrenergic ADs, which, in turn, have greater efficacy than the selective serotonin reuptake inhibitor ADs (serotonergic).[48,49] Thus Algorithm A includes this concept.

Anxiety is very frequently comorbidly associated with pain/depression (discussed below). Most CPPs suffer from sleep abnormalities and a few CPPs have prominent addictive problems. All these have been included in Algorithm A.

Suicidal ideation and suicide attempts are not unusual in the chronic pain population. Because of the frequency of affective disorders and other suicide risk factors[50] in this population, this finding is not unexpected. In addition, the presence of chronic pain may be a suicide risk factor because suicide rates for CPPs may be higher than for non-CPPs.[51] Fishbain[50] has reviewed this literature recently and concluded that besides pain itself, other variables such as longer pain duration, higher pain intensity, neuropathic pain, worker compensation status, significant functional impairment secondary to pain, and chronic abdominal pain may be suicide risk factors. CPPs demonstrating suicidal ideation should be handled as suggested in Table 37-1. Any CPPs complaining of severe pain and depression should be evaluated psychiatrically for suicidal ideation.

▶ TABLE 37-1. PSYCHIATRIC COMORBIDITIES: MEASUREMENT AND TREATMENT APPROACHES

Psychiatric Comorbidities	Measurement Approach	Treatment Approach
1. Affective disorders (major depression, dysthymia, adjustment disorder with depressed mood).	1. Psychiatric examination for DSM-IV[3] criteria for major depression or other forms of depression. 2. Beck Depression Inventory[10,11] (patient rated). 3. Hamilton Depression Scale[10,11] (clinician rated). 4. Carroll Depression Scale[11] (patient rated).	1. Refer to Treatment of Depression Algorithm. 2. Cognitive therapy for depression.
2. Suicide Ideation attempt.	1. Psychiatric examination for DSM-IV[3] diagnosis and need for psychiatric hospitalization. 2. Beck Scale for Suicide Ideation.[10,11]	1. Determine if psychiatric hospitalization is required. 2. If psychiatric hospitalization not required, treat under Algorithm A.
3. Anxiety syndromes (panic disorder, generalized anxiety disorder, adjustment disorder with anxious mood, agoraphobia, obsessive-compulsive disorder, post-traumatic stress disorder, social phobia, obsessive-compulsive spectrum disorder).	1. Psychiatric examination for DSM-IV criteria[3] for anxiety syndromes. 2. State-Trait Anxiety Inventory[10,11] (self-rated) for generalized anxiety disorder. 3. Post-traumatic Chronic Pain Test.[31] 4. Padau Inventory[10,11] for obsessive-compulsive disorder (self-rated).	Refer to Treatment of Anxiety Algorithm B.
4. Pychoactive substance use-related disorders (e.g., opioid dependence/abuse, alcohol dependence/abuse, etc.).	1. History: Cigarette smoking, blackouts, frequent accidents/falls, driver license problems. 2. Physical finding associated with alcohol abuse such as hepatomegaly, etc. 3. Physical findings associated with other types of drug abuse such as eroded nasal septum, cigarette burns, unexplained bruising, track marks, etc. 4. Laboratory findings suggestive of drug abuse: elevated liver function tests, positive toxicology studies (urine or blood) for illicit drugs. 5. Psychiatric examination for DSM diagnosis. 6. Cage Questionnaire[10,11] for alcohol abuse (self-report). 7. Michigan Alcohol Screening Test (MAST).[10,11] 8. Fagerstrom Test for Nicotine Dependence (FTND) (self-report). 9. Drug Abuse Screening Test (DAST).[32] 10. Aberrant drug-related behaviors [33] (Table II) suggestive of addiction.	Refer to Treatment of Addiction Algorithm C.
5. Somatoform disorders A) Pain disorder	Psychiatric examination for DSM-IV diagnosis of pain disorder.	No specific treatment for pain disorder. Treat associated psychiatric comorbidities with emphasis on improving pain.

(Continued)

TABLE 37-1. (CONTINUED)

Psychiatric Comorbidities	Measurement Approach	Treatment Approach

B) Conversion disorder (paralysis, nondermatomal sensory abnormalities, gait abnormalities).	Psychiatric examination for DSM-IV diagnosis of conversion disorder.	Specific treatments utilized:
		A. *Paralysis*[34,35]
		   Physical therapy[34]
		   Suggestion hypnosis.[36]
		   Narcosuggestion [34]
		   Functionally-based rehab program
		    — EMG biofeedback[37]
		    — Functional electrical stimulation[38]
		    — Combination treatment[39]
		    — Methamphetamine[40]
		B. *Nondermatomal sensory abnormalities*
		   Blocks.[41,42]
		C. *Gait abnormalities*
		   Behavior modification[43]
		   Psychotherapy.[44]

6. Intermittent Explosive Disorder and irritability/anger/threatened violence/violence.	1. Psychiatric examination for history of irritability/anger/threats of violence/violence/loss of control, and for the diagnosis of impulse control disorder or other diagnoses associated with violence.[61] 2. State-Trait Anger Expression Inventory (STAX1-2).[10,11]	Refer to Treatment of Potential Violence Algorithm D.
7. Secondary gains and malingering.	1. Psychiatric interview for potential secondary gains. 2. The presence of secondary gains is a requirement for the diagnosis of malingering. However, having a secondary gain does not a malingerer make. There is no way to diagnose malingering except by covert observation.	No treatment approaches exist for malingering as it is rarely diagnosed.
8. Personality disorders.	1. Structured Clinical Interview for DSM-IV Axis II Personality Disorders (SCID-II). 2. Personality Diagnostic Questionnaire for DSM-IV (self-report).	Treatment of Personality Disorders is complex and time consuming. There are no specific psychopharmacological approaches. As such, treatment should only be attempted by psychiatrists and psychologists for this problem.

Psychoactive Substance Use-Related Disorders

The second most common psychiatric comorbidity found in CPPs is that of *psychoactive substance use–related disorders*. Measurement and treatment approaches for this comorbidity are presented in Table 37-1. Fishbain and colleagues[52] reviewed this research area and concluded that the prevalence percentages for the diagnosis of drug abuse, dependence, or addiction in CPPs were in the range of 3.2 to 18.9 percent. A significant percentage of CPPs, 6.4 to 12.5 percent, was reported to abuse illicit drugs (marijuana and cocaine).[52] There was, however, little evidence that addictive behaviors were common.[52] The treatment of this comorbidity in the context of chronic pain revolves around whether that CPP does or does not have an addiction problem. If a particular CPP does have an addiction problem, then the approach to that patient should be as outlined in

Algorithm C. It is to be noted here that if detoxification is required, these patients would do better in multidisciplinary pain facilities than in addiction facilities.[52] This is so because pain can be treated at pain facilities while detoxification is progressing. This is not the case in addiction facilities.

Somatoform Disorders

The third most common psychiatric comorbidity to be found in CPPs is that of *somatoform disorders*. Here both conversion disorder and psychogenic/pain disorder are reported to have high prevalence rates. There are major discrepancies, however, between authors on the prevalences of these two disorders. These discrepancies have been commented on previously and relate to reliability problems with DSM-III criteria for both these diagnoses.[53] Both these diagnoses contain or have contained criteria that require the examiner to make a value judgment about a symptom. Because of these problems, the DSM criteria for psychogenic pain were changed in the DSM-III-R and further changed in the DSM-IV.[53] It is likely, therefore, that the data for the prevalence of these disorders are unreliable. The approach for the diagnosis/measurement and treatment of these two disorders is presented in Table 37-1. It is to be noted that at present there is no definitive psychopharmacologic or other treatment for these two disorders. Since these disorders are often associated with other psychiatric comorbidities, treatment should be directed at these comorbidities in an aggressive fashion in the hope of improving the status of the somatoform disorder.

Anxiety Disorders

The fourth most common group of psychiatric comorbidities to be found in CPPs is that of *anxiety disorders*. These disorders have been reported[54] to occur in the following frequencies: panic disorder, 11 percent; agoraphobia with panic attacks, 2.1 percent; generalized anxiety disorder, 15 to 20 percent; obsessive-compulsive disorder, 1.1 percent; post-traumatic stress disorder, 1.1 percent; adjustment disorder with anxious mood, 42.8 percent; and phobic disorder, 9 percent. Although, as demonstrated earlier, anxiety is common in CPPs, evidence indicates that these reported prevalences are underestimates. The evidence for this statement comes from studies performed on non-CPPs. Katon[55] has reported that 81 percent of panic disorder patients had pain as a presenting complaint. Gilcrist[56] has reported that in general practices low back pain patients are more likely to have a diagnosis of anxiety than non–low back pain patients. Finally, CPPs, when compared with non-CPPs, are more likely to demonstrate avoidance of particular situations [e.g., injections and minor surgery, hospitals, sight of blood, thoughts of injury and illness (blood or injury phobia), being watched or stared at, and speaking or acting to an audience (social phobia)].[57] It is likely, therefore, that anxiety syndromes are comorbidly associated with chronic pain at a greater frequency than reported.

In addition to the preceding diagnoses, CPPs may suffer from subclinical anxiety syndromes, such as subclinical obsessive-compulsive disorder. This now has been termed *obsessive-compulsive spectrum disorder* (OCSD).[58] A number of psychiatric disorders, such as body dysmorphic disorder, anorexia nervosa, binge eating, hypochondriasis, sexual compulsions, pyromania, kleptomania, trichotillomania, compulsive buying, pathologic gambling, and some self-injurious behaviors, appear to demonstrate some obsessive traits and, therefore, are included in OCSD. The OCSD patients use the types of mechanisms and behaviors often noted in OCD. As such, these mechanisms and behaviors make the index disorder, e.g., gambling, worse and more difficult to treat. At issue, then, is whether these types of mechanisms operate in some other somatizing disorder such as pain disorder. The features of OCSD and OCD overlap in many respects, including demographics, repetitive intrusive thoughts or behaviors, comorbidity, and etiology. Most important, it appears that this group of disorders responds preferentially to antiobsessional drugs such as clomipramine and the selective serotonin reuptake inhibitors (SSRIs), e.g., fluvoxamine.[58] There have been no treatment studies using the OCSD concept for pain disorders. However, a number of cases have been published. Fishbain and colleagues[59] have reported on the positive response to antiobsessional agents in three cases of chronic atypical facial pain. Recently, there also has been a report of positive response in two patients with schizophrenia and chronic pain to clomipramine.[60] This limited evidence indicates that perhaps CPPs with "nonspecific" pain and prominent obsessive-compulsive components should be treated with SSRIs at the first opportunity. This concept is then incorporated into the treatment of anxiety in Algorithm B. It is to be noted that Algorithm B also addresses the question of anxiety of such intensity that antianxiety agents are required. These agents do have a place in the treatment of the CPP.

Intermittent explosive disorder (IED) has been reported in up to 9.9 percent of CPPs.[54] This is a disorder characterized by intermittent loss of control episodes that can be associated with violent behavior. Anger/irritability also has been noted to be found frequently in CPPs.[62] Associated psychiatric comorbidities and pain often can increase the levels of anger/irritability.[61] Threats of violence against pain professionals also have been reported, as well as incidents of murder.[61] Thus, threats of violence should be taken seriously, and referral to a psychiatrist and/or psychologist should be made immediately.[61] The treatment of potential violence/anger/irritability is presented in Algorithm D. It is to be noted that uncontrolled pain, other psychiatric comorbidities as contributors to potential violence/anger/irritability, and behavioral situations leading to anger are included as issues in this algorithm.

Secondary gain has been defined as "acceptable or legitimate" interpersonal advantage that results[63] when a patient avoids an activity that is noxious to him or her, or receives support from the environment that is not otherwise forthcoming.[63,64] Examples of these situations include the following: fulfillment of dependency needs, prevention of desertion by the spouse, desire to get away from a bad job situation, desire to be retired before the injury, increased attention, avoidance of coercive activities (e.g., sexual intercourse), increased self-esteem, increased attachment needs, aggressive gratification (revenge), dissatisfaction with impairment rating before settlement (patient believes that he or she should have a 100 percent impairment rating), dissatisfaction with settlement terms, desire for retraining or to go back to school (patient had wanted to increase educational status before the injury), and decrease in home/family responsibilities. Secondary gains occur by unconscious mechanisms.[63] It is, therefore, not clear whether desire for financial compensation should be classified along with the other secondary gains.

In addition to secondary gains, CPPs often demonstrate the following problems:[7,54] (1) The organic diagnosis is often elusive or nonexistent; (2) there is excessive disability versus the medical impairment; (3) nonorganic physical findings are present; (4) CPPs usually are financially dependent on compensation and disability benefits; and (5) CPPs usually are involved in litigation.[63,64] This combination of problems opens CPPs for the accusations that they are malingering their complaints.[65] Malingering is not classified as a mental disorder.[65] The DSM-III-R states that malingering is an "act" and classifies malingering under the V codes that are included for a supplementary classification of factors influencing health status and contact with health services.[65] The DSM-III-R suggests that the possibility of malingering should not be entertained unless there is evidence that physical symptoms are intentionally produced.[65] Rarely is there strong evidence or any evidence for this conclusion. The Institute of Medicine report on pain and disability[65] also has concluded that malingering is rare. The presence of either primary gain or secondary gain or even monetary gain is not evidence of malingering.[63] Although malingering is discussed often in reference to CPPs, there are no or few studies of this problem within the CPP literature.[65] Physicians should not consider this label unless they have definitive proof, such as that obtained by covert observation.

Personality disorders that are diagnosed in Axis II of the DSM-IV are also found frequently in CPPs. Here, studies indicate that the prevalence of personality disorders in CPPs may range from 31 to 59 percent.[54] These rates may be high because of a number of problems pointed out by Fishbain and colleagues.[54] This area

▶ **TABLE 37-2.** SPECTRUM OF ABERRANT DRUG-RELATED BEHAVIORS THAT RAISE CONCERN ABOUT THE POTENTIAL FOR ADDICTION

More Suggestive of Addiction

Selling prescription drugs
Prescription forgery
Stealing of drugs from others
Injecting oral formulations
Obtaining prescription drugs from nonmedical sources
Concurrent abuse of alcohol or illicit drugs
Repeated dose escalation or similar noncompliance despite multiple warnings
Repeated visits to other clinicians or emergency rooms without informing prescriber
Drug-related deterioration in function at work, in the family, or socially.
Repeated resistance to changes in therapy despite evidence of adverse drug effects

Less Suggestive of Addiction

Aggressive complaining
Drug hoarding during periods of reduced symptoms as patient needs more drugs when symptoms increase
Requesting specific drugs.
Openly acquiring similar drugs from other medical sources
Occasional unsanctioned dose escalation or other noncompliance
Unapproved use of the drug to treat another symptom
Reporting psychic effects not intended by the clinician
Resistance to a change in therapy associated with tolerable adverse effects
Intense expressions of anxiety about recurrent symptom

SOURCE: Abstracted from Portenoy.[33]

requires further study. In reference to the types of personality disorders comorbidly associated with chronic pain, no clear, consistent trends are evident. Authors are in conflict as to which personality disorders are found most commonly in CPPs. Personality disorders are important because they often interact with the DSM-IV Axis I diagnoses, e.g., depression, thereby increasing the difficulty of treatment and affecting treatment outcome adversely.[54] It is also possible that the presence of this diagnosis could make pain treatment more difficult. As such, CPPs with this diagnosis should be identified, if possible. Unfortunately, psychiatric treatment for these diagnoses is time-consuming (see Table 37-1). Moreover, the only benefit to the pain clinician from identification of such CPPs would be the partial understanding of why treatment is difficult or perhaps unsuccessful.

▶ SOMATIC COMORBIDITIES

In addition to the psychiatric comorbidities diagnosable by DSM-IV criteria, the CPP also usually demonstrates a great

TABLE 37-3. SOMATIC COMORBIDITIES: MEASUREMENT AND TREATMENT APPROACHES

Somatic Comorbidity	Diagnostic Approach	Treatment Approach
1. Fatigue	History Multidisciplinary Fatigue Intentory[18]	1. Modafinil 2. Multidisciplinary pain facility[19]
2. Sleep	History Pittsburgh Sleep Quality Index[20]	Refer to Algorithm A
3. Headache	History Physical examination	?Migraine, Abortive and preventative Rx Pain facility treatment
4. Memory/concentration	History Psychiatric examination Formal cognitive testing	Modafinil
5. Sexual dysfunction	History KDS-15 Questionnaire[21]	Urology referral Sildenafil
6. Nonorganic physical findings	Physical examination Somatic Amplification Rating Scale[22]	Treat all psychiatric comorbidities plus pain
7. Greater disability versus medical impairment	Functional status history Sickness Impact Profile[23] Functional Assessment Screening Questionnaire[24] Pain and Impairment Relationship Scale[25] Pain Disability Index[26]	Multidisciplinary pain facility
8. Somatization	History Symptom Checklist 90-R[27]	Treat symptomatically
9. Pain behavior	Physical examination UAB Pain Behavior Scale[28]	Multidisciplinary pain facility
10. Irritable bowel symptoms	History Physical examination Fibro Fatigue Scale[30]	Treat symptomatically
11. Autonomic disturbance, such as dizziness, increased sweating, nausea	History Physical examination Fibro Fatigue Scale[30]	Treat symptomatically

number of somatic comorbidities (Table 37-3). These are problems that are not psychiatric but are essentially somatic complaints. However, in the vast majority of CPPs, a treatable disorder usually cannot be found as an explanation for the somatic complaint. Yet these somatic complaints increase the difficulty of treating the CPPs if they interfere with treatment of the main complaint (pain) and other psychiatric comorbidities. Thus, it is important to understand the following: (1) Somatic comorbidities in some CPPs appear to be part of the chronic pain syndrome, whether through predisposition or for other reasons; (2) these comorbidities interfere with treatment of the pain complaint; and (3) the main complaint cannot be treated successfully unless these complaints are at least nominally addressed in a therapeutic fashion. Some of these somatic complaints will be discussed below.

Fatigue

Fatigue is found commonly in CPPs.[19] Recently, Fishbain and colleagues[19] have demonstrated that fatigue is associated with pain and that pain may be etiologically related to fatigue.[19] The mechanism by which pain would be presumed to cause fatigue is unknown.[19] Fatigue may respond to multidisciplinary pain facility treatment.[29] There are no reported psychopharmacologic fatigue treatment trials in CPPs. Modafinil, however, may be a treatment option.

Sleep Problems

Sleep problems are a universal CPP problem. The vast majority of CPPs complain of severe sleep disturbance.[66] The sleep problems are usually a result of pain and may be associated directly with pain intensity.[66] Although the CPP usually can fall asleep, he or she is aroused frequently from sleep by pain.[66] Thus, the sleep is fragmented and of poor quality.[66] The sleep disturbances are not a function of depressed mood.[66] Sleep disturbance is a comorbidity that can interfere significantly with treatment of the CPP. As such, it should be treated aggressively as a target symptom. However, since these sleep disturbances likely are related to pain, the pain also should be treated aggressively. Success in treating pain should have an impact on the sleep disturbance. In treating sleep, the sedating antidepressants should be used as the primary modality. They should be titrated to maximum dose or maximum tolerable dose or to the dose that is effective to initiate 6 to 8 hours of sleep. These concepts are used in treatment Algorithm A.

Headache

Headaches are a significant somatic comorbidity found routinely in CPPs suffering from chronic low back pain and/or neck pain. Fishbain and colleagues[67] reported that severe headaches interfering with function were found in 10.5 percent of consecutive CPPs. These headaches were classified as usually migraine in nature or cervicogenic, although there was significant diagnostic overlap.[67] What was most interesting was that most CPPs with headache, whether migraine or otherwise, had neck-associated symptoms, i.e., neck pain, tender point at the back of the neck, etc. As such, the usual treatment approach to these headaches at our pain facility is two-pronged: (1) treatment of the migraine aspect of the headache, usually with migraine agents; and (2) treatment of the neck aspect of the headache, usually with physical therapy. A recent study[68] has demonstrated that pain facility treatment could be effective for migraine patients.

Concentration and Memory Problems

Concentration and memory problems are another class of somatic comorbidities often encountered in CPPs.[69] These are seen more often in CPPs complaining of neck pain and/or headaches. Impairments are most evident on tests assessing attention, capacity, processing/psychomotor speed, and memory.[69] It is not yet clear whether pain itself interferes with cognitive function or if psychological distress independent of pain affects cognitive function. At present, we do not understand the etiology of this comorbidity well. In addition, because of this, there have been no psychopharmacologic treatment trials specifically directed at this comorbidity. As indicated in Table 37-3, the psychostimulants, such as modafinil, can be an option in these treatment trials.

Sexual Dysfunction

Sexual dysfunction is also a frequent complaint of CPPs.[70,71] It has been claimed that up to 63 percent of men complaining of low back pain were found to be impotent.[71] However, according to the author's experience, very few CPPs actually are truly impotent. The usual complaint is that sexual intercourse is painful, and therefore, the male CPP is unable to maintain an erection. This serves as a disincentive to further sexual intercourse, which, in turn, results in decreased interest in sex. The decreased interest in sex then can lead to marital difficulties. Evidence indicates that there is a significant amount of dissatisfaction with the ability to carry out sexual activity because of pain. Seventy-three percent of CPPs in one survey reported pain-related difficulties with sexual activity.[70] Complaints of sexual dysfunction should be investigated, and CPPs whose symptoms actually indicate a possibility of impotence should be referred to a urologist. Again, as with other CPP comorbidities, we do not understand the etiology of this comorbidity well. As such, there have been no psychopharmacologic or other types of treatment trials for this comorbidity. Sildenafil could be an option for male CPPs complaining of impotence or an inability to maintain erection.

Nonorganic Physical Findings

Another somatic comorbidity often demonstrated in CPPs is that of *nonorganic physical findings*. These are eight physical examination signs found in CPPs with lower back pain or neck pain that have been identified as predominately nonorganic.[72] Waddell and colleagues[72] have demonstrated that a large percentage of patients with chronic low back pain demonstrate these eight nonorganic physical findings. These signs appear to be predictive of treatment outcome,[72] and Waddell and colleagues[72] have suggested that the presence or absence of these signs should be used as a basis for surgical decisions. Most important, these signs correlate with the degree of pain behaviors and, therefore, with a more difficult treatment problem.[72] In addition, the presence of one or more of these signs suggests that a patient is a candidate for the psychiatric diagnosis of conversion disorder, a somatoform disorder. Main and Waddell have stated recently that these signs are not a test of credibility or faking but represent a "psychological yellow flag" indicating that psychological factors need to be considered.[72] Fishbain and colleagues[72] have reviewed this literature recently through an evidence-based structured review process. According to the reviewed evidence, they have concluded the following: (1) Waddell signs do not correlate with psychological distress; (2) Waddell signs do not discriminate organic from nonorganic problems; (3) Waddell signs may represent organic phenomena; (4) Waddell signs are associated with poor treatment outcome; (5) Waddell signs are associated with increased pain levels; and (6) Waddell signs are not associated with secondary gain. Currently, there is no specific treatment for these signs. They appear to improve if pain improves. In addition, because they are associated with significant psychiatric comorbidity, that comorbidity should be targeted aggressively for treatment, as well as the pain.

Because of a number of complex problems outlined previously,[6] CPPs often demonstrate discordance between what the physician thinks they should be able to do, based on the medical impairment, and what they actually do, i.e., their functional status. This then leads to the

CPP being labeled as having secondary gain (discussed earlier) or being a malingerer (discussed earlier). This, however, is not a correct perception because this discordance is the norm for CPPs. If such a discordance is noted, these CPPs should be referred for treatment in a multidisciplinary facility[7] because this problem has been noted to be one of the inclusion criteria for potential referral.

Somatization

Another somatic comorbidity found within CPPs with great regularity is that of *somatization*. This is defined as "a tendency to experience and communicate somatic distress and symptoms that are unaccounted for by pathologic findings and to attribute them to physical illness and to seek medical help for them."[73] The presence of somatization does not necessarily imply that a psychiatric disorder is present because it does not represent either a psychiatric diagnosis or medical diagnosis. As such, this concept does not have any operational criteria and historically has been investigated by a questionnaire or inventories. The questionnaire studies have demonstrated the following: (1) CPPs demonstrate elevated somatization scores (Modified Somatic Perceptions Inventory); (2) CPPs demonstrate elevated hypochondriasis scores (Illness Behavior Questionnaire); (3) CPPs have greater somatization scores versus controls; and (4) somatization may be predictive of pain treatment outcome. However, it is not clear whether these questionnaires are measuring the concept of somatization or simply the comorbidities usually found in CPPs. As such, this comorbidity requires further exploration. There is currently no specific treatment for this group of comorbidities. Treatment, therefore, should be symptomatic.

Pain Behaviors

Another common somatic comorbidity found within CPPs is that of *pain behaviors*. These have been defined as "any and all outputs of the patient that a reasonable observer would characterize as suggestive of pain, such as (but not limited to) posture, facial expression, verbalizing, pain expressions, taking medications, seeking assistance and receiving compensation."[74] Pain behaviors usually are visible on observation or can be elicited by physical examination, which presumably increases pain and thereby generates the pain behavior. Current research on pain behavior can be summarized as follows: (1) The presence of pain behaviors correlates with perceived pain severity, longer pain history, numbers of previous surgeries, and extent of functional impairment; and (2) the presence of significant pain behaviors makes the CPP a more difficult treatment problem with poor outcomes likely,[3] and as pain improves, pain behaviors improve. Pain behaviors were thought originally (when pain facilities started) to be a behavior determined by operant conditioning. Thus, it was thought that pain facilities should modify this behavior. This idea has now been abandoned because it has become clear that such modification is not necessary for improvement. However, significant pain behavior has been identified as an inclusion criterion for possible admission to a multidisciplinary pain facility.[7]

Irritable Bowel Syndrome (IBS) and Autonomic Disturbances

Two final somatic comorbidities also found frequently within CPPs are that of irritable bowel symptoms or a frank diagnosis of *irritable bowel syndrome* (IBS) and *autonomic disturbances* such as dizziness, sweating, nausea, and low blood pressure. IBS has been noted to be found frequently in fibromyalgia, and as such, it would be expected that such patients would be overrepresented in pain facilities. In addition, autonomic disturbances also are encountered frequently in fibromyalgic patients. Thus, patients with these symptoms also should be overrepresented in pain facilities. There is no current information as to why these comorbidities are concentrated in CPPs. It is a clinical impression that autonomic disturbances are seen often in CPPs with greater pain and occur at times of pain elevation. In addition, this group of comorbidities is encountered frequently in CPPs with headaches/neck pain. Finally, the concentration of these comorbidities within CPPs would explain, in part, the finding of increased prevalence of somatic symptoms or somatization as compared with control groups. At present, there is no information on specific treatment approaches for these comorbidities. As such, symptomatic treatment is recommended, if available.

▶ BEHAVIORAL/PSYCHOLOGICAL COMORBIDITIES

In addition to the psychiatric comorbidities diagnosable by the DSM-IV and somatic comorbidities, the CPP will demonstrate a significant number of behavioral/psychological comorbidities. These are summarized in Table 37-4, along with measurement and treatment approaches. These comorbidities also increase the degree of difficulty in treating the CPP because many of these can accentuate the psychiatric comorbidities or interfere with treatment of the main complaint (pain). As such, these comorbidities often have to be addressed in the course of pain treatment. Some of these comorbidities will be discussed below.

Because the etiology of the pain is often elusive, the CPP may begin a search for a diagnosis and thus a cure. Failure to find a cause or a cure can lead to dissatisfaction/anger with the medical system and/or physicians[7] (see Table 37-4, items 1, 2, 3, 6, and 10). This can lead to fantasies of violence against the medical system or physicians. Both these problems should be

▶ **TABLE 37-4.** BEHAVIORAL/PSYCHOLOGICAL COMORBIDITIES: MEASUREMENT AND TREATMENT APPROACHES

Behavioral/Psychological Comorbidities	Measurement Approach	Treatment
1. Organic diagnosis elusive 2. Search for cure 3. Hostility/anger at MD secondary to failure to diagnose or to cure, or involvement in patient's legal situation 4. Anger or grief over employer negligence in injury or for forced job loss 5. Financial issues causing stress 6. Dissatisfaction with medical care either from failure to diagnose/cure or from inability to get perceived necessary medical care (usually as a result of nonauthorization of case by carrier) 7. Anger at compensation/insurance carrier for #6 above or for other issues, such as late payment of benefits, etc. 8. Litigation stress such as dissatisfaction with lawyer, etc. 9. Spousal problems/stress such as perceived nonsupport or blaming or not believing in patient's pain 10. Confusion over conflicting diagnoses or recommendations.	For 1-10 and 15–17: Psychiatric or psychological interview	Counseling for: 1, 2, 3, 4, 5, 6, 7, 8, 9, 10, 11, 12, 13, 14, 15, 16, 17. Specific Counseling Target for: 3, 4, 7. Anger control with evaluation for dangerousness, if necessary Specific medication regimen 3, 4, and 7 targeted for anger control 11. Coping strategies 12. Spouse to decrease over-solicitous behavior 13. Cognitive counseling to decrease fear of pain 14. Cognitive counseling to increase self-efficacy 15. Job stress and perhaps job change
11. Poor coping strategies	Modified Coping Strategies Questionnaire[12]	
12. Spouse over-solicitousness	West-Haven-Yale Multi-Dimensional Pain Inventory[13]	
13. Fear of pain	Pain Anxiety Symptom Scale[14] Tampa Scale for Kinesophobia[15]	
14. Poor self-efficacy	Arthritis Self Efficacy Scale[16]	
15. Pre-injury job stress	Fear Avoidance Beliefs Questionnaire (FABQ)[17]	
16. Childhood victimization	Psychiatric or Psychological Interview	
17. Unrealistic expectations about treatment	Psychiatric or Psychological Interview	

addressed in counseling. Counseling should be oriented toward making the CPP understand why the diagnosis is "elusive" and that he or she may have a chronic problem that he or she may need to learn to control. In addition, the CPP should be made aware of the fact that his or her failure to get well is not his or her fault but is due to a lack of medical knowledge about his or her condition. Such an intervention may limit future searches for cure and decrease anger at the medical system and physicians.

Spouse or significant other problems (see Table 37-4, items 9 and 12) as a consequence of the development of chronic pain have been noted frequently.[4] Pawlicki[75] has pointed out that the family is prone to two psychological tensions because of CPPs' pain, lack of improvement and the diagnosis being elusive: (1) uncertainty about the diagnosis, leading to family anxiety, and (2) uncertainty about the authenticity of the pain. Spouse or family perceptions of nonauthenticity can lead to the perception by CPPs of marital nonsupport and thus to marital/family conflict and eventual dissolution of the marriage. A certain percentage of spouses move in this direction. This probably depends on the previous strength of the marriage and the relationship of the couple. As a consequence of the pain and resulting disability, however, some marital relationships move in the opposite direction. The spouse now becomes very protective rather than rejecting/nonsupportive. Early pain literature had postulated that the oversolicitous spouse could reinforce pain behavior.[75] There is some evidence for this assertion, and as such, multidisciplinary pain facilities usually try to modify this behavior in the spouse.[75] However, it is unclear whether such interventions have an impact on the pain of CPPs. In any case, both these problems are encountered frequently in CPPs and should be addressed psychotherapeutically in an educational fashion.

Some CPPs report childhood victimization. Early pain research postulated that the childhood victimization was etiologically related to the pain of the CPPs.[76] Recent evidence in the only prospective study[76] to date, however, refutes this theory. In this study, there was no difference in pain complaints by adulthood between a victimized cohort and an age/sex-matched nonvictimized control group.[76] There also was no difference in the number of medically unexplainable medical symptoms.[76] CPPs presenting with this comorbidity should be referred for counseling if they feel that at the present time victimization memories are having an impact on their psychological well-being. However, it is unlikely that this counseling will solve their pain complaint.

Problems with the medical/insurance system/litigation system in reference to access to medical care/financial support/litigation often are voiced by CPPs, especially if they are involved in the worker compensation system (see Table 37-4, items 3, 5, 6, 7, and 8). These problems lead to increased stress, including financial stress. They can present as the following specific complaints: (1) compensation checks always arriving late, (2) compensation benefits terminated for no reason, according to the CPP, (3) denial of medical care (doctors, medical procedures, or medical supplies) that the CPP has requested or wants or has been recommended by his or her physician, (4) forcing the CPP to visit a physician he or she does not want to visit, (5) forcing the patient to do a job search when the patient feels that he or she is incapable of going back to work, (6) inability to reach the adjustor, and (7) inability to reach an equitable settlement with the carrier. It is to be noted that these are perceptions of the CPP and may be colored by the litigation process. The author believes that when there is litigation between the worker compensation patient and the carrier, the stress is even greater. Occasionally, these perceptions can interact with the CPP's paranoid tendencies, or this situation can happen to a CPP who has been habitually violent. Under these circumstances, and as discussed earlier, the CPP may begin to harbor fantasies about violent retribution against the insurance carrier or the "insurance physicians."[61] If there are such fantasies or wishes, the CPP should be referred immediately to a psychiatrist.[61]

Many CPPs complain of preinjury job stress.[77] Preinjury job stress appears to be important to pain treatment outcome. In a prospective follow-up study, Fishbain and colleagues[77] indicated that intent to return to work after pain facility treatment was predicted by perceived preinjury job stress. Therefore, in treating the CPP, it is important to understand work-related perceptions, such as those of perceived job dissatisfaction and job stress. These problems should be treated in counseling when encountered.

Recent psychological studies using new constructs have indicated that some CPPs could have unreasonable fear of increased pain,[14,15] poor self-efficacy,[16] and poor coping strategies (see Table 37-4, items 11, 13 and 14). As can be imagined, unreasonable fear of increased pain can interfere with rehabilitation efforts because, generally, any exercise/activation program involves some pain. Thus, this fear can have an impact on treatment outcome. Self-efficacy expectations are beliefs regarding one's capacity to execute the behavior required to produce a certain outcome. Coping strategies are the process of executing a cognitive response to threat. CPPs may have incorrect self-efficacy expectations about their pain and poor coping strategies. Training in the use of CPP coping strategies has been shown to decrease reported pain and to increase pain tolerance and threshold.[78] These three concepts address problem areas where treatment interventions could be targeted.

The final behavior comorbidity is that of unrealistic treatment expectations.[79] Many CPPs wish to be totally healed and to be exactly the same as before the pain problem began, including the performance of now-unreasonable activities. This desire leads to a "search for a cure" (see Table 37-4, item 2) rather than a search for symptom control. Very often in this "search for a cure," the CPP will undergo increasingly riskier treatments that have less and less chance for cure versus a greater and greater chance for further damage. In this quest, there is often physician collusion. Rather than provide realistic/guarded expectations for the success of the treatment offered, the physician may provide inflated claims for the best possible success of the treatment. Thus, some authors[79] have advised that physicians should be cautious about optimistic prognoses because later these may create disappointment and anger. In general, CPPs need to hear realistic information about the limits of the treatment being offered.

▶ CONCLUSIONS

Psychiatric, somatic, and behavioral/psychological comorbidities are commonly associated with chronic pain and can affect the index problem—pain. In addition, these comorbidities can interfere dramatically with treatment or make treatment more difficult, thus affecting treatment outcome. Early recognition and treatment of these comorbidities are mandatory in the successful treatment of CPPs.

APPENDICES 37

▶ **ALGORITHM A.** TREATMENT OF DEPRESSION AS A COMORBIDITY

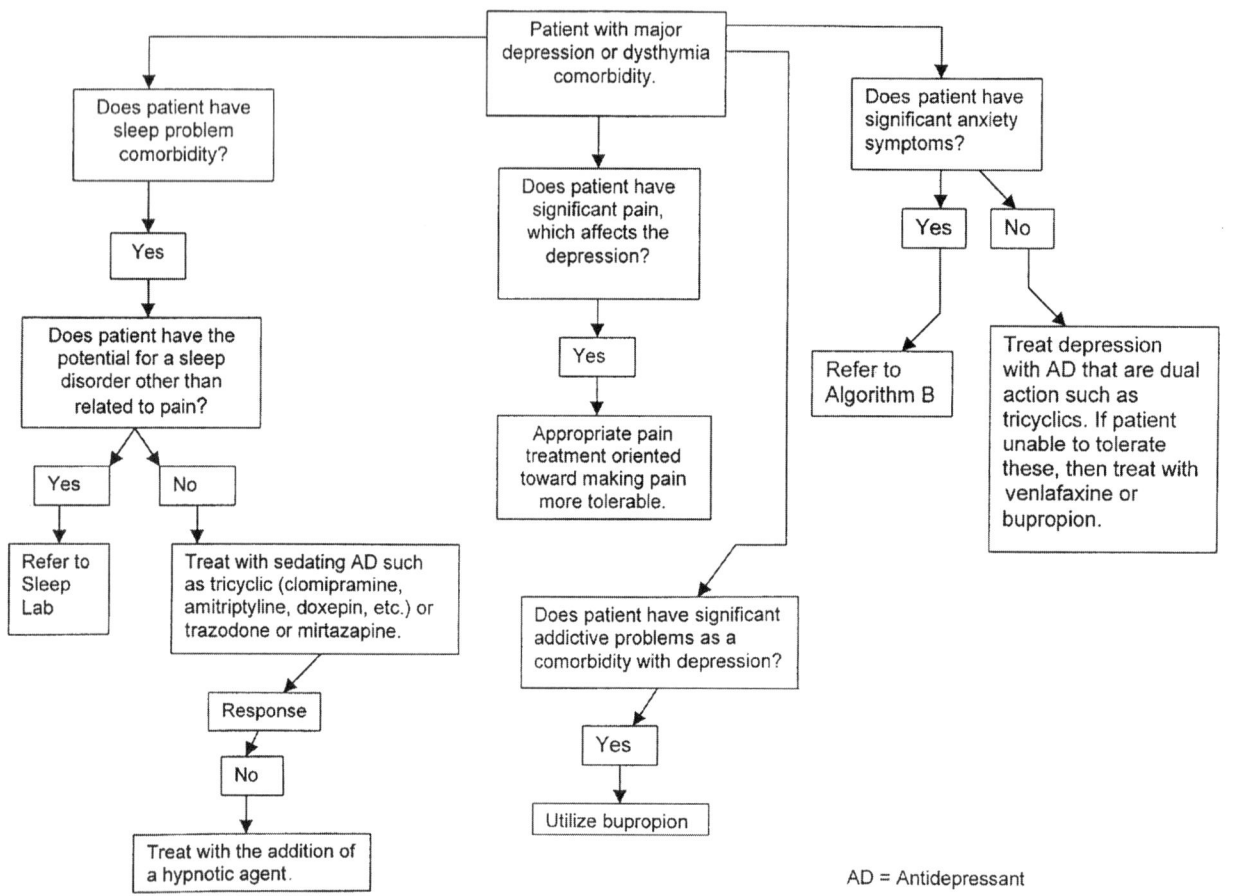

AD = Antidepressant

▶ **ALGORITHM B.** TREATMENT OF ANXIETY AS A COMORBIDITY

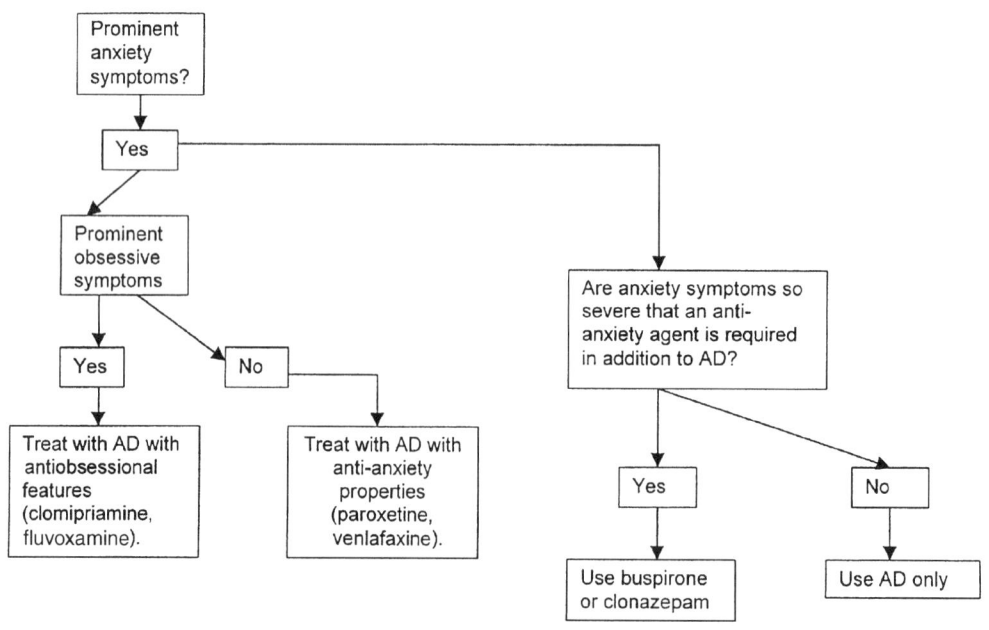

CHAPTER 37 DIAGNOSTIC AND TREATMENT APPROACHES 539

▶ **ALGORITHM C.** TREATMENT OF ADDICTION IN THE CONTEXT OF CHRONIC PAIN

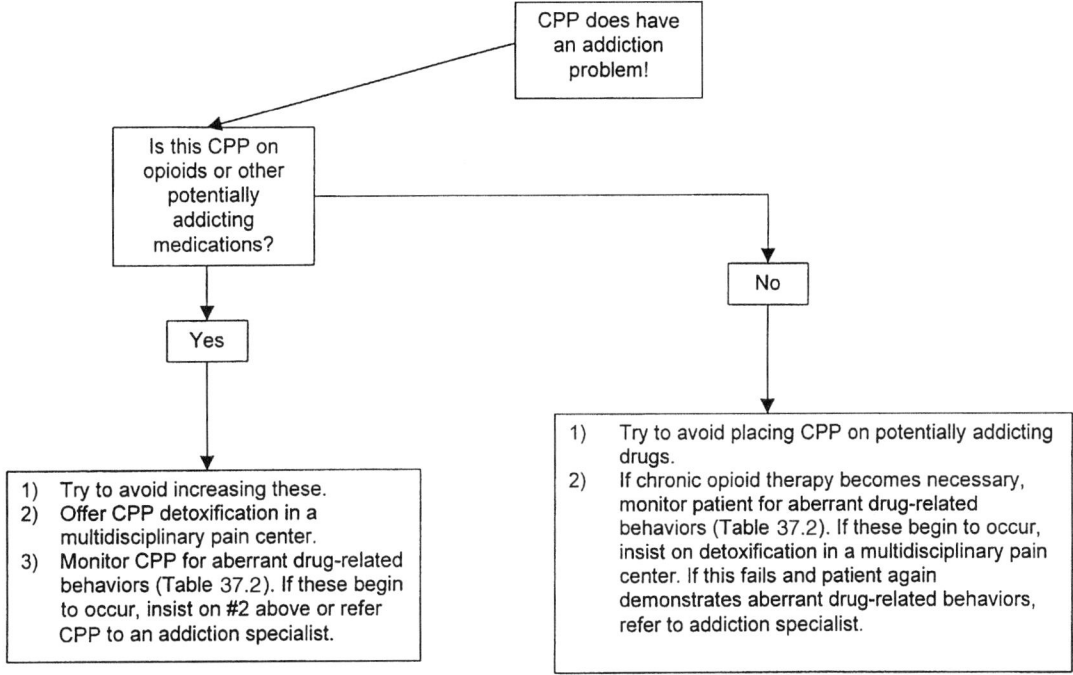

▶ **ALGORITHM D.** TREATMENT OF THREATENED VIOLENCE/ANGER/IRRITABILITY AS A COMORBIDITY

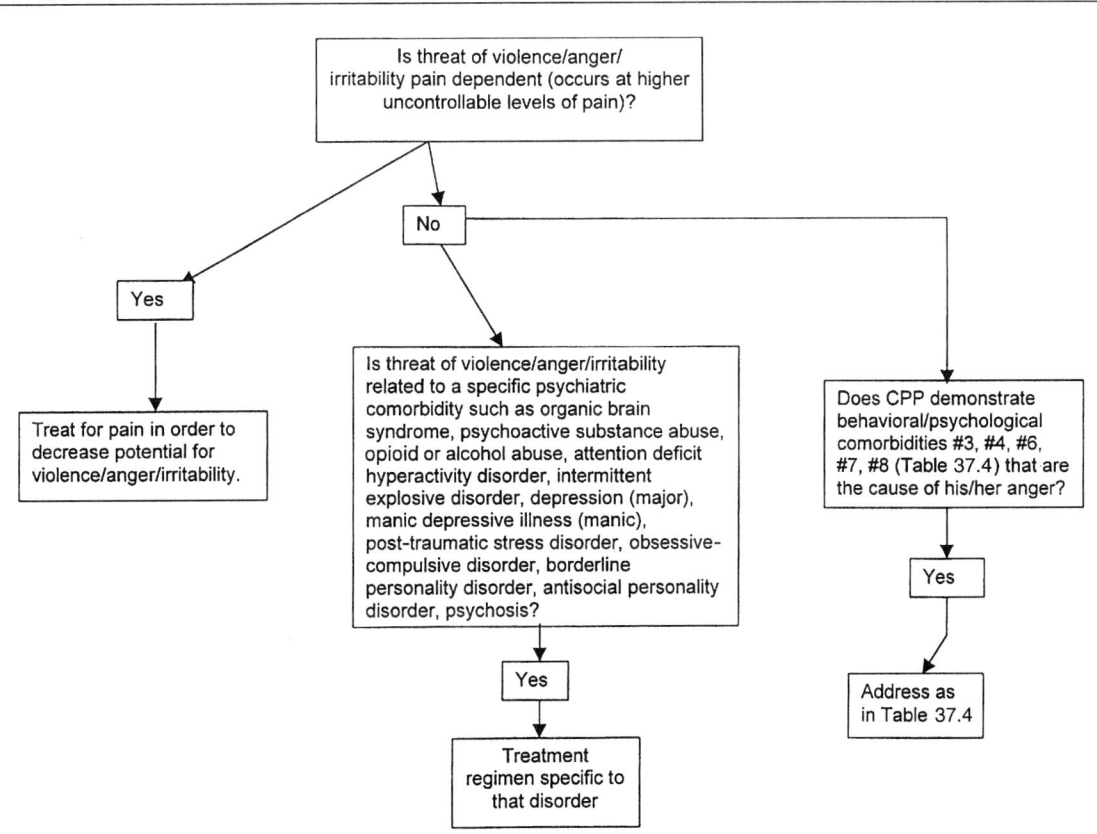

REFERENCES

1. Feinstein A: The pre-therapeutic classification of comorbidity in chronic disease. *J Chron Dis* 1970;13:455.
2. Merikangas KR, Gelernter CS: Comorbidity for alcoholism and depression. *Psychiatr Clin North Am* 1991;13:613–631.
3. American Psychiatric Association: *Diagnostic and Statistical Manual of Mental Disorders,* 4th ed. Washington: American Psychiatric Association, 1994.
4. Fishbain DA, Goldberg M, Meagher BR, Rosomoff H: Male and female chronic pain patients categorized by DSM-III psychiatric diagnostic criteria. *Pain* 1986;26:181–197.
5. Fishbain DA, Cutler RB, Steele-Rosomoff R, et al: The problem-oriented psychiatric examination of the chronic pain patient and its application to the litigation consultation. *Clin J Pain* 1994;10:28–51.
6. Fishbain DA: Pain and psychopathology, in Fogel BS, Schiffer RB, Rao SM (eds), *Neuropsychiatry.* Baltimore: Williams & Wilkins, 1996:443–483.
7. Fishbain DA: Psychiatric and psychological problems associated with chronic pain. *Primary Care Psychiatry* 1997;3:75–81.
8. Romano JM, Turner JA: Chronic pain and depression: Does the evidence support a relationship? *Psychol Bull* 1985;97:18–34.
9. Fishbain DA, Cutler R, Rosomoff HL, et al: Chronic pain-associated depression: Antecedent or consequence of chronic pain? A review. *Clin J Pain* 1997;13:116–137.
10. Ishak WW, Burt T, Sederer LI (eds): *Outcome Measurement in Psychiatry: A Critical Review.* Washington: American Psychiatric Publishing, 2002.
11. Sajatovic M, Ramirez LF: *Rating Scales in Mental Health.* Hudson, OH: Lex-Comp, Inc., 2001.
12. Rosenstiel AK, Keefe FJ: The use of coping strategies in chronic low back pain patients: Relationship to patient characteristics and current adjustment. *Pain* 1983;17:33–44.
13. Bernstein IH, Matthew E, Jaremko, Hinkley BS: On the utility of the West Haven–Yale Multidimensional Pain Inventory. *Spine* 1995;20:956–963.
14. Admundson GJG, Norton GR, Allerdings MD: Fear and avoidance in dysfunctional chronic back pain patients. *Pain* 1997;69:231–236.
15. Vlaeyen JWS, Kolke-Snijders AMJ, Boeren RGB, van Eeek H: Fear of movement/(re)injury in chronic low back pain and its relation to behavioral performance. *Pain* 1995;62:363–372.
16. Lorig K, Chastain RI, Ung E, Large P: Development and evaluation of a scale to measure perceived self-efficacy in people with arthritis. *Arthritis Rheum* 1989;32:37–44.
17. Waddell G, Newton M, Henderson I, et al: A fear-avoidance beliefs questionnaire (FABQ) and the role of fear-avoidance beliefs in chronic low back pain and disability. *Pain* 1993;52:157–168.
18. Schneider RA: Concurrent validity of the Beck Depression Inventory and Multidimensional Fatigue Inventory-20 in assessing fatigue among cancer patients. *Psychol Rep* 1998;82:883–886.
19. Fishbain DA, Cutler B, Cole B, et al: Is pain fatiguing? A structured evidence-based review. *Pain Med* 2003;4:51–62.
20. Buysse DJ, Reynolds CF III, Month TH, et al: The Pittsburgh Sleep Quality Index: A new instrument for psychiatric practice and research. *Psychiatr Res* 1989;28:193–213.
21. Frank E, Anderson MSW, Rubenstein D: Frequency of sexual dysfunction in "normal" couples. *N Engl J Med* 1978;299:111–115.
22. Korbon GA, DeGood DE, Schroeder ME: The development of a somatic amplification rating scale for low back pain. *Spine* 1987;12:787–791.
23. Follick MJ, Smith TW, Ahern DK: The Sickness Impact Profile: A global measure of disability in chronic low back pain. *Pain* 1985;21:67–76.
24. Millard RW: The functional assessment screening questionnaire application for evaluating pain-related disability. *Arch Phys Med Rehabil* 1989;70:303–307.
25. Riley JF, Ahern DK, Follick MJ: Chronic low back pain and functional improvement: Assessing beliefs about the relationship. *Arch Phys Med Rehabil* 1988;69:579–582.
26. Tait RC, Pollard CA, Margolis RB: The Pain Disability Index: Psychometric and validity data. *Arch Phys Med Rehabil* 1987;68:438–441.
27. Derogatis LR: *The SCL-90-R Manual II: Administration Scoring and Procedures.* Towson, MD: Clinical Psychometric Research, 1983.
28. Richards JS, Nepomuceno C, Riles R, Suer Z: Assessing pain behavior: The UAB Pain Behavior Scale. *Pain* 1982;14:393–398.
29. Fishbain DA, Cutler RB, Cole B, et al: Is pain-associated fatigue responsive to multidisciplinary pain facility treatment. *Pain Med* (accepted).
30. Zachrisson O, Regland B, Jahreskog M, et al: A rating scale for fibromyalgia and chronic fatigue syndrome (the FibroFatigue scale). *J Psychosom Res* 2002;52:501–509.
31. Muse M, Frigola G: Development of a quick screening instrument for detecting post-traumatic stress disorder in the chronic pain patient: Construction of the Post-traumatic Chronic Pain Test (PCPT). *Clin J Pain* 1987;2:151–153.
32. Skinner HA: The Drug Abuse Screening Test. *Addict Behav* 1982;7:363–371.
33. Portenoy RK: Opioid therapy for chronic nonmalignant pain: Current status, in Fields HL, Liebeskind JC (eds), *Progress in Pain Research and Management,* Vol. 1: *Pharmacological Approaches to the Treatment of Chronic Pain: New Concepts and Critical Issues.* Seattle: IASP Press, 1994:247–287.
34. Watanabe TK, O'Dell MW, Togliatti TJ: Diagnosis and rehabilitation strategies for patients with hysterical hemiparesis: A report of four cases. *Arch Phys Med Rehabil* 1998;79:709–714.
35. Campo JV, Negrini BJ: Case study: Negative reinforcement and behavioral management of conversion disorder. *J Am Acad Child Adolesc Psychiatry* 2000;39:787–790.
36. Moene FC, Hoogduin KA, Van Dyck R: The inpatient treatment of patients suffering from (motor) conversion symptoms: A description of eight cases. *J Clin Exp Hypnosis* 1998;46:171–190.
37. Fishbain DA, Goldberg M, Khalil TM, et al: The utility of electromyographic biofeedback in the treatment of conversion paralysis. *Am J Psychiatry* 1988;145:1572–1575.
38. Abdel-Moty E, Fishbain DA, Goldberg M, et al: Functional electrical stimulation treatment of postradiculopathy-associated muscle weakness. *Arch Phys Med Rehabil* 1994;75:680–686.
39. Speed J: Behavioral management of conversion disorder: Retrospective study. *Arch Phys Med Rehabil* 1996;77:147–154.

40. Hafeiz HB: Hysterical conversion: A prognostic study. *Br J Psychiatry* 1980;136:548–551.
41. Verdugo RJ, Ochoa JL: Reversal of hypesthesia by nerve block, or placebo: A psychologically mediated sign in chronic pseudoneuropathic pain patients. *J Neurol Neurosurg Psychiatry* 1998;65:196–203.
42. Moriwaki K, Yugae O, Nishioka K, et al: Reduction in the size of tactile hypesthesia and allodynia closely associated with pain relief in patients with chronic pain. *Prog Pain Res Manag* 1994;2:819–830.
43. Teasell RW, Shapiro AP: Strategic-behavioral intervention in the treatment of chronic nonorganic motor disorders. *Am J Phys Med Rehabil* 1994;73:44–50.
44. Sinel M, Eisenberg MS: Two unusual gait disturbances: astasia abasia and camptocormia. *Arch Phys Med Rehabil* 1990;71:1078–1080.
45. Marley S, Williams AC, Black S: A confirmatory factor analysis of the Beck Depression Inventory in chronic pain. *Pain* 2002;99:289–298.
46. Ericsson M, Poston WS, Linder J, et al: Depression predicts disability in long-term chronic pain patients. *Disabil Rehabil* 2002;24:334–340.
47. Bair MJ, Robinson RL, Katon W, Kroenke K: Depression and pain comorbidity: a literature review. *Arch Int Med* 2003;163:2433–2435.
48. Fishbain DA, Cutler B, Rosomoff H, Steele-Rosomoff R: Evidence-based data from animal and human experimental studies on pain relief with antidepressants. *Pain Med* 2000;1:310–316.
49. Fishbain DA, Cutler B, Rosomoff H, Steele-Rosomoff R: Evidence-based data on pain relief with antidepressants. *Ann Med* 2000;32:305–316.
50. Fishbain DA: The association of chronic pain and suicide. *Semin Clin Neuropsychiatr* 1999;4:221–227.
51. Fishbain DA, Goldberg M, Steele-Rosomoff R, et al: Case report: Completed suicide in chronic pain. *Clin J Pain* 1991;7:29–36.
52. Fishbain DA, Steele-Rosomoff R, Rosomoff HL: Drug abuse, dependence, and addiction in chronic pain patients. *Clin J Pain* 1992;8:77–85.
53. Fishbain DA: DSM-IV: Implications and issues for the pain clinician. *Am Pain Soc Bull* 1995;5:6–18.
54. Fishbain DA, Cutler B, Rosomoff H, Steele-Rosomoff R: Comorbidity between psychiatric disorders and chronic pain. *Curr Rev Pain* 1998;2:1–10.
55. Katon W: Panic disorder and somatization: Review of 55 cases. *Am J Med* 1984;77:101–106.
56. Gilcrist JC: Psychiatric and social factors related to low-back pain in general practice. *Rheumatol Rehabil* 1976;18:101–107.
57. Asmudson GJ, Norton GR, Jacobson SJ: Social, blood/injury, and agoraphobic fears in patients with physically unexplained chronic pain: Are they clinically significant? *Anxiety* 1988;2:87–95.
58. Hollander E: Treatment of obsessive-compulsive spectrum disorders with SSRIs. *Br J Psychiatry* 1998;35:7–12.
59. Fishbain DA, Trescott J, Cutler B, et al: Do some chronic pain patients with atypical facial pain overvalue and obsess about their pain? *Psychomatics* 1993;34:355–359.
60. Kurokawa K, Tanino R: Effectiveness of clomipramine for obsessive-compulsive symptoms and chronic pain in two patients with schizophrenia. *J Clin Psychopharmacol* 1997;17:329–330.
61. Fishbain DA, Cutler RB, Rosomoff HL, Steele-Rosomoff R: Risk for violent behavior in patients with chronic pain: Evaluation and management in the pain facility setting. *Pain Med* 2000;1:140–155.
62. Bruehe S, Burns JW, Chung OY, et al: Anger and pain sensitivity in chronic low back pain patients and pain free controls: The role of endogenous opioids. *Pain* 2002;99:223–233.
63. Fishbain DA: Secondary gain concept: Definition problems and its abuse in medical practice. *Am Pain Soc J* 1994;3:264–273.
64. Fishbain DA, Rosomoff HL, Cutler RB, Steele-Rosomoff R: Secondary gain concept: A review of the scientific evidence. *Clin J Pain* 1995;11:6–21.
65. Fishbain DA, Cutler R, Rosomoff HL, et al: Chronic pain disability exaggeration/malingering research and submaximal effort research. *Clin J Pain* 1999;15:244–274.
66. Atkinson JH, Ancoli-Israel S, Slater MA, et al: Subjective sleep disturbance in chronic back pain. *Clin J Pain* 1988;4:225–232.
67. Fishbain DA, Cutler R, Cole B, et al: International Headache Society headache diagnostic patterns in pain facility patients. *Clin J Pain* 2001;17:78–93.
68. Lenstra M, Stewart B, Olazynski WP: Headache pain facility treatment. *Headache* 2002; 42:845–854.
69. Hart RP, Martelli MF, Zasler ND: Chronic pain and neuropsychological functioning. *Neuropsychol Rev* 2000;10: 131–149.
70. Williams AN, Hill P, Gunary R, Cratchley G: Sexual difficulties of chronic pain patients. *Clin J Pain* 2001;17:138–145.
71. LaBan MM, Burk RD, Johnson EW: Sexual impotence in men having low back syndrome. *Arch Phys Med Rehabil* 1966;45:715–723.
72. Fishbain DA, Cole B, Cutler RB, et al: A structured evidence based review on the meaning of non-organic physical signs: Waddell signs. *Pain Med* 2003;4:51–62.
73. Lipowski ZJ: Somatization: The concept and its clinical application. *Am J Psychiatry* 1988;145:1358–1366.
74. Turk DC, Matyas TA: Pain-related behaviors: Communication of pain. *Am Pain Soc J* 1992;1:109–111.
75. Pawlicki RE: A neglected issue: The family of the chronic pain patient. *Am Pain Soc Bull* 1992;1:5–6.
76. Raphael NG, Spatz Widon C, Lange G: Childhood victimization and pain in adulthood: A prospective investigation. *Pain* 2001;92:283–293.
77. Fishbain DA, Cutler R, Rosomoff HL, et al: The prediction of chronic pain patient "intent," "discrepancy with intent" and "discrepancy with non-intent" for return to work post pain facility treatment. *Clin J Pain* 1999;61:165–175.
78. Tan SY: Cognitive and cognitive behavioral methods for pain control: A selective review. *Pain* 1982;12:201–228.
79. Cherkin DC, Deyo RA, Street JH, Barlow W: Predicting poor outcome for back pain seen in primary care using patients' own criteria. *Spine* 1996;21:2900–2907.

PART IV

The Therapeutic Basis of Pain

CHAPTER 38

Nonsteroidal Anti-Inflammatory Medications and Acetaminophen

Brian D. Golden and Steven B. Abramson

Nonsteroidal anti-inflammatory drugs (NSAIDs) are among the most widely prescribed medications throughout the world. Their use is indicated for a variety of painful musculoskeletal conditions, including osteoarthritis (OA) and rheumatoid arthritis (RA), as well as for various acute and chronic pain syndromes, including dysmenorrhea and perioperative pain. Despite a generally favorable benefit-to-risk ratio, long-term NSAID use can be limited by the development of peptic ulcer disease or renal insufficiency, particularly in elderly populations. Both the therapeutic benefit and the potential toxicity of these drugs are, in large measure, due to their capacity to inhibit the synthesis of prostaglandins (PGs) by the cyclooxygenase (COX) enzymes.[1] There are at least two distinct isoforms of prostaglandin H synthase, or cyclooxygenase: COX-1 and COX-2. COX-1 is constitutively expressed in many tissues, where it regulates physiologic functions. COX-2, an inducible form, is not normally expressed by most tissues but is upregulated at sites of inflammation and within certain neoplasms.[2,3] This fundamental paradigm (the basis of the so-called COX hypothesis; see below), although useful to conceptualize the cyclooxygenase isoforms, should be understood to be an oversimplification. For example, COX-2 is expressed constitutively in normal tissue, such as kidney and brain, and such expression may increase in response to physiologic stress (Table 38-1). Conversely, although COX-1 functions as a "housekeeping" enzyme in many tissues, its expression also can be upregulated in disease.[2,3]

Soon after the discovery of the inducible isoform of cyclooxygenase (COX-2), it was hypothesized that drugs that selectively inhibited COX-2 would have the beneficial properties of conventional NSAIDs without the disadvantage of their associated gastrointestinal (GI) side effects.[4] This led to the development and introduction of COX-2-selective agents, the coxibs, that have been studied extensively in multicenter randomized clinical trials over the past decade.[5,6] However, despite the burgeoning amount of scientific and clinical information about the coxibs, fundamental questions remain, particularly with regard to their safety profile, and new ones have emerged that have direct impact on clinical practice. In part driven by these uncertainties, there is still a clinical role for acetaminophen, itself not a member of the NSAID or coxib class and whose mechanism of action is not via cyclooxygenase inhibition, in the treatment of arthritis and pain. This chapter addresses recent advances in our understanding of the biology of the cyclooxygenases and the use and safety of NSAIDs, specific COX-2 inhibitors, and acetaminophen in the treatment of arthritis and pain.

▶ TABLE 38-1. CONSTITUTIVE EXPRESSION OF COX-2*

Tissue	COX-2 Expression	Possible Function(s)
Kidney	Macula densa (juxtaglomerular apparatus) Medullary interstitial cells	Regulation of intravascular volume
Brain	Endothelial cells	Febrile response
	Cortical excitatory neurons	Pain
		Neuronal connectivity
		CNS development
		Learning and memory
Bone	Osteoblasts	Osteoclast differentiation
		Regulation of bone remodeling
Colon cancer	Mucosal epithelium	Adhesion of epithelium to extracellular matrix
		Resistance to apoptosis
Female reproductive system	Ovary	Ovulation (follicular rupture)
	Uterus	Embryo implantation
Gastrointestinal tract	Intestinal epithelium	Mucosal fluid secretion
		Bacterial clearance
	Gastric ulcers	Ulcer healing
Blood vessels	Endothelium	Anti-thrombotic (?)

*See reviews in references.[2,4]

▶ BIOLOGY AND REGULATION OF THE CYCLOOXYGENASES

Until the mid-1980s, it was believed that prostaglandin formation was limited solely by the activation of phospholipases, which released arachidonate from cell membranes as substrate for a constitutively expressed cyclooxygenase enzyme. In 1988, an interleukin 1 (IL-1)–dependent transcriptional upregulation of COX was identified in cultured dermal fibroblasts.[1-3] This suggested the existence of a second, novel COX isoenzyme that was synthesized de novo in the presence of inflammatory stimuli. A major breakthrough occurred by 1992 with the molecular cloning and characterization of a previously unrecognized COX isoform, designated COX-2 (or PGH synthase 2) that, like COX-1, had an apparent molecular weight of 70 kDa.[2,3] COX-1 and COX-2 share approximately 60 percent amino acid sequence homology.[2] The promoter region of the COX-2 gene contains response elements that are sensitive to inflammatory mediators, which accounts for its rapid inducibility, whereas the gene for COX-1 has features consistent with a "housekeeping" gene product involved in physiologic homeostatic processes. Despite these differences between the COX genes, the domains important for enzymatic function for arachidonate metabolism are remarkably similar.[2] The COX-2 substrate access channel is larger than the COX-1 channel, however, primarily due to the presence of a side pocket in the COX-2 molecule.[7] This difference can be used to develop COX-1- or COX-2-specific agents.

The most important difference between the two isoforms is their pattern of tissue expression and regulation.[2-4] COX-1 is constitutively expressed in most tissues, notably platelets, endothelial cells, GI tract, renal microvasculature, glomerulus, and collecting ducts.[2,3] Its expression can increase two- to fourfold under stimulatory conditions and is little affected by glucocorticoids. COX-2, on the other hand, is typically undetectable in most tissues under basal conditions, but its expression in many cell types, including macrophages, fibroblasts, chondrocytes, and epithelial and endothelial cells, is augmented 10- to 80-fold on stimulation by inflammatory cytokines, growth factors, or endotoxin. Moreover, COX-2 expression is inhibited by glucocorticoids.[2]

Since the original discovery of COX-2 in cytokine-stimulated cells, it has become apparent that COX-2 is also expressed in a variety of noninflammatory tissues, including kidney, brain, neoplasms, bone, and cartilage,[2,3] particularly under "physiologic stress" conditions (see Table 38-1). In the kidney, prostaglandins modulate vascular tone homeostasis of salt and water.[2,8] "Constitutive" expression of COX-2 has been detected in the vasculature, cortical macula densa, and medullary interstitial cells of the kidney, and its expression increases with age.[1,8] COX-2 mRNA and protein are increased by salt and water restriction, suggesting that physiologically important renal prostaglandins derive from both COX-1 and COX-2 activity. Therefore, it is not surprising that clinical studies indicate that selective COX-2 agents, like conventional NSAIDs, impair compensated renal function in the setting of congestive heart failure or volume depletion.[1,8,9]

COX-2 also is expressed constitutively in the central nervous system (CNS). Selective COX-2 inhibitors are antipyretic, indicating a role for this isoform in the febrile response. COX-2 mRNA and protein are constitutively expressed in excitatory neurons in the brain of rats, predominantly in forebrain neurons in the cortex and hippocampus, and in dendritic spines, which are intimately involved in synaptic signaling. There is evidence that CNS COX-2 plays a role in learning and memory.[2,3] Moreover, there is evidence for separate central analgesic actions of NSAIDs, mediated by inhibition of CNS COX-2 activity.[10] Finally, with regard to CNS COX expression, there is epidemiologic and observational evidence that NSAIDs reduce the incidence and may provide therapeutic benefit in patients with Alzheimer's disease.[11,12] However, it also should be noted that NSAID therapy has been associated with decreased cognition, particularly in the elderly, including longitudinal memory loss.[13]

▶ INFLAMMATION AND COX-2 EXPRESSION IN ARTHRITIS

Prostaglandins produced by cells within the inflamed joint contribute to the classic inflammatory signs of heat, redness, swelling, and pain. One of the earliest reports of the production of prostaglandins in human arthritis was by Dayer and colleagues,[14] who reported the production of collagenase and prostaglandins by isolated adherent rheumatoid synovial cells. Since the introduction of COX-2-selective agents for the treatment of arthritis, interest has focused on the distribution of the COX-2 isoform in joint tissues. Crofford and colleagues,[15] in an immunohistologic analysis of COX expression in synovium from patients with rheumatoid arthritis (RA), observed extensive and intense intracellular COX-2 staining in mononuclear cells and vascular endothelium, with weaker staining in the synovial lining layer. Siegle and colleagues[16] analyzed sections of synovial tissue from patients with inflammatory arthritides (RA, psoriatic arthritis [PsA], ankylosing spondylitis [AS]) and from patients with OA using COX-1- and COX-2-specific antisera for immunostaining. They found that strong COX-2 immunostaining was present in synovial endothelium, lining cells, fibroblast-like cells, and chondrocytes in samples from patients with inflammatory arthritis, whereas staining of OA synovial tissue was scant. COX-1 staining was confined to synovial lining cells, and no significant differences in staining were apparent among the different arthritides. The detection of COX-1 expression by synovial lining cells was unexpected, and the role of COX-1-derived prostanoids produced by the synovial lining is unknown.

The data from studies by Siegle and colleagues[17] indicate that the intensity of COX-2 staining is much greater in RA than in OA synovial tissue. In contrast to synovium, however, OA cartilage produces a prodigious amount of prostaglandins, comparable with that observed in rheumatoid arthritis. OA cartilage explants cultured ex vivo spontaneously release PGE_2 at levels 50-fold times higher than normal cartilage and 18-fold higher than those produced by normal cartilage stimulated with cytokines and endotoxin.[17] This superinduction of PGE_2 coincides with the upregulation of chondrocyte COX-2 mRNA and protein. The addition of IL-1 antagonists to OA explant cultures inhibits the spontaneous production of PGE_2 by these cultures, indicating that IL-1 derived from chondrocytes induces COX-2 expression and stimulates the production of prostaglandins.[18] Dexamethasone, nonselective NSAIDs, and coxibs inhibit the production of PGE_2 by OA cartilage.

The existing literature is contradictory with respect to the potential effects of eicosanoid overproduction on cartilage metabolism. For example, it has been reported that PGE_2 reverses proteoglycan degradation induced by IL-1 in bovine and human cartilage explants, inhibits IL-1-induced matrix metalloprotease (collagenase, stromelysin) expression in human synovial fibroblast, and enhances collagen type II and proteoglycan synthesis.[18] Conversely, Robinson and colleagues[19] reported in 1978 that prostaglandins produced by rheumatoid explant tissues inhibited cartilage proteoglycan synthesis. In addition, PGE_2 activates metalloproteinases, as has been reported in epithelial cells, human synoviocytes, and human OA cartilage explants.[2] Thus, eicosanoids released by chondrocytes and synovial cells may exert both anabolic and catabolic effects on matrix metabolism, with the net result on cartilage integrity being uncertain at present.

OA is not classified as an inflammatory arthritis. Neutrophils, the cellular hallmark of an acute inflammatory response, do not generally accumulate in OA synovial fluid. Classic signs and symptoms of inflammation—heat, redness, swelling, and pain—are also not typically present. Moreover, as noted earlier, the expression of the induced isoform of cyclooxygenase, COX-2, a marker of inflammatory events in tissue, is not a characteristic feature of OA synovium. COX-2-derived eicosanoids produced by OA chondrocytes embedded within the avascular cartilage are more likely to play an autocrine/paracrine role in modifying cartilage metabolism than in provoking signs and symptoms of inflammation. Based on these observations, the inhibition of COX-2 by NSAIDs, selective or nonselective, would not be expected to be an important component of therapeutic strategies for OA. Indeed, for many patients, NSAIDs are no more effective than simple analgesics, as will be discussed below.

THE USE OF NSAIDs AND COXIBS IN ARTHRITIS AND PAIN

The Place of Anti-Inflammatories in OA Therapy Today

OA, the most common form of joint disease, is a cause of significant pain and disability, particularly in the elderly population.[20] While acetaminophen is effective (see below), a meta-analysis of randomized clinical trials has found that traditional NSAIDs afford significantly greater improvements than acetaminophen in pain at rest and pain on motion,[21] which may result from their ability to treat the underlying inflammation (see above), as well as providing relief from pain. Based on important efficacy and safety studies, it was felt that NSAIDs—given as either a traditional NSAID plus a gastroprotective agent or as monotherapy with a COX-2-selective inhibitor—are a reasonable first-line therapy for patients with moderate or severe pain, especially when inflammation is present, and as second-line therapy in those who gain only little or no relief from acetaminophen.[22] Both U.S. and European guidelines are designed to be flexible and permit development of patient-specific strategies tailored to individual circumstances.

Three COX-2 selective inhibitors (celecoxib, valdecoxib, and rofecoxib) have been approved by the Food and Drug Administration (FDA) for the treatment of OA. The efficacy of COX-2-selective inhibitors has been examined in large, randomized, double-blind trials in adults with OA of the knee or hip. In these studies, celecoxib doses of 100 and 200 mg bid and rofecoxib doses of 12.5 and 25 mg qd were found to provide significant improvements in Western Ontario McMasters University's Osteoarthritis Index (WOMAC) pain end points, including pain walking on a flat surface, night pain, pain at rest, and morning stiffness.[23–30] Importantly, both celecoxib and rofecoxib also have been shown to improve functional status (assessed using WOMAC) and health-related quality of life significantly (assessed using Short-Form 36-item [SF-36] generic questionnaire); improvements were noted within 2 weeks of starting therapy and were maintained for the 6 to 12 weeks of the study.[31,32] Good efficacy and tolerability also have been reported in the elderly.[24,29,32] Celecoxib appears to be equally effective and well tolerated when given either as a 200 mg qd dose or as a 100 mg bid dose.[30]

Comparative trials have shown celecoxib 100 or 200 mg bid to be at least as effective as diclofenac and naproxen in patients with OA of the hip or knee but to have a superior tolerability and safety profile, particularly with regard to GI side effects.[23,26,27,33,34] Rofecoxib 12.5 mg qd has achieved symptomatic improvements comparable with those of diclofenac/misoprostol (Arthrotec),[35] diclofenac,[36] high-dose ibuprofen,[28,37] and nabumetone[29] in comparative trials in patients with OA but was found to cause significantly fewer POBs than these comparator NSAIDs.[38]

Treatment effects beyond symptomatic relief—such as reduced disease progression or true chondroprotection—remain to be proven for most rheumatologic treatments. In particular, the potential disease-modifying effects of NSAIDs need to be clarified. Although in vitro and in vivo animal data support a long-term beneficial effect of NSAIDs in OA, this has not been borne out in the clinic, possibly because traditional NSAIDs usually have to be withdrawn in time because of side effects.[39] In fact, it has been suggested that NSAIDs actually may accelerate structural decline by reducing perfusion of the osteoarthritic joint (via their vasoconstrictive effects, thereby inhibiting repair of joint structures) and also, by relieving joint pain, enabling a greater degree of physical activity, causing increased wear and tear on the affected joint.[39] Long-term indomethacin treatment has been found to have a deleterious effect on cartilage,[40] although this does not appear to be the case with other NSAIDs. Inflammation now is known to play a role in joint pathology and cartilage breakdown, raising the question of whether progression of OA may be slowed if long-term suppression of inflammation could be achieved safely. With the introduction of new COX-2-selective inhibitors with improved long-term safety, particularly with respect to GI effects, it now may be possible to design studies to assess the potential disease-modifying effects of chronic suppression of inflammation without having excessive patient withdrawals.

Analgesic Effects of COX-2-Selective Inhibitors in the Acute Pain Setting

The use of NSAIDs and coxibs in the management of pain and inflammation associated with OA and RA is fairly well established. There are other indications and potential indications for COX-2 inhibition, including various acute and chronic pain syndromes such as dental extraction, primary dysmenorrhea, and potential applications in chronic mechanical low back pain and cancer pain.

Acute pain is one of the most common reasons for patients seeking medical care and is managed most commonly with an NSAID or an opioid analgesic. It is an essentially subjective parameter, and multiple model systems have been developed to try to quantify reliably the degree of pain and the effect of pain-relieving medications. However, while the principal animal models used for screening potential analgesics in the laboratory can differentiate mechanical, visceral, and thermal elements of pain, they predict human pain poorly. Indeed, no animal tests are available to evaluate spontaneous pain (the most common pain presentation in humans) or episodic shooting pains (as in trigeminal neuralgia) or to understand the impact of other factors that affect human perception of pain, such as stress, anxiety, depression, abuse, gender, and cultural issues.

In humans, third molar extraction has emerged as a useful and reproducible model for acute pain and is used widely for assessing the efficacy of potential analgesics. It is believed to be representative of the majority of postoperative pain situations.[41] Its recent refinement through the addition of in situ microdialysis probes means that it is now possible to quantify the relationship between inflammatory mediators and clinical pain and to assess analgesic activity and surrogate end points for COX-1 and COX-2 activity, as well as other measures of inflammation such as edema.[42] The model also permits examination of later hyperalgesia, as well as preventive analgesia. Bilateral extractions and crossover designs permit measurement of more than one inflammatory mediator in the same patient and reduce between-group variables to enhance assay sensitivity. Using this model, celecoxib 200 mg has been shown to provide significant relief from dental extraction pain relative to placebo, although the peak analgesic effect was approximately half that achieved with ibuprofen 600 mg.[42] Ibuprofen consistently suppressed PGE_2 levels (a marker of COX-2 activity) and thromboxane B_2 (TxB_2) levels (a marker of COX-1 activity), whereas celecoxib suppressed PGE_2 at later timepoints only and had no effect on TxB_2 levels, indicating that celecoxib has a COX-1-sparing effect in vivo as well as in in vitro biochemical selectivity assays. Other dental pain studies have found celecoxib 100 to 400 mg to have comparable analgesic efficacy to aspirin 650 mg.[43] Results of early multiple-dose studies of celecoxib in orthopedic and general surgery were inconclusive,[42] and thus the drug did not receive initial FDA approval for the management of acute pain.

Rofecoxib has been shown to provide superior pain relief to celecoxib in postoperative dental pain, postoperative surgical pain, and primary dysmenorrhea.[42] In comparative trials in patients undergoing dental extraction, the analgesic efficacy of rofecoxib 50 to 500 mg was significantly greater than that of celecoxib 200 mg, codeine 60 mg/acetaminophen 600 mg, and placebo,[44,45] and comparable with that of naproxen sodium 550 mg[46] and ibuprofen 400 mg (although significantly fewer subjects in the rofecoxib group needed an additional analgesic within 24 hours of study drug compared with ibuprofen-treated patients).[47,48] In the postoperative setting, rofecoxib 50 mg was found to be of similar efficacy to naproxen sodium 550 mg in relieving orthopedic surgical pain[49] and more effective than celecoxib 200 mg in patients experiencing postoperative pain following spinal fusion surgery.[50] Thus, the published data for rofecoxib provide clear evidence of an analgesic effect in various models of acute pain, as well as in in vitro selectivity for COX-2 inhibition. This, coupled with the fact that it is well tolerated and does not appear to inhibit COX-1-mediated platelet aggregation (see subsequent sections), provided a basis for approval of rofecoxib as the first COX-2-selective inhibitor approved by the FDA for the management of acute pain (up to a maximum of 5 days).

Primary dysmenorrhea, manifested as cramping lower abdominal discomfort, affects as many as 90 percent of premenopausal women, many experiencing associated somatic symptoms as well, including nausea, sweating, mood changes, and low back ache.[51] Although the etiology of primary dysmenorrhea is not completely understood, it is thought to be due to release of uterine prostaglandins, including PGE_2 and PGF_α, during endometrial sloughing. The prostaglandins stimulate myometrial contractions and sensitize nerve endings, causing pain. In fact, dysmenorrheic patients, as compared with normal individuals, have been shown to have higher levels of menstrual fluid prostaglandins, and therapy with NSAIDs is associated with a reduction in those levels and is correlated with symptom relief.[52] Given the role of prostaglandins in the pathophysiology of dysmenorrhea, the clinical efficacy of NSAIDs in this setting, therefore, is not unexpected. Retrospective meta-analyses have confirmed the utility of NSAIDs.[53] In this study, ibuprofen, naproxen, and mefenamic acid were superior to placebo. More recent data confirm the anticipated effectiveness of coxibs in the management of dysmenorrhea. Morrison and colleagues[54] randomized 179 patients to receive either naproxen 550 mg, rofecoxib 25 or 50 mg, or placebo. The study showed that over a 12-hour observation period, the pain relief in the rofecoxib group was comparable with that achieved in the naproxen group; both treatment groups were superior to placebo.

Another emerging indication for the coxib class is in the area of perioperative pain management. By virtue of its virtual lack of platelet inhibition, COX-2-specific inhibition would be expected to provide analgesia without increasing bleeding risk in the perioperative period; moreover, coxibs should reduce the need for use of postoperative opioid analgesics, thereby also reducing complications (Table 38-2). For example, Reuben and colleagues[55] demonstrated that rofecoxib 50 mg given 1 hour prior to arthroscopic surgery of the knee had a prolonged analgesic effect (>13 hours) compared with postoperative dosing (>7 hours), and the overall need for opioids within the first 24 hours in the preoperative group was halved. Of course, certain potential side effects still apply, including renal and cardiovascular concerns[56] (see below).

Central versus Peripheral Analgesic Effects

In response to tissue injury, COX-2 catalyzes the breakdown of arachidonic acid into prostaglandins (e.g., PGE_2)

▶ TABLE 38-2. PERIOPERATIVE USE OF COXIBS

Advantages
  Opioid-sparing effects
    Less ileus
    Less somnolence
    Less hypotension
    Less respiratory depression
    Less nausea/vomiting
    Lower addiction potential
  Virtually eliminates perioperative bleeding
  Reduced risk of gastrointestinal irritation
  Potency approaches opioid equivalence
Disadvantages
  Same renal risks as for traditional NSAIDs
  Uncertainty regarding bone healing in orthopedics
  Uncertainty regarding cardiovascular/ thrombotic risk in susceptible patients

that, in addition to causing vasodilation, also sensitize nociceptive afferent nerve terminals to the effect of other mediators (i.e., cause hyperalgesia). While it had long been thought that the analgesic effects of NSAIDs resulted purely from inhibition of prostaglandin formation in the periphery, it now appears that NSAIDs also have a central component to their analgesic activity.[42] A number of other potential mechanisms of NSAID analgesia have been proposed, including effects mediated by COX-1, leukotriene B_4, and intracellular transcription elements such as peroxisome proliferator-activated receptor γ (PPAR-γ).[57]

Future directions for analgesic research include the use of nitric oxide NSAIDs, the possibility of reducing CNS levels of nitric oxide, inhibiting the vanilloid receptor 1, inhibiting adenosine kinase, activating PPAR-?, and mimicking superoxide dismutase, as well as combinations of complementary-acting analgesics. COX-2-selective inhibitors offer improved safety and permit potentially chronic use for analgesia, but COX-2 inhibition may not be the total answer to pain.[57] The more targeted and mechanistically specific agents become, the more it will be important to use rational drug combinations such as a COX-2 inhibitor and opioids and COX-2 inhibitors and antiepileptic agents.

▶ SAFETY OF COX-2-SELECTIVE NSAIDs VERSUS NONSELECTIVE NSAIDs

Nonselective NSAIDs are associated not only with an increased risk for serious upper GI complications but also with nephrotoxicity, including renal insufficiency, hypertension, peripheral edema, and congestive heart failure.[8,58–60] Epidemiologic studies indicate that the risk of serious upper GI complications is greater in certain patient groups, particularly the elderly.[58–60]

Soon after the discovery of the inducible isoform of cyclooxygenase (COX-2), it was hypothesized that an NSAID that selectively inhibited COX-2 would have the beneficial properties of NSAIDs without the associated GI side effects, the basis of the COX hypothesis. Endoscopic studies have shown that the coxibs are associated with a lower incidence of gastroduodenal ulcers than comparator nonselective NSAIDs.[61–63] Moreover, there is evidence that the coxibs are better tolerated than nonselective NSAIDs with respect to the incidence of nonspecific abdominal pain and dyspepsia.[64] Langman and colleagues,[65] in an analysis of eight double-blind, randomized clinical trials of rofecoxib in patients with OA, found that the cumulative incidence of nonspecific GI adverse events, such as dyspepsia, epigastric pain, nausea, and diarrhea, was less prevalent with this coxib than with comparator NSAIDs over 6 months. However, the magnitude of the difference, although statistically significant ($p = 0.02$), was small (23.5 versus 25.5 percent, respectively) and not *clinically* significant, and the incidence rates converged after 6 months

VIGOR and CLASS

Two phase IV large outcome studies of at least 6 months' duration published in the fall of 2000 indicate that use of coxibs is accompanied by a lower incidence of serious GI toxicity than that seen with comparator NSAIDs.[5,6] The Celecoxib Long-term Arthritis Safety Study (CLASS) compared celecoxib at a dose of 400 mg twice daily with diclofenac 75 mg twice daily and ibuprofen 800 mg three times daily in a study that randomized more than 8000 patients, approximately 75 percent of whom had OA and 25 percent RA. Analysis of data from more than 4400 patients who received treatment for 6 months showed that celecoxib did not significantly reduce the frequency of ulcer complications (0.76 versus 1.45 percent, $p = 0.09$) in comparison with the nonselective NSAIDs,[5] although the difference with respect to the combination of symptomatic ulcers and ulcer complications was statistically significant (2.08 versus 3.54 percent, $p = 0.02$). Furthermore, no superiority of celecoxib relative to the comparator NSAIDs was apparent in subjects who received low-dose aspirin for cardiovascular prophylaxis during the study (approximately 20 percent of all subjects enrolled), and there was no significant difference between treatment groups with respect to either ulcer complications or the combination of ulcer complications and symptomatic ulcers among those who remained on treatment for 13 months.[66]

The Vioxx Gastrointestinal Outcomes Research (VIGOR) trial compared rofecoxib 50 mg once daily with naproxen 500 mg twice daily in nearly 9000 patients with RA, none of whom was taking low-dose aspirin. In this study, the median duration of follow-up was 9 months, and the incidence of confirmed upper GI events was reduced significantly in patients receiving rofecoxib.[6]

The CLASS and VIGOR trials were not designed to demonstrate that the risk of hemorrhage or perforation in patients using coxibs is equivalent to that of subjects receiving placebo. However, the results of these studies, and of other randomized clinical trials that preceded them, make it reasonable to recommend the use of a coxib when an NSAID is indicated for patients who are at increased risk for serious upper GI complications[4] (see Table 38-2). Coadministration of a gastroprotective agent, such as a proton pump inhibitor or misoprostol, has been shown to reduce the incidence of GI adverse events in patients taking a nonselective NSAID.[67-69] The merit of combining a gastroprotective agent with a coxib in at-risk patients is unknown.

Cardiovascular Issues

An unanticipated finding in the VIGOR trial was an apparent increase in the incidence of myocardial infarction in the rofecoxib group relative to the naproxen group. Specifically, 20 of the 4047 subjects in the rofecoxib group (0.5 percent) but only 4 of 4029 subjects in the naproxen group (0.1 percent) had a myocardial infarction during the study. In contrast to the CLASS trial, low-dose aspirin use was an exclusion criterion in the VIGOR study. However, even if subjects who were candidates for secondary cardiovascular prophylaxis (e.g., those with a history of myocardial infarction, cerebral vascular accident, transient ischemic attacks, angina, a coronary artery bypass graft, or angioplasty) were excluded, the incidence of myocardial infarction in the rofecoxib group remained higher than that in the naproxen group (12 of 3877 [0.3 percent] versus 4 of 3878 [0.1 percent], respectively).[66]

Because the absolute number of cardiovascular events in the VIGOR trial was low and the study was not powered to examine the incidence of cardiovascular events, additional studies are needed to determine whether this observation was due to an increased risk for thrombosis imparted by rofecoxib, to a protective effect of the comparator drug, naproxen, or to chance alone. Interpretation of the VIGOR data are also complicated by the fact that the patients enrolled had RA, a disease that may itself increase the risk of coronary thrombosis.[70]

Unopposed COX-1 activity has become the focus of current debate regarding the possible thrombogenic potential of the coxibs. Physiologically, thromboxane derived from platelets promotes vasoconstriction and platelet aggregation, whereas prostacyclin (PGI_2) derived from vascular endothelium has opposite effects, promoting vasodilation and inhibiting platelet aggregation. The consequences of a reduction in prostacyclin production by vascular endothelial cells resulting from selective inhibition of COX-2 in the presence of unopposed thromboxane production mediated by unopposed COX-1 activity in the platelets have led to questions about the cardiovascular safety of coxibs.[1,6] In a canine model of coronary artery thrombosis, celecoxib abolished the prolongation of the time to artery occlusion that resulted from administration of aspirin and the vasodilation that normally occurs in response to the production of prostacyclin by vascular endothelium.[71] Conversely, other data raise the possibility that inhibition of COX-2 could be *beneficial* in atherosclerotic disease. COX-2 is expressed in macrophage-rich areas of atherosclerotic plaques. Pharmacologic inhibition of COX-2 is accompanied by a decrease in the levels of matrix metalloproteinases (MMPs) in the atherosclerotic plaques.[72] It has been suggested that local synthesis of COX-2 by activated macrophages may be associated with acute ischemic syndromes, perhaps as a result of the rupture of plaques induced by the action of MMPs.[72]

Until more information is available, it is prudent to prescribe low-dose aspirin (≤325 mg daily) or another antiplatelet agent in patients treated with coxibs who are at risk for cardiovascular events. Certainly, because they do not inhibit platelet aggregation, it should be recognized that coxibs alone confer no cardiovascular protection.[73] Questions remain, however. For example, does the cardioprotective administration of low-dose aspirin eliminate or weaken the superior gastrointestinal safety profile of the COX-2 inhibitors? If coxibs are thrombogenic, is aspirin prophylaxis adequate to counteract that risk?

To further add to the complexity of the current clinical decision, a new question has emerged regarding the concomitant use of NSAIDs and aspirin; namely, do nonselective NSAIDs, which may compete for aspirin at the COX-1 catalytic site, interfere with its cardioprotective activity? A report by Catella-Lawson and colleagues[74] has indicated that the concomitant administration of ibuprofen (400 mg) but not rofecoxib (25 mg) or diclofenac (75 mg) antagonized the irreversible platelet inhibition (platelet aggregation and TxB_2 production) induced by low-dose aspirin (8 mg). In these studies, acetaminophen (1000 mg) was shown to be a weak nonspecific COX inhibitor but did not antagonize the aspirin effect. Future studies will need to address these unanswered questions regarding the interaction among aspirin, nonselective NSAIDs, and coxibs with respect to both gastrointestinal and cardiovascular safety.

Can the cardiovascular findings in the VIGOR study be generalized to all patients taking COX-2-selective inhibitors? The answer is not known. Although the absolute incidence of myocardial infarction in the VIGOR trial was comparable with that in the CLASS trial, the VIGOR study excluded patients with angina pectoris or symptomatic congestive heart failure, whereas the CLASS trial did not. Hence, patients in the CLASS study may have been at greater risk for myocardial infarction than those in the VIGOR trial. On the other hand, as noted earlier, the VIGOR study excluded subjects taking low-dose aspirin or other antiplatelet agents such as ticlopidine (surrogates for coronary artery disease). Furthermore, as noted, the study population in the VIGOR study consisted entirely of RA patients, many of whom were taking corticosteroids; this group, therefore, may have been inherently more prothrombotic than those in the CLASS study, where the patients were mixed RA and OA. Therefore, several key questions remain unanswered that require further clinical investigation:

- Does rofecoxib increase the risk of thromboses, and if so, is this a property of the class of highly selective COX-2 inhibitors?
- Should risk factors for cardiovascular disease or thrombosis (e.g., hypertension, diabetes, obesity, antiphospholipid antibodies, total knee replacement surgery) be considered relative contraindications to therapy with a selective COX-2 inhibitor?
- Does the use of low-dose aspirin nullify the gastroprotective advantage of COX-2-selective NSAIDs over conventional NSAIDs, as suggested, but not proven, by the CLASS trial?

These key clinical questions cannot be answered unequivocally on the basis of the current evidence and require further intensive study.[75]

Renal Issues

GI complications are not the only risks of NSAID therapy. Other adverse events include alterations in renal function and effects on blood pressure and fluid retention, including increased risk for congestive heart failure. Because renal sodium excretion is at least partially mediated by COX-2, fluid retention and hypertension occur with both nonspecific and COX-2-specific inhibitors.[1] In the CLASS trial, the percentage of patients having the adverse event of peripheral edema was similar for those receiving celecoxib and those receiving nonselective NSAIDs.[5] In the VIGOR trial, the incidence of adverse effects related to renal function in the rofecoxib group was similar to that in naproxen-treated patients.[6] There is no convincing evidence that COX-2 in vascular endothelium, kidney, or elsewhere is inhibited more effectively by coxibs than by conventional NSAIDs. However, it is unclear whether unopposed COX-1 activity resulting from selective COX-2 inhibition will result in yet undetermined toxicity. At present, the data indicate that COX selectivity confers neither an advantage nor a disadvantage with respect to the development of renal side effects.

▶ ACETAMINOPHEN VERSUS ANTI-INFLAMMATORY DRUGS IN OA

Efficacy of Acetaminophen versus NSAIDs

Acetaminophen still has a place in the revised 2000 ACR guidelines as initial therapy for mild to moderate OA pain, up to maximal dosage.[76] The guidelines were amended further to include the addition of other new agents. Acetaminophen has been found to provide symptomatic relief in a number of patients with OA and was advocated in the 1995 guidelines from the ACR and in the European League of Associations of Rheumatology (EULAR) guidelines as first-line therapy for all patients with OA of the knee and hip.[77,78] These recommendations for first-line use of acetaminophen in all OA patients most probably were based on its perceived good tolerability profile and low cost.

In choosing specific pharmacotherapy, it is important to keep in mind that the reduction of joint pain for most patients with OA is modest, on the order of 20 to 30 percent.[79] Therefore, treatment programs should ensure that nonpharmacologic therapy is optimized, including proper exercise, weight loss, and the use of assistive devices as needed.[80–82] Strategies for drug treatment need to take into account the heterogeneity of the disease among patients, which can affect large and small joints, as well as articulations of the axial skeleton. The course of disease also will vary among individuals. Typically, symptomatic OA is characterized by chronic pain, worsened by activity, that can range from mild to severe. Symptoms often deteriorate over time as the structural damage to articular cartilage progresses. As noted earlier, chronic symptoms may be punctuated by acute exacerbations, sudden increases in pain or swelling that may require therapeutic intervention with anti-inflammatory drugs such as intra-articular corticosteroids or brief courses of NSAIDs.

It is generally accepted that the initial drug treatment for most patients with symptomatic OA is simple analgesia using agents such as acetaminophen. This is the formal recommendation in the guidelines published by the European League Against Rheumatism (EULAR) in 2000[83] and the ACR in 1995.[82] The revised ACR guidelines, published in 2000, state that "the prescription of an NSAID merits consideration as an alternative initial therapeutic approach" in patients with severe pain.[81] These revised recommendations have been the subject of debate that has focused on several issues: (1) the evidence whether NSAIDs are more effective than acetaminophen as initial therapy in selected OA populations (e.g., those with inflammation); (2) the evidence that the clinician's criteria for "severe" or inflammatory disease can predict NSAID responders; (3) the uncertainty that the COX-2-selective NSAIDs have a safety profile comparable with that of simple analgesics; and (4) the socioeconomic cost-benefit of a simple analgesic versus the more expensive COX-2-selective agents or coxibs. Despite the merits of the debate, it is noteworthy that since the introduction of the selective COX-2 agents, the use of NSAIDs for the treatment of OA has increased markedly, so that more than 50 percent of all NSAID prescriptions for patients over 65 years of age who are "new" to NSAID therapy are now written for a coxib rather than for a nonselective NSAID.[79]

What is the evidence that informs this debate? Clinical trials evaluating NSAIDs clearly demonstrate a short-term symptomatic effect that is superior to placebo, and most indicate a safer GI profile for coxibs than for nonselective NSAIDs. However, the cost of the coxibs is substantially higher than that of simple analgesics or most nonselective NSAIDs, and concerns for cardiovascular safety have emerged, as discussed earlier, which make it important to evaluate the benefit-to-risk of NSAIDs versus simple analgesia with acetaminophen. In the classic studies by Bradley and Brandt in patients with mild to moderate knee OA, anti-inflammatory doses of ibuprofen (2400 mg/d) were not more effective than either analgesic doses of ibuprofen (1200 mg/d) or acetaminophen (4 g/d).[84,85] Subset analysis, however, did suggest that the anti-inflammatory dose of ibuprofen was superior to acetaminophen in patients with rest pain. Based on this and other studies, the EULAR recommendations state that acetaminophen (or paracetamol) "is the oral analgesic to try first and, if successful, is the preferred long-term oral analgesic."[83,84]

While these and other studies indicate that acetaminophen *can be* as effective as NSAIDs in many patients with OA, there are also data to indicate that NSAIDs provide superior efficacy in subsets of patients.[86,87] A meta-analysis conducted by the North of England Non-Steroidal Anti-Inflammatory Drug Guideline Development Group[88] showed that patients taking NSAIDs had significantly greater improvement in both pain at rest and pain on motion than those taking acetaminophen. Wolfe and colleagues,[86] examining the opinions of patients with hip or knee OA about the effectiveness of their own treatment with NSAIDs and acetaminophen, found that a greater percentage of those surveyed reported that NSAIDs were more effective than acetaminophen than vice versa. Nonetheless, nearly half of those who responded to the survey reported that acetaminophen was at least effective and as satisfactory as the NSAIDs they had received.

For patients with mild to moderate joint pain, some studies have indicated that the difference in efficacy between NSAIDs and acetaminophen is negligible but that differences between these two treatments emerge among patients with more severe symptoms or, possibly, in those with disease of the hip.[87,89] The sum of the evidence suggests that although NSAIDs are more efficacious than acetaminophen in some OA patients, it is reasonable to use acetaminophen as initial treatment, particularly in patients who have no prior experience with either NSAIDs or acetaminophen or who have risk factors associated with NSAID-induced GI adverse events. However, patients who report previous lack of benefit from a full dose of acetaminophen (i.e., 4 g/d) or, as suggested by recent ACR Guidelines,[90] who present with a flare of disease accompanied by the recent onset of signs of inflammation, might rationally be treated with an NSAID. In retrospective analyses, however, neither the presence of a knee effusion nor the severity of knee pain predicted a better response to an anti-inflammatory dose of ibuprofen than to acetaminophen.[85,91] Prospective clinical trials of NSAIDs versus acetaminophen, in which patients are randomized on the basis of signs of inflammation or severity of joint pain, have not been performed.

With regard to head-to-head comparator studies of acetaminophen and coxibs, a recent study by Geba and colleagues[92] in 382 patients with OA of the knee previously treated with NSAIDs or acetaminophen compared rofecoxib and celecoxib with acetaminophen. This was a multicenter, randomized, parallel-group, double-blind trial. The study compared doses of the agents indicated for initial OA management: rofecoxib 12.5 and 25 mg/d, celecoxib 200 mg once per day, and acetaminophen 4000 mg in divided doses. Rofecoxib 25 mg was superior to the other arms across several clinical end points. The authors found that during the first 6 days of therapy, rofecoxib 25 mg/d demonstrated significantly superior efficacy than the other therapies with respect to relief of pain on walking, rest pain, night pain, and morning stiffness. Over the 6-week trial period, patients treated with rofecoxib 25 mg/d also experienced greater improvements in night pain, joint stiffness, physical functioning, and composite pain score compared with patients receiving other therapies.

In any case, clinical guidelines should not be rigid. The judgment of the physician, the effectiveness of nonpharmacologic measures, and observations made during timely follow-up assessments all should factor into therapeutic decisions about the care of the individual patient.

Toxicity of Acetaminophen

While the general safety profile of acetaminophen is excellent, making it one of the most widely-used agents in the world for a variety of indications and patient populations, it is a dose-related hepatotoxin. When consumed in large overdoses in excess of 150 mg/kg per day (>10 g/d)[93] (or rarely at therapeutic doses in susceptible patients), it may cause fulminant hepatic necrosis.[94] Hepatotoxicity is due to accumulation of a toxic and reactive metabolite, N-acetyl-p-benzoquinoneimine (NAPQI), which is formed by cytochrome P450 oxidation of acetaminophen. NAPQI covalently binds to hepatic cell proteins, causing cell necrosis. Under normal circumstances, when acetaminophen is taken at therapeutic doses, glutathione binds to and detoxifies NAPQI by forming mercapturic acid, a stable, renally-excreted metabolite. If the formation of NAPQI exceeds the binding capacity of glutathione, as occurs in the setting of massive overdose and/or concomitant alcohol ingestion (which both depletes glutathione *and* induces cytochrome P450 enzymes), hepatic necrosis may ensue. In alcoholic patients, for this reason, acetaminophen hepatic injury may occur even at usually therapeutic dose ranges.[93,95] However, a systematic review of the effect of therapeutic doses of acetaminophen in patients with chronic alcoholism showed that serious confounding conditions that could have caused hepatotoxicity were present in nearly all reported cases.[96] Furthermore, administration of acetaminophen to chronic alcoholics shortly after their admission to a detoxification center resulted in no greater increase in serum transaminase levels than placebo.[97] Whether patients with nonalcoholic chronic liver disease are at risk for acetaminophen hepatotoxicity is unknown at present. Benson[98] reported that administration of acetaminophen, 4 g/d, to subjects with a variety of chronic liver diseases increased the mean half-life of the drug but did not result in an increase in liver damage. However, these observations reflect an experience in only 20 subjects studied for less than 2 weeks and must be interpreted with caution. No prospective studies examining the risk of hepatotoxicity from chronic administration of therapeutic doses of acetaminophen in patients with underlying liver disease have been performed. The antidote for acetaminophen overdose, acetylcystine, repletes hepatic glutathione stores, but it is only effective if administered promptly, before irreversible liver damage has transpired.

Safety of Acetaminophen versus NSAIDs

The safety profile of acetaminophen is not as clean as originally thought, particularly in patients with excess alcohol consumption and/or reduced hepatic function, and at doses of more than 2 g/d, it has been found to convey the same risk for serious upper GI complications as traditional NSAIDs.[99]

Given the data that suggest that NSAIDs may have superior efficacy in selected OA populations, it should be noted that expert recommendations that favor acetaminophen as initial therapy are influenced by its lower cost and perceived greater safety. While there is little doubt that the economics favor acetaminophen use, two questions have emerged regarding adverse events: (1) Is full-dose acetaminophen (4 g/d) associated with more toxicity than previously appreciated; and (2) does the improved GI safety profile of the coxibs change the benefit-to-risk ratio that previously had favored acetaminophen over nonselective NSAIDs?

An additional area that merits consideration concerns the potential association of analgesic drugs and chronic renal insufficiency. In a recent nationwide, case-controlled study, Swedish investigators demonstrated that regular use of either aspirin or acetaminophen was associated, in a dose-dependent manner, with an increased risk of chronic renal failure.[100] Based on these and prior studies, it is recommended that clinicians should consider carefully the use of aspirin, acetaminophen, and NSAIDs in patients with chronic renal disease.[101]

The most serious adverse events seen with use of nonselective NSAIDs are related to NSAID-associated peptic ulcer disease and its complications (e.g., hemorrhage, perforation, gastric outlet obstruction). Notably, many patients who incurred a serious GI complication from use of an NSAID have not had prior GI symptoms.[59] Furthermore, GI hemorrhage appears to be associated not only with prescription use of aspirin and other NSAIDs but also with over-the-counter (OTC) use of these agents. In a recent study of more than 400 patients undergoing evaluation for upper GI hemorrhage, use of OTC aspirin or a nonaspirin NSAID during the week prior to admission was reported by 35 and 9 percent, respectively, whereas prescription use of aspirin or a nonaspirin NSAID was reported in 6 and 14 percent, respectively.[102] Given the very common use of OTC agents, short-term NSAID use may be a major cause of ulcer-related GI hemorrhage.

Clinical trials that have compared acetaminophen with NSAIDs in the treatment of OA generally have been short term (<12 weeks), using 4 g/d of acetaminophen. In these studies, the risk of adverse GI events associated with the use of nonselective NSAIDs has been greater than that with acetaminophen, resulting in a benefit-to-risk ratio that favored acetaminophen. However, a recent nested case-control study by García Rodríguez and colleagues,[103] using the United Kingdom General Practitioners Research database, indicated that after adjustment for age, sex, ulcer history, smoking, and the use of steroids, anticoagulants, gastroprotective drugs, aspirin, and prescription (but not OTC) NSAIDs, doses of acetaminophen greater than 2 g/d conferred a risk for upper GI complications as great as that with traditional NSAIDs. If this observation is correct, the mechanism underlying upper GI toxicity of high-dose acetaminophen theoretically could be related to its ability to function as a weak inhibitor of COX-1.[74,104] It should be noted that these recent findings by García Rodríguez and colleagues[44] require further investigation. Direct comparisons between COX-2 specific inhibitors and acetaminophen in large outcome trials exceeding 6 months' duration are not available but are needed to determine the most appropriate initial pharmacologic therapy of patients with OA.

▶ SUMMARY AND CONCLUSION

NSAIDs are among the most widely prescribed drugs used for the treatment of OA and pain. While simple analgesics such as acetaminophen are indicated for initial pharmacologic therapy of OA, some patients will obtain insufficient benefit from acetaminophen and will require treatment with NSAIDs for symptomatic improvement. In some instances, symptomatic improvement with an NSAID is greater than with acetaminophen. The use of NSAIDs has been associated with serious adverse events in 1 to 4 percent of patients, including GI hemorrhage and perforation, congestive heart failure, and renal insufficiency. The introduction of selective COX-2 inhibitors has improved the benefit-to-risk ratio in comparison with nonselective COX inhibitors because of the decreased incidence of serious GI adverse events. However, the coxibs offer no apparent advantage over conventional NSAIDs with respect to other toxicities, such as hypertension, fluid retention, and congestive heart failure, presumably because COX-2 is expressed not only at sites of inflammation but also in normal tissue, often in response to conditions of physiologic stress. Because they do not confer the antiplatelet effects associated with COX-1 inhibitors, there is some concern that selective COX-2 inhibitors may increase the risk for cardiovascular thrombotic events. Although short-term clinical pain models support the efficacy of coxibs, large clinical outcome trials will be needed to determine the validity of unresolved safety concerns.

REFERENCES

1. Fitzgerald GA, Patrono C: The coxibs: Selective inhibitors of cyclooxygenase-2. *N Engl J Med* 2001;345:433–442.
2. DuBois RN, et al: Cyclooxygenase in biology and disease. *FASEB J* 1998;12:1063–1073.
3. Smith WL, Langenbach R: Why there are two cyclooxygenase isozymes. *J Clin Invest* 2001;107:1491–1495.
4. Lipsky PE, et al: Analysis of the effect of COX-2 specific inhibitors and recommendations for their use in clinical practice. *J Rheumatol* 2000;27:1338–1340.
5. Silverstein FE, et al: Gastrointestinal toxicity with celecoxib vs nonsteroidal anti-inflammatory drugs for osteoarthritis and rheumatoid arthritis. The CLASS Study: A randomized controlled trial. *JAMA* 2000;284:1247–1255.
6. Bombardier C, et al: A double-blind comparison of rofecoxib and naproxen on the incidence of clinically important upper gastrointestinal events: The VIGOR trial. *N Engl J Med* 2000;343:1520–1528.
7. Fitzgerald GA, Loll P: COX in a crystal ball: Currrent status and future promise of prostaglandin research. *J Clin Invest* 2001;107:1335–1337.
8. Brater DC, et al: Renal effects of COX-2 selective inhibitors. *Am J Nephrol* 2001;21:1–15.
9. Swan SK, et al: Effect of COX-2 inhibition on renal function in elderly persons receiving a low salt diet. *Ann Intern Med* 2000;133:1–9.
10. Cashman JN: The mechanisms of action of NSAIDs in analgesia. *Drugs* 1996;52:13–23.
11. Stewart WF, et al: Risk of Alzheimer's disease and duration of NSAID use. *Neurology* 1997;48:626–632.
12. Bas A, et al: Nonsteroidal anti-inflammatory drugs and the risk of Alzheimer's disease. *N Engl J Med* 2001;345:1515–1521.
13. Saag KG, et al: Nonsteroidal antiinflammatory drugs and cognitive decline in the elderly. *J Rheumatol* 1995;22:2142–2147.
14. Dayer JM, et al: Production of collagenase and prostaglandins by isolated adherent rheumatoid synovial cells. *Proc Natl Acad Sci USA* 1976;73:945–949.
15. Crofford J: COX-2 in synovial tissues. *Osteoarth Cartilage* 1999;7:406–408.
16. Siegle L, et al: Expression of cyclooxygenase 1 and cyclooxygenase 2 in human synovial tissue. *Arthritis Rheum* 1998;41:122–129.
17. Amin AR, et al: Superinduction of cyclooxygenase-2 activity in human osteoarthritis-affected cartilage. *J Clin Invest* 1997;99:1231–1237.
18. Pelletier JP, Martel-Pelletier J, Abramson SB: Osteoarthritis, an inflammatory disease: Potential implication of new therapeutic targets. *Arthritis Rheum* 2001;44:1237–1247.
19. Lippiello L, et al: Involvement of prostaglandins from rheumatoid synovium in inhibition of articular cartilage metabolism. *Arthritis Rheum* 1978;21:909–917.
20. Creamer P, FLores R, Hochberg M: The management of osteoarthritis in older adults. *Clin Geriatr Med* 1988;14:435–454.
21. Eccles M, Freemantle N, Mason J, for the North of England Non-Steroidal Anti-Inflammatory Drug Guideline Development Group: North of England evidence-based guideline development project: Summary guideline for nonsteroidal anti-inflammatory drugs versus basic analgesia in treating the pain of degenerative arthritis. *Br Med J* 1998;317:526–530.
22. Moskowitz R: The role of anti-inflammatory drugs in the treatment of osteoarthritis: A United States viewpoint. *Clin Exp Rheumatol* 2001;19(suppl 25):S3–S8.
23. Bensen W, Fiechtner J, McMillen JI: Treatment of osteoarthritis with celecoxib, a cyclooxygenase-2 inhibitor: A randomized, controlled trial. *Mayo Clin Proc* 1999;74:1095–1105.
24. Detora L, et al: Rofecoxib shows consistent efficacy in osteoarthritis clinical trials, regardless of specific patient demographic and disease factors. *J Rheumatol* 2001;28:2494–2503.

25. Ehrich E, Schnitzer T, McIlwain H: Effect of specific COX-2 inhibition in osteoarthritis of the knee: A 6 week double-blind, placebo-controlled pilot study of rofecoxib. Rofecoxib Osteoarthritis Pilot Study Group. *J Rheumatol* 1999;26:2438–2447.
26. Kivitz A, Moskowitz R, Woods E: Comparative efficacy and safety of celecoxib and naproxen in the treatment of osteoarthritis of the hip. *J Int Med Res* 2001;29:467–479.
27. McKenna F, et al: Celecoxib versus diclofenac in the management of osteoarthritis of the knee. *J Scand J Rheumatol* 2001;30:11–18.
28. Saag K, Heijde D, Fisher C: Rofecoxib, a new cyclooxygenase 2 inhibitor, shows sustained efficacy, comparable with other nonsteroidal anti-inflammatory drugs: A 6-week and a 1-year trial in patients with osteoarthritis. Osteoarthritis Studies Group. *Arch Fam Med* 2000;9:1124–1134.
29. Truitt K, Sperling R, Ettinger WH Jr: A multicenter, randomized, controlled trial to evaluate the safety profile, tolerability, and efficacy of rofecoxib in advanced elderly patients with osteoarthritis. *Aging (Milano)* 2001;13:112–121.
30. Williams G, et al: Comparison of once-daily and twice-daily administration of celecoxib for the treatment of osteoarthritis of the knee. *Clin Ther* 2001;23:213–227.
31. Ehrich E, et al: Effect of rofecoxib therapy on measures of health-related quality of life in patients with osteoarthritis. *Am J Manag Care* 2001;7:609–616.
32. Lisse J, et al: Functional status and health-related quality of life of elderly osteoarthritic patients treated with celecoxib. *J Gerontol* 2001;56:M167–M175.
33. Goldstein J, et al: Reduced incidence of gastroduodenal ulcers with celecoxib, a novel cyclooxygenase-2 inhibitor, compared to naproxen in patients with arthritis. *Am J Gastroenterol* 2001;96:1019–1027.
34. Zhao S, McMillen J, Markenson JA: Evaluation of the functional status aspects of health-related quality of life of patients with osteoarthritis treated with celecoxib. *Pharmacotherapy* 1999;19:1269–1278.
35. Acevedo E, Castaneda O, Ugaz M: Tolerability profiles of rofecoxib (Vioxx) and Arthrotec: A comparison of six weeks treatment in patients with osteoarthritis. *Scand J Rheumatol* 2001;30:19–24.
36. Cannon G, Caldwell J, Holt P: Rofecoxib, a specific inhibitor of cyclooxygenase-2, with clinical efficacy comparable with that of diclofenac sodium: Results of a one-year, randomized clinical trial in patients with osteoarthritis of the knee and hip. Rofecoxib Phase III Protocol 035 Study Group. *Arthritis Rheum* 2000;43:978–987.
37. Day R, et al: A randomized trial of the efficacy and tolerability of the COX-2 inhibitor rofecoxib vs ibuprofen in patients with osteoarthritis. Rofecoxib/Ibuprofen Comparator Study Group. *Arch Intern Med* 2000;160:1781–1787.
38. Langman M, Jensen D, Watson DJ: Adverse upper gastrointestinal effects of rofecoxib compared with NSAIDs. *JAMA* 1999;282:1929–1933.
39. Dougados M: The role of anti-inflammatory drugs in the treatment of osteoarthritis: A European viewpoint. *Clin Exp Rheumatol* 2001;19:S9–S14.
40. Huskisson E, et al on behalf of the LINK Study Group: Effects of anti-inflammatory drugs on the progression of osetoarthritis of the knee. *J Rheumatol* 1995;22:1941–1946.
41. Cooper S: Five studies on ibuprofen for postsurgical dental pain. *Am J Med* 1984;77:70–77.
42. Dionne R, Khan A, Gordon S: Analgesia and COX-2 inhibition. *Clin Exp Rheumatol* 2001;19:S63–70.
43. Clemett D, Goa K: Celecoxib: A review of its use in osteoarthritis, rheumatoid arthritis and acute pain. *Drugs* 2000;59:957–980.
44. Morrison B, et al: Analgesic efficacy of the cyclooxygenase-2-specific inhibitor rofecoxib in post-dental surgery pain: A randomized, controlled trial. *Clin Ther* 1999;21:943–953.
45. Chang D, et al: Rofecoxib versus codeine/acetaminophen in postoperative dental pain: A double-blind, randomized, placebo- and active comparator-controlled clinical trial. *Clin Ther* 2001;23:1446–1455.
46. Fricke J, Morrison B, Christensen S: MK-966 versus naproxen sodium 550-mg in postsurgical dental pain (abstract). *Clin Pharmacol Ther* 1999;645:119.
47. Malmstrom K, et al: Comparison of rofecoxib and celecoxib, two cyclooxygenase-2 inhibitors, in postoperative dental pain: a randomized, placebo- and active-comparator-controlled clinical trial. *Clin Ther* 1999;21:1653–1663.
48. Ehrich E, Dallob A, De Lepeleire I: Characterization of rofecoxib as a cyclooxygenase-2 isoform inhibitor and demonstration of analgesia in the dental pain model. *Clin Pharmacol Ther* 1999;65:336–347.
49. Reicin A, Brown J, Jove M: Efficacy of single-dose and multidose rofecoxib in the treatment of post-orthopedic surgery pain. *Am J Orthop* 2001;30:40–48.
50. Reuben S, Connelly N: Postoperative analgesic effects of celecoxib or rofecoxib after spinal fusion surgery. *Anesth Analg* 2000;91:1221–1225.
51. Coco A: Primary dysmenorrhea (review). *Am Fam Phys* 1999;60:489–496.
52. Chan W, Fuchs F, Powell A: Effects of naproxen sodium on menstrual prostaglandins and primary dysmenorrhea. *Obstet Gynecol* 1983;61:285–291.
53. Zhang W, Po LW: Efficacy of minor analgesics in primary dysmenorrhea: A systematic review. *Br J Obstet Gynaecol* 1998;105:780–789.
54. Morrison B, Daniels S, Kotey P: Rofecoxib, a specific cyclooxygenase-2 inhibitor, in primary dysmenorrhea: a randomized controlled trial. *Obstet Gynecol* 1999; 94:504–508.
55. Reuben S, Bhopatkar S, Maciolek, H: The preemptive analgesic effect of rofecoxib after ambulatory arthroscopic knee surgery. *Anesth Analg*, 2002;94:55–59.
56. Ruoff G, Lema M: Strategies in pain management: New and potential indications for COX-2 specific inhibitors. *J Symptom Pain Manage* 2003;25:21–31.
57. Furst D, Manning D: Future directions in pain management. *Clin Exp Rheumatol* 2001;19:S71–S76.
58. Hernandez-Diaz, García Rodríguez LA: Epidemiological assessment of the safety of conventional nonsteroidal anti-inflammatory drugs. *Am J Med* 2001;110:S20–S27.
59. Singh G, et al: Gastrointestinal tract complications of non-steroidal anti-inflammatory drug treatment in rheumatoid arthritis. *Arch Intern Med* 1996;156:1530–1536.
60. Singh G, Triadafilopoulos G: Epidemiology of NSAID-induced gastrointestinal complications. *J Rheumatol* 1999;26:18–24.

61. Laine L, et al: A randomized trial comparing the effect of rofecoxib, a cyclooxygenase-2 specific inhibitor, with that of ibuprofen on the gastroduodenal mucosa of patients with osteoarthritis. *Gastroenterology* 1999;117:776–783.
62. Hawkey C, et al: Comparison of the effect of rofecoxib (a cyclooxygenase-2 inhibitor), ibuprofen, and placebo on the gastroduodenal mucosa of patients with osteoarthritis: A randomized, a double-blind, placebo-controlled trial. *Arthritis Rheum* 2000;43:370–377.
63. Goldstein JL, et al: Reduced incidence of gastroduodenal ulcers with celecoxib, a novel cyclooxygenase-2 inhibitor, compared to naproxen in patients with arthritis. *Am J Gastroenterol* 2001;96:1019–1027.
64. Feldman M, McMahon AT: Do cyclooxygenase-2 inhibitors provide benefits similar to those of traditional nonsteroidal anti-inflammatory drugs, with less gastrointestinal toxicity? *Ann Intern Med* 2000;132:134–143.
65. Langman MJ, et al: Adverse upper gastrointestinal effects of rofecoxib compared with NSAIDs. *JAMA* 1999;282:1929–1933.
66. FDA Arthritis Advisory Committee Meeting, Gaithersburg, MD, February 7, 2001.
67. Garcia Rodriguez LA, Ruigomez A: Secondary prevention of upper gastrointestinal bleeding associated with maintenance acid-suppressing treatment in patients with peptic ulcer bleed. *Epidemiology* 1999;10:228–232.
68. Yeomans ND, et al: A comparison of omeprazole with ranitidine for ulcers associated with nonsteroidal anti-inflammatory drugs. Acid Suppression Trial: Ranitidine versus Omeprazole for NSAID-Associated Ulcer Treatment (ASTRONAUT) Study Group. *N Engl J Med* 1998;338:719–726.
69. Silverstein FE, et al: Misoprostol reduces serious gastrointestinal complications in patients with rheumatoid arthritis receiving nonsteroidal anti-inflammatory drugs: A randomized, double-blind, placebo-controlled trial. *Ann Intern Med* 1995;123:241–249.
70. Del Rincon I, et al: High incidence of cardiovascular events in a rheumatoid arthritis cohort not explained by traditional cardiac risk factors. *Arthritis Rheum* 2001;44:2737–2745.
71. Hennan JK, et al: Effects of selective cyclooxygenase-2 inhibition on vascular responses and thrombosis in canine coronary arteries. *Circulation* 2001;104:820–825.
72. Cippollone F, et al: Overexpression of functionally coupled cyclooxygenase-2 and prostaglandin E synthase in symptomatic atherosclerotic plaques as a basis of prostaglandin E_2-dependent plaque instability. *Circulation* 2001;104:921–927.
73. Lanas A: Nonsteroidal anti-inflammatory drugs, low-dose aspirin, and potential ways of reducing the risk of complications. *Eur J Gastroenterol Hepatol* 2001;13:623–626.
74. Catella-Lawson F, et al: Cyclooxygenase inhibitors and the antiplatelet effects of aspirin. *N Engl J Med* 2001;345:1809–1817.
75. Catella-Lawson F, Crofford LJ: Cyclooxygenase inhibition and thrombogenicity. *Am J Med* 2001;110:S12–S32.
76. Altman R, et al: Recommendations for the medical management of osteoarthritis of the hip and knee. *Arthritis Res* 2000;43:1905–1915.
77. Hochberg M, Altman R, Brandt KD: Guidelines for the medical management of osteoarthritis: Osteoarthritis of the hip. *Arthritis Rheum* 1995;38:1535–1540.
78. Pendleton A, Arden N, Dougados M: EULAR recommendations for the management of knee osteoarthritis: Report of a task force of the Standing Committee for International Clinical Studies Including Therapeutic Trials (ESCISIT). *Ann Rheum Dis* 2000;59:936–944.
79. Brandt KD, Bradley JD: Should the initial drug used to treat osteoarthritis pain be a nonsteroidal antiinflammatory drug? *J Rheumatol* 2001;28:467–473.
80. Mazzucas A, et al: Effects of self-care education on the health status of inner-city patients with osteoarthritis of the knee. *Arthritis Rheum* 1997;40:1466–1474.
81. American College of Rheumatology Subcommittee on Osteoarthritis Guidelines: Recommendations for the medical management of osteoarthritis of the hip and knee: 2000 update. *Arthritis Rheum* 2000;43:1905–1915.
82. Hochberg MC, et al: Guidelines for the medical management of osteoarthritis: I. Osteoarthritis of the hip. II. Osteoarthritis of the knee. *Arthritis Rheum* 1995;38:1535–1546.
83. Pendleton A, et al: EULAR recommendations for the management of knee osteoarthritis: Report of a task force of the Standing Committee for International Clinical Studies Including Therapeutic Trials (ESCISIT). *Ann Rheum Dis* 2000;59:936–944.
84. Bradley JD, et al: Comparison of an anti-inflammatory dose of ibuprofen, an analgesic dose of ibuprofen, and acetaminophen in the treatment of patients with osteoarthritis of the knee. *N Engl J Med* 1991;325:87–91.
85. Bradley JD, et al: Treatment of knee osteoarthritis: Relationship of clinical features of joint inflammation to the response to a nonsteroidal anti-inflammatory drug or pure analgesic. *J Rheumatol* 1992;19:1950–1954.
86. Wolfe F, Zhao S, Lane N: Preference for nonsteroidal anti-inflammatory drugs over acetaminophen by rheumatic disease patients. *Arthritis Rheum* 2000;43:378–385.
87. Pincus T, et al: A randomized, double-blind, crossover clinical trial of diclofenac plus misoprostol versus acetaminophen in patients with osteoarthritis of the hip or knee. *Arthritis Rheum* 2001;44:1587–1598.
88. Eccles M, Freemantle N, Mason J: North of England Evidence Based Guideline Development Project: Summary guideline for non-steroidal anti-inflammatory drugs versus basic anagesia in treating the pain of degenerative arthritis. *Br Med J* 2001;1998:317–526.
89. Felson DT: The verdict favors nonsteroidal antiinflammatory drugs for the treatment of osteoarthritis and a plea for more evidence on other treatments. *Arthritis Rheum* 2001;44:1477–1480.
90. Altman RD, et al: Recommendations for the medical management of osteoarthritis of the hip and knee. *Arthritis Rheum* 2000;43:1905–1915.
91. Bradley JD, Katz BP, Brandt KD: Severity of knee pain does not predict a better response to an anti-inflammatory dose of ibuprofen than to analgesic therapy in patients with osteoarthritis. *J Rheumatol* 2001;28:1073–1076.

92. Geba G, et al: Vioxx, Acetaminophen, Celecoxib Trial (VACT) Group. Efficacy of rofecoxib, celecoxib, and acetaminophen in osteoarthritis of the knee: A randomized trial. *JAMA* 2002;287:64–71.
93. Tanaka E, Yamazaki K, Misawa S: Update: The clinical importance of acetaminophen hepatotoxicity in nonalcoholic and alcoholic subjects. *J Clin Pharmacol Ther* 2000;25:325–332.
94. Tolman K: Hepatotoxicity of antirheumatic drugs. *J Rheumatol* 1990;17:6–11.
95. McClain CJ, et al: Acetaminophen hepatotoxicity: An update. *Curr Gastroenterol Rep* 1999;1:42–49.
96. Dart RC, Kuffner EK, Rumack BR: Treatment of pain or fever with paracetamol (acetaminophen) in the alcoholic patient: A systematic review. *Am J Med* 2000;7:123–134.
97. Kuffner EK, et al: Effect of maximal daily doses of acetaminophen on the liver of alcoholic patients. *Arch Intern Med* 2001;161:2247–2252.
98. Benson GD: Acetaminophen in chronic liver disease. *Clin Pharmacol Ther* 1983;33:95–101.
99. Rodriguez LG, Hernandez-Diaz S: The risk of upper gastrointestinal complications associated with non-steroidal anti-inflammatory drugs, gluco-corticoids, acetaminophen, and combinations of these agents. *Arthritis Res* 2001;2:98–101.
100. Fored CM, et al: Acetaminophen, aspirin and chronic renal failure. *N Engl J Med* 2001;345:1801–1809.
101. Crofford L: Rational use of analgesic and anti-inflammatory drugs. *N Engl J Med* 2001;345:1844–1846.
102. Wilcox CM, Shalek KA, Cotsonis G: Striking prevalence of over-the-counter nonsteroidal anti-inflammatory drug use in patients with upper gastrointestinal hemorrhage. *Arch Intern Med* 1999;154:42–46.
103. García-Rodríguez LA, Hernandez-Diaz S: The relative risk of upper gastrointestinal complications among users of acetaminophen and non-steroidal anti-inflammatory drugs. *Epidemiology* (in press).
104. Ouellet M, Percival MD: Mechanism of acetaminophen inhibition of cyclooxygenase isoforms. *Arch Biochem Biophys* 2001;387:273–280.

CHAPTER 39

Opioid Therapy

Mellar P. Davis

Among the remedies which it has pleased Almighty God to give man to relieve his sufferings, none is so universal and efficacious as opium.

Sydenham, 1680

Opium, an extract from the poppy, Papaver somniferum, *has five major alkaloids.*[1] *The phenanthrene derivatives consist of morphine, codeine, and thebaine. Although thebaine is not an analgesic, it is the precursor for many of the semisynthetic opioid compounds such as oxycodone.* Opioid *was coined by Acheson to designate drugs that act in a similar fashion to morphine.*[2] *Opioid was the original name given to compounds that were distinctly different from phenanthrene derivatives. Phenanthrene derivatives were called* opiates. *The term* opioid *now has been expanded to include morphine receptor agonists and antagonists that have a wide spectrum of activity and are located throughout the peripheral and central nervous systems.*

▶ OPIOID RECEPTORS

Opioids produce analgesia by binding to three opioid receptors: mu, delta, and kappa (MOR, DOR, and KOR, respectively). Opioid receptors are part of the family of rhodopsin-type G-protein-coupled receptors that initiate cellular activity through activation of a second-messenger cyclic AMP derived via adenylate cyclase. Activation of opioid receptors inhibits adenylate cyclase resulting in decreased calcium flux through voltage-dependent calcium channels. The activation of opioid receptors prevents the release of neuropeptides such as substance P from primary afferent C fibers and hyperpolarize secondary postsynaptic afferent neurons.[3] Opioid receptors are found in the peripheral nerve, dorsal horn, spinal cord, brain stem, along the periaqueductal gray and rostroventral medulla, thalamus, and cortex. In close proximity to opioid receptor sites are the endogenous opioid peptide ligands: enkephalin, endorphin, and dynorphin.[3-5] Receptor function is modulated by G-protein protein kinases. Opioid receptor function will be reduced by non-G-protein-related protein kinases and phospholipases and by induction of extracellular signal-regulated kinases. All these kinases relate to neuroplasticity and, theoretically, opioid analgesic tolerance (reduced analgesia per dose with chronic opioid dosing).[6] Opioid receptor mutations have been described as single nucleotide polymorphisms or exon deletions that reduce receptor density and alter internalization and activation. Such events determine intrinsic efficacy despite similar opioid receptor-binding affinities.[6-10]

▶ OPIOIDS

Clinical differences between various opioids are not only determined by opioid receptor configuration and physiology but also by the physicochemical properties of opioids that determine opioid bioavailability at effect sites. Such properties influence opioid kinetics and dynamics because opioids need to transverse anatomic and physiologic barriers, including the blood-brain barrier and spinal cord, to reach receptor sites (exclusive of peripheral receptors).[11] Such physicochemical differences lead to different rates and degrees of absorption and metabolism by type 1 and type 2 hepatic and extrahepatic cytochrome enzymes, the volume of distribution, redistribution by the enterhepatic recirculation, mobilization from tissue storage sites, and elimination. Lipophilic opioids such as fentanyl and methadone will transverse the lipid based blood-brain barrier more easily than the hydrophilic opioids such as morphine, oxycodone, and hydromorphone. However, owing to a partition coefficient of greater than 2 (particularly for fentanyl), lipophilic opioids become sequestered in myelin and epidural fat or redistribute systematically without directly reaching dorsal gray (Fig. 39-1). The relative lipophilic opioid potency ratio to morphine depends on the route of administration and half-life of the opioid.[11] In addition, some opioids have active metabolites that add to analgesia and/or toxicity. For instance, morphine is derived from codeine, morphine-6-glucuronide is derived from morphine, and desmethyl tramadol is derived from tramadol. The ionization constant (pK_a) and the drug transporter, P-glycoprotein, as well as protein binding, influence drug concentration at effector sites.[11–14] As a result, plasma pharmacokinetics are but a single factor determining analgesia, and plasma levels of opioid are little help in predicting opioid response.

Pharmacogenetics account for large differences in dose-responses among individuals and even larger differences in a single individual among opioids. Polymorphisms have been described for *UGT* regulator genes (particularly intestinal *UGT2B7* for morphine metabolism), the *CYP3A4* regulator gene (responsible for methadone and fentanyl metabolism), and the *CYP2D6* structural gene (responsible for methadone, codeine, tramadol, oxycodone, and hydrocodone metabolism).[15–25] Such polymorphisms are responsible for opioid kinetic differences among individuals, as well as opioid-drug interactions.[19] Opioid receptor pharmacogenetics determine response and non-cross-tolerance among opioids.[16–18,24,25]

▶ EQUIANALGESIA AND OPIOID ROTATION

Clinical equivalence between two opioids focuses on efficacy (analgesia) as an outcome. *Prescribability* of an opioid refers to the average performance of a drug such as morphine given for the first time. Prescribability may be different based on clinical circumstances. Intrinsic and extrinsic factors will influence prescribability and rarely are accounted for in opioid equivalent tables (Table 39-1). The type of pain, organ function, interfering medications, and other factors are among several intrinsic and extrinsic factors that influence equianalgesia.[26] Opioid selection varies based on clinical circumstances. By tradition, morphine is preferred for cancer pain and fentanyl for postoperative pain. Methadone is a logical choice in renal failure. *Switchability* is the transfer of a patient from one opioid to another opioid, also known as *opioid rotation*. Switchability depends on the known equivalence derived from experience.

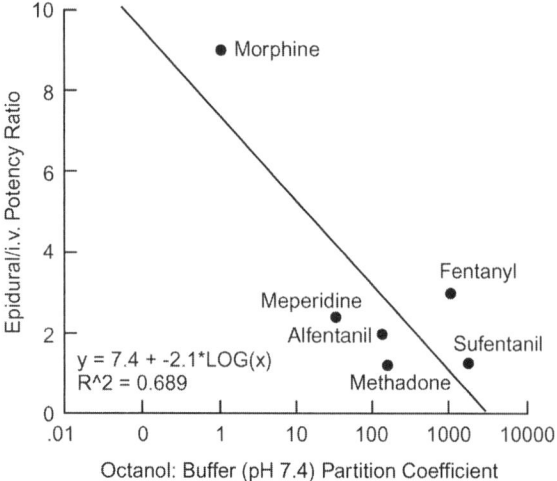

Figure 39-1. Relative potency of intraspinal morphine increases compared with lipophilic opioids. High octanol:buffer partition coefficient opioids quickly exit the spinal canal for epidural fat or remain bound to myelin rather than reaching dorsal horn. *(Reprinted with permission from Cambridge University Press).*[11]

▶ **TABLE 39-1.** SELECTED INTRINSIC AND EXTRINSIC FACTORS

	Prescribing	Switching
Intrinsic Factor		
Genetic polymorphism	Yes	No
Age	Yes	Yes
Gender	Yes	No
Height	Yes	No
Weight	Yes	Yes
Lean body mass	Yes	Yes
Body composition	Yes	Yes
Organ dysfunction	Yes	Yes
Extrinsic Factor		
Concomitant medicine	Yes	Yes
Diet	Yes	Yes
Tobacco	Yes	Yes
Alcohol	Yes	Yes

NOTE: Prescribing refers to the dose selection between populations. Switching refers to changing the dosing conditions in an individual subject. (Reprinted with permission from Elsevier).[26]

▶ TABLE 39-2. EQUIANALGESIA

Drug	Oral	IV	Oral/IV Potency
Morphine	30	10 mg	1:3
Oxycodone	30[1]	—	—
Hydromorphone	7.5[1]	1.5 mg	1:5
Fentanyl	—	125 mcg	—
Codeine	300	—	—
Methadone	Morphine <90 mg/day 1:4 (methadone:morphine) >90 <300 mg/day 1:8 >300 mg/day 1:12[2]	½ oral	1:2
Tramadol	300 mg[3]	—	—

[1]Subject to directional differences depending upon the sequence of administration with wide differences in published potency ratios
[2]Potency of methadone increases and ratio changes with morphine dose
[3]Oral potency of tramadol has been published to be 1/3 of morphine with poor individual predictability
Twycross R[156]
Anderson R[28]
Pereira J[27]
Mercadante S[34]
Quigley C, Wiffen P[165]
Wilder-Smith CH[166]

Equianalgesic tables assume that the rotation is both safe and efficacious[26] (Table 39-2). Intrinsic and extrinsic factors may alter equivalence and lead to inaccuracies, resulting in under- or overdosing with an alternative opioid. The primary indication for opioid rotation is opioid toxicity and incompletely controlled pain. Alternatives to rotation in such situations are *opioid conversion* (to a different route), *opioid sparing* (by the addition of an adjuvant), or simply treating opioid toxicity and maintaining opioid dosing (particularly in the actively dying). Paradoxically, these bioequivalent tables are not used in opioid-naïve individuals for equivalency but for the purpose of non-cross-tolerance among opioids in rotation. Many equianalgesic tables are based on acute pain (i.e., a non-steady-state dosing). The methadone-to-morphine ratio in non-steady-state equianalgesic tables is 10 mg morphine to 10 mg methadone; but not for chronic pain (see Table 39-2). It is assumed that equivalences are the same for age and gender, and there is no pharmacogenetic differences nor differences with organ failure or interfering medications between opioids. It is also assumed that equivalences are directionless when switching from one opioid to another.[27,28] Potencies are assumed to be dose-independent. All these assumptions are incorrect.[27,28] In addition, there is a wide range of estimated dose ratios published among opioids, particularly for those with significant differences in intrinsic efficacy (as occurs between morphine and methadone). Gender differences in pain response to opioids are well described.[29,30] Routes of administration alter dose ratios and are related to bioavailability and lipophilicity. Intraspinal lipophilic opioids become relatively less potent compared to hydrophilic opioids. Hydrophilic opioids such as morphine are retained within the cerebrospinal fluid (CSF). There are several rules of thumb when using equianalgesic tables: (1) Use tables with steady-state equivalence (for chronic pain). (2) Equianalgesic dose ratios or guidelines need to be adapted or modified to the individual patient owing to intrinsic and extrinsic factors. (3) Equianalgesic dosing should be reduced by a third when rotating to an alternative opioid owing to incomplete analgesic cross-tolerance. (4) Be aware of rotating "into" or "out of" interfering drug combinations that will influence opioid kinetics and bioavailability. (5) Methadone is more potent than previously recognized and has a dose-dependent equivalence ratio with morphine and hence requires a unique dosing strategy compared with other opioids. (6) When rotating from high doses of morphine, use 50 percent or less of the equianalgesic dose ratio. (7) Take into account bidirectional differences in opioid dose ratios when rotating between morphine and oxycodone or morphine and hydromorphone.[27,28] (8) Subcutaneous fentanyl-to-subcutaneous morphine potency ratios (inverse of equianalgesic dose ratios) are 80 to 100:1 rather than 150:1, as published by the manufacturer. (9) Monitor frequently when rotating.[27,28,31]

▶ TOLERANCE

Tolerance can be a pharmacodynamic, pharmacokinetic, or psychological process. Tolerance cannot be said to occur if the underlying disease is advancing. Diminutive analgesic response with the same dose is primarily due to neuroadaptation. Tolerance inversely correlates with the number of receptors bound and inversely relates to intrinsic efficacy. Intrinsic efficacy is the opioid receptor reserve or number of left unbound opioid receptors with analgesia. Tolerance to analgesia and toxicity dissociates with time. Fortunately the magnitude of tolerance is different for analgesia and toxicity such that the adverse effects of opioids diminish while analgesia is usually maintained, widening the therapeutic window[8,34] (Fig. 39-2). Intrinsic efficacy, dose, and duration of exposure each play a role in analgesic tolerance. Asymmetric analgesic tolerance and altered equivalence also may be due to kinetic differences that involve absorption, metabolism, and elimination.[34] The use of adjuvant analgesics for which tolerance does not develop curtails opioid dose escalation and secondarily reduces analgesic tolerance.[34] How much tolerance quantitatively plays a role in opioid dose escalation with time is unknown.

▶ OPIOID CHOICES

Opioid preferences depend on the clinical situation (Table 39-3). Fentanyl has kinetics that favor use for acute postoperative pain rather than cancer pain. For this reason, fentanyl has become a popular choice for postoperative pain owing to its short duration of action and rapid onset to analgesia. Sufentanil and afentanil are alternatives. Fentanyl is a second-line agent for cancer pain because it lacks the versatility and half-life of morphine and methadone for chronic pain. Drug delivery by transdermal fentanyl has changed this advantage. Methadone is used rarely postoperatively owing to its long half-life and accumulation. Dosing strategies employing patient-controlled analgesia (PCA) and spinal infusion are used more commonly postoperatively owing to the transient nature of pain and reduced postoperative morbidity and greater patient satisfaction with these two strategies. PCA is used less frequently for cancer pain but is effective in managing incident pain. Around-the-clock oral dosing and provision for breakthrough doses are the key to the management of cancer pain but not postoperative pain. Around-the-clock dosing for chronic noncancer pain is one strategy where dosing to activities and time is also an acceptable approach. Opioid rotation for less than optimal response and associated opioid toxicity and the use of adjuvant analgesics are common to noncancer and chronic cancer pain management and not important in acute pain management.[35] Long-acting opioids, methadone and sustained-release preparations, are typical analgesic choices for daily chronic noncancer pain and cancer pain and are preferred to short-acting and parenteral opioids for compliance reasons. Parenteral administration and short-acting opioids are particularly avoided in patients with psychological dependence and chronic noncancer pain. For those on methadone maintenance programs, methadone is continued in order to avoid withdrawal, and either an additional opioid may be added or methadone may be increased to control pain.[36]

Organ failure alters opioid kinetics and influences analgesic choice. Morphine and methadone kinetics are relatively unaltered in liver failure, but renal failure significantly influences the clearance of the active morphine metabolite morphine-6-glucuronide and increases the risk of morphine toxicity. Methadone does not appear to accumulate in renal failure. Hemodialysis normalizes

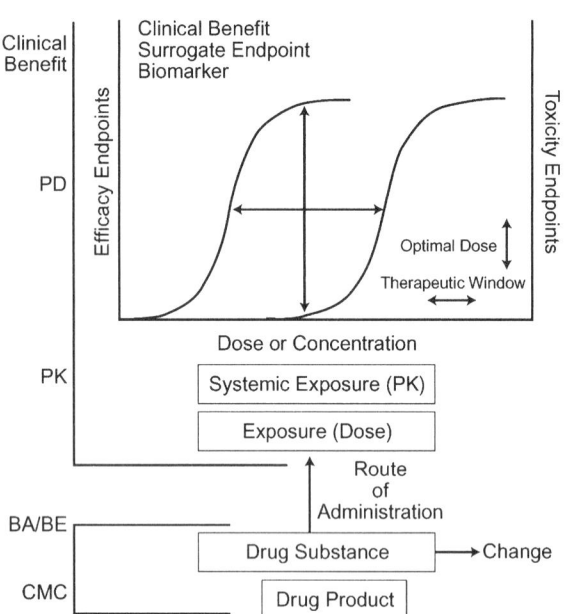

Figure 39-2. Optimal opioid doses occur between efficacy endpoints shaped by pain severity, opioid potency, intrinsic efficacy, and opioid kinetics. Toxicity end points usually are limited by neurotoxicity and, rarely, nausea. The optimal dose will shift based on intrinsic and extrinsic factors that influence individual opioid pharmacology. The relationship between exposure and outcomes defines the optimal dose and therapeutic window. *(Reprinted with permission from Williams RL, Chen ML, Hauck W.[26])*

▶ TABLE 39-3. ADVANTAGES AND DISADVANTAGES OF DIFFERENT ROUTES OF OPIOID ADMINISTRATION

Method of Administration	Advantages	Disadvantages
Oral	• Convenient for staff and patients • Inexpensive • Simple	• Absorption slow and variable • Impractical after surgery due to risk of vomiting and delayed gastric emptying • Problem with first-pass metabolism of morphine
Intramuscular	• Convenient for staff • Inexpensive	• Absorption slow and variable • Uncomfortable for patients
Rectal	• Feasible when oral or parenteral administration not possible • Useful for children	• Absorption slow and variable • Cultural objections in some countries
Intravenous infusion	• Administration simple • Guaranteed absorption	• Risk of respiratory depression and hypoxia • Risk of malnutrition with infusion pump • Requires careful monitoring
IV bolus (titrated effect)	• Enables individualization of therapy • Inexpensive	• Staff training required • Strict monitoring (labor intensive)
Patient-controlled analgesia (PCA)	• High patient satisfaction • Enables individualization of therapy	• Expensive equipment • Risk of malfunction/error • Staff training required • Strict monitoring (labor intensive)
Peripheral nerve blocks	• Excellent analgesia • Superior to other methods • No cardiovascular or respiratory problems	• Requires skilled anesthesiologist

Source: Reprinted with permission, Cambridge University Press.[54]

morphine metabolite kinetics, but peritoneal dialysis does not. Most opioids can be used in the face of renal or hepatic failure but need to be started with low doses and be titrated cautiously with frequent assessments for response and toxicity.

▶ PREEMPTIVE OPIOID THERAPY

Preemptive opioid therapy has been explored as a way of reducing postoperative pain. Uncontrolled acute postoperative pain causes detrimental effects on multiple organ systems.[37] Painful stimulation during an operation causes central nervous system (CNS) sensitization by (1) directly activating high-threshold receptors, (2) producing mechanical and thermal damage to nerves, (3) indirectly activating chemosensitive nociceptors through release of inflammatory mediators, and (4) reducing the threshold of primary nociceptors through release of inflammatory cytokines and growth factors, causing neural depolarization.[38,39] Acute pain after surgery predicts long-term pain; hence, preemptive pain therapy may in fact reduce the development of chronic postoperative pain.[37,40–43] The concept of preemptive analgesia was introduced by Patrick Wall in 1988.[44] Treatment of postoperative pain is seen as a continuum that starts before the surgical incision and extends through surgery until the surgical injury is healed.[38,40] Preemptive opioid benefits include (1) prolonged time to first analgesic dose request after surgery, (2) reduced total analgesic consumed by PCA, (3) pain relief requiring less postoperative opioid, and (4) better recovery and reduced hospitalization.[38] Expected pain severity postoperatively must be significant for preemptive and continuous analgesia to be of value. Large operations, such as abdominal and thoracic surgery, are the ideal setting.[41–43]

Three studies have demonstrated benefits to preemptive morphine prior to hysterectomy.[45–47] Doses of morphine in these studies were 10 mg, 0.15 mg/kg, and 0.15 mg/kg with anesthesia induction, respectively.

Single-bolus doses of short-acting opioids such as fentanyl or alfentanil have marginal benefits.[48] Preoperative short-acting opioids fail to maintain a durable opioid effect during operation necessary to block nociception.[38] Although preemptive epidural fentanyl at 4 μg/kg 15 minutes before surgical incision has been effective, intraoperative infusions of fentanyl are even more effective.[43,49] Opioid-sparing preemptive analgesic combinations with ketamine, an *N*-methyl-D-aspartate (NMDA) receptor blocker, and morphine with epidural doses of 60 and 20 mg, respectively, during abdominal surgery produce longer-lasting analgesia and no differences in adverse effects compared with the usual care.[50] Preemptive intra-articular morphine 3 mg prior to arthroscopic surgery combined with pre- and postoperative surgical bupivacaine and postoperative intra-articular morphine produces a longer duration of analgesia and reduced postoperative opioid consumption than without preemptive therapy.[51] Negative trials with preemptive analgesia or combinations of analgesics have been reported.[52] Some of these negative results are subject to flaws related to patient numbers or trial design.[38] In summary, preoperative intravenous (IV) morphine bolus or continuous infusions of short-acting synthetic opioids started preincision may reduce postoperative pain and have additional benefits.

▶ ACUTE PAIN MANAGEMENT

Acute pain management is centered mainly on postoperative analgesia (see Table 39-5). Progressive improvement in pain control has occurred with the development of PCA techniques with on-demand intravenous opioid. Epidural and intrathecal opioids with or without local anesthetics have been reported to improve postoperative recovery from major surgery.[53–55] A hierarchy of pain management strategies has been proposed by Breivik.[56] Epidural opioids with or without local anesthetics are appropriate for approximately 5 to 15 percent of operations, particularly for those with the greatest risk of cardiopulmonary complications (usually being with major thoracic, abdominal, or orthopedic surgery). Intravenous PCA is required in 10 to 30 percent of patients with low risk for postoperative complications but an anticipated need for 2 days or more of postoperative opioid analgesics. Traditional methods of pain control are intermittent parenteral injections of opioids, the success of which will depend on nursing skills and dedication to pain management. Traditional pain management without assisted devices can be used for uncomplicated surgery with a short expected postoperative pain duration. Traditional pain management is less expensive than PCA and spinal techniques. Comparisons between postoperative PCA technology and traditional methods of pain control have produced mixed results.[57]

Opioid doses differ depending on the type of surgery, the location of the incision, and interindividual differences in pharmacokinetics and pharmacodynamics.[55,58] *Ab libidum* dosing by various routes is the usual practice, which reduces the risk of overdosing but all too frequently leads to inadequate analgesia. Subcutaneous or intravenous opioids are preferred to intramuscular injections, which have no particular advantages and are more painful.[59] An indwelling subcutaneous cannula or butterfly needle reduces the number of needle sticks and improves patient satisfaction. The dosages, intervals, and conversion ratios are the same whether intramuscular, subcutaneous, or intravenous injections are used.

Two groups of patients are undertreated postoperatively. Those on regular doses of opioids preoperatively are at risk of being undertreated postoperatively with *ab libidum* dosing alone without a basal opioid infusion. A protocol of continuous opioids at half the daily preoperative dose and an as-needed increment of one-sixth of the infused dose for titration and rescue can reduce rest pain from 69 to 54 percent and incident pain from 40 to 13 percent compared with as-needed dosing alone in this group of patients.[60] Postoperative patients moved to ICUs are undertreated. In one study, nearly one-half of such patients did not receive opioids at all, and 42 percent had only a single injection per day.[61]

Agonist-Antagonist Opioids, Postoperative Sublingual and Intranasal Analgesia

Agonist-antagonist opioids may be safer than pure mu agonists owing to their ceiling effect for respiratory depression.[3] Buprenorphine, a lipophilic opioid with high mu receptor affinity and rapid sublingual absorption, has favorable characteristics for postoperative analgesia. Potency is 25 to 50 times greater than morphine, and the duration of action is 5 to 6 hours. However, respiratory depression, if it does occur with buprenorphine, is poorly reversed by naloxone.[62,63] Sublingual fentanyl in a candy matrix or as a parenteral solution is readily absorbed transmucosally and has the advantage of a short half-life. Children given 15 to 20 μg/kg of sublingual fentanyl have been treated successfully postoperatively. Facial pruritus and nausea are dose-limiting.[64]

Intranasal fentanyl or sufentanil reduces postoperative pain. Sufentanil 10 to 20 μg or 1.5 to 3 μg/kg or fentanyl 1 μg/kg has been used in children but, theoretically, also could be used in adults.[65–68] Both sufentanil and fentanyl may be given intranasally for the incident pain of dressing changes or with advanced cancer incident pain.[69] Intranasal hydrophilic opioids such as diamorphine and morphine are less effective.[70]

Patient-Controlled Analgesia

Patient satisfaction with analgesia is greater with PCA compared to traditional nurse-controlled *ab libidum* dosing.[71] The advantages of PCA are its ability to deliver doses at 5 to 10 minutes and the patient-centered maintenance of opioid within therapeutic or optimal dose range, thereby improving patient satisfaction.[74] The principles of PCA dosing involve (1) physician-directed opioid titration to pain relief and (2) patient-directed maintenance with small doses at frequent intervals as needed[53] (Fig. 39-4). The first step is intravenous titration, which requires injections of 1 to 2 mg morphine or 10 to 20 µg fentanyl every 4 to 5 minutes until the numerical pain score is 3 (10 is the worst pain and 0 no pain).[74] A 30-minute observational interval for sedation or respiratory depression is then followed by setting the PCA to the titrated dose and assessing every 3 hours.[54] After individual loading doses are titrated, lockout intervals on the PCA device are set at 5- to 10-minute intervals at optimal doses. The median dose of morphine is 1.4 mg (1.6 mg mean) and a lockout interval of 6 minutes (8 minutes mean). Fentanyl doses are 20 µg (23.7 µg mean) and lockout interval 5 minutes (3.5 minutes mean).[58,75] Individualization is the key to safe dosing. Minimal effective doses for postoperative PCA vary widely. Continuous-infusion opioid is an ineffective dosing strategy for acute pain (Tables 39-4 through 39-6).

Pre- and postoperative instructions on use of the PCA device should be given to patients. The goal of therapy is improved function with pain control and not complete eradication of pain. Spouse-controlled PCA is to be discouraged.[76] Vigilance must be constant because malfunction of the PCA device will lead to pain. The addition of a benzodiazepine or barbiturate or performing a regional nerve block while patients are on PCA can lead to respiratory depression.[77,78] Respiratory depression also will occur if renal insufficiency or hypotension due to bleeding develops postoperatively. Both lead to opioid accumulation (particularly morphine's active metabolite morphine-6-glucuronide). Finally, dislodgment of the intravenous catheter may lead to subcutaneous deposits of opioid and delayed respiratory depression.[79,80]

Pharmacokinetics and Opioid Choices for PCA

The duration of opioids analgesia does not correlate with blood concentrations of drugs and varies widely among individuals (Fig. 39-5). The effect delay (the time between the maximum plasma concentrations and development of analgesia) is related to the physicochemical properties of the opioid (see Table 39-6). Lipophilic opioids have a shorter time to analgesia compared with hydrophilic opioids.[53] The effect delay is independent of effective CNS dwell time or duration of opioid action (see Fig. 39-5). Morphine has a long time to effective CNS concentration (i.e., a delayed onset to analgesia) compared with fentanyl and a prolonged CNS dwell time compared with fentanyl, which dampens the fluctuations in effective levels with intermittent doses[53] (see Fig. 39-3 and Table 39-3). Therefore, the kinetics of

▶ TABLE 39-4. RELATIVE CENTRAL NERVOUS SYSTEM CONCENTRATION PROFILES FOR VARIOUS ADMINISTRATION REGIMENS OF OPIOIDS

Drug and Route of Administration	Time of Relative Onset (min)	t_{max} (min)	Relative Duration (min)
Single Intravenous Bolus			
Morphine	6	19	96
Fentanyl	2	4	7
Alfentanil	1	2	2
Constant Intravenous Infusion			
Morphine	420	1250	900
Fentanyl	950	3000	2090
Alfentanil	190	510	340
Single Intramuscular Dose			
Morphine	20	48	110
Single Oral Dose			
Morphine	37	82	139

NOTE: The duration of the relative CNS concentration does not relate to the clinically observed duration of action but is a tool for assessing the suitability of various opioids and regimens for the titration process. Reprinted by permission from Adis International Ltd.[53]
ABBREVIATIONS: t_{max} = time to reach peak concentration following drug administration

▶ **TABLE 39-5.** POSTOPERATIVE ANALGESIA TECHNIQUES

I. *Administration of opioids*
   Intramuscular injection
   Subcutaneous (intermittent bolus injection, continuous infusion)
   Oral (tablets, mixture)
   Patient-controlled analgesia (PCA)
   Rectal
   Intravenous (intermittent bolus, continuous infusion)
   Epidural (intermittent bolus, continuous infusion)
   Sublingual
   Oral transmucosal (Oralet) ("lollipop")
   Transdermal (regular "patch", iontophoresis "patch")
   Intranasal

II. *Administration of nonopioid analgesics*
   Acetaminophen (oral, rectal)
   Nonsteroidal anti-inflammatory drugs (NSAIDs)
       (oral, rectal, IM, IV, intra-articular)
   Dipyrone (Novalgin) (oral, rectal, IM, IV)

III. *Regional Techniques*
   Epidural (local anesthetics and/or opioids, and/or clonidine)
   Spinal (local anesthetics and/or opioids, and/or clonidine)
   Paravertebral
   Peripheral nerve blocks
   Wound infiltration
   Interpleural
   Intra-articular (local anesthetic and/or opioid)

IV. *Nonpharmacologic methods*
   Transcutaneous electrical nerve stimulation (TENS)
   Cryoanalgesia
   Acupuncture

V. *Psychological methods*

Reprinted with permission of Cambridge University Press.[54]

intravenous morphine by PCA confer little improvement over upgraded traditional postoperative analgesia.[53,81] Lipophilic opioids such as fentanyl have a short effect delay time and rapid redistribution (i.e., a short CNS dwell time).[58] However, chronic dosing leads to fentanyl accumulation once peripheral tissue deposit sites in fat and muscle are saturated. This is irrelevant in acute pain management with short duration of fentanyl exposure.[11]

▶ **TABLE 39-6.** MINIMUM EFFECTIVE CONCENTRATIONS (MEC) DURING POSTOPERATIVE PCA

Analgesic	Median	Minimum	Maximum	Variability (%)	
		(ng/ml)		Intrasubject	Intersubject
Sufentanil	0.04	0.01	0.56	80.0	81.0
Buprenorphine	0.38	0.01	6.56	67.9	107.3
Fentanyl	1.16	0.18	8.01	27.2	63.9
Alfentanil	14.87	0.57	99.20	37.0	62.5
Tramadol	287.7	20.20	936.30	38.2	59.1

SOURCES: Lehman, 1993, 1994
Minimum effect opioid concentrations vary significantly between individuals, which illustrates the need for individual titration to effect rather than upon serum concentrations. Reprinted with permission of Cambridge University Press.[58]

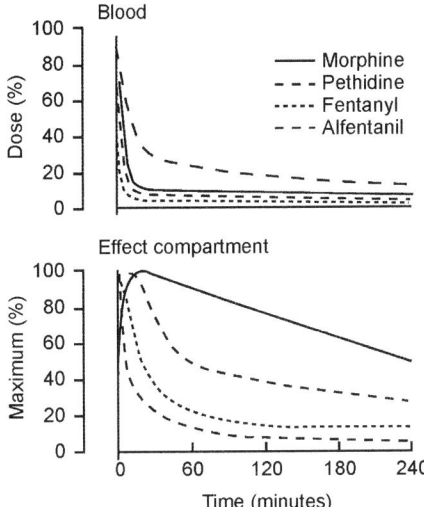

Figure 39-3. The duration-of-effect compartment concentrations of opioids differ more than their blood concentrations. The blood concentrations of various opioids shown in the figure are expressed as a percentage of the dose and are superimposed. The simultaneous-effect compartment concentrations are expressed as a percentage of the peak-effect concentration shown for comparison. The longer the effect delay, the larger is the difference in the time course of the blood and effect compartment concentrations. *(Reprinted with permission from Upton RN, Semple TJ, Macintyre PE).*[53]

As a result, effective doses better correlate with blood concentrations of fentanyl than morphine.[82] Continuous infusions (rather than rapidly titrated injections with a PCA delivery device) delays the effective CNS concentration of opioids (see Table 39-4) and should be used only with PCA devices in those who preoperatively required regular doses of opioids, as previously discussed[53,81,82] (see Table 39-4). The operative procedure may reduce pain significantly, and hence it is difficult to determine the exact dose necessary postoperatively in patients who received chronic opioids preoperatively.

Dose adjustments of opioids by demand-only PCA are managed by interval rather than by amount (see Fig. 39-4). If dose intervals are short, dose levels are effectively increased, and when dose intervals are lengthened, the dose is reduced. Long intervals and large doses produce wider fluctuations in which patients may experience alternating mini-withdrawal and opioid toxicity[53] (Fig. 39-6).

Spinal Opioids

The introduction of spinal opioids into clinical practice occurred in 1979.[54] Analgesia with fewer adverse effects

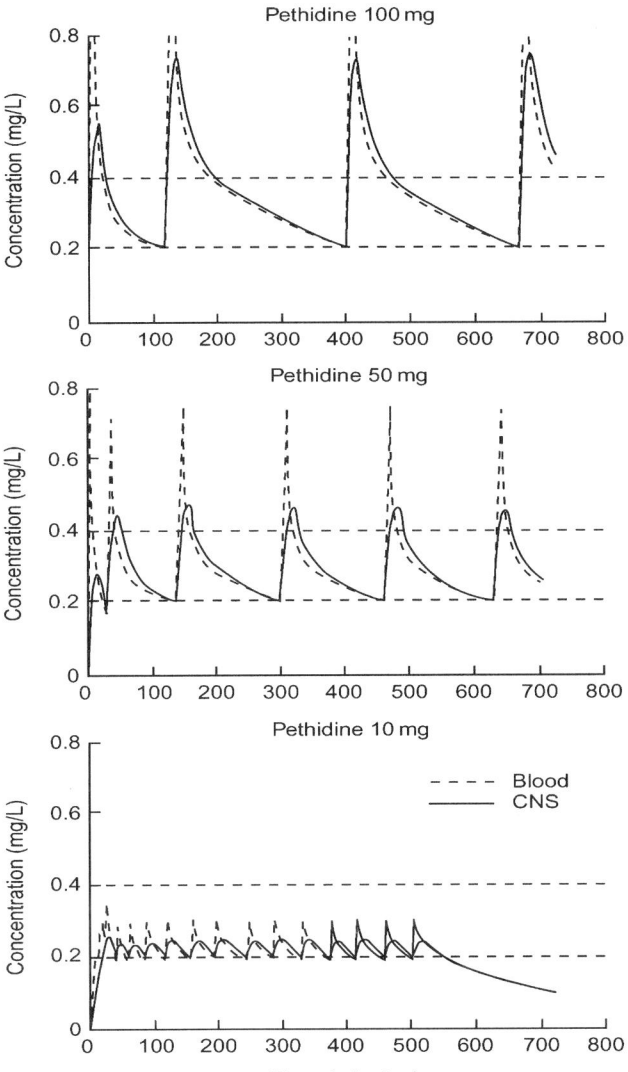

Figure 39-4. The effect of more frequent administration of smaller opioid doses maintains levels in the therapeutic window. Pethidine is not a good choice for acute pain management, but the figure illustrates an excellent principle in opioid dosing. *(Reprinted with permission from Upton RN, Semple TJ, Macintyre PE).*[53]

compared with systemic opioids was the advantage. This segmental analgesia became popular in the management of surgical, obstetric, and nonmalignant chronic pain, as well as cancer pain.[83–85] In a recent review of postoperative spinal analgesia, 41 percent of German hospitals offered epidural analgesia, 57 percent used bolus doses, 20 percent used epidural PCA, and 58 percent used continuous epidural opioids. The popularity of opioids in descending order was sufentanil, morphine, and fentanyl.[86] This differs from country to country. Large variations among individuals by as much as 200 percent of the amount of opioid required for pain control

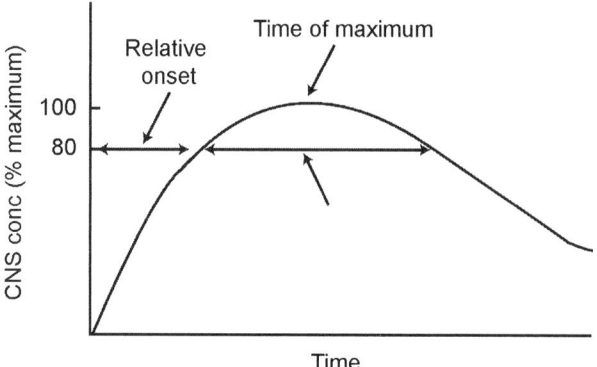

Figure 39-5. The time to relative onset of analgesia is determined in part by lipophilic characteristics. The effective CNS dwell time, as indicated by the arrow, may be relatively independent of the onset to analgesia. Dosing strategies need to take both characteristics into account. *(Reprinted with permission from Upton RN, Semple TJ, Macintyre PE).*[53]

have been described.[87] Clinical benefits are reduced pain, fewer pulmonary complications, and reduced ICU and hospital stays. Unique benefits to spinal opioid analgesia compared with spinal local anesthetics (as analgesics) are the lack of orthostatic hypotension and the lack of sensory, sympathetic, and motor deficits, which allows for early ambulation. The most impressive responses occur with morphine.[54] The patient population that benefits the most consists of high-risk patients with (1) compromised cardiovascular and respiratory function, (2) morbid obesity, and (3) advanced age.[88] Better spinal analgesia is achieved by combining opioids with local anesthetics, but this brings a greater risk for hypertension, motor weakness, urinary retention, and pressure sores owing to loss of skin sensitivity.[54] Epidural opioid PCAs combine feasibility, convenience, and patient-tailored analgesia.[90,91]

Comparisons of spinal opioids to systemic PCA or traditional analgesia have had mixed results.[54] In one study, epidural analgesia was the most commonly practiced analgesia for lower extremity amputations.[92] In a randomized trial, patients undergoing aortic aneurysm repair with epidural PCA had a shorter time to extubation compared with IV PCA but no improvement in length of stay, costs, time in intensive care, gastrointestinal recovery time, or time to ambulation.[93] Epidural morphine improved pain control and reduced mortality compared with the usual practice in a small series of patients undergoing aortic aneurysm repair. Intensive care stays and times to extubation were shortened significantly. Epidural morphine did not improve the overall outcome of gastric, biliary, or colonic surgery in this same series.[94] Continuous epidural morphine for 5 days after thoracoabdominal incision for esophageal cancer improves pain control over IV PCA morphine but does not improve pulmonary function.[95–97] The addition of epidural fentanyl to epidural local anesthetics improves pain control but increases postoperative nausea and vomiting and delays readiness to discharge.[98,99] In a small prospective trial, women undergoing breast reconstruction had lower pain scores with epidural opioids compared with IV PCA and slightly reduced hospital stays (101 versus 126 hours) but no difference in time to ambulation, bowel recovery, time to oral nutrition, nausea, vomiting, or pruritus.[100] In a meta-analysis of postoperative analgesia after thoracotomy, cesarean section, and abdominal, hip, biliary, and coronary surgery, epidural opioids reduced postoperative atelectasis compared with systemic opioids and also reduced postoperative complications. Pulmonary function, as measured by spirometry, was no different.[101,102] Epidural fentanyl reduces bowel function recovery time compared with epidural morphine.[103] Combinations of epidural fentanyl and local anesthetics are slightly better than epidural morphine alone.[104] Continuous-intravenous fentanyl compared with continuous intrathecal fentanyl in pediatric cardiovascular surgery patients did not improve pain control or outcomes.[105] Intrathecal morphine produces better analgesia than fentanyl with or without a cyclooxygenase inhibitor.[106]

Only opioids truly confined regionally provide sedation-free analgesia, and no presently available opioid qualifies.[107] Spinal lipophilic opioids, particularly those with a partition coefficient of greater than 2 (such as fentanyl), have rapid redistribution and systemic absorption into epidural fat and myelin and do not diffuse into the hydrophilic spinal gray matter within the dorsal horn[11] (see Fig. 39-6). Spinal fentanyl probably induces analgesia through systemic absorption. Analgesic plasma concentrations for continuous epidural and continuous intrathecal fentanyl are indistinguishable.

The adverse effects of spinal opioids are similar to those of systemic opioids but also include a higher incidence of pruritus, back pain, headache, cord compression due to hematomas or spinal abscess, meningitis, and nerve root damage. Late-onset respiratory depression can occur with spinal morphine and continuous epidural or intrathecal fentanyl.[108,109] The mechanism for respiratory depression is rostral spread of morphine within the spinal canal and release of fentanyl from tissue storage with systemic recirculation. Therefore, the early-onset analgesia and modest benefits, particularly of epidural and intrathecal lipophilic opioids, may not justify the danger, costs, and inconvenience of spinal administration over parenteral routes, except in high-risk individuals.[54,110,111]

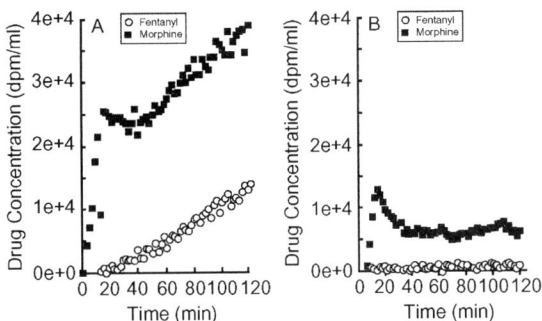

Figure 39-6. Concentration versus time profile of morphine and fentanyl within the spinal cord following simultaneous application of equal doses of both drugs to the surface of the spinal cord. Drug samples from rabbits were obtained using microdialysis probes inserted in the white matter at a depth of 1.5 mm (A) and the gray matter at a depth of 2.5 mm (B). Radiotracer methods were used to measure drug concentrations, which are presented in units of dpm/mL. Note that morphine penetrates the cord much more rapidly and to a greater depth than does fentanyl, which is virtually undetectable in the gray matter. *(Reprinted with permission from Cambridge University Press).*[11]

▶ OPIOIDS IN NONCANCER PAIN

Thirty million individuals in the United States suffer from chronic pain, disabling and severe. Most of these patients do not have cancer.[112] Opioids are used by 0.2 percent of the population, 58 percent of whom have chronic noncancer pain, most commonly back pain, headaches, arthritis, pancreatitis, angina, and phantom limb pain (Table 39-7). Opioids for noncancer pain account for much of the recent rise in morphine consumption.[113]

Half of chronic noncancer pain patients change physicians because their physicians (1) lack the willingness to treat pain adequately, (2) lack concern about pain severity and pain interference with activities, and (3) lack professional knowledge of pain management.[114] Physicians choose to use opioids based on pain behaviors rather than pain severity, the objective pathology, or duration of pain.[115] Physician concerns about prescribing opioids for noncancer pain center around the potential for (1) cognitive and psychomotor deterioration while on opioids, (2) physical dependency, (3) analgesic tolerance, (4) abuse, addiction, and opioid diversion, (5) pain facilitation and reinforcement, and (6) the use of opioids for nonpain purposes such as anxiety.[116]

The equation for opioid use in chronic noncancer pain differs in degree from cancer pain even though there is little to distinguish them on a neurophysiologic and anatomic basis.[115,117] Cancer pain is managed based on a biomedical model, and noncancer pain is managed based on a biopsychosocial model.[117–119] The difference lies in degree and outcome. Chronic noncancer pain management focuses on function rather than pain relief as its primary outcome measure, whereas cancer pain benefits are measured by pain relief more than function.[120,121]

Pain relief has been reported to occur in 50 percent of patients with chronic pain, and pain reduction has been reported in 25 percent with improvement in performance status.[122] On the other hand, two randomized trials have demonstrated modest pain reduction (13 to 29 percent)

▶ TABLE 39-7. GUIDELINES FOR OPIOID THERAPY IN CHRONIC NONMALIGNANT PAIN

- Perform a complete pain and psychosocial history and a physical examination.
- A single physician who sets up an agreement with the patient should be responsible for opioid prescriptions. The agreement should specify the drug regimen, the goals of treatment, possible side effects, and violations that will result in the termination of opioid therapy.
- The opioid analgesic of choice (preferably a sustained-release preparation) should be administered around the clock with an initial titration phase of 3–6 weeks to minimize side effects. Dosing should generally be time contingent rather than pain contingent except during the titration phase where rescue doses for breakthrough pain should be used to help determine the maintenance dose.
- Incremental dosing during the titration phase should result in a graded analgesic response or at least partial pain relief. Failure to realize at least partial analgesia at initial doses may mean that the pain syndrome is unresponsive to opioid treatment.
- The patient should be seen at least monthly for the first few months. At each visit the patient should be assessed for analgesia and opioid-related behavior. All of the information should be documented in the medical record.
- The patient should be reminded that the goal of opioid therapy is to make the pain tolerable and perhaps to improve function as part of a comprehensive treatment program.

Adapted from Moulin DE by permission of Cambridge University Press.[21]

in individuals previously exposed to various opioids who were placed on controlled doses of potent opioids.[116,123] Bias occurs in many of these studies.[124-126] The lack of control groups and the obligated closer attention to the treated groups by pain management staff may improve patient satisfaction with opioid therapy despite continued pain.[127] The benefits of therapy will be influenced by the patient-perceived attitude of the caregiver. A poor pain response may be a reflection of the caregiver's attitude transmitted to the patient. The physician's negative reaction to potential legitimate "drug seeking" individuals certainly will influence analgesic response.[121] Often a skewed population of patients enters pain management programs. Patients usually are referred to pain clinics because of traumatic pain, personality traits, and previously failed trials of therapy. Alternatively, improved function as an outcome may not be realized because of patient deconditioning, physical disabilities, or the affective state of the patient. Groups of patients in analgesic trials have diverse pain patterns with a variety of psychosocial backgrounds.[128]

Because the goal of pain management in chronic noncancer pain is restoration of function, it is unrealistic to consider eradication of pain as a primary outcome. Such a goal leads to patient frustration.[116,129-131] The goal of pain control is not well served by a focused reliance on analgesics alone nor evaluation by unidimensional pain scales. This fosters dependence and passivity and is detrimental to the rehabilitation goal of restoration of function with independence.[115-117,132]

Response to opioids depends on (1) the type of pain; (2) the type of opioid; (3) the dose strategy; (4) the route of administration; (5) the time of follow-up; (6) compliance; (7) the development of tolerance; (8) organ function; (9) interacting medications; and (10) patient characteristics.[133] There is a poor association between the degree of impairment and disability with physical pathology and pain severity. Dysfunctional patients have high pain severity, a high degree of perceived pain interference with activity and affective distress, and low level of control over life events, resulting in a poor expectant outcome with opioid therapy. Interpersonally distressed individuals have a perceived lack of social support and elicit a negative response by others to their pain behavior and, as a result, have a high level of affective distress and a predictably poor response to opioids. Individuals who adaptively cope with pain have low pain interference, low affective distress, and high levels of perceived control over life events and activities with a greater chance of response to opioids.[134-136] Patients who attribute pain to specific traumatic events report higher emotional distress, life interference due to pain, and pain severity than those with insidious onset of pain or spontaneous pain.[137]

Nociceptive pain tends to respond better to opioids than neuropathic pain.[113,120,138-141]

Factors Influencing Opioid Therapy in Noncancer Pain

Physical Dependence

Physical dependence is a physiologic neuroadaptive response that becomes clinically apparent on abrupt opioid withdrawal.[116,129] Physical dependence is not the same as psychological dependence, nor does it lead to psychological dependence. It is rather consistent, although the severity is variably expressed among all patients receiving opioids for more than several days. Patients may "drug seek" to relieve withdrawal symptoms and be mistaken for addicts.[116,129] An "on-off" phenomenon occurs when short-acting opioids are spaced at longer than appropriate intervals. This results in mood and sleep disturbances and increases pain due to the multiple "mini-withdrawal" episodes between doses.[116,128,129] Mini-withdrawal espisodes resolve by switching to sustained-release opioids or transdermal fentanyl or methadone. Acute withdrawal can be managed with clonidine 0.1 mg every 12–24 hours. Reduction of sustained-release preparations by 25 percent every 2 to 3 days will prevent withdrawal. In general, physical dependence is not a major deterrent to opioid therapy.[116]

Opioid Tolerance and Pain Facilitation

Opioid tolerance is not a major problem in most patients. Tolerance to side effects such as sedation, nausea, and reduced cognitive function and psychomotor reaction time is greater with time than to analgesia and increases the therapeutic window, as mentioned earlier.[142] Analgesic tolerance is an increased dose to produce the same analgesia and, like physical dependence, is a neuroadaptive process caused by exposure to opioids. A corollary to increasing dose is a requirement for a stable disease state. Most escalated dose requirements are related to progressive pathology and not analgesic tolerance.[116] Increasing pain on stable doses of opioids needs reassessment for progressive disease; patient characteristics such as delirium, depression, and psychological or spiritual crisis; and drug interactions that accelerate opioid clearance.[116,143-145] Therapeutic options (once assessment is complete) are (1) dose escalation; (2) addition of an adjuvant; (3) opioid rotation to an alternative agent; and (4) opioid conversion.[128,146,147] Some patients' pain may improve with reduced opioid doses or tapered or discontinued opioids.[113] Opioids can induce changes in receptors, as described earlier, that are pronociceptive.[147] This may be related to reduced endogenous opioid release, as well as neuroplasticity.[5] Hyperalgesia has been reported with high doses of morphine. Methadone maintenance is associated with lower pain thresholds. However, analgesic tolerance seldom compromises long-term efficacy, nor does it justify withholding a trial of opioids.[116,120,129]

Pain Reinforcement

Opioids may reinforce a psychological reward through stimulation of dopamine release from the nucleus accumbens.[116,129,148] Primary reinforcement produces euphoria and tranquility. If opioids are made contingent on pain, they act as a reinforcer to the presence of pain.[116,129] The learned association between euphoria with opioids is reinforced by the presence of pain. Pairing opioids with activity or function rather than with pain reduces their reinforcement potential with pain. This has implications for dosing strategies in noncancer pain.[116,129] Around-the-clock dosing also may reduce the reinforcing effects. Euphoria is more of a problem when opioids are taken for nonpain purposes. In the presence of pain, dysphoria is experienced more commonly.[149]

Addiction and Opioid Diversion

Addiction is the "compulsive use of a substance with loss of control resulting in physical, psychological, and social harm to the user and continued use despite harm."[148] This is clearly different from physical dependence and analgesic tolerance. It is estimated that the addiction risk within the noncancer pain population is between 3 and 16 percent. Mere exposure to opioids does not appear to be a significant risk factor for addiction.[116,120] There is a genetic predisposition, as seen in twin studies.[150] A family or personal history of alcoholism or substance abuse disorder is a primary risk factor. Other risk factors include a personality disorder and a lifestyle that engenders major social disruption or chaos.[123] These factors are strong but not absolute predictors and are not an absolute contraindication to a trial of opioids.[129,151] Warning signs of addictive behavior are a (1) preoccupation with the opioid rather than pain relief, (2) drug use without regard to prescribed route, dose, and schedule, (3) "lost" prescriptions, and (4) requests for early refills. Certain opioid preparations can be dissolved and taken intranasally or parenterally. Forging prescriptions is a major indication of substance abuse. Multiple prescribers of the same medication with the use of multiple pharmacies is another strong sign of addiction. Drug diversion differs from abuse. Diverters seek economic gain at the risk of harming others and yet may not be psychologically dependent.[116] Psychological dependence is rare in hospitalized patients who receive opioids for acute pain and in those without a family or personal history of substance abuse.[113] Some authors believe that short-acting opioids produce a reinforcing "high" and have a greater risk than sustained-release opioids. Shorter-acting opioids and parenteral opioids should be discouraged when treating patents with chronic noncancer pain and a substance abuse history.[152–154] Some behaviors associated with psychological dependence are mimicked by pseudoaddiction, which results from undertreatment of pain. Patients with *pseudoaddiction* seek opioids for the purpose of pain relief. The opioid preoccupation and abnormal pain behaviors resolve once pain has been controlled effectively.[129] Physicians can ruin a relationship by assuming that the legitimate need for higher opioid doses or better timing of doses is addictive "opioid seeking" behavior. Opioids sometimes are necessary to control pain in patients with a history of psychological dependence or substance abuse. Opioid maintenance therapy can be an effective component to drug rehabilitation. Providing opioids for addiction therapy requires a special license, but opioids may be used to treat pain in individuals with psychological dependence without a license. However, this should be done in conjunction with an addiction specialist. Characteristics associated with a risk of relapse of psychological dependence while on opioids are (1) a recent history of polysubstance abuse, (2) early prescription abuse in the course of treatment, (3) a prior history of oxycodone abuse, and (4) failure to attend rehabilitation groups.[151] Those less likely to relapse on opioids have (1) a history of alcohol abuse alone, (2) a remote history of polysubstance abuse, (3) continued active participation in rehabilitation, and (4) a stable family and/or social support system.[151] Guidelines to prescribing opioids in individuals with a history of psychological dependence or substance abuse include an opioid educational agreement on chronic opioid therapy that, although not proven in efficacy, at least clarifies expectations and behaviors and may be a springboard into a therapeutic relationship. Opioid agreements should articulate the rationale and expectations of therapy. Restrictions should be spelled out clearly, but the document should not be exhaustive.[152] If patients are on methadone maintenance, maintenance should be continued, with methadone adjusted for pain control, or a second opioid should be added.[36] Long half-life opioids should be used, and rehabilitation therapy should be a requirement for ongoing opioid prescribing.[151] Single-clinician prescribers and single-pharmacy providers limit the potential for diversion and provide an opportunity to measure opioid consumption. Urine testing for compliance is helpful, although short-acting opioids or methadone may be missed unless specifically ordered.[113] Periodic review of pain, opioid side effects, pain behavior, and functional compliance with prescription and rehabilitation is important. A treatment course of opioids is not irreversible.[116,129]

A titration trial of opioids for noncancer pain should extend for at least 4 weeks.[128] If patients are not compliant, fail to obtain pain relief, or worsen functionally while on treatment, the opioid should be discontinued. Despite the best intentions, caring physicians can be duped by the behavior of skilled and manipulative patients who act to deceive them. It is the legitimate role of drug enforcement agencies and regulatory bodies to identify such persons and inform physicians. Negative sanctions against physicians should be avoided.

Opioid Therapy: Cognitive and Psychomotor Function, Driving Ability

Opioid neurotoxicity includes cognitive failure, severe sedation, hallucinations, myoclonus, and hyperalgesia that usually occurs with high doses and prolonged therapy. It is seen with any potent opioid and is not reversed with opioid antagonists but improves with dose reduction. Mild and reversible cognitive impairment and prolongation of psychomotor reaction time occur with acute dosing and improve with time. Tolerance develops rapidly.[113] Temporary mild cognitive impairment occurs with a change in dose but is not related to absolute dose.[152] Patients frequently are less aware of the cognitive changes compared with other side effects.[152] Sustained attention span and psychomotor skills improve at a steady dose level and as pain diminishes.[122] The development of sedation and cognitive impairment on steady doses of opioids is unusual without a predisposing cause, such as organ failure or drug interactions.[123,153] There appears to be no increase in automobile accidents or traumatic injuries to individuals on opioids at steady-state levels. Driving ability is not diminished.[154] Patients should be on steady-state doses for 2 weeks before attempting to drive an automobile.

The elderly have a risk for hip fractures due to falls when started on codeine or propoxyphene.[155] Combinations of opioids and psychotropics produce an even greater risk (relative risk of 2.6). Older patients are more sensitive to opioid side effects. Opioids should be started at lower doses and titrated slowly in the elderly; and other psychotropic agents should be limited.

Opioid Dosing Strategies for Noncancer Pain

The World Health Organization (WHO) analgesic ladder is a guideline governing analgesic choices based on pain severity.[156] It is both a teaching tool for the proper use of opioids and an incentive to make morphine available internationally for those with severe pain. Absent from the guidelines are recommendations for alternative opioids, alternative routes and dosing strategies, and anesthetic techniques such as nerve blocks.[157] The WHO ladder is also incomplete in that it lacks recommendations about nonpharmacologic approaches to pain and rehabilitation.

Opioid prescribing strategies have been proposed and summarized for chronic noncancer pain.[116,128,129] A scheduled respite strategy provides time- or activity-contingent short-acting opioids on an intermittent basis. The focus of dosing is not pain control but function and activity. The advantages are a low risk for tolerance and physical dependence, as well as withdrawal and the "on-off" phenomenon. Pain facilitation would be rare and pain reinforcement minimized. However, individuals would experience only intermittent pain relief.

Short-acting opioids are best used in patients with only occasional pain, perhaps several times a week or month. Pain relief is provided when needed, and the patient remains opioid-free otherwise. However, if opioids are required several times a day, pain reinforcement, physical dependency, and the "on-off" phenomenon are likely to occur, resulting in increasing pain, insomnia, irritability, and increased opioid use.

Finally, sustained-release or long-acting opioids around the clock provide stable blood levels and preemptive pain control. Continuous opioids will be particularly necessary for daily or constant pain. Physical dependency will occur but is not a major problem unless the opioid is withdrawn abruptly. Tolerance is possible. The "on-off" phenomenon and pain reinforcement are minimized owing to the sustained level of opioid and the time-contingent rather than pain-contingent dosing. Sustained-release opioids may be less euphoric and thus may minimize drug-seeking behavior.

Summary of the Management of Chronic Noncancer Pain

Opioids are not first-line analgesics for chronic noncancer pain.

1. The goals of therapy are improved function and quality of life with pain control.
2. Application of opioids requires a biopsychosocial model of pain and multidisciplinary management.
3. Time-contingent titration trials of opioids for 2 to 4 weeks may be necessary to gauge benefit.
4. The risk of addiction is low in patients without a personal or family history of substance abuse or alcoholism.
5. Opioids are not contraindicated in patients with a high risk for addiction but should be instituted with a treatment structure that establishes expected benefits, behaviors, and consequences if there is a failure to comply. The involvement of an addiction specialist may be necessary.

6. Various opioid dosing strategies are based on pain patterns and goals of care. There are pros and cons to each approach.

▶ CANCER PAIN

At diagnosis, 35 percent of cancer patients will have pain, and as cancer advances, 65 to 85 percent of patients will suffer from pain due to their cancer (Table 39-8). Most cancer patients will have more than one pain.[158] Unlike noncancer pain, patients with cancer pain will have a multitude of additional symptoms, including anorexia, weight loss, weakness, nausea, vomiting, dyspnea, depression, and delirium, that will alter the tolerance to opioids and predispose to opioid toxicity. The multiple symptoms may be addressed poorly in clinics that are centered on pain management. The existential nature of a life-limiting illness such as cancer changes the perception of pain severity because it is a persistent reminder of the approaching end to life.

Pain can be treated successfully in 85 to 95 percent of cancer patients with an integrated approach of opioids, adjuvants, surgery, radiation, and nonpharmacologic therapy.[159] Successful pain management is a patient-defined subjective satisfaction with pain relief. There is a paradox in that patients frequently are satisfied with incompletely controlled pain. This is due to a complex combination of low expectations for relief, fear of side effects, previous experiences with opioids in family members, and perceived physician attitude.[160] Patients will sense a reluctance of physicians to prescribe the appropriate opioid dose, or family members fear addiction and side effects and will purposely encourage minimal dosing. Nihilistic expectations for pain relief and the desire to be a "good," noncomplaining patient add to this complexity. On the other hand, physicians who focus on "absolute" analgesia in terms of quantity (derived from a visual analog or numerical scale) without regard to patient goals will miss the mark as patients seek enough relief to remain physically or socially functional but not complete anagesia.

Opioid Therapy in Cancer Pain Management

Three basic pain management principles are (1) modify the source of pain; (2) alter the central perception of pain; and (3) block transmission to the CNS.[161] Opioids are vital to the latter two principles. Principles of opioid prescribing are (1) by mouth; (2) by the clock (as preemptive); (3) by the WHO analgesic ladder for severity (Fig. 39-7); (4) individualized prescribing; and (5) attention to detail.[156] Stepwise control of pain involves (1) pain control at night to promote sleep; (2) pain control at rest and during the day; and (3) acceptable pain relief with activity.[35]

Pain intensity determines analgesic choice and is quantifiable by unidimensional scales. However, assessment of pain severity alone is inadequate. Pain location, quality, radiation, palliating or worsening factors, pattern, and diurnal variations are essential to successful

▶ **TABLE 39-8. PHARMACOKINETICS OF OPIOIDS**

1. Preemptive around-the-clock dosing for continuous pain
2. Provide for breakthrough and incident pain
3. Incremental step ladder response should be pain relief
    a) at night allowing for sleep
    b) during the day and at rest
    c) with activity
4. Identify alternative dosing routes
5. Quantify response in magnitude and duration, and titrate opioids appropriately based upon a percentage of the baseline dose
6. Use preemptive laxatives and anticipate treating nausea
7. Responses are incremental and require regular reassessment
8. Sculpt dosing to pain pattern
9. Be aware of potential confounding factors to opioid kinetics such as interfering drugs, organ failure, age, and gender
10. Recognize dose-limiting side effects such as myoclonus, hallucinations and severe cognitive failure, as well as respiratory depression
11. Identify pain syndromes that are relatively opioid resistant such as colic
12. Identify patient characteristics that will predict poor responses to opioids such as depression, anxiety, delirium, existential crisis, history of drug abuse or methadone maintenance
13. Be aware of equianalgesic doses and factors that influence equivalence
14. Be aware of conversion and rotation ratios between potent opioids
15. When adjusting an analgesic program, change either route, dose or opioid but not simultaneously.

Figure 39-7. World Health Organization three-step analgesic ladder.

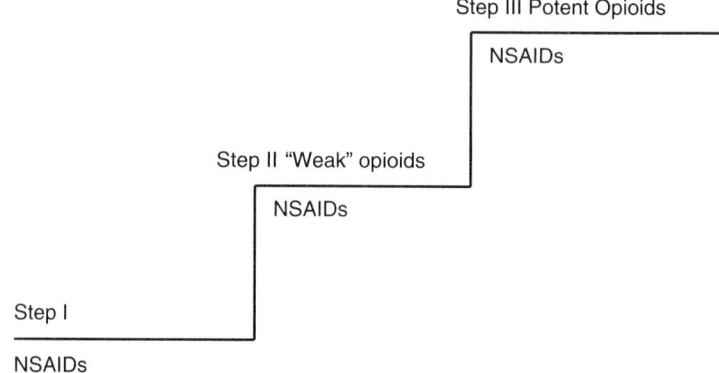

Analgesic Ladder

Nonsteroidal anti-inflammatory drugs are the first step in the WHO ladder (Fig. 39-7). If the predominant pain is neuropathic, tricyclic antidepressants or antiseizure medications may be appropriate.[161] Step II involves "weak" opioids (codeine, tramadol, and dextropropoxyphene), which have a relative ceiling effect and low intrinsic efficacy. Combinations of oxycodone (or other opioids) and NSAIDs have a ceiling effect due to the NSAID. Low doses of a potent opioid may be substituted for step II opioids in countries with potent opioids liberally available. Switching to another weak opioid when severe pain is unrelieved by the first weak opioid usually fails. Patients failing one weak opioid should be switched to a potent opioid and the weak opioid discontinued.[156] Step III agents are morphine, fentanyl, oxycodone, methadone, hydromorphone, and diamorphine. Mixed agonists and antagonists, as well as meperidine, should be avoided.[35,156,161]

▶ MANAGEMENT OF SIDE EFFECTS

Preemptive pain relief, to be successful, requires preemptive management of expected side effects. Laxation in the form of softeners and stimulants (docusate, senna, and bisacodyl) are provided with the first dose of opioids because tolerance does not occur to constipation.[35,157,161] Antiemetics should be available if nausea occurs. Mild nausea usually abates with chronic dosing. Sedation and mild cognitive failure can be observed with expected improvement over several days.[161] There are no predictors for opioid response with opioid dose or plasma levels. In general, younger patients require higher doses. Opioid ceilings are relative to toxicity and not analgesia. The rate-limiting toxicity, neurotoxicity, poorly correlates with dose and is widely variable among individuals.[35,156] Opioids, unlike adjuvant analgesics, do not produce permanent end-organ damage. Responses are logarithmic and related to baseline dose such that dose titration is a percentage of baseline dose rather than an absolute increment.[35]

Opioid Dosing Strategies

Dosing intervals for normal-release formulations of morphine, oxycodone, and hydromorphone are 4 hours. The frail elderly may have extended intervals to 6 hours owing to delayed clearance.[35,157] The onset to action is usually 30 minutes. Sustained-release opioids such as sustained-release morphine and sustained-release oxycodone are prescribed at 12-hour intervals with an onset to analgesia of approximately 2 hours. The onset to analgesia for parenteral morphine is 10 to 15 minutes, and is 6 minutes for fentanyl. The duration of analgesia is 2 to 4 hours for morphine. Hydromorphone approximates morphine.[35,161] Converting from normal-release to a sustained-release oral opioid requires a 2-hour overlap. Converting from a continuous IV or subcutaneous fentanyl to transdermal fentanyl requires a 12-hour overlap. This can be done by incrementally reducing the dose by 50 percent at 6 hours after applying the transdermal patch and discontinuing the IV or subcutaneous infusion at 12 hours.[161]

Dose adjustments of around-the-clock opioids are based on pain severity. Increments of less than 25 percent of the baseline 4-hourly dose usually are ineffective. A 50 percent increment is usual for moderate pain and 100 percent for continued severe pain.[161] The total amount of rescue opioid for nonincident breakthrough pain can be added to the 24-hour total opioid dose if the rescue dosing has been effective and was required frequently (usually more than four per day). Once the around-the-clock dose is titrated to analgesia, the rescue dose for nonincident pain should be readjusted in accord with the new baseline infusion.[35,162]

Adjustment of doses downward may be necessary when side effects occur and pain has been relieved. Pain relief may occur with single-fraction radiation, surgical correction of bone fracture or bowel obstruction, response to an adjuvant analgesic, the development of end-organ failure, or complete cord compression. This can be particularly problematic with sustained-release opioids, methadone, or transdermal fentanyl. Opioids can be reduced by 50 percent on the first day without withdrawal symptoms and thereafter by 50 percent every 1 to 3 days until the around-the-clock dose is discontinued or pain reappears. Unadjusted rescue doses of opioid should be available during opioid withdrawal.[35,161]

In the opioid-naive, a starting dose of morphine is usually 1 mg/h either subcutaneously or intravenously with 2 mg every 2 hours as needed for rescue. In the elderly and frail, 0.5 mg/h is a safe starting parenteral dose.[35] In patients with an unknown opioid history, a fixed dose of 2 to 4 mg parenterally at 2-hour intervals will prevent withdrawal until the correct dose can be determined. Assessment at 2- to 4-hour intervals is important in this circumstance.[35]

Breakthrough Pain

Patients have two time-related components: (1) continuous pain and (2) intermittent pain. Therefore, a provision must be made for both pains. As-needed dosing for continuous pain leads to higher doses of opioids needed to reestablish pain control and exposes patients to alternating opioid toxicity and pain. Sculpting opioid prescribing to the type of pain, either continuous or intermittent or both, will prevent premature opioid rotation and patient frustration. Attention to details of pain history, pattern, and response is required.[35,157,162]

Breakthrough pain can be (1) end-of-dose failure, (2) non-incident, and (3) incident. There is only a general relationship between the around-the-clock dose and the rescue dose, with large interindividual differences for end-of-dose failure and non-incident pain. Increases or decreases in doses should be stepwise, and separate decisions about frequency and dose changes may be necessary regarding continuous and intermittent pain, particularly for incident pain. The opioid changes should be one at a time, either dose, frequency, or route.[35] Recommendations for usual oral rescue opioid dose are (1) the same as the 4-hourly dose; (2) 50 percent of the 4-hourly dose; or (3) 5 to 15 percent of the total daily opioid dose.[35,157,162] Patients with end-of-dose failure usually have suboptimal around-the-clock doses, which simply requires an incremental increase of 30 to 50 percent of the around-the-clock dose and maintenance of the dosing interval. Alternatively, the same dose may be given more frequently, but this risks reduced compliance. In the case of sustained-release morphine or oxycodone, every 8 hours is appropriate, and for transdermal fentanyl, every 60 to 48 hours is appropriate. Rescue doses may be taken at any time during a 4-hour period of the oral normal-release opioid schedule, and the around-the-clock dose should not be deleted nor delayed. Frequent rescue doses (sooner than every 2 hours by vein or 4 hours by mouth) are sometimes required but can be self-defeating, cause excessive side effects, and reduce compliance. This pattern should stimulate a review of the opioid dosing strategy.[35] Incident pain that occurs with activity needs to be anticipated. Opioid titration to incident pain will be independent of the around-the-clock dose. Rescue doses for incident pain may be proportionately larger than the around-the-clock dose.[35] Dose adjustments of the around-the-clock opioid are based on the previous 24-hour requirement for breakthrough pain. Doses for incident pain should not be added to the total 24-hour dose requirement in order to avoid opioid toxicity during rest.[35]

Acute Pain in Cancer Patients

Dosing strategies are similar to guidelines established for acute noncancer pain. Morphine at 1 mg/min is given until analgesia or side effects. If there is no response, then a repeat dose of 1 mg/min for 10 minutes is initiated after a 5-minute rest. This is repeated once again if there is no response. Titration is stopped with the onset of side effects or a maximum dose of 30 mg over 45 minutes.[35] Opioid maintenance infusion is initiated based on the titrated dose and baseline pretitration dose. Hourly doses are one-quarter to one-third of the effective titrated dose added to the baseline dose (if the patient is on chronic opioids). Naloxone should be available.[35] Immediate oral dose conversion has also been done based on the effective intravenous bolus dose. The effective titrated dose is the every-4-hour morphine dose, which is converted to oral by multiplying by 3.[163] Fentanyl 20 μg/min or hydromorphone 0.2 mg/min may be substituted for morphine.[35]

Opioids in the Actively Dying

Pain should be assessed by a gentle and appropriate history. Restlessness may be a sign of pain, but it is important to check the patient for urinary retention or impaction. Opioids should be continued, although an alternative route may be necessary. Family members may attribute the symptoms of dying to opioids. A gentle explanation and reassurance and frequent returns to the bedside are important.[35]

Cancer Pain Poorly Responsive to Opioids

Interindividual differences in response to opioids can vary "1000-fold." There is no particular dose above which a patient will not respond.[164] Opioid dosing limits usually are influenced by the onset of neurotoxicity rather than the lack of analgesia. Many physicians have opioid dosing ceilings that are arbitrary and not based on rational pharmacology. The unfortunate result is either continued pain, premature opioid rotation, or the addition of a second or third potent opioid as a "cocktail." Poorly controlled pain may be the result of neuropathic pain mechanism(s).[164] One of the usual multiple pains cancer patients experience may be more resistant to opioids and become evident because other pains are managed easily. Neuropathic pain, colic, and tenesmus will require the early addition of adjuvant analgesics. Opioid titration, conversion, or rotation are effective strategies.[164] Social, psychological, or spiritual distress can be interpreted easily by patients and physicians as physical pain. Patients under these circumstances will develop troublesome side effects with opioid titration with no perceptible analgesic response. This is a clue that the patient is experiencing "opioid-irrelevant pain."[164]

▶ CONCLUSIONS

Opioids are seminal to the management of moderate to severe acute and chronic pain. Dosing strategies for acute and chronic pain are based on rational pharmacology and evidence-based practice. Opioid dosing strategies differ depending on the type of pain, the type of patient, and the goals of care. Successful management requires an understanding of opioid physicochemical properties and opioid pharmacology with dosing strategies. Close attention to detail and reassessment cannot be overemphasized.

▶ ACKNOWLEDGMENT

I would like to acknowledge the skills of Michele Wells and Becky Phillips in preparing this manuscript.

REFERENCES

1. Pugsley MK: The diverse molecular mechanisms responsible for the actions of opioids on the cardiovascular system. *Pharmacol Ther* 2002;93:51–75.
2. Martin WR: Opioid antagonists. *Pharmacol Rev* 1967;19:463–521.
3. Inturrisi CE: Clinical pharmacology of opioids for pain. *Clin J Pain* 2002;18:S3–13.
4. Roques BP, Noble F, Fournie-Zaluski MC: Endogenous opioid peptides and analgesia, in Stein C (ed), *Opioids in Pain Control*. Cambridge: Cambridge University Press, 1999:21.
5. Heinricher MM, Morgan MM: Supraspinal mechanisms of opioid analgesia, in Stein C (ed), *Opioids in Pain Control*. Cambridge: Cambridge University Press, 1999:46.
6. Law PY, Wong YH, Loh HH: Molecular mechanisms and regulation of opioid receptor signaling. *Annu Rev Pharmacol Toxicol* 2000;40:389–430.
7. Law PY, Wong YH, Loh HH: Mutational analysis of the structure and function of opioid receptors. *Biopoly* 1999;51:440–455.
8. Pasternak GW: Insights into mu opioid pharmacology: The role of mu opioid receptor subtypes. *Life Sci* 2001;68:2213–229.
9. Pasternak GW: Incomplete cross-tolerance and multiple mu opioid peptide receptors. *Trends Pharmacol Sci* 2001;22:67–70.
10. Gaveriaux-Ruff C, Kieffer B: Opioid receptors: Gene structure and function, in Stein C (ed), *Opioids in Pain Control*. Cambridge: Cambridge University Press, 1999:1.
11. Bernard CM: Clinical implications of physiochemical properties of opioids, in Stein C (ed), *Opioids in Pain Control*. Cambridge: Cambridge University Press, 1999:166.
12. Wandel C, Kim R, Wood M, et al: Interaction of morphine, fentanyl, sufentanil, alfentanil, and loperamide with the efflux drug transporter P-glycoprotein. *Anesthesiology* 2002;96:913–920.
13. Fromm MF: P-glycoprotein: A defense mechanism limiting oral bioavailability and CNS accumulation of drugs. *Int J Clin Pharmacol Ther* 2000;38:69–74.
14. Huwyler J, Drewe J, Gutmann H, et al: Modulation of morphine-6-glucuronide penetration into the brain by P-glycoprotein. *Int J Clin Pharmacol Ther* 1998;36:69–70.
15. Mikus G, Somogyi AA, Bochner F, et al: Polymorphic metabolism of opioid narcotic drugs: Possible clinical implications. *Ann Acad Med Singapore* 1991;20(1):9–12.
16. Mikus G: Clinical relevance of drug metabolism polymorphisms. *Ther Umschau* 2000;57:573–578.
17. Flores CM, Mogil JS: The pharmacogentics of analgesia: Toward a genetically based approach to pain management. *Pharmacogenomics* 2001;2:177–194.
18. Burd AL, El-Kouhen R, Erickson LJ, et al: Identification of serine 356 and serine 363 as the amino acids involved in etoporphine-induced down-regulation of the mu-opioid receptor. *J Biol Chem* 1998;273:34488–34495.
19. Caraco Y: Genetic determinants of drug responsiveness and drug interactions. *Ther Drug Monit* 1998;20:517–524.
20. Benet LZ, Izumi T, Zhang Y, et al: Intestinal MDR transport proteins and P-450 enzymes as barriers to oral drug delivery. *J Controlled Release* 1999;62:25–31.
21. Luo G, Cunningham M, Kim S, et al: CYP3A4 induction by drugs: Correlation between a pregnane X receptor reporter gene assay and CYP3A4 expression in human hepatocytes. *Drug Metabol Disp* 2002;30:795–804.
22. Strassburg CP, Kneip S, Topp J, et al: Polymorphic gene regulation and interindividual variation of UPD-glycuronosyltransferase activity in human small intestine. *J Biol Chem* 2000;275:36164–36171.
23. Strassburg CP, Kneip S, Topp J, et al: Polymorphic gene regulation and interindividual variation of UDP-glycuronosyltransferase activity in human small intestine. *J Biol Chem* 2000;275:36164–36171.

24. Befort K, Filliol D, Decaillot FM, et al: A single nucleotide polymorphic mutation in the human mu-opioid receptor severely impairs receptor signaling. *J Biol Chem* 2001;276:3130–3137.
25. Sora I, Elmer G, Funada M, et al: Mu opiate receptor gene dose effects on different morphine actions: Evidence for differential in vivo receptor reserve. *Neuropsychopharmacology* 2001;25:41–54.
26. Williams RL, Chen ML, Hauck WW: Equivalence approaches. *Clin Pharmcol Ther* 2002;72:229–237.
27. Pereira J, Lawlor P, Vigano A, et al: Equianalgesic dose ratios for opioids: A critical review and proposals for long-term dosing. *J Pain Symptom Manage* 2001;22:672–686.
28. Anderson R, Saiers JH, Abram S, et al: Accuracy in equianalgesic dosing: Conversion dilemmas. *J Pain Symptom Manage* 2001;21:397–406.
29. Yue QY, Svensson JO, Alm C, et al: Interindividual and interethnic differences in the demethylation and glucuronidation of codeine. *J Clin Pharmacol* 1989;28:629–637.
30. Bartok RE, Craft RM: Sex differences in opioid antinociception. *J Pharmacol Exp Ther* 1997;282:769–778.
31. Indelicato A, Portenoy RK: Opioid rotation in the management of refractory cancer pain. *J Clin Oncol* 2002;20:348–352.
32. de Leon-Casasola O, Yarussi A: Pathophysiology of opioid tolerance and clinical approach to the opioid tolerant patient. *Curr Rev Pain* 2000;4:203–205.
33. Portenoy RK: Tolerance to opioid analgesics: Clinical aspects. *Cancer Surv* 1994;21:49–65.
34. Mercadante S: Opioid rotation for cancer pain: Rationale and clinical aspects. *Cancer* 1999;86:1856–1866.
35. Walsh D: Pharmacological management of cancer pain. *Semin Oncol* 2000;27:45–63.
36. Manfredi PL, Gonzales GR, Cheville AL, et al: Methadone analgesia in cancer pain patients on chronic methadone maintenance therapy. *J Pain Symptom Manage* 2001;21:169–174.
37. Farris DA, Fiedler MA: Preemptive analgesia applied to postoperative pain management. *AANA J* 2001;69:223–228.
38. Woolf C, Bromley L: Pre-emptive analgesia by opioids, in Stein C (ed), *Opioids in Pain Control*, Cambridge: Cambridge University Press, 1999:212.
39. Kelly DJ, Ahmad M, Brull SJ: Preemptive analgesia: I. Physiological pathways and pharmacological modalities. *Can J Anesth* 2001;48:1000–1010.
40. Katz J: Pre-emptive analgesia: Evidence, current status and future directions. *Eur J Anesthesthesiol Suppl* 1995;10:8–13.
41. Katz J, Jackson M, Kavanagh BP, et al: Acute pain after thoracic surgery predicts long-term post-thoracotomy pain. *Clin J Pain* 1996;12:50–55.
42. Katz J, Kavanagh BP, Sandler AN: Effect of preoperative opioid administration on postoperative pain. *Br J Anaesth* 1992;69:424–425.
43. Katz J, Kavanagh BP, Sandler AN, et al: Preemptive analgesia: Clinical evidence of neuroplasticity contributing to postoperative pain. *Anesthesiology* 1992;77:439–446.
44. Wall PD: The prevention of postoperative pain. *Pain* 1998;33:289–290.
45. Kilickan L, Toker K: The effect of preemptive intravenous morphine on postoperative analgesia and surgical stress response. *Panminerva Med* 2001;43:171–175.
46. Katz J, Clairoux M, Redahan C, et al: High doses alfentanil pre-empts pain after abdominal hysterectomy. *Pain* 1996;68:109–118.
47. Vaida SJ, Ben David B, Somri M, et al: The influence of preemptive spinal anesthesia on postoperative pain. *J Clin Anesth* 2000;12:347.
48. Sabanathan S: Has postoperative pain been eradicated? *Ann R Coll Surg Engl* 1995;77:202–209.
49. Tverskoy M, Braslavsky A, Mazor A, et al: The peripheral effect of fentanyl on postoperative pain. *Anesth Analg* 1998;87:1121–1124.
50. Choe H, Choi YS, Kim YH, et al: Epidural morphine plus ketamine for upper abdominal surgery: Improved analgesia from preincisional versus postincisional administration. *Anesth Analg* 1997;84:560–563.
51. Reuben SS, Sklar J, El-Mansouri M: The preemptive analgesic analgesic effect of intraarticular bupivacaine and morphine after ambulatory arthroscopic knee surgery. *Anesth Analg* 2001;92:923–926.
52. Holthusen H, Backhaus P, Boeminghaus F, et al: Preemptive analgesia: No relevant advantage of preoperative compared with postoperative intravenous administration of morphine, ketamine, and clonidine in patients undergoing transperitoneal tumor nephrectomy. *Reg Anaesth Pain Med* 2002;27:249–253.
53. Upton RN, Semple TJ, Macintyre PE: Pharmacokinetic optimisation of opioid treatment in acute pain therapy. *Clin Pharmacokinet* 1997;33:225–244.
54. Rawal N: Opioids in acute pain, in Stein C (ed), *Opioids in Pain Control*. Cambridge: Cambridge University Press, 1999:247.
55. Hug CC: Intraoperative use of opioids, in Stein C (ed), *Opioids in Pain Control*. Cambridge: Cambridge University Press, 1999:234.
56. Breivik H: Benefits, risks and economics of post-operative pain management programs. *Baillieres Clin Anesthesiol* 1995;9:403–421.
57. Walder B, Schafer M, Henzi I, et al: Efficacy and safety of patient-controlled opioid analgesia for acute postoperative pain: A quantitative systematic review. *Database of Abstracts of Reviews of Effect* 4, 2002.
58. Lehmann KA: Patient-controlled analgesia with opioids, in Stein C (ed), *Opioids in Pain Control*. Cambridge: Cambridge University Press, 1999:270.
59. Cooper IM: Morphine for postoperative analgesia: A comparison of intramuscular and subcutaneous routes of administration. *Anesth Intensive Care* 1996;24:574–578.
60. Peacock JE, Wright BM, Withers MR, et al: Evaluation of a pilot regimen for postoperative pain control in patients receiving oral morphine pre-operatively. *Anesthesia* 2000;55:1208–1212.
61. Bertolini G, Minelli C, Latronico N, et al: The use of analgesic drugs in postoperative patients: The neglected problem of pain control in intensive care units. An observational, prospective, multicenter study in 128 Italian intensive care units. *Eur J Clin Pharmacol* 2002;58:73–77.

62. Thorn SE, Rawal N, Wennhager M: Prolonged respiratory depression caused by sublingual buprenorphine. *Lancet* 1988;1:179–180.
63. Sekar M, Mimpriss TJ: Buprenorphine, benzodiazepines and prolonged respiratory depression. *Anaesthesia* 1987;42:567–568.
64. Feld LH, Champeau MW, van Steennis CA, et al: Preanesthetic medication in children: A comparison of oral transmucosal fentanyl citrate versus placebo. *Anesthesiology* 1989;71:347–347.
65. Toussaint S, Maidl J, Striebel HW: Patient-controlled intranasal analgesia: Effective alternative to intravenous PCA for postoperative pain relief. *Can J Anesth* 2000;47:299–302.
66. Schwagmeier R, Oelmann T, Dannappel T, et al: Patient acceptance of patient-controlled intranasal analgesia. *Anaesthetist* 1996;45:231–234.
67. Streibel HW, Oelmann T, Spies C, et al: Patient-controlled intranasal analgesia: A method for noninvasive postoperative pain management. *Anesth Analg* 1996;83:548–551.
68. Henderson JM, Fisher DM: Intranasal sufentanil. *Can J Anaesth* 1990;37:387.
69. Fine PG, Marcus M, De Boer AJ, et al: An open label study of oral transmucosal fentanyl citrate (OTFC) for the treatment of breakthrough cancer pain. *Pain* 1991;45:149–153.
70. Ward M, Minto G, Alexander-Williams JM: A comparison of patient-controlled analgesia administered by the intravenous or intranasal route during the early postoperative period. *Anesthesia* 2002;57:48–52.
71. Choiniere M, Rittenhouse BE, Perreault S, et al: Efficacy and costs of patient-controlled analgesia versus regularly administered intramuscular opioid therapy. *Anesthesiology* 1998;89:1377–1388.
72. Rawal N: Treating postoperative pain improves outcome. *Minerva Anestesiol* 2001;67:200–205.
73. Ballantyne JC, Carr DB, Chalmers TC, et al: Postoperative patient-controlled analgesia: Meta-analyses of initial randomized control trials. *J Clin Anesth* 1993;5:182–193.
74. Macintyre PE, Ready LB: *Acute Pain Management: A Practical Guide*. London: Saunders, 1996.
75. Ginsberg B, Gil KM, Muir M, et al: The influence of lockout intervals and drug selection on patient-controlled analgesia following gynecological surgery. *Pain* 1995;62:95–100.
76. Lam FY: Patient-controlled analgesia by proxy. *Br J Anesth* 1993;70:113.
77. Etches RC: Patient-controlled analgesia. *Surg Clin North Am* 1999;79:297–312.
78. Etches RC: Respiratory depression associated with patient-controlled. *Can J Anesth* 1994;41:87–90.
79. Dahlstrom B, Tamsen A, Paalzow L, et al: Patient-controlled analgesic therapy: IV. Pharmacokinetics and analgesic plasma concentrations of morphine. *Clin Pharmacokinet* 1982;7:266–279.
80. Fleming BM, Coombs DW: A survey of complications documented in a quality-controlled analysis of patient-controlled analgesia in the postoperative patient. *J Pain Symptom Manage* 1992;7:463–469.
81. Walder B, Schafer M, Henzi I, et al: Efficacy and safety of patient-controlled opioid analgesia for acute postoperative pain: A qualitative systematic review. *Acta Anaesth Scand* 2001;45:795–804.
82. Shafer SL, Gregg KM: Algorithms to rapidly achieve and maintain stable drug concentrations at the site of drug effect with a computer-controlled infusion pump. *J Pharmacokinet Biopharm* 1992;20:147–169.
83. Cousins MJ, Mather LE: Intrathecal and epidural administration of opioids. *Anesthesiology* 1984;61:276–310.
84. Vercautern MP: Epidural PCA: Practical use and misuse. *Acta Anesthesiol Belg* 1999;50:171–176.
85. Vercautren MP: PCA by epidural route (PCAE). *Acta Anesthesiol Belg* 1992;43:33–39.
86. Kampe S, Kiencke P, Krombach J, et al: Current practice in postoperative epidural analgesia: A German survey. *Anesth Analg* 2002;95:1767–1717.
87. Van Boerum DH, Smith JT, Curtin MJ: A comparison of the effects of patient-controlled analgesia with intravenous opioids versus epidural analgesia on recovery after surgery for idiopathic scoliosis. *Spine* 2000;25:2355–2357.
88. Tunman KJ, McCarthy RJ, March RJ, et al: Effects of epidural anesthesia and analgesia on coagulation and outcome after major vascular surgery. *Anesth Analg* 1991;73:696–704.
89. Yeager MP, Glass DD, Neff RK, et al: Epidural anesthesia and analgesia in high-risk surgical patients. *Anesthesiology* 1987;66:729–736.
90. Egan KJ, Ready LB: Patient satisfaction with intravenous PCA or epidural morphine. *Can J Anesth* 1994;41:6–11.
91. Lutz LJ, Lamer TJ: Management of postoperative pain: Review of current techniques and methods. *Mayo Clin Proc* 1990;65:1031–1032.
92. Campbell WB, Marriott S, Eve R, et al: Anesthesia and analgesia for major lower limb amputation. *Cardiovasc Surg* 2000;8:572–575.
93. Norris EJ, Beattie C, Perler BA, et al: Double-masked randomized trial comparing alternate combinations of intraoperative anesthesia and postoperative analgesia in abdominal aortic surgery. *Anesthesiology* 2001;95:1054–1067.
94. Young Park W, Thomas JS, Lee K, et al: Effect of epidural anesthesia and analgesia on perioperative outcome: A randomized, controlled Veterans Affairs Cooperative Study. *Ann Surg* 2001;234:560–571.
95. Flisberg P, Tornebrandt K, Walther B, et al: Pain relief after esophagectomy: Thoracic epidural analgesia is better than parenteral opioids. *J Cardiothorac Vasc Surg* 2001;15:287–287.
96. Ballantyne JC, Carr DB, deFerranti S, et al: The comparative effects of postoperative analgesic therapies on pulmonary outcome: Cumulative meta-analyses of randomized, controlled trials. *Anesth Analg* 1999;86:598–612.
97. Jorgensen H, Wetterslev J, Moiniche S, et al: Epidural local anaesthetics versus opioid-based analgesic regimens on postoperative gastrointestinal paralysis, PONV and pain after abdominal surgery. *Cochrane Database Syst Rev* 2002;4.
98. Lovstad RZ, Stoen R: Postoperative epidural analgesia in children after major orthopaedic surgery: A randomised study of the effect on PONV of two anaesthetic techniques: Low and high dose IV fentanyl and epidural

98. infusions with and without fentanyl. *Acta Anesthesiol Scand* 1977;45:482–488.
99. Finucane BT, Ganapathy S, Carli F, et al: Prolonged epidural infusions of ropivacaine (2 mg/mL) after colonic surgery: The impact of adding fentanyl. *Anesth Analg* 2001;92:1276–1285.
100. Correll DJ, Viscusi ER, Grunwald Z, et al: Epidural analgesia compared with intravenous morphine patient-controlled analgesia: Postoperative outcome measures after mastectomy with immediate TRAM flap breast reconstruction. *Reg Anaesth Pain Med* 2001;26:444–449.
101. Ballantyne JC, Carr DB, deFerranti S, et al: The comparative effects of postoperative analgesic therapies on pulmonary outcome: Cumulative meta-analyses of randomized, controlled trials. *Cochrane Database Abstr Rev Effect* 2002;4.
102. Ballantyne JC, Carr DB, deFerranti S, et al: The comparative effects of postoperative analgesic therapies on pulmonary outcome: Cumulative meta-analyses of randomized, controlled trials. *Anesth Analg* 1998;86:598–612.
103. Vallejo MC, Edwards RP, Shannon KT, et al: Improved bowel function after gynecological surgery with epidural bupivacaine-fentanyl than bupivacaine-morphine infusion. *Can J Anesth* 2000;47:406–411.
104. Reinoso-Barbero F, Saavedra B, Hervilla S, et al: Lidocaine with fentanyl, compared to morphine, marginally improves postoperative epidural analgesia in children. *Can J Anaesth* 2002;49:67–71.
105. Pirat A, Akpek E, Arslan G: Intrathecal versus IV fentanyl in pediatric cardiac anesthesia. *Anesth Analg* 2002;95:1207–1214.
106. McCrory C, Diviney D, Moriarty J, et al: Comparison between bolus intrathecal morphine and an epidurally delivered bupivacaine and fentanyl combination in the management of postthoracotomy pain with or without cyclooxygenase inhibition. *J Cardiothorac Vasc Anesth* 2002;16:607–611.
107. Boswell MV: Lipid soluability and epidural opioid efficacy. *Anesthesiology* 1995;83:427.
108. Weightman WM: Respiratory arrest during epidural infusion of bupivacaine and fentanyl. *Anesth Intensive Care* 1991;19:282–284.
109. Brockway MS, Noble DW, Sharwood-Smith GH, et al: Profound respiratory depression after extradural fentanyl. *Br J Anesth* 1990;64:243–245.
110. Nagle CJ, McQuay HJ: Extradural phethidine. *Br J Anaesth* 1990;65:730–731.
111. Camu F, Debucquoy F: Alfentanil infusion for postoperative pain: A comparison of epidural and intravenous routes. *Anesthesiology* 1992;76:319.
112. Joranson DE, Lietman R: *The McNeil National Pain Study.* New York: Louis and Harris Associates, 1994.
113. Collett BJ: Chronic opioid therapy for non-cancer pain. *Br J Anaesth* 2001;87:133–143.
114. Savage SR: Long-term opioid therapy: Assessment of consequences and risk. *J Pain Symptom Manage* 1996;11:274–286.
115. Turk DC: Clinical effectiveness and cost-effectiveness of treatments for patients with chronic pain. *Clin J Pain* 2002;18:355–365.
116. Savage SR: Opioid use in the management of chronic pain. *Med Clin North Am* 1999;83:761–786.
117. Turk DC: Remember the distinction between malignant and benign pain? Well, forget it. *Clin J Pain* 2002;18:75–76.
118. Turk DC: Okifuji A: Psychological factors in chronic pain: Evolution and revolution. *J Consult Clin Psychol* 2002;70:678–690.
119. Turk DC: Combining somatic and psychosocial treatment for chronic pain patients: perhaps 1 + 1 does = 3. *Clin J Pain* 2001;17:281–283.
120. Portenoy RK: Opioid therapy for chronic nonmalignant pain: A review of the critical issues. *J Pain Symptom Manage* 1996;11:203–217.
121. Moulin DE: Opioids in chronic nonmalignant pain, in Stein C (ed), *Opioids in Pain Control.* Cambridge: Cambridge University Press, 1999:295.
122. Haythornthwaite JA, Menefee LA, Quatrano-Piacentini AL, et al: Outcome of chronic opioid therapy for non-cancer pain. *J Pain Symptom Manage* 1998;15:185–194.
123. Moulin DE, Iezzi A, Amireh R, et al: Randomised trial of oral morphine for chronic non-cancer pain. *Lancet* 1996;347:143–147.
124. Jamison RN, Raymond SA, Slawsby EA, et al: Opioid therapy for chronic noncancer back pain. *Spine* 1998;23:2591–2600.
125. Maruta T, Swanson DW, Finlayson RE: Drug abuse and dependency in patients with chronic pain. *Mayo Clin Proc* 1979;54:241–244.
126. Turner JA, Calsyn DA, Fordyce WE, et al: Drug utilization patterns in chronic pain patients. *Pain* 1982;12:357–363.
127. Passik SD, Kirsh KL: Probing the paradox of patients' satisfaction with inadequate pain managent. *J Pain Symptom Manage* 2002;24:361–362.
128. Breivik H: Opioids in cancer pain and chronic non-cancer pain therapy: Indications and controversies. *Acta Anesthesiol Scand* 2002;45:1059–1966.
129. Savage SR: Opioid therapy of chronic pain: Assessment of consequences. *Acta Anaesthesiol Scand* 1999;43:909–917.
130. Brena SF, Sanders SH: Chronic pain: Fix or manage? *Clin J Pain* 1992;8:170–171.
131. Flor H, Fydrich T, Turk DC: Efficacy of multidisciplinary pain treatment centers: A meta-analytic review. *Pain* 1992;49:221–230.
132. Large RD, Schug SA: Opioids for chronic pain of non-malignant origin: Caring and crippling. *Health Care Analysis* 1995;3:5–11.
133. Dellemijn PL: Opioids in non-cancer pain: A life-time sentence? *Eur J Pain* 2001;5:333–339.
134. McCracken LM, Turk DC: Behavioral and cognitive-behavioral treatment for chronic pain: Outcome, predictors of outcome, and treatment process. *Spine* 2002;27:2564–2573.
135. Turk DC, Sist TC, Okifuji A, et al: Adaptation to metastatic cancer pain, regional/local cancer pain and non-cancer pain: Role of psychological and behavioral factors. *Pain* 1998;74:247–256.
136. Turk DC: The role of psychological factors in chronic pain. *Acta Anesthesiol Scand* 1999;43:885–888.
137. Turk DC: Precipitation of traumatic onset: Compensation status, and physical findings: Impact on pain severity, emotional distress, and disability in chronic pain patients, *J Behav Med* 1996;19:435–453.

138. McQuay HJ: Neuropathic pain: Evidence matters. *Eur J Pain* 2002;6:11–18.
139. McQuay H: Opioids in chronic non-malignant pain: There's too little information on which drugs are effective and when. *Br J Med* 2001;322:1134–1135.
140. Arner S, Meyerson BA: Lack of analgesic effect of opioids on neuropathic and idiopathic forms of pain. *Pain* 1988;33:11–23.
141. Jadad AR, Carroll D, Glynn CJ, et al: Morphine responsiveness of chronic pain: Double-blind randomised crossover study with patient-controlled analgesia. *Lancet* 1992;339:1367–1371.
142. Portenoy RK, Foley KM: Chronic use of opioid analgesics in non-malignant pain: Report of 38 cases. *Pain* 1986;25:171–186.
143. Willweber-Strumpf A, Zenz M, et al: Drug dependence in therapy of chronic pain. *Z Ges Med Grenzg* 1992;47:312–317.
144. Zenz M: Morphine myths: Sedation, tolerance, addiction. *Postgrad Med J* 1991;67:S100–S102.
145. Ytterberg SR, Mahowald ML, Woods SR: Codeine and oxycodone use in patients with chronic rheumatic disease pain. *Arthritis Rheum* 1998;41:1603–1612.
146. Portenoy RK: Tolerance to opioid analgesics: Clinical aspects. *Cancer Surv* 1994;21:49–65.
147. Collin E, Cesselin F: Neurobiological mechanisms of opioid tolerance and dependence. *Clin Neuropharmacol* 1991;14:465–488.
148. Gardner E: Brain reward mechanisms, in Lowinson J, Ruiz P, Millman R (eds), *Substance Abuse: A Comprehensive Text*. Baltimore: Williams & Wilkins, 1992:70–99.
149. Javrik LF, Simpson JH, Guthrie D, et al: Morphine, experimental pain, and psychological reactions. *Psychopharmacologica* 1981;75:124–131.
150. Grove WM, Eckert ED, Heston L, et al: Heritability of substance abuse and antisocial behavior: A study of monozygotic twins reared apart. *Bioll Psychiatr* 1990;27:1293–1304.
151. Dunbar SA, Katz NP: Chronic opioid therapy for nonmalignant pain in patients with a history of substance abuse: Report of 20 cases. *J Pain Symptom Manage* 1996;11:163–171.
152. Kreek MJ: Opiates, opioids and addiction. *Mol Psychiatr* 1996;1:232–254.
153. LaForge KS, Yuferov V, Kreek MJ: Opioid reactor and peptide gene polymorphisms: Potential implications for addictions. *Eur J Pharmacol* 2000;410:249–268.
154. Kreek MJ: Drug addictions: Molecular and cellular endpoints. *Ann NY Acad Sci* 2001;937:27–49.
155. Fishman SM, Bandman TB, Edwards A, et al: The opioid contract in the management of chronic pain. *J Pain Symptom Manage* 1999;18:27–37.
156. Twycross RG: Opioids, in Wall PD, Melzack R (eds): *Textbook of Pain,* 4th ed. New York: Churchill Livingstone, 2001:1187.
157. Hanks GW, de Conno F, Cherny N, et al: Morphine and alternative opioids in cancer pain: The EAPC recommendations. *Br J Cancer* 2001;84:587–593.
158. Grond S, Zech D, Diefenbach C, et al: Prevalence and patter of symptoms in patients with cancer pain: A prospective evaluation of 1635 cancer patients referred to a pain clinic. *J Pain Symptom Manage* 1994;9:372–382.
159. Du Pen SL, Du Pen AR, Polissar N, et al: Implementing guidelines for cancer pain management: Results of a randomized controlled clinical trial. *J Clin Oncol* 1999;17:361–370.
160. Dawson R, Spross JA, Jablonski ES, et al: Probing the paradox of patients' satisfaction with inadequate pain management. *J Pain Symptom Manage* 2002;23:211–220.
161. Levy MH: Drug therapy: Pharmacologic treatment of cancer pain. *N Engl J Med* 1996;335:11245–11232.
162. Portenoy RK, Lesage P: Management of cancer pain. *Lancet* 1999;353:1695–1700.
163. Mercadante S, Villari P, Ferrera P, et al: Rapid titration with intravenous morphine for severe cancer pain and immediate oral conversion. *Cancer* 2002;95:203–208.
164. Hanks GW, Forbes K: Opioid responsiveness. *Acta Anaesthesiol Scand* 1997;41:154–158.
165. Quigley C, Wiffen P: A systematic review of hydromorphone in acute and chronic pain. *J Pain Symptom Manage* 2003;25:169–178.
166. Wilder-Smith CH, Schimke J, Osterwalker B, Senn H: Oral tramadol, a μ-opioid agonist and monoamine reuptake-blocker, and morphine for strong cancer-related pain. *Ann Oncol* 1994;5:141–146.

CHAPTER 40

Antidepressants, Anticonvulsants, and Miscellaneous Agents

Flaminia Coluzzi and Consalvo Mattia

This chapter reviews the contemporary pharmacologic nonopioid analgesic therapies in chronic pain. For a long time, tricyclic antidepressants (TCAs) have been considered the main therapy for postherpetic neuralgia and diabetic polyneuropathy. Subsequently, new antidepressants have been introduced and studied in several neuropathic pain conditions. More recently, the newer anticonvulsants, e.g., gabapentin and its heir pregabalin, have been shown to be effective in managing different chronic pain syndromes.

Recently, new drugs have been evaluated as analgesics. Cannabinoids and N-methly-D-aspartate (NMDA)–receptor antagonists are two of the most interesting emerging therapies. Unfortunately, despite that preclinical results have been promising, solid efficacy data are still lacking for these drugs to be considered new analgesic options. Finally, topical therapies are discussed because of the positive results and optimal safety profile.

▶ ANTIDEPRESSANTS

The comorbidity depression in patients with chronic pain is common and it has been extensively studied. Depression is a virtually universal complication of intractable pain. Some authors reported a prevalence rate approaching 100 percent,[1] which led to a wide use of antidepressants in chronic pain management. Antidepressant drugs not only provide their primary function of mood elevation, but they also may have analgesic properties.

Antidepressants have become a routine adjunctive therapy for most forms of chronic pain. The first report on imipramine in pain management appeared 40 years ago. Since then, TCAs have been used for the treatment of various chronic and neuropathic pain conditions.[2] Their efficacy is now well established, although the multiple mechanisms by which antidepressants produce analgesia are still unclear.[3] Their therapeutic effect in painful conditions can be understood on the basis of common neurotransmitters implicated in both depression and pain. The analgesic properties of these drugs occur independent of their antidepressant action, in lower doses, and with a faster onset of action.

Despite the fact that the older TCAs continue to be utilized for some chronic painful syndromes, a major drawback to their use has been the high incidence of poorly-tolerated side effects. Newer non-TCAs are emerging, and their efficacy as analgesics is being studied.[4]

Mechanisms of Action as Analgesics

The acute physiologic reaction to noxious stimuli involves corticoadrenal and sympathetic responses, resulting in the production of glucocorticoids and increasing the output of the neurogenic amines norepinephrine and serotonin within the neuronal synapse.[5] Experimental studies in animal models of persistent pain have demonstrated analgesic activity with antidepressants. The analgesic efficacy seems to be related mainly to the ability of these drugs to inhibit the presynaptic uptake and storage of neurogenic amines. Serotoninergic and noradrenergic processes are involved in two descending modulatory pathways from the brain stem to the spinal cord that represent the endogenous pain inhibitory mechanisms. The serotonin pathway originates at the level of the midbrain in the periaqueductal gray (PAG) matter and the nucleus raphae magnus, whereas the norepinephrine pathway originates at the level of the locus ceruleus in the medulla.

Apart from neurogenic amine reuptake inhibition, antidepressants have been shown to exert several other pharmacologic actions, and each drug has a unique profile and exhibits diverse pharmacologic properties. Other receptors, such as the histaminic, cholinergic, muscarinic, and cholinergic-nicotinic receptors, are known to be bound by antidepressants, but frequently they are also the cause of intolerable side effects.

Opioid receptors have been involved recently in the hypothetical mechanism of analgesic action of some antidepressants. Both venlafaxine[6] and mirtazapine,[7] two newer antidepressants, have been shown to enhance opioid receptor–mediated analgesia when coadministered with opioid agonists, and their antinociceptive activity seems to be inhibited by selective opioid antagonists.

Antidepressants were found to produce antinociceptive effects by binding to the NMDA-receptor complex and inhibiting NMDA-induced sensitization. TCAs have been found to prevent the elevation of intracellular calcium induced by excitatory amino acid (EAA) in the spinothalamic tract and dorsal horn, confirming their neuroprotective effect. Moreover, antidepressants are supposed to inhibit the uptake of adenosine, a significant mediator of analgesia, and to activate G-protein, resulting in an analgesic effect.[5]

Finally, antidepressants have been demonstrated to have powerful local anesthetic properties by blocking sodium channels in both the central and peripheral nervous systems. This could explain pain relief in terms of reduced firing frequency at ectopic sites.

Several routes of administration have been tested to study the antinociceptive activity of antidepressants in animal models. In the past, the systemic route, such as intraperitoneal, intravenous, and oral administration, was used mainly in animals to mimic oral intake in humans, focusing on central antidepressant actions. Subsequently, antidepressants have been administered either into the cerebral ventricles or spinally, and both these routes produce analgesia, but supraspinal administration is more effective than spinal injection.[8] Most recently, animal studies have shown that antidepressants may produce peripheral analgesia, though the involved mechanisms still remains unclear.

In both the inflammatory pain animal model (the formalin test)[9] and the neuropathic pain animal model (the spinal nerve ligation),[10] amitriptyline has been shown to produce a peripheral antinociceptive action that is clearly due to a local mechanism, since it was not observed following administration of amitriptyline into the contralateral paw. In the neuropathic pain model, amitriptyline was effective in alleviating thermal hyperalgesia but was ineffective against mechanical allodynia.

Peripheral antinociceptive activity has also been shown for desipramine when administrated locally in both inflammatory and neuropathic pain animal models, whereas fluoxetine appeared to have a peripheral antinociceptive action only in the inflammatory pain model.[11]

Tricyclic Antidepressants

For many years, TCAs have been considered the "gold standard" in the treatment of neuropathic pain because their superiority compared either with placebo or with other available drugs has been demonstrated. Several TCAs have been studied. Imipramine, amitriptyline, and clomipramine caused a reuptake inhibition of both serotonin and norepinephrine, whereas desipramine and maprotiline are relatively selective norepinephrine reuptake inhibitors.

Fifteen randomized, placebo-controlled trials have been conducted to analyze the analgesic effects of the TCAs in neuropathic pain. TCAs were shown to have efficacy in postherpetic neuralgia, in painful diabetic neuropathy, in central pain, and in poststroke pain[3] (Table 40-1). Their analgesic effect occurs in the absence of depression and at lower doses than those used to obtain an antidepressant effect.

The number needed to treat (NNT) has been used as an outcome measure to compare and combine several trials. It shows the number of patients needed to treat to produce greater than 50 percent pain relief,[12] and it is used for retrospective evaluation of the analgesic efficacy of drugs. TCAs have been studied extensively in several pathologic conditions. For painful diabetic neuropathy, the NNT for TCAs is 2.4 (2.0–3.0), meaning that two to three patients have to be treated before one patient reports greater than 50 percent pain relief from the drug. The NNT for TCAs is almost identical across different painful conditions, such as postherpetic neuralgia (2.3), painful diabetic polyneuropathy (2.4), and painful peripheral nerve injury (2.5), although it is lower for central pain (1.7).[13]

Unfortunately, TCAs exert their action on a wide range of receptors, leading to a high incidence of

▶ **TABLE 40-1.** TCAs AND SSRIs FOR NEUROPATHIC CHRONIC PAIN: RANDOMIZED, DOUBLE-BLIND, PLACEBO-CONTROLLED TRIALS

TCAs Study	Year	Drug	Dosage mg/day	Pain Diagnosis	n	Outcome
Watson et al.	1982	Amitriptyline	75	Postherpetic neuralgia	24	+
Max et al.	1988	Amitriptyline	150	Postherpetic neuralgia	34	+
Kishore-Kumar et al.	1990	Desipramine	167 (avg)	Postherpetic neuralgia	19	+
Kvinesdal et al.	1984	Imipramine	100	Diabetic neuropathy	12	+
Max et al.	1987	Amitriptyline	25–150	Diabetic neuropathy	29	+
Max et al.	1991	Desipramine	12.5–150	Diabetic neuropathy	20	+
Sindrup et al.	1989	Imipramine	125–200	Diabetic neuropathy	9	+
Sindrup et al.	1990	Clomipramine	50–75	Diabetic neuropathy	19	+
Sindrup et al.	1990	Desipramine	50–200	Diabetic neuropathy	20	+
Sindrup et al.	1990	Imipramine	25–350	Diabetic neuropathy		+
Sindrup et al.	1992	Imipramine	25–350	Diabetic neuropathy	18	+
Panerai et al.	1990	Clomipramine, Nortriptyline	25–100	Central pain		+
Leijon and Boivie	1989	Clomipramine	150	Post-stroke pain	15	+
Vrethem et al.	1997	Amitriptyline	75	Polyneuropathy	33	+
Vrethem et al.	1997	Maprotiline	75	Polyneuropathy	33	+
SSRIs						
Max et al.	1992	Fluoxetine	40	Diabetic neuropathy	46	–
Sindrup et al.	1990	Paroxetine	40	Diabetic neuropathy	19	+
Sindrup et al.	1992	Citalopram	40	Diabetic neuropathy	18	+

SOURCE: *Modified from Lynch ME.*[3]

adverse events. TCAs often are associated with intolerable side effects, and this remains the major challenge in their use. Sedation is a common side effect of TCAs, although agitation, insomnia, hallucinations, and tonic-clonic seizure have been described in some patients.

Apart from interactions with histaminic, acetacholine-muscarinic, and α-adrenergic receptors, the increased availability of serotonin and norepinephrine also can be considered as causes of side effects, because they act both centrally and peripherally in several organs.

Antihistaminic side effects include drowsiness and weight gain. Serotonin also can contribute to increased appetite and weight gain, which may affect the quality of life significantly and reduce compliance. Moreover, serotonin seems to affect different brain activities, including sleeping, learning, sensorial perception, thermoregulation, appetite, sexual behavior, and hormonal secretion.

TCAs block cholinergic-muscarinic receptors, which are located on cells activated by parasympathetic postganglionic neurons. Blurred vision is the most common ophthalmologic side effect, owing to loss of accommodation, especially in patients with glaucoma. Cholinergic blockade also can induce dry mouth, constipation, and urinary retention.

Noradrenergic blockade induces dizziness and decreased blood pressure. In healthy subjects, orthostatic hypotension has been observed owing to α-adrenergic antagonism and β_2-adrenergic-mediated vasodilatation.

The major limitation to TCA therapy is their cardiac effects because these represent a high-risk factor, especially in patients with cardiac disease, such as previous acute myocardial infarction, arrhythmias, or congestive heart failure. TCAs interfere with the conduction system, leading to prolonged QT intervals, rhythm abnormalities, and depressed ST segments on electrocardiograms.

Finally, TCAs can be lethal in accidental or intentional overdoses.

Selective Serotonin Reuptake Inhibitors (SSRIs)

The SSRIs were proposed in the late 1980s as a new class of antidepressants that act by increasing synaptic serotoninergic availability and neurotransmission, with high selectivity for 5-HT receptors and a low or null affinity for others kinds of receptors, e.g. the α-adrenergic, muscarinic, cholinergic, and histaminic receptors. Four major advantages were identified in using SSRIs over TCAs and monoamine oxidase (MAO) inhibitors: (1) the reduced incidence of side effects, resulting in higher tolerability; (2) the ability to administer them once daily without titration; (3) the reduced interaction with sedatives, antiarrhythmics, sympathomimetics, and ETOH; and (4) better tolerance in case of overdose. However, SSRIs may cause the life-threatening central serotonin syndrome, characterized by abdominal pain, fever, diarrhea, sweating, hypertension, elevated mood, altered mental state, delirium, myoclonus, increased motor activity, irritability, and hostility. Other common side effects of SSRIs include agitation, anxiety, sleep disturbance, tremor, sexual dysfunction, and headache.

Despite the proven advantages of SSRIs as antidepressants, when studied in pain states, SSRIs showed to be much less effective than the TCAs. A total of 83 patients have been studied in three randomized, controlled trials examining SSRIs in the treatment of painful diabetic neuropathy (see Table 40-1). Max and colleagues[14] compared fluoxetine, amitriptyline, and desipramine to placebo. There was no significant difference between fluoxetine and placebo in the treatment of painful diabetic neuropathy. TCAs were as effective in nondepressed as in depressed patients, whereas the SSRI fluoxetine was effective only in depressed patients. In two smaller studies, paroxetine[15] and citalopram[16] were found to exhibit a greater analgesic effect than placebo.

Smith analyzed 20 trials available in the literature, examining the analgesic efficacy of SSRIs;[17] among those evaluated, the only positive trials were placebo-controlled. It was shown that antidepressants with both serotonin and norepinephrine reuptake inhibition have greater analgesic effect.

The NNT was calculated to be 6.7 for SSRIs compared with 2.6 for TCAs.[13] In comparative studies between SSRIs and TCAs, the analgesia obtained with TCAs was always superior to that of SSRIs,[18] suggesting that norepinephrine reuptake inhibition is required to obtain an analgesic effect and that serotonin may enhance this mechanism. Recently, it has been demonstrated that in the norepinephrine transporter knock-out mice model enhanced antinociception occurs, likely due to increased activation of the α_2 adrenoreceptor. Moreover, in norepinephrine transporter knock-out mice, morphine treatment produced greater analgesia, supporting the idea that the blockade of the norepinephrine transporter results in potentiated opioid analgesia.[19]

Other Antidepressants

Recently, newer antidepressants have been introduced in clinical use. These new drugs have been classified on the basis of their mechanism of action into three categories:[20]

- *Serotonin and noradrenergic reuptake inhibitors* (SNaRIs), e.g., venlafaxine and nefazodone
- *Noradrenergic and specific serotoninergic antidepressants* (NaSSAs), e.g. mirtazapine
- Bupropion
- Reboxetine

Venlafaxine

Venlafaxine (an SNaRI) is a phenylethylamine antidepressant drug that blocks the synaptic uptake of norepinephrine and serotonin. At low doses, the serotonin reuptake inhibition predominates, and it acts as an SSRI, whereas at higher doses, norepinephrine reuptake inhibition is prominent. Venlafaxine weakly inhibits dopamine reuptake and shows a low affinity for cholinergic-muscarinic, H_1 histaminic, and α-adrenergic receptors.[4] Moreover, venlafaxine was found to induce a dose-dependant opioid-mediated antinociceptive effect that is inhibited by the opioid antagonists naloxone, nor-BNI, and naltrindole. Venlafaxine enhances opioid analgesia when coadministered with selective opioid receptor agonists. Its antinociception seems to be mainly mediated by $kappa_1$, $kappa_2$, and delta opioid receptor subtypes. Finally, venlafaxine's antinociceptive effect is decreased by adrenergic antagonists, whereas it is potentiated significantly by the α_2-adrenergic agonist clonidine.[6]

These results confirm the previous finding of a similarity between venlafaxine and tramadol. Both these agents interact with opioid receptors and inhibit the reuptake of norepinephrine and serotonin. Moreover, these two agents share molecular and pharmacologic features: Both exhibit methoxyphenyl, *N,N*-demethylamino, and hydroxycycloexyl groups; both are metabolized by cytochrome oxidase isoenzyme P450 2D6, and both yield pharmacologically active *O*-desmethyl metabolites.[21]

Venlafaxine is available in two formulations: immediate- and extended-release. The suggested starting dose of venlafaxine is 37.5 mg/d. The maximum dose of 37.5 mg/d can be reached to obtain an antidepressant effect. Dose adjustment is necessary in patients with renal failure, in whom the dose should be reduced by 25 percent, and in dyalitic or hepatic insufficient patients, in whom it should be halved. The extended-release formulation offers the advantage of once-a-day dosing, resulting in enhanced patient compliance.

Venlafaxine is usually well tolerated, and its side effects include nausea, dizziness, somnolence, ejaculatory abnormalities, sweating, dry mouth, and nervousness. At doses of 100 to 300 mg/d, the incidence of hypertension varies from 3 to 7 percent but rises up to 13 percent at greater doses; thus, arterial pressure monitoring is recommended.

Venlafaxine produced pain relief in 12 patients affected by different peripheral neuropathies (postherpetic neuralgia, intercostal neuralgia, atypical facial pain, and multiple sclerosis), but the low number of patients involved does not allow confirmation of therapeutic success.[22]

Venlafaxine was shown to prevent thermal hyperalgesia in a rat model of neuropathic pain (sciatic nerve ligation).[23] Similarly, a human experimental pain model on 16 healthy volunteers showed that at a dose of 37.5 mg every 12 hours, venlafaxine increased the threshold for pain tolerance by single electrical stimulation of the sural nerve.[24] These results suggest a potential analgesic efficacy of venlafaxine for the treatment of chronic neuropathic pain.

In several studies, venlafaxine has been used as a second-choice agent, when other therapies (NSAIDs and other antidepressants) have failed owing to either their inefficacy or intolerable side effects. Both immediate- and sustained-release preparations of venlafaxine at a total dose of 75mg/d were effective in chronic pain management.[25,26]

Venlafaxine was found to be effective in the treatment of postherpetic neuralgia and painful peripheral diabetic neuropathy[27] at the initial dose of 37.5 mg/d extended-release formulation, with progressive increments to a therapeutic dose of 225 mg/d.

The only available randomized controlled trial of venlafaxine, from Sindrup and colleagues, did not show any significant difference between venlafaxine 225 mg daily and imipramine 150 mg daily in relieving pain in polyneuropathy. Both antidepressants were shown more effective than placebo; however, NNT were 5.2 for venlafaxine and 2.7 for imipramine.[28]

Nefazodone

Nefazodone is chemically similar to trazodone. It is a serotonin- and noradrenergic-reuptake inhibitor (SNaRI), significantly weaker than venlafaxine, that antagonizes the action of 5-HT$_2$ receptors; thus, the increased level of serotonin interacts with other receptor subtypes, such as 5-HT$_{1A}$. Moreover, nefazodone shows a weak affinity for α_1- and β-adrenergic receptors and no activity on histaminic, dopaminergic, or muscarinic-cholinergic receptors, resulting in a significantly lower incidence of side effects. The most common side effects are nausea, somnolence, dry mouth, constipation, asthenia, and blurred vision.

In an experimental study on rats, the coadministration of morphine and nefazodone was found to enhance the analgesic effect related to mu$_1$ and mu$_2$ opioid receptors without effects on gastroenteric motility.[29] In humans, only an open-label trial has been published on the analgesic efficacy of nefazodone. The mean dose of 340 mg daily of nefazodone was effective in improving pain, paresthesias, and numbness associated with diabetic neuropathy.[30] *Unfortunately, nefazodone has been associated with jaundice, hepatitis and hepatocellular necrosis in patients receiving therapeutic doses. Because of these severe hepatic adverse events, including liver failure requiring transplantation, nefazodone has been recently removed from the market in Canada and in the U.S.*[31]

Mirtazapine

The mechanism of action of mirtazapine (a NaSSA) as an antidepressant is via dual enhancement of both noradrenergic and serotoninergic activity. First, mirtazapine antagonizes central presynaptic α_2-adrenergic inhibitory autoreceptors, which physiologically prevents further norepinephrine release. Consequently, it facilitates noradrenergic output. Second, mirtazapine increases hippocampal serotonin levels.

Mirtazapine is a strong antagonist of 5-HT$_2$ and 5-HT$_3$ serotoninergic postsynaptic receptors.[32] The antagonism on 5-HT$_2$ receptors facilitates sleep and reduces the incidence of anxiety and agitation, whereas the 5-HT$_3$ antagonism reduces nausea and the need for antiemetic pharmacotherapy, especially in patients on chemotherapy.

In animal models of pain, mirtazapine, similarly to venlafaxine, was found to induce a clear antinociceptive effect. However, the dose-response curve of venlafaxine is linear, whereas mirtazapine shows a therapeutic window effect at doses from 7.5 to 10 mg/kg.[33] Its antinociceptive effect seems to be mediated through serotoninergic, noradrenergic, and opioid mechanisms.[7]

Among the newer antidepressants, involvement of the opioid system occurs only with venlafaxine and mirtazapine. Both these drugs modulate norepinephrine and serotonin release, but NaSSAs (e.g., mirtazapine) differ from SNaRIs (e.g., venlafaxine) because the former does not block the reuptake of biogenic monoamines, as do most of the other antidepressants.[4] The antinociceptive effect of venlafaxine is mediated by mu, delta, kappa$_1$, and kappa$_3$ receptor subtypes, whereas the antinociceptive effect of mirtazapine mainly involves mu and kappa$_3$ opioid receptors.[33]

The most common adverse events are increased appetite and weight gain, but these could offer a benefit in cancer patients.[34] Rare patients with severe neutropenia and symptomatic agranulocytosis were reported; thus, periodic blood monitoring is necessary. Two cases of mirtazapine-induced arthralgia have been described. The mechanism of iatrogenic arthralgia is unknown, but it is possibly related to the similarity with the chemical structure of mianserin, a tetracyclic antidepressant responsible for more than 20 cases of articular adverse effects.[35] Mirtazapine 15 mg/d at bedtime has been used successfully in a patient with chronic back pain after fluoxetine and amitriptyline have been discontinued because of intolerable side effects.[36]

Reboxetine

Reboxetine is a norepinephrine reuptake inhibitor (NeRI), without serotoninergic, dopaminergic, and monoamine oxidase inhibitory properties. Side effects, such as constipation, dry mouth, and impotence, have been observed in 5 percent of treated men. Recently, reboxetine was shown to be more effective than placebo in relieving pain on capsaicin-irritated skin in healthy volunteers.[37]

Bupropion

Bupropion is an old drug, that, unfortunately, showed a high incidence of seizures and other side effects, such as hallucinations, dizziness, and insomnia. In late 1996, a new sustained-release formulation, bupropion SR, was approved; its use is associated with no sexual side effects and a lower incidence of seizure.

Bupropion is classified as a multiple-action antidepressant. It is a dopamine reuptake inhibitor with weak norepinephrine and serotonin reuptake inhibition activity; thus, it has the potential for being an analgesic antidepressant. The first open-label study on bupropion in neuropathic pain showed a consistent reduction in pain after 8 weeks of therapy (150 mg SR bupropion once daily for the first week, followed by 7 weeks of 150 mg SR bupropion twice daily).[38] Recently, in a randomized, double-blind, placebo-controlled crossover study, bupropion SR at 150 to 300 mg/d was shown to be effective for the treatment of neuropathic pain.[39]

In conclusion, even though the newer antidepressants are generally better tolerated than TCAs, further studies are warranted to confirm their role in chronic pain management.[40]

▶ ANTICONVULSANTS

Anticonvulsants are a broad category of drugs that share the ability to suppress CNS epileptic activity. Anticonvulsants not only may prevent or reduce excessive neuronal discharge of seizure foci, but also reduce the spread of neuronal excitation from seizure foci.[41] Although their mechanisms of action are only poorly understood, anticonvulsants have proved efficacious in the treatment of epilepsy.

Growing recognition of the similarities between epilepsy and neuropathic pain has increased interest in the use of anticonvulsants to treat chronic pain. Both disorders may arise from excess or abnormal neuronal activity.[42]

The use of anticonvulsants has a long tradition. They have been used for decades to treat chronic pain, and there is sufficient evidence to establish these drugs as specialized analgesics for neuropathic pain. A number of putative mechanisms of action relate to the pathophysiology of neuropathic pain and may account for the effectiveness of anticonvulsants.[43]

Some drugs have the property of blocking sodium channels in a use-dependent manner but may also exert their effects through other mechanisms. Other drugs do not have sodium channel antagonism but nevertheless seem to be very efficacious. Proposed mechanisms of analgesic action include sodium and calcium channel blockade, augmentation of inhibitory systems, inhibition of excitatory amino acid release, and free-radical scavenging.

Carbamazepine

Carbamazepine is an iminostilbene derivative that is structurally related to the TCAs.

Carbamazepine is primarily a sodium channel modulator that enhances inactivation of voltage-gated sodium channels by reducing high-frequency repetitive firing of action potentials.[44] The importance of sodium channel expression in the development of neuropathic pain is well established. The abnormal accumulation and kinetics of sodium channels, expressed in nociceptors and associated dorsal root ganglia, following a peripheral nerve injury leads to alterations of the electrical properties of the axon membrane. Injured nociceptors undergo a lowering of the depolarization threshold, and ectopic discharges are generated. The sodium channels are divided into tetradoxin-sensitive and tetradoxin-resistant channels. There is evidence that there is a preferential expression and accumulation of tetradotoxin-resistant sodium channels following peripheral nerve injury. The specific nervous system (SNS)–peripheral neuron 3 (PN3) channel (also known as Na(V)1.8) is the best characterized tetradotoxin-resistant sodium channel and relevant to the mechanisms of neuropathic pain. Knock-out SNS-null mutant mice, which only expressed tetradotoxin-sensitive sodium channels, were found to have an increased threshold to noxious mechanical stimuli as well as a delayed development of inflammatory hyperalgesia.[45]

Sodium channel modulators, such as certain antiepileptic drugs, are used for the treatment of neuropathic pain and are most effective in this disorder owing to injury of the peripheral nervous system. The mechanism of action of carbamazepine seems to result from a shift in sodium channels to an inactive state, from which recovery is delayed. This inactivation may be responsible for the reduction in spontaneous activity. The frequency-dependent effect of carbamazepine may explain why this drug reduces tonic discharges arising from nociceptors without affecting normal nerve conduction. The therapeutic effects of sodium channel modulating drugs are due to their ability to prevent the generation of spontaneous ectopic discharges at concentrations lower than those required to block normal impulse generation and propagation. In addition, carbamazepine was shown to modulate L-type calcium channels and increase the release of serotinin and dopaminergic transmission.

Carbamazepine is absorbed slowly, and peak concentrations in plasma are obtained 4 to 8 hours after oral administration. It is significantly bound to plasma proteins, with an extent of about 75 percent. It is metabolized by the hepatic pathway, and inactive compounds are excreted in the urine principally as glucuronides.

Carbamazepine was first used in clinical trials for the treatment of trigeminal neuropathy without any rationale based on defined pharmacologic actions. The effectiveness of carbamazepine led to a number of clinical trials that subsequently established definitive scientific evidence for carbamazepine in the treatment of trigeminal neuralgia.

Three double-blind, placebo-controlled, crossover trials were conducted on trigeminal neuralgia, involving a total of 151 patients.[46–48] Improvement in pain was two- to threefold greater with carbamazepine than with placebo. Response rates for carbamazepine in these studies ranged from 70 to 89 percent with oral doses from 400 to 1000 mg/d after 5 to 14 days of treatment.

A small randomized, double-blind, crossover study compared carbamazepine with placebo in the treatment of painful diabetic neuropathy.[49] Carbamazepine was shown to relieve pain and paresthesias in 28 of the 30 patients studied following a 2-week period of therapy when compared with 20 percent of patients on placebo.

Sixteen patients with severe distal, symmetric, predominantly sensitive diabetic polyneuropathy participated in a randomized, double-blind, crossover study where the efficacy and tolerance of the combination of nortriptyline-fluphenazine versus carbamazepine (CMZ) were compared.[50] No superiority of carbamazepine over the tricyclic-neuroleptic combination was demonstrated.

A small study in 15 patients with central poststroke pain failed to show efficacy of carbamazepine over amitriptyline.[51] Moreover, carbamazepine caused more side effects, and the final dose had to be reduced in 4 patients. Only 1 patient had to be taken off medication, on day 25, owing to drug interaction.

Finally, the analgesic efficacy of carbamazepine, controlled-released morphine, and placebo was compared in 43 patients with peripheral neuropathic pain responsive to spinal cord stimulation (SCS). During brief SCS-free periods, the patients were randomized to receive carbamazepine (600 mg/d) vs placebo or controlled release morphine (90 mg/d) vs placebo. Return of the pain was significantly delayed in the carbamazepine group.[52] Toxic effects of carbamazepine include diplopia, dizziness, nystagmus, dysarthria, lethargy, and leukopenia. Idiosyncratic reactions include apastic anemia, theombiotopemia, and liver toxicity. These reactions require monitoring laboratory tests.[52]

Newer Antiepileptic Drugs

Recently, newer antiepileptic drugs have been introduced in clinical practice for the treatment of neuropathic pain. The newer antiepileptic drugs (AEDs) possess the potential advantages of better tolerability and fewer drug-drug interactions compared with standard treatments or established AEDs.[53] Recent experimental data showed that newer AEDs are effective antihyperalgesic agents in an animal model of neuropathic pain.[54]

Gabapentin/Pregabalin

Gabapentin was developed as a structural GABA analogue described as 1-(aminomethyl)cyclohexanacetic acid and was approved in the United States in 1995 as adjunctive therapy for seizures. Although structurally related to GABA, gabapentin does not have any interaction with either $GABA_A$ or $GABA_B$ receptors and does not affect GABA uptake or metabolism. It does not exhibit any affinity for several other common receptor sites, such as benzodiazepine, glutamate, NMDA, quisqualate, kainate, α- and β-adrenergic, cholinergic-muscarinic or nicotinic, dopamine D_1 and D_2, histamine H_1, serotonin, opioid mu, delta or kappa, or cannabinoid CB_1 receptors. Gabapentin has no effect on voltage-dependent sodium channels. Moreover, it does not affect the cellular uptake of neurotransmitters.

However, recent evidence suggests that both gabapentin and pregabalin, a gabapentin analogue, can modulate N- and P/Q type voltage-gated calcium channels by binding the $\alpha_2\delta$ subunit of the channel, and decrease the intracellular calcium influx.[55,56]

Since gabapentin is excreted unmetabolized, all pharmacologic actions are due to the activity of the parent compound.

The predominant effect of gabapentin was shown on N-type calcium channels, which are expressed in the laminae of the dorsal horn and are important for central modulation of nociceptive input.[57]

Gabapentin was shown to prevent pain-related responses in several animal models of both neuropathic (spinal nerve ligation models) and inflammatory (formalin test) pain.

Although the mechanism of action is still being evaluated, gabapentin is probably the most used anticonvulsant for chronic pain. The success appears to be due to the improved side-effect profile, the ease of monitoring, and evidence of efficacy, as shown by a number of placebo-controlled trials.

There are five randomized, placebo-controlled trials in the literature that established the efficacy of gabapentin for the relief of neuropathic pain. Gabapentin was effective in the treatment of painful diabetic neuropathy, postherpetic neuralgia, and other neuropathic pain syndromes with a NNT of 3.8 (2.4 to 8.7) for diabetic polyneuropathy and 3.2 (2.4 to 5.0) for postherpetic neuralgia. Thus, based on the NNT, gabapentin is comparable with the TCAs.[58]

Gabapentin relieved symptoms of allodynia, burning pain, shooting pain, and hyperesthesia. Pain relief was observed most commonly during the second week of treatment. Moreover, patients had improvements in sleep and mood, and in their quality of life.

As with all anticonvulsants, the most common side effects are related to CNS depression, consisting primarily of somnolence and dizziness. Adverse effects typically were mild to moderate and well tolerated, usually subsiding within approximately 10 days from initiation of treatment.

Two randomized, double-blind, placebo-controlled multicenter studies have been conducted to evaluate the efficacy of gabapentin for postherpetic neuralgia (PHN) management.[59,60] A total of 563 patients (336 in the gabapentin group and 227 in the placebo group) suffering from pain for more than 3 months were involved. In both trials, gabapentin showed a significant difference compared with placebo in reducing PHN pain. A significant reduction in weekly mean pain intensity was seen by week 1 and was maintained to the end of the treatment. Secondary outcomes, including sleep interference, change in overall status, and alterations in quality of life, showed significant improvement with gabapentin.

Gabapentin was found to be effective in diabetic peripheral neuropathy. A total of 165 patients with a 1- to 5-year history of painful diabetic neuropathy were enrolled in a randomized, double-blind, placebo-controlled trial. Gabapentin was titrated from 900 to 3600 mg/d or maximum tolerated dosage. Mean daily pain score was significantly lower ($p < 0.001$) in patients treated with gabapentin compared with placebo. Furthermore, gabapentin was shown to significantly improve sleep, mood, and quality of life.[61]

All the studies included a titration period, during which a maximum tolerated dose was established for each patient. This was followed by a stable dose period. Thus the treatment should be started at a dose of 900 mg/d (300 mg/d on day 1, 600 mg/d on day 2, and 900 mg/d on day 3) administered in three divided doses. Additional titration to 1800 mg/d is recommended for greater efficacy. The effective dose should be individualized according to patient response and tolerability. The dosage of 3600 mg/d or higher may be needed in some patients. Dose adjustment should be applied in patients with renal failure, whereas no adjustment is required for patients with hepatic failure because the drug is excreted unmetabolized.[58]

Recently, gabapentin was shown to be effective in relieving the pain of 18 patients affected with Guillain-Barré syndrome admitted to the intensive care unit for ventilatory support.[62]

A randomized trial in 24 patients with restless legs syndrome showed that gabapentin, with a mean effective dosage of about 1800 mg, improves sensory and motor symptoms and also improves sleep architecture and significantly reduces periodic leg movements during sleep.[63]

In conclusion, ease of use, safe side-effect profile, good tolerability, and no significant interactions with other medications currently make gabapentin the most common first choice agent in the treatment of patients with neuropathic pain.

Further investigation into gabapentin analogues led to the identification of pregabalin.[64] Both gabapentin and pregabalin are 3-alkylated GABA analogues. Pregabalin appears to have a similar pharmacologic profile as gabapentin, with both greater bioavailability and potency. As with gabapentin, pregabalin was shown to be effective in several models of neuropathic pain by blocking the development of thermal hyperalgesia and mechanical allodynia.[65] Pregabalin is currently undergoing a number of clinical trials for painful diabetic neuropathy, fibromyalgia and postherpetic neuralgia.

The first available multicenter, double-blind, placebo-controlled, randomized clinical trial was conducted on 173 patients to evaluate the efficacy and safety of pregabalin in the treatment of postherpetic neuralgia. Patients randomized to pregabalin received either 600 mg/d (creatinine clearance >60 mL/min) or 300 mg/d (creatinine clearance 30 to 60 mL/min). Pregabalin was shown to be safe and significantly more effective than placebo in relieving pain and sleep interference ($p = 0.0001$).[66]

In a recent multicenter, doubleblind, placebo-controlled trial, both 150 and 300 mg/day pregabalin significantly reduced pain and improved sleep and mood disturbances in patients with PHN. Patients with history of failure to respond to previous treatment for PHN with gabapentin at doses ≥1200 mg/day were excluded. The most frequent adverse events were dizziness, somnolence, peripheral edema, weight gain,[67] headache, and dry mouth.

Lamotrigine

Lamotrigine is a phenyltriazine derivative approved in the United States as an add-on therapy for partial complex seizures. Lamotrigine blocks voltage-dependent sodium channels and inhibits glutamate release.

In a 14-week multicenter, randomized, double-blind, placebo-controlled trial in 42 patients with painful HIV-associated neuropathy, lamotrigine at a dose of 300 mg (25 mg/d slowly titrated over 7 weeks to 300 mg/d) was found to reduce pain more significantly than placebo ($p = 0.03$). A large dropout rate was seen in this study: 11 of 20 patients in the drug group; 5 were due to severe rash that forced them to discontinue treatment. In conclusion, lamotrigine was called a "promising" drug for HIV-associated neuropathy, where most other treatments, including amitriptyline, have not shown any benefit compared with placebo.[68]

A randomized, double-blind, placebo-controlled trial of lamotrigine monotherapy (titrated from 25 to 400 mg/d) for painful diabetic neuropathy was conducted in 59 patients over a 6-week period. Twenty-four of 29 patients (83 percent) receiving lamotrigine and 22 of 30 (73 percent) patients receiving placebo completed the study. On the basis of self-recording of pain intensity twice daily with a 0–10 numerical pain scale, the mean daily pain score was reduced from 6.4 ± 0.1 to 4.2 ± 0.1 in the drug-treated group and from 6.5 ± 0.1 to 5.3 ± 0.1 in the placebo group ($p < 0.001$ for lamotrigine doses of 200, 300, and 400 mg). The results on the McGill Pain Questionnaire (MPQ), the Beck Depression Inventory (BDI), and the Pain Disability Index (PDI) remained unchanged in both groups, and in addition, the adverse-events profile was similar. Lamotrigine was concluded to be effective in relieving pain associated with diabetic neuropathy.[69]

A smaller group of 30 patients with central poststroke pain were treated with lamotrigine in a randomized, double-blind, placebo-controlled crossover study during an 8-week treatment period. Lamotrigine 200 mg/d was found to reduce the median pain score to 5 compared with 7 for the placebo group ($p = 0.01$). No significant effect was obtained at lower doses (25, 50, or 100 mg/d). Two patients treated with lamotrigine withdrew from the study because of severe rash.[70] Oral lamotrigine 200 mg/d was concluded to be a well-tolerated and moderately effective treatment for central poststroke pain.

Lamotrigine can cause cutaneous side effects. Maculopapular eruptions have been seen in 3% to 10% of patients. However, serious cutaneous reactions, such as Stevens-Johnson syndrome and toxic epidermal necrolysis have also been observed, especially when the drug was given in association with valproic acid and neuroleptic drugs. The overall incidence of severe rash has been reported to be 3 per 1000 adult patients treated.[71–73]

Oxcarbazepine

Oxcarbazepine, the 10-keto homologue of carbamazepine, is a second-generation AED. Oxcarbazepine has not been evaluated in any randomized, controlled trials in patients with neuropathic pain, but a number of open-label trials in patients with trigeminal neuralgia that was refractory to carbamazepine suggested that oxcarbazepine can be useful and well tolerated.

Oxcarbazepine has been used in the treatment of trigeminal neuralgia at a starting dose of 600 mg/d (300 mg twice daily). The effective dose ranges from 600 to 1200 mg/d, with a maximum of 2400 mg/d. In several trials, oxcarbazepine was shown to be as effective as carbamazepine, with fewer adverse events. Some patients with trigeminal neuralgia obtained pain relief with oxcarbazepine, although previously resistant to carbamazepine. Oxcarbazepine seems to have several advantages when compared to carbamazepine: (1) it is not associated with hematologic or hepatic effects, so there is no need for routine monitoring; (2) it has minimal drug-drug interactions, which makes it safe to use in combination therapy; 3) it does not autoinduce its metabolism, which makes dose titration and achieving steady-state levels relatively simple; and (4) it can be administered in a twice-daily dosing regimen with or without food, thus facilitating patient compliance.[74]

In the rat models of neuropathic pain, oxcarbazepine did not affect mechanical hyperalgesia or tactile allodynia induced by partial sciatic nerve ligation in the rat. However, it produced up to 90 percent reversal of mechanical hyperalgesia in a guinea pig model. Similarly, its active metabolite, monohydroxy derivative, was shown to be active against mechanical hyperalgesia only in a guinea pig model but not in the rat.[54]

Zonisamide

No published randomized, controlled trials of zonisamide in the treatment of neuropathic pain are available in the literature. However, several open-label case series and anecdotal reports have been published. It has been used at a starting dosage of 100 mg at bedtime with progressive increments up to 500 mg/d given twice daily or at bedtime. The most frequently reported side effects were somnolence, dizziness, anorexia, and weight loss.[53]

In a chronic constriction injury model of neuropathic pain, systemic zonisamide was shown to relieve thermal hyperalgesia in a dose-dependent manner. However, mechanical sensitivity relief was observed only at the 100 mg/kg dose, which is sedating in the rat. Zonisamide was considered useful in the treatment of some types of neuropathic pain.[75]

▶ NMDA RECEPTOR ANTAGONISTS

NMDA is an amino acid that binds selectively to a special type of glutamate ionotropic receptor. Glutamate is the most abundant excitatory neurotransmitter in the CNS, with three different receptor subtypes: metabotropic, non-NMDA, and NMDA-receptor subtypes. The NMDA-receptor complex is known to play an important role in synaptic plasticity and in central sensitization. Following peripheral tissue or nerve injury, hyperalgesia and allodynia not only are due to an increase in the sensitivity of primary afferent nociceptors at the site of injury but also depend on NMDA receptor–mediated central changes in synaptic excitability.[76]

NMDA Receptor Antagonists and Analgesia

NMDA receptors are composed of assemblies of NR_1 and NR_2 subunits. Expression of both subunits is required to form functional channels. Every receptor includes a magnesium binding site, which prevents channel opening; a strychnine-insensitive glycine binding site, which is needed for channel opening; a polyamine site (NR_{2B}-selective) that regulates NMDA receptor excitability; and a phencyclidine site located inside the cationic channel. In addition to glutamate, the NMDA receptor requires the binding of a coagonist, glycine, to function. At normal resting membrane potential, NMDA receptors are inactive. The channel pore is blocked by magnesium ions. When a train of impulses arriving at a presynaptic terminal depolarizes the postsynaptic cell, the channel inhibition is released by electrostatic repulsion of magnesium ions, thus allowing NMDA receptor activation, and calcium ions to flow into the cell. Calcium activates several biochemical processes, including membrane phosphorylation, activation of nitric oxide synthase, and activation of immediate-early genes coding for factors that regulate protein synthesis.

The calcium ion influx into the postsynaptic cell is thought to be a crucial signal for the induction of NMDA-receptor-dependent *long-term potentiation*. The increased intracellular calcium activates the enzyme protein kinase C (PKC), which translocates from the cytoplasm to the cell membrane, where it phosphorylates the proteins of the NMDA receptor, causing further release of magnesium ions. Thus, phosphorylation causes an increased responsiveness of the NMDA receptor to glutamate, resulting in central sensitization. When NMDA receptors are sensitized, their hyperresponsiveness includes the neuron responding to non-noxious input from the peripheral nervous system. Nerve injury results in a significant increase in spinal glutamate, which opens the NMDA ionophore channel, causing calcium influx and resulting in spinal windup.[77]

Specific agents, such as ketamine and dextromethorphan, called *NMDA antagonists,* can block the channel in a noncompetitive use-dependent fashion. In animal models of neuropathic pain, NMDA receptor antagonists inhibit various behavioral responses such as allodynia-like responses, thermal hyperalgesia, and tactile allodynia. Despite the promising preclinical data, evidence for clinical efficacy of the NMDA receptor antagonists is still lacking.

Clinical use of NMDA antagonists as analgesics presently is restricted by their side-effect profile. These compounds appear to have a low therapeutic index and poor tolerability at doses that are efficacious clinically. NMDA receptors are ubiquitous in the human CNS. Therefore, antagonists that block NMDA receptors interfere not only with pain perception but also with sensory perception in general and cause memory impairment, psychotomimetic effects, ataxia, and motor incoordination. However, the restricted localization of NR_{2B}-containing receptors in small fiber afferents and in the dorsal horn of the spinal cord suggest that selective antagonists for these subtypes could produce analgesic without the intolerable side effects of the traditional NMDA antagonists. Both preclinical and clinical data indicate that NR_{2B}-selective compounds are better tolerated and retain antinociceptive activity comparable with that of other NMDA receptor antagonists. Moreover, this class of NMDA antagonists could have a broader spectrum of action in several pain conditions, including acute pain.

Peripheral NMDA receptors also offer a very attractive target for NMDA receptor antagonists that do not cross the blood-brain barrier in inflammatory and visceral pain. Such agents might be devoid of CNS side effects at doses producing antinociception at peripheral NMDA receptors.

NMDA Receptor Antagonists and Opioid Tolerance

NMDA receptor antagonists appear to slow or prevent the development of opioid tolerance.[78] In addiction, dextromethorphan was shown to reverse established tolerance even in the continuing presence of morphine and to prevent the expression of withdrawal symptoms in animal models.

Tolerance to opioids shares a common mechanism with NMDA-mediated central sensitization. In fact, opioid tolerance may be considered a change in receptor sensitivity as a result of phosphorylation.[79]

Opioid receptor desensitization, defined as the progressive loss of receptor function under continued exposure to an agonist, includes several mechanisms.[80]

Thus, NMDA antagonists could be considered in combination with opioids in the treatment of chronic pain in order to prevent tolerance and to induce a synergistic effect.[81]

The analogy between the mechanism of opioid tolerance and NMDA-receptor-mediated central sensitization could explain the observed relative opioid resistance of neuropathic pain. The intracellular signaling system activated during NMDA-mediated central sensitization is the same as that engaged by repeated opioid administration. This could explain why neuropathic pain seems to require a higher dose of opioids than nociceptive pain, leading to an increased incidence of related side effects.

A number of NMDA receptor antagonists are available clinically. These include ketamine, dextromethorphan, memantine, amantadine, and D-methadone. NMDA antagonists have been investigated in only a few studies, case reports, and clinical trials in experimental, acute, and chronic pain. Ketamine and dextromethorphan have been the most studied NMDA antagonists.

Ketamine

Ketamine is a noncompetitive NMDA receptor antagonist that binds the phencyclidine site. This drug has been used as an anesthetic for more than 30 years, but analgesic properties where shown when it was used in subanesthetic doses. Side effects of ketamine on awakening from anesthesia include unpleasant dreams, hallucinations, excitation, and emergence delirium. At analgesic doses, the main side effects include visual and auditory disturbances, cognitive impairment, disturbed proprioception, hallucinations, mental disturbances, and feelings of unreality.

Ketamine has been investigated in different models of pain and in some chronic pain syndromes. In 1995, two 3 × 3 crossover studies were conducted by comparing single-dose intravenous ketamine versus alfentanil or placebo (saline). Ketamine was shown to be as effective as alfentanil in diminishing the spread of pinprick hyperalgesia and allodynia produced by experimental intradermal capsaicin in healthy human subjects.[82] Positive observations have been made on the use of ketamine for postherpetic neuralgia, phantom limb pain, spinal cord tumor, and poststroke pain.[76]

Ketamine, similarly to alfentanil, showed a statistically significant analgesic effect over placebo in patients with chronic post-traumatic pain and allodynia distant from the site of injury. However, sedation, nausea, dissociative reactions, dizziness, and visual distortions have been reported as dose-limiting side effects.[83]

Ketamine is available for intravenous or subcutaneous delivery. However, there are some reports of oral delivery (compounded preparations of ketamine). The average oral dose reported to be used empirically has been 100 to 200 mg/d.[84]

Case reports of patients with neuropathic pain syndromes treated with oral ketamine as an analgesic adjuvant have been described. These low doses were effective in controlling pain without intolerable side effects.[85]

Dextromethorphan

Dextromethorphan, an effective over-the-counter cough suppressant, is a noncompetitive NMDA receptor antagonist. Dextromethorphan is metabolized to dextrorphan, which also demonstrates NMDA receptor antagonist activity.

The lower affinity of dextromethorphan and memantine for NMDA receptors, compared with the high affinity of ketamine, has been supposed to reduce the dose-limiting side effects by shortening the duration of receptor occupancy at the phenyclidine site. However, low doses of dextromethorphan (90 mg/d or less) have been shown to be ineffective in relieving chronic neuropathic pain, whereas daily doses of 300 to 400 mg may be necessary to detect a significant analgesic effect.

High-dose oral dextromethophan was shown to be effective in diabetic peripheral neuropathy but not in postherpetic neuralgia. Following a 5-week titration at the maximum tolerated dose, 26 subjects received a mean dextromethorphan dose of 400 mg/d for an additional week. By design, the maximum dose was 960 mg/d; however, all patients developed side effects during the dose titration. Patients with diabetic neuropathy obtained significant reduction of pain with dextromethorphan by a mean of 24 percent compared with placebo ($p = 0.014$), whereas no significant difference was recorded in the postherpetic neuralgia group. For diabetic neuropathy, the mean global pain relief score for dextromethorphan was significantly greater than for placebo ($p = 0.002$), whereas there was no significant effect in postherpetic neuralgia.[86]

Furthermore, preliminary results suggest that a fixed combination of morphine and dextromethorphan provided effective pain control at a lower total daily morphine dose, although pharmacokinetic interaction was not demonstrated.[81]

Conversely, in a recent randomized trial, 100 mg oral dextromethorphan was shown to have no effect on opioid-mediated analgesia when administrated 4 hours prior to an intravenous infusion of morphine 15 mg in patients with chronic pain. Moreover, common adverse effects, such as dizziness, nausea, and sedation, were reported.[87]

In conclusion, NMDA receptor antagonists have been shown to reduce spontaneous pain and hyperalgesia, but they have a narrow therapeutic window that has limited their use because of intolerable side effects, such as psychotomimetic side effects, memory difficulties, and changes in cognitive function due to the wide distribution of NMDA receptors in the brain, spinal cord, cortex, and hippocampus and their involvement in learning processes.

▶ CANNABINOIDS

The first reference to therapeutic use of *Cannabis sativa* dates back 5000 years to ancient Chinese medicine. Shen-Nung, a legendary Chinese emperor who identified hundreds of different plants and discovered their various qualities, wrote the oldest known medical text, entitled the *Pen Ts'ao*, that dates back to 2800 B.C. *Cannabis* was regarded as a superior agent for a wide variety of medical conditions, such as malaria, constipation, gout, rheumatism, and the pain of childbirth.

In 1814, Culpepper published his *Complete Herbal*, which listed all the known medicinal uses of *Cannabis*. In 1839, the physician William O'Shaughnessy introduced the medicinal uses of *Cannabis* to Europe and America by publishing his studies on the antiemetic, anticonvulsive, and antihypnotic effects of *Cannabis* in a paper titled "On the Preparation of the Indian Hemp or Gunja." In 1880, even Reynolds, personal physician to Queen Victoria, described *Cannabis* as a useful drug for the treatment of dysmenorrhea, migraine, neuralgia, convulsions, and insomnia.

Although *Cannabis sativa* has been used for therapeutic purposes for millennia, only recently has the scientific basis of its effects been elucidated. In the last decade, *Cannabis* has been reevaluated as a potential therapeutic agent.

The modern history of *Cannabis* can be summarized in three main steps:

1964: Isolation of Δ^9-tetrahydrocannabinol (Δ-9-THC), its major psychoactive constituent, in more than 60 active ingredients known as *cannabinoids*[88]

1990: Discovery in the human body of a cannabinoid receptor for Δ-9-THC[89]

1992: Isolation of arachidonylethanolamide, the first endocannabinoid, called *anandamide*, from the Sanskrit word *Ananda*, meaning "bliss"[90]

Cannabinoids exert many of their effects by combining with specific receptors in the brain and periphery. Two cannabinoid receptors have been identified and cloned, both members of the G-protein-coupled superfamily.[91]

The central CB_1 receptor is predominantly expressed in the brain, whereas the peripheral CB_2 receptor is found in immune cells. Both receptors are coupled through $G_{i/o}$ proteins, negatively to adenylate cyclase and positively to mitogen-activated protein kinase (MAP). Moreover, CB_1 receptors, when activated, can open potassium channels and close N-type and P/Q-type calcium channels.

The CB_1 receptor was the first to be identified in the murine brain and cloned from rat DNA. It is one of the most abundant receptors in the CNS and is expressed primarily by neurons located in the basal ganglia of the brain, mainly in the substantia nigra, globus pallidus, caudatum nucleus, and putamen. The distribution of CB_1 receptors has been mapped by autoradiography, hybridization, and immunocytochemistry. CB_1 receptors have been identified in the limbic system, including the hippocampus, amygdala, and hypothalamus, in the cerebellum, and in the cortex. Detailed analysis also has found CB_1 receptors in the spinal cord, especially in the dorsolateral funiculus, superficial dorsal horn (in laminae I and II), and lamina X.

The CB_2 receptors were cloned originally from macrophages in the marginal zone of the spleen. CB_2 receptors are generally expressed by immune cells, including B- and T-lymphocytes, natural killer cells, monocytes, and macrophages. CB_2 receptors also are found in the vas deferens and myenteric plexus. Thus, they are termed *peripheral cannabinoid receptors*. As for CB_1, the CB_2 receptor is negatively coupled to AC cyclase via $G_{i/o}$ proteins.[92]

Endocannabinoids

Endogenous ligands for cannabinoid receptors have been identified. These are unsaturated fatty acid amides that are involved in pain and inflammation.

Anandamide was the first discovered endocannabinoid, which was isolated from the porcine brain in 1992.[86] Subsequently, other endogenous agonists, including 2-arachidonilglycerol (2-AG), were isolated, and since then, an increasing list of endocannabinoids has evolved.[93]

Anandamide is an endogenous eicosanoid with moderate affinity for the two cannabinoid receptor subtypes. There is strong evidence that anandamide is not stored but rather is formed "on demand" in postsynaptic neurons in response to depolarization. Anandamide production is initiated by calcium ion influx into the cell, and two mechanisms have been described. The first involves the hydrolysis of a phospholipid precursor called *N*-arachidonyl phosphatidylethanolamine (NAPE) and is catalyzed by a phospholipase D enzyme. The rate-limiting enzyme in anandamide synthesis is the membrane-associated *N*-acyltransferase, which converts phosphatidylethanolamine to NAPE. The second mechanism is de novo synthesis of anandamide from the condensation of arachidonic acid and ethanolamine. Anandamide is inactivated via reuptake and enzymatic hydrolysis by cells. Anandamide is also metabolized by cyclooxygenase and lipooxygenase.[94]

Anandamide exerts its effect via a G-protein-coupled apparatus, leading to the molecular consequences of cannabinoid receptor activation, including inhibition of N-type, P/Q-type, and L-type voltage-gated calcium channels; activation of potassium channels; and inhibition of neurotransmitter release. When bound to cannabinoid receptors, anandamide evokes the traditional tetrad: antinociception, catalepsy, hypothermia, and hypolocomotion.

Anandamide is also a full agonist of the vanilloid VR_1 receptor.[95] The vanilloid receptor was first identified as a target for capsaicin, the irritant ingredient in red pepper. Subsequently, other substances have been discovered that share the ability to bind VR_1, such as resiniferatoxin and anandamide. Anandamide has been defined as an *endovanilloid*, and recent studies have shown that endogenous vanilloid receptor agonists can play a role in the development of neuropathic pain and inflammatory hyperalgesia.[96] Agents that activate CB_1 or VR_1 receptors can modulate pain perception. Both CB_1 and VR_1 receptors are coexpressed in several spinal and dorsal root ganglion neurons and can be considered as targets for new analgesic drugs. In particular, downregulation of VR_1 expression and/or activity could be a promising therapy for the future.

Potential Therapeutic Uses of Cannabinoids

The main therapeutic use of cannabinoids in humans is as an antiemetic, especially in the prevention of nausea and vomiting caused by anticancer chemotherapy.[97] They should be administrated 2 hours before the antineoplastic therapy and repeated 6 hours later.

In patients with cancer or AIDS, Δ-9-THC can be used to increase appetite and treat weight loss. Moreover, cannabinoids have a strong vasodilatation effect, especially for conjunctiva tissue. Thus, they reduce intraocular pressure and could be used for the treatment of glaucoma. THC is also considered a potent bronchodilator, which could be useful in the treatment of asthma.[98]

Finally, several studies have been carried out to assess the therapeutic effectiveness of cannabinoids as analgesics. The major active constituent of cannabis, Δ-9-THC, has been shown to have antinociceptive effects. Analgesic sites of action for cannabinoids have been identified in brain areas, in the spinal cord, and in the periphery.

Cannabinoid-Induced Antinociception

Cannabinoids have been studied widely in several animal pain models. In acute (phasic) pain models, a short noxious heat stimuli (hot plate) or a mechanical pressure (tail flicking) is applied to rodent paw or tail, whereas in chronic (tonic) pain models, longer-lasting signs of pain are induced either by nerve injury (sciatic nerve ligation) or by tissue injury (subcutaneous injection of noxious agents). In acute pain models, Δ-9-THC is the most widely studied cannabinoid. When administrated orally, systematically, or directly into the brain or spinal cord, Δ-9-THC exhibits antinociceptive activity. Moreover, when administered peripherally, cannabinoids depress motor activity and induce hypothermia. However, there is a evidence that CB_1 receptor-mediated antinociception is not a consequence of motor impairment or hypothermia.

Cannabinoids have greater potency in suppressing responses to noxious pressure or thermal stimuli when these stimuli are applied to paws or tail that have been inflamed or made hyperalgesic than to uninflamed/nonhyperalgesic areas. Cannabinoids seem to be effective against inflammatory pain.

When applied on inflamed paws, cannabinoids have a peripheral action, and induce antinociception at lower doses than those obtained with effective concentrations in the CNS.

Cannabinoids selectively modulate the activity of nociceptive neurons in the spinal dorsal horn. Their inhibitory effects are selective for pain-sensitive neurons and have been observed with different noxious stimulations.

The selective CB_1 receptor antagonist SR141716A can prevent the antinociceptive effects of cannabinoid receptor agonists at an appropriately high potency, and antinociceptive responses to Δ9-THC are absent or attenuated in CB_1 knock-out mice. Cannabinoid-mediated antinociception also can be attenuated by pertussis toxin and other agents that impair CB_1 receptor signaling. All these findings support the hypothesis that antinociception induced by cannabinoids is mediated by receptors, in particular the CB_1 receptors.[99]

CB_1 receptors were found in the dorsolateral funiculus, the superficial dorsal horn, and lamina X of the spinal cord. Most CB_1 expression is located on spinal interneurons, which implicates a hypothetical spinal action of cannabinoids as well.[100]

Cannabinoids can interact synergistically with opioid receptor agonists in the production of antinociception. Such synergism occurs when the opioid and the cannabinoid are both administered peripherally, intrathecally, or intracerebroventricularly. It seems to be receptor-related, and it can be blocked by both cannabinoid and opioid receptor antagonists.

Clinical Studies on the Analgesic Efficacy of Cannabinoids

In contrast to the strong preclinical data, good clinical evidence on the efficacy of cannabinoids is lacking.

Unfortunately, the few randomized trials available in the literature that evaluate the analgesic efficacy and safety of cannabinoids seem to have several limitations.[101]

In two postoperative pain trials, intramuscular levonantradol (1.5 to 3.0 mg) was superior to placebo but not more effective than codeine. Unlike these relatively negative analgesic results in nociceptive pain, more interesting are the results obtained in spasticity and neuropathic pain. THC 5 mg was shown to have equivalent analgesia to codeine 50 mg for neuropathic pain and spasticity, but only THC reduced spasticity. All five trials in cancer pain were single-dose trials. Cannabinoid-mediated analgesia was found similar to that of codeine but with dose-limiting adverse effects. Side effects associated with cannabinoids were common in six of the eight trials. Depression of the CNS seems to be the predominant adverse effect that limits their use.

A recent randomized trial in 12 healthy volunteers compared the analgesic effects of THC (20 mg), morphine (30 mg), and a THC-morphine combination (20 mg + 30 mg) by using experimental heat, cold, pressure, and single and repeated transcutaneous electrical stimuli. THC did not significantly reduce pain, whereas the THC-morphine combination completely reversed hyperalgesia and produced a slight additive analgesic effect in the electrical stimulation tests.[102]

In chronic neuropathic pain, 1′,1′-dimethylheptyl-Δ^8-tetrahydrocannabinol-11-oic acid (CT-3), a THC-11-oic acid analogue, at a dose of 40 mg/d was shown to be effective compared with placebo without major adverse effects. Twenty-one patients with chronic neuropathic pain were randomized in a placebo-controlled, double-blind, crossover trial. Three hours after the administration of CT-3, visual analogue scale (VAS) values differed significantly from placebo ($p = 0.02$), whereas, after 8 hours, these differences between the groups were less marked. Dry mouth and tiredness were the most common side effects, significantly more frequent in the CT-3 group ($p = 0.02$). However, no significant differences were recorded in blood tests and electrocardiogram.[103]

The major challenge is the need to reduce or abolish the adverse central effects of cannabinoids. For the purpose of having an acceptable therapeutic index, the spinal and peripheral sites of action should be considered as main targets, although these sites could not act independently from each other.

▶ TOPICAL ANALGESICS

The term *topical analgesic* has been developed to describe analgesics whose mechanism is primarily through reducing pain transmission within the peripheral nervous system. Pain perception involves both peripheral and central nervous system processes.[104]

In both acute and chronic pain, topical analgesics have been shown to be effective. They present several potential advantages compared with systemic analgesic agents. Most available medications for chronic pain have side effects that limit their usefulness, raising the need for newer safe and effective treatments. Unlike systemic analgesics, topical preparations act at peripheral sites of pain generation and do not produce a significant systemic concentration of the analgesic. Thus, they present the advantage of high drug concentration at the site of pain with minimal systemic effects in contrast to transdermal preparations, which must be considered systemic analgesics.

Topical analgesics can be used regularly without significant systemic accumulation, thus reducing the risk of side effects. Sometimes, localized reactions such as rash may occur in some patients, but adverse effects generally are mild and transient. In addition, the lack of relevant systemic effects significantly reduces the risk of drug-drug interactions. For topical analgesics, no dose titration is required, as for systemic agents.[105] However, it should be pointed out that not all topical analgesics are equivalent, and the side-effect profile can be different for various agents.

Lidocaine, and capsaicin are the agents that can be administered topically to produce pain relief in patients suffering from neuropathic pain.

Lidocaine

As a local anesthetic, the mechanism of action of lidocaine is related to its ability to suppress the activity of peripheral sodium channels on excitable membranes within sensory afferents. Consequently, lidocaine reduces pain transmission by preventing the generation and conduction of nerve impulses, especially ectopic and paroxysmal discharges.

Lidocaine administered via a variety of routes has been shown to relieve postherpetic pain. Animal studies have shown that systemic lidocaine can suppress C afferent fiber–evoked activity in the spinal cord and spontaneous activity in damaged peripheral neurons. However, intravenous lidocaine can result in plasma concentrations associated with antiarrhythmic effects. Topical application has been shown to be effective in relieving pain on allodynic skin without serious adverse effects. Topical lidocaine is available in cream (EMLA) and patch (Lidoderm) forms.

EMLA cream is a mixture of local anesthetics consisting of 2.5% lidocaine and 2.5% prilocaine. When applied on skin, it may result in a clear anesthetic effect.

Lidoderm is a soft adhesive patch (10×14 cm) containing 5% lidocaine (700 mg) that provides analgesic relief by blocking neuronal sodium channels. The U.S. Food and Drug Administration (FDA) has approved the lidocaine 5% patch specifically to relieve pain associated with postherpetic neuralgia. Patches should be applied directly to the painful area and worn for no more than 12 hours within any 24-hour period.[106]

Systemic absorption of lidocaine from the patch is minimal. The mean peak lidocaine plasma concentration achieved with the lidocaine 5% patch was 0.13 mg/L after 11 hours, whereas antiarrhythmic effects are obtained with 0.6 mg/L. At low plasma concentrations, lidocaine is 70 percent bound to plasma proteins. Lidocaine is metabolized by hepatic mechanisms and excreted via the kidneys.[107]

Two double-blind, randomized, placebo-controlled clinical trials involving a total of 67 patients demonstrated that the lidocaine 5% patch is efficacious and safe for postherpetic neuralgia.[108,109] Topical lidocaine patches were shown to relieve postherpetic pain more effectively than vehicle topical patches, which are identical except for the absence of lidocaine. Most of the subjects reported moderate or greater pain relief without serious adverse events.

Another randomized, double-blind, vehicle-controlled study conducted in 96 patients with postherpetic neuralgia involving the torso demonstrated that the lidocaine 5% patch alleviates all the most common neuropathic pain qualities, evaluated on the basis of the Neuropathic Pain Scale.[110] Eight specific categories were assessed and rated on a scale of 0 to 10: sharp, hot, dull, cold, skin sensitivity, itchy, deep pain, and surface pain. The lidocaine 5% patch had a significant effect on all the variables, suggesting that patients suffering from a variety of peripheral neuropathic pain conditions could benefit from this topic analgesic. In fact, an open-label study evaluating 16 patients with severe, refractory chronic neuropathic pain other than postherpetic neuralgia, with brush-evoked allodynia on examination, demonstrated that the lidocaine 5% patch can provide pain relief without any significant side effects in a variety of peripheral neuropathic pain conditions, including post-thoracotomy pain, neuroma pain, intercostal neuralgia, and complex regional pain syndrome.[111] Patients involved in this study were refractory to other traditional pharmacologic treatments, such as anticonvulsants, antidepressants, opioids, and mexiletine.

Other open-label studies have been conducted in patients with cervical spine pain, myofascial pain, and chronic low back pain. Adding the lidocaine patch to existing baseline pain medications in 27 patients with moderate to severe myofascial pain with identified trigger points resulted in a statistically significant ($p < 0.05$) mean improvement for average pain intensity, sleep, and quality of life.[107]

In conclusion, literature reports have demonstrated that the lidocaine 5% patch is efficacious, safe, and well tolerated in several neuropathic pain conditions, although it was approved by the FDA only for postherpetic neuralgia. Compared with lidocaine cream, the patch offers the benefits of a targeted peripheral analgesia without apparent loss of sensation under the area of application.

Capsaicin

Capsaicin is a substance found in hot chili peppers and is known as a C-fiber/nociceptor–specific neurotoxin. Topical analgesics containing capsaicin act through their interaction with the vanilloid VR_1 receptors on C fibers. Degeneration of nerve fibers following capsaicin applications has been documented histopathologically; thus, analgesia could be considered a consequence of this neurodegenerative effect. Following prolonged topical use, a significant decrease in the cutaneous density of C fibers occurs.

Although capsaicin has been reported to be beneficial for the treatment of neuropathic pain, topical application of capsaicin is followed by a severe burning sensation in almost 80 percent of patients treated, and EMLA failed to produce a long-lasting attenuation of the pain induced by topical application of 1% capsaicin.[112] Although some positive results have been reported on the effectiveness of capsaicin cream (0.025% and 0.05%) on pain associated with a number of neuropathic pain disorders, the high incidence of side effects and the short duration of benefits highly reduced patient compliance.

A multicenter, controlled, randomized, double-blind clinical trial conducted on 26 subjects showed that capsaicin is ineffective in relieving pain associated with HIV-associated distal symmetric peripheral neuropathy. Subjects receiving capsaicin reported higher current pain scores than did subjects receiving the vehicle.[113]

REFERENCES

1. Fishbain DA: Approaches to treatment decisions for psychiatric comorbidity in management of chronic pain patients. *Med Clin North Am* 1999;3:737–761.
2. MacPherson RD: The pharmacological bases of contemporary pain management. *Pharmacol Ther* 2000;88:163–185.
3. Lynch ME: Antidepressants as analgesics: A review of randomized controlled trials. *J Psychiatr Neurosci* 2001;26:30–36.
4. Mattia C, Coluzzi F: Antidepressants in chronic neuropathic pain. *Mini Rev Med Chem* 2003;3:773–784.
5. Reisner L: Antidepressants for chronic neuropathic pain. *Curr Pain Headache Rep* 2003;7:24–33.
6. Schreiber S, Backer MM, Pick CG: The antinociceptive effect of venlafaxine in mice is mediated through opioid and adrenergic mechanisms. *Neurosci Lett* 1999;273:85–88.
7. Schreiber S, Rigai T, Katz Y, Pick CG: The antinociceptive effect of mirtazapine in mice is mediated through serotoninergic, noradrenergic and opioid mechanism. *Brain Res Bull* 2002;58:601–605.
8. Sawynok J, Esser MJ, Reid AR: Antidepressants as analgesics: An overview of central and peripheral mechanisms of action. *J Psychiatr Neurosci* 2001;26:21–29.
9. Sawynok J, Reid AR, Esser MJ: Peripheral antinociceptive action of amitriptyline in the rat formalin test: Involvement of adenosine. *Pain* 1999;80:45–55.
10. Esser MJ, Sawynok J: Acute amitriptyline in a rat model of neuropathic pain: Differential symptom and route effects. *Pain* 1999;80:643–653.
11. Sawynok J, Esser MJ, Reid AR: Peripheral antinociceptive actions of desipramine and fluoxetine in an inflammatory and neuropathic pain test in the rat. *Pain* 1999;82:149–158.
12. McQuay H, Tramer M, Nye B, et al: A systematic review of antidepressants in neuropathic pain. *Pain* 1996;68:217–227.
13. Sindrup SH, Jensen TS: Efficacy of pharmacological treatments of neuropathic pain: An update and effect related to mechanism of drug action. *Pain* 1999;83:389–400.
14. Max MB, Lynch SA, Muir J, et al: Effects of desipramine, amitriptyline, and fluoxetine on pain in diabetic neuropathy. *N Engl J Med* 1992;326:1250–1256.
15. Sindrup SH, Gram LF, Brosen K, et al: The selective serotonin reuptake inhibitor paroxetine is effective in the treatment of diabetic neuropathy symptoms. *Pain* 1990;42:135–144.
16. Sindrup SH, Bjerre U, Dejgaard A, et al: The selective serotonin reuptake inhibitor citalopram relieves the symptoms of diabetic neuropathy. *Clin Pharmacol Ther* 1992;52:547–552.
17. Smith AJ: The analgesic effects of selective serotonin reuptake inhibitors. *J Psychopharmacol* 1998;12:407–413.
18. Mays TA: Antidepressants in the management of cancer pain. *Curr Pain Headache Rep* 2001;5:227–236.
19. Bohn L, Xu F, Gainetdinov RR, et al: Potentiated opioid analgesia in norepinephrine transporter knockout mice. *J Neurosci* 2000;20:9040–9045.

20. Kent JM: SNaRIs, NaSSAs, and NaRIs: New agents for the treatment of depression. *Lancet* 2000;355:911–918.
21. Markowitz JS, Patrick KS: Venlafaxine-tramadol similarities. *Med Hypoth* 1998;51:167–168.
22. Taylor K, Rowbotham MC: Venlafaxine hydrocloride and chronic pain. *West J Med* 1996;165:147–148.
23. Lang E, Hord AH, Denson D: Venlafaxine hydrochloride (Effexor) relieves thermal hyperalgesia in rats with an experimental mononeuropathy. *Pain* 1996;68:151–155.
24. Enggaard TP, Klitgaard NA, Gram LF, et al: Specific effect of venlafaxine on single and repetitive experimental painful stimuli in humans. *Clin Pharmacol Ther* 2001;69:245–251.
25. Pernia A: Venlafaxine for the treament of neuropathic pain. *J Pain Symptom Manage* 2000;19:408–410.
26. Sumpton JE, Moulin DE: Treatment of neuropathic pain with venlafaxine. *Ann Pharmacother* 2001;35:557–559.
27. Lithner F: Venlafaxine in treatment of severe painful peripheral diabetic neuropathy. *Diabetes Care* 2000;23:1710–1711.
28. Sindrup SH, Bach FW, Madsen C, et al: Venlafaxine versus imipramine in painful polyneuropathy: a randomized, controlled trial. *Neurology* 2003;60:1284–9.
29. Pick CG, Paul D, Eison MS. et al: Potentiation of opioid analgesia by the antidepressant nefazodone. *Eur J Pharmacol* 1992;211:375–381.
30. Goodnick PJ, Breakstone K, Kumar A: Nefazodone in diabetic neuropathy. *Psychosom Med* 2000;62:599–600.
31. Stewart DE: Hepatic adverse reactions associated with nefazodone. *Can J Psychiatry* 2002;47:375–377.
32. De Boer T: The pharmacological profile of mirtazapine. *J Clin Psychiatry* 1996;57:19–25.
33. Schreiber S, Bleich A, Pick CG: Venlafaxine and mirtazapine: Different mechanisms of antidepressant action, common opioid-mediated antinociceptive effects. A possible opioid involment in severe depression? *J Mol Neurosci* 2002;18:143–149.
34. Theobald DE, Kirsh KL, Holtsclaw BA: An open-label crossover trial of mirtazapine (15 and 30 mg) in cancer patients with pain and other distressing symptoms. *J Pain Symptom Manage* 2002;23:442–447.
35. Jolliet P, Veyrac G, Bourin M: First report of mirtazapine-induced arthralgia. *Eur Psychiatry* 2001;16:503–505.
36. Brannon GE, Stone KD: The use of mirtazapine in a patient with chronic pain. *J Pain Symptom Manage* 1999;18:382–385.
37. Schuler P, Seibel K, Chevts V, et al: Analgesic effect of the selective noradrenaline reuptake inhibitor (reboxetine). *Nervenarzt* 2002;73:149–154.
38. Semenchuk MR, Davis B: Efficacy of sustained-release bupropion in neuropathic pain: An open-label study. *Clin J Pain* 2000;16:6–11.
39. Semenchuk MR, Sherman S, Davis B: Double-blind, randomized trial of bupropion SR for the treatment of neuropathic pain. *Neurology* 2001;57:1583–1588.
40. Ansari A: The efficacy of newer antidepressants in the treatment of the chronic pain: A review of current literature. *Harvard Rev Psychiatry* 2000;7:257–277.
41. Rowbotham MC: Neuropathic pain: From basic science to evidence-based treatment, in Giamberardino MA (ed), *IASP Refresher Courses on Pain Management, Pain 2002: An Updated Review*. Refresher Course Syllabus/IASP Scientific Program Committee. Seattle: IASP Press, 2002: 165–173.
42. Dickenson AH, Chapman V: New and old anticonvulsants as analgesics, in Devor M, Rowbotham MC, Wiesenfeld-Hallin Z (eds), *Proceedings of the 9th World Congress on Pain: Progress in Pain Research and Management,* Vol. 16. Seattle: IASP Press, 2000:875–886.
43. Tremont-Lukats IW, Megeff C, Backonja MM: Anticonvulsants for neuropathic pain syndromes: Mechanisms of action and place in therapy. *Drugs* 2000;60:1029–1052.
44. Backonja M: Anticonvulsants for the treatment of neuropathic pain syndromes. *Curr Pain Headache Rep* 2003; 7:39–42.
45. Beydoun A, Backonja M: Mechanistic stratification of anti-neuralgic agents. *J Pain Symptom Manage* 2003;25:S18–S30.
46. Campbell FG, Graham JG, Zilkha KJ: Clinical trial of carbamazepine in trigeminal neuralgia. *J Neurol Neurosurg Psychiatry* 1966;29:265–267.
47. Killian JM, Fromm GH: Carbamazepine in the treatment of neuralgia: Use and side effects. *Arch Neurol* 1968;19:129–136.
48. Nicol CF: A four-year double-blind study of tegretol in facial pain. *Headache* 1969;9:54–57.
49. Rull J, Quibrera R, Gonzalez-Millan, et al: Symptomatic treatment of peripheral diabetic neuropathy with carbamazepine: Double blind crossover trial. *Diabetologia* 1969;5:215–218.
50. Gomez-Peres FJ, Choza R, Rios JM, et al: Nortriptyline-fluphenazine vs carbamazepine in the symptomatic treatment of diabetic neuropathy. *Arch Med Res* 1996;27:525–529.
51. Leijon G, Boivie J: Central post-stoke pain: A controlled trial of amitriptyline and carbamazepine. *Pain* 1989;36:27–36.
52. Harke H, Gretenkort P, Ladleif HU, et al: The response of neuropathic pain and pain in complex regional pain syndrome I to carbamazepine and sustained-release morphine in patients pretreated with spinal cord stimulation: A double-blinded randomized study. *Anesth Analg* 2001;92:488–495.
53. Pappagallo M: Newer antiepileptic drugs: Possible uses in the treatment of neuropathic pain and migraine. *Clin Ther* 2003;25:2506–2538.
54. Fox A, Gentry C, Patel S, et al: Comparative activity of the anti-convulsants oxcarbazepine, carbamazepine, lamotrigine and gabapentin in a model of neuropathic pain in the rat and guinea pig. *Pain* 2003;105:355–362.
55. Matthews EA, Dickenson AH: Effects of spinally delivered N- and P-type voltage-dependent calcium channel antagonists on dorsal horn neuronal responses in a rat model of neuropathy. *Pain* 2001;92:235.
56. Stahl SM: Anticonvulsants and the relief of chronic pain: pregabalin and gabapentin as $\alpha_2\delta$ ligands at voltage-Gated calcium channels. *J Clin Psychiatry* 2004;65:596–597.
57. Maneuf YP, Gonzalez MI, Sutton KS, et al: Cellular and molecular action of the putative GABA-mimetic, gabapentin. *Cell Mol Life Sci* 2003;60:742–750.
58. Backonja M, Glanzman RL: Gabapentin dosing for neuropathic pain: evidence from randomized, placebo-controlled clinical trials. *Clin Ther* 2003;25:81–104.
59. Rowbotham M, Harden N, Stacey B, et al: Gabapentin for the treatment of postherpetic neuralgia: a randomized controlled trial. *JAMA* 1998;280:1837–1842.

60. Rice AS, Maton S: Gabapentin in postherpetic neuralgia: a randomized, double blind, placebo controlled study. *Pain* 2001;94:215–224.
61. Backonja M, Beydoun A, Edward KR, et al: Gabapentin for the symptomatic treatment of painful neuropathy in patients with diabetes mellitus: a randomized controlled trial. *JAMA* 1998;280:1831–1836.
62. Pandey CK, Bose N, Garg G, et al: Gabapentin for the treatment of pain in Guillain-Barré syndrome: a double-blinded, placebo-controlled, crossover study. *Anesth Analg* 2002;95:1719–1723.
63. Garcia-Borreguero D, Larrosa O, De la Llave Y: Treatment of restless legs syndrome with gabapentin: a double-blind, cross-over study. *Neurology* 2002;59:1573–1579.
64. Bryans JS, Wustrow DJ: 3-substituted GABA analogs with central nervous system activity: a review. *Med Res Rev* 1999;19:149–77.
65. Wallin J, Cui JG, Yakhnitsa V, et al: Gabapentin and pregabalin suppress tactile allodynia and potentiate spinal cord stimulation in a model of neuropathy. *Eur J Pain* 2002;6:261–372.
66. Dworkin RH, Corbin AE, Young JP Jr, et al: Pregabalin for the treatment of postherpetic neuralgia: A randomized, placebo-controlled trial. *Neurology* 2003;60:1274–1283.
67. Sabatowski R, Galvez R, Cherry DA, et al: Pregabalin reduces pain and improves sleep and mood disturbances in patients with post-herpetic neuralgia: results of a randomised, placebo-controlled clinical trial. *Pain* 2004;109:26–35.
68. Simpson DM, Olney R, McArthur JC, et al: A placebo-controlled trial of lamotrigine for HIV-associated neuropathy. *Neurology* 2000;54:2115–2119.
69. Eisenberg E, Lurie Y, Braker C, et al: Lamotrigine reduces painful diabetic neuropathy: A randomized, controlled study. *Neurology* 2001;57:505–509.
70. Vestergaard K, Andersen G, Gottrup H, et al: Lamotrigine for central poststroke pain: A randomized, controlled trial. *Neurology* 2001;56:184–190.
71. Yalcin B, Karaduman A: Stevens-Johnson syndrome associated with concomitant use of lamotrigine and valproic acid. *J Am Acad Dermatol* 2000;43:898–899.
72. Sogut A, Yilmaz A, Kilinc M, et al: Suspected lamotrigine-induced toxic epidermal necrolysis. *Acta Neurol Belg* 2003;103:95–98.
73. Rzany B, Correia O, Kelly JP, et al: Risk of Stevens-Johnson syndrome and toxic epidermal necrolysis during first weeks of antiepileptic therapy: a case-control study. Study Group of the International Case Control Study on Severe Cutaneous Adverse Reactions. *Lancet* 1999 353:2190–2194.
74. Carrazana E, Mikoshiba I: Rationale and evidence for the use of oxcarbazepine in neuropathic pain. *J Pain Symptom Manage* 2003;25(5 suppl):S31–S35.
75. Hord AH, Denson DD, Chalfoun AG, et al: The effect of systemic zonisamide (Zonegran) on thermal hyperalgesia and mechanical allodynia in rats with an experimental mononeuropathy. *Anesth Analg* 2003;96:1700–1706.
76. Hewitt DJ: N-Methyl-D-aspartate–enhanced analgesia. *Curr Pain Headache Rep* 2003;7:43–47.
77. Eide PK: Clinical trials of NMDA-receptor antagonists as analgesics, in Devor M, Rowbotham MC, Wiesenfeld-Hallin Z (eds), *Proceedings of the 9th World Congress on Pain: Progress in Pain Research and Management*, Vol 16. Seattle: IASP Press, 2000:817–832.
78. Price DD, Mayer DJ, Mao J, et al: NMDA-receptor antagonists and opioid receptor interactions as related to analgesia and tolerance. *J Pain Symptom Manage* 2000;19:S7–S11.
79. Liu JG, Anand KJS: Protein kinases modulate the cellular adaptations associated with opioid tolerance and dependence. *Brain Res Rev* 2001;38:1–19.
80. Raith K, Hochhaus G: Drugs used in the treatment of opioid tolerance and physical dependence: a review. *Int J Clin Pharmacol Ther* 2004;42:191–203.
81. Katz NP: MorphiDex (MS:DM) double-blind, multiple-dose studies in chronic pain patients. *J Pain Symptom Manage* 2000;19:S37–S41.
82. Park KM, Max MB, Robinovitz E, et al: Effects of intravenous ketamine, alfentanil, or placebo on pain, pinprick hyperalgesia, and allodynia produced by intradermal capsaicin in human subjects. *Pain* 1995;63:163–172.
83. Max MB, Byas-Smith MG, Gacely RH, et al: Intravenous infusion of the NMDA antagonist, ketamine, in chronic posttraumatic pain with allodynia: A double-blind comparison to alfentanil and placebo. *Clin Neuropharmacol* 1995;18:360–368.
84. Hocking G, Cousins MJ: Ketamine in chronic pain management: an evidence-based review. *Anesth Analg* 2003;97:1730–1739.
85. Fitzgibbon EJ, Hall P, Schroder C, et al: Low-dose ketamine as an analgesic adjuvant in difficult pain syndromes: A strategy for conversion from parenteral to oral ketamine. *J Pain Symptom Manage* 2002;23:165–170.
86. Nelson KA, Park KM, Robinovitz E, et al: High-dose dextromethorphan versus placebo in painful diabetic neuropathy and postherpetic neuralgia. *Neurology* 1997;48:1212–1218.
87. Heiskanen T, Hartel B, Dahl ML, et al: Analgesic effects of dextromethorphan and morphine in patients with chronic pain. *Pain* 2002;96:261–267.
88. Gaoni Y, Mechoulam R: Isolation, structure and partial synthesis of an active constituent of hashish. *J Am Chem Soc* 1964;86:1646–1647.
89. Matsuda LA, Lolait SJ, Brownstein MJ, et al: Structure of a cannabinoid receptor and functional expression of the cloned cDNA. *Nature* 1990;346:561–564.
90. Devane WA, Hanus L, Breuer A, et al: Isolation and structure of a brain constituent that binds to the cannabinoid receptor. *Science* 1992;258:1946–1949.
91. Pertwee RG: Pharmacology of cannabinoid CB_1 and CB_2 receptors. *Pharmacol Ther* 1997;74:129–180.
92. Holdcroft A, Hargreaves KM, Rice AS: Cannabinoids and pain modulation in animals and humans, in Devor M, Rowbotham MC, Wiesenfeld-Hallin Z (eds), *Proceedings of the 9th World Congress on Pain: Progress in Pain Research and Management*, Vol. 16. Seattle: IASP Press, 2000:817–832.
93. Stella N, Schweitzer P, Piomelli D: A second endogenous cannabinoid that modulates long-term potentiation. *Nature* 1997;388:733–778.

94. Habayeb OM, Bell SC, Konje JC: Endogenous cannabinoids: Metabolism and their role in reproduction. *Life Sci* 2002;70:1963–1977.
95. Di Marzo V, Bisogno T, De Petrocellis L: Anandamide: Some like it hot. *Trends Pharmacol Sci* 2001;22:346–349.
96. Di Marzo V, Blumberg PM, Szallasi A: Endovanilloid signalling in pain. *Curr Opin Neurobiol* 2002;12:372–379.
97. Ashton CH: Adverse effects of cannabis and cannabinoids. *Br J Anaesth* 1999;83:637–649.
98. Hirst RA, Lambert DG, Notcutt WG: Pharmacology and potential uses of cannabis. *Br J Anaesth* 1998;81:77–84.
99. Pertwee RG: Cannabinoid receptors and pain. *Prog Neurobiol* 2001;63:569–611.
100. Farquhar-Smith WP, Egertova M, Bradbury EJ, et al: Cannabinoid CB(1) receptor expression in rat spinal cord. *Mol Cell Neurosci* 2000;15:510–521.
101. Campbell FA, Tramer MR, Carroll D, et al: Are cannabinoids an effective and safe treatment option in the management of pain? A qualitative systematic review. *Br Med J* 2001;323:1–6.
102. Naef M, Curatolo M, Petersen-Felix S, et al: The analgesic effect of oral delta-9-tetrahydrocannabinol (THC), morphine, and a THC-morphine combination in healthy subjects under experimental pain conditions. *Pain* 2003;105:79–88.
103. Karst M, Salim K, Burstein S, et al: Analgesic effect of the synthetic cannabinoid CT-3 on chronic neuropathic pain: A randomized, controlled trial. *JAMA* 2003;290:1757–1762.
104. Argoff CE: Targeted topical peripheral analgesics in the management of pain. *Curr Pain Headache Rep* 2003;7:34–38.
105. Sawynok J: Topical and peripherally acting analgesics. *Pharmacol Rev* 2003;55:1–20.
106. Comer AM, Lamb HM: Lidocaine patch 5%. *Drugs* 2000;59:245–249.
107. Gammaitoni AR, Alvarez NA, Galer BS: Safety and tolerability of the lidocaine patch 5%, a targeted peripheral analgesic: A review of the literature. *J Clin Pharmacol* 2003;43:111–117.
108. Rowbotham MC, Davies PS, Verkempinck C, et al: Lidocaine patch: Double-blind controlled study of a new treatment method for post-herpetic neuralgia. *Pain* 1996;65:39–44.
109. Galer BS, Rowbotham MC, Perander J, et al: Topical lidocaine patch relieves postherpetic neuralgia more effectively than a vehicle topical patch: Results of an enriched enrollment study. *Pain* 1999;80:533–538.
110. Galer BS, Jensen MP, Ma T, et al: The lidocaine patch 5% effectively treats all neuropathic pain qualities: Results of a randomized, double-blind, vehicle-controlled, 3-week efficacy study with use of the neuropathic pain scale. *Clin J Pain* 2002;18:297–301.
111. Devers A, Galer BS: Topical lidocaine patch relieves a variety of neuropathic pain conditions: An open-label study. *Clin J Pain* 2000;16:205–208.
112. Fuchs PN, Pappagallo M, Meyer RA: Topical EMLA pretreatment fails to decrease the pain induced by 1% topical capsaicin. *Pain* 1999;80:637–642.
113. Paice JA, Ferrans CE, Lashley FR, et al: Topical capsaicin in the management of HIV-associated peripheral neuropathy. *J Pain Symptom Manage* 2000;19:45–52.

CHAPTER 41

Psychological Treatments of Patients with Chronic Pain

Allen H. Lebovits

The psychological intervention with the patient who has chronic pain is an integral part of a multidisciplinary approach to pain management. Interdisciplinary pain programs have been shown to deliver effective and cost-efficient treatments to patients in pain.[1-3] The Commission on Accreditation of Rehabilitation Facilities will not offer accreditation to a pain management facility that does not have a mental health specialist as an integral member of the medical team. The foundation of this is the widely held recognition that the biopsychosocial model of pain, which views pain as a complex interaction of physical, psychological and social factors, is the most accurate conceptual model to use when treating the patient suffering with chronic pain. Diverse factors such as age, ethnicity, level of stress, social support, and coping style alter significantly the perception of pain.

It is now widely accepted that the dichotomous conceptualization of pain as being either medical or "psychogenic" is not in the best interest of the patient suffering with chronic pain; usually chronic pain is a combination of both medical and psychological factors. To be most effective, psychological techniques should be used in conjunction with a comprehensive medical evaluation and treatment—psychological interventions with chronic pain patients should complement, not replace, ongoing medical management. The second prerequisite for initiating psychological treatments is a comprehensive psychological evaluation. The findings from these two approaches should guide the mental health specialist in the choice of which psychological treatment to implement. The American Psychological Association has specified that the psychological treatment of chronic pain is one of 25 areas for which there is empirical validation for psychological intervention.

The mental health practitioner must decide which psychological treatment is optimal for each chronic pain patient. Pain specialists' perceptions of the patient uniformity myths and treatment uniformity myths have shown that it is important to customize treatments for patients with chronic pain.[4] Patients with pain must be evaluated and treated as unique individuals with different personalities, different coping skills, different circumstances of pain onset, and different family and support environments. As such, it would be highly inappropriate to treat pain patients with a cookie-cutter approach that applies the same psychological intervention to all patients.

In 1995, the NIH convened a technology conference to review the behavioral and relaxation interventions for chronic pain and sleep.[1] Their meta-analysis of the literature using the criteria of strong, moderate, fair and weak evidence concluded that there was strong to moderate evidence for the effectiveness of these methods in alleviating chronic pain, in particular:

1. They concluded that there was strong evidence that relaxation methods alleviate a variety of chronic pain syndromes.

2. Regarding hypnosis, the NIH reviewers cited strong evidence for it alleviating chronic cancer pain, and a suggested effectiveness with temporomandibular disorders, tension headaches, irritable bowel syndrome, and oral mucositis.
3. Summarizing the literature on cognitive-behavioral therapy, the conference participants concluded that there was evidence for moderate usefulness and that it was superior to routine care in low back pain and arthritis.
4. Evaluating the literature on biofeedback, the NIH panel felt there was moderate evidence for its effectiveness in pain and noted a particularly effective role in headaches.

The postulated mechanisms of action of these psychological approaches from this conference are that relaxation results in a decrease in sympathetic activity (as evidenced by a decrease in oxygen consumption, heart rate, and blood pressure), hypnosis activates the frontal-limbic attention system to inhibit pain transmission from the thalamic to cortical structures, and cognitive-behavioral therapy decreases depression and anxiety, which share the same brain regions with pain modulation. The NIH reviewers also noted that these methods give the (often previously helpless) pain patient an enhanced sense of self-control over their pain, which facilitates better coping abilities.

▶ COGNITIVE-BEHAVIORAL APPROACH

The most commonly used psychological approach for treating patients with chronic pain is the cognitive-behavioral approach. The general goal of cognitive-behavioral treatment strategies is to assist patients to reconceptualize their belief about pain from it being an uncontrollable medical symptom to a belief that their response to pain can be under their control.[5,6] The cognitive-behavioral approach is based upon the theory that thoughts, emotions, and behavior can influence the pain experience. The pain is not "cured" but the patient may be better able to cope with it. Rather than being passive participants in their treatment, as is the case in standard medical care, patients are required to shift into becoming active partners in their care. A meta-analysis of randomized controlled clinical trials of the cognitive behavioral approach for chronic pain showed significant positive changes in measures of pain experience, mood, coping, pain behavior, and activity level.[7]

The most commonly practiced elements of the cognitive-behavioral approach are (1) education, (2) relaxation (accomplished through many methods including guided imagery, progressive muscle relaxation, meditation, biofeedback, and hypnosis), and (3) cognitive-behavioral therapy (including cognitive restructuring, skills acquisition, and generalization).

Education

The first step is to educate the patient about pain, the mind-body relationship, attitudinal issues, and the expectations and goals of treatment. The American Pain Society and the Agency for Health Care Policy and Research recommend in their clinical practice guidelines that education about pain be an integral component of treatment.[8,9] The education of patients about what to expect regarding pain after procedures or surgery has been demonstrated to significantly improve their satisfaction with pain control.[10] Patients need to receive a rationale to understand their pain, the benefits of a comprehensive approach to pain management, and the specific role of the psychological intervention in the management of their pain. The effectiveness of this step depends on the patient's defensiveness, level of knowledge about the mechanism of pain, and attitudes about the mind-body relationship. The patient needs to be educated about the mind-body relationship according to the patient's ability to understand. Patient education materials, such as instruction sheets and audiotapes, can supplement the clinician's efforts. The patient needs to be reassured that their pain is *real*, that it is not *in their head* (as they may have been told by well meaning but frustrated physicians). The very referral to the mental health specialist can cause patients to become defensive about the *realness* of their pain; this needs to be addressed at the outset of any psychological treatment, otherwise one is faced with a noncooperative patient with minimal motivation.

Additionally, maladaptive attitudinal issues need to be addressed, specifically the fear of addiction and nonadherence to medication or to physical therapy protocols. The generally unwarranted fear of addiction in patients with chronic pain prevents some patients from taking analgesics and adjuvants that might have a significant beneficial impact on their pain, functional level, and quality of life. Additionally, patients may refuse to take specific medications that have "addict" connotations to them, such as methadone and Oxycontin. Cancer patients often worry that their doctors and nurses may not see them as good patients if they complain about their pain;[11] as a result, some patients may take their opioids only when their pain is severe rather than following the recommended fixed schedule. Nonadherence to medication or to physical therapy protocols is a major problem that should be part of the psychoeducational intervention. Patients need to be instructed not to give up on protocols too soon, to allow, for example, the longer-acting tricyclic antidepressants to take effect. A decrease in physician visits following patient education programs has been demonstrated.[12]

One of the most important objectives of patient education is the setting of treatment goals. The patient should be actively involved in this phase. Many patients expect "cure," and quickly! Patients need to be taught that we refer to *pain management* rather than *cure*, and that chronic pain improves slowly, and usually not in a linear manner. Most important, and perhaps most difficult for some pain patients to accept, is the primary treatment goal of *improved function* rather than *pain relief*. The goal of most multidisciplinary pain centers is to return patients to the optimal levels of function that their impairments allow. Most often, behavioral function is the primary goal, followed by emotional function. Important but secondary treatment goals are decreased reliance on medications and lessened utilization of the health care system, as well as reduced subjective pain sensation.

Relaxation

The mainstay of the cognitive-behavioral approach is *relaxation training*, which helps patients to redirect their focus away from pain, reduce autonomic reactivity, reduce muscle tension and enhance their sense of self-control. Relaxation training has been found to be more effective than no treatment at all for chronic pain, but only equally as effective as other self-regulation techniques.[13] Often the initial step of relaxation training is to learn controlled diaphragmatic breathing; it diverts the patient's attention and can induce the relaxation effect by itself. Relaxation training can be accomplished through several techniques used separately or in combination: (1) guided imagery, (2) progressive muscular relaxation, (3) meditation, (4) biofeedback, and (5) hypnosis.

Guided Imagery

A psychotherapist using the *guided imagery* technique asks the patient to focus on a multisensory imaginary scene. Focusing on the different sensory modalities of the scene can make the image more engaging. Typically, the image is elicited from the patient, and the psychotherapist guides the patient through the image, substituting sensations such as warmth or numbness for pain. Patients need to set aside time to practice in a comfortable position without any interruptions. Imagery can work as an effective distraction technique. An alternative use of imagery is to have the patient focus on the pain rather than distract away from it: in this technique, the patient might visualize the pain as a color, for example red, and make it less bright until it turns light pink—corresponding to lower pain intensity.

Progressive Muscular Relaxation

In *progressive muscular relaxation*, patients are taught to alternately tense and relax major muscle groups throughout the body.[14] Only nonpainful muscle groups and body locations are used. Patients learn to recognize and differentiate feelings of tension from relaxation and then apply these skills in situations that are painful. Sixteen muscle groups are initially tensed and relaxed; the number is reduced as the patient becomes more proficient. The patient is instructed to focus on the pleasantness of the relaxation phase.

Meditation

Meditation is defined as "the intentional self-regulation of attention from moment to moment."[15] *Concentration meditation*, involving the focused attention on a point or a mantra differs from *mindfulness meditation*, which emphasizes detached observation, from one moment to the next, of a changing field of objects. The primary advantage of mindfulness meditation is that it can be used to adapt a detached view of the pain sensation, which can lead to an *uncoupling* of the affective from the sensory interpretation of pain. As a result, patients have lower levels of reactivity to pain and exhibit less pain behavior. A study of 51 refractory chronic pain patients going through a mindfulness meditation program showed that 65 percent experienced a reduction of over 33 percent in their pain ratings.[15]

Biofeedback

Biofeedback is a particularly effective modality for teaching chronic pain patients relaxation and self-regulation of physiological processes. Patients learn to modify specific physiological processes based on auditory and/or visual feedback. It is based on the educational paradigm that learning occurs with feedback, which then enables a desired response. In advanced biofeedback systems, body sensors are attached to a computer by a fiber-optic cable and multiple physiological systems can be looked at simultaneously. Ongoing physiological processes (such as muscle tension or surface EMG, temperature, heart rate, sweat gland activity or basal skin response, and breath rate) can be monitored and both visual feedback (using graphs, images, or games) and auditory feedback (using tones or music) are provided. The latest application of biofeedback is neurofeedback, which teaches patients to regulate their EEG activity or brain waves.

The biofeedback protocol typically used with pain patients is (1) an evaluation of their baseline state, (2) an evaluation of their physiological measures while talking and/or thinking of stress, and (3) an evaluation of their physiological state when relaxed. The results for each condition are compared and discussed with the patient. It is often a powerful lesson for the patient to *see* the differences in measurements such as heart rate and muscle tension between when they are stressed and when they are relaxed. Pain syndromes for which biofeedback is most effective include headaches, temporomandibular dysfunction, myofascial pain, irritable bowel syndrome, Raynaud's disease, fibromyalgia, and other pain exacerbated by stress or anxiety. The mechanism of the effectiveness of biofeedback with pain is through:

1. Reduction of general arousal levels, which correlates with a reduction of the central processing of peripheral sensory inputs.
2. Relaxation, which increases pain tolerance and decreases distress.
3. The ability to relax specific muscle spasms.
4. The perception that biofeedback technology is a nonthreatening method of learning about the mind-body relationship.
5. The learning of physiological self-control, which can enhance self-efficacy, and improve both coping skills and hopefulness.

In a comparison of electromyographic (EMG) biofeedback to cognitive-behavioral therapy and to conservative medical intervention, 57 patients with chronic back pain and 21 patients with temporomandibular joint dysfunction (TMD) were evaluated. At 24 months, only the biofeedback group maintained significant reductions in pain severity, interference, affective distress, and use of the health care system.[16]

Hypnosis

Hypnosis is another particularly effective therapeutic technique with pain patients. Hypnosis can induce a state of selective attention focusing, often called *dual awareness*. It not only teaches patients relaxation and a passive disregard of intrusive thoughts, but has the unique feature of introducing specific goals through suggestions. These suggestions enable patients to experience analgesia or reinterpretation of their pain. Patients can experience numbness, for example, instead of pain. Additionally, posthypnotic suggestions allow the patient continued use of the new behavior and assistance in recreating the relaxed state whenever needed following termination of hypnosis. Individuals vary in their hypnotic susceptibility, for largely unknown reasons.

Hypnosis is also used with surgical patients to reduce presurgical fear and, at times, to create hypnoanesthesia when anesthetic agents are contraindicated, when it is desirable for patients to respond, or when their fear of anesthesia is significant. Another utilization of this method is when positive suggestions are made during surgery to reduce postoperative pain and side effects. In a study evaluating intraoperative therapeutic suggestions during ambulatory surgery, 70 patients undergoing elective hernia repair under standard general anesthesia listened continuously during anesthesia to either a therapeutic tape (reassurances regarding a favorable postoperative outcome) or a comparison tape (neutral history of the hospital).[17] The choice of tape was made at random, and the patient, anesthesiologist, surgeon, and nurses were unaware of the tape contents. While there were no between-group differences in postoperative pain scores over time, the therapeutic tape group experienced fewer side effects over the entire postoperative time period; in particular, they had fewer postoperative headaches and less muscular discomfort, nausea and vomiting (the nausea and vomiting difference was limited to the first 90 minutes after the operation) than the comparison tape group.

In a study by Spiegel and Bloom,[18] women with metastatic breast carcinoma pain who were undergoing weekly group therapy with self-hypnosis had significantly lower pain ratings over one year than a control group. A recent review of outcome studies using hypnosis with chronic pain patients concluded that hypnosis is "consistently superior" to no treatment but only equally as effective as other treatments.[13]

Cognitive-Behavioral Therapy

Cognitive-behavioral therapy refers to specific interventions such as skills acquisition, cognitive restructuring, and generalization and maintenance. Many practitioners also put education and relaxation, which are almost always used together, within the domain of cognitive-behavioral therapy. Cognitive-behavioral therapy needs to be modified to the unique needs of each pain patient, specifically being sensitive to the patient's educational and cultural background. A recent review of the literature of the effectiveness of cognitive-behavioral therapy with specific disease states concluded that cognitive-behavioral therapy is *well-established* for rheumatoid arthritis and is *probably efficacious* for patients with osteoarthritis of the knee and for patients with irritable bowel syndrome, but *experimental* for patients with fibromyalgia.[19] Similarly, national guidelines have recommended that cognitive-behavioral therapy should be used to reduce pain and psychological disability in patients with rheumatoid arthritis, as well as to enhance self-efficacy and pain coping.[9]

Skills Acquisition

Patients learn new behaviors and cognitions to improve function, better manage their pain, and improve their coping skills. Patients learn to adapt the more effective *active* coping styles rather than the passive ineffective coping styles such as *catastrophizing, avoidance*, and *denial*. Patients are instructed to practice their newly-learned skills and behaviors in their home and work environments; family and partners can be very helpful in ensuring that patients practice their "homework."

Activity pacing, the scheduling of rest periods so that patients don't overdo it and sabotage their progress, can be very beneficial for many pain patients. Overexertion, as well as resulting in increased pain and the need for prolonged rest, often has negative sequelae such as increased muscle tension and increased use of medications. Teaching patients to moderate their daily activities—by scheduling periods of moderate activity followed by limited rest—can increase their self-confidence.[20] Overly inactive patients are taught to initiate activities in a very limited fashion and to gradually increase activities followed by rest. Patients are also taught to schedule pleasant and enjoyable activities during the day. Additionally, the use of pain diaries to help identify stressful situations, or times of day that exacerbate pain, can help patients regulate their behaviors and emotions to achieve more adaptive pain coping skills.

Cognitive Restructuring

An essential part of cognitive-behavioral therapy is *cognitive restructuring*, which is based on the theory that cognitions determine behavior, affect, and physiology (such as increased muscle tension). Patients learn to identify, challenge, and eventually change self-defeating thoughts (such as "I am worthless"). With this technique, pain patients are taught to identify maladaptive negative thoughts (such as "pain signifies something is terribly wrong", "pain means I need more surgery" or "no one can help me") that pervade their thinking and to replace them with more constructive and adaptive positive thoughts (such as "I can still do many important things"). The maladaptive thoughts often take the form of statements about oneself or one's illness that are negative, overgeneralizing, or catastrophizing. Patients are taught to use their adaptive thoughts when confronted with pain or situations that lead to pain.

Generalization and Maintenance

Unless patients practice, they will relapse when they encounter stressful or difficult situations, which can lead to increased depression and helplessness.[21] To help patients maintain their newly acquired pain-coping skills, training them in problem solving and relapse prevention is important. The generalization and maintenance component of cognitive-behavioral therapy helps patients to identify both difficult situations and the early warning signs of relapse. It teaches patients to rehearse what they have learned, and enables families to reinforce good coping skills during signs of early relapse. Role-playing can be an effective vehicle to identify problematic coping strategies and to practice more adaptive responses to potential situations that lead to pain. Support from a partner can be very influential in ensuring success of the intervention. Many psychologists readily enlist the support of family members to promote generalization and maintenance.

▶ OPERANT THERAPY

Fordyce[22,23] was first to propose that a significant amount of clinical pain consists of overt pain behaviors, which can be influenced by environmental factors (in addition to tissue damage). More specifically, Fordyce proposed that the behavioral expression of pain, *pain behavior*, is the result of positive and negative *reinforcers* from the patient's environment, such as social reinforcement from family and friends, medications from physicians, financial incentives, or the avoidance of activities. Such environmental factors also included environmental stressors such as marital conflict, work demands, or economic difficulties, all of which can can initiate or maintain pain behaviors. *Positive reinforcers*, such as increased attention, which can maintain pain responses, are also considered environmental factors.

Operant therapy modifies pain responses. More specifically, operant therapy applies positive or negative reinforcers to overt pain responses, resulting in behavioral changes in the desired direction. Similarly *overt well behaviors,* such as an increase in adaptive functional behavior or a reduction in pain medication usage, are also reinforced. Reviews of the literature regarding the efficacy of operant and behavioral methods conclude that they have significant effects on overt pain behaviors—such as increased activity, decreased usage of pain medication, and fewer verbal expressions or ratings of pain,[7,24-28] particularly with chronic low back pain patients.

▶ GROUP THERAPY

Group therapy has become a popular form of psychological intervention for the chronic pain patient.[29] A recent meta-analysis of randomized controlled trials of cognitive- behavioral therapy for chronic pain found that most treatments were delivered in groups.[7] Many psychologists will see most of their pain patients in a group rather than individually, because of its effectiveness: group therapy has been shown to significantly reduce pain and improve function in a variety of chronic pain disorders such as osteoarthritis.[30] The advantages of group therapy are that pain patients learn they are not alone in their suffering, the group can be an effective support system, and patients can learn pain coping skills from other patients. Patients will often accept challenges from other patients to improve function more readily than from an individual therapist whom the patient may feel does not understand or appreciate their pain. The major goals of group therapy often are to promote behavior change, to educate patients, and to provide social support.[29] Social support can be influential in reducing psychological disability.

Group therapy typically uses cognitive-behavioral methods to accomplish behavior change, education, and support. Training patients in coping skills helps them to reconceptualize their pain, bringing it more under their control. Additionally, relaxation training, activity-rest cycling (learning pacing), and attention-diversion strategies are techniques often taught in cognitive-behavioral groups.

▶ PSYCHODYNAMIC PSYCHOTHERAPY

The first psychological model of chronic pain was the psychodynamic approach, which emphasized the psychological etiology of pain. Freud viewed pain as a symptomatic expression of an unconscious conflict seeking awareness. Psychodynamic psychotherapy can play an important role in the psychological intervention with pain patients. The integrative psychotherapeutic approach generally integrates behavioral and cognitive strategies within a psychodynamic framework that focuses on developmental issues and interpersonal conflicts.[31] This approach is particularly useful for the *vulnerable* or *pain prone* individuals who may attribute specific psychological meanings to their pain.[32] Psychodynamic themes that are often explored in psychodynamic psychotherapy with individuals experiencing chronic pain include childhood development, early experiences with pain and illness, childhood physical and sexual abuse, unresolved anger, pain as an affect, alexithymia (the inability to communicate affective experiences), pain as punishment, and somatization.[32] Developmental and ongoing relationships are explored, as well as the patient-therapist relationship.[32]

▶ STRESS MANAGEMENT

Because many pain patients report a strong relationship between stress and their pain, stress management interventions can be very helpful. Stress is often defined as "the response of the body to changes in the environment." Stress is, therefore, seen as a physiological response, which can often emerge as pain. The initial step in stress management programs is to identify one's stressors in daily life. This is frequently followed by cognitive-behavioral methods as outlined above, such as relaxation training and cognitive restructuring. Other important stress management interventions that can be particularly helpful to chronic pain patients include:

- Physical exercise on a regular basis. It is usually recommended that exercise be done three times a week for 20–30 minutes. Patients who have been physically inactive need to be cautioned to start out slowly to avoid injury. Chronic pain patients should never initiate a physical exercise program without the guidance of a physiatrist or physical therapist. Swimming is considered one of the best cardiovascular exercises, being particularly good for chronic pain patients as there is limited stress placed on the joints.
- Time management is an important intervention, particularly for "workaholics" or very disorganized patients. Time management involves teaching patients to make daily lists of tasks to be done, prioritizing them with regard to their importance, estimating the amount of time each task takes, and possibly delegating the ones that others can do. If done properly, time management methods can relieve a significant amount of stress for pain patients who often feel overwhelmed.
- Sharing feelings and problems with others such as partners, other patients, or professionals, can be an effective method of relieving stress. Internalizing emotions (keeping them pent up) is generally considered to be unhealthy and has been correlated with a variety of medical conditions including chronic pain. Repressed anger, in particular, has been related to chronic pain, especially low back pain, and has been termed *tension myositis*.[33] Patients with strong support systems have been shown to cope more effectively with stress, which is a reason why, as noted above, group therapy for pain patients can be so helpful.
- The use of humor can be an effective stress reducer. Laughing at one's problems and taking a humorous perspective on difficult situations can reduce stress. Similarly, making time for fun and being involved in recreational activities can be a good distraction and break up the chronicity of stress.

▶ PSYCHOLOGICAL INTERVENTIONS WITH CHILDREN AND ADOLESCENTS

Research on the use of psychological interventions with children and adolescents in pain is less extensive than with adults. Much of the relevant literature has been limited to procedure-related pain, where distraction techniques and the presence of a parent can be very helpful. It is increasingly recognized that psychological interventions suitable for adults may not be appropriate in the pediatric setting. There may be specific psychological interventions for children and adolescents that are particularly effective with chronic pain. Since children often have active imaginations, they are amenable to imagery and relaxation methods. Although cognitive-behavioral methods have been demonstrated to be effective in relieving headaches in children, the evidence for other types of chronic pain has not been as conclusively demonstrated.[34]

▶ BARRIERS TO INTEGRATION OF PSYCHOLOGICAL THERAPIES

Despite the generally accepted efficacy of these methods with pain patients, their relative ease of implementation, and their very low side-effect profile, barriers nevertheless still exist that prevent psychological therapies from being integrated into standard medical care:[1]

1. There still remains an overemphasis on the biomedical model, both in clinical care and in medical education.
2. There is a lack of standardization of psychological techniques.
3. There is a lack of patient compliance in practicing these methods.
4. There is a reluctance by physicians to prescribe psychological methods, due to:
   a. Lack of awareness of the benefits of these techniques;
   b. Concern regarding patient perception that referral reflects mental illness.
5. Inconsistent and poor reimbursement by third-party payers hinders the delivery of services.
6. There are ill-defined credentialing criteria for providers of such services, which creates an unreliability in the delivery of these methods.
7. Psychosocial interventions are time intensive and often necessitate many visits, which can impede physician and patient acceptance.

Integration of psychological interventions with conventional medical methods is essential, particularly in the area of pain. This is highlighted by reports of increased mortality as a result of unresolved pain.[35,36] The clinical practice guideline for the management of cancer pain recommends the early introduction of psychological methods in the course of illness so as to improve the chances of success.[8] Additionally, the success of medical interventions such as surgery and spinal cord implantation, particularly in the area of pain, has been shown to be largely dependent on psychosocial factors.[37] The barriers to the integration and implementation of psychological therapies into the pain management practice hopefully can be overcome with physician and patient education, as well as additional research.[1]

REFERENCES

1. NIH Technology Assessment Panel on Integration of Behavioral and Relaxation Approaches Into the Treatment of Chronic Pain and Insomnia: Integration of behavioral and relaxation approaches into the treatment of chronic pain and insomnia. *JAMA* 1996;276:313–318.
2. Okifuji, A: Interdisciplinary pain management with pain patients: Evidence for its effectiveness. *Sem Pain Med* 2003;110–119.
3. Lebovits AH: Chronic pain: The multidisciplinary approach International Anesthesiology Clinics 1991;29:1–7.
4. Turk, DC: Customizing treatment for chronic pain patients: Who, what and why? *Clin J Pain* 1990;6:255–270.
5. Holzman AD, Turk DC, Kerns RD: The cognitive-behavioral approach to the management of chronic pain, in Holzman AD, Turk DC (eds), *Pain Management—A Handbook of Psychological Treatment Approaches*. New York: Pergamon, 1986:31–50.
6. Bradley LA: Cognitive-behavioral therapy for chronic pain, in Gatchel RJ, Turk DC (eds), *Psychological Approaches to Pain Management*. New York: Guilford Press, 1996:131–147.
7. Morley S, Eccleston C, Williams A: Systematic review and meta-analysis of randomized controlled trials of cognitive-behavioral therapy and behavior therapy for chronic pain in adults, excluding headache. *Pain* 1999;80:1–13.
8. Agency for Health Care Policy and Research. Management of Cancer Pain. Rockville, MD: US Department of Health and Human Services, 1994.
9. American Pain Society: *Guideline for the Management of Pain in Osteoarthritis, Rheumatoid Arthritis, and Juvenile Chronic Arthritis*. Glenview, Illinois: American Pain Society, 2002.
10. Lebovits AH, Zenetos P, O'Neill DK, et al: Satisfaction with epidural and intravenous patient-controlled analgesia. *Pain Med* 2001;2:280–286.
11. Ward SE, Goldberg N, Miller-McCauley V, et al: Patient-related barriers to management of cancer pain. *Pain* 1993; 52:319–324.
12. Lorig K, Laurin J, Holman HR: Arthritis self-management: A study of the effectiveness of patient education for the elderly. *Gerontologist* 1984;24:455–457.
13. Kessler R, Patterson DR, Dane J: Hypnosis and relaxation with pain patients: Evidence for effectiveness. *Sem Pain Med* 2003;1.
14. Lebovits AH, Bassman LE: Psychological aspects of chronic pain management, in Lefkowitz M, Lebovits AH, Wlody D, Rubin S (eds), *A Practical Approach to Pain Management*. Boston: Little Brown, 1996:124–128.
15. Kabat-Zinn J: An outpatient program in behavioral medicine for chronic pain patients based on the practice of mindfulness meditation: Theoretical considerations and preliminary results. *Gen Hosp Psychiatry* 1982;4: 33–47.

16. Flor H, Birbaumer N: Comparison of the efficacy of EMG biofeedback, cognitive-behavior therapy, and conservative medical interventions on the treatment of chronic musculoskeletal pain. *J Consult Clin Psychol* 1993;61:653–658.
17. Lebovits AH, Twersky R, McEwan B: Intraoperative therapeutic suggestions in ambulatory surgery: Are there benefits for postoperative outcome? *Br J Anaesth* 1999;82: 861–866.
18. Spiegel D, Bloom J: Group therapy and hypnosis reduce metastatic breast carcinoma pain. *Psychosom Med* 1983;45: 333–339.
19. Bradley LA, McKendree-Smith NL, Cianfrini LR: Cognitive-behavioral therapy interventions for pain associated with chronic illness: Evidence for their effectiveness. *Sem Pain Med* 2003;1.
20. Hirano PC, Laurent DD, Lorig K: Arthritis patient education studies, 1987–1991: A review of the literature. *Patient Educ Couns* 1994;24:9–54.
21. Keefe FJ, Van Horn Y: Cognitive-behavioral treatment of rheumatoid arthritis pain: Maintaining treatment gains. *Arthritis Care Res* 1993;6:213–222.
22. Fordyce WE: *Behavioral Methods for Chronic Pain and Illness*. St. Louis, MO: Mosby, 1976.
23. Fordyce WE: Learned pain: Pain as behavior, in Loeser J, Butler C, Chapman R, Turk DC (eds), *Bonica's Management of Pain* (3rd). New York: Lippincott, 2000:470–482.
24. Fordyce WE, Roberts AH, Sternbach RA: The behavioral management of chronic pain: A response to critics. *Pain* 1985;22:113–125.
25. Linton SJ: Behavioral remediation of chronic pain: A status report. *Pain* 1986;24:125–141.
26. Keefe FJ, Dunsmore J, Burnett R: Behavioral and cognitive-behavioral approaches to chronic pain: Recent advances and future directions. *J Consult Clin Psychol* 1992;60:528–536.
27. Compas BE, Haaga DA, Keefe FJ, et al: Sampling of empirically supportive psychological treatments for health psychology: Smoking, chronic pain, cancer, and bulimia nervosa. *J Consult Clini Psychol* 1998;66:89–112.
28. Sanders, SH: Operant therapy with pain patients: evidence for its effectiveness. *Sem Pain Med* 2003;1.
29. Keefe FJ, Beaupre PM, Gil KM: Group therapy for patients with chronic pain, in Gatchel RJ, Turk DC (eds), *Psychological Approaches to Pain Management*. New York: Guilford Press, 1996:259–282.
30. Keefe FJ, Caldwell DS, Williams DA et al: Pain coping skills training in the management of osteoarthritic knee pain: A comparative study. *Behav Ther* 1990;21:49–62.
31. Dworkin RH, Grzesiak RC: Chronic pain: On the integration of psyche and soma. in Stricker G, & Gold Jr. (eds), *Comprehensive Handbook of Psychotherapy Integration*. New York: Plenum Press 1993:365–384.
32. Grzesiak RC, Ury GM, Dworkin RH: Psychodynamic psychotherapy with chronic pain patients, in Gatchel RJ, Turk DC (eds), *Psychological Approaches to Pain Management*. New York: Guilford Press, 1996:148–178.
33. Sarno J: *Healing Back Pain: The Mind-Body Connection*. New York: Warner Books, 1991.
34. McGrath PA, Holohan AL: Psychological interventions with children and adolescents: Evidence for their effectiveness in treating chronic pain. *Sem Pain Med* 2003;1.
35. McBeth J, Silman AJ, Macfarlane GJ: Association of widespread body pain with an increased risk of cancer and reduced cancer survival. *Arthritis Rheum* 2003; 48:1686–1692.
36. Liebeskind, JC: Pain can kill. *Pain* 1991;44:3–4.
37. Nelson DV, Kennington M, Novy DM: Psychological selection criteria for implantable spinal cord stimulators. *Pain Forum* 1996;5:93–103.

CHAPTER 42

Principles of Interventional Pain Medicine

Milan P. Stojanovic

Over the last few decades, the role of anesthesia procedures in the treatment of chronic and cancer pain has changed significantly. These changes are multifactorial. Most of the neuroablative procedures aimed at treating chronic pain have been abandoned, because poor outcomes and the potential for complications have been recognized. The neuroablative procedures have been replaced by new, minimally invasive procedures such as spinal cord stimulation. There is a growing use of regional techniques, such as diagnostic facet joint blocks, for diagnostic purposes. The technical aspects of regional procedures have been greatly improved. Many of these procedures are now performed with fluoroscopic guidance, with improved accuracy of medication delivery.

The pain procedures have traditionally been plagued by the lack of clinical outcome studies to support their efficacy. Although we are still in need of randomized controlled trials to support interventional pain procedures, recent well-designed studies have shown very favorable outcomes with many interventional approaches.

New, improved interventional approaches are being used to bridge the gap between the conservative, medical approach and surgical treatments, because of very favorable risk-to-benefit ratios for the majority of minimally invasive interventional treatments.

When considering the use of these procedures, one should always be aware of the potential for false positive results: this is particularly important when performing diagnostic blocks. The reasons for false positive results are the systemic absorption of medications, the regional spread of local anesthetics, and, most commonly, the placebo effect.

This chapter is an overview of the various approaches to minimally invasive treatments for pain. Detailed instructions on how to perform these procedures are found in regional anesthesia manuals, atlases,[1,2] and formal hands-on instructions.[1,2]

▶ MEDICATION CHOICE

Local Anesthetic

Local anesthetic is the most commonly used solution for regional procedures, where it is used for both diagnostic and therapeutic purposes. When used for diagnosis in patients with chronic pain, its role is to determine the location of the peripheral pain generator. This is very useful in cases where other means (such as imaging studies) fail to locate the source of pain, such as in lumbar or cervical facet disease.

The therapeutic value of local anesthetic for the treatment of chronic pain is the potential in certain patients to decrease the "wind up" phenomena in CNS and, therefore, provide pain relief that, in some cases, far exceeds the duration of local anesthetic action. However, this kind of response is very unpredictable and seems to depend on the molecular mechanism of CNS injury. The main rationale for performing these procedures is their favorable risk-to-benefit ratio. The complications from the use of local anesthetic are very rare and they most commonly occur with intravascular absorption of a high dose of local anesthetic, which can lead to seizures or cardiac arrhythmias.

The most commonly used local anesthetics are lidocaine (1%; 0.5%) and bupivacaine (0.25%; 0.5%) with or without epinephrine. The epinephrine prolongs the action of local anesthetics and can also have an antinociceptive effect on its own. The local anesthetics are often mixed with a small amount of steroid solution in order to provide potentially longer pain relief.

Steroids

Steroids have been a well-known treatment for chronic pain. In 1953, Lievre reported for the first time the use of steroids for the treatment of sciatica. Since then, the steroid solution has been used for a variety of purposes, ranging from peripheral nerve blocks to epidural steroid injections.

The mechanism of action of steroids is twofold: (1) they decrease the inflammatory process in injured nerves and, therefore, reduce the sensitization of the peripheral pain generators; and (2) they decrease the electrical discharge in injured nerves and, therefore, decrease pain. Their onset of action is usually delayed, ranging from three days to two weeks. The duration of the action of steroids is unpredictable, providing pain relief from several weeks to several months: it seems that the mechanism of neuropathic pain is responsible for this variation.

Side effects and complications with steroid use are rare. However, keep in mind that systemic absorption of steroids occurs even after peripheral use, causing adrenal suppression for two to four weeks. In diabetic patients, steroids can increase blood glucose values, so the monitoring of blood glucose is warranted if steroids are used. In very rare cases, steroids have been shown to cause manic episodes in patients with bipolar disorder, even after only a single use. It seems prudent to limit the number of steroid injections over the year to a reasonable number in order to prevent systemic effects.

The most commonly used steroid preparations for the treatment of chronic pain are triamcinolone and methylprednisolone.

Neurolytic Agents

The use of neurolytic agents is limited mostly to the treatment of cancer pain. Although patients with chronic pain may initially have favorable responses to neurolytic blocks, very often the pain becomes worse after several weeks or months due to additional nerve damage and central sensitization with neurolytic substance.

Commonly used neurolytic agents are 50 to 100 percent alcohol and 5 to 10 percent phenol. Both agents seem to have similar efficacy with some differences of pain at the injection site (less noxious with phenol), and baricity (hypobaric with alcohol; hyperbaric with phenol).

Hypertonic (10 percent) saline is the only neurolytic agent that is still used for the treatment of some forms of nonmalignant pain, such as epidural lysis of adhesions; however, more outcome studies are needed to fully support this kind of treatment.

Botulinum Toxin

The clinical use of botulinum toxin has become more promising in recent years. Among chronic pain conditions, botulinum toxin may have a significant role in the treatment of myofascial pain syndrome. The botulinum toxin type A is derived from the bacterium *Clostridium botulinum*. This bacterium produces a substance that blocks the release of acetylcholine and most likely decreases vesicle-dependent exocytosis of other neurotransmitters and neuropeptides. Blocked release of acetylcholine leads to muscle relaxation, improved muscle perfusion, and reduction of muscle spindle activity. It possibly inhibits cholinergic interneurons and diminishes central sensitization.

A typical dose of botulinum toxin type A is 25 to 50 units per muscle. A total dose of 300 units should not be exceeded per treatment session, although total doses of up to 800 units have been shown to be safe: the toxic dose estimate is 3000 units. The botulinum toxin type B has a similar mechanism of action, but the doses should be higher than for botulinum toxin type A. The most commonly injected muscles in the cervical area are the trapezius muscle, semispinalis capitis and cervicis, levator scapulae, and splenius capitis and cervicis. For the lower back, piriformis muscle, multifidus muscle, and possibly the paraspinal muscle should be targeted. For optimum results when injections are performed, the mid-belly of the muscle, instead of the actual trigger point, should be targeted.

The duration of action of botulinum toxin type A is up to three months. If injections are repeated more often than every three months, the treatment may reduce its effectiveness. On the other hand, if treatments are performed less often than every three months, the effects of botulinum toxin type A may be potentiated. More studies are needed to fully support this treatment, but early results seem to be very promising.[3]

▶ PERIPHERAL NERVE BLOCKS

Peripheral nerve blocks with local anesthetics can be used as an additional diagnostic modality. Usually, a combination of local anesthetics and steroids is used for this purpose in order to provide patients with potentially prolonged pain relief (of the order of weeks or months).

Although virtually any peripheral nerve can be blocked, some peripheral nerve blocks are used more than others. The most likely reason for this lies in the fact that these peripheral nerves are the most common pain generators.

Occipital Nerve Block

The *greater and lesser occipital nerves* supply the postero-lateral portion of the head and occiput. The greater occipital nerve originates as a medial branch of the C2 nerve root and the lesser occipital nerve may have some innervations from the communicating branch from the C3. The chronic pain condition originating from this nerve is called *occipital neuralgia*.

Greater and lesser occipital nerve blockade may alleviate this painful condition. The greater occipital nerve is blocked at the level of the superior nuchal line, just lateral from the external occipital protuberance using the 25G needle. The lesser occipital nerve is found one-third of the distance from the external occipital protuberance to the mastoid process on the nuchal line.

If occipital nerve blocks do not yield acceptable results, the C2 nerve root block or occipital nerve stimulator are possible treatment options.

Ilioinguinal and Genitofemoral Nerve Block

The *ilioinguinal nerve* originates from the first lumbar ramus. The nerve emerges from the lateral border of the psoas major muscle and makes its way to the inguinal canal just below the spermatic cord. The genitofemoral nerve, its genital branch, enters the inguinal canal and supplies the scrotal skin and the cremaster muscle.

Ilioinguinal and genitofemoral neuralgias have a variety of causes but are often seen after surgical interventions such as inguinal herniorrhaphy.

These nerve blocks may reduce the chronic pain in the inguinal and scrotal region.

Lateral Femoral Cutaneous Nerve Block

The *lateral cutaneous* is a sensory nerve that originates from the dorsal branches from the second and third lumbar ventral rami. The chronic pain from the lateral femoral cutaneous nerve is better known as *meralgia paresthetica*. Although its etiology is not clear, it seems that a significant impact on pain is due to mechanical irritation of this nerve.

The classic pain distribution is in the lateral thigh. Many times the disease can be overlooked and the pain misdiagnosed as lumbar radiculopathy.

The diagnostic-therapeutic blockade of this nerve is performed 2 cm medial and 2 cm caudad from the anterior superior iliac spine.

▶ SYMPATHETIC BLOCKS

Sympathetically maintained pain (SMP) is a subgroup of chronic pain conditions that are identified by their response to sympathetic blocks. It seems that the pathophysiology of chronic pain in SMP includes up regulation of peripheral alpha receptors, and that increased sympathetic tone is not present. Still, the most common tool used for the diagnosis of SMP is a sympathetic blockade. The most accurate test for the diagnosis of SMP, intravenous phentolamine test, lacks sensitivity and is not commonly used due to the poor availability of phentolamine.

More commonly used sympathetic blocks are: stellate ganglion block, lumbar sympathetic block, ganglion impar block, and intravenous regional block.

Stellate Ganglion Block

The *stellate ganglion* is usually composed of the inferior cervical and first thoracic ganglion. It is located between C7 and T1 vertebral levels. The ganglion lies lateral to the longus colli muscle, anterior to the transverse process, and posterior to the vertebral artery. It provides supply to the upper extremity through the C7-T1 gray communicating rami. Aberrant supply to the upper extremity from T2-3 gray communicating rami (Kuntz's nerves) occurs only occasionally, resulting in no pain relief from the "classic approach" to the stellate ganglion block.

The stellate ganglion block provides interruption of sympathetic fibers supplying the ipsilateral upper extremity and the face. Besides a diagnostic-therapeutic purpose in patients with SMP, the stellate ganglion block is used for the treatment of Raynaud's disease and other conditions causing impaired circulation to the arm.

The stellate ganglion block is usually performed in the supine position by identifying Chassaignac's tubercle at C6 level. Although a "blind" approach is most commonly used in practice, the confirmation of needle placement by fluoroscopic guidance and the administration of contrast media can be very informative and can possibly decrease the incidence of potentially serious complications such as intravascular or intrathecal injection. Other complications and side effects from stellate ganglion blockade are pneumothorax, recurrent laryngeal nerve block (hoarseness), and phrenic nerve block. Temperature increase in ipsilateral extremity is a sign of a successful block. Horner's syndrome (*ptosis, miosis, anhidrosis*) is a very common side effect of stellate ganglion block.

If prolonged pain relief from the stellate ganglion block is achieved (from seven days to a few weeks), a repeated block is warranted. If the pain relief is present only for the duration of the local anesthetics, it is not justifiable to repeat the stellate ganglion block: in that case, another treatment solution should be sought.

Although neurolytic blocks with radiofrequency (and sometimes with phenol) are clinically utilized, more studies are needed to support this treatment modality.

Recent literature indicates that sympathetic blocks may have significant positive predictive value in outcomes to spinal cord stimulation.[4] This means if the patient has even short-lasting pain relief from the stellate ganglion block, there is a very good chance of a favorable response to a spinal cord stimulator trial and permanent implant.

Lumbar Sympathetic Block

Lumbar sympathetic block (LSB) is equivalent to a stellate ganglion block for lower extremity pain. It serves a diagnostic-therapeutic role for a lower extremity SMP.

The indications for LSB are very similar to those for the stellate ganglion block. The lumbar sympathetic chain is blocked at L3 or L2 level in the anterolateral portion of the vertebral body. The new technique for LSB uses an oblique fluoroscopic view and advancement of the needle in co-axial view, which significantly shortens the time and discomfort of the procedure. Complications with LSB are very rare, and they most commonly involve only transient weakness that occurs immediately after the LSB because of leakage of local anesthetic to adjacent nerve roots. Genitofemoral neuralgia is a potential complication of LSB: although its incidence has been reported as 15%, it seems that clinically it may be much lower.

"Permanent" LSB with radiofrequency lesioning or phenol can be performed; however, there is a lack of studies supporting its efficacy. As for the stellate ganglion block, a good response to an LSB seems to be an excellent predictor for outcomes with spinal cord stimulation.

Celiac Plexus Block

The *celiac plexus* is located just anterior to the diaphragmatic crura and aorta. The nerve supply to the celiac ganglion is mixed from preganglionic splanchnic nerves (greater T5-T10; lesser T10-T11; least T11-T12), vagal parasympathetic fibers, postganglionic sympathetic fibers, and phrenic nerve sensory supply.

The most common and effective use of the neurolytic celiac plexus block is to treat malignant pain (pancreatic cancer, malignant spread of the retroperitoneum, and upper abdomen pain). The neurolytic substances administered to celiac plexus are alcohol and phenol. The use of celiac blockade for nonmalignant pain is more controversial: only anecdotal evidence currently supports the use of the celiac plexus block for these conditions (such as chronic abdominal pain and chronic pancreatitis).

▶ EPIDURAL STEROID INJECTIONS

Epidural steroid injections (ESI) are one of the most common regional techniques for the treatment of chronic pain. The first time that an epidural injection was used to treat low back pain was in 1901 by Cathelin,[5] and the first report of epidural steroid use was in 1953 when Lievre and associates reported the use of epidural hydrocortisone in 20 patients.[6]

Over 40 studies are published on the use of epidural steroids for the treatment of low back pain and lumbar and cervical radiculopathy, with reported success rates in the range of 18% to 90%. However, the number of randomized controlled studies is low, with a majority of the published reports suffering from some methodological flaws. The most common criticisms of these studies include: (1) the lack of narrow patient selection criteria; (2) short follow-up periods; and (3) lack of fluoroscopic guidance.

However, it is clear that the ESIs have a very low risk-to-benefit ratio, and that many patients benefit from these procedures. ESI bridges the gap between conservative treatments and surgery. Johnson and colleagues reported only four minor complications out of a total of 5334 ESIs done at various spinal levels under fluoroscopic guidance.[7]

The Role of Fluoroscopy in ESI

In recent years, it has become clear that fluoroscopy improves the accuracy of medication placement when ESIs are performed. Although literature linking improved outcomes to the use of fluoroscopy is lacking, many studies support fluoroscopy when performing ESI.[8]

It seems that when lumbar ESI are performed without fluoroscopic guidance, the accuracy of locating the epidural space is only 70 percent. In cervical levels and with a caudal approach, the success rates of the nonfluoroscopic approach to ESI are less than 50 percent.[9] Even when the needle is positioned in the epidural space, the initial medication spread is unilateral in 50 percent of the cases.

Since intravascular injection has been shown to occur in 11 percent of cases with certain ESI approaches, and is not recognized in up to 50 percent of cases, it is important to use fluoroscopy with contrast administration to assure adequate needle placement.

With all these factors combined, it is clear that fluoroscopy and epidurography can improve medication delivery when ESIs are performed. When fluoroscopy is used, the lateral view should be chosen for the final needle position and contrast spread confirmation.

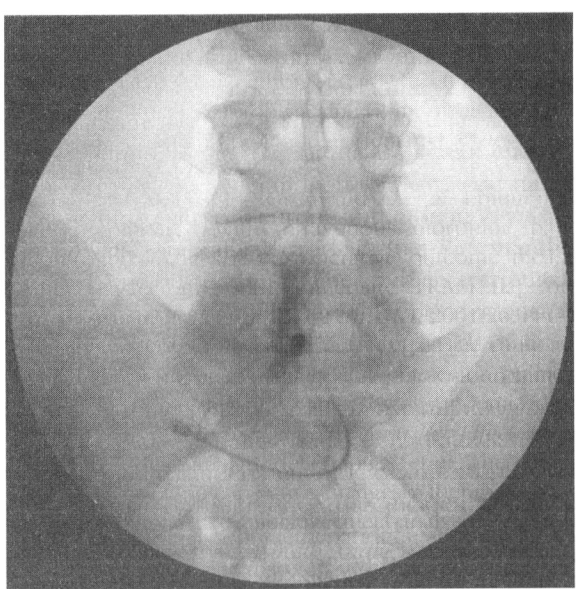

Figure 42-1. Interlaminar (translaminar) epidural steroid injection and contrast spread as seen in anteroposterior fluoroscopic view.

Technical Approaches

Interlaminar (Translaminar) Approach

Traditionally, ESIs were performed via the interlaminar approach.[10] This approach is simple and reliable in patients with preserved anatomy of the epidural space (Fig. 42-1; Fig. 42-1a). In many institutions, this approach is used as the initial approach to the ESI. However, in patients with previous laminectomy or epidural scar tissue, other approaches may be better suited.

Transforaminal Approach

Recently, the potential advantages of transforaminal approaches were shown in several studies, which had high success rates in patient outcomes (75–84 percent). It seems that this approach is particularly useful for lumbar levels in patients with previous back surgery—the surgery can result in epidural scar formation, limiting the epidural contrast and medication spread. The transforaminal approach can assure accurate medication delivery to the site of pathology (Fig. 42-2).

When the interlaminar approach is used, the ventral epidural space (where most of the pathology is located) is reached in just over 25 percent of the cases.[11] The transforaminal approach provides reliable medication spread ventrally.

While the lumbar transforaminal approach is safe, this approach should be used with caution in cervical levels, because several cases of cervical spinal cord infarction have been reported. The exact etiology of this is not clear, although it seems that interrupted vascular supply to the spinal cord may play a significant role.

Caudal Approach

The caudal approach is one of the early approaches used for ESIs. However, recently it has been criticized because it needs a diluted solution of steroids, therefore diminishing the amount of steroids reaching the site of pathology. In many cases, it seems more prudent to use a more

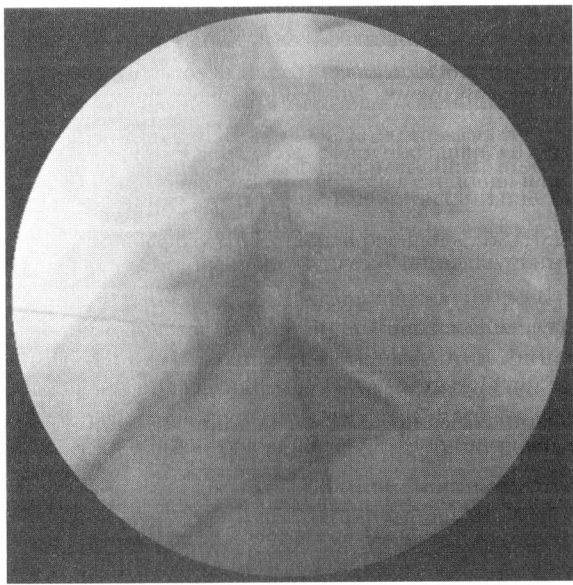

Figure 42-1A. Interlaminar (translaminar) epidural steroid injection and contrast spread as seen in lateral fluoroscopic view.

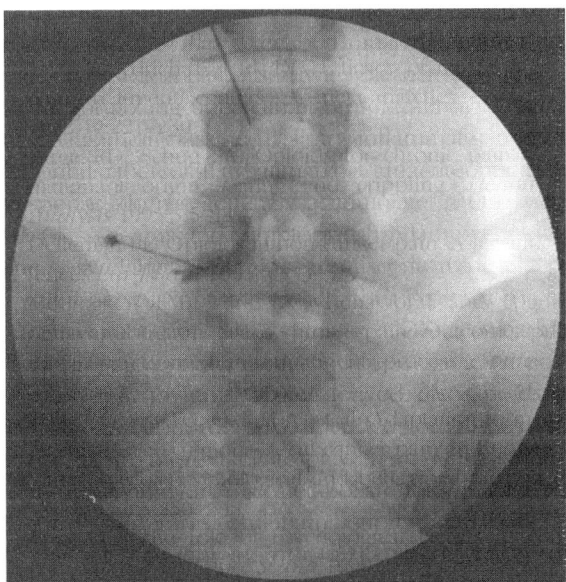

Figure 42-2. Transforaminal epidural steroid injection and contrast spread as seen in anteroposterior fluoroscopic view.

concentrated steroid solution and administer it closer to the site of pathology.

▶ FACET JOINT BLOCKS AND RADIOFREQUENCY LESIONING

Diagnostic Facet-Medial Branch Blocks

A growing body of evidence shows that facet joint pain is a significant cause of pain and suffering. Studies have shown that the prevalence of cervical facet joint pain is 54 to 60 percent, whereas lumbar facet joints cause pain in 15 to 40 percent of patients with chronic low back pain.[12,13] None of the data from history and physical examination are helpful to make a diagnosis of lumbar facet joint pain. However, pain on back extension, tenderness over the facet joints and radiation patterns (not below the knee) suggest lumbar facet joint pain. The facet pain can often coexist with another source of pain (such as diskogenic pain, radiculopathy, or myofascial pain). The best diagnostic test for facet joint pain is the diagnostic medial branch facet joint block.

The purpose of the diagnostic medial branch block is to block the nerve supply to the facet joints. Usually two separate blocks are needed to establish the diagnosis. Under fluoroscopic guidance, the needle is placed over 3–4 medial branches at a time (and L5 dorsal ramus) and 0.3cc of local anesthetics (1% lidocaine or 1:1 mix of 1% lidocaine and 0.5% bupivacaine) is administered. If more than 50 percent pain relief (some experience 90 percent) is provided for at least 4–6 hours, the diagnosis of facet joint pain is very likely. Despite the two positive diagnostic medial branch blocks, there is still a possibility for false-positive results. False positive results occur due to: (1) leakage of local anesthetics to other structures such as disk and nerve root; (2) relief of the myofascial component of the pain by anesthetizing the skin and muscles; and (3) a placebo response.

When performing lumbar medial branch blocks, the needle is positioned under fluoroscopic guidance so that the tip of the needle rests at the junction between the superior articular process (SAP) and the transverse process (eye of the "Scotty dog") in oblique view. The final needle position is always confirmed in the anterior-posterior view so that the tip of the needle lies just at the lateral margin of the SAP and caudally from the superior margin of the transverse process. A similar approach to that for cervical medial branch blocks is used.

Traditionally, for medial branch blocks, a separate needle is used for each level to be blocked. A new approach uses a *single needle technique*, which can possibly reduce the rate of false positive response and is less noxious to the patient.

The intra-articular facet joint blocks with steroids have not been shown to provide good pain relief and are rarely used; instead, the patients who experienced good pain relief with two diagnostic medial branch blocks may benefit from the radiofrequency lesioning of medial branches.

It has been recognized recently that, in many cases, neck pain and occipital headaches are caused by cervical facet disease.[14] In a similar way to that used in the lumbar and thoracic regions, a diagnostic medial branch block followed by radiofrequency can provide adequate pain relief.

Radiofrequency Lesioning for Facet Pain

Radiofrequency lesioning was first performed in 1974 for facet joint denervation. This technique involves the placement of cannulae under fluoroscopic guidance in a similar fashion as in diagnostic medial branch blocks. Once the cannulae are in adequate positions, sensory and motor testing is performed in order to confirm placement and to assure that the cannulae are not close to the nerve root (Fig. 42-3; Fig. 42-4). A small amount (0.5cc) of local anesthetic is placed at each target site. At this point, heating at 80 degrees Celsius for 90 seconds is performed at each level. After an initial increase in pain for two to three days, pain relief takes place. Various studies have demonstrated pain relief and improved outcomes with radiofrequency lesioning for facet pain. However, in certain patients, there is a

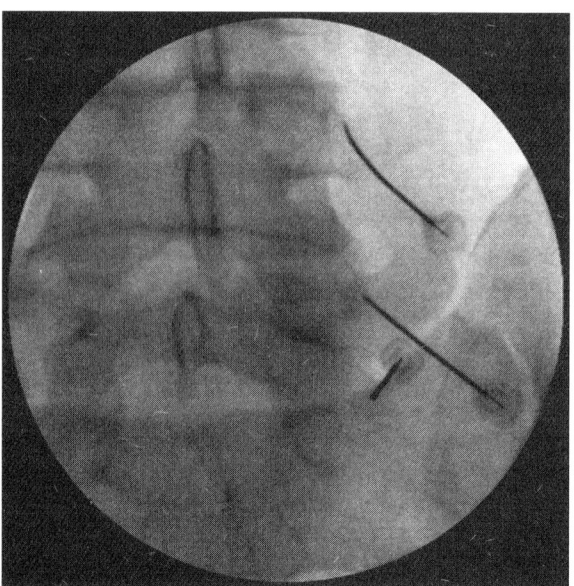

Figure 42-3. Lumbar facet radiofrequency lesioning with cannulae positioned at right L4, L5 medial branches and L5 dorsal ramus as seen in anteroposterior fluoroscopic view.

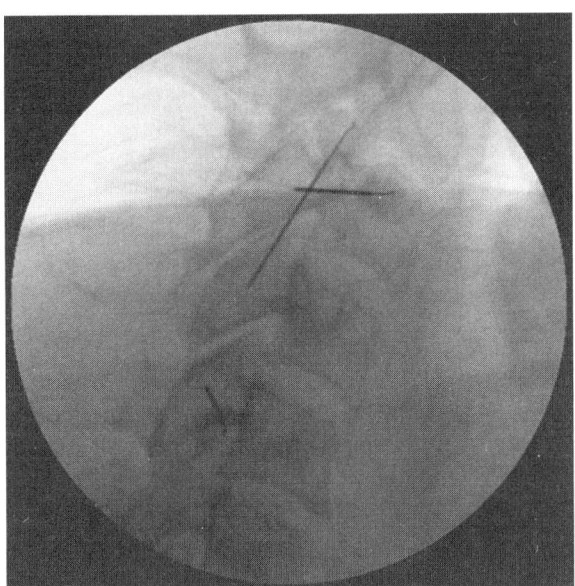

Figure 42-4. Cervical facet radiofrequency lesioning with cannulae positioned at right C3, C4 and C5 medial branches as seen in lateral fluoroscopic view.

recurrence of pain at 8–12 months after treatment due to regeneration of the lesioned nerves; a repeat procedure in those cases is warranted. Many clinical studies support this form of treatment for facet joint pain in cervical and lumbar levels.[15]

Recently, a new *pulse* mode of radiofrequency has been introduced. It uses a maximum temperature of 42 degrees Celsius for 120 seconds for each nerve. Although the exact mechanism of action is unclear, it seems that this treatment modality "stuns" the nerve endings, most likely by modulation of the neurotransmitters. The initial results of this treatment modality are encouraging, but more studies are needed to further support its role.

▶ DISKOGENIC PAIN AND DISKOGRAPHY

Pathophysiology and Patient Presentation

In diskogenic pain, annular lamellae develop microfractures, and annular nociceptors become sensitized with a decrease in their firing thresholds. Furthermore, the damaged disk promotes growth of nerve fibers along radial tears into the inner annulus. Histological studies have revealed a two-fold increase of nerve fibers expressing substance P in annulus fibrosus of patients suffering from diskogenic low back pain.

Most patients present with low back pain limited to the back area (axial pain), or with radiation to one or both lower extremities. Pain is increased by prolonged sitting or standing. Physical exam results can be normal, including those of the straight leg raising test.

Diskogenic pain can present with normal magnetic resonance imaging (MRI) results. However, certain MRI findings are highly suggestive of diskogenic disease: (1) decreased disk signal intensity on T2-weighted MRI images; (2) *high intensity zone* (HIZ), a high T2-weighted signal within the annulus; or (3) a *bulging* or *protruding* disk.

Provocation diskography remains a gold standard for the diagnosis of diskogenic pain.

Diskography

Concordant pain with low pressure or low volume diskography is the most important diagnostic finding. During diskography, separate needles are placed into the suspected problem disk and two control disks, under fluoroscopic guidance (Fig. 42-5). The contrast media is then injected into one disk at a time, with the patient being blinded to the timing of the injection. The diskography is positive when concordant pain is produced with injection. Concordant pain is sought with less than 30 PSI above opening pressure, or less than 1.25cc of contrast administered into the disk. The low back (or lower extremity) pain reproduced under these conditions is considered to be diskogenic in origin. The disk disruption and leakage of dye through the annular tear is usually seen with the onset of pain. Disk disruption alone, without reproduction of the patient's pain, is an insufficient finding for the diagnosis of diskogenic pain. The use of

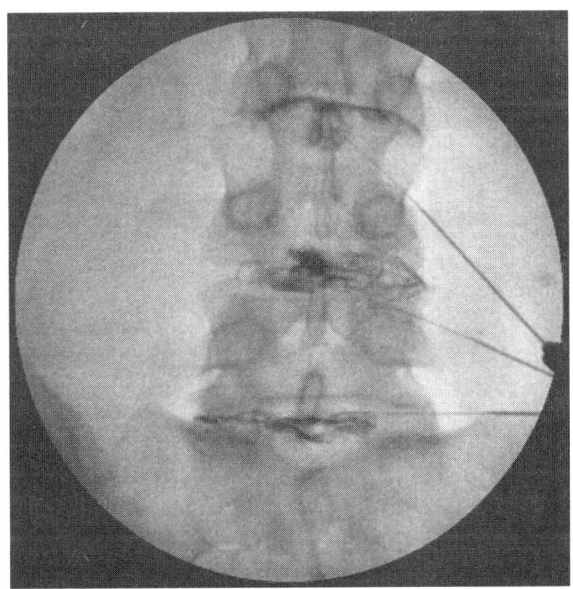

Figure 42-5. Lumbar diskography with needles positioned at the L3–4, L4–5 and L5–S1 levels with intradiskal contrast spread as seen in anteroposterior fluoroscopic view.

postdiskography CT is a helpful, but not necessary, diagnostic tool. The most serious complication of diskography is diskitis. Although rare, it is very resistant to treatment due to limited blood supply to the disk. The intradiskal and intravenous antibiotic administration minimizes its occurrence. Diskography has been shown to be a safe procedure, not producing any damage to the disk.[16]

Diskogenic Low Back Pain: Treatment Options

There are several treatment options in a patient presenting with diskogenic pain. A conservative therapy such as McKenzie exercises or dynamic lumbar stabilization exercises can be helpful for some patients.

New treatment options for diskogenic pain are promising.[17] Intradiskal Electrothermal Therapy (IDET) (Fig. 42-6) is supported by many outcome studies and recently a new form of radiofrequency (RF) denervation of the annulus has been introduced.

Surgical approaches (anterior and posterior lumbar fusion, and titanium cages) are the most commonly used treatment for diskogenic low back pain, with success rates in the 50 to 85 percent range. Unfortunately, surgery is expensive and carries a risk of complications, so may be reserved for severe morphological disk damage where the regeneration of annulus fibrosus is unlikely.

Intradiskal Electrothermal Therapy (IDET)

IDET is a minimally invasive approach for the treatment of diskogenic low back pain, and is performed under local anesthesia. It involves percutaneously threading a flexible catheter into the disk tissue in a fluoroscopically guided procedure. The catheter is composed of thermal resistive coil, which enables its distal part to be heated to the desired temperature. The technique for approaching the disk for IDET procedure is similar to diskography. The final position of the catheter is such that the end of the catheter is placed circumferentially around the inner surface of the posterior annulus.

Once the catheter is in a satisfactory position, as confirmed by AP and lateral fluoroscopy, the distal part of the catheter is gradually heated. The increments in temperature are automatically achieved, with a target temperature of 80–90°C for 4–6 minutes. For optimum results, it is important to maintain a maximum temperature of at least 80°C. The actual annular tissue temperature is up to 15°C lower than the temperature of the catheter tip. Comprehensive patient and cadaver temperature mapping studies have shown the safety of reaching such a high temperature as long as the catheter tip is located within the disk tissue.[18]

The primary mechanism of IDET action may involve thermal modification of collagen fibers. Post-treatment histology in human cadavers demonstrates increased collagen density, moderate fibroplasias of capsule with the capsule being thicker than control, increased vasculature, and plump and active endothelial cells. The secondary putative mechanism of IDET action involves the destruction of sensitized nociceptors in the annular wall. It is important to point out that patient recovery is gradual. The *healing* process reaches its peak in four months after the IDET procedure; during that period it is particularly important to ensure that patients limit their physical activity with a carefully structured rehabilitation program.

Over 10 clinical outcome studies have been published on IDET in the last five years, showing an average 70 percent improvement in pain and function in followups up to two years.[19] Preliminary results of a recent randomized, blinded, placebo-controlled study suggest that IDET is superior to placebo.[20] At the same time, complication rates of this treatment have been very low. More studies are needed to identify the outcome predictors of IDET. It seems that decrease of more than half in disk height and morbid obesity are clear contraindications for this treatment.[21]

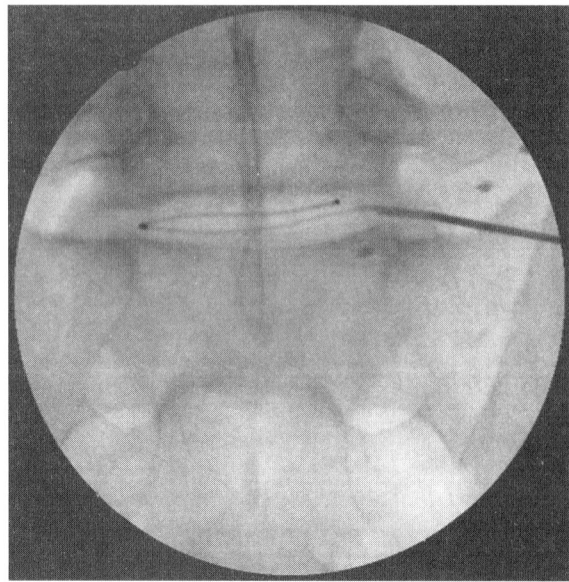

Figure 42-6. IDET procedure with catheter positioned intradiskally at the L5–S1 lumbar level as seen in anteroposterior fluoroscopic view.

▶ NEUROSTIMULATION

Even in 600 BC, the electrical power of the torpedo fish was used for medical purposes.[22] The first report of electrical stimulation of the human brain dates back to 1874.[22] In 1967, the *spinal cord stimulation* (SCS) technique was introduced by Shealy and associates, based on Melzack and Wall's gate theory of pain. Since then, the various forms of neurostimulation have become very common treatments of chronic pain disorders. Neurostimulation is minimally invasive, reversible and its efficacy is widely documented in the literature.[23] It can provide better pain relief than many other treatment modalities in carefully selected patients.[24]

Spinal Cord Stimulation (SCS)

With SCS the stimulating electrodes are placed in the epidural space, connected to the subcutaneous *internalized pulse generator* (IPG) and the internal or external power source (Figs. 42-7, 42-8). An SCS screening trial is performed prior to permanent SCS implantation.

The initial SCS techniques involved open intrathecal implantation of electrodes via laminotomy. With the improvement of hardware quality, the procedure has become much simpler, allowing for SCS trials with favorable risk to benefit ratios. The alternative term for SCS is *dorsal column stimulation*, since SCS electrodes stimulate dorsal columns of the spinal cord.

Mechanism of Action

The Melzak and Wall theory proposes that stimulation of A-beta fibers modulates the dorsal horn gate and, therefore, reduces the nociceptive input from the periphery. Recent studies suggest that other mechanisms may better explain the SCS action.[25]

It has been shown that intrathecal administration of the GABA-agonist baclofen enhances the antinociceptive action of SCS in both animals and humans.[26] Similar

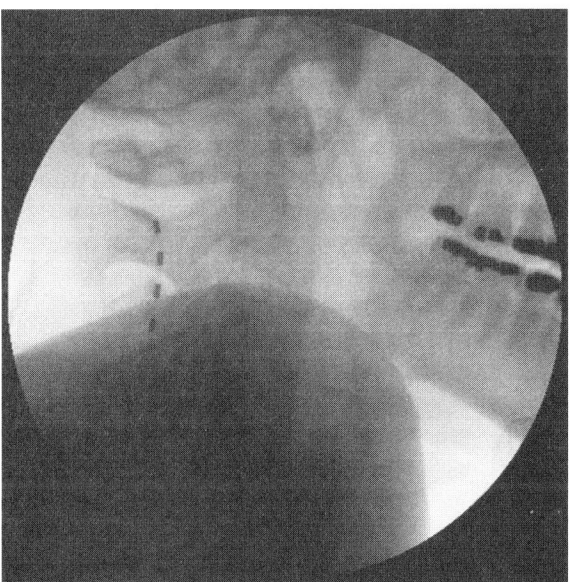

Figure 42-7. Spinal cord stimulator lead, placed in cervical epidural space with its tip positioned at the C2 vertebral level as seen in lateral fluoroscopic view.

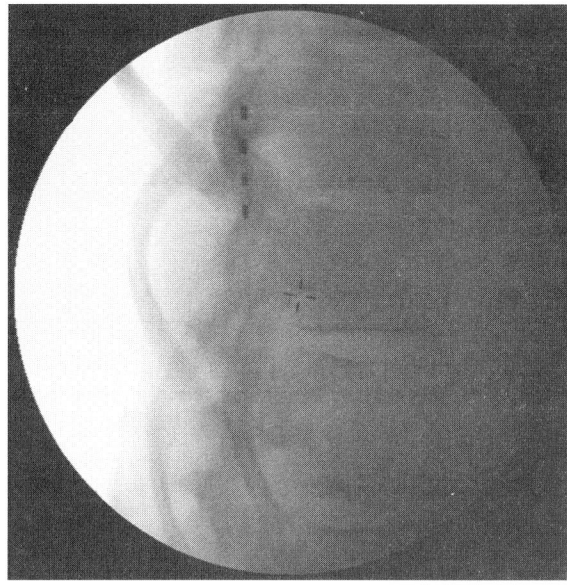

Figure 42-8. Spinal cord stimulator lead, placed in lumbar epidural space with its tip positioned at the T9 vertebral level as seen in lateral fluoroscopic view.

findings point to the possible role of adenosine as a mediator in the SCS action. It seems that the SCS may have a role in activating the descending inhibitory pathways originating in the periaqueductal gray (PAG).

The mechanism of SCS action in patients with peripheral ischemic pain may differ. The SCS was shown to suppress the sympathetic activity via adrenoreceptors and increase the release of calcitonin gene-related peptide (CGRP). In patients with myocardial ischemia, the SCS mediates in the redistribution of the coronary blood flow from regions with normal perfusion in favor of regions with impaired myocardial perfusion.

Although the exact mechanism of SCS action is unclear, it seems very likely that various mechanisms may play a role in its effectiveness.

Indications for SCS Treatment

Patients with lumbar and cervical radiculopathy who are not surgical candidates, and patients with Failed Back Surgery Syndrome (FBSS) are the best candidates for SCS. Reported success rates in these conditions vary from 12 to 88 percent. In a recent study, Turner et al. performed a systematic review of literature related to SCS and FBSS, which revealed that an average of 59 percent of patients had at least 50 percent pain relief and improved functional status.[27] Many studies suggest that patients with a radiating pain pattern to the leg seem to respond better to SCS than patients with isolated axial low back pain.

Many studies support the use of SCS for Complex Regional Pain Syndrome (CRPS) [previously, reflex sympathetic dystrophy (RSD)].[4] A recent publication suggests that CRPS patients with sympathetically maintained pain respond significantly better to SCS than those patients with sympathetically-independent pain. The effectiveness of pain relief with SCS in CRPS varies from 50 to 91 percent.

Certain subgroups of patients with peripheral neuropathy may respond well to SCS. It seems that SCS can be an effective treatment for diabetic neuropathy, providing analgesia and potentially salvaging the affected limb, but possibly with increased rates of infection with the SCS implantation.[28] Although a recent study showed excellent results with SCS in postherpetic neuralgia patients, more studies are needed to fully support the SCS efficacy.

Myocardial ischemia and anginal pain refractory to pharmacologic and surgical interventions respond well to SCS. These patients have demonstrated an increase in exercise capacity, reduction in anginal complaints, decreased use of short-acting nitrates and improved quality of life.[29] The fear of potential increase in myocardial damage, and cardiac arrhythmia does not seem to be justified.

In patients with peripheral vascular disease, SCS may decrease pain, increase the peripheral blood flow, promote ulcer healing and potentially contribute to limb salvage. Many European studies firmly support the role of SCS for peripheral ischemic pain.[30]

Stimulation Trial

Before proceeding with permanent SCS implantation, a stimulation trial is warranted. The SCS trial can positively predict a long-term outcome in 50 to 70 percent of cases. The epidural space is identified by the loss of resistance technique. The SCS lead is inserted into the epidural space, under continuous fluoroscopic guidance. Once an adequate lead position is obtained, the trial stimulation is performed with a goal that paresthesias provide 70 to 80 percent overlap with the patient's pain location. The lead contains four to eight electrodes, and various electrode combinations can be used to achieve optimal results. The typical trial time is three to five days, allowing the patient to adequately assess the effectiveness of SCS prior to potentially receiving a permanent implant. At the end of the trial, the percutaneous lead is removed.

Permanent SCS Implant

The successful percutaneous trial is followed, at a separate date, by the placement of a permanent lead and internal pulse generator. The permanent SCS hardware consists of an SCS lead, an extension cable, and an internal pulse generator (IPG). The entire procedure is performed under local anesthesia in the operating room. The lead is inserted in a similar fashion as in the trial, and is anchored to the fascia and then connected to the IPG, which is usually placed either in the abdominal or gluteal area. In certain instances, the SCS lead can be placed via open laminotomy for better effectiveness and decreased chance of migration. For the first six to eight weeks following permanent SCS implantation patients should avoid any extreme activity, in order to prevent lead migration and to allow for epidural scar tissue formation. The epidural scar tissue prevents undesirable migration of the electrodes.

After the permanent implant, patients are able to adjust the parameters of stimulation via a small remote control. If more complex reprogramming and SCS analysis is needed, a separate physician-controlled external programmer is used. In case of inadequate stimulation, the physician can change the polarity and number of functioning electrodes, and other stimulation parameters, in order to provide better stimulation coverage. The SCS batteries have to be changed every three to six years, which requires a brief visit to the operating room.

Complications with SCS are rare and are, in most cases, limited to equipment failure, infection, and subcutaneous hematoma. SCS and pacemakers can be combined in the same patient, but with great caution because of potential interference between the devices and inhibition of the cardiac pacemaker if they are used simultaneously.

Peripheral Nerve Stimulation

Peripheral nerve stimulation (PNS) involves placing the electrodes close to the affected nerve. This modality seems to be particularly effective in patients with CRPS 2 and sympathetic components of pain, with the best indications being the post-therapeutic neuritis and extremity burns. The published success rates for PNS vary from 24 to 62 percent.[31] The mechanism of action of PNS includes loss of sensory perception in A-δ fiber distribution and suppression of C-fiber activity.

The most common implantation sites of PNS are the median, ulnar, radial, posterior tibial and common peroneal nerves. After surgical dissection and exposure of the chosen peripheral nerve, the stimulator electrode is placed over the nerve and covered by a fascial flap. After a successful trial, the pulse generator is implanted in the subcutaneous pocket and connected to the stimulator lead in a similar fashion as the SCS.

For the occipital neuralgia, the PNS electrode is placed subcutaneously around the C1 spinous process. If adequate pain relief is obtained, an electrode is sutured to the underlying fascia and tunneled to the IPG site. In a long-term follow-up study, two-thirds of patients with occipital neuralgia had greater than 75 percent pain relief and one-third had greater than 50 percent pain relief.[32]

Sacral nerve root stimulation (SNRS) may provide particularly good response in patients with interstitial cystitis. It involves inserting four epidural leads in a retrograde manner through S2, S3, and S4 foramina with the lead tip facing anteriorly. Sacral nerve stimulation is an FDA-approved treatment for various forms of voiding dysfunction.

Deep Brain Stimulation

When deep brain stimulation (DBS) is used for pain control, the stimulating electrodes are positioned in the problem brain areas, under local anesthesia, via stereotaxic technique. The electrodes can be positioned to stimulate the brain areas involved in processing noxious stimulation. The most common target areas are the sensory thalamic nuclei (VPM and VPL) and periaqueductal and periventricular gray regions (PAG and PVG). While the proposed action of DBS in VPM/VPL stimulation is interruption of pain pathways, PAG/PVG stimulation seems to activate descending analgesia pathways.

While immediate results are encouraging, long-term failure due to tolerance does occur. The indications for DBS are various forms of neuropathic pain, low back pain, and cancer pain; however, many of these patients might be better suited to SCS. It seems that DBS should be reserved for only those patients who have failed all other treatment modalities.

▶ INTRATHECAL MEDICATION DELIVERY FOR PAIN

The intrathecal route provides targeted delivery of medications and avoids the side effects encountered by systemic administration. The device for intrathecal administration is a surgically-implanted intrathecal catheter connected to a subcutaneous pump that contains a reservoir for medication. The reservoir is percutaneously refilled with medication every two to three months, depending on the infusion rate. Patients with implanted pumps can safely undergo magnetic resonance imaging (MRI).

The alternative to an implanted pump, the epidural route, is more costly due to maintenance of the external system, and is frequently less convenient for the patient, and, therefore, should be reserved for an anticipated use of less than three months.

Several outcome studies strongly support the use of intrathecal drug delivery in carefully selected patient populations.[33] The use of intrathecal pumps has significantly reduced the number of neurosurgical ablative procedures for pain.

Patient Selection

The best candidates for intrathecal pump implants are cancer patients with life expectancies of more than three months. Only patients who do not tolerate other routes of administration (oral or intravenous) due to side effects (such as nausea or constipation) and who initially responded to opioids should be considered for the intrathecal trial. Cancer patients with a neuropathic component of pain, not responding to conservative treatments, may be considered for intrathecal administration of local anesthetic.

Patients with nonmalignant pain may be candidates for an intrathecal pump; however, these patients should be selected carefully. The long-term need for opioid administration should be kept in mind and strict preimplant expectations should be set with the patient.

Medication Selection

All the medications used for intrathecal administration should be free of preservatives. Morphine is the only medication approved by the Food and Drug Administration (FDA) for intrathecal delivery via the implanted system. However, other opioids can be useful in patients that cannot tolerate morphine: these include fentanyl, sufentanil, and hydromorphone. The opioids are delivered to the intrathecal space via a surgically implanted subcutaneous pump that contains a reservoir for medication.

Other groups of medications, including local anesthetics, clonidine, and baclofen, have recently been used for intrathecal delivery.[34] These medications can be used alone or can be combined with opioids. This approach seems to be very promising due to numerous receptors involved in nociceptive transmission.

Future medications that could prove beneficial for intrathecal administration are NMDA antagonists (ketamine), A1 adenosine receptor agonists (adenosine), calcium channel antagonists (ziconotide), and somastatin agonists (octreotide).

Screening Methods

Prior to a permanent implant, patients should undergo a trial in order to assess their suitability for the intrathecal pump implant. Oral opioids should be discontinued or decreased before the trial. The screening can be performed in several ways: (1) intrathecal catheter placement and continous administration of opioids at 1/300 of oral daily dose; (2) epidural catheter placement and administration of opioids at 1/30 of oral daily dose; or (3) use of a one-time intrathecal bolus.

Medication Dosage

If the patient is opioid-naïve, morphine should be started at 0.2 mg/day and gradually titrated to effect. In opioid-tolerant patients, the initial intrathecal dose should be less than the standard conversion dose. For breakthrough pain, oral short-acting opioids should be used.

Opioid conversion dosage from other routes of administration is as follows:

- intrathecal to epidural = 1:10
- intrathecal to intravenous = 1:100
- intrathecal to oral = 1:300

Local anesthetics may be added to opioids if indicated for the neuropathic component of pain. Bupivacaine is the most commonly used medication and it is used in the concentration of 30–40 mg/ml (3–4%). The intrathecal dose for bupivacaine varies from 2–30 mg/day, although doses of over 300 μg/day have been reported in literature.[35] Clonidine (alpha-adrenergic agonists) is FDA approved for epidural administration; its intrathecal dose is 50–900 μg/day. It has been used for the treatment of spasticity, and it seems to have significant antinociceptive action as well. Care should be taken when clonidine is administered, since hypotension can occur.

Side Effects and Complications

Hypersensitivity to intrathecal morphine may result in paradoxical lowering of pain thresholds with allodynia and increased pain: in these cases, opioid rotation should be considered. The most common side effects are respiratory depression, pruritus, nausea, vomiting, urinary retention, reduced libido, edema with weight gain, and constipation. Respiratory depression is much more frequent in opioid-naïve patients and very rarely occurs in opioid-tolerant cases.

Prolonged use of opioids may lead to significant endocrine dysfunction with more than 10 percent of patients developing growth hormone or adrenal insufficiency.[36] Routine hormonal monitoring should be considered in these patients.

Rare surgical complications include granuloma formation, infection seroma, meningitis, bleeding and postdural puncture headache. Pump malfunction, catheter kinking, disconnection, dislodgement, breaks and migration can occur. In these cases, the withdrawal symptoms and loss of analgesia are signs of inadequate drug delivery and warrant further investigation. Because of the pump safety mechanism, spontaneous overdose of intrathecal medication from the pump does not occur; however, there is a possibility of overdose during the pump refilling and testing.

REFERENCES

1. Waldman SD: *Interventional Pain Management*. Philadelphia: W.B. Saunders Company, 2001.
2. Fenton DS, Czervionke LF: *Image-Guided Spine Intervention*. Philadelphia: W.B. Saunders, 2003.
3. Lang AM: Botulinum toxin type A therapy in chronic pain disorders. *Arch Phys Med Rehabil*. 2003;84:S69–S73.
4. Hord D, Cohen S, Ahmed S, Chang Y, Vallejo R, Cosgrove GR, Stojanovic MP: Does sympathetic block predict success in complex regional pain syndrome patients undergoing spinal cord stimulation? *Neurosurgery* 2003:53:626-632; discussion 632–633.
5. Cathelin MF: Mode d' action de la cocaine injecté dons l'espace epidural par le procedé de canal sacre. *C R Soc Biol* 1901;53:478–479.
6. Lievre JA, Block MH, Pean G, Uno J: L'hydrocortisone en injection local. *Rev Rhum Mal Osteoartic* 1953;20:300–301.
7. Johnson BA, Schellhas KP, Pollei SR: Epidurography and therapeutic epidural injections: technical considerations and experience with 5334 cases. *AJNR Am J Neuroradiol* 1999;20:697–705.
8. Cluff R, Abdel-Kader M, Cohen SP, et al: The technical aspects of epidural steroid injections: A national survey. *Anesthesia & Analgesia* 2002;95:403–408.
9. Stojanovic MP, Vu T, Caneris O, et al: The role of fluoroscopy in cervical epidural steroid injections: An analysis of contrast dispersal patterns. *Spine* 2002;27:509–514.
10. Koes B, Scholten R, Mens J, Bouter LM: Efficacy of epidural steroid injections for low-back pain and sciatica: a systematic review of randomized clinical trials. *Pain* 1995;63:279–288.
11. Tomczak R, Seeling W, Rieber A, Sokiranski R, Rilinger N, Brambs HJ: Epidurography: comparison with CT, spiral CT and MR epidurography Rofo. *Fortschritte auf dem Gebiete der Rontgenstrahlen und der Neuen Bildgebenden Verfahren*. 1996;165:123–129.
12. Dreyer SJ, Dreyfuss PH: Low back pain and the zygapophysial (facet) joints. *Arch Phys Med Rehabil* 1996; 77:290–300.
13. Schwarzer AC, Wang S-C, Bogduk N, et al: Prevalence and clinical features of lumbar zygapophyseal joint pain: a study in an Australian population with chronic low back pain. *Ann Rheum Dis* 1995;54:100–106.
14. Fukui S, Ohseto K, Shiotani M, et al: Referred pain distribution of the cervical zygapophyseal joints and cervical dorsal rami. *Pain* 1996;68:79–83.
15. van Kleef M, Brandse GA, Kessels A, et al: Randomized trial of radiofrequency lumbar facet denervation for chronic low back pain. *Spine* 1999;15:24:1937–1942.
16. Tehranzadeh J: Discography 2000. *Radiol Clin North Am* 1998;36:463–495.
17. Pauza KJ, Howell S, Dreyfuss, et al: A randomized, placebo-controlled trial of intradiscal electrothermal therapy for the treatment of discogenic low back pain. *Spine J* 2004;4:27–35.
18. Saal JA, Saal JS: Intradiscal electrothermal treatment for chronic discogenic low back pain: a prospective outcome study with minimum 1-year follow-up. *Spine* 2000; 25:2622–2627.
19. Wong: Intradiscal electrothermal therapy (IDET). *JBR-BTR* 2003;86:297–299.
20. Pauza KJ, Howell S, Dreyfuss, et al: A randomized, placebo-controlled trial of intradiscal electrothermal therapy for the treatment of discogenic low back pain. *Spine J* 2004;4:27–35.
21. Heary RF: Intradiscal electrothermal annuloplasty: The IDET procedure. *J Spinal Disord* 2001;14:353–360.

22. Melzack R, Wall P: Pain mechanism: a new theory. *Science* 1965;150:951–979.
23. Shealy C, Mortimer J, Reswick J: Electrical inhibition of pain by stimulation of the dorsal columns: preliminary report. *Anesth Analg* 1967;46:489–491.
24. Stojanovic M: Stimulation methods for neuropathic pain control. *Cur Pain Headache Rev* 2001:131–137.
25. Meyerson B, Linderoth B: Mechanisms of spinal cord stimulation in neuropathic pain. *Neurolog Res* 2000;22:285–292.
26. Meyerson BA, Cui J-G, Yakhnitsa V, et al: Modulation of spinal pain mechanisms by spinal cord stimulation and the potential role of adjuvant pharmacotherapy. *Stereotact Funct Neurosurg* 1997;68:129–140.
27. Turner JA, Loeser JD, Bell KG: Spinal cord stimulation for chronic low back pain: A systematic literature synthesis. *Neurosurg* 1995;37:1088–1096.
28. Torrens JK, Stanley PJ, Ragunathan PL, et al: Risk of infection with electrical spinal-cord stimulation [letter; comment]. *Lancet* 1997;349:729.
29. Jessurun GA, Ten Vaarwerk IA, DeJongste MJ, et al: Sequelae of spinal cord stimulation for refractory angina pectoris. Reliability and safety profile of long-term clinical application. *Coron Artery Dis* 1997;8:33–38.
30. Claeys L: Spinal cord stimulation for peripheral vascular disease: a critical review-European studies. Pain Digest 1999; 9:337–341.
31. Campbell J, Long D: Peripheral nerve stimulation in the treatment of intractable pain. *J Neurosurg* 1976;45:692–699.
32. Weiner RL, Reed KL: Peripheral neurostimulation for control of intractable occipital neuralgia. *Neuromodulation* 1999:217–221.
33. Rainov NG, Heidecke V, Burkert W: Long-term intrathecal infusion of drug combinations for chronic back and leg pain. *J Pain Symptom Manage* 2001;22:862–871.
34. Deer TR, Caraway DL, Kim CK, Dempsey CD, Stewart CD, McNeil KF: Clinical experience with intrathecal bupivacaine in combination with opioid for the treatment of chronic pain related to failed back surgery syndrome and metastatic cancer pain of the spine. *Spine J* 2002;2:274–278.
35. Rainov NG, Heidecke V, Burkert W: Long-term intrathecal infusion of drug combinations for chronic back and leg pain. *J Pain Symptom Manag* 2001;22:862–871.
36. Paice J, Penn R, Shott S: Intraspinal morphine for chronic pain: a retrospective multicenter study. *J Pain Symptom Manage* 1996;11;71–80.

CHAPTER 43

Neurostimulatory Techniques in Pain Medicine

Ashwini D. Sharan, John Birkness, and Ali R. Rezai

The therapeutic use of electrical stimulation for pain relief is an ancient art. The Egyptians and the Greeks used electric eels to apply shock therapy and the Romans applied the torpedo fish to treat maladies such as cephalgia and arthralgia.[1-3]

Subsequently, modern human knowledge developed, the anatomy of the pain tracts was elucidated, and it became evident that electrical stimulation of the nervous system could be predictably used for therapeutic benefits. In the mid-1900s, neurosurgeons routinely applied electrical stimulation to the brain to treat and to study movement, and psychiatric and pain disorders. In fact, in 1954, Heath and Pool both reported on implantation of temporary electrodes in the septum pellucidum to treat patients with schizophrenia and pain from metastatic carcinoma.[4,5] Many other targets have since been enthusiastically explored for the treatment of patients with chronic pain, including the thalamus, caudate, cingulate, and the periaqueductal grey. With the renewed interest in chronic deep brain stimulation, it is hoped that physicians may again study, and offer patients an opportunity to interrupt, the pain circuits in the deep cerebral targets.

The further use of central nervous system stimulation developed with the introduction of the gate theory for pain control by Melzack and Wall.[6] They noted that stimulation of large myelinated fibers of peripheral nerves resulting in paresthesias blocked the activity in small nociceptive projections. Shealy applied this knowledge in 1967 by inserting the first dorsal column stimulator in a human suffering from terminal metastatic cancer.[7,8] The therapeutic use of electrical stimulation developed further when Shealy, in collaboration with Long, prompted Hagfers and Maurer to independently produce the first two solid-state transcutaneous electrical nerve stimulators (TENS)[7,8] (actually, TENS were originally developed as a screening tool for spinal cord stimulators). Subsequently, electrodes have been implanted via a laminectomy in the subarachnoid space, between the two layers of the dura or in the epidural space, both dorsal or ventral to the spinal cord, and later the percutaneous technique was introduced.[9-14]

In 1991, Tsubokawa et al. reported on their initial experience in 12 patients treated for deafferentation pain.[15] The basis for the Tsubokawa description originally stemmed from experimental work performed by the same group.[15-17] Tsubokawa observed that thalamic hyperactivity could be reduced by chronic sensorimotor cortex stimulation. Since then, many other groups have independently reported on the treatment of chronic pain conditions via cortical targets.[18-22]

Chronic pain conditions are becoming an increasing problem, with growing costs generally requiring a multidisciplinary approach. The central nervous system, at some point, clearly becomes involved in the processing of these painful conditions, with an integration of complex changes in neurophysiology and behavior. Many ablative techniques have been employed in the past to interrupt these signals; however, the results were often temporary and pain tended to recur.

The more modern approach suggests that modulation of the nervous system elements may be a more resilient method for treating such chronic pain disorders. As well, many of these pain conditions are dynamic and evolving, and as such, need a similar treatment modality. Neurostimulation allows therapeutic dosing of electrical current in a variety of pulse forms, amplitudes, pulse widths, and frequencies, to affect a particular system. Furthermore, it is not destructive, it is reversible, and it can be remotely adjusted and programmed over time; these are clear advantages over previous surgical therapies. This chapter gives an overview of spinal cord stimulation (SCS), motor cortex stimulation (MCX), and deep brain stimulation (DBS) for chronic pain disorders.

▶ SPINAL CORD STIMULATION

SCS has been utilized for a variety of pain conditions, the most common being complex regional pain syndrome (CRPS), failed back surgery syndrome (FBSS), nerve root injury, and angina. SCS is particularly indicated with any type of *neuropathic* pain. Indications have been extended to include the treatment of intractable pain due to other causes, including ischemic peripheral vascular disease pain, cervical neuritis pain, spinal cord injury pain, postherpetic neuralgia, neurogenic thoracic outlet syndrome, and temporomandibular joint syndrome refractory to multiple surgical interventions.

Although a large body of work has been published, the exact mechanisms of action of SCS remain unclear. The computer modeling work of Coburn and, more recently, of Holsheimer and Strujik, have shed some light, at least theoretically, on the distribution of the electrical fields within the spinal structures.[23–28] It is clear that stimulation on the dorsal aspect of the epidural space creates complex electrical fields that affect a large number of structures. We do not know whether activation afferents within the peripheral nerve, dorsal columns, or supra-lemniscal pathways share an equivalent mechanism of action. Additionally, there may be antidromic action potentials passing caudally in the dorsal columns to activate spinal segmental mechanisms in the dorsal horns, as well as action potentials ascending in the dorsal columns activating cells in the brain stem, which, in turn, might activate descending inhibition. At the chemical level, animal studies suggest that the SCS triggers the release of serotonin, substance P, and GABA within the dorsal horn.[29–31]

General Pain

Several studies have been published over the last 30 years that assess the general clinical efficacy of spinal cord stimulators in the management of chronic benign pain.[32–37] The general experience is that, in properly selected patients, SCS will produce at least 50 percent pain relief in 50 to 60 percent of the implanted patients. Interestingly, with the proper follow-up care, these results can be maintained over several years.

Kumar retrospectively reviewed a 15-year experience with SCS with a mean follow-up period of 5.5 years in 235 patients.[37] In their series, 59 percent of patients experienced satisfactory relief on follow-up and 47 patients were gainfully employed. He reported that the results were better in patients with failed back syndrome (FBSS), reflex sympathetic dystrophy (RSD), and peripheral vascular disease of the lower limbs. Patients with cauda equina injury, phantom pain, and paraplegic pain did not respond as well. He also noted better results with multipolar systems. An additional observation, particularly in patients with FBBS, included better outcome in patients who were directed to SCS implantation without excessively long delays following the "failed" surgical procedures.

North et al. reviewed the experience with SCS at Johns Hopkins between 1972 and 1990 via a survey.[35] Of 320 tested patients, 78 percent underwent permanent implantation (171 were included in the questionnaire and mean follow-up was 7 years). Fifty-two percent of the patients experienced over 50 percent relief from their pain and almost 60 percent of them reported a reduction in drug usage. Two additional important findings were a statistical correlation between pain relief and paresthesia coverage, and a decrease in complications related to hardware failure with programmable multichannel devices.

Lang has published similar results.[34] He reported on 200 patients with a mean follow-up of 44 months, with a majority of patients having FBSS. In 42 percent of patients, the pain was manageable by the SCS alone. In Spiegelman's series, 43 patients with an average follow-up of 13 months underwent implantation of a Medtronic Resume electrode (Medtronic Inc., Minneapolis, MN) under general anesthesia.[36] In all cases, the electrode was externalized and a trial stimulation period was given. Seventy percent of the patients were implanted, resulting in more than 60 percent pain relief, and statistical significance was noted in the reduction in opioid medication intake. Koeze also reported on 26 patients with two-year follow-up, and in 58 percent of these cases, narcotic drug usage was reduced.[32]

Failed Back Surgery Syndrome (FBSS)

Failed back surgery syndrome remains vaguely defined. This syndrome has included pain localized to the center of the lower lumbar area, pain in the buttocks, or diffuse lower extremity pain from a host of sources including arachnoiditis, epidural fibrosis, radiculitis, microinstability, and recurrent disk herniations. SCS is accepted in the treatment of leg pain, but its use for relief of pain in the lower lumbar area is not widespread. Further, most existing literature does not report on how effective the overlap is between the paresthesia and the pain regions. Additionally, no group has been able to demonstrate the efficacy of SCS against a placebo treatment. There have been three recent prospective series on the effects of SCS on FBSS.[38–41]

Barolat et al. prospectively enrolled patients with only low back pain, or with low back pain greater than or equal in severity to their leg pain.[41] These patients underwent implantation with a multi-lead paddle electrode by ANS (Advanced Neuromodulation Systems, Inc., Plano, TX) and were followed with a *visual analog scale* (VAS), *Oswestry questionnaire*, and *Sickness Impact Profile* (SIP). The study demonstrated a 69 percent successful reduction in back pain and 88 percent successful reduction in leg pain at one-year follow-up; there was significant improvement in the VAS, Oswestry, and SIP scores. Burchiel et al., in 1996, reported on 70 patients with a one-year follow-up and showed successful management of pain in 55 percent of patients.[40] Medication usage and work status was unchanged in this study. North et al. performed the first prospective, randomized comparison of SCS with any other treatment modality with a six-month crossover arm in the study, where 51 patients with FBSS consented to randomization. The study demonstrated significant difference between the patients who opted for crossover from SCS to reoperation versus the opposite. The study concluded that SCS is a viable alternative to reoperation for FBSS.

Other well-conducted retrospective reviews on SCS for FBSS exist.[32,33,42–45] A few studies have further reported on the cost-effectiveness of SCS in FBSS.[46–48]

Complex Regional Pain Syndrome (CRPS)

Complex regional pain syndrome is a term used for reflex sympathetic dystrophy (RSD), causalgia, sympathetically maintained pain (SMP), and related problems. Many patients do not present with all four cardinal signs or symptoms—pain, swelling, stiffness, and discoloration of the extremity—and not all series closely follow the IASP diagnostic criterion. Nevertheless, the literature tends to support the use of SCS in CRPS.

Two prospective trials have been reported with SCS and CRPS.[49,50] Kemler et al. reported on 54 randomized patients, where 36 patients were assigned to receive SCS and physical therapy and 18 received only physical therapy.[49] At six months, patients who received SCS had significant reduction in pain and improvement in health-related quality of life. In 1999, Oakley and Weiner reported on 19 patients with CRPS prospectively followed to assess the efficacy of SCS with an average follow-up of 7.9 months.[50] All patients received at least partial relief, 30 percent received full relief, and there were significant improvements in VAS and SIP scores.

Many retrospective studies exist on the clinical efficacy of SCS for CRPS.[51–55] The largest study was published by Bennet et al.[53]; they studied 101 patients with CRPS separated into two groups. This study seemed to demonstrate improved results with multiple arrays, and concluded that sometimes frequencies in excess of 250 Hz were necessary to maintain pain control.

Implementation of SCS in patients with CRPS can be difficult. The possibility of aggravating the original pain or causing a new pain or allodynia at the implanted hardware site is significant. The pain often spreads to multiple body parts and we have used multiple electrodes in the cervical and low thoracic area to independently provide coverage to the upper and lower extremities, since the voltage requirements between the upper and lower extremities will differ.

Angina

The role of SCS in the management of refractory angina pectoris seems to be very promising and many studies have uniformly shown good results.[56–62] However, the mechanisms of action of SCS in relief of anginal pain are unclear. It is not definitively known whether the pain relief is a result of direct depression of the nociceptive signals in the spinal cord or whether it is a secondary gain from a reduction in the ischemia.[63,64]

The first report on the antianginal effects of SCS on refractory angina was published by Murphy and Giles in 1987.[65] Since then, the role of SCS in refractory angina has been scientifically studied in a prospective manner. The ESBY Study (electrical stimulation versus coronary artery bypass surgery-CABG) has provided some useful statistical data.[61,66] One hundred and four patients were randomized to either SCS (53) or CABG (51). There was no significant difference in pain relief between the two groups. The CABG group demonstrated increased exercise capacity and less ST-segment depression on maximum workloads but also noted a higher mortality (8 vs. 1) and cerebrovascular morbidity. The number of anginal episodes in both groups was statistically decreased.

The study concluded that when the effects on ischemia, on morbidity, and on mortality are factored in the decision making, SCS seemed a reasonable alternative for patients with an increased risk of surgical complications. Additional reports have found that both the pain and the health aspects of quality of life improved significantly after three months of SCS and were maintained after one year.[67] Further, SCS does not appear[68] to conceal the symptoms of an acute myocardial infarction.[69]

Technique

A good therapeutic response often begins with the delivery of a pleasant paresthesia in all the pain-affected dermatomes.[68,70–72] Because of this, an understanding of the spinal dermatomal anatomy assists in the implantation of the SCS electrodes.[68, 70–72]

The distribution of the paresthesia is determined not only by the longitudinal position of the electrode(s) in the spinal canal, but also by its mediolateral location. The following is a general description of various body areas that are frequently targets for spinal cord stimulation.[71,72]

An electrode at C2 covers the ipsilateral posterior occipital area as well as the angle of the jaw. To obtain complete coverage of the hand, the electrode should optimally be placed at C4–5. Selective coverage of the ulnar aspect of the hand can be obtained with electrodes at C7–T1. An electrode in the lower thoracic or upper lumbar area will have a greater than 70 percent chance of stimulating foot fibers, but L5 or S1 root stimulation is likely to present a more consistent approach to target the foot. The perineum is extremely difficult to stimulate. Alternative strategies may include electrodes placed at T11–12–L1 or electrodes placed in the sacral foramina. The low back area remains difficult to stimulate without intervening chest or abdominal wall stimulation. In our experience, the best location is at about T9–10 with an electrode placed strictly in the midline. Law showed that the low back fibers may be more selectively activated by a matrix of closely spaced electrodes at the T9–10 spine level.[71,72]

SCS implantation can be performed with either monitored (local anesthetic and intravenous sedation) or general anesthesia. The distribution of the stimulation-induced paresthesia is vital, and, therefore, it seems intuitive that testing a patient who is awake would yield superior results, because it allows for immediate feedback regarding the stimulation-induced paresthesia. Even when implantation is performed under general anesthesia, the patient can be awakened for the testing phase, after placement of the electrodes. This latter concept is an important one, as extreme lateral placement of an SCS electrode is often perceived by the patient as unpleasant and painful. When the whole procedure is performed under general anesthesia, the implanting physician must rely on either the radiographic correlation or on evoked motor or sensory responses to assure proper electrode positioning.[73] It has been our experience that motor-evoked responses in the cervical area have a higher degree of correlation with the sensory paresthesia than those evoked in the thoracic area.

The implanting physician must try to avoid any discomfort to the patient during the procedure. Generous infiltration of the skin and the periosteum with long-acting anesthetic agents minimizes the requirements for intravenous sedation. Excessive use of intravenous benzodiazepines should be avoided, since patients might remain confused as they emerge. Propofol (Diprivan, Stuart Pharmaceuticals, Wilmington, DE), an intravenous hypnotic agent which acts promptly and whose effects last only a few minutes when discontinued, has proven to be a very useful drug for this procedure; the patient wakes up promptly and lucidly as the drug is discontinued. In our institution, SCS implantation with plate electrodes is usually performed with the patient in a lateral decubitus position after the anesthesiologist places a laryngeal mask airway (LMA).

Today's technology allows the physician to deliver effective stimulation to the spinal cord via two types of electrodes: (1) an electrode that can be inserted percutaneously; and (2) an electrode that requires an open technique and direct visualization for insertion.

Most contemporary electrodes are either *quadripolar* or *octopolar* (referring to the number of contacts present on the electrode). The general trend is to utilize one or two quadripolar electrodes for limb pain and one or two octopolar electrodes for axial pain. The chief differences between these systems are the number of contacts, the length of the contacts, and the spacing between them. In addition, insertion of multiple parallel electrodes permits construction of different configuration matrices that can create variably focused electrical fields. Inherently, percutaneous electrodes are more flexible, thus allowing them to be inserted through a Touhy needle, although some require the presence of a stylet for insertion.

Percutaneous electrodes have several advantages. First and foremost, they can be inserted through a needle, avoiding surgical dissection of the paraspinal muscles and boney removal. Second, they can be easily advanced over several segments in the epidural space, allowing the testing of several spinal cord levels and optimizing electrode position. Finally, there is the added benefit of performing a trial stimulation to assess candidacy for a permanent implant, and temporary percutaneous electrodes can easily be removed in the implanting physician's office. Permanent percutaneous implantation technique is similar except that a surgical incision is required to anchor the electrode in place, extension wires must be tunneled a few inches away from the insertion site, and a return trip to the operating room is necessary for removal or internalization of the electrode.

There are also some disadvantages to percutaneous electrodes: they have a tendency to migrate and the stimulation may be more susceptible to postural changes, the electrical field generated by percutaneous electrodes is fairly narrow (compared to that of plate electrodes) and power requirements are higher, and these electrodes must be placed under fluoroscopic guidance.

Plate-type electrodes may sometimes be referred to as *ribbon electrodes*, *paddle electrodes*, or *laminotomy electrodes*. The most commonly utilized plate electrodes are quadripolar leads. The Medtronic Resume (Medtronic Inc., Minneapolis, MN) and the ANS Lamitrode (Advanced Neuromodulation Systems, Plano, TX) both have four linearly arranged contacts in one paddle. The Medtronic Symmix (Medtronic Inc., Minneapolis, MN) has four contacts arranged in a diamond pattern, which may facilitate bilateral lower extremity stimulation. The ANS Peritrode (Advanced Neuromodulation Systems, Plano, TX) consists of two smaller paddles, each with two contacts. These have the advantage of allowing the surgeon to place the paddles in two different locations or with two different orientations. Both Medtronic and ANS offer a double quadripolar electrode, with two columns of four contacts on the same paddle. ANS also manufactures the Lamitrode 8 and Lamitrode 88 (Advanced Neuromodulation Systems, Plano, TX) electrodes, which have eight contacts or two columns of eight contacts, respectively. These electrodes should reduce implanting physician variability in relation to cranial caudal orientation. Bear in mind that with the increasing number of contacts, there is a significant increase in power consumption, and the complexity of programming rises in even greater magnitude. However, there is some data to suggest that implantation of complex arrays have a significant advantage with changing symptomatology over time.[74]

The choice of one versus the other is most often dictated by the individual implanting physician's preferences and patterns of practice.

Electrical stimulation currently consists of rectangular pulses delivered to the epidural space through implanted electrodes via either a constant voltage or a constant current system. Two types of system are available: a completely implantable pulse generator (IPG) and a radiofrequency (RF) coupled pulse generator with an implantable receiver.

An IPG provides stimulation with fine resolution increments of 0.05 V and varying rates. The IPG contains a lithium battery, the life span of which varies with usage and parameters, but under average usage it lasts between 2.5 and 4.5 years. Replacement of the IPG requires a surgical (but outpatient) procedure. Further, activation and control of the IPG occurs through an external transcutaneous telemetry device. The IPG can alternatively be turned on or off through a small external magnet. Three commonly used IPG systems exist: (1) the Itrel 3 pulse generator (Medtronic Inc., Minneapolis, MN) is a unichannel device, which can power up to four contacts; (2) the Synergy (Medtronic Inc., Minneapolis, MN); and (3) the Genesis (Advanced Neuromodulation Systems, Plano, TX) both allow programming of two independent stimulation channels.

RF-driven systems contain a subcutaneously implanted receiver and an external transmitter, which must be worn in order to obtain stimulation. An antenna applied to the skin connects the transmitter to the receiver transcutaneously. In the postoperative period, obtaining adequate contact between the receiver and the antenna may be hampered by swelling at the site. Even further inconvenience may be added to an individual with handicapped motor function in the upper extremities or in a patient with reflex sympathetic dystrophy (RSD), who might not tolerate an antenna taped to the skin. Also, the equipment cannot be worn while swimming or showering, and severe perspiration, as with exercise and physical therapy, might make proper contact of the antenna more problematic. Most importantly, RF-driven systems can be customized to deliver more power than the corresponding lithium-powered IPG systems. These systems also allow the patient greater control over the stimulation parameters, since pulse width, cycling, and, more importantly, electrode combinations can be accessed via the transmitter. Additionally, RF systems can deliver stimulation with rates up to 1,400 Hz. Available RF-driven systems currently include the Matrix (Medtronic Inc., Minneapolis, MN) and the Renew (Advanced Neuromodulation Systems, Plano, TX).

The use of an IPG versus an RF system should be tailored to the individual patient. The chief advantage of the RF system is the ability to sustain larger power requirements over time. Rechargeable battery technologies are being developed that will allow transcutaneous recharging of the batteries, obviating the need for operative replacements, and which may ultimately make radiofrequency systems obsolete.

Implantation of a SCS system is only one of the factors involved in the successful implementation of the modality. After implantation, three variables have to be established: the perception threshold, the discomfort threshold, and the usage range. When the stimulator has been implanted, all the combinations must be tested extensively in order to find the ones that have the best electrical characteristics and provide the broadest coverage of the painful area. It has been clearly demonstrated that unless the paresthesia covers the area of pain, no pain relief will occur. This testing process can be time-consuming. The concept of a computer-controlled, patient-interactive stimulation has been promoted by Jay Law and Richard North for several years.[72,75] The CCSTIM system recently developed by ANS provides an optimal graphical interface between the implanter, the patient, and the stimulation device. It is not unusual for optimal stimulation parameters not to be established until four to five months have elapsed since the implantation.

▶ MOTOR CORTEX STIMULATION

Many chronic pain conditions may become refractory to conventional medical therapies. Tsubokawa et al. in 1991 first reported on deafferentation pain treated with epidural motor cortex stimulation.[15] A chronic stimulating electrode was placed epidurally such that stimulation of the underlying cortex produced motor contractions in the painful region. Since then, multiple groups from around the world have reported on their success with motor cortex stimulation (MCX).[18–22]

The mechanisms of MCX are currently under investigation but it is hoped that stimulation of the cortical structures leads to activation of non-nociceptive sensory neurons that are believed to exert an inhibitory effect on their nociceptive counterparts. This type of interaction may be present at multiple levels of the somatosensory pathway along the peripheral and central nervous systems. In deafferentation pain, the flow of afferent nervous impulses mediating noxious stimuli may be interrupted, resulting in the development of aberrant connections. While these interactions are thought to be disrupted at the level of the lesion, it is proposed that they are preserved rostrally. Thus, stimulating non-nociceptive sensory neurons at a rostral cortical level may bypass the aberrant connections and exert an inhibitory effect on the nociceptive system.

The methodology previously applied to electric field modeling in the spinal cord and subdural cortical arrays has also been applied to a model of an extradural electrode array.[76–85] Similar to SCS, the model demonstrated that the cerebrospinal fluid (CSF) has a large shunting effect on the effectiveness of the stimulus current. Older patients, with atrophic brains (and wider CSF spaces as a result) may not be amenable to epidural placement, and may likely require subdural strip electrode placement.[86]

Technique

The primary motor cortex is situated immediately anterior to the central sulcus as established by Broadman. The somatotopical representation of the homunculus on the precentral gyrus has been previously demonstrated, and the basic strategy employed by the surgeon should be to implant an electrode on the precentral gyrus corresponding to the region on the homunculus targeted yielding the lowest motor threshold contraction. Some recent electrical potential modeling has suggested that activation within the depths of the central sulcus may not be the primary site of electrical activation and stimulation of the motor cortex as a goal refers instead to the *anatomical placement* of the electrode and not necessarily the physiological target.[86]

The target may be localized in a variety of ways. Modern imaging, including magnetic resonance imaging (MRI) and computed tomography (CT), allows visualization of the central sulcus. In order to utilize CT, a curved reconstruction following the cortical surface will provide superior localization of the intended target than a straight reconstruction.[87] Similarly, a high resolution MRI scan may be used to localize the central sulcus.[88] Alternatively, functional MRI techniques may need to be employed in certain patient populations. PET scan and fMRI provide a functional roadmap. This may be particularly valuable in assessing for displacement or reorganization of the motor cortex region in patients with poststroke pain and phantom limb syndrome.

The surgical goal should be to allow accurate placement of an incision so that the creation of a craniotomy will permit placement of an electrode over the area corresponding to the somatotopy of the painful region. A more extensive discussion of the technique can be found in a variety of sources.[86,89]

The most commonly utilized electrode is the 3986-Resume TL (Medtronic, Minneapolis, MN). It is the thinnest four-contact paddle electrode; it has a thickness of 1.4 mm, and the paddle dimensions are 7 mm by 44 mm. Many different placement strategies exist: Our preference is to place two paddle electrodes parallel to the central sulcus with one directly over the sulcus and the other immediately adjacent anteriorly; others prefer to place the electrode perpendicular to the sulcus. The new Resume circular electrode (Medtronic, Inc., Minneapolis, MN, USA) may resolve the issue of electrode orientation. Nevertheless, the implanter must plan a strategy that will allow for strategic reprogramming should there be adaptation to the stimulation over time; the presence of multiple electrode arrays definitely allows this.

Epidural placement is the standard procedure today, mainly because it is perceived to be safer and less invasive. However, with epidural placement of the electrode, the stimulation voltage must penetrate the CSF, which requires a higher intensity. This can be a limiting issue with patients with brain atrophy, such as the elderly.

Subdural placement of electrode may eliminate this variable and requires less intensity, but the risks of a subdural hematoma, CSF leak risk, and possible electrode scarring makes its use less favorable. For a patient to be a candidate for MCX, it is essential that cortical stimulation be possible in the homuncular region corresponding to the painful area. Patients who have suffered large cortical strokes and have significant encephalomalacia in the corresponding region of cortex may not be optimal candidates.

Most patients experienced pain relief relatively soon after adjustment of the electrical parameters.

However, analysis of the electrical parameters used to achieve therapeutic benefits has not been revealing. Many patients report changes in their therapeutic benefit in the initial postoperative period, which may be related to postoperative fibrosis. Alternatively, we have also had patients who have lost therapeutic benefit after six months to over a year, and with either changing the contacts for stimulation, incorporating cycling modes, or even repositioning the electrodes, we have been able to reestablish the therapeutic benefit of cortical stimulation.

As is true with most therapies available for these different chronic pain conditions, the results vary tremendously and depend largely on the differing definitions of success. MCX has been used best consistently for trigeminal neuropathic pain and poststroke or central pain. Nguyen et al. reported on 22 patients with trigeminal pain where 13 obtained marked improvement and 5 obtained satisfactory improvement; only 4 were not improved.[89-91] Ebel et al. reported sustained good to excellent relief in 3 of 7 patients with trigeminal pain over time.[92] Meyerson et al. have reported between 60 and 90 percent pain relief in five patients with trigeminal neuropathic pain.[93] Tsubokawa initially reported on the treatment of central pain with MCX with 8 of 12 patients having continued effect after one year of therapy.[15] Many others have reported on their experience with similar patients. Katayama et al. noted satisfactory results in 2 of 3 patients with brainstem infarcts.[94] Mertens et al. noted 60 percent excellent or good relief in poststroke pain.[19] An analysis of the literature by Nguyen et al. has revealed 52 percent success (82 of 159 patients) for central pain.[89] This therapy is young, but with standardization of the technique, it will likely prove its merits over time.

▶ DEEP BRAIN STIMULATION

As previously mentioned, Heath and Pool in 1954 both reported on implantation of temporary electrodes in the septum pellucidum to treat patients with schizophrenia and pain from metastatic carcinoma.[4,5] The technique of DBS was then described in the ventroposterolateral thalamic nucleus by Mazars et al. in 1960.[95] Other targets were subsequently explored. PVG/PAG stimulation was described by multiple groups.[96-98]

The exact mechanism of pain modulation by DBS is unknown. PAG/PVG stimulation is, however, felt to result in stimulation of an opiod dependent pathway. Thalamic stimulation may modulate the abnormal firing patterns in the thalamic neurons secondary to deafferentation.

Currently, the two most common targets for DBS for pain are the periventricular grey matter and the ventrocaudalis thalamus. As a generalization, patients with neuropathic pain should undergo paresthesia-producing stimulation with implantation in Vc thalamus, whereas those patients with nociceptive pain should undergo PVG/PAG stimulation. Many patients will inevitably have mixed components of nociceptive and neuropathic pain and thus both Vc and PVG/PAG stimulation trials may be indicated. With PVG/PAG stimulation, patients may feel a pleasant warmth, a diffuse burning, or even anxiety. The internal capsule and medial lemniscus has also been used successfully.

Implantation of a DBS lead for pain begins with application of a stereotactic head frame. The patient then undergoes either a CT and/or MRI on the morning of surgery. Either direct or indirect targeting methods can be employed. Indirect targeting requires identification of the AC-PC (anterior and posterior commissural points). The target for the PVG is 2–5 mm anterior to the PC, 2 mm lateral to the medial wall of the third ventricle, and at the level of the PC. The PAG is 2–3 mm lateral to the midline, 1–2 mm lateral to the aqueduct, and 2–3 mm inferior to the AC-PC line. A burr hole is created in the contralateral hemisphere; intraoperative microelectrode recording is often helpful in delineating the somatotopy of the target, but in cases of lesional pain macrostimulation may be more helpful. Once the target is identified, a quadripolar electrode is placed and secured with a burr hole fastener device. The electrode is externalized for a period of trial stimulation, and, if the period demonstrates adequate pain relief, the electrode and extension can be later internalized and connected to an IPG.

Since the therapy of DBS for pain has been practiced for decades, many case series have been published.[98-106] It seems that patients with nociceptive pain (as in cancer pain and FBSS) respond best to DBS. Cancer pain has been reported to respond to PVG/PAG with pain reduction in 25 to 100 percent of cases, and in FBSS with pain reduction in 30 to 80 percent. Additionally, brachial plexus injuries, peripheral neuropathies, and phantom limb pain appear to respond to Vc DBS. Again, spinal cord injury and postherpetic neuralgia pain are difficult to treat.

REFERENCES

1. Stillings D: A survey of the history of electrical stimulation for pain to 1900. *Med Instrum* 1975;9:255–259.
2. Kellaway P: The part played by electric fish in the early history of bioelectricity and electrotherapy. *Bull Hist Med* 1946;20:112–137.
3. Schiller F: The history of algology, algotherapy, and the role of inhibition. *Hist Phil Life Sci* 1990;12:27–50.
4. Heath R: *Studies in Schizophrenia: A Multidisciplinary Approach to Mind-Brain Relationships.* Cambridge, MA: Harvard University Press, 1954.
5. Pool J: Psychosurgery in older people: *J Am Geriat Soc* 1954;2:456–465.
6. Melzack R, Wall P: Pain mechanisms: a new theory. *Science* 1965;150:971–979.

7. Shealy CN: Dorsal column stimulation: optimization of application. *Surgical Neurology*. 1975;4:142–145.
8. Shealy C, Cady R: Historical perspective of pain management. *Pain Management: A Practical Guide for Clinicians*. FL: St. Lucie Press, 1998:7–15.
9. Hoppenstein R: Electrical stimulation of the ventral and dorsal columns of the spinal cord for relief of chronic intractable pain. *Surg Neurol* 1975;4:187–194.
10. Larson S, Sances A, Cusick J, Meyer G, Swiontek T: A comparison between anterior and posterior implant systems. *Surg Neurol* 1975;4:180–186.
11. Larson SJ, Sances A, Jr., Riegel DH, Meyer GA, Dallmann DE, Swiontek T: Neurophysiological effects of dorsal column stimulation in man and monkey. *J Neurosurgery* 1974; 41:217–223.
12. Lazorthes Y, Verdie J, Arbus L: Stimulation analgesique medullaire anterieure et posterieure par technique d'implantation percutanee. *Acta Neurochir* 1978;40:253–276.
13. Dooley DM: Spinal cord stimulation. *AORN Journal* 1976;23:1209–1212.
14. Dooley D: *Percutaneous Electrical Stimulation of the Spinal Cord*. Bal Harbour, Florida: Assoc Neurol Surg, 1975.
15. Tsubokawa T, Katayama Y, Yamamoto T, et al: Chronic motor cortex stimulation for the treatment of central pain. *Acta Neurochir Suppl* (wein) 1991;52:137–139.
16. Hirayama T, Tsubokawa T, Katayama Y, Yamamoto Y, Koyama S: Chronic changes in activity of thalamic relay neurons following spinothalamic tractotomy in cat: Effects of motor cortex stimulation. *Pain* 1990;5 Suppl:273.
17. Yamamoto T, Katayama Y, Tsubokawa T, et al: Usefulness of the morphine/thiamylal test for the treatment of deafferentation pain. *Pain Res* (Tokyo) 1991;6:143–146.
18. Katayama Y, Fukaya C, Yamamoto T: Poststroke pain control by chronic motor cortex stimulation: neurological characteristics predicting a favorable response. *J Neurosurg* 1998;89:585–591.
19. Mertens P, Nuti C, Sindou M, et al: Precentral cortex stimulation for the treatment of central neuropathic pain. *Stereotact Funct Neurosurg* 1999;73:122–125.
20. Saitoh Y, Shibata M, Sanada Y, Mashimo T: Motor cortex stimulation for phantom limb pain. *Lancet* 1999;353: 212.
21. Nguyen JP, Lefaucher JP, Le Guerinel C, et al: Motor cortex stimulation in the treatment of central and neuropathic pain. *Arch Med Res* 2000;31:263–265.
22. Smith H, Joint C, Schlugman D, Nandi D, Stein J, Aziz T: Motor cortex stimulation in neuropathic pain. *Neurosurgical Focus* 2002;11:Article 2.
23. Coburn B: Electrical stimulation of the spinal cord: Two-dimensional finite element analysis with particular reference to epidural electrodes. *Med Biol Eng Comput* 1980;18: 573–584.
24. Coburn B: A theoretical study of epidural electrical stimulation of the spinal cord—Part II: Effects on long myelinated fibers. *IEEE Trans Biomed Eng* 1985;32:978–986.
25. Coburn B, Sin WK: A theoretical study of epidural electrical stimulation of the spinal cord—Part I: Finite element analysis of stimulus fields. *IEEE Trans Biomed Eng* 1985;32:971–977.
26. Holsheimer J, Struijk JJ: How do geometric factors influence epidural spinal cord stimulation? A quantitative analysis by computer modeling. *Stereotact Funct Neurosurg* 1991;56:234–249.
27. Holsheimer J, Barolat G, Struijk JJ, He J: Significance of the spinal cord position in spinal cord stimulation. *Acta Neurochir Suppl* 1995;64:119–124.
28. Holsheimer J, Wesselink WA: Effect of anode-cathode configuration on paresthesia coverage in spinal cord stimulation. *Neurosurgery* 1997;41:654–9;discussion 659–660.
29. Linderoth B, Stiller CO, Gunasekera L, O'Connor WT, Ungerstedt U, Brodin E: Gamma-aminobutyric acid is released in the dorsal horn by electrical spinal cord stimulation: an in vivo microdialysis study in the rat. *Neurosurgery* 1994;34:484–8;discussion 488–489.
30. Linderoth B: Dorsal column stimulation and pain: experimental studies of putative neurochemical and neurophysiological mechanisms, in *Kongl Carolinska Medico Chirurgiska Institutet*, Stockholm, 1992.
31. Foreman RD, Beall JE, Coulter JD, Willis WD: Effects of dorsal column stimulation on primate spinothalamic tract neurons. *J Neurophysiol* 1976;39:534–546.
32. Koeze TH, Williams AC, Reiman S: Spinal cord stimulation and the relief of chronic pain. *J Neurol Neurosurg Psychiatr* 1987;50:1424–1429.
33. North RB, Kidd DH, Zahurak M, James CS, Long DM: Spinal cord stimulation for chronic, intractable pain: experience over two decades. *Neurosurgery* 1993;32:384–394;discussion 394–395.
34. Lang P: The treatment of chronic pain by epidural spinal cord stimulation—a 15-year follow-up; present status. *Axon* 1997;18:71–73.
35. North RB, Wetzel FT: Spinal cord stimulation for chronic pain of spinal origin: a valuable long-term solution. *Spine* 2002;27:2584–2591;discussion 2592.
36. Spiegelmann R, Friedman W: Spinal cord stimulation: a contemporary series. *Neurosurgery* 1991;28:65–71.
37. Kumar K, Toth C, Nath RK, Laing P: Epidural spinal cord stimulation for treatment of chronic pain—some predictors of success: A 15-year experience. *Surg Neurol* 1998;50:110–20;discussion 120–121.
38. North R, Kidd D, Piantadosi S: Spinal cord stimulation versus reoperation for failed back surgery syndrome: a prospective, randomized study design. *Acta Neurochirurg Suppl* 1995;64:106–108.
39. North R, Kidd D, Lee M, Piantadosi S: A prospective, randomized study of spinal cord stimulation versus reoperation for failed back surgery syndrome: initial results. *Stereotact Funct Neurosurg* 1994;62:267–272.
40. Burchiel K, Anderson V, Brown F, Fessler R, Friedman W, Pelofsky: Prospective, multicenter study of spinal cord stimulation for relief of chronic back and extremity pain. *Spine* 1996;21:2786–2794.
41. Barolat G, Oakley J, Law J, Notth R, Ketcik B, Sharan A: Epidural spinal cord stimulation with multiple electrode paddle lead is effective in treating intractable low back pain. *Neuromodulation* 2001;4:59–66.
42. Turner J, Loeser J, Bell K: Spinal cord stimulation for chronic low back pain: a systematic literature synthesis. *Neurosurgery* 1995;37:1088–1095.

43. Meglio M, Cioni B: Spinal cord stimulation in low back and leg pain. *Stereotact Funct Neurosurg* 1994;62:263–266.
44. De La Porte C: Spinal cord stimulation in failed back surgery syndrome. *Pain* 1993;52:55–61.
45. LeDoux M, Langford K: Spinal cord stimulation for the failed back syndrome. *Spine* 1993;18:191–194.
46. North R, Ewend M, Lawton M, Kidd D, Piantadosi S: Failed back surgery syndrome: 5-year follow-up after spinal cord stimulator implantation. *Neurosurgery* 1991;28:692–699.
47. Devulder J, De Laat: Spinal cord stimulation: a valuable treatment for chronic failed back surgery patients. *J Pain Symptom Manage* 1997;13:296–301.
48. Bell G: Cost-effectiveness analysis of spinal cord stimulation in treatment of failed back surgery syndrome. *J Pain Symptom Manage* 1997;13:286–295.
49. Kemler M, Barendse G, van Kleef M: Spinal cord stimulation in patients with chronic reflex sympathetic dystrophy. *N Engl J Med* 2000;343:618–624.
50. Oakley J, Weiner R: Spinal cord stimulation for complex regional pain syndrome: a prospective study of 19 patients at two centers. *Neuromodulation* 1999;2:47–50.
51. Barolat G, Schwartzma R, Woo R: Epidural spinal cord stimulation in the management of reflex sympatheitc dystrophy. *Stereotact Funct Neurosurg* 1989;53:29–39.
52. Kemler M, Barendse G, van Kleef M: Pain relief in complex regional pain syndrome due to spinal cord stimulation does not depend on vasodilation. *Anesthesiology* 2000;92:1653–1660.
53. Bennet D, Alo K, Oakley J, Feler C: Spinal cord stimulation for complex regional pain syndrome I [RSD]: a retrospective multicenter experience from 1995 to 1998 of 101 patients. *Neuromodulation* 1999;2:202–210.
54. Kemler M, Barendse GvK: Electrical spinal cord stimulation in reflex sympathetic dystrophy: retrospective analysis of 23 patients. *J Neurosurg* 1999;90:79–83.
55. Kumar K, Nath R, Toth C: Spinal cord stimulation is effective in the management of reflex sympathetic dystrophy. *Neurosurgery* 1997;40:503–509.
56. Augustinsson L: Spinal cord electrical stimulation in severe angina pectoris: surgical technique, intraoperative physiology, complications, and side effects. *Pace* 1989;12:693–694.
57. DeJongste M, Haaksma J, Hautvast R, Hillege H, Meyler J: Effects of spinal cord stimulation on daily life myocardial ischemia in patients with severe coronary artery disease: A prospective ambulatory ECG study. *Br Heart J* 1994;71:413–418.
58. DeJongste M, Hautvast R, Hillege H, Lie K: Efficacy of spinal cord stimulation as an adjuvant therapy of angina pectoris: A prospective randomized study. *J Am Coll Cardiol* 1994;23:1592–1597.
59. DeJongste M: Spinal cord stimulation for ischemic heart disease. *Neurol Res* 2000;22:293–298.
60. Mannheimer C, Augustinsson L, Carlsson C, Manhem K, Wilhelmsson C: Epidural spinal electrical stimulation in severe angina pectoris. *Br Heart J* 1988;59:56–61.
61. Mannheimer C, Eliasson: Electrical stimulation versus coronary artery bypass surgery in severe angina pectoris: the ESBY study. *Circulation* 1998;97:1157–1163.
62. Sanderson J, Brooksby P, Waterhouse D, Palmer R, Neubauer K: Epidural spinal electrical stimulation for severe angina: a study of its effects on symptoms, exercise tolerance and degree of ischaemia. *Eur Heart J* 1992;13:628–633.
63. Meller S, Gebhart G: A critical review of the afferent pathways and the potential chemical mediators involved in cardiac pain. *Neurosc* 1992;48:501–524.
64. Thamer V, Deussen A, Schipke J, Tolle T, Heush G: Pain and myocardial ischemia: the role of the sympathetic system. *Bas Res Card* 1990;85:253–266.
65. Murphy D, Giles K: Dorsal column stimulation for pain relief from intractable angina pectoris. *Pain* 1987;28:365–368.
66. Norrsell H, Pilhall: Effects of spinal cord stimulation and coronary artery bypass grafting on myocardial ischemia and heart rate variability: further results from the ESBY study. *Cardiology* 2000;94:12–18.
67. Vulink N, Overgaauw D, Jesserun G, et al: The effects of spinal cord stimulation on quality of life in patients with therapeutically chronic refractory angina pectoris. *Neuromodulation* 1999;2:33–40.
68. Barolat G: Experience with 509 plate electrodes implanted epidurally from C1 to L1. *Stereotact Funct Neurosurg* 1993;61:60–79.
69. Andersen C, Hole P, Oxhoj H: Will SCS treatment for angina pectoris pain conceal myocardial infarction? *Abstracts of the First Meeting of the International Neuromodulation Society*, Rome 1992:39.
70. Barolat G, Massaro G, He J, Zeme S, Ketcik B: Mapping of sensory responses to epidural stimulation of the intraspinal neural structures in man. *J Neurosurg* 1993;78:233–239.
71. Law J: Spinal stimulation: Statistical superiority of monophasic stimulation of narrowly separated, longitudinal bipoles having rostral cathodes. *Appl Neurophys* 1983;46:129–137.
72. Law J: A new method for targeting a spinal stimulator: Quantitatively paired comparisons. *Appl Neurophys* 1987;50:436.
73. Iacono R, Boswell M, Guthkelch A: Placement of spinal cord stimulators using spinal anaesthesia and monitored by cortical evoked potentials obtained from spinal cord stimulation. *Pain* 1990;5:S234.
74. Sharan A, Cameron T, Barolat G: Evolving patterns of spinal cord stimulation in patients implanted for intractable low back and leg pain. *Neuromodulation* 2002;5:167–179.
75. North R, Fowler K, Nigrin D, Szymanski R, Piantadosi S: Automated "pain drawing" analysis by computer-controlled, patient-interactive neurological stimulation system. *Pain* 1992;50:51–57.
76. Grill WM: Electrical properties of implant encapsulation tissue. *Ann Biomed Eng* 1994;22:23–33.
77. Holsheimer JJ: MR assessment of the normal position of the spinal cord in the spinal canal. *Am J Neuroradiol* 1994;15:951–959.
78. Ranck JB Jr: Which elements are excited in electrical stimulation of mammalian central nervous system: A review. *Brain Res* 1975;98:417–440.
79. Rattay F: Analysis of the electrical excitation of CNS neurons. *IEEE Trans Biomed Eng* 1998;45:766–772.
80. Rushton WAH: The effect upon the threshold for nervous excitation of the length of nerve exposed, and the angle between current and nerve. *J Physiol* 1927;63:357–377.

81. Testerman R: Neural electrode configuration design: A design comparison via generated electric fields in spinal cord models. *Medtronic Report* 2001, April 13.
82. Testerman R: Electric field model of a cortical electrode array. *Medtronic Report* 2001, June 1.
83. Testerman R: Electric field model of a cortical electrode array (Supplement). *Medtronic Report* 2001, June 11.
84. Warman EN: Modeling the effects of electric fields on nerve fibers: Determination of excitation thresholds. *IEEE Trans Biomed Eng* 1992;39:1244–1254.
85. Wesselink WA: Analysis of current density and related parameters in spinal cord stimulation. *IEEE Trans Rehab Eng* 1998;6:200–207.
86. Sharan A, Rosenow J, Turbay M, Testerman R, Rezai A: Precentral stimulation for chronic pain, in Henderson J, (ed), *Neurosurg Clinic N Amer* Vol. 14, 2003:1–7.
87. Lee U, Bastos A, Alonso-Vanegas M, Morris A, Olivier A: Topographic analysis of the gyral patterns of the central area. *Stereotact Funct Neurosurg* 1998;70:38–51.
88. Herregodts P, Stadnik T, De Ridder F, D'Haens J: Cortical stimulation for central neuropathic pain: 3-D surface MRI for easy determination of the motor cortex. *Acta Neurochir Suppl* 1995;64:132–135.
89. Nguyen J, Lefaucheur J, Keravel Y: Motor cortex stimulation, in Burchiel K (ed), *Electrical stimulation and the relief of pain*. New York: Elsevier Science B.V., 2003:197–209.
90. Nguyen J, Keravel Y, Feve A, et al: Treatment of deafferentation pain by chronic stimulation of the motor cortex: report of a series of 20 cases. *Acta Neurochir Suppl* 1997;68:54–60.
91. Nguyen JP: Motor cortex stimulation, *10th World Congress on Pain*, San Diego, 2002.
92. Ebel H, Rust D, Tronnier V, et al: Chronic precentral stimulation in trigeminal neuropathic pain. *Acta Neurochir Suppl (wein)* 1996;138:1300–1306.
93. Meyerson B, Lindblom U, Linderoth B, et al: Motor cortex stimulation as treatment of trigeminal neuropathic pain. *Acta Neurochir Suppl (Wein)* 1993;58:150–153.
94. Katayama Y, Tsubokawa T, Yamamoto T: Chronic motor cortex stimulation for central deafferentation pain: experience with bulbar pain secondary to Wallenberg syndrome. *Stereotact Funct Neurosurg* 1994;62:295–299.
95. Mazars G, Roge R, Mazars Y: Stimulation of the spinothalamic fasciculus and their bearing on the pathophysiology of pain. *Revue neurologique* 1960;103:136–138.
96. Richardson D, Akil H: Pain reduction by electrical brain stimulation in man. Part 1: Acute administration in periaqueductal and periventricular sites. *J Neurosurg* 1977;47:178–183.
97. Richardson D, Akil H: Pain reduction by electrical brain stimulation in man. Part 2: Chronic self-administration in the periventricular gray matter. *J Neurosurg* 1977;47:184–194.
98. Hosobuchi Y, Adams J, Linchitz R: Pain relief by electrical stimulation of the central gray matter in humans and its reversal by naloxone. *Science* 1977;197.
99. Richardson D, Akil H: Long-term results of periventricular gray self-stimulation. *Neurosurgery* 1977;1:199–202.
100. Young R, Kroening R, Fulton W, et al: Electrical stimulation of the brain in treatment of chronic pain. *J Neurosurg* 1985;62:389–396.
101. Turnball I, Shulman R, Woodhusst W: Thalamic stimulation for neuropathic pain. *J Neurosurg* 1980; 52:486–493.
102. Meyerson B, Boethius J, Carlsson A: Alleviation of malignant pain by electrical stimulation in the periventricular region: Pain relief as related to stimulation sites. *Adv Pain Res Ther* 1979;3:525–533.
103. Levy R, Lamb S, Adams J: Treatment of chronic pain by deep brain stimulation: long-term follow-up and review of the literature. *Neurosurg* 1987;21:885–893.
104. Kumar K, Toth C, Nath R: Deep brain stimulation for intractable pain: a 15-year experience. *Neurosurg* 1997;40:736–746.
105. Tasker R, Vilela Filho O: Deep brain stimulation for neuropathic pain. *Stereotact Funct Neurosurg* 1995;65:122–124.
106. Dieckmann G, Witzmann A: Initial and long-term results of deep brain stimulation for chronic intractable pain. *Appl Neurophys* 1982;45:167–172.

CHAPTER 44

Neurosurgical Approaches to the Treatment of Pain

James N. Campbell and Daniel M. Sciubba

Neurosurgical procedures provide important approaches for the treatment of chronic pain. In fact, pain is one of the most common symptoms leading to neurosurgical consultation, forcing neurosurgeons by default to become pain specialists. Many of the specialized pain treatments are highly effective. One example applies to trigeminal neuralgia, one of the most agonizing of neuropathic pain states. Neurosurgeons effectively apply both ablative and decompressive techniques to treat this disorder.

The cliché, if you have a hammer everything looks like a nail, unfortunately applies to medicine and, more specifically, to pain care. In other words, specialists treat pain using the tools provided to them within their field of medical training. The neurologist and oncologist may be most interested in medicinal approaches to a pain problem. The anesthesiologist, on the other hand, may consider nerve blocks more efficient at alleviating pain. However, there are relatively few neurosurgeons with a primary interest in pain as a subspecialty. As a consequence, pain-neurosurgery is often not discussed as an option for the patient with moderate to severe pain, possibly to the detriment of the patient.

In this review, not all pain problems that present to a general neurosurgeon's clinic will be discussed. Rather, conditions appropriate for the neurosurgical pain specialist will be presented, highlighting specific clinical problems and the possible options recommended by a pain-neurosurgeon. As will be borne out in this review, the neurosurgeon may control pain via: selective stimulation of parts of the nervous system, directed drug delivery, decompression of nervous structures, reconstructive procedures, and ablative procedures. The indications for and expected outcomes of such surgical procedures will be covered here. The treatment of trigeminal neuralgia is covered elsewhere in this volume.

▶ BACK AND NECK PAIN

The most common complaint for the pain specialist, and one of the leading health care problems in medicine today, is back pain. Axial spine pain may result from one of three surgically treatable pathophysiological mechanisms: (1) compression; (2) instability; and (3) malalignment. If practical and feasible, surgical correction directed at one or more of these mechanisms may be undertaken. The decision whether or not to undergo surgery is based not only on analysis of the relative risks and benefits, but also on patient preference and surgeon experience and skill.

Decompressive procedures for back and neck pain are well established and quite successful in the properly selected patient. Procedures include laminectomies, discectomies, and foraminotomies. Each is intended to decompress stenosis of the spinal canal and/or transverse spinal foramina. In addition, the allograft or autograft with and without the aid of rigid internal fixation (pedicle screws, lateral mass screws, or anterior plating with interbody grafts) has been used to stabilize and/or realign the bony spinal column, in order to address axial spine pain. Fusion of the anterior column (interbody fusion) may be required along with a posterior fusion in some instances. Surgery may fail because the surgical objectives were not accomplished (e.g., pseudarthrosis) necessitating further surgery. Alternatively, surgery may fail because the plan or concept of the surgery was flawed. In cases where surgery has failed or has an unfavorable risk-to-benefit ratio, the interventional pain specialist may offer other surgical options.

Dorsal column stimulation (DCS) is a very important tool to consider when patients have intractable back and leg pain problems not amenable to straightforward surgery.[31] In addition, DCS may prove more effective in treating pain attributable to lumbar arachnoiditis. *Intrathecal drug delivery* in medically refractory cases is also an option. Deer et al. have shown that combination treatment of bupivacaine with opioids may give rise to better outcomes than treatment with opioids alone.[11] Intrathecal bupivacaine in combination with an opioid is an option for the treatment of chronic pain related to failed back surgery syndrome and metastatic cancer pain of the spine. Drawbacks to such treatment include catheter complications, long-term analgesic tolerance, and the need for return visits to the clinician for prescription refills. Ongoing technical improvements, and the availability of more choices for intrathecal therapy, may make such therapy more attractive in the near future.

▶ NERVE INJURY PAIN

Four mechanisms of nerve injury pain have been identified:[7] (1) central changes, (2) abnormal neural discharge from the nerve injury site, (3) abnormal discharges arising from the intact nociceptors that share the innervation territory of the injured nerve, and (4) abnormal discharges that arise from the dorsal root ganglion. A given patient may have pain due to one or more of these mechanisms. If the cause is central, peripheral nerve operations have little likelihood of helping the patients. Fortunately for many patients, peripheral mechanisms dominate.

Traumatic neuromas form at the site of nerve injury as a consequence of unsuccessful nerve regeneration. Laboratory studies indicate that the level of hyperalgesia after nerve transection varies with whether or not a nerve is successfully regenerating.[21] Moreover, once the nerve has regenerated to its target, the level of hyperalgesia drops even further. Thus, one goal in surgically approaching nerve injury pain is to foster an environment favorable to nerve regeneration. Such an environment can often be created successfully when there is a large nerve injury. In these cases, the neuroma may be excised with subsequent joining of the healthy ends of the nerve via nerve graft repair. In the case of small cutaneous nerves, however, nerve repair may be impractical. To approach this problem, neuroma pathophysiology must first be explained further.

Neuromas may be the source of pain in two ways: ectopic excitability and spontaneous activity. *Ectopic excitability* refers to the property whereby the axons that innervate the neuroma may be excited by mechanical stimuli. Thus, the location of the neuroma is of pathophysiological significance. For instance, if the injured nerve is adherent to moving structures such as tendons via scar formation, then normal motion of an extremity may induce neural activity in nociceptive fibers and lead to pain. Removal of such a neuroma may only lead to formation of a new neuroma with the same inherent proclivity to induce pain. Therefore, "removal" of a neuroma is a misnomer; the surgeon may only be able to "relocate" the neuroma. Such *neuroma relocation surgery* is predicated on ectopic mechanosensitivity being an important mechanism for the pain. The surgical goal is to relocate the neuroma away from scar tissue, away from pressure points, and away from moving structures such as tendons or joints. Thus, to treat a neuroma involving the infrapatellar branch of the saphenous nerve, for example, the surgeon may consider severing the saphenous nerve in the adductor canal (mid-thigh level).[6]

Nerve entrapment is a very common source of pain that may be treated surgically via various release operations, yielding success in general. However, pain may persist after such procedures due to intrinsic nerve damage caused by the initial entrapment. It is well known that pain and the degree of nerve entrapment are not necessarily correlated. Accordingly, severe pain may be seen in mild cases of nerve entrapment, and no pain may be present in severe cases of nerve entrapment. Genetic differences likely play an important role in the liability for pain. It is, therefore, a mistake to not consider entrapment the cause of serious pain simply because the entrapment is *mild*.

The majority of nerve entrapments are easily identified, and surgery to address entrapments is likewise fairly easy to accomplish. *The key to addressing entrapment problems is diagnostic acumen and awareness.* If pain fails to get better after nerve entrapment surgery, the clinicians should entertain two diagnostic possibilities: inaccurate diagnosis or inadequate surgery.

Clinicians also should always entertain the possibility of a nerve entrapment as a cause of otherwise unexplained pain. The use of electromyogram and nerve conduction studies to diagnose nerve entrapments is not always reliable due to sensitivity and specificity issues.

▶ **TABLE 44-1.** EXAMPLES OF PAINFUL NERVE ENTRAPMENTS

Symptom	Entrapment
Shoulder pain	Suprascapular nerve entrapment
Shoulder and arm pain	Thoracic outlet syndrome
Forearm pain	Radial tunnel syndrome Pronator syndrome
Groin pain	Entrapment of the ilioinguinal nerve
Calf and medial thigh pain	Entrapment of saphenous nerve at Hunter's canal
Foot pain	Anterior tarsal tunnel syndrome

Table 44-1 lists some diagnoses, sometimes overlooked, that should be considered for certain symptoms. Two keys to making these diagnoses are looking for point tenderness over the entrapment site and looking at the effects of selective nerve blocks proximal to the nerve entrapment. For example, a patient may have unexplained forearm and hand pain. A diligent examination may disclose focal point tenderness over the brachioradialis muscle six to seven centimeters below the lateral epicondyle. This is the point where the posterior interosseus branch of the radial nerve enters the arcade of Fröhse (a fibrous ring formed by the two heads of the supinator muscle). A proximal block of the radial nerve will eliminate the pain entirely for the duration of the anesthetic block if the diagnosis of radial tunnel syndrome is correct. Surgical release of the nerve at the arcade of Fröhse may dramatically relieve the patient's pain.

Anesthetic nerve blocks in general are used quite commonly but are not very effective as a long-term treatment of pain. One exception is the case of sympathetically maintained pain. Mixing steroids with an anesthetic may confer temporary benefit if given at the entrapment site (e.g., as with carpal tunnel syndrome), but is of no use if given to an area remote from the entrapment. Despite limitations in therapeutic efficacy, nerve blocks should be considered as a useful adjunctive test for diagnosing the origin of pain, specifically for both neuromas and putative nerve entrapments. If the site of the neuroma is the source of pain, then anesthesia applied directly to the neuroma or entrapment, or proximal to the neuroma or entrapment along the course of the involved nerve, should relieve the pain for the duration of the anesthetic action.

In some situations, nerve injury pain persists despite attempts to foster nerve regeneration, to relocate neuromas, or to free entrapped nerves. In these cases implantation of a *peripheral nerve stimulator* may be useful. These devices work similarly to dorsal column stimulators, except that the electrodes are placed on the involved nerve. [8,32]

Finally, further ablative surgery may be useful in addressing nerve injury pain when all other techniques have been exhausted. For example, groin and testicular pain may persist after inguinal herniorrhaphy. In these operations, branches of the lumbar plexus—in particular the ilioinguinal and genitofemoral nerves—may be injured. Proximal resection of these nerves in the retroperitoneum or revision of the hernia repair are generally first-line techniques to treat this pain. If such procedures fail, patients are instructed to undergo nerve blocks of T11, T12, L1, and L2 to determine the major dermatome responsible for the pain. If the patient responds well to a T12 block, then an intradural dorsal root rhizotomy or a dorsal root ganglionectomy of T12 and the bordering dermatomes, T11 and L1, may be undertaken: These last two techniques will be discussed below. In some cases of nerve injury, efferent effects of the sympathetic nervous system drive the pain (sympathetically-maintained pain, see below).

▶ GANGLIONECTOMY AND DORSAL ROOT RHIZOTOMY

Neurosurgeons may offer dorsal root ganglionectomies or dorsal root rhizotomies for a variety of conditions. These are not first-line therapies generally, but may be offered when other simpler techniques have failed (e.g., medical therapy, dorsal column stimulation, or peripheral neuroma relocation). Although long-term efficacy remains an issue for such individuals, most data indicate that at least some patients report sustained pain relief. [45,42]

A *dorsal root rhizotomy* is a preganglionic lesion, leaving the peripheral projection of the dorsal root ganglion intact. The nerve fibers in the peripheral nerve remain connected to the dorsal root ganglion and thus survive, though connectivity with the central nervous system is lost. With dorsal root ganglionectomy, the cell bodies of the peripheral nerve fibers are removed, and so the nerve fibers undergo wallerian degeneration in both directions (the dorsal root and the peripheral nerve). Otherwise the procedure is the same as a dorsal rhizotomy. Preference for ganglionectomy over rhizotomy stemmed largely from evidence suggesting the existence of *ventral* root pain afferents. [10] It was found that the ventral root afferents had cell bodies in the dorsal root ganglia, however. Thus, a ganglionectomy procedure would offer a more complete interruption of afferent input to the spinal cord than a dorsal root rhizotomy by inducing degeneration in both the ventral and dorsal roots. However, subsequent animal studies indicated that these ventral root afferents likely loop back and eventually still make their way to the spinal cord through the dorsal root.[40] However, no studies to date have compared outcomes head-to-head for these two approaches. Hosobuchi reported three patients who, after failing to achieve pain relief with a dorsal rhizotomy, underwent a dorsal root ganglionectomy of the same roots and achieved lasting pain relief.[18]

Some recent work has raised concern that ganglionectomy may invoke a further mechanism for neuropathic pain, however. Inevitably with nerve transection (including ganglionectomy) there exists a region where intact fibers share the region denervated as a result of the nerve injury. Several lines of evidence indicate that these intact nociceptors become sensitized and spontaneously active. A dorsal rhizotomy is a preganglionic lesion, leaving the peripheral projection of the dorsal root ganglion cell intact: therefore, there is no peripheral wallerian degeneration. This may decrease the sensitization that would otherwise occur in the intact nociceptors that enter the spinal cord through the adjacent intact roots.

Only empirical trials comparing the two procedures can ultimately resolve the issue of which, if either, of the procedures is better. In any case, current theory suggests that the dorsal rhizotomy may be the better option.

▶ POST-THORACOTOMY PAIN

Pain following thoracotomy, a fairly common postoperative morbidity, is a further example of nerve injury pain. Retractors used in chest surgery routinely crush the intercostal nerve, particularly on the rostral side[35] and it is not uncommon for the nerve to be severed or caught up in sutures by the end of the case. The discussion above regarding surgical approaches to nerve injury applies. Usually much of the post-thoracotomy pain will abate with time, but unfortunately the pain persists for many patients. Proximal resection of the nerve involved may solve the problem but a new neuroma will form that could become the new pain generator. At this point, the neurosurgeon may offer rhizotomy or ganglionectomy. However, pain may recur even after intervals of over a year. The mechanism is unknown but could relate to the mechanisms that create phantom pain following deafferentation. One mechanism may relate to the rich plexus formation that occurs between nerves. Overlapping innervation territory between nerves is such that numbness may be barely detectable after resection of one intercostal nerve. Arguably, lesioning the spinal root (or ganglion) above and below the root corresponding to the painful dermatome may help circumvent the problem of recurrent pain. Many surgeons, therefore, advocate sectioning of three roots (or ganglia), even though the pain may appear to be served by one nerve. Dorsal column stimulation is generally not useful for post-thoracotomy pain, perhaps because the dorsal column representation of the intercostal nerves is too small.

▶ HEADACHE AND OCCIPITAL NEURALGIA

Migraine is the most common cause of headache, but in medically refractory cases, consideration should be given to pathology at the level of the C2 and C3 nerve roots, namely occipital neuralgia. This condition presents with symptoms that may overlap with those of migraine. The prototypic patient describes sharp lancinating pain from the upper neck to an area over the posterior scalp with associated tenderness at the superior nuchal line. The pain may be unilateral or bilateral. The diagnosis may be investigated by doing selective blocks of the C2 and C3 nerve roots. If there is complete relief of pain, the patient may be a candidate for either a C2–3 rhizotomy or a C2–3 ganglionectomy (as described previously).[23] Implantation of subcutaneous suboccipital electrodes represents an alternative to treating this form of headache, and has been tried as a less invasive alternative. In the majority of cases, the problem is idiopathic. An occult nerve root entrapment may be the actual lesion in these "idiopathic" cases. In other situations there may be an underlying anatomic abnormality such as C1–2 arthritis that leads to root compression.

▶ THE DREZ OPERATION

Neurosurgeons have a long tradition of trying to cut afferent pathways to eliminate pain. Whereas one might think that such approaches would work, they do so inconsistently. One of the explanations postulated is that the signaling for pain is displaced to the next processing center for nociceptive information, namely the *dorsal horn* of the spinal cord. Nashold developed an operation initially aimed at lesioning the tract of Lissauer. Later it was realized that the lesioning actually destroyed the lamina I-V. The operation came to be known as the DREZ operation (for *dorsal root entry zone*). For certain conditions, the operation works extremely well, but for other indications the results are disappointing. *As a general rule, the operation works best if the lesion is preganglionic. The results are disappointing for postganglionic lesions (peripheral nerve).* The DREZ operation has been done also in the dorsal horn equivalent for the face, namely the *nucleus caudalis*.[5]

The *Nashold* technique involves placement of a microelectrode 2 mm into the dorsolateral sulcus of the spinal cord, a location that grossly approximates the DREZ. The electrode is then heated to about 75°C for 15 seconds to create a lesion. As many as 100 longitudinal lesions may be necessary to ablate the targeted area.[27] Sindou described a similar procedure in which a small incision is made in the dorsolateral sulcus and the superficial dorsal horn is lesioned with a bipolar cautery. The results of surgery with the two techniques appear comparable, but no group has done a comparative study.[13] In the following sections, we consider conditions where the DREZ operation has a role in treatment.

Lesions of the Brachial Plexus

One of the classic causes of severe neuropathic pain is an injury to the brachial plexus in which the dorsal roots are torn from the spinal cord. Though considered a brachial plexus injury, in some sense this is actually an injury to the spinal cord dorsal horn. The injury consists of a stretch injury of the plexus that is severe enough to literally tear the roots away from the spinal cord. The diagnosis should be suspected in brachial plexus lesions accompanied by serious pain. In essence, this is a form of *phantom pain* (even though the limb is still there). The diagnosis is further supported by findings of meningoceles seen on myelography or MRI scanning. The patients have clinical evidence of a proximal injury with evidence of serratus anterior denervation, rhomboid denervation, and/or a Horner's syndrome. At surgery the surgeon sees absence of the dorsal roots and evidence of scarring in the area where the dorsal roots normally enter the spinal cord. A classical presentation is for the patient to complain of pain in the hand even though there is no feeling in the hand. In addition to the ongoing pain there are episodes of lightening-like pain. The pain probably results from epileptiform discharge in the dorsal horn stemming from the injury to the dorsal horn as a result of the avulsion.

Morbidity with the DREZ operation is secondary to inadvertent ablation of adjacent spinal tracts. Corticospinal tract damage may be associated with ipsilateral leg spasticity and/or weakness. Posterior column damage may be associated with sensory deficits in the ipsilateral leg. Furthermore, the cervical laminectomy required for surgical exposure may lead to instability and spinal deformity. The reported incidence of significant enduring complications is in the range of 5 percent or less in the hands of most experienced surgeons.[28] In our own series of 10 patients, there were no significant complications: all patient reported pain relief, with a mean pain relief of 85 percent.[9]

Avulsion lesions also occur at the lumbar plexus and sacral plexus, but such conditions often are initially misdiagnosed. Patients typically present after severe trauma with extensive pelvic fractures complaining of pain in an area denervated by a lumbar or lumbosacral plexus lesion. The myelogram or MRI scan may reveal meningoceles in one or more of the avulsed roots. The pain from these lesions responds well to the DREZ operation, which should be considered the treatment of choice.[25]

Spinal Cord Injury Pain

Neuropathic pain is a common sequela of spinal cord injury. Two somewhat distinct pain patterns are evident: one is segmental, corresponding to the level of the injury, and the other is infralesional, occurring at locations below the injury. This infralesional pain may occur in complete spinal cord transections, and in a sense is a form of phantom pain. Medical management, dorsal column stimulation, and intrathecal delivery of opioids all typically offer little relief for this type of pain. Nashold and colleagues maintained that the DREZ operation helped with segmental pain, but believed that the DREZ operation offered little for the distal *phantom* pain. However, Edgar et al.[12] (and later Falci et al.)[14] determined that the DREZ lesioning may indeed relieve distal pain as well.

Electrophysiological recordings guided the lesion extent in these studies. Quite likely the extent of lesioning performed by Falci, Edgar, and colleagues was more extensive then the lesions performed by Nashold. How the dorsal horn in the upper thoracic area, for example, becomes involved in signaling pain in distal parts of the body is unexplained. Nevertheless, the DREZ operation ought to be considered as an option in the treatment of pain from spinal cord injury—particularly in cases of complete spinal cord transection in the thoracic area, where the threat of further neurological damage from DREZ lesioning is not really an issue.

Postherpetic Neuralgia

Postherpetic neuralgia, another neuropathic pain condition, may also respond to DREZ surgery. Friedman and Bullitt[15] noted significant pain relief in 91 percent of patients directly after surgery; however, long-term follow-up showed significant pain relief in only 25 percent of patients. In addition, the complication rate was not trivial. The results, however, may be better for the face. The nucleus caudalis separates out from the main sensory nucleus, *nucleus oralis*, and *nucleus interpolaris* so that it is possible to lesion the second order *pain* neurons while leaving tactile sensibility intact. Bernard and colleagues reported pain relief in 71 percent of patients with postherpetic neuralgia,[4] and suggested that favorable results tended to correlate with a lesser preoperative sensory deficit, with pain restricted to trigeminal distributions, and with pain of a burning, lancinating, or penetrating quality. Ataxia from lesioning the spinocerebellar tract is a potential complication.

Lesions of the Cauda Equina

DREZ surgery may relieve pain in patients with injury to the cauda equina, such as gunshot wounds; these operations are associated with minimal morbidity, as roots have already been interrupted.[38] This follows the rule that the

DREZ operation should be considered for lesions proximal to the dorsal root ganglia.

► COMPLEX REGIONAL PAIN SYNDROME AND THE ROLE OF SURGICAL SYMPATHECTOMY

Complex regional pain syndrome (CRPS) denotes a clinical syndrome characterized by pain that is out of proportion to the inciting event (e.g., trauma). Frequently the patient has tactile pain (allodynia) and other manifestations of hyperalgesia (pain to cooling and punctate mechanical stimuli). Other manifestations include swelling, marked skin temperature disparity with the opposite side, abnormal sweating, and nail or hair growth abnormalities. In perhaps a minority of these patients, the sympathetic nervous system maintains the painful state. The empiric designation *sympathetically maintained pain* (SMP) denotes this dependency. All of the pain may be sympathetically maintained, or the patient may have both SMP and *sympathetically independent pain* (SIP). The patient may (CRPS type 2) or may not (CRPS type 1) have an underlying nerve injury.

The surgeon and pain clinician have three vital roles to play in approaching the patient with CRPS:

1. Establish the presence or absence of SMP.
2. Look for alternative diagnoses (e.g., occult nerve entrapments).
3. Palliate (e.g., medical therapy, dorsal column stimulation, PT aimed at avoiding tissue contractures).

The fortunate patient with CRPS has SMP, because SMP can be treated, whereas CRPS by itself has no direct treatment except for "tincture of time" and symptomatic therapy. As a rule all patients with SMP have cooling hyperalgesia.[16] Therefore, at the bedside, the clinician may be alerted to the possibility of SMP by examining the patient to see if there is hyperalgesia to cooling stimuli. Cooling hyperalgesia is a sensitive but nonspecific marker of SMP; in other words, if the patient does not have cooling hyperalgesia, it may not be necessary to pursue the presence of SMP by doing sympathetic blocks. A further important point is that patients may have SMP without having many of the classic features of CRPS, such as edema, skin temperature changes, and trophic skin changes.

Mechanism of SMP

The mechanism of SMP has been elucidated through recent studies. Sato and Perl first determined that nociceptors develop chemical sensitivity to norepinephrine after a nerve injury.[39] Subsequent studies in primates confirmed that intact nociceptors that share the innervation territory of injured nerve develop adrenergic sensitivity. The α-1 adrenergic receptor in primates appears to be the major culprit.[2] These studies indicate that SMP is not a disorder of the sympathetic nervous system *per se*, but rather a receptor disorder, whereby nociceptors acquire abnormal catechol sensitivity. This has been confirmed in human studies that have shown that a physiological concentration of norepinephrine injected into the skin of SMP patients—rendered pain free via a sympathetic ganglion block—evokes pain. Norepinephrine does not cause pain when injected into normal volunteers.[1]

Diagnosis

The distinction between SMP and SIP is made empirically, based on the response to a sympathetic block. To achieve a sympathetic block, local anesthesic is injected along the sympathetic chain. The patient with SMP has pain relief at least for the duration of the anesthetic action. In some cases, one or more sympathetic ganglion blocks may exert a long-term benefit, and, therefore, may be considered therapeutic as well as diagnostic.

Shortcomings of a sympathetic ganglion block include pain from the injection itself, the need for fluoroscopy, and the possibility of both false positives and false negatives. False positives occur because of nonspecific blockade of nearby somatic roots, systemic effects of the local anesthetic, and placebo effects. False negatives occur primarily when the anesthetic fails to reach the target. A systemic infusion of the short-acting α-adrenergic blocking agent phentolamine provides an alternative to a sympathetic ganglion block, and provides a safe, painless means of blocking the sympathetic nervous system with placebo control. Patients are pretreated with propranolol and receive a rapid infusion of phentolamine (in blinded fashion) while in the recumbent position, at a dose sufficient to raise the distal limb temperature to near core temperature. Patients have sensory testing every five minutes to determine relief from hyperalgesia to mechanical and cooling stimuli.[33,34,44]

Surgical Treatment

Patients with SMP who fail to have lasting benefit from sympathetic blocks are candidates for *surgical sympathectomy*. Treatment of upper extremity SMP involves T2-4 sympathectomies via a thoraco-endoscopic approach. SMP in one lower extremity may require bilateral lumbar sympathectomies. Though patients may have protracted postoperative recovery periods, overall favorable results have been noted.[3] Others have had less good fortune with sympathectomies.[24] Our experience indicates good results in carefully selected patients.

Evidence suggests that patients who test positive for SMP are particularly good candidates for treatment with dorsal column stimulation.[17] Performance of a surgical sympathectomy does not preclude use of dorsal column stimulation, and vice versa.

► CANCER PAIN: CORDOTOMY, MIDLINE MYELOTOMY, AND OTHER ABLATIVE TECHNIQUES

Cordotomy is an effective treatment for unilateral pelvic and leg pain due to cancer. By sectioning the anterolateral quadrant of the spinal cord, interruption occurs in the *spinothalamic tract* (STT) with subsequent loss of contralateral pain and temperature sensation. The procedure can be done as a percutaneous radiofrequency ablation at C1–2 or via a laminectomy. Because the STT is organized such that sacral fibers are lateral and the more rostral body parts are represented medially, treatment of pain in the leg is comparably easier than treatment of pain in the arm. In addition, cordotomy appears to be more effective in addressing intermittent shooting pain than steady burning pain.[41] Unfortunately, benefit tends to subside with time, and thus its use in treating chronic pain receives little attention.

Bilateral cordotomy may be performed for bilateral pain. However, high bilateral cordotomies (C1–2) carry the risk of leading to respiratory insufficiency, and thus should not be done. In addition, sexual function and bladder control may be adversely affected by a bilateral cordotomy. On the other hand, a unilateral percutaneous C1–2 cordotomy combined with an open thoracic cordotomy on the contralateral side has been shown to achieve immediate benefit in nearly 100 percent of patients with bilateral pain.[36] This benefit reduced to 37 percent at 5 to 10 years follow-up. However, a long-term success rate of 37 percent is worth considering for some patients with otherwise intractable symptoms.

Intrathecal delivery of opioids (and possibly bupivacaine) augments medical therapies and reduces the need for ablative techniques for cancer pain, particularly for caudal pain. Nevertheless, these medical therapies are not a panacea, and ablative techniques still have a role. Various ablative procedures are also useful in treating cancer pain. As noted above, the DREZ operation may be useful for pain from head and neck cancer, with the advantage that tactile sensibility is preserved. Dorsal root rhizotomies may be beneficial for patients with chest wall pain. It has been hypothesized that malignancies may induce pain through somatic and visceral mechanisms.

Willis and colleagues identified a separate pathway for transmission of visceral nociceptive stimuli in the midline on the dorsal side of the spinal cord in animals.[46] As a result, *midline myelotomy* has been advocated as a way of treating visceral pain associated with cancer. Nauta and colleagues performed punctate midline myelotomies in humans with abdominal/pelvic pain due to cancer,

and reported encouraging results.[29] Despite the need for a thoracic laminectomy, this operation is technically simple, and neurological morbidity is minimal.

Ligation of the thecal sac at S2 is yet another simple procedure that may provide relief in patients suffering from severe pelvic pain with loss of bladder and bowel function.

Deep Brain Stimulation and Motor Cortex Stimulation

Deep brain stimulation (DBS) has yet to find itself in the mainstream of pain treatment options. Moreover, recent interest has shifted to *motor cortex stimulation* (MCX, see below). Nonetheless, favorable results are still periodically reported. Given the large impact that DBS is having on the treatment of movement disorders, it is likely that, in time, DBS may find its niche in pain treatment. The sites for stimulation are typically the periventricular gray and the sensory thalamus (ventroposterolateral thalamus). Nandi et al. reported on successful use of DBS in eight patients with pain of central origin.[26] Kumar et al.[20] reported that 60 percent of patients with failed back syndrome had meaningful benefit with follow-up over two years. Additionally, refractory cluster headache has been successfully treated with stimulation in the posterior hypothalamus.[22]

Cortical Stimulation for Facial Pain, Stroke Pain, and Other Indications

Tsubokawa and colleagues reported the remarkable finding that motor cortex stimulation may relieve pain in neuropathic pain conditions. Since then there have been many publications that have corroborated this finding.[43] The mechanism by which stimulation in this area relieves pain is unclear, but long-term results are impressive in this difficult-to-treat group of patients. Katayama et al.[19] stimulated the spinal cord, thalamus, and motor cortex for treatment of poststroke pain in 45 patients, and demonstrated progressively improved efficacy as stimulation was moved up the neural axis (7%, 25%, and 48%, respectively). Dorsal column stimulation is not an option for treatment of facial pain, and thus the advent of motor cortex stimulation represents a new way to control pain in the facial area as well.[30] Saitoh et al.[37] reported 14 patients with intractable neuropathic pain where electrodes were placed in the interhemispheric fissure for leg pain and within the central sulcus for arm pain. Success was achieved in nine of the patients.

► CONCLUSIONS

Neurosurgical procedures offer effective means to treat acute and chronic pain. A large menu of possible therapeutic interventions exists. Electrical stimulation, CNS drug delivery, ablative procedures, reconstruction, stabilization, and decompression all have a role in the treatment of pain disorders. Knowledge of these options, including their efficacy, morbidity, and technical nuances, is a prerequisite to the modern practice of pain medicine.

REFERENCES

1. Ali Z, Raja SN, Wesselmann U, et al: Intradermal injection of norepinephrine evokes pain in patients with sympathetically-maintained pain. *Pain* 2000;88:161.
2. Ali Z, Ringkamp M, Hartke TV, et al: Uninjured C-fiber nociceptors develop spontaneous activity and alpha adrenergic sensitivity following L6 spinal nerve ligation in the monkey. *J Neurophysiol* 1999;81:455.
3. Bandyk DF, Johnson BL, Kirkpatrick AF, et al: Surgical sympathectomy for reflex sympathetic dystrophy syndromes. *J Vasc Surg* 2002;35:269.
4. Bernard EJ Jr, Nashold BS Jr, Caputi F, et al: Nucleus caudalis DREZ lesions for facial pain. *Br J Neurosurg* 1987;1:81.
5. Bullard DE, Nashold BS Jr: The caudalis DREZ for facial pain. *Stereotact Funct Neurosurg* 1997;68:168.
6. Burchiel KJ, Johans TJ, Ochoa J: The surgical treatment of painful traumatic neuromas. *J Neurosurg* 1993;78:714.
7. Campbell JN: Nerve lesions and the generation of pain. *Muscle Nerve* 2001;24:1261.
8. Campbell JN, Long DM: Peripheral nerve stimulation in the treatment of intractable pain. *J Neurosurg* 1976;45:692.
9. Campbell JN, Solomon CT, James CS: The Hopkins experience with lesions of the dorsal horn (Nashold's operation) for pain from avulsion of the brachial plexus. *Appl Neurophysiol* 1988;51:170.
10. Coggeshall RE, Applebaum ML, Frazen M, et al: Unmyelinated axons in human ventral roots, a possible explanation for the failure of dorsal rhizotomy to relieve pain. *Brain* 1975;98:157.
11. Deer TR, Caraway DL, Kim CK, et al: Clinical experience with intrathecal bupivacaine in combination with opioid for the treatment of chronic pain related to failed back surgery syndrome and metastatic cancer pain of the spine. *Spine J* 2002;2:274.
12. Edgar RE, Best LG, Quail PA, et al: Computer-assisted DREZ microcoagulation: post-traumatic spinal deafferentation pain. *J Spinal Disord* 1993;6:48.
13. Emery E, Blondet E, Mertens P, et al: Microsurgical DREZotomy for pain due to brachial plexus avulsion: long-term results in a series of 37 patients. *Stereotact Funct Neurosurg* 1997;68:155.
14. Falci S, Best L, Bayles R, et al: Dorsal root entry zone microcoagulation for spinal cord injury-related central pain: operative intramedullary electrophysiological guidance and clinical outcome. *J Neurosurg* 2002;97:193.
15. Friedman AH, Bullitt E: Dorsal root entry zone lesions in the treatment of pain following brachial plexus avulsion, spinal cord injury and herpes zoster. *Appl Neurophysiol* 1988;51:164.
16. Frost SA, Raja SN, Campbell JN, et al: Does hyperalgesia to cooling stimuli characterize patients with sympathetically maintained pain (reflex sympathetic dystrophy)?, in Dubner R, Gebhart GF, Bond MR (eds), *Proceedings of the Vth World Congress on Pain*. Amsterdam, Elsevier Science Publishers BV, 1988;151.
17. Hord ED, Cohen SP, Cosgrove GR, et al: The predictive value of sympathetic block for the success of spinal cord stimulation. *Neurosurgery* 2003;53:626.
18. Hosobuchi Y: The majority of unmyelinated afferent axons in human ventral roots probably conduct pain. *Pain* 1980;8:167.
19. Katayama Y, Yamamoto T, Kobayashi K, et al: Motor cortex stimulation for post-stroke pain: comparison of spinal cord and thalamic stimulation. *Stereotact Funct Neurosurg* 2001;77:183.
20. Kumar K, Toth C, Nath RK: Deep brain stimulation for intractable pain: a 15-year experience. *Neurosurgery* 1997;40:736.
21. Lancelotta MP, Sheth RN, Meyer RA, et al: Severity and duration of hyperalgesia in rat varies with type of nerve lesion. *Neurosurgery* 2003;53:1200.
22. Leone M, Franzini A, Broggi G, et al: Hypothalamic deep brain stimulation for intractable chronic cluster headache: a 3-year follow-up. *Neurolog Sci* 2003;24 Supplement 2:S143–S145.
23. Lozano AM, Vanderlinden G, Bachoo R, et al: Microsurgical C-2 ganglionectomy for chronic intractable occipital pain. *J Neurosurg* 1998;89:359.
24. Mailis A, Furlan A: Sympathectomy for neuropathic pain. *Cochrane Database of Systematic Reviews* 2003;CD002918.
25. Moossy JJ, Nashold BS Jr, Osborne D, et al: Conus medullaris nerve root avulsions. *J Neurosurg* 1987;66:835.
26. Nandi D, Aziz T, Carter H, et al: Thalamic field potentials in chronic central pain treated by periventricular gray stimulation—a series of eight cases. *Pain* 2003;101:97.
27. Nashold BS Jr: Current status of the DREZ operation: 1984. *Neurosurgery* 1984;15:942.
28. Nashold BS, Nashold J: The DREZ operation, in Tindall GT, Cooper PR, Barrow DL (eds), *The Practice of Neurosurgery*. Baltimore: Williams and Wilkins, 1996;p 3129.
29. Nauta HJ, Soukup VM, Fabian RH, et al: Punctate midline myelotomy for the relief of visceral cancer pain. *J Neurosurg* 2000;92:125.
30. Nguyen JP, Keravel Y, Feve A, et al: Treatment of deafferentation pain by chronic stimulation of the motor cortex: report of a series of 20 cases. *Acta Neurochirurg Supp (Wien)* 1997;68:54.
31. North RB, Wetzel FT: Spinal cord stimulation for chronic pain of spinal origin: a valuable long-term solution. *Spine* 2002;27:2584.
32. Novak CB, Mackinnon SE: Outcome following implantation of a peripheral nerve stimulator in patients with chronic nerve pain. *Plast Reconstr Surg* 2000;105:1967.
33. Raja SN, Treede R-D, Davis KD, et al: Systemic alpha-adrenergic blockade with phentolamine: a diagnostic test for sympathetically maintained pain. *Anesthesiology* 1991;74:691.
34. Raja SN, Turnquist JL, Meleka S, et al: Monitoring adequacy of α-adrenoceptor blockade following systemic phentolamine administration. *Pain* 1996;64:197.

35. Rogers ML, Henderson L, Mahajan RP, et al: Preliminary findings in the neurophysiological assessment of intercostal nerve injury during thoracotomy. *Eur J Cardiothorac Surg* 2002;21:298.
36. Rosomoff HL, Papo I, Loser JH: Neurosurgical operations on the spinal cord, in Bonica J (ed), *The Management of Pain,* 2nd ed. Philadelphia: Lea and Febiger, 1990:2067.
37. Saitoh Y, Kato A, Ninomiya H, et al: Primary motor cortex stimulation within the central sulcus for treating deafferentation pain. *Acta Neurochirurg Suppl* 2003;87:149.
38. Sampson JH, Cashman RE, Nashold BS Jr, et al: Dorsal root entry zone lesions for intractable pain after trauma to the conus medullaris and cauda equina. *J Neurosurg* 1995; 82:28.
39. Sato J, Perl ER: Adrenergic excitation of cutaneous pain receptors induced by peripheral nerve injury. *Science* 1991;251:1608.
40. Shin HK, Kim J, Nam SC, et al: Spinal entry route for ventral root afferent fibers in the cat. *Exp Neurol* 1986;94:714.
41. Tasker RR, DeCarvalho GT, Dolan EJ: Intractable pain of spinal cord origin: clinical features and implications for surgery. *J Neurosurg* 1992;77:373.
42. Taub A, Robinson F, Taub E: Dorsal root ganglionectomy for intractable monoradicular sciatica: A series of 61 patients. *Stereotact Funct Neurosurg* 1995;65:106.
43. Tsubokawa T, Katayama Y, Yamamoto T, et al: Chronic motor cortex stimulation for the treatment of central pain. *Acta Neurochirurg Suppl (Wien)* 1991;52:137.
44. Wehnert Y, Muller B, Larsen B, et al: Sympathetically-maintained pain (SMP): phentolamine test vs sympathetic nerve blockade: Comparison of two diagnostic methods. *Orthopade* 2002;31:1076.
45. Wilkinson HA, Chan AS: Sensory ganglionectomy: theory, technical aspects, and clinical experience. *J Neurosurg* 2001;95:61.
46. Willis WD, Al Chaer ED, Quast MJ, et al: A visceral pain pathway in the dorsal column of the spinal cord. *Proc Nat Acad Sci USA* 1999;96:7675.

CHAPTER 45

Principles of Palliative Care for the Pain Physician

Karla S. Hayes and E. Daniela Hord

Palliative medicine has developed into a recognized discipline that is patient-focused, family-oriented, and relationship-centered. The goal is to enhance patient and family quality of life while minimizing suffering throughout the duration of a life-challenging illness. This chapter focuses on palliative care of terminally ill patients, with particular emphasis on cancer pain. Pain is a common symptom in cancer patients, and numerous pain syndromes have been described in this population. Among cancer patients, approximately 30 percent present with pain at the time of diagnosis, but the incidence increases up to 90 percent during the terminal phase of the illness. Unrelieved pain can be incapacitating. It can preclude a satisfying quality of life when it gets to the point where it interferes with activities of daily living. Persistent pain interferes with the ability to eat, sleep, think, and interact with others. It is also strongly associated with heightened psychological distress.[1] Pain management is just part of a comprehensive plan needed to address the many factors affecting the quality of a cancer patient's life. Palliative care seeks to optimize quality of life throughout the course of an illness by carefully attending to the many physical, spiritual, and psychosocial needs of the patient and family.

▶ SPECIFIC ASPECTS OF PALLIATIVE CARE

Symptom Management

Control of pain and other physical symptoms, of psychological issues, and of existential distress are the major goals of palliative care. Pain usually does not occur in isolation, particularly in patients with advanced disease. In studies of cancer and AIDS populations, multiple symptoms have always been found.[2,3] Portenoy and colleagues reported that among 246 patients receiving active treatment, independent of tumor type, 40 to 80 percent experienced pain, drowsiness, dry mouth, insomnia, lack of energy, or psychological distress. The mean number of symptoms was 11.5 per patient.[2] Multiple surveys of cancer patients have found a high prevalence of pain, fatigue, weakness, dyspnea, delirium, nausea, vomiting, adjustment disorders, depression, and anxiety. Other symptoms to be aware of are constipation, bowel or bladder incontinence, urinary retention, dysphagia, anorexia, pruritis, and stomatitis. The frequency and severity of these symptoms tends to increase as the illness progresses. Treatment of all physical symptoms is necessary for the overall comfort of the patient. Unfortunately, too many patients suffer from unrelieved symptoms: Patients and families may accept symptoms as inevitable and, therefore, not complain; physicians may fail to ask about specific symptoms, especially those that are not readily apparent, such as insomnia, confusion, or constipation; and even when symptoms are recognized, one or more often become inadequately treated.

Care Guidelines

A recent study identified five domains that, from the patient's perspective, are considered critical to assuring quality end-of-life care: (1) receiving adequate pain and symptom management; (2) avoiding inappropriate prolongation of dying; (3) achieving a sense of control; (4) relieving burden; (5) and strengthening relationships with loved ones.[4] Palliative care is not simply a matter of controlling pain or other physical symptoms; competent care of the dying patient requires a dedicated multidisciplinary team with expertise in the delivery of palliative care services, some of which will include addressing spiritual needs and psychosocial issues.

Suffering

In addition to meticulous attention to symptom assessment and management, it is very important to appreciate that the suffering of the terminally ill patient is distinct from his or her disease or other distressing symptoms. Suffering is a complex phenomenon caused by a variety of problems, including pain and other factors such as physical symptoms, psychological distress, lack of family or social support, limited financial resources, or lack of function. Suffering was defined by Cassel[5] as "the state of severe distress associated with events that threaten the intactness of the person." It is extremely important to address suffering and related phenomena when assessing pain; treating only pain might be unsuccessful in a patient whose suffering is caused predominantly by other factors.

Psychosocial Care

In palliative medicine, psychosocial issues are considered to be just as important as physical symptoms: In fact, the psychological and social stressors that confront cancer patients and their families can be even more significant than the physical symptoms. Pain is not simply related to tissue damage. Pain is a complex of subtle multidimensional processes that involve perception, nociception, perceived or ascribed meaning, effects, somatic variables, and learned sociocultural responses.[6,7] The study of psychological factors contributing to the etiology and treatment of cancer pain dates as far back as the 1950s.

A vital aspect of successful cancer pain management is being able to understand the patient's perspective. It is important to make time to ask about the patient's beliefs and expectations at the beginning of the treatment process, and then to review these attitudes at appropriate intervals, since they will likely be dynamic in nature. Despite advances in cancer treatment, patients and family members still tend to view cancer as a death sentence and they often react with hopelessness and despair when given the diagnosis. Despite the significant distress experienced at the time of diagnosis, the majority of patients gradually adjust during the six-month period following the diagnosis.[8] One of the best predictors of positive adaptation is the psychological state of the patient prior to the initiation of the therapeutic process.[9] Brief screening techniques that identify depression, anxiety, and other signs of distress should be incorporated into the care of the patient. Patients who are identified as having high levels of distress should be referred to a mental health professional for treatment. For some patients, ongoing mental health services are necessary, whereas other patients will require assistance only at critical points during their illness. Clinical practice suggests that virtually all patients can benefit from some type of psychosocial intervention during the illness process, especially near the end of life.

Spiritual Care

Patients will try to find ways of maintaining their *sense of self* when they are dying. Spiritual pain describes the anguish that can occur when patients cannot come to terms with their illness. Spiritual care is based on accepting each patient or family member and their attitudes, hopes, and fears. Spiritual needs are universal and distinct from religious beliefs. The spiritual dimension is the realm in which people search for meaning in their lives, and where they develop their own beliefs and values. It is influenced by a person's life experiences, as well as cultural factors. People who cope best with their illness and the associated suffering seem to do so by finding some sort of meaning in the suffering. Spirituality can assist a patient in finding hope and meaning in the midst of despair. Naturally, the spiritual needs of each patient will vary. Although some needs are common to all, the recognition of an individual's particular needs is essential for effective care.

Cancer Pain

The most common pain syndromes associated with cancer are listed in Table 45-1, and can be categorized as acute versus chronic, and neurological versus non-neurological. However, there is a large variety of chronic pain syndromes described in cancer patients including, but not limited to, the syndromes listed in the table. The etiology of the pain in most cases is direct tumor involvement. Examples include compression of neural structures, bone invasion, vascular obstruction, and mucous membrane ulceration. Cancer-induced syndromes such as paraneoplastic syndromes, postherpetic neuralgia, GI/GU spasms or dysfunction, and bedsores, account for less than 10 percent of the pain reported in cancer patients. Acute pain caused by diagnostic or therapeutic procedures accounts for 20 percent of the pain reported, which would include postoperative pain as well as side

TABLE 45-1. PAIN SYNDROMES ASSOCIATED WITH CANCER

Acute neurological pain syndromes	Post-lumbar puncture headache
	Opioid headache
	Spinal opioid hyperalgesia syndrome
	Epidural injection pain
	Chemotherapy-induced painful peripheral neuropathy
	Chemotherapy-induced headache (post intrathecal MTX, post-systemic L-asparaginase, post-transretinoic acid therapy)
	Post-radiation therapy early-induced brachial plexopathy
	Acute herpetic neuralgia
	Subacute radiation myelopathy
	Headache from increased intracranial pressure
Acute non-neurological pain syndromes	Postoperative pain
	Injection pain
	Chemotherapy intravenous infusion-induced pain
	Hepatic artery chemotherapy-induced pain
	Intraperitoneal chemotherapy pain
	Mucositis
	Acute radiation enteritis and proctocolitis
	Immunotherapy-related pain
	Procedure-related pain (bone marrow biopsy, paracentesis, thoracentesis, pleurodesis, tumor embolization)
	Pathologic fractures pain
	Acute obstruction of bowel or perforation
Chronic neurological pain syndromes	Atlanto-axial destruction
	Odontoid fractures
	Epidural compression
	Headache related to intracerebral tumors, leptomeningeal metastases, or base of skull metastases
	Facial pain related to glossopharingeal neuralgia or trigeminal neuralgia
	Tumor-related plexopathy
	Tumor-related PHN
	Paraneoplastic painful peripheral neuropathy
	Post-chemotherapy painful peripheral neuropathy
	Plexopathy associated with intra-arterial infusion
	Radiation-induced plexopathy
	Chronic radiation myelopathy
	Postsurgical pain syndromes (postmastectomy syndrome, post-thoracotony syndrome, postradical neck dissection syndrome, postnephrectomy syndrome, stump pain and phantom pain)
Chronic non-neurological pain syndromes	Multifocal bone pain related to multiple bony mets
	Pain caused by tumor invasion of the joints or soft tissue
	Hepatic distension pain syndrome
	Midline retroperitoneal pain syndrome
	Chronic intestinal obstruction pain
	Peritoneal carcinomatosis pain
	Malignant pelvic floor myalgia
	Ureteric obstruction
	Postchemotherapy avascular necrosis of femoral or humeral head
	Painful gynecomastia post hormonal therapy for prostate cancer
	Postmastectomy pain syndrome
	Postradical neck dissection pain
	Post-thoracotomy pain
	Postop frozen shoulder
	Phantom pain syndromes
	Stump pain
	Postsurgical pelvic floor myalgia
	Radiation-induced chronic pelvic pain
	Painful lymphedema

effects from chemotherapy or radiation treatment. Pain that is unrelated to the malignancy or its treatment accounts for less than 10 percent of pain complaints.[10]

▶ PAIN MANAGEMENT

Assessment

A comprehensive assessment of cancer pain is the first important step toward optimal pain relief. As mentioned earlier in this chapter, the undertreatment of cancer pain continues to be a common problem. Rates of undertreatment may be as high as 40 percent.[11] The undertreatment of cancer pain can be attributed to the combined effects of deficiencies in clinician practice, patient underreporting and noncompliance, and system-wide impediments to optimal analgesic therapy.[12] Physicians can improve this situation by continually reassuring the patients about the safety of analgesic therapy and by encouraging pain reporting by those who may otherwise minimize symptoms because of stoicism, a desire to be liked, or a concern about distracting the physician from the disease.[13]

Undertreatment is also likely if patients with persistent pain do not receive analgesics at all. For example, in a recent study on institutionalized elderly patients, 26 percent of patients with daily pain were not treated.[14] Lack of treatment was associated with age greater than 85 years, minority race, impaired cognition, and the need for multiple other medications.

Inadequate assessment is thought to be one of the primary causes of this problem. In a study to evaluate the correlation between patient and clinician evaluation of pain severity, Grossman et al. found that when patients rated their pain as moderate to severe, oncology fellows failed to appreciate the severity of the pain in 73 percent of cases.[15]

A pain assessment should provide the clinician with sufficient information to estimate the severity of pain, form a clinical impression regarding the etiology of the pain, determine the need for further diagnostic studies, and formulate therapeutic recommendations that take into account the patient's overall medical and psychosocial status. The special challenges associated with the assessment of cancer pain include the entirely subjective nature of the pain, the complex multisystem involvement in patients with advanced malignancies, and the ever-changing clinical situation in this patient population.[16] A proper assessment includes not only an accurate investigation of the pain itself, but also an evaluation of the impact of the pain on the patient's life. Cancer patients frequently have more than one pain, so each should be evaluated separately. The assessment should review the pain characteristics of quality, intensity, distribution, and temporal relationships. The quality of the pain often suggests its pathophysiology. The intensity and quality of pain may help characterize the pain mechanism and underlying syndrome. The intensity should be measured using validated pain assessment scales and results should be serially documented in the patient's medical record.

Pain Mechanisms

It may be useful to distinguish a patient's pain as being *nociceptive* versus *neuropathic*.

Nociceptive pain is that which is perceived to originate from tissue damage in a somatic or visceral structure. Nociceptive pain that arises from somatic structures is usually well localized and described as *sharp, aching,* or *throbbing*. Visceral pain, on the other hand, tends to be more diffuse, with its character depending on the involved visceral structure. Nociceptive pains, especially somatic pains, usually respond well to opioid analgesics or to neurolytic blocks.

Neuropathic pain is due to abnormalities of the peripheral or central nervous system. It is often sustained by aberrant somatosensory processing either at the periphery, the dorsal root ganglion, or in the brain. Neuropathic pain can be described as *burning, pins and needles, lancinating,* or an uncomfortable sense of *numbness*. The response of neuropathic pain to opioid analgesics is less predictable than for nociceptive pain. The usual treatment adds tricyclic antidepressants or antiepileptic medications to opioid analgesics. Sympathetic nerve blocks have also been shown to be useful in some cases.[17]

Treatment

In advanced disease, opioids are the mainstay of pain management. Nearly 85 percent of patients with cancer pain can be well controlled with conventional oral medications. Numerous studies have demonstrated that opioid-based pharmacotherapy administered according to simple guidelines can provide adequate pain relief to more than three-quarters of patients with cancer pain.[18–21] Although this favorable experience does not eliminate the need for NSAIDs and adjuvant analgesics, as well as numerous other analgesic strategies, it has led to a consensus view that pharmacotherapy should be the main approach to the management of cancer pain, and that opioid therapy is the first-line if pain is moderate or severe.[22,23] The long-term effectiveness and safety of opioid therapy justifies a parallel approach to the management of chronic pain associated with other progressive incurable illnesses. Side effects of opioids should be aggressively treated.

Nevertheless, interventional pain procedures should be seriously considered when the pain does not respond to opioids, or intolerable side effects limit the use of opioids. The most common interventional approaches are antineoplastic therapy, neurolytic blocks, neurostimulatory techniques, intrathecal pumps and cordotomy. These more invasive therapies should provide pain relief to an additional 10 percent of patients, leaving only a small fraction of cancer patients with inadequate relief.

Crescendo pain in cancer patients, as well as in all other types of palliative care patients, should be managed as an emergency, with rapid escalating doses of parenteral opioids—patient-controlled analgesia being an efficient method of administration. Methadone should be considered as a therapeutic option. Close monitoring is advisable when titrating opioids rapidly and when practicing opioid rotation.[24,25]

Treatment of chronic cancer pain should follow the WHO guidelines, with some specific issues. Malignant bone pain refractory to treatment with a combination of opioid and NSAID can be managed with radiation therapy if focal or if associated with impending fracture. If the pain is multifocal, bisphosphonate compounds[26,27] or calcitonin[28,29] should be considered. A new treatment option to relieve pain in patients with osteoblastic metastatic lesions is radiopharmaceutical treatment; Samarium SM-153 lexidronan, is administered as a single injection, with the objective of targeting bone lesions due to metastatic cancer.

The Need for Alternative Routes of Opioid Delivery

Because of asthenia, dysphagia, muscle weakness, nausea and vomiting, the oral route often is not possible for patients with advanced disease.[30] There are several other routes for the delivery of opioids, allowing management at home for most of the patients.

The transdermal route for pain control is available for fentanyl. The fentanyl patch offers a 72-hour dosing interval, and is most useful if the pain syndrome is relatively stable. When compared to controlled-release morphine in cancer patients, the transdermal fentanyl patch was found to provide greater overall satisfaction,[31] and fewer significant side effects.[32] The main disadvantage is that the smallest patch (25 µg per hour) is roughly equivalent to 75 mg oral morphine in 24 hours, so patients who require smaller doses should not be started on this patch.

The subcutaneous route for pain control is usually considered when a terminally ill patient is treated at home, or the peripheral vein access is difficult. A 25-gauge butterfly needle is used, which is usually changed every week. The bioavailability of the opioids delivered subcutaneously is similar to that for intravenous opioids.[33] Patient-controlled analgesia, with or without a basal rate, can also be used. The transmucosal route is available for fentanyl (Actiq) as lozenges, and is recommended when faster onset of action is needed or if nausea is limiting oral intake of other types of opioids. Some opioids, such as oxycodone 20 mg/ml (Oxyfast), have good sublingual absorption.

The rectal route allows effective absorption of several opioids. Few opioids are available as suppositories, but the pharmacist can prepare the desired opioid of the specified strength.

Summary

Preliminary pain assessment data are obtained, including a brief history, physical examination, laboratory data, and pain intensity information. Patient comfort is maximized with prompt administration of analgesics. This facilitates the evaluation process. The World Health Organization's analgesic ladder provides a framework for analgesic prescription:

1. Mild pain is usually treated with nonopioids such as aspirin, acetaminophen, and/or nonsteroidal anti-inflammatory drugs (NSAIDS) while the opioid analgesics are used for moderate to severe pain.
2. The oral route of administration is preferred for virtually all patients.
3. Analgesics with short half-lives are used to facilitate rapid dose escalation and prompt relief of pain.
4. Analgesics are given *around the clock* rather than *as needed*.
5. Once baseline opioid requirements are determined, sustained-release opioid preparations can be used to reduce the number of pills taken each day.
6. PRN medications are available for breakthrough pain.
7. A prophylactic bowel regimen is initiated with all opioid therapy. Nausea and vomiting are treated with aggressive antiemetic therapy. These may become less prominent after the first several days of opioid therapy.
8. The dose of each analgesic is maximized before considering changes in drugs or routes of administration.
9. Changes in analgesics or routes of administration are generally based on the development of significant toxicities that are not responsive to usual measures, rather than the total administered dose.
10. Equianalgesic tables are used to approximate opioid dose conversions.
11. Serial pain intensity ratings are obtained and documented in the medical record. A general goal is to get the patient below a 50 percent level of pain.

► CONCLUSION

Given that cancer pain—and, by extrapolation, pain in terminally ill patients—can be adequately managed in almost all circumstances, there is no reason why its deleterious impact on the patient's physical, psychological, and spiritual resources should be allowed to be an ongoing issue. Pain management in these patients should be based on the principles of palliative care, in conjunction with multisymptom management.

There are numerous negative consequences of unrelieved pain, but no known benefits. If the primary treating physician or consultant does not feel comfortable treating a cancer patient's pain, then it is imperative that they refer the patient to a pain management specialist.

REFERENCES

1. Ferrell BR: The impact of pain on quality of life: A decade of research. *Nurs Clin North Am* 1995;30:609–24.
2. Portenoy RK, Thaler HT, Kornblith AB, et al: Symptom prevalence, characteristics and distress in a cancer population. *Qual Life Res* 1994;3:183–189.
3. Vogl D, Rosenfeld B, Breitbart W, et al: Symptom prevalence, characteristics, and distress in AIDS outpatients. *J Pain Symptom Manage* 1999;18:253–262.
4. Singer PA, Martin DK, Kelner M: Quality end-of-life care: patients' perspectives. *JAMA* 1999;281:163–168.
5. Cassel EJ: The nature of suffering and the goals of medicine. *N Engl J Med* 1982;75:639–645.
6. Cherny NI, Coyle N, Foley KM: Suffering in the advanced cancer patient: a definition and taxonomy. *J Palliat Care* 1994;10:57–70.
7. Melzack R, Wall PD: Pain mechanisms: a theory. *Science* 1965;150:971.
8. Weisman AD, Worden JW: The existential plight in cancer: significance for the first 100 days. *Int J Psychiatry Med* 1977;7:1.
9. Carlsson M, Mamrin E: Psychological and psychosocial aspects of breast cancer treatments. *Cancer Nurs* 1994;17:418.
10. Abeloff MD: *Clinical Oncology*, 2nd edition. London: Churchill Livingstone, 2000:539–40.
11. Cleeland CS, Gonin R, Hatfield AK, et al: Pain and its treatment in outpatients with metastatic cancer. *N Engl J Med* 1994;330;29:73–83.
12. Pargeon KL, Hailey BJ: Barriers to effective cancer pain management: a review of the literature. *J Pain Symptom Manage* 1999;18:358–368.
13. Ward SE, Goldberg N, Miller-McCauley V, et al: Patient-related barriers to management of cancer. *Pain* 1993;52: 319–324.
14. Bernabei R, Gambassi G, Lapane K, et al: Management of pain in elderly patients with cancer. *JAMA* 1998; 279:1877–1882.
15. Grossman SA, Sheidler VR, et al: Correlation of patient and caregiver ratings of cancer pain. *J Pain Symptom Manage* 1991;6:53–57.
16. Portenoy RK, Coyle N: Controversies in the long-term management of analgesic therapy in patients with advanced cancer. *J Polliat Care* 1991;7(2):13–24.
17. Cherny NI, Arbit E, Jain S: Invasive techniques in the management of cancer pain. *Hematol Oncol Clin North Am* 1996;10:121–137.
18. Schug SA, Zech D, Dorr U: Cancer pain management according to WHO analgesic guidelines. *J Pain Symptom Manage* 1990;5:27–32.
19. Takeda F: Results of field testing in Japan of the WHO draft interim guidelines on relief of cancer pain. *Pain Clinic* 1986;1:83–89.
20. Ventafridda V, Tamburini M, Caraceni A, et al: A validation study of the WHO method for cancer pain relief. *Cancer* 1987;59:850–856.
21. Grond S, Zech D, Schug SA, et al: Validation of World Health Organization guidelines for cancer pain relief during the last days and hours of life. *J Pain Symptom Manage* 1991;6:411–412.
22. World Health Organization: *Cancer Pain Relief*, 2nd edition, with a Guide to Opioid Availability. Geneva: World Health Organization, 1996.
23. Jacox A, Carr DB, Payne R: *Management of Cancer Pain*. AHCPR Publication No. 94-0592: Clinical Practice Guideline No. 9. Rockville, MD, U.S. Department of Health and Human Services, Public Health Service, March 1994.
24. Thomas Z, Bruera E: Use of methadone in a highly tolerant patient receiving parenteral hydromorphone. *J Pain Symptom Manage* 1995;10:315–317.
25. Hagen NA, Elmwood T, Scott E: Cancer pain emergencies: A protocol for management. *J Pain Symptom Manage* 1997;14: 45–50.
26. Ernst DS, MacDonald RN, Paterson AH, et al: A double blind crossover trial of intravenous clodronate in metastatic bone pain. *J Pain Symptom Manage* 1992;7:4–11.
27. Hortobagyi GN, Theriault RL, Porter L, et al: Efficacy of pamidronate in reducing skeletal complications in patients with breast cancer and lytic metastases. *N Engl J Med* 1996;335:1785–1791.
28. Roth A, Kolaric K: Analgesic activity of calcitonin in patients with painful osteolytic metastasis of breast cancer. Results of a randomized study. *Oncology* 1986;43:283–287.
29. Serdengecti S, Serdengecti K, Derman, et al: Salmon calcitonin in the treatment of bone metastases. *Int J Clin Pharmacol Res* 1986;6:151–155.
30. Bruera E, Higginson I, Neuman CM: Palliative care, in Loeser LD (ed), *Bonica's Management of Pain, 3rd ed.* Philadelphia: Lippincott, Williams & Wilkins, 2001:754–762.
31. Ahmedzai S, Brooks D: Transdermal fentanyl versus sustained-release oral morphine in cancer pain: Preference, efficacy, and quality of life. The TTS-Fentanyl Comparative Trial Group. *J Pain Symptom Manage* 1997;13:254–261.
32. Payne R, Mathias SD, Pasta DJ, et al: Quality of life and cancer pain: satisfaction and side effects with transdermal fentanyl versus oral morphine. *J Clin Oncol* 1998; 16:1588–1593.
33. Moulin DE, Kreeft JH, Murray-Parsons N, et al: Comparison of continuous subcutaneous and intravenous hydromorphone infusions for the management of cancer pain. *Lancet* 1991;337:465–468.

CHAPTER 46

Physical Therapy and Rehabilitation

Maureen J. Simmonds

Pain and physical dysfunction are central to rehabilitation. Both are complex multidimensional problems and the human element adds to the complexity. Research in the last couple of decades has led to a more complete and comprehensive understanding of the factors that influence pain and the impact of pain. This knowledge has had a profound influence on rehabilitation as the traditional practice-led, biomedical, disease-based approach has shifted toward one that is person- (patient or client) centered, best-evidence, activity driven and biopsychosocially based.

Regardless of the practice environment (i.e. primary, tertiary, or community) and the specific medical condition (e.g., musculoskeletal or neurological) of the patient, the best-evidence model of rehabilitation practice is one in which a knowledgeable and empathetic practitioner employs a holistic approach, empowers the patient, and bases the assessment and management on sound science. Essentially the approach recognizes and values the art (e.g., the charisma and empathy of the therapist and the therapeutic relationship), as well as the science (e.g., mechanisms) of the therapeutic intervention. The model also recognizes the individual nature, the past experiences and the future expectations of patients. Thus patients' beliefs and bodies are assessed and managed.

A best-evidence rehabilitation model utilizes standardized assessment and outcome measures to guide, refine, and prove the effectiveness of specific types of interventions. Activity and exercise are central components of this model, which promotes general health and well-being as well as improving function. Finally, a best-evidence model encourages and facilitates patient self-management and is time-limited. This chapter presents and critically discusses (1) a personcentered, best-evidence, and activity-driven biopsychosocial model of rehabilitation for patients with pain; (2) assessment methods and measures; and (3) principles of management.

▶ MODELS OF REHABILITATION

In the last decade, a number of theories and lines of evidence have converged and resulted in the current person-centered, best-evidence and activity-driven biopsychosocial model of rehabilitation. Historically, the biomedical disease model formed the basic construct on which most physical therapies and rehabilitation were based. This traditional disease-based model provided a useful construct that explained many of the signs and symptoms that patients with specific diseases presented with. However, the model did not adequately explain the wide variation in symptom presentation (e.g., pain) and explained less about the impact of symptoms (e.g., disability), especially in chronic conditions. Acknowledgement of the limitations of the biomedical model of disease and an understanding of the strong influence of psychological (e.g., fear of injury and activity) and social factors (e.g., employment

Figure 46-1. Patient-centered Biopsychosocial Model.

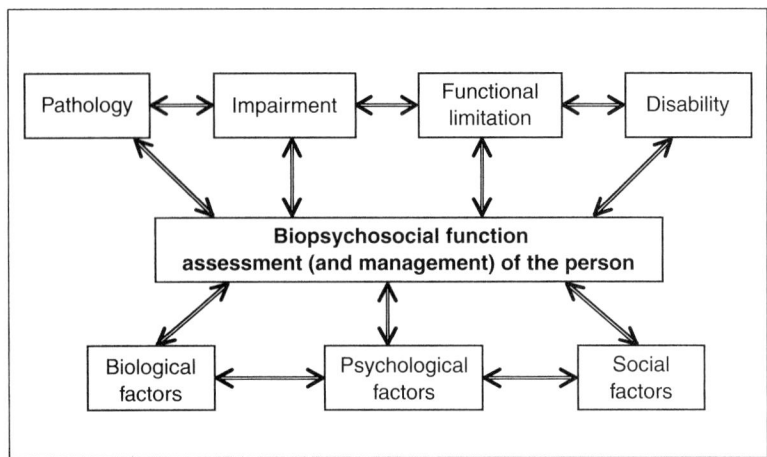

difficulties) on symptoms and symptom impact (e.g., distress and dysfunction) has led to an expanded model of practice, the biopsychosocial model.[1-3]

Application of this expanded conceptual model is exemplified by a holistic approach. For example, the use of psychosocial screening questions, questionnaires, and functional tests as part of the assessment, and the use of cognitive-behavioral principles, exercise quotas, or graded exposure to activities are part of the intervention.

A second conceptual model that provides a practice framework for rehabilitation is the *disablement model*.[4] Like the previous model, the disablement model has expanded to incorporate the improved understanding of pain and disability phenomena. The disablement model is based on the World Health Organization (WHO) disease model.[5] The latter model illustrates a progression from pathology through impairment to disability. The WHO defines *impairment* as any loss or abnormality of psychological, physiological, or anatomical structure or function (e.g., decreased range of motion or strength). *Disability* is defined by WHO as any restriction or lack (resulting from an impairment) of ability to perform an activity in the manner or within the range considered normal (e.g., the inability to work). Nagi[6] recognized the need for a concept that bridged impairment and disability, and proposed the term *functional limitation*. *Functional limitation* is the compromised ability to perform activities of daily living (ADLs). Nagi's model is as follows.

Pathology → Impairment
→ Functional Limitation → Disability

A shortcoming of this model is that the implied unidirectional and linear progression from pathology to disability is not accurate. Rather, each of the constructs is complex and is influenced by a myriad of factors. For example, a similar back injury (pathology) can be a minor inconvenience for someone with good coping skills and a *flexible*, relatively sedentary, work environment (minimal disability). The same type of injury can lead to a downward spiral of distress and disability for an individual who is very anxious about the injury, catastrophizes inappropriately about the implications of the injury, and who has a heavy manual occupation with few employment options (major disability). The second shortcoming of Nagi's model is the implied unidirectional progression from pathology to disability. Although it is difficult to separate pathological processes (e.g., arthritis) from pathological consequences (disability), there is a clear bi-directional influence. For example, joint stiffness and muscle weakness were long thought to be disease expressions of osteoarthritis—the condition. And, obesity was thought to be a predictor of osteoarthritis or at least a co-morbid problem. It is now known that these problems are consequences of inactivity (which may be secondary to arthritis)[7] rather than part of the disease process per se. Inactivity may also be due to inadequately managed pain and inaccurate beliefs about the harmful effects of exercise on arthritic joints. Research has shown that judicious exercise and activity are not harmful to joints but actually promote health and wellness of the person with the arthritic joints.[8,9] Figures 46-1 and 46-2 illustrate the models of rehabilitation discussed above with examples of their application in practice.

▶ ASSESSMENT AND OUTCOME

Central to the management of patients with pain is education and activity; both components start and are interwoven with the initial interview and assessment. However, a summary of key education and activity aspects should be discussed with the patient as a separate *stand alone* activity to enable full discussion and ensure the patient's full understanding of the message.

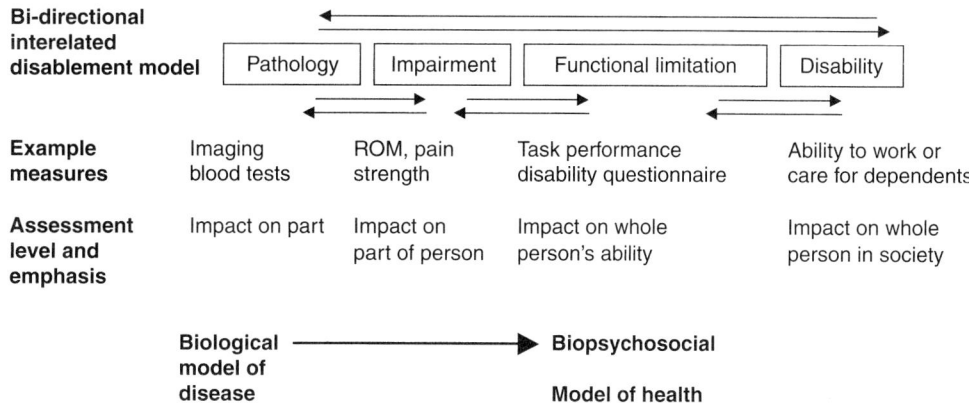

Figure 46-2. Disablement Model with examples of assessment tests and their level of emphasis.

Clinical Interview: Understanding the Patient with Pain and Movement Dysfunction

Regardless of the specific disease and the type of facility (in- or out-patient and primary or tertiary care), pain and movement dysfunction are the most common reasons for physical therapy referral.[10] Optimum management is predicated on a sound and comprehensive understanding of each of the problems, the relationships between the problems, the impact of the problems on the person, and the impact of the person on the problems. In a patient-centered approach, it is vital to understand the patient's beliefs about the problem and its potential impact, and also how they believe it should be managed (e.g., rest or exercise).

The clinical interview provides the first opportunity to establish a supportive but empowering therapeutic relationship. The quality of this relationship is important. Probably the biggest single influence on treatment outcome is the physical therapist's ability to engage, educate, and empower the patient to adopt and maintain a healthy active lifestyle. In a recent study,[33] treatment outcomes were used to identify expert physical therapists. Therapists classified as expert had a patient-centered approach to care, characterized by collaborative clinical reasoning and promotion of patient empowerment. Not surprisingly, years of experience did not distinguish expertise.

Individuals with pain will have a variety of personal theories for their pain that are shaped by personal experience, culture, and education. The therapist must elicit the patients' theories during the interview and help correct them if appropriate. This may be difficult because complex and erroneous beliefs about pain—that may include notions of blame and responsibility—are frequently difficult to change.

Erroneous beliefs and inappropriate pain behaviors are often shaped in part by medical practitioners who have provided erroneous and/or incomplete explanations or advice.[12] Explanations from an authority figure, imparted with an air of certainty, are difficult to displace and interfere with subsequent treatment. For example, patients told that they have a "disintegrating spine" and that they "should rest or avoid bending" will understandably have anxieties about embarking on a treatment program that involves activity and flexibility exercises. Such apparently contradictory medical advice can add to feelings of distress and anxiety. Likewise, unhelpful and perhaps exaggerated pain behaviors can result from patients believing they have to *prove* they have a problem. This can result from medical practitioners implying that there is nothing wrong with the patient because the "tests were normal" or the "injury has healed." Explanations that clarify the purpose and limitations of specific tests (e.g., the fact that imaging tests cannot image pain, and that persistent pain does not mean persistent injury) should help to decrease patient distress and disability. Exaggerated pain behaviors are also used to communicate distress and suffering. Eliciting, understanding, and (where appropriate) addressing the source of distress may help decrease distress-related pain behaviors.

In addition to establishing patients' pain beliefs during the clinical interview, it is worth eliciting their treatment beliefs, preferences, and goals. Nonspecific (placebo) treatment effects are significantly influenced by treatment expectations (both patients' and practitioners'). Fears, anxieties, and enthusiasm will influence their participation and adherence to treatment, and thus the outcome from treatment.

Setting treatment goals is easy to discuss but difficult to do. Physical therapists believe that it is important to include patients in goal-setting and that treatment outcomes will improve if they do so. Likewise patients have indicated that participation is important for them. However, research has shown that both groups need education in goal-setting activities.[13] The clinical interview initiates the goal setting process and the mutually agreed treatment plan completes the process. Common sense and best evidence suggest that goals must be mutually agreed upon, reasonable, realistic, achievable, but challenging. They must also have timelines for expected achievement.

In a study designed to determine whether physical therapists involved patients in goal setting, Baker and colleagues[13] videotaped 22 physical therapists as they conducted an initial assessment of 73 patients. The frequency of therapists' attempts to include patients in goal setting was measured using the *Participation Method Assessment Instrument* (PMAI). Despite a stated belief by therapists of the importance of mutually agreed goals, only 9 of the 22 therapists introduced goal-setting with the patient, and only 3 explained the patient's role to the patient. Of the 22 therapists, 15 explicitly stated treatment goals, but these were their own. Clearly, what therapists intend to do (and how they believe they practice) may not be what they actually do. In the UK, the Physiotherapy Pain Association has published *Pain Management Standards*.[14] Standard 3 states that "there is written evidence that individual goals are *agreed with the patient* and incorporated into the overall treatment plan." The issue may be with how *agreed* is reached. *Agreed* may simply reflect a lack of disagreement with the therapists' suggested goals and a socially acceptable response. Perhaps that is acceptable because, regardless of this lack of involvement, patients claimed that they were satisfied. It may be that expressions of satisfaction were based on a positive relationship with the therapist and may have had nothing to do with the goal setting exercise. Some patients probably respect, appreciate, and indeed expect professional advice. Subjects in Baker's study[13] were elderly and perhaps comfortable with a more authoritarian medical relationship. In contrast and in a cardiac rehabilitation center, Moore and colleagues[15] reported that perceived arbitrary goal setting by therapists was a source of patient dissatisfaction.

Tailoring treatment, including goal setting, to the patients' needs and preferences is clearly important. Setting goals with patient participation does not require more time if education and discussion is woven into the assessment. Greater goal attainment, increased patient satisfaction, higher functional gains, and increased adherence to treatment is an expected and stated outcome, but further research is necessary to test the process and determine whether the expected outcomes are realized.

A final purpose of the clinical interview is to identify patients at risk of serious pathology or poor outcome, and to refer or manage them as appropriate. Patients at risk of poor outcome and the development of chronic disability can be identified early and on the basis of psychosocial factors.[16] Predictive factors (see Table 46-1) include patient variables (e.g., high levels of distress at the onset of the problem), social variables (e.g., workplace issues), and therapeutic relationship variables (e.g., inadequate and unclear explanations of the problem). Gatchel et al.[17] evaluated the clinical effectiveness of an early intervention in high-and low-risk patients with acute low back pain. High-risk patients were randomly assigned to one of two groups: a functional restoration early intervention group ($n = 22$) or a nonintervention group ($n = 48$). A group of low-risk subjects ($n = 54$) who did not receive any early intervention was also evaluated. High-risk subjects who received early intervention displayed fewer indices of chronic pain disability on a wide range of work, health care utilization, medication use, and self-report pain variables, relative to the high-risk subjects who did not receive such early intervention. In addition, the high-risk nonintervention group displayed significantly more symptoms of chronic pain disability on these variables relative to the initially low-risk subjects. Cost-comparison savings data revealed greater cost savings associated with the early intervention group versus the no early intervention group. Clearly, tailoring treatment to individual patients on the basis of psychosocial as well as biomedical characteristics provides both human and economic benefits, whereas tailoring treatments solely on biomedical factors seems destined to contribute to human and economic costs. A caveat is the potential biasing effect of psychosocial risk factors on the therapeutic milieu. Expected poor outcomes can be actualized as a self-fulfilling prophecy.[18]

Standardized Clinical Assessment

Traditionally, standard clinical assessments were largely based on a narrow biomedical model. Impairment tests

▶ **TABLE 46-1.** RISK FACTORS FOR CHRONIC PAIN AND DISABILITY

Personal Factors
• Misunderstanding that pain is directly related to injury
• High level of distress at onset of the problem
• Catastrophizing about the problem and the implications
• Passive approach, i.e., expecting others to cure the problem
• Significant anger about the pain and initiating injury
• High level of fear or avoidance of activity

Social Factors
• Compensation or litigation
• Family reinforcement of inactivity
• Work issues, e.g., financial disincentives, difficulties with supervisor

Therapeutic Relationship Factors
• Unclear explanations
• Different diagnostic labels from different practitioners
• Inadequate examination and assessment
• Inappropriate and/or ineffective treatments
• Reinforcement of patient passivity
• Unrealistic predictions of treatment outcome

SOURCE: *Adapted from Main and Watson 2002.*

such as range of motion, strength, and specific joint movements have been described in great detail, are entrenched in current practice, and have apparent clinical utility. Yet many are unreliable and the constrained manner of testing may have little to do with the patient's functional ability, especially when the problem is chronic or recurrent. The expanded biopsychosocial conceptual model recognizes salient factors beyond anatomical and biomechanical aspects that can affect the functional ability of the person as a whole. Multimethod, multimeasure assessment approaches are necessary to optimally understand and, therefore, manage the problem. A comprehensive assessment using standardized, reliable, and valid tools should not be long or arduous for the patient or the clinician. Simple methods with clinical utility are available. However, it is important to use complementary methods of assessment such as patient self-reports and clinician-measured performance tests.[19,20] The assessment should provide answers to the following questions: What can (and does) the patient do, why? and why not? What are the barriers to improved performance and increased activity (e.g., lack of pain control, strength, range of motion, confidence, understanding, or motivation)? The answers to these questions provide the guide to treatment.

Self-Report Measures

Many standardized self-report assessment measures are available and in clinical use for measuring constructs such as pain, disability, and quality of life. The measures range from simple unidimensional or global questionnaires that are specific and take a few minutes to complete, to complex multidimensional questionnaires that can sample a wide range of activities, thoughts, and interference with social roles, and that are useful for patients with a range of conditions. Given that the questionnaires under consideration have similar measurement properties (reliability, validity, and responsiveness) the choice of which questionnaire to use in a given situation depends on the reason for its use and what the clinical investigator wants to know. Investigators may be more interested in what the patient *can* do, *does* do, *did* do, *wants* to do, *thinks or feels* they can do, and importantly *why?* or *why not?* The assessment should guide treatment, so it important that the information is not just collected but is actually used to determinate the type of treatment, to guide and refine it, and also to evaluate its effectiveness.

An advantage of generic questionnaires is that the same tool can be used for a wide variety of patient groups allowing comparisons across groups. A disadvantage is that they may not adequately assess some of the activities that tend to be compromised by a specific disease. Some investigators have attempted to address this problem by modifying a general questionnaire. For example, Wu and colleagues modified the generic SF-36 to more comprehensively assess patients with HIV/AIDS.[21] In contrast, Cella and colleagues[22] addressed the problem by designing one general questionnaire that is used to assess all patients and a set of subquestionnaires used in addition to the general questionnaire to better assess a particular problem.[22,23] For example, the *Functional Assessment of Cancer Therapy Scale* (FACT-G) is a general quality-of-life questionnaire that contains physical activity items and can be completed by all patients with cancer. Patients with a particular problem (e.g., breast cancer) would then also complete a subquestionnaire, *FACT-Breast Cancer*, that measures specific breast cancer related difficulties.

Other types of questionnaires are *symptom* specific as opposed to *condition* specific. For example, the *Brief Pain Inventory* (BPI)[24] and *Brief Fatigue Inventory* (BFI)[25] assess the magnitude of a particular symptom and also how much that symptom, pain or fatigue, interferes with activity. However, the questions regarding activity interference are very general and are not judged against any standard criteria. For example, using the BPI, patients may report that pain interferes with walking, and they may score this as a 5 on a 0–10 scale. However, this 5 doesn't differentiate between the walking ability of patients who can walk between the bed and commode and patients who can walk in the community.

Finally, other questionnaires assess the individual's attitudes toward, or confidence in, their ability to perform specific movements or activities. For example, the *Fear Avoidance Questionnaire*[3] and *Tampa Scale of Kinesiphobia*[26] measure the individual's movement-related fears or anxieties whereas self-efficacy questionnaires address the patient's confidence in being able to perform certain activities or manage their pain.[27,28] Again, the important point for the clinician is that the data gathered from the questionnaires is used to guide and determine the effectiveness of treatment. For example, individuals with back pain who have a high level of fear related to activity should benefit more from carefully graded exposure to exercise and activity, compared to those with a similar level of impairment but a low level of fear. The latter group may do well with a self-management exercise program.

Although standardized questionnaires of physical function have clinical utility, they may not be a valid reflection of a patient's actual functional status,[29] especially when an external reference is unavailable. Self-report measures are more closely associated with other self-report measures than with physical capabilities observed or measured by clinicians.[30] Patient self-reports and clinician-measured tests of the same functions are moderately correlated at best,[31,32] hence the need to use both methods. The moderate correlations suggest that self-reports and physical performance are differentially influenced by current mood, attitudes, expectancies, memory, reporting biases, and situational demands.[31] The strength of correlation is also due to individual patients using idiosyncratic criteria to make self-report judgments on their compromised functional ability or perceived difficulties. In contrast, physical performance measures use standard criteria (time and/or distance) and *measure* rather than *judge* performance ability.

Another reason for using both self-report and physical performance is that self-reports of function may be inaccurate. Inaccuracy of reporting is not intentional nor is it confined to patients. Most individuals, including physicians,[33] have difficulty estimating the time it takes to complete a task, their tolerance for different activities, or the distance they can walk.

Physical Performance Measures

Contemporary clinical physical assessments in rehabilitation now usually include direct tests of functional movement. Several independent research teams have developed simple performance task batteries for use in different patient populations, especially those with chronic diseases and clusters of symptoms such as pain and fatigue (e.g., spinal dysfunction, chronic pain, cancer, and HIV/AIDS).[34-38]

Physical performance tests usually comprise a series of tasks that challenge physical function in different positions (e.g., lying, sitting, and standing) and challenge different components of physical function (e.g., fine and gross motor function, velocity of motion, strength, flexibility, endurance, coordination, and balance) in a functionally meaningful way. Disease and injury, and the symptoms of disease and injury, lead to a decline in task performance that is evidenced by reduced movement velocity and endurance.[35,37,38] In the clinical environment, performance ability is easily measured on the simple basis of time taken, number of repetitions completed, or distance reached or walked.

Tables 46-2a and 46-2b present examples of performance test batteries that have been evaluated in individuals with different conditions. Clinicians and researchers can use the whole battery or select specific tests based on the patient's movement problems. The reliabilities of all of the tests are well established (ICC values range between .7 and .9). Discriminant validity is strong, the tests differentiate patients from healthy individuals ($p < .0001$ to $p < .05$). Together with a measure of pain, specific performance tests are significant predictors of disability[37,38] and quality of life.[39] Factor analysis has shown that the test batteries (cancer and back pain) primarily load on one of two factors. One factor essentially represents speed and coordination and the second represents endurance, strength, and balance.[38,40]

The robust reliability across patient groups (intra-, inter- and test-retest) is largely a result of the simplicity of the tests, both for the patients to perform and the practi-

▶ **TABLE 46-2A.** BACK PAIN PERFORMANCE BATTERY

Task	Procedure	Measure
Repeated Sit-to-Stand	Subjects rise to standing and return to sitting as quickly as possible five times. After a brief pause the task is repeated.	The average of the two task times is recorded.
Repeated Trunk Flexion	Subjects are timed as they bend forward to the limit of their range and return to the upright position as fast as tolerated five times. After a brief pause the task is repeated.	The average of the two task times is recorded.
Loaded Reach	Subjects stand next to a wall on which a meter rule is mounted horizontally at shoulder height. They hold a weight that is 5% of their body weight (up to maximum of 5 kg) at shoulder height and close to the body and then reach forward.	Maximum distance reached in centimeters is recorded.
50-Foot Walk	Subjects walk 25 feet, turn around and walk back to start as fast as they can.	Time taken is recorded.
Five-Minute Walk	Subjects walk as far and as fast as they can for 5 minutes.	Distance walked is recorded.
360° Rollover	Subjects lie supine on a treatment bed. They roll over 360° as fast as they can. After a brief pause, they roll 360° in the opposite direction.	The time to complete a rollover in both directions is summed and recorded.

SOURCE: *Simmonds et al, 1998.*[37]

▶ TABLE 46-2B. PHYSICAL PERFORMANCE BATTERY FOR USE IN CANCER AND HIV/AIDS

Task	Procedure	Measure
Coin Test	Subjects sit at a table. They are timed as they pick up four coins and place them in a cup. (They are required to pick up each coin individually).	Time taken is recorded.
Belt Tie	Subjects sit in a standard chair. They are timed as they wrap a bandage (approximately 4 feet long) around their waists and tie it in front of them.	Time taken is recorded.
Sock Test	Subjects sit in a standard chair. They are timed as they put on one loose-fitting sock.	Time taken is recorded.
Repeated Sit-to-Stand	Subjects rise to standing and return to sitting as quickly as possible twice. After a brief pause the task is repeated.	The average of the two task times is recorded.
Repeated Reach-Up	Subjects stand facing a wall and reach up as high as they can with both hands. A mark is placed on the wall at the reach distance. Subjects then reach up and return their hands to their sides three times, as fast as they can. After a brief pause the task is repeated.	The average of the two task times is recorded.
Forward Reach	Subjects stand sideways next to a wall on which a meter rule is mounted horizontally at shoulder height. They then reach forward as far as they can.	Maximum distance reached in centimeters is recorded.
Pen Pick up from Floor	Subjects stand and a pen is placed on the floor directly in front of the subject's feet. They are timed as they bend down and pick up the pen as fast as they can.	Time taken is recorded.
50-Foot Walk	Subjects walk 25 feet, turn around and walk back to start as fast as they can.	Time taken is recorded.
Six-Minute Walk	Subjects walk as far and as fast as they can for 6 minutes. (They are allowed to sit and rest as necessary during the 6-minute period.)	Distance walked is recorded.

Source: *(HIV/AIDS Simmonds, Novy et al. 2002)(Simmonds 2002).*

tioner to administer and measure. The tests are basic activities (or components of activities) that are done many times throughout the day. Thus, it is relatively easy to determine the impact (burden) of the condition on the patient.

Most tests require the use of only a simple stopwatch and a method of measuring distance easily. Reach distance can be easily measured using a tape measure. Walk distance can be easily tested over premeasured and marked known distances. For portability of testing in different inpatient, outpatient, and community environments, a simple surveyor's wheel can be easily and cheaply obtained.

Compared to healthy individuals, the physical abilities of patients are compromised across performance tests. However, the magnitude of compromise is influenced in a task-specific manner based on the physical demands of the task and factors such as usual activity level, age, gender, disease/disorder, and specific symptoms (e.g., pain and fatigue). For example, we tested the performance ability of 100 outpatients with HIV/AIDS and not—surprisingly—found it to be lower than age-equivalent healthy norms. Patients took up to four times as long to perform the repeated sit-to-stand task and walked 75 percent of the distance compared to age equivalent norms.[41] Considering how often individuals rise from sitting during the day, this performance decrement places a significant time and energy burden on patients. In Simmonds' study, 50 percent of the patients with HIV/AIDS had pain at the time of testing and the majority of those had pain that was moderate to severe (mean 6.3/10 ±2.4). Fatigue was a problem for 98 percent of the patients (mean 5.4/10 ±2.3). Comparisons across pain (present or absent) and fatigue (mild or severe) subsamples showed that pain influenced performance to a greater extent than fatigue. Patients had particular difficulty with tests that required whole body movements in the upright position. These tests are obviously more physically demanding and require more energy, effort, and endurance; hence, the decrement in performance is of relatively high magnitude.

The fact that individuals with a disease or disorder move more slowly than normal healthy individuals is a fundamental and robust finding. Preliminary results show that patients move more slowly than pain-free individuals across speed conditions (i.e., self-selected slow, preferred, and fastest speed) and that anticipated pain influences movement speed. Patients tend to overestimate movement-related pain with fast movements, and underestimate pain with slow movements.[42] Thus, they may "choose" to move slower in an effort to avoid pain, and that pattern may persist. Unfortunately, slow movements are inefficient in terms of the time they take and the relative energy they consume. The relatively high physiological cost can be partly explained by the relatively high levels of muscle activity used during the performance of a specific task. For example, we compared movement and muscle activity during sit-to-stand and forward reach tasks in patients and healthy pain-free control subjects. We found that patients used more muscle activity and for a longer time than healthy pain-free subjects.[43] Figure 46-3 illustrates the difference in magnitude and duration of muscle activity in a group of patients with back pain compared to an age-matched group of healthy individuals during performance of a sit-to-stand task.

The relatively higher energy cost associated with moving while impaired has been well documented. It is also evident that the energy cost of movement is matched by patients' judgments of perceived effort during task performance. It is interesting that during performance of a timed distance walk, patients with a benign musculoskeletal problem

Figure 46-3. Muscle activity during a sit-to-stand test (*a*) and loaded reach forward test (*b*) in age- and gender-matched groups of individuals with and without low back pain (*n* = 60).

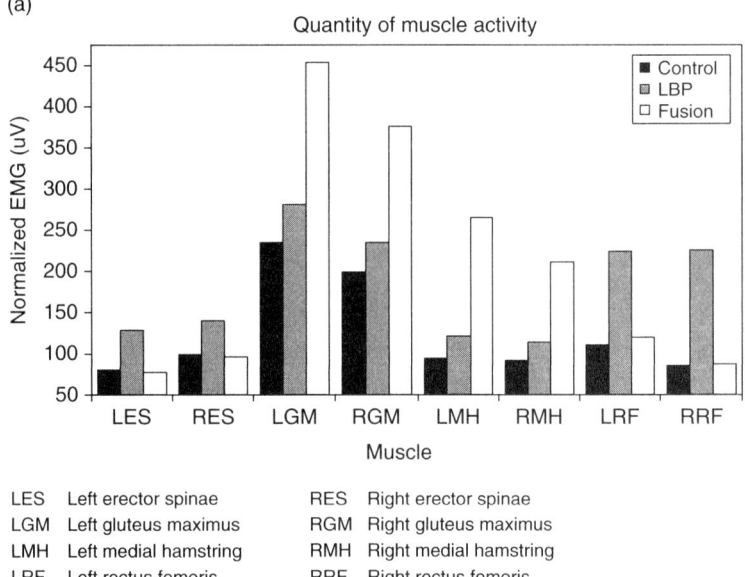

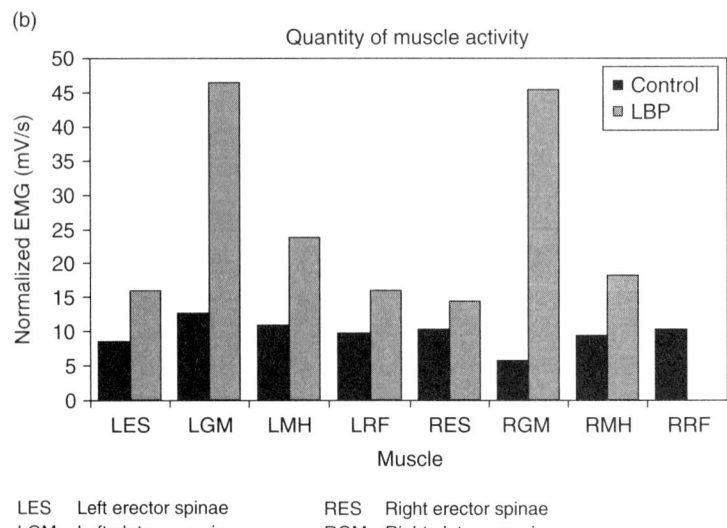

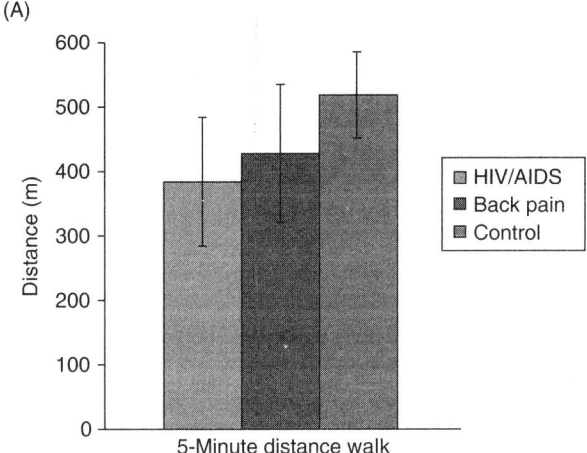

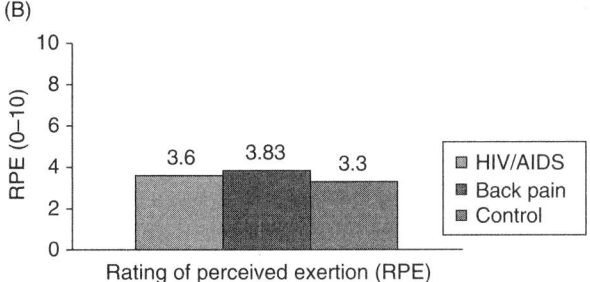

Figure 46-4. A. Shows difference in walking performance between subjects with back pain, HIV/AIDS and healthy control individuals. B. There was no significant difference in perceived effort.

(back pain), a systemic disease (HIV/AIDS) and healthy individuals selected a walking speed that reflected a similar level of perceived effort ($\cong 3/10$)[37,41] (Figure 46-4). However, for an equivalent amount of effort, there were significant differences in the distances that different patient groups were able to walk. It is plausible that patients' predictions of effort, as well as pain, influence their activity participation decisions.

It is important to remember that physical performance is not simply a test of physical function. Rather, an individual's physical performance is influenced by a variety of factors in each of the biopsychosocial domains. Clinicians need to identify and potentially manage those factors that have the greatest influence on the *patient's* performance, and these may not be physical. Fear of activity is known to influence performance, and graded exposure to the feared activity can improve performance in individuals with significant levels of fear. Education can also improve performance. For example, Moseley[44] showed that patients' physical impairment measures improved following a pain education intervention (that did not permit activity).

In summary, the assessment should include patient self-report questionnaires and clinician-measured performance tests. The purpose of the assessment is to identify key barriers to enhanced activity and to manage them as appropriate. The barriers may be primarily biological, such as pain and weakness. Therefore, the treatment should be primarily biological, such as pain management with heat and muscle strengthening exercises designed to improve activity. The primary barriers to activity may be psychological, for example, inaccurate beliefs about pain or fear of re-injury. Therefore, the treatment should focus on education and graded exposure to the feared activity. Finally, it is humbling to consider that the primary barrier to activity may be social—poor therapeutic relationship.

▶ INTERVENTION

Rehabilitation strategies must be individualized to reduce pain and optimize improvement in physical function. Activity, activity-related goal setting, and pacing of activity, play key roles in rehabilitation. Modalities (heat, cold, TENS, and manual therapy) also play a limited role. They can be used on a time-limited basis to help reduce pain in order to improve engagement in activity.

Numerous studies have demonstrated the benefits of physical exercise and activity for health promotion, disease prevention, disability management, and overall quality of life. A consequence of pain, regardless of its genesis, is a temporary or permanent reduction in activity. Physical inactivity has a detrimental effect on general health as well as psychosocial well-being. Therefore, for patients with pain, maintenance of (or early return to) activity is a fundamental aim of management. What is less clear is whether any specific exercise or activity regimen will facilitate resumption of normal activity, and if so, whether the specific mechanisms of effects are biological, psychological, sociological, or all of the above. Unfortunately, and regardless of proven health benefits, many individuals have difficulty initiating and/or maintaining exercise programs or even assuming a more active lifestyle. To state the obvious, the beneficial effects of exercise or activity will only accrue if the exercise or activity is done and is maintained; hence the value of simple activities such as walking.

Physical activity is an umbrella term that includes concepts such as fitness, exercise, training, and conditioning. Essentially any bodily movement that increases energy expenditure above the resting level is physical activity. Exercise is frequently used interchangeably with physical activity. However, exercise and exercise training is purposeful activity specifically designed to improve or maintain a particular component of physical fitness (e.g., flexibility, strength, or endurance—cardiovascular or musculoskeletal).

Although there is evidence-based consensus on the beneficial effects of exercise and activity for patients with pain, there is less supporting evidence of the beneficial effects of any specific type of exercise. For example, in the case of *low back pain* (LBP), a variety of different treatment regimens and specific types and intensity of exercise are used in clinical practice with the aim of preventing and/or treating LBP and its consequences. However, most regimens were based on a biomedical impairment model and thus focused on improving specific trunk strength, endurance and/or flexibility, and aerobic fitness. The implied rationale of this assessment and management approach is that trunk muscles are weak, the spine is stiff, and the patient is unfit. However, trunk strength and mobility predict neither LBP nor disability, and patients with LBP are no less fit than most healthy individuals.[45,46] Nevertheless, Jette et al.[10] reported that endurance exercises were included in 52 percent of treatment plans of 739 individuals with LBP and resulted in improved outcomes compared to patients who had a less active intervention. Thus, the intervention seems appropriate and effective but the rationale and implied mechanisms of effect are obviously incomplete.

A meaningful review of the literature on exercise for pain is difficult. Interventions are often inadequately described in terms of type, intensity, or duration; patients are heterogeneous or inadequately described; and outcome measures vary from specific *impairment measures* (e.g., pain reduction) to *social measures* such as return to work. Although the latter is a very important outcome, it is influenced more by education, job skills, and unemployment rate than by pain severity.

A number of randomized controlled trials and systematic reviews have addressed exercise and activity in individuals with acute and chronic pain, and many have been conducted in back pain patients. It seems clear that exercise and activity is beneficial for individuals with pain as it can reduce the impact of pain and enhance the sense of well-being. It is less clear whether the type, intensity, frequency, or duration of exercise or activity is important. Based on current evidence, it seems that *the most effective exercise or activity regimen is that which is done.* The UK guidelines[47] recommend advising patients to continue ordinary activity. However, this advice should also be accompanied by education, reassurance, and specific instructions about self-management of the problem.

Based on a systematic review, and acknowledging the limited evidence, the Paris Task Force[48] recommended exercise programs (for LBP) that combine strength training, stretching and/or fitness. A subsequent Cochrane review by van Tulder reported that in acute LBP, when exercise therapy is compared to inactive treatment and other active treatments, there was no difference between treatment types.[49] Effectiveness was judged on the basis of reduction in pain intensity, increase in self-report of functional status, overall improvement, and return to work.

Clearly there are difficulties with the literature that reflect the complexity of the problems of pain and the impact of pain on disability—apparent contradictions between *expert consensus* and the results of randomized controlled trials. The outcome of individual patients may be "lost" in the group data analyzed from randomized controlled trials but "remembered" in expert consensus recommendations. The way in which responses are elicited and the context in which data are recorded can affect reported outcomes, potentially leading to erroneous decisions about what interventions benefit patients. Finally, although the evidence for the effectiveness of specific exercise in acute pain is weak, evidence for the harmful effect of rest is strong.

The consensus recommendations from the Paris Task Force for *acute* LBP are as follows:

1. Low stress aerobic exercise can prevent debilitation due to inactivity and may help return patients to the highest level of function appropriate to their circumstances. *Strength of evidence: Limited (at least one adequate scientific study).*
2. Aerobic (endurance) exercise programs, which minimally stress the back (walking, biking, or swimming) can be started during the first two weeks for most patients with acute low back problems. *Strength of evidence: No research basis.*
3. Conditioning exercises for trunk muscles (especially back extensors), gradually increased, are helpful for patients with acute back pain. *Strength of evidence: Limited (at least one adequate scientific study).*
4. Back-specific exercises on machines provide no benefit over traditional exercise. *Strength of evidence: No research basis.*
5. Stretching exercises (of back muscles) are not recommended. *Strength of evidence: No research basis.*
6. Exercises using quotas yield better outcomes than exercises using pain as a guide to progression. *Strength of evidence: Limited (at least one adequate scientific study).*

In the management of chronic pain, a number of randomized, controlled trials and systematic reviews have reported beneficial effects of exercise and activity. The Paris Task Force reviewed 10 scientifically rigorous, randomized, controlled trials of exercise for chronic LBP. Patients did better than control subjects in 7 of the 10 trials. However, the regimens were characterized by variability of type, intensity, and duration. Moreover, some treatment programs included additional components such as education and/or behavior modification.

A variety of factors—personal (physical and psychological), social, and environmental—account for the difficulties in engaging and retaining normally sedentary individuals in exercise programs. Individuals with chronic diseases are thought to face additional physical and psychological barriers to activity (e.g., pain, fatigue, weakness, and anxiety about their disease and its short- and long-term impact).

However, for nondistressed individuals with pain, barriers and motivators related to physical activity are similar to those in the general population (e.g., lack of time, inclement weather, and family commitments). Pain problems appear to be both barriers to, and motivators of, activity.[50] This suggests that it is not the pain *per se* that is influencing activity participation; rather it is the individual's interpretation of the symptom that is influencing their behavior.

In a qualitative study, Keen and colleagues interviewed 27 individuals who were participants in a randomized, controlled trial of a progressive exercise program. They reported that some individuals believed that being more physically active helped ease their back pain and made them feel better, and were worried about stopping exercise for fear that their back pain would return. Others did not exercise on a regular basis but resumed exercise when reminded to by their backache. Still others avoided physical activity for fear of an aggravation of their LBP. Although all subjects identified the avoidance of some physical activity (e.g., lifting and gardening), not all were fearful or anxious about such activity. It appears that those individuals in the latter category reported that their confidence was restored over time through (1) reassurance and advice from health professionals, (2) modifying the way an activity was done (e.g., less vigorous); and (3) a progressive exercise program.[50] It appears that a change in behavior led to a change in belief about the ability to be active.

Participation is essential if exercise benefits are to accrue. Friedrich and colleagues[51] conducted a double-blind, randomized study and evaluated the effect of a motivation program on exercise compliance (adherence) and disability. Ninety-three patients with LBP were randomly assigned to either a standard exercise program ($n = 49$) group or a combined exercise and motivation group ($n = 44$). The exercise program consisted of an individual, submaximal, gradually increased training session. Each patient was prescribed 10 sessions that each lasted about 25 minutes. The specific exercises were aimed at "improving spinal mobility, as well as trunk and lower limb muscle length, force, endurance, and coordination, thereby restoring normal function." Flexibility exercises for the trunk and lower limbs preceded strengthening exercises for the trunk. The motivation program consisted of five sessions that included counseling, providing information about LBP and exercise, reinforcement, forming a treatment contract between patient and therapist, and keeping an exercise diary.

The combined exercise and motivation group increased the rate of attendance and reduced disability and pain in the short term (4 and 12 months). However, there was no difference in exercise adherence in the long term. Long-term adherence to exercise is an acknowledged problem in the general population and is no different in those with pain. In the case of LBP, when exercise benefits may not be immediate or even apparent and recurrence of LBP is inevitable, it is no surprise to find that adherence to exercise is problematic. For some individuals, encouraging and facilitating them to have a more active lifestyle, and demystifying and demedicalizing the problem may be more beneficial than prescribing a special exercise program for their "medical" problem that they don't do anyway.

Frost et al.[52] tested a supervised general fitness program. Cognitive-behavioral principles and a normal model of human behavior rather than a disease model was followed. Participants were encouraged to compare themselves to sports participants who had been laid off from training and who needed to get back to their previous activity level. Eighty-one individuals with chronic LBP participated and were assigned to one of two groups (exercise and control). Both groups were taught exercises and attended an educational program. The exercise group also attended eight sessions of a supervised fitness program that extended over four weeks. Participants were encouraged to improve their own performance record (not compete with others) and to complete an activity diary. A mean reduction of 7.7 percent (pain and disability) was obtained in the exercise group compared to a 2.4 percent reduction in the control group. This difference was statistically significant and was maintained at the two-year follow-up. However, the authors note that the confidence interval of the differences between groups was large, indicating a wide variation in treatment effect.

It is not surprising that in studies comparing relatively active and relatively passive interventions, the active intervention appears more effective. However, in a recent Volvo award-winning study, Mannion and colleagues[53] compared three active therapies for chronic low back pain. The 148 subjects were randomly assigned to (1) an active physiotherapy program; (2) a muscle reconditioning program using training devices; and (3) a low-impact aerobics program. Subjects attended their program twice a week for three months. All programs led to reductions in pain and disability that were maintained at six months. There were no differences between groups, suggesting a lack of treatment specificity.

In summary, it appears that the specific type of exercise, or exercise regimen, is much less important than once thought. Although the notion may be anathema to traditional-thinking clinicians entrenched in a narrow, structurally focused biomedical (impairment) model, it is less surprising to those who recognize the biopsychosocial nature of chronic LBP. Exercises targeted at a specific biological or structural impairment may effect changes in impairment but the actual impairment may contribute relatively little to the individual's pain problem.

Although speculative at present, it is plausible to suggest that exercise is beneficial for those with chronic pain because it improves mood and confidence and changes the perception of self as disabled. Thus, the primary benefits of exercise or activity for individuals with chronic LBP are central rather than peripheral or structural. The key for therapists is to identify and address patient-specific barriers to activity using a biopsycho- social framework.

REFERENCES

1. Waddell G: A new clinical model for the treatment of low back pain. *Sine* 1987;12:632–644.
2. Vlaeyen J, Linton S: Fear-avoidance and its consequences in chronic musculoskeletal pain: a state of the art. *Pain* 2000;85:317–322.
3. Waddell G, et al: A fear avoidance beliefs questionnaire (FABQ) and the role of fear avoidance beliefs in chronic low back pain and disability. *Pain* 1993;52:157–168.
4. American Physical Therapy Association: *Guide to Physical Therapy Practice, 2nd ed*. Alexandria, VA;2000.
5. World Health Organization: *International Classification of Impairments, Disabilities and Handicaps (ICIDH)*. Geneva;1980.
6. Nagi SZ: Disability concepts revisited: Implications for prevention. In: Pope A, Tarlou A (eds.): *Disability in America: Toward a National Agenda for Prevention*. Washington: National Academy Press;1991:309–327.
7. Jordan J, et al: Systemic risk factors for osteartharitis. In: Felson DT, Osteoarthritis: New insights. Part 1: The disease and it's risk factors. *Ann Intern Med* 2000;133:637–639.
8. Minor M, Kay D (eds.): Arthritis. In: *Exercise Management for Persons with Chronic Diseases and Disabilities*, Champaign, IL:Human Kinetics;1997;149–154.
9. Thomas K, et al: Home based exercise programme for knee pain and knee osteoarthritis: Randomised controlled trial. *BMJ* 2002;325:752.
10. Jette D, Jette A: Physical therapy and health outcomes in patients with spinal impairments. *Phys Ther* 1996;76:930–941.
11. Resnik L, Hart D: Using clinical outcomes to identify expert physical therapists. *Phys Ther* 2003;83:990–1002.
12. Pither C, Nicholas M: The identification of iatrogenic factors in the development of chronic pain syndromes: abnormal treatment behaviour? In: *Proceedings of the VIth World Congress on Pain*. New York: Elsevier Science, 1991.
13. Baker S, et al: Patient participation in physical therapy goal setting. *Phys Ther* 2001;81:1118–1126.
14. Muncy H: *Standards for physiotherapists working in pain management programmes*. Chartered Society of Physiotherapists. 1999.
15. Moore S, Kramer F: Women's and men's preferences for cardiac rehabilitation program features. *J Cardiopulm Rehabil* 1996;16:163–168.
16. Linton S: A review of psychological risk factors in back and neck pain. *Spine* 2000;25:1148–1156.
17. Gatchel RJ, et al:, Treatment- and cost-effectiveness of early intervention for acute low-back pain patients: a one-year prospective study. *J Occup Rehabil,* 2003;13:1–9.
18. Simmonds M, Kumar S, Lechelt E: Does knowledge of patient's workers compensation status influence clinical judgements? *J Occup Rehabil* 1996;6:93–107.
19. Simmonds M: Physical function in patients with cancer. Psychometric characteristics and clinical usefulness of a physical performance test battery. *J Pain Symptom Manage* 2002;24:404–414.
20. Lee C, et al: A comparison of self-report and clinician measured physical function among patients with low back pain. *Arch Phys Med Rehabil* 2000;82:227–231.
21. Wu AW, et al: Applications of the Medical Outcomes Study health-related quality of life measures in HIV/AIDS. *Qual Life Res* 1997;6:531–554.
22. Cella D, et al: The Functional Assessment of Cancer Therapy Scale: Development and validation of the general measure. *J Clin Oncol,* 1993;11:570–579.
23. Cella, D: Factors influencing quality of life in cancer patients: anemia and fatigue. *Semin Oncol* 1998;25:43–46.
24. Cleeland C: Measurement of pain by subjective report. In: Chapman C, Loeser J (eds.): *Issues in Pain Measurement*, New York: Raven Press; 1989:391–403.
25. Mendoza TR, et al: The rapid assessment of fatigue severity in cancer patients. *Cancer* 1999;85:1186–1196.
26. Kori S, Miller R, Todd D: Kinesiophobia: a new view of chronic pain behavior. *Pain Management* 1990; Jan/Feb:35–43.
27. Estlander A, et al: Anthropometric variables, self-efficacy beliefs, and pain and disability ratings on the isokinetic performance of low back pain patients. *Spine* 1994;19:941–947.
28. Anderson K, et al: Development and initial validation of a scale to measure self-efficacy beliefs in patients with chronic pain. *Pain* 1995;3:77–84.
29. Fordyce W, et al: Pain measurement and pain behavior. *Pain* 1984;18:53–69.
30. Deyo R, Centor R: Assessing the responsiveness of functional scales to clinical change: an analogy to diagnostic test performance. *J Chronic Dis* 1986; 39:897–906.
31. Lee CE, Simmonds MJ, Novy DM, Jones S: Self-reports and clinician-measured physical function among patients with low back pain: a comparison. *Arch Phys Med Rehabil* 2001;82.
32. Fisher K, Johnston M: Validation of the Oswestry Low Back Pain Disability Questionnaire, its sensitivity as a measure of change following treatment and its relationship with other aspects of the chronic pain experience. *Physiother Theo Pract* 1997;13:67–80.
33. Sharrack B, Hughes R: Reliability of distance estimation by doctors and patients: cross sectional study. *BMJ* 1997; 315:1652–1654.
34. Gronblad M, Hurri H, Kouri J-P: Relationships between spinal mobility, physical performance tests, pain intensity and disability assessments in chronic low back pain patients. *Scand J Rehab Med* 1997;29:17–24.
35. Harding V, et al: The development of a battery of measures for assessing physical functioning of chronic pain patients. *Pain* 1994;58:367–375.

36. Lin Y, Davey R, Cochrane T: Tests for physical function of the elderly with knee and hip osteoarthritis. *Scand J Med Sci Sports* 2001;11:280–286.
37. Simmonds MJ, et al: Psychometric characteristics and clinical usefulness of physical performance tests in patients with low back pain. *Spine* 1998;23:2412–2421.
38. Simmonds MJ: Physical function in patients with cancer: Psychometric characteristics and clinical usefulness of a physical performance test battery. *J Pain Symptom Manage* 2002;24(4):404–415.
39. Simmonds MJ, Sandoval R, Lee JQ: Pain and physical performance measures as predictors of quality of life in ambulatory patients with HIV/AIDS. *Proceedings of the 21st Annual Meeting of the American Pain Society, Baltimore, MD,* 2002;Glenview, IL: American Pain Society.
40. Novy DM, Simmonds MJ, Lee CE: Physical performance tasks: what are the underlying constructs? *Arch Phys Med Rehabil* 2002;83:44–47.
41. Simmonds MJ, Novy DM, Sandoval R: The differential influence of pain and fatigue on physical performance and health status in ambulatory patients with HIV. *Clin J Pain* 2002.
42. Simmonds MJ, Rebelo R: Self-selected speed of movement during a repeated sit-to-stand task in individuals with and without LBP. In: *EFIC—Pain in Europe IV* 2003; Prague, Czech Republic.
43. Simmonds MJ: Physical performance: What are the measures and what do they mean? In: Max M (ed.): *Pain Clinical Update*; Seattle: IASP Press;1999.
44. Moseley GL: Evidence for a direct relationship between cognitive and physical change during an education intervention in people with chronic low back pain. *Eur J Pain* 2004;8:39–45.
45. Verbunt JA, et al: Disuse and deconditioning in chronic low back pain: concepts and hypotheses on contributing mechanisms. *Eur J Pain* 2003;7:9–21.
46. Wittink H, et al: Deconditioning in patients with chronic low back pain: Fact or fiction? *Spine* 2000;25:2221–2228.
47. CSAG (Clinical Standards Advisory Group): *Back Pain: Report of a CSAG Committee on Back Pain*, London,1994.
48. Abenhaim L, et al: The role of activity in the therapeutic management of back pain. Report of the International Paris Task Force on Back Pain. *Spine* 2000;25:1S–33S.
49. van Tulder MW, Malmivaara A, Esmail R, and Koes BW: Exercise therapy for low back pain. *Cochrane Data base of Systemic Reviews*. Oxford:Update Software; 2000;(2):CD 000 335.
50. Keen S, et al: Individuals with low back pain: how do they view physical activity? *J Fam Pract* 1999;16:39–45.
51. Friedrich M, et al: Combined exercise and motivation program: effect on the compliance and level of disability of patients with chronic low back pain: a randomized controlled trial. *Arch Phys Med Rehabil* 1998;79:475–487.
52. Frost H, et al: A fitness programme for patients with chronic low back pain: 2-year follow-up of a randomized controlled trial. *Pain* 1998;75:273–279.
53. Mannion A, et al: A randomized clinical trial of three active therapies for chronic low back pain. *Spine* 1999; 24:2435–2448.

INDEX

Page numbers followed by italic *f* or *t* indicate figures or tables, respectively.

Ablative procedures. *See* Neurosurgical ablation
Acetaminophen
   for cancer pain, 476
   in children, 231*t*
   for osteoarthritis, 552–553
   safety, 554–555
   toxicity, 553–554
Acid-sensing ion channels, 332
Acquired immunodeficiency syndrome (AIDS), 350–351, 350*t*
Activity pacing, 602–603
Acupuncture
   for cervicogenic headache, 417
   for complex regional pain syndromes, 371
   for migraine, 395
Acute brachial plexus neuropathy. *See* Parsonage-Turner syndrome
Acute pain
   in cancer patients, 575
   characteristics, 196*t*
   coxibs for, 548–549
   from medical procedures, 234–235, 235*t*
   patient-controlled analgesia for, 522–254
   in patients with addictive disorders, 263–264
   peripheral nerve blocks for, 519–522
Acute small-fiber sensory neuropathy, 348
Addiction. *See also* Substance abuse
   acute pain treatment in, 263–264
   in chronic pain patients, 266–268, 267*f*, 571–572
   definitions, 258, 291
   legal issues, 268–269
   life-threatening diseases and, 265
   pain treatment barriers in, 262–263
   physical findings, 262, 262*t*
   risk, 260–261
   treatment, 539*f*
Adenosine receptors, 38, 332
Adolescent Pediatric Pain Tool (APPT), 203, 203*t*
Affective disorders, 528, 529*t*
A fibers
   anatomy, 5, 6*f*
   chemically invoked sensation in, 24
   heat sensitive, 22
   mechanically sensitive, 23
   neuropeptide Y and, 43
   type I, 22
   type II, 22
Aftersensation, 213, 213*t*
AIDS. *See* Acquired immunodeficiency syndrome (AIDS)
Allodynia
   in central neuropathic pain, 303
   definition, 40, 213, 290
   imaging, 153
   mechanisms, 26–27, 214*t*
Alloesthesia, 129
α_2-adrenergic antagonists, 334
α_2-adrenergic receptors, 37
Alzheimer's disease, 254
American Nurses Association, 176
American Pain Society, 277*t*
AMPA receptors, 33
Amyloidosis, 328
Amyotrophic lateral sclerosis, 347–348
Analgesics. *See also specific drugs and drug categories*
   adjuvant, 233*t*, 334–335, 476–477
   imaging studies, 154
   non-opioid. *See* Non-opioids
   opioid. *See* Opioids
   overuse, 383–384, 386
   patient-controlled. *See* Patient-controlled analgesia
   stepladder, 336–337, 336*t*
   topical, 594–595
Anandamide, 593–594. *See also* Cannabinoids
Anemia, 499
Anesthesia dolorosa, 129
Aneurysm, aortic, 488
Angina pectoris, 484–486, 623–624
Anosodiaphoria, 129
Anterior pretectal nucleus, 12–13
Anterior spinal artery syndrome, 136
Antiarrhythmics, 309
Anticonvulsants
   for cancer pain, 476–477
   for central neuropathic pain, 308–309
   in children, 233*t*
   for chronic daily headache, 387
   mechanisms of action, 586
   for myofascial pain syndrome, 509
   for peripheral neuropathic pain, 333
   types of, 586–589
Antidepressants
   for cancer pain, 476
   for central neuropathic pain, 308
   in children, 233*t*
   for chronic daily headache, 387
   clinical trials, 583*t*
   for complex regional pain syndromes, 369
   for fibromyalgia, 506
   mechanisms of action, 581–582

Antidepressants *(Continued)*
    for migraine, 396–397
    for peripheral neuropathic pain, 334
    for spinal pain, 434–435
    tricyclic, 582–583, 583t
    types of, 582–586
Antiemetics, 396, 593
Antiepileptics. *See* Anticonvulsants
Antihistamines, 477
Anti-inflammatory drugs. *See* Coxibs; Nonsteroidal anti-inflammatory drugs (NSAIDs)
Anxiety disorders, 215–216, 528, 529–530t, 531–532, 538f
Anxiolytics, 233t
Aortic aneurysm, 488
Aortic arc syndrome. *See* Takayasu's arteritis
Aortic dissection, 488
Aortic regurgitation, 486
APPT. *See* Adolescent Pediatric Pain Tool (APPT)
Arachnoid, 130–131
Arterial diseases, 487–490
Arthritis. *See* Osteoarthritis; Rheumatoid arthritis
Aspirin, 551
Assisted suicide, 176–177
Atherosclerosis, 487
Aura, migraine, 113–114
Autoimmune diseases, 326–327. *See also specific diseases*
Autonomic nervous system
    anatomy, 105–106, 106f, 149–150
    nociception and, 159–160
    pathophysiology, 106–109

Baclofen
    for cancer pain, 477
    intrathecal, 310
    for peripheral neuropathic pain, 335
    for trigeminal neuralgia, 406
BDI. *See* Beck Depression Inventory (BDI)
BDNF. *See* Brain-derived neurotrophic factor (BDNF)
Beck Depression Inventory (BDI), 216
Behavioral rating scales, 205
Behavioral therapy. *See* Cognitive-behavioral therapy
Benzodiazepines, 335, 477
Beta-blockers, 397
Biliary colic, 495–496
Biliary tract disease, 495–496
Biofeedback, 601–602
Bisphosphonates, 335, 477
Body outline drawing, 197f
Bone scintigraphy, 367–368, 367f
Bone tumors, 482
Botulinum toxin
    for migraine, 397
    for regional anesthesia, 608
    for spinal pain, 436–437
    for trigger point injections, 509
Bowel regimens, 472t
BPI. *See* Brief Pain Inventory (BPI)
Brachial plexus
    anatomy, 140, 140f
    neuropathies, 141–144, 635
Bradykinin receptors, 54
Brain-derived neurotrophic factor (BDNF), 82
Brain stem, 87
Breakthrough pain, 575
Breast cancer surgery, 296, 297t
B2 receptors. *See* Bradykinin receptors

Brief Pain Inventory (BPI), 203t, 204, 211
Brown-Sequard syndrome, 136, 137f
Buprenorphine, 264, 269
Bupropion, 586. *See also* Antidepressants
Burger's disease. *See* Thromboangiitis obliterans
Burning mouth syndrome, 409–410
Butterbur root, 394

CAGE-AID screen, 262, 262t
Calcitonin, 477
Calcitonin gene-related peptide (CGRP), 32
Calcium channel blockers, 39, 55
Calcium-regulating drugs, 370
Campylobacter jejuni infection, 348
Cancer pain
    adjuvant analgesics for, 476–477
    assessment, 468–469, 468t, 644
    diagnosis, 469–470
    interventional approaches, 477–478, 478t
    mechanisms, 644
    non-opioids for, 476–477
    nonresponsive to opioids, 575–576, 645
    opioids for, 471–476, 472–474t, 562–563, 573–574, 573t, 644–645
    pathophysiology, 467
    peripheral neuropathic, 326
    pharmacologic treatment, 470–471, 574t
    surgical treatment, 637
    syndromes, 642, 643t
Cannabinoid receptors, 56, 333
Cannabinoids
    for central neuropathic pain, 310
    efficacy, 593–594
    history of use, 591–592
    for peripheral neuropathic pain, 335
    therapeutic uses, 593
Capsaicin
    efficacy, 595
    nociceptors and, 24
    for peripheral neuropathic pain, 334
    pharmacology, 35, 595
    secondary hyperalgesia and, 26
Carbamazepine, 586–587. *See also* Anticonvulsants
    for central neuropathic pain, 308
    side effects, 406
    for trigeminal neuralgia, 405–406
Cardiac pain, 484–485
Cardiovascular disease, 484–490
Carpal tunnel syndrome, 142
Causalgia, 359. *See also* Complex regional pain syndrome (CRPS)
Cavernous sinus, 117
Celecoxib. *See* Coxibs
Celiac plexus, 149, 610
Center for Epidemiological Studies Depression Scale (CES-D), 216
Central cord syndrome, 136, 138f
Central neuropathic pain
    antidepressants for, 308
    antiepileptics for, 308–309
    cannabinoids for, 310
    definition, 301–302
    diagnosis, 304
    epidemiology, 302, 302t
    etiology, 302, 304–305, 304f
    intrathecal therapy for, 310
    mechanisms, 305–306
    in multiple sclerosis, 347

NMDA receptor antagonists for, 310
opioids for, 309
in Parkinson's disease, 352
pharmacologic treatment, 306–311, 307t
surgical treatment, 311–312
symptoms, 303
treatment strategy, 312
vs. peripheral neuropathic pain, 312–313, 312t, 322f
Central poststroke pain syndrome. *See* Poststroke pain
Central sensitization. *See* Sensitization
Cervical plexus, 137–139, 139f
Cervical root avulsion syndromes, 140
Cervicogenic headache, 415–418
CES-D. *See* Center for Epidemiological Studies Depression Scale (CES-D)
C fibers
anatomy, 5, 6–7f
chemically invoked sensation in, 24
heat sensitive, 21–22
mechanically insensitive, 23
mechanically sensitive, 23
CGRP. *See* Calcitonin gene-related peptide (CGRP)
Charcot-Marie-Tooth disease, 329–330
Charcot's joint, 346
Cheiralgia paresthetica, 144
Chemically invoked sensation, 24
Children
adjuvant analgesics for, 233t
chronic pain in, 237t, 238–239
complex regional pain syndromes in, 365, 371
drug therapy, 231–233t
Guillain-Barré syndrome in, 348
nondrug therapy, 231–234
non-opioids for, 231t
opioids for, 232t
pain assessment in, 228–230, 230f
pain experience of, 186–187, 225–226
pain from medical procedures, 234–235, 235t
pain management in, 231–233t, 231–234
plasticity of pain in, 226–228, 227f
recurrent pain syndromes in, 235–238
Cholangitis, 496
Cholecystitis, 496
Chronic daily headache
classification, 379–382
comorbidity, 382
diagnostic tests, 384–385
differential diagnosis, 384
epidemiology, 379
pathogenesis, 383–384
pathophysiology, 382–383, 385f
prevention, 388
prognosis, 388
treatment, 385–388
Chronic pain
active addiction and, 266–268, 267f
assessment. *See* Pain assessment
behavioral comorbidities, 535–537, 536t
characteristics, 196t
classification, 292–297, 294–297t
definition, 291–292
imaging, 152–153
management. *See* Pain management
mechanisms, 209–210
opioids for, 569–572, 569t
psychiatric comorbidities, 528–532, 529–530t, 538–539f

psychological factors, 215–217. *See also* Constructivism; Psychological evaluation
somatic comorbidities, 532–535, 533t
Chronic paroxysmal hemicrania, 392
Churg-Strauss vasculitis, 327
Clinical practice guidelines, 279, 279f, 280–281t
Clonidine
for cancer pain, 477
intrathecal, 310
for peripheral neuropathic pain, 334
Cluster headache
diagnosis, 392–394
pathophysiology, 116–118, 392
treatment, 397–398
C mechanoheat-sensitive fibers, 21–22. *See also* C fibers
CMHs. *See* C mechanoheat-sensitive fibers
C2 nerve. *See* Occipital nerve
Codeine, 232t, 472t, 474. *See also* Opioids
Co-enzyme Q_{10}, 394
Cogan's syndrome, 134–135
Cognitive-behavioral therapy, 600–603
Cold pain sensation, 23–24
Complex regional pain syndrome (CRPS)
bone scintigraphy, 367–368, 367f
in children, 365, 371
clinical features, 363–365, 364t, 365f
complementary medicine, 371
complications, 372
diagnostic criteria, 295–296, 296t, 365–366, 636
diagnostic tests, 366–368
differential diagnosis, 368
genetics, 363
incidence, 359–360
inciting events, 359, 360t
neurostimulation for, 371, 623
pathophysiology, 360–363, 636
peripheral neuropathic pain in, 330–331
pharmacologic therapy, 369–370
physical examination, 366
physical therapy, 371
prevention, 371
prognosis, 372
psychological therapy, 371
socioeconomics, 360
spinal drug application, 371
surgical treatment, 636–637
therapeutic algorithm, 371
Concentration problems, 533t, 534
Confounders, 179
Connective tissue diseases, 326–327. *See also specific diseases*
Connectivity, 154
Constructivism, 162–166, 163f
Cordotomy, 637
Coronary artery bypass graft, 141
Cor pulmonale, 487
Cortex, somatosensory areas, 127–129, 128–129f
Cortical silent period (CSP), 115
Cortical stimulation, 637
Corticosteroids
for cancer pain, 477
in children, 233t
for complex regional pain syndromes, 370
epidural injection, 610–611, 611f
for migraine, 396
for regional anesthesia, 608
for spinal pain, 437

COX-1, 545–546
COX-2
  in arthritis, 547
  expression, 545–547, 546t
  inhibitors. See Coxibs
Coxibs
  actions, 545, 549–550
  for acute pain, 548–549
  for dysmenorrhea, 549
  for osteoarthritis, 548
  outcome studies, 550–551
  for perioperative pain, 549, 550t
  safety, 550
  side effects, 551–552
Cranial nerve function, 201, 201t
Cranial pain, 130–131
Crohn's disease, 493–492
Crossed afterdischarge, 83
Cross-sectional studies, 179
Cryoglobulinemia, 328
Cryptogenic disorders, 330–331. See also specific diseases
CSP. See Cortical silent period (CSP)
Cyclooxygenase. See COX-1; COX-2
Cyr-Wartman screen, 262, 262t
Cytokines, 39–40, 57. See also Tumor necrosis factor-α

Deafferentation, 291
Deep brain stimulation, 617, 627, 637
Deep tendon reflexes, 213, 215t
Defense response, 160
Dejerine-Roussy syndrome, 129
Delta opioid receptor clone (DOR-1), 62t, 66
Dementia, 188–189
Dependence, 258, 291, 570. See also Substance abuse
Depression
  in chronic pain patients, 215, 528, 529t
  in pelvic pain patients, 461
  treatment, 538f
Descending facilitation, 45–46
Descending nociceptive inhibitory system, 13, 124–126, 126f
Dextromethorphan, 477, 591. See also NMDA receptor antagonists
Diabetes mellitus, 324, 349, 588
Diclofenac, 231t
Diffuse noxious inhibitory control (DNIC) system, 11
Diffusion tensor magnetic resonance imaging (DTI), 154
Disability, 216, 650t
Disinhibition theory, 305–306
Diskography, 613–614, 613f
Distal sensorimotor polyneuropathy (DSP), 349, 350
Divalproex sodium, 397
Diverticulitis, 494
DNIC system. See Diffuse noxious inhibitory control (DNIC) system
DOR-1. See Delta opioid receptor clone (DOR-1)
Dorsal column-medial lemniscal pathway
  anatomy, 123–124, 124–126f
  decussation, 136, 137f
Dorsal columns disease, 136
Dorsal horn
  anatomy, 4, 5f, 7–9, 121–122, 123f
  effector systems, 32–33
  neurons, 9, 40, 121–122, 123f
  in peripheral nerve lesions, 83
  sensory processing in, 84–87, 86f
  transmitter-receptor pairs, 32–33
Dorsal root ganglia (DRGs)
  anatomy, 121, 122f
  basket formation in, 107
  rhizotomy, 633–634
Double effect, 176, 176f
DREZ operation, 634–635
DRGs. See Dorsal root ganglia (DRGs)
DTI. See Diffusion tensor magnetic resonance imaging (DTI)
Duodenal ulcer, 492
Dura mater, 130
Dynorphins
  in inflammatory pain, 44
  in neuropathic pain, 44–45
  pharmacology, 43–44
  structure, 62, 62t
Dysafferentation, 291
Dysmenorrhea, 549
Dyspepsia, 491

EAAs. See Excitatory amino acids (EAAs)
Ear pain. See Otalgia
Effector systems, 32–33, 35. See also Neurotransmitters; Receptors
Elderly patients. See also Alzheimer's disease; Parkinson's disease
  atypical presentation of pain, 252–253
  laboratory studies of pain, 245–249, 246–248f
  pain clinic referrals, 249, 249f
  pain experience, 187–189
  pain prevalence, 250–251f, 250–252, 252t
  presbyalgos in, 245, 253t, 254
Electrodes
  for motor cortex stimulation, 626
  for spinal cord stimulation, 624–625
Endocannabinoids, 592–593
Endometriosis, 455–456
Endomorphins, 87
Endorphins, 62, 62t
Enkephalins, 61–62, 62t
Entrapment neuropathy(ies), 329, 632–633, 633t
Epidural adhesiolysis, 438–439
Epidural analgesia, 522–524, 567–569
Epidural injections, 437–438, 610–611, 611f
Erb-Duchenne's palsy, 140
Ergots, 396
Erythermalgia, 489
Erythromelalgia, 489
Esophageal pain, 490
Ethics, 171–172
Euthanasia, 176–177
Evidence-based guidelines, 279, 280–281t
Evoked pain, 303, 363. See also Allodynia; Hyperalgesia
Evoked release, 46–47
Excitatory amino acids (EAAs), 32. See also Glutamate
Eye pain. See Ophthalmodynia
Eye pathology, 135–136

Fabry's disease, 330
Facet joint blocks, 612
Facial pain
  anatomy, 132–135f
  atypical, 134, 410
  syndromes, 131–135
Failed back surgery syndrome (FBSS), 623
Familial amyloid polyneuropathy, 330
Fascia iliaca nerve block, 522
Fatigue, 533, 533t
Femoral nerve block, 521–522, 609

Fentanyl. *See also* Opioids
  for cancer pain, 472t, 475
  in children, 232t
  intranasal, 564
  in patient-controlled epidural analgesia, 523
  for postoperative pain, 562
Feverfew, 394
Fibromyalgia
  clinical presentation, 504–505, 505t
  diagnosis, 505–506, 505t, 510
  treatment, 506–507, 506–507t, 510
  vs. myofascial pain syndrome, 513–514
First pain, 22
Flowmetry, 357
Fluoroscopy, 610
FMRI. *See* Functional magnetic resonance imaging (fMRI)
Forebrain, 87
Free-radical scavengers, 370
Functional magnetic resonance imaging (fMRI), 151–152
Funicular pain, 136

GABA, 85
GABA agonists, 335, 369
Gabapentin, 587–588. *See also* Anticonvulsants
  for central neuropathic pain, 309
  for complex regional pain syndromes, 369
  for peripheral neuropathic pain, 333
GABA receptors, 38, 85, 86f
GAL. *See* Galanin (GAL)
Galanin (GAL), 32, 83
γ-aminobutyric acid. *See* GABA
Ganglionectomy, 633–634
Gasserian ganglion. *See* Semilunar ganglion
Gastric pain, 490–492
Gastric tumors, 492
Gastric ulcer, 491–492
Gene splicing, 66–67, 66f
Geniculate neuralgia, 135
Genitofemoral nerve, 147, 609
Giant cell arteritis. *See* Temporal arteritis
GIRK channels. *See* G-protein coupled inwardly rectifying K$^+$ (GIRK) channels
Glazer protocol, 459, 459t
Glossopharyngeal neuralgia, 135, 331, 407–408
Glutamate, 5–7, 32
Glutamate receptors. *See also* NMDA receptors
  G-proteins and, 38
  metabotropic, 34
  in peripheral neuropathic pain, 332
  in spinothalamic tract, 10, 10f
Gluteal nerve, 149
Glycine receptors, 38
Gold standard, 179
G-protein coupled inwardly rectifying K$^+$ (GIRK) channels, 38
G-protein coupling, 33, 36, 38
Gradenigo syndrome, 132
Graphic rating scales, 205, 205f
Gray matter, 6f, 8, 8f
Group therapy, 603
Growth factors, 39–40, 56–57
Guided imagery, 601
Guillain-Barré syndrome, 326, 348

Headache. *See also* Chronic daily headache; Cluster headache; Migraine
  anatomical sources, 131
  cervicogenic, 415–418
  in chronic pain patients, 533t, 534
  chronic tension-type, 380–381
  in multiple sclerosis, 347
  new persistent daily, 381
  rebound, 392
  recurrent, 235–238, 236–237t
  sinus, 136
Health care providers, 173
Health care quality assurance. *See* Quality, health care
Heart disease, 183, 484–485, 486. *See also specific conditions*
Heat pain, 21–22, 24
Hemicrania continua, 381–382
Hemopathies, 499–500
Hemophilia, 500
Henoch-Schönlein purpura, 489
Hepatic disease, 324, 495
Hepatitis, 325, 495
Herpes zoster infection. *See* Postherpetic neuralgia
Histamine, 24
HIV infection. *See* Human immunodeficiency virus (HIV) infection
5-HT$_2$ antagonists, 387
Human immunodeficiency virus (HIV) infection, 325
Human T-cell lymphotrophic virus infection, 325
Humerus, 143, 144
Hydrocodone, 474. *See also* Opioids
Hydromorphone. *See also* Opioids
  for cancer pain, 472t, 475
  in children, 232t
Hydroxyurea, 499
Hyperalgesia
  definition, 24, 213, 213t
  imaging, 153
  nociceptor sensitization in, 25
  primary, 24–25
  punctate, 26–27
  secondary, 25–26
  stroking. *See* Allodynia
  viscerovisceral, 498
Hypertrophic cardiomyopathy, 486
Hypnosis, 602
Hypnotics, 233t
Hypothalamus, 116–117
Hypothyroidism, 325

IASP. *See* International Association in the Study of Pain (IASP)
Ibuprofen, 231t
IDET. *See* Intradiskal electrothermal annuloplasty (IDET)
Idiopathic brachial plexitis. *See* Parsonage-Turner syndrome
Idiopathic cervical dystonia. *See* Torticollis
Iliohypogastric nerve, 147
Ilioinguinal nerve, 147, 609
Imaging. *See also specific conditions*
  of analgesic effects, 154
  in chronic pain, 152–153
  connectivity, 154
  in headaches, 393
  in hyperalgesia, 153–154
  in pain assessment, 151–152
  reliability, 152
  sensitivity, 152
  technologies, 151
Incidence, 179
Infections, 325–328
Inflammatory bowel disease, 493–494
Inflammatory pain
  descending facilitation and, 45–46

Inflammatory pain *(Continued)*
  neuropeptide release and, 46–47
  neuropeptide Y and, 43
  sodium channels in, 41
  vs. neuropathic pain, 40
Infraclavicular nerve block, 521
Inotropic effector systems, 33
Intercostal neuralgia, 144
Interictal disturbances, 114–116
Interleukin-6, 57
Interleukin-β, 57
Intermittent explosive disorder, 530t, 531–532, 539f
International Association in the Study of Pain (IASP)
  pain classification, 292–297, 294–297t
  pain subtypes, 291–292
  terminology, 289–291
Interneurons, 6f, 9
Interosseous nerve, 142, 144
Interscalene nerve block, 520–521
Intestinal tumors, 494
Intraarticular blocks, 437
Intradiskal electrothermal annuloplasty (IDET), 439, 614, 614f
Intradiskal therapies, 439
Intrathecal pump
  for central neuropathic pain, 310
  medications, 617–618
  patient selection, 617
  for peripheral neuropathic pain, 335–336
  psychological evaluation, 222
  side effects, 618
Irritable bowel syndrome, 456, 494–495, 535
Ischemia, 483–485

Kainate receptors, 33
Kappa opioid receptor clone (KOR-1), 62t, 66
Ketamine, 591. *See also* NMDA receptor antagonists
Kidney disease, 497–499
Klumpke's paralysis, 140
KOR-1. *See* Kappa opioid receptor clone (KOR-1)

Lamotrigine, 308, 588–589. *See also* Anticonvulsants
LANSS. *See* Leeds Assessment of Neuropathic Symptoms and Signs (LANSS)
Laparoscopy, 455
Lasegue sign, 213
Leeds Assessment of Neuropathic Symptoms and Signs (LANSS), 72, 212
Leg spasms, 347
Leprosy, 325
Levorphanol, 472t, 475
Lhermitte's phenomenon, 136
Lidocaine. *See also* Antiarrhythmics
  for cancer pain, 477
  for central neuropathic pain, 309
  clinical trials, 594–595
  intrathecal, 310
  mechanisms of action, 594
  transdermal, 594–595
  for trigger point injections, 509
Lissauer's tract, 8
Liver disease. *See* Hepatic disease
Local anesthetics
  for central neuropathic pain, 309
  for complex regional pain syndromes, 370
  in patient-controlled epidural analgesia, 522–523
  for peripheral neuropathic pain, 334
  for regional anesthesia, 607–608
  for spinal pain, 436–437
Locus ceruleus, 126, 126f
Low back pain. *See* Lumbar pain
Lower trunk lesions, 141
Lumbar disk herniation, 444
Lumbar pain. *See also* Spinal pain
  assessment, 652t
  clinical algorithm, 443f
  epidemiology, 421
  imaging, 152
  neurosurgical approaches, 631–632
  pathophysiology, 613
  physical therapy, 656–657
  socioeconomic status and, 181–182
  treatment, 442–443, 614, 614f, 631–632
Lumbar puncture, 394
Lumbar sympathetic block, 610
Lumbosacral plexus
  anatomy, 145–146, 146f
  neuropathies, 145–149
  radiculopathies, 145
Lung tumors, 490
Lyme disease, 145, 325–326

Magnesium, 394
Magnetic resonance imaging (MRI), functional. *See* Functional magnetic resonance imaging (fMRI)
Malingering, 530t, 532
Mandibular nerve, 134, 401
Manipulative therapy, 417
Mastectomy, 144
Maxillary nerve, 133–134, 401
McGill Pain Questionnaire (MPQ), 203t, 204, 211, 221
Mechanically insensitive afferents (MIAs), 23. *See also* C fibers
Mechanically sensitive nociceptive afferents (MSAs), 23. *See also* A fibers; C fibers
Mechanical pain sensation, 23–25
Medial nerve, 141–142
Medical procedures, in children, 234–235, 235t
Meditation, 601
Medulla, 45–46, 87
Mees' lines, 322
Memorial Pain Assessment Card (MPAC), 203t, 204–205
Memory problems, 533t, 534
Meninges, 130–131
Mental status examination, 200, 200t
Meperidine, 472t, 474. *See also* Opioids
Meralgia paresthetica, 147
Methadone. *See also* Opioids
  acute pain treatment and, 264
  for cancer pain, 472t, 475–476
  in children, 232t
  legal issues, 268
Mexiletine, 309, 477. *See also* Antiarrhythmics
MIAs. *See* Mechanically insensitive afferents (MIAs)
Microvascular angina, 486
Microvascular decompression, 407
MIDAS. *See* Migraine Disability Assessment Scale (MIDAS)
Migraine
  aura mechanisms, 113–114
  chronic, 380, 392
  diagnosis, 392–394
  disability assessment, 393
  interictal disturbances, 114–116
  neurosurgical approaches, 634

nonpharmacologic treatment, 394–395
pain mechanism, 114
pathophysiology, 113–116, 391–392
pharmacologic treatment, 395–396
prophylactic therapy, 396–397
status, 392
transformed, 380. *See also* Chronic daily headache
Migraine Disability Assessment Scale (MIDAS), 393, 398
Mirtazapine, 585. *See also* Antidepressants
Mitral stenosis, 486
Mitral valve prolapse, 486
Monoclonal proteins, 327–328
Mononeuropathy multiplex, 351
Mood, assessment of, 215–216
MOR-1. *See* Mu opioid receptor clone (MOR-1)
Moral agency, 172
Moral integrity, 172–173
Morbidity, demographics, 245
Morphine. *See also* Opioids
　for cancer pain, 472t, 474–475
　in children, 232t
　intrathecal, 310, 617
　in patient-controlled epidural analgesia, 523
Mortality, demographics, 244–245
Morton's metatarsalgia, 149
Motor cortex stimulation
　for cancer pain, 637
　in central neuropathic pain, 311
　technique, 626–627
Motor examination, 213
Movement disorders, 352–353
MPAC. *See* Memorial Pain Assessment Card (MPAC)
MPI. *See* Multidimensional Pain Inventory (MPI)
MPQ. *See* McGill Pain Questionnaire (MPQ)
MSAs. *See* Mechanically sensitive nociceptive afferents (MSAs)
Multidimensional Pain Inventory (MPI), 216, 221, 297
Multiple myeloma, 500
Multiple sclerosis, 346–347
Mu opioid receptor clone (MOR-1), 62t, 66–67, 67f
Murphy's sign, 496
Muscle pain, 483–484, 503–504
Muscle relaxants, 387
Muscular dystrophy(ies), 351
Muscular ischemia, 483–484
Musculocutaneous nerve, 141
Myelopathic pain, 136
Myeloproliferative disorders, 500
Myelotomy, 637
Myocardial infarction, 484
Myofascial pain syndrome
　clinical presentation, 507–508, 516
　diagnosis, 508, 508t, 515–516
　epidemiology, 514
　historical background, 514
　pathophysiology, 507, 514–515
　treatment, 508–510t, 508–511, 516–517
Myofascial trigger point. *See* Trigger point(s)
Myopathy(ies), 351–352, 483
Myositis
　eosinophilic, 351
　orbital. *See* Orbital pseudotumor

Naproxen, 231t
Na$_v$1.8, 41–42, 80. *See also* Sodium channels
Neck, innervation, 139f
Neck pain. *See also* Spinal pain

clinical algorithm, 441f
epidemiology, 421–422
neurosurgical approaches, 631–632
surgical treatment, 440
Nefazodone, 585. *See also* Antidepressants
Nerve block(s)
　for complex regional pain syndromes, 370
　equipment, 519–520
　fascia iliaca, 522
　femoral, 521–522
　infraclavicular, 521
　interscalene, 520–521
　for pelvic pain, 463
　for spinal pain, 437
　technique, 520
Nerve fibers. *See* A fibers; C fibers
Nerve injuries. *See also* Peripheral neuropathic pain
　brain stem in, 87
　cell bodies in, 80
　changes in intact nerves following, 77–79, 78f
　forebrain in, 87
　ligand-gated ion channels in, 80–82
　neuroma site, 79–80, 81f
　neurosurgical approaches, 632–633
　phenotypic shifts following, 82–83
　sensory processing in, 84–86, 86f
　spinal glia activation in, 86–87
　structural changes following, 83–84
Nerve(s)
　injured. *See* Nerve injuries
　primary afferent. *See* Primary afferent nerve(s)
Nervous intermedius neuralgia, 408
Neuralgia(s). *See also specific syndromes*
　of cutaneous nerve of the thigh, 149
　geniculate, 135
　glossopharyngeal. *See* Glossopharyngeal neuralgia
　intercostal, 144
　mandibular, 134
　maxillary, 133–134
　nervous intermedius, 408
　occipital, 138
　postherpetic, 144
　Raeder's paratrigeminal, 131–132
　spinal accessory, 135
　superior laryngeal, 408
　thoracic, 144–145
　trigeminal. *See* Trigeminal neuralgia
Neuralgic amyotrophy. *See* Parsonage-Turner syndrome
Neurokinin A (NKA), 32
Neurokinin receptors, 34
Neuroleptics, 477
Neurolytic agents, 608
Neuromas, 79–80, 81f, 632
Neurons, 9–10
　dorsal horn, 9, 40
　injured. *See* Nerve injuries
　postsynaptic dorsal column, 11, 12f
　projection, 9–10
　summation, 84
　windup, 84
Neuropathic pain. *See also* Nerve injuries; Neuropathy(ies)
　central. *See* Central neuropathic pain
　definition, 290
　descending facilitation and, 46
　diagnosis, 217
　etiologies, 212, 212t

Neuropathic pain *(Continued)*
  imaging, 152–153, 217
  neuropeptide release and, 47
  neuropeptide Y and, 43
  peripheral. *See* Peripheral neuropathic pain
  sodium channels in, 41–42, 79
  in spinal cord injury. *See* Spinal cord injury
  symptoms, 214t
  vs. inflammatory pain, 40
Neuropathic Pain Scale, 212
Neuropathy(ies). *See also* Nerve injuries; Neuropathic pain
  brachial plexus, 141–144
  entrapment, 329
  hereditary, 329–330
  inferior gluteal, 149
  inflammatory demyelinating, 326
  interosseous, 142, 144
  lower extremity, 146–149
  lumbosacral plexus, 145–149
  median nerve, 141–142
  musculocutaneous nerve, 141
  porphyric, 330
  pudendal, 149
  radial, 143–144
  superior gluteal, 149
  suprascapular, 141
  tibial, 148
  toxic, 329
  trigeminal sensory, 408
  ulnar, 143
Neuropeptides, evoked release, 46–47
Neuropeptide Y (NPY), 42–43, 82
Neuropeptide Y (NPY) receptors, 37
Neurostimulation, 614–617, 615f
Neurosurgical ablation, 311–312, 336
Neurotoxin TTX, 40–41
Neurotransmitters, 5–7, 32–33. *See also specific substances*
NF-kappa B. *See* Nuclear factor (NF)-kappa B
Nitric oxide (NO), 35
NKA. *See* Neurokinin A (NKA)
NMDA/NOS cascade, 40
NMDA receptor antagonists
  analgesia and, 590
  for central neuropathic pain, 310
  for complex regional pain syndromes, 370
  opioid tolerance and, 590–591
  for peripheral neuropathic pain, 335
  types of, 590–591
NMDA receptors, 33–34. *See also* Glutamate receptors
  in central sensitization, 84–85
  dynorphin and, 44
Nociception
  autonomic effects, 159–160
  central affective consequences, 160–162
  in children, 225–226
  defense response, 160
  impacts of, 159
  vs. pain, 158–159
Nociceptive pathways
  anatomy, 4f
  ascending, 10–13
  cortical, 12–13
  descending inhibitory, 13, 124–126, 126f
  dorsal columns-medial lemniscal. *See* Dorsal column-medial lemniscal pathway
  integration with visceral signals, 105–106, 106f
  limbic, 12–13
  postsynaptic dorsal column, 11, 12f
  spinohypothalamic, 12–13
  spinomesencephalic. *See* Spinomesencephalic tract
  spinoreticular. *See* Spinoreticular system
  spinothalamic. *See* Spinothalamic tract
  visceral pain, 12f
Nociceptor(s). *See also* A fibers; C fibers
  anatomy, 4–7, 6–8f, 121–122, 122t
  cold pain sensation and, 23–24
  definition, 291
  heat stimuli and, 21–22
  in hyperalgesia, 24–27
  mechanical stimuli and, 23
Non-opioids, 231t, 476–477
Nonorganic physical findings, 533t, 534–535
Nonsteroidal anti-inflammatory drugs (NSAIDs). *See also* Coxibs
  for cancer pain, 476
  for dysmenorrhea, 549
  for osteoarthritis, 548, 552–553
  for peripheral neuropathic pain, 335
  safety, 550, 554–555
  for spinal pain, 434
Noradrenergic system, 126
Notalgia paresthetica, 145
Noxious input transmission, 4–5, 6f
NPY receptors. *See* Neuropeptide Y (NPY) receptors
NSAIDs. *See* Nonsteroidal anti-inflammatory drugs (NSAIDs)
Nuclear factor (NF)-kappa B, 333
Nucleus proprius, 122. *See also* Dorsal horn
Nursing home patients, 188–189

Obsessive-compulsive spectrum disorder, 531
Obturator nerve, 147
Occipital nerve
  blockade, 418, 609
  neuralgia, 138, 634
  radiculopathy, 138
Odds ratio, 179
OFF cells, 45–46
ON cells, 45–46
Operant therapy, 603
Ophthalmic nerve, 401
Ophthalmodynia, 131
Ophthalmoplegias, 131
Opioid peptides, 61–62, 62t. *See also specific substances*
Opioid receptors
  classification, 62, 62t
  clones, 66–67
  mechanism of action, 559
  molecular structure, 65–67
  at peripheral afferent terminals, 36–37
  postsynaptic, 37
  presynaptic, 36
  variability in effects, 259–260
Opioids. *See also specific drugs*
  abuse potential, 259–260, 266t, 267f
  actions, 62–63, 65–67
  addiction, 266–268, 267f, 571–572
  bowel regimen, 472t
  for cancer pain, 471–476, 472–474t
  for central neuropathic pain, 309
  in children, 232t
  clinical use, 64–65
  cognitive effects, 572
  for complex regional pain syndromes, 369

dependence, 258, 291, 570
dependence risk assessment, 65, 222, 260–262, 262t, 265, 266t, 571
dosing strategy, 572, 574–575
in dying patients, 575
epidural, 567–569, 569f
equianalgesic dosing, 560–561, 561t
ethical issues, 176–177
imaging studies, 154
for migraine, 396
for myofascial pain syndrome, 510, 511
for patient-controlled analgesia, 522–523, 564–567, 565–566t
for peripheral neuropathic pain, 333–334
pharmacokinetics, 573t
pharmacology, 559–560, 560f
preemptive therapy, 563–564
prescribability factors, 560t
reversal, 65
rotation, 473–474f, 560–561, 560–561t
routes of administration, 563t
selection, 64, 562–563
side effect management, 574–575
for spinal pain, 435
structure, 62–63, 63f
subcutaneous, 645
tolerance, 64–65, 471, 562, 562f, 570–571
transdermal, 645
withdrawal symptoms, 263–264
Optic neuritis, 131, 347
Orbital pseudotumor, 132, 352
Osteoarthritis, 482–483, 547–548
Osteomyelitis, 481–482
Osteoporosis, 482
Otalgia, 134
Outcome variables, 275–276
Oxcarbazepine, 308, 406, 589. *See also* Anticonvulsants
Oxycodone, 232t, 472t, 474. *See also* Opioids

Paget's disease of bone, 482
Pain. *See also specific diseases and locations*
  acute. *See* Acute pain
  age-related differences, 249–252, 249–252f
  asymbolia to, 129
  central. *See* Central neuropathic pain
  in children. *See* Children
  chronic. *See* Chronic pain
  classification, 292–297, 294–297t
  constructivist model. *See* Constructivism
  coping strategies, 216–217
  definitions, 195, 289–290
  in demented patients, 188–189
  in elderly patients. *See* Elderly patients
  evoked. *See* Evoked pain
  first, 22
  higher processing of, 13–14
  imaging, 151–155
  inflammatory. *See* Inflammatory pain
  neuropathic. *See* Neuropathic pain; Neuropathy(ies)
  patient education, 600–601
  perception of, 22. *See also* Pain experience
  perioperative, 549, 550t
  persisting beyond healing, 264–265, 265f, 267f
  postoperative, 563–567, 565–567t
  psychological aspects, 215–217. *See also* Constructivism; Psychological evaluation
  radicular, 136–137. *See also specific nerves and syndromes*
  referred. *See* Referred pain
  second, 22
  in skeletal muscle, 483–484
  spontaneous. *See* Spontaneous pain
  subtypes, 291–292
  sympathetically maintained. *See* Sympathetically maintained pain (SMP)
  terminology, 213, 213t, 290–291
  transmission of, 7–10, 32–36. *See also* Effector systems; Nociceptive pathways
  vs. nociception, 158–159
Pain assessment. *See also specific conditions and types of pain*
  characteristics of pain, 210–211
  in children, 228–230, 230f
  goals, 209
  history taking, 200, 200t, 210–212
  initial, 199–200, 199t
  measurement tools, 202–206, 203t, 205–206f, 205–206t, 211
  mental status examination, 200, 200t
  ongoing, 202
  physical examination, 200–202, 212–215
  psychological factors, 215–217
Pain behaviors, 533t, 535
Pain Disability Index, 216
Pain experience
  affective dimension, 196t, 197
  age-related differences, 245–249, 246–248f. *See also* Children; Elderly patients
  behavioral dimension, 196t, 198
  characteristics, 196t
  in children. *See* Children
  cognitive dimension, 196t
  culture-related differences, 185–186
  in elderly patients. *See* Elderly patients
  ethnicity-related differences, 184–186
  physiologic dimension, 196, 196t
  race-related differences, 184–186
  sensory dimension, 196–197, 196t, 197f
  sex-related differences, 183–184
  sociocultural dimension, 198
  socioeconomic status and, 181–182
Pain management. *See also specific conditions and types of pain*
  certification, 276
  in children. *See* Children
  clinical practice guidelines, 279, 279f, 280–281t
  cognitive-behavioral approach, 600–603
  conflict management in, 175
  culture-related differences, 185–186
  decision making, 173–175
  in elderly patients. *See* Elderly patients
  ethical issues, 173–178, 174f
  ethnicity-related differences, 184–186
  evidence-based guidelines, 279, 280–281t
  group therapy, 603
  insufficient, 175–176
  moral agency in, 172
  moral integrity and, 172–173
  operant therapy, 603
  outcome objectives, 275–276, 281t
  psychotherapy, 603–604
  public accountability, 279, 282–283t, 283
  quality assurance, 276
  quality improvement, 276–279, 277–278t
  race-related differences, 184–186
  sex-related differences, 184
  stress management, 604
Pain transmission, 7–9

Palliative care, 641–642
Pancoast's syndrome, 490
Pancoast syndrome, 144
Pancreatic tumors, 497
Pancreatitis, 496–497
PAR2. *See* Proteinase-activated receptor (PAR2)
Parkinson's disease, 253–254, 352
Parsonage-Turner syndrome, 140–141, 331, 348
Patient-controlled analgesia (PCA)
   epidural, 522–524, 568
   intravenous, 524
   postoperative, 564–567, 565–566t, 567–568f
Pediatric patients. *See* Children
Pelvic pain
   assessment, 454–455, 454t
   bladder-related, 456–457
   bowel-related, 456
   diagnostic tests, 455
   epidemiology, 454
   gynecologic causes, 455–456
   musculoskeletal causes, 457–458, 458t
   nerve-related, 457
   neurobiology, 453–454
   physical therapy, 458–460, 459t
   psychological factors, 460–462
   treatment, 462–463
Peptides, 32, 39, 61–62, 62t. *See also specific substances*
Percutaneous disk decompression, 439
Percutaneous rhizotomy, 407
Pericarditis, 486–487
Perioperative pain, 549, 550t
Peripheral nerve blocks, 519–522, 608–609
Peripheral nerve stimulation, 616–617
Peripheral neuropathic pain. *See also specific conditions*
   adjuvant analgesics for, 334–335
   analgesic stepladder, 336–337, 336t
   animal models, 76
   assessment, 72–75, 74–75t, 321–324. *See also* Pain assessment
   diagnosis, 323–324, 323t
   diversity of, 71–72
   entrapment neuropathy and, 329
   etiologies, 322t, 348t
   genetic factors, 76
   intrathecal pumps in, 335–336
   invasive treatment, 335–336
   ligand-gated ion channels in, 79–82
   mechanisms, 74–75t, 76–77, 88t
   monoclonal proteins and, 327–328
   neurotrophic factors, 77–78, 78f
   pathophysiology, 331–333
   pharmacologic treatment, 333–335
   physical examination, 322–323
   sympathetic mechanisms, 75t, 88t
   treatment, 88t
   tumor necrosis factor-α in, 73, 78, 78f
   vitamin deficiencies and, 328–329
   vs. central neuropathic pain, 312–313, 312t, 322f
Peroneal nerve, 149
Personality disorders, 530t, 532
PET. *See* Positron-emission tomography (PET)
Phalen's test, 215
Phantom limb pain, 165, 635
Phenytoin, 406
Phlegmasia alba dolens, 489
Phlegmasia coerulea dolens, 489
Physical dependence, 258, 291. *See also* Substance abuse

Physical therapy
   clinical assessment, 650–652
   clinical interview, 648–650, 650t
   for complex regional pain syndrome (CRPS), 371
   interventions, 655–657
   for pelvic pain, 458–460, 459t
   performance measures, 652–653t, 652–655, 654–655f
   for spinal pain, 435–436
Pia mater, 131
Piriformis syndrome, 148
Pituitary apoplexy, 132
Placebos, 177–176
Pleurisy, 490
Polyarteritis nodosa, 327, 488
Polycystic kidney disease, 498
Polymyalgia rheumatica, 483
Polymyositis, 351
Polyneuropathy, 330
Polyradiculopathy, 351
Pontine strokes, 129
Porphyric neuropathy, 330
Positron-emission tomography (PET), 151–152
Posterolateral cord syndrome. *See* Dorsal columns disease
Postherpetic neuralgia
   incidence, 325
   neurosurgical approaches, 635
   ophthalmic, 408–409
   symptoms, 144
Postmastectomy pain, 144
Postoperative pain, 563–567, 565–567t
Poststroke pain, 304, 304f, 343–344
Post-thoracotomy pain, 144, 634
Potassium channels
   in injured nerve fibers, 81–82
   in neuropathic pain, 332
   therapeutic use, 56
Pregabalin, 333
Presbyalgos, 245, 253t, 254. *See also* Elderly patients
Prevalence rate, 180
Primary afferent nerve(s)
   in sensory processing, 4f, 5, 6f
   sympathetic terminals and, 107–108, 108f
   terminal distribution, 6–7f
   in visceral pain, 98–100
Prinzmetal's variant angina, 486
Progressive polyradiculopathy, 351
Projection neurons, 9–10
Pronator teres syndrome, 142
Propoxyphene, 474. *See also* Opioids
Prostaglandins, 35, 41, 547
Proteinase-activated receptor (PAR2), 54
Proton-gated ion channels, 55
Pseudoaddiction, 259. *See also* Substance abuse
Psychological evaluation
   clinical interview, 219–221
   in complex regional pain syndromes, 368
   in myofascial pain syndrome, 510
   in pelvic pain, 460–462
   prior to treatment, 221–222
   purpose, 219
   testing, 221
Psychostimulants, 233t
Psychotherapy, 603–604
Pudendal nerve, 149
Purine (P2X) receptors, 56, 332

P2X receptors. *See* Purine (P2X) receptors
Pyelonephritis, 498

Qigong, 371
QSART. *See* Quantitative sudomotor axon reflex test (QSART)
QST. *See* Quantitative sensory testing (QST)
Quality, health care
    definition, 274–276, 274f, 275t
    historical influences, 273–274
    improvement measures, 276–277
Quality assurance, 276
Quantitative sensory testing (QST), 217, 368
Quantitative sudomotor axon reflex test (QSART), 367

Radial nerve, 143–144
Radiculopathy(ies). *See also specific syndromes*
    cervical, 138–139, 440
    lumbosacral, 145, 443–444, 444f
    occipital, 138
    thoracic, 144–145, 441
Radiofrequency lesioning
    for spinal pain, 437
    technique, 612–613, 612–613f
    for trigeminal neuralgia, 407
Radiopharmaceuticals, 477
Raeder's paratrigeminal neuralgia, 131–132
Raphe spinal pathway, 13
Raynaud's phenomenon, 489
Reboxetine, 585–586. *See also* Antidepressants
Receptors
    adenosine, 38
    α_2-adrenergic. *See* α_2-adrenergic receptors
    AMPA, 33
    excitatory, 35–36
    GABA, 38
    glutamate. *See* Glutamate receptors
    glycine, 38
    inhibitory, 38–39
    kainate, 33
    neurokinin, 34
    neuropeptide Y, 37
    NMDA, 33–34
    opioid. *See* Opioid receptors
    purinergic, 35
    serotonin. *See* Serotonin receptors
    spinal inotropic, 33–34
    vanilloid, 35
Referred pain
    in biliary tract disease, 495–496
    in cardiac disease, 485
    in esophageal disease, 490
    in gastric disease, 491–492
    in intestinal disease, 493
    in myofascial pain syndrome, 507–508, 508t
    in pancreatic disease, 496–497
    pathophysiology, 484
    patterns, 131
    in ureteral colic, 498
    in visceral diseases, 484
Reflex sympathetic dystrophy, 295, 359. *See also* Complex regional pain syndrome (CRPS)
Rehabilitation, 647–648, 648–649f. *See also* Physical therapy
Relaxation techniques, 601–602
Reliability, 180
Renal failure, 324–325
Renal infarction, 498

Renal tumors, 499
Renal vein thrombosis, 498
Respiratory disease, 490
Response sensitization, 161
Restless leg syndrome, 352–353, 588
Rheumatoid arthritis, 327, 482–483, 547
Riboflavin, 394
Rofecoxib. *See* Coxibs
Rostroventromedial medulla, 45–46, 87

Sarcoidosis, 327
Scalloped primary afferent nerve endings, 8. *See also* Primary afferent nerve(s)
Schemata, 164–165
Sciatic nerve, 147–148
Secondary gains, 530t, 532
Second pain, 22
Sedatives, 233t
Selective serotonin reuptake inhibitors (SSRIs), 583–584, 583t. *See also* Antidepressants
Self-medication, 259. *See also* Substance abuse
Semilunar ganglion, 131
Semmes-Weinstein examination, 349
Senescence, 244. *See also* Elderly patients
Sensitization
    central, 26, 84–85, 210, 322f, 382–383
    definition, 24
    nociceptor, 382
    peripheral, 209–210, 322f
    response, 161
Sensory deficit, 303–304, 364
Sensory neuron specific receptors (SNSRs), 54
Sensory processing, 4f
Sensory testing, 73, 212–213
Serotonergic system, 124
Serotonin receptors, 37–38, 54
Sexual dysfunction, 533t, 534
Sickle cell disease, 499
Sinus headache, 136
Sjögren syndrome, 326–327, 352
Sleep disturbances, 506, 533–534, 533t
Slipping rib syndrome, 144
SNSRs. *See* Sensory neuron specific receptors (SNSRs)
Socioeconomic status, 181–182
Sodium channel blockers, 39, 369
Sodium channels
    in inflammatory pain, 41
    in injured nerve fibers, 80–81
    $Na_v1.8$, 41–42
    in neuropathic pain, 41–42, 79, 332
    in primary afferent neurons, 40
    subtypes, 40–41
    therapeutic use, 55–56
SOM. *See* Somatostatin (SOM)
Somatization, 533t, 535
Somatoform disorders, 529–530t, 531
Somatostatin (SOM), 32
SP. *See* Substance P (SP)
Spasmolytics, 434
Sphenopalatine ganglion, 131
Spinal accessory nerve neuralgia, 135
Spinal cord. *See also* Nociceptive pathways; Primary afferent nerve(s)
    anatomy, 5f
    autonomic dysreflexia, 106–107
    dorsal horn. *See* Dorsal horn

Spinal cord *(Continued)*
   hemisection, 136
   myelopathy, 136
Spinal cord injury
   pain classification, 293–295, 295t, 344
   pain syndromes, 304, 304f, 344–345, 635
Spinal cord stimulation
   for central neuropathic pain, 311
   for chronic benign pain, 622–623
   for complex regional pain syndromes, 371
   for failed back surgery syndrome, 623
   indications, 615–616
   mechanisms of action, 615
   psychological evaluation, 222
   technique, 615–616, 615f, 624–625
Spinal glia, 86–87
Spinal inotropic receptors, 33–34
Spinal interneurons, 6f, 9
Spinal pain. *See also* Lumbar pain; Neck pain
   assessment, 432, 432t
   biomechanics of injury, 428–429
   diagnostic strategies, 433
   diskogenic, 424
   epidemiology, 421–422
   facet joint, 425
   interventional techniques, 437–439
   intradiskal therapies, 439
   local anesthetics for, 436–437
   mechanical, 429, 429t
   muscular, 426
   neurophysiology, 426–427
   nonmechanical, 429, 430t
   pathoanatomy, 423–424
   pathophysiology, 424–426
   patient history, 429, 429t
   pharmacologic therapy, 434–435
   physical examination, 429–432, 430t
   physical factors, 427–428, 428t
   physical therapy, 435–436
   psychological factors, 427, 428t
   psychosocial factors, 428, 428t
   radicular, 424–425
   sacroiliac joint, 425–426
   surgical treatment, 439–444
   treatment strategies, 433
Spinal stenosis, 444–445
Spine, 422–423
Spinomesencephalic tract, 123, 124f
Spinoreticular system, 11–12, 123, 124f
Spinothalamic tract, 6f, 8f, 9–11, 10f, 122–123, 122f
Spinotrigeminal nucleus, 9
Spontaneous pain, 303, 363
Spurling's test, 215
Statins, 352
Status migraine, 392. *See also* Migraine
Stellate ganglion block, 609–610
Steroids. *See* Corticosteroids
Stiff person syndrome, 353
Stimulus interaction, 22
Straight-leg raising test, 213, 431
Stress management, 604
Stroke(s), 129. *See also* Poststroke pain
Stroking hyperalgesia. *See* Allodynia
Substance abuse
   assessment for potential, 261–262, 262t, 532t
   in chronic pain patients, 529t, 530, 539f
   diagnosis, 258t
   etiology, 258
   prevalence, 257–258
   terminology, 258–259
Substance P (SP), 32
Substantia gelatinosa, 8–9. *See also* Dorsal horn
Suicide risk, 528, 529t
Summation, 84, 213
Superior laryngeal neuralgia, 408
Superior orbital fissure syndrome, 132
Suprascapular nerve, 141
Sural nerve, 148
Sympathectomy, surgical, 370–371, 636–637
Sympathetically maintained pain (SMP), 72, 361–362, 636
Sympathetic nerve blocks, 370, 609–610
Sympathetic nerves, 149–150
Sympathetic terminals, 107–108, 108f
Syncope, 106–107
Syringomyelia
   pathophysiology, 138f, 345
   in spinal cord injury, 345
   surgical treatment, 346
   symptoms, 136, 304, 345–346
Systemic lupus erythematosus, 327

Takayasu's arteritis, 488
Temporal arteritis
   pathophysiology, 487–488
   polymyalgia rheumatica and, 483
   symptoms, 327, 488
Temporomandibular joint (TMJ) syndrome, 136, 410–411
Tender points, 505, 505t
Tennis elbow, 144
TENS. *See* Transcutaneous electrical nerve stimulation (TENS)
Thalamus
   anatomy, 126, 127f
   pain, 129, 344
Thalidomide, 335
Thermography, 367
Thermometry, 367
Thermoregulatory sweat test (TST), 367
Thiamine deficiency, 328
Thoracic nerves, 144–145
Thoracic outlet syndrome, 141, 141f
Thromboangiitis obliterans, 488
Thrombophlebitis, 489–490
Tibial nerve, 148
Tic douloureux, 133
Tinel's sign, 322
Tinel's test, 215
Tizanidine, 334, 509
TMJ syndrome. *See* Temporomandibular joint (TMJ) syndrome
TMS. *See* Transcranial magnetic stimulation (TMS)
TNF-α *See* Tumor necrosis factor-α (TNF-α)
Tolerance, 259, 291. *See also* Substance abuse
Tolosa Hunt syndrome, 132
Torticollis, 353
Toxic neuropathy, 329
Tramadol, 476
Transcranial magnetic stimulation (TMS), 115
Transcutaneous electrical nerve stimulation (TENS), 417
Transient receptor potential (TRP) channels, 54–55, 79
Transmitter-receptor pairs, 32–33
Tricyclic antidepressants, 582–583, 583t. *See also* Antidepressants
Trigeminal nerve, 132–135f, 401–404, 402f
Trigeminal neuralgia, 133

afferent fibers in, 83
diagnostic tests, 411–412, 411*f*
etiology, 404–405, 406*f*
pathophysiology, 405, 405*f*
peripheral neuropathic pain in, 331
pharmacologic treatment, 405–406, 589
surgical treatment, 406–407
symptoms, 404
Trigeminal sensory neuropathy, 408
Trigger point(s)
definition, 513
injections, 417–418, 509, 509*t*, 517
in myofascial pain syndrome, 507, 507*t*, 508–509
Triptans, 395
TrkA receptor. *See* Tyrosine kinase (TrkA) receptor
TRP channels. *See* Transient receptor potential (TRP) channels
TST. *See* Thermoregulatory sweat test (TST)
Tumor necrosis factor-α (TNF-α)
in nerve injury, 333
in neuropathic pain, 73, 78, 78*f*, 335
pharmacology, 57
Tyrosine kinase (TrkA) receptor, 332

Ulcerative colitis, 493–494
Ulnar nerve, 142–143
Upper trunk lesions, 140–141
Ureteral colic, 498–499
Ureteral tumors, 499

Vagal afferents, 108–109
Valdecoxib. *See* Coxibs
Validity, 180
Vanilloid receptors, 35, 332
Varicella-zoster virus, 325
Vasculitis, 351, 489
Vasoactive intestinal polypeptide (VIP), 32

Venlafaxine, 584–585. *See also* Antidepressants
Venous pain, 489
Ventral horn, 4, 5*f*
Verbal descriptor scales, 205
VGSCs. *See* Sodium channels
VIP. *See* Vasoactive intestinal polypeptide (VIP)
Visceral pain
animal models, 97–98
characteristics, 95–96, 484
hypersensitivity syndromes, 96
nerve pathway, 12*f*
neurobiology, 98–101
vagal afferents in, 108–109
vs. superficial pain, 95–96
Visual analogue scale, 205–206, 206*f*, 206*t*
Vitamin deficiencies, 328–329
Voltage-gated ion channels, 39. *See also* Calcium channel blockers; Sodium channels

Waddell's signs, 215, 431, 534
Wallenberg syndrome
pathophysiology, 130*f*, 409
poststroke pain in, 344
prognosis, 409
symptoms, 129–130, 304
Warm fibers, 21
Wartenberg syndrome, 144
Wegener's granulomatosis, 327
Windup, 84
Wisconsin Brief Pain Inventory. *See* Brief Pain Inventory (BPI)
Withdrawal symptoms, 263–264
Writhing test, 97

Ziconotide, 310
Zollinger-Ellison syndrome, 491–492
Zonisamide, 589. *See also* Anticonvulsants